SIGNALS
and
SYSTEMS

A MATLAB® Integrated Approach

OKTAY ALKIN

SIGNALS
and
SYSTEMS

A MATLAB® Integrated Approach

CRC Press
Taylor & Francis Group
Boca Raton   London   New York

CRC Press is an imprint of the
Taylor & Francis Group, an **informa** business

CRC Press
Taylor & Francis Group
6000 Broken Sound Parkway NW, Suite 300
Boca Raton, FL 33487-2742

© 2014 by Taylor & Francis Group, LLC
CRC Press is an imprint of Taylor & Francis Group, an Informa business

No claim to original U.S. Government works

Printed on acid-free paper
Version Date: 20140206

International Standard Book Number-13: 978-1-4665-9853-9 (Hardback)

Library of Congress Cataloging-in-Publication Data

Alkin, Oktay.
   Signals and systems : a MATLAB integrated approach / Oktay Alkin.
     pages cm
   Includes bibliographical references and index.
   ISBN 978-1-4665-9853-9 (hardback)
   1. Signal processing--Data processing. 2. System analysis--Data processing. 3. MATLAB. I. Title.

TK5102.9.A447 2014
621.382'2028553--dc23   2013036873

**Visit the Taylor & Francis Web site at
http://www.taylorandfrancis.com**

**and the CRC Press Web site at
http://www.crcpress.com**

to my mother
Tayyibe Alkin

and to the memory of my father
Alptekin Alkin

Contents

Preface

The subject of signals and systems is an integral component of any undergraduate program in electrical engineering. It also has applications in computer engineering, mechanical engineering, aerospace engineering and bioengineering. It encompasses analysis, design and implementation of systems as well as problems involving signal-system interaction. A solid background in signals and systems is essential for the student to be able to venture into fields such as signal processing, image processing, communications and control systems.

This textbook was written with the goal of providing a modern treatment of signals and systems at the undergraduate level. It covers both continuous-time and discrete-time signals and systems, and can be used as the main textbook for a one-semester introductory course in signals and systems or for a two-semester course sequence. It can also be used for self study. Writing style is student-friendly, starting with the basics of each topic and advancing gradually. Proofs and derivations that may safely be skipped at a first reading are color coded. Also, important concepts and conclusions are highlighted to stand out. No prior signals and systems knowledge is assumed. The level of presentation is appropriate for a second or third year engineering student with differential calculus background.

There are many excellent textbooks available for use in undergraduate level courses in the area of signals and systems. Some have matured over long periods of time, decades rather than just years, and have gained great popularity and traction among instructors as well as students. Consequently, the writing of a new textbook in signals and systems is a difficult task as one must consider the questions of what new ideas the book would employ, and what value it would add to the teaching of the subject. This textbook resulted from the author's efforts over the past couple of decades in trying to find ways to incorporate software into the teaching of the material not only as a computational tool but also as a pedagogical one. It utilizes MATLAB software due to its popularity in the engineering community and its availability for a variety of operating systems. Software use is integrated into the material at several levels:

1. **Interactive programs:** Graphical user interface based MATLAB programs allow key concepts of signals and systems to be visualized. We are all visual learners; we tend to remember an interesting scene in a movie much better than something we hear on the radio or something we read in a textbook. Taking this one step further, if we are also able to control the scene in a movie through our actions, we may remember the cause-effect relationships between our actions and the results they create even better. This is perhaps the main reason why children can become very proficient in video games. A large number of interactive programs are available for download with this

textbook. Some allow a static figure in the text to come alive on a computer where the student can change key parameters and observe the results of such changes to understand cause-effect relationships. Some take a solved example in the text and expand it in an open-ended manner to allow the student to set up "what if" scenarios. Some programs provide animations for concepts that are difficult to teach. Examples of this group include linear convolution, Fourier series representation of signals, periodic convolution, three-dimensional visualizations for Laplace and z transforms, and steady-state response of systems.

2. **MATLAB code for solved examples:** Most of the solved examples in each chapter have associated MATLAB listings available for download. These listings include detailed comments, and are useful in a variety of ways: They help the student check his or her work against a computer solution. They reinforce good coding practices. They also allow the student to experiment by changing parameter values and running the code again.

3. **MATLAB exercises:** In addition to the code listings associated with solved examples, there is a section at the end of each chapter which contains stand-alone MATLAB exercises that take the student through exploration and/or simulation of a concept by developing the necessary code step by step. Intermediate steps are explained in detail, and good coding practices are enforced throughout. Exercises are designed to help the student become more proficient with MATLAB while working on problems in the context of signals and systems. Furthermore, MATLAB exercises are synchronized with the coverage of the material. At specific points in the text the student is referred to MATLAB exercises relevant to the topic being discussed. The goal is not just to provide cookbook style quick solutions to problems, but rather to develop additional insight and deeper understanding of the material through the use of software.

4. **MATLAB based problems and projects:** In addition to traditional end-of-chapter problems, the textbook contains problems that require MATLAB solutions and project ideas that are MATLAB-based. These can be used by instructors as the basis of computer assignments.

5. MATLAB coverage of the book is integrated into the help browser of the MATLAB application. This allows the student to have the textbook and a computer running MATLAB side by side while studying. Additionally, it gives the instructor the freedom to display MATLAB exercises on a projector while lecturing, without the need to type and run any code, if preferred.

While software is an integral part of the textbook, and one that is meant to distinguish it from other works in the same field, it "never gets in the way". If desired, one can ignore all MATLAB-related content and use it as a traditional textbook with which to teach or learn signals and systems. The coverage of the theory is not cluttered with code segments. Instead, all MATLAB exercises, problems and projects are presented in their own sections at the end of each chapter, with references provided within the narrative of the chapter. Apart from the software use, the textbook includes 287 solved examples and 350 traditional end-of-chapter problems.

Organization of the Material

Chapter 1 deals with mathematical modeling of continuous-time and discrete-time signals. Basic building blocks for both continuous-time and discrete-time signals are presented as well as mathematical operations applied to signals. Classification methods for signals are discussed; definitions of signal energy and signal power are given. The idea of impulse decomposition of a signal is presented in preparation for the discussion of convolution operation in later chapters. The concept of a phasor is introduced for continuous-time sinusoidal signals. The chapter includes MATLAB exercises that focus on techniques for generating and graphing signals, developing functions for basic building blocks, and simulating signal operations.

In Chapter 2 the concept of a continuous-time system is introduced. Simplifying assumptions of linearity and time invariance that are useful in building a framework for analysis and design of systems are discussed. The chapter proceeds with the use of constant-coefficient linear differential equations for describing the behavior of linear systems. Rather than assuming prior background in solving differential equations or simply referring the student to math textbooks, differential equation solution methods are explained at a level sufficient for working with linear and time-invariant systems. If the students have already been exposed to a course on differential equations, corresponding sections of this chapter could be skipped or reviewed quickly. Representation of a differential equation by a block diagram is also briefly discussed. The concepts of impulse response and convolution are developed, and their link to the differential equation of the system is shown. Definitions of stability and causality as well as the time-domain conditions for a system to achieve them are given. MATLAB exercises are provided for testing linearity and time invariance of continuous-time systems and for obtaining approximate numerical solutions to differential equations.

Chapter 3 provides a treatment of time-domain analysis for discrete-time systems, and parallels the coverage of Chapter 2. After the introduction of linearity and time invariance from a discrete-time system perspective, it proceeds with the analysis of discrete-time systems by means of their difference equations. Solution methods for difference equations are summarized for the sake of completeness. Representation of a difference equation by a block diagram is discussed. Impulse response and convolution concepts are developed for discrete-time systems, and their relationship to the difference equation of the system is shown. Stability and causality concepts are also detailed for discrete-time systems. The chapter includes MATLAB exercises that focus on software implementation of discrete-time systems from difference equations or by using discrete-time convolution.

Chapter 4 is on Fourier analysis of continuous-time signals and systems. It begins with the analysis of periodic continuous-time signals in terms of their frequency content. First, the idea of finding the best approximation to a periodic signal through the use of a few trigonometric functions is explored in order to build intuition. Afterwards trigonometric, exponential and compact variants of the Fourier series are discussed. The Fourier transform for non-periodic signals is then introduced by generalizing the exponential Fourier series representation of a periodic signal with infinitely large period. A discussion of Parseval's theorem is provided leading to the concepts of energy and power spectral density for deterministic signals. System function concept is introduced for continuous-time systems. Response of linear and time-invariant systems to both periodic and non-periodic input signals is studied. The chapter includes a number of MATLAB exercises dealing with finite-harmonic approximations to periodic signals and the problem of graphing system functions.

The development of Chapter 5 mirrors that of Chapter 4 for discrete-time signals. Analysis of periodic discrete-time signals through the use of discrete-time Fourier series (DTFS) is presented. Afterwards the discrete-time Fourier transform (DTFT) is developed by generalizing the discrete-time Fourier series. The relationship between the DTFS coefficients of a periodic signal and the DTFT of a single isolated period of it is emphasized to highlight the link between the indices of the DTFS coefficients and the angular frequencies of the DTFT spectrum. Energy and power spectra concepts for deterministic discrete-time signals are discussed. DTFT-based system function is introduced. Response of linear and time-invariant systems to both periodic and non-periodic input signals is studied. An introduction to the discrete Fourier transform (DFT) is provided through a gentle transition from the DTFS. The chapter provides MATLAB exercises dealing with DTFS computations, periodic convolution, steady state response of linear time-invariant systems to discrete-time sinusoidal input signals, the relationship between DFT and DTFT, the use of the DFT for approximating the DTFT, circular time shifting and reversal, symmetry properties of the DFT, convolution via DFT, and use of the DFT in approximating continuous Fourier series and/or transform.

Chapter 6 is dedicated to the topic of sampling. First the concept of impulse sampling a continuous-time signal is explored. The relationship between the Fourier transforms of the original signal and its impulse-sampled version is explored, and the Nyquist sampling criterion is derived. DTFT of the discrete-time signal obtained through sampling is related to the Fourier transform of the original continuous-time signal. The aliasing phenomenon is explained through special emphasis given to the sampling of sinusoidal signals. Practical forms of sampling such as natural sampling and zero-order hold sampling are discussed. Reconstruction of a continuous signal from its discrete-time version by means of zero- and first-order hold as well as bandlimited interpolation is studied. The chapter concludes with a discussion of resampling of discrete-time signals and the concepts of decimation and interpolation. Several MATLAB exercises are provided with focus on the sampling of sinusoids, zero- and first-order sampling and reconstruction, spectral relationships in sampling, and resampling of discrete-time signals.

In Chapter 7 the Laplace transform is discussed as an analysis tool for continuous-time signals and systems. Its relationship to the continuous-time Fourier transform is explored. The bulk of the chapter focuses on the more general, bilateral, variant of the Laplace transform. The convergence characteristics of the transform as well as the region of convergence concept are introduced. Next the fundamental properties of the Laplace transform are detailed, and solved examples of utilizing those properties are given. The method of finding the inverse Laplace transform through the use of partial fraction expansion is explained. The chapter proceeds with application of the Laplace transform to the analysis of linear time-invariant systems. Special emphasis is placed on making the connections between the s-domain system function, the differential equation and the impulse response. Graphical interpretation of pole-zero diagrams for the purpose of determining the frequency response of CTLTI systems is illustrated. The chapter also provides a discussion of causality and stability concepts as they relate to the s-domain system function. Characteristics of allpass systems and inverse systems are outlined, and an introduction to Bode plots is given. Direct-form, cascade and parallel simulation structures for CTLTI systems are derived from the s-domain system function. Finally, the unilateral version of the Laplace transform is introduced as a practical tool for use in solving constant-coefficient linear differential equations with specified initial conditions. The chapter includes a number of MATLAB exercises that deal with three-dimensional visualization of the Laplace transform, computation of the Fourier transform from the Laplace transform, graphing pole-zero diagrams, residue calculations for partial fraction expansion, symbolic calculations for the Laplace transform, computation of the frequency response from pole-zero layout, the use of MATLAB's system objects, Bode plots, and solution of a differential equation using the unilateral Laplace transform.

Chapter 8 is dedicated to the study of the z-transform, and its structure is similar to that of Chapter 7 for Laplace transform. The z-transform is introduced as a more general version of the discrete-time Fourier transform (DTFT) studied in Chapter 5. The relationship between the z-transform and the DTFT is illustrated. Convergence characteristics of the z-transform are explained and the importance of the region of convergence concept is highlighted. Fundamental properties of the z-transform are detailed along with solved examples of their use. The problem of finding the inverse z-transform using the inversion integral, partial fraction expansion and long division is discussed. Examples are provided with detailed solutions in using the partial fraction expansion and long division to compute the inverse transform for all possible types of the region of convergence. Application of the z-transform to the analysis of linear time-invariant systems is treated in detail. Connections between the z-domain system function, the difference equation and the impulse response are explored to provide insight. Graphical interpretation of pole-zero diagrams for the purpose of determining the frequency response of DTLTI systems is illustrated. Causality and stability concepts are discussed from the perspective of the z-domain system function. Characteristics of allpass systems and inverse systems are outlined. Direct-form, cascade and parallel implementation structures for DTLTI systems are derived from the z-domain system function. The unilateral version of the z-transform is introduced as a practical tool for use in solving constant-coefficient linear difference equations with specified initial conditions. The chapter includes a number of MATLAB exercises that deal with three-dimensional visualization of the z-transform, computation of the DTFT from the z-transform, graphing pole-zero diagrams, relationship between convolution and polynomial multiplication, partial fraction expansion and long division, computation of the frequency response from pole-zero layout, calculations for the development of block diagrams, implementing DTLTI systems using cascaded second-order sections, and solution of a difference equation using the unilateral z-transform.

Chapter 9 provides an introductory treatment of state-space analysis of both continuous-time and discrete-time systems. Analysis of continuous-time systems is studied in Section 9.1, followed by similar analysis of discrete-time systems in Section 9.2. This structure allows appropriate material to be chosen based on the contents of a particular course. Within each of the first two sections the concept of state variables is developed by starting with a differential (or difference) equation and expressing it in terms of first order differential or difference equations. Methods of obtaining state space models from a physical description of the system or from a system function are also detailed as applicable. Solution of the state-space model is given for both continuous-time and discrete-time systems. Section 9.3 is dedicated to the problem of using state-space models for obtaining a discrete-time system that approximates the behavior of a continuous-time system. MATLAB exercises for this chapter deal with obtaining a state-space model from the system function, diagonalization of the state matrix, computation of the state transition matrix, solving the homogeneous state equation, symbolic computation of the state transition matrix, obtaining the system function from the state-space model, and discretization of the state-space model.

Chapter 10 focuses on the application of the material presented so far to the analysis and design of continuous-time and discrete-time filters. The discussion begins with an introduction to distortionless transmission and ideal filter behavior. An introductory treatment of analog filter design using Butterworth, Chebyshev and inverse Chebyshev approximation functions is given. While most introductory courses on signals and systems skip analog filter design all together, it is helpful for the student to have a basic understanding of the ideas behind analog filter design to appreciate the use of analog prototypes in discrete-time filters, and to understand the interplay of parameters such as filter order, critical frequencies, ripple and attenuation. The discussion of analog filters is followed by a discussion of design methods for IIR and FIR discrete-time filters. MATLAB exercises included in this chapter focus on using MATLAB for analog filter design using Butterworth and Chebyshev approximations, frequency transformations from lowpass to other filter types, IIR design using impulse invariance and bilinear transformation, and FIR filter design using either Fourier series method or Parks-McClellan method. Instead of using MATLAB's do-it-all "Filter Design and Analysis Tool", emphasis is placed on using the primitive functions for step-by step design of filters in a way that parallels the discussion in the chapter.

Chapter 11 provides an introductory treatment of another application of the material in the area of communication systems, namely amplitude modulation. It begins by introducing the idea of modulation and the need for it. Generation of amplitude modulated signals and their analysis in the frequency domain using the techniques developed in earlier chapters are discussed. Techniques for demodulation of amplitude-modulated signals are summarized. Two variants of amplitude modulation, namely double sideband suppressed carrier modulation and single sideband modulation, are briefly introduced. The chapter contains MATLAB exercises that cover computation and graphing of the AM signal, computation of the exponential Fourier series representation of the tone-modulated AM signal, simulation of the switching modulator and the square-law modulator, and simulation of the envelope detector.

A number of appendices are provided for complex numbers and Euler's formula, various mathematical relations, proofs of orthogonality for the basis functions used in Fourier analysis, partial fraction expansion, and a brief review of matrix algebra.

To the Instructor

The organization of the material allows the textbook to be used in a variety of circumstances and for a variety of needs.

- The material can be covered in its entirety in a two-semester sequence.

- A one-semester course on continuous-time signals and systems can be based on Sections 1 through 3 of Chapter 1 followed by Chapters 2, 4, 7. The sections of Chapter 2 that deal with solving linear constant coefficient differential equations may be skipped or quickly reviewed if the students had a prior course on differential equations. Selected material from Chapters 9 (Section 9.1), 10 (Sections 10.1 through 10.4) and 11 may be used based on time availability and curricular needs.

- A one-semester course on discrete-time signals and systems can be based on Section 4 of Chapter 1 followed by Chapters 3, 5, 6 and 8. Selected material from Chapters 9 (Sections 9.3 and 9.4), 10 (Section 10.5) and 11 may be used based on time availability and curricular needs.

Supplementary Materials

The following supplementary materials are available to the instructors who adopt the textbook for classroom use:

- A solutions manual in pdf format that contains solutions to the problems at the end of each chapter, including the MATLAB problems

- Presentation slides for lectures in pdf format

- Presentation slides in pdf format for the solutions of individual problems, suitable for use in lectures.

- Image files in encapsulated postscript format for the figures in the book

The following supplementary materials are available to all users of the textbook:

- A downloadable archive containing MATLAB code files for interactive GUI-based programs, exercises and examples in the textbook as well as the files needed for integrating them into MATLAB help browser

Downloadable materials are available from the publisher's website and from

www.signalsandsystems.org

Acknowledgments

I am indebted to many of my faculty colleagues who have reviewed parts of the manuscript and gave me ideas and suggestions for improvements. Their feedback has helped shape the contents of the book. I'm also grateful to my students in ECE 436 who have used parts of the material at various stages of development, and provided feedback.

The expertise of the project team at CRC press have been invaluable. I would like to thank Nora Konopka for being a wonderful editor to work with. Thanks are also due to Charlotte Byrnes and Jessica Vakili for their patience and encouragement throughout the project.

Professor Ronald C. Houts, my advisor from the days of graduate school, has developed and nurtured my early interest in the field of signal processing, and for that I'm grateful.

Finally, I would like to thank my sweetheart, Esin, for putting up with me and supporting me through the writing of this text.

Chapter 1

Signal Representation and Modeling

Chapter Objectives

- Understand the concept of a *signal* and how to work with mathematical models of signals.

- Discuss fundamental signal types and signal operations used in the study of signals and systems. Experiment with methods of simulating continuous- and discrete-time signals with MATLAB.

- Learn various ways of classifying signals and discuss symmetry properties.

- Explore characteristics of sinusoidal signals. Learn *phasor* representation of sinusoidal signals, and how phasors help with analysis.

- Understand the decomposition of signals using unit-impulse functions of appropriate type.

- Learn energy and power definitions.

1.1 Introduction

Signals are part of our daily lives. We work with a wide variety of signals on a day-to-day basis whether we realize it or not. When a commentator in a radio broadcast studio speaks into a microphone, the sound waves create changes in the air pressure around the microphone. If we happen to be in the same studio, membranes in our ears detect the changes in the air pressure, and we perceive them as sound. Pressure changes also cause vibrations on the membrane within the microphone. In response, the microphone produces a time-varying electrical voltage in such a way that the variations of the electrical voltage mimic the variations of air pressure in the studio. Thus, the microphone acts as a *transducer* which converts an *acoustic signal* to an *electrical signal*. This is just one step in the process of radio broadcast. The electrical signal produced by a microphone is modified and enhanced in a variety of ways, and its strength is boosted. It is then applied to a transmitting antenna which converts it to an *electromagnetic signal* that is suitable for transmission over the air for long distances. These electromagnetic waves fill the air space that we live in. An electromagnetic signal picked up by the antenna of a radio receiver is converted back to the form of an electrical voltage signal which is further processed within the radio receiver and sent to a loudspeaker. The loudspeaker converts this electrical voltage signal to vibrations which recreate the air pressure variations similar to those that started the whole process at the broadcast studio.

The broadcast example discussed above includes acoustic, electrical and electromagnetic signals. Examples of different physical signals can also be found around us. Time variations of quantities such as force, torque, speed and acceleration can be taken as examples of mechanical signals. For example, in an automobile, a tachometer measures the speed of the vehicle as a signal, and produces an electrical voltage the strength of which is proportional to the speed at each time instant. Subsequently, the electrical signal produced by the tachometer can be used in a speed control system to regulate the speed of the vehicle.

Consider a gray-scale photograph printed from a film negative obtained using an old film camera. Each point on the photograph has a shade of gray ranging from pure black to pure white. If we can represent shades of gray numerically, say with values ranging from 0.0 to 1.0, then the photograph can be taken as a signal. In this case, the strength of each point is not a function of time, but rather a function of horizontal and vertical distances from a reference point such as the bottom left corner of the photograph. If a digital camera is to be used to capture an image, the resulting signal is slightly different. The sensor of a digital camera is made up of photo-sensitive cells arranged in a rectangular grid pattern. Each cell measures the intensity of light it receives, and produces a voltage that is proportional to it. Thus, the signal that represents a digital image also consists of light intensity values as a function of horizontal and vertical distances. The difference is that, due to the finite number of cells on the sensor, only the distance values that are multiples of cell width and cell height are meaningful. Light intensity information only exists for certain values of the distance, and is undefined between them.

Examples of signals can also be found outside engineering disciplines. Financial markets, for example, rely on certain economic indicators for investment decisions. One such indicator that is widely used is the Dow Jones Industrial Average (DJIA) which is computed as a weighted average of the stock prices of 30 large publicly owned companies. Day-to-day variations of the DJIA can be used by investors in assessing the health of the economy and in making buying or selling decisions.

1.2 Mathematical Modeling of Signals

In the previous section we have discussed several examples of signals that we encounter on a day-to-day basis. These signals are in one or another physical form, some are electrical signals, some are in the form of air pressure variations over time, some correspond to time variations of a force or a torque in a mechanical system, and some represent light intensity and/or color as a function of horizontal and vertical distance from a reference point. There are other physical forms as well. In working with signals, we have one or both of the following goals in mind:

1. *Understand the characteristics of the signal in terms of its behavior in time and in terms of the frequencies it contains (signal analysis).*
   For example, consider the electrical signal produced by a microphone. Two people uttering the same word into the microphone do not necessarily produce the same exact signal. Would it be possible to recognize the speaker from his or her signal? Would it be possible to identify the spoken word?

2. *Develop methods of creating signals with desired characteristics (signal synthesis).*
   Consider an electronic keyboard that synthesizes sounds of different musical instruments such as the piano and the violin. Naturally, to create the sound of either instrument we need to understand the characteristics of the sounds these instruments make. Other examples are the synthesis of human speech from printed text by text-to-speech software and the synthesis of test signals in a laboratory for testing and calibrating specialized equipment.

In addition, signals are often used in conjunction with systems. A signal is applied to a system as *input*, and the system responds to the signal by producing another signal called the *output*. For example, the signal from an electric guitar can be applied (connected via cable) to a system that boosts its strength or enhances the sound acoustically. In problems that involve the interaction of a signal with a system we may have additional goals:

3. *Understand how a system responds to a signal and why (system analysis).*
   What characteristics of the input signal does the system retain from its input to its output? What characteristics of the input signal are modified by the system, and in what ways?

4. *Develop methods of constructing a system that responds to a signal in some prescribed way (system synthesis).*
   Understanding different characteristics of signals will also help us in the design of systems that affect signals in desired ways. An example is the design of a system to enhance the sound of a guitar and give it the "feel" of being played in a concert hall with certain acoustic properties. Another example is the automatic speed control system of an automobile that is required to accelerate or decelerate automobile to the desired speed within prescribed constraints.

In analyzing signals of different physical forms we need to develop a uniform framework. It certainly would not be practical to devise different analysis methods for different types of physical signals. Therefore, we work with *mathematical models* of physical signals.

Most of the examples utilized in our discussion so far involved signals in the form of a physical quantity varying over time. There are exceptions; some signals could be in the form of variations of a physical quantity over a spatial variable. Examples of this are the distribution of force along the length of a steel beam, or the distribution of light intensity at different points in an image. To simplify the discussion, we will focus our attention on signals that are functions of time. The techniques that will be developed are equally applicable to signals that use other independent variables as well.

The mathematical model for a signal is in the form of a formula, function, algorithm or a graph that approximately describes the time variations of the physical signal.

1.3   Continuous-Time Signals

Consider $x(t)$, a mathematical function of time chosen to approximate the strength of the physical quantity at the time instant t. In this relationship t is the independent variable, and x is the dependent variable. The signal $x(t)$ is referred to as a *continuous-time* signal or an *analog* signal.   An example of a continuous-time signal is shown in Fig. 1.1.

Figure 1.1 – A continuous-time signal.

Fig. 1.2 illustrates the mathematical model for a 25 ms segment of the voice signal that corresponds to the vowel "o" in the word "hello". Similarly, a 25 ms segment of the sound from a violin playing the note A3 is shown in Fig. 1.3.

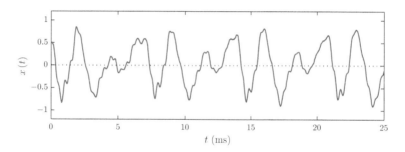

Figure 1.2 – A segment from the vowel "o" of the word "hello".

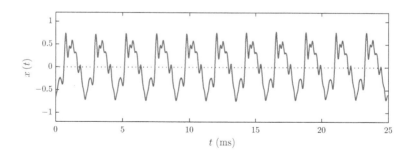

Figure 1.3 – A segment from the sound of a violin.

Some signals can be described analytically. For example, the function

$$x(t) = 5\sin(12t)$$

describes a sinusoidal signal. Sometimes it is convenient to describe a signal model by patching together multiple signal segments, each modeled with a different function.   An example of this is

$$x\left(t\right) = \begin{cases} e^{-3t} - e^{-6t}\ , & t \geq 0 \\ 0\ , & t < 0 \end{cases}$$

Software resources: See MATLAB Exercise 1.1.

1.3.1 Signal operations

In this section we will discuss time-domain operations that are commonly used in the course of working with continuous-time signals. Some are simple arithmetic operators. Examples of these are addition and multiplication of signals in the time domain, addition of a constant offset to a signal, and multiplication of a signal with a constant gain factor. Other signal operations that are useful in defining signal relationships are time shifting, time scaling and time reversal. More advanced signal operations such as convolution and sampling will be covered in their own right in later parts of this textbook.

Arithmetic operations

Addition of a constant offset A to the signal $x\left(t\right)$ is expressed as

$$g\left(t\right) = x\left(t\right) + A \tag{1.1}$$

At each time instant the amplitude of the result $g\left(t\right)$ is equal to the amplitude of the signal $x\left(t\right)$ plus the constant offset value A. A practical example of adding a constant offset to a signal is the removal of the dc component at the output of a class-A amplifier circuit through capacitive coupling. Fig. 1.4 illustrates the addition of positive and negative offset values to a signal.

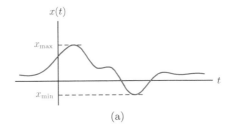

(a)

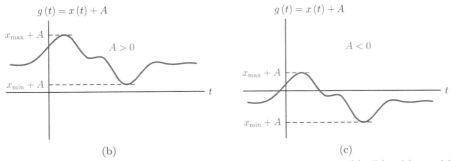

(b)   (c)

Figure 1.4 – Adding an offset A to signal $x\left(t\right)$: (a) Original signal $x\left(t\right)$, (b) $g\left(t\right) = x\left(t\right) + A$ with $A > 0$, (c) $g\left(t\right) = x\left(t\right) + A$ with $A < 0$.

A signal can also be multiplied with a constant gain factor:

$$g(t) = B\,x(t) \tag{1.2}$$

The result of this operation is a signal $g(t)$, the amplitude of which is equal to the product of the amplitude of the signal $x(t)$ and the constant gain factor B at each time instant. This is illustrated in Fig. 1.5.

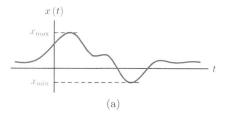

(a)

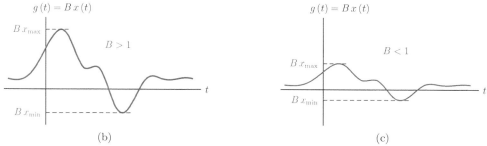

(b)   (c)

Figure 1.5 – Multiplying signal $x(t)$ with a constant gain factor: (a) Original signal $x(t)$, (b) $g(t) = B\,x(t)$ with $B > 1$, (c) $g(t) = B\,x(t)$ with $B < 1$.

A practical example of Eqn. (1.2) is a constant-gain amplifier used for amplifying the voltage signal produced by a microphone.

Addition of two signals is accomplished by adding the amplitudes of the two signals at each time instant. For two signals $x_1(t)$ and $x_2(t)$, the sum

$$g(t) = x_1(t) + x_2(t) \tag{1.3}$$

is computed at each time instant $t = t_1$ by adding the values $x_1(t_1)$ and $x_2(t_2)$. This is illustrated in Fig. 1.6.

Signal adders are useful in a wide variety of applications. As an example, in an audio recording studio, signals coming from several microphones can be added together to create a complex piece of music.

Multiplication of two signals is carried out in a similar manner. Given two signals $x_1(t)$ and $x_2(t)$, the product

$$g(t) = x_1(t)\,x_2(t) \tag{1.4}$$

is a signal, the amplitude of which at any time instant $t = t_1$ is equal to the product of the amplitudes of the signals $x_1(t)$ and $x_2(t)$ at the same time instant, that is, $g(t_1) = x_1(t_1)\,x_2(t_1)$. This is illustrated in Fig. 1.7.

One example of a practical application of multiplying two signals is in modulating the amplitude of a sinusoidal carrier signal with a message signal. This will be explored when we discuss amplitude modulation in Chapter 11.

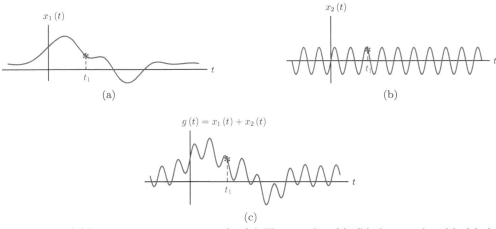

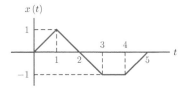

Figure 1.6 – Adding continuous-time signals: (a) The signal $x_1(t)$, (b) the signal $x_2(t)$, (c) the sum signal $g(t) = x_1(t) + x_2(t)$.

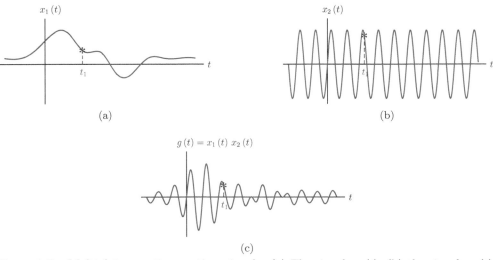

Figure 1.7 – Multiplying continuous-time signals: (a) The signal $x_1(t)$, (b) the signal $x_2(t)$, (c) the product signal $g(t) = x_1(t)\,x_2(t)$.

Example 1.1: **Constant offset and gain**

Consider the signal shown in Fig. 1.8. Sketch the signals

a. $g_1(t) = 1.5\,x(t) - 1$
b. $g_2(t) = -1.3\,x(t) + 1$

Figure 1.8 – The signal $x(t)$ for Example 1.1.

Solution:

a. Multiplying the signal with a constant gain factor of 1.5 causes all signal amplitudes to be scaled by 1.5. The intermediate signal $g_{1a}(t) = 1.5\,x(t)$ is shown in Fig. 1.9(a). In the next step, the signal $g_1(t) = g_{1a}(t) - 1$ is obtained by lowering the graph of the the signal $g_{1a}(t)$ by 1 at all time instants. The result is shown in Fig. 1.9(b).

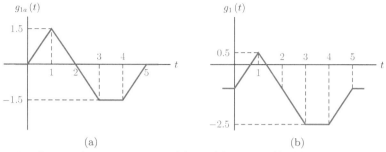

(a)   (b)

Figure 1.9 – Signals for Example 1.1: (a) $g_{1a}(t) = 1.5\,x(t)$, (b) $g_1(t) = 1.5\,x(t) - 1$.

b. We will proceed in a similar fashion. In this case the gain factor is negative, so the signal amplitudes will exhibit sign change. The intermediate signal $g_{2a}(t) = -1.3\,x(t)$ is shown in Fig. 1.10(a). In the next step, the signal $g_2(t) = g_{2a}(t) + 1$ is obtained by raising the graph of the the signal $g_{2a}(t)$ by 1 at all time instants. The result is shown in Fig. 1.10(b).

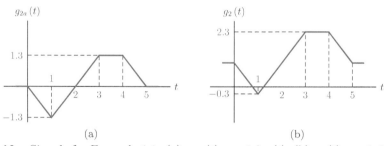

(a)   (b)

Figure 1.10 – Signals for Example 1.1: (a) $g_{2a}(t) = -1.3\,x(t)$, (b) $g_2(t) = -1,3\,x(t) + 1$.

Software resources: See MATLAB Exercise 1.2.

Interactive Demo: `sop_demo1`

The demo program "`sop_demo1.m`" is based on Example 1.1. It allows experimentation with elementary signal operations involving constant offset and constant gain factor. Specifically, the signal $x(t)$ shown in Fig. 1.8 is used as the basis of the demo program, and the signal

$$g(t) = B\,x(t) + A$$

is computed and graphed. The parameters A and B can be adjusted through the use of slider controls.

Software resources:
`sop_demo1.m`

Example 1.2: **Arithmetic operations with continuous-time signals**

Two signals $x_1(t)$ and $x_2(t)$ are shown in Fig. 1.11. Sketch the signals

a. $g_1(t) = x_1(t) + x_2(t)$
b. $g_2(t) = x_1(t) \, x_2(t)$

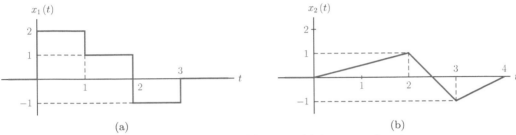

(a)   (b)

Figure 1.11 – Signals $x_1(t)$ and $x_2(t)$ for Example 1.2.

Solution: The following analytical descriptions of the signals $x_1(t)$ and $x_2(t)$ can be inferred from the graphical descriptions given in Fig. 1.11:

$$x_1(t) = \begin{cases} 2, & 0 < t < 1 \\ 1, & 1 < t < 2 \\ -1, & 2 < t < 3 \\ 0, & \text{otherwise} \end{cases} \quad \text{and} \quad x_2(t) = \begin{cases} \frac{1}{2}t, & 0 < t < 2 \\ -2t + 5, & 2 < t < 3 \\ t - 4, & 3 < t < 4 \\ 0, & \text{otherwise} \end{cases}$$

a. The addition of the two signals can be accomplished by adding the values in each segment separately. The analytical form of the sum is

$$g_1(t) = \begin{cases} \frac{1}{2}t + 2, & 0 < t < 1 \\ \frac{1}{2}t + 1, & 1 < t < 2 \\ -2t + 4, & 2 < t < 3 \\ t - 4, & 3 < t < 4 \\ 0, & \text{otherwise} \end{cases} \tag{1.5}$$

and is shown graphically in Fig. 1.12(a).

b. The product of the two signals can also computed for each segment, and is obtained as

$$g_2(t) = \begin{cases} t, & 0 < t < 1 \\ \frac{1}{2}t, & 1 < t < 2 \\ 2t - 5, & 2 < t < 3 \\ 0, & \text{otherwise} \end{cases} \tag{1.6}$$

This result is shown in Fig. 1.12(b).

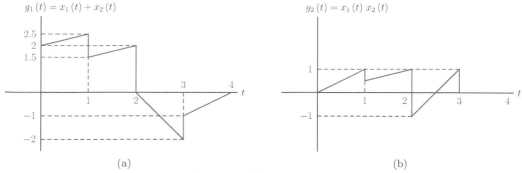

(a)   (b)
Figure 1.12 – Signals $g_1(t)$ and $g_2(t)$ for Example 1.2.

Software resources:
ex_1_2.m

Time shifting

A *time shifted* version of the signal $x(t)$ can be obtained through

$$g(t) = x(t - t_d) \tag{1.7}$$

where t_d is any positive or negative constant. Fig. 1.13 illustrates the relationship described by Eqn. (1.7). In part (a) of Fig. 1.13 the amplitude of $x(t)$ at the time instant $t = t_1$ is marked with a star-shaped marker. Let that marker represent a special event that takes place within the signal $x(t)$. Substituting $t = t_1 + t_d$ in Eqn. (1.7) we obtain

$$g(t_1 + t_d) = x(t_1) \tag{1.8}$$

Thus, in the signal $g(t)$, the same event takes place at the time instant $t = t_1 + t_d$. If t_d is positive, $g(t)$ is a delayed version of $x(t)$, and t_d is the amount of time delay. A negative t_d, on the other hand, corresponds to advancing the signal in time by an amount equal to $-t_d$.

Time scaling

A *time scaled* version of the signal $x(t)$ is obtained through the relationship

$$g(t) = x(at) \tag{1.9}$$

where a is a positive constant. Depending on the value of a, the signal $g(t)$ is either a compressed or an expanded version of $x(t)$. Fig. 1.14 illustrates this relationship. We will use a logic similar to the one employed in explaining the time shifting behavior: In part (a) of Fig. 1.14, let the star-shaped marker at the time instant $t = t_1$ represent a special event in the signal $x(t)$. Substituting $t = t_1/a$ in Eqn. (1.9) we obtain

$$g\left(\frac{t_1}{a}\right) = x(t_1) \tag{1.10}$$

The corresponding event in $g(t)$ takes place at the time instant $t = t_1/a$. Thus, a scaling parameter value of $a > 1$ results in the signal $g(t)$ being a compressed version of $x(t)$. Conversely, $a < 1$ leads to a signal $g(t)$ that is an expanded version of $x(t)$.

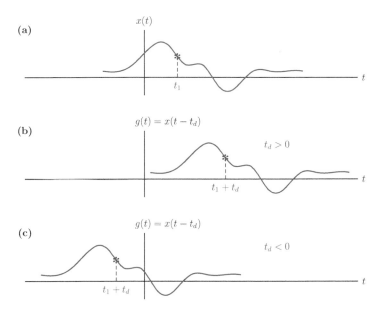

Figure 1.13 – Time shifting a signal: (a) Original signal $x(t)$, (b) time shifted signal $g(t)$ for $t_d > 0$, (c) time shifted signal $g(t)$ for $t_d < 0$.

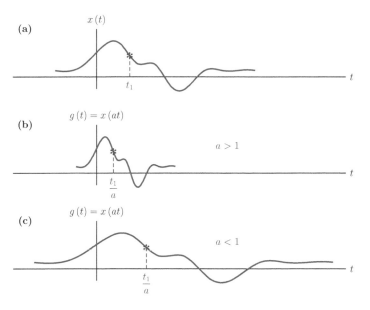

Figure 1.14 – Time scaling a signal: (a) Original signal $x(t)$, (b) time scaled signal $g(t)$ for $a > 0$, (c) time scaled signal $g(t)$ for $a < 0$.

Time reversal

A *time reversed* version of the signal $x(t)$ is obtained through

$$g(t) = x(-t) \tag{1.11}$$

which is illustrated in Fig. 1.15.

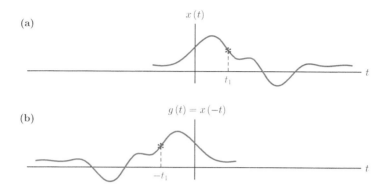

Figure 1.15 – Time reversing a signal: (a) Original signal $x(t)$, (b) time reversed signal $g(t)$.

An event that takes place at the time instant $t = t_1$ in the signal $x(t)$ is duplicated at the time instant $t = -t_1$ in the signal $g(t)$. Graphically this corresponds to folding or flipping the signal around the vertical axis.

Interactive Demo: sop_demo2

The demo program "sop_demo2.m" is based on the discussion above, and allows experimentation with elementary signal operations, namely shifting, scaling and reversal of a signal as well as addition and multiplication of two signals. The desired signal operation is selected from the drop-down list control. When applicable, slider control for the time delay parameter t_d or the time scaling parameter a becomes available to facilitate adjustment of the corresponding value. Also, the star-shaped marker used in Figs. 1.13 through 1.15 is also shown on the graphs for $x(t)$ and $g(t)$ when it is relevant to the signal operation chosen.
Software resources:
sop_demo2.m

Example 1.3: **Basic operations for continuous-time signals**

Consider the signal $x(t)$ shown in Fig. 1.16. Sketch the following signals:

  a. $g(t) = x(2t - 5)$,
  b. $h(t) = x(-4t + 2)$.

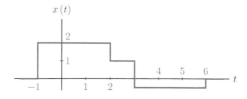

Figure 1.16 – The signal $x(t)$ for Example 1.3.

Solution:

a. We will obtain $g(t)$ in two steps: Let an intermediate signal be defined as $g_1(t) = x(2t)$, a time scaled version of $x(t)$, shown in Fig. 1.17(b). The signal $g(t)$ can be expressed as

$$g(t) = g_1(t - 2.5) = x(2[t - 2.5]) = x(2t - 5)$$

and is shown in Fig. 1.17(c).

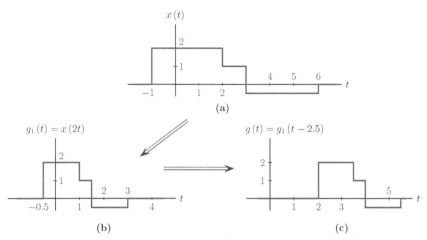

Figure 1.17 – (a) The intermediate signal $g_1(t)$, and (b) the signal $g(t)$ for Example 1.3.

b. In this case we will use two intermediate signals: Let $h_1(t) = x(4t)$. A second intermediate signal $h_2(t)$ can be obtained by time shifting $h_1(t)$ so that

$$h_2(t) = h_1(t + 0.5) = x(4[t + 0.5]) = x(4t + 2)$$

Finally, $h(t)$ can be obtained through time reversal of $h_2(t)$:

$$h(t) = h_2(-t) = x(-4t + 2)$$

The steps involved in sketching $h(t)$ are shown in Fig. 1.18(a)–(d).

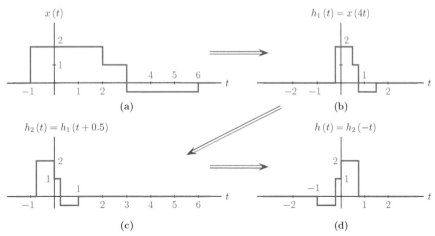

Figure 1.18 – (a) The intermediate signal $h_1(t)$, (b) the intermediate signal $h_2(t)$, and (c) the signal $h(t)$ for Example 1.3.

Integration and differentiation

Integration and differentiation operations are used extensively in the study of linear systems. Given a continuous-time signal $x(t)$, a new signal $g(t)$ may be defined as its time derivative in the form

$$g(t) = \frac{dx(t)}{dt} \tag{1.12}$$

A practical example of this is the relationship between the current $i_C(t)$ and the voltage $v_C(t)$ of an ideal capacitor with capacitance C as given by

$$i_C(t) = C\frac{dv_C(t)}{dt} \tag{1.13}$$

For example, if the voltage in the shape of a periodic triangular waveform shown in Fig. 1.19(a) is applied to an ideal capacitor, the current that flows through the capacitor would be proportional to its derivative, resulting in a square-wave signal as shown in Fig. 1.19(b).

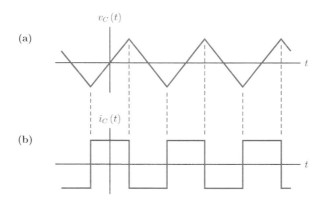

Figure 1.19 – (a) A periodic triangular waveform $v_C(t)$ used as the voltage of a capacitor, and (b) the periodic square-wave current signal $i_C(t)$ that results.

Similarly, a signal can be defined as the integral of another signal in the form

$$g(t) = \int_{-\infty}^{t} x(\lambda)\, d\lambda \tag{1.14}$$

The relationship between the current $i_L(t)$ and the voltage $v_L(t)$ of an ideal inductor can serve as an example of this. Specifically we have

$$i_L(t) = \frac{1}{L}\int_{-\infty}^{t} v_L(\lambda)\, d\lambda \tag{1.15}$$

If the voltage in the shape of a periodic square-wave shown in Fig. 1.20(a) is applied to an ideal inductor, the current that flows through it would be proportional to its integral, resulting in a periodic triangular waveform as shown in Fig. 1.20(b).

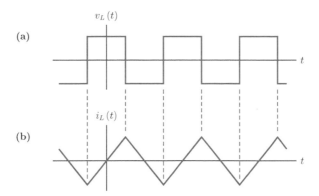

Figure 1.20 – (a) A periodic square-wave signal $v_L(t)$ used as the voltage of an inductor, and (b) the periodic triangular current signal $i_L(t)$ that results.

1.3.2   Basic building blocks for continuous-time signals

There are certain basic signal forms that can be used as building blocks for describing signals with higher complexity. In this section we will study some of these signals. Mathematical models for more advanced signals can be developed by combining these basic building blocks through the use of the signal operations described in Section 1.3.1.

Unit-impulse function

We will see in the rest of this chapter as well as the chapters to follow that the unit-impulse function plays an important role in mathematical modeling and analysis of signals and linear systems. It is defined by the following equations:

$$\delta(t) = \begin{cases} 0\,, & \text{if } t \neq 0 \\ \text{undefined}\,, & \text{if } t = 0 \end{cases} \tag{1.16}$$

and

$$\int_{-\infty}^{\infty} \delta(t)\, dt = 1 \tag{1.17}$$

Note that Eqn. (1.16) by itself represents an incomplete definition of the function $\delta(t)$ since the amplitude of it is defined only when $t \neq 0$, and is undefined at the time instant $t = 0$. Eqn. (1.17) fills this void by defining the area under the function $\delta(t)$ to be unity. The impulse function cannot be graphed using the common practice of graphing amplitude as a function of time, since the only amplitude of interest occurs at a single time instant $t = 0$, and it has an undefined value. Instead, we use an arrow to indicate the location of that undefined amplitude, as shown in Fig. 1.21.

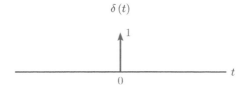

Figure 1.21 – Unit-impulse function.

In a number of problems in this text we will utilize shifted and scaled versions of impulse function. The function $a\,\delta\,(t - t_1)$ represents a unit-impulse function that is time shifted by t_1 and scaled by constant scale factor a. It is described through the equations

$$a\,\delta\,(t - t_1) = \begin{cases} 0\,, & \text{if } t \neq t_1 \\ \text{undefined}\,, & \text{if } t = t_1 \end{cases} \tag{1.18}$$

and

$$\int_{-\infty}^{\infty} a\,\delta\,(t - t_1)\,dt = a \tag{1.19}$$

and is shown in Fig. 1.22.

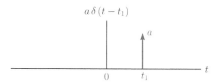

Figure 1.22 – Scaled and time shifted impulse function.

In Figs. 1.21 and 1.22 the value displayed next to the up arrow is not an amplitude value. Rather, it represents the area of the impulse function.

In Eqns. (1.17) and (1.19) the integration limits do not need to be infinite; any set of integration limits that includes the impulse would produce the same result. With any time increment $\Delta t > 0$ Eqn. (1.17) can be written in the alternative form

$$\int_{-\Delta t}^{\Delta t} \delta\,(t)\,dt = 1 \tag{1.20}$$

and Eqn. (1.19) can be written in the alternative form

$$\int_{t_1 - \Delta t}^{t_1 + \Delta t} a\,\delta\,(t - t_1)\,dt = a \tag{1.21}$$

How can a function that has zero amplitude everywhere except at one point have a non-zero area under it? One way of visualizing how this might work is to start with a rectangular pulse centered at the origin, with a duration of a and an amplitude of $1/a$ as shown in Fig. 1.23(a). Mathematically, the pulse $q\,(t)$ is defined as

$$q\,(t) = \begin{cases} \dfrac{1}{a}\,, & |t| < \dfrac{a}{2} \\ 0\,, & |t| > \dfrac{a}{2} \end{cases} \tag{1.22}$$

The area under the pulse is clearly equal to unity independent of the value of a. Now imagine the parameter a getting smaller, resulting in the pulse becoming narrower. As the pulse becomes narrower, it also becomes taller as shown in Fig. 1.23(b). The area under the pulse remains unity. In the limit, as the parameter a approaches zero, the pulse approaches an impulse, that is

$$\delta\,(t) = \lim_{a \to 0}\,[q\,(t)] \tag{1.23}$$

This is illustrated in Fig. 1.23(c).

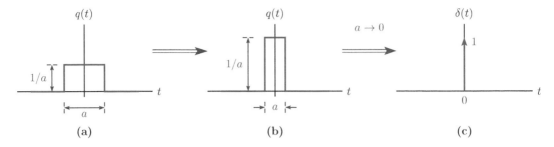

Figure 1.23 – Obtaining a unit impulse as the limit of a pulse with unit area. (a) The pulse $q(t)$, (b) a narrower and taller version obtained by reducing a, (c) unit-impulse function obtained as the limit.

It is also possible to obtain a unit impulse as the limit of other functions besides a rectangular pulse. Another example of a function that approaches a unit impulse in the limit is the *Gaussian* function. Consider the limit

$$\delta(t) = \lim_{a \to 0} \left[\frac{1}{a\sqrt{\pi}} e^{-t^2/a^2} \right] \qquad (1.24)$$

The Gaussian function inside the limit has unit area, and becomes narrower as the parameter a becomes smaller. This is illustrated in Fig. 1.24.

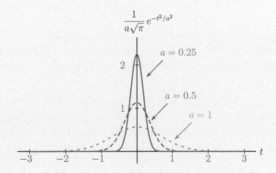

Figure 1.24 – Obtaining a unit impulse as the limit of a Gaussian pulse.

A third function that yields a unit impulse in the limit is a *squared sinc pulse*. It is also possible to define a unit impulse as the limit case of the squared sinc pulse:

$$\delta(t) = \lim_{a \to 0} \left[\frac{a}{(\pi t)^2} \sin^2 \left(\frac{\pi t}{a} \right) \right] \qquad (1.25)$$

The convergence of this limit to a unit impulse is illustrated in Fig. 1.25.

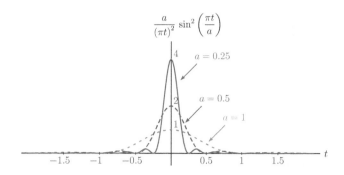

Figure 1.25 – Obtaining a unit impulse as the limit of a squared sinc pulse.

Interactive Demo: imp_demo

The demo program "imp_demo.m" illustrates the construction of a unit-impulse function as the limit of functions with unit area, namely a rectangular pulse, a Gaussian pulse and a squared sinc pulse. Refer to Eqns. (1.23), (1.24) and (1.25) as well as the corresponding Figs. 1.23, 1.24 and 1.25 for details.

The parameter a used in changing the shape of each of these functions can be varied by using a slider control. Observe how each function exhibits increased similarity to the unit impulse as the value of a is reduced.

Software resources:

imp_demo.m

The impulse function has two fundamental properties that are useful. The first one, referred to as the *sampling property*[1] of the impulse function, is stated as

$$f(t)\,\delta(t - t_1) = f(t_1)\,\delta(t - t_1) \tag{1.26}$$

where $f(t)$ is any function of time that is continuous at $t = t_1$. Intuitively, since $\delta(t - t_1)$ is equal to zero for all $t \neq t_1$, the only value of the function $f(t)$ that should have any significance in terms of the product $f(t)\,\delta(t - t_1)$ is its value at $t = t_1$. The sampling property of the unit-impulse function is illustrated in Fig. 1.26.

A consequence of Eqn. (1.26) is the *sifting property* of the impulse function which is expressed as

$$\int_{-\infty}^{\infty} f(t)\,\delta(t - t_1)\,dt = f(t_1) \tag{1.27}$$

The integral of the product of a function $f(t)$ and a time shifted unit-impulse function is equal to the value of $f(t)$ evaluated at the location of the unit impulse. The sifting property expressed by Eqn. (1.27) is easy to justify through the use of the sampling property.

---

[1]   A thorough mathematical study of singularity functions such as the unit-impulse function is quite advanced, and is well beyond the scope of this text. From a theoretical standpoint, one could set up conditions under which the sampling property given by Eqn. (1.26) may not hold. However, it will be valid and useful for our purposes in formulating signal models for engineering applications.

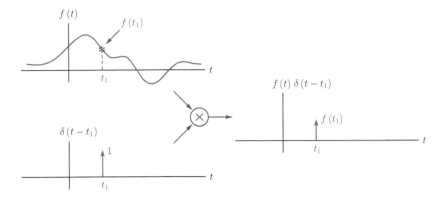

Figure 1.26 – Illustration of the sampling property of the unit-impulse function.

Substituting Eqn. (1.26) into Eqn. (1.27) we obtain

$$\int_{-\infty}^{\infty} f(t)\,\delta(t-t_1)\,dt = \int_{-\infty}^{\infty} f(t_1)\,\delta(t-t_1)\,dt$$

$$= f(t_1) \int_{-\infty}^{\infty} \delta(t-t_1)\,dt$$

$$= f(t_1) \tag{1.28}$$

As discussed above, the limits of the integral in Eqn. (1.27) do not need to be infinitely large for the result to be valid; any integration interval that includes the impulse will work. Given a time increment $\Delta t > 0$, the following is also valid:

$$\int_{t_1-\Delta t}^{t_1+\Delta t} f(t)\,\delta(t-t_1)\,dt = f(t_1) \tag{1.29}$$

It should be clear from the foregoing discussion that the unit-impulse function is a mathematical idealization. No physical signal is capable of occurring without taking any time and still providing a non-zero result when integrated. Therefore, no physical signal that utilizes the unit-impulse function as a mathematical model can be obtained in practice. This realization naturally prompts the question *"why do we need the unit-impulse signal?"* or the question *"how is the unit-impulse signal useful?"* The answer to these questions lies in the fact that the unit-impulse function is a fundamental building block that is useful in approximating events that occur quickly. The use of the unit-impulse function allows us to look at a signal as a series of events that occur momentarily.

Consider the sifting property of the impulse function which was described by Eqn. (1.27). Through the use of the sifting property, we can focus on the behavior of a signal at one specific instant in time. By using the sifting property repeatedly at every imaginable time instant we can analyze a signal in terms of its behavior as a function of time. This idea will become clearer when we discuss *impulse decomposition of a signal* in the next section which will later become the basis of the development leading to the convolution operation in Chapter 2. Even though the impulse signal is not obtainable in a practical sense, it can be approximated using a number of functions; the three signals we have considered in Figs. 1.23, 1.24 and 1.25 are possible candidates when the parameter a is reasonably small.

Unit-step function

The unit-step function is useful in situations where we need to model a signal that is turned on or off at a specific time instant. It is defined as follows:

$$u(t) = \begin{cases} 1, & t > 0 \\ 0, & t < 0 \end{cases} \tag{1.30}$$

The function $u(t)$ is illustrated in Fig. 1.27.

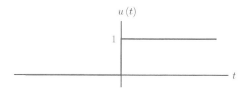

Figure 1.27 – Unit-step function.

A time shifted version of the unit-step function can be written as

$$u(t - t_1) = \begin{cases} 1, & t > t_1 \\ 0, & t < t_1 \end{cases} \tag{1.31}$$

and is shown in Fig. 1.28.

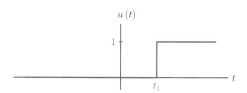

Figure 1.28 – Time shifted unit-step function.

Consider a sinusoidal signal given by

$$x(t) = \sin(2\pi f_0 t) \tag{1.32}$$

which oscillates for all values of t. If we need to represent a sinusoidal signal that is switched on at time $t = 0$, we can use the unit-step function with Eqn. (1.32), and write

$$x(t) = \sin(2\pi f_0 t)\, u(t) = \begin{cases} \sin(2\pi f_0 t), & t > 0 \\ 0, & t < 0 \end{cases}$$

Alternatively, the signal can be switched on at an arbitrary time instant $t = t_1$ through the use of a time shifted unit-step function:

$$x(t) = \sin(2\pi f_0 t)\, u(t - t_1) = \begin{cases} \sin(2\pi f_0 t), & t > t_1 \\ 0, & t < t_1 \end{cases}$$

Fig. 1.29 illustrates this use of the unit-step function for switching a signal on at a specified time instant.

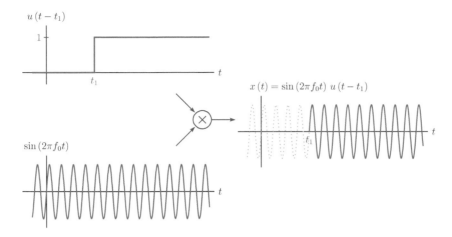

Figure 1.29 – The use of the unit-step function for switching a signal on at a specified time.

Interactive Demo: stp_demo1

This demo program illustrates the use of the unit-step function for switching a signal on at a specified time. It is based on Fig. 1.29. The delay parameter t_1 for the unit-step function can be varied by using the slider control.
Software resources:
stp_demo1.m

Conversely, suppose the sinusoidal signal has been on for a long time, and we would like to turn it off at time instant $t = t_1$. This could be accomplished by multiplication of the sinusoidal signal with a time reversed and shifted unit-step function in the form

$$x\left(t\right) = \sin\left(2\pi f_0 t\right) u\left(-t + t_1\right) = \begin{cases} \sin\left(2\pi f_0 t\right), & t < t_1 \\ 0, & t > t_1 \end{cases}$$

and is illustrated in Fig. 1.30.

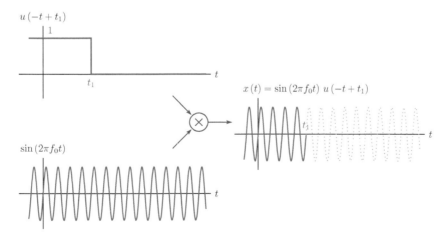

Figure 1.30 – The use of the unit-step function for switching a signal off at a specified time.

This demo program illustrates the use of the unit-step function for switching a signal off at a specified time. It is based on Fig. 1.30. As in "stp_demo1.m", the delay parameter t_1 for the unit-step function can be varied by using the slider control.

Software resources:

stp_demo2.m

The relationship between the unit-step function and the unit-impulse function is important. The unit-step function can be expressed as a *running integral* of the unit-impulse function:

$$u(t) = \int_{-\infty}^{t} \delta(\lambda) \, d\lambda \tag{1.33}$$

In Eqn. (1.33) we have changed the independent variable to λ to avoid any confusion with the upper limit t of integration. Recall that the integral of the unit-impulse function equals unity provided that the impulse is within the integration limits. This is indeed the case in Eqn. (1.33) if the upper limit of the integral is positive, that is, if $t = t_0 > 0$. Conversely, if $t = t_0 < 0$, then the integration interval stops short of the location of the impulse, and the result of the integral is zero. This is illustrated in Fig. 1.31.

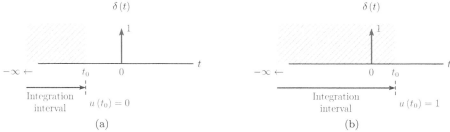

Figure 1.31 – Obtaining a unit step through the running integral of the unit impulse: (a) $t = t_0 < 0$, (b) $t = t_0 > 0$.

In Fig. 1.23 and Eqns. (1.22) and (1.23) we have represented the unit-impulse function as the limit of a rectangular pulse with unit area. Fig. 1.32 illustrates the running integral of such a pulse. In the limit, the result of the running integral approaches the unit-step function as $a \to 0$.

Figure 1.32 – (a) Rectangular pulse approximation to a unit impulse, (b) running integral of the rectangular pulse as an approximation to a unit step.

The unit-impulse function can be written as the derivative of the unit-step function, that is,

$$\delta(t) = \frac{du}{dt} \tag{1.34}$$

consistent with Eqn. (1.33).

Unit-pulse function

We will define the unit-pulse function as a rectangular pulse with unit width and unit amplitude, centered around the origin. Mathematically, it can be expressed as

$$\Pi(t) = \begin{cases} 1, & |t| < \frac{1}{2} \\ 0, & |t| > \frac{1}{2} \end{cases} \tag{1.35}$$

and is shown in Fig. 1.33.

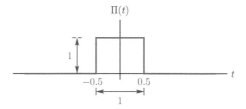

Figure 1.33 – Unit-pulse function.

The unit-pulse function can be expressed in terms of time shifted unit-step functions as

$$\Pi(t) = u\left(t + \tfrac{1}{2}\right) - u\left(t - \tfrac{1}{2}\right) \tag{1.36}$$

The steps in constructing a unit-pulse signal from unit-step functions are illustrated in Fig. 1.34. It is also possible to construct a unit-pulse signal using unit-impulse functions. Let

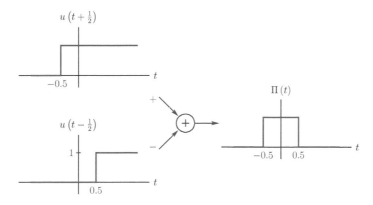

Figure 1.34 – Constructing a unit-pulse signal using unit-step functions.

us substitute Eqn. (1.33) into Eqn. (1.36) to obtain

$$\Pi\left(t\right) = \int_{-\infty}^{t+1/2} \delta\left(\lambda\right) d\lambda - \int_{-\infty}^{t-1/2} \delta\left(\lambda\right) d\lambda$$

$$= \int_{t-1/2}^{t+1/2} \delta\left(\lambda\right) d\lambda \tag{1.37}$$

The result in Eqn. (1.37) can be interpreted as follows: For the result of the integral to be equal to unity, the impulse needs to appear between the integration limits. This requires that the lower limit be negative and the upper limit be positive. Thus we have

$$\int_{t-1/2}^{t+1/2} \delta\left(\lambda\right) d\lambda = \begin{cases} 1, & \text{if} \quad t - \frac{1}{2} < 0 \text{ and } t + \frac{1}{2} > 0 \\ 0, & \text{otherwise} \end{cases}$$

$$= \begin{cases} 1, & -\frac{1}{2} < t < \frac{1}{2} \\ 0, & \text{otherwise} \end{cases} \tag{1.38}$$

which matches the definition of the unit-pulse signal given in Eqn. (1.35). A graphical illustration of Eqn. (1.37) is given in Fig. 1.35.

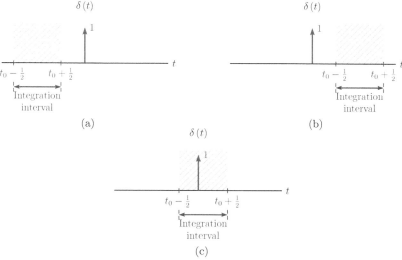

Figure 1.35 – Constructing a unit pulse by integrating a unit impulse: (a) $t = t_0 < -1/2$, (b) $t = t_0 > 1/2$, (c) $-1/2 < t < 1/2$.

Unit-ramp function

The unit-ramp function is defined as

$$r\left(t\right) = \begin{cases} t, & t \geq 0 \\ 0, & t < 0 \end{cases} \tag{1.39}$$

It has zero amplitude for $t < 0$, and unit slope for $t \geq 0$. This behavior is illustrated in Fig. 1.36.

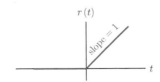

Figure 1.36 – Unit-ramp function.

An equivalent definition of the unit-ramp function can be written using the product of a linear signal $g(t) = t$ and the unit-step function as

$$r(t) = t\,u(t) \tag{1.40}$$

as illustrated in Fig. 1.37.

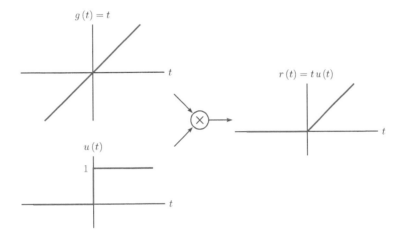

Figure 1.37 – Constructing a unit-ramp function.

Alternatively, the unit-ramp function may be recognized as the integral of the unit-step function:

$$r(t) = \int_{-\infty}^{t} u(\lambda)\, d\lambda \tag{1.41}$$

as shown in Fig. 1.38.

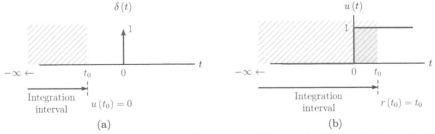

Figure 1.38 – Obtaining a unit-ramp function as the running integral of a unit step.

We will find time shifted and time scaled versions of the unit-ramp function useful in constructing signals that have piecewise linear segments.

Unit-triangle function

The unit-triangle function is defined as

$$\Lambda\left(t\right) = \begin{cases} t+1, & -1 \le t < 0 \\ -t+1, & 0 \le t < 1 \\ 0, & \text{otherwise} \end{cases} \tag{1.42}$$

and is shown in Fig. 1.39.

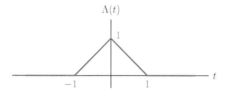

Figure 1.39 – Unit-triangle function.

Unit-triangle function can be expressed in terms of shifted and scaled unit-ramp functions as

$$\Lambda\left(t\right) = r\left(t+1\right) - 2\,r\left(t\right) + r\left(t-1\right) \tag{1.43}$$

In order to understand Eqn. (1.43) we will observe the slope of the unit-triangle waveform in each segment. The first term $r\left(t+1\right)$ has a slope of 0 for $t < -1$, and a slope of 1 for $t > -1$. The addition of the term $-2r\left(t\right)$ changes the slope to -1 for $t > 0$. The last term, $r\left(t-1\right)$, makes the slope 0 for $t > 1$. The process of constructing a unit triangle from shifted and scaled unit-ramp functions is illustrated in Fig. 1.40.

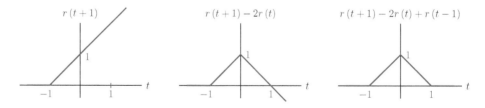

Figure 1.40 – Construction of a unit triangle using unit-ramp functions.

Interactive Demo: `wav_demo1`

The waveform explorer program "`wav_demo1.m.m`" allows experimentation with the basic signal building blocks discussed in Section 1.3.2. The signal $x\left(t\right)$ that is graphed is constructed in the form

$$x\left(t\right) = x_1\left(t\right) + x_2\left(t\right) + x_3\left(t\right) + x_4\left(t\right) + x_5\left(t\right)$$

Each term $x_i\left(t\right),\ \ i = 1,\ldots,5$ in Eqn. (1.44) can be a scaled and time shifted step, ramp, pulse or triangle function. Additionally, step and ramp functions can be time reversed;

pulse and triangle functions can be time scaled. If fewer than five terms are needed for $x(t)$, unneeded terms can be turned off.
Software resources:
`wav_demo1.m`

Sinusoidal signals

The general form of a sinusoidal signal is

$$x(t) = A\cos(\omega_0 t + \theta) \tag{1.44}$$

where A is the *amplitude* of the signal, and ω_0 is the *radian frequency* which has the unit of *radians per second*, abbreviated as rad/s. The parameter θ is the *initial phase angle* in radians. The radian frequency ω_0 can be expressed as

$$\omega_0 = 2\pi f_0 \tag{1.45}$$

where f_0 is the *frequency* in Hz. This signal is shown in Fig. 1.41.

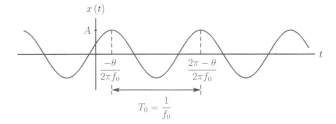

Figure 1.41 – The sinusoidal signal $x(t) = A\cos(2\pi f_0 t + \theta)$.

The amplitude parameter A controls the peak value of the signal. The phase θ affects the locations of the peaks:

- If $\theta = 0$, the cosine waveform has a peak at the origin $t_0 = 0$. Other positive or negative peaks of the waveform would be at time instants that are solutions the equation
$$2\pi f_0 t_k = k\pi, \quad k: \text{integer} \tag{1.46}$$
which leads to $t_k = k/(2f_0)$.

- With a different value of θ, the first peak is shifted to $t_0 = -\theta/(2\pi f_0)$, the time instant that satisfies the equation $2\pi f_0 t_0 + \theta = 0$. Other peaks of the waveform are shifted to time instants $t_k = (k\pi - \theta)/(2\pi f_0)$ which are solutions to the equation
$$2\pi f_0 t_k + \theta = k\pi, \quad k: \text{integer} \tag{1.47}$$

Finally, the frequency f_0 controls the repetition of the waveform. Let's refer to the part of the signal from one positive peak to the next as one *cycle* as shown in Fig. 1.41. When expressed in Hz, f_0 is the number cycles (or periods) of the signal per second. Consequently, each cycle or period spans $T_0 = 1/f_0$ seconds.

Interactive Demo: `sin_demo1`

This demo program illustrates the key properties of a sinusoidal signal based on the preceding discussion. Refer to Eqn. (1.44) and Fig. 1.41. Slider controls allow the amplitude A, the frequency f_0 and the phase θ to be varied.

1. Observe the value of the peak amplitude of the waveform changing in response to changes in parameter A.

2. Pay attention to the reciprocal relationship between the frequency f_0 and the period T_0.

3. Set $f_0 = 250$ Hz, and $\theta = 0$ degrees. Count the number of complete cycles over 40 ms in the time interval -20 ms $\leq t \leq 20$ ms. How does the number of full cycles correspond to the value of f_0?

4. When $\theta = 0$, the cosine waveform has a peak at $t = 0$. Change the phase θ slightly in the negative direction by moving the slider control to the left, and observe the middle peak of the waveform move to the right. Observe the value of this *time-delay* caused by the change in θ, and relate it to Eqn. (1.47).

Software resources:
`sin_demo1.m`

1.3.3   Impulse decomposition for continuous-time signals

Sifting property of the unit-impulse function is given by Eqn. (1.27). Using this property, a signal $x(t)$ can be expressed in the form

$$x(t) = \int_{-\infty}^{\infty} x(\lambda)\, \delta(t - \lambda)\, d\lambda \tag{1.48}$$

We will explore this relationship in a bit more detail to gain further insight. Suppose we wanted to approximate the signal $x(t)$ using a series of rectangles that are each Δ wide. Let each rectangle be centered around the time instant $t_n = n\Delta$ where n is the integer index, and also let the height of each rectangle be adjusted to the amplitude of the signal $x(t)$ at the mid-point of the rectangle. Our approximation of $x(t)$ would be in the form

$$\hat{x}(t) = \sum_{n=-\infty}^{\infty} x(n\Delta)\, \Pi\left(\frac{t - n\Delta}{\Delta}\right) \tag{1.49}$$

which is graphically depicted in Fig. 1.42.

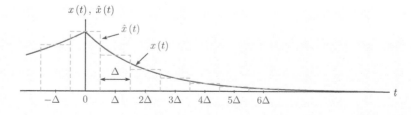

Figure 1.42 – Approximating a signal with rectangular pulses.

Let's multiply and divide the term in the summation on the right side of Eqn. (1.49) by Δ to obtain

$$\hat{x}(t) = \sum_{n=-\infty}^{\infty} x(n\Delta)\, \frac{1}{\Delta}\, \Pi\left(\frac{t-n\Delta}{\Delta}\right) \Delta \tag{1.50}$$

Furthermore, we will define $p_n(t)$ as

$$p_n(t) = \frac{1}{\Delta}\, \Pi\left(\frac{t-n\Delta}{\Delta}\right) \tag{1.51}$$

Clearly, $p_n(t)$ is a rectangular pulse of width Δ and height $(1/\Delta)$, and it is centered around the time instant $(n\Delta)$. It has unit area independent of the value of Δ. Substituting Eqn. (1.51) into Eqn. (1.50) leads to

$$\hat{x}(t) = \sum_{n=-\infty}^{\infty} x(n\Delta)\, p_n(t)\, \Delta \tag{1.52}$$

We know from Eqns. (1.22) and (1.23) and Fig. 1.23 that the unit-impulse function $\delta(t)$ can be obtained as the limit of a rectangular pulse with unit area by making it narrower and taller at the same time. Therefore, the pulse $p_n(t)$ turns into an impulse if we force $\Delta \to 0$.

$$\lim_{\Delta \to 0} [p_n(t)] = \lim_{\Delta \to 0} \left[\frac{1}{\Delta}\, \Pi\left(\frac{t-n\Delta}{\Delta}\right)\right] = \delta(t) \tag{1.53}$$

Thus, if we take the limit of Eqn. (1.52) as $\Delta \to 0$, the approximation becomes an equality:

$$x(t) = \lim_{\Delta \to 0} [\hat{x}(t)]$$

$$= \lim_{\Delta \to 0} \left[\sum_{n=-\infty}^{\infty} x(n\Delta)\, p_n(t)\, \Delta\right] \tag{1.54}$$

In the limit[2] we get $p_n(t) \to \delta(t)$, $n\Delta \to \lambda$ and $\Delta \to d\lambda$. Also, the summation in Eqn. (1.54) turns into an integral, and we obtain Eqn. (1.48). This result will be useful as we derive the convolution relationship in the next chapter.

Interactive Demo: `id_demo.m`

The demo program "`id_demo.m`" illustrates the derivation of impulse decomposition as outlined in Eqns. (1.48) through (1.54). It shows the the progression of $\hat{x}(t)$ as defined by Eqn. (1.49) into $x(t)$ as the value of Δ approaches zero. A slider control allows the value of Δ to be adjusted.

Software resources:
`id_demo.m`

---

[2]   It should be noted that we are not aiming for a mathematically rigorous proof here, but rather some insight. We are assuming that $x(t)$ is a well behaving function that allows the limit in Eqn. (1.54).

1.3.4 Signal classifications

In this section we will summarize various methods and criteria for classifying the types of continuous-time signals that will be useful in our future discussions.

Real vs. complex signals

A real signal is one in which the amplitude is real-valued at all time instants. For example, if we model an electrical voltage or current with a mathematical function of time, the resulting signal $x(t)$ is obviously real-valued. In contrast, a complex signal is one in which the signal amplitude may also have an imaginary part. A complex signal may be written in *Cartesian form* using its real and imaginary parts as

$$x(t) = x_r(t) + x_i(t) \tag{1.55}$$

or in *polar form* using its magnitude and phase as

$$x(t) = |x(t)| \, e^{j\angle x(t)} \tag{1.56}$$

The two forms in Eqns. (1.55) and (1.56) can be related to each other through the following set of equations:

$$|x(t)| = \left[x_r^2(t) + x_i^2(t)\right]^{1/2} \tag{1.57}$$

$$\angle x(t) = \tan^{-1}\left[\frac{x_i(t)}{x_r(t)}\right] \tag{1.58}$$

$$x_r(t) = |x(t)| \, \cos\left(\angle x(t)\right) \tag{1.59}$$

$$x_i(t) = |x(t)| \, \sin\left(\angle x(t)\right) \tag{1.60}$$

In deriving Eqns. (1.59) and (1.60) we have used Euler's formula.[3]

Even though we will mostly focus on the use of real-valued signals in our discussion in the rest of this text, there will be times when the use of complex signal models will prove useful.

Periodic vs. non-periodic signals

A signal is said to be *periodic* if it satisfies

$$x(t + T_0) = x(t) \tag{1.61}$$

at all time instants t, and for a specific value of $T_0 \neq 0$. The value T_0 is referred to as the *period* of the signal. An example of a periodic signal is shown in Fig. 1.43.

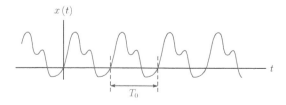

Figure 1.43 – Example of a signal that is periodic.

---

[3] $e^{\pm ja} = \cos(a) \pm j \sin(a)$.

Applying the periodicity definition of Eqn. (1.61) to $x\left(t + T_0\right)$ instead of $x(t)$ we obtain

$$x\left(t + 2T_0\right) = x\left(t + T_0\right) \tag{1.62}$$

which, when combined with Eqn. (1.61), implies that

$$x\left(t + 2T_0\right) = x\left(t\right) \tag{1.63}$$

and, through repeated use of this process, we can show that

$$x\left(t + kT_0\right) = x\left(t\right) \tag{1.64}$$

for any integer value of k. Thus we conclude that, if a signal is periodic with period T_0, then it is also periodic with periods of $2T_0, 3T_0, \ldots, kT_0, \ldots$ where k is any integer. When we refer to the period of a periodic signal, we will imply the smallest positive value of T_0 that satisfies Eqn. (1.61). In cases when we discuss multiple periods for the same signal, we will refer to T_0 as the *fundamental period* in order to avoid ambiguity.

The *fundamental frequency* of a periodic signal is defined as the reciprocal of its fundamental period:

$$f_0 = \frac{1}{T_0} \tag{1.65}$$

If the fundamental period T_0 is expressed in seconds, then the corresponding fundamental frequency f_0 is expressed in Hz.

Example 1.4: **Working with a complex periodic signal**

Consider a signal defined by

$$x\left(t\right) = x_r\left(t\right) + x_i\left(t\right)$$

$$= A\,\cos\left(2\pi f_0 t + \theta\right) + j\,A\,\sin\left(2\pi f_0 t + \theta\right)$$

Using Eqns. (1.57) and (1.58), polar complex form of this signal can be obtained as $x\left(t\right) = |x\left(t\right)|\,e^{j\angle x(t)}$, with magnitude and phase given by

$$|x\left(t\right)| = \left[\left[\,A\,\cos\left(2\pi f_0 t + \theta\right)\right]^2 + \left[A\,\sin\left(2\pi f_0 t + \theta\right)\right]^2\right]^{1/2} = A \tag{1.66}$$

and

$$\angle x(t) = \tan^{-1}\left[\frac{\sin\left(2\pi f_0 t + \theta\right)}{\cos\left(2\pi f_0 t + \theta\right)}\right] = \tan^{-1}\left[\tan\left(2\pi f_0 t + \theta\right)\right] = 2\pi f_0 t + \theta \tag{1.67}$$

respectively. In deriving the results in Eqns. (1.66) and (1.67) we have relied on the appropriate trigonometric identities.[4] Once the magnitude $|x\left(t\right)|$ and the phase $\angle x\left(t\right)$ are obtained, we can express the signal $x\left(t\right)$ in polar complex form:

$$x\left(t\right) = |x\left(t\right)|\,e^{j\angle x(t)} = A\,e^{j\,\left(2\pi f_0 t + \theta\right)} \tag{1.68}$$

The components of the Cartesian and polar complex forms of the signal are shown in Fig. 1.44. The real and imaginary parts of $x\left(t\right)$ have a 90 degree phase difference between them. When the real part of the signal goes through a peak, the imaginary part goes through zero and vice versa. The phase of $x\left(t\right)$ was found in Eqn. (1.66) to be a linear function of the

---

[4] $\cos^2\left(a\right) + \sin^2\left(a\right) = 1$, and $\tan\left(a\right) = \sin\left(a\right)/\cos\left(a\right)$.

time variable t. The broken-line appearance of the phase characteristic in Fig. 1.44 is due to normalization of the phase angles to keep them in the range $-\pi \leq \measuredangle x\left(t\right) < \pi$.

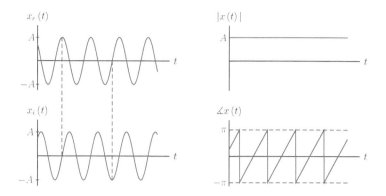

Figure 1.44 – Cartesian and polar representations of the signal $x\left(t\right)$ in Example 1.4.

Software resources:
ex_1_4.m

Interactive Demo: cexp_demo

This demo is based on Example 1.4 and Fig. 1.44. It allows experimentation with the four graphs in Fig. 1.44 while varying the parameters A, f_0, and θ. Three slider controls at the top of the graphical user interface window can be used for varying the three parameters within their prescribed ranges: $0.5 \leq A \leq 5$, $50\text{Hz} \leq f_0 \leq 500\text{Hz}$, and $-180° \leq \theta \leq 180°$ respectively. Moving a slider to the right increases the value of the parameter it controls, and moving it to the left does the opposite. It is also possible to type the desired value of a parameter into the edit field associated with its slider control. Vary the parameters and observe the following:

  1. As the amplitude A is increased, the graphs for $x_r\left(t\right)$, $x_i\left(t\right)$ and $\left|x\left(t\right)\right|$ scale accordingly, however, the phase $\measuredangle x\left(t\right)$ is not affected.
  2. Changing θ causes a time-delay in $x_r\left(t\right)$ and $x_i\left(t\right)$.

Software resources:
cexp_demo.m

Software resources: See MATLAB Exercise 1.4.

Example 1.5: **Periodicity of continuous-time sinusoidal signals**

Consider two continuous-time sinusoidal signals

$$x_1\left(t\right) = A_1 \sin\left(2\pi f_1 t + \theta_1\right)$$

and

$$x_2\left(t\right) = A_2 \sin\left(2\pi f_2 t + \theta_2\right)$$

Determine the conditions under which the sum signal

$$x(t) = x_1(t) + x_2(t)$$

is also periodic. Also, determine the fundamental period of the signal $x(t)$ as a function of the relevant parameters of $x_1(t)$ and $x_2(t)$.

Solution: Based on the discussion above, we know that the individual sinusoidal signals $x_1(t)$ and $x_2(t)$ are periodic with fundamental periods of T_1 and T_2 respectively, and the following periodicity properties can be written:

$$x_1(t + m_1 T_1) = x_1(t) , \quad T_1 = 1/f_1 \tag{1.69}$$

$$x_2(t + m_2 T_2) = x_2(t) , \quad T_2 = 1/f_2 \tag{1.70}$$

where m_1 and m_2 are arbitrary integer values. In order for the sum signal $x(t)$ to be periodic with period T_0, we need the relationship

$$x(t + T_0) = x(t) \tag{1.71}$$

to be valid at all time instants. In terms of the signals $x_1(t)$ and $x_2(t)$ this requires

$$x_1(t + T_0) + x_2(t + T_0) = x_1(t) + x_2(t) \tag{1.72}$$

Using Eqns. (1.69) and (1.70) with Eqn. (1.72) we reach the conclusion that Eqn. (1.71) would hold if two integers m_1 and m_2 can be found such that

$$T_0 = m_1 T_1 = m_2 T_2 \tag{1.73}$$

or, in terms of the frequencies involved,

$$\frac{1}{f_0} = \frac{m_1}{f_1} = \frac{m_2}{f_2} \tag{1.74}$$

The sum of two sinusoidal signals $x_1(t)$ and $x_2(t)$ is periodic provided that their frequencies are integer multiples of a frequency f_0, i.e., $f_1 = m_1 f_0$ and $f_2 = m_2 f_0$. The period of the sum signal is $T_0 = 1/f_0$. The frequency f_0 is the fundamental frequency of $x(t)$.

The reasoning above can be easily extended to any number of sinusoidal signals added together. Consider a signal $x(t)$ defined as

$$x(t) = \sum_{n=1}^{N} A_n \sin(2\pi f_n t + \theta_n) \tag{1.75}$$

The signal $x(t)$ is periodic provided that the frequencies of all of its components are integer multiples of a fundamental frequency f_0, that is,

$$f_n = m_n f_0 , \quad n = 1, \ldots, N \tag{1.76}$$

This conclusion will be significant when we discuss the Fourier series representation of periodic signals in Chapter 4.

Example 1.6: **More on the periodicity of sinusoidal signals**

Discuss the periodicity of the signals

  a. $\qquad x\left(t\right)=\sin\left(2\pi\,1.5\,t\right)+\sin\left(2\pi\,2.5\,t\right)$

  b. $\qquad y\left(t\right)=\sin\left(2\pi\,1.5\,t\right)+\sin\left(2\pi\,2.75\,t\right)$

Solution:

a. For this signal, the fundamental frequency is $f_0 = 0.5$ Hz. The two signal frequencies can be expressed as

$$f_1 = 1.5 \text{ Hz } = 3f_0 \quad \text{and} \quad f_2 = 2.5 \text{ Hz } = 5f_0$$

The resulting fundamental period is $T_0 = 1/f_0 = 2$ seconds. Within one period of $x\left(t\right)$ there are $m_1 = 3$ full cycles of the first sinusoid and $m_2 = 5$ cycles of the second sinusoid. This is illustrated in Fig. 1.45.

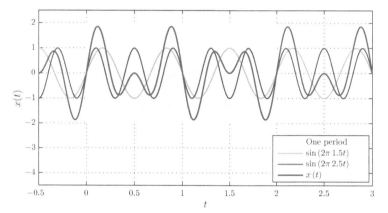

Figure 1.45 – Periodicity of $x\left(t\right)$ of Example 1.6.

b. In this case the fundamental frequency is $f_0 = 0.25$ Hz. The two signal frequencies can be expressed as

$$f_1 = 1.5 \text{ Hz } = 6f_0 \quad \text{and} \quad f_2 = 2.75 \text{ Hz } = 11f_0$$

The resulting fundamental period is $T_0 = 1/f_0 = 4$ seconds. Within one period of $x\left(t\right)$ there are $m_1 = 6$ full cycles of the first sinusoid and $m_2 = 11$ cycles of the second sinusoid. This is illustrated in Fig. 1.46.

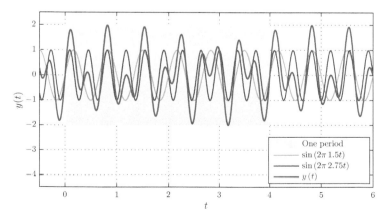

Figure 1.46 – Periodicity of $y\left(t\right)$ of Example 1.6.

Software resources:
ex_1_6a.m
ex_1_6b.m

Interactive Demo: sin_demo2

This demo program allows exploration of the periodicity of multi-tone signals. A signal of the type given by Eqn. (1.75) can be graphed using up to four frequencies. The beginning and the end of one period of the signal is indicated on the graph with dashed vertical lines.
Software resources:
sin_demo2.m

Deterministic vs. random signals

Deterministic signals are those that can be described completely in analytical form in the time domain. The signals that we worked with in this chapter have all been examples of deterministic signals. Random signals, on the other hand, are signals that occur due to random phenomena that cannot be modeled analytically. An example of a random signal is the speech signal converted to the form of an electrical voltage waveform by means of a microphone. A very short segment from a speech signal was shown in Fig. 1.2. Other examples of random signals are the vibration signal recorded during an earthquake by a seismograph, the noise signal generated by a resistor due to random movements and collisions of its electrons and the thermal energy those collisions produce. A common feature of these signals is the fact that the signal amplitude is not known at any given time instant, and cannot be expressed using a formula. Instead, we must try to model the statistical characteristics of the random signal rather than the signal itself. These statistical characteristics studied include average values of the signal amplitude or squared signal amplitude, distribution of various probabilities involving signal amplitude, and distribution of normalized signal energy or power among different frequencies. Study of random signals is beyond the scope of this text, and the reader is referred to one of the many excellent texts available on the subject.

1.3.5   Energy and power definitions

Energy of a signal

With physical signals and systems, the concept of *energy* is associated with a signal that is applied to a load. The signal source delivers the energy which must be dissipated by the load. For example, consider a voltage source with voltage $v(t)$ connected to the terminals of a resistor with resistance R as shown in Fig. 1.47(a). Let $i(t)$ be the current that flows through the resistor. If we wanted to use the voltage $v(t)$ as our basis in energy calculations, the total energy dissipated in the resistor would be

$$E = \int_{-\infty}^{\infty} v(t)\, i(t)\, dt = \int_{-\infty}^{\infty} \frac{v^2(t)}{R}\, dt \qquad (1.77)$$

Alternatively, consider the arrangement in Fig. 1.47(b) where a current source with a time-varying current $i(t)$ is connected to the terminals of the same resistor. In this case, the energy dissipated in the resistor would be computed as

$$E = \int_{-\infty}^{\infty} v(t)\, i(t)\, dt = \int_{-\infty}^{\infty} R\, i^2(t)\, dt \qquad (1.78)$$

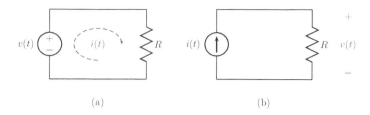

Figure 1.47 – Energy dissipation in a load when (a) a voltage source is used, and (b) a current source is used.

In the context of this text we use mathematical signal models that are independent of the physical nature of signals that are being modeled. A signal such as $x(t)$ could represent a voltage, a current, or some other time-varying physical quantity. It would be desirable to have a definition of signal energy that is based on the mathematical model of the signal alone, without any reference to the load and to the physical quantity from which the mathematical model may have been derived. If we don't know whether the function $x(t)$ represents a voltage or a current, which of the two equations discussed above should we use for computing the energy produced by the signal? Comparing Eqns. (1.77) and (1.78) it is obvious that, if the resistor value is chosen to be $R = 1\ \Omega$, then both equations would produce the same numerical value:

$$E = \int_{-\infty}^{\infty} \frac{v^2(t)}{(1)}\, dt = \int_{-\infty}^{\infty} (1)\, i^2(t)\, dt \tag{1.79}$$

This is precisely the approach we will take. In order to simplify our analysis, and to eliminate the need to always pay attention to the physical quantities that lead to our mathematical models, we will define the *normalized energy* of a real-valued signal $x(t)$ as

$$E_x = \int_{-\infty}^{\infty} x^2(t)\, dt \tag{1.80}$$

provided that the integral in Eqn. (1.80) can be computed. If the signal $x(t)$ is complex-valued, then its normalized energy is computed as

$$E_x = \int_{-\infty}^{\infty} |x(t)|^2\, dt \tag{1.81}$$

where $|x(t)|$ is the norm of the complex signal. In situations where normalized signal energy is not computable, we may be able to compute the *normalized power* of the signal instead.

Example 1.7: **Energy of a right-sided exponential signal**

Compute the normalized energy of the right-sided exponential signal

$$x(t) = A\, e^{-\alpha t}\, u(t)$$

where $\alpha > 0$.

Solution: The signal $x(t)$ is shown in Fig. 1.48.   Notice how the signal amplitude starts with a value of $x(0) = A$ and declines to $x(1/\alpha) = A/e$ after $1/\alpha$ seconds elapse. At the time instant $t = 2/\alpha$ seconds, the amplitude is further down to $x(2/\alpha) = A/e^2$.   Thus,

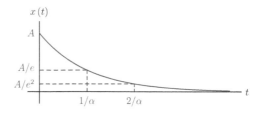

Figure 1.48 – The signal $x(t)$ for Example 1.7.

the parameter α controls the rate at which the signal amplitude decays over time. The normalized energy of this signal can be computed by application of Eqn. (1.80) as

$$E_x = \int_0^\infty A^2\, e^{-2\alpha t}\, dt = A^2\, \frac{e^{-2\alpha t}}{-2\alpha} \bigg|_0^\infty = \frac{A^2}{2\alpha} \tag{1.82}$$

The restriction $\alpha > 0$ was necessary since, without it, we could not have evaluated the integral in Eqn. (1.82).

Interactive Demo: exp_demo

This demo program is based on Example 1.7, Fig. 1.48, and Eqn. (1.82). Parameters A and α of the right-sided exponential signal $x(t)$ can be controlled with the slider controls available.

1. The shape of the right-sided exponential depends on the parameter α. Observe the key points with amplitudes A/e and A/e^2 as the parameter α is varied.
2. Observe how the displayed signal energy E_x relates to the shape of the exponential signal.

Software resources:
exp_demo.m

Time averaging operator

In preparation for defining the power in a signal, we need to first define the time average of a signal. We will use the operator $\langle \ldots \rangle$ to indicate time average. If the signal $x(t)$ is periodic with period T_0, its time average can be computed as

$$\langle x(t) \rangle = \frac{1}{T_0} \int_{-T_0/2}^{T_0/2} x(t)\, dt \tag{1.83}$$

If the signal $x(t)$ is non-periodic, the definition of time average in Eqn. (1.83) can be generalized with the use of the limit operator as

$$\langle x(t) \rangle = \lim_{T \to \infty} \left[\frac{1}{T} \int_{-T/2}^{T/2} x(t)\, dt \right] \tag{1.84}$$

One way to make sense out of Eqn. (1.84) is to view the non-periodic signal $x(t)$ as though it is periodic with an infinitely large period so that we never get to see the signal pattern repeat itself.

Example 1.8: **Time average of a pulse train**

Compute the time average of a periodic pulse train with an amplitude of A and a period of $T_0 = 1$ s, defined by the equations

$$x(t) = \begin{cases} A, & 0 < t < d \\ 0, & d < t < 1 \end{cases}$$

and

$$x(t + k T_0) = x(t + k) = x(t) \text{ for all } t, \text{ and all integers } k$$

The signal $x(t)$ is shown in Fig. 1.49.

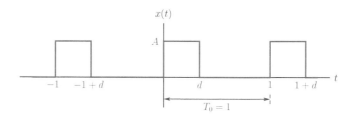

Figure 1.49 – The periodic pulse-train signal $x(t)$ for Example 1.8.

Solution: The parameter d is the width of the pulse within each period. Since the length of one period is $T_0 = 1$ s, the parameter d can also be taken as the fraction of the period length in which the signal amplitude equals A units. In this context, d is referred to as the *duty cycle* of the pulse train. To compute the time average of $x(t)$ we need to apply Eqn. (1.83) over one period:

$$\langle x(t) \rangle = \int_0^1 x(t)\, dt = \int_0^d (A)\, dt + \int_d^1 (0)\, dt = A\,d$$

As expected, the time average of the pulse-train signal $x(t)$ is proportional to both the pulse amplitude A and the duty cycle d.

Interactive Demo: `tavg_demo`

This demo program illustrates the time averaging of pulse train $x(t)$ as shown in Fig. 1.49 of Example 1.8 with amplitude A and duty cycle d. Values of both parameters can be adjusted using the slider controls provided. Observe how the pulse train changes when the duty cycle is varied. Also pay attention to the dashed red line that shows the value of the time average $\langle x(t) \rangle$.
Software resources:
`tavg_demo.m`

Power of a signal

Consider again the simple circuits in Fig. 1.47 that were used in defining normalized signal energy. Using $v(t)$ as the voltage across the terminals of the load and $i(t)$ as the current flowing through the load, the *instantaneous power* dissipated in the load resistor would be

$$p_{\text{inst}}(t) = v(t)\, i(t) \tag{1.85}$$

If the load is chosen to have a value of $R = 1 \ \Omega$, the *normalized instantaneous power* can be defined as

$$p_{\text{norm}}(t) = x^2(t) \tag{1.86}$$

where $x(t)$ could represent either the voltage or the current. Often, a more useful concept is the *normalized average power* defined as the time average of $p_{\text{norm}}(t)$, that is

$$P_x = \langle x^2(t) \rangle \tag{1.87}$$

Note that this definition will work for both periodic and non-periodic signals. For a periodic signal, Eqn. (1.87) can be used with Eqn. (1.83) to yield

$$P_x = \frac{1}{T_0} \int_{-T_0/2}^{T_0/2} x^2(t) \ dt \tag{1.88}$$

For a non-periodic signal, Eqn. (1.84) can be substituted into Eqn. (1.87), and we have

$$P_x = \lim_{T \to \infty} \left[\frac{1}{T} \int_{-T/2}^{T/2} x^2(t) \ dt \right] \tag{1.89}$$

In our discussion, we are assuming that the signal $x(t)$ is real-valued. If we need to consider complex-valued signals, then the definition of signal power can be generalized by using the squared norm of the signal in the time averaging operation:

$$P_x = \left\langle |x(t)|^2 \right\rangle \tag{1.90}$$

Recall that the squared norm of a complex signal is computed as

$$|x(t)|^2 = x(t) \ x^*(t) \tag{1.91}$$

where $x^*(t)$ is the complex conjugate of $x(t)$. Thus, the power of a periodic complex signal is

$$P_x = \frac{1}{T_0} \int_{-T_0/2}^{T_0/2} |x(t)|^2 \ dt \tag{1.92}$$

and the power of a non-periodic complex signal is

$$P_x = \lim_{T \to \infty} \left[\frac{1}{T} \int_{-T/2}^{T/2} |x(t)|^2 \ dt \right] \tag{1.93}$$

Example 1.9: **Power of a sinusoidal signal**

Consider the signal

$$x(t) = A \sin(2\pi f_0 t + \theta)$$

which is not limited in time. It is obvious that the energy of this signal can not be computed since the integral in Eqn. (1.80) would not yield a finite value. On the other hand, signal power can be computed through the use of Eqn. (1.88). The period of the signal is $T_0 = 1/f_0$, therefore

$$P_x = f_0 \int_{-1/2f_0}^{1/2f_0} A^2 \sin^2(2\pi f_0 t + \theta) \ dt \tag{1.94}$$

Using the appropriate trigonometric identity,[5] we can write Eqn. (1.94) as

$$P_x = f_0 \int_{-1/2f_0}^{1/2f_0} \frac{A^2}{2} \, dt - f_0 \int_{-1/2f_0}^{1/2f_0} \frac{A^2}{2} \cos\left(4\pi f_0 t + 2\theta\right) dt \qquad (1.95)$$

The second integral in Eqn. (1.95) evaluates to zero, since we are integrating a cosine function with the frequency $2f_0$ over an interval of $T_0 = 1/f_0$. The cosine function has two full cycles within this interval. Thus, the normalized average power in a sinusoidal signal with an amplitude of A is

$$P_x = f_0 \int_{-1/2f_0}^{1/2f_0} \frac{A^2}{2} \, dt = \frac{A^2}{2} \qquad (1.96)$$

This is a result that will be worth remembering.

Example 1.10: **Right-sided exponential signal revisited**

Let us consider the right-sided exponential signal of Example 1.7 again. Recall that the analytical expression for the signal was

$$x(t) = A \, e^{-\alpha t} \, u(t)$$

Using Eqn. (1.89) it can easily be shown that, for $\alpha > 0$, the power of the signal is $P_x = 0$. On the other hand, it is interesting to see what happens if we allow $\alpha = 0$. In this case, the signal $x(t)$ becomes a step function:

$$x(t) = A \, u(t)$$

The normalized average signal power is

$$P_x = \lim_{T \to \infty} \left[\frac{1}{T} \int_{-T/2}^{T/2} x(t) \, dt \right]$$

$$= \lim_{T \to \infty} \left[\frac{1}{T} \int_{-T/2}^{0} (0) \, dt + \frac{1}{T} \int_{0}^{T/2} A^2 \, dt \right]$$

$$= \lim_{T \to \infty} \left[\frac{A^2}{2} \right] = \frac{A^2}{2}$$

and the normalized signal energy becomes infinitely large, i.e., $E_x \to \infty$.

Energy signals vs. power signals

In Examples 1.7 through 1.10 we have observed that the concept of signal energy is useful for some signals and not for others. Same can be said for signal power. Based on our observations, we can classify signals encountered in practice into two categories:

- Energy signals are those that have finite energy and zero power, i.e., $E_x < \infty$, and $P_x = 0$.

- Power signals are those that have finite power and infinite energy, i.e., $E_x \to \infty$, and $P_x < \infty$.

---

[5] $\sin^2(a) = \frac{1}{2}\left[1 - \cos(2a)\right]$.

Just as a side note, all voltage and current signals that can be generated in the laboratory or that occur in the electronic devices that we use in our daily lives are energy signals. A power signal is impossible to produce in any practical setting since doing so would require an infinite amount of energy. The concept of a power signal exists as a mathematical idealization only.

RMS value of a signal

The *root-mean-square (RMS)* value of a signal $x(t)$ is defined as

$$X_{RMS} = \left[\left\langle x^2(t) \right\rangle \right]^{1/2} \tag{1.97}$$

The time average in Eqn. (1.97) is computed through the use of either Eqn. (1.83) or Eqn. (1.84), depending on whether $x(t)$ is periodic or not.

Example 1.11: **RMS value of a sinusoidal signal**

In Example 1.9, we have found the normalized average power of the sinusoidal signal $x(t) = A \sin(2\pi f_0 t + \theta)$ to be $P_x = \langle x^2(t) \rangle = A^2/2$. Based on Eqn. (1.97), the RMS value of this signal is

$$X_{RMS} = \frac{A}{\sqrt{2}} \tag{1.98}$$

as illustrated in Fig. 1.50.

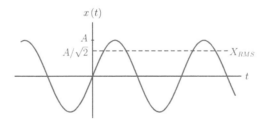

Figure 1.50 – RMS value of a sinusoidal signal.

Example 1.12: **RMS value of a multi-tone signal**

Consider the two-tone signal given by

$$x(t) = a_1 \cos(2\pi f_1 t + \theta_1) + a_2 \cos(2\pi f_2 t + \theta_2)$$

where the two frequencies are distinct, i.e., $f_1 \neq f_2$. Determine the RMS value and the normalized average power of this signal.

Solution: Computation of both the RMS value and the normalized average power for the signal $x(t)$ requires the computation of the time average of $x^2(t)$. The square of the signal $x(t)$ is

$$\begin{aligned} x^2(t) = {} & a_1^2 \cos^2(2\pi f_1 t + \theta_1) + a_2^2 \cos^2(2\pi f_2 t + \theta_2) \\ & + 2\,a_1 a_2 \cos(2\pi f_1 t + \theta_1) \cos(2\pi f_2 t + \theta_2) \end{aligned} \tag{1.99}$$

Using the trigonometric identities[6] Eqn. (1.99) can be written as

$$x^2(t) = \frac{a_1^2}{2} + \frac{a_2^2}{2} + \frac{a_1^2}{2}\cos\left(2\pi\left[2f_1\right]t + 2\theta_1\right) + \frac{a_2^2}{2}\cos\left(2\pi\left[2f_2\right]t + 2\theta_2\right)$$
$$+ a_1 a_2 \cos\left(2\pi\left[f_1 + f_2\right]t + \theta_1 + \theta_2\right) + a_1 a_2 \cos\left(2\pi\left[f_1 - f_2\right]t + \theta_1 - \theta_2\right)$$

We can now apply the time averaging operator to $x^2(t)$ to yield

$$\langle x^2(t)\rangle = \frac{a_1^2}{2} + \frac{a_2^2}{2} + \frac{a_1^2}{2}\left\langle \cos\left(2\pi\left[2f_1\right]t + 2\theta_1\right)\right\rangle + \frac{a_2^2}{2}\left\langle \cos\left(2\pi\left[2f_2\right]t + 2\theta_2\right)\right\rangle$$
$$+ a_1 a_2 \left\langle \cos\left(2\pi\left[f_1 + f_2\right]t + \theta_1 + \theta_2\right)\right\rangle + a_1 a_2 \left\langle \cos\left(2\pi\left[f_1 - f_2\right]t + \theta_1 - \theta_2\right)\right\rangle \tag{1.100}$$

The time average of each of the cosine terms in Eqn. (1.100) is equal to zero, and we obtain

$$\langle x^2(t)\rangle = \frac{a_1^2}{2} + \frac{a_2^2}{2} \tag{1.101}$$

Thus, the RMS value of the signal $x(t)$ is

$$X_{RMS} = \sqrt{\frac{a_1^2}{2} + \frac{a_2^2}{2}} \tag{1.102}$$

and its normalized average power is

$$P_x = \frac{a_1^2}{2} + \frac{a_2^2}{2} \tag{1.103}$$

The generalization of these results to signals containing more than two sinusoidal components is straightforward. If $x(t)$ contains M sinusoidal components with amplitudes $a_1, a_2, \ldots, a_M$, and distinct frequencies $f_1, f_2, \ldots, f_M$, its RMS value and normalized average power are

$$X_{RMS} = \sqrt{\frac{a_1^2}{2} + \frac{a_2^2}{2} + \ldots + \frac{a_M^2}{2}} \tag{1.104}$$

and

$$P_x = \frac{a_1^2}{2} + \frac{a_2^2}{2} + \ldots + \frac{a_M^2}{2} \tag{1.105}$$

respectively.

1.3.6 Symmetry properties

Some signals have certain symmetry properties that could be utilized in a variety of ways in the analysis. More importantly, a signal that may not have any symmetry properties can still be written as a linear combination of signals with certain symmetry properties.

---

[6] $\cos^2(a) = \frac{1}{2} + \frac{1}{2}\cos(2a)$ and $\cos(a)\cos(b) = \frac{1}{2}\cos(a+b) + \frac{1}{2}\cos(a-b)$.

Even and odd symmetry

A real-valued signal is said to have *even symmetry* if it has the property

$$x\left(-t\right) = x\left(t\right) \tag{1.106}$$

for all values of t. A signal with even symmetry remains unchanged when it is time reversed. Fig. 1.51(a) shows an example of a signal with even symmetry property. Similarly, a real-valued signal is said to have *odd symmetry* if it has the property

$$x\left(-t\right) = -x\left(t\right) \tag{1.107}$$

for all t. Time reversal has the same effect as negation on a signal with odd symmetry. This is illustrated in Fig. 1.51(b).

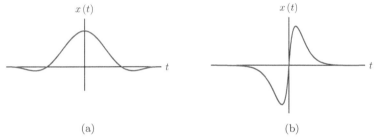

(a)   (b)

Figure 1.51 – (a) Even signal, (b) odd signal.

As a specific example the signal $x\left(t\right) = \cos\left(\omega t\right)$ has even symmetry since $\cos\left(-\omega t\right) = \cos\left(\omega t\right)$. Similarly, the signal $x\left(t\right) = \sin\left(\omega t\right)$ has odd symmetry since $\sin\left(-\omega t\right) = -\sin\left(\omega t\right)$.

Decomposition into even and odd components

It is always possible to split a real-valued signal into two components in such a way that one of the components is an even function of time, and the other is an odd function of time. Consider the following representation of the signal $x\left(t\right)$:

$$x\left(t\right) = x_e\left(t\right) + x_o\left(t\right) \tag{1.108}$$

where the two components $x_e\left(t\right)$ and $x_o\left(t\right)$ are defined as

$$x_e\left(t\right) = \frac{x\left(t\right) + x\left(-t\right)}{2} \tag{1.109}$$

and

$$x_o\left(t\right) = \frac{x\left(t\right) - x\left(-t\right)}{2} \tag{1.110}$$

The definitions of the two signals $x_e\left(t\right)$ and $x_o\left(t\right)$ guarantee that they always add up to $x\left(t\right)$. Furthermore, symmetry properties $x_e(-t) = x_e\left(t\right)$ and $x_o(-t) = -x_o\left(t\right)$ are guaranteed by the definitions of the signals $x_e\left(t\right)$ and $x_o\left(t\right)$ regardless of any symmetry properties the signal $x\left(t\right)$ may or may not possess. Symmetry properties of $x_e\left(t\right)$ and $x_o\left(t\right)$ can be easily verified by forming the expressions for $x_e\left(-t\right)$ and $x_o\left(-t\right)$ and comparing them to Eqns. (1.109) and (1.110).

Let us use Eqn. (1.109) to write $x_e(-t)$:

$$x_e(-t) = \frac{x(-t) + x(t)}{2} = x_e(t) \qquad (1.111)$$

Similarly, $x_o(-t)$ can be found from Eqn. (1.110) as

$$x_o(-t) = \frac{x(-t) - x(t)}{2} = -x_o(t) \qquad (1.112)$$

We will refer to $x_e(t)$ and $x_o(t)$ as the *even component* and the *odd component* of $x(t)$ respectively.

Example 1.13: **Even and odd components of a rectangular pulse**

Determine the even and the odd components of the rectangular pulse signal

$$x(t) = \Pi\left(t - \tfrac{1}{2}\right) = \begin{cases} 1, & 0 < t < 1 \\ 0, & \text{elsewhere} \end{cases}$$

which is shown in Fig. 1.52.

Figure 1.52 – The rectangular pulse signal of Example 1.13.

Solution: The even component of the signal $x(t)$ as computed from Eqn. (1.109) is

$$x_e(t) = \frac{\Pi\left(t - \tfrac{1}{2}\right) + \Pi\left(-t - \tfrac{1}{2}\right)}{2} = \frac{1}{2}\Pi(t/2)$$

Its odd component is computed using Eqn. (1.110) as

$$x_o(t) = \frac{1}{2}\left[\Pi\left(t - \tfrac{1}{2}\right) - \Pi\left(-t - \tfrac{1}{2}\right)\right]$$

The two components of $x(t)$ are graphed in Fig. 1.53.

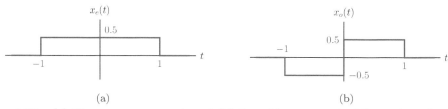

(a)   (b)

Figure 1.53 – (a) The even component, and (b) the odd component of the rectangular pulse signal of Example 1.13.

Example 1.14: **Even and odd components of a sinusoidal signal**

Consider the sinusoidal signal

$$x(t) = 5 \cos(10t + \pi/3) \tag{1.113}$$

What symmetry properties does this signal have, if any? Determine its even and odd components.

Solution: The signal $x(t)$ is shown in Fig. 1.54. It can be observed that it has neither even symmetry nor odd symmetry.

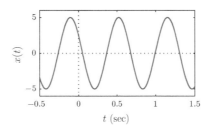

Figure 1.54 – The signal of Example 1.14.

We will use Eqns. (1.109) and (1.110) in decomposing this signal into its even and odd components. The even component is

$$x_e(t) = \frac{5}{2} \cos(10t + \pi/3) + \frac{5}{2} \cos(-10t + \pi/3) \tag{1.114}$$

Using the appropriate trigonometric identity,[7] Eqn. (1.114) can be written as

$$x_e(t) = 5 \cos(10t) \cos(\pi/3) = 2.5 \cos(10t)$$

Similarly, the odd component of $x(t)$ is

$$x_o(t) = \frac{5}{2} \cos(10t + \pi/3) - \frac{5}{2} \cos(-10t + \pi/3) \tag{1.115}$$

which, through the use of the same trigonometric identity, can be written as

$$x_o(t) = -5 \sin(10t) \sin(\pi/3) = -4.3301 \sin(10t)$$

Even and odd components of $x(t)$ are shown in Fig. 1.55.

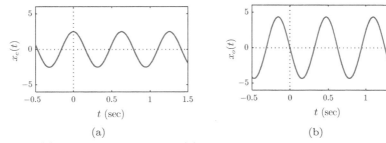

(a)   (b)

Figure 1.55 – (a) Even component, and (b) odd component of the signal of Example 1.14.

---

[7] $\cos(a + b) = \cos(a) \cos(b) - \sin(a) \sin(b)$.

Software resources:
ex_1_14.m

Symmetry properties for complex signals

Even and odd symmetry definitions given by Eqns. (1.106) and (1.107) for real-valued signals can be extended to work with complex-valued signals as well. A complex-valued signal is said to be *conjugate symmetric* if it satisfies

$$x(-t) = x^*(t) \tag{1.116}$$

for all t. For a signal with conjugate symmetry, time reversal has the same effect as complex conjugation. Similarly, a complex-valued signal is said to be *conjugate antisymmetric* if it satisfies

$$x(-t) = -x^*(t) \tag{1.117}$$

for all t. If a signal is conjugate antisymmetric, time reversing it causes the signal to be conjugated and negated simultaneously.

It is easy to see from Eqns. (1.116) and (1.106) that, if the signal is real-valued, that is, if $x(t) = x^*(t)$, the definition of conjugate symmetry reduces to that of even symmetry. Similarly, conjugate antisymmetry property reduces to odd symmetry for a real valued $x(t)$ as revealed by a comparison of Eqns. (1.117) and (1.107).

Example 1.15: **Symmetry of a complex exponential signal**

Consider the complex exponential signal

$$x(t) = A e^{j\omega t}, \quad A: \text{real} \tag{1.118}$$

What symmetry property does this signal have, if any?

Solution: Time reversal of $x(t)$ results in

$$x(-t) = A e^{-j\omega t} \tag{1.119}$$

Complex conjugate of the signal $x(t)$ is

$$x^*(t) = \left(A e^{j\omega t}\right)^* = A e^{-j\omega t} \tag{1.120}$$

Since $x(-t) = x^*(t)$, we conclude that the signal $x(t)$ is conjugate symmetric.

Decomposition of complex signals

It is possible to express any complex signal $x(t)$ as the sum of two signals

$$x(t) = x_E(t) + x_O(t) \tag{1.121}$$

such that $x_E(t)$ is conjugate symmetric, and $x_O(t)$ is conjugate antisymmetric. The two components of the decomposition in Eqn. (1.121) are computed as

$$x_E(t) = \frac{x(t) + x^*(-t)}{2} \tag{1.122}$$

and

$$x_O(t) = \frac{x(t) - x^*(-t)}{2} \tag{1.123}$$

Definitions of $x_E(t)$ and $x_O(t)$ ensure that they always add up to $x(t)$. Furthermore, the two components are guaranteed to have the conjugate symmetry and conjugate antisymmetry properties respectively.

By conjugating the first component $x_E(t)$ we get

$$x_E^*(t) = \left[\frac{x(t) + x^*(-t)}{2} \right]^* = \frac{x^*(t) + x(-t)}{2} = x_E(-t)$$

proving that $x_E(t)$ is conjugate symmetric. Conjugation of the other component $x_O(t)$ yields

$$x_O^*(t) = \left[\frac{x(t) - x^*(-t)}{2} \right]^* = \frac{x^*(t) - x(-t)}{2} = -x_O(-t)$$

proving that $x_O(t)$ is conjugate antisymmetric.

1.3.7 Graphical representation of sinusoidal signals using phasors

In certain situations we find it convenient to represent a sinusoidal signal with a complex quantity referred to as a *phasor*. Use of the phasor concept is beneficial in analyzing linear systems in which multiple sinusoidal signals may exist with the same frequency but with differing amplitude and phase values. Consider again a sinusoidal signal in the form

$$x(t) = A \cos\left(2\pi f_0 t + \theta\right) \tag{1.124}$$

with three adjustable parameters, namely the amplitude A, the phase θ, and the frequency f_0. Using Euler's formula,[8] it is possible to express the sinusoidal signal in Eqn. (1.124) as the real part of a complex signal in the form

$$x(t) = \mathrm{Re}\left\{ A\, e^{j(2\pi f_0 t + \theta)} \right\} \tag{1.125}$$

The exponential function in Eqn. (1.125) can be factored into two terms, yielding

$$x(t) = \mathrm{Re}\left\{ A\, e^{j\theta}\, e^{j2\pi f_0 t} \right\} \tag{1.126}$$

Let the complex variable $\boldsymbol{X}$ be defined as

$$\boldsymbol{X} \triangleq A\, e^{j\theta} \tag{1.127}$$

The complex variable $\boldsymbol{X}$ can be viewed as a vector with norm A and angle θ. Vector interpretation of $\boldsymbol{X}$ is illustrated in Fig. 1.56.

---

[8] $e^{ja} = \cos(a) + j\sin(a)$.

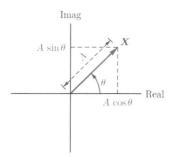

Figure 1.56 – Complex variable $\boldsymbol{X}$ shown as a vector.

Using $\boldsymbol{X}$ in Eqn. (1.126), we obtain

$$x\left(t\right) = \text{Re}\left\{\boldsymbol{X}\, e^{j2\pi f_0 t}\right\} \tag{1.128}$$

In Eqn. (1.128) the exponential term $e^{j2\pi f_0 t}$ represents rotation. It is also complex-valued, and has unit norm. Its angle $2\pi f_0 t$ is a linear function of time. Over an interval of one second, the angle of the exponential term grows by $2\pi f_0$ radians, corresponding to f_0 revolutions around the origin. The expression in brackets in Eqn. (1.128) combines the vector $\boldsymbol{X}$ with this exponential term. The combination represents a vector with norm A and initial phase angle θ that rotates at the rate of f_0 revolutions per second. We will refer to this rotating vector as a *phasor*. At any time instant, the amplitude of the signal $x\left(t\right)$ equals the real part of the phasor. This relationship is illustrated in Fig. 1.57.

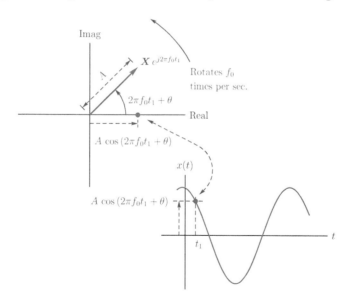

Figure 1.57 – Phasor representation of the signal $x\left(t\right)$.

Interactive Demo: `phs_demo`

The demo program in "`phs_demo.m`" provides a graphical user interface for experimenting with phasors.

The graph window in the upper right section depicts the complex plane. One or two phasors can be visible; the first phasor is shown in blue, and the second phasor is shown in red. The real part of each phasor is shown with thick dashed lines of the same color as the phasor.

The graph window at the bottom shows the time-domain representation of the signal that corresponds to each phasor, using the same color scheme. The sum of the two sinusoidal signals is shown in black. The real part of each phasor, displayed with thick dashed lines on the complex plane, is duplicated on the time graph to show its relationship to the respective time-varying signal.

The norm, phase angle and rotation rate of each phasor can be specified by entering the values of the parameters into fields marked with the corresponding color, blue or red. The program accepts rotation frequencies in the range $0.1 \leq f \leq 10$ Hz. Magnitude values can be from 0.5 to 3. Initial phase angles must naturally be in the range $-180 \leq \theta \leq 180$ degrees.

Visibility of a phasor and its associated signals can be turned on and off using the checkbox controls next to the graph window. The time variable can be advanced by clicking on the arrow button next to the slider control. As t is increased, each phasor rotates counterclockwise at its specified frequency.

Software resources:
phs_demo.m

Software resources: See MATLAB Exercise 1.5.

1.4 Discrete-Time Signals

Discrete-time signals are not defined at all time instants. Instead, they are defined only at time instants that are integer multiples of a fixed time increment T, that is, at $t = nT$. Consequently, the mathematical model for a discrete-time signal is a function $x[n]$ in which independent variable n is an integer, and is referred to as the *sample index*. The resulting signal is an indexed sequence of numbers, each of which is referred to as a *sample* of the signal. Discrete-time signals are often illustrated graphically using *stem* plots, an example of which is shown in Fig. 1.58.

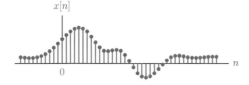

Figure 1.58 – A discrete-time signal.

As a practical example of a discrete-time signal, daily closing values of the Dow Jones Industrial Average for the first three months of 2003 are shown in Fig. 1.59. In this case the time interval T corresponds to a day, and signal samples are indexed with integers corresponding to subsequent days of trading.

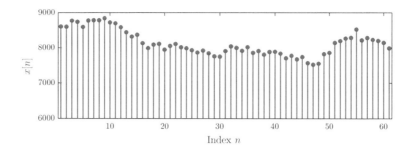

Figure 1.59 – Dow Jones Industrial Average for the first three months of 2003.

Sometimes discrete-time signals are also modeled using mathematical functions. For example

$$x[n] = 3 \sin[0.2n]$$

is a discrete-time sinusoidal signal. In some cases it may be more convenient to express a signal in a tabular form. A compact way of tabulating a signal is by listing the significant signal samples between a pair of braces, and separating them with commas:

$$x[n] = \{\, 3.7,\, 1.3,\, -1.5,\, 3.4,\, 5.9 \,\} \qquad (1.129)$$

The up-arrow indicates the position of the sample index $n = 0$, so we have $x[-1] = 3.7$, $x[0] = 1.3$, $x[1] = -1.5$, $x[2] = 3.4$, and $x[3] = 5.9$.

If the significant range of signal samples to be tabulated does not include $n = 0$, then we specify which index the up-arrow indicates. For example

$$x[n] = \{\, 1.1,\, 2.5,\, 3.7,\, 3.2,\, 2.6 \,\} \qquad (1.130)$$
$$n=5$$

indicates that $x[4] = 1.1$, $x[5] = 2.5$, $x[6] = 3.7$, $x[7] = 3.2$, and $x[8] = 2.6$.

For consistency in handling discrete-time signals we will assume that any discrete-time signal $x[n]$ has an infinite number of samples for $-\infty < n < \infty$. If a signal is described in tabular form as shown in Eqns. (1.129) and (1.130), any samples not listed will be taken to have zero amplitudes.

In a discrete-time signal the time variable is discrete, yet the amplitude of each sample is continuous. Even if the range of amplitude values may be limited, any amplitude within the prescribed range is allowed.

If, in addition to limiting the time variable to the set of integers, we also limit the amplitude values to a discrete set, the resulting signal is called a *digital* signal. In the simplest case there are only two possible values for the amplitude of each sample, typically indicated by "0" and "1". The corresponding signal is called a *binary* signal. Each sample of a binary signal is called a *bit* which stands for *binary digit*. Alternatively each sample of a digital signal could take on a value from a set of M allowed values, and the resulting digital signal is called an *M-ary signal*.

Software resources: See MATLAB Exercise 1.6.

1.4.1   Signal operations

Basic signal operations for continuous-time signals were discussed in Section 1.3.1. In this section we will discuss corresponding signal operations for discrete-time signals. Arithmetic operations for discrete-time signals bear strong resemblance to their continuous-time counterparts. Discrete-time versions of time shifting, time scaling and time reversal operations will also be discussed. Technically the use of the word *"time"* is somewhat inaccurate for these operations since the independent variable for discrete-time signals is the sample index n which may or may not correspond to time. We will, however, use the established terms of time shifting, time scaling and time reversal while keeping in mind the distinction between t and n. More advanced signal operations such as convolution will be covered in later chapters.

Arithmetic operations

Consider a discrete-time signal $x[n]$. A constant offset value can be added to this signal to obtain

$$g[n] = x[n] + A \tag{1.131}$$

The offset A is added to each sample of the signal $x[n]$. If we were to write Eqn. (1.131) for specific values of the index n we would obtain

$$\vdots$$
$$g[-1] = x[-1] + A$$
$$g[0] = x[0] + A$$
$$g[1] = x[1] + A$$
$$\vdots$$

and so on. This is illustrated in Fig. 1.60.

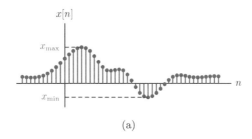

(a)

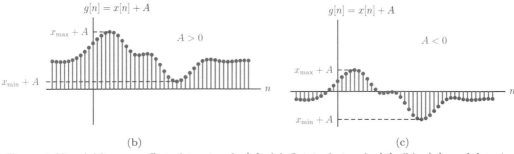

(b)   (c)

Figure 1.60 – Adding an offset A to signal $x[n]$: (a) Original signal $x[n]$, (b) $g[n] = x[n] + A$ with $A > 0$, (c) $g[n] = x[n] + A$ with $A < 0$.

Multiplication of the signal $x[n]$ with gain factor B is expressed in the form

$$g[n] = B\,x[n] \tag{1.132}$$

The value of each sample of the signal $g[n]$ is equal to the product of the corresponding sample of $x[n]$ and the constant gain factor B. For individual index values we have

$$\vdots$$
$$g[-1] = B\,x[-1]$$
$$g[0] = B\,x[0]$$
$$g[1] = B\,x[1]$$
$$\vdots$$

and so on. This is illustrated in Fig. 1.61.

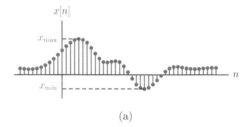

(a)

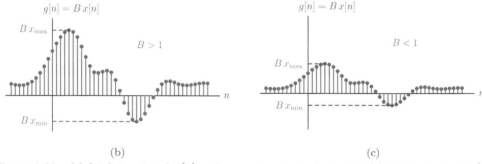

(b)   (c)

Figure 1.61 – Multiplying signal $x[n]$ with a constant gain factor B: (a) Original signal $x[n]$, (b) $g[n] = B\,x[n]$ with $B > 1$, (c) $g[n] = B\,x[n]$ with $B < 1$.

Addition of two discrete-time signals is accomplished by adding the amplitudes of the corresponding samples of the two signals. Let $x_1[n]$ and $x_2[n]$ be the two signals being added. The sum

$$g[n] = x_1[n] + x_2[n] \tag{1.133}$$

is computed for each value of the index n as

$$\vdots$$
$$g[-1] = x_1[-1] + x_2[-1]$$
$$g[0] = x_1[0] + x_2[0]$$
$$g[1] = x_1[1] + x_2[1]$$
$$\vdots$$

This is illustrated in Fig. 1.62.

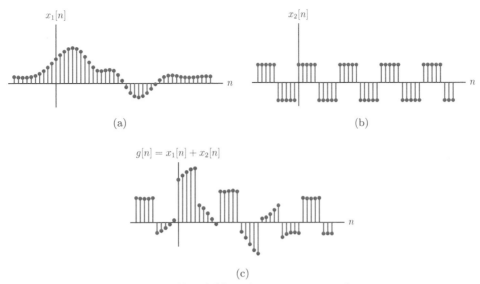

(a)   (b)

(c)

Figure 1.62 – Adding discrete-time signals.

Two discrete-time signals can also be multiplied in a similar manner. The product of two signals $x_1[n]$ and $x_2[n]$ is

$$g[n] = x_1[n] \, x_2[n] \qquad (1.134)$$

which can be written for specific values of the index as

$$\vdots$$
$$g[-1] = x_1[-1] \, x_2[-1]$$
$$g[0] = x_1[0] \, x_2[0]$$
$$g[1] = x_1[1] \, x_2[1]$$
$$\vdots$$

and is shown in Fig. 1.63.

Time shifting

Since discrete-time signals are defined only for integer values of the sample index, time shifting operations must utilize integer shift parameters. A time shifted version of the signal $x[n]$ is obtained as

$$g[n] = x[n - k] \qquad (1.135)$$

where k is any positive or negative integer. The relationship between $x[n]$ and $g[n]$ is illustrated in Fig. 1.64.

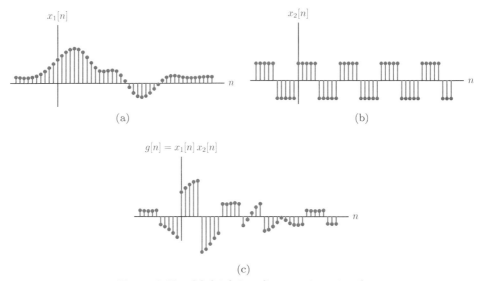

(a)   (b)

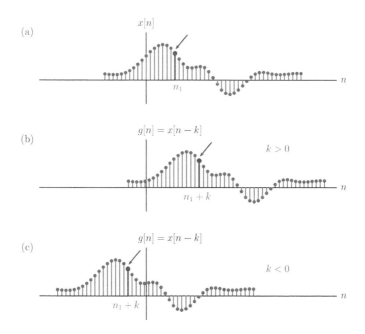

(c)

Figure 1.63 – Multiplying discrete-time signals.

Figure 1.64 – Time shifting a discrete-time signal.

In part (a) of Fig. 1.64 the sample of $x[n]$ at index n_1 is marked with a thicker stem. Let that sample correspond to a special event in signal $x[n]$. Substituting $n = n_1 + k$ in Eqn. (1.135) we have

$$g[n_1 + k] = x[n_1] \tag{1.136}$$

It is clear from Eqn. (1.136) that the event that takes place in $x[n]$ at index $n = n_1$ takes place in $g[n]$ at index $n = n_1 + k$. If k is positive, this corresponds to a delay by k samples. Conversely, a negative k implies an advance.

Time scaling

For discrete-time signals we will consider time scaling in the following two forms:

$$g[n] = x[kn] , \quad k: \text{integer} \tag{1.137}$$

and

$$g[n] = x[n/k] , \quad k: \text{integer} \tag{1.138}$$

Let us first consider the form in Eqn. (1.137). The relationship between $x[n]$ and $g[n] = x[kn]$ is illustrated in Fig. 1.65 for $k = 2$ and $k = 3$.

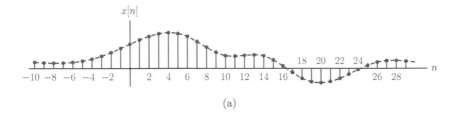

(a)

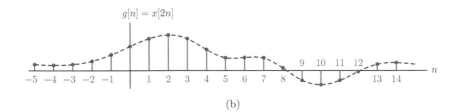

(b)

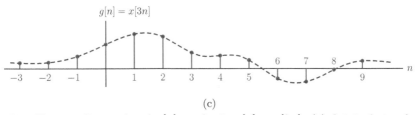

(c)

Figure 1.65 – Time scaling a signal $x[n]$ to obtain $g[n] = x[kn]$: (a) Original signal $x[n]$, (b) $g[n] = x[2n]$, (c) $g[n] = x[3n]$.

It will be interesting to write this relationship for several values of the index n. For $k = 2$ we have

$$\ldots \quad g[-1] = x[-2] , \quad g[0] = x[0] , \quad g[1] = x[2] , \quad g[2] = x[4] , \quad \ldots$$

which suggests that $g[n]$ retains every other sample of $x[n]$, and discards the samples between them. This relationship is further illustrated in Fig. 1.66.

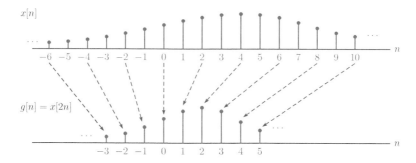

Figure 1.66 – Illustration of time scaling (downsampling) a discrete-time signal by a factor of 2.

For $k = 3$, samples of $g[n]$ are

$$\ldots \quad g[-1] = x[-3] \,, \quad g[0] = x[0] \,, \quad g[1] = x[3] \,, \quad g[2] = x[6] \,, \quad \ldots$$

In this case every third sample of $x[n]$ is retained, and the samples between them are discarded, as shown in Fig. 1.67.

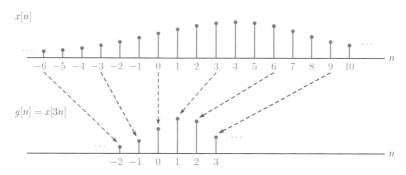

Figure 1.67 – Illustration of time scaling (downsampling) a discrete-time signal by a factor of 3.

This raises an interesting question: Does the act of discarding samples lead to a loss of information, or were those samples redundant in the first place? The answer depends on the characteristics of the signal $x[n]$, and will be explored when we discuss *downsampling* and *decimation* in Chapter 6.

An alternative form of time scaling for a discrete-time signal was given by Eqn. (1.138). Consider for example, the signal $g[n]$ defined based on Eqn. (1.138) with $k = 2$:

$$g[n] = x[n/2] \tag{1.139}$$

Since the index of the signal on the right side of the equal sign is $n/2$, the relationship between $g[n]$ and $x[n]$ is defined only for values of n that make $n/2$ an integer. We can write

$$\ldots \quad g[-2] = x[-1] \,, \quad g[0] = x[0] \,, \quad g[2] = x[1] \,, \quad g[4] = x[2] \,, \quad \ldots$$

The sample amplitudes of the signal $g[n]$ for odd values of n are not linked to the signal $x[n]$ in any way. For the sake of discussion let us set those undefined sample amplitudes equal to zero. Thus, we can write a complete description of the signal $g[n]$

$$g[n] = \begin{cases} x[n/2] \,, & \text{if } n/2 \text{ is integer} \\ 0 \,, & \text{otherwise} \end{cases} \tag{1.140}$$

Fig. 1.68 illustrates the process of obtaining $g[n]$ from $x[n]$ based on Eqn. (1.140). The relationship between $x[n]$ and $g[n]$ is known as *upsampling*, and will be discussed in more detail in Chapter 6 in the context of *interpolation*.

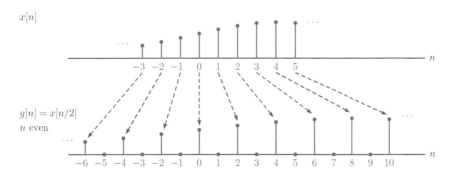

Figure 1.68 – Illustration of time scaling (upsampling) a discrete-time signal by a factor of 2.

Time reversal

A time reversed version of the signal $x[n]$ is

$$g[n] = x[-n] \tag{1.141}$$

An event that takes place at index value $n = n_1$ in the signal $x[n]$ takes place at index value $n = -n_1$ in the signal $g[n]$. Graphically this corresponds to folding or flipping the signal $x[n]$ around $n = 0$ axis as illustrated in Fig. 1.69.

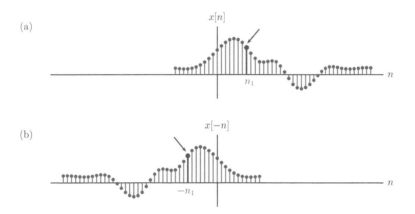

Figure 1.69 – Time reversal of a discrete-time signal.

1.4.2 Basic building blocks for discrete-time signals

In this section we will look at basic discrete-time signal building blocks that are used in constructing mathematical models for discrete-time signals with higher complexity. We will see that many of the continuous-time signal building blocks discussed in Section 1.3.2 have discrete-time counterparts that are defined similarly, and that have similar properties. There are also some fundamental differences between continuous-time and discrete-time versions of the basic signals, and these will be indicated throughout our discussion.

Unit-impulse function

The discrete-time unit-impulse function is defined by

$$\delta[n] = \left\{ \begin{array}{ll} 1, & n = 0 \\ 0, & n \neq 0 \end{array} \right. \tag{1.142}$$

and is shown graphically in Fig. 1.70.

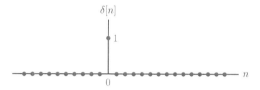

Figure 1.70 – Discrete-time unit-impulse signal.

As is evident from the definition in Eqn. (1.142) and Fig. 1.70, the discrete-time unit-impulse function does not have the complications associated with its continuous-time counterpart. The signal $\delta[n]$ is unambiguously defined for all integer values of the sample index n.

Shifted and scaled versions of the discrete-time unit-impulse function are used often in problems involving signal-system interaction. A unit-impulse function that is scaled by a and time shifted by n_1 samples is described by

$$a\,\delta[n - n_1] = \left\{ \begin{array}{ll} a, & n = n_1 \\ 0, & n \neq n_1 \end{array} \right. \tag{1.143}$$

and is shown in Fig. 1.71.

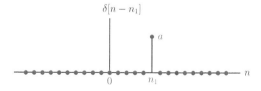

Figure 1.71 – Scaled and time shifted discrete-time unit-impulse signal.

The fundamental properties of the continuous-time unit-impulse function discussed in Section 1.3.2 can be readily adapted to its discrete-time counterpart. The *sampling property* of the discrete-time unit-impulse function is expressed as

$$x[n]\,\delta[n - n_1] = x[n_1]\,\delta[n - n_1] \tag{1.144}$$

It is important to interpret Eqn. (1.144) correctly: $x[n]$ and $\delta[n - n_1]$ are both infinitely long discrete-time signals, and can be graphed in terms of the sample index n as shown in Fig. 1.72.

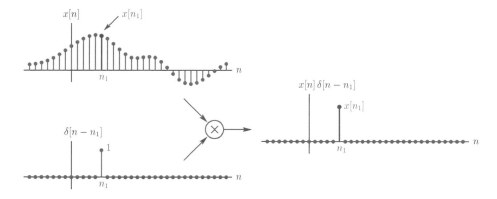

Figure 1.72 – Illustration of the sampling property of the discrete-time unit-impulse signal.

The claim in Eqn. (1.144) is easy to justify: If the two signals $x[n]$ and $\delta[n - n_1]$ are multiplied on a sample-by-sample basis, the product signal is equal to zero for all but one value of the sample index n, and the only non-zero amplitude occurs for $n = n_1$. Mathematically we have

$$x[n]\,\delta[n - n_1] = \begin{cases} x[n_1]\,, & n = n_1 \\ 0\,, & n \neq n_1 \end{cases} \qquad (1.145)$$

which is equivalent to the right side of Eqn. (1.144).

The *sifting property* for the discrete-time unit-impulse function is expressed as

$$\sum_{n=-\infty}^{\infty} x[n]\,\delta[n - n_1] = x[n_1] \qquad (1.146)$$

which easily follows from Eqn. (1.144). Substituting Eqn. (1.144) into Eqn. (1.146) we obtain

$$\sum_{n=-\infty}^{\infty} x[n]\,\delta[n - n_1] = \sum_{n=-\infty}^{\infty} x[n_1]\,\delta[n - n_1]$$

$$= x[n_1] \sum_{n=-\infty}^{\infty} \delta[n - n_1]$$

$$= x[n_1] \qquad (1.147)$$

where we have relied on the sum of all samples of the impulse signal being equal to unity. The result of the summation in Eqn. (1.146) is a scalar, the value of which equals sample n_1 of the signal $x[n]$.

Unit-step function

The discrete-time unit-step function can also be defined in a way similar to its continuous-time version:

$$u[n] = \begin{cases} 1\,, & n \geq 0 \\ 0\,, & n < 0 \end{cases} \qquad (1.148)$$

The function $u[n]$ is shown in Fig. 1.73.

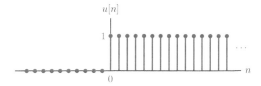

Figure 1.73 – Discrete-time unit-step signal.

As in the case of the discrete-time unit-impulse function, this function also enjoys a clean definition without any of the complications associated with its continuous-time counterpart. Eqn. (1.148) provides a complete definition of the discrete-time unit-step function for all integer values of the sample index n. A time shifted version of the discrete-time unit-step function can be written as

$$u[n - n_1] = \begin{cases} 1 , & n \geq n_1 \\ 0 , & n < n_1 \end{cases} \tag{1.149}$$

and is illustrated in Fig. 1.74.

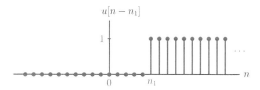

Figure 1.74 – Time shifted discrete-time unit-step signal.

Recall that a continuous-time unit-impulse could be obtained as the first derivative of the continuous-time unit-step. An analogous relationship exists between the discrete-time counterparts of these signals. It is possible to express a discrete-time unit-impulse signal as the *first difference* of the discrete-time unit-step signal:

$$\delta[n] = u[n] - u[n - 1] \tag{1.150}$$

This relationship is illustrated in Fig. 1.75.

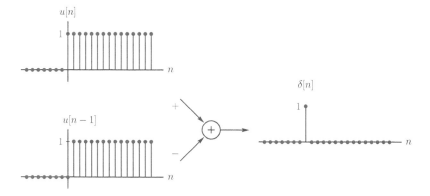

Figure 1.75 – Obtaining a discrete-time unit-impulse from a discrete-time unit-step through first difference.

Conversely, a unit-step signal can be constructed from unit-impulse signals through a *running sum* in the form

$$u[n] = \sum_{k=-\infty}^{n} \delta[k] \tag{1.151}$$

This is analogous to the running integral relationship between the continuous-time versions of these signals, given by Eqn. (1.33). In Eqn. (1.151) we are adding the samples of a unit-step signal $\delta[k]$ starting from $k = -\infty$ up to and including the sample for $k = n$. If n, the upper limit of the summation, is zero or positive, the summation includes the only sample with unit amplitude, and the result is equal to unity. If $n < 0$, the summation ends before we reach that sample, and the result is zero. This is shown in Fig. 1.76.

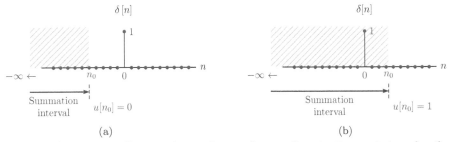

Figure 1.76 – Obtaining a discrete-time unit-step from a discrete-time unit-impulse through a running sum: (a) $n = n_0 < 0$, (b) $n = n_0 > 0$.

An alternative approach for obtaining a unit-step from a unit-impulse is to use

$$u[n] = \sum_{k=0}^{\infty} \delta[n - k] \tag{1.152}$$

in which we add the signals $\delta[n], \delta[n-1], \delta[n-2], \ldots, \delta[n-k]$ to construct a unit-step signal. This is an example of impulse decomposition that will be discussed in Section 1.4.3.

Unit-ramp function

The discrete-time version of the unit-ramp function is defined as

$$r[n] = \begin{cases} n, & n \geq 0 \\ 0, & n < 0 \end{cases} \tag{1.153}$$

and is shown in Fig. 1.77.

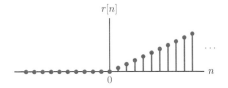

Figure 1.77 – Discrete-time unit-ramp function.

The definition in Eqn. (1.153) can be written in a more compact form as the product of the linear signal $g[n] = n$ and the unit-step function. We can write $r[n]$ as

$$r[n] = n\, u[n] \tag{1.154}$$

which is illustrated in Fig. 1.78.

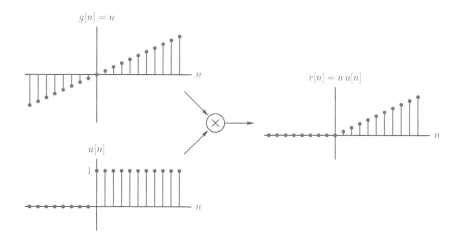

Figure 1.78 – Obtaining a discrete-time unit ramp.

Alternatively, the discrete-time unit-ramp function can be expressed as a running summation applied to the discrete-time unit-step function in the form

$$r[n] = \sum_{k=-\infty}^{n-1} u[k] \tag{1.155}$$

By trying out the summation in Eqn. (1.155) for a few different values of the index n it is easy to see that it produces values consistent with the definition of the unit-ramp function given by Eqn. (1.154). This is analogous to the relationship between the continuous-time versions of these signals where a running integral is used for obtaining a continuous-time unit-ramp from a continuous-time unit-step as stated in Eqn. (1.41). The process of obtaining a discrete-time unit-ramp through a running sum is illustrated in Fig. 1.79.

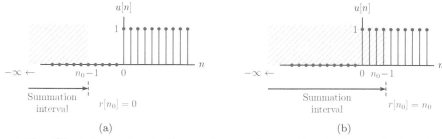

Figure 1.79 – Obtaining a discrete-time unit-ramp from a discrete-time unit-step through a running sum: (a) $n = n_0 < 0$, (b) $n = n_0 > 0$.

Sinusoidal signals

A discrete-time sinusoidal signal is in the general form

$$x[n] = A \cos{(\Omega_0 n + \theta)} \qquad (1.156)$$

where A is the *amplitude*, Ω_0 is the *angular frequency* in radians, and θ is the phase angle which is also in radians. The angular frequency Ω_0 can be expressed as

$$\Omega_0 = 2\pi F_0 \qquad (1.157)$$

The parameter F_0, a dimensionless quantity, is referred to as the *normalized frequency* of the sinusoidal signal. Fig. 1.80 illustrates discrete-time sinusoidal signals for various values of Ω_0.

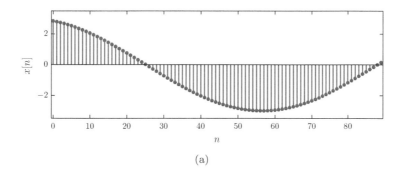

(a)

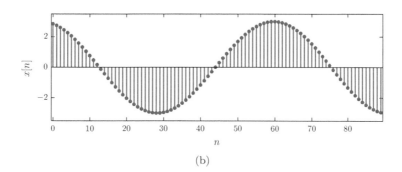

(b)

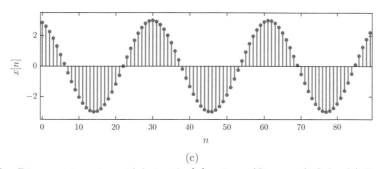

(c)

Figure 1.80 – Discrete-time sinusoidal signal $x[n] = 3 \cos{(\Omega_0 n + \pi/10)}$ for (a) $\Omega_0 = 0.05$ rad, (b) $\Omega_0 = 0.1$ rad, (c) $\Omega_0 = 0.2$ rad.

At this point we will note a fundamental difference between a discrete-time sinusoidal signal and its continuous-time counterpart:

1. A continuous-time sinusoidal signal was defined by Eqn. (1.44). Its parameter ω_0 is in rad/s since it appears next to the time variable t, and the product $\omega_0 t$ must be in radians to qualify as the argument of a trigonometric function.

2. In contrast, the parameter Ω_0 is in radians since it appears next to a dimensionless index parameter n, and the product $\Omega_0 n$ must be in radians. For this reason, Ω_0 is referred to as the angular frequency of the discrete-time sinusoidal signal.

Even though the word "frequency" is used, Ω_0 is not really a frequency in the traditional sense, but rather an angle. In support of this assertion, we will see in later parts of this text that values of Ω_0 outside the range $-\pi \leq \Omega_0 < \pi$ are mathematically indistinguishable from values within that range.

Similar reasoning applies to the parameter F_0: It is not really a frequency but a dimensionless quantity that can best be thought of as a "percentage". In order to elaborate further on the meanings of the parameters Ω_0 and F_0, we will consider the case of obtaining a discrete-time sinusoidal signal from a continuous-time sinusoidal signal in the form

$$x_a(t) = A \cos(\omega_0 t + \theta) \tag{1.158}$$

Let us evaluate the amplitude of $x_a(t)$ at time instants that are integer multiples of T_s, and construct a discrete-time signal with the results.

$$\begin{aligned} x[n] &= x_a(nT_s) \\ &= A \cos(\omega_0 T_s n + \theta) \\ &= A \cos(2\pi f_0 T_s n + \theta) \end{aligned} \tag{1.159}$$

This is illustrated in Fig. 1.81.

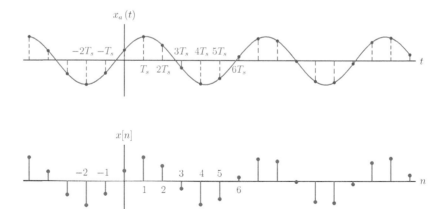

Figure 1.81 – Obtaining a discrete-time sinusoidal signal from a continuous-time sinusoidal signal.

Since the signal $x_a(t)$ is evaluated at intervals of T_s, the number of samples taken per unit time is

$$f_s = \frac{1}{T_s} \tag{1.160}$$

Substituting Eqn. (1.160) into Eqn. (1.159) we obtain

$$x[n] = A \cos\left(2\pi \left[\frac{f_0}{f_s}\right] n + \theta\right) = A \cos\left(2\pi F_0 n + \theta\right) \tag{1.161}$$

which reveals the meaning of the normalized frequency F_0:

$$F_0 = \frac{f_0}{f_s} \tag{1.162}$$

The act of constructing a discrete-time signal by evaluating a continuous-time signal at uniform intervals is called *sampling*, and will be discussed in more detail in Chapter 6. The parameters f_s and T_s are referred to as the *sampling rate* and the *sampling interval* respectively. Eqn. (1.162) suggests that the normalized frequency F_0 is essentially the frequency f_0 of the continuous-time sinusoid expressed as a percentage of the sampling rate.

1.4.3 Impulse decomposition for discrete-time signals

Consider an arbitrary discrete-time signal $x[n]$. Let us define a new signal $x_k[n]$ by using the k-th sample of the signal $x[n]$ in conjunction with a time shifted unit-impulse function as

$$x_k[n] = x[k]\,\delta[n-k] = \begin{cases} x[k], & n = k \\ 0, & n \neq k \end{cases} \tag{1.163}$$

The signal $x_k[n]$ is a scaled and time shifted impulse signal, the only non-trivial sample of which occurs at index $n = k$ with an amplitude of $x[k]$. If we were to repeat Eqn. (1.163) for all possible values of k, we would obtain an infinite number of signals $x_k[n]$ for $k = -\infty, \dots, \infty$. In each of these signals there would only be one non-trivial sample the amplitude of which equals the amplitude of the corresponding sample of $x[n]$. For example, consider the signal

$$x[n] = \{\, 3.7,\ \underset{\uparrow}{1.3},\ -1.5,\ 3.4,\ 5.9 \,\}$$

The signals $x_k[n]$ for this case would be as follows:

$$\vdots$$

$$
\begin{aligned}
x_{-1}[n] &= \{\, 3.7, \quad \underset{\uparrow}{0}, \quad\ 0, \quad\ \ 0, \quad 0 \,\} \\
x_0[n] &= \{\, 0, \quad \underset{\uparrow}{1.3}, \quad 0, \quad\ \ 0, \quad 0 \,\} \\
x_1[n] &= \{\, 0, \quad \underset{\uparrow}{0}, \quad -1.5, \quad 0, \quad 0 \,\} \\
x_2[n] &= \{\, 0, \quad \underset{\uparrow}{0}, \quad\ \ 0, \quad 3.4, \quad 0 \,\} \\
x_3[n] &= \{\, 0, \quad \underset{\uparrow}{0}, \quad\ \ 0, \quad\ \ 0, \quad 5.9 \,\}
\end{aligned}
$$

$$\vdots$$

The signal $x[n]$ can be reconstructed by adding these components together.

$$x[n] = \sum_{k=-\infty}^{\infty} x_k[n] = \sum_{k=-\infty}^{\infty} x[k]\,\delta[n-k] \tag{1.164}$$

Eqn. (1.164) represents an impulse decomposition of the discrete-time signal $x[n]$. We will use it to derive the convolution relationship in Chapter 3.

1.4.4 Signal classifications

Discussion in this section will parallel that of Section 1.3.4 for continuous-time signals.

Real vs. complex signals

A discrete-time signal may be real or complex valued. A complex signal can be written in *Cartesian form* using its real and imaginary parts as

$$x[n] = x_r[n] + x_i[n] \tag{1.165}$$

or in *polar form* using its magnitude and phase as

$$x[n] = |x[n]| \, e^{j\angle x[n]} \tag{1.166}$$

The equations to convert from one complex form to the other are identical to those for continuous-time complex signals.

Periodic vs. non-periodic signals

A discrete-time signal is said to be periodic if it satisfies

$$x[n] = x[n + N] \tag{1.167}$$

for all values of the integer index n and for a specific value of $N \neq 0$. The parameter N is referred to as the *period* of the signal. An example of a periodic discrete-time signal is shown in Fig. 1.82.

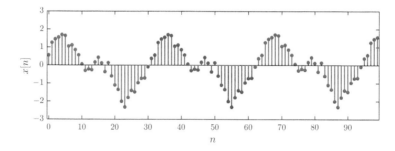

Figure 1.82 – Example of a discrete-time signal that is periodic.

A discrete-time signal that is periodic with a period of N samples is also periodic with periods of $2N, 3N, \ldots, kN$ for any positive integer k. Applying the periodicity definition of Eqn. (1.167) to $x[n + N]$ instead of $x[n]$ yields

$$x[n + N] = x[n + 2N] \tag{1.168}$$

Substituting Eqn. (1.168) into Eqn. (1.167) we obtain

$$x[n] = x[n + 2N] \tag{1.169}$$

and, through repeated use of this process, we can show that

$$x[n] = x[n + kN] \tag{1.170}$$

The smallest positive value of N that satisfies Eqn. (1.167) is called the *fundamental period*. To avoid ambiguity when we refer to the period of a signal, fundamental period will be implied unless we specifically state otherwise. The *normalized fundamental frequency* of a discrete-time periodic signal is the reciprocal of its fundamental period, i.e.,

$$F_0 = \frac{1}{N} \tag{1.171}$$

Periodicity of discrete-time sinusoidal signals

The general form of a discrete-time sinusoidal signal $x[n]$ was given by Eqn. (1.161). For $x[n]$ to be periodic, it needs to satisfy the periodicity condition given by Eqn. (1.167). Specifically we need

$$
\begin{aligned}
A \cos\left(2\pi F_0 n + \theta\right) &= A \cos\left(2\pi F_0 \left[n + N\right] + \theta\right) \\
&= A \cos\left(2\pi F_0 n + 2\pi F_0 N + \theta\right)
\end{aligned} \tag{1.172}
$$

For Eqn. (1.172) to hold, the arguments of the cosine functions must differ by an integer multiple of 2π. This requirement results in

$$2\pi F_0 N = 2\pi k \tag{1.173}$$

and consequently

$$N = \frac{k}{F_0} \tag{1.174}$$

for the period N. Since we are dealing with a discrete-time signal, there is the added requirement that the period N obtained from Eqn. (1.174) must be an integer value. Thus, the discrete-time sinusoidal signal defined by Eqn. (1.161) is periodic provided that Eqn. (1.174) yields an integer value for N. The fundamental period of the sinusoidal signal is then obtained by using the smallest integer value of k, if any, that results in N being an integer.

It should be obvious from the foregoing discussion that, contrary to a continuous-time sinusoidal signal always being periodic, a discrete-time sinusoidal signal may or may not be periodic. The signal will not be periodic, for example, if the normalized frequency F_0 is an irrational number so that no value of k produces an integer N in Eqn. (1.174).

Example 1.16: Periodicity of a discrete-time sinusoidal signal

Check the periodicity of the following discrete-time signals:

a. $x[n] = \cos\left(0.2n\right)$
b. $x[n] = \cos\left(0.2\pi n + \pi/5\right)$
c. $x[n] = \cos\left(0.3\pi n - \pi/10\right)$

Solution:

a. The angular frequency of this signal is $\Omega_0 = 0.2$ radians which corresponds to a normalized frequency of

$$F_0 = \frac{\Omega_0}{2\pi} = \frac{0.2}{2\pi} = \frac{0.1}{\pi}$$

This results in a period

$$N = \frac{k}{F_0} = 10\pi k$$

Since no value of k would produce an integer value for N, the signal is not periodic.

b. In this case the angular frequency is $\Omega_0 = 0.2\pi$ radians, and the normalized frequency is $F_0 = 0.1$. The period is

$$N = \frac{k}{F_0} = \frac{k}{0.1} = 10k$$

For $k = 1$ we have $N = 10$ samples as the fundamental period. The signal $x[n]$ is shown in Fig. 1.83.

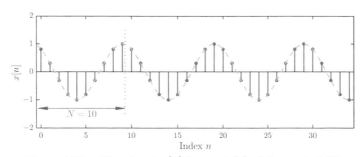

Figure 1.83 – The signal $x[n]$ for part (b) of Example 1.16.

c. For this signal the angular frequency is $\Omega_0 = 0.3\pi$ radians, and the corresponding normalized frequency is $F_0 = 0.15$. The period is

$$N = \frac{k}{F_0} = \frac{k}{0.15}$$

The smallest positive integer k that would result in an integer value for the period N is $k = 3$. Therefore, the fundamental period is $N = 3/0.15 = 20$ samples. The signal $x[n]$ is shown in Fig. 1.84.

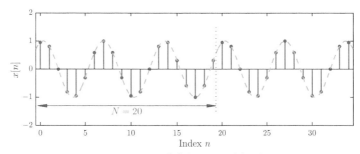

Figure 1.84 – The signal $x[n]$ for part (c) of Example 1.16.

It is interesting to observe from Fig. 1.84 that the period of $N = 20$ samples corresponds to three full cycles of the continuous-time sinusoidal signal from which $x[n]$ may have been derived (see the outline shown in the figure). This is due to the fact that $k = 3$ and, based on Eqn. (1.173), the argument of the cosine function is advanced by

$$2\pi F_0 N = 2\pi k = 6\pi \quad \text{radians}$$

after one period of 20 samples.

Software resources:
ex_1_16a.m
ex_1_16b.m
ex_1_16c.m

Example 1.17: **Periodicity of a multi-tone discrete-time sinusoidal signal**

Comment on the periodicity of the two-tone discrete-time signal

$$x[n] = 2\cos(0.4\pi n) + 1.5\sin(0.48\pi n) \tag{1.175}$$

Solution: The signal is in the form

$$x[n] = x_1[n] + x_2[n] \tag{1.176}$$

with

$$x_1[n] = 2\cos(\Omega_1 n) \ , \quad \Omega_1 = 0.4\pi \text{ rad} \tag{1.177}$$

and

$$x_2[n] = 1.5\sin(\Omega_2 n) \ , \quad \Omega_2 = 0.48\pi \text{ rad} \tag{1.178}$$

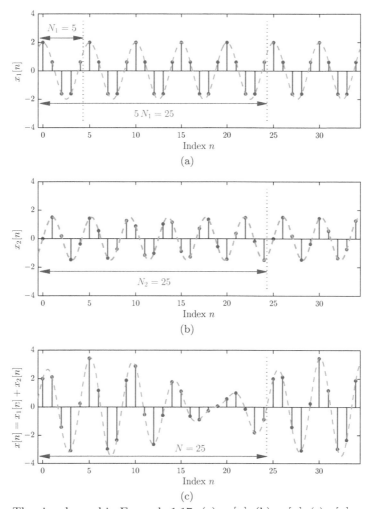

Figure 1.85 – The signals used in Example 1.17: (a) $x_1[n]$, (b) $x_2[n]$, (c) $x[n] = x_1[n] + x_2[n]$.

Corresponding normalized frequencies are $F_1 = 0.2$ and $F_2 = 0.24$. The period of each component can be found as follows:

$$\text{For } x_1[n]: \quad N_1 = \frac{k_1}{F_1} = \frac{k_1}{0.2}, \quad k_1 = 1, \quad N_1 = 5$$

$$\text{For } x_2[n]: \quad N_2 = \frac{k_2}{F_2} = \frac{k_2}{0.24}, \quad k_2 = 6, \quad N_2 = 25$$

Thus the fundamental period for $x_1[n]$ is $N_1 = 5$ samples, and the fundamental period for $x_2[n]$ is $N_2 = 25$ samples. The period of the total signal $x[n]$ is $N = 25$ samples. Within the period of $N = 25$ samples, the first component $x_1[n]$ completes 5 cycles since $N = 5N_1$. The second component completes 6 cycles since $k_2 = 6$ to get an integer value for N_2. This is illustrated in Fig. 1.85.

Software resources:

ex_1_17.m

Deterministic vs. random signals

Deterministic signals are signals that can be described completely in analytical form in the time domain. Random signals are signals that cannot be modeled analytically. They can be analyzed in terms of their statistical properties. Study of random signals is beyond the scope of this text, and the reader is referred to one of the many excellent texts available on the subject.

1.4.5 Energy and power definitions

Energy of a signal

The energy of a real-valued discrete-time signal is computed as

$$E_x = \sum_{n=-\infty}^{\infty} x^2[n] \tag{1.179}$$

provided that it can be computed. If the signal under consideration is complex-valued, its energy is given by

$$E_x = \sum_{n=-\infty}^{\infty} \left| x[n] \right|^2 \tag{1.180}$$

Time averaging operator

We will use the operator $\langle \ldots \rangle$ to indicate time average. If the signal $x[n]$ is periodic with period N, its time average can be computed as

$$\langle x[n] \rangle = \frac{1}{N} \sum_{n=0}^{N-1} x[n] \tag{1.181}$$

For a signal $x[n]$ that is non-periodic, the definition of time average in Eqn. (1.181) can be generalized with the use of the limit operator as

$$\langle x[n] \rangle = \lim_{M \to \infty} \left[\frac{1}{2M+1} \sum_{n=-M}^{M} x[n] \right] \tag{1.182}$$

Power of a signal

The average power of a real-valued discrete-time signal is computed as

$$P_x = \left\langle x^2[n] \right\rangle \tag{1.183}$$

This definition works for both periodic and non-periodic signals. For a periodic signal, Eqn. (1.181) can be used with Eqn. (1.183) to yield

$$P_x = \frac{1}{N} \sum_{n=0}^{N-1} x^2[n] \tag{1.184}$$

For a non-periodic signal, Eqn. (1.182) can be substituted into Eqn. (1.183) to yield

$$P_x = \lim_{M \to \infty} \left[\frac{1}{2M+1} \sum_{n=-M}^{M} x^2[n] \right] \tag{1.185}$$

Eqns. (1.184) and (1.185) apply to signals that are real-valued. If the signal under consideration is complex, then Eqns. (1.184) and (1.185) can be generalized by using the squared norm of the signal $x[n]$, i.e.,

$$P_x = \left\langle \left| x[n] \right|^2 \right\rangle \tag{1.186}$$

Thus, the power of a periodic complex signal is

$$P_x = \frac{1}{N} \sum_{n=0}^{N-1} \left| x[n] \right|^2 \tag{1.187}$$

and the power of a non-periodic complex signal is

$$P_x = \lim_{M \to \infty} \left[\frac{1}{2M+1} \sum_{n=-M}^{M} \left| x[n] \right|^2 \right] \tag{1.188}$$

Energy signals vs. power signals

As with continuous-time signals, it is also possible to classify discrete-time signals based on their energy and power:

- Energy signals are those that have finite energy and zero power, i.e., $E_x < \infty$, and $P_x = 0$.

- Power signals are those that have finite power and infinite energy, i.e., $E_x \to \infty$, and $P_x < \infty$.

1.4.6 Symmetry properties

Definitions of the symmetry properties given in Section 1.3.6 for continuous-time signals can easily be adapted to discrete-time signals. In addition, discrete-time signals without symmetry can be expressed in terms of symmetric components.

Even and odd symmetry

A real-valued discrete-time signal is even symmetric if it satisfies

$$x[-n] = x[n] \tag{1.189}$$

for all integer values of the sample index n. In contrast, a discrete-time signal with odd symmetry satisfies

$$x[-n] = -x[n] \tag{1.190}$$

for all n. A signal $x[n]$ that is even symmetric remains unchanged when it is reversed in time. If $x[n]$ has odd symmetry, time reversing it causes the signal to be negated. Examples of discrete-time signals with even and odd symmetry are shown in Fig. 1.86(a) and (b).

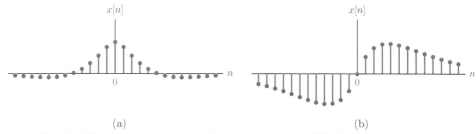

(a)   (b)

Figure 1.86 – (a) Discrete-time signal with even symmetry, (b) discrete-time signal with odd symmetry.

Decomposition into even and odd components

As in the case of continuous-time signals, any real-valued discrete-time signal $x[n]$ can be written as the sum of two signals, one with even symmetry and one with odd symmetry. The signal $x[n]$ does not have to have any symmetry properties for this to work. Consider the following representation of the signal $x[n]$:

$$x[n] = x_e[n] + x_o[n] \tag{1.191}$$

The signal $x_e[n]$ is the even component defined as

$$x_e[n] = \frac{x[n] + x[-n]}{2} \tag{1.192}$$

and $x_o[n]$ is the odd component defined as

$$x_o[n] = \frac{x[n] - x[-n]}{2} \tag{1.193}$$

It can easily be verified that $x_e[-n] = x_e[n]$ and $x_o[-n] = -x_o[n]$. Furthermore, the signals $x_e[n]$ and $x_o[n]$ always add up to $x[n]$ owing to the way they are defined.

Symmetry properties for complex signals

Even and odd symmetry definitions given by Eqns. (1.189) and (1.190) for a real-valued signal $x[n]$ can be extended to apply to complex-valued signals as well. A complex-valued signal $x[n]$ is conjugate symmetric if it satisfies

$$x[-n] = x^*[n] \tag{1.194}$$

for all n. Similarly, $x[n]$ is conjugate antisymmetric if it satisfies

$$x[-n] = -x^*[n] \tag{1.195}$$

for all n. If $x[n]$ is conjugate symmetric, time reversing it has the same effect as conjugating it. For a conjugate antisymmetric $x[n]$, time reversal is equivalent to conjugation and negation applied simultaneously.

If the signal $x[n]$ is real-valued, then its complex conjugate is equal to itself, that is, $x^*[n] = x[n]$. In this case the definition of conjugate symmetry reduces to that of even symmetry as can be seen from Eqns. (1.194) and (1.189). Similarly, conjugate antisymmetry property reduces to odd symmetry for a real-valued $x[n]$.

Decomposition of complex signals

Any complex signal $x[n]$ can be expressed as the sum of two signals of which one is conjugate symmetric and the other is conjugate antisymmetric. The component $x_E[n]$ defined as

$$x_E[n] = \frac{x[n] + x^*[-n]}{2} \tag{1.196}$$

is always conjugate symmetric, and the component $x_O[n]$ defined as

$$x_O[n] = \frac{x[n] - x^*[-n]}{2} \tag{1.197}$$

is always conjugate antisymmetric. The relationship

$$x[n] = x_E[n] + x_O[n] \tag{1.198}$$

holds due to the way $x_E[n]$ and $x_O[n]$ are defined.

1.5   Further Reading

[1] A.B. Carlson and P.B. Crilly. *Communication Systems*. McGraw-Hill, 2009.

[2] R.P. Kanwal. *Generalized Functions: Theory and Applications*. Birkhèauser, 2004.

[3] P. Prandoni and M. Vetterli. *Signal Processing for Communications*. Communication and Information Sciences. Taylor & Francis, 2008.

[4] A. Spanias, T. Painter, and V. Atti. *Audio Signal Processing and Coding*. Wiley, 2006.

[5] M. Tohyama. *Sound and Signals*. Signals and Communication Technology. Springer, 2011.

[6] U. Zölzer. *Digital Audio Signal Processing*. Wiley, 2008.

MATLAB Exercises

MATLAB Exercise 1.1: **Computing and graphing continuous-time signals**

Often we have the need to simulate mathematical models of signals on a computer. Sometimes this is done to check the results of hand calculations involving a signal, or to gain graphical insight into signal behavior. At other times we may need a numerical approx-imation to the mathematical model of a signal for use in a simulation study involving a signal-system interaction problem. In this exercise we will explore methods of computing and graphing continuous-time signal models in MATLAB.

Caveat: Since any row or column vector must have a finite-number of elements, a completely accurate representation of a continuous-time signal is not possible in MATLAB. The best we can do is to approximate a continuous-time signal by constructing a vector that holds amplitudes of the signal at equally-spaced time instants. If the vector used for this purpose has a sufficient number of amplitude values, and if the time spacing of these amplitude values is chosen carefully,[9] the result can still be a reasonably good approximation to a continuous-time signal. In graphing such a signal, the signal amplitudes are connected with straight lines to give the appearance of a continuous-time signal.

a. Compute the signal
$$x_1(t) = 5\sin(12t)$$

at 500 points in the time interval $0 \le t \le 5$, and graph the result.

Solution: Type the following three lines into MATLAB command window to obtain the graph in Fig. 1.87.

```
>>  t = linspace(0,5,500);
>>  x1 = 5*sin(12*t);
>>  plot(t,x1);
```

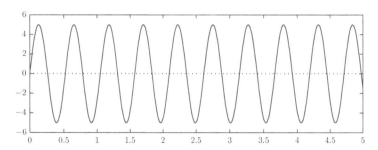

Figure 1.87 – Graph obtained in MATLAB Exercise 1.1 part (a).

b. Compute and graph the signal
$$x_2(t) = \begin{cases} e^{-3t} - e^{-6t}, & t \ge 0 \\ 0, & t < 0 \end{cases}$$

in the time interval $0 \le t \le 5$ seconds, using a time increment of $\Delta t = 0.01$ seconds.

Solution: This signal can be computed and graphed with the following statements:

```
>>  t = [0:0.01:5];
>>  x2 = exp(-3*t)-exp(-6*t);
>>  plot(t,x2);
```

The first line demonstrates an alternative method of constructing a time vector. The result is shown in Fig. 1.88.

---

[9] A more formal method of determining the appropriate time spacing for signal amplitude values will be given when we discuss the subject of *sampling* in Chapter 6.

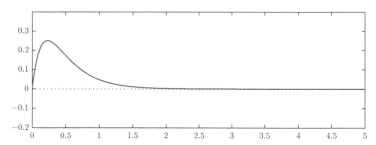

Figure 1.88 – Graph obtained in MATLAB Exercise 1.1 part (b).

c. Compute and graph the signal

$$x_2(t) = \begin{cases} e^{-3t} - e^{-6t}, & t \geq 0 \\ 0, & t < 0 \end{cases}$$

in the time interval $-2 \leq t \leq 3$ seconds, using a time increment of $\Delta t = 0.01$ seconds. Solution: The signal we need to graph is the same as the signal in part (b). The slight complication presented by this exercise is due the fact that we need to graph the signal starting at a negative time instant, and signal amplitudes are equal to zero for negative values of t. As a result, two separate descriptions of the signal need to be utilized, one for $t < 0$ and the other for $t \geq 0$. In part (b) we avoided this problem by graphing the signal only for positive values of t and thus using a single analytical expression to compute all signal amplitudes.

The simplest method of dealing with this problem is to use the logic operator ">=" which stands for *"greater than or equal to"*. The following lines produce the graph shown in Fig. 1.89.

```
>>  t = [-2:0.01:3];
>>  w = (t>=0);
>>  x2 = (exp(-3*t)-exp(-6*t)).*w;
>>  plot(t,x2);
```

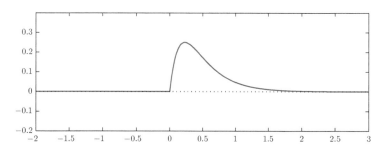

Figure 1.89 – Graph obtained in MATLAB Exercise 1.1 part (c).

Common mistake: It is important to use the *element-by-element multiplication* operator ".*" and not the *vector multiplication* operator "*" in the third line above. The expression for computing the vector "x2" involves two vectors of the same size, namely "(exp(-3*t)-exp(-6*t))" and "w". The operator ".*" creates a vector of the same size by multiplying the corresponding elements of the two vectors. The operator "*", on the other hand, is used for computing the *scalar product* of two vectors, and would not work in this case.

d. Compute and graph the signal

$$x_3(t) = \begin{cases} e^{-3t} - e^{-6t}, & 0 \leq t \leq 1 \\ 0, & \text{otherwise} \end{cases}$$

in the time interval $-2 \leq t \leq 3$ seconds, using a time increment of $\Delta t = 0.01$ seconds.
Solution: This signal is similar to the one in part (c). The only difference is that $x_3(t) = 0$ for $t > 1$. The logical test needs to be modified to reflect this. Using the *logical and* operator "&" we can type the following lines into MATLAB command window to compute and graph $x_3(t)$:

```
>>  t = [-2:0.01:3];
>>  x3 = (exp(-3*t)-exp(-6*t)).*((t>=0)&(t<=1));
>>  plot(t,x3);
```

The resulting graph is shown in Fig. 1.90.

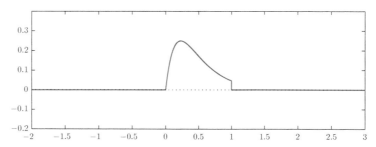

Figure 1.90 – Graph obtained in MATLAB Exercise 1.1 part (d).

Software resources:
matex_1_1a.m
matex_1_1b.m
matex_1_1c.m
matex_1_1d.m

MATLAB Exercise 1.2: Describing signals using piecewise linear segments

Consider the signal $x(t)$ obtained in Example 1.1 and shown in Fig. 1.8. A signal such as this one can also be described by specifying the endpoints of linear segments and interpolating between them. MATLAB function `interp1(..)` is used for this purpose.

The endpoints of the signal under consideration are

$$(t_p, x_p) = \{(-1, 0), (0, 0), (1, 1), (3, -1), (4, -1), (5, 0), (6, 0)\}$$

The code below computes the signal for $-1 \leq t \leq 6$ and graphs the result.

```
>>  tp = [-1,0,1,3,4,5,6];
>>  xp = [0,0,1,-1,-1,0,0];
>>  t = [-1:0.01:6];
>>  x = interp1(tp,xp,t,'linear');
>>  plot(t,x,'b-',tp,xp,'ro'); grid;
```

Caution: The function `interp1(..)` requires that all t_p values be distinct, that is, no two points could have the same time value. As a result, signals that have discontinuities (vertical edges) cannot be graphed using this approach.

Software resources:
`matex_1_2.m`

MATLAB Exercise 1.3: Signal operations for continuous-time signals

In MATLAB Exercise 1.1 we have used vectors for representing mathematical models for continuous-time signals. This approach is often useful for simulating signal-system interaction problems, and will be utilized extensively in later parts of this textbook. In this exercise, however, we will consider alternative approaches for representing signals in MATLAB for the purpose of experimenting with signal operations such as time shifting, time scaling and time reversal.

Consider again the signal $x(t)$ used in Example 1.3 and shown in Fig. 1.16. Analytically $x(t)$ can be written in the form

$$x(t) = \begin{cases} 2, & -1 \leq t < 2 \\ 1, & 2 \leq t < 3 \\ -0.5, & 3 \leq t \leq 6 \end{cases}$$

The following three lines of code create a vector "x" to represent this signal and graph it.

```
>>  t = [-10:0.01:10];
>>  x = 2*((t>=-1)&(t<2))+1*((t>=2)&(t<3))-0.5*((t>=3)&(t<=6));
>>  plot(t,x);
```

Since "x" is a vector, there is no easy way to compute and graph the signals $g(t) = x(2t - 5)$ and $h(t) = x(-4t + 2)$ that were graphed in Example 1.3. Instead of using a vector, we will change our strategy, and use a MATLAB function to represent the signal. Create an ASCII text file named "`sigx.m`" in the current directory and place the following two lines in it:

```
function x = sigx(t)
   x = 2*((t>=-1)&(t<2))+1*((t>=2)&(t<3))-0.5*((t>=3)&(t<=6));
```

The result is a MATLAB function `sigx(..)` that returns the value of the signal $x(t)$ at any specified time instant, or even at a set of specified time instants. A vector "g" that holds the amplitude values for the signal $g(t) = x(2t - 5)$ can be computed and graphed with the following set of statements:

```
>>  t = [-10:0.01:10];
>>  g = sigx(2*t-5);
>>  plot(t,g);
```

The graph that results should match Fig. 1.17(c). Similarly, a vector "h" with amplitude values of the signal $h(t) = x(-4t + 2)$ can be computed and graphed with the statements

```
>>  t = [-10:0.01:10];
>>  h = sigx(-4*t+2);
>>  plot(t,h);
```

The graph produced should match Fig. 1.18(c).

Use of the function `sigx(..)` allows us to write signal operations in a natural form. One slight drawback is that it requires the creation of an ASCII file "`sigx.m`" on disc with the function code in it. This may not always be desirable, especially if the signal $x(t)$ will only be used for one exercise and will not be needed again. An alternative method is to create an *anonymous function* that does not need to be saved into a file. An anonymous function is created for the current MATLAB session, and stays in memory as long as MATLAB remains active. It is discarded when MATLAB is closed. The following statement creates an anonymous function with the name "**sx**" that works the same way as the function `sigx(..)`:

```
>>   sx = @(t) 2*((t>=-1)&(t<2))+1*((t>=2)&(t<3))-0.5*((t>=3)&(t<=6));
```

Signals $g(t)$ and $h(t)$ can now be computed and graphed using

```
>>   plot(t,sx(2*t-5));
>>   plot(t,sx(-4*t+2));
```

Software resources:
`matex_1_3a.m`
`matex_1_3b.m`
`matex_1_3c.m`
`matex_1_3d.m`
`sigx.m`

MATLAB Exercise 1.4: **Creating periodic signals**

MATLAB has a number of functions for creating specific periodic waveforms. In this exercise we will explore functions `square(..)` and `sawtooth(..)`. Afterwards, we will present a simple technique for creating arbitrary periodic waveforms.

a.   One period of a square-wave signal is defined as

$$
x(t) = \begin{cases} 1, & 0 < t < \dfrac{T}{2} \\ -1, & \dfrac{T}{2} < t < T \end{cases}
$$

MATLAB function `square(..)` produces a square wave signal with period equal to $T = 2\pi$ by default, so we need to use time scaling operation to obtain it with any other value of T. The code below will compute and graph a square wave with period $T = 1$ s for the time interval $-1 < t < 10$ s.

```
>>   t = [-1:0.01:10];
>>   x = square(2*pi*t);
>>   plot(t,x);
```

b.   A square-wave signal with a duty cycle of $0 < d \le 1$ is defined as

$$
x(t) = \begin{cases} 1, & 0 < t < Td \\ -1, & Td < t < T \end{cases}
$$

The following code will compute and graph a square-wave signal with period $T = 1$ s and duty cycle $d = 0.2$ (or 20 percent).

```
>>   x = square (2*pi*t,20);
>>   plot(t,x);
```

c. The function `sawtooth(..)` generates a sawtooth waveform with period $T = 2\pi$. Its definition for one period is

$$x(t) = t/T \quad \text{for} \ 0 < t < T$$

The code below can be used for obtaining a sawtooth signal with period $T = 1.5$ s.

```
>>   x = sawtooth (2*pi*t/1.5);
>>   plot(t,x);
```

d. A signal $x(t)$ that is periodic with period $T = 2.5$ s is defined through the following:

$$x(t) = \begin{cases} t, & 0 \leq t < 1 \\ e^{-5(t-1)}, & 1 \leq t < 2.5 \end{cases} \quad \text{and} \quad x(t + 2.5k) = x(t)$$

The following can be used for graphing this signal in the time interval $-2 < t < 12$ s.

```
>>   t = [-2:0.01:12];
>>   x1 = @(t) t.*((t>=0)&(t<1))+exp(-5*(t-1)).*((t>=1)&(t<2.5));
>>   x = x1(mod(t,2.5));
>>   plot(t,x); grid;
```

The strategy is straightforward: We first create an anonymous function "x1" to describe one period of the signal. In the next line we use that anonymous function with the argument `mod(t,2.5)` to cause periodic repetition.

Software resources:
`matex_1_4a.m`
`matex_1_4b.m`
`matex_1_4c.m`
`matex_1_4d.m`

MATLAB Exercise 1.5: Functions for basic building blocks

It is possible to extend MATLAB by developing custom functions for performing tasks that are encountered often. In this exercise we will develop a few simple functions for the computation of basic signal building blocks described in Section 1.3.2.

The unit-step function can be implemented by creating a text file with the name "`ss_step.m`" and the following two lines in it:

```
1    function x = ss_step(t)
2       x = 1*(t>=0);
```

It returns the amplitude values for the unit-step function evaluated at each time instant specified by the vector "`t`". The unit-ramp function can be implemented with the following two lines saved into a text file named "`ss_ramp.m`":

```
1    function x = ss_ramp(t)
2       x = t.*(t>=0);
```

Functions for computing a unit pulse and a unit triangle can be developed based on the functions `ss_step(..)` and `ss_ramp(..)`. We will obtain the unit pulse by applying Eqn. (1.36) as follows:

```
1    function x = ss_pulse(t)
2      x = ss_step(t+0.5)-ss_step(t-0.5);
```

Similarly, a unit triangle can be obtained by the application of Eqn. (1.43):

```
1    function x = ss_tri(t)
2      x = ss_ramp(t+1)-2*ss_ramp(t)+ss_ramp(t-1);
```

Software resources:

matex_1_5.m

ss_step.m

ss_ramp.m

ss_pulse.m

ss_tri.m

MATLAB Exercise 1.6: Computing and graphing discrete-time signals

In previous exercises we have graphed continuous-time signals through the use of the plot(..) function which connects dots with straight lines so that the visual effect is consistent with a signal defined at all time instants. When we work with discrete-time signals we will use a *stem plot* to emphasize the discrete-time nature of the signal.

a. Compute and graph the signal

$$x_1[n] = \{\, 1.1, \underset{\substack{\uparrow \\ n=5}}{2.5}, 3.7, 3.2, 2.6 \,\} \tag{1.199}$$

for the range of the sample index $4 \leq n \leq 8$.

Solution: Type the following three lines into MATLAB command window to obtain the graph shown in Fig. 1.91.

```
>> n = [4:8];
>> x1 = [1.1, 2.5, 3.7, 3.2, 2.6];
>> stem(n,x1);
```

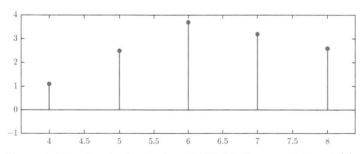

Figure 1.91 – Graph obtained in MATLAB Exercise 1.6 part (a).

b. Compute and graph the signal

$$x_2[n] = \sin(0.2n)$$

for the index range $n = 0, 1, \ldots, 99$.

Solution: This signal can computed and graphed with the following statements:

```
>>   n = [0:99];
>>   x2 = sin(0.2*n);
>>   stem(n,x2);
```

The result is shown in Fig. 1.92.

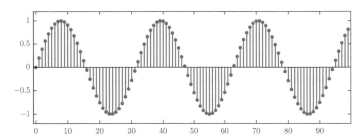

Figure 1.92 – Graph obtained in MATLAB Exercise 1.6 part (b).

c. Compute and graph the signal

$$x_3[n] = \begin{cases} \sin(0.2n) \, , & n = 0, \ldots, 39 \\ 0 \, , & \text{otherwise} \end{cases}$$

for the interval $n = -20, \ldots, 59$.

Solution: The signal $x_3[n]$ to be graphed is similar to the signal of part (b), but is a finite-length signal that equals zero for $n < 0$ and for $n > 39$. As in MATLAB Exercise 1.1 we will solve this problem by using logic operators on the index vector "**n**". Type the following statements into the command window to compute and graph the signal $x_3[n]$.

```
>>   n = [-20:59];
>>   x3 = sin(0.2*n).*((n>=0)&(n<=39));
>>   stem(n,x3);
```

The result is shown in Fig. 1.93.

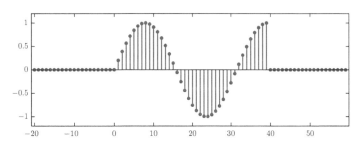

Figure 1.93 – Graph obtained in MATLAB Exercise 1.6 part (c).

Software resources:
matex_1_6a.m
matex_1_6b.m
matex_1_6c.m

MATLAB Exercise 1.7: Periodic extension of a discrete-time signal

Sometimes a periodic discrete-time signal is specified using samples of one period. Let $x[n]$ be a length-N signal with samples in the interval $n = 0, \ldots, N - 1$. Let us define the signal

$\tilde{x}[n]$ as the periodic extension of $x[n]$ so that

$$\tilde{x}[n] = \sum_{m=-\infty}^{\infty} x[n+mN]$$

In this exercise we will develop a MATLAB function to periodically extend a discrete-time signal. The function ss_per(..) given below takes a vector "x" that holds one period (N samples) of $x[n]$. The second argument "idx" is a vector that holds the indices at which the periodic extension signal should be evaluated. The samples of the periodic signal $\tilde{x}[n]$ are returned in the vector "xtilde".

```
1    function xtilde = ss_per(x,idx)
2        N = length(x);    % Period of the signal.
3        n = mod(idx,N);   % Modulo indexing.
4        nn = n+1;         % MATLAB indices start with 1.
5        xtilde = x(nn);
6    end
```

Consider a length-5 signal $x[n]$ given by

$$x[n] = n, \quad \text{for } n = 0, 1, 2, 3, 4$$

The periodic extension $\tilde{x}[n]$ can be computed and graphed for $n = -15, \ldots, 15$ with the following statements:

```
>>   x = [0,1,2,3,4]
>>   n = [-15:15]
>>   xtilde = ss_per(x,n)
>>   stem(n,xtilde)
```

What if we need to compute and graph a time reversed version $\tilde{x}[-n]$ in the same interval? That can also be accomplished easily using the following statements:

```
>>   xtilde = ss_per(x,-n)
>>   stem(n,xtilde)
```

Software resources:
matex_1_7.m
ss_per.m

Problems

1.1. Sketch and label each of the signals defined below:

a. $x_a(t) = \begin{cases} 0, & t < 0 \text{ or } t > 4 \\ 2, & 0 < t < 1 \\ 1, & 1 < t < 2 \\ t-1, & 2 < t < 3 \\ 2, & 3 < t < 4 \end{cases}$

b. $x_b(t) = \begin{cases} 0, & t < 0 \text{ or } t > 5 \\ t, & 0 < t < 2 \\ -t+4, & 2 < t < 3 \\ -2t+7, & 3 < t < 4 \\ t-5, & 4 < t < 5 \end{cases}$

1.2.   Consider the signals shown in Fig. P.1.2. For each signal write the analytical description in segmented form similar to the descriptions of the signals in Problem 1.1.

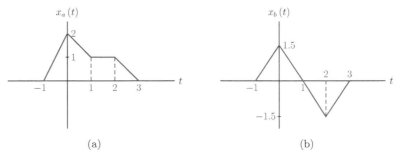

(a)   (b)

Figure P. 1.2

1.3.   Using the two signals $x_a(t)$ and $x_b(t)$ given in Fig. P.1.2, compute and sketch the
signals specified below:

a. $g_1(t) = x_a(t) + x_b(t)$
b. $g_2(t) = x_a(t)\, x_b(t)$
c. $g_3(t) = 2x_a(t) - x_b(t) + 3$

1.4.   For the signal $x(t)$ shown in Fig. P.1.4, compute the following:

a. $g_1(t) = x(-t)$
b. $g_2(t) = x(2t)$
c. $g_3(t) = x\left(\dfrac{t}{2}\right)$
d. $g_4(t) = x(-t + 3)$
e. $g_5(t) = x\left(\dfrac{t-1}{3}\right)$
f. $g_6(t) = x(4t - 3)$
g. $g_7(t) = x\left(1 - \dfrac{t}{3}\right)$

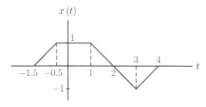

Figure P. 1.4

1.5.   Consider the signal
$$x(t) = \left(e^{-t} - e^{-2t}\right) u(t)$$

Determine and sketch the following signals derived from $x(t)$ through signal operations:

a. $g_1(t) = x(2t - 1)$
b. $g_2(t) = x(-t + 2)$

c. $g_3(t) = x(-3t + 5)$

d. $g_4(t) = x\left(\dfrac{t-1}{3}\right)$

I.6. Let b be a positive constant. Show that

$$\delta(bt) = \frac{1}{b}\delta(t)$$

Hint: Start with Eqn. (1.23) that expresses the impulse function as the limit of a pulse $q(t)$ with height $1/a$ and unit area. Apply time scaling to $q(t)$ and then take the limit as $a \to 0$.

I.7. Consider again Eqn. (1.23) that expresses the impulse function as the limit of a pulse $q(t)$ with height $1/a$ and area equal to unity. Show that, for small values of a, we have

$$\int_{-\infty}^{\infty} f(t)\, q(t - t_1)\, dt \approx f(t_1)$$

where $f(t)$ is any function that is continuous in the interval $t_1 - a/2 < t < t_1 + a/2$. Afterwards, by taking the limit of this result, show that the sifting property of the impulse function holds.

I.8. Sketch each of the following functions.

a. $\delta(t) + \delta(t-1) + \delta(t-2)$

b. $\delta(t-1) + u(t)$

c. $e^{-t}\,\delta(t-1)$

d. $e^{-t}\,[u(t-1) - u(t-2)]$

e. $\displaystyle\sum_{n=0}^{\infty} e^{-t}\,\delta(t - 0.1n)$

I.9. Sketch each of the following functions in the time interval $-1 \le t \le 5$. Afterwards use the waveform explorer program "wav_demo1.m" to check your results.

a. $u(t) + u(t-1) - 3u(t-2) + u(t-3)$

b. $r(t) - 2r(t-2) + r(t-3)$

c. $u(t) + r(t-2) - u(t-3) - r(t-4)$

d. $2\Pi\left(\dfrac{t-1}{2}\right) - \Pi\left(\dfrac{t-2}{1.5}\right) + 2\Pi(t-3)$

e. $\Lambda(t) + 2\Lambda(t-1) + 1.5\Lambda(t-3) - \Lambda(t-4)$

I.10. Express each of the signals shown in Fig. P.1.10 using scaled and time shifted unit-step functions. Afterwards use the waveform explorer program "wav_demo1.m" to check your results.

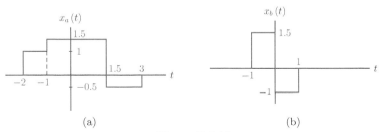

(a)   (b)

Figure P. 1.10

1.11. Consider again the signals in Fig. P.1.10. Express each signal using scaled and time shifted unit-pulse functions. Afterwards use the waveform explorer program "`wav_demo1.m`" to check your results.

1.12. Express the signal $x(t)$ shown in Fig. P.1.12 using

  a. Unit-ramp functions
  b. Unit-triangle functions

In each case, check your results using the waveform explorer program "`wav_demo1.m`".

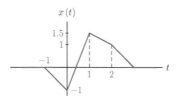

Figure P. 1.12

1.13. Suppose that in an experiment we collect data in the form of (t_i, x_i) pairs for $i = 1, \ldots, N$, and would like to construct a continuous-time signal $x(t)$ by linear interpolation, that is, by connecting the points with straight line segments as shown in Fig. P.1.13(a). The only restriction is that all t_i values are distinct; no two data points have the same time value. Let us define a *skewed triangle function* in the form

$$\Lambda_s(t, a, b) = \begin{cases} 0, & t < -a \text{ or } t > b \\ \dfrac{t}{a} + 1, & -a < t < 0 \\ -\dfrac{t}{b} + 1, & 0 < t < b \\ 0, & \end{cases}$$

which is shown in Fig. P.1.13(b). Devise a method of expressing the linearly interpolated signal $x(t)$ in terms of the skewed triangle functions.

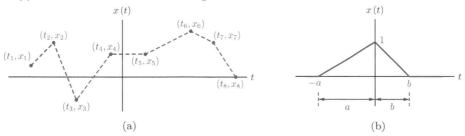

(a)   (b)

Figure P. 1.13

1.14. Time derivative of the unit-impulse function $\delta(t)$ is called a *doublet*. Given a function $f(t)$ that is continuous at $t = 0$ and also has continuous first derivative at $t = 0$, show that

$$\int_{-\infty}^{\infty} f(t)\, \delta'(t)\, dt = -f'(0) = -\left. \frac{df(t)}{dt} \right|_{t=0}$$

Hint: Use integration by parts, and then apply the sifting property of the impulse function.

1.15. Signum function is defined as

$$\text{sgn}(t) = \begin{cases} 1, & t > 0 \\ -1, & t < 0 \end{cases}$$

Express the function $\text{sgn}(t)$ in terms of the unit-step function.

1.16. Using Euler's formula, prove the following identities:

a. $\cos(a) = \dfrac{1}{2} e^{ja} + \dfrac{1}{2} e^{-ja}$

b. $\sin(a) = \dfrac{1}{2j} e^{ja} - \dfrac{1}{2j} e^{-ja}$

c. $\dfrac{d}{da}[\cos(a)] = -\sin(a)$

d. $\cos(a+b) = \cos(a)\cos(b) - \sin(a)\sin(b)$

e. $\sin(a+b) = \sin(a)\cos(b) + \cos(a)\sin(b)$

f. $\cos^2(a) = \dfrac{1}{2} + \dfrac{1}{2}\cos(2a)$

1.17. Using the definition of periodicity, determine if each signal below is periodic or not. If the signal is periodic, determine the fundamental period and the fundamental frequency.

a. $x(t) = 3\cos(2t + \pi/10)$
b. $x(t) = 2\sin(\sqrt{20}t)$
c. $x(t) = 3\cos(2t + \pi/10)\,u(t)$
d. $x(t) = \cos^2(3t - \pi/3)$
e. $x(t) = e^{-|t|}\cos(2t)$

f. $x(t) = \displaystyle\sum_{k=-\infty}^{\infty} e^{-(t-kT_s)} u(t - kT_s)$

g. $x(t) = e^{j(2t+\pi/10)}$
h. $x(t) = e^{(-1+j2)t}$

1.18. Determine the periodicity of each of the multi-tone signals given below. For those that are periodic, determine the fundamental period and the fundamental frequency.

a. $x(t) = 2\cos(5t + \pi/10) + 3\sin(5\pi t)$
b. $x(t) = \cos(10\pi t) - 3\sin(30\pi t + \pi/4)$
c. $x(t) = \sin(\sqrt{2}t) + \sin(2t)$
d. $x(t) = \cos(45\pi t) + \cos(55\pi t)$

1.19. A signal $g(t)$ is defined in terms of another signal $x(t)$ as

$$g(t) = A\,x(t), \quad A \neq 0$$

a. If $x(t)$ is an energy signal with normalized energy equal to E_x, show that $g(t)$ is also an energy signal. Determine the normalized energy of $g(t)$ in terms of E_x and A.
b. If $x(t)$ is a power signal with normalized average power equal to P_x, show that $g(t)$ is also a power signal. Determine the normalized average power of $g(t)$ in terms of P_x and A.

1.20. A signal $g(t)$ is defined in terms of another signal $x(t)$ as

$$g(t) = Ax(t) + B, \quad A, B \neq 0$$

a. If $x(t)$ is an energy signal with normalized energy equal to E_x, is $g(t)$ is an energy signal or a power signal?

b. Determine either the normalized energy or the normalized average power of $g(t)$ as appropriate, in terms of E_x, A and B.

1.21. Determine the normalized energy of each of the signals given below.

a. $x(t) = e^{-2|t|}$

b. $x(t) = e^{-2t} u(t)$

c. $x(t) = e^{-2t} \cos(5t) u(t)$

Hint: Express the cosine term in $x(t)$ using Euler's formula prior to squaring and integrating.

1.22. Determine the normalized energy of each of the signals shown in Fig. P.1.2.

1.23. Determine the normalized average power of each of the periodic signals shown in Fig. P.1.23.

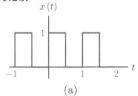
(a)

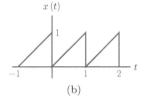
(b)

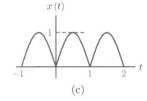
(c)

Figure P. 1.23

1.24. In Example 1.12 the RMS value of a multi-tone signal was derived for the case of distinct frequencies. Now consider the signal

$$x(t) = 2\cos(2\pi f_1 t) + 3\sin(2\pi f_1 t) - 6\cos(2\pi f_2 t)$$

where the first two terms have the same frequency. Determine the RMS value of the signal.

1.25. Identify which of the signals in Fig. P.1.25 are even, which ones are odd, and which signals are neither even nor odd.

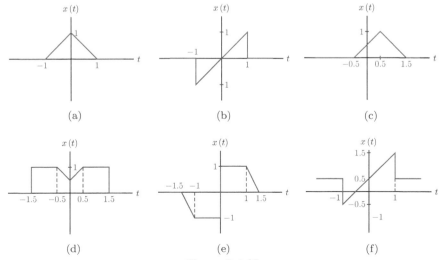

Figure P. 1.25

1.26.   Consider a signal $x(t)$. Show that

a.   If $x(t)$ is even, then

$$\int_{-\lambda}^{\lambda} x(t)\,dt = 2\int_{0}^{\lambda} x(t)\,dt$$

b.   If $x(t)$ is odd, then

$$\int_{-\lambda}^{\lambda} x(t)\,dt = 0$$

1.27.   Let $x(t) = x_1(t)\,x_2(t)$. Show that

a.   If both $x_1(t)$ and $x_2(t)$ are even, then $x(t)$ is even.
b.   If both $x_1(t)$ and $x_2(t)$ are odd, then $x(t)$ is even.
c.   If both $x_1(t)$ is even and $x_2(t)$ is odd, then $x(t)$ is odd.

1.28.   Determine the conjugate symmetric and conjugate antisymmetric components of the signal $x(t) = e^{(-2+j10\pi)t}$. Sketch each component in polar form, that is, in terms of its magnitude and phase.

1.29.   For each of the signals listed below, find the even and odd components $x_e(t)$ and $x_o(t)$. In each case, sketch the original signal and its two components.

a. $x(t) = e^{-5t}\sin(t)\,u(t)$
b. $x(t) = e^{-3|t|}\cos(t)$
c. $x(t) = e^{-3|t|}\sin(t)$
d. $x(t) = \left(t\,e^{-3t} + 2\right)u(t)$
e. $x(t) = e^{-2|t-1|}$

1.30.   Determine and sketch even and odd components of the signal $x(t)$ used in Problem 1.12 and shown in Fig. P.1.12.

1.31.   Transform each of the sinusoidal signals given below into phasor form.

a. $x(t) = 3\cos(200\pi t)$
b. $x(t) = 7\sin(100\pi t)$
c. $x(t) = 2\sin(10\pi t) - 5\cos(10\pi t)$

1.32.   Using an operating frequency of $f_0 = 10$ Hz, express the phasors given below as time-domain signals.

a. $\boldsymbol{X} = 5\,e^{j14°}$
b. $\boldsymbol{X} = 2\,e^{j28°} + 3\,e^{j18°}$
c. $\boldsymbol{X} = 2\,e^{j28°} - 3\,e^{j18°}$

1.33.   For the signal $x[n]$ shown in Fig. P.1.33, sketch the following signals.

a. $g[n] = x[n-3]$
b. $g[n] = x[2n-3]$
c. $g[n] = x[-n]$
d. $g[n] = x[2-n]$
e. $g[n] = \begin{cases} x[n/2], & \text{if } n/2 \text{ is integer} \\ 0, & \text{otherwise} \end{cases}$

f. $g[n] = x[n]\,\delta[n]$
g. $g[n] = x[n]\,\delta[n - 3]$
h. $g[n] = x[n]\,\{u[n + 2] - u[n - 2]\}$

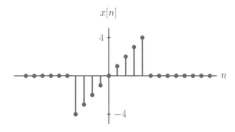

Figure P. 1.33

1.34. Consider the sinusoidal discrete-time signal

$$x[n] = 5\cos\left(\frac{3\pi}{23}n + \frac{\pi}{4}\right)$$

Is the signal periodic? If yes, determine the fundamental period.

1.35. Carefully sketch each signal described below.

a. $x[n] = \{\underset{\uparrow}{1},\ -1,\ 3,\ -2,\ 5\,\}$
b. $x[n] = (0.8)^{n}\,u[n]$
c. $x[n] = u[n] - u[n - 10]$
d. $x[n] = r[n] - 2r[n - 5] + r[n - 10]$

1.36. Determine the normalized energy of each signal described in Problem 1.35.

1.37. Consider a signal $x[n]$. Show that

a. If $x[n]$ is even, then $\displaystyle\sum_{n=-M}^{M} x[n] = x[0] + 2\sum_{n=1}^{M} x[n]$

b. If $x[n]$ is odd, then $\displaystyle\sum_{n=-M}^{M} x[n] = 0$

1.38. Let $x[n] = x_1[n]\,x_2[n]$. Show that

a. If both $x_1[n]$ and $x_2[n]$ are even, then $x[n]$ is even.
b. If both $x_1[n]$ and $x_2[n]$ are odd, then $x[n]$ is even.
c. If $x_1[n]$ is even and $x_2[n]$ is odd, then $x[n]$ is odd.

1.39. Determine even and odd components of the signal $x[n]$ shown in Fig. P.1.39.

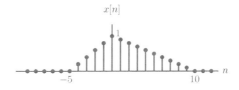

Figure P. 1.39

MATLAB Problems

1.40. Write a MATLAB script to compute and graph the signals $x_a(t)$ and $x_b(t)$ described in Problem 1.1. Use an anonymous function for each signal based on the analytical description given (see MATLAB Exercise 1.1). Graph each signal in the interval $-1 \leq t \leq 6$ with a time increment of $\Delta t = 0.01$. Label the axes appropriately.

1.41.

a. Consider the signals $x_a(t)$ and $x_b(t)$ described in Problem 1.2. Using the `interp1(..)` function of MATLAB, compute each signal in the time interval $-2 \leq t \leq 4$ with a time increment of $\Delta t = 0.01$ (see MATLAB Exercise 1.2).

b. Afterwards, compute the signals $g_1(t)$, $g_2(t)$ and $g_3(t)$ from the signals $x_a(t)$ and $x_b(t)$ as specified in Problem 1.3. Graph each signal in the same time interval as in part (a). Carefully label the axes of each graph.

1.42. Refer to the signals $g_1(t)$ and $g_2(t)$ found in Example 1.2 and shown in Fig. 1.12. Develop a MATLAB script to verify the analytical forms of these signals given by Eqns. (1.5) and (1.6). Your script should

a. Compute samples of each signal in the time interval $-1 \leq t \leq 4$ using a time increment of $\Delta t = 0.01$ s.

b. Graph the results using the "`subplot`" function.

Ensure that the computations in part (a) correspond directly to Eqns. (1.5) and (1.6).

1.43. Refer to the signal $x(t)$ used in Problem 1.4 and shown in Fig. P.1.4.

a. Express the signal $x(t)$ through the use of a MATLAB anonymous function that utilizes the function `interp1(..)`.

b. Using MATLAB, compute and graph each of the signals for parts (a) through (g) of Problem 1.4 in the time interval $-10 < t < 10$ s, using a time increment of $\Delta t = 0.02$ s. Compare MATLAB graphs to those obtained manually in Problem 1.4.

1.44. Refer to the signal $x(t)$ used in Problem 1.5.

a. Express the signal $x(t)$ through the use of a MATLAB anonymous function.

b. Using MATLAB, compute and graph each of the signals for parts (a) through (d) of Problem 1.5 in the time interval $-10 < t < 10$ s, using a time increment of $\Delta t = 0.02$ s. Compare MATLAB graphs to those obtained manually in Problem 1.5.

1.45. Using MATLAB functions `ss_step(..)`, `ss_ramp(..)`, `ss_pulse(..)` and `ss_tri(..)` developed in MATLAB Exercise 1.5, graph each of the signals specified in Problem 1.9 parts (a) through (e) in the time interval $-2 < t < 8$ s, using a time increment of $\Delta t = 0.01$ s. Compare MATLAB graphs to those obtained manually in Problem 1.9.

1.46. Consider the signal $x(t)$ used in Problem 1.12 and graphed in Fig. P.1.12.

   a. Express this signal through an anonymous MATLAB function that utilizes the function `ss_ramp(..)`, and graph the result in an appropriate time interval.
   b. Repeat part (a) using the function `ss_tri(..)` instead of the function `ss_ramp(..)`.
   c. In MATLAB, compute and graph even and odd components of the signal $x(t)$. Code Eqns. (1.109) and (1.110) directly in MATLAB for this purpose. Compare your MATLAB graphs to those obtained manually in Problem 1.30.

1.47. Consider the discrete-time signal $x[n]$ used in Problem 1.33 and graphed in Fig. P.1.33.

   a. Express this signal through an anonymous MATLAB function that utilizes the function `ss_ramp(..)`, and graph the result for index range $n = -10, \ldots, 10$.
   b. Express each of the signals in parts (a) through (h) of Problem 1.33 in MATLAB, and graph the results. Use functions `ss_step(..)` and `ss_ramp(..)` as needed.

1.48. Consider the discrete-time signal $x[n]$ used in Problem 1.39 and graphed in Fig. P.1.39.

   a. Express this signal through an anonymous MATLAB function that utilizes the function `ss_ramp(..)`, and graph the result for index range $n = -10, \ldots, 10$.
   b. Write a script to compute and graph even and odd components of this signal.

MATLAB Projects

1.49. In Problem 1.13 a method was developed for linearly interpolating between data points (t_i, x_i) for $i = 1, \ldots, N$ using the *skewed triangle function* shown in Fig. P.1.13b.

   a. Write a MATLAB function `ss_trs(..)` to compute values of the skewed triangle function at time instants specified in vector "`t`". The function should have the syntax

```
x = ss_trs(t,a,b)
```

     where "`t`" is a row vector. Arguments "`a`" and "`b`" are both positive scalars. Test your function to compute and graph the functions $\Lambda_s(t, 3, 2)$, $\Lambda_s(t, 1, 4)$, and $\Lambda_s(t, 2.7, 1.3)$ in the time interval $-5 \le t \le 5$ using a time increment $\Delta t = 0.01$.
   b. Next, write a MATLAB function `ss_interp(..)` to interpolate between a set of data points. The function should have the syntax

```
x = ss_interp(tvec,xvec,t)
```

     The arguments "`tvec`" and "`xvec`" are row vectors of equal length. The vector "`tvec`" holds $t_1, \ldots, t_N$, and the vector "`xvec`" holds $x_1, \ldots, x_N$. The returned vector "`x`" holds values of the interpolated signal at time instants specified in the vector "`t`".

c. Write a MATLAB script to test the function `ss_interp(..)` with the set of data points

$$(t_i, x_i) = (-4, 0.5), (-3, 1.5), (-2, -1), (-0.5, 1), (1, 1), (3, 2), (4, 1.5), (5, 0)$$

Graph the interpolated signal in the time interval $-4 \leq t \leq 5$ using a time increment of $\Delta t = 0.01$.

1.50. In some digital systems such as inexpensive toys, sinusoidal signals at specific frequencies may be generated through table look-up, thus eliminating the need for the system to include the capability to compute trigonometric functions. Samples corresponding to exactly one period of the sinusoidal signal are computed ahead of time and stored in memory. When the signal is needed, the contents of the look-up table are played back repeatedly. Consider a sinusoidal signal

$$x_a = \cos\left(2\pi f_1 t\right)$$

with $f_1 = 1336$ Hz. Let $x_1[n]$ be a discrete-time sinusoidal signal obtained by evaluating $x_a(t)$ at intervals of 125 µs corresponding to a *sampling rate* of 8000 samples per second.

$$x_1[n] = x_a\left(125 \times 10^{-6}n\right) = x_a\left(\frac{n}{8000}\right)$$

a. Determine the fundamental period N_1 for the signal $x_1[n]$.

b. Write a MATLAB script to accomplish the following:

- Create a vector "x1" that holds exactly one period of the signal.
- By repeating the vector "x1" as many times as needed, create a vector "x" that holds 8000 samples of the sinusoidal signal that corresponds to a time duration of 1 second.
- Play back the resulting sound using the `sound(..)` function of MATLAB. If the fundamental period was computed properly in part (a) then the repeated sections of the signal should fit together seamlessly, and you should hear a clean tone.

c. Modify the script in part (b) so that the number of samples used in creating the vector "x1" is 10 less than the correct value found in part (a). Create a vector "x" by repeating this imperfect vector "x1" as many times as necessary. Listen to the audio playback of the vector "x", and comment on the difference. Also, graph the resulting signal. Using the zoom tool, zoom into the transition area between two repetitions, and observe the flaw in the transition that is due to incorrect period length.

d. Repeat parts (a) through (c) using the sinusoidal signal

$$x_b = \cos\left(2\pi f_b t\right)$$

with $f_1 = 852$ Hz, and the corresponding discrete-time signal

$$x_2[n] = x_b\left(125 \times 10^{-6}n\right) = x_b\left(\frac{n}{8000}\right)$$

e. Consider the dual-tone signal

$$x[n] = x_1[n] + x_2[n]$$

Using the table look-up method with the vectors "x1" and "x2" obtained above, generate a vector that holds 8000 samples of the dual-tone signal. Play back the resulting vector and comment.

1.51.   In this project the concept of *dual-tone multi-frequency (DTMF)* signaling will be explored. As the name implies, DTMF signals are mixtures of two sinusoids at distinct frequencies. They are used in communications over analog telephone lines. A particular version of DTMF signaling is utilized in dialing a number with push-button telephone handsets, a scheme known as *touch tone dialing*. When the caller dials a number, the DTMF generator produces a dual-tone signal for each digit dialed. The synthesized signal is in the form

$$x_k(t) = \sin(2\pi f_1 t) + \sin(2\pi f_2 t) , \qquad 0 \le t \le T_d$$

Frequency assignments for the digits on a telephone keypad are shown in Fig. P.1.51.

f_1 \ f_2	1209 Hz	1336 Hz	1477 Hz
697 Hz	1	2	3
770 Hz	4	5	6
852 Hz	7	8	9
941 Hz	*	0	#

Figure P. 1.51

The goal of this project is to develop a DTMF synthesizer function for MATLAB.

a. Develop a function named `ss_dtmf1(..)` to produce the signal for one digit. The syntax of the function should be

```
x = ss_dtmf1 ( n , t )
```

The first argument "`n`" is the digit for which the DTMF signal is to be generated. Let values $n = 0$ through $n = 9$ represent the corresponding keys on the keypad. Map the remaining two keys "`*`" and "`#`" to values $n = 10$ and $n = 11$ respectively. Finally, the value $n = 12$ should represent a pause, that is, a silent period. The vector "`t`" contains the time instants at which the DTMF signal $x(t)$ is evaluated and returned in vector "`x`".

b. Develop a function named `ss_dtmf(..)` with the syntax

```
x = ss_dtmf ( number , dt , nd , np )
```

The arguments for the function `ss_dtmf(..)` are defined as follows:

number: The phone number to be dialed, entered as a vector. For example, to dial the number 555-1212, the vector "`number`" would be entered as

```
number = [5,5,5,1,2,1,2]
```

dt: The time increment Δt to be used in computing the amplitudes of the DTMF signal.

nd: Parameter to control the duration of the DTMF signal for each digit. The duration of each digit should be

$$T_d = n_d \, \Delta t$$

np: Parameter to control the duration of pause between consecutive digits. The duration of pause should be

$$T_p = n_p \, \Delta t$$

The function ss_dtmf(..) should use the function ss_dtmf1(..) to produce the signals $x_k(t)$ for each digit (and the pauses between digits) and append them together to create the signal $x(t)$.

c. Write a script to test the function ss_dtmf(..) with the number 555-1212. Use a time-increment of 125 microseconds corresponding to 8000 values per second. The duration of each digit should be 200 milliseconds with 80 millisecond pauses between digits.

d. Play back the resulting signal $x(t)$ using the sound(..) function.

Chapter 2

Analyzing Continuous-Time Systems in the Time Domain

Chapter Objectives

- Develop the notion of a *continuous-time system*.

- Learn simplifying assumptions made in the analysis of systems. Discuss the concepts of *linearity* and *time invariance*, and their significance.

- Explore the use of differential equations for representing continuous-time systems.

- Develop methods for solving differential equations to compute the output signal of a system in response to a specified input signal.

- Learn to represent a differential equation in the form of a block diagram that can be used as the basis for simulating a system.

- Discuss the significance of the *impulse response* as an alternative description form for linear and time-invariant systems.

- Learn how to compute the output signal for a linear and time-invariant system using *convolution*. Understand the graphical interpretation of the steps involved in carrying out the convolution operation.

- Learn the concepts of causality and stability as they relate to physically realizable and usable systems.

2.1   Introduction

In this chapter, we will begin our examination of the system concept. An overly simplified and rather broad definition of a system may be given as follows:

In general, a system is any physical entity that takes in a set of one or more physical signals and, in response, produces a new set of one or more physical signals.

Consider a microphone that senses the variations in air pressure created by the voice of a singer, and produces a small electrical signal in the form of a time-varying voltage. The microphone acts as a system that facilitates the conversion of an acoustic signal to an electrical signal. Next, consider an amplifier that is connected to the output terminals of the microphone. It takes the small-amplitude electrical signal from the microphone, and produces a larger-scale replica of it suitable for use with a loudspeaker. Finally a loudspeaker that is connected to the output terminals of the amplifier converts the electrical signal to sound. We can view each of the physical entities, namely the microphone, the amplifier and the loudspeaker, as individual systems. Alternatively, we can look at the combination of all three components as one system that consists of three subsystems working together.

Another example is the transmission of images from a television studio to the television sets in our homes. The system that achieves the goal of bringing sound and images from the studio to our homes is a rather complex one that consists of a large number of subsystems that carry out the tasks of 1) converting sound and images to electrical signals at the studio, 2) transforming electrical signals into different formats for the purposes of enhancement, encoding and transmission, 3) transmitting electrical or electromagnetic signals over the air or over a cable connection, 4) receiving electrical or electromagnetic signals at the destination point, 5) processing electrical signals to convert them to formats suitable for a television set, and 6) displaying the images and playing the sound on the television set. Within such a large-scale system some of the physical signals are continuous-time signals while others are discrete-time or digital signals.

Not all systems of interest are electrical. An automobile is an example of a large-scale system that consists of numerous mechanical, electrical and electromechanical subsystems working together. Other examples are economic systems, mass transit systems, ecological systems and computer networks. In each of these examples, large-scale systems are constructed as collections of interconnected smaller-scale subsystems that work together to achieve a prescribed goal.

In this textbook we will not attempt a complete treatment of any particular large-scale system. Instead, we will focus our efforts on developing techniques for understanding, analyzing and designing subsystems with accurate and practical models based on laws of physics and mathematical transforms. In general, a system can be viewed as any physical entity that defines the cause-effect relationships between a set of signals known as *inputs* and another set of signals known as *outputs*. The input signals are *excitations* that drive the system, and the output signals are the *responses* of the system to those excitations.

In Chapter 1 we have discussed basic methods of mathematically modeling continuous-time and discrete-time signals as functions of time.

The mathematical model of a system is a function, formula or algorithm (or a set of functions, formulas, algorithms) to approximately recreate the same cause-effect relationship between the mathematical models of the input and the output signals.

If we focus our attention on single-input/single-output systems, the interplay between the system and its input and output signals can be graphically illustrated as shown in Fig. 2.1.

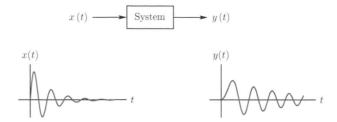

Figure 2.1 – Continuous-time signal-system interaction.

If the input and the output signals are modeled with mathematical functions $x\,(t)$ and $y\,(t)$ respectively, then the system itself needs to be modeled as a function, formula, or algorithm of some sort that converts the signal $x\,(t)$ into the signal $y\,(t)$. Sometimes we may have the need to design a system to achieve a particular desired effect in the conversion of the input signal $x\,(t)$ to the output signal $y\,(t)$. For example, an amplifier increases the power level of a signal that may be too weak to be heard or transmitted, while preserving its information content. A frequency-selective filter can be used for either removing or boosting certain frequencies in a signal. A speech recognition system can analyze the signal $x\,(t)$ in an attempt to recognize certain words, with the goal of performing certain tasks such as connecting the caller to the right person via an automated telephone menu system.

In contrast with systems that we design in order to achieve a desired outcome, sometimes the effect of a particular system on the input signal is not a desired one, but one that we are forced to accept, tolerate, or handle. For example, a wire connection is often used so that a signal that exists at a particular location could be duplicated at a distant remote location. In such a case, we expect the wire connection to be transparent within the scheme of transmitting the signal from point A to point B; however, that is usually not the case. The physical conductors used for the wire connection are far from ideal, and cause changes in the signal as it travels from A to B. The characteristics of the connection itself represent a system that is more of a nuisance than anything else. We may need to analyze this system, and find ways to compensate for the undesired effects of this system if we are to communicate successfully.

The relationship between the input and the output signals of a continuous-time system will be mathematically modeled as

$$y(t) = \text{Sys}\,\{x(t)\} \tag{2.1}$$

where the operator $\text{Sys}\,\{\ldots\}$ represents the transformation applied to $x\,(t)$. This transformation can be anything from a very simple one to a very complicated one. Consider, for example, a system that amplifies its input signal by a constant gain factor K to yield an output signal

$$y\,(t) = K\,x\,(t) \tag{2.2}$$

or one that delays its input signal by a constant time delay τ to produce

$$y\,(t) = x\,(t - \tau) \tag{2.3}$$

or a system that produces an output signal that is proportional to the square of the input signal as in

$$y\,(t) = K\,\left[x\,(t)\right]^{2} \tag{2.4}$$

In all three examples above, we have system definitions that could be expressed in the form of simple functions. This is not always the case. More interesting systems have more complicated definitions that cannot be reduced to a simple function of the input signal, but must rather be expressed by means of an algorithm, a number of functional relationships interconnected in some way, or a differential equation.

In deriving valid mathematical models for physical systems, we rely on established laws of physics that are applicable to individual components of a system. For example, consider a resistor in an electronic circuit. In mathematically modeling a resistor we use Ohm's law which states that the voltage between the terminals of a resistor is proportional to the current that flows through it. This leads to the well-known relationship $v = Ri$ between the resistor voltage and the resistor current. On the other hand, laboratory experiments indicate that the resistance of an actual carbon resistor varies as a function of temperature, a fact which is neglected in the mathematical model based on Ohm's law. Thus, Ohm's law provides a simplification of the physical relationship involved. Whether this simplification is acceptable or not depends on the specifics of the circuit in which the resistor is used. How significant are the deviations of the mathematical model of the resistor from the behavior of the actual resistor? How significant are the temperature changes that cause these deviations? How sensitive is the circuit to the variations in the value of R? Answers to these and similar questions are used for determining if the simple mathematical model is appropriate, or if a more sophisticated model should be used. Modeling of a system always involves some simplification of the physical relationships between input and output signals. This is necessary in order to obtain mathematical models that are practical for use in understanding system behavior. Care should be taken to avoid oversimplification and to ensure that the resulting model is a reasonably accurate approximation of reality.

Two commonly used simplifying assumptions for mathematical models of systems are *linearity* and *time invariance* which will be the subjects of Section 2.2. Section 2.3 focuses on the use of differential equations for representing continuous-time systems. Restrictions that must be placed on differential equations in order to model linear and time-invariant systems will be discussed in Section 2.4. Methods for solving linear constant-coefficient differential equations will be discussed in Section 2.5. Derivation of block diagrams for simulating continuous-time linear and time-invariant systems will be the subject of Section 2.6. In Section 2.7 we discuss the significance of the impulse response and its use in the context of the convolution operator for determining the output signal of a system. Concepts of causality and stability of systems are discussed in Sections 2.8 and 2.9 respectively.

2.2   Linearity and Time Invariance

In most of this textbook we will focus our attention on a particular class of systems referred to as *linear and time-invariant* systems. Linearity and time invariance will be two important properties which, when present in a system, will allow us to analyze the system using well-established techniques of the linear system theory. In contrast, the analysis of systems that are not linear and time-invariant tends to be more difficult, and often relies on methods that are specific to the types of systems being analyzed.

2.2.1   Linearity in continuous-time systems

A system is said to be *linear* if the mathematical transformation $y(t) = \text{Sys}\{x(t)\}$ that governs the input-output relationship of the system satisfies the following two equations for any two input signals $x_1(t)$, $x_2(t)$ and any arbitrary constant gain factor α_1.

Conditions for linearity:

$$\text{Sys}\{x_1(t) + x_2(t)\} = \text{Sys}\{x_1(t)\} + \text{Sys}\{x_2(t)\} \tag{2.5}$$

$$\text{Sys}\{\alpha_1\, x_1(t)\} = \alpha_1\, \text{Sys}\{x_1(t)\} \tag{2.6}$$

The condition in Eqn. (2.5) is the *additivity rule* which can be stated as follows: The response of a linear system to the sum of two signals is the same as the sum of individual responses to each of the two input signals. The condition in Eqn. (2.6) is the *homogeneity rule*. Verbally stated, scaling the input signal of a linear system by a constant gain factor causes the output signal to be scaled with the same gain factor. The two criteria given by Eqns. (2.5) and (2.6) can be combined into one equation which is referred to as the *superposition principle*.

Superposition principle:

$$\text{Sys}\{\alpha_1\, x_1(t) + \alpha_2\, x_2(t)\} = \alpha_1\, \text{Sys}\{x_1(t)\} + \alpha_2\, \text{Sys}\{x_2(t)\} \tag{2.7}$$

A continuous-time system is linear if it satisfies the superposition principle stated in Eqn. (2.7) for any two arbitrary input signals $x_1(t)$, $x_2(t)$ and any two arbitrary constants α_1 and α_2. The superposition principle applied to two signals can be expressed verbally as follows: The response of the system to a weighted sum of two input signals is equal to the same weighted sum of the responses of the system to individual input signals. This concept is very important, and is graphically illustrated in Fig. 2.2.

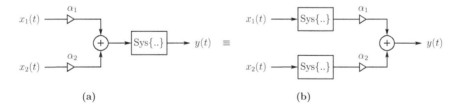

(a)   (b)

Figure 2.2 – Illustration of Eqn. (2.7). The two configurations shown are equivalent if the system under consideration is linear.

Furthermore, if superposition works for the weighted sum of any two input signals, it can be proven by induction that it also works for an arbitrary number of input signals. Mathematically we have

$$\text{Sys}\left\{\sum_{i=1}^{N} \alpha_i\, x_i(t)\right\} = \sum_{i=1}^{N} \alpha_i\, \text{Sys}\{x_i(t)\} \tag{2.8}$$

The generalized form of the superposition principle can be expressed verbally as follows: The response of a linear system to a weighted sum of N arbitrary signals is equal to the same weighted sum of the individual responses of the system to each of the N signals. Let $y_i(t)$ be the response to the input component $x_i(t)$ alone, that is $y_i(t) = \text{Sys}\{x_i(t)\}$ for

$i = 1, \ldots, N$. Superposition principle implies that

$$y(t) = \mathrm{Sys}\left\{ \sum_{i=1}^{N} \alpha_i \, x_i(t) \right\} = \sum_{i=1}^{N} \alpha_i \, y_i(t) \tag{2.9}$$

This is graphically illustrated in Fig. 2.3.

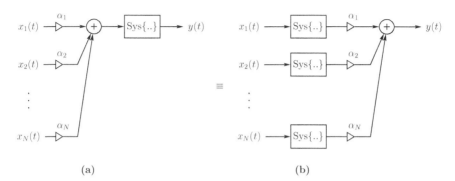

(a)   (b)

Figure 2.3 – Illustration of Eqn. (2.7). The two configurations shown are equivalent if the system under consideration is linear.

Testing a system for linearity involves checking whether the superposition principle holds true with two arbitrary input signals. Example 2.1 will demonstrate this procedure.

Example 2.1: **Testing linearity of continuous-time systems**

Four different systems are described below through their input-output relationships. For each, determine if the system is linear or not:

a. $y(t) = 5 \, x(t)$
b. $y(t) = 5 \, x(t) + 3$
c. $y(t) = 3 \, [x(t)]^2$
d. $y(t) = \cos(x(t))$

Solution:

a. If two input signals $x_1(t)$ and $x_2(t)$ are applied to the system individually, they produce the output signals $y_1(t) = 5 \, x_1(t)$ and $y_2(t) = 5 \, x_2(t)$ respectively. Let the input signal be $x(t) = \alpha_1 \, x_1(t) + \alpha_2 \, x_2(t)$. The corresponding output signal is found using the system definition:

$$
\begin{aligned}
y(t) =& 5 \, x(t) \\
     =& 5 \, [\alpha_1 \, x_1(t) + \alpha_2 \, x_2(t)] \\
     =& \alpha_1 \, [5 \, x_1(t)] + \alpha_2 \, [5 \, x_2(t)] \\
     =& \alpha_1 \, y_1(t) + \alpha_2 \, y_2(t)
\end{aligned}
$$

Superposition principle holds; therefore this system is linear.

b. If two input signals $x_1(t)$ and $x_2(t)$ are applied to the system individually, they produce the output signals $y_1(t) = 5 \, x_1(t) + 3$ and $y_2(t) = 5 \, x_2(t) + 3$ respectively.

We will again use the combined input signal $x(t) = \alpha_1 x_1(t) + \alpha_2 x_2(t)$ for testing. The corresponding output signal for this system is

$$\begin{aligned} y(t) &= 5\,x(t) + 3 \\ &= 5\,\alpha_1 x_1(t) + 5\,\alpha_2 x_2(t) + 3 \end{aligned}$$

The output signal $y(t)$ cannot be expressed in the form $y(t) = \alpha_1 y_1(t) + \alpha_2 y_2(t)$. Superposition principle does not hold true in this case. The system in part (b) is not linear.

c. Using two input signals $x_1(t)$ and $x_2(t)$ individually, the corresponding output signals produced by this system are $y_1(t) = 3\,[x_1(t)]^2$ and $y_2(t) = 3\,[x_2(t)]^2$ respectively. Applying the linear combination $x(t) = \alpha_1 x_1(t) + \alpha_2 x_2(t)$ to the system produces the output signal

$$\begin{aligned} y(t) &= 3\,[\alpha_1 x_1(t) + \alpha_2 x_2(t)]^2 \\ &= 3\,\alpha_1^2\,[x_1(t)]^2 + 6\,\alpha_1\alpha_2\,x_1(t)\,x_2(t) + 3\,\alpha_2^2\,[x_2(t)]^2 \end{aligned}$$

It is clear that this system is not linear either.

d. The test signals $x_1(t)$ and $x_2(t)$ applied to the system individually produce the output signals $y_1(t) = \cos[x_1(t)]$ and $y_2(t) = \cos[x_2(t)]$ respectively. Their linear combination $x(t) = \alpha_1 x_1(t) + \alpha_2 x_2(t)$ produces the output signal

$$y(t) = \cos[\alpha_1 x_1(t) + \alpha_2 x_2(t)]$$

This system is not linear either.

Software resources: See MATLAB Exercise 2.1.

2.2.2 Time invariance in continuous-time systems

Another important concept in the analysis of systems is time invariance. A system is said to be *time-invariant* if its behavior characteristics do not change in time. Consider a continuous-time system with the input-output relationship

$$\mathrm{Sys}\,\{x(t)\} = y(t)$$

If the input signal applied to a time-invariant system is time-shifted by τ seconds, the only effect of this delay should be to cause an equal amount of time shift in the output signal, but to otherwise leave the shape of the output signal unchanged. If that is the case, we would expect the relationship

$$\mathrm{Sys}\,\{x(t - \tau)\} = y(t - \tau)$$

to also be valid.

Condition for time-invariance:

$$\mathrm{Sys}\,\{x(t)\} = y(t) \quad \text{implies that} \quad \mathrm{Sys}\,\{x(t - \tau)\} = y(t - \tau) \qquad (2.10)$$

This relationship is depicted in Fig. 2.4.

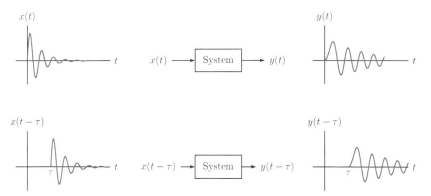

Figure 2.4 – Illustration of time invariance.

Alternatively, the relationship described by Eqn. (2.10) can be characterized by the equivalence of the two system configurations shown in Fig. 2.5.

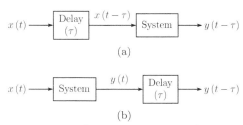

Figure 2.5 – Another interpretation of time-invariance. The two configurations shown are equivalent for a time-invariant system.

Testing a system for time invariance involves checking whether Eqn. (2.10) holds true for any arbitrary input signal. This procedure will be demonstrated in Example 2.2.

Example 2.2: **Testing time invariance of continuous-time systems**

Three different systems are described below through their input-output relationships. For each, determine whether the system is time-invariant or not:

    a. $y(t) = 5\, x(t)$
    b. $y(t) = 3\, \cos(x(t))$
    c. $y(t) = 3\, \cos(t)\, x(t)$

Solution:

    a. For this system, if the input signal $x(t)$ is delayed by τ seconds, the corresponding output signal would be

$$\text{Sys}\{x(t-\tau)\} = 5\, x(t-\tau) = y(t-\tau)$$

    and therefore the system is time-invariant.

    b. Let the input signal be $x(t-\tau)$. The output of the system is

$$\text{Sys}\{x(t-\tau)\} = 3\, \cos(x(t-\tau)) = y(t-\tau)$$

    This system is time-invariant as well.

c. Again using the delayed input signal $x(t - \tau)$ we obtain the output

$$\text{Sys}\{x(t - \tau)\} = 3\cos(t)\, x(t - \tau) \neq y(t - \tau)$$

In this case the system is not time-invariant since the time-shifted input signal leads to a response that is not the same as a similarly time-shifted version of the original output signal.

Before we leave this example we will use this last part of the problem as an opportunity to address a common source of confusion. The question may be raised as to whether the term t in the argument of the cosine function should be replaced with $t - \tau$ as well. In other words, should we have written

$$\text{Sys}\{x(t - \tau)\} \stackrel{?}{=} 3\cos(t - \tau)\, x(t - \tau)$$

which would have led to the conclusion that the system under consideration might be time-invariant? The answer is no. The term $\cos(t)$ is part of the system definition and not part of either the input or the output signal. Therefore we cannot include it in the process of time shifting input and output signals.

Software resources:	See MATLAB Exercise 2.2.

Example 2.3: **Using linearity property**

A continuous-time system with input-output relationship $y(t) = \text{Sys}\{x(t)\}$ is known to be linear. Whether the system is time-invariant or not is not known. Assume that the responses of the system to four input signals $x_1(t)$, $x_2(t)$ $x_3(t)$ and $x_4(t)$ shown in Fig. 2.6 are known.

Discuss how the information provided can be used for finding the response of this system to the signal $x(t)$ shown in Fig. 2.7.

Solution: Through a first glance at the four input signals given we realize that $x(t)$ is a time-shifted version of $x_1(t)$, that is, $x(t) = x_1(t - 0.5)$. As a result, we may be tempted to conclude that $y(t) = y_1(t - 0.5)$, however, this would be the wrong approach since the system is not necessarily time-invariant. We do not know for a fact that the input signal $x_1(t - 0.5)$ produces the response $y_1(t - 0.5)$. For the same reason, we cannot base our solution on the idea of constructing the input signal from $x_2(t)$ and $x_3(t)$ as

$$x(t) = x_2(t) + x_3(t - 1.5) \tag{2.11}$$

since it involves time shifting the signal $x_3(t)$, and we do not know the response of the system to $x_3(t - 1.5)$. We need to find a way to construct the input signal from input signals given, using only scaling and addition operators, but without using any time shifting. A possible solution is to write $x(t)$ as

$$x(t) = 0.6\, x_2(t) + 0.8\, x_4(t) \tag{2.12}$$

Since the system is linear, the output signal is

$$y(t) = 0.6\, y_2(t) + 0.8\, y_4(t) \tag{2.13}$$

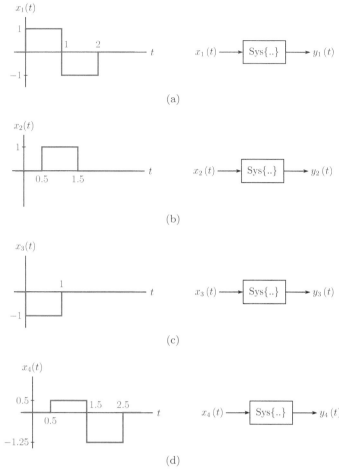

(a)

(b)

(c)

(d)

Figure 2.6 – Input-output pairs for Example 2.3.

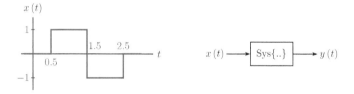

Figure 2.7 – Input signal for Example 2.3.

2.2.3 CTLTI systems

In the rest of this textbook, we will work with continuous-time systems that are both linear and time-invariant. A number of time- and frequency-domain analysis and design techniques will be developed for such systems. For simplicity, we will use the acronym *CTLTI* to refer to *continuous-time linear and time-invariant* systems.

2.3   Differential Equations for Continuous-Time Systems

One method of representing the relationship established by a system between its input and output signals is a differential equation that approximately describes the interplay of the physical quantities within the system. Such a differential equation typically involves the input and the output signals as well as various derivatives of either or both. Following is an example:

$$\frac{d^2y}{dt^2} + 3x(t)\frac{dy}{dt} + y(t) - 2x(t) = 0$$

If we need to model a mechanical system with a differential equation, physical relationships between quantities such as mass, force, torque and acceleration may be used. In the case of an electrical circuit we use the physical relationships that exist between various voltages and currents in the circuit, as well as the properties of the circuit components that link those voltages and currents. For example, the voltage between the leads of an inductor is known to be approximately proportional to the time rate of change of the current flowing through the inductor. Using functions $v_L(t)$ and $i_L(t)$ as mathematical models for the inductor voltage and the inductor current respectively, the mathematical model for an *ideal inductor* is

$$v_L(t) = L\frac{di_L(t)}{dt}$$

Similarly, the current that flows through a capacitor is known to be proportional to the rate of change of the voltage between its terminals, leading to the mathematical model of an *ideal capacitor* as

$$i_C(t) = C\frac{dv_C(t)}{dt}$$

where $v_C(t)$ and $i_C(t)$ are the mathematical models for the capacitor voltage and the capacitor current respectively.

Figure 2.8 – Mathematical models for (a) ideal inductor, (b) ideal capacitor.

Ideal inductor and the ideal capacitor represent significant simplification of real-life versions of these devices with their physical voltage and current quantities. The following examples will demonstrate the process of obtaining a differential equation from the physical description of a system.

Example 2.4: Differential equation for simple RC circuit

Consider the simple first-order RC circuit shown in Fig. 2.9. The input signal $x(t)$ is the voltage applied to a series combination of a resistor and a capacitor, and the output signal is the voltage $y(t)$ across the terminals of the capacitor.

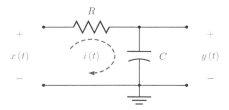

Figure 2.9 – RC circuit for Example 2.4.

Even though the RC circuit is a very simple example of a system, it will prove useful in discussing fundamental concepts of linear systems theory, and will be used as the basis of several examples that will follow. The techniques that we will develop using the simple RC circuit of Fig. 2.9 as a backdrop will be applicable to the solution of more complex problems involving a whole host of other linear systems.

It is known from circuit theory that both the resistor and the capacitor carry the same current $i(t)$. Furthermore, the voltage drop across the terminals of the resistor is

$$v_R(t) = R\,i(t) \tag{2.14}$$

and the capacitor voltage $y(t)$ is governed by the equation

$$i(t) = C\,\frac{dy(t)}{dt} \tag{2.15}$$

Using Eqns. (2.14) and (2.15) the relationship between the input and the output signals can be written as

$$RC\,\frac{dy(t)}{dt} + y(t) = x(t) \tag{2.16}$$

Multiplying both sides of Eqn. (2.16) by $1/RC$, we have

$$\frac{dy(t)}{dt} + \frac{1}{RC}\,y(t) = \frac{1}{RC}\,x(t) \tag{2.17}$$

Eqn. (2.17) describes the behavior of the system through its input-output relationship. If $x(t)$ is specified, $y(t)$ can be determined by solving Eqn. (2.17) using either analytical solution methods or numerical approximation techniques. The differential equation serves as a complete description of the system in this case. For example, the response of the system to the sinusoidal input signal $x(t) = \sin(2\pi f_0 t)$ can be found by solving the differential equation

$$\frac{dy(t)}{dt} + \frac{1}{RC}\,y(t) = \frac{1}{RC}\,\sin(2\pi f_0 t)$$

If we are interested in finding how the same system responds to a single isolated pulse with unit amplitude and unit duration, we would solve the following differential equation instead:

$$\frac{dy(t)}{dt} + \frac{1}{RC}\,y(t) = \frac{1}{RC}\,\Pi(t)$$

Example 2.5: **Another RC circuit**

Find a differential equation between the input and the output signals of the circuit shown in Fig. 2.10.

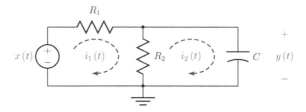

Figure 2.10 – Circuit for Example 2.5.

Solution: Using the two mesh currents $i_1(t)$ and $i_2(t)$, and applying Kirchhoff's voltage law (KVL) we obtain the following two equations:

$$-x(t) + R_1 i_1(t) + R_2 [i_1(t) - i_2(t)] = 0 \qquad (2.18)$$

$$R_2 [i_2(t) - i_1(t)] + y(t) = 0 \qquad (2.19)$$

We also know that the current that runs through the capacitor is proportional to the rate of change of its voltage. Therefore

$$i_2(t) = C \frac{dy(t)}{dt} \qquad (2.20)$$

Substituting Eqn. (2.20) into Eqn. (2.19) and solving for $i_1(t)$ yields

$$i_1(t) = C \frac{dy(t)}{dt} + \frac{1}{R_2} y(t) \qquad (2.21)$$

Finally, using Eqns. (2.20) and (2.21) in Eqn. (2.18) results in

$$-x(t) + R_1 C \frac{dy(t)}{dt} - \frac{R_1 + R_2}{R_2} y(t) = 0$$

which can be rearranged to produce the differential equation we seek:

$$\frac{dy(t)}{dt} + \frac{R_1 + R_2}{R_1 R_2 C} y(t) = \frac{1}{R_1 C} x(t) \qquad (2.22)$$

The order of a differential equation is determined by the highest order derivative that appears in it. In Examples 2.4 and 2.5 we obtained first-order differential equations; therefore, the systems they represent are also of first order. In the next example we will work with a circuit that will yield a second-order differential equation and, consequently, a second-order system.

Example 2.6: **Differential equation for RLC circuit**

Find a differential equation between the input signal $x(t)$ and the output signal $y(t)$ to serve as a mathematical model for the series RLC circuit shown in Fig. 2.11.

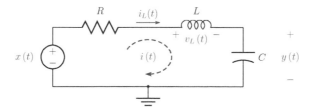

Figure 2.11 – Circuit for Example 2.6.

Solution: Applying KVL around the main loop we get

$$-x\left(t\right) + R\,i\left(t\right) + v_L\left(t\right) + y\left(t\right) = 0 \tag{2.23}$$

with $x\left(t\right)$ and $y\left(t\right)$ representing the input and the output signals of the system respectively. The inductor voltage is proportional to the time derivative of the current:

$$v_L\left(t\right) = L\,\frac{di\left(t\right)}{dt} \tag{2.24}$$

Similarly, realizing that the loop current $i\left(t\right)$ is proportional to the time derivative of the capacitor voltage $y\left(t\right)$, we can write

$$i\left(t\right) = C\,\frac{dy\left(t\right)}{dt} \tag{2.25}$$

Differentiating both sides of Eqn. (2.25) with respect to time yields

$$\frac{di\left(t\right)}{dt} = C\,\frac{d^2y\left(t\right)}{dt^2} \tag{2.26}$$

and substituting Eqn. (2.26) into Eqn. (2.24) we obtain

$$v_L\left(t\right) = LC\,\frac{d^2y\left(t\right)}{dt^2} \tag{2.27}$$

Finally, substituting Eqns. (2.25) and (2.27) into Eqn. (2.23) leads to the differential equation for the RLC circuit:

$$-x\left(t\right) + RC\,\frac{dy\left(t\right)}{dt} + LC\,\frac{d^2y\left(t\right)}{dt^2} + y\left(t\right) = 0$$

Rearranging terms, we have

$$\frac{d^2y\left(t\right)}{dt^2} + \frac{R}{L}\,\frac{dy\left(t\right)}{dt} + \frac{1}{LC}\,y\left(t\right) = \frac{1}{LC}\,x\left(t\right) \tag{2.28}$$

Thus, the RLC circuit of Fig. 2.11 leads to a second-order differential equation.

2.4   Constant-Coefficient Ordinary Differential Equations

In Section 2.2 we have discussed the properties of a class of systems known as linear and time-invariant. Linearity and time-invariance assumptions allow us to develop a robust set of methods and techniques for analyzing and designing systems. We will see in later parts of this text that, by limiting our focus to systems that are both linear and time-invariant (for which we use the acronym CTLTI), we will be able to use the convolution operation and the system function concept for the computation of the output signal.

In Examples 2.4, 2.5 and 2.6 we have explored methods of finding differential equations for several electrical circuits. In general, CTLTI systems can be modeled with *ordinary differential equations* that have *constant coefficients*. An ordinary differential equation is one that does not contain partial derivatives. The differential equation that represents a

CTLTI system contains the input signal $x(t)$, the output signal $y(t)$ as well as simple time derivatives of the two, namely

$$\frac{d^k y(t)}{dt^k}, \quad k = 0, \ldots, N$$

and

$$\frac{d^k x(t)}{dt^k}, \quad k = 0, \ldots, M$$

A general constant-coefficient differential equation representing a CTLTI system is therefore in the form

$$a_N \frac{d^N y(t)}{dt^N} + a_{N-1} \frac{d^{N-1} y(t)}{dt^{N-1}} + \ldots + a_1 \frac{dy(t)}{dt} + a_0 y(t) =$$

$$b_M \frac{d^M x(t)}{dt^M} + b_{M-1} \frac{d^{M-1} x(t)}{dt^{M-1}} + \ldots + b_1 \frac{dx(t)}{dt} + b_0 x(t) \qquad (2.29)$$

or it can be expressed in closed summation form.

Constant-coefficient ordinary differential equation:

$$\sum_{k=0}^{N} a_k \frac{d^k y(t)}{dt^k} = \sum_{k=0}^{M} b_k \frac{d^k x(t)}{dt^k} \qquad (2.30)$$

The order of the differential equation (and therefore the order of the system) is the larger of N and M. As an example, the differential equation given by Eqn. (2.17) for the circuit of Fig. 2.9 fits the standard form of Eqn. (2.30) with $N = 1$, $M = 0$, $a_1 = 1$, $a_0 = 1/RC$ and $b_0 = 1/RC$. Similarly, the differential equation given by Eqn. (2.28) for the circuit of Fig. 2.11 also conforms to the standard format of Eqn. (2.30) with $N = 2$, $M = 0$, $a_2 = 1$, $a_1 = R/L$, $a_0 = 1/LC$ and $b_0 = 1/LC$.

In general, a constant-coefficient ordinary differential equation in the form of Eqn. (2.30) has a family of solutions. In order to find a unique solution for $y(t)$, initial values of the output signal and its first $N - 1$ derivatives need to be specified at a time instant $t = t_0$. We need to know

$$y(t_0), \quad \left.\frac{dy(t)}{dt}\right|_{t=t_0}, \quad \ldots, \quad \left.\frac{d^{N-1} y(t)}{dt^{N-1}}\right|_{t=t_0}$$

to find the solution for $t > t_0$.

Example 2.7: **Checking linearity and time invariance of a differential equation**

Determine whether the first-order constant-coefficient differential equation

$$\frac{dy(t)}{dt} + a_0 y(t) = b_0 x(t) \qquad (2.31)$$

represents a CTLTI system.

Solution: In order to check the linearity of the system described by the first-order differential equation of Eqn. (2.31) we will assume that two input signals $x_1(t)$ and $x_2(t)$ produce the responses $y_1(t)$ and $y_2(t)$ respectively. The input-output pairs must satisfy the differential equation. Therefore we have

$$\frac{dy_1(t)}{dt} + a_0 \, y_1(t) = b_0 \, x_1(t) \tag{2.32}$$

and

$$\frac{dy_2(t)}{dt} + a_0 \, y_2(t) = b_0 \, x_2(t) \tag{2.33}$$

Now let a new input signal be constructed as a linear combination of $x_1(t)$ and $x_2(t)$ in the form

$$x_3(t) = \alpha_1 \, x_1(t) + \alpha_2 \, x_2(t) \tag{2.34}$$

where α_1 and α_2 are two arbitrary constants. If the system under consideration is linear, then with $x_3(t)$ as the input signal the solution of the differential equation must be

$$y_3(t) = \alpha_1 \, y_1(t) + \alpha_2 \, y_2(t) \tag{2.35}$$

Does $y(t) = y_3(t)$ satisfy the differential equation when the input is equal to $x(t) = x_3(t)$? Using $y_3(t)$ in the differential equation we obtain

$$\frac{dy_3(t)}{dt} + a_0 \, y_3(t) = \frac{d}{dt} \left[\alpha_1 \, y_1(t) + \alpha_2 \, y_2(t) \right] + a_0 \left[\alpha_1 \, y_1(t) + \alpha_2 \, y_2(t) \right] \tag{2.36}$$

Rearranging terms on the right side of Eqn. (2.36) we have

$$\frac{dy_3(t)}{dt} + a_0 \, y_3(t) = \alpha_1 \left[\frac{dy_1(t)}{dt} + a_0 \, y_1(t) \right] + \alpha_2 \left[\frac{dy_2(t)}{dt} + a_0 \, y_2(t) \right] \tag{2.37}$$

and substituting Eqns. (2.32) and (2.33) into Eqn. (2.37) results in

$$\begin{aligned} \frac{dy_3(t)}{dt} + a_0 \, y_3(t) &= \alpha_1 \left[b_0 \, x_1(t) \right] + \alpha_2 \left[b_0 \, x_2(t) \right] \\ &= b_0 \left[\alpha_1 \, x_1(t) + \alpha_2 \, x_2(t) \right] \\ &= b_0 \, x_3(t) \end{aligned} \tag{2.38}$$

Signals $x_3(t)$ and $y_3(t)$ as a pair satisfy the differential equation. As a result, we may be tempted to conclude that the corresponding system is linear, however, we will take a cautious approach and investigate a bit further: Let us begin by recognizing that the differential equation we are considering is in the same form as the one obtained in Example 2.4 for the simple RC circuit. As a matter of fact it would be identical to it with $a_0 = b_0 = 1/RC$. In the simple RC circuit, the output signal is the voltage across the terminals of the capacitor. What would happen if the capacitor is initially charged to V_0 volts? Would Eqn. (2.35) still be valid? Let us write Eqn. (2.35) at time $t = t_0$:

$$y_3(t_0) = \alpha_1 \, y_1(t_0) + \alpha_2 \, y_2(t_0) \tag{2.39}$$

On the other hand, because of the initial charge of the capacitor, any solution found must start with the same initial value $y(t_0) = V_0$ and continue from that point on. We must have

$$y_1(t_0) = V_0 \, , \quad y_2(t_0) = V_0 \, , \quad y_3(t_0) = V_0$$

It is clear that Eqn. (2.39) can only be satisfied if $V_0 = 0$, that is, if the capacitor has no initial charge. Therefore, the differential equation in Eqn. (2.31) represents a CTLTI system if and only if the initial value of the output signal is equal to zero.

Another argument to convince ourselves that a system with non-zero initial conditions cannot be linear is the following: Based on the second condition of linearity given by Eqn. (2.6), a zero input signal must produce a zero output signal (just set $\alpha_1 = 0$). A system with a non-zero initial state $y(t_0) = V_0$ produces a non-zero output signal even if the input signal is zero, and therefore cannot be linear.

Next we need to check the system described by the differential equation in Eqn. (2.31) for time invariance. If we replace the time variable t with $(t - \tau)$ we get

$$\frac{dy(t - \tau)}{dt} + a_0\, y(t - \tau) = b_0\, x(t - \tau)$$

Delaying the input signal by τ causes the output signal to also be delayed by τ. Therefore, the system is time-invariant.

In Example 2.7 we verified that the first-order constant-coefficient differential equation corresponds to a CTLTI system provided that the system is initially relaxed. It is also a straightforward task to prove that the general constant-coefficient differential equation given by Eqn. (2.30) corresponds to a CTLTI system if all initial conditions are equal to zero.

Assuming that two input signals $x_1(t)$ and $x_2(t)$ produce the output signals $y_1(t)$ and $y_2(t)$ respectively, we will check and see if the input signal $x_3(t) = \alpha_1\, x_1(t) + \alpha_2\, x_2(t)$ leads to the output signal $y_3(t) = \alpha_1\, y_1(t) + \alpha_2\, y_2(t)$. Through repeated differentiation it can be shown that

$$\frac{d^k x_3(t)}{dt^k} = \frac{d^k}{dt^k}\{\alpha_1\, x_1(t) + \alpha_2\, x_2(t)\}$$

$$= \alpha_1\, \frac{d^k x_1(t)}{dt^k} + \alpha_2\, \frac{d^k x_2(t)}{dt^k} \tag{2.40}$$

for $k = 0, \ldots, M$, and similarly,

$$\frac{d^k y_3(t)}{dt^k} = \frac{d^k}{dt^k}\{\alpha_1\, y_1(t) + \alpha_2\, y_2(t)\}$$

$$= \alpha_1\, \frac{d^k y_1(t)}{dt^k} + \alpha_2\, \frac{d^k y_2(t)}{dt^k} \tag{2.41}$$

for $k = 0, \ldots, N$. Substituting Eqn. (2.41) into the left side of Eqn. (2.30) yields

$$\sum_{k=0}^{N} a_k\, \frac{d^k y_3(t)}{dt^k} = \sum_{k=0}^{N} a_k\left\{\alpha_1\, \frac{d^k y_1(t)}{dt^k} + \alpha_2\, \frac{d^k y_2(t)}{dt^k}\right\}$$

$$= \alpha_1 \sum_{k=0}^{N} a_k\, \frac{d^k y_1(t)}{dt^k} + \alpha_2 \sum_{k=0}^{N} a_k\, \frac{d^k y_2(t)}{dt^k} \tag{2.42}$$

Since we have assumed that $[x_1(t), y_1(t)]$ and $[x_2(t), y_2(t)]$ are solution pairs and therefore satisfy the differential equation, we have

$$\sum_{k=0}^{N} a_k\, \frac{d^k y_1(t)}{dt^k} = \sum_{k=0}^{M} b_k\, \frac{d^k x_1(t)}{dt^k} \tag{2.43}$$

and

$$\sum_{k=0}^{N} a_k \frac{d^k y_2(t)}{dt^k} = \sum_{k=0}^{M} b_k \frac{d^k x_2(t)}{dt^k} \tag{2.44}$$

Using Eqns. (2.43) and (2.44) in Eqn. (2.42) it can be shown that

$$\sum_{k=0}^{N} a_k \frac{d^k y_3(t)}{dt^k} = \sum_{k=0}^{M} b_k \left\{ \alpha_1 \frac{d^k x_1(t)}{dt^k} + \alpha_2 \frac{d^k x_2(t)}{dt^k} \right\}$$

$$= \sum_{k=0}^{M} b_k \frac{d^k x_3(t)}{dt^k} \tag{2.45}$$

where, in the last step, we have used the result found in Eqn. (2.40). As an added condition, Eqn. (2.41) must also be satisfied at $t = t_0$ for all derivatives in the differential equation, requiring

$$\left. \frac{d^k y(t)}{dt^k} \right|_{t=t_0} = 0 \tag{2.46}$$

for $k = 0, \dots, N-1$.

The differential equation

$$\sum_{k=0}^{N} a_k \frac{d^k y(t)}{dt^k} = \sum_{k=0}^{M} b_k \frac{d^k x(t)}{dt^k}$$

represents a linear system provided that all initial conditions are equal to zero:

$$y(t_0) = 0, \quad \left. \frac{dy(t)}{dt} \right|_{t=t_0} = 0, \quad \dots, \quad \left. \frac{d^{N-1} y(t)}{dt^{N-1}} \right|_{t=t_0} = 0$$

Time invariance of the corresponding system can also be proven easily by replacing T with $t - \tau$ in Eqn. (2.30):

$$\sum_{k=0}^{N} a_k \frac{d^k y(t-\tau)}{dt^k} = \sum_{k=0}^{M} b_k \frac{d^k x(t-\tau)}{dt^k} \tag{2.47}$$

2.5   Solving Differential Equations

One method of determining the output signal of a system in response to a specified input signal is to solve the corresponding differential equation. In later parts of this text we will study alternative methods of accomplishing the same task. The use of these alternative methods will also be linked to the solution of the differential equation to provide further insight into linear system behavior.

In this section we will present techniques for solving linear constant-coefficient differential equations. Two distinct methods will be presented; one that can only be used with a first-order differential equation, and one that can be used with any order differential equation in the standard form of Eqn. (2.30).

2.5.1   Solution of the first-order differential equation

The first-order differential equation

$$\frac{dy(t)}{dt} + \alpha\, y(t) = r(t) \tag{2.48}$$

represents a first-order CTLTI system. This could be the differential equation of the RC circuit considered in Example 2.4 with $\alpha = 1/RC$ and $r(t) = (1/RC)\, x(t)$. In this section we will formulate the general solution of this differential equation for a specified initial value $y(t_0)$.

Let the function $f(t)$ be defined as

$$f(t) = e^{\alpha t}\, y(t) \tag{2.49}$$

Differentiating $f(t)$ with respect to time yields

$$\begin{aligned}
\frac{df(t)}{dt} &= e^{\alpha t}\frac{dy(t)}{dt} + \alpha\, e^{\alpha t}\, y(t)\\[2mm]
&= e^{\alpha t}\left[\frac{dy(t)}{dt} + \alpha\, y(t)\right]
\end{aligned} \tag{2.50}$$

The expression in square brackets is recognized as $r(t)$ from Eqn. (2.48), therefore

$$\frac{df(t)}{dt} = e^{\alpha t}\, r(t) \tag{2.51}$$

Integrating both sides of Eqn. (2.51) over time, we obtain $f(t)$ as

$$f(t) = f(t_0) + \int_{t_0}^{t} e^{\alpha \tau}\, r(\tau)\, d\tau \tag{2.52}$$

Using Eqn. (2.49)

$$f(t_0) = e^{\alpha t_0}\, y(t_0) \tag{2.53}$$

the solution for $f(t)$ becomes

$$f(t) = e^{\alpha t_0}\, y(t_0) + \int_{t_0}^{t} e^{\alpha \tau}\, r(\tau)\, d\tau \tag{2.54}$$

The solution for $y(t)$ is found through the use of this result in Eqn. (2.49):

$$\begin{aligned}
y(t) &= e^{-\alpha t}\, f(t)\\[2mm]
&= e^{-\alpha(t-t_0)}\, y(t_0) + e^{-\alpha t}\int_{t_0}^{t} e^{\alpha \tau}\, r(\tau)\, d\tau\\[2mm]
&= e^{-\alpha(t-t_0)}\, y(t_0) + \int_{t_0}^{t} e^{-\alpha(t-\tau)}\, r(\tau)\, d\tau
\end{aligned} \tag{2.55}$$

Thus, Eqn. (2.55) represents the complete solution of the first-order differential equation in Eqn. (2.48), and is practical to use as long as the right-side function $r(t)$ allows easy evaluation of the integral.

The differential equation

$$\frac{dy(t)}{dt} + \alpha y(t) = r(t), \qquad y(t_0): \text{specified} \tag{2.56}$$

is solved as

$$y(t) = e^{-\alpha(t-t_0)} y(t_0) + \int_{t_0}^{t} e^{-\alpha(t-\tau)} r(\tau) \, d\tau \tag{2.57}$$

We may be tempted to question the significance of the solution found in Eqn. (2.55) especially since it is only applicable to a first-order differential equation and therefore a first-order system. The result we have found will be very useful for working with higher-order systems, however, when we study the state-space description of a CTLTI system in Chapter 9.

Example 2.8: **Unit-step response of the simple RC circuit**

Consider the simple RC circuit that was first introduced in Example 2.4. Let the element values be $R = 1 \ \Omega$ and $C = 1/4$ F. Assume the initial value of the output at time $t = 0$ is $y(0) = 0$. Determine the response of the system to an input signal in the form of a unit-step function, i.e., $x(t) = u(t)$.

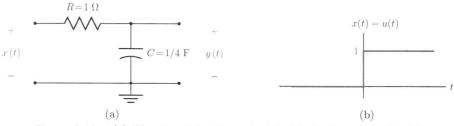

(a)   (b)

Figure 2.12 – (a) The circuit for Example 2.8, (b) the input signal $x(t)$.

Solution: The differential equation of the circuit was obtained in Eqn. (2.17) of Example 2.4. Using the unit-step input signal specified, it can be written as

$$\frac{dy(t)}{dt} + \frac{1}{RC} y(t) = \frac{1}{RC} u(t)$$

Applying Eqn. (2.57) with $\alpha = 1/RC$, $t_0 = 0$ and $r(t) = (1/RC) u(t)$ we have

$$y(t) = \int_0^t e^{-(t-\tau)/RC} \frac{1}{RC} u(t) \, d\tau = \frac{e^{-t/RC}}{RC} \int_0^t e^{\tau/RC} \, d\tau = 1 - e^{-t/RC} \tag{2.58}$$

for $t \geq 0$. In compact form, the result in Eqn. (2.58) is

$$y(t) = \left(1 - e^{-t/RC}\right) u(t) \tag{2.59}$$

If we now substitute the specified values of R and C into Eqn. (2.59) we obtain the output signal

$$y(t) = \left(1 - e^{-4t}\right) u(t)$$

which is graphed in Fig. 2.13.

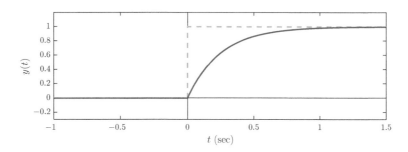

Figure 2.13 – The output signal $y(t)$ for Example 2.8.

Software resources:
ex_2_8.m

Example 2.9: **Pulse response of the simple RC circuit**

Determine the response of the RC circuit of Example 2.4 to a rectangular pulse signal

$$x(t) = A \, \Pi(t/w) \tag{2.60}$$

shown in Fig. 2.14. Element values for the circuit are $R = 1 \, \Omega$ and $C = 1/4$ F. The initial value of the output signal at time $t = -w/2$ is $y(-w/2) = 0$.

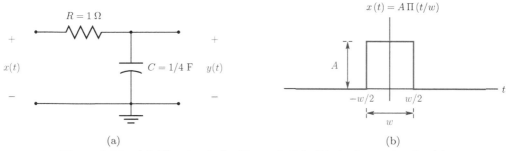

Figure 2.14 – (a) The circuit for Example 2.9, (b) the input signal $x(t)$.

Solution: The differential equation of the circuit is obtained from Eqn. (2.17) with the substitution of specified element values and the input signal:

$$\frac{dy(t)}{dt} + 4\, y(t) = 4A \, \Pi(t/w)$$

The solution is in the form of Eqn. (2.57) with $\alpha = 4$, $y(-w/2) = 0$ and $r(t) = 4A \, \Pi(t/w)$:

$$y(t) = \int_{-w/2}^{t} e^{-4(t-\tau)} \, 4A \, \Pi(\tau/w) \, d\tau \tag{2.61}$$

The integral in this result can be evaluated using two possible ranges of the variable t.

Case 1: $-\dfrac{w}{2} < t \le \dfrac{w}{2}$

In this range of t we have $\Pi\left(t/w\right) = 1$, and the response in Eqn. (2.61) simplifies to

$$y\left(t\right) = 4A \int_{-w/2}^{t} e^{-4(t-\tau)}\,d\tau$$

which can be evaluated as

$$y\left(t\right) = 4A\,e^{-4t} \int_{-w/2}^{t} e^{4\tau}\,d\tau = A\left[1 - e^{-2w}\,e^{-4t}\right] \tag{2.62}$$

Case 2: $t > \dfrac{w}{2}$

In this case we have $\Pi\left(t/w\right) = 1$ for $-w/2 \le t \le w/2$ and $\Pi\left(t/w\right) = 0$ for $t > w/2$. The result in Eqn. (2.61) simplifies to

$$y\left(t\right) = 4A \int_{-w/2}^{w/2} e^{-4(t-\tau)}\,d\tau$$

which leads to the response

$$y\left(t\right) = 4A\,e^{-4t} \int_{-w/2}^{w/2} e^{4\tau}\,d\tau = A\,e^{-4t}\left[e^{2w} - e^{-2w}\right] \tag{2.63}$$

The complete response of the system can be expressed by combining the results found in Eqns. (2.62) and (2.63) as

$$y\left(t\right) = \begin{cases} A\left[1 - e^{-2w}\,e^{-4t}\right], & -\dfrac{w}{2} < t \le \dfrac{w}{2} \\[2mm] A\,e^{-4t}\left[e^{2w} - e^{-2w}\right], & t > \dfrac{w}{2} \end{cases} \tag{2.64}$$

The signal $y\left(t\right)$ is shown in Fig. 2.15 for $A = 1$ and $w = 1$.

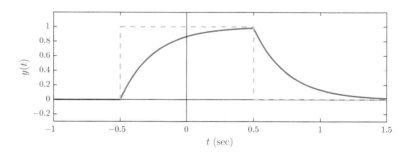

Figure 2.15 – The output signal $y\left(t\right)$ for Example 2.9.

Software resources:
ex_2_9.m

Example 2.10: **Pulse response of the simple RC circuit revisited**

Rework the problem in Example 2.9 by making use of the unit-step response found in Example 2.8 along with linearity and time-invariance properties of the RC circuit.

Solution: The pulse signal used as input in Example 2.9 can be expressed as the difference of two unit-step signals in the form, i.e.,

$$x\left(t\right) = A\,\Pi\left(t/w\right) = A\,u\left(t + \frac{w}{2}\right) - A\,u\left(t - \frac{w}{2}\right)$$

This is illustrated in Fig. 2.16.

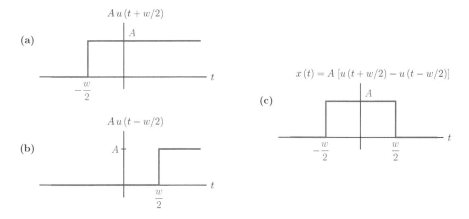

Figure 2.16 – Constructing a pulse from time-shifted step functions.

Since the system under consideration is linear, the response to $x\left(t\right)$ is

$$\text{Sys}\left\{x\left(t\right)\right\} = \text{Sys}\left\{A\,u\left(t + \frac{w}{2}\right)\right\} - \text{Sys}\left\{A\,u\left(t - \frac{w}{2}\right)\right\}$$

$$= A\,\text{Sys}\left\{u\left(t + \frac{w}{2}\right)\right\} - A\,\text{Sys}\left\{u\left(t - \frac{w}{2}\right)\right\} \tag{2.65}$$

Furthermore, since the system is time-invariant, its responses to time-shifted unit-step functions can be found from the unit-step response that was computed in Example 2.8. It was determined that

$$\text{Sys}\left\{u\left(t\right)\right\} = \left(1 - e^{-4t}\right) u\left(t\right)$$

Using the time invariance property of the system we have

$$\text{Sys}\left\{u\left(t + \frac{w}{2}\right)\right\} = \left[1 - e^{-4(t+w/2)}\right] u\left(t + \frac{w}{2}\right) \tag{2.66}$$

and

$$\text{Sys}\left\{u\left(t - \frac{w}{2}\right)\right\} = \left[1 - e^{-4(t-w/2)}\right] u\left(t - \frac{w}{2}\right) \tag{2.67}$$

Substituting Eqns. (2.66) and (2.67) into Eqn. (2.65) we find the pulse response of the RC circuit as

$$\text{Sys}\left\{x\left(t\right)\right\} = A\,\left[1 - e^{-4(t+w/2)}\right] u\left(t + \frac{w}{2}\right) - A\,\left[1 - e^{-4(t-w/2)}\right] u\left(t - \frac{w}{2}\right) \tag{2.68}$$

which is in agreement with the result found in Example 2.9. The steps used in the solution are illustrated in Fig. 2.17.

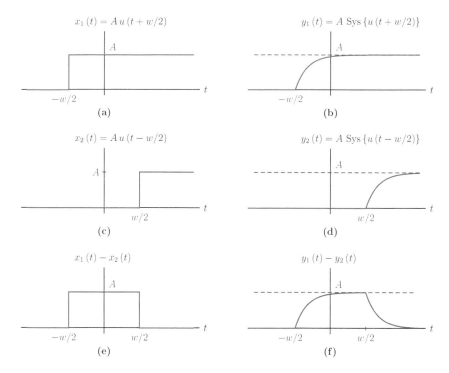

Figure 2.17 – The steps employed in the solution of Example 2.10.

Software resources:
ex_2_10.m

Interactive Demo: rc_demo1.m

The demo "rc_demo1.m" is based on Example 2.10, Eqn. (2.65) and Fig. 2.17. The pulse response of the RC circuit is obtained from its step response using superposition. The input pulse applied to the RC circuit is expressed as the difference of two step signals, and the output signal is computed as the difference of the individual responses to step functions. Circuit parameters R and C as well as the pulse width w can be varied using slider controls.

Software resources:
rc_demo1.m

Software resources: See MATLAB Exercise 2.3.

2.5.2 Solution of the general differential equation

The solution method discussed in the previous section applies to a first-order differential equation only, although, in Chapter 9, we will discuss extension of this method to higher-order differential equations through the use of state variables. In our efforts to solve the general constant-coefficient differential equation in the form given by Eqn. (2.30) we will consider two separate components of the output signal $y(t)$ as follows:

$$y(t) = y_h(t) + y_p(t) \tag{2.69}$$

The first term, $y_h(t)$, is the solution of the *homogeneous differential equation* found by ignoring the input signal, that is, by setting $x(t)$ and all of its derivatives equal to zero in Eqn. (2.30) for all values of t:

$$\sum_{k=0}^{N} a_k \frac{d^k y(t)}{dt^k} = 0 \qquad (2.70)$$

As a mathematical function, $y_h(t)$ is the *homogeneous solution* of the differential equation. From the perspective of the output signal of a system, $y_h(t)$ is called the *natural response* of the system. As one of the components of the output signal, the homogeneous solution of the differential equation or, equivalently, the natural response of the system to which it corresponds, $y_h(t)$ depends on the structure of the system as well as the initial state of the system. It does not depend, however, on the input signal applied to the system. It is the part of the response that is produced by the system due to a release of the energy stored within the system. Recall the circuits used in Examples 2.4 through 2.6. Some circuit elements such as capacitors and inductors are capable of storing energy which could later be released under certain circumstances. If, at some point in time, a capacitor or an inductor with stored energy is given a chance to release this energy, the circuit could produce a response through this release even when there is no external input signal being applied.

When we discuss the stability property of CTLTI systems in later sections of this chapter we will see that, for a stable system, $y_h(t)$ tends to gradually disappear in time. Because of this, it is also referred to as the *transient response* of the system.

In contrast, the second term $y_p(t)$ in Eqn. (2.69) is part of the solution that is due to the input signal $x(t)$ being applied to the system. It is referred to as the *particular solution* of the differential equation. It depends on the input signal $x(t)$ and the internal structure of the system, but it does not depend on the initial state of the system. It is the part of the response that remains active after the homogeneous solution $y_h(t)$ gradually becomes smaller and disappears. When we study the system function concept in later chapters of this text we will link the particular solution of the differential equation to the *steady-state response* of the system, that is, the response to an input signal that has been applied for a long enough time for the transient terms to die out.

2.5.3 Finding the natural response of a continuous-time system

Computation of the natural response of a CTLTI system requires solving the homogeneous differential equation. Before we tackle the problem of finding the homogeneous solution for the general constant-coefficient differential equation, we will consider the first-order case. The first-order homogeneous differential equation is in the form

$$\frac{dy(t)}{dt} + \alpha\, y(t) = 0 \qquad (2.71)$$

where α is any arbitrary constant. The homogeneous differential equation in Eqn. (2.71) has many possible solutions. In order to find a unique solution from among many, we also need to know the initial value of $y(t)$ at some time instant $t = t_0$. We will begin by writing Eqn. (2.71) in the alternative form

$$\frac{dy(t)}{dt} = -\alpha\, y(t) \qquad (2.72)$$

It is apparent from Eqn. (2.72) that the solution we are after is a function $y(t)$ the time derivative of which is proportional to itself. An exponential function of time has that property, so we can make an educated guess for a solution in the form

$$y(t) = c\,e^{st} \tag{2.73}$$

Trying this guess[1] in the homogeneous differential equation we have

$$\frac{d}{dt}\left[c\,e^{st}\right] + \alpha\,c\,e^{st} = 0 \tag{2.74}$$

which yields

$$s\,c\,e^{st} + \alpha\,c\,e^{st} = 0 \tag{2.75}$$

By factoring out the common term $c\,e^{st}$, Eqn. (2.75) becomes

$$c\,e^{st}\,(s + \alpha) = 0 \tag{2.76}$$

There are two ways to make the left side of Eqn. (2.76) equal zero:

1.　　Select $c\,e^{st} = 0$

2.　　Select $(s + \alpha) = 0$

The former choice leads to the trivial solution $y(t) = 0$, and is obviously not very useful. Also, as a solution it is only valid when the initial value of the output signal is $y(0) = 0$ as we cannot satisfy any other initial value with this solution. Therefore we must choose the latter, and use $s = -\alpha$. Substituting this value of s into Eqn. (2.73) we get

$$y(t) = c\,e^{-\alpha t} \tag{2.77}$$

where the constant c must be determined based on the desired initial value of $y(t)$ at $t = t_0$. We will explore this in the next two examples.

Example 2.11: **Natural response of the simple RC circuit**

Consider again the RC circuit of Example 2.4 shown in Fig. 2.18 with the element values $R = 1\ \Omega$ and $C = 1/4$ F. Also, let the input terminals of the circuit be connected to a battery that supplies the circuit with an input voltage of 5 V up to the time instant $t = 0$.

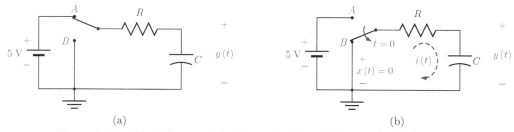

(a)　　　　　　　　　　　　　　　　　(b)

Figure 2.18 – The RC circuit for Example 2.11: (a) for $t < 0$, (b) for $t \geq 0$.

---

[1]　　Another way to justify the guess utilized in Eqn. (2.73) is to write Eqn. (2.71) in the form

$$\frac{dy}{y} = -\alpha\,dt$$

Integrating both sides leads to

$$\int \frac{dy}{y} = -\int \alpha\,dt \quad \Rightarrow \quad \ln(y) = -\alpha t + K \quad \Rightarrow \quad y(t) = e^{K}e^{-\alpha t} = c\,e^{-\alpha t}$$

Assuming the battery has been connected to the circuit for a long time before $t = 0$, the capacitor voltage has remained at a steady-state value of 5 V. Let the switch be moved from position A to position B at $t = 0$ ensuring that $x(t) = 0$ for $t \geq 0$. The initial value of capacitor voltage at time $t = 0$ is $y(0) = 5$ V. Find the output signal as a function of time.

Solution: Since $x(t) = 0$ for $t > 0$, the output signal is $y(t) = y_h(t)$, and we are trying to find the homogeneous solution. Substituting the specified parameter values, the homogeneous differential equation is found as

$$\frac{dy(t)}{dt} + 4\,y(t) = 0$$

We need $(s + 4) = 0$, and the corresponding homogeneous solution is in the form

$$y_h(t) = c\,e^{-st} = c\,e^{-4t}$$

for $t \geq 0$. The initial condition $y_h(0) = 5$ must be satisfied. Substituting $t = 0$ into the homogeneous solution, we get

$$y_h(0) = c\,e^{-4(0)} = c = 5$$

Using the value found for the constant c, the natural response of the circuit is

$$y_h(t) = 5\,e^{-4t}, \quad \text{for } t \geq 0$$

The natural response can be expressed in a more compact form through the use of the unit-step function:

$$y_h(t) = 5\,e^{-4t}\,u(t) \tag{2.78}$$

This solution is shown in Fig. 2.19.

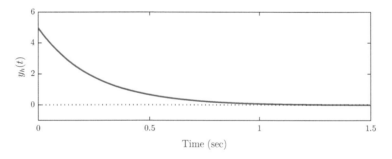

Figure 2.19 – The natural response of the circuit in Example 2.11.

Next we will check this solution against the homogeneous differential equation to verify its validity. The first derivative of the output signal is

$$\frac{dy_h(t)}{dt} = \frac{d}{dt}\left[5\,e^{-4t}\right] = -20\,e^{-4t}$$

Using $y_h(t)$ and $dy_h(t)/dt$ in the homogeneous differential equation we have

$$\frac{dy_h(t)}{dt} + 4\,y_h(t) = -20\,e^{-4t} + 4\left[5\,e^{-4t}\right] = 0$$

indicating that the solution we have found is valid.

Software resources:
ex_2_11.m

Example 2.12: Changing the start time in Example 2.11

Rework the problem in Example 2.11 with one minor change: The initial value of the output signal is specified at the time instant $t = -0.5$ seconds instead of at $t = 0$, and its value is $y(-0.5) = 10$. Physically this would correspond to using a 10 V battery instead of the 5 V battery shown in Fig. 2.18(a), and moving the switch from position A to position B at time instant $t = -0.5$ seconds instead of $t = 0$. These differences are illustrated in Fig. 2.20.

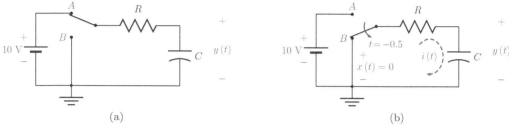

(a)   (b)

Figure 2.20 – The RC circuit for Example 2.12: (a) for $t < -0.5$, (b) for $t \geq -0.5$.

Solution: The general form of the solution found in Example 2.11 is still valid, that is,

$$y_h(t) = c\,e^{-4t}$$

To satisfy $y_h(-0.5) = 10$ we need

$$y_h(-0.5) = c\,e^{-4(-0.5)} = c\,e^2 = 10$$

and therefore

$$c = \frac{10}{e^2} = 1.3534$$

The homogeneous solution is

$$y_h(t) = 1.3534\,e^{-4t}, \quad \text{for } t \geq -0.5$$

or, using the unit step function

$$y_h(t) = 1.3534\,e^{-4t}\,u(t+0.5) \tag{2.79}$$

If the initial condition is specified at a time instant other than $t = 0$, then the solution we find using the procedure outlined starts at that time instant as shown in Fig. 2.21.

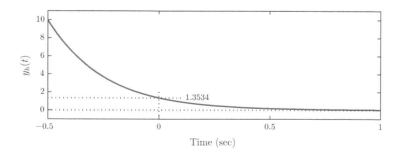

Figure 2.21 – The natural response of the circuit in Example 2.12.

Software resources:
ex_2_12.m

We are now ready to solve the general homogeneous differential equation in the form

$$\sum_{k=0}^{N} a_k \frac{d^k y(t)}{dt^k} = 0 \qquad (2.80)$$

If we use the same initial guess

$$y(t) = c e^{st} \qquad (2.81)$$

for the solution, various derivatives of $y(t)$ will be

$$\frac{dy(t)}{dt} = s c e^{st}, \quad \frac{d^2 y(t)}{dt^2} = s^2 c e^{st}, \quad \frac{d^3 y(t)}{dt^3} = s^3 c e^{st}, \quad \cdots$$

Through repeated differentiation it can be shown that

$$\frac{d^k y(t)}{dt^k} = s^k c e^{st} \qquad (2.82)$$

We need to determine which values of the parameter s would lead to valid solutions for the homogeneous differential equation. Substituting Eqn. (2.82) into Eqn. (2.80) results in

$$\sum_{k=0}^{N} a_k s^k c e^{st} = 0 \qquad (2.83)$$

Since the term $c e^{st}$ is independent of the summation index k, it can be factored out to yield

$$c e^{st} \sum_{k=0}^{N} a_k s^k = 0 \qquad (2.84)$$

requiring one of the following conditions to be satisfied for $y(t) = c e^{st}$ to be a solution of the differential equation:

1. $c e^{st} = 0$
   This leads to the trivial solution $y(t) = 0$ for the homogeneous equation, and is therefore not very interesting. Also, since this solution would leave us with no adjustable parameters, it cannot satisfy any non-zero initial conditions.

2. $\sum_{k=0}^{P} a_k s^k = 0$
   This is called the *characteristic equation* of the system. Values of s that are the solutions of the characteristic equation can be used in exponential functions as solutions of the homogeneous differential equation.

The characteristic equation:

$$\sum_{k=0}^{N} a_k s^k = 0 \qquad (2.85)$$

The characteristic equation is found by starting with the homogeneous differential equation, and replacing the k-th derivative of the output signal $y(t)$ with s^k.

To obtain the characteristic equation, substitute:

$$\frac{d^k y(t)}{dt^k} \quad \rightarrow \quad s^k$$

Let us write the characteristic equation in open form:

$$a_N s^N + a_{N-1} s^{N-1} + \ldots + a_1 s + a_0 = 0 \tag{2.86}$$

The polynomial of order N on the left side of Eqn. (2.86) is referred to as the *characteristic polynomial*, and is obtained by simply replacing each derivative in the homogeneous differential equation with the corresponding power of s. Let the roots of the characteristic polynomial be $s_1, s_2, \ldots, s_N$ so that Eqn. (2.86) can be written as

$$a_N (s - s_1)(s - s_2) \ldots (s - s_N) = 0 \tag{2.87}$$

Using any of the roots of the characteristic polynomial, we can construct a signal

$$y_k(t) = c_k e^{s_k t}, \quad k = 1, \ldots, N \tag{2.88}$$

that will satisfy the homogeneous differential equation. The general solution of the homogeneous differential equation is obtained as a linear combination of all valid terms in the form of Eqn. (2.88) as

$$y_h(t) = c_1 e^{s_1 t} + c_2 e^{s_2 t} + \ldots + c_N e^{s_N t} = \sum_{k=1}^{N} c_k e^{s_k t} \tag{2.89}$$

The unknown coefficients $c_1, c_2, \ldots, c_N$ are determined from the initial conditions. The exponential terms $e^{s_k t}$ in the homogeneous solution given by Eqn. (2.89) are called the *modes of the system*. In later chapters of this text we will see that the roots s_k of the characteristic polynomial of the system will be identical to the *poles* of the system function and to the *eigenvalues* of the state matrix.

Example 2.13: **Time constant concept**

We will again refer to the RC circuit introduced in Example 2.4 and shown in Fig. 2.9. The differential equation governing the behavior of the circuit was found in Eqn. (2.17). The characteristic equation of the system is found as

$$s + \frac{1}{RC} = 0$$

This is a first-order system, and its only mode is $e^{-t/RC}$. If the initial value of the output signal is $y(0) = V_0$, the homogeneous solution of the differential equation (or the natural response of the system) is

$$y_h(t) = V_0 e^{-t/RC} u(t) \tag{2.90}$$

Let parameter τ be defined as $\tau = RC$, and the natural response can be written as

$$y_h(t) = V_0 e^{-t/\tau} u(t) \tag{2.91}$$

For this type of a system with only one mode, the parameter τ is called the *time constant* of the system. Based on Eqn. (2.91) the time constant τ represents the amount it takes for the natural response to decay down to $1/e$ times or, equivalently, 36.8 percent of its initial value. This is illustrated in Fig. 2.22.

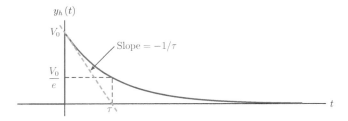

Figure 2.22 – Illustration of the time constant for a first-order system.

It is evident from Fig. 2.22 that the line that is tangent to $y_h(t)$ at $t = 0$ has a slope of $-1/\tau$, that is,

$$\left. \frac{dy_h(t)}{dt} \right|_{t=0} = -1/\tau$$

and it intersects the time axis at $t = \tau$. The amplitude of the output signal at $t = \tau$ is $y_h(\tau) = V_0/e$.

Software resources:
ex_2_13.m

Interactive Demo: rc_demo2.m

The demo "rc_demo2.m" is based on Example 2.13, Eqn. (2.91) and Fig. 2.22. The natural response of the RC circuit is computed and graphed for a specified initial value. Circuit parameters R and C and the initial value V_0 of the output signal at time $t = 0$ can be varied using slider controls.

  1. Reduce the time constant τ by reducing either R or C and observe its effect on how long it takes the natural response to become negligibly small. Can you develop a rule-of-thumb on how many time constants it takes for the natural response to become negligible?
  2. As the time constant is changed, observe the slope of the natural response at time $t = 0$. The greater the slope, the faster the natural response disappears.

Software resources:
rc_demo2.m

Example 2.14: **Natural response of second-order system**

The differential equation for the RLC circuit used in Example 2.6 and shown in Fig. 2.11 was given by Eqn. (2.28). Let the element values be $R = 5\ \Omega$, $L = 1$ H and $C = 1/6$ F. At time $t = 0$, the initial inductor current is $i(0) = 2$ A and the initial capacitor voltage is $y(0) = 1.5$ V. No external input signal is applied to the circuit, therefore $x(t) = 0$. Determine the output voltage $y(t)$.

Solution: Without an external input signal, the particular solution is zero, and the total solution of the differential equation includes only the homogeneous solution, that is, $y(t) = y_h(t)$. The homogeneous differential equation is

$$\frac{d^2 y(t)}{dt^2} + 5\frac{dy(t)}{dt} + 6\,y(t) = 0$$

The characteristic equation of the system is

$$s^2 + 5s + 6 = 0$$

with solutions $s_1 = -2$ and $s_2 = -3$. The homogeneous solution of the differential equation is in the form

$$y_h(t) = c_1\,e^{-2t} + c_2\,e^{-3t} \qquad (2.92)$$

for $t \geq 0$. The unknown coefficients c_1 and c_2 need to be adjusted so that the specified initial conditions are satisfied. Evaluating $y_h(t)$ for $t = 0$ we obtain

$$y_h(0) = c_1 e^{-2(0)} + c_2 e^{-3(0)} = c_1 + c_2 = 1.5 \qquad (2.93)$$

Using the initial value of the inductor current we have

$$i(0) = C\left.\frac{dy_h(t)}{dt}\right|_{t=0} = 2$$

from which the initial value of the first derivative of the output signal can be obtained as

$$\left.\frac{dy_h(t)}{dt}\right|_{t=0} = \frac{i(0)}{C} = \frac{2}{1/6} = 12$$

Differentiating the solution found in Eqn. (2.92) and imposing the initial value of $dy_h(t)/dt$ leads to

$$\left.\frac{dy_h(t)}{dt}\right|_{t=0} = \left.\left[-2c_1 e^{-2t} - 3c_2 e^{-3t}\right]\right|_{t=0}$$
$$= -2c_1 - 3c_2 = 12 \qquad (2.94)$$

The coefficients c_1 and c_2 can now be determined by solving Eqns. (2.93) and (2.94) simultaneously, resulting in

$$c_1 = 16.5\,, \quad \text{and} \quad c_2 = -15$$

and the natural response of the system is

$$y_h(t) = 16.5\,e^{-2t} - 15\,e^{-3t}\,, \quad t \geq 0 \qquad (2.95)$$

which is graphed in Fig. 2.23.

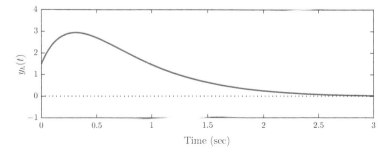

Figure 2.23 – The natural response of the second-order system in Example 2.14.

Software resources:
ex_2_14.m

In Example 2.14 the two roots of the characteristic polynomial turned out to be both real-valued and different from each other. As a result we were able to express the solution of the homogeneous equation in the standard form of Eqn. (2.89). We will now review several possible scenarios for the types of roots obtained:

Case 1: All roots are distinct and real-valued.

In this case the homogeneous solution is

$$y_h(t) = \sum_{k=1}^{N} c_k e^{s_k t} \tag{2.96}$$

as discussed above. If a real root s_k is negative, the corresponding term $c_k e^{s_k t}$ decays exponentially over time. Alternatively, if s_k is positive, the corresponding term grows exponentially over time, and may cause the output signal to become very large without bound. Fig. 2.24 illustrates these two possibilities.

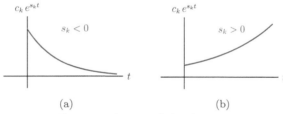

Figure 2.24 – Terms corresponding to real roots of the characteristic equation: (a) $s_k < 0$, (b) $s_k > 0$.

Case 2: Characteristic polynomial has complex-valued roots.

Since the coefficients of the differential equation, and consequently the coefficients of the characteristic polynomial, are all real-valued, any complex roots must appear in the form of conjugate pairs. Therefore, if $s_{1a} = \sigma_1 + j\omega_1$ is a complex root of the characteristic polynomial, then its complex conjugate $s_{1b} = s_{1a}^* = \sigma_1 - j\omega_1$ must also be a root. Let the part of the homogeneous solution that is due to these two roots be

$$\begin{aligned} y_{h1}(t) &= c_{1a} e^{s_{1a} t} + c_{1b} e^{s_{1b} t} \\ &= c_{1a} e^{(\sigma_1 + j\omega_1)t} + c_{1b} e^{(\sigma_1 - j\omega_1)t} \end{aligned} \tag{2.97}$$

where the coefficients c_{1a} and c_{1b} are to be determined from the initial conditions. Continuing with the reasoning we started above, since the coefficients of the differential equation are real-valued, the homogeneous solution $y_h(t)$ must also be real. Furthermore, $y_{h1}(t)$, the part of the solution that is due to the complex conjugate pair of roots we are considering, must also be real. This in turn implies that the coefficients c_{1a} and c_{1b} must form a complex conjugate pair as well. Let us write the two coefficients in polar complex form as

$$c_{1a} = |c_1| e^{j\theta_1}, \quad \text{and } c_{1b} = |c_1| e^{-j\theta_1} \tag{2.98}$$

Substituting Eqn. (2.98) into Eqn. (2.97) leads to the result

$$y_{h1}(t) = |c_1|\; e^{\sigma_1 t}\, e^{j(\omega_1 t + \theta_1)} + |c_1|\; e^{\sigma_1 t}\, e^{-j(\omega_1 t + \theta_1)}$$

$$= 2\,|c_1|\; e^{\sigma_1 t} \cos\left(\omega_1 t + \theta_1\right) \tag{2.99}$$

a. A pair of complex conjugate roots for the characteristic polynomial leads to a solution component in the form of a cosine signal multiplied by an exponential signal.

b. The oscillation frequency of the cosine signal is determined by ω_1, the imaginary part of the complex roots.

c. The real part of the complex roots, σ_1, impacts the amplitude of the solution. If $\sigma_1 < 0$, then the amplitude of the cosine signal decays exponentially over time. In contrast, if $\sigma_1 > 0$, the amplitude of the cosine signal grows exponentially over time. These two possibilities are illustrated in Fig. 2.25.

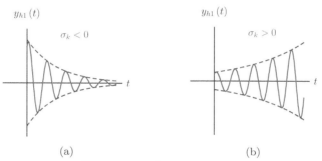

(a)   (b)

Figure 2.25 – Terms corresponding to pair of complex conjugate roots of the characteristic equation: (a) $\sigma_1 < 0$, (b) $\sigma_1 > 0$.

Using the appropriate trigonometric identity[2] it is also possible to write Eqn. (2.99) as

$$y_{h1}(t) = 2\,|c_1| \cos\left(\theta_1\right) e^{\sigma_1 t} \cos\left(\omega_1 t\right) - 2\,|c_1| \sin\left(\theta_1\right) e^{\sigma_1 t} \sin\left(\omega_1 t\right)$$

$$= d_1\, e^{\sigma_1 t} \cos\left(\omega_1 t\right) + d_2\, e^{\sigma_1 t} \sin\left(\omega_1 t\right) \tag{2.100}$$

Case 3: Characteristic polynomial has some multiple roots.

Consider again the factored version of the characteristic equation first given by Eqn. (2.87):

$$a_N\left(s - s_1\right)\left(s - s_2\right)\ldots\left(s - s_N\right) = 0$$

What if the first two roots are equal, that is, $s_2 = s_1$? If we used the approach we have employed so far, we would have a natural response in the form

$$y_h(t) = c_1\, e^{s_1 t} + c_2\, e^{s_2 t} + \text{other terms}$$

$$= c_1\, e^{s_1 t} + c_2\, e^{s_1 t} + \text{other terms}$$

$$= \left(c_1 + c_2\right) e^{s_1 t} + \text{other terms}$$

$$= \bar{c}_1\, e^{s_1 t} + \text{other terms} \tag{2.101}$$

---

[2] $\cos\left(a + b\right) = \cos\left(a\right) \cos\left(b\right) - \sin\left(a\right) \sin\left(b\right)$.

The problem with the response in Eqn. (2.101) is that we have lost one of the co-efficients. For a homogeneous differential equation of order N we need to satisfy N initial conditions at some time instant $t = t_0$, namely

$$y(t_0) , \quad \frac{dy(t)}{dt}\bigg|_{t=t_0} , \quad \ldots, \quad \frac{d^{N-1}y(t)}{dt^{N-1}}\bigg|_{t=t_0}$$

To satisfy N initial conditions we need as many adjustable parameters $c_1, \ldots, c_N$ sometimes referred to as N *degrees of freedom*. Losing one of the coefficients creates a problem for our ability to satisfy N initial conditions. In order to gain back the coefficient we have lost, we need an additional term for the two roots at $s = s_1$. A solution in the form

$$y_h(t) = c_{11} e^{s_1 t} + c_{12} t e^{s_1 t} + \text{other terms} \qquad (2.102)$$

will work for this purpose. In general, a root of multiplicity r requires r terms in the homogeneous solution. If the characteristic polynomial has a factor $(s - s_1)^r$ then the resulting homogeneous solution will be

$$y_h(t) = c_{11} e^{s_1 t} + c_{12} t e^{s_1 t} + \ldots + c_{1r} t^{r-1} e^{s_1 t} + \text{other terms} \qquad (2.103)$$

Example 2.15: **Natural response of second-order system revisited**

Consider again the RLC circuit which was first used in Example 2.6 and shown in Fig. 2.11. At time $t = 0$, the initial inductor current is $i(0) = 0.5$ A, and the initial capacitor voltage is $y(0) = 2$ V. No external input signal is applied to the circuit, therefore $x(t) = 0$. Determine the output voltage $y(t)$ if

  a. the element values are $R = 2 \ \Omega$, $L = 1$ H and $C = 1/26$ F,
  b. the element values are $R = 6 \ \Omega$, $L = 1$ H and $C = 1/9$ F.

Solution: Since no external input signal is applied to the circuit, the output signal is equal to the natural response, i.e., $y(t) = y_h(t)$, which we will obtain by solving the homogeneous differential equation. Two initial conditions are specified. The first one is that $y_h(0) = 2$. Using the specified initial value of the inductor current we have

$$i(0) = C \frac{dy_h(t)}{dt}\bigg|_{t=0} = 0.5$$

which leads to the initial value for the first derivative of the output signal as

$$\frac{dy_h(t)}{dt}\bigg|_{t=0} = \frac{0.5}{C}$$

Now we are ready to find the solutions for the two sets of component values given in parts (a) and (b):

  a. Using the specified component values, the homogeneous differential equation is

$$\frac{d^2y(t)}{dt^2} + 2\frac{dy(t)}{dt} + 26\,y(t) = 0$$

and the characteristic equation is

$$s^2 + 2s + 26 = 0$$

The roots of the characteristic equation are $s_1 = -1 + j5$ and $s_2 = -1 - j5$. The natural response is therefore in the form given by Eqn. (2.100) with $\sigma_1 = -1$ and $\omega_1 = 5$ rad/s.

$$y_h(t) = d_1 e^{-t} \cos(5t) + d_2 e^{-t} \sin(5t)$$

Now we can impose the initial conditions. The first one is straightforward:

$$y_h(0) = d_1 = 2$$

For the specified capacitance value $C = 1/26$ F, the initial value of $dy_h(t)/dt$ is

$$\left. \frac{dy_h(t)}{dt} \right|_{t=0} = 13$$

To impose the specified initial value of $dy_h(t)/dt$ we will first differentiate the homogeneous solution to obtain

$$\frac{dy_h(t)}{dt} = -d_1 e^{-t} \cos(5t) - 5d_1 e^{-t} \sin(5t) - d_2 e^{-t} \sin(5t) + 5d_2 e^{-t} \cos(5t)$$

Evaluating this derivative at $t = 0$ we have

$$\left. \frac{dy_h(t)}{dt} \right|_{t=0} = -d_1 + 5d_2 = 13$$

Since we previously found that $d_1 = 2$, we need $d_2 = 3$, and the natural response of the circuit is

$$y_h(t) = 2 e^{-t} \cos(5t) + 3 e^{-t} \sin(5t) \tag{2.104}$$

for $t \geq 0$. This solution is shown in Fig. 2.26(a).

b. For this case the homogeneous differential equation becomes

$$\frac{d^2 y(t)}{dt^2} + 6 \frac{dy(t)}{dt} + 9 y(t) = 0$$

and the characteristic equation is

$$s^2 + 6s + 9 = 0, \quad \text{or} \quad (s+3)^2 = 0$$

Since both roots of the characteristic equation are at $s_1 = -3$ we will use a homogeneous solution in the form

$$y_h(t) = c_{11} e^{-3t} + c_{12} t e^{-3t}$$

for $t \geq 0$. Imposing the initial value $y(0) = 2$ yields

$$c_{11} = 2$$

for the first coefficient. For the specified capacitance value of $C = 1/9$ F the initial value of the derivative of the output signal is

$$\left. \frac{dy_h(t)}{dt} \right|_{t=0} = 4.5$$

To satisfy the initial condition on $dy_h\left(t\right)/dt$ we will differentiate the homogeneous solution and evaluate the result at $t=0$ which yields

$$\left.\frac{dy_h\left(t\right)}{dt}\right|_{t=0} = \left.\frac{d}{dt}\left[c_{11}\,e^{-3t}+c_{12}\,te^{-3t}\right]\right|_{t=0}$$

$$= \left.\left[-3c_{11}\,e^{-3t}+c_{12}\,e^{-3t}-3c_{12}\,te^{-3t}\right]\right|_{t=0}$$

$$= -3\,c_{11}+c_{12}=4.5$$

and leads to the result

$$c_{12}=10.5$$

Therefore the natural response for this case is

$$y_h\left(t\right)=2\,e^{-3t}+10.5\,te^{-3t} \tag{2.105}$$

which is shown in Fig. 2.26(b).

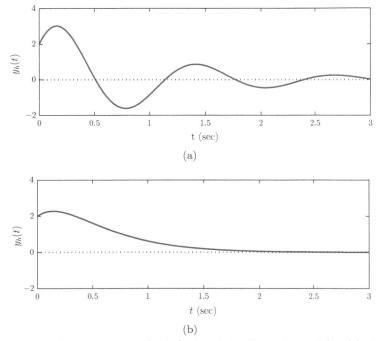

(a)

(b)

Figure 2.26 – Natural responses on the RC circuit in Example 2.15 for (a) characteristic equation roots $s_{1,2}=-1\pm j2$, and (b) characteristic equation roots $s_1=s_2=-3$.

Software resources:
ex_2_15a.m
ex_2_15b.m

Interactive Demo: nr_demo1

The interactive demo program **nr_demo1.m** illustrates different types of homogeneous solutions for a second-order continuous-time system based on the roots of the characteristic polynomial. Recall that the three possibilities were explored above, namely distinct real roots, complex conjugate roots and multiple roots.

In the demo program, the two roots can be specified using slider controls, and the corresponding natural response can be observed. If the roots are both real, then they can be controlled independently. If complex values are chosen, however, then the two roots move simultaneously to keep their complex conjugate relationship. The locations of the two roots s_1 and s_2 are marked on the complex plane. The differential equation, the characteristic equation and the analytical solution for the natural response are displayed and updated as the roots are moved.

1. Start with two real roots that are both negative, say $s_1 = -1$ and $s_2 = -0.5$. Set initial conditions as

$$y_h (t) = 0, \qquad \left. \frac{dy_h (t)}{dt} \right|_{t=0} = 1$$

   Observe the shape of the natural response.

2. Gradually move s_2 to the right, bringing it closer to the vertical axis. Observe the changes in the shape of the natural response. What happens when the root reaches the vertical axis, that is, when $s_2 = 0$?

3. Keep moving s_2 to the right until $s_2 = 0.3$. How does the natural response change when a real root moves into the positive territory?

4. Set the two roots to be $s_{1,2} = -0.3 \pm j1.2$ and observe the natural response.

5. Gradually increase the imaginary parts of the two roots toward $\pm j4$ while keeping their real parts fixed at -0.3. How does the change in the imaginary parts affect the natural response?

6. Gradually move the real parts of the two roots to the right while keeping the imaginary parts equal to $\pm j4$. How does this affect the natural response? What happens when the roots cross over to the right side of the vertical axis?

Software resources:
nr_demo1.m

2.5.4   Finding the forced response of a continuous-time system

In the preceding discussion we have concentrated our efforts on finding the homogeneous solution of the constant-coefficient differential equation which corresponds to the natural response $y_h (t)$ of the system under consideration when no external input signal is applied to it. To complete the solution process, we also need to determine the particular solution for a specified input signal $x (t)$ that is applied to the system. To find the particular solution, we start with an educated guess about the form of the solution we seek, and then adjust its parameters so that the result satisfies the differential equation. The form of the particular solution should include the input signal $x (t)$ and all of its derivatives that are linearly independent, assuming $x (t)$ has a finite number of linearly independent derivatives. For example, if the input signal is $x (t) = \cos (at)$, we need to construct a particular solution that includes the terms $\cos (at)$ and $\sin (at)$ in the form

$$y_p (t) = k_1 \cos (at) + k_2 \sin (at) \qquad (2.106)$$

with parameters k_1 and k_2 to be determined from the differential equation. On the other hand, if the input signal is $x (t) = t^3$, then we need a particular solution that includes the terms t^3, t^2, t, and a constant term in the form

$$y_p(t) = k_3 t^3 + k_2 t^2 + k_1 t + k_0 \tag{2.107}$$

Table 2.1 lists some of the common types of input signals and the forms of particular solutions to be used for them.

Input signal	Particular solution
K (constant)	k_1
$K\, e^{at}$	$k_1\, e^{at}$
$K\, \cos(at)$	$k_1 \cos(at) + k_2 \sin(at)$
$K\, \sin(at)$	$k_1 \cos(at) + k_2 \sin(at)$
$K\, t^n$	$k_n t^n + k_{n-1} t^{n-1} + \ldots + k_1 t + k_0$

Table 2.1 – Choosing a particular solution for various input signals.

The coefficients of the particular solution are determined from the differential equation by assuming all initial conditions are equal to zero (recall that the particular solution does not depend on the initial conditions of the differential equation or the initial state of the system). The specified initial conditions of the differential equation are imposed in the subsequent step for determining the unknown coefficients of the homogeneous solution, not the particular solution.

Now we have all the tools we need to determine the forced response of the system for a specified input signal. The following is a summary of the procedure to be used:

1. Write the homogeneous differential equation. Find the characteristic equation by replacing derivatives of the output signal with corresponding powers of s.
2. Solve for the roots of the characteristic equation and write the homogeneous solution in the form of Eqn. (2.96). If some of the roots are complex conjugate pairs, then use the form in Eqn. (2.97) for them. If there are some multiple roots, use the procedure outlined in Eqn. (2.103) for them. Leave the homogeneous solution in parametric form with undetermined coefficients; do not attempt to compute the coefficients $c_1, c_2, \ldots$ of the homogeneous solution yet.
3. Find the form of the particular solution by either picking the appropriate form of it from Table 2.1, or by constructing it as a linear combination of the input signal and its time derivatives. (This latter approach requires that the input signal have a finite number of linearly independent derivatives.)
4. Try the particular solution in the non-homogeneous differential equation and determine the coefficients $k_1, k_2, \ldots$ of the particular solution. At this point the particular solution should be uniquely determined. However, the coefficients of the homogeneous solution are still undetermined.
5. Add the homogeneous solution and the particular solution together to obtain the total solution. Compute the necessary derivatives of the total solution. Impose the necessary initial conditions on the total solution and its derivatives. Solve the resulting set of equations to determine the coefficients $c_1, c_2, \ldots$ of the homogeneous solution.

These steps for finding the forced solution of a differential equation will be illustrated in the next example.

Example 2.16: **Forced response of the first-order system for sinusoidal input**

Consider once again the RC circuit of Fig. 2.9 with the element values of $R = 1 \ \Omega$ and $C = 1/4$ F. The initial value of the output signal is $y(0) = 5$. Determine the output signal in response to a sinusoidal input signal in the form

$$x(t) = A \cos(\omega t)$$

with amplitude $A = 20$ and radian frequency $\omega = 8$ rad/s.

Solution: Using the specified component values, the non-homogeneous differential equation of the circuit under consideration is

$$\frac{dy(t)}{dt} + 4\,y(t) = 4\,x(t)$$

and we have found in Example 2.11 that the homogeneous solution is in the form

$$y_h(t) = c\,e^{-4t}$$

for $t \geq 0$. We will postpone the task of determining the value of the constant c until after we find the particular solution. Using Table 2.1, the form of the particular solution we seek is

$$y_p(t) = k_1 \cos(\omega t) + k_2 \sin(\omega t) \tag{2.108}$$

Differentiating Eqn. (2.108) with respect to time yields

$$\frac{dy_p(t)}{dt} = -\omega k_1 \sin(\omega t) + \omega k_2 \cos(\omega t) \tag{2.109}$$

The particular solution $y_p(t)$ must satisfy the non-homogeneous differential equation. Substituting Eqns. (2.108) and (2.109) along with the specified input signal $x(t)$ into the differential equation we obtain

$$-\omega k_1 \sin(\omega t) + \omega k_2 \cos(\omega t) + 4\left[k_1 \cos(\omega t) + 4k_2 \sin(\omega t)\right] = A \cos(\omega t)$$

which can be written in a more compact form as

$$(4k_1 + \omega k_2 - A) \cos(\omega t) + (4k_2 - \omega k_1) \sin(\omega t) = 0 \tag{2.110}$$

Eqn. (2.110) must be satisfied for all values of t, therefore we must have

$$4k_1 + \omega k_2 - A = 0 \tag{2.111}$$

and

$$4k_2 - \omega k_1 = 0 \tag{2.112}$$

Eqns. (2.111) and (2.112) can be solved simultaneously to yield

$$k_1 = \frac{4A}{16 + \omega^2} \quad \text{and} \quad k_2 = \frac{A\omega}{16 + \omega^2}$$

The forced solution is obtained by adding the homogeneous solution and the particular solution together:

$$y(t) = y_h(t) + y_f(t)$$

$$= ce^{-4t} + \frac{4A}{16 + \omega^2} \cos(\omega t) + \frac{A\omega}{16 + \omega^2} \sin(\omega t) \qquad (2.113)$$

Let us now substitute the numerical values $A = 20$ and $\omega = 8$ rad/s. The output signal becomes

$$y(t) = ce^{-4t} + \cos(8t) + 2\sin(8t)$$

Finally, we will impose the initial condition $y(0) = 5$ to obtain

$$y(0) = 5 = c + \cos(0) + 2\sin(0)$$

which yields $c = 4$. Therefore, the complete solution is

$$y(t) = 4e^{-4t} + \cos(8t) + 2\sin(8t) \qquad (2.114)$$

for $t \geq 0$. The solution found in Eqn. (2.114) has two fundamentally different components, and can be written in the form

$$y(t) = y_t(t) + y_{ss}(t) \qquad (2.115)$$

The first term

$$y_t(t) = 4e^{-4t} \qquad (2.116)$$

represents the part of the output signal that disappears over time, that is,

$$\lim_{t \to \infty} \{y_t(t)\} = 0$$

It is called the *transient* component of the output signal. The second term

$$y_{ss}(t) = \cos(8t) + 2\sin(8t) \qquad (2.117)$$

is the part of the output signal that remains after the transient term disappears. Therefore it is called the *steady-state response* of the system. Transient and steady-state components as well as the complete response are shown in Fig. 2.27.

Before we leave this example, one final observation is in order: The homogeneous solution for the same circuit was found in Example 2.11 as

$$y_h(t) = 5e^{-4t} \qquad (2.118)$$

with the input signal $x(t)$ set equal to zero. In this example we have used a sinusoidal input signal, and the transient part of the response obtained in Eqn. (2.116) does not match the homogeneous solution for the zero-input case; the two differ by a scale factor. This justifies our decision to postpone the computation of the constant c until the very end. The initial condition is specified for the total output signal $y(t)$ at time $t = 0$; therefore, we must first add the homogeneous solution and the particular solution, and only after doing that we can impose the specified value of $y(0)$ on the solution to determine the constant c.

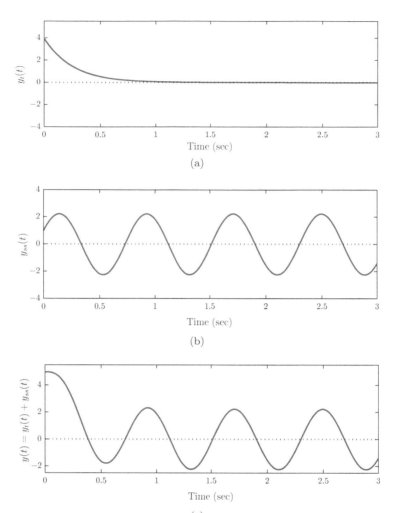

Figure 2.27 – Computation of the output signal of the circuit in Example 2.16: (a) transient component, (b) steady-state component, (c) the complete output signal.

We will revisit this example when we study system function concepts in Chapter 7.
Software resources:
ex_2_16.m

Interactive Demo: fr_demo1

The interactive demo program fr_demo1.m is based on Example 2.16. It allows experimentation with parameters of the problem. The amplitude A is fixed at $A = 20$ so that the input signal is

$$x(t) = 20 \cos(\omega t)$$

The resistance R, the capacitance C, the radian frequency ω and the initial output value $y(0)$ can be varied using slider controls. The effect of parameter changes on the transient response $y_t(t)$, the steady-state response $y_{ss}(t)$ and the total forced response $y_t(t) + y_{ss}(t)$ can be observed.

1. Start with the settings $R = 5 \ \Omega$, $C = 10$ F, $\omega = 4$ rad/s and $y(0) = 0$. Since the initial value of the output signal is zero, the resulting system is CTLTI. In addition, the initial value of the input signal is also zero, and therefore the output signal has no transient component. Observe the peak amplitude value of the steady-state component. Confirm that it matches with what was found in Example 2.16.

2. Now gradually increase the radian frequency ω up to $\omega = 12$ rad/s and observe the change in the peak amplitude of the steady-state component of the output. Compare with the result obtained in Eqn. (2.113).

3. Set parameter values as $R = 1 \ \Omega$, $C = 0.3$ F, $\omega = 12$ rad/s and $y(0) = 4$ V. The time constant is $\tau = RC = 0.3$ s. Comment on how long it takes for the transient component of the output to become negligibly small so that the output signal reaches its steady-state behavior. (As a rule of thumb, 4 to 5 time constants are sufficient for the output signal to be considered in steady state.)

4. Gradually increase the value of R to $R = 3 \ \Omega$ and observe the changes in the transient behavior.

Software resources:
fr_demo1.m

2.6 Block Diagram Representation of Continuous-Time Systems

Up to this point we have studied methods for analyzing continuous-time systems represented in the time domain by means of constant-coefficient ordinary differential equations. Given a system description based on a differential equation, and an input signal $x(t)$, the output signal $y(t)$ can be determined by solving the differential equation.

In some cases we may need to realize a continuous-time system that has a particular differential equation. Alternately, it may be desired to simulate a continuous-time system on a digital computer in an approximate sense. In problems that involve the realization or the simulation of a continuous-time system, we start with a block diagram representation of the differential equation. Once a block diagram is obtained, it may either be realized using circuits that approximate the behavior of each component of the block diagram, or simulated on a computer using code segments that simulate the behavior of each component.

In general, the problem of converting a differential equation to a block diagram has multiple solutions that are functionally equivalent even though they may look different. In this section we will present one particular technique for obtaining a block diagram from a constant-coefficient ordinary differential equation. Alternative methods of constructing block diagrams will be presented in Chapter 7 in the context of obtaining a block diagram from a system function.

Block diagrams for continuous-time systems are constructed using three types of components, namely constant-gain amplifiers, signal adders and integrators. These components are shown in Fig. 2.28. The technique for finding a block diagram from a differential equation is best explained with an example. Consider a third-order differential equation in the form

$$\frac{d^3 y}{dt^3} + a_2 \frac{d^2 y}{dt^2} + a_1 \frac{dy}{dt} + a_0 \, y = b_2 \frac{d^2 x}{dt^2} + b_1 \frac{dx}{dt} + b_0 \, x \qquad (2.119)$$

which is in the standard form of Eqn. (2.30) with $N = 3$ and $M = 2$. The differential equation is shown in compact notation with the understanding that $x(t)$, $y(t)$ and all of their derivatives are functions of time. Also, for convenience, we have chosen the coefficient

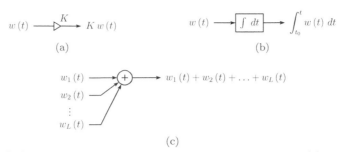

Figure 2.28 – Block diagram components for continuous-time systems: (a) constant-gain amplifier, (b) integrator, (c) signal adder.

of the highest derivative of $y(t)$ to be $a_N = 1$ (which, in this case, happens to be a_3). In cases where $a_N \neq 1$ both sides of Eqn. (2.119) can be divided by a_N to satisfy this requirement.

As the first step for finding a block diagram for this differential equation, we will introduce an intermediate variable $w(t)$. This new variable will be used in place of $y(t)$ in the left side of the differential equation in Eqn. (2.119), and the result will be set equal to $x(t)$ to yield

$$\frac{d^3 w}{dt^3} + a_2 \frac{d^2 w}{dt^2} + a_1 \frac{dw}{dt} + a_0 \, w = x \tag{2.120}$$

The differential equation in Eqn. (2.120) is relatively easy to implement in the form of a block diagram. Rearranging terms in Eqn. (2.120) we obtain

$$\frac{d^3 w}{dt^3} = x - a_2 \frac{d^2 w}{dt^2} - a_1 \frac{dw}{dt} - a_0 \, w \tag{2.121}$$

One possible implementation is shown in Fig. 2.29.

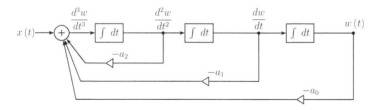

Figure 2.29 – The block diagram for Eqn. (2.121).

We will now show that the output signal $y(t)$ can be expressed in terms of the intermediate variable $w(t)$ as

$$y = b_2 \frac{d^2 w}{dt^2} + b_1 \frac{dw}{dt} + b_0 \, w \tag{2.122}$$

This is essentially the right side of the differential equation in Eqn. (2.119) in which $x(t)$ was replaced with $w(t)$. Together, Eqns. (2.120) and (2.122) are equivalent to Eqn. (2.119).

The proof for Eqn. (2.122) will be given starting with the right side of Eqn. (2.119) and expressing the terms in it through the use of Eqn. (2.120). We have the following relationships:

$$b_0 \, x = b_0 \left(\frac{d^3 w}{dt^3} + a_2 \frac{d^2 w}{dt^2} + a_1 \frac{dw}{dt} + a_0 \, w \right) \tag{2.123}$$

$$b_1 \frac{dx}{dt} = b_1 \left(\frac{d^4 w}{dt^4} + a_2 \frac{d^3 w}{dt^3} + a_1 \frac{d^2 w}{dt^2} + a_0 \frac{dw}{dt} \right) \tag{2.124}$$

$$b_2 \frac{d^2 x}{dt^2} = b_2 \left(\frac{d^5 w}{dt^5} + a_2 \frac{d^4 w}{dt^4} + a_1 \frac{d^3 w}{dt^3} + a_0 \frac{d^2 w}{dt^2} \right) \tag{2.125}$$

Adding Eqns. (2.123) through (2.125) we get

$$b_0 \, x + b_1 \frac{dx}{dt} + b_2 \frac{d^2 x}{dt^2} = \left(b_0 \frac{d^3 w}{dt^3} + b_1 \frac{d^4 w}{dt^4} + b_2 \frac{d^5 w}{dt^5} \right)$$

$$+ a_2 \left(b_0 \frac{d^2 w}{dt^2} + b_1 \frac{d^3 w}{dt^3} + b_2 \frac{d^4 w}{dt^4} \right)$$

$$+ a_1 \left(b_0 \frac{dw}{dt} + b_1 \frac{d^2 w}{dt^2} + b_2 \frac{d^3 w}{dt^3} \right)$$

$$+ a_0 \left(b_0 \, w + b_1 \frac{dw}{dt} + b_2 \frac{d^2 w}{dt^2} \right)$$

$$= \frac{d^3 y}{dt^3} + a_2 \frac{d^2 y}{dt^2} + a_1 \frac{dy}{dt} + a_0 \, y \tag{2.126}$$

proving that Eqns. (2.120) and (2.122) are indeed equivalent to Eqn. (2.119). Thus, the output $y(t)$ can be obtained through the use of Eqn. (2.122).

Since the derivatives on the right-side of Eqn. (2.122) are readily available in the block diagram of Fig. 2.29, we will simply add the required connections to it to arrive at the completed block diagram shown in Fig. 2.30.

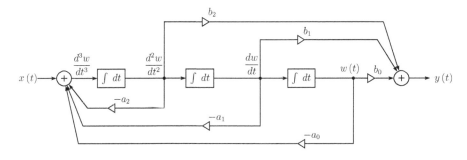

Figure 2.30 – The completed block diagram for Eqn. (2.119).

Even though we have used a third-order differential equation to demonstrate the process of constructing a block diagram, the extension of the technique to a general constant-coefficient differential equation is straightforward. In Fig. 2.30 the feed-forward gains of the block diagram are the right-side coefficients $b_0, b_1, \ldots, b_M$ of the differential equation. Feedback gains of the block diagram are the negated left-side coefficients $-a_0, -a_1, \ldots, -a_{N-1}$ of the differential equation. Recall that we must have $a_N = 1$ for this to work.

Imposing initial conditions

It is also possible to incorporate initial conditions into the block diagram. In the first step, initial values of $y(t)$ and its first $N-1$ derivatives need to be converted to corresponding initial values of $w(t)$ and its first $N-1$ derivatives. Afterwards, appropriate initial value can be imposed on the output of each integrator as shown in Fig. 2.31.

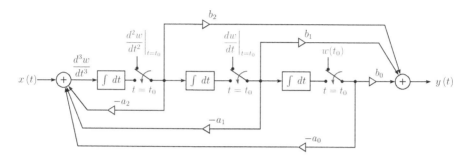

Figure 2.31 – Incorporating initial conditions into a block diagram.

Example 2.17: Block diagram for continuous-time system

Construct a block diagram to solve the differential equation

$$\frac{d^3y}{dt^3} + 5\frac{d^2y}{dt^2} + 17\frac{dy}{dt} + 13\,y = x + 2\frac{dx}{dt}$$

with the input signal $x(t) = \cos(20\pi t)$ and subject to initial conditions

$$y(0) = 1\,, \quad \left.\frac{dy}{dt}\right|_{t=0} = 2\,, \quad \left.\frac{d^2y}{dt^2}\right|_{t=0} = -4$$

Solution: Using the intermediate variable $w(t)$ as outlined in the preceding discussion we obtain the following pair of differential equations equivalent to the original differential equation:

$$\frac{d^3w}{dt^3} + 5\frac{d^2w}{dt^2} + 17\frac{dw}{dt} + 13\,w = x \tag{2.127}$$

$$y = w + 2\frac{dw}{dt} \tag{2.128}$$

Initial conditions specified in terms of the values of y, dy/dt and d^2y/dt^2 at time instant $t = 0$ need to be expressed in terms of the integrator outputs w, dw/dt and d^2w/dt^2 at time instant $t = 0$. For this purpose we will start by writing Eqn. (2.128) at time $t = 0$:

$$y(0) = 1 = w(0) + 2\left.\frac{dw}{dt}\right|_{t=0} \tag{2.129}$$

By differentiating both sides of Eqn. (2.128) and evaluating the result at $t = 0$ we obtain

$$\left.\frac{dy}{dt}\right|_{t=0} = 2 = \left.\frac{dw}{dt}\right|_{t=0} + 2\left.\frac{d^2w}{dt^2}\right|_{t=0} \tag{2.130}$$

Differentiating Eqn. (2.128) twice and evaluating the result at $t = 0$ yields

$$\frac{d^2y}{dt^2}\bigg|_{t=0} = -4 = \frac{d^2w}{dt^2}\bigg|_{t=0} + 2\ \frac{d^3w}{dt^3}\bigg|_{t=0} \tag{2.131}$$

Initial value of d^3w/dt^3 needed in Eqn. (2.131) is obtained from Eqn. (2.127) as

$$\frac{d^3w}{dt^3}\bigg|_{t=0} = x\,(0) - 5\ \frac{d^2w}{dt^2}\bigg|_{t=0} - 17\ \frac{dw}{dt}\bigg|_{t=0} - 13\,w\,(0) \tag{2.132}$$

Substitution of Eqn. (2.132) into Eqn. (2.131) results in

$$\frac{d^2y}{dt^2}\bigg|_{t=0} = -4 = -9\ \frac{d^2w}{dt^2}\bigg|_{t=0} - 34\ \frac{dw}{dt}\bigg|_{t=0} - 26\,w\,(0) + 2\,x\,(0) \tag{2.133}$$

In addition we know that $x\,(0) = 1$. Solving Eqns. (2.129), (2.130) and (2.133) the initial values of integrator outputs are obtained as

$$w\,(0) = \frac{-71}{45}\,, \qquad \frac{dw}{dt}\bigg|_{t=0} = \frac{58}{45}\,, \qquad \frac{d^2w}{dt^2}\bigg|_{t=0} = \frac{16}{45}$$

The block diagram can be constructed as shown in Fig. 2.32.

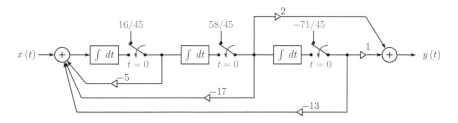

Figure 2.32 – Block diagram for Example 2.17.

2.7   Impulse Response and Convolution

In previous sections of this chapter we have explored the use of differential equations for describing the time-domain behavior of continuous-time systems, and have concluded that a CTLTI system can be completely described by means of a constant-coefficient ordinary differential equation. An alternative description of a CTLTI system can be given in terms of its *impulse response* $h\,(t)$ which is simply the forced response of the system under consideration when the input signal is a unit impulse. This is illustrated in Fig. 2.33.

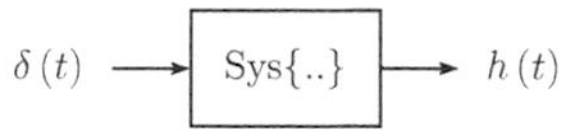

$$\delta\,(t) \longrightarrow \boxed{\text{Sys\{..\}}} \longrightarrow h\,(t)$$

Figure 2.33 – Computation of the impulse response for a CTLTI system.

It will be shown later in this chapter that the impulse response also constitutes a complete description of a CTLTI system. Consequently, the response of such a system to any arbitrary input signal $x(t)$ can be uniquely determined from the knowledge of its impulse response.

It should be noted that our motivation for exploring additional description forms for CTLTI systems is not due to any deficiency or shortcoming of the differential equation of the system. We know that the differential equation is sufficient for the solution of any signal-system interaction problem. The impulse response $h(t)$ does not provide any new information or capability beyond that provided by the differential equation. What we gain through it is an alternative means of working with signal-system interaction problems which can sometimes be more convenient, and which can provide additional insight into system behavior. In the following sections we will discuss how the impulse response of a CTLTI system can be obtained from the underlying differential equation. The reverse operation is also possible, and will be discussed in later chapters.

2.7.1 Finding impulse response of a CTLTI system

When we worked on computing the forced response of a system from its differential equation, we relied on the entries in Table 2.1 to find a particular solution. First, the form of the particular solution appropriate for the type of input signal under consideration is obtained from the table. The particular solution is then tested against the differential equation, and values of any unknown coefficients are computed. Afterwards the particular solution is combined with the homogeneous solution to form the forced response of the system, and initial conditions are imposed to determine the values of any remaining coefficients.

In determining the impulse response of a system from its differential equation, we run into a roadblock: There is no entry in Table 2.1 that corresponds to an input signal in the form $x(t) = \delta(t)$. For a first-order differential equation we can use the technique outlined in Section 2.5.1 and obtain the impulse response through the use of Eqn. (2.55) with $t_0 = 0$, $y(t_0) = 0$ and $x(\tau) = \delta(\tau)$. However, this approach is not applicable to a higher-order differential equation. We will therefore take a slightly different approach in finding the impulse response of a CTLTI system:

1. Use a unit-step function for the input signal, and compute the forced response of the system using the techniques introduced in previous sections. This is the *unit-step response* of the system.
2. Differentiate the unit-step response of the system to obtain the impulse response, i.e.,

$$h(t) = \frac{dy(t)}{dt} \tag{2.134}$$

This idea relies on the fact that differentiation is a linear operator. Given that

$$y(t) = \text{Sys}\{x(t)\} \tag{2.135}$$

we have

$$\frac{dy(t)}{dt} = \frac{d}{dt}\left[\text{Sys}\{x(t)\}\right] = \text{Sys}\left\{\frac{dx(t)}{dt}\right\} \tag{2.136}$$

By choosing the input signal to be a unit-step function, that is, $x(t) = u(t)$, and also recalling that $du(t)/dt = \delta(t)$, we obtain

$$\text{Sys}\{\delta(t)\} = \text{Sys}\left\{\frac{du(t)}{dt}\right\} = \frac{d}{dt}\left[\text{Sys}\{u(t)\}\right] \tag{2.137}$$

It is important to remember that the relationship expressed by Eqn. (2.137) is only valid for a CTLTI system. This is not a serious limitation at all, since we will rarely have a reason to compute an impulse response unless the system is linear and time-invariant.

Example 2.18: **Impulse response of the simple RC circuit**

Determine the impulse response of the first-order RC circuit of Example 2.4, shown in Fig. 2.9, first in parametric form, and then using element values $R = 1\ \Omega$ and $C = 1/4$ F. Assume the system is initially relaxed, that is, there is no initial energy stored in the system. (Recall that this is a necessary condition for the system to be CTLTI.)

Solution: We will solve this problem using two different methods. The differential equation for the circuit was given by Eqn. (2.17). Since we have a first-order differential equation, we can use the solution method that led to Eqn. (2.55) in Section 2.5.1. Letting $\alpha = 1/RC$, $t_0 = 0$ and $y(0) = 0$ we have

$$y(t) = \int_0^t e^{-(t-\tau)/RC} \frac{1}{RC} x(\tau)\, d\tau \tag{2.138}$$

If the input signal is chosen to be a unit-impulse signal, then the output signal becomes the impulse response of the system. Setting $x(t) = \delta(t)$ leads to

$$h(t) = \int_0^t e^{-(t-\tau)/RC} \frac{1}{RC} \delta(\tau)\, d\tau \tag{2.139}$$

Using the sifting property of the unit-impulse function, we obtain

$$h(t) = \frac{1}{RC} e^{-t/RC} u(t)$$

Now we will obtain the same result using the more general method developed in Section 2.7.1. Recall that the unit-step response of the system was found in Example 2.8 Eqn. (2.59) as

$$y(t) = \left(1 - e^{-t/RC}\right) u(t)$$

Differentiating Eqn. (2.140) with respect to time we obtain the impulse response:

$$h(t) = \frac{dy(t)}{dt} = \frac{d}{dt}\left[\left(1 - e^{-t/RC}\right) u(t)\right] = \frac{1}{RC} e^{-t/RC} u(t)$$

which is the same result obtained in Eqn. (2.140). With the substitution of the specified element values, the impulse response becomes

$$h(t) = 4\,e^{-4t}\,u(t) \tag{2.140}$$

which is graphed in Fig. 2.34.

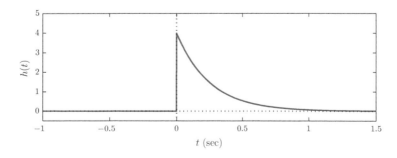

Figure 2.34 – Impulse response of the system in Example 2.18.

Software resources:
ex_2_18.m

Example 2.19: Impulse response of a second-order system

Determine the impulse response of the RLC circuit that was used in Example 2.6 and shown in Fig. 2.11. Use element values $R = 2\ \Omega$, $L = 1$ H and $C = 1/26$ F.

Solution: The homogeneous solution for the system was found in part (a) of Example 2.15 to be

$$y_h\,(t) = d_1\,e^{-t}\cos\,(5t) + d_2\,e^{-t}\sin\,(5t)$$

We will begin by finding the unit-step response. The form of the particular solution is obtained from Table 2.1 as

$$y_p\,(t) = k_1$$

Testing the particular solution with the differential equation leads to the conclusion that $k_1 = 1$, and the complete solution is

$$
\begin{aligned}
y\,(t) =& y_h\,(t) + y_p\,(t) \\
=& d_1\,e^{-t}\cos\,(5t) + d_2\,e^{-t}\sin\,(5t) + 1
\end{aligned}
$$

Since we are determining the impulse response, we will naturally assume that the system is CTLTI, and is therefore initially relaxed. This requires

$$y\,(0) = d_1 + 1 = 0 \tag{2.141}$$

and

$$
\begin{aligned}
\frac{dy_h\,(t)}{dt}\bigg|_{t=0} &= \left[(-d_1 + 5d_2)\,e^{-t}\cos\,(5t) + (-5d_1 - d_2)\,e^{-t}\sin\,(5t)\right]\big|_{t=0} \\
&= -d_1 + 5d_2 = 0 \tag{2.142}
\end{aligned}
$$

Solving Eqns. (2.141) and (2.141) yields $d_1 = -1$ and $d_2 = -0.2$. Therefore, the unit-step response of the system is

$$y\,(t) = y_h\,(t) + y_p\,(t) = -e^{-t}\cos\,(5t) - (0.2)\,e^{-t}\sin\,(5t) + 1$$

for $t \geq 0$. The impulse response is found by differentiating the unit-step response with respect to time, and is

$$
\begin{aligned}
h\,(t) =& \frac{dy\,(t)}{dt} \\
=& e^{-t}\cos\,(5t) + 5\,e^{-t}\sin\,(5t) + (0.2)\,e^{-t}\sin\,(5t) - e^{-t}\cos\,(5t) \\
=& (5.2)\,e^{-t}\sin\,(5t) \tag{2.143}
\end{aligned}
$$

for $t \geq 0$. This result is graphed in Fig. 2.35.

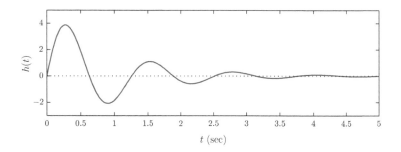

Figure 2.35 – Impulse response of the system in Example 2.19.

Software resources:
ex_2_19.m

2.7.2   Convolution operation for CTLTI systems

In Section 1.3.3 we have formulated the decomposition of an arbitrary signal into impulse
functions. Specifically, we have established that a signal $x(t)$ can be expressed through a
decomposition in the form

$$x(t) = \int_{-\infty}^{\infty} x(\lambda)\, \delta(t-\lambda)\, d\lambda \tag{2.144}$$

Using Eqn. (2.144) the output signal $y(t)$ of a system can be written in terms of its input
signal $x(t)$ as

$$y(t) = \text{Sys}\{x(t)\} = \text{Sys}\left\{ \int_{-\infty}^{\infty} x(\lambda)\, \delta(t-\lambda)\, d\lambda \right\} \tag{2.145}$$

Assuming the system under consideration is linear, we will take liberty[3] to swap the order
of integration and system transformation, and write the output signal as

$$y(t) = \int_{-\infty}^{\infty} \text{Sys}\{x(\lambda)\, \delta(t-\lambda)\}\, d\lambda \tag{2.146}$$

The justification for this action can be given by considering the integral of Eqn. (2.144) as
the limit case of an infinite sum as demonstrated in Chapter 1, Eqn. (1.54).

For the sake of discussion let us assume that we know the system under consideration to
be linear, but not necessarily time-invariant. The response of a linear system to an impulse
signal scaled by $x(\lambda)$ and time-shifted by λ is

$$\text{Sys}\{x(\lambda)\, \delta(t-\lambda)\} = x(\lambda)\, \text{Sys}\{\delta(t-\lambda)\}$$

Eqn. (2.146) can therefore be written as

$$y(t) = \int_{-\infty}^{\infty} x(\lambda)\, \text{Sys}\{\delta(t-\lambda)\}\, d\lambda \tag{2.147}$$

---

[3]   Eqn. (2.146) essentially represents the superposition principle applied to an infinite sum of terms
$x(\lambda)\,\delta(t-\lambda)$. There are studies that conclude that one may find particular signals and circumstances
under which the transition from Eqn. (2.145) to Eqn. (2.146) would be problematic. For the types of signals
and systems encountered in engineering practice, however, Eqn. (2.146) will generally be valid.

Using Eqn. (2.147), the response of a linear system to any arbitrary input signal $x(t)$ can be computed, provided that we know the responses of the system to impulse signals time-shifted by all possible values. Thus, the knowledge necessary for determining the output of a linear system in response to an arbitrary signal $x(t)$ is

$$\left\{ h_\lambda(t) = \text{Sys}\left\{\delta(t-\lambda)\right\}, \quad \text{all } \lambda \right\}$$

While Eqn. (2.147) provides us with a viable method of determining system output $y(t)$ for any input signal $x(t)$, the vast amount of knowledge that we need to possess about the system makes this highly impractical.

For the result in Eqn. (2.147) to be of practical use, the prerequisite knowledge about the system must be reduced to a manageable level. Let the *impulse response* of the system be defined as

$$h(t) = \text{Sys}\left\{\delta(t)\right\} \tag{2.148}$$

If, in addition to being linear, the system is also known to be time-invariant, then the response of the system to any shifted impulse signal can be derived from the knowledge of $h(t)$ alone, that is,

$$h_\lambda(t) = \text{Sys}\left\{\delta(t-\lambda)\right\} = h(t-\lambda) \tag{2.149}$$

consistent with the definition of time invariance given by Eqn. (2.10). Now the output signal can be written as

$$y(t) = \int_{-\infty}^{\infty} x(\lambda)\, h(t-\lambda)\, d\lambda \tag{2.150}$$

Eqn. (2.150) is known as the *convolution integral* for continuous-time signals. The output signal $y(t)$ of a CTLTI system is obtained by *convolving* the input signal $x(t)$ and the impulse response $h(t)$ of the system. This relationship is expressed in compact notation as

$$y(t) = x(t) * h(t) \tag{2.151}$$

where the symbol $*$ represents the *convolution operator*. We will show later in this section that the convolution operator is commutative, that is, the relationship in Eqn. (2.150) can also be written in the alternative form

$$y(t) = h(t) * x(t)$$
$$= \int_{-\infty}^{\infty} h(\lambda)\, x(t-\lambda)\, d\lambda \tag{2.152}$$

The roles of $h(t)$ and $x(t)$ can be swapped without affecting the end result.

Continuous-time convolution summary:

$$y(t) = x(t) * h(t) = \int_{-\infty}^{\infty} x(\lambda)\, h(t-\lambda)\, d\lambda \tag{2.153}$$

$$= h(t) * x(t) = \int_{-\infty}^{\infty} h(\lambda)\, x(t-\lambda)\, d\lambda \tag{2.154}$$

In computing the convolution integral it is helpful to sketch the signals. The graphical steps involved in computing the convolution of two signals $x(t)$ and $h(t)$ at a specific time-instant t can be summarized as follows:

1. Sketch the signal $x(\lambda)$ as a function of the independent variable λ. This corresponds to a simple name change on the independent variable, and the graph of the signal $x(\lambda)$ appears identical to the graph of the signal $x(t)$. (See Fig. 2.36.)

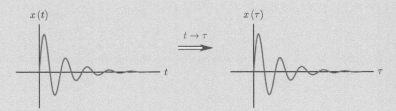

Figure 2.36 – Obtaining $x(\lambda)$ for the convolution integral.

2. For one specific value of t, sketch the signal $h(t-\lambda)$ as a function of the independent variable λ. This task can be broken down into two steps as follows:

   a. Sketch $h(-\lambda)$ as a function of λ. This step amounts to time-reversal of the signal $h(\lambda)$.

   b. In $h(\lambda)$ substitute $\lambda \to \lambda - t$. This step yields

$$h(-\lambda)\Big|_{\lambda \to \lambda - t} = h(t-\lambda) \tag{2.155}$$

   and amounts to time-shifting $h(-\lambda)$ by t.

   See Fig. 2.37 for an illustration of the steps for obtaining $h(t-\lambda)$.

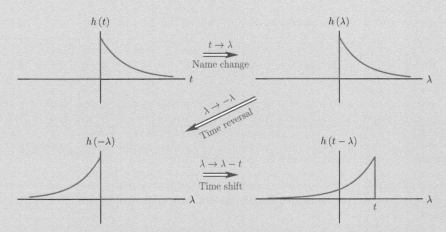

Figure 2.37 – Obtaining $h(t-\lambda)$ for the convolution integral.

3. Multiply the two signals sketched in 1 and 2 to obtain the product $f(\lambda) = x(\lambda)\,h(t-\lambda)$.

4. Compute the area under the product $f(\lambda) = x(\lambda)\,h(t-\lambda)$ by integrating it over the independent variable λ. The result is the value of the output signal at the specific time instant t.

5. Repeat steps 1 through 4 for all values of t that are of interest.

The next several examples will illustrate the details of the convolution operation for CTLTI systems.

Example 2.20: **Unit-step response of RC circuit revisited**

The unit-step response of the simple RC circuit was found in Example 2.8 using the direct solution method discussed in Section 2.5.1. Solve the same problem using the convolution operation.

Solution: In Example 2.18 we found the impulse response of the RC circuit to be

$$h(t) = \frac{1}{RC} e^{-t/RC} u(t)$$

Based on the convolution integral of Eqn. (2.150), the two functions we need are $x(\lambda)$ and $h(t - \lambda)$, both as functions of λ. Fig. 2.38 shows these functions for the two cases of $t \leq 0$ and of $t > 0$.

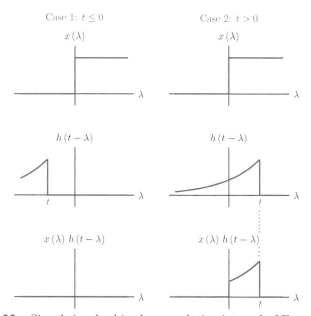

Figure 2.38 – Signals involved in the convolution integral of Example 2.20.

It is obvious that, for $t \leq 0$ these two functions do not overlap anywhere, and therefore the integrand $x(\lambda) h(t - \lambda)$ of the convolution integral is zero for all values of λ. Thus, an easy conclusion is

$$y(t) = 0, \quad \text{for } t \leq 0$$

If $t > 0$, the two functions $x(\lambda)$ and $h(t - \lambda)$ overlap for values of λ in the interval $(0, t)$. In this interval, we have $x(\lambda) = 1$ and $h(t - \lambda) = \frac{1}{RC} e^{-(t-\lambda)/RC}$. Therefore, the output signal is

$$y(t) = \int_0^t \frac{1}{RC} e^{-(t-\lambda)/RC} d\lambda = 1 - e^{-t/RC}, \quad \text{for} \quad t > 0$$

Using a unit-step function to account for both segments of the solution found, the general solution valid for all t can be written as

$$y(t) = \left(1 - e^{-t/RC}\right) u(t)$$

which agrees with the result found in Example 2.8 by solving the differential equation directly.

Interactive Demo: `conv_demo1.m`

The demo program "`conv_demo1.m`" is based on Example 2.20, and allows visualization of the convolution operation. Signals $x(\lambda)$, $h(t-\lambda)$ and the product $[x(\lambda)\, h(t-\lambda)]$ are graphed on the left side of the demo screen. The time variable can be advanced by means of a slider control. As the time t is varied, graphs for the factor $h(t-\lambda)$ and the integrand $[x(\lambda)\, h(t-\lambda)]$ are updated. The area under the integrand $[x(\lambda)\, h(t-\lambda)]$ is shaded to correspond to the result of the integral for that particular value of t.

Software resources:

`conv_demo1.m`

Example 2.21: **Pulse response of RC circuit revisited**

Using convolution, determine the response of the simple RC circuit of Example 2.4 to a unit-pulse input signal $x(t) = \Pi(t)$.

Solution: As in the previous example, it will be useful to sketch the functions involved in the convolution integral, namely $x(\lambda)$ and $h(t-\lambda)$. In doing so, we will consider three distinctly different possibilities for the time variable t as shown in Fig. 2.39.

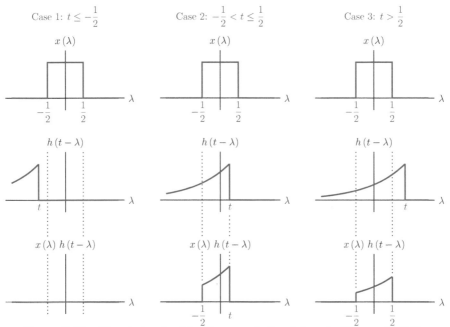

Figure 2.39 – Signals involved in the convolution integral of Example 2.21.

The convolution integral will be evaluated separately for each of the three cases. It is suggested that the reader follow the development below while using the interactive demo program "`conv_demo2.m`" to observe the overlaps of $x(\lambda)$ and $h(t-\lambda)$ in each of the three cases.

Case 1: $t \leq -\frac{1}{2}$

There is no overlap between the two functions in this case, therefore their product equals zero for all values of λ. As a result, the output signal is

$$y(t) = 0 , \quad \text{for} \quad t \leq -\frac{1}{2} \tag{2.156}$$

Case 2: $-\frac{1}{2} < t \leq \frac{1}{2}$

For this case, the two functions overlap in the range $-\frac{1}{2} < \lambda \leq t$. Setting integration limits accordingly, we can compute the output signal as

$$y(t) = \int_{-1/2}^{t} \frac{1}{RC} e^{-(t-\lambda)/RC} \, d\lambda$$

$$= \left(1 - e^{-(t+1/2)/RC}\right) , \quad \text{for} \quad -\frac{1}{2} < t \leq \frac{1}{2}$$

Case 3: $t > \frac{1}{2}$

For this case, the overlap of the two functions will occur in the range $-\frac{1}{2} < \lambda \leq \frac{1}{2}$. Setting integration limits to the endpoints of this new range, we have

$$y(t) = \int_{-1/2}^{1/2} \frac{1}{RC} e^{-(t-\lambda)/RC} \, d\lambda$$

$$= e^{-t/RC} \left(e^{1/2RC} - e^{-1/2RC}\right) , \quad \text{for} \quad t > \frac{1}{2}$$

We have computed the output signal in each of the three distinct time intervals we have identified, mainly before the start of the pulse, during the pulse, and after the end of the pulse. Putting these three partial solutions together, the complete solution for the output signal is obtained as

$$y(t) = \begin{cases} 0 , & t \leq -\frac{1}{2} \\ \left(1 - e^{-(t+1/2)/RC}\right) , & -\frac{1}{2} < t \leq \frac{1}{2} \\ e^{-t/RC} \left(e^{1/2RC} - e^{-1/2RC}\right) , & t > \frac{1}{2} \end{cases} \tag{2.157}$$

and is shown in Fig. 2.40.

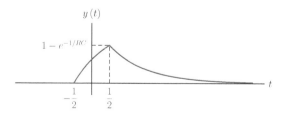

Figure 2.40 – Convolution result for Example 2.21.

It is easy to see that the output signal found in Eqn. (2.157) becomes identical to the result obtained in Eqn. (2.64) of Example 2.9 if we set $A = 1$, $w = 1$, $R = 1 \, \Omega$ and $C = 1/4$ F.

Software resources:
ex_2_21.m

Interactive Demo: conv_demo2.m

The demo program "conv_demo2.m" is based on Example 2.21, Fig. 2.39 and Fig. 2.40. It allows visualization of the convolution operation. The graphical user interface is similar to that of "conv_demo1.m" except it is adapted to use the signals of Example 2.21.
Software resources:
conv_demo2.m

Example 2.22: **A more involved convolution problem**

Consider a system with the impulse response

$$h(t) = e^{-t} \left[u(t) - u(t-2) \right]$$

Let the input signal applied to this system be

$$x(t) = \Pi(t - 0.5) - \Pi(t - 1.5)$$

$$= \begin{cases} 1, & 0 \leq t < 1 \\ -1, & 1 \leq t < 2 \\ 0, & \text{otherwise} \end{cases}$$

Determine the output signal $y(t)$ using convolution.

Solution: The functions $x(\lambda)$ and $h(t - \lambda)$ involved in the convolution integral are shown in Fig. 2.41.

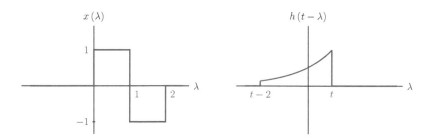

Figure 2.41 – The functions $x(\lambda)$ and $h(t - \lambda)$ for the convolution problem of Example 2.22.

For this problem we will need to distinguish between six different regions for the time variable t as shown in Fig. 2.42. The demo program "conv_demo3.m" can be used along with the solution outlined below to observe the overlapped regions in time.

Case I: $t \leq 0$

The functions $x(\lambda)$ and $h(t - \lambda)$ do not overlap for $t \leq 0$, and therefore their product equals zero for all values of λ. As a result, the output signal is

$$y(t) = 0, \quad \text{for} \quad t \leq 0$$

Case 2: $0 < t \leq 1$

In this case the two functions overlap for the range $0 < \lambda \leq t$. The output signal is

$$y(t) = \int_0^t (1) \, e^{-(t-\lambda)} \, d\lambda$$

$$= 1 - e^{-t} , \quad \text{for} \quad 0 < t \leq 1$$

Case 3: $1 < t \leq 2$

In this case the two functions overlap for the range $0 < \lambda \leq t$. In the interval $0 < \lambda \leq 1$ we have $x(\lambda) = 1$, and the resulting integrand is positive. On the other hand, in the interval $1 < \lambda \leq t$ we have $x(\lambda) = -1$, and the integrand is negative. As a result, two integrals will be needed for the solution:

$$y(t) = \int_0^1 (1) \, e^{-(t-\lambda)} \, d\lambda + \int_1^t (-1) \, e^{-(t-\lambda)} \, d\lambda$$

$$= -1 + 4.4366 \, e^{-t} , \quad \text{for} \quad 1 < t \leq 2$$

Case 4: $2 < t \leq 3$

In this case the overlap region is $t - 2 < \lambda \leq 2$. Furthermore, the left edge of $h(t - \lambda)$ is in the interval $0 < t - 2 \leq 1$. In the interval $t - 2 < \lambda \leq 1$ the input signal is $x(\lambda) = 1$, and the resulting integrand $[x(\lambda) \, h(t - \lambda)]$ is positive. In the remainder of the overlap interval, that is, for $1 < \lambda \leq 2$ we have $x(\lambda) = -1$, and the integrand is negative. Again, two integrals will be formed:

$$y(t) = \int_{t-2}^1 (1) \, e^{-(t-\lambda)} \, d\lambda + \int_1^2 (-1) \, e^{-(t-\lambda)} \, d\lambda$$

$$= -0.1353 - 1.9525 \, e^{-t} , \quad \text{for} \quad 2 < t \leq 3$$

Case 5: $3 < t \leq 4$

In this case the overlap region is $t - 2 < \lambda \leq 2$. The left edge of $h(t - \lambda)$ is in the interval $1 < t - 2 \leq 2$. The output signal is

$$y(t) = \int_{t-2}^2 (-1) \, e^{-(t-\lambda)} \, d\lambda$$

$$= 0.1353 - 7.3891 \, e^{-t} , \quad \text{for} \quad 3 < t \leq 4$$

Case 6: $t > 4$

The functions $x(\lambda)$ and $h(t - \lambda)$ do not overlap for $t > 4$, and therefore the output signal is

$$y(t) = 0 , \quad \text{for} \quad t > 4$$

The complete output signal is graphed in Fig. 2.43 by combining the results found for all six regions of time.

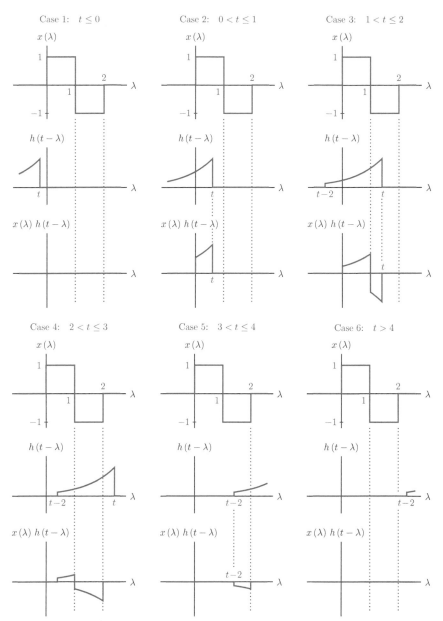

Figure 2.42 – Signals involved in the convolution integral of Example 2.22.

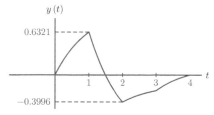

Figure 2.43 – Convolution result for Example 2.22.

Software resources:
ex_2_22.m

Interactive Demo: conv_demo3.m

The demo program "conv_demo3.m" is based on Example 2.22. It facilitates visualization of the overlapping regions between the functions $x(\lambda)$ and $h(t-\lambda)$ as the time variable t is varied. The graphical user interface is similar to the user interfaces of "conv_demo1.m" and "conv_demo2.m".
Software resources:
conv_demo3.m

We have stated in Eqn. (2.152) that the convolution operator is commutative. The proof is straightforward, and will now be given.

Let us apply the variable change $t - \lambda = \gamma$ to the integral in Eqn. (2.150), and write it as

$$y(t) = x(t) * h(t)$$
$$= \int_{\infty}^{-\infty} x(t-\gamma) \, h(\gamma) \, (-d\gamma) \tag{2.158}$$

where we have also used the two consequences of the variable change employed, namely

1. $d\lambda = -d\gamma$
2. $\lambda \to \pm\infty$ means $\gamma \to \mp\infty$

Swapping the integration limits in Eqn. (2.158) and negating the integral to compensate for it, the output signal can now be written as

$$y(t) = \int_{-\infty}^{\infty} x(t-\gamma) \, h(\gamma) \, d\gamma$$
$$= h(t) * x(t) \tag{2.159}$$

thus proving the commutative property of the convolution operator.

The following example will illustrate the use of the alternative form of the convolution integral based on the commutative property.

Example 2.23: **Using alternative form of convolution**

Find the unit-step response of the RC circuit with impulse response

$$h(t) = \frac{1}{RC} \, e^{-t/RC} \, u(t)$$

using the alternative form of the convolution integral given by Eqn. (2.159).

Solution: Using the convolution integral of Eqn. (2.159), the two functions we need are $h(\gamma)$ and $x(t-\gamma)$, both as functions of γ. Fig. 2.44 shows these functions for the two cases of $t \leq 0$ and of $t > 0$. For $t \leq 0$ these two functions do not overlap anywhere, and therefore the integrand $[h(\gamma) \, x(t-\gamma)]$ of the convolution integral is zero for all values of γ. We therefore conclude that

$$y(t) = 0 , \quad \text{for } t \leq 0 \tag{2.160}$$

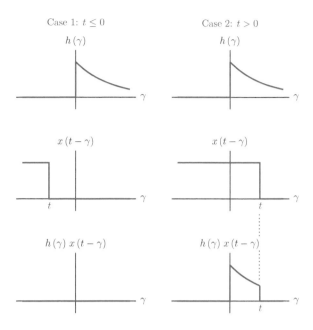

Figure 2.44 – Signals involved in the convolution integral of Example 2.23.

For values of $t > 0$, the two functions $h(\gamma)$ and $x(t-\gamma)$ overlap for values of γ in the interval $(0, t)$. In this interval, we have $x(t-\gamma) = 1$ and $h(\gamma) = \frac{1}{RC} e^{-\gamma/RC}$. Therefore, the output signal is

$$y(t) = \int_0^t \frac{1}{RC} e^{-\gamma/RC} \, d\gamma = 1 - e^{-t/RC} \, , \quad \text{for} \quad t > 0$$

Using a unit-step function to account for both segments of the solution found, the general solution valid for all t is

$$y(t) = \left(1 - e^{-t/RC} \right) u(t)$$

which agrees with the result found in Example 2.20.

Interactive Demo: conv_demo4.m

The demo program "conv_demo4.m" is based on Example 2.23. It facilitates visualization of the overlapping regions between the functions $h(\gamma)$ and $x(t-\gamma)$ as the time variable t is varied. The graphical user interface is identical to the user interface of "conv_demo1.m".
Software resources:
conv_demo4.m

2.8   Causality in Continuous-Time Systems

Causality is an important feature of physically realizable systems. A system is said to be causal if the current value of the output signal depends only on current and past values of the input signal, but not on its future values. For example, a continuous-time system with input-output relationship

$$y(t) = x(t) + x(t - 0.01) + x(t - 0.02) \tag{2.161}$$

is causal since the output signal can be computed based on current and past values of the input signal. Conversely, the system given by

$$y(t) = x(t) + x(t - 0.01) + x(t + 0.01) \tag{2.162}$$

is non-causal since the computation of the output signal requires anticipation of a future value of the input signal.

Can a non-causal system be realized?

The answer depends on what type of realization we seek. Consider the system represented by Eqn. (2.161). For the sake of discussion, suppose that the signal $x(t)$ is the output of a microphone in a recording studio where a live band is playing. The signal is delayed by 0.01 seconds and 0.02 seconds in two separate systems, and the outputs of these delay systems are added to the original signal $x(t)$ to produce the signal $y(t)$ in *real time*, that is, while the input signal is being played. This type of signal processing is called *real-time processing*, and is illustrated in Fig. 2.45.

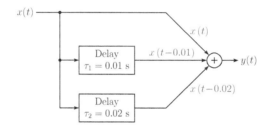

Figure 2.45 – Causal system given by Eqn. (2.161).

   Now consider the system in Eqn. (2.162). Clearly, this system cannot be realized for real-time processing since the value $x(t + 0.01)$ is not available at the time instant t when the output is computed. If we must realize this system (some non-causal systems may have desired properties in other respects) then we must realize it in *post processing mode*. The signals $x(t)$, $x(t - 0.01)$, and $x(t + 0.01)$ are each recorded. The output signal is then computed from these recordings after the fact, e.g., after the band finishes playing the song. Another way to look at this type of processing is to consider a new signal $w(t)$ defined as

$$w(t) = y(t - 0.01) \tag{2.163}$$

and write the system description in terms of the new, delayed, output signal $w(t)$ in the form

$$w(t) = x(t - 0.01) + x(t - 0.02) + x(t) \tag{2.164}$$

The signal $w(t)$ can be computed in real time, and we can listen to it while the band is playing. However, it represents a delayed version of the actual output signal $y(t)$ that is of interest to us. In essence we are doing post processing. Some non-causal systems can be made causal through the addition of delays as we have done in Eqn. (2.163). However, we will see that this is not always possible.

For CTLTI systems the causality property can be related to the impulse response of the system. Recall that the output signal $y(t)$ of the system is equal to the convolution of its input signal $x(t)$ and its impulse response $h(t)$. We will use the alternative form of the convolution integral given by

$$y(t) = h(t) * x(t)$$
$$= \int_{-\infty}^{\infty} h(\lambda)\, x(t-\lambda)\, d\lambda \tag{2.165}$$

For the system under consideration to be causal, computation of $y(t)$ should not require any future values of the input signal. Consequently, the term $x(t-\lambda)$ in the integral should not contain references to any future values. We need

$$t - \lambda \leq t \tag{2.166}$$

and equivalently $\lambda \geq 0$. Thus, the integrand in Eqn. (2.166) should not have any non-zero values for $\lambda < 0$. This in turn requires that the impulse response of the system be equal to zero for all negative values of its argument, i.e.,

$$h(t) = 0 \quad \text{for all} \quad t < 0 \tag{2.167}$$

This result intuitively makes sense. The impulse response of a causal CTLTI system cannot possibly start before $t = 0$ since doing so would require the system to anticipate the impulse to occur at $t = 0$. Using Eqn. (2.167), the convolution relationship in Eqn. (2.165) can be written in the right-sided form

$$y(t) = \int_{0}^{\infty} h(\lambda)\, x(t-\lambda)\, d\lambda \tag{2.168}$$

2.9   Stability in Continuous-Time Systems

A system is said to be *stable* in the *bounded-input bounded-output (BIBO)* sense if any bounded input signal is guaranteed to produce a bounded output signal.

An input signal $x(t)$ is said to be bounded if an upper bound B_x exists such that

$$\left| x(t) \right| < B_x < \infty \tag{2.169}$$

for all values of t. It doesn't matter how large the upper bound B_x may be, as long as it is finite. For the system to be stable a finite upper bound B_y must exist for the output signal in response to any input signal bounded as described by Eqn. (2.169).

For stability of a continuous-time system:

$$\left| x(t) \right| < B_x < \infty \quad \text{implies that} \quad \left| y(t) \right| < B_y < \infty \tag{2.170}$$

If the system under consideration is CTLTI, then we would like to relate the stability condition given by Eqn. (2.170) to the impulse response of the system as well as its differential equation. Derivation of the necessary condition follows:

The output signal of a CTLTI system is found from its input signal and impulse response through the use of the convolution integral

$$y(t) = \int_{-\infty}^{\infty} h(\lambda) x(t - \lambda) \, d\lambda \tag{2.171}$$

The absolute value of the output signal is

$$|y(t)| = \left| \int_{-\infty}^{\infty} h(\lambda) x(t - \lambda) \, d\lambda \right| \tag{2.172}$$

Absolute value of an integral is less than or equal to the integral of the absolute value of the integrand, so we can write the inequality

$$|y(t)| \leq \int_{-\infty}^{\infty} |h(\lambda) x(t - \lambda)| \, d\lambda \tag{2.173}$$

The integrand in Eqn. (2.173) can be expressed as

$$|h(\lambda) x(t - \lambda)| = |h(\lambda)| \, |x(t - \lambda)| \tag{2.174}$$

and the inequality in Eqn. (2.173) becomes

$$|y(t)| \leq \int_{-\infty}^{\infty} |h(\lambda)| \, |x(t - \lambda)| \, d\lambda \tag{2.175}$$

Since $|x(t - \tau)| < B_x$ we can write Eqn. (2.175) as

$$|y(t)| \leq \int_{-\infty}^{\infty} |h(\lambda)| \, B_x \, d\lambda \tag{2.176}$$

or, equivalently

$$|y(t)| \leq B_x \int_{-\infty}^{\infty} |h(\lambda)| \, d\lambda < B_y < \infty \tag{2.177}$$

Eqn. (2.177) implies that we need

$$\int_{-\infty}^{\infty} |h(\lambda)| \, d\lambda < \infty \tag{2.178}$$

For a CTLTI system to be stable, its impulse response must be *absolute integrable*.

Example 2.24: **Stability of a first-order continuous-time system**

Evaluate the stability of the first-order CTLTI system described by the differential equation

$$\frac{dy(t)}{dt} + a \, y(t) = x(t)$$

where a is a real-valued constant.

Solution: The impulse response of the system in question is

$$h(t) = e^{-at} u(t)$$

as can be easily verified through the use of Eqn. (2.55) with $r(t) = \delta(t)$, $t_0 = 0$ and $y(0) = 0$. (We take the initial value to be zero since the system is specified to be CTLTI.) Now we can check to see if the requirement given by Eqn. (2.178) is satisfied.

$$\int_{-\infty}^{\infty} |h(\lambda)| \, d\lambda = \int_{0}^{\infty} e^{-a\lambda} \, d\lambda = \frac{1}{a} \quad \text{provided that} \quad a > 0 \qquad (2.179)$$

Thus the system is stable if $a > 0$. (Verify that the integral in Eqn. (2.179) cannot be evaluated if $a \leq 0$.) It is also interesting to recognize that the characteristic equation that corresponds to the differential equation of the system is

$$s + a = 0$$

and its only solution is at $s_1 = -a$.

The stability of a CTLTI system can also be associated with the modes of the differential equation that governs the behavior of the system. In Example 2.24 the system proved to be stable for $a > 0$ which meant that the only root of the characteristic polynomial had to be negative. Furthermore, we have seen in Section 2.5.3 that the locations of the roots of the characteristic polynomial control the type of the transient behavior of the system. In the case of real roots, a negative real root leads to a decaying exponential signal while a positive real root produces a growing exponential signal. In the case of a pair of complex conjugate roots, if the real part of the pair is negative, the resulting signal is oscillatory with exponentially decaying amplitude. If the real part of the conjugate pair is positive, then the oscillatory response has a growing amplitude. Summarizing both cases we can conclude that, for a causal system, characteristic polynomial roots in the left half of the complex plane are associated with stable behavior, and those in the right half of the complex plane are associated with unstable behavior. For a causal CTLTI system to be stable, all roots of the characteristic polynomial must be in the left half of the complex plane.

A side note: In describing the associations between the roots of the characteristic polynomial and stability, we referred to a *causal* CTLTI system. If we were to consider an *anti-causal* system, the impulse response of which proceeds in the negative direction toward $t \to -\infty$, then the associations described above would have to be reversed. In that case roots in the right half of the complex plane would lead to stable behavior.

Why is stability important for a CTLTI system?

An unstable CTLTI system is capable of producing an unbounded output signal in response to at least some bounded input signals. What is the practical significance of this? Consider a battery-operated amplifier circuit that may be unstable due to a design flaw. We certainly don't expect the voltages in a battery-operated device to be infinitely large, and it does not happen. What may happen is that the output voltage may go up to the maximum level it can physically attain (probably close to the power supply voltage) and stay there, causing circuit components to saturate, and causing the system to cease its useful linear operation. Another example would be an electromechanical system such as an electric motor. If it is unstable as a system, its speeds may reach levels that are too high for its normal operation, and cause physical damage to the motor.

2.10   Approximate Numerical Solution of a Differential Equation

For a system modeled by means of a differential equation, the output signal can be determined by solving the differential equation with the specified input signal and initial conditions. In Section 2.5 we have explored time-domain methods for solving linear differential equations with constant coefficients. Transform-domain methods for accomplishing the same task will be presented in Chapter 7.

Sometimes we may run into differential equations for which analytical solutions may be difficult or impossible to obtain. The differential equation may have nonlinear terms, or the analytical definition of the input signal applied to the system may not be simple or practical enough for obtaining an analytical solution. In those cases, we may have to rely on approximate step-by-step solutions that are obtained on a computer by the use of an iterative numerical algorithm. A number of numerical algorithms for finding approximate solutions to differential equations are documented in the literature, ranging from very simple to very sophisticated. A thorough review of these algorithms is also well beyond the scope of this text, and can be found in a number of excellent textbooks on numerical analysis. In this section we will briefly discuss the simplest of the numerical solution methods, the *Euler method*, which works reasonably well for our purposes as long as some basic precautions are taken.

Consider the RC circuit that was analyzed in Examples 2.8 and 2.9 using analytical solution techniques. The first-order differential equation that governs the operation of the circuit is

$$\frac{dy(t)}{dt} + \frac{1}{RC} y(t) = \frac{1}{RC} x(t) \tag{2.180}$$

where $x(t)$ and $y(t)$ represent the input and the output signals respectively. The first step will be to rearrange the terms of the differential equation so that the derivative of the output signal is left by itself on the left side of the equal sign:

$$\frac{dy(t)}{dt} = -\frac{1}{RC} y(t) + \frac{1}{RC} x(t) \tag{2.181}$$

Eqn. (2.181) is in the general form

$$\frac{dy(t)}{dt} = g[t, y(t)] \tag{2.182}$$

where $g(\ldots)$ represents some function of t and $y(t)$. For this particular circuit we have

$$g[t, y(t)] = -\frac{1}{RC} y(t) + \frac{1}{RC} x(t) \tag{2.183}$$

We know from basic calculus that one way of approximating a derivative is through the use of finite differences. At the time instant $t = t_0$, the derivative on the left side of Eqn. (2.182) can be approximated as

$$\left. \frac{dy(t)}{dt} \right|_{t=t_0} \approx \frac{y(t_0 + T) - y(t_0)}{T} \tag{2.184}$$

provided that the time step T is sufficiently small. This is illustrated in Fig. 2.46. The slope of the tangent passing through the point $[t_0, y(t_0)]$ represents the true value of the derivative $dy(t)/dt$ at $t = t_0$. The slope of the chord passing through the points $[t_0, y(t_0)]$ and $[t_0 + T, y(t_0 + T)]$ is the approximate value of the same.

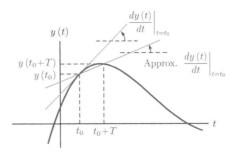

Figure 2.46 – Approximating the first derivative using a finite difference.

Substituting Eqn. (2.184) into Eqn. (2.182) and rearranging the terms we obtain

$$y(t_0 + T) \approx y(t_0) + T\,g\,[t_0, y(t_0)] \tag{2.185}$$

which gives us the iteration formula for the Euler method. The solution starts with a known initial value of the output signal at the time instant $t = t_0$, typically $t = 0$. Suppose $y(0)$ is given. We use it to approximate $y(T)$ as

$$y(T) \approx y(0) + T\,g\,[0, y(0)] \tag{2.186}$$

For the RC circuit example, Eqn. (2.186) becomes

$$y(T) \approx y(0) + T\left[-\frac{1}{RC}\,y(0) + \frac{1}{RC}\,x(0)\right]$$

Having obtained an approximate value for $y(T)$, we can use it to approximate the solution at the time instant $t = 2T$ as

$$y(2T) \approx y(T) + T\,g\,[T, y(T)]$$
$$= y(T) + T\left[-\frac{1}{RC}\,y(T) + \frac{1}{RC}\,x(T)\right]$$

The knowledge of an approximate value for $y(2T)$ can then be used for finding an approximate value for $y(3T)$ as

$$y(3T) \approx y(2T) + T\,g\,[2T, y(2T)]$$
$$= y(2T) + T\left[-\frac{1}{RC}\,y(2T) + \frac{1}{RC}\,x(2T)\right]$$

and so on. For this approximation technique to work in a reasonably accurate manner, the step size T needs to be chosen to be small enough. As far as the system is concerned, a good rule of thumb is for the step size T to be 10 percent of the time constant or smaller. In addition, time variations of the input signal must also be considered in selecting the step size. The Euler method may suffer from numerical instability problems if the step size T is not chosen small enough. A variant known as the *backward Euler method* improves the stability of the solution. The discussion of sampling in Chapter 6 will further clarify this issue.

Although the Euler method as presented in this section only applies to a first-order differential equation, we will be able to generalize it to apply to higher-order systems when we discuss state-space models in Chapter 9.

MATLAB Exercises 2.4 and 2.5 provide examples of finding numerical solutions in MAT-LAB using the Euler method or the more sophisticated methods MATLAB implements.

Software resources: See MATLAB Exercises 2.4 and 2.5.

2.11   Further Reading

[1] S.C. Chapra. *Applied Numerical Methods with MATLAB for Engineers and Scientists.* McGraw-Hill, 2008.

[2] S.C. Chapra and R.P. Canale. *Numerical Methods for Engineers.* McGraw-Hill, 2010.

[3] T.L. Harman, J.B. Dabney, and N. Richert. *Advanced Engineering Mathematics Using MATLAB V.4.* PWS BookWare Companion Series. PWS Publishing Company, 1997.

[4] P.V. O'Neil. *Advanced Engineering Mathematics.* Cengage Learning, 2011.

[5] Frank L. Severance. *System Modeling and Simulation: An Introduction.* John Wiley and Sons, 2001.

[6] C.R. Wylie and L.C. Barrett. *Advanced Engineering Mathematics.* McGraw-Hill, 1995.

MATLAB Exercises

MATLAB Exercise 2.1: Testing linearity of continuous-time systems

In Example 2.1 we have checked four different systems for linearity. In each case we started with two arbitrary signals $x_1(t)$ and $x_2(t)$. We have formulated the response of each system to the combined signal

$$x(t) = \alpha_1\, x_1(t) + \alpha_2\, x_2(t)$$

with arbitrary constants α_1 and α_2, and determined if the response to the combined signal matched the expected response

$$y(t) = \alpha_1\, y_1(t) + \alpha_2\, y_2(t)$$

for a linear system. In this exercise, we will simulate the four systems considered in Example 2.1, and test them using signals generated in MATLAB.

Let us begin by creating a vector "t" of time instants, and generating vectors "x1" and "x2" to correspond to two test signals $x_1(t)$ and $x_2(t)$ with the following commands:

```
>>  t = [0:0.01:5];
>>  x1 = cos(2*pi*5*t);
>>  x2 = exp(-0.5*t);
```

Let $\alpha_1 = 2$ and $\alpha_2 = 1.25$. The vector "x" will hold amplitudes of the combined input signal $x(t) = \alpha_1\, x_1(t) + \alpha_2\, x_2(t)$ at the time instants listed in vector "t". To compute and graph $x(t)$ type the following set of commands:

```
>>   alpha1 = 2;
>>   alpha2 = 1.25;
>>   x = alpha1*x1+alpha2*x2;
>>   plot(t,x);
```

We are now ready to simulate the system under consideration. One method for doing this is to develop a MATLAB function for each system. For example, the system in part (a) of Example 2.1 can be simulated using the function `sys_a(..)` defined with the following code:

```
1   function y = sys_a(x)
2     y = 5*x;
```

These two lines must be placed into a file named "`sys_a.m`" and saved at a location visible to MATLAB. An alternative is to use anonymous functions that were introduced in MATLAB Exercise 1.3. An anonymous function for the system in part (a) of Example 2.1 can be entered by typing

```
>>   sys_a = @(x) 5*x;
```

directly into the command window. This is the preferred approach in situations where the function in question may be needed only briefly, and we may not want to save it to disk. Once the function is defined, the response expected from a linear system and the actual response of our system can be obtained by typing the following lines:

```
>>   y1 = sys_a(x1);
>>   y2 = sys_a(x2);
>>   y_exp = alpha1*y1+alpha2*y2;
>>   y_act = sys_a(x);
```

The vector "`y_exp`" holds the response expected from a linear system, and the vector "`y_act`" holds the output signal produced by the system in response to the combined input signal in vector "`x`".

The MATLAB script listed below represents a complete program that tests the system in part (a) of Example 2.1 for linearity.

```
1    % Script: matex_2_1.m
2    %
3    t = [0:0.01:4];              % Create a time vector.
4    x1 = cos(2*pi*5*t);          % Test signal 1.
5    x2 = exp(-0.5*t);            % Test signal 2.
6    alpha1 = 2;                  % Set parameters alpha1
7    alpha2 = 1.25;               %    and alpha2.
8    x = alpha1*x1+alpha2*x2;     % Combined signal.
9    % Define anonymous functions for the systems in Example 2-1.
10   sys_a = @(x) 5*x;
11   sys_b = @(x) 5*x+3;
12   sys_c = @(x) 3*x.*x;
13   sys_d = @(x) cos(x);
14   % Test the system in part (a) of Example 2-1.
15   y1 = sys_a(x1);
16   y2 = sys_a(x2);
17   y_exp = alpha1*y1+alpha2*y2;  % Expected response for a linear system.
18   y_act = sys_a(x);             % Actual response.
19   clf;                          % Clear figure.
20   subplot(1,2,1);
21   plot(t,y_exp);                % Graph expected response.
22   title('y_{exp} = \alpha_1 y_1 + \alpha_2 y_2')
```

```
23    xlabel('t (sec)');
24    ylabel('Amplitude');
25    subplot(1,2,2);
26    plot(t,y_act);                    % Graph  actual  response.
27    title('y_{act} = Sys_a\{\alpha_1 x_1 + \alpha_2 x_2\}')
28    xlabel('t (sec)');
29    ylabel('Amplitude');
```

The graph produced by the script is shown in Fig. 2.47.

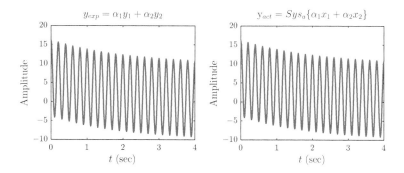

Figure 2.47 – Signals $y_{exp}(t)$ and $y_{act}(t)$ for MATLAB Exercise 2.1.

The two signals seem to be identical, supporting our earlier conclusion that the system in part (a) is linear. A word of caution is in order here: For a system to be linear, superposition principle must hold for *any two* signals $x_1(t)$ and $x_2(t)$ and not just the two signals we generated above. If we only had the graphs in Fig. 2.47 without our earlier analysis of Example 2.1, we would not be able to claim that the system is linear; the best we could do would be to say that the system *may be* linear. Nonlinearity of a system is easier to prove; we only need to show one case where superposition fails.

The systems in parts (b) through (d) of Example 2.1 can be simulated by editing the code above and changing the anonymous function used for the system.

Software resources:

matex_2_1.m

MATLAB Exercise 2.2: Testing time invariance of continuous-time systems

In Example 2.2 we have checked three systems for time invariance. In each case an arbitrary test signal was used with the system in its original form $x(t)$ and in time-shifted form $x(t-\tau)$. If the system under consideration is time-invariant we need

$$\mathrm{Sys}\{x(t-\tau)\} = y(t-\tau)$$

where $y(t)$ is the response to the original signal $x(t)$.

In this exercise, we will simulate the three systems considered in Example 2.2, and test them using signals generated in MATLAB. Recall that, in MATLAB Exercise 2.1, anonymous functions were used for simulating the systems, and MATLAB vectors were used for simulating of the input signals. For the purpose of this exercise we will find it more convenient to use an anonymous function for the input signal as well, since time shifts are easier to represent with functions rather than vectors.

Let the test signal $x(t)$ be

$$x(t) = e^{-0.5t} u(t)$$

We will begin by creating a vector "**t**" of time instants, and developing an anonymous function "**x**" for the test signal $x(t)$ with the following commands:

```
>>  t = [0:0.01:10];
>>  x = @(t) exp(-0.5*t).*(t>=0);
>>  plot(t,x(t),t,x(t-2));
```

We are now ready to simulate the system under consideration. For the sake of discussion, let's pick the system in part (c) of Example 2.2 and express it with an anonymous function as

```
>>  sys_c = @(x) 3*cos(t).*x;
```

directly into the command window. We are now ready to compute the responses of the system to the signals $x(t)$ and $x(t-2)$ as

```
>>  y1 = sys_c(x(t));
>>  y2 = sys_c(x(t-2));
```

Vectors "**y1**" and "**y2**" are vectors that hold the responses of the system under consideration. The two responses can be graphed on the same coordinate system with the statement

```
>>  plot(t,y1,'b-',t,y2,'r:');
```

The MATLAB script listed below represents a complete program that tests the system in part (c) of Example 2.2 for time-invariance.

```
1   % Script matex_2_2.m
2   %
3   t = [0:0.01:10];                % Create a time vector.
4   x = @(t) exp(-0.5*t).*(t>=0);   % Anonymous function for test signal.
5   % Define anonymous functions for the systems in Example 2-2.
6   sys_a = @(x) 5*x;
7   sys_b = @(x) 3*cos(x);
8   sys_c = @(x) 3*cos(t).*x;
9   % Test the system in part (c) of Example 2-2.
10  y1 = sys_c(x(t));
11  y2 = sys_c(x(t-2));
12  clf;                            % Clear figure.
13  plot(t,y1,'b-',t,y2,'r:');      % Graph the two responses.
14  title('Responses to x(t) and x(t-2)')
15  xlabel('t (sec)');
16  ylabel('Amplitude');
17  legend('Sys\{x(t)\}','Sys\{x(t-2)\}');
```

The graph produced by the script is shown in Fig. 2.48.

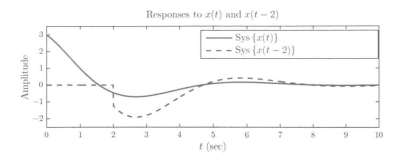

Figure 2.48 – Signals $\text{Sys}\{x(t)\}$ and $\text{Sys}\{x(t-2)\}$ for MATLAB Exercise 2.2.

The two signals seem to be different in shape, supporting our earlier conclusion that the system in part (c) is not time-invariant. The systems in parts (a) and (b) of Example 2.2 can be simulated by editing the code above and changing the anonymous function used for the system.

Software resources:
`matex_2_2.m`

MATLAB Exercise 2.3: **Using linearity to determine the response of the RC circuit**

Refer to Examples 2.8, 2.9 and 2.10. In Example 2.8 the unit-step response of the simple RC circuit was determined by solving the governing differential equation. The circuit was assumed to be initially relaxed, resulting in $y(0) = 0$. In Example 2.9 we have determined the response of the same circuit to a pulse signal, again by solving the differential equation. Example 2.10 explored an alternative approach to finding the response to the pulse signal; namely using the response to a unit-step and the superposition principle. (Superposition could be used since the circuit is initially relaxed and therefore linear and time-invariant.)

In this exercise we will explore this concept using MATLAB. The response of the circuit to a unit-step signal was found in Example 2.8 to be

$$y_u(t) = \text{Sys}\{u(t)\} = \left(1 - e^{-4t}\right) u(t)$$

which can be expressed by means of an anonymous function as

```
yu = @(t) (1-exp(-4*t)).*(t>=0);
```

The script listed below uses the unit-step response for determining and graphing the response of the system to the pulse signal

$$x_1(t) = \Pi(t) = u(t+0.5) - u(t-0.5)$$

```
1   % Script: matex_2_3a.m
2   %
3   % Anonymous function for unit-step response.
4   yu = @(t) (1-exp(-4*t)).*(t>=0);
5   t = [-5:0.01:5];              % Vector of time instants.
6   y1 = yu(t+0.5)-yu(t-0.5);     % Compute response to x1(t).
7   plot(t,y1);
```

This idea can be extended to input signals with higher complexity. Consider the signal

$$x_2\left(t\right) = u\left(t\right) - 2\,u\left(t - 1\right) + u\left(t - 2\right)$$

which is shown in Fig. 2.49.

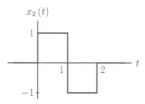

Figure 2.49 – Signal $x_2\left(t\right)$ for MATLAB Exercise 2.3.

```
1   % Script: matex_2_3b.m
2   %
3   % Anonymous function for unit-step response.
4   yu = @(t) (1-exp(-4*t)).*(t>=0);
5   t = [-5:0.01:5];              % Vector of time instants.
6   y2 = yu(t)-2*yu(t-1)+yu(t-2); % Compute response to x2(t)].
7   plot(t,y2);
```

Software resources:
matex_2_3a.m
matex_2_3b.m

MATLAB Exercise 2.4: Numerical solution of the RC circuit using Euler method

In this exercise we will use the Euler method to find an approximate numerical solution for the RC circuit problem of Example 2.8, and compare it to the exact solution that was found. The RC circuit and the input signal are shown in Fig. 2.50.

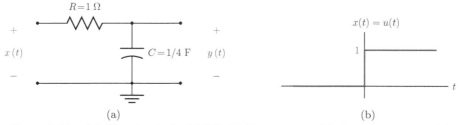

Figure 2.50 – (a) The circuit for MATLAB Exercise 2.4, (b) the input signal $x\left(t\right)$.

For the specified input signal, the differential equation of the circuit is

$$\frac{dy\left(t\right)}{dt} + 4\,y\left(t\right) = 4\,u\left(t\right)$$

With the circuit initially relaxed, that is, with $y\left(0\right) = 0$, the exact solution for the output signal was found to be

$$y\left(t\right) = \left(1 - e^{-4t}\right)u\left(t\right) \tag{2.187}$$

In order to use the Euler method for approximate solution, we will write the differential equation in the form

$$\frac{dy(t)}{dt} = g(t, y(t))$$

with the right-side function given by

$$g(t, y(t)) = -4\,y(t) + 4\,u(t) \tag{2.188}$$

The Euler method approximation $\hat{y}(t)$ to the solution is based on the iteration formula

$$\begin{aligned}\hat{y}((k+1)T_s) &= \hat{y}(kT_s) + T_s\,g(kT_s, \hat{y}(kT_s)) \\ &= \hat{y}(kT_s) + T_s\,(-4\,\hat{y}(kT_s) + 4\,u(kT_s))\end{aligned} \tag{2.189}$$

The MATLAB script given below computes and graphs exact and approximate solutions, using a step size of $T_s = 0.1$ s. In addition, the percent error of the approximate solution is computed as

$$\varepsilon(kT_s) = \frac{\hat{y}(kT_s) - y(kT_s)}{y(kT_s)} \times 100$$

and graphed for $k = 1, 2, \ldots, 10$ to span a time interval of 1 s.

```
1   % Script: matex_2_4.m
2   %
3   Ts = 0.1;              % Time increment
4   t = [0:Ts:1];          % Vector of time instants
5   % Compute the exact solution.
6   y = 1-exp(-4*t);       % Eqn.(2.186)
7   % Compute the approximate solution using Euler method.
8   yhat = zeros(size(t));
9   yhat(1) = 0;           % Initial value.
10  for k = 1:length(yhat)-1,
11    g = -4*yhat(k)+4;            % Eqn.(2.188)
12    yhat(k+1) = yhat(k)+Ts*g;   % Eqn.(2.189)
13  end;
14  % Graph exact and approximate solutions.
15  clf;
16  subplot(211);
17  plot(t,y,'-',t,yhat,'ro'); grid;
18  title('Exact and approximate solutions for RC circuit');
19  xlabel('Time (sec)');
20  ylabel('Amplitude');
21  legend('Exact solution','Approximate solution',...
22    'Location','SouthEast');
23  % Compute and graph the percent approximation error.
24  err_pct = (yhat-y)./y*100;
25  subplot(212);
26  plot(t(2:length(t)),err_pct(2:length(t)),'ro'); grid
27  title('Percent approximation error');
28  xlabel('Time (sec)');
29  ylabel('Error (%)');
```

The loop that runs from line 10 to 13 of the code computes the approximate solution step by step. The graphs produced by the MATLAB script above are given in Fig. 2.51. The largest approximation error seems to be around 21 percent, and it goes down in time as the output signal settles close to its final value.

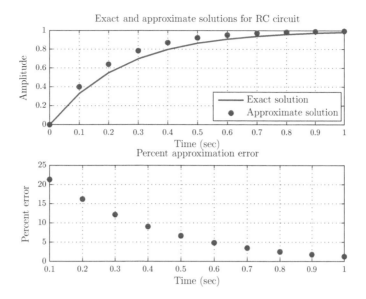

Figure 2.51 – Actual and approximate solutions for the RC circuit and the percent error for $\Delta t = 0.1$ s.

The time step that we used in the code above was $T_s = 0.1$ seconds which is 40 percent of the time constant of $\tau = 0.25$ s. For better accuracy a smaller time step is needed. Euler method is a first-order approximation method, and is the simplest of all numerical methods for solving differential equations. Fig. 2.52 shows the results obtained with the same script, by simply changing the step size to $T_s = 0.02$ seconds or 8 percent of the time constant. In this case, the largest error between the exact and approximate solutions is on the order of 4 percent.

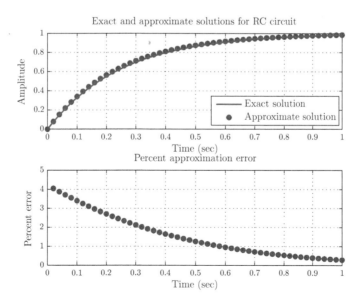

Figure 2.52 – Actual and approximate solutions for the RC circuit and the percent error for $\Delta t = 0.02$ s.

Another way to reduce the approximation error is to use a more sophisticated higher-order approximation method as we will see in the next exercise.

Software resources:

matex_2_4.m

MATLAB Exercise 2.5: Improved numerical solution of the RC circuit

For numerical solution of differential equations MATLAB has a number of functions that are based on more sophisticated algorithms than the Euler method we have used in MATLAB Exercise 2.4. In this exercise we will use one of those functions, namely the function ode45(..). The function ode45(..) uses the *4-th order Runge-Kutta technique*, a higher-order method than the Euler method of the previous example. For a given step size, approximation errors for the 4-th order Runge-Kutta technique are generally smaller than those encountered with the Euler method.

Before we can use the function ode45(..), we need to write a function to compute the right side $g(t, y(t))$ of the differential equation. The code for such a function rc1(..) is given below:

```
1   function ydot = rc1(t,y)
2     ydot = -4*y+4;
3   end
```

The function rc1(..) takes two arguments, t and y, and returns the corresponding value of $g(t, y(t))$. It is used by the function ode45(..) in computing the approximate solution. The script listed below computes an approximate solution in the time interval $0 \leq t \leq 1$ s using a time step of $T = 0.1$ s, and compares it to the exact solution.

```
1    % Script: matex_2_5a.m
2    %
3    t = [0:0.1:1];       % Vector of time instants
4    % Compute the exact solution.
5    y = 1-exp(-4*t);     % Eqn.(2.187)
6    % Compute the approximate solution using ode45().
7    [t,yhat] = ode45(@rc1,t,0);
8    % Graph exact and approximate solutions.
9    clf;
10   subplot(211);
11   plot(t,y,'-',t,yhat,'ro'); grid;
12   title('Exact and approximate solutions for RC circuit');
13   xlabel('Time (sec)');
14   ylabel('Amplitude');
15   legend('Exact solution','Approximate solution',...
16     'Location','SouthEast');
17   % Compute and graph the percent approximation error.
18   err_pct = (yhat-y')./y'*100;
19   subplot(212);
20   plot(t(2:max(size(t))),err_pct(2:max(size(t))),'ro'); grid
21   title('Percent approximation error');
22   xlabel('Time (sec)');
23   ylabel('Percent error');
```

Graphs produced by the script **matex_2_5a.m** are shown in Fig. 2.53. Even though the step size is quite large, the accuracy of the solution obtained is much better than that obtained in the previous exercise (compare Fig. 2.53 to Fig. 2.51).

If the right-side function $g(t, y(t))$ is simple enough to be expressed on one line, an anonymous function can be used in place of "rc1.m". The script listed below uses an anonymous function "rc2" to solve the same problem.

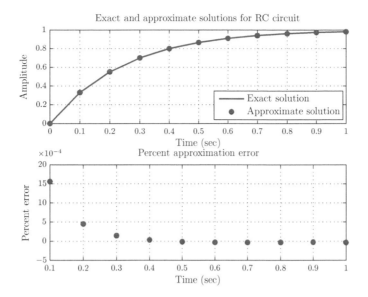

Figure 2.53 – Actual and approximate solutions for the RC circuit and the percent error with the `ode45(..)` function.

```
1   % Script: matex_2_5b.m
2   %
3   t = [0:0.1:1];        % Vector of time instants
4   % Compute the exact solution.
5   y = 1-exp(-4*t);      % Eqn.(2.187)
6   % Compute the approximate solution using ode45().
7   rc2 = @(t,y) -4*y+4;
8   [t,yhat] = ode45(rc2,t,0);
9   % Graph exact and approximate solutions.
10  clf;
11  subplot(211);
12  plot(t,y,'-',t,yhat,'ro'); grid;
13  title('Exact and approximate solutions for RC circuit');
14  xlabel('Time (sec)');
15  ylabel('Amplitude');
16  legend('Exact solution','Approximate solution',...
17    'Location','SouthEast');
18  % Compute and graph the percent approximation error.
19  err_pct = (yhat-y')./y'*100;
20  subplot(212);
21  plot(t(2:max(size(t))),err_pct(2:max(size(t))),'ro'); grid
22  title('Percent approximation error');
23  xlabel('Time (sec)');
24  ylabel('Percent error');
```

Software resources:
matex_2_5a.m
rc1.m
matex_2_5b.m

Problems

2.1.   A number of systems are specified below in terms of their input-output relationships. For each case, determine if the system is linear and/or time-invariant.

    a. $y(t) = |x(t)| + x(t)$

    b. $y(t) = t\,x(t)$

    c. $y(t) = e^{-t}\,x(t)$

    d. $y(t) = \displaystyle\int_{-\infty}^{t} x(\lambda)\, d\lambda$

    e. $y(t) = \displaystyle\int_{t-1}^{t} x(\lambda)\, d\lambda$

    f. $y(t) = (t+1)\displaystyle\int_{-\infty}^{t} x(\lambda)\, d\lambda$

2.2.   Consider the cascade combination of two systems shown in Fig. P.2.2(a).

               (a)   (b)

Figure P. 2.2

    a. Let the input-output relationships of the two subsystems be given as

$$\text{Sys}_1\{x(t)\} = 3\,x(t) \quad \text{and} \quad \text{Sys}_2\{w(t)\} = w(t-2)$$

    Write the relationship between $x(t)$ and $y(t)$.

    b. Let the order of the two subsystems be changed as shown in Fig. P.2.2(b). Write the relationship between $x(t)$ and $\bar{y}(t)$. Does changing the order of two subsystems change the overall input-output relationship of the system?

2.3.   Repeat Problem 2.2 with the following sets of subsystems:

    a. $\text{Sys}_1\{x(t)\} = 3\,x(t) \quad \text{and} \quad \text{Sys}_2\{w(t)\} = t\,w(t)$

    b. $\text{Sys}_1\{x(t)\} = 3\,x(t) \quad \text{and} \quad \text{Sys}_2\{w(t)\} = w(t) + 5$

2.4.   Find a differential equation between the input voltage $x(t)$ and the output voltage $y(t)$ for the circuit shown in Fig. P.2.4. At $t=0$ the initial values are

$$i_L(0) = 1 \text{ A}, \qquad v_C(0) = 2 \text{ V}$$

Express the initial conditions for $y(t)$ and $dy(t)/dt$.

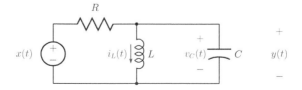

Figure P. 2.4

2.5.   Find a differential equation between the input voltage $x(t)$ and the output voltage $y(t)$ for the circuit shown in Fig. P.2.5. At $t = 0$ the initial values are

$$v_1(0) = 2 \text{ V}, \qquad v_2(0) = -2 \text{ V}$$

Express the initial conditions for $y(t)$ and $dy(t)/dt$.

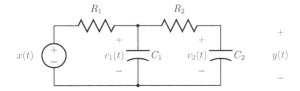

Figure P. 2.5

2.6.   One of the commonly used examples for modeling dynamic systems with differential equations is the so-called *predator-prey* problem which is stated as follows:

Suppose that the following set of rules were experimentally determined to govern the populations of predators and prey that live in the same environment:

- The number of prey grows at a rate proportional to its current value if there are no predators.

- The number of predators decreases at a rate proportional to its current value if there is no prey.

- The number of encounters between predator and prey is proportional to the product of the current populations of each.

- Each encounter between the predator and the prey increases the number of predators and decreases the number of prey.

Designate $x(t)$ as the number of prey at time t, and $y(t)$ as the number of predators at time t. Using parametric proportion constants as needed, write two differential equations to model the system. Are the equations linear or nonlinear? Justify your answer.

2.7.   Consider the differential equation for the RC circuit in Fig. P.2.12:

$$\frac{dy(t)}{dt} + 4y(t) = 4x(t)$$

Let the input signal be a unit step, that is, $x(t) = u(t)$. Using the first-order differential equation solution technique discussed in Section 2.5.1 find the solution $y(t)$ for $t \geq 0$ subject to each initial condition specified below:

    a. $y(0) = 0$
    b. $y(0) = 5$
    c. $y(0) = 1$
    d. $y(0) = -1$
    e. $y(0) = -3$

2.8.   Solve each of the first-order differential equations given below for the specified input signal and subject to the specified initial condition. Use the first-order solution technique discussed in Section 2.5.1.

a. $\dfrac{dy\,(t)}{dt} + 4\,y\,(t) = x\,(t)\ ,\qquad x\,(t) = u\,(t)\ ,\qquad y\,(0) = -1$

b. $\dfrac{dy\,(t)}{dt} + 2\,y\,(t) = 2\,x\,(t)\ ,\qquad x\,(t) = u\,(t) - u\,(t-5)\ ,\qquad y\,(0) = 2$

c. $\dfrac{dy\,(t)}{dt} + 5\,y\,(t) = 3\,x\,(t)\ ,\qquad x\,(t) = \delta\,(t)\ ,\qquad y\,(0) = 0.5$

d. $\dfrac{dy\,(t)}{dt} + 5\,y\,(t) = 3\,x\,(t)\ ,\qquad x\,(t) = t\,u\,(t)\ ,\qquad y\,(0) = -4$

e. $\dfrac{dy\,(t)}{dt} + y\,(t) = 2\,x\,(t)\ ,\qquad x\,(t) = e^{-2t}\,u\,(t)\ ,\qquad y\,(0) = -1$

2.9.   For each homogeneous differential equation given below, find the characteristic equation and show that it only has simple real roots. Find the homogeneous solution for $t \ge 0$ in each case subject to the initial conditions specified.

a. $\dfrac{d^2y\,(t)}{dt^2} + 3\,\dfrac{dy\,(t)}{dt} + 2\,y\,(t) = 0\ ,\quad y\,(0) = 3\ ,\quad \left.\dfrac{dy\,(t)}{dt}\right|_{t=0} = 0$

b. $\dfrac{d^2y\,(t)}{dt^2} + 4\,\dfrac{dy\,(t)}{dt} + 3\,y\,(t) = 0\ ,\quad y\,(0) = -2\ ,\quad \left.\dfrac{dy\,(t)}{dt}\right|_{t=0} = 1$

c. $\dfrac{d^2y\,(t)}{dt^2} - y\,(t) = 0\ ,\quad y\,(0) = 1\ ,\quad \left.\dfrac{dy\,(t)}{dt}\right|_{t=0} = -2$

d. $\dfrac{d^3y\,(t)}{dt^3} + 6\,\dfrac{d^2y\,(t)}{dt^2} + 11\,\dfrac{dy\,(t)}{dt} + 6\,y\,(t) = 0,\ y\,(0) = 2,\ \left.\dfrac{dy\,(t)}{dt}\right|_{t=0} = -1,\ \left.\dfrac{d^2y\,(t)}{dt^2}\right|_{t=0} = 1$

2.10.   For each homogeneous differential equation given below, find the characteristic equation and show that at least some of its roots are complex. Find the homogeneous solution for $t \ge 0$ in each case subject to the initial conditions specified.

a. $\dfrac{d^2y\,(t)}{dt^2} + 3\,y\,(t) = 0\ ,\quad y\,(0) = 2\ ,\quad \left.\dfrac{dy\,(t)}{dt}\right|_{t=0} = 0$

b. $\dfrac{d^2y\,(t)}{dt^2} + 2\,\dfrac{dy\,(t)}{dt} + 2\,y\,(t) = 0\ ,\quad y\,(0) = -2\ ,\quad \left.\dfrac{dy\,(t)}{dt}\right|_{t=0} = -1$

c. $\dfrac{d^2y\,(t)}{dt^2} + 4\,\dfrac{dy\,(t)}{dt} + 13\,y\,(t) = 0\ ,\quad y\,(0) = 5\ ,\quad \left.\dfrac{dy\,(t)}{dt}\right|_{t=0} = 0$

d. $\dfrac{d^3y\,(t)}{dt^3} + 3\,\dfrac{d^2y\,(t)}{dt^2} + 4\,\dfrac{dy\,(t)}{dt} + 2\,y\,(t) = 0,\ y\,(0) = 1,\ \left.\dfrac{dy\,(t)}{dt}\right|_{t=0} = 0,\ \left.\dfrac{d^2y\,(t)}{dt^2}\right|_{t=0} = -2$

2.11.   For each homogeneous differential equation given below, find the characteristic equation and show that it has multiple-order roots. Find the homogeneous solution for $t \ge 0$ in each case subject to the initial conditions specified.

a. $\dfrac{d^2y\,(t)}{dt^2} + 2\,\dfrac{dy\,(t)}{dt} + y\,(t) = 0\ ,\quad y\,(0) = 1\ ,\quad \left.\dfrac{dy\,(t)}{dt}\right|_{t=0} = 0$

b. $\dfrac{d^3y\,(t)}{dt^3}+7\,\dfrac{d^2y\,(t)}{dt^2}+16\,\dfrac{dy\,(t)}{dt}+12\,y\,(t)=0,\ y\,(0)=0,\ \left.\dfrac{dy\,(t)}{dt}\right|_{t=0}=-2,\ \left.\dfrac{d^2y\,(t)}{dt^2}\right|_{t=0}=1$

c. $\dfrac{d^3y\,(t)}{dt^3}+6\,\dfrac{d^2y\,(t)}{dt^2}+12\,\dfrac{dy\,(t)}{dt}+8\,y\,(t)=0,\ y\,(0)=-1,\ \left.\dfrac{dy\,(t)}{dt}\right|_{t=0}=0,\ \left.\dfrac{d^2y\,(t)}{dt^2}\right|_{t=0}=3$

2.12. Consider the simple RC circuit used in Example 2.8 and shown in Fig. P.2.12(a).

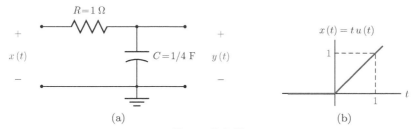

(a)   (b)

Figure P. 2.12

The differential equation between the input voltage and the output voltage was found to be

$$\frac{dy\,(t)}{dt}+4\,y\,(t)=4\,x\,(t)$$

Assuming the circuit is initially relaxed, compute and sketch the response to the unit-ramp signal $x\,(t)=r\,(t)=t\,u\,(t)$ shown in Fig. P.2.12(b).

2.13. For each first-order differential equation given below, find the solution for $t\ge 0$ with the specified input signal and subject to the specified initial value. Use the general solution technique outlined in Section 2.5.4.

a. $\dfrac{dy\,(t)}{dt}+4\,y\,(t)=x\,(t)\,,\qquad x\,(t)=u\,(t)\,,\qquad y\,(0)=-1$

b. $\dfrac{dy\,(t)}{dt}+2\,y\,(t)=2\,x\,(t)\,,\qquad x\,(t)=\sin\,(2t)+2\,\cos\,(t)\,,\qquad y\,(0)=2$

c. $\dfrac{dy\,(t)}{dt}+5\,y\,(t)=3\,x\,(t)\,,\qquad x\,(t)=t\,u\,(t)\,,\qquad y\,(0)=-4$

d. $\dfrac{dy\,(t)}{dt}+y\,(t)=2\,x\,(t)\,,\qquad x\,(t)=e^{-2t}\,u\,(t)\,,\qquad y\,(0)=-1$

2.14. For each differential equation given below, find the solution for $t\ge 0$ with the specified input signal and subject to the specified initial value. Use the general solution technique outlined in Section 2.5.4.

a. $\dfrac{d^2y\,(t)}{dt^2}+3\,\dfrac{dy\,(t)}{dt}+2\,y\,(t)=x(t)\,,\quad x\,(t)=u\,(t)\,,\quad y\,(0)=3\,,\quad \left.\dfrac{dy\,(t)}{dt}\right|_{t=0}=0$

b. $\dfrac{d^2y\,(t)}{dt^2}+4\,\dfrac{dy\,(t)}{dt}+3\,y\,(t)=x(t)\,,\quad x\,(t)=(t+1)\,u\,(t)\,,\quad y\,(0)=-2\,,\quad \left.\dfrac{dy\,(t)}{dt}\right|_{t=0}=1$

c. $\dfrac{d^2y\,(t)}{dt^2}+3\,y\,(t)=x(t)\,,\quad x\,(t)=u\,(t)\,,\quad y\,(0)=2\,,\quad \left.\dfrac{dy\,(t)}{dt}\right|_{t=0}=0$

d. $\dfrac{d^2y\,(t)}{dt^2}+2\,\dfrac{dy\,(t)}{dt}+y\,(t)=x(t)\,,\quad x\,(t)=e^{-2t}\,u\,(t)\,,\quad y\,(0)=1\,,\quad \left.\dfrac{dy\,(t)}{dt}\right|_{t=0}=0$

2.15. A system is described by the differential equation

$$\frac{d^2 y(t)}{dt^2} + 4 \frac{dy(t)}{dt} + 3 y(t) = \frac{dx(t)}{dt} - 2 x(t)$$

and has the initial conditions

$$y(0) = -2, \quad \left. \frac{dy(t)}{dt} \right|_{t=0} = 1$$

Using the technique outlined in Section 2.6 draw a block diagram to simulate this system. Incorporate the initial conditions into the block diagram.

2.16. Draw a block diagram for each DTLTI system described below by a differential equation.

a. $\dfrac{d^2 y(t)}{dt^2} + 4 \dfrac{dy(t)}{dt} + 13 y(t) = x(t)$

b. $\dfrac{d^3 y(t)}{dt^3} + 3 \dfrac{d^2 y(t)}{dt^2} + 4 \dfrac{dy(t)}{dt} + 2 y(t) = \dfrac{d^2 x(t)}{dt^2} + 6 \dfrac{dx(t)}{dt} + 3 x(t)$

c. $\dfrac{d^3 y(t)}{dt^3} + 6 \dfrac{d^2 y(t)}{dt^2} + 12 \dfrac{dy(t)}{dt} + 8 y(t) = \dfrac{d^2 x(t)}{dt^2} + 4 \dfrac{dx(t)}{dt} + 2 x(t)$

2.17. Two CTLTI systems with impulse responses $h_1(t)$ and $h_2(t)$ are connected in cascade as shown in Fig. P.2.17(a).

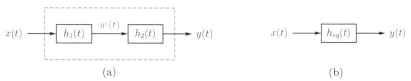

(a)   (b)

Figure P. 2.17

a. Determine the impulse response $h_{eq}(t)$ of the equivalent system, as shown in Fig. P.2.17(b) in terms of $h_1(t)$ and $h_2(t)$.
Hint: Use convolution to express $w(t)$ in terms of $x(t)$. Afterwards use convolution again to express $y(t)$ in terms of $w(t)$.
b. Let $h_1(t) = h_2(t) = \Pi(t - 0.5)$ where $\Pi(t)$ is the unit pulse. Determine and sketch $h_{eq}(t)$ for the equivalent system.
c. With $h_1(t)$ and $h_2(t)$ as specified in part (b), let the input signal be a unit step, that is, $x(t) = u(t)$. Determine and sketch the signals $w(t)$ and $y(t)$.

2.18. Two CTLTI systems with impulse responses $h_1(t)$ and $h_2(t)$ are connected in parallel as shown in Fig. P.2.18(a).

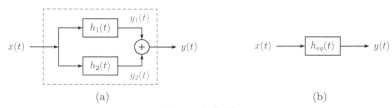

(a)   (b)

Figure P. 2.18

a. Determine the impulse response $h_{eq}(t)$ of the equivalent system, as shown in Fig. P.2.18(b) in terms of $h_1(t)$ and $h_2(t)$.
Hint: Use convolution to express the signals $y_1(t)$ and $y_2(t)$ in terms of $x(t)$. Afterwards express $y(t)$ in terms of $y_1(t)$ and $y_2(t)$.

b. Let $h_1(t) = e^{-t}u(t)$ and $h_2(t) = -e^{-3t}u(t)$. Determine and sketch $h_{eq}(t)$ for the equivalent system.

c. With $h_1(t)$ and $h_2(t)$ as specified in part (b), let the input signal be a unit-step, that is, $x(t) = u(t)$. Determine and sketch the signals $y_1(t)$, $y_2(t)$ and $y(t)$.

2.19.   Three CTLTI systems with impulse responses $h_1(t)$, $h_2(t)$ and $h_3(t)$ are connected as shown in Fig. P.2.19(a).

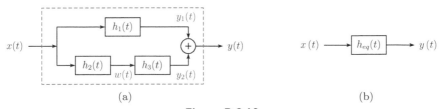

(a)   (b)

Figure P. 2.19

a. Determine the impulse response $h_{eq}(t)$ of the equivalent system, as shown in Fig. P.2.19(b) in terms of $h_1(t)$, $h_2(t)$ and $h_3(t)$.

b. Let $h_1(t) = e^{-t}u(t)$, $h_2(t) = \delta(t-2)$, and $h_3(t) = e^{-2t}u(t)$. Determine and sketch $h_{eq}(t)$ for the equivalent system.

c. With $h_1(t)$, $h_2(t)$ and $h_3(t)$ as specified in part (b), let the input signal be a unit step, that is, $x(t) = u(t)$. Determine and sketch the signals $w(t)$, $y_1(t)$, $y_2(t)$ and $y(t)$.

2.20.   Consider the CTLTI system shown in Fig. P.2.20(a).

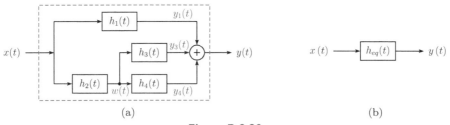

(a)   (b)

Figure P. 2.20

a. Express the impulse response of the system as a function of the impulse responses of the subsystems.

b. Let

$$h_1(t) = e^{-t}u(t)$$

$$h_2(t) = h_3(t) = u(t) - u(t-1)$$

and

$$h_4(t) = \delta(t-1)$$

Determine the impulse response $h_{eq}(t)$ of the equivalent system.

c. Let the input signal be a unit-step, that is, $x(t) = u(t)$. Determine and sketch the signals $w(t)$, $y_1(t)$, $y_3(t)$ and $y_4(t)$.

2.21. Let $y(t)$ be the response of a CTLTI system to an input signal $x(t)$. Starting with the definition of the convolution integral, show that the response of the system to the derivative $dx(t)/dt$ is equal to $dy(t)/dt$.

2.22. Using the convolution integral given by Eqns. (2.153) and (2.154) prove each of the relationships below:

a. $x(t) * \delta(t) = x(t)$

b. $x(t) * \delta(t - t_0) = x(t - t_0)$

c. $x(t) * u(t - 2) = \int_{-\infty}^{t-2} x(\lambda)\, d\lambda$

d. $x(t) * u(t - t_0) = \int_{-\infty}^{t-t_0} x(\lambda)\, d\lambda$

e. $x(t) * \Pi\left(\dfrac{t - t_0}{T}\right) = \int_{t-t_0-T/2}^{t-t_0+T/2} x(\lambda)\, d\lambda$

2.23. The impulse response of a CTLTI system is

$$h(t) = \delta(t) - \delta(t - 1)$$

Determine sketch the response of this system to the triangular waveform shown in Fig. P.2.23.

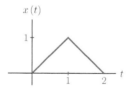

Figure P. 2.23

2.24. A CTLTI system has the impulse response

$$h(t) = \delta(t) + 0.5\,\delta(t - 1) + 0.3\,\delta(t - 2) + 0.2\,\delta(t - 3)$$

Determine sketch the response of this system to the exponential input signal

$$x(t) = e^{-t}\, u(t)$$

shown in Fig. P.2.24.

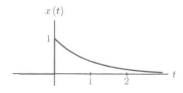

Figure P. 2.24

2.25. Let $x(t)$ be a signal limited in the range $(-\tau, \tau)$, that is

$$x(t) = 0 \quad \text{for } |t| > \tau$$

Show that a periodic extension of $x(t)$ can be obtained by convolving it with a periodic impulse train as follows:

$$\tilde{x}(t) = x(t) * \sum_{n=-\infty}^{\infty} \delta(t - nT_s)$$

where the period is $T_s \geq 2\tau$.

2.26. For each pair of signals $x(t)$ and $h(t)$ given below, find the convolution $y(t) = x(t) * h(t)$. In each case sketch the signals involved in the convolution integral and determine proper integration limits.

    a. $x(t) = u(t)$, $h(t) = e^{-2t} u(t)$
    b. $x(t) = u(t)$, $h(t) = \left[e^{-t} - e^{-2t}\right] u(t)$
    c. $x(t) = u(t-2)$, $h(t) = e^{-2t} u(t)$
    d. $x(t) = u(t) - u(t-2)$, $h(t) = e^{-2t} u(t)$
    e. $x(t) = e^{-t} u(t)$, $h(t) = e^{-2t} u(t)$

2.27. For each pair of signals $x(t)$ and $h(t)$ given below, find the convolution $y(t) = x(t) * h(t)$ first graphically, and then analytically using the convolution integral.

    a. $x(t) = \Pi\left(\dfrac{t-2}{4}\right)$, $h(t) = u(t)$

    b. $x(t) = 3\,\Pi\left(\dfrac{t-2}{4}\right)$, $h(t) = e^{-t} u(t)$

    c. $x(t) = \Pi\left(\dfrac{t-2}{4}\right)$, $h(t) = \Pi\left(\dfrac{t-2}{4}\right)$

    d. $x(t) = \Pi\left(\dfrac{t-2}{4}\right)$, $h(t) = \Pi\left(\dfrac{t-3}{6}\right)$

2.28. Let $x(t)$ and $h(t)$ both be finite-duration signals such that, given parameter values $t_1 < t_2$ and $t_3 < t_4$, we have

$$x(t) = 0 \quad \text{for } t < t_1 \text{ or } t > t_2$$
$$h(t) = 0 \quad \text{for } t < t_3 \text{ or } t > t_4$$

Show that the convolution of $x(t)$ and $h(t)$ is also a finite duration signal:

$$y(t) = x(t) * h(t) = 0 \quad \text{for } t < t_5 \text{ or } t > t_6$$

Determine the limits t_5 and t_6 in terms of t_1, t_2, t_3 and t_4.

2.29. In Section 2.9 it was shown that, for a CTLTI system to be stable in BIBO sense, the impulse response must be absolute integrable, that is,

$$\int_{-\infty}^{\infty} |h(\lambda)| \, d\lambda < \infty$$

Suppose a particular system does not satisfy this condition. Prove that the system cannot be stable by finding a bounded input signal $x(t)$ that produces an unbounded output signal $y(t)$.

2.30.   The system shown in Fig. P.2.30 represents addition of echos to the signal $x(t)$:

$$y(t) = x(t) + \alpha_1 x(t - \tau_1) + \alpha_2 x(t - \tau_2)$$

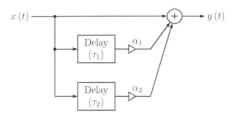

Figure P. 2.30

Comment on the system's

a. Linearity
b. Time invariance
c. Causality
d. Stability

2.31.   For each system described below find the impulse response. Afterwards determine if the system is causal and/or stable.

a. $y(t) = \text{Sys}\{x(t)\} = \displaystyle\int_{-\infty}^{t} x(\lambda)\, d\lambda$

b. $y(t) = \text{Sys}\{x(t)\} = \displaystyle\int_{t-T}^{t} x(\lambda)\, d\lambda, \qquad T > 0$

c. $y(t) = \text{Sys}\{x(t)\} = \displaystyle\int_{t-T}^{t+T} x(\lambda)\, d\lambda, \qquad T > 0$

MATLAB Problems

2.32.   Refer to Problem 2.2. The cascade combination of the two subsystems will be tested using the input signal

$$x(t) = e^{-t} \cos(2t)\, u(t)$$

Write a MATLAB script to do the following:

a. Create an anonymous anonymous function "**x**" to compute the input signal at all time instants specified in a vector "**t**".
b. Implement the cascade system as shown in Fig. P.2.2(a). Compute the signals $w(t)$ and $y(t)$ in the time interval $-1 \le t \le 5$ s using a time increment of 0.01 s.
c. Implement the cascade system in the alternative form shown in Fig. P.2.2(b). Compute the signals $\bar{w}(t)$ and $\bar{y}(t)$ in the time interval $-1 \le t \le 5$ s using a time increment of 0.01 s. Compare $y(t)$ and $\bar{y}(t)$.

2.33.   Repeat the steps of Problem 2.32 using the subsystems described in Problem 2.3.

2.34.   Consider the differential equations given in Problem 2.8. Using the Euler method, find a numerical solution for each differential equation with the specified input signal and initial condition. Refer to MATLAB Exercise 2.4 for an example of doing this. For each case use a step size T_s equal to 10 percent of the time constant, and obtain the numerical solution in the time interval $0 \leq t \leq 2$ s. Compare the numerical solution to the exact solution found in Problem 2.32.

Hint: Part (c) may be tricky due to the unit-impulse input signal. Approximate $\delta(t)$ using a pulse with unit area and a width equal to the step size.

2.35. Repeat Problem 2.34 using MATLAB function `ode45(..)` instead of the Euler method.

2.36. Refer to Problem 2.12 in which the unit-ramp response of the simple RC circuit of Fig. P.2.12(a) was determined. Since the circuit is assumed to be initially relaxed, it is linear and time-invariant. As a result, the unit-ramp response obtained in Problem 2.12 can be used with superposition for finding the response of the circuit to signals with higher complexity. Consider the trapezoidal input signal shown in Fig. P.2.36.

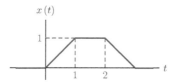

Figure P. 2.36

Develop a script to do the following:

a. Create an anonymous function `xr(..)` to compute the unit ramp function at time instants in a specified vector. You may wish to review the script developed in MATLAB Exercise 2.3 as an example.

b. Create an anonymous function `yr(..)` to compute the unit-ramp response of the RC circuit at time instants in a specified vector.

c. Express the input signal $x(t)$ in terms of scaled and time shifted unit ramp functions, and compute it in the time interval $-1 \leq t \leq 5$ seconds with an increment of 0.01 seconds.

d. Express the response $y(t)$ of the circuit in terms of scaled and time shifted versions of the unit-ramp response, and compute it in the same time interval with the same increment.

e. Graph the input and the output signals on the same coordinate system.

2.37. Refer again to the simple RC circuit shown in Fig. P.2.12(a). In MATLAB Exercise 2.3 an anonymous function was developed for computing the unit-step response of the circuit. In Problem 2.36 a similar anonymous function was developed for computing its unit-ramp response. Consider the input signal shown in Fig. P.2.37.

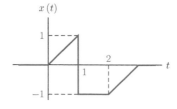

Figure P. 2.37

Develop a MATLAB script to do the following:

a. Express the input signal $x(t)$ using a combination of scaled and time shifted unit-step and unit-ramp functions.

b. Use the anonymous function `yu(..)` from MATLAB Exercise 2.3 and the anonymous function `yr(..)` from Problem 2.36. Express the output signal $y(t)$ of the circuit in terms of scaled and time shifted versions of the unit-step response and the unit-ramp response, and compute it in the time interval $-1 \leq t \leq 5$ seconds with an increment of 0.01 seconds.

c. Graph the input and the output signals on the same coordinate system.

2.38.   Write a script to verify the solution to Problem 2.23. Express the triangular waveform $x(t)$ using an anonymous function. Afterwards compute and graph the output signal $y(t)$ in the time interval $-1 \leq t \leq 5$ s.

2.39.   Write a script to verify the solution to Problem 2.24. Express the input signal $x(t) = e^{-t} u(t)$ using an anonymous function. Afterwards compute and graph the output signal $y(t)$ in the time interval $-1 \leq t \leq 7$ s.

MATLAB Projects

2.40.   Consider a general second-order homogeneous differential equation

$$\frac{d^2 y(t)}{dt^2} + a_1 \frac{dy(t)}{dt} + a_0\, y(t) = 0$$

with initial conditions

$$y(0) = p_1, \qquad \left. \frac{dy(t)}{dt} \right|_{t=0} = p_2$$

The coefficients a_1 and a_0 of the differential equation and the initial values p_1 and p_2 are all real-valued, and will be left as parameters.

a. Let s_1 and s_2 be the roots of the characteristic polynomial. On paper, solve the homogeneous differential equation for the three possibilities for the roots:

1. The roots are real and distinct.
2. The roots are a complex conjugate pair.
3. The two roots are real and equal.

Find the solution $y(t)$ as a function of the roots s_1 and s_2 as well as the initial values p_1 and p_2.

b. Develop a MATLAB function `ss_diff2solve(..)` with the syntax

```
y = ss_diff2solve(a1,a0,p1,p2,t)
```

The vector "t" contains the time instants at which the solution should be computed. The returned vector "y" holds the solution at the time instants in vector "t" so that the solution can be graphed with

```
plot(t,y)
```

Your function should perform the following steps:

1. Form the characteristic polynomial and find its roots.
2. Determine which of the three categories the roots fit (simple real roots, complex conjugate pair or multiple roots).
3. Compute the solution accordingly.

c. Test the function `ss_diff2solve(..)` with the homogeneous differential equations in Problems 2.9a, b, c, Problems 2.10a, b, c, and Problem 2.11a.

2.41. Refer to the discussion in Section 2.10 regarding numerical solution of a differential equation using the Euler method. A modified version of this method is the backward Euler method which uses the approximation

$$\left.\frac{dy\,(t)}{dt}\right|_{t=t_0} \approx \frac{y\,(t_0) - y\,(t_0 - T)}{T}$$

instead of Eqn. (2.184). With this modification Eqn. (2.185) becomes

$$y\,(t_0) \approx y\,(t_0 - T) + T\,g\,[t_0, y\,(t_0)]$$

Consider the RC circuit problem the numerical solution of which was explored in MATLAB Exercise 2.4.

a. Start with the script developed in MATLAB Exercise 2.4. Modify it so that the step size is $T = 0.7$ s and the approximate solution is computed and graphed for $0 < t \leq 7$ s.
b. Develop a new script that uses the backward Euler method to solve the same problem. To accomplish this, start with the original differential equation and approximate the derivative on the left side with the backward difference. Simplify the resulting equation and use it in the code. Compute and graph the approximate solution using a step size of $T = 0.7$ s as in part (a).
c. Compare the results obtained in parts (a) and (b). Notice that the step size T used in both scripts is very large compared to the time constant. How does the performance of the backward Euler method compare to that of the Euler method?

2.42. Backward Euler method can be extended to apply to second-order differential equations using the approximation

$$\left.\frac{d^2y\,(t)}{dt^2}\right|_{t=t_0} \approx \frac{y\,(t_0) - 2\,y\,(t_0 - T) + y\,(t_0 - 2T)}{T^2}$$

for the second derivative. Using this approximation, develop a script to compute and graph the approximate solution of the homogeneous differential equation

$$\frac{d^2y\,(t)}{dt^2} + 2\,\frac{dy\,(t)}{dt} + 26\,y\,(t) = 0$$

subject to the initial conditions

$$y\,(0) = 2\,,\qquad \left.\frac{dy\,(t)}{dt}\right|_{t=0} = 13$$

Experiment with the choice of the step size T. Compare your approximate solutions to the exact solution found in Example 2.15 part (a).

Chapter 3

Analyzing Discrete-Time Systems in the Time Domain

Chapter Objectives

- Develop the notion of a *discrete-time system*.

- Learn simplifying assumptions made in the analysis of discrete-time systems. Discuss the concepts of *linearity* and *time invariance*, and their significance.

- Explore the use of difference equations for representing discrete-time systems.

- Develop methods for solving difference equations to compute the output signal of a system in response to a specified input signal.

- Learn to represent a difference equation in the form of a block diagram that can be used as the basis for simulating or realizing a system.

- Discuss the significance of the *impulse response* as an alternative description form for linear and time-invariant systems.

- Learn how to compute the output signal for a linear and time-invariant system using *convolution*. Understand the graphical interpretation of the steps involved in carrying out the convolution operation.

- Learn the concepts of causality and stability as they relate to physically realizable and usable systems.

3.1   Introduction

The definition of a discrete-time system is similar to that of its continuous-time counterpart:

In general, a discrete-time system is a mathematical formula, method or algorithm that defines a cause-effect relationship between a set of discrete-time input signals and a set of discrete-time output signals.

Signal-system interaction involving a single-input single-output discrete-time system is illustrated in Fig. 3.1.

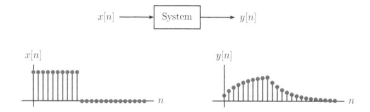

Figure 3.1 – Discrete-time signal-system interaction.

The input-output relationship of a discrete-time system may be expressed in the form

$$y[n] = \mathrm{Sys}\left\{x[n]\right\} \tag{3.1}$$

where $\mathrm{Sys}\left\{\ldots\right\}$ represents the transformation that defines the system in the time domain. A very simple example is a system that simply multiplies its input signal by a constant gain factor K

$$y[n] = K\,x[n] \tag{3.2}$$

or a system that delays its input signal by m samples

$$y[n] = x[n - m] \tag{3.3}$$

or one that produces an output signal proportional to the square of the input signal

$$y[n] = K\,\left[x[n]\right]^2 \tag{3.4}$$

A system with higher complexity can be defined using a difference equation that establishes the relationship between input and output signals.

Two commonly used simplifying assumptions used in studying mathematical models of systems, namely *linearity* and *time invariance*, will be the subject of Section 3.2. Section 3.3 deals with the issue of deriving a difference equation model for a discrete-time system, and Section 3.4 discusses the characteristics of constant-coefficient linear difference equations. Solution methods for constant-coefficient linear difference equations are presented in Section 3.5. Block diagrams for realizing discrete-time systems are introduced in Section 3.6. In Section 3.7 we discuss the significance of the impulse response for discrete-time systems, and its use in the context of the convolution operator for determining the output signal. Concepts of causality and stability of systems are discussed in Sections 3.8 and 3.9 respectively.

3.2   Linearity and Time Invariance

In our discussion of continuous-time systems in Chapter 2 we have relied heavily on the simplifying assumptions of linearity and time invariance. We have seen that these two assumptions allow a robust set of analysis methods to be developed. The same is true for the analysis of discrete-time systems.

3.2.1 Linearity in discrete-time systems

Linearity property will be very important as we analyze and design discrete-time systems. Conditions for linearity of a discrete-time system are:

Conditions for linearity:

$$\text{Sys}\left\{x_1[n] + x_2[n]\right\} = \text{Sys}\left\{x_1[n]\right\} + \text{Sys}\left\{x_2[n]\right\} \tag{3.5}$$

$$\text{Sys}\left\{\alpha_1\, x_1[n]\right\} = \alpha_1\, \text{Sys}\left\{x_1[n]\right\} \tag{3.6}$$

Eqns. (3.5) and (3.6), referred to as the *additivity rule* and the *homogeneity rule* respectively, must be satisfied for any two discrete-time signals $x_1[n]$ and $x_2[n]$ as well as any arbitrary constant α_1. These two criteria can be combined into one equation known as the *superposition principle.*

Superposition principle:

$$\text{Sys}\left\{\alpha_1\, x_1[n] + \alpha_2\, x_2[n]\right\} = \alpha_1\, \text{Sys}\left\{x_1[n]\right\} + \alpha_2\, \text{Sys}\left\{x_2[n]\right\} \tag{3.7}$$

Verbally expressed, Eqn. (3.7) implies that the response of the system to a weighted sum of any two input signals is equal to the same weighted sum of the individual responses of the system to each of the two input signals. Fig. 3.2 illustrates this concept.

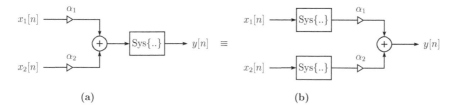

(a)   (b)

Figure 3.2 – Illustration of Eqn. (3.7). The two configurations shown are equivalent if the system under consideration is linear.

A generalization of the principle of superposition for the weighted sum of N discrete-time signals is expressed as

$$\text{Sys}\left\{\sum_{i=1}^{N} \alpha_i\, x_i[n]\right\} = \sum_{i=1}^{N} \alpha_i\, \text{Sys}\left\{x_i[n]\right\} \tag{3.8}$$

The response of a linear system to a weighted sum of N arbitrary signals is equal to the same weighted sum of individual responses of the system to each of the N signals. Let $y_i[n]$ be the response to the input term $x_i[n]$ alone, that is $y_i[n] = \text{Sys}\left\{x_i[n]\right\}$ for $i = 1, \ldots, N$.

Superposition principle implies that

$$y[n] = \text{Sys}\left\{\sum_{i=1}^{N}\alpha_i\,x_i[n]\right\} = \sum_{i=1}^{N}\alpha_i\,y_i[n] \tag{3.9}$$

This is illustrated in Fig. 3.3.

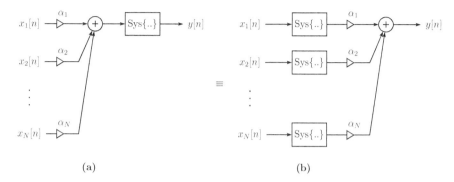

(a)   (b)

Figure 3.3 – Illustration of Eqn. (3.7). The two configurations shown are equivalent if the system under consideration is linear.

Example 3.1: **Testing linearity in discrete-time systems**

For each of the discrete-time systems described below, determine whether the system is linear or not:

a. $y[n] = 3\,x[n] + 2x[n-1]$
b. $y[n] = 3\,x[n] + 2\,x[n+1]\,x[n-1]$
c. $y[n] = a^{-n}\,x[n]$

Solution:

a. In order to test the linearity of the system we will think of its responses to the two discrete-time signals $x_1[n]$ and $x_2[n]$ as

$$y_1[n] = \text{Sys}\{x_1[n]\} = 3\,x_1[n] + 2\,x_1[n-1]$$

and

$$y_2[n] = \text{Sys}\{x_2[n]\} = 3\,x_2[n] + 2\,x_2[n-1]$$

The response of the system to the linear combination signal $x[n] = \alpha_1\,x_1[n] + \alpha_2\,x_2[n]$ is computed as

$$\begin{aligned}
y[n] &= \text{Sys}\{\alpha_1\,x_1[n] + \alpha_2\,x_2[n]\}\\
&= 3\,(\alpha_1\,x_1[n] + \alpha_2\,x_2[n]) + 2\,(\alpha_1\,x_1[n-1] + \alpha_2\,x_2[n-1])\\
&= \alpha_1\,(3\,x_1[n] + 2\,x_1[n-1]) + \alpha_2\,(3\,x_2[n] + 2\,x_2[n-1])\\
&= \alpha_1\,y_1[n] + \alpha_2\,y_2[n]
\end{aligned}$$

Superposition principle holds, and therefore the system in question is linear.

b. Again using the test signals $x_1[n]$ and $x_2[n]$ we have

$$y_1[n] = \text{Sys}\,\{x_1[n]\} = 3\,x_1[n] + 2\,x_1[n+1]\,x_1[n-1]$$

and

$$y_2[n] = \text{Sys}\,\{x_2[n]\} = 3\,x_2[n] + 2\,x_2[n+1]\,x_2[n-1]$$

Use of the linear combination signal $x[n] = \alpha_1\,x_1[n] + \alpha_2\,x_2[n]$ as input to the system yields the output signal

$$
\begin{aligned}
y[n] =& \text{Sys}\,\{\alpha_1\,x_1[n] + \alpha_2\,x_2[n]\} \\
=& 3\,(\alpha_1\,x_1[n] + \alpha_2 x_2[n]) \\
& + 2\,(\alpha_1\,x_1[n+1] + \alpha_2\,x_2[n+1])\,(\alpha_1\,x_1[n-1] + \alpha_2\,x_2[n-1])
\end{aligned}
$$

In this case the superposition principle does not hold true. The system in part (b) is therefore not linear.

c. The responses of the system to the two test signals are

$$y_1[n] = \text{Sys}\,\{x_1[n]\} = a^{-n}\,x_1[n]$$

and

$$y_2[n] = \text{Sys}\,\{x_2[n]\} = a^{-n}\,x_2[n]$$

and the response to the linear combination signal $x[n] = \alpha_1\,x_1[n] + \alpha_2\,x_2[n]$ is

$$
\begin{aligned}
y[n] =& \text{Sys}\,\{\alpha_1\,x_1[n] + \alpha_2\,x_2[n]\} \\
=& a^{-n}\,(\alpha_1\,x_1[n] + \alpha_2\,x_2[n]) \\
=& \alpha_1\,a^{-n}\,x_1[n] + \alpha_2\,a^{-n}\,x_2[n] \\
=& \alpha_1\,y_1[n] + \alpha_2\,y_2[n]
\end{aligned}
$$

The system is linear.

3.2.2 Time invariance in discrete-time systems

The definition of the time invariant property for a discrete-time system is similar to that of a continuous-time system. Let a discrete-time system be described with the input-output relationship $y[n] = \text{Sys}\,\{x[n]\}$. For the system to be considered *time-invariant*, the only effect of time-shifting the input signal should be to cause an equal amount of time shift in the output signal.

Condition for time invariance:

$$\text{Sys}\,\{x[n]\} = y[n] \quad \text{implies that} \quad \text{Sys}\,\{x[n-k]\} = y[n-k] \qquad (3.10)$$

This relationship is depicted in Fig. 3.4.

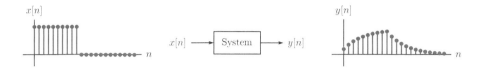

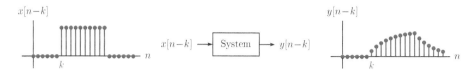

Figure 3.4 – Illustration of time-invariance for a discrete-time system.

Alternatively, the time-invariant nature of a system can be characterized by the equivalence of the two configurations shown in Fig. 3.5.

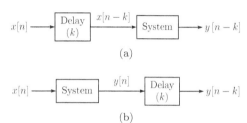

Figure 3.5 – Another interpretation of time-invariance. The two configurations shown are equivalent for a time-invariant system.

Example 3.2: **Testing time invariance in discrete-time systems**

For each of the discrete-time systems described below, determine whether the system is time-invariant or not:

a. $y[n] = y[n-1] + 3\,x[n]$
b. $y[n] = x[n]\,y[n-1]$
c. $y[n] = n\,x[n-1]$

Solution:

a. We will test the time-invariance property of the system by time-shifting both the input and the output signals by the same number of samples, and see if the input–output relationship still holds. Replacing the index n by $n-k$ in the arguments of all input and output terms we obtain

$$\text{Sys}\,\{x[n-k]\} = y[n-k-1] + 3\,x[n-k] = y[n-k]$$

The input–output relationship holds, therefore the system is time-invariant.
b. Proceeding in a similar fashion we have

$$\text{Sys}\,\{x[n-k]\} = x[n-k]\,y[n-k-1] = y[n-k]$$

This system is time-invariant as well.

c. Replacing the index n by $n - k$ in the arguments of all input and output terms yields

$$\text{Sys}\,\{x[n-k]\} = n\,x[n-k-1] \neq y[n-k]$$

This system is clearly not time-invariant since the input–output relationship no longer holds after input and output signals are time-shifted.

Should we have included the factor n in the time shifting operation when we wrote the response of the system to a time-shifted input signal? In other words, should we have written the response as

$$\text{Sys}\,\{x[n-k]\} \overset{?}{=} (n-k)\,x[n-k-1]$$

The answer is no. The factor n that multiplies the input signal is part of the system definition and not part of either the input or the output signal. Therefore we cannot include it in the process of time-shifting input and output signals.

3.2.3 DTLTI systems

Discrete-time systems that are both linear and time-invariant will play an important role in the rest of this textbook. We will develop time- and frequency-domain analysis and design techniques for working with such systems. To simplify the terminology, we will use the acronym *DTLTI* to refer to *discrete-time linear and time-invariant* systems.

3.3 Difference Equations for Discrete-Time Systems

In Section 2.3 we have discussed methods of representing continuous-time systems with differential equations. Using a similar approach, discrete-time systems can be modeled with difference equations involving current, past, or future samples of input and output signals. We will begin our study of discrete-time system models based on difference equations with a few examples. Each example will start with a verbal description of a system and lead to a system model in the form of a difference equation. Some of the examples will lead to systems that will be of fundamental importance in the rest of this text while other examples lead to nonlinear or time-varying systems that we will not consider further. In the sections that follow we will focus our attention on difference equations for DTLTI systems, and develop solution techniques for them using an approach that parallels our study of differential equations for CTLTI systems.

Example 3.3: **Moving-average filter**

A length-N *moving average filter* is a simple system that produces an output equal to the arithmetic average of the most recent N samples of the input signal. Let the discrete-time output signal be $y[n]$. If the current sample index is 100, the current output sample $y[100]$ would be equal to the arithmetic average of the current input sample $x[100]$ and $(N-1)$ previous input samples. Mathematically we have

$$y[100] = \frac{x[100] + x[99] + \ldots + x[100 - (N-1)]}{N}$$

$$= \frac{1}{N} \sum_{k=0}^{N-1} x[100 - k]$$

The general expression for the length-N moving average filter is obtained by expressing the n-th sample of the output signal in terms of the relevant samples of the input signal as

$$y[n] = \frac{x[n] + x[n-1] + \ldots + x[n-(N-1)]}{N}$$

$$= \frac{1}{N} \sum_{k=0}^{N-1} x[n-k] \tag{3.11}$$

Eqn. (3.11) is a difference equation describing the input-output relationship of the moving average filter as a discrete-time system. The operation of the length-N moving average filter is best explained using the analogy of a window, as illustrated in Fig. 3.6.

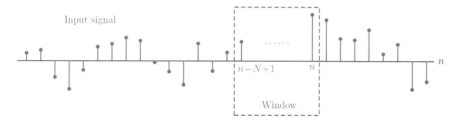

Figure 3.6 – Length-N moving average filter.

Suppose that we are observing the input signal through a window that is wide enough to hold N samples of the input signal at any given time. Let the window be stationary, and let the input signal $x[n]$ move to the left one sample at a time, similar to a film strip. The current sample of the input signal is always the rightmost sample visible through the window. The current output sample is the arithmetic average of the input samples visible through the window.

Moving average filters are used in practical applications to smooth the variations in a signal. One example is in analyzing the changes in a financial index such as the Dow Jones Industrial Average. An investor might use a moving average filter on a signal that contains the values of the index for each day. A 50-day or a 100-day moving average window may be used for producing an output signal that disregards the day-to-day fluctuations of the input signal and focuses on the slowly varying trends instead. The degree of smoothing is dependent on N, the size of the window. In general, a 100-day moving average is smoother than a 50-day moving average.

Let $x[n]$ be the signal that holds the daily closing values of the index for the calendar year 2003. The 50-day moving averages are computed by

$$y_1[n] = \frac{1}{50} \sum_{k=0}^{49} x[n-k]$$

and 100-day moving averages are computed by

$$y_2[n] = \frac{1}{100} \sum_{k=0}^{99} x[n-k]$$

Fig. 3.7 shows the daily values of the index for the calendar year 2003 as a discrete-time signal as well as the outputs produced by 50-day and 100-day moving average filters. Note

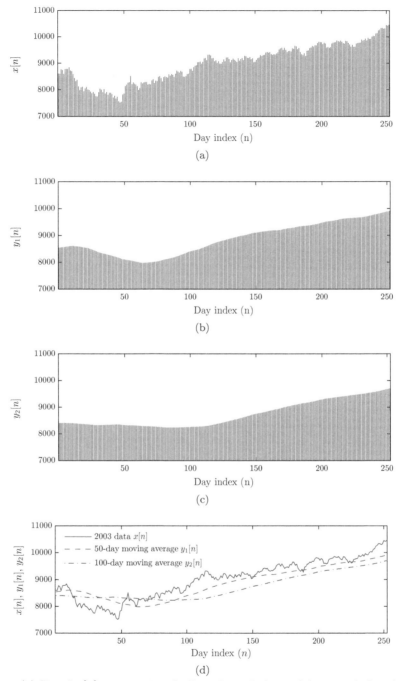

Figure 3.7 – (a) Signal $x[n]$ representing the Dow Jones Industrial Average daily values for the calendar year 2003, (b) the signal $y_1[n]$ holding 50-day moving average values for the index, (c) the signal $y_2[n]$ holding 100-day moving average values for the index, (d) signals $x[n]$, $y_1[n]$ and $y_2[n]$ depicted as line graphs on the same set of axes for comparison.

that, in the computation of the output signal samples $y_1[0], \ldots, y_1[48]$ as well as the output signal samples $y_2[0], \ldots, y_2[98]$, the averaging window had to be supplied with values from 2002 data for the index. For example, to compute the 50-day moving average on January 1 of 2003, previous 49 samples of the data from December and part of November of 2002 had to be used.

We will develop better insight for the type of smoothing achieved by moving average filters when we discuss frequency domain analysis methods for discrete-time systems.

Interactive Demo: `ma_demo1`

The interactive demo program `ma_demo1.m` illustrates the length-N moving average filter discussed in Example 3.3. Because of the large number of samples involved, the two discrete-time signals are shown through line plots as opposed to stem plots. The first plot is the Dow Jones Industrial Average data for the year 2003 which is preceded by partial data from the year 2002. The second plot is the smoothed output signal of the moving average filter. The current sample index n and the length N of the moving average filter can each be specified through the user interface controls. The current sample index is marked on each plot with a red dot. Additionally, the window for the moving average filter is shown on the first plot, superimposed with the input data. The green horizontal line within the window as well as the green arrows indicate the average amplitude of the samples that fall within the window.

1. Increment the sample index and observe how the window slides to the right each time. Observe that the rightmost sample in the window is the current sample, and the window accommodates a total of N samples.
2. For each position of the window, the current output sample is the average of all input samples that fall within the window. Observe how the position of the window relates to the value of the current output sample.
3. The length of the filter, and consequently the width of the window, relates to the degree of smoothing achieved by the moving average filter. Vary the length of the filter and observe the effect on the smoothness of the output signal.

Software resources:
`ma_demo1.m`

Example 3.4: **Length-2 moving-average filter**

A length-2 moving average filter produces an output by averaging the current input sample and the previous input sample. This action translates to a difference equation in the form

$$y[n] = \frac{1}{2}\,x[n] + \frac{1}{2}\,x[n-1] \tag{3.12}$$

and is illustrated in Fig. 3.8. The window through which we look at the input signal accommodates two samples at any given time, and the current input sample is close to the right edge of the window. As in the previous example, we will assume that the window is stationary, and the system structure shown is also stationary. We will imagine the input signal moving to the left one sample at a time like a film strip. The output signal in the lower part of the figure also acts like a film strip and moves to the left one sample at a time, in sync with the input signal. If we wanted to write the input-output relationship of the length-2 moving average filter for several values of the index n, we would have

$$n = 0 : \qquad y[0] = \frac{1}{2} x[0] + \frac{1}{2} x[-1]$$

$$n = 1 : \qquad y[1] = \frac{1}{2} x[1] + \frac{1}{2} x[0]$$

$$n = 2 : \qquad y[2] = \frac{1}{2} x[2] + \frac{1}{2} x[1]$$

and so on.

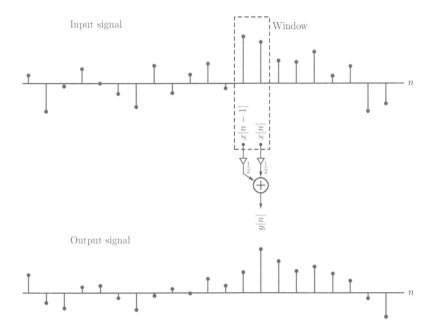

Figure 3.8 – Illustration of length-2 moving average filter.

Interactive Demo: `ma_demo2`

The interactive demo program `ma_demo2.m` illustrates a length-2 moving average filter discussed in Example 3.4. The input and output signals follow the film strip analogy, and can be moved by changing the current index through the user interface. Input signal samples that fall into the range of the length-2 window are shown in red color. Several choices are available for the input signal for experimentation.

1. Set the input signal to a unit-impulse, and observe the output signal of the length-2 moving average filter. What is the *impulse response* of the system?

2. Set the input signal to a unit-step, and observe the output signal of the system. Pay attention to the output samples as the input signal transitions from a sample amplitude of 0 to 1. The smoothing effect should be most visible during this transition.

Software resources:
`ma_demo2.m`

Example 3.5: **Length-4 moving average filter**

A length-4 moving average filter produces an output by averaging the current input sample and the previous three input samples. This action translates to a difference equation in the form

$$y_{[n]} = \frac{1}{4}\,x[n] + \frac{1}{4}\,x[n-1] + \frac{1}{4}\,x[n-2] + \frac{1}{4}\,x[n-3] \tag{3.13}$$

and is illustrated in Fig. 3.9.

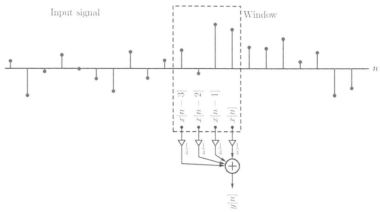

Figure 3.9 – Illustration of length-4 moving average filter.

For a few values of the index we can write the following set of equations:

$$n = 0: \quad y[0] = \frac{1}{4}\,x[0] + \frac{1}{4}\,x[-1] + \frac{1}{4}\,x[-2] + \frac{1}{4}\,x[-3]$$

$$n = 1: \quad y[1] = \frac{1}{4}\,x[1] + \frac{1}{4}\,x[0] + \frac{1}{4}\,x[-1] + \frac{1}{4}\,x[-2]$$

$$n = 2: \quad y[2] = \frac{1}{4}\,x[2] + \frac{1}{4}\,x[1] + \frac{1}{4}\,x[0] + \frac{1}{4}\,x[-1]$$

Interactive Demo: `ma_demo3`

The interactive demo program `ma_demo3.m` illustrates a length-4 moving average filter discussed in Example 3.5. Its operation is very similar to that of the program `ma_demo2.m`.

1. Set the input signal to a unit-impulse, and observe the output signal of the length-4 moving average filter. What is the *impulse response* for this system?

2. Set the input signal to a unit-step, and observe the output signal of the system. Pay attention to the output samples as the input signal transitions from a sample amplitude of 0 to 1. How does the result differ from the output signal of the length-2 moving average filter in response to a unit-step?

Software resources:

`ma_demo3.m`

Software resources: See MATLAB Exercises 3.1 and 3.2.

Example 3.6: **Exponential smoother**

Another method of smoothing a discrete-time signal is through the use of an *exponential smoother* which employs a difference equation with feedback. The current output sample is computed as a mix of the current input sample and the previous output sample through the equation

$$y[n] = (1 - \alpha)\, y[n-1] + \alpha\, x[n] \qquad (3.14)$$

The parameter α is a constant in the range $0 < \alpha < 1$, and it controls the degree of smoothing. According to Eqn. (3.14), the current output sample $y[n]$ has two contributors, namely the current input sample $x[n]$ and the previous output sample $y[n-1]$. The contribution of the current input sample is proportional to α, and the contribution of the previous output sample is proportional to $1-\alpha$. Smaller values of α lead to smaller contributions from each input sample, and therefore a smoother output signal. Writing the difference equation given by Eqn. (3.14) for several values of the sample index n we obtain

$$n = 0: \quad y[0] = (1 - \alpha)\, y[-1] + \alpha\, x[0]$$
$$n = 1: \quad y[1] = (1 - \alpha)\, y[0] + \alpha\, x[1]$$
$$n = 2: \quad y[2] = (1 - \alpha)\, y[1] + \alpha\, x[2]$$

and so on. Since the difference equation in Eqn. (3.14) is linear with constant coefficients, the exponential smoother would be an example of a DTLTI system provided that it is initially relaxed which, in this case, implies that $y[-1] = 0$. Fig. 3.10 illustrates the application of the linear exponential smoother to the 2003 Dow Jones Industrial Average data for $\alpha = 0.1$ and $\alpha = 0.2$.

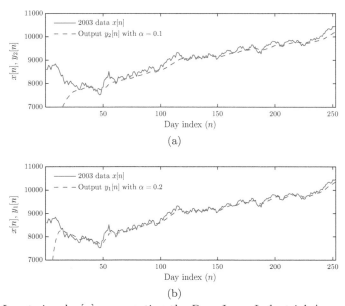

Figure 3.10 – Input signal $x[n]$ representing the Dow Jones Industrial Average daily values for the calendar year 2003 compared to the output of the linear exponential smoother with (a) $\alpha = 0.1$, and (b) $\alpha = 0.2$.

Software resources: See MATLAB Exercise 3.3.

Interactive Demo: es_demo

The interactive demo program es_demo.m illustrates the operation of the exponential smoother discussed in Example 3.6. The input and output signals follow the film strip analogy, and can be moved by changing the current index through the user interface. The smoothing parameter α can also be varied. Several choices are available for the input signal for experimentation.

1. Set the input signal to a unit impulse, and observe the output signal of the linear exponential smoother. What is the length of the impulse response for this system?

2. Vary the parameter α and observe its effect on the response to a unit impulse.

3. Set the input signal to Dow Jones Industrial Average data and observe the smoothing effect as the parameter α is varied.

Software resources:
es_demo.m

Example 3.7: **Loan payments**

A practical everyday use of a difference equation can be found in banking: borrowing money for the purchase of a house or a car. The scenario is familiar to all of us. We borrow the amount that we need to purchase that dream house or dream car, and then pay back a fixed amount for each of a number of periods, often measured in terms of months. At the end of each month the bank will compute our new balance by taking the balance of the previous month, increasing it by the monthly interest rate, and subtracting the payment we made for that month. Let $y[n]$ represent the amount we owe at the end of the n-th month, and let $x[n]$ represent the payment we make in month n. If c is the monthly interest rate expressed as a fraction (for example, 0.01 for a 1 percent monthly interest rate), then the loan balance may be modeled as the output signal of a system with the following difference equation as shown in Fig. 3.11:

$$y[n] = (1 + c) \, y[n-1] - x[n] \tag{3.15}$$

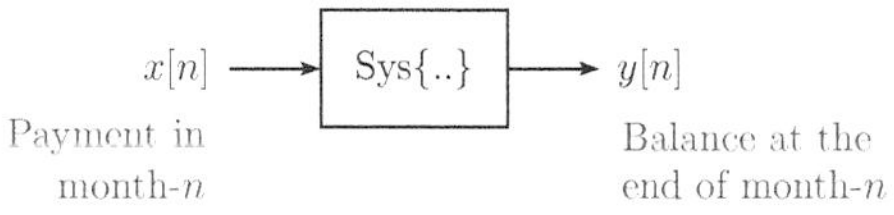

Figure 3.11 – System model for loan balance

Expressing the input-output relationship of the system through a difference equation allows us to analyze it in a number of ways. Let A represent the initial amount borrowed in month $n - 0$, and let the monthly payment be equal to B for months $n - 1, 2, \ldots$. One method of finding $y[n]$ would be to solve the difference equation in Eqn. (3.15) with the input signal

$$x[n] = B \, u[n-1] \tag{3.16}$$

and the initial condition $y[0] = A$. Alternatively we can treat the borrowed amount as a negative payment in month 0, and solve the difference equation with the input signal

$$x[n] = -A\,\delta[n] + B\,u[n-1] \tag{3.17}$$

and the initial condition $y[-1] = 0$.

In later parts of this text, as we develop the tools we need for analysis of systems, we will revisit this example. After we learn the techniques for solving linear constant-coefficient difference equations, we will be able to find the output of this system in response to an input given by Eqn. (3.17).

Example 3.8: **A nonlinear dynamics example**

An interesting example of the use of difference equations is seen in chaos theory and its applications to nonlinear dynamic systems. Let the output $y[n]$ of the system represent the population of a particular kind of species in a particular environment, for example, a certain type of plant in the rain forest. The value $y[n]$ is normalized to be in the range $0 < y[n] < 1$ with the value 1 corresponding to the maximum capacity for the species which may depend on the availability of resources such as food, water, direct sunlight, etc. In the *logistic growth* model, the population growth rate is assumed to be proportional to the remaining capacity $1 - y[n]$, and population change from one generation to the next is given by

$$y[n] = r\,(1 - y[n-1])\,y[n-1] \tag{3.18}$$

where r is a constant. This is an example of a nonlinear system that does not have an input signal. Instead, it produces an output signal based on its initial state.
Software resources:
ex_3_8.m

Example 3.9: **Newton-Raphson method for finding a root of a function**

In numerical analysis, one of the simplest methods for finding the real roots of a well-behaved function is the Newton-Raphson technique. Consider a function $u = f(w)$ shown in Fig. 3.12. Our goal is to find the value of w for which $u = f(w) = 0$

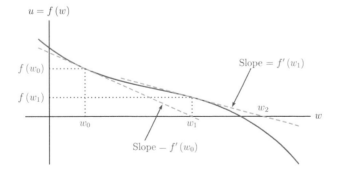

Figure 3.12 – Newton Raphson root finding technique.

Starting with an initial guess $w = w_0$ for the solution, we draw the line that is tangent to the function at that point, as shown in Fig. 3.12. The value $w = w_1$ where the tangent line intersects the real axis is our next, improved, guess for the root. The slope of the tangent line is $f'(w_0)$, and it passes through the point $(w_0, f(w_0))$. Therefore, the equation of the tangent line is

$$u - f(w_0) = f'(w_0)(w - w_0) \tag{3.19}$$

At the point where the tangent line intersects the horizontal axis we have $u = 0$ and $w = w_1$. Substituting these values into Eqn. (3.19) we have

$$-f(w_0) = f'(w_0)(w_1 - w_0) \tag{3.20}$$

and solving for w_1 results in

$$w_1 = w_0 - \frac{f(w_0)}{f'(w_0)} \tag{3.21}$$

Thus, from an initial guess w_0 for the solution, we find a better guess w_1 through the use of Eqn. (3.21). Repeating this process, an even closer guess can be found as

$$w_2 = w_1 - \frac{f(w_1)}{f'(w_1)} \tag{3.22}$$

and so on. The technique can be modeled as a discrete-time system. Let the output signal $y[n]$ represent successive guesses for the root, that is, $y[n] = w_n$. The next successive guess can be obtained from the previous one through the difference equation

$$y[n] = y[n-1] - \frac{f(y[n-1])}{f'(y[n-1])} \tag{3.23}$$

As an example of converting this procedure to a discrete-time system, let the function be

$$u = f(w) = w^2 - A$$

where A is any positive real number. We can write

$$f'(w) = 2w , \quad \text{and} \quad \frac{f(w)}{f'(w)} = \frac{w}{2} - \frac{A}{2w} \tag{3.24}$$

The difference equation in Eqn. (3.23) becomes

$$y[n] = y[n-1] - \frac{y[n-1]}{2} + \frac{A}{2y[n-1]}$$

$$= \frac{1}{2}\left(y[n-1] + \frac{A}{y[n-1]}\right) \tag{3.25}$$

Obviously Eqn. (3.25) represents a nonlinear difference equation. Let us use this system to iteratively find the square root of 10 which we know is $\sqrt{10} = 3.162278$. The function the root of which we are seeking is $f(w) = w^2 - 10$. Starting with an initial guess of $y[0] = 1$ we have

$$y[1] = \frac{1}{2}\left(y[0] + \frac{10}{y[0]}\right) = \frac{1}{2}\left(1 + \frac{10}{1}\right) = 5.5$$

The next iteration produces

$$y[2] = \frac{1}{2}\left(y[1] + \frac{10}{y[1]}\right) = \frac{1}{2}\left(5.5 + \frac{10}{5.5}\right) = 3.659091$$

Continuing in this fashion we obtain $y[3] = 3.196005$, $y[4] = 3.162456$, and $y[5] = 3.162278$ which is accurate as the square root of 10 up to the sixth decimal digit.
Software resources:
ex_3_9.m

3.4   Constant-Coefficient Linear Difference Equations

DTLTI systems can be modeled with constant-coefficient linear difference equations. A linear difference equation is one in which current, past and perhaps even future samples of the input and the output signals can appear as linear terms. Furthermore, a constant coefficient linear difference equation is one in which the linear terms involving input and output signals appear with coefficients that are constant, independent of time or any other variable.

The moving-average filters we have explored in Examples 3.3, 3.4 and 3.5 were described by constant-coefficient linear difference equations:

$$\text{Length-2}: \quad y[n] = \frac{1}{2}\,x[n] + \frac{1}{2}\,x[n-1]$$

$$\text{Length-4}: \quad y[n] = \frac{1}{4}\,x[n] + \frac{1}{4}\,x[n-1] + \frac{1}{4}\,x[n-2] + \frac{1}{4}\,x[n-3]$$

$$\text{Length-10}: \quad y[n] = \frac{1}{10}\sum_{k=0}^{9} x[n-k]$$

A common characteristic of these three difference equations is that past or future samples of the output signal do not appear on the right side of any of them. In each of the three difference equations the output $y[n]$ is computed as a function of current and past samples of the input signal. In contrast, reconsider the difference equation for the exponential smoother

$$y[n] = (1-\alpha)\,y[n-1] + \alpha\,x[n]$$

or the difference equation for the system that computes the current balance of a loan

$$y[n] = (1+c)\,y[n-1] - x[n]$$

Both of these systems also have constant-coefficient linear difference equations (we assume that parameters α and c are constants). What sets the last two systems apart from the three moving average filters above is that they also have *feedback* in the form of past samples of the output appearing on the right side of the difference equation. The value of $y[n]$ depends on the past output sample $y[n-1]$.

Examples 3.8 and 3.9, namely the logistic growth model and the Newton-Raphson algorithm for finding a square root, utilized the difference equations

$$y[n] = r\,(1 - y[n-1])\,y[n-1]$$

and

$$y[n] = \frac{1}{2}\left(y[n-1] + \frac{A}{y[n-1]}\right)$$

Both of these difference equations are nonlinear since they contain nonlinear terms of $y[n-1]$.

A general constant-coefficient linear difference equation representing a DTLTI system is in the form

$$a_0\,y[n] + a_1\,y[n-1] + \ldots + a_{N-1}\,y[n-N+1] + a_N\,y[n-N] =$$

$$b_0\,b[n] + b_1\,x[n-1] + \ldots + b_{M-1}\,x[n-M+1] + b_M\,x[n-M] \qquad (3.26)$$

or in closed summation form as shown below.

Constant-coefficient linear difference equation:

$$\sum_{k=0}^{N} a_k \, y[n-k] = \sum_{k=0}^{M} b_k \, x[n-k] \tag{3.27}$$

The order of the difference equation (and therefore the order of the system it represents) is the larger of N and M. For example, the length-2 moving average filter discussed in Example 3.4 is a first-order system. Similarly, the orders of the length-4 and the length-10 moving average filters of Examples 3.5 and 3.3 are 3 and 9 respectively.

A note of clarification is in order here: The general form we have used in Eqns. (3.26) and (3.27) includes current and past samples of $x[n]$ and $y[n]$ but no future samples. This is for practical purposes only. The inclusion of future samples in a difference equation for the computation of the current output would not affect the linearity and the time invariance of the system represented by that difference equation, as long as the future samples also appear as linear terms and with constant coefficients. For example, the difference equation

$$y[n] = y[n-1] + x[n+2] - 3x[n+1] + 2x[n] \tag{3.28}$$

is still a constant-coefficient linear difference equation, and it may still correspond to a DTLTI system. We just have an additional challenge in computing the output signal through the use of this difference equation: We need to know future values of the input signal. For example, computation of $y[45]$ requires the knowledge of $x[46]$ and $x[47]$ in addition to other terms. We will explore this further when we discuss the causality property later in this chapter.

Example 3.10: **Checking linearity and time-invariance of a difference equation**

Determine whether the first-order constant-coefficient linear difference equation in the form

$$a_0 y[n] + a_1 y[n-1] = b_0 x[n]$$

represents a DTLTI system.

Solution: Our approach will be similar to that employed in Example 2.7 of Chapter 2 for a continuous-time system. Assume that two input signals $x_1[n]$ and $x_2[n]$ produce the corresponding output signals $y_1[n]$ and $y_2[n]$ respectively. Each of the signal pairs $x_1[n] \leftrightarrow y_1[n]$ and $x_2[n] \leftrightarrow y_2[n]$ must satisfy the difference equation, so we have

$$a_0 y_1[n] + a_1 y_1[n-1] = b_0 x_1[n] \tag{3.29}$$

and

$$a_0 y_2[n] + a_1 y_2[n-1] = b_0 x_2[n] \tag{3.30}$$

Let a new input signal be constructed as a linear combination of $x_1[n]$ and $x_2[n]$ as

$$x_3[n] = \alpha_1 \, x_1[n] + \alpha_2 \, x_2[n] \tag{3.31}$$

For the system described by the difference equation to be linear, its response to the input signal $x_3[n]$ must be

$$y_3[n] = \alpha_1 \, y_1[n] + \alpha_2 \, y_2[n] \tag{3.32}$$

and the input-output signal pair $x_3[n] \leftrightarrow y_3[n]$ must also satisfy the difference equation. Substituting $y_3[n]$ into the left side of the difference equation yields

$$a_0 y_3[n] + a_1 y_3[n-1] = a_0 \left(\alpha_1 \, y_1[n] + \alpha_2 \, y_2[n] \right) + a_1 \left(\alpha_1 \, y_1[n-1] + \alpha_2 \, y_2[n-1] \right) \quad (3.33)$$

By rearranging the terms on the right side of Eqn. (3.33) we get

$$a_0 y_3[n] + a_1 y_3[n-1] = \alpha_1 \left(a_0 \, y_1[n] + a_1 \, y_1[n-1] \right) + \alpha_2 \left(a_0 \, y_2[n] + a_1 \, y_2[n-1] \right) \quad (3.34)$$

Substituting Eqns. (3.29) and (3.30) into Eqn. (3.34) leads to

$$\begin{aligned} a_0 y_3[n] + a_1 y_3[n-1] &= \alpha_1 \left(b_0 \, x_1[n] \right) + \alpha_2 \left(b_0 \, x_2[n] \right) \\ &= b_0 \left(\alpha_1 \, x_1[n] + \alpha_2 \, x_2[n] \right) \\ &= b_0 \, x_3[n] \end{aligned} \quad (3.35)$$

proving that the input-output signal pair $x_3[n] \leftrightarrow y_3[n]$ satisfies the difference equation.

Before we can claim that the difference equation in question represents a linear system, we need to check the initial value of $y[n]$. We know from previous discussion that a difference equation like the one given by Eqn. (3.29) can be solved iteratively starting at a specified value of the index $n = n_0$ provided that the value of the output sample at index $n = n_0 - 1$ is known. For example, let $n_0 = 0$, and let $y[n_0 - 1] = y[-1] = A$. Starting with the specified value of $y[-1]$ we can determine $y[0]$ as

$$y[0] = \left(\frac{a_1}{a_0} \right) A + b_0 \, x_0$$

Having determined the value of $y[0]$ we can find $y[1]$ as

$$\begin{aligned} y[1] &= \left(-\frac{a_1}{a_0} \right) y[0] + b_0 \, x_1 \\ &= \left(-\frac{a_1}{a_0} \right) \left[\left(-\frac{a_1}{a_0} \right) A + b_0 \, x_0 \right] + b_0 \, x_1 \end{aligned}$$

and continue in this fashion. Clearly the result obtained is dependent on the initial value $y[-1] = A$. Since the $y_1[n]$, $y_2[n]$ and $y_3[n]$ used in the development above are all solutions of the difference equation for input signals $x_1[n]$, $x_2[n]$ and $x_3[n]$ respectively, they must each satisfy the specified initial condition, that is,

$$y_1[-1] = A \, , \quad y_2[-1] = A \, , \quad y_3[-1] = A$$

In addition, the linearity condition in Eqn. (3.32) must be satisfied for all values of the index n including $n = -1$:

$$y_3[-1] = \alpha_1 \, y_1[-1] + \alpha_2 \, y_2[-1]$$

Thus we are compelled to conclude that the system represented by Eqn. (3.29) is linear only if $y[-1] = 0$. In the general case, we need $y[n_0 - 1] = 0$ if the solution is to start at index $n = n_0$.

Our next task is to check the time-invariance property of the system described by the difference equation. If we replace the index n with $n - m$, Eqn. (3.29) becomes

$$a_0 y[n-m] + a_1 y[n-m-1] = b_0 x[n-m] \quad (3.36)$$

Delaying the input signal $x[n]$ by m samples causes the output signal $y[n]$ to be delayed by the same amount. The system is time-invariant.

In Example 3.10 we have verified that the first-order constant-coefficient linear difference equation corresponds to a DTLTI system provided that its initial state is zero. We are now ready to generalize that result to the constant-coefficient linear difference equation of any order. Let the two input signals $x_1[n]$ and $x_2[n]$ produce the output signals $y_1[n]$ and $y_2[n]$ respectively. The input-output signal pairs $x_1[n] \leftrightarrow y_1[n]$ and $x_2[n] \leftrightarrow y_2[n]$ satisfy the difference equation, so we can write

$$\sum_{k=0}^{N} a_k\, y_1[n-k] = \sum_{k=0}^{M} b_k\, x_1[n-k] \tag{3.37}$$

and

$$\sum_{k=0}^{N} a_k\, y_2[n-k] = \sum_{k=0}^{M} b_k\, x_2[n-k] \tag{3.38}$$

To test linearity of the system we will construct a new input signal as a linear combination of $x_1[n]$ and $x_2[n]$:

$$x_3[n] = \alpha_1\, x_1[n] + \alpha_2\, x_2[n] \tag{3.39}$$

If the system described by the difference equation is linear, its response to the input signal $x_3[n]$ must be

$$y_3[n] = \alpha_1\, y_1[n] + \alpha_2\, y_2[n] \tag{3.40}$$

We will test the input-output signal pair $x_3[n] \leftrightarrow y_3[n]$ through the difference equation. Substituting $y_3[n]$ into the left side of the difference equation yields

$$\sum_{k=0}^{N} a_k\, y_3[n-k] = \sum_{k=0}^{N} a_k\, \left(\alpha_1\, y_1[n-k] + \alpha_2\, y_2[n-k]\right) \tag{3.41}$$

Rearranging the terms on the right side of Eqn. (3.41) and separating them into two separate summations yields

$$\sum_{k=0}^{N} a_k\, y_3[n-k] = \alpha_1 \sum_{k=0}^{N} a_k\, y_1[n-k] + \alpha_2 \sum_{k=0}^{N} a_k\, y_2[n-k] \tag{3.42}$$

The two summations on the right side of Eqn. (3.42) can be substituted with their equivalents from Eqns. (3.37) and (3.38), resulting in

$$\sum_{k=0}^{N} a_k\, y_3[n-k] = \alpha_1 \sum_{k=0}^{M} b_k\, x_1[n-k] + \alpha_2 \sum_{k=0}^{M} b_k\, x_2[n-k] \tag{3.43}$$

Finally, we will combine the two summations on the right side of Eqn. (3.43) back into one summation to obtain

$$\sum_{k=0}^{N} a_k\, y_3[n-k] = \sum_{k=0}^{M} b_k\, \left(\alpha_1\, x_1[n-k] + \alpha_2\, x_2[n-k]\right)$$

$$= \sum_{k=0}^{M} b_k\, x_3[n-k] \tag{3.44}$$

We conclude that the input-output signal pair $x_3[n] \leftrightarrow y_3[n]$ also satisfies the difference equation. The restriction discussed in Example 3.10 regarding the initial conditions will be applicable here as well. If we are interested in finding a unique solution for $n \geq n_0$, then the initial values

$$y[n_0 - 1] \,, \quad y[n_0 - 2] \,, \ldots, \quad y[n_0 - N]$$

are needed. The linearity condition given by Eqn. (3.40) must be satisfied for all values of n including index values $n = n_0 - 1, n_0 - 2, \ldots, n_0 - N$. Consequently, the system that corresponds to the difference equation in Eqn. (3.27) is linear only if all the initial conditions are zero, that is,

$$y[n_0 - k] = 0$$

for $k = 1, \ldots, N$. Next we need to check for time invariance. Replacing the index n with $n - m$ in Eqn. (3.27) we get

$$\sum_{k=0}^{N} a_k \, y[n - m - k] = \sum_{k=0}^{M} b_k \, x[n - m - k] \tag{3.45}$$

indicating that the input-output signal pair $x_[n - m] \leftrightarrow y[n - m]$ also satisfies the difference equation. Thus, the constant-coefficient linear difference equation is time-invariant.

3.5   Solving Difference Equations

The output signal of a discrete-time system in response to a specified input signal can be determined by solving the corresponding difference equation. In some of the examples of Section 3.3 we have already experimented with one method of solving a difference equation, namely the *iterative method*. Consider again the difference equation for the exponential smoother of Example 3.6. By writing the difference equation for each value of the index n we were able to obtain the output signal $y[n]$ one sample at a time. Given the initial value $y[-1]$ of the output signal, its value for $n = 0$ is found by

$$y[0] = (1 - \alpha) \, y[-1] + \alpha \, x[n]$$

Setting $n = 1$ we obtain

$$y[1] = (1 - \alpha) \, y[0] + \alpha \, x[1]$$

Repeating for $n = 2$ leads to

$$y[2] = (1 - \alpha) \, y[1] + \alpha \, x[2]$$

and we can continue in this fashion indefinitely. The function ss_expsmoo(..) developed in MATLAB Exercise 3.3 is an implementation of the iterative solution of this difference equation.

The iterative solution method is not limited to DTLTI systems; it can also be used for solving the difference equations of nonlinear and/or time-varying systems. Consider, for example, the nonlinear difference equation for the logistic growth model of Example 3.8. For a specified parameter value r and initial value $y[-1]$, the output at $n = 0$ is

$$y[0] = r \, (1 - y[-1]) \, y[-1]$$

Next, $y[1]$ is computed from $y[0]$ as

$$y[1] = r \, (1 - y[0]) \, y[0]$$

and so on. Fig. 3.13 shows the first 50 samples of the solution obtained in this fashion for $r = 3.1$ and $y[-1] = 0.3$.

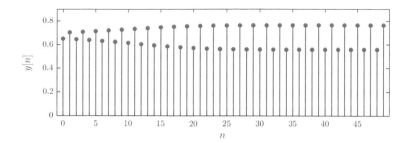

Figure 3.13 – First 50 samples of the iterative solution of the difference equation for the logistic growth model.

Iterative solution of a difference equation can be also used as the basis of implementing a discrete-time system on real-time signal processing hardware. One shortcoming of this approach, however, is the lack of a complete analytical solution. Each time we iterate through the difference equation we obtain one more sample of the output signal, but we do not get an expression or a formula for computing the output for an arbitrary value of the index n. If we need to know $y[1527]$, we must iteratively compute samples $y[0]$ through $y[1526]$ first.

Software resources: See MATLAB Exercise 3.4.

In the rest of this section we will concentrate our efforts on developing an analytical method for solving constant-coefficient linear difference equations. Analytical solution of nonlinear and/or time-varying difference equations is generally difficult or impossible, and will not be considered further in this text.

The solution method we are about to present exhibits a lot of similarities to the method developed for solving constant-coefficient ordinary differential equations in Section 2.5. We will recognize two separate components of the output signal $y[n]$ in the form

$$y[n] = y_h[n] + y_p[n] \tag{3.46}$$

The term $y_h[n]$ is the solution of the *homogeneous difference equation* found by setting $x[n] = 0$ in Eqn. (3.27) for all values of n:

$$\sum_{k=0}^{N} a_k \, y[n-k] = 0 \tag{3.47}$$

Thus, $y_h[n]$ is the signal at the output of the system when no input signal is applied to it. As in the continuous-time case, we will refer to $y_h[n]$ as the *homogeneous solution* of the difference equation or, equivalently, as the *natural response* of the system to which it corresponds. It depends on the structure of the system which is expressed through the set of coefficients a_i for $i = 0, \ldots, N$. Furthermore, it depends on the initial state of the system that is expressed through the output samples $y[n_0 - 1], y[n_0 - 2], \ldots, y[n_0 - N]$. (Recall that n_0 is the beginning index for the solution; usually we will use $n_0 = 0$.) When we discuss the stability property of DTLTI systems in Section 3.9 we will discover that, for a stable system, $y_h[n]$ approaches zero for large positive and negative values of the index n.

The second term $y_p[n]$ in Eqn. (3.46) is the part of the solution that is due to the input signal $x[n]$ applied to the system. It is referred to as the *particular solution* of the difference equation. It depends on both the input signal $x[n]$ and the internal structure of the system. It is independent of the initial state of the system. The combination of the homogeneous solution and the particular solution is referred to as the *forced solution* or the *forced response*.

3.5.1   Finding the natural response of a discrete-time system

We will begin the discussion of the solution method for solving the homogeneous equation by revisiting the linear exponential smoother first encountered in Example 3.6.

Example 3.11: **Natural response of exponential smoother**

Determine the natural response of the exponential smoother defined in Example 3.6 if $y[-1] = 2$.

Solution: The difference equation for the exponential smoother was given in Eqn. (3.6). The homogeneous difference equation is found by setting $x[n] = 0$:

$$y[n] = (1 - \alpha)\, y[n - 1] \tag{3.48}$$

The natural response $y_h[n]$ yet to be determined must satisfy the homogeneous difference equation. We need to start with an educated guess for the type of signal $y_h[n]$ must be, and then adjust any relevant parameters. Therefore, looking at Eqn. (3.48), we ask the question: "What type of discrete-time signal remains proportional to itself when delayed by one sample?" A possible answer is a signal in the form

$$y_h[n] = c\, z^n \tag{3.49}$$

where z is a yet undetermined constant. Delaying $y[n]$ of Eqn. (3.49) by one sample we get

$$y_h[n - 1] = c\, z^{n-1} = z^{-1} c\, z^n = z^{-1} y[n] \tag{3.50}$$

Substituting Eqns. (3.49) and (3.50) into the homogeneous difference equation yields

$$c\, z^n = (1 - \alpha)\, z^{-1} c\, z^n$$

or, equivalently

$$c\, z^n \left[1 - (1 - \alpha)\, z^{-1} \right] = 0$$

which requires one of the following conditions to be true for all values of n:

a. $c\, z^n = 0$
b. $\left[1 - (1 - \alpha)\, z^{-1} \right] = 0$

We cannot use the former condition since it leads to the trivial solution $y[n] = c\, z^n = 0$, and is obviously not very useful. Furthermore, the initial condition $y[-1] = 2$ cannot be satisfied using this solution. Therefore we must choose the latter condition and set $z = (1 - \alpha)$ to obtain

$$y_h[n] = c\, (1 - \alpha)^n \, , \quad \text{for } n \geq 0$$

The constant c is determined based on the desired initial state of the system. We want $y[-1] = 2$, so we impose it as a condition on the solution found in Eqn. (3.51):

$$y_h[-1] = c(1-\alpha)^{-1} = 2$$

This yields $c = 2(1-\alpha)$ and

$$y_h[n] = 2(1-\alpha)(1-\alpha)^n = 2(1-\alpha)^{n+1}$$

The natural response found is shown in Fig. 3.14 for $\alpha = 1$ and $\alpha = 0.2$.

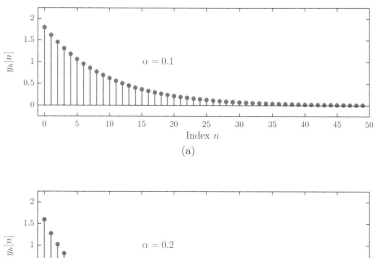

(a)

(b)

Figure 3.14 – The natural response of the linear exponential smoother for (a) $\alpha = 0.1$, and (b) $\alpha = 0.2$.

Software resources:
ex_3_11.m

Let us now consider the solution of the general homogeneous difference equation in the form

$$\sum_{k=0}^{N} a_k\, y[n-k] = 0 \qquad (3.51)$$

Let us start with the same initial guess that we used in Example 3.11:

$$y_h[n] = c\,z^n \qquad (3.52)$$

Shifted versions of the prescribed homogeneous solution are

$$y_h[n-1] = c\,z^{n-1} = z^{-1}c\,z^n$$

$$y_h[n-2] = c\,z^{n-2} = z^{-2}c\,z^n$$

$$y_h[n-3] = c\,z^{n-3} = z^{-3}c\,z^n$$

which can be expressed in the general form

$$y_h[n-k] = c\,z^{n-k} = z^{-k}c\,z^n \tag{3.53}$$

Which value (or values) of z can be used in the homogeneous solution? We will find the answer by substituting Eqn. (3.53) into Eqn. (3.51):

$$\sum_{k=0}^{N} a_k\, z^{-k} c\, z^n = 0 \tag{3.54}$$

The term $c\,z^n$ is independent of the summation index k, and can be factored out to yield

$$c\,z^n \sum_{k=0}^{N} a_k\, z^{-k} = 0 \tag{3.55}$$

There are two ways to satisfy Eqn. (3.55):

1. $c\,z^n = 0$
    This leads to the trivial solution $y[n] = 0$ for the homogeneous equation, and is therefore not very interesting. Also, we have no means of satisfying any initial conditions with this solution other than $y[-i] = 0$ for $i = 1, \ldots, N$.

2. $\sum_{k=0}^{N} a_k\, z^{-k} = 0$
    This is called the *characteristic equation* of the system. Values of z that are the solutions of the characteristic equation can be used in exponential functions as solutions of the homogeneous difference equation.

The characteristic equation:

$$\sum_{k=0}^{N} a_k\, z^{-k} = 0 \tag{3.56}$$

The characteristic equation for a DTLTI system is found by starting with the homogeneous difference equation and replacing delayed versions of the output signal with the corresponding negative powers of the complex variable z.

To obtain the characteristic equation, substitute:

$$y[n-k] \rightarrow z^{-k} \tag{3.57}$$

The characteristic equation can be written in open form as

$$a_0 + a_1 z^{-1} + \ldots + a_{N-1} z^{-N+1} + a_N z^{-N} = 0 \tag{3.58}$$

If we want to work with non-negative powers of z, we could simply multiply both sides of the characteristic equation by z^P to obtain

$$a_0 z^N + a_1 z^{N-1} + \ldots + a_{N-1} z^1 + a_N = 0 \tag{3.59}$$

The polynomial on the left side of the equal sign in Eqn. (3.59) is the *characteristic polynomial* of the DTLTI system. Let the roots of the characteristic polynomial be $z_1, z_2, \ldots, z_N$ so that Eqn. (3.59) can be written as

$$a_0 \left(a - z_1\right) \left(a - z_2\right) \ldots \left(a - z_N\right) = 0 \qquad (3.60)$$

Any of the roots of the characteristic polynomial can be used in a signal in the form

$$y_i[n] = c_i z_i^n, \quad i = 1, \ldots, N \qquad (3.61)$$

which satisfies the homogeneous difference equation:

$$\sum_{k=0}^{N} a_k y_i[n - k] = 0 \quad \text{for } i = 1, \ldots, N \qquad (3.62)$$

Furthermore, any linear combination of all valid terms in the form of Eqn. (3.61) satisfies the homogeneous difference equation as well, so we can write

$$y_h[n] = c_1 z_1^n + c_2 z_2^n + \ldots + c_N z_N^n = \sum_{k=1}^{N} c_k z_k^n \qquad (3.63)$$

The coefficients $c_1, c_2, \ldots, c_N$ are determined from the initial conditions. The exponential terms z_i^n in the homogeneous solution given by Eqn. (3.63) are the *modes of the system*. In later parts of this text we will see that the modes of a DTLTI system correspond to the *poles of the system function* and the *eigenvalues of the state matrix*.

Example 3.12: **Natural response of second-order system**

A second-order system is described by the difference equation

$$y[n] - \frac{5}{6} y[n - 1] + \frac{1}{6} y[n - 2] = 0$$

Determine the natural response of this system for $n \geq 0$ subject to initial conditions

$$y[-1] = 19, \quad \text{and} \quad y[-2] = 53$$

Solution: The characteristic equation is

$$z^2 - \frac{5}{6} z + \frac{1}{6} = 0$$

with roots $z_1 = 1/2$ and $z_2 = 1/3$. Therefore the homogeneous solution of the difference equation is

$$y_h[n] = c_1 \left(\frac{1}{2}\right)^n + c_2 \left(\frac{1}{3}\right)^n$$

for $n \geq 0$. The coefficients c_1 and c_2 need to be determined from the initial conditions. We have

$$y_h[-1] = c_1 \left(\frac{1}{2}\right)^{-1} + c_2 \left(\frac{1}{3}\right)^{-1}$$

$$= 2\,c_1 + 3\,c_2 = 19 \qquad (3.64)$$

and

$$
\begin{aligned}
y_h[-2] &= c_1 \left(\frac{1}{2}\right)^{-2} + c_2 \left(\frac{1}{3}\right)^{-2} \\
&= 4\,c_1 + 9\,c_2 = 53
\end{aligned}
\tag{3.65}
$$

Solving Eqns. (3.64) and (3.65) yields $c_1 = 2$ and $c_2 = 5$. The natural response of the system is

$$
y_h[n] = 2\left(\frac{1}{2}\right)^n u[n] + 5\left(\frac{1}{3}\right)^n u[n]
$$

Software resources:
ex_3_12.m

In Example 3.12 the characteristic equation obtained from the homogeneous difference equation had two distinct roots that were both real-valued, allowing the homogeneous solution to be written in the standard form of Eqn. (3.63). There are other possibilities as well. Similar to the discussion of the homogeneous differential equation in Section 2.5.3 we will consider three possible scenarios:

Case 1: All roots are distinct and real-valued.

This leads to the homogeneous solution

$$
y[n] = \sum_{k=1}^{N} c_k\, z_k^n
\tag{3.66}
$$

for $n \geq n_0$ as we have seen in Example 3.12. The value of the real root z_k determines the type of contribution made to the homogeneous solution by the term $c_k\, z_k^n$. If $|z_k| < 1$ then z_k^n decays exponentially over time. Conversely, $|z_k| > 1$ leads to a term z_k^n that grows exponentially. A negative value for z_k causes the corresponding term in the homogeneous solution to have alternating positive and negative sample amplitudes. Possible forms of the contribution z_k^n are shown in Fig. 3.15.

Case 2: Characteristic polynomial has complex-valued roots.

Since the difference equation and its characteristic polynomial have only real-valued coefficients, any complex roots of the characteristic polynomial must appear in conjugate pairs. Therefore, if

$$
z_{1a} = r_1\, e^{j\Omega_1}
$$

is a complex root, then its conjugate

$$
z_{1b} = z_{1a}^{*} = r_1\, e^{-j\Omega_1}
$$

must also be a root. Let the part of the homogeneous solution that is due to these two roots be

$$
\begin{aligned}
y_{h1}[n] &= c_{1a}\, z_{1a}^n + c_{1b}\, z_{1b}^n \\
&= c_{1a}\, r_1^n\, e^{j\Omega_1 n} + c_{1b}\, r_1^n\, e^{-j\Omega_1 n}
\end{aligned}
\tag{3.67}
$$

The coefficients c_{1a} and c_{1b} are yet to be determined from the initial conditions. Since the coefficients of the difference equation are real-valued, the solution $y_h[n]$ must also

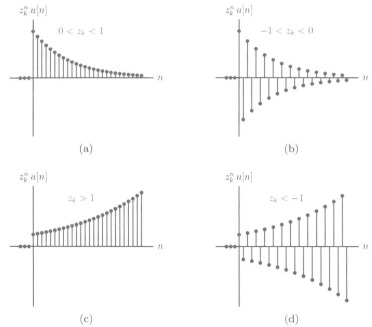

Figure 3.15 – The term $z_k^n\, u[n]$ for (a) $0 < z_k < 1$, (b) $-1 < z_k < 0$, (c) $z_k > 1$, (d) $z_k < -1$.

be real. Furthermore, $y_{h1}[n]$, the part of the solution that is due to the complex conjugate pair of roots we are considering must also be real. This implies that the coefficients c_{1a} and c_{1b} must form a complex conjugate pair. We will write the two coefficients in polar complex form as

$$c_{1a} = |c_1|\, e^{j\theta_1}\,, \quad \text{and } c_{1b} = |c_1|\, e^{-j\theta_1} \tag{3.68}$$

Substituting Eqn. (3.68) into Eqn. (3.67) we obtain

$$\begin{aligned} y_{h1}[n] &= |c_1|\, r_1^n\, e^{j(\Omega_1 n + \theta_1)} + |c_1|\, r_1^n\, e^{-j(\Omega_1 n + \theta_1)} \\ &= 2\,|c_1|\, r_1^n\, \cos\left(\Omega_1 n + \theta_1\right) \end{aligned} \tag{3.69}$$

The contribution of a complex conjugate pair of roots to the solution is in the form of a cosine signal multiplied by an exponential signal. The oscillation frequency of the discrete-time cosine signal is determined by Ω_1. The magnitude of the complex conjugate roots, r_1, impacts the amplitude behavior. If $r_1 < 1$, then the amplitude of the cosine signal decays exponentially over time. If $r_1 > 1$ on the other hand, the amplitude of the cosine signal grows exponentially over time. These two possibilities are illustrated in Fig. 3.16.

With the use of the the appropriate trigonometric identity[1] Eqn. (3.69) can also be written in the alternative form

$$\begin{aligned} y_{h1}[n] &= 2\,|c_1|\, \cos\left(\theta_1\right)\, r_1^n\, \cos\left(\Omega_1 n\right) - 2\,|c_1|\, \sin\left(\theta_1\right)\, r_1^n\, \sin\left(\Omega_1 n\right) \\ &= d_1\, r_1^n\, \cos\left(\Omega_1 n\right) + d_2\, r_1^n\, \sin\left(\Omega_1 n\right) \end{aligned} \tag{3.70}$$

---

[1] $\cos\left(a + b\right) = \cos\left(a\right)\cos\left(b\right) - \sin\left(a\right)\sin\left(b\right)$.

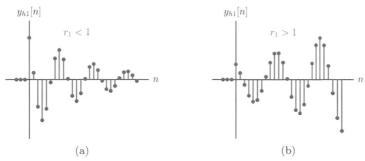

Figure 3.16 – Terms corresponding to a pair of complex conjugate roots of the characteristic equation: (a) $r_1 < 0$, (b) $r_1 > 1$.

Case 3: Characteristic polynomial has some multiple roots.

Consider again the factored version of the characteristic equation first given by Eqn. (3.60):

$$a_0 (z - z_1) (z - z_2) \ldots (z - z_N) = 0$$

What if the first two roots are equal, that is, $z_2 = z_1$? If we were to ignore the fact that the two roots are equal, we would have a natural response in the form

$$
\begin{aligned}
y_h[n] &= c_1 z_1^n + c_2 z_2^n + \text{other terms} \\
&= c_1 z_1^n + c_2 z_1^n + \text{other terms} \\
&= (c_1 + c_2) z_1^n + \text{other terms} \\
&= \tilde{c}_1 z_1^n + \text{other terms}
\end{aligned}
\tag{3.71}
$$

The equality of two roots leads to loss of one of the coefficients that we will need in order to satisfy the initial conditions. In order to gain back the coefficient we have lost, we need an additional term for the two roots at $z = z_1$, and it can be obtained by considering a solution in the form

$$y_h[n] = c_{11} z_1^n + c_{12} n z_1^n + \text{other terms} \tag{3.72}$$

In general, a root of multiplicity r requires r terms in the homogeneous solution. If the characteristic polynomial has a factor $(z - z_1)^r$, the resulting homogeneous solution is

$$y_h[n] = c_{11} z_1^n + c_{12} n z_1^n + \ldots + c_{1r} n^{r-1} z_1^n + \text{other terms} \tag{3.73}$$

Example 3.13: **Natural response of second-order system revisited**

Determine the natural response of each of the second-order systems described by the difference equations below:

a. $y[n] - 1.4 \, y[n-1] + 0.85 \, y[n-2] = 0$

with initial conditions $y[-1] = 5$ and $y[-2] = 7$.

b. $y[n] - 1.6 \, y[n-1] + 0.64 \, y[n-2] = 0$

with initial conditions $y[-1] = 2$ and $y[-2] = -3$.

Solution:

a. The characteristic equation is

$$z^2 - 1.4\,z + 0.85 = 0$$

which can be solved to yield

$$z_{1,2} = 0.7 \pm j0.6$$

Thus, the roots of the characteristic polynomial form a complex conjugate pair. They can be written in polar complex form as

$$z_{1,2} = 0.922\,e^{\pm j0.7086}$$

which leads us to a homogeneous solution in the form

$$y_h[n] = d_1\,(0.922)^n\,\cos\,(0.7086n) + d_2\,(0.922)^n\,\sin\,(0.7086n)$$

for $n \geq 0$. The coefficients d_1 and d_2 need to be determined from the initial conditions. Evaluating $y_h[n]$ for $n = -1$ and $n = -2$ we have

$$
\begin{aligned}
y_h[-1] &= d_1\,(0.922)^{-1}\,\cos\,(-0.7086) + d_2\,(0.922)^{-1}\,\sin\,(-0.7086) \\
&= 0.6923\,d_1 - 0.5385\,d_2 = 5
\end{aligned}
\tag{3.74}
$$

and

$$
\begin{aligned}
y_h[-2] &= d_1\,(0.922)^{-2}\,\cos\,(-1.4173) + d_2\,(0.922)^{-2}\,\sin\,(-1.4173) \\
&= 0.1893\,d_1 - 0.7456\,d_2 = 7
\end{aligned}
\tag{3.75}
$$

Solving Eqns. (3.74) and (3.75) yields $d_1 = 1.05$ and $d_2 = -5.8583$. The natural response of the system is

$$y_h[n] = 1.05\,(0.922)^n\,\cos\,(0.7086\,n)\,u[n] - 5.8583\,(0.922)^n\,\sin\,(0.7086\,n)\,u[n]$$

and is graphed in Fig. 3.17.

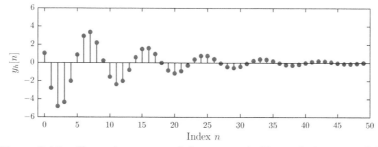

Figure 3.17 – Natural response of the system in Example 3.13 part (a).

b. For this system the characteristic equation is

$$z^2 - 1.6\,z + 0.64 = 0$$

The roots of the characteristic polynomial are $z_1 = z_2 = 0.8$. Therefore we must look for a homogeneous solution in the form of Eqn. (3.72):

$$y_h[n] = c_1\,(0.8)^n + c_2\,n\,(0.8)^n$$

for $n \geq 0$. Imposing the initial conditions at $n = -1$ and $n = -2$ we obtain

$$y_h[-1] = c_1 \, (0.8)^{-1} + c_2 \, (-1) \, (0.8)^{-1}$$

$$= 1.25 \, c_1 - 1.25 \, c_2 = 2 \tag{3.76}$$

and

$$y_h[-2] = c_1 \, (0.8)^{-2} + c_2 \, (-2) \, (0.8)^{-2}$$

$$= 1.5625 \, c_1 - 3.125 \, c_2 = -3 \tag{3.77}$$

Eqns. (3.76) and (3.77) can be solved to obtain the coefficient values $c_1 = 5.12$ and $c_2 = 3.52$. The natural response is

$$y_h[n] = 5.12 \, (0.8)^n + 3.52 \, n \, (0.8)^n$$

and is graphed in Fig. 3.18.

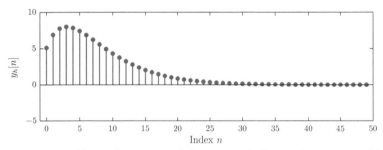

Figure 3.18 – Natural response of the system in Example 3.13 part (b).

Software resources:
ex_3_13a.m
ex_3_13b.m

Interactive Demo: nr_demo2

The interactive demo program **nr_demo2.m** illustrates different types of homogeneous solutions for a second-order discrete-time system based on the roots of the characteristic polynomial. Three possible scenarios were explored above, namely distinct real roots, complex conjugate roots and identical real roots.

In the demo program, the two roots can be specified in terms of their norms and angles using slider controls, and the corresponding natural response can be observed. If the roots are both real, then they can be controlled independently. If the roots are complex, then they move simultaneously to keep their complex conjugate relationship. The locations of the two roots z_1 and z_2 are marked on the complex plane. A circle with unit radius that is centered at the origin of the complex plane is also shown. The difference equation, the characteristic equation and the analytical solution for the natural response are displayed and updated as the roots are moved.

1. Start with two complex conjugate roots as given in part (a) of Example 3.13. Set the roots as

$$z_{1,2} = 0.922 \, e^{\pm 40.5998^\circ}$$

Set the initial values as they were set in the example, that is, $y[-1] = 5$ and $y[-2] = 7$. The natural response displayed should match the result obtained in Example 3.13.

2. Gradually increase the norm of the first root. Since the roots are complex, the second root will also change to keep the complex conjugate relationship of the roots. Observe the natural response as the norm of the complex roots become greater than unity and cross over the circle to the outside. What happens when the roots cross over?

3. Bring the norm of the roots back to $z_{1,2} = 0.922$. Gradually decrease the angle of z_1. The angle of z_2 will also change. How does this impact the shape of the natural response?

4. Set the angle of the z_1 equal to zero, so that the angle of z_2 also becomes zero, and the roots can be moved individually. Set the norms of the two roots as

$$|z_1| = 0.8 \quad \text{and} \quad |z_2| = 0.5$$

and observe the natural response.

5. Gradually increase $|z_1|$ and observe the changes in the natural response, especially as the root moves outside the circle.

Software resources:
nr_demo2.m

3.5.2   Finding the forced response of a discrete-time system

In the preceding section we focused our efforts on determining the homogeneous solution of the constant-coefficient linear difference equation or, equivalently, the natural response $y_h[n]$ of the system when no external input signal exists. As stated in Eqn. (3.46), the complete solution is the sum of the homogeneous solution with the particular solution that corresponds to the input signal applied to the system. The procedure for finding a particular solution for a difference equation is similar to that employed for a differential equation. We start with an educated guess about the form of the particular solution we seek, and then adjust the values of its parameters so that the difference equation is satisfied. The form of the particular solution picked should include the input signal $x[n]$ as well as the delayed input signals $x[n - k]$ that differ in form. For example, if the input signal is $x[n] = K \cos(\Omega_0 n)$, then we assume a particular solution in the form

$$y_p[n] = k_1 \cos(\Omega_0 n) + k_2 \sin(\Omega_0 n) \tag{3.78}$$

Both cosine and sine terms are needed since $x[n - 1]$ is in the form

$$\begin{aligned} x[n - 1] &= K \cos(\Omega_0 [n - 1]) \\ &= K \cos(\Omega_0) \cos(\Omega_0 n) + K \sin(\Omega_0) \sin(\Omega_0 n) \end{aligned}$$

Other delays of $x[n]$ do not produce any terms that differ from these. If the input signal is in the form $x[n] = n^m$ then the delays of $x[n]$ would contain the terms $n^{m-1}, n^{m-2}, \ldots, n^1, n^0$, and the particular solution is in the form

$$y_p[n] = k_m n^m + k_{m-1} n^{m-1} + \ldots + k_1 n + k_0 \tag{3.79}$$

Table 3.1 lists some of the common types of input signals and the forms of particular solutions to be used for them.

Input signal	Particular solution
K (constant)	k_1
$K\,e^{an}$	$k_1\,e^{an}$
$K\,\cos\left(\Omega_0 n\right)$	$k_1\,\cos\left(\Omega_0 n\right) + k_2\,\sin\left(\Omega_0 n\right)$
$K\,\sin\left(\Omega_0 n\right)$	$k_1\,\cos\left(\Omega_0 n\right) + k_2\,\sin\left(\Omega_0 n\right)$
$K\,n^m$	$k_m n^m + k_{m-1} n^{m-1} + \ldots + k_1 n + k_0$

Table 3.1 – Choosing a particular solution for various discrete-time input signals.

The unknown coefficients k_i of the particular solution are determined from the difference equation by assuming all initial conditions are zero (recall that the particular solution does not depend on the initial conditions of the difference equation, or the initial state of the system). Initial conditions of the difference equation are imposed in the subsequent step for determining the unknown coefficients of the homogeneous solution, not the coefficients of the particular solution. The procedure for determining the complete forced solution of the difference equation is summarized below:

1. Write the homogeneous difference equation, and then find the characteristic equation by replacing delays of the output signal with corresponding negative powers of the complex variable z.

2. Solve for the roots of the characteristic equation and write the homogeneous solution in the form of Eqn. (3.66). If some of the roots appear as complex conjugate pairs, then use the form in Eqn. (3.70) for those roots. If there are any multiple roots, use the procedure outlined in Eqn. (3.73). Leave the homogeneous solution in parametric form with undetermined coefficients; do not attempt to compute the coefficients $c_1, c_2, \ldots$ of the homogeneous solution yet.

3. Find the form of the particular solution by either picking the appropriate form of it from Table 3.1, or by constructing it as a linear combination of the input signal and its delays. (This latter approach requires that delays of the input signal produce a finite number of distinct signal forms.)

4. Try the particular solution in the non-homogeneous difference equation and determine the coefficients $k_1, k_2, \ldots$ of the particular solution. At this point the particular solution should be uniquely determined. However, the coefficients of the homogeneous solution are still undetermined.

5. Add the homogeneous solution and the particular solution together to obtain the total solution. Impose the necessary initial conditions and determine the coefficients $c_1, c_2, \ldots$ of the homogeneous solution.

The next two examples will illustrate these steps.

Example 3.14: **Forced response of exponential smoother for unit-step input**

Find the forced response of the exponential smoother of Example 3.6 when the input signal is a unit-step function, and $y[-1] = 2.5$.

Solution: In Example 3.11 the homogeneous solution of the difference equation for the exponential smoother was determined to be in the form

$$y_h[n] = c\,(1-\alpha)^n$$

For a unit-step input, the particular solution is in the form

$$y_p[n] = k_1$$

The particular solution must satisfy the difference equation. Substituting $y_p[n]$ into the difference equation we get

$$k_1 = (1-\alpha)\,k_1 + \alpha$$

and consequently $k_1 = 1$. The forced solution is the combination of homogeneous and particular solutions:

$$\begin{aligned}y[n] &= y_h[n] + y_p[n]\\ &= c\,(1-\alpha)^n + 1\end{aligned}$$

The constant c needs to be adjusted to satisfy the specified initial condition $y[-1] = 2.5$.

$$y[-1] = c\,(1-\alpha)^{-1} + 1 = 2.5$$

results in $c = 1.5\,(1-\alpha)$, and the forced response of the system is

$$\begin{aligned}y_h[n] &= 1.5\,(1-\alpha)\,(1-\alpha)^n + 1\\ &= 1.5\,(1-\alpha)^{n+1} + 1\;, \quad \text{for } n \geq 0\end{aligned}$$

This signal is shown in Fig. 3.19 for $\alpha = 0.1$.

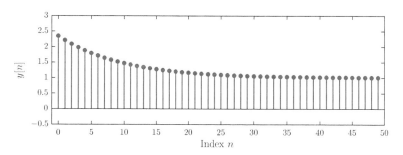

Figure 3.19 – Forced response of the system in Example 3.14.

Software resources:
ex_3_14.m

Example 3.15: **Forced response of exponential smoother for sinusoidal input**

Find the forced response of the exponential smoother of Example 3.6 when the input signal is a sinusoidal function in the form

$$x[n] = A\,\cos\,(\Omega n)$$

Use parameter values $A = 20$, $\Omega = 0.2\pi$, and $\alpha = 0.1$. The initial value of the output signal is $y[-1] = 2.5$.

Solution: Recall that the difference equation of the exponential smoother is

$$y[n] = (1 - \alpha) \, y[n - 1] + \alpha \, x[n]$$

The homogeneous solution is in the form

$$y_h[n] = c \, (1 - \alpha)^n$$

For a sinusoidal input signal, the form of the appropriate particular solution is obtained from Table 3.1 as

$$y_p[n] = k_1 \cos (\Omega n) + k_2 \sin (\Omega n) \tag{3.80}$$

The particular solution must satisfy the difference equation of the exponential smoother. Therefore we need

$$y_p[n] - (1 - \alpha) \, y_p[n - 1] = \alpha \, x[n] \tag{3.81}$$

The term $y_p[n - 1]$ is needed in Eqn. (3.81). Time-shifting both sides of Eqn. (3.80)

$$y_p[n - 1] = k_1 \cos (\Omega[n - 1]) + k_2 \sin (\Omega[n - 1]) \tag{3.82}$$

Using the appropriate trigonometric identities[2] Eqn. (3.82) can be written as

$$y_p[n - 1] = [k_1 \cos (\Omega) - k_2 \sin (\Omega)] \cos (\Omega n) + [k_1 \sin (\Omega) + k_2 \cos (\Omega)] \sin (\Omega n) \tag{3.83}$$

Let us define

$$\beta = 1 - \alpha$$

to simplify the notation. Substituting Eqns. (3.80) and (3.83) along with the input signal $x[n]$ into the difference equation in Eqn. (3.81) we obtain

$$[k_1 - \beta \, k_1 \cos (\Omega) + \beta \, k_2 \sin (\Omega)] \cos (\Omega n)$$
$$+ [k_2 - \beta \, k_1 \sin (\Omega) - \beta \, k_2 \cos (\Omega)] \sin (\Omega n)$$
$$= \alpha A \cos (\Omega n) \tag{3.84}$$

Since Eqn. (3.84) must be satisfied for all values of the index n, coefficients of $\cos (\Omega n)$ and $\sin (\Omega n)$ on both sides of the equal sign must individually be set equal to each other. This leads to the two equations

$$k_1 \, [1 - \beta \cos (\Omega)] + k_2 \, \beta \sin (\Omega) = A \tag{3.85}$$

and

$$-k_1 \, \beta \sin (\Omega) + k_2 \, [1 - \beta \cos (\Omega)] = 0 \tag{3.86}$$

Eqns. (3.85) and (3.86) can be solved for the unknown coefficients k_1 and k_2 to yield

$$k_1 = \frac{\alpha A \, [1 - \beta \cos (\Omega)]}{1 - 2 \, \beta \cos (\Omega) + \beta^2} \quad \text{and} \quad k_2 = \frac{\alpha A \, \beta \sin (\Omega)}{1 - 2 \, \beta \cos (\Omega) + \beta^2} \tag{3.87}$$

Now the forced solution of the system can be written by combining the homogeneous and particular solutions as

$$y[n] = y_h[n] + y_p[n]$$
$$= c \, \beta^n + \frac{\alpha A \, [1 - \beta \cos (\Omega)]}{1 - 2 \, \beta \cos (\Omega) + \beta^2} \cos (\Omega n) + \frac{\alpha A \, \beta \sin (\Omega)}{1 - 2 \, \beta \cos (\Omega) + \beta^2} \sin (\Omega n) \tag{3.88}$$

---

[2] $\cos (a + b) = \cos (a) \cos (b) - \sin (a) \sin (b)$ and $\sin (a + b) = \sin (a) \cos (b) + \cos (a) \sin (b)$.

Using the specified parameter values of $A = 20$, $\Omega = 0.2\pi$, $\alpha = 0.1$ and $\beta = 0.9$, the coefficients k_1 and k_2 are evaluated to be

$$k_1 = 1.5371 \quad \text{and} \quad k_2 = 2.9907$$

and the forced response of the system is

$$y[n] = c\,(0.9)^n + 1.5371\,\cos\,(0.2\pi n) + 2.9907\,\sin\,(0.2\pi n)$$

We need to impose the initial condition $y[-1] = 2.5$ to determine the remaining unknown coefficient c. For $n = -1$ the output signal is

$$y[-1] = c\,(0.9)^{-1} + 1.5371\,\cos\,(-0.2\pi) + 2.9907\,\sin\,(-0.2\pi)$$

$$= 1.1111\,c - 0.5144 = 2.5$$

Solving for the coefficient c yields to $c = 2.7129$. The forced response can now be written in complete form:

$$y[n] = 2,7129\,(0.9)^n + 1.5371\,\cos\,(0.2\pi n) + 2.9907\,\sin\,(0.2\pi n) \tag{3.89}$$

for $n \geq 0$. The forced response consists of two components. The first term in Eqn. (3.89) is the *transient response*

$$y_t[n] = 2,7129\,(0.9)^n \tag{3.90}$$

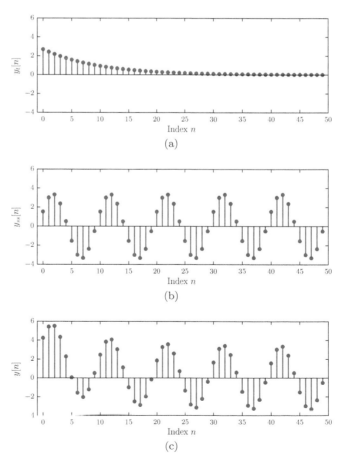

Figure 3.20 – The signals obtained in Example 3.15: (a) $y_t[n]$, (b) and $y_{ss}[n]$, (c) $y[n]$.

which is due to the initial state of the system. It disappears over time. The remaining terms in Eqn. (3.89) represent the *steady-state response* of the system:

$$y_{ss}[n] = 1.5371 \cos(0.2\pi n) + 2.9907 \sin(0.2\pi n) \tag{3.91}$$

Signals $y[n]$, $y_t[n]$ and $y_{ss}[n]$ are shown in Fig. 3.20.
Software resources:
ex_3_15.m

Interactive Demo: fr_demo2

The interactive demo program fr_demo2.m is based on Example 3.15, and allows experimentation with parameters of the problem. The amplitude A is fixed at $A = 20$ so that the input signal is

$$x[n] = 20 \cos(\Omega n)$$

The exponential smoother parameter α, the angular frequency Ω and the initial output value $y[-1]$ can be varied using slider controls. The effect of parameter changes on the transient response $y_t[n]$, the steady-state response $y_{ss}[n]$ and the total forced response $y_t[n] + y_{ss}[n]$ can be observed.

1. Start with the settings $\alpha = 0.1$, $\Omega = 0.2\pi = 0.62832$ radians, and $y[-1] = 2.5$. Observe the peak amplitude value of the steady-state component. Confirm that it matches with what was found in Example 3.15, Fig. 3.20.
2. Now gradually increase the angular frequency Ω up to $\Omega = 0.5\pi = 1.5708$ radians, and observe the change in the peak amplitude of the steady-state component of the output. Compare with the result obtained in Eqn. (3.88).
3. Set parameter values back to $\alpha = 0.1$, $\Omega = 0.2\pi = 0.62832$ radians, and $y[-1] = 2.5$. Pay attention to the transient response, and how many samples it takes for it to become negligibly small.
4. Gradually decrease the value of α toward $\alpha = 0.05$ and observe the changes in the transient behavior. How does the value of α impact the number of samples it takes for the output signal to reach steady state?

Software resources:
fr_demo2.m

3.6   Block Diagram Representation of Discrete-Time Systems

A discrete-time system can also be represented with a block diagram, and multiple solutions exist that are functionally equivalent. In this section we will discuss just one particular technique for obtaining a block diagram, and the discussion of other techniques will be deferred until Chapter 8.

Block diagrams are useful for discrete-time systems not only because they provide additional insight into the operation of a system, but also because they allow implementation of the system on a digital computer. We often use the block diagram as the first step in developing the computer code for implementing a discrete-time system. An example of this is given in MATLAB Exercise 3.5.

Three types of operators are utilized in the constant-coefficient linear difference equation of Eqn. (3.27): multiplication of a signal by a constant gain factor, addition of two signals, and time shift of a signal. Consequently, the fundamental building blocks for use in block diagrams of discrete-time systems are constant-gain amplifier, signal adder and one-sample delay element as shown in Fig. 3.21.

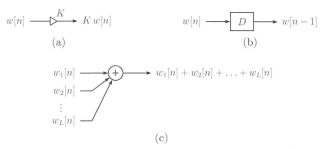

Figure 3.21 – Block diagram components for discrete-time systems: (a) constant-gain amplifier, (b) one-sample delay, (c) signal adder.

We will begin the discussion with a simple third-order difference equation expressed as

$$y[n] + a_1\, y[n-1] + a_2\, y[n-2] + a_3\, y[n-3] = b_0\, x[n] + b_1\, x[n-1] + b_2\, x[n-2] \quad (3.92)$$

This is a constant-coefficient linear difference equation in the standard form of Eqn. (3.27) with parameter values $N = 3$ and $M = 2$. Additionally, the coefficient of the term $y[n]$ is chosen to be $a_0 = 1$. This is not very restricting since, for the case of $a_0 \neq 1$, we can always divide both sides of the difference equation by a_0 to satisfy this condition. It can be shown that the following two difference equations that utilize an intermediate signal $w(t)$ are equivalent to Eqn. (3.92):

$$w[n] + a_1\, w[n-1] + a_2\, w[n-2] + a_3\, w[n-3] = x[n] \quad (3.93)$$

$$y[n] = b_0\, w[n] + b_1\, w[n-1] + b_2\, w[n-2] \quad (3.94)$$

The proof is straightforward. The terms on the right side of Eqn. (3.92) can be written using Eqn. (3.93) and its time-shifted versions. The following can be written:

$$b_0\, x[n] = b_0\, \big(w[n] + a_1\, w[n-1] + a_2\, w[n-2] + a_3\, w[n-3]\big) \quad (3.95)$$

$$b_1\, x[n-1] = b_1\, \big(w[n-1] + a_1\, w[n-2] + a_2\, w[n-3] + a_3\, w[n-4]\big) \quad (3.96)$$

$$b_2\, x[n-2] = b_2\, \big(w[n-2] + a_1\, w[n-3] + a_2\, w[n-4] + a_3\, w[n-5]\big) \quad (3.97)$$

We are now in a position to construct the right side of Eqn. (3.92) using Eqns. (3.95) through (3.97):

$$b_0\, x[n] + b_1\, x[n-1] + b_2\, x[n-2] =$$
$$b_0\, \big(w[n] + a_1\, w[n-1] + a_2\, w[n-2] + a_3\, w[n-3]\big)$$
$$+ b_1\, \big(w[n-1] + a_1\, w[n-2] + a_2\, w[n-3] + a_3\, w[n-4]\big)$$
$$+ b_2\, \big(w[n-2] + a_1\, w[n-3] + a_2\, w[n-4] + a_3\, w[n-5]\big) \quad (3.98)$$

Rearranging the terms of Eqn. (3.98) we can write it in the form

$$b_0\,x[n]+b_1\,x[n-1]+b_2\,x[n-2] =$$

$$\big(b_0\,w[n]+b_1\,w[n-1]+b_2\,w[n-2]\big)$$

$$+\,a_1\,\big(b_0\,w[n-1]+b_1\,w[n-2]+b_2\,w[n-3]\big)$$

$$+\,a_2\,\big(b_0\,w[n-2]+b_1\,w[n-3]+b_2\,w[n-4]\big)$$

$$+\,a_3\,\big(b_0\,w[n-3]+b_1\,w[n-4]+b_2\,w[n-5]\big) \qquad (3.99)$$

and recognizing that the terms on the right side of Eqn. (3.99) are time-shifted versions of $y[n]$ from Eqn. (3.94) we obtain

$$b_0\,x[n]+b_1\,x[n-1]+b_2\,x[n-2] = y[n]+a_1\,y[n-1]+a_2\,y[n-2]+a_3\,y[n-3] \quad (3.100)$$

which is the original difference equation given by Eqn. (3.92). Therefore, Eqns. (3.93) and (3.94) form an equivalent representation of the system described by Eqn. (3.92).

One possible block diagram implementation of the difference equation in Eqn. (3.93) is shown in Fig. 3.22. It takes the discrete-time signal $x[n]$ as input, and produces the intermediate signal $w[n]$ as output.

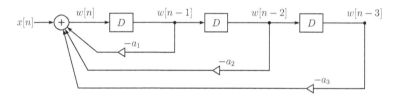

Figure 3.22 – The block diagram for Eqn. (3.94).

Three delay elements are used for obtaining the samples $w[n-1]$, $w[n-2]$ and $w[n-3]$. The intermediate signal $w[n]$ is then obtained via an adder that adds scaled versions of the three past samples of $w[n]$. Keep in mind that our ultimate goal is to obtain the signal $y[n]$ which is related to the intermediate signal $w[n]$ through Eqn. (3.94). The computation of $y[n]$ requires the knowledge of $w[n]$ as well as its two past samples $w[n-1]$ and $w[n-2]$, both of which are available in the block diagram of Fig. 3.22. The complete block diagram for the system is obtained by adding the necessary connections to the block diagram in Fig. 3.22, and is shown in Fig. 3.23.

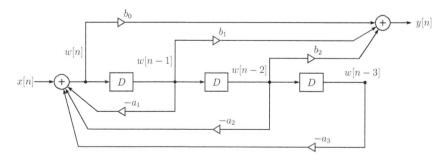

Figure 3.23 – The completed block diagram for Eqn. (3.92).

The development above was based on a third-order difference equation, however, the extension of the technique to a general constant-coefficient linear difference equation is straightforward. In Fig. 3.23 the feed-forward gains of the block diagram are the right-side coefficients $b_0, b_1, \ldots, b_M$ of the difference equation in Eqn. (3.92). Feedback gains of the block diagram are the negated left-side coefficients $-a_1, -a_2, \ldots, -a_N$ of the difference equation. Recall that we must have $a_0 = 1$ for this to work.

Imposing initial conditions

Initial conditions can easily be incorporated into the block diagram. The third-order difference equation given by Eqn. (3.92) would typically be solved subject to initial values specified for $y[-1]$, $y[-2]$ and $y[-3]$. For the block diagram we need to determine the corresponding values of $w[-1]$, $w[-2]$ and $w[-3]$ through the use of Eqns. (3.93) and (3.94). This will be illustrated in Example 3.16.

Example 3.16: **Block diagram for discrete-time system**

Construct a block diagram to solve the difference equation

$$y[n] - 0.7\,y[n-1] - 0.8\,y[n-2] + 0.84\,y[n-3] = 0.1\,x[n] + 0.2\,x[n-1] + 0.3\,x[n-2]$$

with the input signal $x[n] = u[n]$ and subject to initial conditions

$$y[-1] = 0.5\,, \quad y[-2] = 0.3\,, \quad y[-3] = -0.4$$

Solution: Using the intermediate variable $w[n]$ as outlined in the preceding discussion we can write the following pair of difference equations that are equivalent to the original difference equation:

$$w[n] - 0.7\,w[n-1] - 0.8\,w[n-2] + 0.84\,w[n-3] = x[n] \tag{3.101}$$

$$y[n] = 0.1\,w[n] + 0.2\,w[n-1] + 0.3\,w[n-2] \tag{3.102}$$

The block diagram can now be constructed as shown in Fig. 3.24.

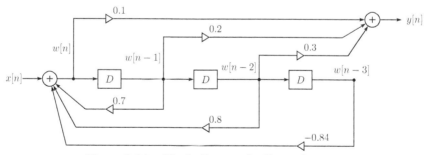

Figure 3.24 – Block diagram for Example 3.16.

Initial conditions specified in terms of the values of $y[-1]$, $y[-2]$ and $y[-3]$ need to be translated to corresponding values of $w[-1]$, $w[-2]$ and $w[-3]$. Writing Eqn. (3.102) for $n = -1, -2, -3$ yields the following three equations:

$$y[-1] = 0.1\,w[-1] + 0.2\,w[-2] + 0.3\,w[-3] = 0.5 \tag{3.103}$$

$$y[-2] = 0.1\,w[-2] + 0.2\,w[-3] + 0.3\,w[-4] = 0.3 \tag{3.104}$$

$$y[-3] = 0.1\,w[-3] + 0.2\,w[-4] + 0.3\,w[-5] = -0.4 \tag{3.105}$$

These equations need to be solved for $w[-1]$, $w[-2]$ and $w[-3]$. The two additional unknowns, namely $w[-4]$ and $w[-5]$, need to be obtained in other ways. Writing Eqn. (3.101) for $n = -1$ yields

$$w[-1] - 0.7\,w[-2] - 0.8\,w[-3] + 0.84\,w[-4] = x[-1] \tag{3.106}$$

Since $x[n] = u[n]$ we know that $x[-1] = 0$, therefore $w[-4]$ can be expressed as

$$w[-4] = -1.1905\,w[-1] + 0.8333\,w[-2] + 0.9524\,w[-3] \tag{3.107}$$

which can be substituted into Eqn. (3.104) to yield

$$y[-2] = -0.3571\,w[-1] + 0.35\,w[-2] + 0.4857\,w[-3] = 0.3 \tag{3.108}$$

Similarly, writing Eqn. (3.101) for $n = -2$ yields

$$w[-2] - 0.7\,w[-3] - 0.8\,w[-4] + 0.84\,w[-5] = x[-2] \tag{3.109}$$

We know that $x[-2] = 0$, therefore

$$
\begin{aligned}
w[-5] = &-1.1905\,w[-2] + 0.8333\,w[-3] + 0.9524\,w[-4] \\
= &-1.1905\,w[-2] + 0.8333\,w[-3] \\
&+ 0.9524\left(-1.1905\,w[-1] + 0.8333\,w[-2] + 0.9524\,w[-3]\right) \\
= &-1.1338\,w[-1] - 0.3968\,w[-2] + 1.7404\,w[-3] \tag{3.110}
\end{aligned}
$$

Substituting Eqns. (3.107) and (3.110) into Eqn. (3.105) we obtain

$$y[-3] = -0.5782\,w[-1] + 0.0476\,w[-2] + 0.8126\,w[-3] = -0.4 \tag{3.111}$$

Eqns. (3.103), (3.108) and (3.111) can be solved simultaneously to determine the initial conditions in terms of $w[n]$ as

$$w[-1] = 1.0682\,, \quad w[-2] = 1.7149\,, \quad w[-3] = 0.1674 \tag{3.112}$$

In the block diagram of Fig. 3.24 the outputs of the three delay elements should be set equal to these values before starting the simulation.

Interactive Demo: **dgm_demo1**

The interactive demo program **dgm_demo1.m** illustrates the solution of the difference equation of Example 3.16 through the use of the block diagram constructed and shown in Fig. 3.24. Two choices are given for the input signal $x[n]$: a unit-step signal and a periodic sawtooth signal both of which have $x[-1] = x[-2] = 0$. Numerical values of the node variables are shown on the block diagram. For $n = 0$, initial values $w[n-1] = w[-1]$, $w[n-2] = w[-2]$ and $w[n-3] = w[-3]$ are shown on the diagram as computed in Example 3.16.

Incrementing the sample index n by clicking the button to the right of the index field causes the node values to be updated in an animated fashion, illustrating the operation of the block diagram. Additionally, input and output samples of the system are shown as stem plots with the current samples indicated in red color.

Software resources:

dgm_demo1.m

Software resources: See MATLAB Exercise 3.5.

3.7   Impulse Response and Convolution

A constant-coefficient linear difference equation is sufficient for describing a DTLTI system such that the output signal of the system can be determined in response to any arbitrary input signal. However, we will often find it convenient to use additional description forms for DTLTI systems. One of these additional description forms is the impulse response which is simply the forced response of the system under consideration when the input signal is a unit impulse. This is illustrated in Fig. 3.25.

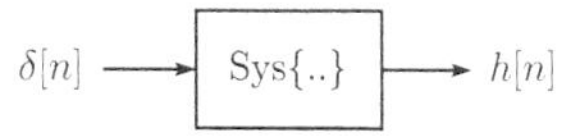

$$\delta[n] \longrightarrow \boxed{\text{Sys}\{..\}} \longrightarrow h[n]$$

Figure 3.25 – Computation of the impulse response for a DTLTI system.

The impulse response also constitutes a complete description of a DTLTI system. The response of a DTLTI system to any arbitrary input signal $x[n]$ can be uniquely determined from the knowledge of its impulse response.

In the next section we will discuss how the impulse response of a DTLTI system can be obtained from its difference equation. The reverse is also possible, and will be discussed in later chapters.

3.7.1   Finding impulse response of a DTLTI system

Finding the impulse response of a DTLTI system amounts to finding the forced response of the system when the forcing function is a unit impulse, i.e., $x[n] = \delta[n]$. In the case of a difference equation with no feedback, the impulse response is found by direct substitution of the unit impulse input signal into the difference equation. If the difference equation has feedback, then finding an appropriate form for the particular solution may be a bit more difficult. This difficulty can be overcome by finding the unit-step response of the system as an intermediate step, and then determining the impulse response from the unit-step response.

The problem of determining the impulse response of a DTLTI system from the governing difference equation will be explored in the next two examples.

Example 3.17: **Impulse response of moving average filters**

Find the impulse response of the length-2, length-4 and length-N moving average filters discussed in Examples 3.3, 3.4 and 3.5.

Solution: Let us start with the length-2 moving average filter. The governing difference equation is

$$y[n] = \frac{1}{2}\, x[n] + \frac{1}{2}\, x[n-1]$$

Let $h_2[n]$ denote the impulse response of the length-2 moving average filter (we will use the subscript to indicate the length of the window). It is easy to compute $h_2[n]$ is by setting

$x[n] = \delta[n]$ in the difference equation:

$$h_2[n] = \text{Sys}\{\delta[n]\} = \frac{1}{2}\delta[n] + \frac{1}{2}\delta[n-1]$$

The result can also be expressed in tabular form as

$$h_2[n] = \{\, 1/2,\ 1/2 \,\} \qquad\qquad (3.113)$$
$$\uparrow$$

Similarly, for a length-4 moving average filter with the difference equation

$$y[n] = \frac{1}{4}x[n] + \frac{1}{4}x[n-1] + \frac{1}{4}x[n-2] + \frac{1}{4}x[n-3]$$

the impulse response is

$$h_4[n] = \text{Sys}\{\delta[n]\} = \frac{1}{4}\delta[n] + \frac{1}{4}\delta[n-1] + \frac{1}{4}\delta[n-2] + \frac{1}{4}\delta[n-3]$$

or in tabular form

$$h_4[n] = \{\, 1/4,\ 1/4,\ 1/4,\ 1/4 \,\} \qquad\qquad (3.114)$$
$$\uparrow$$

These results are easily generalized to a length-N moving average filter. The difference equation of the length-N moving average filter is

$$y[n] = \sum_{k=0}^{N-1} x[n-k] \qquad\qquad (3.115)$$

Substituting $x[n] = \delta[n]$ into Eqn. (3.115) we get

$$h_N[n] = \text{Sys}\{\delta[n]\} = \sum_{k=0}^{N-1} \delta[n-k] \qquad\qquad (3.116)$$

The result in Eqn. (3.116) can be written in alternative forms as well. One of those alternative forms is

$$h_N[n] = \begin{cases} \dfrac{1}{N}\,, & n = 0,\ldots,N-1 \\ 0\,, & \text{otherwise} \end{cases}$$

and another one is

$$h_N[n] = \frac{1}{N}\left(u[n] - u[n-N]\right)$$

Example 3.18: **Impulse response of exponential smoother**

Find the impulse response of the exponential smoother of Example 3.6 with $y[-1] = 0$.

Solution: With the initial condition $y[-1] = 0$, the exponential smoother described by the difference equation in Eqn. (3.14) is linear and time-invariant (refer to the discussion in Section 3.4). These two properties will allow us to use superposition in finding its impulse response. Recall that the unit-impulse function can be expressed in terms of unit-step functions as

$$\delta[n] = u[n] - u[n-1]$$

As a result, the impulse response of the linear exponential smoother can be found through the use of superposition in the form

$$
\begin{aligned}
h[n] = \text{Sys}\{\delta[n]\} &= \text{Sys}\{u[n] - u[n-1]\} \\
&= \text{Sys}\{u[n]\} - \text{Sys}\{u[n-1]\}
\end{aligned} \tag{3.117}
$$

We will first find the response of the system to a unit-step signal. The homogeneous solution of the difference equation at hand was already found in Example 3.11 as

$$
y_h[n] = c\,(1-\alpha)^n
$$

For a unit-step input, the particular solution is in the form

$$
y_p[n] = k_1
$$

Using this particular solution in the difference equation we get

$$
k_1 = (1-\alpha)\,k_1 + \alpha
$$

which leads to $k_1 = 1$. Combining the homogeneous and particular solutions, the forced solution is found to be in the form

$$
\begin{aligned}
y[n] &= y_h[n] + y_p[n] \\
&= c\,(1-\alpha)^n + 1
\end{aligned} \tag{3.118}
$$

with coefficient c yet to be determined. If we now impose the initial condition $y[-1] = 0$ on the result found in Eqn. (3.118) we get

$$
y[-1] = c\,(1-\alpha)^{-1} + 1 = 0
$$

and consequently

$$
c = -\,(1-\alpha)
$$

Thus, the unit-step response of the linear exponential smoother is

$$
y[n] = \text{Sys}\{u[n]\} = 1 - (1-\alpha)\,(1-\alpha)^n
$$

for $n \geq 0$. In compact notation, this result can be written as

$$
y[n] = \left[1 - (1-\alpha)^{n+1}\right] u[n] \tag{3.119}
$$

Since the system is time-invariant, its response to a delayed unit-step input is simply a delayed version of $y[n]$ found in Eqn. (3.119):

$$
\begin{aligned}
\text{Sys}\{u[n-1]\} &= y[n-1] \\
&= [1 - (1-\alpha)^n]\,u[n-1]
\end{aligned} \tag{3.120}
$$

The impulse response of the linear exponential smoother is found using Eqns. (3.119) and (3.120) as

$$
h[n] = y[n] - y[n-1] = \alpha\,(1-\alpha)^n\,u[n] \tag{3.121}
$$

and is shown in Fig. 3.26 for $\alpha = 0.1$.

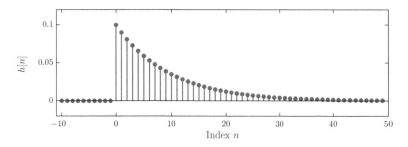

Figure 3.26 – Impulse response of the linear exponential smoother of Example 3.18 for $\alpha = 0.1$.

Software resources:
ex_3_18.m

3.7.2   Convolution operation for DTLTI systems

The development of the convolution operation for DTLTI systems will be based on the impulse decomposition of a discrete-time signal. It was established in Section 1.4.3 that any arbitrary discrete-time signal can be written as a sum of scaled and shifted impulse signals, leading to a decomposition in the form

$$x[n] = \sum_{k=-\infty}^{\infty} x[k]\,\delta[n-k] \tag{3.122}$$

Time-shifting a unit-impulse signal by k samples and multiplying it by the amplitude of the k-th sample of the signal $x[n]$ produces the signal $x_k[n] = x[k]\,\delta[n-k]$. Repeating this process for all integer values of k and adding the resulting signals leads to Eqn. (3.122).

If $x[n]$ is the input signal applied to a system, the output signal can be written as

$$y[n] = \text{Sys}\left\{x[n]\right\} = \text{Sys}\left\{\sum_{k=-\infty}^{\infty} x[k]\,\delta[n-k]\right\} \tag{3.123}$$

The input signal is the sum of shifted and scaled impulse functions. If the system under consideration is linear, then using the additivity rule given by Eqn. (3.5) we can write

$$y[n] = \sum_{k=-\infty}^{\infty} \text{Sys}\left\{x[k]\,\delta[n-k]\right\} \tag{3.124}$$

Furthermore, using the homogeneity rule given by Eqn. (3.6), Eqn. (3.124) becomes

$$y[n] = \sum_{k=-\infty}^{\infty} x[k]\,\text{Sys}\left\{\delta[n-k]\right\} \tag{3.125}$$

For the sake of discussion, let us assume that we know the system under consideration to be linear, but not necessarily time-invariant. The response of the linear system to any arbitrary input signal $x[n]$ can be computed through the use of Eqn. (3.125), provided that we already know the response of the system to impulse signals shifted in by all possible delay amounts. The knowledge necessary for determining the output of a linear system in response to an arbitrary signal $x[n]$ is

$$\left\{ \text{Sys}\{\delta[n-k]\} \;,\quad \text{all } k \right\}$$

Eqn. (3.125) provides us with a viable, albeit impractical, method of determining system output $y[n]$ for any input signal $x[n]$. The amount of the prerequisite knowledge that we must possess about the system to be able to use Eqn. (3.125) diminishes its usefulness. Things improve, however, if the system under consideration is also time-invariant.

Let the *impulse response* of the system be defined as

$$h[n] = \text{Sys}\{\delta[n]\} \tag{3.126}$$

If, in addition to being linear, the system is also known to be time-invariant, then the response of the system to any shifted impulse signal can be derived from the knowledge of $h[n]$ through

$$\text{Sys}\{\delta[n-k]\} = h[n-k] \tag{3.127}$$

consistent with the definition of time invariance given by Eqn. (3.10). This reduces the prerequisite knowledge to just the impulse response $h[n]$, and we can compute the output signal as

$$y[n] = \sum_{k=-\infty}^{\infty} x[k]\, h[n-k] \tag{3.128}$$

Eqn. (3.128) is known as the *convolution sum* for discrete-time signals. The output signal $y[n]$ of a DTLTI system is obtained by *convolving* the input signal $x[n]$ and the impulse response $h[n]$ of the system. This relationship is expressed in compact notation as

$$y[n] = x[n] * h[n] \tag{3.129}$$

where the symbol $*$ represents the *convolution operator*. We will show later in this section that the convolution operator is commutative, that is, the relationship in Eqn. (3.128) can also be written in the alternative form

$$y[n] = h[n] * x[n]$$
$$= \sum_{k=-\infty}^{\infty} h[k]\, x[n-k] \tag{3.130}$$

by swapping the roles of $h[n]$ and $x[n]$ without affecting the end result.

Discrete-time convolution summary:

$$y[n] = x[n] * h[n] = \sum_{k=-\infty}^{\infty} x[k]\, h[n-k]$$

$$= h[n] * x[n] = \sum_{k=-\infty}^{\infty} h[k]\, x[n-k]$$

Example 3.19: **A simple discrete-time convolution example**

A discrete-time system is described through the impulse response

$$h[n] = \{ \underset{\underset{n=0}{\uparrow}}{4} , 3, 2, 1 \}$$

Use the convolution operation to find the response of the system to the input signal

$$x[n] = \{ \underset{\underset{n=0}{\uparrow}}{-3} , 7, 4 \}$$

Solution: Consider the convolution sum given by Eqn. (3.128). Let us express the terms inside the convolution summation, namely $x[k]$ and $h[n-k]$, as functions of k.

$$x[k] = \{ \underset{\underset{k=0}{\uparrow}}{-3} , 7, 4 \}$$

$$h[-k] = \{ 1, 2, 3, \underset{\underset{k=0}{\uparrow}}{4} \}$$

$$h[n-k] = \{ 1, 2, 3, \underset{\underset{k=n}{\uparrow}}{4} \}$$

In its general form both limits of the summation in Eqn. (3.128) are infinite. On the other hand, $x[k] = 0$ for negative values of the summation index k, so setting the lower limit of the summation to $k = 0$ would have no effect on the result. Similarly, the last significant sample of $x[k]$ is at index $k = 2$, so the upper limit can be changed to $k = 2$ without affecting the result as well, leading to

$$y[n] = \sum_{k=0}^{2} x[k]\, h[n-k] \tag{3.131}$$

If $n < 0$, we have $h[n-k] = 0$ for all terms of the summation in Eqn. (3.131), and the output amplitude is zero. Therefore we will only concern ourselves with samples for which $n \geq 0$. The factor $h[n-k]$ has significant samples in the range

$$0 \leq n - k \leq 3$$

which can be expressed in the alternative form

$$n - 3 \leq k \leq n$$

The upper limit of the summation in Eqn. (3.131) can be set equal to $k = n$ without affecting the result, however, if $n > 2$ then we should leave it at $k = 2$. Similarly, the lower limit can be set to $k = n - 3$ provided that $n - 3 > 0$, otherwise it should be left at $k = 0$. So, a compact version of the convolution sum adapted to the particular signals of this example would be

$$y[n] = \sum_{k=\max(0,n-3)}^{\min(2,n)} x[k]\, h[n-k] , \quad \text{for } n \geq 0 \tag{3.132}$$

where the lower limit is the larger of $k = 0$ and $k = n-3$, and the upper limit is the smaller of $k = 2$ and $k = n$. We will use this result to compute the convolution of $x[n]$ and $h[n]$.

For $n = 0$:

$$y[0] = \sum_{k=0}^{0} x[k]\, h[0-k]$$

$$= x[0]\, h[0] = (-3)\,(4) = -12$$

For $n = 1$:

$$y[1] = \sum_{k=0}^{1} x[k]\, h[1-k]$$

$$= x[0]\, h[1] + x[1]\, h[0]$$
$$= (-3)\,(3) + (7)\,(4) = 19$$

For $n = 2$:

$$y[2] = \sum_{k=0}^{2} x[k]\, h[2-k]$$

$$= x[0]\, h[2] + x[1]\, h[1] + x[2]\, h[0]$$
$$= (-3)\,(2) + (7)\,(3) + (4)\,(4) = 31$$

For $n = 3$:

$$y[3] = \sum_{k=0}^{2} x[k]\, h[3-k]$$

$$= x[0]\, h[3] + x[1]\, h[2] + x[2]\, h[1]$$
$$= (-3)\,(1) + (7)\,(2) + (4)\,(3) = 23$$

For $n = 4$:

$$y[4] = \sum_{k=1}^{2} x[k]\, h[4-k]$$

$$= x[1]\, h[3] + x[2]\, h[2]$$
$$= (7)\,(1) + (4)\,(2) = 15$$

For $n = 5$:

$$y[5] = \sum_{k=2}^{2} x[k]\, h[5-k]$$

$$= x[2]\, h[3] = (4)\,(1) = 4$$

Thus the convolution result is

$$y[n] = \{\, \underset{\underset{n=0}{\uparrow}}{-12}\,,\, 19,\, 31,\, 23,\, 15,\, 4 \,\}$$

Example 3.20: **Simple discrete-time convolution example revisited**

Rework the convolution problem of Example 3.19 with the following modifications applied to the two signals:

$$h[n] = \{ \underset{\underset{n=N_2}{\uparrow}}{4} , 3, 2, 1 \}$$

and

$$x[n] = \{ \underset{\underset{n=N_1}{\uparrow}}{-3} , 7, 4 \}$$

Assume the starting indices N_1 and N_2 are known constants.

Solution: We need to readjust limits of the summation index. The function $x[k]$ has significant samples in the range

$$N_1 \leq k \leq N_1 + 2 \tag{3.133}$$

For the function $h[n-k]$, the significant range of the index k is found from the inequality

$$N_2 \leq n - k \leq N_2 + 3$$

the terms of which can be rearranged to yield

$$n - N_2 - 3 \leq k \leq n - N_2 \tag{3.134}$$

Using the inequalities in Eqns. (3.133) and (3.134) the convolution sum can be written as

$$y[n] = \sum_{k=K_1}^{K_2} x[k] \, h[n-k] \tag{3.135}$$

with the limits

$$K_1 = \max \left(N_1, n - N_2 - 3 \right) , \quad \text{and} \quad K_2 = \min \left(N_1 + 2, n - N_2 \right) , \tag{3.136}$$

For example, suppose we have $N_1 = 5$ and $N_2 = 7$. Using Eqn. (3.135) with the limits in Eqn. (3.136) we can write the convolution sum as

$$y[n] = \sum_{k=\max(5,n-10)}^{\min(7,n-7)} x[k] \, h[n-k] \tag{3.137}$$

For the summation to contain any significant terms, the lower limit must not be greater than the upper limit, that is, we need

$$\max \left(5, n - 10 \right) \leq \min \left(7, n - 7 \right)$$

As a result, the leftmost significant sample of $y[n]$ will occur at index $n = 12$. In general, it can be shown (see Problem 3.23 at the end of this chapter) that the leftmost significant sample will be at the index $n = N_1 + N_2$. The sample $y[12]$ is computed as

$$y[12] = \sum_{k=5}^{5} x[k] \, h[12-k]$$

$$= x[5] \, h[7] = (-3) \, (4) = -12$$

Other samples of $y[n]$ can be computed following the procedure demonstrated in Example 3.19 and yield the same pattern of values with the only difference being the starting index. The complete solution is

$$y[n] = \{\,-\underset{\underset{n=12}{\uparrow}}{12}\,,\,19,\,31,\,23,\,15,\,4\,\}$$

Software resources: See MATLAB Exercise 3.7.

In computing the convolution sum it is helpful to sketch the signals. The graphical steps involved in computing the convolution of two signals $x[n]$ and $h[n]$ at a specific index value n can be summarized as follows:

1. Sketch the signal $x[k]$ as a function of the independent variable k. This corresponds to a simple name change on the independent variable, and the graph of the signal $x[k]$ appears identical to the graph of the signal $x[n]$. (See Fig. 3.27.)

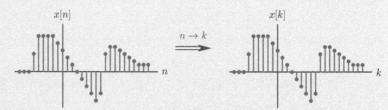

Figure 3.27 – Obtaining $x[k]$ for the convolution sum.

2. For one specific value of n, sketch the signal $h[n-k]$ as a function of the independent variable k. This task can be broken down into two steps as follows:

   a. Sketch $h[-k]$ as a function of k. This step amounts to time-reversal of the signal $h[k]$.

   b. In $h[-k]$ substitute $k \rightarrow k-n$. This step yields

$$h[-k]\Big|_{k \rightarrow k-n} = h[n-k] \qquad\qquad (3.138)$$

   and amounts to time-shifting $h[-k]$ by n samples.

   See Fig. 3.28 for an illustration of the steps for obtaining $h[n-k]$.

3. Multiply the two signals sketched in 1 and 2 to obtain the product $x[k]\,h[n-k]$.

4. Sum the sample amplitudes of the product $x[k]\,h[n-k]$ over the index k. The result is the amplitude of the output signal at the index n.

5. Repeat steps 1 through 4 for all values of n that are of interest.

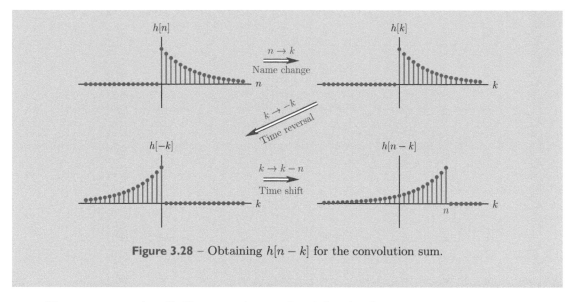

Figure 3.28 – Obtaining $h[n - k]$ for the convolution sum.

The next example will illustrate the graphical details of the convolution operation for DTLTI systems.

Example 3.21: **A more involved discrete-time convolution example**

A discrete-time system is described through the impulse response

$$h[n] = (0.9)^n \, u[n]$$

Use the convolution operation to find the response of a system to the input signal

$$x[n] = u[n] - u[n - 7]$$

Signals $h[n]$ and $x[n]$ are shown in Fig. 3.29.

(a)   (b)

Figure 3.29 – The signals for Example 3.21: (a) Impulse response $h[n]$, (b) input signal $x[n]$.

Solution: Again we will find it useful to sketch the functions $x[k]$, $h[n - k]$ and their product before we begin evaluating the convolution result. Such a sketch is shown in Fig. 3.30. It reveals three distinct possibilities for the overlap of $x[k]$ and $h[n - k]$. The convolution sum needs to be set up for each of the three regions of index n.

Case I: $n < 0$

There is no overlap between the two functions in this case, therefore their product equals zero for all values of k. The output signal is

$$y[n] = 0, \quad \text{for} \quad n < 0$$

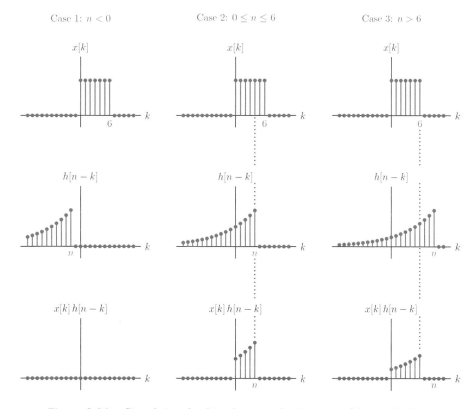

Figure 3.30 – Signals involved in the convolution sum of Example 3.21.

Case 2: $0 \leq n \leq 6$

For this case, the two functions overlap for the range of the index $0 \leq k \leq n$. Setting summation limits accordingly, the output signal can be written as

$$y[n] = \sum_{k=0}^{n} (1) \, (0.9)^{n-k} \qquad (3.139)$$

The expression in Eqn. (3.139) can be simplified by factoring out the common term $(0.9)^{n}$ and using the geometric series formula (see Appendix C) to yield

$$y[n] = (0.9)^{n} \sum_{k=0}^{n} (0.9)^{-k}$$

$$= (0.9)^{n} \, \frac{1 - (0.9)^{-(n+1)}}{1 - (0.9)^{-1}}$$

$$= -9 \left[(0.9)^{n} - \frac{1}{0.9} \right] , \quad \text{for} \quad 0 \leq n \leq 6 \qquad (3.140)$$

Case 3: $n > 6$

For $n > 6$ the overlap of the two functions will occur in the range $0 \leq k \leq 6$, so the summation limits need to be adjusted.

$$y[n] = \sum_{k=0}^{6} (1)\,(0.9)^{n-k} \tag{3.141}$$

Again factoring out the $(0.9)^n$ term and using the geometric series formula we get

$$y[n] = (0.9)^n \sum_{k=0}^{6} (0.9)^{-k}$$

$$= (0.9)^n \frac{1 - (0.9)^{-7}}{1 - (0.9)^{-1}}$$

$$= 9.8168\,(0.9)^n \,, \quad \text{for} \quad n > 6 \tag{3.142}$$

Thus, we have computed the output signal in each of the three distinct intervals we have identified. Putting these three partial solutions together, the complete solution for the output signal is obtained as

$$y[n] = \begin{cases} 0\,, & n < 0 \\ -9\left[0.9^n - \dfrac{1}{0.9}\right]\,, & 0 \le n \le 6 \\ 9.8168\,(0.9)^n\,, & n > 6 \end{cases} \tag{3.143}$$

and is shown in Fig. 3.31.

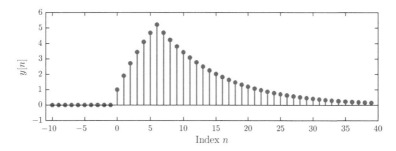

Figure 3.31 – Convolution result for Example 3.21.

Software resources:
ex_3_21.m

Interactive Demo: conv_demo5.m

The demo program "conv_demo5.m" is based on the discrete-time convolution problem in Example 3.21. It facilitates visualization of the overlapping samples between the functions $x[k]$ and $h[n-k]$ as the index k is varied. The impulse response used in the demo program is in the general form

$$h[n] = a^n\,u[n]$$

and can be made identical to the impulse response used in Example 3.21 by setting $a = 0.9$.
Software resources:
conv_demo5.m

Software resources: See MATLAB Exercise 3.8.

3.8   Causality in Discrete-Time Systems

Causality is an important feature of physically realizable systems. A system is said to be causal if the current value of the output signal depends only on current and past values of the input signal, but not on its future values. For example, a discrete-time system defined by the relationship

$$y[n] = y[n-1] + x[n] - 3\,x[n-1] \tag{3.144}$$

is causal, and one defined by

$$y[n] = y[n-1] + x[n] - 3\,x[n+1] \tag{3.145}$$

is non-causal. Causal systems can be implemented in *real-time processing* mode where a sample of the output signal is computed in response to each incoming sample of the input signal. On the order hand, implementation of non-causal systems may only be possible in *post processing* mode where the entire input signal must be observed and recorded before processing can begin.

In the case of a DTLTI system, the causality property can easily be related to the impulse response. Recall that the output signal $y[n]$ of a DTLTI system can be computed as the convolution of its input signal $x[n]$ with its impulse response $h[n]$ as

$$y[n] = h[n] * x[n] = \sum_{k=-\infty}^{\infty} h[k]\,x[n-k] \tag{3.146}$$

For the system under consideration to be causal, the computation of $y[n]$ should not require any future samples of the input signal. Thus, the term $x[n-k]$ in the summation should not contain index values in the future. This requires

$$n - k \leq n \quad \Longrightarrow \quad k \geq 0 \tag{3.147}$$

The product $h[k]\,x[n-k]$ should not have any non-zero values for $k < 0$. Therefore, the impulse response $h[n]$ should be equal to zero for all negative index values:

$$h[n] = 0 \quad \text{for all} \quad n < 0 \tag{3.148}$$

For a causal DTLTI system, the convolution relationship in Eqn. (3.146) can be written in the right-sided form by setting the lower limit of the summation index equal to zero:

$$y[n] = \sum_{k=0}^{\infty} h[k]\,x[n-k] \tag{3.149}$$

3.9   Stability in Discrete-Time Systems

A system is said to be *stable* in the *bounded-input bounded-output (BIBO)* sense if any bounded input signal is guaranteed to produce a bounded output signal.

A discrete-time input signal $x[n]$ is said to be bounded if an upper bound B_x exists such that

$$|x[n]| < B_x < \infty \tag{3.150}$$

for all values of the integer index n. A discrete-time system is stable if a finite upper bound B_y exists for the output signal in response to any input signal bounded as described in Eqn. (3.150).

For stability of a discrete-time system:

$$|x[n]| < B_x < \infty \quad \text{implies that} \quad |y[n]| < B_y < \infty \tag{3.151}$$

If the system under consideration is DTLTI, it is possible to relate the stability condition given by Eqn. (3.151) to the impulse response of the system as well as its difference equation. Derivation of the necessary condition follows:

The output signal of a DTLTI system is found from its input signal and impulse response through the use of the convolution sum expressed as

$$y[n] = \sum_{k=-\infty}^{\infty} h[k]\, x[n-k] \tag{3.152}$$

The absolute value of the output signal is

$$|y[n]| = \left| \sum_{k=-\infty}^{\infty} h[k]\, x[n-k] \right| \tag{3.153}$$

Absolute value of a sum is less than or equal to the sum of the absolute values, so an inequality involving $y[n]$ can be written as

$$|y[n]| \leq \sum_{k=-\infty}^{\infty} |h[k]\, x[n-k]| \tag{3.154}$$

The summation term in Eqn. (3.154) can be expressed as

$$|h[k]\, x[n-k]| = |h[k]|\, |x[n-k]| \tag{3.155}$$

and the inequality in Eqn. (3.154) becomes

$$|y[n]| \leq \sum_{k=-\infty}^{\infty} |h[k]|\, |x[n-k]| \tag{3.156}$$

Replacing the term $|x[n-k]|$ in Eqn. (3.156) with B_x makes the term on the right side of the inequality even greater, so we have

$$|y[n]| \leq \sum_{k=-\infty}^{\infty} |h[k]|\, B_x \tag{3.157}$$

or, by factoring out the common factor B_x

$$|y[n]| \leq B_x \sum_{k=-\infty}^{\infty} |h[k]| < \infty \tag{3.158}$$

Eqn. (3.158) leads us to the conclusion

$$\sum_{k=-\infty}^{\infty} |h[k]| < \infty \qquad (3.159)$$

For a DTLTI system to be stable, its impulse response must be *absolute summable*.

Example 3.22: **Stability of a length-2 moving-average filter**

Comment on the stability of the length-2 moving-average filter described by the difference equation

$$y[n] = \frac{1}{2} x[n] + \frac{1}{2} x[n-1]$$

Solution: We will approach this problem in two different ways. First, let's check directly to see if any arbitrary bounded input signal is guaranteed to produce a bounded output signal. The absolute value of the output signal is

$$|y[n]| = \left| \frac{1}{2} x[n] + \frac{1}{2} x[n-1] \right| \qquad (3.160)$$

Since absolute value of a sum is less than or equal to sum of absolute values, we can derive the following inequality from Eqn. (3.160):

$$|y[n]| \leq \frac{1}{2} |x[n]| + \frac{1}{2} |x[n-1]| \qquad (3.161)$$

Furthermore, since we assume $|x[n]| < B_x$ for all n, replacing $|x[n]|$ and $|x[n-1]|$ terms on the right side of the inequality in Eqn. (3.161) does not affect the validity of the inequality, so we have

$$|y[n]| \leq \frac{1}{2} B_x + \frac{1}{2} B_x = B_x \qquad (3.162)$$

proving that the output signal $y[n]$ is bounded as long as the input signal $x[n]$ is.

An alternative way to attack the problem would be to check the impulse response for the absolute summability condition in Eqn. (3.159). The impulse response of the length-2 moving-average filter is

$$h[n] = \frac{1}{2} \delta[n] + \frac{1}{2} \delta[n-1] \qquad (3.163)$$

which is clearly summable in an absolute sense.

Example 3.23: **Stability of the loan balance system**

Comment on the stability of the loan balance system discussed in Example 3.7.

Solution: Recall that the governing difference equation is

$$y[n] = (1+c)\, y[n-1] - x[n]$$

where c is a positive constant interest rate. This system can be analyzed in terms of its stability using the same approach that we have used in the previous example. However, we will take a more practical approach and analyze the system from a layperson's perspective.

What if we borrowed the money and never made a payment? The loan balance would keep growing each period. If the initial value of the output signal is $y[0] = A$ and if $x[n] = 0$ for $n \geq 1$, then we would have

$$y[1] = (1 + c)\, A$$

$$y[2] = (1 + c)^2\, A$$

$$\vdots$$

$$y[n] = (1 + c)^n\, A$$

which indicates that the system is unstable since we were able to find at least one bounded input signal that leads to an unbounded output signal.

It is also worth noting that the characteristic equation is

$$z - (1 + c) = 0$$

and its only solution is at $z_1 = 1 + c$.

The stability of a DTLTI system can also be associated with the modes of the difference equation that governs its behavior. We have seen in Section 3.5.1 that, for a causal DTLTI system, real and complex roots z_k of the characteristic polynomial for which $|z_k| \geq 1$ are associated with unstable behavior. Thus, for a causal DTLTI system to be stable, the magnitudes of all roots of the characteristic polynomial must be less than unity. If a circle is drawn on the complex plane with its center at the origin and its radius equal to unity, all roots of the characteristic polynomial must lie inside the circle for the corresponding causal DTLTI system to be stable.

A side note: In describing the associations between the roots of the characteristic polynomial and stability, we referred to a *causal* DTLTI system. If we were to consider an *anti-causal* system, the impulse response of which proceeds in the negative direction toward $n \to -\infty$, then the associations described above would have to be reversed. In that case roots outside the unit circle of the complex plane would lead to stable behavior.

Why is stability important for a DTLTI system?

An unstable DTLTI system is one that is capable of producing an unbounded output signal in response to at least some bounded input signals. What is the practical significance of this? Since we are dealing with numerical algorithms implemented via software on a computer, an unstable algorithm may lead to an output signal containing numbers that keep growing until they reach magnitudes that can no longer be handled by the number representation conventions of the processor or the operating system. When this occurs, the software that implements the system ceases to function in the proper way.

3.10   Further Reading

[1] S. Elaydi. *An Introduction to Difference Equations.* Springer Undergraduate Texts in Mathematics and Technology. Springer, 2005.

[2] W.G. Kelley and A.C. Peterson. *Difference Equations: An Introduction With Applications.* Harcourt/Academic Press, 2001.

[3] V. Lakshmikantham. *Theory of Difference Equations Numerical Methods and Applications*. Marcel Dekker, 2002.

[4] R.E. Mickens. *Difference Equations: Theory and Applications*. Van Nostrand Reinhold Company, 1990.

MATLAB Exercises

MATLAB Exercise 3.1: Writing functions for moving average filters

In this exercise we will develop MATLAB functions for implementing the length-2 and length-4 moving average filters that were discussed in Examples 3.4 and 3.5 respectively. As stated before, our purpose is not the development of fastest and most efficient functions. MATLAB already includes some very powerful functions for implementing a wide variety of discrete-time systems, and those could certainly be used for the simulation of moving average filters as well. The use of two of these built-in MATLAB functions, namely `conv(..)` and `filter(..)`, will be illustrated in later exercises. In this exercise, our purpose in developing our own functions for simulating moving average filters is to provide further insight into their operation. Therefore, we will sacrifice execution speed and efficiency in favor of a better understanding of coding a difference equation for a moving average filter.

Before developing any MATLAB function, it would be a good idea to consider how that function would eventually be used. Our development of the MATLAB code in this exercise will parallel the idea of *real-time processing* where the input and the output signals exist in the form of "data streams" of unspecified length. Algorithmically, the following steps will be executed:

1. Pick the *current input sample* from the input stream.
2. Process the *current input sample* through the moving average filter to compute the *current output sample*.
3. Put the *current output sample* into the output stream.
4. Repeat steps 1 through 3.

Let us begin by developing a function named "`ss_movavg2`" to implement step 2 above for a length-2 moving average filter. The function will be used with the syntax

```
y = ss_movavg2(x)
```

where "x" represents the *current input sample*, and "y" is the *current output sample*. Once developed, the function can be placed into the loop described above to simulate a length-2 moving average filter with any input data stream.

The difference equation of a length-2 moving average filter is

$$y[n] = \frac{1}{2}x[n] + \frac{1}{2}x[n-1]$$

A local variable named "xnm1" will be used for keeping track of the previous sample $x[n-1]$. The following two lines would compute the output sample and update the previous input sample for future use.

```
y = (x+xnm1)/2;
xnm1 = x;
```

Herein lies our first challenge: In MATLAB, local variables created within functions are discarded when the function returns. The next time the same function is called, its local variable "**xnm1**" would not have the value previously placed in it. The solution is to declare the variable "**xnm1**" to be a *persistent* variable, that is, a variable that retains its value between function calls. Following is a listing of the completed function ss_movavg2(..):

```
1    function y=ss_movavg2(x)
2      persistent xnm1;
3      if isempty(xnm1)
4        xnm1 = 0;
5      end;
6      y = (x+xnm1)/2;
7      xnm1 = x;
```

Another detail is that the first time a persistent variable is created (the first time the function ss_movavg2(..) is called), it is created as an empty matrix. Lines 3 through 5 of the code above set the variable "**xnm1**" equal to zero if it happens to be empty.

The function ss_movavg4(..) can be developed in a similar manner. Consider Eqn. (3.13). In addition to $x[n-1]$ we also need to keep track of two input samples prior to that, namely $x[n-2]$ and $x[n-3]$. Therefore we will define two additional local variables with the names "**xnm2**" and "**xnm3**", and declare them to be persistent as well. The code for the function ss_movavg4(..) is listed below.

```
1    function y = ss_movavg4(x)
2      persistent xnm1 xnm2 xnm3;
3      if isempty(xnm1)
4        xnm1 = 0;
5      end;
6      if isempty(xnm2)
7        xnm2 = 0;
8      end;
9      if isempty(xnm3)
10       xnm3 = 0;
11     end;
12     y = (x+xnm1+xnm2+xnm3)/4;
13     xnm3 = xnm2;
14     xnm2 = xnm1;
15     xnm1 = x;
```

In lines 13 through 15 of the code we perform bookkeeping, and prepare the three local variables for the next call to the function. The value $x[n]$ we have now will be the "previous sample" $x[n-1]$ the next time this function is called, and similar logic applies to the other two variables as well. Therefore the values of {"x", "xnm1", and "xnm2"} need to be moved into {"xnm1", "xnm2", and "xnm3"} respectively. In doing this, the order of assignments is critical. Had we used the order

```
13     xnm1 = x;
14     xnm2 = xnm1;
15     xnm3 = xnm2;
```

all three variables would have ended up (incorrectly) with the same value.

A final detail regarding persistent variables is that sometimes we may need to clear them from memory. Suppose we want to use the function ss_movavg2(..) with another data stream, and need to clear any persistent variables left over from a previous simulation. The following command accomplishes that:

```
>> clear ss_movavg2
```

Software resources:

ss_movavg2.m
ss_movavg4.m

MATLAB Exercise 3.2: Testing the functions written in MATLAB Exercise 3.1

In this exercise we will develop the code for testing the moving average filtering functions developed in Exercise 3.1. Daily values of the Dow Jones Industrial Average data for the years 2001 through 2003 are available in MATLAB data file "djia.mat" in MATLAB variables "x2001", "x2002" and "x2003". The following script computes the length-4 moving average for the 2003 data, and graphs both the input and the output signals.

```
1   % Script matex_3_2a.m
2   %
3   load 'djia.mat';        % Load the input data stream.
4   output = [];            % Create an empty output stream.
5   clear ss_movavg4;       % Clear persistent variables.
6   nsamp = length(x2003);  % Number of samples in the input stream.
7   for n=1:nsamp
8     x = x2003(n);         % "x" is the current input sample.
9     y = ss_movavg4(x);    % "y" is the current output sample.
10    output = [output,y];  % Append "y" to the output stream.
11  end;
12  % Graph input and output signals.
13  clf;
14  plot([1:nsamp],x2003,[1:nsamp],output);
15  axis([1,252,7500,10500]);
```

Note that the function ss_movavg4(..) begins with its persistent variables "xnm1", "xnm2", "xnm3" each set equal to 0. As a result, the first three samples of the output will be inaccurate. This problem can be alleviated by prepending the last three samples of the year 2002 data to the year 2003 data before we begin processing.

The MATLAB script above can be easily modified to simulate a length-2 moving average filter by substituting function ss_movavg2(..) in place of function ss_movavg4(..).

Software resources:

matex_3_2a.m
matex_3_2b.m
djia.mat

MATLAB Exercise 3.3: Writing and testing a function for the exponential smoother

Consider the exponential smoother introduced in Example 3.6 with the difference equation

$$y[n] = (1 - \alpha)\, y[n-1] + \alpha\, x[n]$$

We will implement the exponential smoother in MATLAB and test it with the Dow Jones Industrial Average data using the same algorithmic approach as in earlier exercises:

1. Pick the *current input sample* from the input stream.
2. Process the *current input sample* through the moving average filter to compute the *current output sample*.
3. Put the *current output sample* into the output stream.
4. Repeat steps 1 through 3.

A function named "ss_expsmoo" will be developed for implementing step 2 of the algorithm. The syntax of the function is

```
y = ss_expsmoo(x,alpha)
```

where "x" represents the *current input sample*, and "y" is the *current output sample*. The parameter "alpha" belongs to the exponential smoother. A persistent variable named "ynm1" is utilized for keeping track of the *previous output sample* $y[n-1]$. The code for the function ss_expsmoo(..) is listed below:

```
1   function y = ss_expsmoo(x,alpha)
2     persistent ynm1;
3     if isempty(ynm1)
4       ynm1 = 0;
5     end;
6     y = (1-alpha)*ynm1+alpha*x;
7     ynm1 = y;
```

The following script computes the output of the exponential smoother for the 2003 Dow Jones data as input, and graphs both the input and the output signals.

```
1   % Script: matex_3_3.m
2   %
3   load 'djia.mat';        % Load the input data stream.
4   output = [];            % Create an empty output stream.
5   clear ss_expsmoo;       % Clear persistent variables.
6   nsamp = length(x2003);  % Number of samples in the input stream.
7   alpha = 0.1;            % Parameter for exponential smoother.
8   for n=1:nsamp
9     x = x2003(n);         % "x" is the current input sample.
10    y = ss_expsmoo(x,alpha);   % "y" is the current output sample.
11    output = [output,y];  % Append "y" to the output stream.
12  end;
13  % Graph input and output signals.
14  clf;
15  plot([1:nsamp],x2003,[1:nsamp],output);
16  axis([1,252,7500,10500]);
```

Software resources:
matex_3_3.m
ss_expsmoo.m

MATLAB Exercise 3.4: Iteratively solving a difference equation

It was shown in Section 3.5 that one way to solve a difference equation is to iterate through it one sample index at a time. In this exercise we will explore this concept in MATLAB, using the difference equation for the loan balance problem introduced in Example 3.7. While the specific problem at hand is quite simplistic, it will help us highlight some of the challenges encountered in adapting a signal-system interaction problem for MATLAB coding, in particular with the vector indexing scheme of MATLAB.

The system under consideration accepts monthly payment data $x[n]$ as input and produces an output signal $y[n]$ that represents the balance at the end of each month. The difference equation for the system was given by Eqn. (3.15) and will be repeated here:

$$\text{Eqn. (3.15):} \qquad y[n] = (1+c)\, y[n-1] + x[n]$$

Let the amount borrowed be $10,000$ which we will place into $y[0]$ as the initial value of the output signal. Let us also assume that 100 will be paid each month starting with month 1. Thus, the input signal would be

$$x[n] = 100\, u[n-1]$$

Using a monthly interest rate of 0.5 percent, meaning $c = 0.5/100 = 0.005$, we would like to compute the output signal for $n = 1, \ldots, 18$. Following statements create two vectors "xvec" and "yvec":

```
>>  xvec = [0,100*ones(1,18)];
>>  yvec = [10000,zeros(1,18)];
```

The length-19 vector "xvec" contains payment data. Its first element is zero, and all other elements are equal to 100. The vector "yvec" holds samples of the output signal. Its first element is initialized to 10000, the loan amount, and other elements are set equal to zero as place holders to be modified later.

One of the difficulties encountered in representing discrete-time signals with vectors is that MATLAB vectors do not have an element with index 0; the first element of a vector always has an index of 1. When we translate the difference equation of the system into code, we need to remember that

"xvec(1)" corresponds to $x[0]$,
"xvec(2)" corresponds to $x[1]$,

$$\vdots$$

"yvec(1)" corresponds to $y[0]$,
"yvec(2)" corresponds to $y[1]$,

$$\vdots$$

and so on. Fortunately, this is not a difficult problem to overcome. The method we will use in this exercise is not the most elegant, but it is one that will keep us continuously aware of the indexing differences between MATLAB vectors and our signals. The script listed below iterates through the difference equation. Notice how the MATLAB variable "offset" is used in line 9 to deal with the indexing problem.

```
1   % Script: matex_3_4a.m
2   %
3   xvec = [0,100*ones(1,18)];      % Vector to hold input signal.
4   yvec = [10000,zeros(1,18)];     % Vector to hold output signal.
5   c = 0.005;                      % Interest rate.
6   offset = 1;                     % Offset to fix index issues.
7   % Start the loop to compute the output signal.
8   for n=1:18
9     yvec(offset+n)=(1+c)*yvec(offset+n-1)-xvec(offset+n);
10  end;
11  % Display the output signal.
12  tmp = [[0:18]',yvec'];
13  disp(tmp);
```

The last two lines

```
12  tmp = [[0:18]',yvec'];
13  disp(tmp);
```

create a 19 by 2 matrix "tmp" in which the first column contains the indices from 0 to 19, and the second column contains the corresponding sample amplitudes of the output signal. A better looking tabulated display of the results can be produced with the use of the fprintf(..) function. The modified script shown below tabulates the results a bit more cleanly, and also graphs the output signal $y[n]$.

```
1   % Script: matex_3_4b.m
2   %
3   xvec = [0,100*ones(1,18)];      % Vector to hold input signal.
4   yvec = [10000,zeros(1,18)];     % Vector to hold output signal.
5   c = 0.005;                      % Interest rate.
6   offset = 1;                     % Offset to fix index issues.
7   fprintf(1,'Index   Input   Output\n');   % Print header.
8   % Start the loop to compute and print the output signal.
9   for n=1:18
10      yvec(offset+n)=(1+c)*yvec(offset+n-1)-xvec(offset+n);
11      fprintf(1,'%5d   %5.2f   %5.2f\n',n,xvec(offset+n),yvec(offset+n));
12  end;
13  % Graph the output signal.
14  stem([0:18],yvec);
```

Software resources:
matex_3_4a.m
matex_3_4b.m

MATLAB Exercise 3.5: Implementing a discrete-time system from its block diagram

In this exercise we will implement the discrete-time system for which a block diagram was obtained in Example 3.16, Fig. 3.24. A MATLAB function named "sys1" will be developed with the syntax

```
y = sys1(x)
```

where "x" represents the *current input sample*, and "y" is the *current output sample*. The function sys1(..) is used in a loop structure for processing samples from an input data stream one sample at a time.

Let the outputs $w[n-1]$, $w[n-2]$ and $w[n-3]$ of the three delay elements from left to right be represented by MATLAB variables "wnm1", "wnm2" and "wnm3" respectively. These are the persistent variables in sys1(..). When the function is first called, the initial values obtained in Example 3.16 are placed into the persistent variables. A listing of the completed function sys1(..) is given below:

```
1   function y = sys1(x)
2     persistent wnm1 wnm2 wnm3;
3     if isempty(wnm1)
4       wnm1 = 1.0682;   % Initial value w[-1]
5     end;
6     if isempty(wnm2)
7       wnm2 = 1.7149;   % Initial value w[-2]
8     end;
9     if isempty(wnm3)
10      wnm3 = 0.1674;   % Initial value w[-3]
11    end;
12    wn = x+0.7*wnm1+0.8*wnm2-0.84*wnm3;   % Eqn.(3.101)
13    y = 0.1*wn+0.2*wnm1+0.3*wnm2;         % Eqn.(3.102)
14    % Prepare for the next call to the function.
15    wnm3 = wnm2;
16    wnm2 = wnm1;
17    wnm1 = wn;
```

The script listed below can be used for testing the function sys1(..) with a unit-step input signal and graphing the resulting output signal.

```
1    % Script: matex_3_5.m
2    %
3    input = ones(1,50);     % Input data stream.
4    output = [];            % Create an empty output stream.
5    clear sys1;             % Clear persistent variables.
6    nsamp = length(input);  % Number of samples in the input stream.
7    for n=1:nsamp
8      x = input(n);         % "x" is the current input sample.
9      y = sys1(x);          % "y" is the current output sample.
10     output = [output,y];  % Append "y" to the output stream.
11   end;
12   % Graph the output signal.
13   clf;
14   stem([0:nsamp-1],output);
15   title('The output signal');
16   xlabel('Index n');
17   ylabel('y[n]');
```

The graph produced by this script is shown in Fig. 3.32. Compare it to the result displayed in the interactive demo program dgm_demo1.m.

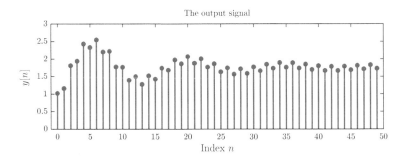

Figure 3.32 – Unit-step response of the system in MATLAB Exercise 3.5.

Software resources:
matex_3_5.m
sys1.m

MATLAB Exercise 3.6: Discrete-time system of MATLAB Exercise 3.5 revisited

Consider again the system that was implemented in MATLAB Exercise 3.5. In this exercise we will implement the same system using the function filter(..) that is part of MATLAB. Recall that the system in question was introduced in Example 3.16. Its difference equation is

$$y[n] - 0.7\,y[n-1] - 0.8\,y[n-2] + 0.84\,y[n-3] = 0.1\,x[n] + 0.2\,x[n-1] + 0.3\,x[n-2]$$

The unit-step response of the system is to be determined subject to the initial conditions

$$y[-1] = 0.5\,, \quad y[-2] = 0.3\,, \quad y[-3] = -0.4$$

The function filter(..) implements the difference equation

$$a_0\,y[n] + a_1\,y[n-1] + \ldots + a_{N-1}\,y[n-N+1] + a_N\,y[n-N] =$$
$$b_0\,b[n] + b_1\,x[n-1] + \ldots + b_{M-1}\,x[n-M+1] + b_M\,x[n-M] \qquad (3.164)$$

using a *transposed direct-form II* implementation which is different than the block diagram used in MATLAB Exercise 3.5. In both Example 3.16 and MATLAB Exercise 3.5 we have converted the initial output values $y[-1]$, $y[-2]$, and $y[-3]$ to values $w[-1]$, $w[-2]$ and $w[-3]$ for the block diagram in Fig. 3.24. This step will need to be revised when using the function `filter(..)`.

The syntax of the function `filter(..)` is

```
y = filter(b,a,x,zi)
```

Meanings of parameters are explained below:

"b":	Vector of feed-forward coefficients $b_0, b_1, \ldots, b_M$
"a":	Vector of feedback coefficients $a_0, a_1, \ldots, a_N$
"x":	Vector containing samples of the input signal
"y":	Vector containing samples of the output signal
"zi":	Vector containing initial values of the delay elements for the *transposed direct-form II* block diagram

The vector "zi" is obtained from the initial values of input and output samples through a call to the function `filtic(..)`. The script listed below computes and graphs the output signal of the system using the two MATLAB functions mentioned. It can easily be verified that the output signal obtained is identical to that of MATLAB Exercise 3.5.

```
1   % Script: matex_3_6.m
2   %
3   a = [1,-0.7,-0.8,0.84];        % Feedback coefficients.
4   b = [0.1,0.2,0.3];             % Feed-forward coefficients,
5   y_init = [0.5,0.3,-0.4];       % y[-1], y[-2], and y[-3].
6   x_init = [0,0,0];              % x[-1], x[-2], and x[-3].
7   inp = ones(1,50);              % Unit-step input signal.
8   zi = filtic(b,a,y_init,x_init);
9   out = filter(b,a,inp,zi);
10  % Graph the output signal.
11  clf;
12  stem([0:49],out);
13  title('The output signal');
14  xlabel('Index n');
15  ylabel('y[n]');
```

Software resources:
matex_3_6.m

MATLAB Exercise 3.7: Convolution using MATLAB

In Example 3.19 we have discussed the convolution operation applied to two finite-length signals, namely

$$h[n] = \{ \underset{\underset{n=0}{\uparrow}}{4} , 3, 2, 1 \}$$

and

$$x[n] = \{ - \underset{\underset{n=0}{\uparrow}}{3} , 7, 4 \}$$

MATLAB has a very convenient function `conv(..)` for performing the convolution operation on two vectors that hold samples of the signals $h[n]$ and $x[n]$. The following set of statements create two vectors "h" and "x" corresponding to the signals we are considering, and then compute their convolution:

```
>>  h = [4,3,2,1];
>>  x = [-3,7,4];
>>  y = conv(h,x)

y =
   -12     19     31     23     15      4
```

The convolution operator is commutative, that is, the roles of $h[n]$ and $x[n]$ can be reversed with no effect on the result. Therefore, the statement

```
>>  y = conv(x,h)
```

produces exactly the same result.

The MATLAB vector "y" produced by the function conv(..) starts with index 1, a characteristic of all MATLAB vectors. Consequently, an attempt to graph it with a statement like

```
>>  stem(y);
```

results in the stem graph shown in Fig. 3.33.

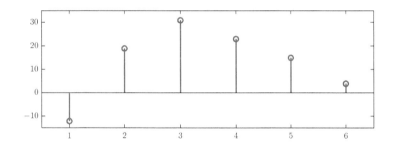

Figure 3.33 – The graph of the convolution result with wrong indices.

A correctly indexed stem graph can be obtained by the statements

```
>>  n = [0:5];
>>  stem(n,y);
```

and is shown in Fig. 3.34.

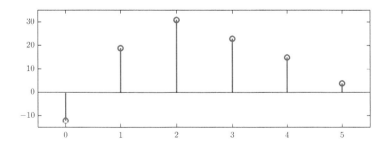

Figure 3.34 – The graph with corrected indices.

What if the starting indices of the two signals $h[n]$ and $x[n]$ are not at $n = 0$ but are specified as arbitrary values N_h and N_x respectively? (See Example 3.20.) We could simply adjust the vector "n" to accommodate the change. Instead, we will use this as an opportunity to develop a *wrapper function* that utilizes the built-in conv(..) function of MATLAB, but also automates the process of generating the appropriate index vector "n" to go with it. The listing for the function ss_conv(..) is given below:

```
1    function [y,n] = ss_conv(h,x,Nh,Nx)
2       y = conv(h,x);
3       Ny = length(y);       % Number of samples in y[n].
4       nfirst = Nh+Nx;       % Correct index for the first sample
5                             %   in vector "y".
6       nlast = nfirst+Ny-1;  % Correct index for the last sample
7                             %   in vector "y".
8       n = [nfirst:nlast];
```

Parameters "Nh" and "Nx" are the correct indices that should be associated with the first samples of vectors "h" and "x" respectively. Example usage of the function ss_conv(..) is given below:

```
>>   [y,n] = ss_conv(h,x,5,7)
>>   stem(n,y);
```

Software resources:
matex_3_7.m
ss_conv.m

MATLAB Exercise 3.8: Implementing a moving average filter through convolution

Consider again the problem of applying a moving-average filter to a signal for the purpose of producing a smoother version of it at the filter output. Suppose the signal $x[n]$ contains samples of the Dow Jones Industrial average data for the calendar year 2003, and we would like to apply a length-50 moving-average filter to it. While it is possible to write a function to implement a length-50 moving-average filter in a way similar to the development of the functions ss_movavg2(..) and ss_movavg4(..) in MATLAB Exercise 3.1, that would be greatly impractical due to the fact that we need to perform bookkeeping on 49 prior input samples at any given time. Instead, we will solve the problem using convolution.

The impulse response of the length-50 moving-average filter is

$$h[n] = \frac{1}{50} \left(u[n] - u[n-50] \right)$$

$$= \begin{cases} 1/50, & 0 \le n \le 49 \\ 0, & \text{otherwise} \end{cases}$$

which can be represented with a MATLAB vector obtained with the statement

```
>>   h = 1/50*ones(1,50);
```

The following set of statements statements load the MATLAB data file "djia.mat" into memory, and list the contents of MATLAB workspace so far.

```
>>   load 'djia.mat';
>>   whos
```

```
Name          Size                    Bytes   Class       Attributes

h             1x50                      400   double
x2001         1x248                    1984   double
x2002         1x252                    2016   double
x2003         1x252                    2016   double
```

We are assuming that we have started a fresh session for this exercise; otherwise there may be additional variables listed above. The data for the calendar year is in vector "x2003", and has 252 samples. Thus, we have an input signal $x[n]$ for $n = 0, \ldots, 251$. Its convolution with the impulse response in vector "h" can be obtained with the statements

```
>>   x = x2003;
>>   y2003 = conv(h,x);
```

There is one practical problem with this result. Recall the discussion in Example 3.3 about observing the input signal through a window that holds as many samples as the length of the filter, and averaging the window contents to obtain the output signal. For the first 49 samples of the output signal, the window is only partially full. (Refer to Fig. 3.6.) In the computation of $y[0]$, the only meaningful sample in the window is $x[0]$, and all samples to the left of it are zero-valued. The next output sample $y[n]$ will be based on averaging two meaningful samples $x[1]$ and $x[0]$ along with 48 zero-amplitude samples, and so on. As a result, the first meaningful sample of the output signal (meaningful in the sense of a moving average) will occur at index $n = 49$ when, for the first time, the window is full.

To circumvent this problem and obtain meaningful moving-average values for all 2003 data, we will prepend the signal $x[n]$ with an additional set of 49 samples borrowed from the last part of the data for calendar year 2002. The following MATLAB statement accomplishes that:

```
>>   x = [x2002(204:252),x2003];
```

The resulting vector "x" represents the signal $x[n]$ that starts with index $n = -49$. We can now convolve this signal with the impulse response of the length-50 moving-average filter to obtain the output signal $y[n]$ which will also begin at index $n = -49$. The statement

```
>>   y = conv(h,x);
```

produces a vector "y" with 350 elements. Its first 49 elements correspond to signal samples $y[-49]$ through $y[-1]$, and represent averages computed with incomplete data due to empty slots on the left side of the window. In addition, the last 49 elements of the vector "y" corresponding to signal samples $y[252]$ through $y[300]$ also represent averages computed with incomplete data, in this case with empty slots appearing on the right side of the window. We will obtain the smoothed data for the index range $n = 0, \ldots, 251$ by discarding the first and the last 49 samples of the vector "y".

```
>>   y = y(50:301);
>>   plot([0:251],x2003,':',[0:251],y,'-');
```

The resulting graph is shown in Fig. 3.35.

Software resources:
matex_3_8.m

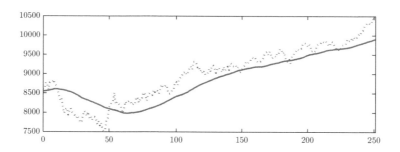

Figure 3.35 – Input and output signals of length-50 moving average filter.

Problems

3.1. A number of discrete-time systems are specified below in terms of their input-output relationships. For each case determine if the system is linear and/or time-invariant.

  a. $y[n] = x[n]\, u[n]$
  b. $y[n] = 3\, x[n] + 5$
  c. $y[n] = 3\, x[n] + 5\, u[n]$
  d. $y[n] = n\, x[n]$
  e. $y[n] = \cos(0.2\pi n)\, x[n]$
  f. $y[n] = x[n] + 3\, x[n-1]$
  g. $y[n] = x[n] + 3\, x[n-1]\, x[n-2]$

3.2. A number of discrete-time systems are described below. For each case determine if the system is linear and/or time-invariant.

  a. $y[n] = \displaystyle\sum_{k=-\infty}^{n} x[k]$

  b. $y[n] = \displaystyle\sum_{k=0}^{n} x[k]$

  c. $y[n] = \displaystyle\sum_{k=n-2}^{n} x[k]$

  d. $y[n] = \displaystyle\sum_{k=n-2}^{n+2} x[k]$

3.3. Consider the cascade combination of two systems shown in Fig. P.3.3(a).

$x[n] \longrightarrow \boxed{\text{Sys}_1\{\ \}} \xrightarrow{\ w[n]\ } \boxed{\text{Sys}_2\{\ \}} \longrightarrow y[n]$
(a)

$x[n] \longrightarrow \boxed{\text{Sys}_2\{\ \}} \xrightarrow{\ \bar{w}[n]\ } \boxed{\text{Sys}_1\{\ \}} \longrightarrow \bar{y}[n]$
(b)

Figure P. 3.3

  a. Let the input-output relationships of the two subsystems be given as

$$\text{Sys}_1\{x[n]\} = 3\, x[n] \quad \text{and} \quad \text{Sys}_2\{w[n]\} = w[n-2]$$

  Write the relationship between $x[n]$ and $y[n]$.

b. Let the order of the two subsystems be changed as shown in Fig. P.3.3(b). Write the relationship between $x[n]$ and $\bar{y}[n]$. Does changing the order of two subsystems change the overall input-output relationship of the system?

3.4. Repeat Problem 3.3 with the following sets of subsystems:

a. $\text{Sys}_1\{x[n]\} = 3\,[n]$ and $\text{Sys}_2\{w[n]\} = n\,w[n]$
b. $\text{Sys}_1\{x[n]\} = 3\,x[n]$ and $\text{Sys}_2\{w[n]\} = w[n] + 5$

3.5. The response of a linear and time-invariant system to the input signal $x[n] = \delta[n]$ is given by

$$\text{Sys}\{\delta[n]\} = \{\ \underset{\underset{n=0}{\uparrow}}{2}\ ,\ 1,\ -1\ \}$$

Determine the response of the system to the following input signals:

a. $x[n] = \delta[n] + \delta[n-1]$
b. $x[n] = \delta[n] - 2\,\delta[n-1] + \delta[n-2]$
c. $x[n] = u[n] - u[n-5]$
d. $x[n] = n\,(u[n] - u[n-5])$

3.6. Consider a system that is known to be linear but not necessarily time-invariant. Its responses to three impulse signals are given below:

$$\text{Sys}\{\delta[n]\} = \{\ \underset{\underset{n=0}{\uparrow}}{1}\ ,\ 2,\ 3\ \}$$

$$\text{Sys}\{\delta[n-1]\} = \{\ \underset{\underset{n=1}{\uparrow}}{3}\ ,\ 3,\ 2\ \}$$

$$\text{Sys}\{\delta[n-2]\} = \{\ \underset{\underset{n=2}{\uparrow}}{3}\ ,\ 2,\ 1\ \}$$

For each of the input signals listed below, state whether the response of the system can be determined from the information given. If the response can be determined, find it. If not, explain why it cannot be done.

a. $x[n] = 5\,\delta[n-1]$
b. $x[n] = 3\,\delta[n] + 2\,\delta[n-1]$
c. $x[n] = \delta[n] - 2\,\delta[n-1] + 4\,\delta[n-2]$
d. $x[n] = u[n] - u[n-3]$
e. $x[n] = u[n] - u[n-4]$

3.7. The discrete-time signal

$$x[n] = \{\ \underset{\underset{n=0}{\uparrow}}{1.7},\ 2.3,\ 3.1,\ 3.3,\ 3.7,\ 2.9,\ 2.2,\ 1.4,\ 0.6,\ -0.2,\ 0.4\ \}$$

is used as input to a length-2 moving average filter. Determine the response $y[n]$ for $n = 0, \ldots, 9$. Use $x[-1] = 0$.

3.8. Consider again the signal $x[n]$ specified in Problem 3.7. If it is applied to a length-4 moving average filter, determine the output signal $y[n]$ for $n = 0, \ldots, 9$. Use $x[-1] = x[-2] = x[-3] = 0$.

3.9. The signal $x[n]$ specified in Problem 3.7 is used as input to an exponential smoother with $\alpha = 0.2$. The initial value of the output signal is $y[-1] = 1$. Determine $y[n]$ for $n = 0, \ldots, 9$.

3.10. Consider the difference equation model for the loan payment system explored in Example 3.7. The balance at the end of month-n is given by

$$y[n] = (1+c)\, y[n-1] - x[n]$$

where c is the monthly interest rate and $x[n]$ represents the payment made in month-n. Let the borrowed amount be $\$10,000$ to be paid back at the rate of $\$250$ per month and with a monthly interest rate of 0.5 percent, that is, $c = 0.005$. Determine the monthly balance for $n = 1, \ldots, 12$ by iteratively solving the difference equation.
Hint: Start with the initial balance of $y[0] = 10000$ and model the payments with the input signal

$$x[n] = 250\, u[n-1]$$

3.11. Refer to Example 3.9 in which a nonlinear difference equation was found for computing a root of a function. The idea was adapted to the computation of the square root of a positive number A by searching for a root of the function

$$f(w) = w^2 - A$$

By iteratively solving the difference equation given by Eqn. (3.25), approximate the square root of

a. $A = 5$
b. $A = 17$
c. $A = 132$

In each case carry out the iterations until the result is accurate up to the fourth digit after the decimal point.

3.12. For each homogeneous difference equation given below, find the characteristic equation and show that it only has simple real roots. Find the homogeneous solution for $n \geq 0$ in each case subject to the initial conditions specified.
Hint: For part (e) use MATLAB function `roots(..)` to find the roots of the characteristic polynomial.

a. $y[n] + 0.2\, y[n-1] - 0.63\, y[n-2] = 0$, $\quad y[-1] = 5$, $\quad y[-2] = -3$
b. $y[n] + 1.3\, y[n-1] + 0.4\, y[n-2] = 0$, $\quad y[-1] = 0$, $\quad y[-2] = 5$
c. $y[n] - 1.7\, y[n-1] + 0.72\, y[n-2] = 0$, $\quad y[-1] = 1$, $\quad y[-2] = 2$
d. $y[n] - 0.49\, y[n-2] = 0$, $\quad y[-1] = -3$, $\quad y[-2] = -1$
e. $y[n] + 0.6\, y[n-1] - 0.51\, y[n-2] - 0.28\, y[n-3] = 0$,
    $y[-1] = 3$, $\quad y[-2] = 2$, $\quad y[-3] = 1$

3.13. For each homogeneous difference equation given below, find the characteristic equation and show that at least some of its roots are complex. Find the homogeneous solution for $n \geq 0$ in each case subject to the initial conditions specified.
Hint: For part (d) use MATLAB function `roots(..)` to find the roots of the characteristic polynomial.

a. $y[n] - 1.4\, y[n-1] + 0.85\, y[n-2] = 0$, $\quad y[-1] = 2$, $\quad y[-2] = -2$
b. $y[n] - 1.6\, y[n-1] + y[n-2] = 0$, $\quad y[-1] = 0$, $\quad y[-2] = 3$
c. $y[n] + y[n-2] = 0$, $\quad y[-1] = 3$, $\quad y[-2] = 2$
d. $y[n] - 2.5\, y[n-1] + 2.44\, y[n-2] - 0.9\, y[n-3] = 0$,
    $y[-1] = 1$, $\quad y[-2] = 2$, $\quad y[-3] = 3$

3.14. For each homogeneous difference equation given below, find the characteristic equation and show that it has multiple-order roots. Find the homogeneous solution for $n \geq 0$ in each case subject to the initial conditions specified.
Hint: For parts (c) and (d) use MATLAB function `roots(..)` to find the roots of the characteristic polynomial.

    a. $y[n] - 1.4\,y[n-1] + 0.49\,y[n-2] = 0$, $y[-1] = 1$, $y[-2] = 1$
    b. $y[n] + 1.8\,y[n-1] + 0.81\,y[n-2] = 0$, $y[-1] = 0$, $y[-2] = 2$
    c. $y[n] - 0.8\,y[n-1] - 0.64\,y[n-2] + 0.512\,y[n-3] = 0$,
        $y[-1] = 1$, $y[-2] = 1$, $y[-3] = 2$
    d. $y[n] + 1.7\,y[n-1] + 0.4\,y[n-2] - 0.3\,y[n-3] = 0$, $y[-1] = 1$, $y[-2] = 2$, $y[-3] = 1$

3.15. Solve each difference equation given below for the specified input signal and initial conditions. Use the general solution technique outlined in Section 3.5.2.

    a. $y[n] = 0.6\,y[n-1] + x[n]$, $x[n] = u[n]$, $y[-1] = 2$
    b. $y[n] = 0.8\,y[n-1] + x[n]$, $x[n] = 2\sin(0.2n)$, $y[-1] = 1$
    c. $y[n] - 0.2\,y[n-1] - 0.63\,y[n-2] = x[n]$, $x[n] = e^{-0.2n}$, $y[-1] = 0$, $y[-2] = 3$
    d. $y[n] + 1.4\,y[n-1] + 0.85\,y[n-2] = x[n]$, $x[n] = u[n]$, $y[-1] = -2$, $y[-2] = 0$
    e. $y[n] + 1.6\,y[n-1] + 0.64\,y[n-2] = x[n]$, $x[n] = u[n]$, $y[-1] = 0$, $y[-2] = 1$

3.16. Consider the exponential smoother explored in Examples 3.6 and 3.14. It is modeled with the difference equation

$$y[n] = (1 - \alpha)\,y[n-1] + \alpha\,x[n]$$

Let $y[-1] = 0$ so that the system is linear.

    a. Let the input signal be a unit step, that is, $x[n] = u[n]$. Determine the response of the linear exponential smoother as a function of α.
    b. Let the input signal be a unit ramp, that is, $x[n] = n\,u[n]$. Determine the response of the linear exponential smoother as a function of α.

3.17. Consider a first-order differential equation in the form

$$\frac{dy_a(t)}{dt} + A\,y_a(t) = A\,x_a(t)$$

The derivative can be approximated using the backward difference

$$\frac{dy_a(t)}{dt} \approx \frac{y_a(t) - y_a(t - T)}{T}$$

where T is the step size.

    a. Using the approximation for the derivative in the differential equation, express $y_a(t)$ in terms of $y_a(t - T)$ and $x_a(t)$.
    b. Convert the differential equation to a difference equation by defining discrete-time signals

$$x[n] = x_a(nT)$$
$$y[n] = y_a(nT)$$
$$y[n-1] = y_a(nT - T)$$

    Show that the resulting difference equation corresponds to an exponential smoother. Determine its parameter α in terms of A and T.

3.18. Construct a block diagram for each difference equation given below.

   a. $y[n] + 0.2\,y[n-1] - 0.63\,y[n-2] = x[n] + x[n-2]$
   b. $y[n] - 2.5\,y[n-1] + 2.44\,y[n-2] - 0.9\,y[n-3] = x[n] - 3\,x[n-1] + 2\,x[n-2]$
   c. $y[n] - 0.49\,y[n-2] = x[n] - x[n-1]$
   d. $y[n] + 0.6\,y[n-1] - 0.51\,y[n-2] - 0.28\,y[n-3] = x[n] - 2\,x[n-2]$

3.19. Two DTLTI systems with impulse responses $h_1[n]$ and $h_2[n]$ are connected in cascade as shown in Fig. P.3.19(a).

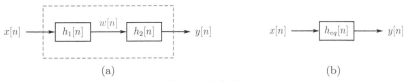

(a)   (b)

Figure P. 3.19

   a. Determine the impulse response $h_{eq}[n]$ of the equivalent system, as shown in Fig. P.3.19(b) in terms of $h_1[n]$ and $h_2[n]$.
   Hint: Use convolution to express $w[n]$ in terms of $x[n]$. Afterwards use convolution again to express $y[n]$ in terms of $w[n]$.
   b. Let $h_1[n] = h_2[n] = u[n] - u[n-5]$. Determine and sketch $h_{eq}[n]$ for the equivalent system.
   c. With $h_1[n]$ and $h_2[n]$ as specified in part (b), let the input signal be a unit step, that is, $x[n] = u[n]$. Determine and sketch the signals $w[n]$ and $y[n]$.

3.20. Two DTLTI systems with impulse responses $h_1[n]$ and $h_2[n]$ are connected in parallel as shown in Fig. P.3.20(a).

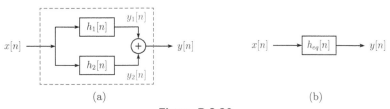

(a)   (b)

Figure P. 3.20

   a. Determine the impulse response $h_{eq}[n]$ of the equivalent system, as shown in Fig. P.3.20(b) in terms of $h_1[n]$ and $h_2[n]$.
   Hint: Use convolution to express the signals $y_1[n]$ and $y_2[n]$ in terms of $x[n]$. Afterwards express $y[n]$ in terms of $y_1[n]$ and $y_2[n]$.
   b. Let $h_1[n] = (0.9)^n\,u[n]$ and $h_2[n] = (-0.7)^n\,u[n]$. Determine and sketch $h_{eq}[n]$ for the equivalent system.
   c. With $h_1[n]$ and $h_2[n]$ as specified in part (b), let the input signal be a unit step, that is, $x[n] = u[n]$. Determine and sketch the signals $y_1[n]$, $y_2[n]$ and $y[n]$.

3.21. Three DTLTI systems with impulse responses $h_1[n]$, $h_2[n]$ and $h_3[n]$ are connected as shown in Fig. P.3.21(a).

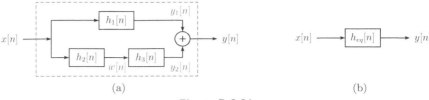

Figure P. 3.21

a. Determine the impulse response $h_{eq}[n]$ of the equivalent system, as shown in Fig. P.3.21(b) in terms of $h_1[n]$, $h_2[n]$ and $h_3[n]$.

b. Let $h_1[n] = e^{-0.1n} u[n]$, $h_2[n] = \delta[n-2]$, and $h_3[n] = e^{-0.2n} u[n]$. Determine and sketch $h_{eq}[n]$ for the equivalent system.

c. With $h_1[n]$, $h_2[n]$ and $h_3[n]$ as specified in part (b), let the input signal be a unit step, that is, $x[n] = u[n]$. Determine and sketch the signals $w[n]$, $y_1[n]$, $y_2[n]$ and $y[n]$.

3.22. Consider the DTLTI system shown in Fig. P.3.22(a).

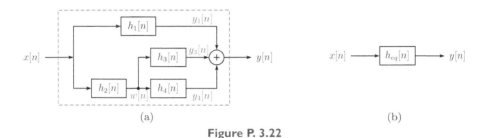

Figure P. 3.22

a. Express the impulse response of the system as a function of the impulse responses of the subsystems.

b. Let

$$h_1[n] = e^{-0.1n} u[n]$$

$$h_2[n] = h_3[n] = u[n] - u[n-3]$$

and

$$h_4[n] = \delta[n-2]$$

Determine the impulse response $h_{eq}[n]$ of the equivalent system.

c. Let the input signal be a unit-step, that is, $x[n] = u[n]$. Determine and sketch the signals $w[n]$, $y_1[n]$, $y_3[n]$ and $y_4[n]$.

3.23.   Consider two finite-length signals $x[n]$ and $h[n]$ that are equal to zero outside the intervals indicated below:

$$x[n] = 0 \quad \text{if } n < N_{x1} \text{ or } n > N_{x2}$$

$$h[n] = 0 \quad \text{if } n < N_{h1} \text{ or } n > N_{h2}$$

In other words, significant samples of $x[n]$ are in the index range $N_{x1}, \ldots, N_{x2}$, and the significant samples of $h[n]$ are in the index range $N_{h1}, \ldots, N_{h2}$.

Let $y[n]$ be the convolution of $x[n]$ and $h[n]$. Starting with the convolution sum given by Eqn. (3.128), determine the index range of significant samples $N_{y1}, \ldots, N_{y2}$ for $y[n]$.

3.24. Let $y[n]$ be the convolution of two discrete-time signals $x[n]$ and $h[n]$, that is

$$y[n] = x[n] * h[n]$$

Show that time shifting either $x[n]$ or $h[n]$ by m samples causes the $y[n]$ to be time shifted by m samples also. Mathematically prove that

$$x[n - m] * h[n] = y[n - m]$$

and

$$x[n] * h[n - m] = y[n - m]$$

3.25. Determine the impulse response of each system described below. Afterwards determine whether the system is causal and/or stable.

a. $\quad y[n] = \mathrm{Sys}\{x[n]\} = \displaystyle\sum_{k=-\infty}^{n} x[k]$

b. $\quad y[n] = \mathrm{Sys}\{x[n]\} = \displaystyle\sum_{k=-\infty}^{n} e^{-0.1\,(n-k)}\, x[k]$

c. $\quad y[n] = \mathrm{Sys}\{x[n]\} = \displaystyle\sum_{k=0}^{n} x[k] \qquad \text{for } n \geq 0$

d. $\quad y[n] = \mathrm{Sys}\{x[n]\} = \displaystyle\sum_{k=n-10}^{n} x[k]$

e. $\quad y[n] = \mathrm{Sys}\{x[n]\} = \displaystyle\sum_{k=n-10}^{n+10} x[k]$

MATLAB Problems

3.26. Refer to the homogeneous difference equations in Problem 3.12. For each one, develop a MATLAB script to iteratively solve it for $n = 0, \ldots, 19$ using the initial conditions specified. Compare the results of iterative solutions to the results obtained analytically in Problem 3.12.

3.27. Refer to Problem 3.3. The cascade combination of the two subsystems will be tested using the input signal

$$x[n] = (0.95)^n \, \cos\,(0.1\pi n)\, u[n]$$

Write a script to do the following:

a. Create an anonymous anonymous function "x" to compute the input signal at all index values specified in a vector "n".
b. Implement the cascade system as shown in Fig. P.3.3(a). Compute the signals $w[n]$ and $y[n]$ for $n = 0, \ldots, 29$.
c. Implement the cascade system in the alternative form shown in Fig. P.3.3(b). Compute the signals $\bar{w}[n]$ and $\bar{y}[n]$ for $n = 0, \ldots, 29$. Compare $y[n]$ and $\bar{y}[n]$.

3.28. Repeat the steps of Problem 3.27 using the subsystems described in Problem 3.4.

3.29. Refer to Problem 3.1. Linearity and time invariance of the systems listed will be tested using the two input signals

$$x_1[n] = n \, e^{-0.2n} \, (u[n] - u[n-20]) \, ;$$

and

$$x_2[n] = \cos(0.05\pi n) \, (u[n] - u[n-20]) \, ;$$

Develop a script to do the following:

a. Express the two input signals by means of two anonymous functions "x1" and "x2". Each anonymous function should take a single argument "n" which could either be a scalar or a vector of index values.
b. Express each of the systems described in Problem 3.1 as an anonymous function. Name the anonymous functions "sys1" through "sys5". Each should take two arguments "n" and "x".
c. Compute the response of each system to

$$\begin{aligned}
x[n] &= x_1[n] \\
x[n] &= x_2[n] \\
x[n] &= x_1[n] + x_2[n] \\
x[n] &= 5 \, x_1[n] - 3 \, x_2[n]
\end{aligned}$$

Identify which systems fail the linearity test.
d. Compute the response of each system to

$$\begin{aligned}
x[n] &= x_1[n-1] \\
x[n] &= x_2[n-3]
\end{aligned}$$

Identify which systems fail the time-invariance test.

3.30.

a. Develop a function ss_lbal(..) to iteratively solve the loan balance difference equation explored in Problem 3.10. It should have the following interface:

```
bal = ss_lbal(A,B,c,n)
```

The arguments are:

A : Amount of the loan; this is also the balance at $n = 0$
B : Monthly payment amount
c : Monthly interest rate
n : Index of the month in which the balance is sought
bal : Computed balance after n months

b. Write a script to test the function with the values specified in Problem 3.10.

3.31. Refer to Problem 3.11.

a. Develop a MATLAB function ss_sqrt(..) that iteratively solves the nonlinear difference equation given by Eqn. (3.25) to approximate the square root of a positive number A. The syntax of the function should be

```
y = ss_sqrt(A,y_init,tol)
```

The argument "A" represents the positive number the square root of which is being sought, "y_init" is the initial value $y[-1]$, and "tol" is the tolerance limit ε. The function should return when the difference between two consecutive output samples is less than the tolerance limit, that is,

$$|y[n] - y[n-1]| \le \varepsilon$$

b. Write a script to test the function ss_sqrt(..) with the values $A = 5$, $A = 17$ and $A = 132$.

3.32. Refer to Problem 3.16 in which the response of the linear exponential smoother to unit-step and unit-ramp signals were found as functions of the parameter α. Consider the input signal shown in Fig. P.3.32.

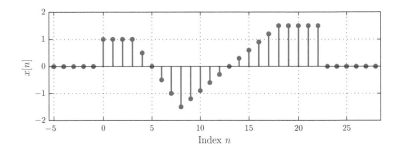

Figure P. 3.32

Develop a script to do the following:

a. Create an anonymous function yu(..) to compute the unit-step response of the linear exponential smoother. It should take two arguments, "alpha" and "n", and should return the unit-step response evaluated at each index in the vector "n".

b. Create an anonymous function yr(..) to compute the unit-ramp response of the linear exponential smoother. Like the function yu(..), the function yr(..) should also take two arguments, "alpha" and "n", and should return the unit-ramp response evaluated at each index in the vector "n".

c. Express the input signal $x[n]$ using a combination of scaled and time shifted unit-step and unit-ramp functions.

d. Use the anonymous functions yu(..) and yr(..) to compute and graph the output signal $y[n]$. Try with $\alpha = 0.1$, 0.2, 0.3 and compare.

MATLAB Projects

3.33. Consider the logistic growth model explored in Example 3.8. The normalized value of the population at index n is modeled using the nonlinear difference equation

$$y[n] = r\,(1 - y[n-1])\,y[n-1]$$

where r is the growth rate. This is an example of a *chaotic system*. Our goal in this project is to simulate the behavior of a system modeled by the nonlinear difference equation given, and to determine the dependency of the population growth on the initial value $y[0]$ and the growth rate r.

a. Develop a function called `ss_lgrowth(..)` with the syntax

```
y = ss_lgrowth (y_init ,r ,N)
```

where "`y_init`" is the initial value of the population at $n = 0$, "`r`" is the growth rate, and "`N`" is the number of iterations to be carried out. The vector returned by the function should have the values

$$[\, y[0], y[1], y[2], \ldots, y[N] \,]$$

1. Write a script to compute $y[n]$ for $n = 1, \ldots, 30$ with specified values $y[0]$ and r.
2. With the growth rate set at $r = 1.5$, compute and graph $y[n]$ for the cases of $y[0] = 0.1$, $y[0] = 0.3$, $y[0] = 0.5$, $y[0] = 0.7$. Does the population reach equilibrium? Comment on the dependency of the population at equilibrium on the initial value $y[0]$.
3. Repeat the experiment with $r = 2, 2.5, 2.75, 3$. For each value of r compute and graph $y[n]$ for the cases of $y[0] = 0.1$, $y[0] = 0.3$, $y[0] = 0.5$, $y[0] = 0.7$. Does the population still reach equilibrium? You should find that beyond a certain value of r population behavior should become oscillatory. What is the critical value of r?

3.34. Consider the second-order homogeneous difference equation

$$y[n] + a_1 \, y[n-1] + a_2 \, y[n-2] = 0$$

with initial conditions $y[-1] = p_1$ and $y[-2] = p_2$. The coefficients a_1 and a_2 of the difference equation and the initial values p_1 and p_2 are all real-valued, and will be left as parameters.

a. Let z_1 and z_2 be the roots of the characteristic polynomial. On paper, solve the homogeneous difference equation for the three possibilities for the roots:

  1. The roots are real and distinct.
  2. The roots are a complex conjugate pair.
  3. The two roots are real and equal.

  Find the solution $y[n]$ as a function of the roots z_1 and z_2 as well as the initial values p_1 and p_2.

b. Develop a MATLAB function `ss_de2solve(..)` with the syntax

```
y = ss_de2solve (a1 ,a2 ,p1 ,p2 ,n)
```

The vector "`n`" contains the index values for which the solution should be computed. The returned vector "`y`" holds the solution at the index values in vector "`n`" so that the solution can be graphed with

```
stem (n,y)
```

Your function should perform the following steps:

  1. Form the characteristic polynomial and find its roots.
  2. Determine which of the three categories the roots fit (simple real roots, complex conjugate pair or multiple roots).
  3. Compute the solution accordingly.

   c. Test the function `ss_de2solve(..)` with the homogeneous difference equations in Problems 3.12a, b, c, d, Problems 3.13a, b, c, and Problem 3.14a and b.

3.35. One method of numerically approximating the integral of a function is the *trapezoidal integration* method. Consider the function $x_a(t)$ as shown in Fig. P.3.35.

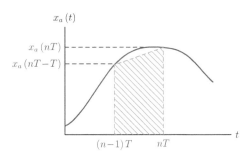

Figure P. 3.35

The area under the function from $t = (n-1)T$ to $t = nT$ may be approximated with the area of the shaded trapezoid:

$$\int_{(n-1)T}^{nT} x_a(t)\, dt \approx \frac{T}{2}\left[x_a(nT - T) + x_a(nT)\right]$$

Let

$$y_a(t) = \int_{-\infty}^{t} x_a(\lambda)\, d\lambda$$

It follows that

$$y_a(nT) \approx y_a(nT - T) + \frac{T}{2}\left[x_a(nT - T) + x_a(nT)\right]$$

If we define discrete-time signals $x[n]$ and $y[n]$ as

$$x[n] = x_a(nT)\,, \qquad \text{and} \qquad y[n] = y_a(nT)$$

then $y[n-1] = y_a(nT - T)$, and we obtain the difference equation

$$y[n] = y[n-1] + \frac{T}{2}\left(x[n] + x[n-1]\right)$$

which is the basis of a trapezoidal integrator.

   a. Develop a function `ss_integ(..)` with the following interface:

        `val = ss_integ(xa,lim1,lim2,y_init,k)`

   The arguments are:

`xa` :	Handle to an anonymous function for $x_a(t)$
`lim1` :	Lower limit of the integral
`lim2` :	Upper limit of the integral
`y_init` :	Initial value of the integral
`k` :	Number of time steps from `lim1` to `lim2`
`val` :	Trapezoidal approximation to the integral

b. Write a script to test the function with the integrand

$$x_a(t) = e^{-t} \sin(t)$$

and compare its approximate integral with the correct result given by

$$\int x_a(t)\, dt = \frac{1}{2} - \frac{1}{2} e^{-t} \cos(t) - \frac{1}{2} e^{-t} \sin(t)$$

Hint: Set the lower limit of the integral to $t = 0$.

Chapter 4

Fourier Analysis for Continuous-Time Signals and Systems

Chapter Objectives

- Learn techniques for representing continuous-time periodic signals using orthogonal sets of periodic basis functions.

- Study properties of exponential, trigonometric and compact Fourier series, and conditions for their existence.

- Learn the Fourier transform for non-periodic signals as an extension of Fourier series for periodic signals.

- Study properties of the Fourier transform. Understand energy and power spectral density concepts.

- Explore frequency-domain characteristics of CTLTI systems. Understand the system function concept.

- Learn the use of frequency-domain analysis methods for solving signal-system interaction problems with periodic and non-periodic input signals.

4.1 Introduction

In Chapters 1 and 2 we have developed techniques for analyzing continuous-time signals and systems from a time-domain perspective. A continuous-time signal can be modeled as a function of time. A CTLTI system can be represented by means of a constant-coefficient linear differential equation, or alternatively by means of an impulse response. The output signal of a CTLTI system can be determined by solving the corresponding differential equation or by using the convolution operation.

In Chapter 1 we have also discussed the idea of viewing a signal as a combination of simple signals acting as "building blocks". Examples of this were the use of unit-step, unit-ramp, unit-pulse and unit-triangle functions for constructing signals (see Section 1.3.2) and the use of the unit-impulse function for impulse decomposition of a signal (see Section 1.3.3).

Another especially useful set of building blocks is a set in which each member function has a unique frequency. Representing a signal as a linear combination of single-frequency building blocks allows us to develop a *frequency-domain* view of a signal that is particularly useful in understanding signal behavior and signal-system interaction problems. If a signal can be expressed as a superposition of single-frequency components, knowing how a linear and time-invariant system responds to each individual component helps us understand overall system behavior in response to the signal. This is the essence of frequency-domain analysis.

In Section 4.2 we discuss methods of analyzing periodic continuous-time signals in terms of their frequency content. Frequency-domain analysis methods for non-periodic signals are presented in Section 4.3. In Section 4.4 representation of signal energy and power in the frequency domain is discussed. System function concept is introduced in Section 4.5. The application of frequency-domain analysis methods to the analysis of CTLTI systems is discussed in Sections 4.6 and 4.7.

4.2   Analysis of Periodic Continuous-Time Signals

Most periodic continuous-time signals encountered in engineering problems can be expressed as linear combinations of sinusoidal *basis functions*, the more technical name for the so-called "building blocks" we referred to in the introductory section. The idea of representing periodic functions of time in terms of trigonometric basis functions was first realized by French mathematician and physicist Jean Baptiste Joseph Fourier (1768-1830) as he worked on problems related to heat transfer and propagation. The basis functions in question can either be individual sine and cosine functions, or they can be in the form of complex exponential functions that combine sine and cosine functions together.

Later in this section we will study methods of expressing periodic continuous-time signals in three different but equivalent formats, namely the *trigonometric Fourier series (TFS)*, the *exponential Fourier series (EFS)* and the *compact Fourier series (CFS)*. Before we start a detailed study of the mathematical theory of Fourier series, however, we will find it useful to consider the problem of *approximating* a periodic signal using a few trigonometric functions. This will help us build some intuitive understanding of frequency-domain methods for analyzing periodic signals.

4.2.1   Approximating a periodic signal with trigonometric functions

It was established in Section 1.3.4 of Chapter 1 that a signal $\tilde{x}(t)$ which is periodic with period T_0 has the property

$$\tilde{x}(t + T_0) = \tilde{x}(t) \tag{4.1}$$

for all t. Furthermore, it was shown through repeated use of Eqn. (4.1) that a signal that is periodic with period T_0 is also periodic with kT_0 for any integer k.

In working with periodic signals in this chapter, we will adopt the convention of using the tilde (˜) character over the name of the signal as a reminder of its periodicity. For the sake of discussion let us consider the square-wave signal $\tilde{x}(t)$ with a period of T_0 as shown in Fig. 4.1.

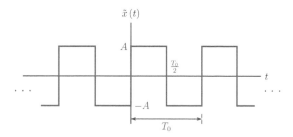

Figure 4.1 – Periodic square-wave signal.

Suppose that we wish to approximate this signal using just one trigonometric function. The first two questions that need to be answered are:

1. Should we use a sine or a cosine?
2. How should we adjust the parameters of the trigonometric function?

The first question is the easier one to answer. The square-wave signal $\tilde{x}(t)$ shown in Fig. 4.1 has odd symmetry signified by $\tilde{x}(-t) = -\tilde{x}(t)$ (see Section 1.3.6 of Chapter 1). Among the two choices for a trigonometric function, the sine function also has odd symmetry, that is, $\sin(-a) = -\sin(a)$ for any real-valued parameter a. On the other hand, the cosine function has even symmetry since $\cos(-a) = \cos(a)$. Therefore it would intuitively make sense to choose the sine function over the cosine function. Our approximation would be in the form

$$\tilde{x}(t) \approx b_1 \sin(\omega t) \tag{4.2}$$

Since $\tilde{x}(t)$ has a fundamental period of T_0, it would make sense to pick a sine function with the same fundamental period, so that

$$\sin(\omega(t + T_0)) = \sin(\omega t) \tag{4.3}$$

For Eqn. (4.3) to work we need $\omega T_0 = 2\pi$ and consequently

$$\omega = \frac{2\pi}{T_0} = \omega_0 = 2\pi f_0 \tag{4.4}$$

where we have defined the frequency f_0 as the reciprocal of the period, that is, $f_0 = 1/T_0$. Let $\tilde{x}^{(1)}(t)$ represent an approximate version of the signal $\tilde{x}(t)$, so that

$$\tilde{x}^{(1)}(t) = b_1 \sin(\omega_0 t) \tag{4.5}$$

In Eqn. (4.5) we have used the superscripted signal name $\tilde{x}^{(1)}$ to signify the fact that we are using only one trigonometric function in this approximation. Our next task is to determine the value of the coefficient b_1. How should b_1 be chosen that Eqn. (4.5) represents the best approximation of the given type possible for the actual signal $\tilde{x}(t)$? There is a bit of subjectivity in this question since we have not yet defined what the "best approximation" means for our purposes.

Let us define the *approximation error* as the difference between the square-wave signal and its approximation:

$$\tilde{\varepsilon}_1(t) = \tilde{x}(t) - \tilde{x}^{(1)}(t) = \tilde{x}(t) - b_1 \sin(\omega_0 t) \tag{4.6}$$

The subscript used on the error signal $\tilde{\varepsilon}_1(t)$ signifies the fact that it is the approximation error that results when only one trigonometric function is used. $\tilde{\varepsilon}_1(t)$ is also periodic with period T_0. One possible method of choosing the best value for the coefficient b_1 would be to choose the value that makes the normalized average power of $\tilde{\varepsilon}_1(t)$ as small as possible.

Recall that the normalized average power in a periodic signal was defined in Chapter 1 Eqn. (1.88). Adapting it to the error signal $\tilde{\varepsilon}_1(t)$ we have

$$P_\epsilon = \frac{1}{T_0} \int_0^{T_0} [\tilde{\varepsilon}_1(t)]^2 \, dt \tag{4.7}$$

This is also referred to as the *mean-squared error (MSE)*. For simplicity we will drop the constant scale factor $1/T_0$ in front of the integral in Eqn. (4.7), and minimize the *cost function*

$$J = \int_0^{T_0} [\tilde{\varepsilon}_1(t)]^2 \, dt = \int_0^{T_0} [\tilde{x}(t) - b_1 \sin(\omega_0 t)]^2 \, dt \tag{4.8}$$

instead. Minimizing J is equivalent to minimizing P_ϵ since the two are related by a constant scale factor. The value of the coefficient b_1 that is optimum in the sense of producing the smallest possible value for MSE is found by differentiating the cost function with respect to b_1 and setting the result equal to zero.

$$\frac{dJ}{db_1} = \frac{d}{db_1} \left[\int_0^{T_0} \left[\tilde{x}(t) - b_1 \sin(\omega_0 t) \right]^2 \, dt \right] = 0$$

Changing the order of integration and differentiation leads to

$$\frac{dJ}{db_1} = \int_0^{T_0} \frac{d}{db_1} \left[\tilde{x}(t) - b_1 \sin(\omega_0 t) \right]^2 \, dt = 0$$

Carrying out the differentiation we obtain

$$\int_0^{T_0} 2 \left[\tilde{x}(t) - b_1 \sin(\omega_0 t) \right] \left[-\sin(\omega_0 t) \right] \, dt = 0$$

or equivalently

$$-\int_0^{T_0} \tilde{x}(t) \sin(\omega_0 t) \, dt + b_1 \int_0^{T_0} \sin^2(\omega_0 t) \, dt = 0 \tag{4.9}$$

It can be shown that the second integral in Eqn. (4.9) yields

$$\int_0^{T_0} \sin^2(\omega_0 t) \, dt = \frac{T_0}{2}$$

Substituting this result into Eqn. (4.9) yields the optimum choice for the coefficient b_1 as

$$b_1 = \frac{2}{T_0} \int_0^{T_0} \tilde{x}(t) \sin(\omega_0 t) \, dt \tag{4.10}$$

For the square-wave signal $\tilde{x}(t)$ in Fig. 4.1 we have

$$b_1 = \frac{2}{T_0} \int_0^{T_0/2} (A) \sin(\omega_0 t) \, dt + \frac{2}{T_0} \int_{T_0/2}^{T_0} (-A) \sin(\omega_0 t) \, dt = \frac{4A}{\pi} \tag{4.11}$$

The best approximation to $\tilde{x}(t)$ using only one trigonometric function is

$$\tilde{x}^{(1)}(t) = \frac{4A}{\pi} \sin(\omega_0 t) \tag{4.12}$$

and the approximation error is

$$\tilde{\varepsilon}_1(t) = \tilde{x}(t) - \frac{4A}{\pi} \sin(\omega_0 t) \tag{4.13}$$

The signal $\tilde{x}(t)$, its single-frequency approximation $\tilde{x}^{(1)}(t)$ and the approximation error $\tilde{\varepsilon}_1(t)$ are shown in Fig. 4.2.

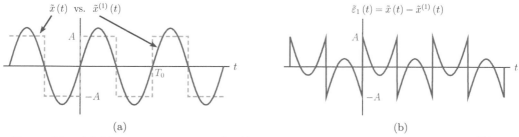

(a)   (b)

Figure 4.2 – (a) The square-wave signal $\tilde{x}(t)$ and its single-frequency approximation $\tilde{x}^{(1)}(t)$, (b) the approximation error $\tilde{\varepsilon}_1(t)$.

In the next step we will try a two-frequency approximation to $\tilde{x}(t)$ and see if the approximation error can be reduced. We know that the basis function $\sin(2\omega_0 t) = \sin(4\pi f_0 t)$ is also periodic with the same period T_0; it just has two full cycles in the interval $(0, T_0)$ as shown in Fig. 4.3.

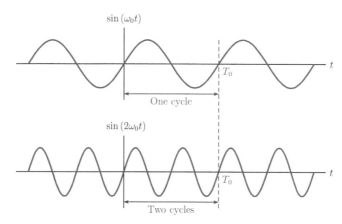

Figure 4.3 – Signals $\sin(\omega_0 t)$ and $\sin(2\omega_0 t)$.

Let the approximation using two frequencies be defined as

$$\tilde{x}^{(2)}(t) = b_1 \sin(\omega_0 t) + b_2 \sin(2\omega_0 t) \tag{4.14}$$

The corresponding approximation error is

$$\begin{aligned} \tilde{\varepsilon}_2(t) &= \tilde{x}(t) - \tilde{x}^{(2)}(t) \\ &= \tilde{x}(t) - b_1 \sin(\omega_0 t) - b_2 \sin(2\omega_0 t) \end{aligned} \tag{4.15}$$

The cost function will be set up in a similar manner:

$$J = \int_0^{T_0} [\tilde{x}(t) - b_1 \sin(\omega_0 t) - b_2 \sin(2\omega_0 t)]^2 \, dt \tag{4.16}$$

Differentiating J with respect to first b_1 and then b_2 we obtain

$$\frac{\partial J}{\partial b_1} = \int_0^{T_0} 2 [\tilde{x}(t) - b_1 \sin(\omega_0 t) - b_2 \sin(2\omega_0 t)] [-\sin(\omega_0 t)] \, dt$$

and

$$\frac{\partial J}{\partial b_2} = \int_0^{T_0} 2 [\tilde{x}(t) - b_1 \sin(\omega_0 t) - b_2 \sin(2\omega_0 t)] [-\sin(2\omega_0 t)] \, dt$$

Setting the partial derivatives equal to zero leads to the following two equations:

$$\int_0^{T_0} \tilde{x}(t) \sin(\omega_0 t) \, dt = b_1 \int_0^{T_0} \sin^2(\omega_0 t) \, dt + b_2 \int_0^{T_0} \sin(2\omega_0 t) \sin(\omega_0 t) \, dt = 0 \tag{4.17}$$

$$\int_0^{T_0} \tilde{x}(t) \sin(2\omega_0 t) \, dt = b_1 \int_0^{T_0} \sin(\omega_0 t) \sin(2\omega_0 t) \, dt + b_2 \int_0^{T_0} \sin^2(2\omega_0 t) \, dt = 0 \tag{4.18}$$

We have already established that

$$\int_0^{T_0} \sin^2(\omega_0 t) \, dt = \frac{T_0}{2}$$

Similarly it can be shown that

$$\int_0^{T_0} \sin^2(2\omega_0 t) \, dt = \frac{T_0}{2} \quad \text{and} \quad \int_0^{T_0} \sin(\omega_0 t) \sin(2\omega_0 t) \, dt = 0$$

and Eqns. (4.17) and (4.18) can now be solved for the optimum values of the coefficients b_1 and b_2, resulting in

$$b_1 = \frac{2}{T_0} \int_0^{T_0} \tilde{x}(t) \sin(\omega_0 t) \, dt \tag{4.19}$$

and

$$b_2 = \frac{2}{T_0} \int_0^{T_0} \tilde{x}(t) \sin(2\omega_0 t) \, dt \tag{4.20}$$

It is interesting to note that the expression for b_1 given in Eqn. (4.19) for the two-frequency approximation problem is the same as that found in Eqn. (4.10) for the single-frequency approximation problem. This is a result of the *orthogonality* of the two sine functions $\sin(\omega_0 t)$ and $\sin(2\omega_0 t)$, and will be discussed in more detail in the next section.

Using the square-wave signal $\tilde{x}(t)$ in Eqns. (4.19) and (4.20) we arrive at the following optimum values for the two coefficients:

$$b_1 = \frac{4A}{\pi} \quad \text{and} \quad b_2 = 0 \tag{4.21}$$

Interestingly, the optimum contribution from the sinusoidal term at radian frequency $2\omega_0$

turned out to be zero, resulting in

$$\tilde{x}^{(2)}\left(t\right) = \frac{4A}{\pi}\ \sin\left(\omega_0 t\right) + (0)\ \sin\left(2\omega_0 t\right) = \tilde{x}^{(1)}\left(t\right) \tag{4.22}$$

for this particular signal $\tilde{x}\left(t\right)$.

It can be shown (see Problem 4.1 at the end of this chapter) that a three-frequency approximation

$$\tilde{x}^{(3)}\left(t\right) = b_1\ \sin\left(\omega_0 t\right) + b_2\ \sin\left(2\omega_0 t\right) + b_3\ \sin\left(3\omega_0 t\right) \tag{4.23}$$

has the optimum coefficient values

$$b_1 = \frac{4A}{\pi}\ ,\quad b_2 = 0\ ,\quad \text{and}\quad b_3 = \frac{4A}{3\pi} \tag{4.24}$$

The signal $\tilde{x}\left(t\right)$, its three-frequency approximation $\tilde{x}^{(3)}\left(t\right)$ and the approximation error $\tilde{\epsilon}_3\left(t\right)$ are shown in Fig. 4.4.

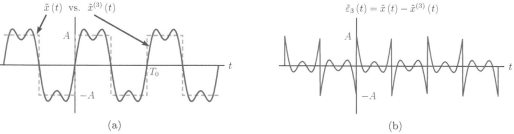

Figure 4.4 – (a) The square-wave signal $\tilde{x}\left(t\right)$ and its three-frequency approximation $\tilde{x}^{(3)}\left(t\right)$, (b) the approximation error $\tilde{\varepsilon}_3\left(t\right)$.

Some observations are in order:

1. Based on a casual comparison of Figs. 4.2(b) and 4.4(b), the normalized average power of the error signal $\tilde{\varepsilon}_3\left(t\right)$ seems to be less than that of the error signal $\tilde{\varepsilon}_1\left(t\right)$. Consequently, $\tilde{x}^{(3)}\left(t\right)$ is a better approximation to the signal than $\tilde{x}^{(1)}\left(t\right)$.

2. On the other hand, the peak value of the approximation error seems to be $\pm A$ for both $\tilde{\varepsilon}_1\left(t\right)$ and $\tilde{\varepsilon}_3\left(t\right)$. If we were to try higher-order approximations using more trigonometric basis functions, the peak approximation error would still be $\pm A$ (see Problem 4.4 at the end of this chapter). This is due to fact that, at each discontinuity of $\tilde{x}\left(t\right)$, the value of the approximation is equal to zero independent of the number of trigonometric basis functions used. In general, the approximation at a discontinuity will yield the average of the signal amplitudes right before and right after the discontinuity. This leads to the well-known *Gibbs phenomenon*, and will be explored further in Section 4.2.6.

Interactive Demo: `appr_demo1`

The demo program in "`appr_demo1.m`" provides a graphical user interface for experimenting with approximations to the square-wave signal of Fig. 4.1 using a specified number of trigonometric functions. Parameter values $A = 1$ and $T_0 = 1$ s are used. On the left side, slider controls allow parameters b_i to be adjusted freely for $i = 1, \ldots, 7$. The approximation

$$\tilde{x}^{(m)}\left(t\right) = \sum_{k=1}^{m} b_k\ \sin\left(k\omega_0 t\right)$$

and the corresponding approximation error

$$\tilde{\varepsilon}_m(t) = \tilde{x}(t) - \tilde{x}^{(m)}(t)$$

are computed and graphed. The value of m may be controlled by setting unneeded coefficients equal to zero. For example $\tilde{x}^{(5)}(t)$, the approximation with 5 trigonometric terms, may be explored by simply setting $b_6 = b_7 = 0$ and adjusting the remaining coefficients. In addition to graphing the signals, the program also computes the value of the cost function

$$J = \int_0^{T_0} [\tilde{\varepsilon}_m(t)]^2 \, dt$$

1. With all other coefficients reset to zero, adjust the value of b_1 until J becomes as small as possible. How does the best value of b_1 correspond to the result found in Eqn. (4.12)? Make a note of the smallest value of J obtained.
2. Keeping b_1 at the best value obtained, start adjusting b_2. Can J be further reduced through the use of b_2?
3. Continue in this manner adjusting the remaining coefficients one at a time. Observe the shape of the approximation error signal and the value of J after each adjustment.

Software resources:
`appr_demo1.m`

4.2.2   Trigonometric Fourier series (TFS)

We are now ready to generalize the results obtained in the foregoing discussion about approximating a signal using trigonometric functions. Consider a signal $\tilde{x}(t)$ that is periodic with fundamental period T_0 and associated fundamental frequency $f_0 = 1/T_0$. We may want to represent this signal using a linear combination of sinusoidal functions in the form

$$\begin{aligned} \tilde{x}(t) =\,& a_0 + a_1 \cos(\omega_0 t) + a_2 \cos(2\omega_0 t) + \ldots + a_k \cos(k\omega_0 t) \ldots \\ & + b_1 \sin(\omega_0 t) + b_2 \sin(2\omega_0 t) + \ldots + b_k \sin(k\omega_0 t) + \ldots \end{aligned} \tag{4.25}$$

or, using more compact notation

$$\tilde{x}(t) = a_0 + \sum_{k=1}^{\infty} a_k \cos(k\omega_0 t) + \sum_{k=1}^{\infty} b_k \sin(k\omega_0 t) \tag{4.26}$$

where $\omega_0 = 2\pi f_0$ is the fundamental frequency in rad/s. Eqn. (4.26) is referred to as the *trigonometric Fourier series (TFS)* representation of the periodic signal $\tilde{x}(t)$, and the sinusoidal functions with radian frequencies of $\omega_0, 2\omega_0, \ldots, k\omega_0$ are referred to as the *basis functions*. Thus, the set of basis functions includes

$$\phi_k(t) = \cos(k\omega_0 t) , \qquad k = 0, 1, 2, \ldots, \infty \tag{4.27}$$

and

$$\psi_k(t) = \sin(k\omega_0 t) , \qquad k = 1, 2, \ldots, \infty \tag{4.28}$$

Using the notation established in Eqns. (4.27) and (4.28), the series representation of the signal $\tilde{x}(t)$ given by Eqn. (4.25) can be written in a more generalized fashion as

$$\tilde{x}(t) = a_0 + \sum_{k=1}^{\infty} a_k \, \phi_k(t) + \sum_{k=1}^{\infty} b_k \, \psi_k(t) \tag{4.29}$$

We will call the frequencies that are integer multiples of the fundamental frequency the *harmonics*. The frequencies $2\omega_0, 3\omega_0, \ldots, k f_0$ are the second, the third, and the k-th harmonics of the fundamental frequency respectively. The basis functions at harmonic frequencies are all periodic with a period of T_0. Therefore, considering our "building blocks" analogy, the signal $\tilde{x}(t)$ which is periodic with period T_0 is represented in terms of building blocks (basis functions) that are also periodic with the same period. Intuitively this makes sense.

Before we tackle the problem of determining the coefficients of the Fourier series representation, it is helpful to observe some properties of harmonically related sinusoids. Using trigonometric identities it can be shown that

$$\int_{t_0}^{t_0+T_0} \cos(m\omega_0 t) \cos(k\omega_0 t) \, dt = \left\{ \begin{array}{ll} T_0/2, & m = k \\ 0, & m \neq k \end{array} \right. \tag{4.30}$$

This is a very significant result. Two cosine basis functions at harmonic frequencies $m\omega_0$ and $k\omega_0$ are multiplied, and their product is integrated over one full period $(t_0, t_0 + T_0)$. When the integer multipliers m and k represent two different harmonics of the fundamental frequency, the result of the integral is zero. A non-zero result is obtained only when $m = k$, that is, when the two cosine functions in the integral are the same. A set of basis functions $\{\cos(k\omega_0 t), \ k = 0, \ldots, \infty\}$ that satisfies Eqn. (4.30) is said to be an *orthogonal* set. Similarly it is easy to show that

$$\int_{t_0}^{t_0+T_0} \sin(m\omega_0 t) \sin(k\omega_0 t) \, dt = \left\{ \begin{array}{ll} T_0/2, & m = k \\ 0, & m \neq k \end{array} \right. \tag{4.31}$$

meaning that the set of basis functions $\{\sin(k\omega_0 t), \ k = 1, \ldots, \infty\}$ is orthogonal as well. Furthermore it can be shown that the two sets are also orthogonal to each other, that is,

$$\int_{t_0}^{t_0+T_0} \cos(m\omega_0 t) \sin(k\omega_0 t) \, dt = 0 \tag{4.32}$$

for any combination of the two integers m and k (even when $m = k$). In Eqns. (4.30) through (4.32) the integral on the left side of the equal sign can be started at any arbitrary time instant t_0 without affecting the result. The only requirement is that integration be carried out over one full period of the signal.

Detailed proofs of orthogonality conditions under a variety of circumstances are given in Appendix D.

We are now ready to determine the unknown coefficients $\{a_k; \ k = 0, 1, \ldots, \infty\}$ and $\{b_k; \ k = 1, \ldots, \infty\}$ of Eqns. (4.25) and (4.26). Let us first change summation indices in Eqn. (4.26) from k to m, then multiply both sides of it with $\cos(k\omega_0 t)$ and integrate over one full period:

$$\begin{aligned} \int_{t_0}^{t_0+T_0} \tilde{x}(t) \cos(k\omega_0 t) \, dt = {} & \int_{t_0}^{t_0+T_0} a_0 \cos(k\omega_0 t) \, dt \\ & + \int_{t_0}^{t_0+T_0} \left[\sum_{m=1}^{\infty} a_m \cos(m\omega_0 t) \right] \cos(k\omega_0 t) \, dt \\ & + \int_{t_0}^{t_0+T_0} \left[\sum_{m=1}^{\infty} b_m \sin(m\omega_0 t) \right] \cos(k\omega_0 t) \, dt \end{aligned} \tag{4.33}$$

Swapping the order of integration and summation in Eqn. (4.33) leads to

$$
\int_{t_0}^{t_0+T_0} \tilde{x}(t) \cos(k\omega_0 t) \, dt = a_0 \int_{t_0}^{t_0+T_0} \cos(k\omega_0 t) \, dt
$$

$$
+ \sum_{m=1}^{\infty} a_m \left[\int_{t_0}^{t_0+T_0} \cos(m\omega_0 t) \cos(k\omega_0 t) \, dt \right]
$$

$$
+ \sum_{m=1}^{\infty} b_m \left[\int_{t_0}^{t_0+T_0} \sin(m\omega_0 t) \cos(k\omega_0 t) \, dt \right] \qquad (4.34)
$$

Let us consider the three terms on the right side of Eqn. (4.34) individually:

1. Let $k > 0$ (we will handle the case of $k = 0$ separately). The first term on the right side of Eqn. (4.34) evaluates to zero since it includes, as a factor, the integral of a cosine function over a full period.

2. In the second term we have a summation. Each term within the summation has a factor which is the integral of the product of two cosines over a span of T_0. Using the orthogonality property observed in Eqn. (4.30), it is easy to see that all terms in the summation disappear with the exception of one term for which $m = k$.

3. In the third term we have another summation. In this case, each term within the summation has a factor which is the integral of the product of a sine function and a cosine function over a span of T_0. Using Eqn. (4.32) we conclude that all terms of this summation disappear.

Therefore, Eqn. (4.34) simplifies to

$$
\int_{t_0}^{t_0+T_0} \tilde{x}(t) \cos(k\omega_0 t) \, dt = a_k \int_{t_0}^{t_0+T_0} \cos^2(k\omega_0 t) \, dt = a_k \frac{T_0}{2}
$$

which can be solved for the only remaining coefficient a_k to yield

$$
a_k = \frac{2}{T_0} \int_{t_0}^{t_0+T_0} \tilde{x}(t) \cos(k\omega_0 t) \, dt \,, \quad \text{for} \quad k = 1, \dots, \infty \qquad (4.35)
$$

Similarly, by multiplying both sides of Eqn. (4.26) with $\sin(k\omega_0 t)$ and repeating the procedure used above, it can be shown that b_k coefficients can be computed as (see Problem 4.6 at the end of this chapter)

$$
b_k = \frac{2}{T_0} \int_{t_0}^{t_0+T_0} \tilde{x}(t) \sin(k\omega_0 t) \, dt \,, \quad \text{for} \quad k = 1, \dots, \infty \qquad (4.36)
$$

Finally, we need to compute the value of the constant coefficient a_0. Integrating both sides of Eqn. (4.26) over a full period, we obtain

$$
\int_{t_0}^{t_0+T_0} \tilde{x}(t) \, dt = \int_{t_0}^{t_0+T_0} a_0 \, dt \qquad (4.37)
$$

$$
+ \int_{t_0}^{t_0+T_0} \left[\sum_{k=1}^{\infty} a_k \cos(k\omega_0 t) \right] dt + \int_{t_0}^{t_0+T_0} \left[\sum_{k=1}^{\infty} b_k \sin(k\omega_0 t) \right] dt
$$

Again changing the order of integration and summation operators, Eqn. (4.36) becomes

$$\int_{t_0}^{t_0+T_0} \tilde{x}(t)\,dt \;=\; \int_{t_0}^{t_0+T_0} a_0\,dt \qquad (4.38)$$

$$+ \sum_{k=1}^{\infty} a_k \left[\int_{t_0}^{t_0+T_0} \cos\left(k\omega_0 t\right)\,dt \right] + \sum_{k=1}^{\infty} b_k \left[\int_{t_0}^{t_0+T_0} \sin\left(k\omega_0 t\right)\,dt \right]$$

Every single term within each of the two summations will be equal to zero due to the periodicity of the sinusoidal functions being integrated. This allows us to simplify Eqn. (4.38) to

$$\int_{t_0}^{t_0+T_0} \tilde{x}(t)\,dt = \int_{t_0}^{t_0+T_0} a_0\,dt = a_0 T_0$$

which can be solved for a_0 to yield

$$a_0 = \frac{1}{T_0} \int_{t_0}^{t_0+T_0} \tilde{x}(t)\,dt \qquad (4.39)$$

A close examination of Eqn. (4.39) reveals that the coefficient a_0 represents the time average of the signal $\tilde{x}(t)$ as defined in Eqn. (1.83) of Chapter 1. Because of this, a_0 is also referred to as the *average value* or the *dc component* of the signal.

Combining the results obtained up to this point, the TFS expansion of a signal can be summarized as follows:

Trigonometric Fourier series (TFS):

1. **Synthesis equation:**

$$\tilde{x}(t) = a_0 + \sum_{k=1}^{\infty} a_k \cos\left(k\omega_0 t\right) + \sum_{k=1}^{\infty} b_k \sin\left(k\omega_0 t\right) \qquad (4.40)$$

2. **Analysis equations:**

$$a_k = \frac{2}{T_0} \int_{t_0}^{t_0+T_0} \tilde{x}(t)\cos\left(k\omega_0 t\right)\,dt, \quad \text{for} \quad k = 1,\ldots,\infty \qquad (4.41)$$

$$b_k = \frac{2}{T_0} \int_{t_0}^{t_0+T_0} \tilde{x}(t)\sin\left(k\omega_0 t\right)\,dt, \quad \text{for} \quad k = 1,\ldots,\infty \qquad (4.42)$$

$$a_0 = \frac{1}{T_0} \int_{t_0}^{t_0+T_0} \tilde{x}(t)\,dt \qquad \text{(dc component)} \qquad (4.43)$$

Example 4.1: Trigonometric Fourier series of a periodic pulse train

A pulse-train signal $\tilde{x}(t)$ with a period of $T_0 = 3$ seconds is shown in Fig. 4.5. Determine the coefficients of the TFS representation of this signal.

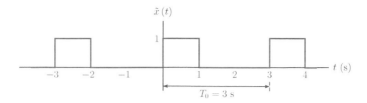

Figure 4.5 – The periodic pulse train used in Example 4.1.

Solution: In using the integrals given by Eqns. (4.41), (4.42), and (4.43), we can start at any arbitrary time instant t_0 and integrate over a span of 3 seconds. Applying Eqn. (4.43) with $t_0 = 0$ and $T_0 = 3$ seconds, we have

$$a_0 = \frac{1}{3}\left[\int_0^1 (1)\, dt + \int_1^3 (0)\, dt\right] = \frac{1}{3}$$

The fundamental frequency is $f_0 = 1/T_0 = 1/3$ Hz, and the corresponding value of ω_0 is

$$\omega_0 = 2\pi f_0 = \frac{2\pi}{3}\quad \text{rad/s.}$$

Using Eqn. (4.41), we have

$$a_k = \frac{2}{3}\left[\int_0^1 (1)\cos(2\pi kt/3)\, dt + \int_1^3 (0)\cos(2\pi kt/3)\, dt\right]$$

$$= \frac{\sin(2\pi k/3)}{\pi k}, \qquad \text{for } k = 1, 2, \ldots, \infty$$

Finally, using Eqn. (4.42), we get

$$b_k = \frac{2}{3}\left[\int_0^1 (1)\sin(2\pi kt/3)\, dt + \int_1^3 (0)\sin(2\pi kt/3)\, dt\right]$$

$$= \frac{1 - \cos(2\pi k/3)}{\pi k}, \qquad \text{for } k = 1, 2, \ldots, \infty$$

Using these coefficients in the synthesis equation given by Eqn. (4.40), the signal $x(t)$ can now be expressed in terms of the basis functions as

$$\tilde{x}(t) = \frac{1}{3} + \sum_{k=1}^{\infty}\left(\frac{\sin(2\pi k/3)}{\pi k}\right)\cos(2\pi kt/3) + \sum_{k=1}^{\infty}\left(\frac{1 - \cos(2\pi k/3)}{\pi k}\right)\sin(2\pi kt/3) \qquad (4.44)$$

Example 4.2: **Approximation with a finite number of harmonics**

Consider again the signal $\tilde{x}(t)$ of Example 4.1. Based on Eqns. (4.25) and (4.26), it would theoretically take an infinite number of cosine and sine terms to obtain an accurate representation of it. On the other hand, values of coefficients a_k and b_k are inversely proportional to k, indicating that the contributions from the higher order terms in Eqn. (4.44) will decline in significance. As a result we may be able to neglect high order terms and still obtain a reasonable approximation to the pulse train. Approximate the periodic pulse train of Example 4.1 using (a) the first 4 harmonics, and (b) the first 10 harmonics.

Solution: Recall that we obtained the following in Example 4.1:

$$a_0 = \frac{1}{3}, \qquad a_k = \frac{\sin\left(2\pi k/3\right)}{\pi k}, \qquad b_k = \frac{1 - \cos\left(2\pi k/3\right)}{\pi k}$$

These coefficients have been numerically evaluated for up to $k = 10$, and are shown in Table 4.1.

k	a_k	b_k
0	0.3333	
1	0.2757	0.4775
2	−0.1378	0.2387
3	0.0	0.0
4	0.0689	0.1194
5	−0.0551	0.0955
6	0.0	0.0
7	0.0394	0.0682
8	−0.0345	0.0597
9	0.0	0.0
10	0.0276	0.0477

Table 4.1 – TFS coefficients for the pulse train of Example 4.2.

Let $\tilde{x}^{(m)}(t)$ be an approximation to the signal $\tilde{x}(t)$ utilizing the first m harmonics of the fundamental frequency:

$$\tilde{x}^{(m)}(t) = a_0 + \sum_{k=1}^{m} a_k \cos\left(k\omega_0 t\right) + \sum_{k=1}^{m} b_k \sin\left(k\omega_0 t\right) \tag{4.45}$$

Using $m = 4$, we have

$$\tilde{x}^{(4)}(t) = 0.3333 + 0.2757 \cos\left(2\pi t/3\right) - 0.1378 \cos\left(4\pi t/3\right) + 0.0689 \cos\left(8\pi t/3\right)$$

$$+ 0.4775 \sin\left(2\pi t/3\right) + 0.2387 \sin\left(4\pi t/3\right) + 0.1194 \sin\left(8\pi t/3\right)$$

A similar but lengthier expression can be written for the case $m = 10$ which we will skip to save space. Fig. 4.6 shows two approximations to the original pulse train using the first 4 and 10 harmonics respectively.

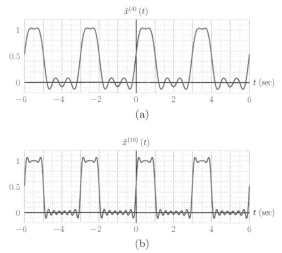

Figure 4.6 – Approximation to the pulse train of Example 4.1 using (a) 4 harmonics, (b) 10 harmonics.

Interactive Demo:　　`tfs_demo1`

The demo program in "`tfs_demo1.m`" provides a graphical user interface for computing finite-harmonic approximations to the pulse train of Examples 4.1 and 4.2. Values of the TFS coefficients a_k and b_k for the pulse train are displayed in the spreadsheet-style table on the left.

Selecting an integer value m from the drop-down list labeled "Largest harmonic to include in approximation" causes the finite-harmonic approximation $\tilde{x}^{(m)}(t)$ to be computed and graphed. At the same time, checkboxes next to the coefficient pairs $\{a_k, b_k\}$ that are included in the approximation are automatically checked.

Alternatively coefficient sets $\{a_k, b_k\}$ may be included or excluded arbitrarily by checking the box to each set of coefficients on or off. When this is done, the graph displays the phrase "free format" as well as a list of the coefficient indices included.

1. Use the drop-down list to compute and graph various finite-harmonics approximations. Observe the similarity of the approximated signal to the original pulse train as larger values of m are used.
2. Use the free format approach for observing the individual contributions of individual $\{a_k, b_k\}$ pairs. (This requires checking one box in the table with all others unchecked.)

Software resources:
`tfs_demo1.m`

Software resources:　　　See MATLAB Exercises 4.1 and 4.2.

Example 4.3: **Periodic pulse train revisited**

Determine the TFS coefficients for the periodic pulse train shown in Fig. 4.7.

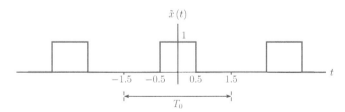

Figure 4.7 – The periodic pulse train used in Example 4.3.

Solution: This is essentially the same pulse train we have used in Example 4.1 with one minor difference: The signal is shifted in the time domain so that the main pulse is centered around the time origin $t = 0$. As a consequence, the resulting signal is an even function of time, that is, it has the property $\tilde{x}(-t) = \tilde{x}(t)$ for $-\infty < t < \infty$.

Let us take one period of the signal to extend from $t_0 = -1.5$ to $t_0 + T_0 = 1.5$ seconds. Applying Eqn. (4.43) with $t_0 = -1.5$ and $T_0 = 3$ seconds, we have

$$a_0 = \frac{1}{3} \int_{t=-0.5}^{0.5} (1) \, dt = \frac{1}{3}$$

Using Eqn. (4.41) yields

$$a_k = \frac{2}{3} \int_{-0.5}^{0.5} (1) \cos\left(2\pi kt/3\right) dt = \frac{2 \sin\left(2\pi k/3\right)}{\pi k}$$

and using Eqn. (4.42)

$$b_k = \frac{2}{3} \int_{-0.5}^{0.5} (1) \sin\left(2\pi kt/3\right) dt = 0$$

Thus, $\tilde{x}(t)$ can be written as

$$\tilde{x}(t) = \frac{1}{3} + \sum_{k=1}^{\infty} \left(\frac{2 \sin\left(\pi k/3\right)}{\pi k} \right) \cos\left(k\omega_0 t\right) \tag{4.46}$$

where the fundamental frequency is $f_0 = 1/3$ Hz. In this case, the signal is expressed using only the $\cos\left(k\omega_0 t\right)$ terms of the TFS expansion. This result is intuitively satisfying since we have already recognized that $\tilde{x}(t)$ exhibits even symmetry, and therefore it can be represented using only the even basis functions $\{ \cos\left(k\omega_0 t\right), \ k = 0, 1, \ldots, \infty \}$, omitting the odd basis functions $\{ \sin\left(k\omega_0 t\right), \ k = 1, 2, \ldots, \infty \}$.
Software resources:
ex_4_3.m

Interactive Demo: tfs_demo2

The demo program in "tfs_demo2.m" is based on Example 4.3, and computes finite-harmonic approximations to the periodic pulse train with even symmetry as shown in Fig. 4.7. It shares the same graphical user interface as in the program "tfs_demo1.m" with the only difference being the even symmetry of the pulse train used. Values of TFS coefficients a_k and b_k are displayed in the spreadsheet-style table on the left. Observe that $b_k = 0$ for all k as we have determined in Example 4.3.
Software resources:
tfs_demo2.m

Example 4.4: **Trigonometric Fourier series for a square wave**

Determine the TFS coefficients for the periodic square wave shown in Fig 4.8.

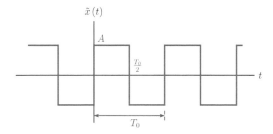

Figure 4.8 – The square-wave signal used in Example 4.4.

Solution: This is a signal with odd symmetry, that is, $\tilde{x}(-t) = -\tilde{x}(t)$ for $-\infty < t < \infty$. Intuitively we can predict that the constant term a_0 and the coefficients a_k of the even basis functions $\{\cos(k\omega_0 t),\ k = 1,\ldots,\infty\}$ in the TFS representation of $\tilde{x}(t)$ should all be equal to zero, and only the odd terms of the series should have significance. Applying Eqn. (4.41) with integration limits $t_0 = -T_0/2$ and $t_0 + T_0 = T_0/2$ seconds, we have

$$a_k = \frac{2}{T_0}\left[\int_{-T_0/2}^{0}(-A)\cos(2\pi kt/T_0)\,dt + \int_{t=0}^{T_0/2}(A)\cos(2\pi kt/T_0)\,dt\right] = 0$$

as we have anticipated. Next, we will use Eqn. (4.43) to determine a_0:

$$a_0 = \frac{1}{T_0}\left[\int_{-T_0/2}^{0}(-A)\,dt + \int_{0}^{T_0/2}(A)\,dt\right] = 0$$

Finally, using Eqn. (4.42), we get the coefficients of the sine terms:

$$b_k = \frac{2}{T_0}\left[\int_{-T_0/2}^{0}(-A)\sin(2\pi kt/T_0)\,dt + \int_{0}^{T_0/2}(A)\sin(2\pi kt/T_0)\,dt\right]$$

$$= \frac{2A}{\pi k}\left[1 - \cos(\pi k)\right]$$

Using the identity

$$\cos(\pi k) = (-1)^k$$

the result found for b_k can be written as

$$b_k = \begin{cases} \dfrac{4A}{\pi k}, & k\ \text{odd} \\[2mm] 0, & k\ \text{even} \end{cases} \qquad (4.47)$$

Compare the result in Eqn. (4.47) to the coefficients b_1 through b_3 we have computed in Eqn. (4.24) in the process of finding the optimum approximation to the square-wave signal $\tilde{x}(t)$ using three sine terms. Table 4.2 lists the TFS coefficients up to the 10-th harmonic.

k	a_k	b_k
0	0.0	
1	0.0	1.2732
2	0.0	0.0
3	0.0	0.4244
4	0.0	0.0
5	0.0	0.2546
6	0.0	0.0
7	0.0	0.1819
8	0.0	0.0
9	0.0	0.1415
10	0.0	0.0

Table 4.2 – TFS coefficients for the square-wave signal of Example 4.4.

Finite-harmonic approximations $\tilde{x}^{(m)}(t)$ of the signal $\tilde{x}(t)$ for $m = 3$ and $m = 9$ are shown in Fig. 4.9(a) and (b).

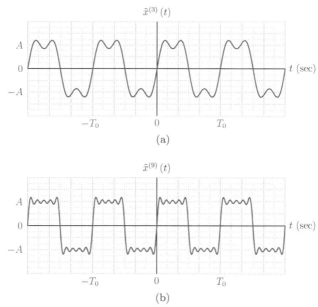

Figure 4.9 – Approximation to the square-wave signal of Example 4.4 using (a) $m = 3$, and (b) $m = 9$.

Software resources:
ex_4_4a.m
ex_4_4b.m

Interactive Demo: tfs_demo3

The demo program in "tfs_demo3.m" is based on the TFS representation of a square-wave signal as discussed in Example 4.4. It computes finite-harmonic approximations to the square-wave signal with odd symmetry as shown in Fig. 4.9. It extends the user interface of the programs "tfs_demo1.m" and "tfs_demo2.m" by allowing the amplitude A and the period T_0 to be varied through the use of slider controls. Observe the following:

1. When the amplitude A is varied, the coefficients b_k change proportionally, as we have determined in Eqn. (4.47).

2. Varying the period T_0 causes the fundamental frequency f_0 to also change. The coefficients b_k are not affected by a change in the period. The finite-harmonic approximation to the signal changes as a result of using a new fundamental frequency with the same coefficients.

Software resources:
tfs_demo3.m

4.2.3   Exponential Fourier series (EFS)

Fourier series representation of the periodic signal $\tilde{x}(t)$ in Eqn. (4.26) can also be written in alternative forms. Consider the use of complex exponentials $\left\{ e^{jk\omega_0 t}, \ n = -\infty, \ldots, \infty \right\}$ as basis functions so that the signal $\tilde{x}(t)$ is expressed as a linear combination of them in the form

$$\tilde{x}(t) = \sum_{k=-\infty}^{\infty} c_k \, e^{jk\omega_0 t} \tag{4.48}$$

where the coefficients c_k are allowed to be complex-valued even though the signal $\tilde{x}(t)$ is real. This is referred to as the *exponential Fourier series (EFS)* representation of the periodic signal. Before we consider this idea for an arbitrary periodic signal $\tilde{x}(t)$ we will study a special case.

Single-tone signals:

Let us first consider the simplest of periodic signals: a single-tone signal in the form of a cosine or a sine waveform. We know that Euler's formula can be used for expressing such a signal in terms of two complex exponential functions:

$$\begin{aligned}
\tilde{x}(t) &= A \cos(\omega_0 t + \theta) \\
&= \frac{A}{2} e^{j(\omega_0 t + \theta)} + \frac{A}{2} e^{-j(\omega_0 t + \theta)} \\
&= \frac{A}{2} e^{j\theta} e^{j\omega_0 t} + \frac{A}{2} e^{-j\theta} e^{-j\omega_0 t}
\end{aligned} \tag{4.49}$$

Comparing Eqn. (4.49) with Eqn. (4.48) we conclude that the cosine waveform can be written in the EFS form of Eqn. (4.48) with coefficients

$$c_1 = \frac{A}{2} e^{j\theta}, \quad c_{-1} = \frac{A}{2} e^{-j\theta}, \quad \text{and} \quad c_k = 0 \text{ for all other } k \tag{4.50}$$

If the signal under consideration is $\tilde{x}(t) = A \sin(\omega_0 t + \theta)$, a similar representation can be obtained using Euler's formula:

$$\begin{aligned}
\tilde{x}(t) &= A \sin(\omega_0 t + \theta) \\
&= \frac{A}{2j} e^{j(\omega_0 t + \theta)} - \frac{A}{2j} e^{-j(\omega_0 t + \theta)}
\end{aligned} \tag{4.51}$$

Using the substitutions

$$\frac{1}{j} = -j = e^{-j\pi/2} \quad \text{and} \quad -\frac{1}{j} = j = e^{j\pi/2}$$

Eqn. (4.51) can be written as

$$\tilde{x}(t) = \frac{A}{2} e^{j(\theta - \pi/2)} e^{j\omega_0 t} + \frac{A}{2} e^{-j(\theta - \pi/2)} e^{-j\omega_0 t} \tag{4.52}$$

Comparison of Eqn. (4.52) with Eqn. (4.48) leads us to the conclusion that the sine waveform in Eqn. (4.52) can be written in the EFS form of Eqn. (4.48) with coefficients

$$c_1 = \frac{A}{2} e^{j(\theta - \pi/2)}, \quad c_{-1} = \frac{A}{2} e^{-j(\theta - \pi/2)}, \quad \text{and} \quad c_k = 0 \text{ for all other } k \tag{4.53}$$

The EFS representations of the two signals are shown graphically, in the form of a *line spectrum*, in Fig. 4.10(a) and (b).

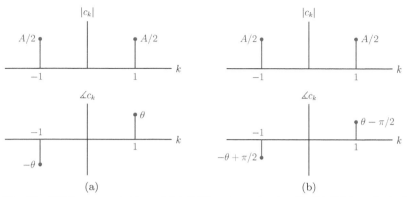

Figure 4.10 – The EFS representation of (a) $\tilde{x}(t) = A \cos(\omega_0 t + \theta)$, (b) $\tilde{x}(t) = A \sin(\omega_0 t + \theta)$.

The magnitudes of the coefficients are identical for the two signals, and the phases differ by $\pi/2$. This is consistent with the relationship

$$A \sin(\omega_0 t + \theta) = A \cos(\omega_0 t + \theta - \pi/2) \tag{4.54}$$

The general case:

We are now ready to consider the EFS representation of an arbitrary periodic signal $\tilde{x}(t)$. In order for the series representation given by Eqn. (4.48) to be equivalent to that of Eqn. (4.26) we need

$$c_0 = a_0 \tag{4.55}$$

and

$$c_{-k} e^{-jk\omega_0 t} + c_k e^{jk\omega_0 t} = a_k \cos(k\omega_0 t) + b_k \sin(k\omega_0 t) \tag{4.56}$$

for each value of the integer index k. Using Euler's formula in Eqn. (4.56) yields

$$[c_{-k} + c_k] \cos(k\omega_0 t) + [-jc_{-k} + jc_k] \sin(k\omega_0 t) = a_k \cos(k\omega_0 t) + b_k \sin(k\omega_0 t) \tag{4.57}$$

Since Eqn. (4.57) must be satisfied for all t, we require that coefficients of sine and cosine terms on both sides be identical. Therefore

$$c_k + c_{-k} = a_k \tag{4.58}$$

and

$$j(c_k - c_{-k}) = b_k \tag{4.59}$$

Solving Eqns. (4.58) and (4.59) for c_k and c_{-k} we obtain

$$c_k = \frac{1}{2}(a_k - jb_k) \tag{4.60}$$

and

$$c_{-k} = \frac{1}{2}(a_k + jb_k) \tag{4.61}$$

for $k = 1, \ldots, \infty$. Thus, EFS coefficients can be computed from the knowledge of TFS coefficients through the use of Eqns. (4.60) and (4.61).

Comparison of Eqns. (4.60) and (4.61) reveals another interesting result: For a real-valued signal $\tilde{x}(t)$, positive and negative indexed coefficients are complex conjugates of each other, that is,

$$c_k = c_{-k}^* \tag{4.62}$$

What if we would like to compute the EFS coefficients of a signal without first having to obtain the TFS coefficients? It can be shown (see Appendix D) that the exponential basis functions also form an orthogonal set:

$$\int_{t_0}^{t_0+T_0} e^{jm\omega_0 t}\, e^{-jk\omega_0 t}\, dt = \begin{cases} T_0\,, & m = k \\ 0\,, & m \neq k \end{cases} \tag{4.63}$$

EFS coefficients can be determined by making use of the orthogonality of the basis function set.

Let us first change the summation index in Eqn. (4.48) from k to m. Afterwards we will multiply both sides with $e^{-jk\omega_0 t}$ to get

$$\tilde{x}(t)\, e^{-jk\omega_0 t} = e^{-jk\omega_0 t} \sum_{m=-\infty}^{\infty} c_m\, e^{jm\omega_0 t}$$

$$= \sum_{m=-\infty}^{\infty} c_m\, e^{jm\omega_0 t}\, e^{-jk\omega_0 t} \tag{4.64}$$

Integrating both sides of Eqn. (4.64) over one period of $\tilde{x}(t)$ and making use of the orthogonality property in Eqn. (4.63), we obtain

$$\int_{t_0}^{t_0+T_0} \tilde{x}(t)\, e^{-jk\omega_0 t}\, dt = \int_{t_0}^{t_0+T_0} \left[\sum_{m=-\infty}^{\infty} c_m\, e^{jm\omega_0 t}\, e^{-jk\omega_0 t} \right] dt$$

$$= \sum_{m=-\infty}^{\infty} c_m \left[\int_{t_0}^{t_0+T_0} e^{jm\omega_0 t}\, e^{-jk\omega_0 t}\, dt \right]$$

$$= c_k\, T_0 \tag{4.65}$$

which we can solve for the coefficient c_k as

$$c_k = \frac{1}{T_0} \int_{t_0}^{t_0+T_0} \tilde{x}(t)\, e^{-jk\omega_0 t}\, dt \tag{4.66}$$

In general, the coefficients of the EFS representation of a periodic signal $\tilde{x}(t)$ are complex-valued. They can be graphed in the form of a line spectrum if each coefficient is expressed in polar complex form with its magnitude and phase:

$$c_k = |c_k|\, e^{j\theta_k} \tag{4.67}$$

Magnitude and phase values in Eqn. (4.67) are computed by

$$|c_k| = \left[(\operatorname{Re}\{c_k\})^2 + (\operatorname{Im}\{c_k\})^2 \right]^{1/2} \tag{4.68}$$

and

$$\theta_k = \angle c_k = \tan^{-1}\left(\frac{\operatorname{Im}\{c_m\}}{\operatorname{Re}\{c_m\}} \right) \tag{4.69}$$

respectively. Example 4.5 will illustrate this. If we evaluate Eqn. (4.66) for $k = 0$ we obtain

$$c_0 = \frac{1}{T_0} \int_{t_0}^{t_0+T_0} \tilde{x}(t)\, dt \tag{4.70}$$

The right side of Eqn. (4.70) is essentially the definition of the time average operator given by Eqn. (1.83) in Chapter 1. Therefore, c_0 is the dc value of the signal $\tilde{x}(t)$.

Exponential Fourier series (EFS):

1. Synthesis equation:

$$\tilde{x}(t) = \sum_{k=-\infty}^{\infty} c_k\, e^{jk\omega_0 t} \qquad (4.71)$$

2. Analysis equation:

$$c_k = \frac{1}{T_0} \int_{t_0}^{t_0+T_0} \tilde{x}(t)\, e^{-jk\omega_0 t}\, dt \qquad (4.72)$$

Example 4.5: Exponential Fourier series for periodic pulse train

Determine the EFS coefficients of the signal $\tilde{x}(t)$ of Example 4.3, shown in Fig. 4.7, through direct application of Eqn. (4.72).

Solution: Using Eqn. (4.72) with $t_0 = -1.5$ s and $T_0 = 3$ s, we obtain

$$c_k = \frac{1}{3} \int_{-0.5}^{0.5} (1)\, e^{-j2\pi kt/3}\, dt = \frac{\sin(\pi k/3)}{\pi k}$$

For this particular signal $\tilde{x}(t)$, the EFS coefficients c_k are real-valued. This will not always be the case. The real-valued result for coefficients $\{c_k\}$ obtained in this example is due to the even symmetry property of the signal $\tilde{x}(t)$. (Remember that, in working with the same signal in Example 4.3, we found $b_k = 0$ for all k. As a result, we have $c_k = c_{-k} = a_k/2$ for all k.)

Before the coefficients $\{c_k\}$ can be graphed or used for reconstructing the signal $\tilde{x}(t)$ the center coefficient c_0 needs special attention. Both the numerator and the denominator of the expression we derived in Eqn. (4.73) become zero for $k = 0$. For this case we need to use L'Hospital's rule which yields

$$c_0 = \frac{\frac{d}{dk}\left[\sin(\pi k/3)\right]}{\frac{d}{dk}[\pi k]}\Bigg|_{k=0} = \frac{(\pi/3)\cos(\pi k/3)}{\pi}\Bigg|_{k=0} = \frac{1}{3}$$

The signal $\tilde{x}(t)$ can be expressed in terms of complex exponential basis functions as

$$\tilde{x}(t) = \sum_{k=-\infty}^{\infty} \left(\frac{\sin(\pi k/3)}{\pi k}\right) e^{j2\pi kt/3} \qquad (4.73)$$

A line graph of the set of coefficients c_k is useful for illustrating the make-up of the signal $\tilde{x}(t)$ in terms of its harmonics, and is shown in Fig. 4.11.

Note that this is not quite in the magnitude-and-phase form of the line spectrum discussed earlier. Even though the coefficients $\{c_k\}$ are real-valued, some of the coefficients are negative. Consequently, the graph in Fig. 4.11 does not qualify to be the magnitude of the spectrum. Realizing that $e^{j\pi} = -1$, any negative-valued coefficient $c_k < 0$ can be expressed as

$$c_k = |c_k|\, e^{j\pi}$$

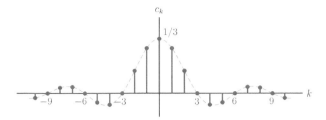

Figure 4.11 – The line spectrum for the pulse train of Example 4.5.

using a phase angle of π radians to account for the negative multiplier. The line spectrum is shown in Fig. 4.12 in its proper form.

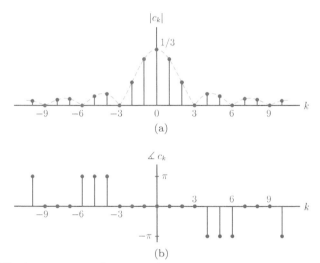

Figure 4.12 – The line spectrum for Example 4.5 in proper magnitude and phase format.

Interactive Demo: efs_demo1

The demo program in "efs_demo1.m" is based on Example 4.5, and provides a graphical user interface for computing finite-harmonic approximations to the pulse train of Fig. 4.7. Values of the EFS coefficients c_k are displayed in Cartesian format in the spreadsheet-style table on the left.

Selecting an integer value m from the drop-down list labeled "Largest harmonic to include in approximation" causes the finite-harmonic approximation $\tilde{x}^{(m)}(t)$ to be computed and graphed. At the same time, checkboxes next to the coefficients c_k included in the finite-harmonic approximation are automatically checked.

Alternatively, arbitrary coefficients may be included or excluded by checking or unchecking the box next to each coefficient. When this is done, the graph displays the phrase "free format" as well as a list of indices of the coefficients included in computation.

1. Use the drop-down list to compute and graph various finite-harmonics approximations. Observe the similarity of the approximated signal to the original pulse train as larger values of m are used.

2. Use the free format approach for observing the individual contributions of individual c_k coefficients by checking only one box in the table with all others unchecked.

Software resources:
efs_demo1.m

| Software resources: | See MATLAB Exercise 4.3. |

Example 4.6: **Periodic pulse train revisited**

In Example 4.5 the EFS coefficients of the pulse train in Fig. 4.7 were found to be purely real. As discussed, this is due to the even symmetry of the signal. In this example we will remove this symmetry to see how it impacts the EFS coefficients. Consider the pulse train shown in Fig. 4.13.

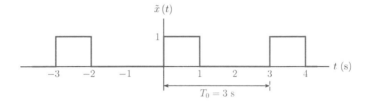

Figure 4.13 – The periodic pulse train for Example 4.6.

Using Eqn. (4.66) with $t_0 = 0$ and $T_0 = 3$ seconds we obtain

$$c_k = \frac{1}{3} \int_0^1 (1) \, e^{-j2\pi kt/3} \, dt = \frac{-1}{j2\pi k} \left[e^{-j2\pi k/3} - 1 \right]$$

After some simplification, it can be shown that real and imaginary parts of the coefficients can be expressed as

$$\text{Re}\{c_k\} = \frac{\sin(2\pi k/3)}{2\pi k} \quad \text{and} \quad \text{Im}\{c_k\} = \frac{\cos(2\pi k/3) - 1}{2\pi k}$$

Contrast these results with the TFS coefficients determined in Example 4.1 for the same signal. TFS representation of the signal $\tilde{x}(t)$ was given in Eqn. (4.44). Recall that the EFS coefficients are related to TFS coefficients by Eqns. (4.60) and (4.61). Using Eqns. (4.68) and (4.69), magnitude and phase of c_k can be computed as

$$|c_k| = \frac{1}{\sqrt{2}\pi |k|} \sqrt{1 - \cos(2\pi k/3)} \tag{4.74}$$

and

$$\theta_k = \tan^{-1}\left(\frac{\cos(2\pi k/3) - 1}{\sin(2\pi k/3)} \right) \tag{4.75}$$

The expression for magnitude can be further simplified. Using the appropriate trigonometric identity[1] we can write

$$\cos(2\pi k/3) = 2\cos^2(\pi k/3) - 1 = 1 - 2\sin^2(\pi k/3)$$

---
[1] $\cos(2a) = 2\cos^2(a) - 1 = 1 - 2\sin^2(a)$.

Substituting this result into Eqn. (4.74) yields

$$|c_k| = \frac{1}{3} \, |\mathrm{sinc}\,(k/3)|$$

where we have used the sinc function defined as

$$\mathrm{sinc}\,(\alpha) \triangleq \frac{\sin(\pi\alpha)}{\pi\alpha}$$

The line spectrum is graphed in Fig. (4.14).

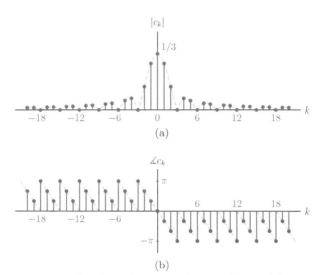

Figure 4.14 – Line spectrum for the pulse train of Example 4.6: (a) magnitude, (b) phase.

Software resources:
ex_4_6.m

Interactive Demo: efs_demo2

The demo program in "efs_demo2.m" is based on Example 4.6 in which we have removed the even symmetry of the pulse train of Example 4.5. If the periodic waveforms used in Examples 4.5 and 4.6 are denoted as $\tilde{x}_1(t)$ and $\tilde{x}_2(t)$ respectively, the relationship between them may be written as

$$\tilde{x}_2(t) = \tilde{x}_1(t - t_d) = \tilde{x}_1(t - 0.5)$$

The demo provides a graphical user interface for adjusting the time delay and observing its effects on the exponential Fourier coefficients as well as the corresponding finite-harmonic approximation.

1. Observe that the magnitude spectrum c_k does not change with changing time-delay.
2. A delay of $t_d = 0.5$ s creates the signal in Example 4.6. Observe the phase of the line spectrum.
3. On the other hand, a delay of $t_d = 0$ creates the signal in Example 4.5 with even symmetry. In this case the phase should be either 0 or 180 degrees.

Software resources:
efs_demo2.m

Software resources: See MATLAB Exercise 4.4.

Example 4.7: **Effects of duty cycle on the spectrum**

Consider the pulse train depicted in Fig. 4.15.

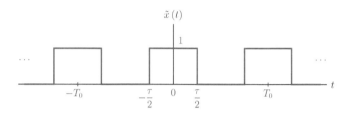

Figure 4.15 – Periodic pulse train for Example 4.7.

Each pulse occupies a time duration of τ within a period length of T_0. The *duty cycle* of a pulse train is defined as the ratio of the pulse-width to the period, that is,

$$d \triangleq \frac{\tau}{T_0} \tag{4.76}$$

The duty cycle plays an important role in pulse-train type waveforms used in electronics and communication systems. Having a small duty cycle means increased blank space between individual pulses, and this can be beneficial in certain circumstances. For example, we may want to take advantage of the large gap between pulses, and utilize the target system to process other signals using a strategy known as *time-division multiplexing*. On the other hand, a small duty cycle does not come without cost as we will see when we graph the line spectrum. Using Eqn. (4.76) with $t_0 = -\tau/2$ we obtain

$$c_k = \frac{1}{T_0} \left[\int_{-\tau/2}^{\tau/2} (1)\, e^{-j2\pi kt/T_0}\, dt \right] = \frac{\sin(\pi k d)}{\pi k}$$

The result in Eqn. (4.77) can be written in a more compact form as

$$c_k = d \operatorname{sinc}(kd) \tag{4.77}$$

In Eqn. (4.77) values of coefficients c_k depend only on the duty cycle and not on the period T_0. On the other hand, the period T_0 impacts actual locations of the coefficients on the frequency axis, and consequently the frequency spacing between successive coefficients. Since the fundamental frequency is $f_0 = 1/T_0$, the coefficient c_k occurs at the frequency $kf_0 = k/T_0$ in the line spectrum.

Magnitudes of coefficients c_k are shown in Fig. 4.16 for duty cycle values of $d = 0.1$, $d = 0.2$, and $d = 0.3$ respectively.

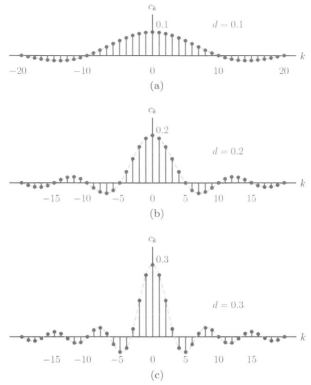

Figure 4.16 – Line spectra for the pulse train of Example 4.7 with duty cycles (a) $d = 0.1$, (b) $d = 0.2$, and (c) $d = 0.3$.

We observe that smaller values of the duty cycle produce increased high-frequency content as the coefficients of large harmonics seem to be stronger for $d = 0.1$ compared to the other two cases.

Software resources:

ex_4_7.m

Example 4.8: **Spectrum of periodic sawtooth waveform**

Consider a periodic sawtooth signal $\tilde{x}(t)$, with a period of $T_0 = 1$ s, defined by

$$\tilde{x}(t) = at\,, \quad 0 < t < 1 \quad \text{and} \quad \tilde{x}(t+1) = \tilde{x}(t)$$

and shown in Fig. 4.17. The parameter a is a real-valued constant. Determine the EFS coefficients for this signal.

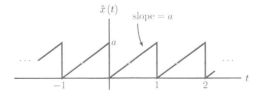

Figure 4.17 – Periodic sawtooth waveform of Example 4.8.

Solution: The fundamental frequency of the signal is

$$f_0 = \frac{1}{T_0} = 1 \text{ Hz}, \qquad \omega_0 = \frac{2\pi}{T_0} = 2\pi \text{ rad/s}$$

Using Eqn. (4.72) we obtain

$$c_k = \int_0^1 at\, e^{-j2\pi kt}\, dt$$

$$= a \left[\frac{t\, e^{-j2\pi kt}}{-j2\pi k} \bigg|_0^1 + \int_0^1 \frac{e^{-j2\pi kt}}{j2\pi k}\, dt \right]$$

$$= a \left[-\frac{e^{-j2\pi k}}{j2\pi k} + \frac{e^{-j2\pi k} - 1}{4\pi^2 k^2} \right] \tag{4.78}$$

where we have used *integration by parts*.[2] Using Euler's formula on Eqn. (4.78), real and imaginary parts of the EFS coefficients are obtained as

$$\text{Re}\{c_k\} = a \left[\frac{\sin(2\pi k)}{2\pi k} + \frac{\cos(2\pi k) - 1}{4\pi^2 k^2} \right] \tag{4.79}$$

and

$$\text{Im}\{c_k\} = a \left[\frac{\cos(2\pi k)}{2\pi k} - \frac{\sin(2\pi k)}{4\pi^2 k^2} \right] \tag{4.80}$$

respectively. These two expressions can be greatly simplified by recognizing that $\sin(2\pi k) = 0$ and $\cos(2\pi k) = 1$ for all integers k. For $k = 0$, the values of $\text{Re}\{c_k\}$ and $\text{Im}\{c_k\}$ need to be resolved by using L'Hospital's rule on Eqns. (4.79) and (4.80). Thus, the real and imaginary parts of c_k are

$$\text{Re}\{c_k\} = \begin{cases} \dfrac{a}{2}, & k = 0 \\ 0, & k \neq 0 \end{cases} \tag{4.81}$$

$$\text{Im}\{c_k\} = \begin{cases} 0, & k = 0 \\ \dfrac{a}{2\pi k}, & k \neq 0 \end{cases} \tag{4.82}$$

Combining Eqns. (4.81) and (4.82) the magnitudes of the EFS coefficients are obtained as

$$|c_k| = \begin{cases} \dfrac{|a|}{2}, & k = 0 \\ \left| \dfrac{a}{2\pi k} \right|, & k \neq 0 \end{cases}$$

For the phase angle we need to pay attention to the sign of the parameter a. If $a \geq 0$, we have

$$\theta_k = \begin{cases} 0, & k = 0 \\ \pi/2, & k > 0 \\ -\pi/2, & k < 0 \end{cases}$$

If $a < 0$, then θ_k needs to be modified as

$$\theta_k = \begin{cases} \pi, & k = 0 \\ -\pi/2, & k > 0 \\ \pi/2, & k < 0 \end{cases}$$

---

[2] $\int_a^b u\, dv = uv\, \big|_a^b - \int_a^b v\, du.$

Magnitude and phase spectra for $\tilde{x}(t)$ are graphed in Fig. 4.18(a) and (b) with the assumption that $a > 0$.

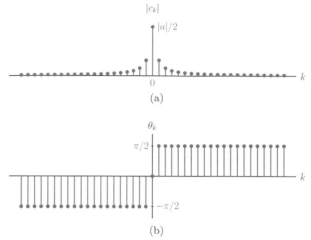

(a)

(b)

Figure 4.18 – Line spectrum for the periodic sawtooth waveform of Example 4.8: (a) magnitude, (b) phase.

Software resources:
ex_4_8a.m
ex_4_8b.m

Example 4.9: **Spectrum of multi-tone signal**

Determine the EFS coefficients and graph the line spectrum for the multi-tone signal

$$\tilde{x}(t) = \cos\left(2\pi\left[10f_0\right]t\right) + 0.8\cos\left(2\pi f_0 t\right)\cos\left(2\pi\left[10f_0\right]t\right)$$

shown in Fig. 4.19.

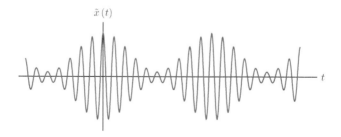

Figure 4.19 – Multi-tone signal of Example 4.9.

Solution: Using the appropriate trigonometric identity[3] the signal $\tilde{x}(t)$ can be written as

$$\tilde{x}(t) = \cos\left(2\pi\left[10f_0\right]t\right) + 0.4\cos\left(2\pi\left[11f_0\right]t\right) + 0.4\cos\left(2\pi\left[9f_0\right]t\right)$$

Applying Euler's formula, we have

$$\begin{aligned}
\tilde{x}(t) =& 0.5\,e^{j2\pi(10f_0)t} + 0.5\,e^{-j2\pi(10f_0)t} \\
&+ 0.2\,e^{j2\pi(11f_0)t} + 0.2\,e^{-j2\pi(11f_0)t} + 0.2\,e^{j2\pi(9f_0)t} + 0.2\,e^{-j2\pi(9f_0)t}
\end{aligned} \qquad (4.83)$$

---

[3] $\cos(a)\cos(b) = \frac{1}{2}\cos(a+b) + \frac{1}{2}\cos(a-b)$.

By inspection of Eqn. (4.83) the significant EFS coefficients for the signal $\tilde{x}(t)$ are found as

$$
\begin{aligned}
c_9 &= c_{-9} = 0.2, \\
c_{10} &= c_{-10} = 0.5, \\
c_{11} &= c_{-11} = 0.2
\end{aligned}
$$

and all other coefficients are equal to zero. The resulting line spectrum is shown in Fig. 4.20.

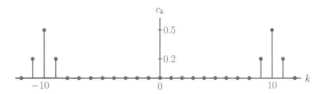

Figure 4.20 – Line spectrum for the multi-tone signal of Example 4.9.

Software resources:
ex_4_9.m

Example 4.10: **Spectrum of half-wave rectified sinusoidal signal**

Determine the EFS coefficients and graph the line spectrum for the half-wave periodic signal $\tilde{x}(t)$ defined by

$$
\tilde{x}(t) = \begin{cases} \sin(\omega_0 t), & 0 \le t < T_0/2 \\ 0, & T_0/2 \le t < T_0 \end{cases}, \quad \text{and} \quad \tilde{x}(t + T_0) = \tilde{x}(t) \tag{4.84}
$$

and shown in Fig. 4.21.

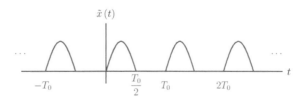

Figure 4.21 – Half-wave rectified sinusoidal signal of Example 4.10.

Solution: Using the analysis equation given by Eqn. (4.72), the EFS coefficients are

$$
c_k = \frac{1}{T_0} \int_0^{T_0/2} \sin(\omega_0 t) \, e^{-jk\omega_0 t} \, dt
$$

In order to simplify the evaluation of the integral in Eqn. (4.85) we will write the sine function using Euler's formula, and obtain

$$
c_k = \frac{1}{j2T_0} \int_0^{T_0/2} \left[e^{j\omega_0 t} - e^{-j\omega_0 t} \right] e^{-jk\omega_0 t} \, dt
$$

$$
= \frac{1}{j2T_0} \left. \left(\frac{e^{-j\omega_0(k-1)}}{-j\omega_0(k-1)} \right) \right|_{t=0}^{T_0/2} - \frac{1}{j2T_0} \left. \left(\frac{e^{-j\omega_0(k+1)}}{-j\omega_0(k+1)} \right) \right|_{t=0}^{T_0/2}
$$

which can be simplified to yield

$$c_k = \frac{1}{4\pi(k-1)}\left[e^{-j\pi(k-1)} - 1\right] - \frac{1}{4\pi(k+1)}\left[e^{-j\pi(k+1)} - 1\right] \tag{4.85}$$

We need to consider even and odd values of k separately.

Case 1: k odd and $k \neq \mp 1$

If k is odd, then both $k-1$ and $k+1$ are non-zero and even. We have $e^{-j\pi(k-1)} = e^{-j\pi(k+1)} = 1$, and therefore $c_k = 0$.

Case 2: $k = 1$

If $k = 1$, c_1 has an indeterminate form which can be resolved through the use of L'Hospital's rule:

$$c_1 = \left.\frac{-j\pi\,e^{-j\pi(k-1)}}{4\pi}\right|_{k=1} = -\frac{j}{4}$$

Case 3: $k = -1$

Similar to the case of $k = 1$ an indeterminate form is obtained for $k = -1$. Through the use of L'Hospital's rule the coefficient c_{-1} is found to be

$$c_{-1} = \left.\frac{j\pi\,e^{-j\pi(k+1)}}{4\pi}\right|_{k=-1} = \frac{j}{4}$$

Case 4: k even

In this case both $k-1$ and $k+1$ are odd, and we have $e^{-j\pi(k-1)} = e^{-j\pi(k+1)} = -1$.

$$e^{-j\pi(k-1)} - 1 = e^{-j\pi(k+1)} - 1 = -2$$

Using this result in Eqn. (4.85) we get

$$c_k = -\frac{1}{2\pi(k-1)} + \frac{1}{2\pi(k+1)} = \frac{-1}{\pi(k^2-1)}$$

Combining the results of the four cases, the EFS coefficients are

$$c_k = \begin{cases} 0, & k \text{ odd and } k \neq \mp 1 \\ -j/4, & k = 1 \\ j/4, & k = -1 \\ \dfrac{-1}{\pi(k^2-1)}, & k \text{ even} \end{cases}$$

The resulting line spectrum is shown in Fig. 4.22.

Software resources:

ex_4_10a.m

ex_4_10b.m

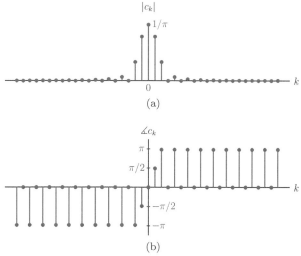

Figure 4.22 – Line spectrum for the half-wave sinusoidal signal of Example 4.10.

4.2.4   Compact Fourier series (CFS)

Yet another form of the Fourier series representation of a periodic signal is the *compact Fourier series (CFS)* expressed as

$$\tilde{x}(t) = d_0 + \sum_{k=1}^{\infty} d_k \cos(k\omega_0 t + \phi_k) \tag{4.86}$$

Using the appropriate trigonometric identity[4] Eqn. (4.86) can be written as

$$\tilde{x}(t) = d_0 + \sum_{k=1}^{\infty} \left[d_k \cos(k\omega_0 t) \cos(\phi_k) - d_k \sin(k\omega_0 t) \sin(\phi_k) \right]$$

$$= d_0 + \sum_{k=1}^{\infty} d_k \cos(k\omega_0 t) \cos(\phi_k) - \sum_{k=1}^{\infty} d_k \sin(k\omega_0 t) \sin(\phi_k)$$

Recognizing that this last equation for $\tilde{x}(t)$ is in a format identical to the TFS expansion of the same signal given by Eqn. (4.26), the coefficients of the corresponding terms must be equal. Therefore the following must be true:

$$a_0 = d_0 \tag{4.87}$$

$$a_k = d_k \cos(\phi_k), \quad k = 1, \ldots, \infty \tag{4.88}$$

$$b_k = -d_k \sin(\phi_k), \quad k = 1, \ldots, \infty \tag{4.89}$$

Solving for d_k and ϕ_k from Eqns. (4.87), (4.88) and (4.89) we obtain

$$d_k = \sqrt{\{a_k^2 + b_k^2\}}, \quad k = 1, \ldots, \infty \tag{4.90}$$

---

[4] $\cos(a+b) = \cos(a)\cos(b) - \sin(a)\sin(b)$.

and

$$\phi_k = -\tan^{-1}\left(\frac{b_k}{a_k}\right) \ , \qquad k = 1, \ldots, \infty \tag{4.91}$$

with $d_0 = a_0$, and $\phi_0 = 0$. CFS coefficients can also be obtained from the EFS coefficients by using Eqns. (4.88) and (4.89) in conjunction with Eqns. (4.60) and (4.61), and remembering that $c_{-k} = c_k^*$ for real-valued signals:

$$d_k = 2\left[(\text{Re}\{c_k\})^2 + (\text{Im}\{c_k\})^2\right]^{1/2} = 2\,|c_k| \ , \qquad k = 1, \ldots, \infty \tag{4.92}$$

and

$$\phi_k = \tan^{-1}\left(\frac{\text{Im}\{c_k\}}{\text{Re}\{c_k\}}\right) = \theta_k \ , \qquad k = 1, \ldots, \infty \tag{4.93}$$

In the use of Eqn. (4.93), attention must be paid to the quadrant of the complex plane in which the coefficient c_k resides.

Example 4.11: **Compact Fourier series for a periodic pulse train**

Determine the CFS coefficients for the periodic pulse train that was used in Example 4.6. Afterwards, using the CFS coefficients, find an approximation to $\tilde{x}(t)$ using $m = 4$ harmonics.

Solution: CFS coefficients can be obtained from the EFS coefficients found in Example 4.6 along with Eqns. (4.92) and (4.93):

$$d_0 = c_0 \qquad \text{and} \qquad d_k = 2\,|c_k| = \frac{2}{3}\,|\operatorname{sinc}(k/3)| \ , \qquad k = 1, \ldots, \infty$$

and

$$\phi_k = \tan^{-1}\left(\frac{\cos(2\pi k/3) - 1}{\sin(2\pi k/3)}\right) \ , \qquad k = 1, \ldots, \infty$$

Table 4.3 lists values of some of the compact Fourier series coefficients for the pulse train $x(t)$.

k	d_k	ϕ_k (rad)
0	0.3333	0.0000
1	0.5513	−1.0472
2	0.2757	−2.0944
3	0.0000	0.0000
4	0.1378	−1.0472
5	0.1103	−2.0944
6	0.0000	0.0000
7	0.0788	−1.0472
8	0.0689	−2.0944
9	0.0000	0.0000
10	0.0551	−1.0472

Table 4.3 – Compact Fourier series coefficients for the waveform in Example 4.11.

Let $\tilde{x}^{(m)}(t)$ be an approximation to the signal $\tilde{x}(t)$ utilizing the first m harmonics of the fundamental frequency, that is,

$$\tilde{x}^{(m)}(t) = d_0 + \sum_{k=1}^{m} d_k \cos(k\omega_0 t + \phi_k) \tag{4.94}$$

Using $m = 4$ and $f_0 = 1/3$ Hz we have

$$x^{(4)}(t) = 0.3333 + 0.5513 \cos\left(2\pi\left(\tfrac{1}{3}\right)t - 1.0472\right)$$

$$+ 0.2757 \cos\left(2\pi\left(\tfrac{2}{3}\right)t - 2.0944\right) + 0.1378 \cos\left(2\pi\left(\tfrac{4}{3}\right)t - 1.0472\right) \qquad (4.95)$$

The individual terms in Eqn. (4.95) as well as the resulting finite-harmonic approximation are shown in Fig. 4.23.

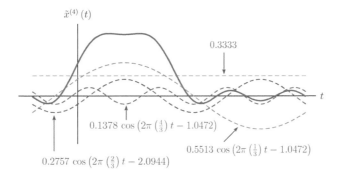

Figure 4.23 – The contributing terms in Eqn. (4.95) and the the resulting finite-harmonic approximation.

4.2.5   Existence of Fourier series

For a specified periodic signal $\tilde{x}(t)$, one question that needs to be considered is the existence of the Fourier series: Is it always possible to determine the Fourier series coefficients? The answer is well-known in mathematics. A periodic signal can be uniquely expressed using sinusoidal basis functions at harmonic frequencies provided that the signal satisfies a set of conditions known as *Dirichlet conditions* named after the German mathematician Johann Peter Gustav Lejeune Dirichlet (1805–1859). A thorough mathematical treatment of Dirichlet conditions is beyond the scope of this text. For the purpose of the types of signals we will encounter in this text, however, it will suffice to summarize the three conditions as follows:

1. The signal $\tilde{x}(t)$ must be integrable over one period in an absolute sense, that is

$$\int_0^{T_0} |\tilde{x}(t)|\, dt < \infty \qquad (4.96)$$

   Any periodic signal in which the amplitude values are bounded will satisfy Eqn. (4.96). In addition, periodic repetitions of singularity functions such as a train of impulses repeated every T_0 seconds will satisfy it as well.

2. If the signal $\tilde{x}(t)$ has discontinuities, it must have at most a finite number of them in one period. Signals with an infinite number of discontinuities in one period cannot be expanded into Fourier series.

3. The signal $\tilde{x}(t)$ must have at most a finite number of minima and maxima in one period. Signals with an infinite number of minima and maxima in one period cannot be expanded into Fourier series.

Most signals we encounter in engineering applications satisfy the existence criteria listed above. Consequently, they can be represented in series expansion forms given by Eqns. (4.26), (4.48) or (4.86). At points where the signal $\tilde{x}(t)$ is continuous, its Fourier series expansion converges perfectly. At a point of discontinuity, however, the Fourier series expansion (TFS, EFS or CFS representation of the signal) yields the average value obtained by approaching the discontinuity from opposite directions. If the signal $\tilde{x}(t)$ has a discontinuity at $t = t_0$, and if $\{c_k\}$ are the EFS coefficients for it, we have

$$\sum_{k=-\infty}^{\infty} c_k\, e^{jk\omega_0 t_0} = \frac{1}{2} \lim_{t \to t_0^-} [\tilde{x}(t)] + \frac{1}{2} \lim_{t \to t_0^+} [\tilde{x}(t)] \tag{4.97}$$

This is illustrated in Fig. 4.24.

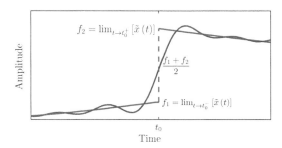

Figure 4.24 – Convergence of Fourier series at a discontunuity.

4.2.6   Gibbs phenomenon

Let us further explore the issue of convergence of the Fourier series at a discontinuity. Consider the periodic square wave shown in Fig. 4.25.

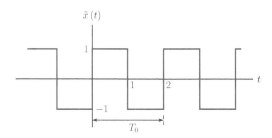

Figure 4.25 – The square wave with period $T_0 = 2$ s.

The TFS coefficients of this signal were found in Example 4.4, and are repeated below:

$$a_k = 0 , \qquad \text{and} \qquad b_k = \begin{cases} \dfrac{4}{\pi k} , & k \text{ odd} \\[2mm] 0 , & k \text{ even} \end{cases}$$

Finite-harmonic approximation to the signal $\tilde{x}(t)$ using m harmonics is

$$\tilde{x}^{(m)}(t) = \sum_{\substack{k=1 \\ k \text{ odd}}}^{m} \left(\frac{4}{\pi k} \right) \sin(\pi t)$$

and the approximation error that occurs when m harmonics is used is

$$\tilde{\varepsilon}^{(m)}\left(t\right) = \tilde{x}\left(t\right) - \tilde{x}^{(m)}\left(t\right)$$

Several finite-harmonic approximations to $\tilde{x}\left(t\right)$ are shown in Fig. 4.26 for $m = 1, 3, 9, 25$. The approximation error is also shown for each case.

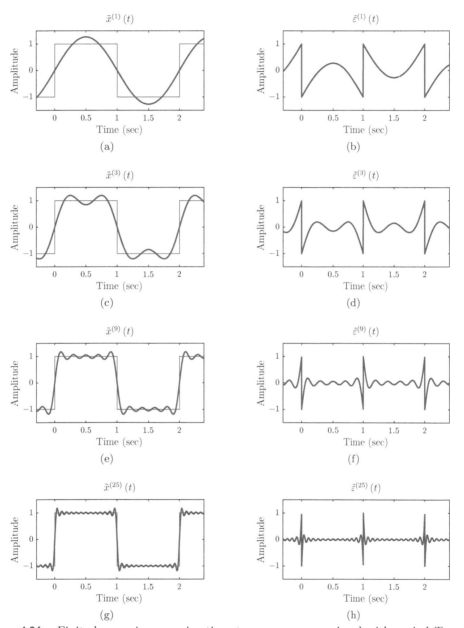

Figure 4.26 – Finite-harmonic approximations to a square-wave signal with period $T_0 = 2$ s and the corresponding approximation errors.

We observe that the approximation error is non-uniform. It seems to be relatively large right before and right after a discontinuity, and it gets smaller as we move away from discontinuities. The Fourier series approximation overshoots the actual signal value at one side of each discontinuity, and undershoots it at the other side. The amount of overshoot or undershoot is about 9 percent of the height of the discontinuity, and cannot be reduced by increasing m. This is a form of the *Gibbs phenomenon* named after American scientist Josiah Willard Gibbs (1839-1903) who explained it in 1899. An enlarged view of the approximation and the error right around the discontinuity at $t = 1$ s are shown in Fig. 4.27 for $m = 25$. A further enlarged view of the approximation error corresponding to the shaded area in Fig. 4.27(b) is shown in Fig. 4.28. Notice that the positive and negative lobes around the discontinuity have amplitudes of about ± 0.18 which is 9 percent of the amplitude of the discontinuity.

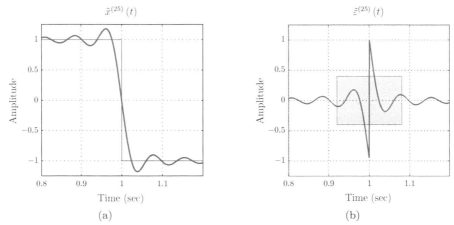

Figure 4.27 – Enlarged view of the finite-harmonic approximation and the approximation error for $m = 25$.

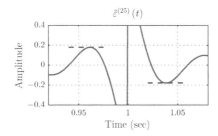

Figure 4.28 – Further enlarged view of the approximation error for $m = 25$.

One way to explain the reason for the Gibbs phenomenon would be to link it to the inability of sinusoidal basis functions that are continuous at every point to approximate a discontinuity in the signal.

4.2.7 Properties of Fourier series

Some fundamental properties of the Fourier series will be briefly explored in this section using the exponential (EFS) form of the series. A more thorough discussion of these properties will be given in Section 4.3.5 for the Fourier transform.

Linearity

For any two signals $\tilde{x}(t)$ and $\tilde{y}(t)$ periodic with $T_0 = 2\pi/\omega_0$ and with their respective series expansions

$$\tilde{x}(t) = \sum_{k=-\infty}^{\infty} c_k\, e^{jk\omega_0 t}$$

$$\tilde{y}(t) = \sum_{k=-\infty}^{\infty} d_k\, e^{jk\omega_0 t}$$

and any two constants α_1 and α_2, it can be shown that the following relationship holds:

Linearity of the Fourier series:

$$\alpha_1\, \tilde{x}(t) + \alpha_2\, \tilde{y}(t) = \sum_{k=-\infty}^{\infty} [\alpha_1\, c_k + \alpha_2\, d_k]\, e^{jk\omega_0 t} \qquad (4.98)$$

Symmetry of Fourier series

Conjugate symmetry and conjugate antisymmetry properties were defined for discrete-time signals in Section 1.4.6 of Chapter 1. Same definitions apply to Fourier series coefficients as well. EFS coefficients c_k are said to be *conjugate symmetric* if they satisfy

$$c_{-k} = c_k^* \qquad (4.99)$$

for all k. Similarly, the coefficients form a *conjugate antisymmetric* set if they satisfy

$$c_{-k} = -c_k^* \qquad (4.100)$$

for all k.

  If the signal $\tilde{x}(t)$ is real-valued, it can be shown that its EFS coefficients form a conjugate symmetric set. Conversely, if the signal $\tilde{x}(t)$ is purely imaginary, its EFS coefficients form a conjugate antisymmetric set.

Symmetry of Fourier series:

$$\tilde{x}(t): \text{Real}, \quad \text{Im}\{\tilde{x}(t)\} = 0 \quad \text{implies that} \quad c_{-k} = c_k^* \qquad (4.101)$$

$$\tilde{x}(t): \text{Imag}, \quad \text{Re}\{\tilde{x}(t)\} = 0 \quad \text{implies that} \quad c_{-k} = -c_k^* \qquad (4.102)$$

Polar form of EFS coefficients

Recall that the EFS coefficients can be written in polar form as

$$c_k = |c_k|\, e^{j\theta_k} \qquad (4.103)$$

If the set $\{c_k\}$ is conjugate symmetric, the relationship in Eqn. (4.99) leads to

$$|c_{-k}|\, e^{j\theta_{-k}} = |c_k|\, e^{-j\theta_k} \qquad (4.104)$$

using the polar form of the coefficients. The consequences of Eqn. (4.104) are obtained by equating the magnitudes and the phases on both sides.

Conjugate symmetric coefficients: $c_{-k} = c_k^*$

$$\text{Magnitude:}\qquad |c_{-k}| = |c_k| \qquad\qquad (4.105)$$

$$\text{Phase:}\qquad \theta_{-k} = -\,\theta_k \qquad\qquad (4.106)$$

It was established in Eqn. (4.101) that the EFS coefficients of a real-valued $\tilde{x}(t)$ are conjugate symmetric. Based on the results in Eqns. (4.105) and (4.106) the magnitude spectrum is an even function of k, and the phase spectrum is an odd function.

Similarly, if the set $\{c_k\}$ is conjugate antisymmetric, the relationship in Eqn. (4.100) reflects on polar form of c_k as

$$|c_{-k}|\;e^{j\theta_{-k}} = -\,|c_k|\;e^{-j\theta_k} \qquad\qquad (4.107)$$

The negative sign on the right side of Eqn. (4.107) needs to be incorporated into the phase since we could not write $|c_{-k}| = -\,|c_k|$ (recall that magnitude must to be non-negative). Using $e^{\mp j\pi} = -1$, Eqn. (4.107) becomes

$$|c_{-k}|\;e^{j\theta_k} = |c_k|\;e^{-j\theta_k}\;e^{\mp j\pi}$$

$$= |c_k|\;e^{-j(\theta_k \pm \pi)} \qquad\qquad (4.108)$$

The consequences of Eqn. (4.108) are summarized below.

Conjugate antisymmetric coefficients: $c_{-k} = -c_k^*$

$$\text{Magnitude:}\qquad |c_{-k}| = |c_k| \qquad\qquad (4.109)$$

$$\text{Phase:}\qquad \theta_{-k} = -\,\theta_k \pm \pi \qquad\qquad (4.110)$$

A purely imaginary signal $\tilde{x}(t)$ leads to a set of EFS coefficients with conjugate anti-symmetry. The corresponding magnitude spectrum is an even function of k as suggested by Eqn. (4.109). The phase is neither even nor odd.

Fourier series for even and odd signals

If the real-valued signal $\tilde{x}(t)$ is an even function of time, the resulting EFS spectrum c_k is real-valued for all k.

$$\tilde{x}(-t) = \tilde{x}(t),\ \text{all}\ t\quad \text{implies that}\quad \text{Im}\,\{c_k\} = 0\,,\text{all}\ k \qquad\qquad (4.111)$$

Conversely it can also be proven that, if the real-valued signal $\tilde{x}(t)$ has odd-symmetry, the resulting EFS spectrum is purely imaginary.

$$\tilde{x}(-t) = -\tilde{x}(t),\ \text{all}\ t\quad \text{implies that}\quad \text{Re}\,\{c_k\} = 0\,,\text{all}\ k \qquad\qquad (4.112)$$

Time shifting

For a signal with EFS expansion

$$\tilde{x}(t) = \sum_{k=-\infty}^{\infty} c_k \, e^{jk\omega_0 t}$$

it can be shown that

$$\tilde{x}(t-\tau) = \sum_{k=-\infty}^{\infty} \left[c_k \, e^{-jk\omega_0 \tau} \right] e^{jk\omega_0 t} \qquad (4.113)$$

The consequence of time shifting $\tilde{x}(t)$ is multiplication of its EFS coefficients by a complex exponential function of frequency.

4.3   Analysis of Non-Periodic Continuous-Time Signals

In Section 4.2 of this chapter we have discussed methods of representing periodic signals by means of harmonically related basis functions. The ability to express a periodic signal as a linear combination of standard basis functions allows us to use the superposition principle when such a signal is used as input to a linear and time-invariant system. We must also realize, however, that we often work with signals that are not necessarily periodic. We would like to have similar capability when we use non-periodic signals in conjunction with linear and time-invariant systems. In this section we will work on generalizing the results of the previous section to apply to signals that are not periodic. These efforts will lead us to the *Fourier transform* for continuous-time signals.

4.3.1   Fourier transform

In deriving the Fourier transform for non-periodic signals we will opt to take an intuitive approach at the expense of mathematical rigor. Our approach is to view a non-periodic signal as the limit case of a periodic one, and make use of the exponential Fourier series discussion of Section 4.2. Consider the non-periodic signal $x(t)$ shown in Fig. 4.29.

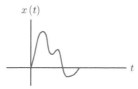

Figure 4.29 – A non-periodic signal $x(t)$.

What frequencies are contained in this signal? What kind of a specific mixture of various frequencies needs to be assembled in order to construct this signal from a standard set of basis functions?

We already know how to represent periodic signals in the frequency domain. Let us construct a periodic extension $\tilde{x}(t)$ of the signal $x(t)$ by repeating it at intervals of T_0.

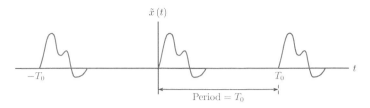

Figure 4.30 – Periodic extension $\tilde{x}(t)$ of the signal $x(t)$.

This is illustrated in Fig. 4.30. The periodic extension $\tilde{x}(t)$ can be expressed as a sum of time-shifted versions of $x(t)$ shifted by all integer multiples of T_0 to yield

$$\tilde{x}(t) = \ldots + x(t + T_0) + x(t) + x(t - T_0) + x(t - 2T_0) + \ldots \tag{4.114}$$

The selected period T_0 should be sufficiently large so as not to change the shape of the signal by causing overlaps. Putting Eqn. (4.114) in summation form we obtain

$$\tilde{x}(t) = \sum_{k=-\infty}^{\infty} x(t - kT_0) \tag{4.115}$$

Since $\tilde{x}(t)$ is periodic, it can be analyzed in the frequency domain by using the methods developed in the previous section. The EFS representation of a periodic signal was given by Eqn. (4.71) which is repeated here:

$$\text{Eqn. (4.71):} \quad \tilde{x}(t) = \sum_{k=-\infty}^{\infty} c_k\, e^{jk\omega_0 t}$$

The coefficients c_k of the EFS expansion of the signal $\tilde{x}(t)$ are computed as

$$\text{Eqn. (4.72):} \quad c_k = \frac{1}{T_0} \int_{-T_0/2}^{T_0/2} \tilde{x}(t)\, e^{-jk\omega_0 t}\, dt$$

We have used the starting point $t_0 = -T_0/2$ for the integral in Eqn. (4.116) so that the lower and upper integration limits are symmetric. Fundamental frequency is the reciprocal of the period, that is,

$$f_0 = \frac{1}{T_0}$$

and the fundamental radian frequency is

$$\omega_0 = 2\pi f_0 = \frac{2\pi}{T_0}$$

Realizing that $\tilde{x}(t) = x(t)$ within the span $-T_0/2 < t < T_0/2$ of the integral, let us write Eqn. (4.116) as

$$c_k = \frac{1}{T_0} \int_{-T_0/2}^{T_0/2} x(t)\, e^{-jk\omega_0 t}\, dt \tag{4.116}$$

If the period T_0 is allowed to become very large, the periodic signal $\tilde{x}(t)$ would start to look more and more similar to $x(t)$. In the limit we would have

$$\lim_{T_0 \to \infty} [\tilde{x}(t)] = x(t) \tag{4.117}$$

As the period T_0 becomes very large, the fundamental frequency f_0 becomes very small. In the limit, as we force $T_0 \to \infty$, the fundamental frequency becomes infinitesimal:

$$T_0 \to \infty \quad \text{implies that} \quad \Delta f = \frac{1}{T_0} \to 0 \text{ and } \Delta\omega = \frac{2\pi}{T_0} \to 0 \tag{4.118}$$

In Eqn. (4.118) we have switched to the notation Δf and $\Delta\omega$ instead of f_0 and ω_0 to emphasize the infinitesimal nature of the fundamental frequency. Applying this change to the result in Eqn. (4.116) leads to

$$c_k = \frac{1}{T_0} \int_{-T_0/2}^{T_0/2} x(t) \, e^{-jk\,\Delta\omega\, t} \, dt \tag{4.119}$$

where c_k is the contribution of the complex exponential at the frequency $\omega = k\,\Delta\omega$. Because of the large T_0 term appearing in the denominator of the right side of Eqn. (4.119), each individual coefficient c_k is very small in magnitude, and in the limit we have $c_k \to 0$. In addition, successive harmonics $k\,\Delta\omega$ are very close to each other due to infinitesimally small $\Delta\omega$. Let us multiply both sides of Eqn. (4.119) by T_0 to obtain

$$c_k\, T_0 = \int_{-T_0/2}^{T_0/2} x(t) \, e^{-jk\,\Delta\omega t} \, dt \tag{4.120}$$

If we now take the limit as $T_0 \to \infty$, and substitute $\omega = k\,\Delta\omega$ we obtain

$$X(\omega) = \lim_{T_0 \to \infty} [c_k\, T_0]$$
$$= \lim_{T_0 \to \infty} \left[\int_{-T_0/2}^{T_0/2} x(t) \, e^{-jk\,\Delta\omega t} \, dt \right]$$
$$= \int_{-\infty}^{\infty} x(t) \, e^{-j\omega t} \, dt \tag{4.121}$$

The function $X(\omega)$ is the *Fourier transform* of the non-periodic signal $x(t)$. It is a continuous function of frequency as opposed to the EFS line spectrum for a periodic signal that utilizes only integer multiples of a fundamental frequency. We can visualize this by imagining that each harmonic of the fundamental frequency comes closer and closer to its neighbors, finally closing the gaps and turning into a continuous function of ω.

At the start of the discussion it was assumed that the signal $x(t)$ has finite duration. What if this is not the case? Would we be able to use the transform defined by Eqn. (4.121) with infinite-duration signals? The answer is yes, as long as the integral in Eqn. (4.121) converges.

Before discussing the conditions for the existence of the Fourier transform, we will address the following question: What does the transform $X(\omega)$ mean? In other words, how can we use the new function $X(\omega)$ for representing the signal $x(t)$? Recall that the Fourier series coefficients c_k for a periodic signal $\tilde{x}_p(t)$ were useful because we could construct the signal from them using Eqn. (4.116). Let us apply the limit operator to Eqn. (4.116):

$$x(t) = \lim_{T_0 \to \infty} [\tilde{x}(t)] = \lim_{T_0 \to \infty} \left[\sum_{k=-\infty}^{\infty} c_k \, e^{jk\omega_0 t} \right] \tag{4.122}$$

If we multiply and divide the term inside the summation by T_0 we obtain

$$x(t) = \lim_{T_0 \to \infty} [\tilde{x}(t)] = \lim_{T_0 \to \infty} \left[\sum_{k=-\infty}^{\infty} c_k T_0 \, e^{jk\omega_0 t} \left(\frac{1}{T_0} \right) \right] \tag{4.123}$$

We know that

$$\frac{1}{T_0} = \frac{\omega_0}{2\pi}$$

and for large values of T_0 we can write

$$\omega_0 \to \Delta\omega , \quad \text{and} \quad \frac{1}{T_0} \to \frac{\Delta\omega}{2\pi}$$

In the limit, Eqn. (4.123) becomes

$$x(t) = \lim_{T_0 \to \infty} [\tilde{x}(t)] = \lim_{T_0 \to \infty} \left[\sum_{k=-\infty}^{\infty} c_k T_0\, e^{jk\,\Delta\omega t} \left(\frac{\Delta\omega}{2\pi} \right) \right] \tag{4.124}$$

Also realizing that

$$\lim_{T_0 \to \infty} [c_k T_0] = X(k\,\Delta\omega) = X(\omega)$$

and changing the summation to an integral in the limit we have

$$x(t) = \frac{1}{2\pi} \int_{-\infty}^{\infty} X(\omega)\, e^{j\omega t}\, d\omega \tag{4.125}$$

Eqn. (4.125) explains how the transform $X(\omega)$ can be used for constructing the signal $x(t)$. We can interpret the integral in Eqn. (4.125) as a continuous sum of complex exponentials at harmonic frequencies that are infinitesimally close to each other. Thus, $x(t)$ and $X(\omega)$ represent two different ways of looking at the same signal, one by observing the amplitude of the signal at each time instant and the other by considering the contribution of each frequency component to the signal.

In summary, we have derived a transform relationship between $x(t)$ and $X(\omega)$ through the following equations:

Fourier transform for continuous-time signals:

1. **Synthesis equation: (Inverse transform)**

$$x(t) = \frac{1}{2\pi} \int_{-\infty}^{\infty} X(\omega)\, e^{j\omega t}\, d\omega \tag{4.126}$$

2. **Analysis equation: (Forward transform)**

$$X(\omega) = \int_{-\infty}^{\infty} x(t)\, e^{-j\omega t}\, dt \tag{4.127}$$

Often we will use the Fourier transform operator $\mathcal{F}$ and its inverse $\mathcal{F}^{-1}$ in a shorthand notation as

$$X(\omega) = \mathcal{F}\{x(t)\}$$

for the *forward transform*, and

$$x(t) = \mathcal{F}^{-1}\{X(\omega)\}$$

for the *inverse transform*. An even more compact notation for expressing the relationship between $x(t)$ and $X(\omega)$ is in the form

$$x(t) \xleftrightarrow{\mathcal{F}} X(\omega) \tag{4.128}$$

In general, the Fourier transform, as computed by Eqn. (4.127), is a complex function of ω. It can be written in Cartesian complex form as

$$X(\omega) = X_r(\omega) + jX_i(\omega)$$

or in polar complex form as

$$X(\omega) = |X(\omega)|\, e^{j\angle X(\omega)}$$

Sometimes it will be more convenient to express the Fourier transform of a signal in terms of the frequency f in Hz rather than the radian frequency ω in rad/s. The conversion is straightforward by substituting $\omega = 2\pi f$ and $d\omega = 2\pi\, df$ in Eqns. (4.126) and (4.127) which leads to the following equations:

Fourier transform for continuous-time signals (using f instead of ω):

1. Synthesis equation: (Inverse transform)

$$x(t) = \int_{-\infty}^{\infty} X(f)\, e^{j2\pi ft}\, df \qquad (4.129)$$

2. Analysis equation: (Forward transform)

$$X(f) = \int_{-\infty}^{\infty} x(t)\, e^{-j2\pi ft}\, dt \qquad (4.130)$$

Note the lack of the scale factor $1/2\pi$ in front of the integral of the inverse transform when f is used. This is consistent with the relationship $d\omega = 2\pi\, df$.

4.3.2 Existence of Fourier transform

The Fourier transform integral given by Eqn. (4.127) may or may not converge for a given signal $x(t)$. A complete theoretical study of convergence conditions is beyond the scope of this text, but we will provide a summary of the practical results that will be sufficient for our purposes. Let $\hat{x}(t)$ be defined as

$$\hat{x}(t) = \frac{1}{2\pi} \int_{-\infty}^{\infty} X(\omega)\, e^{j\omega t}\, d\omega \qquad (4.131)$$

and let $\varepsilon(t)$ be the error defined as the difference between $x(t)$ and $\hat{x}(t)$:

$$\varepsilon(t) = x(t) - \hat{x}(t) = x(t) - \frac{1}{2\pi} \int_{-\infty}^{\infty} X(\omega)\, e^{j\omega t}\, d\omega \qquad (4.132)$$

For perfect convergence of the transform at all time instants, we would naturally want $\varepsilon(t) = 0$ for all t. However, this is not possible at time instants for which $x(t)$ exhibits discontinuities. German mathematician Johann Peter Gustav Lejeune Dirichlet showed that the following set of conditions, referred to as *Dirichlet conditions* are sufficient for the convergence error $\varepsilon(t)$ to be zero at all time instants except those that correspond to discontinuities of the signal $x(t)$:

1. The signal $x(t)$ must be integrable in an absolute sense, that is

$$\int_{-\infty}^{\infty} |x(t)| \, dt < \infty \tag{4.133}$$

2. If the signal $x(t)$ has discontinuities, it must have at most a finite number of them in any finite time interval.
3. The signal $x(t)$ must have at most a finite number of minima and maxima in any finite time interval.

From a practical perspective, all signals we will encounter in our study of signals and systems will satisfy the second and third conditions regarding the number of discontinuities and the number of extrema respectively. The absolute integrability condition given by Eqn. (4.133) ensures that the result of the integral in Eqn. (4.126) is equal to the signal $x(t)$ at all time instants except discontinuities. At points of discontinuities, Eqn. (4.126) yields the average value obtained by approaching the discontinuity from opposite directions. If the signal $x(t)$ has a discontinuity at $t = t_0$, we have

$$\hat{x}(t_0) = \frac{1}{2} \lim_{t \to t_0^-} [x(t)] + \frac{1}{2} \lim_{t \to t_0^+} [x(t)] \tag{4.134}$$

An alternative approach to the question of convergence is to require

$$\int_{-\infty}^{\infty} |\varepsilon(t)|^2 \, dt = \int_{-\infty}^{\infty} |x(t) - \hat{x}(t)|^2 \, dt = 0 \tag{4.135}$$

which ensures that the normalized energy in the error signal $\varepsilon(t)$ is zero even if the error signal itself is not equal to zero at all times. This condition ensures that the transform $X(\omega)$ as defined by Eqn. (4.127) is finite. It can be shown that, the condition stated by Eqn. (4.135) is satisfied provided that the signal $x(t)$ is square integrable, that is

$$\int_{-\infty}^{\infty} |x(t)|^2 \, dt < \infty \tag{4.136}$$

We have seen in Chapter 1 (see Eqn. (1.81)) that a signal that satisfies Eqn. (4.136) is referred to as an *energy signal*. Therefore, all energy signals have Fourier transforms. Furthermore, we can find Fourier transforms for some signals that do not satisfy Eqn. (4.136), such as periodic signals, if we are willing to accept the use of the impulse function in the transform. Another example of a signal that is not an energy signal is the unit-step function. It is neither absolute integrable nor square integrable. We will show, however, that a Fourier transform can be found for the unit-step function as well if we allow the use of the impulse function in the transform.

4.3.3 Developing further insight

The interpretation of the Fourier transform for a non-periodic signal as a generalization of the EFS representation of a periodic signal is fundamental. We will take the time to apply this idea step by step to an isolated pulse in order to develop further insight into the Fourier transform. Consider the signal $x(t)$ shown in Fig. 4.31, an isolated rectangular pulse centered at $t = 0$ with amplitude A and width τ.

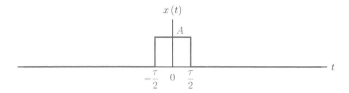

Figure 4.31 – An isolated rectangular pulse.

The analytical definition of $x(t)$ is

$$x(t) = A\,\Pi(t/\tau) = \begin{cases} A\,, & |t| < \tau/2 \\ 0\,, & |t| > \tau/2 \end{cases}$$

Let the signal $x(t)$ be extended into a pulse train $\tilde{x}(t)$ with a period T_0 as shown in Fig. 4.32.

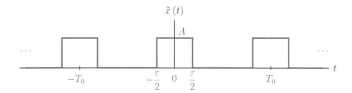

Figure 4.32 – Periodic extension of the isolated rectangular pulse into a pulse train.

By adapting the result found in Example 4.7 to the problem at hand, the EFS coefficients of the signal $\tilde{x}(t)$ can be written as

$$c_k = Ad\,\text{sinc}\,(kd) \tag{4.137}$$

where d is the duty cycle of the pulse train, and is given by

$$d = \frac{\tau}{T_0} \tag{4.138}$$

Substituting Eqn. (4.138) into Eqn. (4.137)

$$c_k = \frac{A\tau}{T_0}\,\text{sinc}\,(k\tau/T_0) \tag{4.139}$$

Let us multiply both sides of Eqn. (4.139) by T_0 and write the scaled EFS coefficients as

$$c_k T_0 = A\tau\,\text{sinc}\,(kf_0\tau) \tag{4.140}$$

where we have also substituted $1/T_0 = f_0$ in the argument of the sinc function. The outline (or the envelope) of the scaled EFS coefficients $c_k T_0$ is a sinc function with a peak value of $A\tau$ at $k = 0$. The first zero crossing of the sinc-shaped envelope occurs at

$$kf_0\tau = 1 \quad \Longrightarrow \quad k = \frac{1}{f_0\tau} \tag{4.141}$$

which may or may not yield an integer result. Subsequent zero crossings occur for

$$k f_0 \tau = 2, 3, 4, \ldots$$

or equivalently for

$$k = \frac{2}{f_0 \tau} \, , \; \frac{3}{f_0 \tau} \, , \; \frac{4}{f_0 \tau} \, , \; \ldots$$

The coefficients $c_k T_0$ are graphed in Fig. 4.33(a) for $A = 1$, $\tau = 0.1$ s and $T_0 = 0.25$ s corresponding to a fundamental frequency of $f_0 = 4$ Hz. Spectral lines for $k = 1$, $k = 2$ and $k = 3$ represent the strength of the frequency components at $f_0 = 4$ Hz, $2 f_0 = 8$ Hz and $3 f_0 = 12$ Hz respectively. For the example values given, the first zero crossing of the sinc envelope occurs at $k = 2.5$, between the spectral lines for $k = 2$ and $k = 3$.

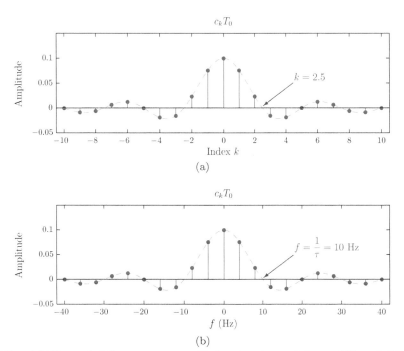

Figure 4.33 – (a) The scaled line spectrum $c_k T_0$ for $A = 1$, $\tau = 0.1$ s and $T_0 = 0.25$ s, (b) the scaled line spectrum as a function of frequency f.

In Fig. 4.33(b) the same line spectrum is shown with actual frequencies f on the horizontal axis instead of the integer index k. The spectral line for index k now appears at the frequency $f = k f_0 = 4k$ Hz. For example, the spectral lines for $k = 1, 2, 3$ appear at frequencies $f = 4, 8, 12$ Hz respectively. The first zero crossing of the sinc envelope is at the frequency

$$f - \frac{1}{\tau} = 10 \text{ Hz}$$

and subsequent zero crossings appear at frequencies

$$f = \frac{2}{\tau} \, , \; \frac{3}{\tau} \, , \; \frac{4}{\tau} \, , \; \ldots = 20, 30, 40, \ldots \text{ Hz}$$

What would happen if the period T_0 is gradually increased while keeping the pulse amplitude A and the pulse width τ unchanged? The spectral lines in 4.33(b) are at integer multiples of the fundamental frequency f_0. Since $f_0 = 1/T_0$, increasing the period T_0 would cause f_0 to become smaller, and the spectral lines to move inward, closer to each other. The zero crossings of the sinc envelope would not change, however, since the locations of the zero crossings depend on the pulse width τ only. Figs. 4.34(a)–(c) show the line spectra for $T_0 = 5, 10, 20$ s respectively. In Fig. 4.34(a), the first zero crossing of the sinc envelope is still at $1/\tau = 10$ Hz, yet the frequency spacing of the spectral lines is now reduced to 2 Hz; therefore the first zero crossing coincides with the 5-th harmonic. In Fig. 4.34(b), the frequency spacing of the spectral lines is further reduced to 1 Hz and, consequently, the first zero crossing of the sinc envelope coincides with the 10-th harmonic, and is still at 10 Hz.

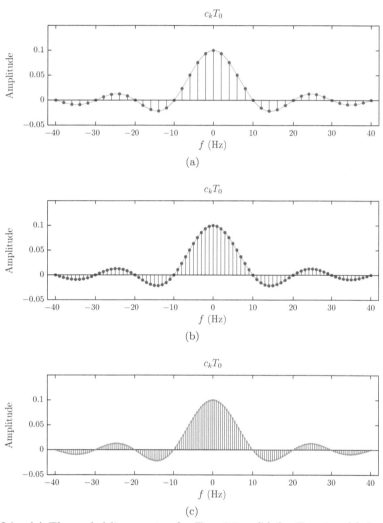

Figure 4.34 – (a) The scaled line spectra for $T_0 = 0.5$ s, (b) for $T_0 = 1$ s, (c) for $T_0 = 2$ s.

We can conclude that, as the period becomes infinitely large, the distances between adjacent spectral lines become infinitesimally small, and they eventually converge to a continuous function in the shape of the sinc envelope. It appears that the Fourier transform of the isolated pulse with amplitude A and width τ is

$$X(f) = A\tau \operatorname{sinc}(f\tau) \tag{4.142}$$

or, using the radian frequency variable ω,

$$X(\omega) = A\tau \, \text{sinc}\left(\frac{\omega\tau}{2\pi}\right) \qquad (4.143)$$

Interactive Demo: `ft_demo1.m`

This demo program is based on the concepts introduced in the discussion above along with Figs. 4.31 through 4.34. The periodic pulse train with amplitude A, pulse width τ and period T_0 is graphed. Its line spectrum based on the EFS coefficients is also shown. The spectrum graph includes the outline (or the envelope) of the EFS coefficients. Compare these graphs to Figs. 4.32 and 4.34. The amplitude is fixed at $A = 5$ since it is not a significant parameter for this demo. The period T_0 and the pulse width τ may be varied by adjusting the slider controls, allowing us to duplicate the cases in Fig. 4.34.

1. Increase the pulse period and observe the fundamental frequency change, causing the spectral lines of the line spectrum to move inward. Pay attention to how the outline remains fixed while this is occurring. Recall that the fundamental frequency is inversely proportional to the period; however, the outline is only a function of the pulse width.
2. With the signal period fixed, increase the pulse width and observe the changes in the outline. Notice how the locations of the spectral lines remain unchanged, but the heights of the spectral lines get adjusted to conform to the new outline.
3. For large values of the signal period observe how the outline approaches the Fourier transform of a single isolated pulse.

Software resources:
`ft_demo1.m`

4.3.4   Fourier transforms of some signals

In this section we will work on examples of determining the Fourier transforms of some fundamental signals.

Example 4.12: **Fourier transform of a rectangular pulse**

Using the forward Fourier transform integral in Eqn. (4.127), find the Fourier transform of the isolated rectangular pulse signal

$$x(t) = A\,\Pi\left(\frac{t}{\tau}\right)$$

shown in Fig. 4.35.

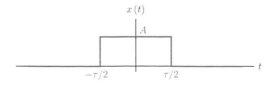

Figure 4.35 – Isolated pulse with amplitude A and width τ for Example 4.12.

Solution: Recall that this is the same isolated pulse the Fourier transform of which was determined in Section 4.3.3 as the limit case of the EFS representation of a periodic pulse train. In this example we will take a more direct approach to obtain the same result through the use of the Fourier transform integral in Eqn. (4.127):

$$X(\omega) = \int_{-\tau/2}^{\tau/2} (A) \, e^{-j\omega t} \, dt = A \left. \frac{e^{-j\omega t}}{-j\omega} \right|_{-\tau/2}^{\tau/2} = \frac{2A}{\omega} \sin\left(\frac{\omega\tau}{2}\right)$$

In order to use the sinc function, the result in Eqn. (4.144) can be manipulated and written in the form

$$X(\omega) = A\tau \frac{\sin(\omega\tau/2)}{(\omega\tau/2)} = A\tau \operatorname{sinc}\left(\frac{\omega\tau}{2\pi}\right) \tag{4.144}$$

In this case it will be easier to graph the transform in terms of the independent variable f instead of ω. Substituting $\omega = 2\pi f$ into Eqn. (4.46) we obtain

$$X(f) = A\tau \operatorname{sinc}(f\tau) \tag{4.145}$$

The spectrum $X(f)$, shown in Fig. 4.36, is purely real owing to the fact that the signal $x(t)$ exhibits even symmetry. The peak value of the spectrum is $A\tau$, and occurs at the frequency $f = 0$. The zero crossings of the spectrum occur at frequencies that satisfy $f\tau = k$ where k is any non-zero integer.

Figure 4.36 – Transform of the pulse in Example 4.12.

Software resources:
ex_4_12.m

The particular spectrum obtained in Example 4.12 is of special interest especially in digital systems. For example, in digital communications, rectangular pulses such as the one considered in Example 4.12 may be used for representing binary 0's or 1's. Let us observe the relationship between the pulse width and the shape of the spectrum:

1. Largest values of the spectrum occur at frequencies close to $f = 0$. Thus, low frequencies seem to be the more significant ones in the spectrum, and the significance of frequency components seems to decrease as we move further away from $f = 0$ in either direction.

2. The zero crossings of the spectrum occur for values of f that are integer multiples of $1/\tau$ for multiplier values $k = \pm1, \pm2, \ldots, \pm\infty$. As a result, if the pulse width is

decreased, these zero crossings move further away from the frequency $f = 0$ resulting in the spectrum being *stretched out* in both directions. This increases the relative significance of large frequencies. Narrower pulses have frequency spectra that expand to higher frequencies.

3. If the pulse width is increased, zero crossings of the spectrum move inward, that is, closer to the frequency $f = 0$ resulting in the spectrum being squeezed in from both directions. This decreases the significance of large frequencies, and causes the spectrum to be concentrated more heavily around the frequency $f = 0$. Wider pulses have frequency spectra that are more concentrated at low frequencies.

Fig. 4.37 illustrates the effects of changing the pulse width on the frequency spectrum.

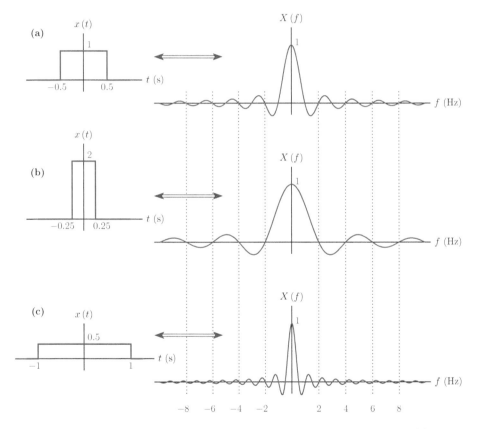

Figure 4.37 – Effects of changing the pulse width on the frequency spectrum: (a) $A = 1$ and $\tau = 1$, (b) $A = 2$ and $\tau = 0.5$, (c) $A = 0.5$ and $\tau = 2$.

In all three cases illustrated in Fig. 4.37, the product of the pulse width and the pulse amplitude is fixed, i.e., $A\tau = 1$. Consequently, the peak of the frequency spectrum is equal to unity in each case. The placements of the zero crossings of the spectrum depend on the pulse width. In Fig. 4.37(a), the zero crossings are 1 Hz apart, consistent with the pulse width of $\tau = 1$ s. In parts (b) and (c) of the figure, the spacing between adjacent zero crossings is 2 Hz and $\frac{1}{2}$ Hz, corresponding to pulse widths of $\tau = 0.5$ s and $\tau = 2$ s respectively.

This demo program is based on the concepts introduced in Example 4.12 and Figs. 4.35, 4.36, and 4.37. The amplitude A and the width τ of the rectangular pulse signal $x(t)$ may be varied by using the two slider controls or by typing values into the corresponding edit fields. The Fourier transform $X(f)$ is computed and graphed as the pulse parameters are varied.

1. Pay particular attention to the relationship between the pulse width τ and the locations of the zero crossings of the spectrum.
2. Observe how the concentrated nature of the spectrum changes as the pulse width is made narrower.
3. Also observe how the height of the spectrum at $f = 0$ changes with pulse amplitude and pulse width.

Software resources:
ft_demo2.m

Example 4.13: **Fourier transform of a rectangular pulse revisited**

Using the forward Fourier transform integral in Eqn. (4.127), find the Fourier transform of the isolated rectangular pulse given by

$$x(t) = A\,\Pi\left(\frac{t - \tau/2}{\tau}\right)$$

which is shown in Fig. 4.38.

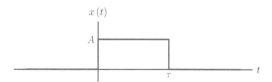

Figure 4.38 – Isolated pulse with amplitude A and width τ for Example 4.13.

Solution: The signal $x(t)$ is essentially a time-shifted version of the pulse signal used in Example 4.12. Using the Fourier transform integral in Eqn. (4.127) we obtain

$$X(\omega) = \int_0^\tau (A)\, e^{-j\omega t}\, dt = A \left. \frac{e^{-j\omega t}}{-j\omega} \right|_0^\tau = \frac{A}{-j\omega}\left[e^{-j\omega\tau} - 1\right] \qquad (4.146)$$

At this point we will use a trick which will come in handy in similar situations in the rest of this text as well. Let us write the part of Eqn. (4.146) in square brackets as follows:

$$\left[1 - e^{j\omega\tau}\right] = \left[e^{-j\omega\tau/2} - e^{j\omega\tau/2}\right] e^{-j\omega\tau/2} \qquad (4.147)$$

The use of Euler's formula on the result of Eqn. (4.147) yields

$$\left[1 - e^{j\omega\tau}\right] = -j\,2\sin(\omega\tau/2)\, e^{-j\omega\tau/2} \qquad (4.148)$$

Substituting Eqn. (4.148) into Eqn. (4.146), the Fourier transform of $x(t)$ is found as

$$X(\omega) = \frac{2A}{\omega} \sin\left(\frac{\omega\tau}{2}\right) e^{-j\omega\tau/2}$$

$$= A\tau \operatorname{sinc}\left(\frac{\omega\tau}{2\pi}\right) e^{-j\omega\tau/2} \tag{4.149}$$

The transform found in Eqn. (4.149) is complex-valued, and is best expressed in polar form. In preparation for that, we will find it convenient to write the transform in terms of the frequency variable f through the substitution $\omega = 2\pi f$:

$$X(f) = A\tau \operatorname{sinc}(f\tau) e^{-j\pi f\tau} \tag{4.150}$$

We would ultimately like to express the transform of Eqn. (4.150) in the form

$$X(f) = |X(f)| \, e^{j\theta(f)}$$

Let the functions $B(f)$ and $\beta(f)$ be defined as

$$B(f) = A\tau \operatorname{sinc}(f\tau)$$

and

$$\beta(f) = -\pi f\tau$$

so that

$$X(f) = B(f) \, e^{j\beta(f)}$$

The functions $B(f)$ and $\beta(f)$ are shown in Fig. 4.39.

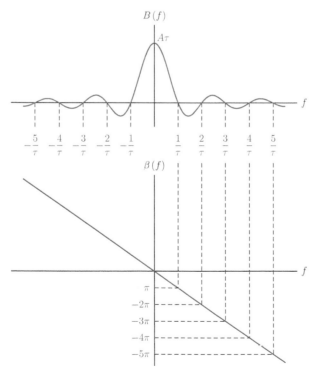

Figure 4.39 – The spectrum $X(f)$ in Example 4.13 expressed in the form $X(f) = B(f) \, e^{j\beta(f)}$.

A quick glance at Fig. 4.39 reveals the reason that keeps us from declaring the functions $B(f)$ and $\beta(f)$ as the magnitude and the phase of the transform $X(f)$: The function $B(f)$ could be negative for some values of f, and therefore cannot be the magnitude of the transform. However, $|X(\omega)|$ can be written in terms of $B(f)$ by paying attention to its sign. In order to outline the procedure to be used, we will consider two separate example intervals of the frequency variable f:

Case I: $-1/\tau < f < 1/\tau$

This is an example of a frequency interval in which $B(f) \geq 0$, and

$$B(f) = |B(f)|$$

Therefore, we can write the transform as

$$X(f) = B(f) e^{j\beta(f)} = |B(f)| e^{j\beta(f)}$$

resulting in the magnitude

$$|X(f)| = |B(f)| = B(f)$$

and the phase

$$\theta(f) = \beta(f)$$

for the transform.

Case 2: $1/\tau < f < 2/\tau$

For this interval we have $B(f) < 0$. The function $B(f)$ can be expressed as

$$B(f) = -|B(f)| = |B(f)| e^{\pm j\pi}$$

Using this form of $B(f)$ in the transform leads to

$$X(f) = |B(f)| e^{\pm j\pi} e^{j\beta(f)}$$

which results in the magnitude

$$|X(f)| = |B(f)| = -B(f)$$

and the phase

$$\theta(f) = \beta(f) \pm \pi$$

for the transform $X(f)$.

Comparison of the results in the two cases above leads us to the following generalized expressions for the magnitude and the phase of the transform $X(f)$ as

$$|X(f)| = |B(f)| \tag{4.151}$$

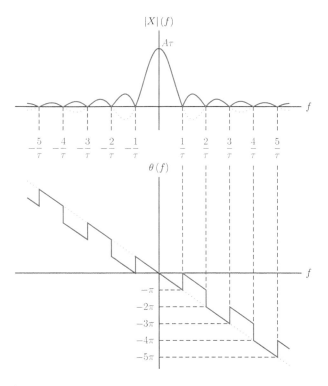

Figure 4.40 – Adjustment of magnitude and phase of the transform in Example 4.13.

and

$$\theta\left(f\right) = \begin{cases} \beta\left(f\right) & \text{if } B\left(f\right) \geq 0 \\ \beta\left(f\right) \pm \pi & \text{if } B\left(f\right) > 0 \end{cases} \tag{4.152}$$

In summary, for frequencies where $B\left(f\right)$ is negative, we add $\pm\pi$ radians to the phase term to account for the factor (-1). This is illustrated in Fig. 4.40.

Phase values outside the interval $(-\pi, \pi)$ radians are indistinguishable from the corresponding values within the interval since we can freely add any integer multiple of 2π radians to the phase. Because of this, it is customary to fit the phase values inside the interval $(-\pi, \pi)$, a practice referred to as *phase wrapping*. Fig. 4.41 shows the phase characteristic after the application of phase wrapping.

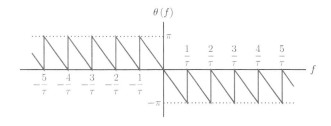

Figure 4.41 – Phase $\theta\left(f\right)$ of the transform in Example 4.13 after phase wrapping.

Software resources:
ex_4_13.m

Example 4.14: **Transform of the unit-impulse function**

The unit-impulse function was defined in Section 1.3.2 of Chapter 1. The Fourier transform of the unit-impulse signal can be found by direct application of the Fourier transform integral along with the sifting property of the unit-impulse function.

$$\mathcal{F}\left\{\delta\left(t\right)\right\} = \int_{-\infty}^{\infty} \delta\left(t\right) e^{-j\omega t}\, dt = e^{-j\omega t}\big|_{t=0} = 1 \tag{4.153}$$

In an effort to gain further insight, we will also take an alternative approach to determining the Fourier transform of the unit-impulse signal. Recall that in Section 1.3.2 we expressed the unit-impulse function as the limit case of a rectangular pulse with unit area. Given the pulse signal

$$q\left(t\right) = \frac{1}{a}\,\Pi\left(\frac{t}{a}\right)$$

the unit-impulse function can be expressed as

$$\delta\left(t\right) = \lim_{a\to 0}\left\{q\left(t\right)\right\}$$

If the parameter a is gradually made smaller, the pulse $q\left(t\right)$ becomes narrower and taller while still retaining unit area under it. In the limit, the pulse $q\left(t\right)$ becomes the unit impulse function $\delta\left(t\right)$. Using the result obtained in Example 4.12 with $A = 1/a$ and $\tau = a$, the Fourier transform of $q\left(t\right)$ is

$$Q\left(f\right) = \mathcal{F}\left\{q(t)\right\} = \operatorname{sinc}\left(fa\right) \tag{4.154}$$

Using an intuitive approach, we may also conclude that the Fourier transform of the unit-impulse function is an expanded, or stretched-out, version of the transform $Q\left(f\right)$ in Eqn. (4.154), i.e.,

$$\mathcal{F}\left\{\delta\left(t\right)\right\} = \lim_{a\to 0}\left\{Q\left(f\right)\right\} = \lim_{a\to 0}\left\{\operatorname{sinc}\left(fa\right)\right\} = 1$$

This conclusion is easy to justify from the general behavior of the function $\operatorname{sinc}\left(fa\right)$. Zero crossings on the frequency axis appear at values of f that are integer multiples of $1/a$. As the value of the parameter a is reduced, the zero crossings of the sinc function move further apart, and the function stretches out or flattens around its peak at $f = 0$. In the limit, it approaches a constant of unity. This behavior is illustrated in Fig. 4.42.

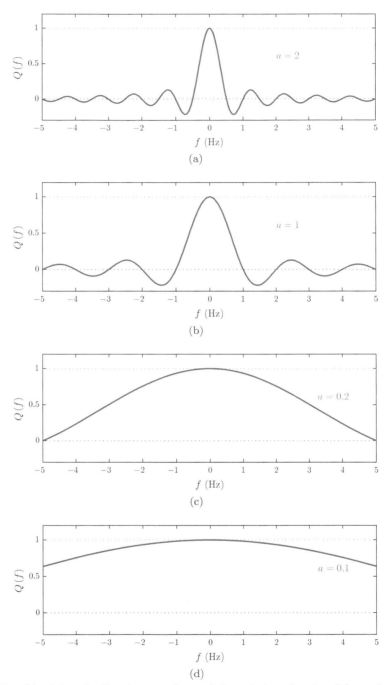

Figure 4.42 – Obtaining the Fourier transform of the unit-impulse signal from the transform of a rectangular pulse: (a) $a = 2$, (b) $a = 1$, (c) $a = 0.2$, (d) $a = 0.1$.

Interactive Demo: ft_demo3.m

The demo program "ft_demo3.m" illustrates the relationship between the Fourier transforms of the unit impulse and the rectangular pulse with unit area as discussed in the preceding section. A rectangular pulse with width equal to a and height equal to $1/a$ is shown along with its sinc-shaped Fourier transform. The pulse width may be varied through the use of a slider control. If we gradually reduce the width of the pulse, it starts to look more and more like a unit-impulse signal, and it becomes a unit-impulse signal in the limit. Intuitively it would make sense for the transform of the pulse to turn into the transform of the unit impulse. The demo program allows us to experiment with this concept.

1. Start with $a = 2$ as in Fig. 4.42(a) and observe the spectrum of the pulse.
2. Gradually reduce the pulse width and compare the spectrum to parts (b) through (d) of Fig. 4.42.

Software resources:
ft_demo3.m

Example 4.15: Fourier transform of a right-sided exponential signal

Determine the Fourier transform of the right-sided exponential signal

$$x(t) = e^{-at} u(t)$$

with $a > 0$ as shown in Fig. 4.43.

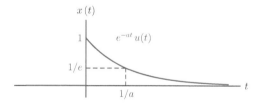

Figure 4.43 – Right-sided exponential signal for Example 4.15.

Solution: Application of the Fourier transform integral of Eqn. (4.127) to $x(t)$ yields

$$X(\omega) = \int_{-\infty}^{\infty} e^{-at} u(t) e^{-j\omega t} \, dt$$

Changing the lower limit of integral to $t = 0$ and dropping the factor $u(t)$ results in

$$X(\omega) = \int_{0}^{\infty} e^{-at} e^{-j\omega t} \, dt = \int_{0}^{\infty} e^{-(a+j\omega)t} \, dt = \frac{1}{a + j\omega}$$

This result in Eqn. (4.155) is only valid for $a > 0$ since the integral could not have been evaluated otherwise. The magnitude and the phase of the transform are

$$|X(\omega)| = \left| \frac{1}{a + j\omega} \right| = \frac{1}{\sqrt{a^2 + \omega^2}}$$

$$\theta(\omega) = \angle X(\omega) = -\tan^{-1}\left(\frac{\omega}{a}\right)$$

Magnitude and phase characteristics are graphed in Fig. 4.44.

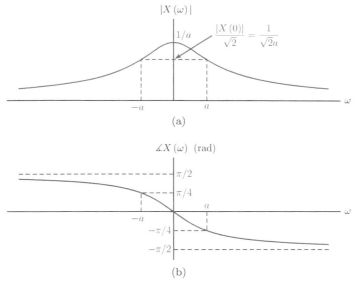

(a)

(b)

Figure 4.44 – (a) The magnitude, and (b) the phase of the transform of the right-sided exponential signal in Example 4.15.

Some observations are in order: The magnitude of the transform is an even function of ω since

$$|X(-\omega)| = \frac{1}{\sqrt{a^2 + (-\omega)^2}} = \frac{1}{\sqrt{a^2 + \omega^2}} = X(\omega)$$

The peak magnitude occurs at $\omega = 0$ and its value is $|X(0)| = 1/a$. At the radian frequency $\omega = a$ the magnitude is equal to $|X(a)| = 1/(\sqrt{2}a)$ which is $1/\sqrt{2}$ times its peak value. On a logarithmic scale this corresponds to a drop of 3 decibels (dB). In contrast with the magnitude, the phase of the transform is an odd function of ω since

$$\theta(-\omega) = -\tan^{-1}\left(\frac{-\omega}{a}\right) = \tan^{-1}\left(\frac{\omega}{a}\right) = -\theta(\omega)$$

At frequencies $\omega \to \pm\infty$ the phase angle approaches

$$\lim_{\omega \to \pm\infty}[\theta(\omega)] = \mp\frac{\pi}{2}$$

Furthermore, at frequencies $\omega = \pm a$ the phase angle is

$$\theta(\pm a) = \mp\frac{\pi}{4}$$

We will also compute real and imaginary parts of the transform to write its Cartesian form representation. Multiplying both the numerator and the denominator of the result in Eqn. (4.155) by $(a - j\omega)$ we obtain

$$X(\omega) = \frac{1}{(a + j\omega)}\frac{(a - j\omega)}{(a - j\omega)} = \frac{a - j\omega}{a^2 + \omega^2}$$

from which real and imaginary parts of the transform can be extracted as

$$X_r(\omega) = \frac{a}{a^2 + \omega^2} \tag{4.155}$$

and

$$X_i(\omega) = \frac{-\omega}{a^2 + \omega^2} \tag{4.156}$$

respectively. $X_r(\omega)$ and $X_i(\omega)$ are shown in Fig. 4.45(a) and (b).

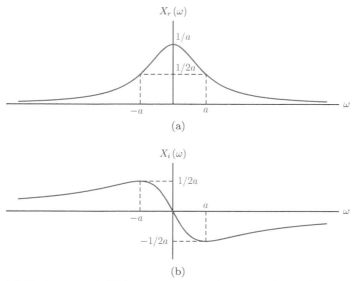

(a)

(b)

Figure 4.45 – (a) Real part and (b) imaginary part of the transform of the right-sided exponential signal in Example 4.15.

In this case the real part is an even function of ω, and the imaginary part is an odd function.

Software resources:

ex_4_15a.m

ex_4_15b.m

___

Interactive Demo: ft_demo4.m

This demo program is based on Example 4.15. The right-sided exponential signal

$$x(t) = e^{-at} u(t)$$

and its Fourier transform are graphed. The parameter a may be modified through the use of a slider control.

Observe how changes in the parameter a affect the signal $x(t)$ and the transform $X(\omega)$. Pay attention to the correlation between the width of the signal and the width of the transform. Does the fundamental relationship resemble the one observed earlier between the rectangular pulse and its Fourier transform?

Software resources:

ft_demo4.m

___

Example 4.16: **Fourier transform of a two-sided exponential signal**

Determine the Fourier transform of the two-sided exponential signal given by

$$x(t) = e^{-a|t|}$$

where a is any non-negative real-valued constant. The signal $x\,(t)$ is shown in Fig. 4.46.

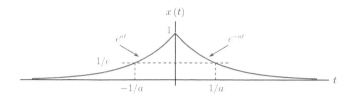

Figure 4.46 – Two-sided exponential signal $x\,(t)$ for Example 4.16.

Solution: Applying the Fourier transform integral of Eqn. (4.127) to our signal we get

$$X\,(\omega) = \int_{-\infty}^{\infty} e^{-a|t|}\, e^{-j\omega t}\, dt$$

Splitting the integral into two halves yields

$$X\,(\omega) = \int_{-\infty}^{0} e^{-a|t|}\, e^{-j\omega t}\, dt + \int_{0}^{\infty} e^{-a|t|}\, e^{-j\omega t}\, dt$$

Recognizing that

$$t \leq 0 \qquad \Rightarrow \qquad e^{-a|t|} = e^{at}$$
$$t \geq 0 \qquad \Rightarrow \qquad e^{-a|t|} = e^{-at}$$

the transform is

$$X\,(\omega) = \int_{-\infty}^{0} e^{at} e^{-j\omega t}\, dt + \int_{0}^{\infty} e^{-at} e^{-j\omega t}\, dt$$

$$-\frac{1}{a - j\omega} + \frac{1}{a + j\omega}$$

$$= \frac{2a}{a^2 + \omega^2} \tag{4.157}$$

$X\,(\omega)$ is shown in Fig. 4.47.

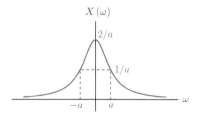

Figure 4.47 – Transform of the two-sided exponential signal of Example 4.16.

The transform $X\,(\omega)$ found in Eqn. (4.157) is purely real for all values of ω. This is a consequence of the signal $x\,(t)$ having even symmetry, and will be explored further as we look at the symmetry properties of the Fourier transform in the next section. In addition

to being purely real, the transform also happens to be non-negative in this case, resulting in a phase characteristic that is zero for all frequencies.
Software resources:
ex_4_16.m

Interactive Demo: ft_demo5.m

This demo program is based on Example 4.16. The two-sided exponential signal

$$x\left(t\right) = e^{-a|t|}$$

and its Fourier transform are graphed. The transform is purely real for all ω owing to the even symmetry of the signal $x\left(t\right)$. The parameter a may be modified through the use of a slider control.

Observe how changes in the parameter a affect the signal $x\left(t\right)$ and the transform $X\left(\omega\right)$. Pay attention to the correlation between the width of the signal and the width of the transform.
Software resources:
ft_demo5.m

Example 4.17: **Fourier transform of a triangular pulse**

Find the Fourier transform of the triangular pulse signal given by

$$x\left(t\right) = A\,\Lambda\left(\frac{t}{\tau}\right) = \begin{cases} A + At/\tau\,, & -\tau < t < 0 \\ A - At/\tau\,, & 0 \le t < \tau \\ 0\,, & |t| \ge \tau \end{cases}$$

where $\Lambda\left(t\right)$ is the unit-triangle function defined in Section 1.3.2 of Chapter 1. The signal $x\left(t\right)$ is shown in Fig. 4.48.

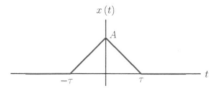

Figure 4.48 – Triangular pulse signal of Example 4.17.

Solution: Using the Fourier transform integral on the signal $x\left(t\right)$, we obtain

$$X\left(\omega\right) = \int_{-\tau}^{0}\left(A + \frac{At}{\tau}\right)e^{-j\omega t}\,dt + \int_{0}^{\tau}\left(A - \frac{At}{\tau}\right)e^{-j\omega t}\,dt$$

$$= A\int_{-\tau}^{0}e^{-j\omega t}\,dt + \frac{A}{\tau}\int_{-\tau}^{0}t\,e^{-j\omega t}\,dt + A\int_{0}^{\tau}e^{-j\omega t}\,dt - \frac{A}{\tau}\int_{0}^{\tau}t\,e^{-j\omega t}\,dt$$

$$= A\int_{-\tau}^{\tau}e^{-j\omega t}\,dt + \frac{A}{\tau}\int_{-\tau}^{0}t\,e^{-j\omega t}\,dt - \frac{A}{\tau}\int_{0}^{\tau}t\,e^{-j\omega t}\,dt \qquad (4.158)$$

The last two integrals in Eqn. (4.158) are similar in form to entry (B.16) of the table of indefinite integrals in Appendix B, repeated here for convenience:

$$\text{Eqn. (B.16):} \qquad \int t\, e^{at}\, dt = \frac{1}{a^2}\, e^{at}\, [\,at - 1\,]$$

Setting $a = -j\omega$ results in

$$\int t\, e^{-j\omega t}\, dt = \frac{1}{\omega^2}\, e^{-j\omega t}\, [\,j\omega t + 1\,]$$

and the Fourier transform is

$$X(\omega) = A\left(\frac{e^{-j\omega t}}{-j\omega}\right)\Bigg|_{-\tau}^{\tau} + \frac{A}{\omega^2 \tau}\left(e^{-j\omega t}\,[\,j\omega t + 1\,]\right)\Bigg|_{-\tau}^{0} - \frac{A}{\omega^2 \tau}\left(e^{-j\omega t}\,[\,j\omega t + 1\,]\right)\Bigg|_{0}^{\tau}$$

$$= \frac{2A \sin(\omega\tau)}{\omega} + \frac{2A}{\omega^2 \tau}\,[\,1 - \omega\tau \sin(\omega\tau) - \cos(\omega\tau)\,]$$

$$= \frac{A}{\omega^2 \tau}\,[\,1 - \cos(\omega\tau)\,]$$

Using the appropriate trigonometric identity[5] it can be shown that

$$X(\omega) = \frac{4A}{\omega^2 \tau}\, \sin^2\left(\frac{\omega\tau}{2}\right) \qquad\qquad (4.159)$$

It would be convenient to use the sinc function with the result in Eqn. (4.159). Recognizing that

$$\text{sinc}\left(\frac{\omega\tau}{2\pi}\right) = \frac{\sin(\omega\tau/2)}{(\omega\tau/2)} = \frac{2}{\omega\tau}\, \sin\left(\frac{\omega\tau}{2}\right)$$

the transform in Eqn. (4.159) can be written as

$$X(\omega) = A\tau\, \text{sinc}^2\left(\frac{\omega\tau}{2\pi}\right)$$

If the transform is written in terms of f instead of ω, we obtain

$$X(f) = A\tau\, \text{sinc}^2(f\tau)$$

Thus, the Fourier transform of a triangular pulse is proportional the square of the sinc function. Contrast this with the Fourier transform of a rectangular pulse which is proportional to the sinc function as we have seen in Example 4.12. The transform of the triangular pulse, shown in Fig. 4.49, seems to be concentrated more heavily around low frequencies compared to the transform of the rectangular pulse.

---

[5] $\sin^2(a) = \frac{1}{2}\,[\,1 - \cos(a)\,]$.

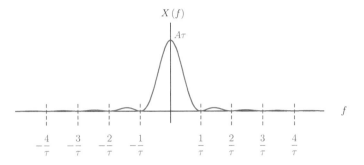

Figure 4.49 – Frequency spectrum of the triangular pulse signal of Example 4.17.

We will have occasion to look at the transform of a triangular pulse in more depth when we study the properties of the Fourier transform in the next section. Specifically, a much simpler method of obtaining $X(\omega)$ for a triangular pulse will be presented after studying the convolution property of the Fourier transform.

Software resources:

ex_4_17.m

Interactive Demo: ft_demo6.m

This demo program is based on Example 4.17. The triangular pulse signal with peak amplitude A and width 2τ is shown along with its Fourier transform as determined in Eqn. (4.160). The transform is purely real for all f owing to the even symmetry of the triangular pulse signal $x(t)$. The parameters A and τ may be modified through the use of slider controls.

Observe how parameter changes affect the signal $x(t)$ and the transform $X(\omega)$. Pay attention to the correlation between the width of the signal and the width of the transform.

Software resources:

ft_demo6.m

Example 4.18: Fourier transform of the signum function

Determine the Fourier transform of the signum function defined as

$$ x(t) = \operatorname{sgn}(t) = \begin{cases} -1, & t < 0 \\ 1, & t > 0 \end{cases} \tag{4.160} $$

and shown graphically in Fig. 4.50.

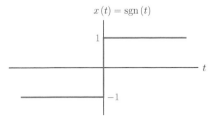

Figure 4.50 – The signum function.

Solution: If we were to apply the Fourier transform integral directly to this signal, we would obtain

$$X(\omega) = \int_{-\infty}^{0} (-1)\, e^{-j\omega t}\, dt + \int_{0}^{\infty} (1)\, e^{-j\omega t}\, dt$$

in which the two integrals cannot be evaluated. Instead, we will define an intermediate signal $p(t)$ as

$$p(t) = \begin{cases} -e^{at}, & t < 0 \\ e^{-at}, & t > 0 \end{cases}$$

where $a \geq 0$. We will first determine the Fourier transform of this intermediate signal, and then obtain the Fourier transform of the signum function from this intermediate result. The Fourier transform of $p(t)$ is found as

$$P(\omega) = \int_{-\infty}^{0} \left(-e^{at}\right)\, e^{-j\omega t}\, dt + \int_{0}^{\infty} \left(e^{-at}\right)\, e^{-j\omega t}\, dt$$

$$= \left(\frac{-e^{at}\, e^{-j\omega t}}{a - j\omega}\right)\Bigg|_{-\infty}^{0} + \left(\frac{e^{-at}\, e^{-j\omega t}}{-a - j\omega}\right)\Bigg|_{0}^{\infty}$$

$$= \frac{-j\,2\omega}{a^2 + \omega^2}$$

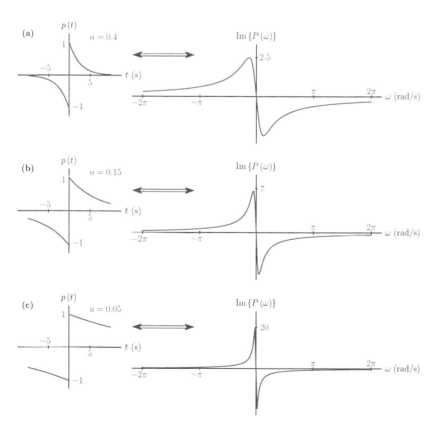

Figure 4.51 – Signum function as a limit case of $p(t)$ of Eqn. (4.161): (a) $p(t)$ for $a = 0.4$ and its transform, (b) $p(t)$ for $a = 0.15$ and its transform, (c) $p(t)$ for $a = 0.05$ and its transform.

The result is purely imaginary for all ω due to the odd symmetry of $p(t)$. As the value of a is reduced, the intermediate signal $p(t)$ starts to look more and more similar to the signum function, and becomes the signum function in the limit as $a \to 0$. The transition of $p(t)$ to $\mathrm{sgn}(t)$ is illustrated in Fig. 4.51 for $a = 0.4, 0.15$, and 0.04.

Setting $a = 0$ in the transform $P(\omega)$, we obtain the Fourier transform of the original signal $x(t)$:

$$X(\omega) = \mathcal{F}\left\{\mathrm{sgn}(t)\right\} = \lim_{a \to 0} \left[\frac{-j\,2\omega}{a^2 + \omega^2} \right] = \frac{2}{j\omega}$$

The magnitude of the transform is

$$|X(\omega)| = \frac{2}{|\omega|}$$

and the phase is

$$\theta(\omega) = \angle X(\omega) = \begin{cases} \pi/2\,, & \omega < 0 \\ -\pi/2\,, & \omega > 0 \end{cases}$$

Magnitude and phase characteristics are shown in Fig. 4.52.

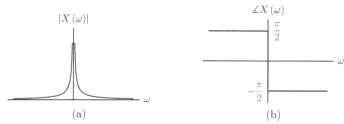

(a) (b)

Figure 4.52 – (a) Magnitude and (b) phase of the Fourier transform of $\mathrm{sgn}(t)$.

Software resources:
ex_4_18.m

Interactive Demo: ft_demo7.m

This demo program is based on Example 4.18, and allows experimentation with the progression of the signal

$$p(t) = \begin{cases} -e^{at}\,, & t < 0 \\ e^{-at}\,, & t > 0 \end{cases}$$

and its corresponding Fourier transform $P(\omega)$ into the signal

$$x(t) = \mathrm{sgn}(t)$$

and its associated transform $X(\omega)$ as the parameter a is made smaller. The parameter a may be adjusted through the use of a slider control. The program displays a graph of the signal $p(t)$ and a graph of the imaginary part of transform $P(\omega)$. The real part of the transform is always equal to zero for this particular signal, and is therefore not shown.

Software resources:
ft_demo7.m

4.3.5   Properties of the Fourier transform

In this section we will explore some of the fundamental properties of the Fourier transform that provide further insight into its use for understanding and analyzing characteristics of signals. We will also see that some of these properties greatly simplify the computation of the Fourier transform for certain types of signals.

Linearity

Fourier transform is a linear operator. For any two signals $x_1(t)$ and $x_2(t)$ with their respective transforms

$$x_1(t) \xleftrightarrow{\mathcal{F}} X_1(\omega)$$

$$x_2(t) \xleftrightarrow{\mathcal{F}} X_2(\omega)$$

and any two constants α_1 and α_2, it can be shown that the following relationship holds:

Linearity of the Fourier transform:

$$\alpha_1 x_1(t) + \alpha_2 x_2(t) \xleftrightarrow{\mathcal{F}} \alpha_1 X_1(\omega) + \alpha_2 X_2(\omega) \qquad (4.161)$$

Proof: Using the forward transform equation given by Eqn. (4.127) with the time domain signal $[\alpha_1 x_1(t) + \alpha_2 x_2(t)]$ leads to:

$$\mathcal{F}\{\alpha_1 x_1(t) + \alpha_2 x_2(t)\} = \int_{-\infty}^{\infty} [\alpha_1 x_1(t) + \alpha_2 x_2(t)] \, e^{-j\omega t} \, dt$$

$$= \int_{-\infty}^{\infty} \alpha_1 x_1(t) \, e^{-j\omega t} \, dt + \int_{-\infty}^{\infty} \alpha_2 x_2(t) \, e^{-j\omega t} \, dt$$

$$= \alpha_1 \int_{-\infty}^{\infty} x_1(t) \, e^{-j\omega t} \, dt + \alpha_2 \int_{-\infty}^{\infty} x_2(t) \, e^{-j\omega t} \, dt$$

$$= \alpha_1 \mathcal{F}\{x_1(t)\} + \alpha_2 \mathcal{F}\{x_2(t)\} \qquad (4.162)$$

Duality

Consider the transform pair

$$x(t) \xleftrightarrow{\mathcal{F}} X(\omega)$$

The transform relationship between $x(t)$ and $X(\omega)$ is defined by the inverse and forward Fourier transform integrals given by Eqns. (4.126) and (4.127) and repeated here for convenience:

$$\text{Eqn. (4.126):} \quad x(t) = \frac{1}{2\pi} \int_{-\infty}^{\infty} X(\omega) \, e^{j\omega t} \, d\omega$$

$$\text{Eqn. (4.127):} \quad X(\omega) = \int_{-\infty}^{\infty} x(t) \, e^{-j\omega t} \, dt$$

Suppose that we swap the roles of the independent variables to construct a new time-domain signal $X(t)$ and a new frequency-domain function $x(\omega)$. What would be the relationship between $X(t)$ and $x(\omega)$?

The expressions for the inverse transform in Eqn. (4.126) and the forward transform in Eqn. (4.127) are interestingly similar. The main differences between the two integrals are the sign of the exponential term and the scale factor $1/2\pi$ for the analysis equation. This similarity leads to the *duality principle* of the Fourier transform. Let us change the integration variable of Eqn. (4.163) from ω to λ, and write it as

$$x\left(t\right) = \frac{1}{2\pi} \int_{-\infty}^{\infty} X\left(\lambda\right) e^{j\lambda t} \, d\lambda \tag{4.163}$$

If $\omega \to \pm\infty$ then $\lambda \to \pm\infty$, so we don't need to change the limits of integration. Now, from a purely mathematical perspective, we will change the name of the independent variable from t to $-\omega$, that is, we will evaluate the function on the left side of the integral in Eqn. (4.163) for $t = -\omega$, and write the result as

$$x\left(-\omega\right) = \frac{1}{2\pi} \int_{-\infty}^{\infty} X\left(\lambda\right) e^{-j\lambda\omega} \, d\lambda \tag{4.164}$$

Changing the integration variable from λ to t yields

$$x\left(-\omega\right) = \frac{1}{2\pi} \int_{-\infty}^{\infty} X\left(t\right) e^{-jt\omega} \, dt \tag{4.165}$$

and by multiplying both sides of Eqn. (4.165) by 2π we obtain

$$2\pi\, x\left(-\omega\right) = \int_{-\infty}^{\infty} X\left(t\right) e^{-jt\omega} \, dt \tag{4.166}$$

Eqn. (4.166) has the same general form as Eqn. (4.127) with substitutions $x\left(t\right) \to X\left(t\right)$ and $X\left(\omega\right) \to 2\pi\, x\left(-\omega\right)$. Thus, the following relationship must also be true:

$$X\left(t\right) \overset{\mathcal{F}}{\longleftrightarrow} 2\pi\, x\left(-\omega\right) \tag{4.167}$$

The time-domain function $X\left(t\right)$ and the frequency-domain function $2\pi\, x\left(-\omega\right)$ form a Fourier transform pair. Thus, if $X\left(\omega\right)$ is the Fourier transform of $x\left(t\right)$, then $2\pi\, x\left(-\omega\right)$ is the Fourier transform of $X\left(t\right)$.

Duality property:

$$x\left(t\right) \overset{\mathcal{F}}{\longleftrightarrow} X\left(\omega\right) \quad \text{implies that} \quad X\left(t\right) \overset{\mathcal{F}}{\longleftrightarrow} 2\pi\, x\left(-\omega\right) \tag{4.168}$$

We may find it more convenient to express the duality property using the frequency f instead of the radian frequency ω since this approach eliminates the 2π factor from the equations:

Duality property (using f instead of ω):

$$x\left(t\right) \overset{\mathcal{F}}{\longleftrightarrow} X\left(f\right) \quad \text{implies that} \quad X\left(t\right) \overset{\mathcal{F}}{\longleftrightarrow} x\left(-f\right) \tag{4.169}$$

Duality property will be useful in deriving some of the other properties of the Fourier transform.

Example 4.19: **Fourier transform of the** sinc **function**

Find the Fourier transform of the signal

$$x(t) = \text{sinc}(t)$$

Solution: The Fourier transform of a rectangular pulse was found in Example 4.12 as a sinc function. Specifically we obtained

$$\mathcal{F}\left\{A\Pi\left(\frac{t}{\tau}\right)\right\} = A\tau\,\text{sinc}\left(\frac{\omega\tau}{2\pi}\right) \tag{4.170}$$

This relationship can be used as a starting point in finding the Fourier transform of $x(t) = \text{sinc}(t)$. Let $\tau = 2\pi$ so that the argument of the sinc function in Eqn. (4.170) becomes ω:

$$\mathcal{F}\left\{A\Pi\left(\frac{t}{2\pi}\right)\right\} = 2\pi A\,\text{sinc}(\omega) \tag{4.171}$$

In the next step, let $A = 1/2\pi$ to obtain

$$\mathcal{F}\left\{\frac{1}{2\pi}\Pi\left(\frac{t}{2\pi}\right)\right\} = \text{sinc}(\omega) \tag{4.172}$$

Applying the duality property given by Eqn. (4.169) to the transform pair in Eqn. (4.172) leads to the result

$$\mathcal{F}\{\text{sinc}(t)\} = \Pi\left(\frac{-\omega}{2\pi}\right) = \Pi\left(\frac{\omega}{2\pi}\right) \tag{4.173}$$

Using f instead of ω in Eqn. (4.173) yields a relationship that is easier to remember:

$$\mathcal{F}\{\text{sinc}(t)\} = \Pi(f) \tag{4.174}$$

The Fourier transform of $x(t) = \text{sinc}(t)$ is a unit pulse in the frequency domain. The signal $x(t)$ and its transform are graphed in Fig. 4.53(a) and (b).

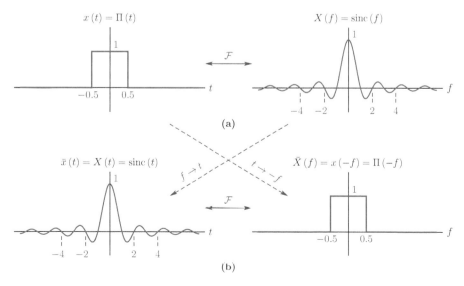

(a)

(b)

Figure 4.53 – (a) The function $x(t) = \Pi(t)$ and its Fourier transform $X(f)$, (b) the function $\bar{x}(t) = \text{sinc}(t)$ and its Fourier transform $\bar{X}(f)$.

Example 4.20: **Transform of a constant-amplitude signal**

Find the Fourier transform of the constant-amplitude signal

$$x(t) = 1, \quad \text{all } t$$

Solution: Finding the Fourier transform of $x(t)$ by direct application of Eqn. (4.127) is not possible since the resulting integral

$$X(\omega) = \int_{-\infty}^{\infty} x(t)\, e^{-j\omega t}\, dt = \int_{-\infty}^{\infty} (1)\, e^{-j\omega t}\, dt$$

could not be evaluated. On the other hand, in Example 4.14 we have concluded that the Fourier transform of the unit-impulse signal is a constant for all frequencies, that is

$$\mathcal{F}\{\delta(t)\} = 1, \quad \text{all } \omega \tag{4.175}$$

Applying the duality property to the transform pair in Eqn. (4.175) we obtain

$$\mathcal{F}\{1\} = 2\pi\, \delta(-\omega) = 2\pi\, \delta(\omega)$$

If f is used instead of ω in the transform, then we have

$$\mathcal{F}\{1\} = \delta(f)$$

This example also presents an opportunity for two important observations:

1. The signal $x(t) = 1$ does not satisfy the existence conditions discussed in Section 4.3.2; it is neither absolute integrable nor square integrable. Therefore, in a strict sense, its Fourier transform does not converge. By allowing the impulse function to be used in the transform, we obtain a function $X(\omega)$ that has the characteristics of a Fourier transform, and that can be used in solving problems in the frequency domain.
2. Sometimes we express the Fourier transform in terms of the radian frequency ω; at other times we use the frequency f. In general, the conversion from one format to another is just a matter of using the relationship $\omega = 2\pi f$, so that

$$X(\omega) = \left. X(f) \right|_{f=\omega/2\pi}, \qquad X(f) = \left. X(\omega) \right|_{\omega=2\pi f}$$

One exception to this rule is seen when the transform contains a singularity function. Compare Eqns. (4.176) and (4.176) in the example above. $X(\omega)$ has a 2π factor that $X(f)$ does not have. This behavior is consistent with scaling of the impulse function (see Problem 1.6 of Chapter 1).

Example 4.21: **Another example of using the duality property**

Using the duality property, find the Fourier transform of the signal

$$x(t) = \frac{1}{3 + 2t^2}$$

which is graphed in Fig. 4.54.

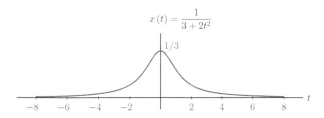

Figure 4.54 – The signal $x(t) = 1/\left(3 + 2t^2\right)$ of Example 4.21.

Solution: While it is possible to solve this problem by a direct application of the Fourier transform integral given by Eqn. (4.127), it would be difficult to evaluate the integral unless we resort to the use of integration tables. Instead, we will simplify the solution by using another transform pair that we already know, and adapting that transform pair to our needs. In Example 4.16 we have found that

$$e^{-a|t|} \quad \overset{\mathcal{F}}{\longleftrightarrow} \quad \frac{2a}{a^2 + \omega^2}$$

where the form of the transform bears a similarity to the time-domain expression for the signal $x(t)$. In both functions the numerator is constant, and the denominator has a second-order term next to a constant term. We will make use of this similarity through the use of the duality property of the Fourier transform. The duality property allows us to write

$$\frac{2a}{a^2 + t^2} \quad \overset{\mathcal{F}}{\longleftrightarrow} \quad 2\pi\, e^{-a|\omega|} \tag{4.176}$$

as another valid transform pair. By taking steps that will preserve the validity of this transform pair, we will work toward making the time-domain component of the relationship in Eqn. (4.176) match the signal $x(t)$ given in the problem statement. The signal $x(t)$ has a $2t^2$ term in its denominator. Let us multiply both the numerator and the denominator of the time-domain component of Eqn. (4.176) by 2:

$$\frac{4a}{2a^2 + 2t^2} \quad \overset{\mathcal{F}}{\longleftrightarrow} \quad 2\pi\, e^{-a|\omega|}$$

To match the constant term in the denominator of $x(t)$ we will choose

$$2a^2 = 3 \quad \Longrightarrow \quad a = \sqrt{\frac{3}{2}}$$

and obtain

$$\frac{2\sqrt{6}}{3 + 2t^2} \quad \overset{\mathcal{F}}{\longleftrightarrow} \quad 2\pi\, e^{-\sqrt{\frac{3}{2}}\,|\omega|} \tag{4.177}$$

Afterwards, by scaling both sides of Eqn. (4.177) with $2\sqrt{6}$ we get

$$\frac{1}{3 + 2t^2} \quad \overset{\mathcal{F}}{\longleftrightarrow} \quad \frac{\pi}{\sqrt{6}}\, e^{-\sqrt{\frac{3}{2}}\,|\omega|}$$

The transform found in Eqn. (4.178) is graphed in Fig. 4.55 using f as the independent variable instead of ω.

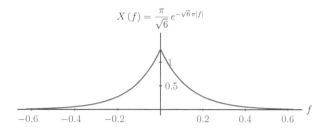

Figure 4.55 – Fourier transform of the function $x(t)$ of Example 4.21.

Software resources:
ex_4_21.m

Example 4.22: **Signum function revisited**

In Example 4.18 the Fourier transform of the signal $x(t) = \text{sgn}(t)$ was found to be

$$X(f) = \mathcal{F}\{\text{sgn}(t)\} = \frac{2}{j\omega}$$

Determine the time-domain signal, the Fourier transform of which is

$$X(\omega) = -j\,\text{sgn}(\omega)$$

Solution: This is another problem that could be simplified by the use of the duality property. Starting with the already known transform pair

$$\text{sgn}(t) \;\overset{\mathcal{F}}{\longleftrightarrow}\; \frac{2}{j\omega}$$

and applying the duality property as expressed by Eqn. (4.168) we conclude that the following must also be a valid transform pair:

$$\frac{2}{jt} \;\overset{\mathcal{F}}{\longleftrightarrow}\; 2\pi\,\text{sgn}(-\omega)$$

Recognizing that the sgn function has odd symmetry, that is, $\text{sgn}(-\omega) = -\text{sgn}(\omega)$,

$$\frac{2}{jt} \;\overset{\mathcal{F}}{\longleftrightarrow}\; -2\pi\,\text{sgn}(\omega) \tag{4.178}$$

Multiplying both sides of Eqn. (4.178) by $j/2\pi$, we have

$$\frac{1}{\pi t} \;\overset{\mathcal{F}}{\longleftrightarrow}\; -j\,\text{sgn}(\omega) \tag{4.179}$$

The signal we are seeking is

$$x(t) = \frac{1}{\pi t}$$

which is graphed in Fig. 4.56.

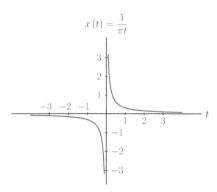

Figure 4.56 – The signal $x(t) = \dfrac{1}{\pi t}$ found in Example 4.22.

It is interesting to look at the magnitude and the phase of the transform $X(\omega)$ in Eqn. (4.178). The transform can be written as

$$X(\omega) = -j\,\mathrm{sgn}(\omega) = \begin{cases} -j\,, & \omega > 0 \\ j\,, & \omega < 0 \end{cases}$$

Realizing that $j = e^{j\pi/2}$ we can write

$$X(\omega) = \begin{cases} e^{-j\pi/2}\,, & \omega > 0 \\ e^{j\pi/2}\,, & \omega < 0 \end{cases}$$

The magnitude and the phase of $X(\omega)$ are shown in Fig. 4.57.

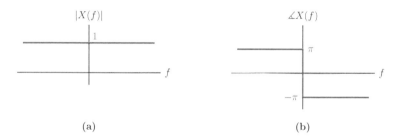

Figure 4.57 – (a) The magnitude and (b) the phase of $X(\omega)$ given by Eqn. (4.178).

The magnitude of the transform is equal to unity for all frequencies. The phase of the transform is $-\pi/2$ radians for positive frequencies and $+\pi/2$ radians for negative frequencies.

Example 4.23: **Fourier transform of the unit-step function**

Determine the Fourier transform of the unit-step function $x(t) = u(t)$.

Solution: The unit-step function does not satisfy the conditions for the existence of the Fourier transform. An attempt to find the transform by direct application of Eqn. (4.127) would result in

$$\mathcal{F}\{u(t)\} = \int_{-\infty}^{\infty} u(t)\,e^{-j\omega t}\,dt = \int_{0}^{\infty} e^{-j\omega t}\,dt$$

which does not converge. Instead, we will try to make use of some of the earlier results found along with the linearity property of the Fourier transform. The unit-step function can be expressed in terms of the signum function and a constant offset as

$$u(t) = \frac{1}{2} + \frac{1}{2} \operatorname{sgn}(t)$$

This is illustrated in Fig. 4.58.

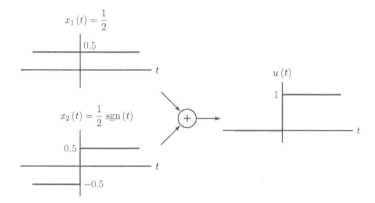

Figure 4.58 – Obtaining the unit-step function as the sum of the signum function and a constant offset.

Using the linearity of the Fourier transform we obtain

$$\mathcal{F}\{u(t)\} = \mathcal{F}\left\{\frac{1}{2} + \frac{1}{2} \operatorname{sgn}(t)\right\}$$

$$= \frac{1}{2} \mathcal{F}\{1\} + \frac{1}{2} \mathcal{F}\{\operatorname{sgn}(t)\} \qquad (4.180)$$

The transform of a constant signal was found in Example 4.20:

$$\mathcal{F}\{1\} = 2\pi \, \delta(\omega)$$

Furthermore, in Example 4.18 we found

$$\mathcal{F}\{\operatorname{sgn}(t)\} = \frac{2}{j\omega}$$

Using these two results in Eqn. (4.180), Fourier transform of the unit-step function is

$$\mathcal{F}\{u(t)\} = \pi \, \delta(\omega) + \frac{1}{j\omega} \qquad (4.181)$$

Fig. 4.59 shows the unit-step signal and the magnitude of its transform.

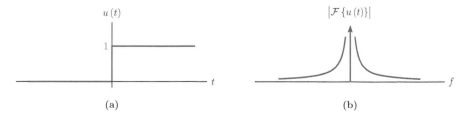

(a) (b)

Figure 4.59 – (a) The unit-step function, and (b) the magnitude of its Fourier transform.

If the transform of the unit-step function is needed in terms of f instead of ω, the result in Eqn. (4.181) needs to be modified as

$$\mathcal{F}\{u(t)\} = \frac{1}{2}\delta(f) + \frac{1}{j2\pi f} \tag{4.182}$$

Symmetry of the Fourier Transform

Conjugate symmetry and conjugate antisymmetry properties were defined for time-domain signals in Section 1.3.6 of Chapter 1. Same definitions apply to a frequency-domain transform as well: A transform $X(\omega)$ is said to be *conjugate symmetric* if it satisfies

$$X^*(\omega) = X(-\omega) \tag{4.183}$$

for all ω. Similarly, a transform $X(\omega)$ is said to be *conjugate antisymmetric* if it satisfies

$$X^*(\omega) = -X(-\omega) \tag{4.184}$$

for all ω. A transform does not have to be conjugate symmetric or conjugate antisymmetric. If it has any symmetry properties, however, they can be explored for simplifying the analysis in the frequency domain.

   If the signal $x(t)$ is real-valued, it can be shown that its Fourier transform $X(\omega)$ is conjugate symmetric. Conversely, if the signal $x(t)$ is purely imaginary, its transform is conjugate antisymmetric.

Symmetry of the Fourier transform:

$$x(t): \text{Real}, \ \text{Im}\{x(t)\} = 0 \quad \text{implies that} \quad X^*(\omega) = X(-\omega) \tag{4.185}$$

$$x(t): \text{Imag}, \ \text{Re}\{x(t)\} = 0 \quad \text{implies that} \quad X^*(\omega) = -X(-\omega) \tag{4.186}$$

Proof:

1. **Real $x(t)$:**
   Let us begin by conjugating both sides of the forward transform integral given by Eqn. (4.127):

$$X^*(\omega) = \left[\int_{-\infty}^{\infty} x(t)\, e^{-j\omega t}\, dt\right]^* = \int_{-\infty}^{\infty} x^*(t)\, e^{j\omega t}\, dt \tag{4.187}$$

   Since $x(t)$ is real valued, $x^*(t) = x(t)$, and we have

$$X^*(\omega) = \int_{-\infty}^{\infty} x(t)\, e^{j\omega t}\, dt \tag{4.188}$$

   In the next step we will derive $X(-\omega)$ from Eqn. (4.127) with the substitution $\omega \to -\omega$:

$$X(-\omega) = \int_{-\infty}^{\infty} x(t)\, e^{j\omega t}\, dt \tag{4.189}$$

   Comparing Eqns. (4.188) and (4.189) we conclude that $X^*(\omega) = X(-\omega)$ when $x(t)$ is real-valued.

2. Imaginary $x(t)$:
In this case the complex conjugate of the transform is

$$X^*(\omega) = \int_{-\infty}^{\infty} x^*(t) \, e^{j\omega t} \, dt = -\int_{-\infty}^{\infty} x(t) \, e^{j\omega t} \, dt \qquad (4.190)$$

since $x^*(t) = -x(t)$ for an imaginary signal. Comparison of Eqns. (4.189) and (4.190) leads to the conclusion that $X^*(\omega) = -X(-\omega)$ when $x(t)$ is purely imaginary.

Cartesian and polar forms of the transform

A complex transform $X(\omega)$ can be written in polar form as

$$X(\omega) = |X(\omega)| \, e^{j\Theta(\omega)} \qquad (4.191)$$

and in Cartesian form as

$$X(\omega) = X_r(\omega) + jX_i(\omega) \qquad (4.192)$$

If $X(\omega)$ is conjugate symmetric, the relationship in Eqn. (4.183) can be written as

$$|X(-\omega)| \, e^{j\Theta(-\omega)} = |X(\omega)| \, e^{-j\Theta(\omega)} \qquad (4.193)$$

using the polar form of the transform, and

$$X_r(-\omega) + jX_i(-\omega) = X_r(\omega) - jX_i(\omega) \qquad (4.194)$$

using its Cartesian form. The consequences of Eqns. (4.193) (4.194) can be obtained by equating the magnitudes and the phases on both sides of Eqn. (4.193) and by equating real and imaginary parts on both sides of Eqn. (4.194). The results can be summarized as follows:

Conjugate symmetric transform: $X(-\omega) = X^*(\omega)$

$$\begin{aligned}
\text{Magnitude:} \quad & |X(-\omega)| = |X(\omega)| & (4.195) \\
\text{Phase:} \quad & \Theta(-\omega) = -\Theta(\omega) & (4.196) \\
\text{Real part:} \quad & X_r(-\omega) = X_r(\omega) & (4.197) \\
\text{Imag. part:} \quad & X_i(-\omega) = -X_i(\omega) & (4.198)
\end{aligned}$$

We have already established the fact that the transform of a real-valued $x(t)$ is conjugate symmetric. The results in Eqns. (4.195) and (4.196) suggest that, for such a transform, the magnitude is an even function of ω, and the phase is an odd function. Furthermore, based on Eqns. (4.197) and (4.198), the real part of the transform exhibits even symmetry while its imaginary part exhibits odd symmetry.

Similarly, if $X(\omega)$ is conjugate antisymmetric, the relationship in Eqn. (4.183) reflects on polar form of $X(\omega)$ as

$$|X(-\omega)| \, e^{j\Theta(-\omega)} = -|X(\omega)| \, e^{-j\Theta(\omega)} \qquad (4.199)$$

The negative sign on the right side of Eqn. (4.199) needs to be incorporated into the phase since we could not write $|X(-\omega)| = -|X(\omega)|$ (recall that magnitude needs to be a non-negative function for all ω). Using $e^{\mp j\pi} = -1$, Eqn. (4.199) can be written as

$$|X(-\omega)| \, e^{j\Theta(-\omega)} = |X(\omega)| \, e^{-j\Theta(\omega)} e^{\mp j\pi}$$

$$= |X(\omega)| \, e^{-j[\Theta(\omega)\pm\pi]} \qquad (4.200)$$

Conjugate antisymmetry property of the transform can also be expressed in Cartesian form as

$$X_r(-\omega) + jX_i(-\omega) = -X_r(\omega) + jX_i(\omega) \qquad (4.201)$$

The consequences of Eqns. (4.199) and (4.201) are given below.

Conjugate antisymmetric transform: $\quad X(-\omega) = -X^*(\omega)$

$$\text{Magnitude:} \quad |X(-\omega)| = |X(\omega)| \qquad (4.202)$$

$$\text{Phase:} \quad \Theta(-\omega) = -\Theta(\omega) \pm \pi \qquad (4.203)$$

$$\text{Real part:} \quad X_r(-\omega) = -X_r(\omega) \qquad (4.204)$$

$$\text{Imag. part:} \quad X_i(-\omega) = X_i(\omega) \qquad (4.205)$$

A purely imaginary signal leads to a Fourier transform with conjugate antisymmetry. For such a transform the magnitude is still an even function of ω as suggested by Eqn. (4.202). The phase is neither even nor odd. The real part is an odd function of ω, and the imaginary part is an even function.

Example 4.24: **Symmetry properties for the transform of right-sided exponential signal**

The Fourier transform of the right-sided exponential signal $x(t) = e^{-at} u(t)$ was found in Example 4.15 to be

$$X(f) = \mathcal{F}\{e^{-at} u(t)\} = \frac{1}{a + j\omega}$$

Elaborate on the symmetry properties of the transform.

Solution: Since the signal $x(t)$ is real-valued, its transform must be conjugate-symmetric. The complex conjugate of the Fourier transform is

$$X^*(\omega) = \left(\frac{1}{a + j\omega}\right)^* = \frac{1}{a - j\omega}$$

and the folded version of the transform, obtained by replacing ω with $-\omega$ in Eqn. (4.206), is

$$X(-\omega) = \left(\frac{1}{a + j\omega}\right)\bigg|_{\omega \to -\omega} = \frac{1}{a - j\omega}$$

Thus we verify that $X^*(\omega) = X(-\omega)$, and the transform is conjugate symmetric. The magnitude and the phase of the transform are

$$|X(\omega)| = \left|\frac{1}{a + j\omega}\right| = \frac{1}{\sqrt{a^2 + \omega^2}}$$

and

$$\Theta\left(\omega\right) = \angle X\left(\omega\right) = -\tan^{-1}\left(\frac{\omega}{a}\right)$$

respectively. We observe that the magnitude is an even function of frequency, that is,

$$\left|X\left(-\omega\right)\right| = \frac{1}{\sqrt{a^2 + \left(-\omega\right)^2}} = \frac{1}{\sqrt{a^2 + \omega^2}} = \left|X\left(\omega\right)\right|$$

consistent with the conclusion reached in Eqn. (4.195). On the other hand, the phase has odd symmetry since

$$\Theta\left(-\omega\right) = -\tan^{-1}\left(\frac{-\omega}{a}\right) = \tan^{-1}\left(\frac{\omega}{a}\right) = -\Theta\left(\omega\right)$$

consistent with Eqn. (4.196). By multiplying both the numerator and the denominator of the transform in Eqn. (4.206) by $(a - j\omega)$ we have

$$X\left(\omega\right) = \frac{(1)}{(a+j\omega)}\frac{(a-j\omega)}{(a-j\omega)} = \frac{a}{a^2+\omega^2} - j\frac{\omega}{a^2+\omega^2} \tag{4.206}$$

Real and imaginary parts of the transform can be found as

$$X_r\left(\omega\right) = \mathrm{Re}\left\{X\left(\omega\right)\right\} = \frac{a}{a^2+\omega^2} \tag{4.207}$$

and

$$X_i\left(\omega\right) = \mathrm{Im}\left\{X\left(\omega\right)\right\} = \frac{-\omega}{a^2+\omega^2} \tag{4.208}$$

It is a trivial matter to show that $X_r\left(\omega\right)$ is an even function of ω, and $X_i\left(\omega\right)$ is an odd function.

Software resources:
ex_4_24a.m
ex_4_24b.m

Transforms of even and odd signals

If the real-valued signal $x\left(t\right)$ is an even function of time, the resulting transform $X\left(\omega\right)$ is real-valued for all ω.

$$x\left(-t\right) = x\left(t\right), \text{ all } t \quad \text{implies that} \quad \mathrm{Im}\left\{X\left(\omega\right)\right\} = 0, \text{ all } \omega \tag{4.209}$$

Proof: If the imaginary part of the transform $X\left(\omega\right)$ is zero, then the transform must remain unchanged when it is conjugated. Thus, another way to express Eqn. (4.209) is by

$$X^*\left(\omega\right) = X\left(\omega\right) \tag{4.210}$$

This is the relationship we will prove. Let us conjugate both sides of the Fourier transform integral in Eqn. (4.127) to obtain

$$X^*\left(\omega\right) = \left[\int_{-\infty}^{\infty} x\left(t\right) e^{-j\omega t}\, dt\right]^* = \int_{-\infty}^{\infty} x^*\left(t\right) e^{j\omega t}\, dt = \int_{-\infty}^{\infty} x\left(t\right) e^{j\omega t}\, dt \tag{4.211}$$

where we have used the fact that $x^*(t) = x(t)$ for a real-valued signal. Changing the integration variable from t to $-\lambda$, recognizing that $dt = -d\lambda$, and adjusting the integration limits accordingly we obtain

$$X^*(\omega) = -\int_{\infty}^{-\infty} x(-\lambda) e^{-j\omega\lambda} d\lambda \tag{4.212}$$

For a signal with even symmetry we have $x(-\lambda) = x(\lambda)$, and the integral in Eqn. (4.212) becomes

$$X^*(\omega) = -\int_{\infty}^{-\infty} x(\lambda) e^{-j\omega\lambda} d\lambda \tag{4.213}$$

Finally, swapping the two integration limits and negating the integral to compensate for it results in

$$X^*(\omega) = \int_{-\infty}^{\infty} x(\lambda) e^{-j\omega\lambda} d\lambda = X(\omega) \tag{4.214}$$

Conversely it can also be proven that, if the real-valued signal $x(t)$ has odd-symmetry, the resulting Fourier transform is purely imaginary. This can be mathematically stated as follows:

$$x(-t) = -x(t), \text{ all } t \quad \text{implies that} \quad \text{Re}\{X(\omega)\} = 0, \text{ all } \omega \tag{4.215}$$

Conjugating a purely imaginary transform is equivalent to negating it, so an alternative method of expressing the relationship in Eqn. (4.215) is

$$X^*(\omega) = -X(\omega) \tag{4.216}$$

The procedure for proving Eqn. (4.216) is very similar to the one given above for the transform of an even symmetric signal. (See Problem 4.25 at the end of this chapter.)

Example 4.25: **Transform of a two-sided exponential signal**

The two-sided exponential signal $x(t) = e^{-a|t|}$ was shown in Fig. 4.46, and has even symmetry. Its Fourier transform was found in Example 4.16 as

$$X(\omega) = \frac{2a}{a^2 + \omega^2}$$

and was graphed in Fig. 4.47. As expected, the transform is real-valued for all ω.

Example 4.26: **Transform of a pulse with odd symmetry**

Consider the signal

$$x(t) = \begin{cases} -1, & -1 < t < 0 \\ 1, & 0 < t < 1 \\ 0, & t < -1 \quad \text{or} \quad t > 1 \end{cases}$$

shown in Fig. 4.60. Determine the Fourier transform $X(\omega)$ and show that it is purely imaginary.

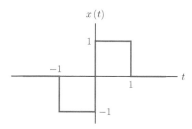

Figure 4.60 – The pulse of Example 4.26 with odd symmetry.

Solution: The transform can be found by direct use of the forward transform integral.

$$X(\omega) = \int_{-1}^{0} (-1) \, e^{-j\omega t} \, dt + \int_{0}^{1} (1) \, e^{-j\omega t} \, dt$$

$$= \frac{j2}{\omega} \left[\cos(\omega) - 1 \right] \qquad (4.217)$$

As expected, the transform is purely imaginary. Its magnitude, phase and imaginary part are graphed in Fig. 4.61.

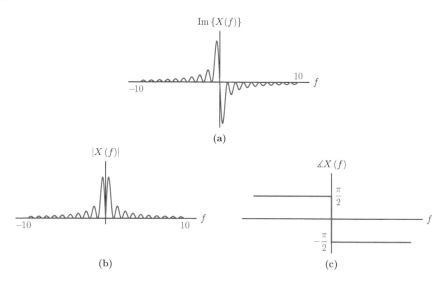

Figure 4.61 – The Fourier transform found in Eqn. (4.217) for the odd symmetric pulse of Example 4.26: (a) imaginary part, (b) magnitude, (c) phase.

Software resources:
ex_4_26.m

Time shifting

For a transform pair

$$x(t) \overset{\mathcal{F}}{\longleftrightarrow} X(\omega)$$

it can be shown that

$$x(t-\tau) \xleftrightarrow{\mathcal{F}} X(\omega) e^{-j\omega\tau} \tag{4.218}$$

Proof: Applying the Fourier transform integral in Eqn. (4.127) to $x(t-\tau)$ we obtain

$$\mathcal{F}\{x(t-\tau)\} = \int_{-\infty}^{\infty} x(t-\tau) e^{-j\omega t}\, dt \tag{4.219}$$

Let $\lambda = t - \tau$ in the integral of Eqn. (4.219) so that

$$\mathcal{F}\{x(t-\tau)\} = \int_{-\infty}^{\infty} x(\lambda) e^{-j\omega(\lambda+\tau)}\, d\lambda \tag{4.220}$$

The exponential function in the integral of Eqn. (4.220) can be written as a product of two exponential functions to obtain

$$\mathcal{F}\{x(t-\tau)\} = \int_{-\infty}^{\infty} x(\lambda) e^{-j\omega\lambda} e^{-j\omega\tau}\, d\lambda$$

$$= e^{-j\omega\tau} \int_{-\infty}^{\infty} x(\lambda) e^{-j\omega\lambda}\, d\lambda$$

$$= e^{-j\omega\tau} X(\omega) \tag{4.221}$$

The consequence of shifting, or delaying, a signal in time is multiplication of its Fourier transform by a complex exponential function of frequency.

Example 4.27: **Time shifting a rectangular pulse**

The transform of a rectangular pulse with amplitude A, width τ and center at $t = 0$ was found in Example 4.12 as

$$A\,\Pi\left(\frac{t}{\tau}\right) \xleftrightarrow{\mathcal{F}} A\tau\,\operatorname{sinc}\left(\frac{\omega\tau}{2\pi}\right)$$

In Example 4.13 the transform of a time shifted version of this pulse was determined by direct application of the Fourier transform integral as

$$A\,\Pi\left(\frac{t-\tau/2}{\tau}\right) \xleftrightarrow{\mathcal{F}} A\tau\,\operatorname{sinc}\left(\frac{\omega\tau}{2\pi}\right) e^{-j\omega\tau/2}$$

The result in Eqn. (4.149) could easily have been obtained from the result in Eqn. (4.144) through the use of the time shifting property.

Example 4.28: **Time shifting a two-sided exponential signal**

Consider again the two-sided exponential signal used in earlier examples. A time shifted version of it is

$$x(t) = e^{-a|t-\tau|}$$

The signal $x(t)$ is shown in Fig 4.62 with the assumption that $a > 0$.

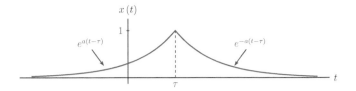

Figure 4.62 – The time-shifted two-sided exponential signal of Example 4.28.

In Example 4.16 its Fourier transform was determined to be

$$\mathcal{F}\left\{ e^{-a\,|t|} \right\} = \frac{2a}{a^2 + \omega^2}$$

Using the time shifting property of the Fourier transform we obtain

$$X\left(\omega\right) = \mathcal{F}\left\{ e^{-a\,|t-\tau|} \right\} = \frac{2a\,e^{-j\omega\tau}}{a^2 + \omega^2}$$

The transform $X\left(\omega\right)$ is shown in Fig. 4.63.

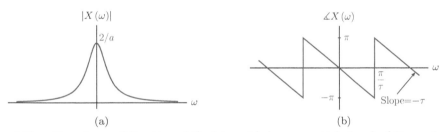

(a)   (b)

Figure 4.63 – Transform of the time-shifted two-sided exponential signal of Example 4.28:
(a) magnitude, (b) phase.

Compare this with the transform of the signal of Example 4.16. Note that time shifting the signal affects only the phase of the frequency spectrum, and not its magnitude.

Software resources:
ex_4_28.m

Frequency shifting

For a transform pair

$$x\left(t\right) \overset{\mathcal{F}}{\longleftrightarrow} X\left(\omega\right)$$

it can be shown that

$$x\left(t\right) e^{j\omega_0 t} \overset{\mathcal{F}}{\longleftrightarrow} X\left(\omega - \omega_0\right) \tag{4.222}$$

Proof: By directly applying the Fourier transform integral in Eqn. (4.127) to the signal $x(t) e^{j\omega_0 t}$ we obtain

$$\mathcal{F}\left\{x(t) e^{j\omega_0 t}\right\} = \int_{-\infty}^{\infty} x(t) e^{j\omega_0 t} e^{-j\omega t} dt$$

$$= \int_{-\infty}^{\infty} x(t) e^{-j(\omega-\omega_0)t} dt$$

$$= X(\omega - \omega_0) \tag{4.223}$$

An alternative proof can be obtained by using the duality principle in conjunction with the time shifting property. While it isn't necessarily shorter or simpler than the proof above, it nevertheless serves as another demonstration of the use of duality principle, and will therefore be given here. We start with the following two relationships:

$$x(t) \overset{\mathcal{F}}{\longleftrightarrow} X(\omega) \tag{4.224}$$

$$x(t - \tau) \overset{\mathcal{F}}{\longleftrightarrow} X(\omega) e^{-j\omega\tau} \tag{4.225}$$

If $x(t)$ and $X(\omega)$ are a valid transform pair in Eqn. (4.224), then it follows through the use of the time-shifting property that Eqn. (4.225) is also valid. Applying the duality principle to each of the transform pairs in Eqns. (4.224) and (4.225) we obtain the following two relationships:

$$X(t) \overset{\mathcal{F}}{\longleftrightarrow} 2\pi x(-\omega) \tag{4.226}$$

$$X(t) e^{-jt\tau} \overset{\mathcal{F}}{\longleftrightarrow} 2\pi x(-\omega - \tau) \tag{4.227}$$

Let $\tilde{x}(t) = X(t)$ and $\tilde{X}(\omega) = 2\pi x(-\omega)$. Also using the substitution $\omega_0 = -\tau$, the two transforms in Eqns. (4.226) and (4.227) can be rewritten as

$$\tilde{x}(t) \overset{\mathcal{F}}{\longleftrightarrow} \tilde{X}(\omega) \tag{4.228}$$

$$\tilde{x}(t) e^{j\omega_0 t} \overset{\mathcal{F}}{\longleftrightarrow} \tilde{X}(\omega - \omega_0) \tag{4.229}$$

Modulation property

Modulation property is an interesting consequence of the frequency shifting property combined with the linearity of the Fourier transform. As its name implies, it finds significant use in modulation techniques utilized in analog and digital communications. We will have the occasion to use it when we study amplitude modulation in Chapter 11.

For a transform pair

$$x(t) \overset{\mathcal{F}}{\longleftrightarrow} X(\omega)$$

it can be shown that

$$x(t)\cos(\omega_0 t) \overset{\mathcal{F}}{\longleftrightarrow} \frac{1}{2}\left[X(\omega - \omega_0) + X(\omega + \omega_0)\right] \tag{4.230}$$

and

$$x(t)\sin(\omega_0 t) \overset{\mathcal{F}}{\longleftrightarrow} \frac{1}{2}\left[X(\omega - \omega_0) e^{-j\pi/2} + X(\omega + \omega_0) e^{j\pi/2}\right] \tag{4.231}$$

Multiplication of a signal by a cosine waveform causes its spectrum to be shifted in both directions by the frequency of the cosine waveform, and to be scaled by $\frac{1}{2}$. Multiplication of the signal by a sine waveform causes a similar effect with an added phase shift of $-\pi/2$ radians for positive frequencies and $\pi/2$ radians for negative frequencies.

Proof: Using Euler's formula, the left side of the relationship in Eqn. (4.230) can be written as

$$x\left(t\right)\cos(\omega_0 t)=\frac{1}{2}x\left(t\right)e^{\omega_0 t}+\frac{1}{2}x\left(t\right)e^{-\omega_0 t} \tag{4.232}$$

The desired proof is obtained by applying the frequency shifting property to the terms on the right side of Eqn. (4.232). Using Eqn. (4.222) we obtain

$$x\left(t\right)e^{j\omega_0 t}\stackrel{\mathcal{F}}{\longleftrightarrow}X\left(\omega-\omega_0\right) \tag{4.233}$$

and

$$x\left(t\right)e^{-j\omega_0 t}\stackrel{\mathcal{F}}{\longleftrightarrow}X\left(\omega+\omega_0\right) \tag{4.234}$$

which could be used together in Eqn. (4.232) to arrive at the result in Eqn. (4.230).

The proof of Eqn. (4.231)is similar, but requires one additional step. Again using Euler's formula, let us write the left side of Eqn. (4.231) as

$$x\left(t\right)\sin(\omega_0 t)=\frac{1}{2j}x\left(t\right)e^{j\omega_0 t}-\frac{1}{2j}x\left(t\right)e^{-j\omega_0 t} \tag{4.235}$$

Realizing that

$$\frac{1}{j}=-j=e^{-j\pi/2}\quad\text{and}\quad\frac{-1}{j}=j=e^{j\pi/2}$$

Eqn. (4.235) can be rewritten as

$$x\left(t\right)\sin(\omega_0 t)=\frac{1}{2}x\left(t\right)e^{j\omega_0 t}e^{-j\pi/2}+\frac{1}{2}x\left(t\right)e^{-j\omega_0 t}e^{j\pi/2} \tag{4.236}$$

The proof of Eqn. (4.231) can be completed by using Eqns. (4.233) and (4.234) on the right side of Eqn. (4.236).

Example 4.29: **Modulated pulse**

Find the Fourier transform of the modulated pulse given by

$$x\left(t\right)=\begin{cases}\cos\left(2\pi f_0 t\right), & |t|<\tau\\ 0, & |t|>\tau\end{cases}$$

Solution: Let $p\left(t\right)$ be defined as a rectangular pulse centered around the time origin with a pulse width of 2τ and a unit amplitude, i.e.,

$$p\left(t\right)=\Pi\left(\frac{t}{2\tau}\right)$$

Using $p\left(t\right)$, the signal $x\left(t\right)$ can be expressed in a more compact form as

$$x\left(t\right)=p\left(t\right)\cos\left(2\pi f_0 t\right)$$

This is illustrated in Fig. 4.64.

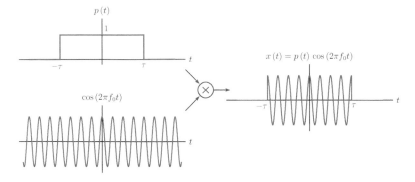

Figure 4.64 – Expressing the modulated pulse signal of Example 4.29 as the product of a rectangular pulse and a sinusoidal signal.

The transform of the pulse $p(t)$ is

$$P(f) = \mathcal{F}\{p(t)\} = 2\tau \operatorname{sinc}(2\tau f) \tag{4.237}$$

We will apply the modulation property to this result to obtain the transform we seek:

$$
\begin{aligned}
X(\omega) &= \frac{1}{2} P(f - f_0) + \frac{1}{2} P(f + f_0) \\
&= \tau \operatorname{sinc}\left(2\tau(f + f_0)\right) + \tau \operatorname{sinc}\left(2\tau(f - f_0)\right)
\end{aligned}
\tag{4.238}
$$

Construction of this spectrum is illustrated in Fig. 4.65.

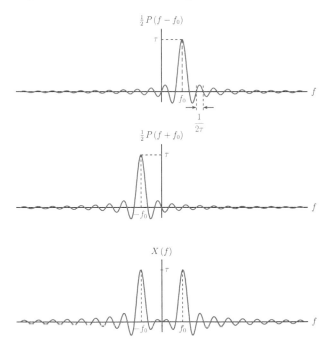

Figure 4.65 – Transform of the modulated pulse of Example 4.29 obtained by adding frequency-shifted versions of $P(\omega)$.

Software resources:
ex_4_29.m

Interactive Demo: ft_demo8.m

This demo program is based on Example 4.29. It allows experimentation with modulated pulse signal and its Fourier transform. Pulse width parameter τ and the frequency f_0 may be varied using slider controls. Both the modulated pulse signal and its transform are graphed.

1. What is the effect of changing the pulse width on the spectrum?
2. How does the frequency f_0 affect the shape of the spectrum?

Software resources:
ft_demo8.m

Time and frequency scaling

For a transform pair

$$x\left(t\right) \overset{\mathcal{F}}{\longleftrightarrow} X\left(\omega\right)$$

it can be shown that

$$x\left(at\right) \overset{\mathcal{F}}{\longleftrightarrow} \frac{1}{\left|a\right|} X\left(\frac{\omega}{a}\right) \tag{4.239}$$

The parameter a is any non-zero and real-valued constant.

Proof: The Fourier transform of $x\left(at\right)$ is

$$\mathcal{F}\left\{x\left(at\right)\right\} = \int_{-\infty}^{\infty} x\left(at\right) e^{-j\omega t}\, dt \tag{4.240}$$

A new independent variable λ will be introduced through the variable change $at = \lambda$. Let us substitute

$$t = \frac{\lambda}{a} \quad \text{and} \quad dt = \frac{d\lambda}{a} \tag{4.241}$$

in the integral of Eqn. (4.241). We need to consider the cases of $a > 0$ and $a < 0$ separately. If $a > 0$, then $t \to \pm\infty$ implies $\lambda \to \pm\infty$. Therefore, the limits of the integral remain unchanged under the variable change, and we have

$$\mathcal{F}\left\{x\left(at\right)\right\} = \frac{1}{a} \int_{-\infty}^{\infty} x\left(\lambda\right) e^{-j\omega\lambda/a}\, d\lambda$$

$$= \frac{1}{a} X\left(\frac{\omega}{a}\right), \quad a > 0 \tag{4.242}$$

If $a < 0$, however, the same substitution leads to

$$\mathcal{F}\left\{x\left(at\right)\right\} = \frac{1}{a} \int_{\infty}^{-\infty} x\left(\lambda\right) e^{-j\omega\lambda/a}\, d\lambda \tag{4.243}$$

since $t \to \pm\infty$ now leads to $\lambda \to \mp\infty$. Swapping the lower and upper limits of the integral and negating the result to compensate for it yields

$$\mathcal{F}\left\{x\left(\lambda\right)\right\} = -\frac{1}{a} \int_{-\infty}^{\infty} x\left(\lambda\right) e^{-j\omega\lambda/a} \, d\lambda$$

$$= -\frac{1}{a} X\left(\frac{\omega}{a}\right), \quad a < 0 \tag{4.244}$$

It is possible to combine Eqns. (4.242) and (4.244) into one compact expression by letting the scale factor in front of the integral be expressed as $1/\left|a\right|$ so that it is valid for both $a > 0$ and $a < 0$. Thus we obtain the desired proof:

$$\mathcal{F}\left\{x\left(at\right)\right\} = \frac{1}{\left|a\right|} X\left(\frac{\omega}{a}\right) \tag{4.245}$$

This property is significant in signal processing. The implication of Eqn. (4.239) is that, if the time variable is scaled by a factor a, the frequency variable is scaled by the same factor, but in the opposite direction.

Example 4.30: **Working with scaled and shifted pulses**

Determine the Fourier transform of the signal

$$x\left(t\right) = \begin{cases} -1, & -0.5 < t < 1.5 \\ 0.75, & 1.5 < t < 5.5 \end{cases}$$

which is shown in Fig 4.66.

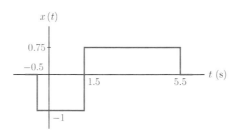

Figure 4.66 – The signal $x\left(t\right)$ of Example 4.30.

Solution: It certainly is possible to use the Fourier transform integral of Eqn. (4.127) directly on the signal $x\left(t\right)$. It would be more convenient, however, to make use of the properties of the Fourier transform. The signal $x\left(t\right)$ can be written as the sum of two rectangular pulses:

1. One with amplitude of $A_1 = -1$, width $\tau_1 = 2$ s, center at $t_{d1} = 0.5$ s

2. One with amplitude of $A_2 = 0.75$, width $\tau_2 = 4$ s, center at $t_{d2} = 3.5$

Using the unit-pulse function $\Pi(t)$ we have

$$x(t) = A_1 \Pi\left(\frac{t - t_{d1}}{\tau_1}\right) + A_2 \Pi\left(\frac{t - t_{d2}}{\tau_2}\right)$$

$$= -\Pi\left(\frac{t - 0.5}{2}\right) + 0.75\,\Pi\left(\frac{t - 3.5}{4}\right)$$

The Fourier transform of the unit-pulse function is

$$\mathcal{F}\{\Pi(t)\} = \operatorname{sinc}\left(\frac{\omega}{2\pi}\right)$$

Using the scaling property of the Fourier transform, the following two relationships can be written:

$$\Pi\left(\frac{t}{2}\right) \xleftrightarrow{\mathcal{F}} 2\operatorname{sinc}\left(\frac{\omega}{\pi}\right)$$

$$\Pi\left(\frac{t}{4}\right) \xleftrightarrow{\mathcal{F}} 4\operatorname{sinc}\left(\frac{2\omega}{\pi}\right)$$

In addition, the use of the time-shifting property leads to

$$\Pi\left(\frac{t - 0.5}{2}\right) \xleftrightarrow{\mathcal{F}} 2\operatorname{sinc}\left(\frac{\omega}{\pi}\right) e^{-j0.5\,\omega} \tag{4.246}$$

$$\Pi\left(\frac{t - 3.5}{4}\right) \xleftrightarrow{\mathcal{F}} 4\operatorname{sinc}\left(\frac{2\omega}{\pi}\right) e^{-j3.5\,\omega} \tag{4.247}$$

Using Eqns. (4.246) and (4.247) the Fourier transform of $x(t)$ is

$$X(\omega) = -2\operatorname{sinc}\left(\frac{\omega}{\pi}\right) e^{-j0.5\,\omega} + 3\operatorname{sinc}\left(\frac{2\omega}{\pi}\right) e^{-j3.5\,\omega}$$

which is graphed in Fig. 4.67.

Software resources:
ex_4_30.m

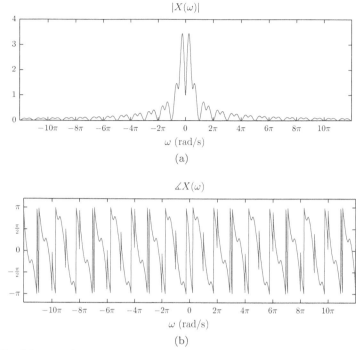

Figure 4.67 – Magnitude and phase of the spectrum of the signal $x(t)$ of Example 4.30.

Interactive Demo: fd_demo9.m

The demo program "fd_demo9.m" is based on Example 4.30. It illustrates the use of time scaling and time shifting properties of the Fourier transform for computing the transform of a signal that can be expressed as the sum of two rectangular pulses. The signal to be constructed is in the form

$$x(t) = A_1 \, \Pi\left(\frac{t - t_{d1}}{\tau_1}\right) + A_2 \, \Pi\left(\frac{t - t_{d2}}{\tau_2}\right)$$

Parameters A_1, A_2, τ_1 τ_2, t_{d1} and t_{d2} can be varied using slider controls. The individual spectrum of each pulse is displayed along with the spectrum of the signal $x(t)$.
Software resources:
fd_demo9.m

Differentiation in the time domain

For a given transform pair

$$x(t) \; \overset{\mathcal{F}}{\longleftrightarrow} \; X(\omega)$$

it can be shown that

$$\frac{d^n}{dt^n}\left[x(t)\right] \overset{\mathcal{F}}{\longleftrightarrow} (j\omega)^n \, X(\omega) \tag{4.248}$$

If we choose to use f instead or ω, then

$$\frac{d^n}{dt^n}\left[x\left(t\right)\right] \overset{\mathcal{F}}{\longleftrightarrow} \left(j2\pi f\right)^n X\left(f\right) \tag{4.249}$$

Proof: We will first prove the relationship in Eqn. (4.248) for $n = 1$. Once this is done, proof for $n > 1$ is easily obtained by repeating the process. The proof for Eqn. (4.249) is very similar, and will be left as an exercise (see Problem 4.28 at the end of this chapter).

Consider the inverse Fourier transform relationship given by Eqn. (4.126). Differentiating both sides of it with respect to t yields

$$\frac{d}{dt}\left[x\left(t\right)\right] = \frac{d}{dt}\left[\frac{1}{2\pi}\int_{-\infty}^{\infty} X\left(\omega\right)e^{j\omega t}\,d\omega\right] \tag{4.250}$$

Swapping the order of differentiation and integration on the right side of Eqn. (4.250) we can write

$$\frac{d}{dt}\left[x\left(t\right)\right] = \frac{1}{2\pi}\int_{-\infty}^{\infty}\frac{d}{dt}\left[X\left(\omega\right)e^{j\omega t}\right]\,d\omega$$

$$= \frac{1}{2\pi}\int_{-\infty}^{\infty}\left[j\omega X\left(\omega\right)\right]e^{j\omega t}\,d\omega$$

$$= \mathcal{F}^{-1}\left\{j\omega X\left(\omega\right)\right\} \tag{4.251}$$

which leads us to the conclusion

$$\frac{d}{dt}\left[x\left(t\right)\right] \overset{\mathcal{F}}{\longleftrightarrow} j\omega X\left(\omega\right) \tag{4.252}$$

Thus, differentiation of the signal with respect to the time variable corresponds to multiplication of the transform with $j\omega$. Differentiating both sides of Eqn. (4.251) one more time yields

$$\frac{d}{dt}\left[\frac{d}{dt}\left[x\left(t\right)\right]\right] = \frac{d}{dt}\left[\frac{1}{2\pi}\int_{-\infty}^{\infty} j\omega X\left(\omega\right)e^{j\omega t}\,df\right]$$

$$= \frac{1}{2\pi}\int_{-\infty}^{\infty}\frac{d}{dt}\left[j\omega X\left(\omega\right)e^{j\omega t}\right]\,df$$

$$= \frac{1}{2\pi}\int_{-\infty}^{\infty}\left[\left(j\omega\right)^2 X\left(\omega\right)\right]e^{j\omega t}\,df$$

$$= \mathcal{F}^{-1}\left\{\left(j\omega\right)^2 X\left(\omega\right)\right\} \tag{4.253}$$

or in compact form

$$\frac{d^2}{dt^2}\left[x\left(t\right)\right] \overset{\mathcal{F}}{\longleftrightarrow} \left(j\omega\right)^2 X\left(\omega\right) \tag{4.254}$$

It is obvious that each differentiation of the signal $x(t)$ will bring up another $(j\omega)$ factor in the corresponding transform. Differentiating the signal $x(t)$ for n times corresponds to multiplying the transform by $(j\omega)^n$.

Example 4.31: **Triangular pulse revisited**

Recall that the Fourier transform of the triangular pulse signal

$$x(t) = A\Lambda\left(\frac{t}{\tau}\right)$$

was found in Example 4.17 through direct application of the Fourier transform integral of Eqn. (4.127). Rework the problem using the differentiation-in-time property.

Solution: Instead of working with the signal $x(t)$ directly, we will find it convenient to work with its derivative

$$w(t) = \frac{dx(t)}{dt}$$

shown in Fig. 4.68.

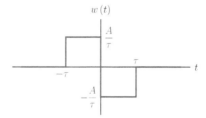

$w(t)$

Figure 4.68 – The intermediate signal $w(t)$ for Example 4.31.

Using the unit-pulse function along with time-scaling and time-shifting operations, $w(t)$ can be expressed as

$$w(t) = \frac{A}{\tau}\left[\Pi\left(\frac{t+\tau/2}{\tau}\right) - \Pi\left(\frac{t-\tau/2}{\tau}\right)\right] \tag{4.255}$$

Using the variable f instead of ω for convenience, the transform $W(f)$ can be computed from Eqn. (4.255) as

$$\begin{aligned} W(f) &= A\,\text{sinc}(f\tau)\,e^{j2\pi f(\tau/2)} - A\,\text{sinc}(f\tau)\,e^{-j2\pi f(\tau/2)} \\ &= A\,\text{sinc}(f\tau)\left[e^{j2\pi f(\tau/2)} - e^{-j2\pi f(\tau/2)}\right] \\ &= 2j\,A\,\text{sinc}(f\tau)\,\sin(\pi f\tau) \end{aligned} \tag{4.256}$$

The real part of $W(f)$ is equal to zero; its imaginary part is shown in Fig. 4.69. Frequency-domain equivalent of the relationship in Eqn. (4.256) is

$$W(f) = (j2\pi f)\,X(f) \tag{4.257}$$

Solving Eqn. (4.257) for $X(f)$ yields

$$X(f) = \frac{W(f)}{j2\pi f} = A\tau\,\text{sinc}^2(f\tau)$$

which is in agreement with the answer found in Example 4.17.

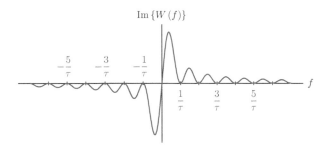

Figure 4.69 – Spectrum of the intermediate signal $w(t)$ for Example 4.31.

Example 4.32: **Fourier transform of a trapezoidal pulse**

Determine the Fourier transform of the trapezoidal pulse $x(t)$ shown in Fig. 4.70

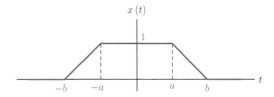

Figure 4.70 – Trapezoidal pulse of Example 4.32.

Solution: Analytical description of this signal could be given in the form

$$x(t) = \begin{cases} 0, & t < -b \\ (t+b)/(b-a), & -b \le t < -a \\ 1, & -a \le t < a \\ (-t+b)/(b-a), & a \le t < b \\ 0, & t > b \end{cases}$$

Once again we will use the intermediate signal $w(t)$ obtained by differentiating $x(t)$ with respect to t:

$$w(t) = \frac{dx(t)}{dt} = \begin{cases} 1/(b-a), & -b < t < -a \\ -1/(b-a), & a \le t < b \\ 0, & \text{otherwise} \end{cases}$$

The signal $w(t)$ is shown in Fig. 4.71.

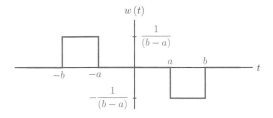

Figure 4.71 – The intermediate signal $w(t)$ for Example 4.32.

Finding the Fourier transform of $w(t)$ is much easier compared to a direct evaluation of the Fourier transform integral for $x(t)$. For notational convenience, let two new parameters be defined as

$$\tau = b - a \quad \text{and} \quad \lambda = \frac{a+b}{2} \tag{4.258}$$

The parameter τ is the width of each of the two pulses that make up $w(t)$, and these pulses are centered at $\pm\lambda$. The derivative $w(t)$ can be written using the unit-pulse function as

$$w(t) = \frac{1}{\tau}\Pi\left(\frac{t+\lambda}{\tau}\right) - \frac{1}{\tau}\Pi\left(\frac{t-\lambda}{\tau}\right)$$

and the corresponding transform $W(f)$ is

$$
\begin{aligned}
W(f) &= \operatorname{sinc}(f\tau)\, e^{j2\pi f\lambda} - \operatorname{sinc}(f\tau)\, e^{-j2\pi f\lambda} \\
&= \operatorname{sinc}(f\tau)\left[e^{j2\pi f\lambda} - e^{-j2\pi f\lambda}\right] \\
&= 2j\,\operatorname{sinc}(f\tau)\,\sin(2\pi f\lambda)
\end{aligned}
\tag{4.259}
$$

The transform of $x(t)$ can now be found by scaling this result:

$$X(f) = \frac{W(f)}{j2\pi f} = 2\lambda\,\operatorname{sinc}(f\tau)\,\operatorname{sinc}(2f\lambda)$$

The transform $X(f)$ is graphed in Fig. 4.72 for various values of the parameters τ and λ.

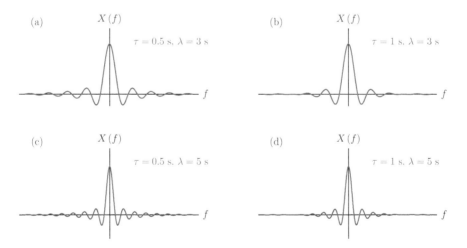

Figure 4.72 – The transform $X(f)$ for Example 4.32.

Having found the solution for $X(f)$ we will speculate on the differentiation in time property a bit further. What if we had differentiated $w(t)$ one more time to obtain a new signal $v(t)$? Let

$$v(t) = \frac{dw(t)}{dt} = \frac{d^2x(t)}{dt^2}$$

The signal $v(t)$ consists of some shifted and scaled impulses, and is shown in Fig. 4.73.

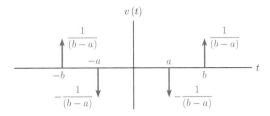

Figure 4.73 – The intermediate signal $v(t)$ for Example 4.32.

The transform of $v(t)$ would be related to the transform of $x(t)$ by

$$X(f) = \frac{V(f)}{(j2\pi f)^2} = -\frac{V(f)}{4\pi^2 f^2} \tag{4.260}$$

Analytically, $v(t)$ can be expressed as

$$v(t) = \frac{1}{b-a}\delta(t+b) - \frac{1}{b-a}\delta(t+a) - \frac{1}{b-a}\delta(t-a) + \frac{1}{b-a}\delta(t-b)$$

Knowing that $\mathcal{F}\{\delta(t)\} = 1$, and utilizing the time-shifting property of the Fourier transform, $V(f)$ is found to be

$$V(f) = \frac{1}{b-a}e^{j2\pi fb} - \frac{1}{b-a}e^{j2\pi fa} - \frac{1}{b-a}e^{-j2\pi fa} + \frac{1}{b-a}e^{-j2\pi fb}$$

$$= \frac{2}{b-a}\left[\cos(2\pi fb) - \cos(2\pi fa)\right] \tag{4.261}$$

Based on Eqn. (4.258), we have

$$a = \lambda - \frac{\tau}{2} \quad \text{and} \quad b = \lambda + \frac{\tau}{2}$$

Substituting a and b into Eqn. (4.261) we obtain

$$V(f) = \frac{2}{\tau}\left[\cos\left(2\pi f\left(\lambda + \frac{\tau}{2}\right)\right) - \cos\left(2\pi f\left(\lambda - \frac{\tau}{2}\right)\right)\right] \tag{4.262}$$

Using the appropriate trigonometric identity,[6] Eqn. (4.262) becomes

$$V(f) = -\frac{4}{\tau}\sin(\pi f\tau)\sin(2\pi f\lambda)$$

and the transform $X(f)$ is obtained by using Eqn. (4.260) as

$$X(f) = \frac{1}{\pi^2 f^2\tau}\sin(\pi f\tau)\sin(2\pi f\lambda)$$

$$= 2\lambda\operatorname{sinc}(f\tau)\operatorname{sinc}(2f\lambda)$$

which matches the result found earlier.

---

[6] $\cos(a+b) = \cos(a)\cos(b) - \sin(a)\sin(b)$.

Software resources:
ex_4_32.m

Interactive Demo: fd_demo10

The demo program "fd_demo10.m" illustrates the steps involved in Example 4.32 for finding the Fourier transform of a trapezoidal pulse through the use of the differentiation in time property. The signals $x(t)$ and $w(t)$ are displayed on the left side of the graphical user interface window. On the right, the transforms are displayed. The parameters τ and λ may be adjusted using the two slider controls provided. Based on Eqn. (4.259) the real part of $W(f)$ is zero, and therefore only its imaginary part is shown. On the other hand, the transform $X(f)$ is purely real as given by Eqn. (4.260). We graph $X(f)/\lambda$ so that the peak magnitude of the resulting graph is fixed, and we can concentrate on the placement of zero crossings as the parameters τ and λ are varied.

It is also interesting to note that, for $\lambda = \tau/2$, the trapezoidal pulse $x(t)$ becomes identical to the triangular pulse of Example 4.31. In the demo program it is possible to see this by setting $\lambda = 0.5$ and $\tau = 1$. The corresponding transforms $W(f)$ and $X(f)/\lambda$ should also agree with the results of Example 4.31.

Software resources:
fd_demo10.m

Differentiation in the frequency domain

For a transform pair

$$x(t) \overset{\mathcal{F}}{\longleftrightarrow} X(\omega) \tag{4.263}$$

it can be shown that

$$(-jt)^n \, x(t) \overset{\mathcal{F}}{\longleftrightarrow} \frac{d^n}{d\omega^n}[X(\omega)] \tag{4.264}$$

If we choose to use f instead or ω, then

$$(-j2\pi t)^n \, x(t) \overset{\mathcal{F}}{\longleftrightarrow} \frac{d^n}{df^n}[X(f)] \tag{4.265}$$

Proof: The proof is straightforward and is similar to the proof of Eqn. (4.248). It will be carried out for Eqn. (4.264). The very similar proof of Eqn. (4.265) will be left as an exercise (see Problems 4.29 and 4.30 at the end of this chapter).

Let us differentiate both sides of the Fourier transform integral in Eqn. (4.127) with respect to ω:

$$\frac{d}{d\omega}\Big[X(\omega)\Big] = \frac{d}{d\omega}\left[\int_{-\infty}^{\infty} x(t) \, e^{-j\omega t} \, dt\right] \tag{4.266}$$

Changing the order of integration and differentiation on the right side of Eqn. (4.266) yields

$$\frac{d}{d\omega}\left[X\left(\omega\right)\right] = \int_{-\infty}^{\infty} \frac{d}{d\omega}\left[x\left(t\right)e^{-j\omega t}\right]dt$$

$$= \int_{-\infty}^{\infty}(-jt)\,x\left(t\right)e^{-j\omega t}\,dt$$

$$= \mathcal{F}\left\{(-jt)\,x\left(t\right)\right\} \tag{4.267}$$

to prove Eqn. (4.264) for $n = 1$. Proof for $n > 1$ can easily be obtained by repeated use of this procedure.

Alternatively, we might see the similarity between the expressions of the differentiation properties in time and frequency domains, and prove the latter from the knowledge of the former through the use of the duality principle. Let's start with the following two relationships:

$$x\left(t\right) \xleftrightarrow{\mathcal{F}} X\left(\omega\right) \tag{4.268}$$

$$\frac{d^n}{dt^n}\left[x\left(t\right)\right] \xleftrightarrow{\mathcal{F}} (j\omega)^n\,X\left(\omega\right) \tag{4.269}$$

If $x\left(t\right)$ and $X\left(f\right)$ are a valid transform pair in Eqn. (4.268), then Eqn. (4.269) is also valid by the differentiation in time property. Application of the duality principle to each of the transform pairs in Eqns. (4.268) and (4.269) leads to the following two relationships:

$$X\left(t\right) \xleftrightarrow{\mathcal{F}} 2\pi\,x\left(-\omega\right) \tag{4.270}$$

$$(jt)^n\,X\left(t\right) \xleftrightarrow{\mathcal{F}} 2\pi \frac{d^n}{d\left(-\omega\right)^n}\left[x\left(-\omega\right)\right] \tag{4.271}$$

Let $\bar{x}\left(t\right) = X\left(t\right)$ and $\bar{X}\left(\omega\right) = 2\pi\,x\left(-\omega\right)$. The two transforms in Eqns. (4.270) and (4.271) can be rewritten as

$$\bar{x}\left(t\right) \xleftrightarrow{\mathcal{F}} \bar{X}\left(\omega\right) \tag{4.272}$$

$$(jt)^n\,\bar{x}\left(t\right) \xleftrightarrow{\mathcal{F}} \frac{d^n}{d\left(-\omega\right)^n}\left[\bar{X}\left(\omega\right)\right] \tag{4.273}$$

Finally, multiplication of both sides of the transform pair in Eqn. (4.273) by $(-1)^n$ yields

$$(-jt)^n\,\bar{x}\left(t\right) \xleftrightarrow{\mathcal{F}} \frac{d^n}{d\omega^n}\left[\bar{X}\left(\omega\right)\right] \tag{4.274}$$

which completes the proof. If Eqn. (4.272) represents a valid transform pair, then so does Eqn. (4.274).

Example 4.33: **Pulse with trapezoidal spectrum**

Find the signal $x\left(t\right)$ the Fourier transform of which is the trapezoidal function given by

$$X\left(f\right) = \begin{cases} 1\,, & |f| \leq f_0\left(1 - r\right) \\ \dfrac{1}{2r}\left[-\dfrac{|f|}{f_0} + 1 + r\right]\,, & f_0\left(1 - r\right) < |f| \leq f_0\left(1 + r\right) \\ 0\,, & |f| > f_0\left(1 + r\right) \end{cases}$$

and shown in Fig. 4.74. Real-valued parameter r is adjustable in the range $0 < r < 1$.

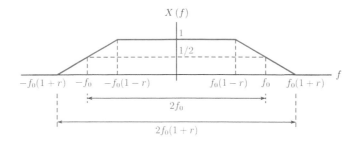

Figure 4.74 – The trapezoidal spectrum $X(f)$ for Example 4.33.

Solution: The answer can easily be found by applying the duality property to the result obtained in Eqn. (4.259) of Example 4.32 with the adjustments $a = f_0(1 - r)$, $b = f_0(1 + r)$, $\lambda = f_0$ and $\tau = 2f_0 r$. (See Problem 4.32 at the end of this chapter.) Instead of using that approach, however, we will use this example as an opportunity to apply the differentiation in frequency property. Let

$$W(f) = \frac{dX(f)}{df}$$

By carrying out the differentiation we obtain $W(f)$ as

$$W(f) = \frac{1}{2f_0 r} \Pi\left(\frac{f + f_0}{2f_0 r}\right) - \frac{1}{2f_0 r} \Pi\left(\frac{f - f_0}{2f_0 r}\right)$$

which is shown in Fig. 4.75.

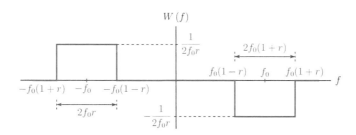

Figure 4.75 – The intermediate spectrum $W(f)$ for Example 4.33.

The Fourier transform of the sinc function is (see Example 4.19)

$$\mathcal{F}\{\operatorname{sinc}(t)\} = \Pi(f)$$

Using the time and frequency scaling property of the Fourier transform given by Eqn. (4.239) we obtain

$$\mathcal{F}\{\operatorname{sinc}(2f_0 r t)\} = \frac{1}{2f_0 r} \Pi\left(\frac{f}{2f_0 r}\right) \tag{4.275}$$

The two frequency-domain pulses that make up the transform $W(f)$ are simply frequency shifted and frequency scaled versions of the right side of Eqn. (4.275). Therefore, the inverse transform of $W(f)$ is

$$w(t) = \text{sinc}(2f_0\,rt)\,e^{-j2\pi f_0 t} - \text{sinc}(2f_0\,rt)\,e^{j2\pi f_0 t}$$
$$= \text{sinc}(2f_0\,rt)\left[e^{-j2\pi f_0 t} - e^{j2\pi f_0 t}\right]$$
$$= -\,2j\,\sin(2\pi f_0 t)\,\text{sinc}(2f_0\,rt)$$

Since $w(t) = -j2\pi t\, x(t)$, we can write $x(t)$ as

$$x(t) = 2f_0\,\text{sinc}(2f_0 t)\,\text{sinc}(2f_0\,rt) \tag{4.276}$$

Fig. 4.76 shows the signal $x(t)$ for various values of the parameter r.

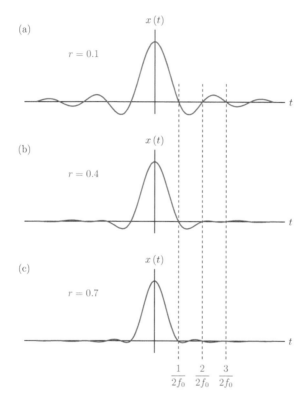

Figure 4.76 – The time-domain signal $x(t)$ with the trapezoidal spectrum $X(f)$ used in Example 4.33: (a) $r = 0.1$, (b) $r = 0.4$, (c) $r = 0.7$.

The signal $x(t)$ with a trapezoidal spectrum exhibits some interesting properties:

1. Zero crossings of the signal $x(t)$ occur at uniform time intervals of $1/(2f_0)$ regardless of the value of r.
2. Larger values of r result in a reduction in the oscillatory behavior of the signal.

These properties are very significant in communication systems in the context of *pulse shaping*. A signal with the properties listed above can be used for representing binary or M-ary symbols while eliminating the interference of those symbols with each other, a phenomenon known as *intersymbol interference (ISI)*. Harry Nyquist has shown that any signal the spectrum of which has certain symmetry properties will exhibit uniformly spaced zero crossings. The trapezoidal spectrum shown in Fig. 4.74 is one example.

Software resources:
ex_4_33.m

Interactive Demo: ft_demo11

The demo program "ft_demo11.m" expands on the concepts explored in Example 4.33. The trapezoidal spectrum $X(f)$ with parameters f_0 and r and its corresponding time-domain signal $x(t)$ are shown. Values of parameters f_0 and r may be adjusted using slider controls. The zero crossings of the signal $x(t)$ are shown with marker lines. Specifically observe the following:

- As the parameter f_0 is changed, the zero crossings of the signal coincide with integer multiples of $1/(2f_0)$. (You may need to zoom in to observe that zero crossings are always uniformly spaced.)

- Vary the parameter r and observe its effect on the locations of zero crossings. Does it seem to have an effect?

- The one-sided bandwidth of the trapezoidal spectrum is $f_0(1+r)$. If f_0 is fixed, bandwidth can be controlled by the parameter r. Observe the effect of increasing the bandwidth of the spectrum on the shape of the signal $x(t)$.

Software resources:
ft_demo11.m

Convolution property

For two transform pairs

$$x_1(t) \xleftrightarrow{\mathcal{F}} X_1(\omega) \quad \text{and} \quad x_2(t) \xleftrightarrow{\mathcal{F}} X_2(\omega) \tag{4.277}$$

it can be shown that

$$x_1(t) * x_2(t) \xleftrightarrow{\mathcal{F}} X_1(\omega)\, X_2(\omega) \tag{4.278}$$

Proof: The proof is obtained by direct application of the Fourier transform integral in Eqn. (4.127) to the convolution of signals $x_1(t)$ and $x_2(t)$:

$$\mathcal{F}\{x_1(t) * x_2(t)\} = \int_{-\infty}^{\infty} \left[\int_{-\infty}^{\infty} x_1(\lambda)\, x_2(t-\lambda)\, d\lambda \right] e^{-j\omega t}\, dt \tag{4.279}$$

Interchanging the order of two integrals and rearranging terms we obtain

$$\mathcal{F}\{x_1(t) * x_2(t)\} = \int_{\infty}^{\infty} \left[\int_{-\infty}^{\infty} x_1(\lambda)\, x_2(t-\lambda)\, e^{-j\omega t}\, dt \right] d\lambda$$

$$= \int_{-\infty}^{\infty} x_1(\lambda) \left[\int_{-\infty}^{\infty} x_2(t-\lambda)\, e^{-j\omega t}\, dt \right] d\lambda \tag{4.280}$$

In Eqn. (4.280), we will recognize the expression in square brackets as the Fourier transform of the time shifted signal $x_2(t - \lambda)$. The use of time shifting property of the Fourier transform given by Eqn. (4.218) yields

$$\int_{-\infty}^{\infty} x_2(t - \lambda)\, e^{-j\omega t}\, dt = \mathcal{F}\left\{x_2(t - \lambda)\right\} = X_2(\omega)\, e^{-j\omega\lambda} \tag{4.281}$$

Substituting Eqn. (4.281) into Eqn. (4.280) we get

$$\mathcal{F}\left\{x_1(t) * x_2(t)\right\} = \int_{-\infty}^{\infty} x_1(\lambda)\left[X_2(\omega)\, e^{-j\omega\lambda}\right] d\lambda$$

$$= \left[\int_{-\infty}^{\infty} x_1(\lambda)\, e^{-j\omega\lambda}\, d\lambda\right] X_2(\omega)$$

$$= X_1(\omega)\, X_2(\omega) \tag{4.282}$$

Multiplication of two signals

For two transform pairs

$$x_1(t) \xleftrightarrow{\mathcal{F}} X_1(\omega) \quad \text{and} \quad x_2(t) \xleftrightarrow{\mathcal{F}} X_2(\omega) \tag{4.283}$$

it can be shown that

$$x_1(t)\, x_2(t) \xleftrightarrow{\mathcal{F}} \frac{1}{2\pi} X_1(\omega) * X_2(\omega) \tag{4.284}$$

If we choose to use f instead of ω, then

$$x_1(t)\, x_2(t) \xleftrightarrow{\mathcal{F}} X_1(f) * X_2(f) \tag{4.285}$$

Proof: The right side of Eqn. (4.284) can be written as

$$\frac{1}{2\pi} X_1(\omega) * X_2(\omega) = \frac{1}{2\pi} \int_{-\infty}^{\infty} X_1(\lambda)\, X_2(\omega - \lambda)\, d\lambda \tag{4.286}$$

Applying the inverse Fourier transform to both sides of Eqn. (4.286) yields

$$\mathcal{F}^{-1}\left\{\frac{1}{2\pi} X_1(\omega) * X_2(\omega)\right\} = \frac{1}{2\pi} \int_{-\infty}^{\infty} \left[\frac{1}{2\pi} \int_{-\infty}^{\infty} X_1(\lambda)\, X_2(\omega - \lambda)\, d\lambda\right] e^{j\omega t}\, d\omega \tag{4.287}$$

Changing the order of the two integrals on the right side of Eqn. (4.287), and arranging terms, we obtain

$$\mathcal{F}^{-1}\left\{\frac{1}{2\pi} X_1(\omega) * X_2(\omega)\right\} = \frac{1}{2\pi} \int_{-\infty}^{\infty} X_1(\lambda)\left[\frac{1}{2\pi} \int_{-\infty}^{\infty} X_2(\omega - \lambda)\, e^{j\omega t}\, d\omega\right] d\lambda \tag{4.288}$$

Recall from the frequency shifting property of the Fourier transform, given by Eqn. (4.222), that $X_2(\omega - \lambda)$ is the Fourier transform of the signal $x_2(t)$ multiplied by a complex exponential. The expression in square brackets on the right side of Eqn. (4.288) represents the inverse Fourier transform of $X_2(\omega - \lambda)$, and evaluates to

$$\frac{1}{2\pi} \int_{-\infty}^{\infty} X_2(\omega - \lambda) e^{j\omega t} d\omega = x_2(t) e^{j\lambda t} \qquad (4.289)$$

Substituting Eqn. (4.289) into Eqn. (4.288) and rearranging the terms on the right side, we have

$$\mathcal{F}^{-1}\left\{ \frac{1}{2\pi} X_1(\omega) * X_2(\omega) \right\} = \frac{1}{2\pi} \int_{-\infty}^{\infty} X_1(\lambda) \left[x_2(t) e^{j\lambda t} \right] d\lambda$$

$$= x_2(t) \left[\frac{1}{2\pi} \int_{-\infty}^{\infty} X_1(\lambda) e^{j\lambda t} d\lambda \right]$$

$$= x_1(t)\, x_2(t) \qquad (4.290)$$

Example 4.34: **Transform of a truncated sinusoidal signal**

A sinusoidal signal that is time-limited in the interval $-\tau < t < \tau$ is given by

$$x(t) = \begin{cases} \cos(2\pi f_0 t), & -\tau < t < \tau \\ 0, & \text{otherwise} \end{cases}$$

Determine the Fourier transform of this signal using the multiplication property.

Solution: The signal $x(t)$ is the same as the modulated pulse signal used in Example 4.29 and shown in Fig. 4.64. In Example 4.29 the transform was determined using the modulation property. Let $x_1(t)$ and $x_2(t)$ be defined as

$$x_1(t) = \cos(2\pi f_0 t) \quad \text{and} \quad x_2(t) = \Pi\left(\frac{t}{2\tau} \right)$$

The signal $x(t)$ can be written as the product of these two signals:

$$x(t) = x_1(t)\, x_2(t) = \cos(2\pi f_0 t)\, \Pi\left(\frac{t}{2\tau} \right)$$

It will be more convenient to use f rather than ω in this case. Fourier transforms of the signals $x_1(t)$ and $x_2(t)$ are

$$X_1(f) = \frac{1}{2}\delta(f + f_0) + \frac{1}{2}\delta(f - f_0)$$

and

$$X_2(f) = 2\tau \operatorname{sinc}(2\tau f)$$

respectively. Using Eqn. (4.285), the Fourier transform of the product $x(t) = x_1(t) \, x_2(t)$ is

$$X(f) = X_1(f) * X_2(f)$$
$$= \tau \operatorname{sinc}\left(2\tau\left(f + f_0\right)\right) + \tau \operatorname{sinc}\left(2\tau\left(f - f_0\right)\right)$$

which matches the result found in Example 4.29.
Software resources:
ex_4_34.m

Interactive Demo: ft_demo12

The demo program "ft_demo12.m" is based on Example 4.34, and illustrates the multiplication property of the Fourier transform. On the left side of the screen, the signals $x_1(t)$ and $x_2(t)$ of Example 4.34 as well as their product $[x_1(t) \, x_2(t)]$ are graphed. The right side of the screen displays the corresponding spectra $X_1(f)$, $X_2(f)$, and $[X_1(f) * X_2(f)]$. Parameters τ and f_0 may be adjusted using slider controls.
Software resources:
ft_demo12.m

Integration

For a transform pair

$$x(t) \overset{\mathcal{F}}{\longleftrightarrow} X(\omega) \tag{4.291}$$

it can be shown that

$$\int_{-\infty}^{t} x(\lambda)\, d\lambda \overset{\mathcal{F}}{\longleftrightarrow} \frac{X(\omega)}{j\omega} + \pi X(0)\, \delta(\omega) \tag{4.292}$$

Proof: Recall that in Example 4.23 we have found the Fourier transform of a unit-step function to be

$$U(\omega) = \mathcal{F}\{u(t)\} = \pi \delta(\omega) + \frac{1}{j\omega}$$

Now consider the convolution of the signal $x(t)$ with a unit-step signal, i.e.,

$$x(t) * u(t) = \int_{-\infty}^{\infty} x(\lambda)\, u(t - \lambda)\, d\lambda \tag{4.293}$$

We know that

$$u(t - \lambda) = \begin{cases} 1, & \lambda < t \\ 0, & \lambda > t \end{cases} \tag{4.294}$$

Using Eqn. (4.294) to modify the integration limits in Eqn. (4.293) we obtain

$$x(t) * u(t) = \int_{-\infty}^{t} x(\lambda)\, d\lambda \tag{4.295}$$

The right side of Eqn. (4.295) is the function the Fourier transform of which we are seeking. We also know from Eqn. (4.279) that

$$\mathcal{F}\left\{x\left(t\right) * u\left(t\right)\right\} = X\left(\omega\right) U\left(\omega\right) \tag{4.296}$$

Thus we have

$$\mathcal{F}\left\{\int_{-\infty}^{t} x(\lambda)\, d\lambda\right\} = \pi\, X\left(\omega\right) \delta\left(\omega\right) + \frac{X\left(\omega\right)}{j\omega} \tag{4.297}$$

Using the sampling property of the impulse function, Eqn. (4.297) becomes

$$\mathcal{F}\left\{\int_{-\infty}^{t} x(\lambda)\, d\lambda\right\} = \pi\, X\left(0\right) \delta\left(\omega\right) + \frac{X\left(\omega\right)}{j\omega} \tag{4.298}$$

Table 4.4 contains a summary of key properties of the Fourier transform. Table 4.5 lists some of the fundamental Fourier transform pairs.

Property	Signal	Transform
Linearity	$\alpha\, x_1\left(t\right) + \beta\, x_2\left(t\right)$	$\alpha\, X_1\left(\omega\right) + \beta\, X_2\left(\omega\right)$
Duality	$X\left(t\right)$	$2\pi\, x\left(-\omega\right)$
Conjugate symmetry	$x\left(t\right)$ real	$X^*\left(\omega\right) = X\left(-\omega\right)$
		Magnitude: $\quad\left\lvert X\left(-\omega\right)\right\rvert = \left\lvert X\left(\omega\right)\right\rvert$
		Phase: $\quad\Theta\left(-\omega\right) = -\Theta\left(\omega\right)$
		Real part: $\quad X_r\left(-\omega\right) = X_r\left(\omega\right)$
		Imaginary part: $X_i\left(-\omega\right) = -X_i\left(\omega\right)$
Conjugate antisymmetry	$x\left(t\right)$ imaginary	$X^*\left(\omega\right) = -X\left(-\omega\right)$
		Magnitude: $\quad\left\lvert X\left(-\omega\right)\right\rvert = \left\lvert X\left(\omega\right)\right\rvert$
		Phase: $\quad\Theta\left(-\omega\right) = -\Theta\left(\omega\right) \mp \pi$
		Real part: $\quad X_r\left(-\omega\right) = -X_r\left(\omega\right)$
		Imaginary part: $X_i\left(-\omega\right) = X_i\left(\omega\right)$
Even signal	$x\left(-t\right) = x\left(t\right)$	$\operatorname{Im}\left\{X\left(\omega\right)\right\} = 0$
Odd signal	$x\left(-t\right) = -x\left(t\right)$	$\operatorname{Re}\left\{X\left(\omega\right)\right\} = 0$
Time shifting	$x(t - \tau)$	$X\left(\omega\right) e^{-j\omega\tau}$
Frequency shifting	$x\left(t\right) e^{j\omega_0 t}$	$X\left(\omega - \omega_0\right)$
Modulation property	$x\left(t\right) \cos(\omega_0 t)$	$\frac{1}{2}\left[X\left(\omega - \omega_0\right) + X\left(\omega + \omega_0\right)\right]$
Time and frequency scaling	$x\left(at\right)$	$\dfrac{1}{\lvert a\rvert} X\left(\dfrac{\omega}{a}\right)$
Differentiation in time	$\dfrac{d^n}{dt^n}\left[x\left(t\right)\right]$	$\left(j\omega\right)^n X\left(\omega\right)$
Differentiation in frequency	$\left(-jt\right)^n x\left(t\right)$	$\dfrac{d^n}{d\omega^n}\left[X\left(\omega\right)\right]$
Convolution	$x_1\left(t\right) * x_2\left(t\right)$	$X_1\left(\omega\right) X_2\left(\omega\right)$
Multiplication	$x_1\left(t\right) x_2\left(t\right)$	$\dfrac{1}{2\pi} X_1\left(\omega\right) * X_2\left(\omega\right)$
Integration	$\displaystyle\int_{-\infty}^{t} x(\lambda)\, d\lambda$	$\dfrac{X\left(\omega\right)}{j\omega} + \pi\, X\left(0\right) \delta\left(\omega\right)$
Parseval's theorem	$\displaystyle\int_{-\infty}^{\infty} \left\lvert x\left(t\right)\right\rvert^2 dt = \dfrac{1}{2\pi}\int_{-\infty}^{\infty} \left\lvert X\left(\omega\right)\right\rvert^2 d\omega$	

Table 4.4 – Fourier transform properties.

Name	Signal	Transform
Rectangular pulse	$x\left(t\right) = A\,\Pi\left(t/\tau\right)$	$X\left(\omega\right) = A\tau\,\operatorname{sinc}\left(\dfrac{\omega\tau}{2\pi}\right)$
Triangular pulse	$x\left(t\right) = A\,\Lambda\left(t/\tau\right)$	$X\left(\omega\right) = A\tau\,\operatorname{sinc}^2\left(\dfrac{\omega\tau}{2\pi}\right)$
Right-sided exponential	$x\left(t\right) = e^{-at}\,u\left(t\right)$	$X\left(\omega\right) = \dfrac{1}{a+j\omega}$
Two-sided exponential	$x\left(t\right) = e^{-a\lvert t\rvert}$	$X\left(\omega\right) = \dfrac{2a}{a^2+\omega^2}$
Signum function	$x\left(t\right) = \operatorname{sgn}\left(t\right)$	$X\left(\omega\right) = \dfrac{2}{j\omega}$
Unit impulse	$x\left(t\right) = \delta\left(t\right)$	$X\left(\omega\right) = 1$
Sinc function	$x\left(t\right) = \operatorname{sinc}\left(t\right)$	$X\left(\omega\right) = \Pi\left(\dfrac{\omega}{2\pi}\right)$
Constant-amplitude signal	$x\left(t\right) = 1,\ \text{all } t$	$X\left(\omega\right) = 2\pi\,\delta\left(\omega\right)$
	$x\left(t\right) = \dfrac{1}{\pi t}$	$X\left(\omega\right) = -j\,\operatorname{sgn}\left(\omega\right)$
Unit-step function	$x\left(t\right) = u\left(t\right)$	$X\left(\omega\right) = \pi\,\delta\left(\omega\right) + \dfrac{1}{j\omega}$
Modulated pulse	$x\left(t\right) = \Pi\left(\dfrac{t}{\tau}\right)\cos\left(\omega_0 t\right)$	$X\left(\omega\right) = \dfrac{\tau}{2}\operatorname{sinc}\left(\dfrac{\left(\omega-\omega_0\right)\tau}{2\pi}\right) +$
		$\dfrac{\tau}{2}\operatorname{sinc}\left(\dfrac{\left(\omega+\omega_0\right)\tau}{2\pi}\right)$

Table 4.5 – Some Fourier transform pairs.

4.3.6 Applying Fourier transform to periodic signals

In developing frequency-domain analysis methods in the previous sections of this chapter we have distinguished between periodic and non-periodic continuous-time signals. Periodic signals were analyzed using various forms of the Fourier series such as TFS, EFS or CFS. In contrast, Fourier transform was used for analyzing non-periodic signals. While this distinction will be appropriate when we work with one type of signal or the other, there may be times when we need to mix periodic and non-periodic signals within the same system. An example of this occurs in amplitude modulation where a non-periodic *message signal* may be multiplied with a periodic *carrier signal*. Another example is the use of a periodic signal as input to a system the impulse response of which is non-periodic. In such situations it would be convenient to use the Fourier transform for periodic signals as well. In general a periodic signal is a power signal that does not satisfy the existence conditions for the Fourier transform. It is neither absolute integrable nor square integrable. On the other hand, we have seen in Example 4.20 that a Fourier transform can be found for a constant-amplitude signal that does not satisfy the existence conditions, as long as we are willing to accept singularity functions in the transform. The next two examples will expand on this idea. Afterwards we will develop a technique for converting the EFS representation of a periodic continuous-time signal to a Fourier transform that is suitable for use in solving problems.

Example 4.35: **Fourier transform of complex exponential signal**

Determine the transform of the complex exponential signal $x(t) = e^{j\omega_0 t}$.

Solution: The transform of the constant unit-amplitude signal was found in Example 4.20 to be

$$\mathcal{F}\{1\} = 2\pi\,\delta(\omega)$$

Using this result along with the frequency shifting property of the Fourier transform given by Eqn. (4.222), the transform of the complex exponential signal is

$$\mathcal{F}\{e^{j\omega_0 t}\} = 2\pi\,\delta(\omega - \omega_0) \tag{4.299}$$

This is illustrated in Fig. 4.77.

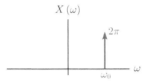

Figure 4.77 – The transform of complex exponential signal $x(t) = e^{j\omega_0 t}$.

Example 4.36: **Fourier transform of sinusoidal signal**

Determine the transform of the sinusoidal signal $x(t) = \cos(\omega_0 t)$.

Solution: The transform of the constant unit-amplitude signal was found in Example 4.20 to be

$$\mathcal{F}\{1\} = 2\pi\,\delta(\omega)$$

Using this result along with the modulation property of the Fourier transform given by Eqn. (4.230), the transform of the sinusoidal signal is

$$\mathcal{F}\{\cos(\omega_0 t)\} = \pi\,\delta(\omega - \omega_0) + \pi\,\delta(\omega + \omega_0) \tag{4.300}$$

This transform is shown in Fig. 4.78.

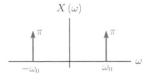

Figure 4.78 – The transform of sinusoidal signal $x(t) = \cos(\omega_0 t)$.

In Examples 4.35 and 4.36 we were able to obtain the Fourier transforms of two periodic signals, namely a complex exponential signal and a sinusoidal signal. Both transforms were possible through the use of impulse functions in the transform. The idea can be generalized to apply to any periodic continuous-time signal that has an EFS representation.

The EFS synthesis equation for a periodic signal $\tilde{x}(t)$ was given by Eqn. (4.71) which is repeated here for convenience:

$$\text{Eqn. (4.71):} \quad \tilde{x}(t) = \sum_{k=-\infty}^{\infty} c_k \, e^{jk\omega_0 t}$$

Using the Fourier transform analysis equation given by Eqn. (4.127), the transform of $\tilde{x}(t)$ is

$$X(\omega) = \int_{-\infty}^{\infty} \tilde{x}(t) \, e^{-j\omega t} \, dt \tag{4.301}$$

$$= \int_{-\infty}^{\infty} \left[\sum_{k=-\infty}^{\infty} c_k \, e^{jk\omega_0 t} \right] e^{-j\omega t} \, dt \tag{4.302}$$

Interchanging the order of the integral and the summation in Eqn. (4.301) and rearranging terms we obtain

$$X(\omega) = \sum_{k=-\infty}^{\infty} c_k \left[\int_{-\infty}^{\infty} e^{jk\omega_0 t} \, e^{-j\omega t} \, dt \right] \tag{4.303}$$

In Eqn. (4.303) the expression in square brackets is the Fourier transform of the signal $e^{jk\omega_0 t}$ which, using the result obtained in Example 4.35, can be written as

$$\int_{-\infty}^{\infty} e^{jk\omega_0 t} \, e^{-j\omega t} \, dt = 2\pi \, \delta(\omega - k\omega_0) \tag{4.304}$$

Using Eqn. (4.304) in Eqn. (4.303), the transform of the periodic signal $\tilde{x}(t)$ with EFS coefficients $\{c_k\}$ is obtained as

$$X(\omega) = \sum_{k=-\infty}^{\infty} 2\pi c_k \, \delta(\omega - k\omega_0) \tag{4.305}$$

The Fourier transform for a periodic continuous-time signal is obtained by converting each EFS coefficient c_k to an impulse with area equal to $2\pi c_k$ and placing it at the radian frequency $\omega = k\omega_0$. This process is illustrated in Fig. 4.79.

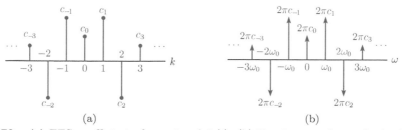

Figure 4.79 – (a) EFS coefficients for a signal $\tilde{x}(t)$, (b) Fourier transform obtained through Eqn. (4.305).

Example 4.37: Fourier transform of periodic pulse train

Determine the Fourier transform of the periodic pulse train with duty cycle $d = \tau/T_0$ as shown in Fig. 4.80.

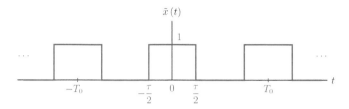

Figure 4.80 – Periodic pulse train for Example 4.37.

Solution: The EFS coefficients for $\tilde{x}(t)$ were determined in Example 4.7 as

$$c_k = d \operatorname{sinc}(kd)$$

Using Eqn. 4.305 the Fourier transform is

$$X(\omega) = \sum_{k=-\infty}^{\infty} 2\pi d \operatorname{sinc}(kd)\, \delta(\omega - k\omega_0) \tag{4.306}$$

where $\omega_0 = 1/T_0$ is the fundamental radian frequency.

4.4   Energy and Power in the Frequency Domain

In this section we will discuss a very important theorem of Fourier series and transform known as *Parseval's theorem* which can be used as the basis of computing energy or power of a signal from its frequency domain representation. Afterwards energy and power spectral density concepts will be introduced.

4.4.1   Parseval's theorem

For a periodic power signal $\tilde{x}(t)$ with period of T_0 and EFS coefficients $\{c_k\}$ it can be shown that

$$\frac{1}{T_0} \int_{t_0}^{t_0+T_0} |\tilde{x}(t)|^2\, dt = \sum_{k=-\infty}^{\infty} |c_k|^2 \tag{4.307}$$

For a non-periodic energy signal $x(t)$ with a Fourier transform $X(f)$, the following holds true:

$$\int_{-\infty}^{\infty} |x(t)|^2\, dt = \int_{-\infty}^{\infty} |X(f)|^2\, df \tag{4.308}$$

The left side of Eqn. (4.307) represents the normalized average power in a periodic signal as we have derived in Eqn. (1.92). The left side of Eqn. (4.308) represents the normalized signal energy as derived in Eqn. (1.81).   The relationships given by Eqns. (4.307) and (4.308) relate signal energy or signal power to the frequency-domain representation of the signal. They are two forms of *Parseval's theorem*.

Proofs: We will begin with the proof of Eqn. (4.307). We know from Eqn. (4.48) that

$$\tilde{x}(t) = \sum_{k=-\infty}^{\infty} c_k \, e^{j\omega_0 t} \tag{4.309}$$

The left side of Eqn. (4.307) can be written as

$$\frac{1}{T_0} \int_{t_0}^{t_0+T_0} |\tilde{x}(t)|^2 \, dt = \frac{1}{T_0} \int_{t_0}^{t_0+T_0} \left[\sum_{k=-\infty}^{\infty} c_k \, e^{jk\omega_0 t} \right] \left[\sum_{m=-\infty}^{\infty} c_m \, e^{jm\omega_0 t} \right]^* dt$$

$$= \frac{1}{T_0} \int_{t_0}^{t_0+T_0} \left[\sum_{k=-\infty}^{\infty} c_k \, e^{jk\omega_0 t} \right] \left[\sum_{m=-\infty}^{\infty} c_m^* \, e^{-jm\omega_0 t} \right] dt \tag{4.310}$$

By rearranging the order of the two summations and the integral in Eqn. (4.310), we have

$$\frac{1}{T_0} \int_{t_0}^{t_0+T_0} |\tilde{x}(t)|^2 \, dt = \frac{1}{T_0} \sum_{k=-\infty}^{\infty} c_k \sum_{m=-\infty}^{\infty} c_m^* \int_{t_0}^{t_0+T_0} e^{j(k-m)\omega_0 t} \, dt \tag{4.311}$$

Using orthogonality of exponential basis functions (see Eqn. (4.63)), we have

$$\int_{t_0}^{t_0+T_0} e^{j(k-m)\omega_0 t} \, dt = \begin{cases} T_0, & k = m \\ 0, & k \neq m \end{cases} \tag{4.312}$$

As a result, the inner summation in Eqn. (4.311) evaluates to $c_k^* T_0$, and we have

$$\frac{1}{T_0} \int_{t_0}^{t_0+T_0} |\tilde{x}(t)|^2 \, dt = \frac{1}{T_0} \sum_{k=-\infty}^{\infty} c_k c_k^* T_0 = \sum_{k=-\infty}^{\infty} |c_k|^2$$

which proves Eqn. (4.307). The proof for Eqn. (4.308) is quite similar. We know that for an energy signal $x(t)$ the inverse Fourier transform equation is

$$x(t) = \int_{-\infty}^{\infty} X(f) \, e^{j2\pi ft} \, dt$$

The normalized energy of $x(t)$ can be written as

$$\int_{-\infty}^{\infty} |x(t)|^2 \, dt = \int_{-\infty}^{\infty} x(t) \, x^*(t) \, dt$$

$$= \int_{-\infty}^{\infty} x(t) \left[\int_{-\infty}^{\infty} X(f) \, e^{j2\pi ft} \, df \right]^* dt$$

$$= \int_{-\infty}^{\infty} x(t) \left[\int_{-\infty}^{\infty} X^*(f) \, e^{-j2\pi ft} \, df \right] dt \tag{4.313}$$

By interchanging the order of the two integrals in Eqn. (4.313) and rearranging terms we get

$$\int_{-\infty}^{\infty} |x(t)|^2 \, dt = \int_{-\infty}^{\infty} X^*(f) \left[\int_{-\infty}^{\infty} x(t) \, e^{-j2\pi ft} \, dt \right] df$$

$$= \int_{-\infty}^{\infty} X^*(f) \, X(f) \, df$$

$$= \int_{-\infty}^{\infty} |X(f)|^2 \, df \tag{4.314}$$

which proves Eqn. (4.308).

4.4.2 Energy and power spectral density

Examining the two statements of Parseval's theorem expressed by Eqns. (4.307) and (4.308), we reach the following conclusions:

1. In Eqn. (4.307) the left side corresponds to the normalized average power of the signal $\tilde{x}(t)$, and therefore the summation on the right side must also represent power. The term $|c_k|^2$ corresponds to the power of the frequency component at $f = kf_0$, that is, the power in the k-th harmonic only. Suppose we construct a new function $S_x(f)$ by starting with the line spectrum c_k and replacing each spectral line with an impulse function scaled by $|c_k|^2$ and time shifted to the frequency kf_0, i.e.,

$$S_x(f) = \sum_{k=-\infty}^{\infty} |c_k|^2 \, \delta(f - kf_0) \tag{4.315}$$

In the next step we will integrate $S_x(f)$ to obtain

$$\int_{-\infty}^{\infty} S_x(f) \, df = \int_{-\infty}^{\infty} \left[\sum_{k=-\infty}^{\infty} |c_k|^2 \, \delta(f - kf_0) \right] df$$

$$= \sum_{k=-\infty}^{\infty} \left[\int_{-\infty}^{\infty} |c_k|^2 \, \delta(f - kf_0) \, df \right] \tag{4.316}$$

Using the sifting property of the unit-impulse function Eqn. (4.316) becomes

$$\int_{-\infty}^{\infty} S_x(f) \, df = \sum_{k=-\infty}^{\infty} |c_k|^2 \tag{4.317}$$

Finally, substituting Eqn. (4.317) into Eqn. (4.307) we have

$$\frac{1}{T_0} \int_{t_0}^{t_0+T_0} |\tilde{x}(t)|^2 \, dt = \int_{-\infty}^{\infty} S_x(f) \, df \tag{4.318}$$

In Eqn. (4.318) the normalized average power of the signal $x(t)$ is computed by integrating the function $S_x(f)$ over all frequencies. Consequently, the function $S_x(f)$ is the *power spectral density* of the signal $x(t)$.

If we use the radian frequency variable ω instead of f, Eqn. (4.318) becomes

$$\frac{1}{T_0} \int_{t_0}^{t_0+T_0} |\tilde{x}(t)|^2 \, dt = \frac{1}{2\pi} \int_{-\infty}^{\infty} S_x(\omega) \, d\omega \qquad (4.319)$$

As an interesting by-product of Eqn. (4.318), the power of $x(t)$ that is within a specific frequency range can be determined by integrating $S_x(f)$ over that frequency range. For example, the power contained at frequencies in the range $(-f_0, f_0)$ can be computed as

$$P_x \text{ in } (-f_0, f_0) = \int_{-f_0}^{f_0} S_x(f) \, df \qquad (4.320)$$

2. In Eqn. (4.308) the left side is the normalized signal energy for the signal $x(t)$, and the right side must be the same. The integrand $|X(f)|^2$ is therefore the *energy spectral density* of the signal $x(t)$. Let the function $G_x(f)$ be defined as

$$G_x(f) = |X(f)|^2 \qquad (4.321)$$

Substituting Eqn. (4.321) into Eqn. (4.308), the normalized energy in the signal $x(t)$ can be expressed as

$$\int_{-\infty}^{\infty} |x(t)|^2 \, dt = \int_{-\infty}^{\infty} G_x(f) \, df$$

If ω is used in place of f, Eqn. (4.322) becomes

$$\int_{-\infty}^{\infty} |x(t)|^2 \, dt = \frac{1}{2\pi} \int_{-\infty}^{\infty} G_x(\omega) \, d\omega$$

3. In Eqn. (4.308) we have expressed Parseval's theorem for an energy signal, and used it to lead to the derivation of the energy spectral density in Eqn. (4.321). We know that some non-periodic signals are power signals, therefore their energy cannot be computed. The example of one such signal is the unit-step function. The power of a non-periodic signal was defined in Eqn. (1.93) which will be repeated here:

$$\text{Eqn. (1.93):} \quad P_x = \lim_{T \to \infty} \left[\frac{1}{T} \int_{-T/2}^{T/2} |x(t)|^2 \, dt \right]$$

In order to write the counterpart of Eqn. (4.308) for a power signal, we will first define a *truncated* version of $x(t)$ as

$$x_T(t) = \begin{cases} x(t), & -T/2 < t < T/2 \\ 0, & \text{otherwise} \end{cases} \qquad (4.322)$$

Let $X_T(f)$ be the Fourier transform of the truncated signal $x_T(t)$:

$$X_T(f) = \mathcal{F}\{x_T(t)\} = \int_{-T/2}^{T/2} x_T(t) \, e^{-j2\pi ft} \, dt \qquad (4.323)$$

Now Eqn. (4.308) can be written in terms of the truncated signal and its transform as

$$\int_{-T/2}^{T/2} |x_T(t)|^2 \, dt = \int_{-\infty}^{\infty} |X_T(f)|^2 \, df \qquad (4.324)$$

Scaling both sides of Eqn. (4.324) by $(1/T)$ and taking the limit as T becomes infinitely large, we obtain

$$\lim_{T \to \infty} \left[\frac{1}{T} \int_{-T/2}^{T/2} |x_T(t)|^2 \, dt \right] = \lim_{T \to \infty} \left[\frac{1}{T} \int_{-\infty}^{\infty} |X_T(f)|^2 \, df \right] \tag{4.325}$$

The left side of Eqn. (4.325) is the average normalized power in a non-periodic signal as we have established in Eqn. (1.93). Therefore, the power spectral density of $x(t)$ is

$$S_x(f) = \lim_{T \to \infty} \left[\frac{1}{T} |X_T(f)|^2 \right] \tag{4.326}$$

Example 4.38: **Power spectral density of a periodic pulse train**

The EFS coefficients for the periodic pulse train $\tilde{x}(t)$ shown in Fig. 4.81 were found in Example 4.5 to be

$$c_k = \frac{1}{3} \text{ sinc}(k/3)$$

Determine the power spectral density for $x(t)$. Also find the total power, the dc power, the power in the first three harmonics, and the power above 1 Hz.

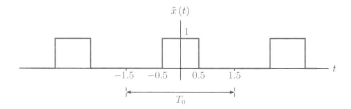

Figure 4.81 – Periodic pulse train used in Example 4.38.

Solution: The period of the signal is $T_0 = 3$ s, and therefore the fundamental frequency is $f_0 = \frac{1}{3}$ Hz. The power spectral density $S_x(f)$ can be written using Eqn. (4.315) as

$$S_x(f) = \sum_{k=-\infty}^{\infty} \left| \frac{1}{3} \text{ sinc}(k/3) \right|^2 \delta(f - k/3)$$

which is graphed in Fig. 4.82.

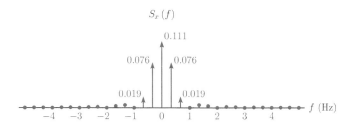

Figure 4.82 – Power spectral density for Example 4.38.

The total power in the signal $x(t)$ can be easily determined using the left side of Eqn. (4.307), i.e.,

$$P_x = \frac{1}{T_0} \int_{t_0}^{t_0+T_0} |\tilde{x}(t)|^2 \, dt = \frac{1}{3} \int_{-0.5}^{0.5} (1)^2 \, dt = 0.3333$$

The dc power in the signal $x(t)$ is

$$P_{dc} = |c_0|^2 = (0.3333)^2 = 0.1111$$

The power in the fundamental frequency is

$$P_1 = |c_{-1}|^2 + |c_1|^2 = 2|c_1|^2 = 2(0.0760) = 0.1520$$

Power values for the second and third harmonics can be found similarly as

$$P_2 = |c_{-2}|^2 + |c_2|^2 = 2|c_2|^2 = 2(0.0190)^2 = 0.0380$$

and

$$P_3 = |c_{-3}|^2 + |c_3|^2 = 2|c_3|^2 = 2(0)^2 = 0$$

The third harmonic is at frequency $f = 1$ Hz. Thus, the power above 1 Hz can be determined by subtracting the power values for the dc and the first three harmonics from the total, i.e.,

$$\begin{aligned}
P_{hf} &= P_x - P_{dc} - P_1 - P_2 - P_3 \\
&= 0.3333 - 0.1111 - 0.1520 - 0.0380 - 0 \\
&= 0.0322
\end{aligned}$$

Example 4.39: **Power spectral density of a sinusoidal signal**

Find the power spectral density of the signal $\tilde{x}(t) = 5\cos(200\pi t)$.

Solution: Using Euler's formula, the signal in question can be written as

$$x(t) = \frac{5}{2} e^{-j200\pi t} + \frac{5}{2} e^{j200\pi t}$$

from an inspection of which we conclude that the only significant coefficients in the EFS representation of the signal are

$$c_{-1} = c_1 = \frac{5}{2}$$

with all other coefficients equal to zero. The fundamental frequency is $f_0 = 100$ Hz. Using Eqn. (4.315), the power spectral density is

$$\begin{aligned}
S_x(f) &= \sum_{n=-\infty}^{\infty} |c_n|^2 \delta(f - 100n) \\
&= |c_{-1}|^2 \delta(f + 100) + |c_1|^2 \delta(f - 100) \\
&= \frac{25}{4} \delta(f + 100) + \frac{25}{4} \delta(f - 100)
\end{aligned}$$

which is shown in Fig. 4.83

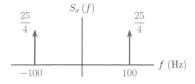

Figure 4.83 – Power spectral density for Example 4.39.

Example 4.40: **Energy spectral density of a rectangular pulse**

Determine the energy spectral density of the rectangular pulse

$$x\left(t\right) = 20\,\Pi\left(100t\right)$$

Solution: It was determined in Example 4.12 that

$$A\,\Pi\left(t/\tau\right) \overset{\mathcal{F}}{\longleftrightarrow} A\tau\,\text{sinc}\left(f\tau\right)$$

Thus the Fourier transform of the pulse given above is

$$X\left(f\right) = 0.2\,\text{sinc}\left(0.01f\right)$$

and the energy spectral density is

$$G_x\left(f\right) = \left|X\left(f\right)\right|^2 = 0.04\,\text{sinc}^2\left(0.01f\right)$$

This result is shown in Fig. 4.84.

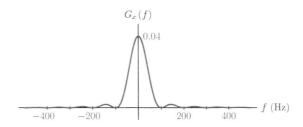

Figure 4.84 – Energy spectral density for Example 4.40.

Example 4.41: **Energy spectral density of the sinc function**

Determine the energy spectral density of the signal

$$x\left(t\right) = \text{sinc}\left(10t\right)$$

Afterwards, compute (a) the total energy in the sinc pulse, and (b) the energy in the sinc pulse at frequencies up to 3 Hz.

Solution: The spectrum of a sinc pulse is a rectangle. Through the use of the duality property of the Fourier transform, $X\left(f\right)$ is found to be

$$X\left(f\right) = \frac{1}{10}\,\Pi\left(\frac{f}{10}\right)$$

The energy spectral density is

$$G_x\left(f\right) = \frac{1}{100}\,\Pi\left(\frac{f}{10}\right)$$

which is shown in Fig. 4.85. The total energy in the signal $x\left(t\right)$ is

$$E_x = \int_{-\infty}^{\infty} G_x\left(f\right)\,df = \int_{-5}^{5}\left(\frac{1}{100}\right)df = 0.1$$

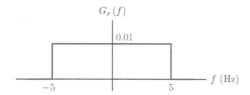

Figure 4.85 – Energy spectral density for Example 4.41.

The energy at frequencies up to 3 Hz is found by integrating $G_x(f)$ in the interval -3 Hz $< f < 3$ Hz:

$$E_x \text{ in } (-3, 3 \text{ Hz}) = \int_{-3}^{3} G_x(f) \, df = \int_{-3}^{3} \left(\frac{1}{100} \right) df = 0.06$$

Fig. 4.86 illustrates the use of $G_x(f)$ for computing the signal energy.

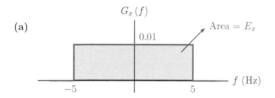

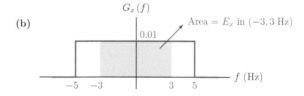

Figure 4.86 – Use of energy spectral density for computing signal energy: (a) total area under $G_x(f)$ corresponding to total energy of the signal, (b) partial area under $G_x(f)$ corresponding to signal energy in frequencies up to 3 Hz.

4.4.3   Autocorrelation

The energy spectral density $G_x(f)$ or the power spectral $S_x(f)$ density for a signal can be computed through direct application of the corresponding equations derived in Section 4.4.2; namely Eqn. (4.321) for an energy signal, and either Eqn. (4.315) or Eqn. (4.326) for a power signal, depending on its type. In some circumstances it is also possible to compute both from the knowledge of the *autocorrelation function* which will be defined in this section. Let $x(t)$ be a real-valued signal.

For an energy signal $x(t)$ the *autocorrelation function* is defined as

$$r_{xx}(\tau) = \int_{-\infty}^{\infty} x(t) \, x(t + \tau) \, dt \qquad (4.327)$$

For a periodic power signal $\tilde{x}(t)$ with period T_0, the corresponding definition of the auto-correlation function is

$$\tilde{r}_{xx}(\tau) = \langle \tilde{x}(t)\, \tilde{x}(t+\tau) \rangle$$

$$= \frac{1}{T_0} \int_{-T_0/2}^{T_0/2} \tilde{x}(t)\, \tilde{x}(t+\tau)\, dt \qquad (4.328)$$

The triangle brackets indicate time average. The autocorrelation function for a periodic signal is also periodic as signified by the tilde ($\tilde{\ }$) character used over the symbol r_{xx} in Eqn. (4.328). Finally, for a non-periodic power signal, the corresponding definition is

$$r_{xx}(\tau) = \langle x(t)\, x(t+\tau) \rangle$$

$$= \lim_{T_0 \to \infty} \left[\frac{1}{T_0} \int_{-T_0/2}^{T_0/2} x(t)\, x(t+\tau)\, dt \right] \qquad (4.329)$$

Even though we refer to $r_{xx}(\tau)$ as a *function*, we will often treat it as if it is a continuous-time signal, albeit one that uses a different independent variable, τ, than the signal $x(t)$ for which it is computed. The variable τ simply corresponds to the time shift between the two copies of $x(t)$ used in the definitions of Eqns. (4.327) through (4.329).

It can be shown that, for an energy signal, the energy spectral density is the Fourier transform of the autocorrelation function, that is,

$$\mathcal{F}\{r_{xx}(\tau)\} = G_x(f) \qquad (4.330)$$

Proof: Let us begin by applying the Fourier transform definition to the autocorrelation function $r_{xx}(\tau)$ treated as a continuous-time signal:

$$\mathcal{F}\{r_{xx}(\tau)\} = \int_{-\infty}^{\infty} r_{xx}(\tau)\, e^{-j2\pi f\tau}\, d\tau \qquad (4.331)$$

$$= \int_{-\infty}^{\infty} \left[\int_{-\infty}^{\infty} x(t)\, x(t+\tau)\, dt \right] e^{-j2\pi f\tau}\, d\tau \qquad (4.332)$$

The two integrals in Eqn. (4.331) can be rearranged to yield

$$\mathcal{F}\{r_{xx}(\tau)\} = \int_{-\infty}^{\infty} x(t) \left[\int_{-\infty}^{\infty} x(t+\tau)\, e^{-j2\pi f\tau}\, d\tau \right] dt \qquad (4.333)$$

Realizing that the inner integral in Eqn. (4.333) is equal to

$$\int_{-\infty}^{\infty} x(t+\tau)\, e^{-j2\pi f\tau}\, d\tau = \mathcal{F}\{x(t+\tau)\} = e^{j2\pi ft} X(f) \qquad (4.334)$$

Eqn. (4.333) becomes

$$\mathcal{F}\{r_{xx}(\tau)\} = X(f) \int_{-\infty}^{\infty} x(t)\, e^{j2\pi ft}\, dt \qquad (4.335)$$

Since $x(t)$ is real

$$\int_{-\infty}^{\infty} x(t)\, e^{j2\pi ft}\, dt = X^*(f) \tag{4.336}$$

and we obtain

$$\mathcal{F}\{r_{xx}(\tau)\} = X(f)\, X^*(f) = |X(f)|^2 \tag{4.337}$$

which completes the proof.

The definition of the autocorrelation function for a periodic signal $\tilde{x}(t)$ can be used in a similar manner in determining the power spectral density of the signal.

Let the signal $\tilde{x}(t)$ and the autocorrelation function $\tilde{r}_{xx}(\tau)$, both periodic with period T_0, have the EFS representations given by

$$\tilde{x}(t) = \sum_{k=-\infty}^{\infty} c_k\, e^{j2\pi k f_0 t} \tag{4.338}$$

and

$$\tilde{r}_{xx}(\tau) = \sum_{k=-\infty}^{\infty} d_k\, e^{j2\pi k f_0 \tau} \tag{4.339}$$

respectively. It can be shown that the EFS coefficients $\{c_k\}$ and $\{d_k\}$ are related by

$$d_k = |c_k|^2 \tag{4.340}$$

and the power spectral density is the Fourier transform of the autocorrelation function, that is,

$$S_x(f) = \mathcal{F}\{\tilde{r}_{xx}(\tau)\} \tag{4.341}$$

Proof: Using the EFS analysis equation given by Eqn. (4.72) with the definition of the autocorrelation function in Eqn. (4.328) leads to

$$
\begin{aligned}
d_k &= \frac{1}{T_0} \int_{-T_0/2}^{T_0/2} \tilde{r}_{xx}(\tau)\, e^{-j2\pi k f_0 \tau}\, d\tau \\
&= \frac{1}{T_0} \int_{-T_0/2}^{T_0/2} \left[\frac{1}{T_0} \int_{-T_0/2}^{T_0/2} \tilde{x}(t)\, \tilde{x}(t+\tau)\, dt \right] e^{-j2\pi k f_0 \tau}\, d\tau
\end{aligned}
\tag{4.342}
$$

Rearranging the order of the two integrals Eqn. (4.342) can be written as

$$d_k = \frac{1}{T_0} \int_{-T_0/2}^{T_0/2} \tilde{x}(t) \left[\frac{1}{T_0} \int_{-T_0/2}^{T_0/2} \tilde{x}(t+\tau)\, e^{-j2\pi k f_0 \tau}\, d\tau \right] dt \tag{4.343}$$

Using the time shifting property of the EFS given by Eqn. (4.113) the expression in square brackets in Eqn. (4.343) is

$$\frac{1}{T_0} \int_{-T_0/2}^{T_0/2} \tilde{x}(t+\tau)\, e^{-j2\pi k f_0 \tau}\, d\tau = e^{j2\pi k f_0 t}\, c_k \tag{4.344}$$

Substituting Eqn. (4.344) into Eqn. (4.343) yields

$$d_k = \frac{1}{T_0} \int_{-T_0/2}^{T_0/2} \tilde{x}(t) \, e^{j2\pi k f_0 t} \, c_k \, dt$$

$$= c_k \frac{1}{T_0} \int_{-T_0/2}^{T_0/2} \tilde{x}(t) \, e^{j2\pi k f_0 t} \, dt \qquad (4.345)$$

Recognizing that

$$\frac{1}{T_0} \int_{-T_0/2}^{T_0/2} \tilde{x}(t) \, e^{j2\pi k f_0 t} \, dt = c_k^*$$

since $\tilde{x}(t)$ is real, we conclude

$$d_k = c_k \, c_k^* = |c_k|^2 \qquad (4.346)$$

Using the technique developed in Section 4.3.6 the Fourier transform of $\tilde{r}_{xx}(\tau)$ is

$$\mathcal{F}\{\tilde{r}_{xx}(\tau)\} = \sum_{k=-\infty}^{\infty} d_k \, \delta(f - k f_0) \qquad (4.347)$$

Using Eqn. (4.345) in Eqn. (4.347) results in

$$\mathcal{F}\{\tilde{r}_{xx}(\tau)\} = 2\pi \sum_{k=-\infty}^{\infty} |c_k|^2 \, \delta(f - k f_0) \qquad (4.348)$$

which is identical to the expression given by Eqn. (4.315) for the power spectral density of a periodic signal.

A relationship similar to the one expressed by Eqn. (4.341) applies to random processes that are *wide-sense stationary*, and is known as the Wiener-Khinchin theorem. It is one of the fundamental theorems of random signal processing.

Example 4.42: **Power spectral density of a sinusoidal signal revisited**

Find the autocorrelation function for the signal

$$\tilde{x}(t) = 5 \cos(200\pi t)$$

Afterwards determine the power spectral density from the autocorrelation function.

Solution: Using Eqn. (4.328) with $T_0 = 0.01$ s we have

$$\tilde{r}_{xx}(\tau) = \frac{1}{0.01} \int_{-0.005}^{0.005} 25 \cos(200\pi t) \cos(200\pi [t + \tau]) \, dt$$

Using the appropriate trigonometric identity[7] the integral is evaluated as

$$\tilde{r}_{xx}(\tau) = \frac{25}{2} \cos(200\pi \tau)$$

Power spectral density is the Fourier transform of the autocorrelation function:

$$S_x(f) = \mathcal{F}\{\tilde{r}_{xx}(\tau)\} = \frac{25}{4} \delta(f + 100) + \frac{25}{4} \delta(f - 100)$$

which is in agreement with the earlier result found in Example 4.39.

---

[7] $\cos(a)\cos(b) = \frac{1}{2}\cos(a + b) + \frac{1}{2}\cos(a - b).$

Properties of the autocorrelation function

The autocorrelation function as defined by Eqns. (4.327), (4.328) and (4.329) has a number of important properties that will be listed here:

1. $r_{xx}(0) \geq |r_{xx}(\tau)|$ for all τ.

   To see why this is the case, we will consider the non-negative function $[x(t) \mp x(t+\tau)]^2$. The time average of this function must also be non-negative, so we can write

   $$\left\langle [x(t) \mp x(t+\tau)]^2 \right\rangle \geq 0$$

   or equivalently

   $$\left\langle x^2(t) \right\rangle \mp \left\langle 2\,x(t)\,x(t+\tau) \right\rangle + \left\langle x^2(t+\tau) \right\rangle \geq 0$$

   which implies that

   $$r_{xx}(0) \mp 2\,r_{xx}(\tau) + r_{xx}(0) \geq 0$$

   which is the same as property 1.

2. $r_{xx}(-\tau) = r_{xx}(\tau)$ for all τ, that is, the autocorrelation function has even symmetry. Recall that the autocorrelation function is the inverse Fourier transform of the power spectral density, i.e., $r_{xx}(\tau) = \mathcal{F}^{-1}\{S_x(f)\}$. We know from the symmetry properties of the Fourier transform that, since $S_x(f)$ is purely real, $r_{xx}(\tau)$ must be an even function of τ.

3. If the signal $x(t)$ is periodic with period T, then its autocorrelation function $\tilde{r}_{xx}(\tau)$ is also periodic with the same period, that is, if $x(t) = x(t+kT)$ for all integers k, then $\tilde{r}_{xx}(\tau) = \tilde{r}_{xx}(\tau + kT)$ for all integers k.

   This property easily follows from the time-average based definition of the autocorrelation function given by Eqn. (4.328). The time average of a periodic signal is also periodic with the same period.

4.5   System Function Concept

In time-domain analysis of systems in Chapter 2 we have relied on two distinct description forms for CTLTI systems:

1. A linear constant-coefficient differential equation that describes the relationship between the input and the output signals

2. An impulse response which can be used with the convolution operation for determining the response of the system to an arbitrary input signal

In this section, the concept of *system function* will be introduced as the third method for describing the characteristics of a system.

The system function is simply the Fourier transform of the impulse response:

$$H(\omega) = \mathcal{F}\{h(t)\} = \int_{-\infty}^{\infty} h(t)\, e^{-j\omega t}\, dt \tag{4.349}$$

Recall that the impulse response is only meaningful for a system that is both linear and time-invariant, since the convolution operator could not be used otherwise. Consequently, the system function concept is valid for linear and time-invariant systems only.

In general, $H(\omega)$ is a complex function of ω, and can be written in polar form as

$$H(\omega) = |H(\omega)|\, e^{j\Theta(\omega)} \tag{4.350}$$

Example 4.43: **System function for the simple RC circuit**

The impulse response of the RC circuit shown in Fig. 4.87 was found in Example 2.18 to be

$$h(t) = \frac{1}{RC}\, e^{-t/RC}\, u(t)$$

Determine the system function.

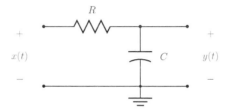

Figure 4.87 – The RC circuit for Example 4.43.

Solution: Taking the Fourier transform of the impulse response we obtain

$$H(\omega) = \int_0^\infty \frac{1}{RC}\, e^{-t/RC}\, e^{-j\omega t}\, dt = \frac{1}{1+j\omega RC}$$

To simplify notation we will define $\omega_c = 1/RC$ and write the system function as

$$H(\omega) = \frac{1}{1+j(\omega/\omega_c)} \tag{4.351}$$

The magnitude and the phase of the system function are

$$|H(\omega)| = \frac{1}{\sqrt{1+(\omega/\omega_c)^2}}$$

and

$$\Theta(\omega) = \angle H(\omega) = -\tan^{-1}(\omega/\omega_c)$$

respectively. The results found in Eqns. (4.352) and (4.352) are shown in Fig. 4.88.

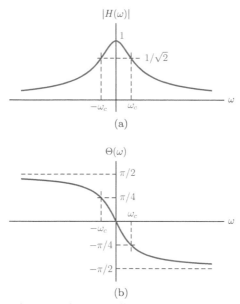

Figure 4.88 – The system function for the RC circuit in Example 4.22: (a) magnitude, (b) phase.

The value of the system function at the frequency ω_c is $H(\omega_c) = 1/(1+j)$, and the corresponding magnitude is $|H(\omega_c)| = 1/\sqrt{2}$. Thus, ω_c represents the frequency at which the magnitude of the system function is 3 decibels below its peak value at $\omega = 0$, that is,

$$20 \log_{10}\left[\frac{|H(\omega_c)|}{|H(0)|}\right] = 20 \log_{10}\left[\frac{1}{\sqrt{2}}\right] \approx -3 \text{ dB}$$

The frequency ω_c is often referred to as the *3-dB cutoff frequency* of the system.
Software resources:
ex_4_43.m

Interactive Demo: sf_demo1.m

The demo program "sf_demo1.m" illustrates the system function concept for the simple RC circuit of Example 4.43. The demo program allows circuit parameters R and C to be varied using slider controls, and displays the effect of parameter changes on the magnitude and the phase of the system function.

1. Observe how the two variable parameters are related to the 3-dB cutoff frequency f_c and consequently the *bandwidth* of the system function. One particular bandwidth measure called the *3-dB bandwidth* is defined as the frequency range in which the magnitude of the system function remains within 3-dB of its peak value, or greater than $1/\sqrt{2}$ in this case.
2. Pay attention to the value of the phase of the system function at the frequency $f = f_c$.

Software resources:
sf_demo1.m

Software resources: See MATLAB Exercise 4.5.

Obtaining the system function from the differential equation

Sometimes we need to find the system function for a CTLTI system described by means of a differential equation. Two properties of the Fourier transform will be useful for this purpose; the convolution property and the time differentiation property.

1. Since the output signal is computed as the convolution of the impulse response and the input signal, that is, $y(t) = h(t) * x(t)$, the corresponding relationship in the frequency domain is $Y(\omega) = H(\omega) X(\omega)$. Consequently, the system function is equal to the ratio of the output transform to the input transform:

$$H(\omega) = \frac{Y(\omega)}{X(\omega)} \tag{4.352}$$

2. Using the time differentiation property the transforms of the individual terms in the differential equation are

$$\frac{d^k y(t)}{dt^k} \xleftrightarrow{\mathcal{F}} (j\omega)^k Y(\omega) \qquad k = 0, 1, \ldots \tag{4.353}$$

$$\frac{d^k x(t)}{dt^k} \xleftrightarrow{\mathcal{F}} (j\omega)^k X(\omega) \qquad k = 0, 1, \ldots \tag{4.354}$$

The system function is obtained from the differential equation by first transforming both sides of the differential equation through the use of Eqns. (4.353) and (4.354), and then using Eqn. (4.352).

Example 4.44: **Finding the system function from the differential equation**

Determine the system function for a CTLTI system described by the differential equation

$$\frac{d^2 y(t)}{dt^2} + 2 \frac{dy(t)}{dt} + 26 y(t) = x(t)$$

Solution: Taking the Fourier transform of both sides of the differential equation we obtain

$$(j\omega)^2 Y(\omega) + 2(j\omega) Y(\omega) + 26 Y(\omega) = X(\omega)$$

or

$$\left[(26 - \omega^2) + j2\omega\right] Y(\omega) = X(\omega)$$

The system function is

$$H(\omega) = \frac{Y(\omega)}{X(\omega)} = \frac{1}{(26 - \omega^2) + j2\omega}$$

4.6   CTLTI Systems with Periodic Input Signals

In earlier sections of this chapter we have developed methods for representing periodic signals as linear combinations of sinusoidal or complex exponential basis functions. Using the TFS representation we have

$$\text{Eqn. (4.26):} \qquad \tilde{x}(t) = a_0 + \sum_{k=1}^{\infty} a_k \cos(k\omega_0 t) + \sum_{k=1}^{\infty} b_k \sin(k\omega_0 t)$$

Alternatively, using the EFS representation

$$\text{Eqn. (4.48):} \quad \tilde{x}(t) = \sum_{k=-\infty}^{\infty} c_k \, e^{jk\omega_0 t}$$

If a periodic signal is used as input to a CTLTI system, the superposition property can be utilized for finding the output signal. The response of the CTLTI system to the periodic signal can be determined as a linear combination of its responses to the individual basis functions in the TFS or EFS representation. We will analyze this concept in two steps; first for a complex exponential input signal, and then for a sinusoidal signal.

4.6.1　Response of a CTLTI system to complex exponential signal

Consider a CTLTI system with impulse response $h(t)$, driven by an input signal in the form of a complex exponential function of time

$$\tilde{x}(t) = e^{j\omega_0 t} \tag{4.355}$$

The response $y(t)$ of the system can be found through the use of the convolution relationship as we have derived in Section 2.7.2 of Chapter 2.

$$\begin{aligned} y(t) &= h(t) * \tilde{x}(t) \\ &= \int_{-\infty}^{\infty} h(\lambda)\, \tilde{x}(t-\lambda)\, d\lambda \end{aligned} \tag{4.356}$$

Using the signal $\tilde{x}(t)$ given by Eqn. (4.355) in Eqn. (4.356) we get

$$y(t) = \int_{-\infty}^{\infty} h(\lambda)\, e^{j\omega_0(t-\lambda)}\, d\lambda \tag{4.357}$$

or, equivalently

$$y(t) = e^{j\omega_0 t} \int_{-\infty}^{\infty} h(\lambda)\, e^{-j\omega_0 \lambda}\, d\lambda \tag{4.358}$$

The integral in Eqn. (4.358) should be recognized as the system function evaluated at the specific radian frequency $\omega = \omega_0$. Therefore

$$y(t) = e^{j\omega_0 t} H(\omega_0) \tag{4.359}$$

The development in Eqns. (4.357) through (4.359) is based on the inherent assumption that the Fourier transform of $h(t)$ exists. This in turn requires the corresponding CTLTI system to be stable. Any natural response the system may have exhibited at one point would have disappeared a long time ago. Consequently, the response found in Eqn. (4.359) is the *steady-state response* of the system.

For a CTLTI system driven by a complex exponential input signal, we have the following important relationship:

Response to complex exponential input:

$$y(t) = \text{Sys}\left\{e^{j\omega_0 t}\right\} = e^{j\omega_0 t} H(\omega_0)$$

$$= |H(\omega_0)| e^{j[\omega_0 t + \Theta(\omega_0)]} \tag{4.360}$$

1. The response of a CTLTI system to a complex exponential input signal is a complex exponential output signal with the same frequency ω_0.
2. The effect of the system on the complex exponential input signal is to
   (a) Scale its amplitude by an amount equal to the magnitude of the system function at $\omega = \omega_0$
   (b) Shift its phase by an amount equal to the phase of the system function at $\omega = \omega_0$

4.6.2   Response of a CTLTI system to sinusoidal signal

Let the input signal to the CTLTI system under consideration be a sinusoidal signal in the form

$$\tilde{x}(t) = \cos(\omega_0 t) \tag{4.361}$$

The response of the system in this case will be determined by making use of the results of the previous section.

We will use Euler's formula to write the input signal using two complex exponential functions as

$$\tilde{x}(t) = \cos(\omega_0 t) = \frac{1}{2} e^{j\omega_0 t} + \frac{1}{2} e^{-j\omega_0 t} \tag{4.362}$$

This representation of $\tilde{x}(t)$ allows the results of the previous section to be used. The output signal can be written using superposition:

$$y(t) = \frac{1}{2} \text{Sys}\left\{e^{j\omega_0 t}\right\} + \frac{1}{2} \text{Sys}\left\{e^{-j\omega_0 t}\right\}$$

$$= \frac{1}{2} e^{j\omega_0 t} H(\omega_0) + \frac{1}{2} e^{-j\omega_0 t} H(-\omega_0)$$

$$= \frac{1}{2} e^{j\omega_0 t} |H(\omega_0)| e^{j\Theta(\omega_0)} + \frac{1}{2} e^{-j\omega_0 t} |H(-\omega_0)| e^{j\Theta(-\omega_0)} \tag{4.363}$$

If the impulse response $h(t)$ is real-valued, the result in Eqn. (4.363) can be further simplified. Recall from Section 4.3.5 that, for real-valued $h(t)$, the transform $H(\omega)$ is conjugate symmetric, resulting in

$$|H(-\omega_0)| = |H(\omega_0)| \quad \text{and} \quad \Theta(-\omega_0) = -\Theta(\omega_0) \tag{4.364}$$

Using these relationships, Eqn. (4.363) becomes

$$y(t) = \frac{1}{2} |H(\omega_0)| e^{j[\omega_0 t + \Theta(\omega_0)]} + \frac{1}{2} |H(\omega_0)| e^{-j[\omega_0 t + \Theta(\omega_0)]}$$

$$= |H(\omega_0)| \cos(\omega_0 t + \Theta(\omega_0)) \tag{4.365}$$

For a CTLTI system driven by a cosine input signal, we have the following important relationship:

Response to cosine input:

$$y(t) = \text{Sys}\{\cos(\omega_0 t)\} = |H(\omega_0)| \cos(\omega_0 t + \Theta(\omega_0)) \qquad (4.366)$$

1. When a CTLTI system is driven by single-tone input signal at frequency ω_0, its output signal is also a single-tone signal at the same frequency.
2. The effect of the system on the input signal is to
   (a) Scale its amplitude by an amount equal to the magnitude of the system function at $\omega = \omega_0$
   (b) Shift its phase by an amount equal to the phase of the system function at $\omega = \omega_0$

Example 4.45: Steady-state response of RC circuit for single-tone input

Consider the RC circuit of Example 4.43. Let the component values be chosen to yield a 3-dB cutoff frequency of $\omega_c = 160\pi$ rad/s, or equivalently $f_c = 80$ Hz. Let the input signal be in the form

$$\tilde{x}(t) = 5\cos(2\pi f t)$$

Compute the steady-state output signal for the cases $f_1 = 20$ Hz, $f_2 = 100$ Hz, $f_3 = 200$ Hz, and $f_4 = 500$ Hz.

Solution: For the system under consideration the system function is

$$H(f) = \frac{1}{1 + j(f/80)} \qquad (4.367)$$

The magnitude and the phase of the system function were determined in Example 4.43 Eqns. (4.352) and (4.352). Using f instead of ω we have

$$|H(f)| = \frac{1}{\sqrt{1 + (f/80)^2}} \qquad \text{and} \qquad \Theta(f) = -\tan^{-1}(f/80)$$

as shown in Fig. 4.89. For the input frequency of $f_1 = 20$ Hz, the magnitude and the phase of the system function are

$$|H(20)| = \frac{1}{\sqrt{1 + (20/80)^2}} = 0.9701 \qquad (4.368)$$

and

$$\Theta(20) = -\tan^{-1}(20/80) = -0.245 \text{ radians} \qquad (4.369)$$

The impact of the system on the 20 Hz input signal is amplitude scaling by a factor of 0.9701 and phase shift by -0.245 radians. The corresponding steady-state output signal is

$$y_1(t) = (5)(0.9701)\cos(40\pi t - 0.245)$$
$$= 4.8507\cos(40\pi(t - 0.0019)) \qquad (4.370)$$

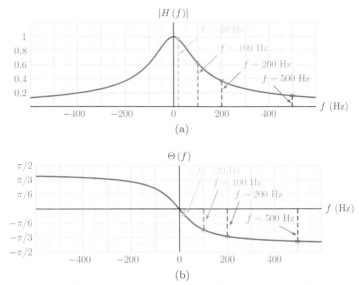

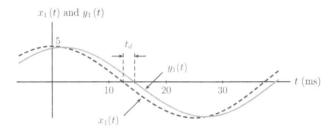

Figure 4.89 – The system function for the system in Example 4.45: (a) magnitude, (b) phase.

The phase shift of -0.245 radians translates to a time-delay of about $t_d = 1.9$ ms. The steady-state output signal $y_1(t)$ is shown in Fig. 4.90 along with the input signal for comparison.

Figure 4.90 – The 20 Hz input signal and the corresponding output signal for Example 4.45.

Calculations in Eqns. (4.368) through 4.370) can easily be repeated for the other three frequency choices, but will be skipped here to save space. The results are summarized in Table 4.6 for all four frequencies. The frequency-discriminating nature of the system is evident.

f (Hz)	$|H(f)|$	$\Theta(f)$ (rad)	t_d (ms)
20	0.9701	-0.2450	1.95
100	0.6247	-0.8961	1.43
200	0.3714	-1.1903	0.94
500	0.1580	-1.4121	0.45

Table 4.6 – Amplitude scaling, phase offset and time delay values for Example 4.45.

Input-output signal pairs for $f_2 = 100$ Hz, $f_3 = 200$ Hz, and $f_4 = 500$ Hz are shown in Fig. 4.91.

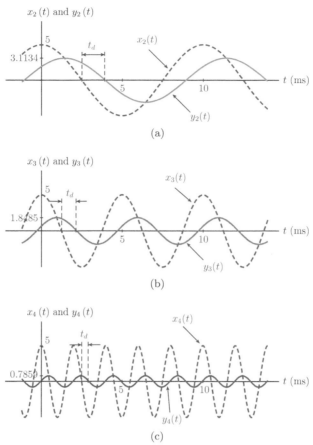

Figure 4.91 – The input-output signal pairs for Example 4.45: (a) $f_2 = 100$ Hz, (b) $f_3 = 200$ Hz, and (c) $f_4 = 500$ Hz.

Software resources:
ex_4_45a.m
ex_4_45b.m
ex_4_45c.m

Interactive Demo: sf_demo2.m

The demo program "sf_demo2.m" parallels the development of Example 4.45. Magnitude, phase and time-delay characteristics of the system function under consideration are shown. The first slider control allows the value of f_c to be adjusted. With the choice of $f_c = 80$ Hz the transfer function of Example 4.45 is obtained. The second slider control specifies the frequency of the input signal which is marked on the magnitude, phase and time-delay graphs. At the bottom of the screen, input and output signals are shown along with delay markers that illustrate the delay caused by the system.

1. Verify the numerical results of Example 4.45 by using $f_c = 80$ Hz, and setting the signal frequency equal to $f_1 = 20$ Hz, $f_2 = 100$ Hz, $f_3 = 200$ Hz, and $f_4 = 500$ Hz. Compare input and output signals to those in Figs. 4.90 and 4.91.

2. With the signal frequency fixed at $f_2 = 100$ Hz, change the 3-dB bandwidth of the system and observe its effect on the output signal in terms of peak amplitude and time delay.

Software resources:
sf_demo2.m

4.6.3 Response of a CTLTI system to periodic input signal

Using the development in the previous sections, we are now ready to consider the use of a general periodic signal $\tilde{x}(t)$ as input to a CTLTI system. Let $\tilde{x}(t)$ be a signal that satisfies the existence conditions for the Fourier series. Using the EFS representation of the signal, the response of the system is

$$\text{Sys}\{\tilde{x}(t)\} = \text{Sys}\left\{\sum_{k=-\infty}^{\infty} c_k\, e^{jk\omega_0 t}\right\} \tag{4.371}$$

Let us use the linearity of the system to write the response as

$$\text{Sys}\{\tilde{x}(t)\} = \sum_{k=-\infty}^{\infty} \text{Sys}\{c_k\, e^{jk\omega_0 t}\} = \sum_{k=-\infty}^{\infty} c_k\, \text{Sys}\{e^{jk\omega_0 t}\} \tag{4.372}$$

The response of the system to an exponential basis function at frequency $\omega = k\omega_0$ is given by

$$\text{Sys}\{e^{jk\omega_0 t}\} = e^{jk\omega_0 t}\, H(k\omega_0) \tag{4.373}$$

Using Eqn. (4.373) in Eqn. (4.372), the response of the system to $\tilde{x}(t)$ is found as

$$\text{Sys}\{\tilde{x}(t)\} = \sum_{k=-\infty}^{\infty} c_k\, H(k\omega_0)\, e^{jk\omega_0 t} \tag{4.374}$$

Two important observations should be made based on Eqn. (4.374):

1. For a CTLTI system driven by a periodic input signal, the output signal is also periodic with the same period.
2. If the EFS coefficients of the input signal are $\{c_k\, ;\, k = 1,\ldots,\infty\}$ then the EFS coefficients of the output signal are $\{c_k\, H(k\omega_0)\, ;\, k = 1,\ldots,\infty\}$.

Example 4.46: RC circuit with pulse-train input

Consider again the RC circuit used in Example 4.45 with system function given by Eqn. (4.368). Let the input signal be a pulse train with period $T_0 = 50$ ms and duty cycle $d = 0.2$ as shown in Fig. 4.92. Determine the output signal in steady state.

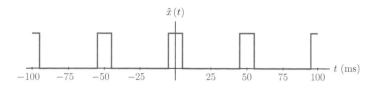

Figure 4.92 – The input signal for Example 4.46.

Solution: The EFS coefficients were determined in Example 4.7 for the type of periodic pulse train shown in Fig. 4.92. Using the specified duty cycle the coefficients are

$$c_k = 0.2 \, \text{sinc} \, (0.2k) \tag{4.375}$$

The fundamental frequency is $f_0 = 1/T_0 = 20$ Hz, and the EFS representation of the signal $\tilde{x}(t)$ is

$$\tilde{x}(t) = \sum_{k=-\infty}^{\infty} 0.2 \, \text{sinc} \, (0.2k) \, e^{j40\pi kt} \tag{4.376}$$

Let $\{d_k \; ; \; k = -\infty, \ldots, \infty\}$ be the EFS coefficients of the output signal $\tilde{y}(t)$. Using Eqn. (4.374) we have

$$d_k = c_k \, H \, (kf_0) = \frac{c_k}{1 + j \, (20k/80)} \, , \qquad k = -\infty, \ldots, \infty \tag{4.377}$$

The EFS representation of the output signal is

$$\tilde{y}(t) = \sum_{k=-\infty}^{\infty} \left(\frac{c_k}{1 + j \, (20k/80)} \right) e^{j40\pi kt} \tag{4.378}$$

Using the expressions for the magnitude and the phase of the system function, Eqn. (4.377) can be written as

$$|d_k| = |c_k| \, |H \, (kf_0)| = \frac{|c_k|}{\sqrt{1 + (20k/80)^2}} \tag{4.379}$$

and

$$\angle d_k = \angle c_k + \Theta \, (kf_0) = \angle c_k - \tan^{-1} (20k/80) \tag{4.380}$$

The relationship between the system function $H(f)$ and the line spectra c_k and d_k is illustrated in Fig. 4.93. An approximation to the output signal $\tilde{y}(t)$ using terms up to the 75-th harmonic is shown in Fig. 4.94.

Software resources:
ex_4_46a.m
ex_4_46b.m

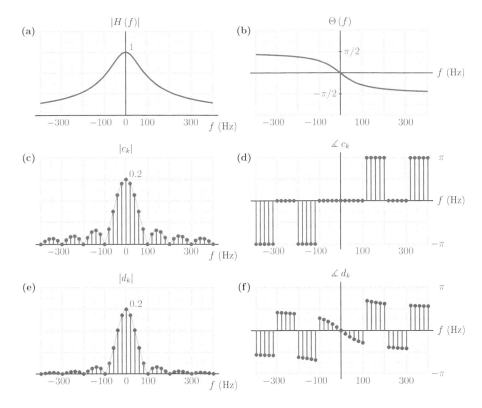

Figure 4.93 – System function and line spectra for Example 4.46: (a) the magnitude of the system function, (b) the phase of the system function, (c) the magnitude of the input spectrum, (d) the phase of the input spectrum, (e) the magnitude of the output spectrum, (f) the phase of the output spectrum.

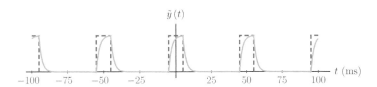

Figure 4.94 – The response of the system in Example 4.5 to pulse train with period $T_0 = 20$ ms and duty cycle $d = 0.2$.

Interactive Demo: `sf_demo3.m`

The demo program "`sf_demo3.m`" is based on Example 4.46, and illustrates the use of the system function with the exponential Fourier series coefficients of a periodic signal. The system under consideration is the RC circuit that we have used extensively in this chapter. The input signal is a periodic pulse train the period and the duty cycle of which may be adjusted using the slider controls. Continuing with the notation established in Example 4.46, we will let c_k and d_k represent the line spectra of the input and the output signals respectively. In the demo program, only the magnitude spectra $|c_k|$ and $|d_k|$ are shown due to screen size constraints.

1. Observe the effect of the period T_0 on the placement of spectral lines in the frequency spectra of input and output signals. The line spectra $|c_k|$ and $|d_k|$ are displayed not in terms of the index k but rather in terms of the actual frequencies kf_0.

2. Observe how the system function $H(f)$ effects the line spectrum for the output signal. Compare the line spectra for the input and the output signals. The more different the two line spectra are, the more different the two time domain signals $\tilde{x}(t)$ and $\tilde{y}(t)$ will be.

Software resources:
sf_demo3.m

4.7 CTLTI Systems with Non-Periodic Input Signals

Let us consider the case of using a non-periodic signal $x(t)$ as input to a CTLTI system. We have established in Section 2.7.2 of Chapter 2 that the output of a CTLTI system is equal to the convolution of the input signal with the impulse response, that is

$$
\begin{aligned}
y(t) &= h(t) * x(t) \\
&= \int_{-\infty}^{\infty} h(\lambda)\, x(t-\lambda)\, d\lambda
\end{aligned}
\tag{4.381}
$$

Let us assume that

1. The system is stable ensuring that $H(\omega)$ converges
2. The input signal has a Fourier transform

We have seen in Section 4.3.5 of this chapter that the Fourier transform of the convolution of two signals is equal to the product of individual transforms:

$$
Y(\omega) = H(\omega)\, X(\omega)
\tag{4.382}
$$

The output transform is the product of the input transform and the system function.

Writing each transform involved in Eqn. (4.382) in polar form using its magnitude and phase we obtain the corresponding relationships:

$$
|Y(\omega)| = |H(\omega)|\, |X(\omega)|
\tag{4.383}
$$

$$
\angle Y(\omega) = \angle X(\omega) + \Theta(\omega)
\tag{4.384}
$$

1. The magnitude of the output spectrum is equal to the product of the magnitudes of the input spectrum and the system function.
2. The phase of the output spectrum is found by adding the phase characteristics of the input spectrum and the system function.

Example 4.47: **Pulse response of RC circuit revisited**

Consider again the RC circuit used Example 4.43. Let $f_c = 1/RC = 80$ Hz. Determine the Fourier transform of the response of the system to the unit-pulse input signal $x(t) = \Pi(t)$.

Solution: In Example 2.21 of Chapter 2 the output signal $y(t)$ was found through the use of convolution operation. In this example we will approach the same problem from a system function perspective. The system function of the RC circuit under consideration was found in Example 4.43 to be

$$\text{Eqn. (4.351):} \quad H(f) = \frac{1}{1 + j\left(f/f_c\right)}$$

The transform of the input signal is

$$X(f) = \text{sinc}(f)$$

Using $f_c = 80$ Hz, the transform of the output signal is

$$Y(f) = H(f)\,X(f)$$

$$= \frac{1}{1 + j\left(\dfrac{f}{80}\right)}\,\text{sinc}(f) \tag{4.385}$$

Magnitude and phase of the transform $Y(\omega)$ can be found using Eqns. (4.383) and (4.384). The magnitude is

$$|Y(f)| = \frac{1}{\sqrt{1 + \left(\dfrac{f}{80}\right)^2}}\,|\text{sinc}(f)| \tag{4.386}$$

and the phase is

$$\angle Y(f) = -\tan^{-1}\left(\frac{f}{80}\right) + \angle\left[\text{sinc}(f)\right] \tag{4.387}$$

as illustrated in Fig. 4.95. With a bit of work it can be shown that $Y(f)$ found above is indeed the Fourier transform of the signal $y(t)$ found in Example 2.21 using the convolution integral.

Software resources:
ex_4_47.m

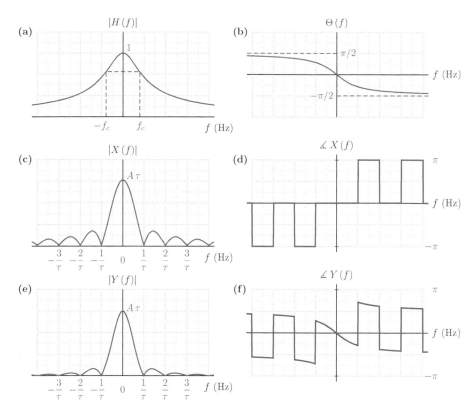

Figure 4.95 – Determining the Fourier transform of the output signal of a linear system: (a) the magnitude of the system function, (b) the phase of the system function, (c) the magnitude of the input spectrum, (d) the phase of the input spectrum, (e) the magnitude of the output spectrum, (f) the phase of the output spectrum.

Interactive Demo: `sf_demo4.m`

The demo program "`sf_demo4.m`" is based on Example 4.47, and illustrates the use of the system function in conjunction with the Fourier transform of a pulse. The transform of the output signal is the product of the system function with the transform of the input signal as we have established in Eqns. (4.385) through (4.387). The input signal $x(t)$ is a single pulse with unit amplitude and width equal to τ, and it is centered around $t = 0$. The user interface of the demo program displays magnitudes of the three transforms involved, namely $X(f)$, $H(f)$ and $Y(f)$. Time domain representations of the input and the output signals are also shown. Phase characteristics are not displayed in the demo due to screen space constraints. System parameter f_c and input signal parameter τ may be varied using slider controls, and the effect of these variations may be observed on magnitude spectra as well as the output signal.

1. Observe how the changes to the system function affect the magnitude spectrum of the output signal.
2. Specifically, how does the bandwidth of the system impact the shape of the output pulse. How does the bandwidth of the system function relate to the level of similarity between the input and the output signals?

3. Vary the duration τ of the input pulse. How does the pulse duration relate to the level of similarity between the input and the output signals?

Software resources:
sf_demo4.m

4.8 Further Reading

[1] R.J. Beerends. *Fourier and Laplace Transforms*. Cambridge University Press, 2003.

[2] R.N. Bracewell. *The Fourier Transform and Its Applications*. Electrical Engineering Series. McGraw-Hill, 2000.

[3] E.A. Gonzalez-Velasco. *Fourier Analysis and Boundary Value Problems*. Elsevier Science, 1996.

[4] R. Haberman. *Applied Partial Differential Equations: With Fourier Series and Boundary Value Problems*. Pearson Education, 2012.

[5] K.B. Howell. *Principles of Fourier Analysis*. Studies in Advanced Mathematics. Taylor & Francis, 2010.

MATLAB Exercises

MATLAB Exercise 4.1: Computing finite-harmonic approximation to pulse train

Develop a MATLAB script to compute and graph the finite-harmonic approximation to the periodic pulse train $\tilde{x}(t)$ of Example 4.1 for $m = 4$.

Solution: Consider the listing given below:

```
1   % Script matex_4_1.m
2   %
3   m = 4;                % Number of harmonics to be used.
4   t = [-7:0.01:7];      % Create a vector of time instants.
5   f0 = 1/3;             % Fundamental frequency is f0 = 1/3 Hz.
6   omega0 = 2*pi*f0;
7   x = 1/3;              % Recall that a0=1/3.
8                         % We will start by setting x(t)=a0.
9   % Start the loop to compute the contribution of each harmonic.
10  for k = 1:m,
11    ak = 1/(pi*k)*sin(k*omega0);
12    bk = 1/(pi*k)*(1-cos(k*omega0));
13    x = x+ak*cos(k*omega0*t)+bk*sin(k*omega0*t);
14  end;
15  % Graph the resulting approximation to x(t).
16  clf;
17  plot(t,x); grid;
18  title('Approximation using m=4');
19  xlabel('Time (sec)');
20  ylabel('Amplitude');
```

The main part of the code is the loop construct between lines 10 and 14. Before entering the loop, the vector "x" holds a zero-order approximation to the signal, that is, $\tilde{x}^{(0)}(t) = a_0 = 1/3$. After the first pass through the loop, the vector "x" holds samples of the signal

$$\tilde{x}^{(1)}(t) = a_0 + a_1 \cos(2\pi f_0 t) + b_1 \sin(2\pi f_0 t)$$

After the second pass through the loop, the vector "x" now holds

$$\tilde{x}^{(2)}(t) = a_0 + a_1 \cos(2\pi f_0 t) + b_1 \sin(2\pi f_0 t)$$
$$+ a_2 \cos(4\pi f_0 t) + b_2 \sin(4\pi f_0 t)$$

and so on. The graph produced by this listing is shown in Fig. 4.96.

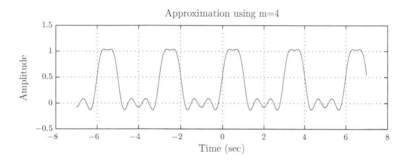

Figure 4.96 – Graph produced by the code in MATLAB Exercise 4.1.

Software resources:
matex_4_1.m

MATLAB Exercise 4.2: Computing multiple approximations to pulse train

MATLAB listing given below computes all approximations to the periodic pulse train $\tilde{x}(t)$ of Example 4.1 for $m = 1, \ldots, 5$. The approximations $\tilde{x}^{(1)}(t)$, $\tilde{x}^{(2)}(t)$, $\tilde{x}^{(4)}(t)$, $\tilde{x}^{(5)}(t)$, are graphed on the same coordinate system for comparison. The approximation $\tilde{x}^{(3)}(t)$ is not graphed since $a_3 = b_3 = 0$ based on Table 4.1, and therefore $\tilde{x}^{(3)}(t) = \tilde{x}^{(2)}(t)$.

```
1    % Script: matex_4_2.m
2    %
3    m = 5;                   % Maximum number of harmonics to be used.
4    t = [-3:0.005:3];        % Create a vector of time instants.
5    f0 = 1/3;                % Fundamental frequency is f0=1/3 Hz.
6    omega0 = 2*pi*f0;
7    x = 1/3*ones(size(t));   % Recall that a0=1/3.
8                             % We will start by setting x(t)=a0.
9    xmat = x;                % Row vector containing dc value of 1/3.
10   % Start the loop to compute the contribution of each harmonic.
11   for k = 1:m,
12      ak = 1/(pi*k)*sin(k*omega0);
13      bk = 1/(pi*k)*(1-cos(k*omega0));
14      x = x+ak*cos(k*omega0*t)+bk*sin(k*omega0*t);
15      xmat = [xmat;x]; % Append another row to matrix 'xmat'.
16   end;
17   % Graph the results.
18   clf;
19   plot(t,xmat(2,:),'b--',t,xmat(3,:),'g-',t,xmat(5,:),'r-',...
20      t,xmat(6,:),'m-'); grid;
```

```
21    title('Approximation to x(t) using m=1,...,5');
22    xlabel('Time (sec)');
23    ylabel('Amplitude');
24    legend('x^{(1)}(t)','x^{(2)}(t)','x^{(4)}(t)','x^{(5)}(t)');
```

On line 9 the matrix "xmat" is initialized as a row vector holding the constant (zero-order) approximation to $\tilde{x}(t)$. The loop construct between lines 11 and 16 is a slightly modified version of the corresponding construct in MATLAB Exercise 4.1. At each pass through the loop a new row is added to the matrix "xmat" as the next approximation to the periodic signal. When the script is finished, the rows of the matrix "xmat" correspond to finite-harmonic approximations as follows:

$$\text{row } 1 \Rightarrow \tilde{x}^{(0)}(t)$$
$$\text{row } 2 \Rightarrow \tilde{x}^{(1)}(t)$$
$$\text{row } 3 \Rightarrow \tilde{x}^{(2)}(t)$$
$$\vdots$$

The graph produced by this script is shown in Fig. 4.97.

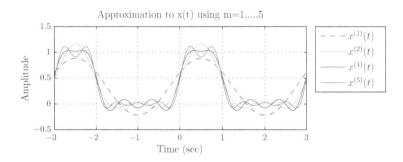

Figure 4.97 – Graph produced by the code in MATLAB Exercise 4.2.

Software resources:
matex_4_2.m

MATLAB Exercise 4.3: Graphing the line spectrum in Example 4.5

The line spectrum found in Example 4.5 can be graphed in MATLAB using a direct implementation of the general expression found in Eqn. (4.73) for the EFS coefficients. The code given below computes and graphs the spectrum in the index range $-25 \le k \le 25$.

```
1    % Script: matex_4_3a.m
2    %
3    k = [-25:25];        % Create a vector of index values.
4    k = k+eps;           % Avoid division by zero.
5    ck = sin(pi*k/3)./(pi*k);    % Compute the EFS coefficients.
6    % Graph the line spectrum.
7    clf;
8    stem(k,ck);
9    axis([-25.5,25.5,-0.1,0.4]);
10   xlabel('Index k');
11   ylabel('Coefficient');
```

The resulting line spectrum graph is shown in Fig. 4.98.

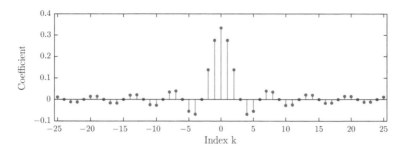

Figure 4.98 – The line spectrum for MATLAB Exercise 4.3.

Recall that the general expression for c_k as given by Eqn. (4.73) produced an indefinite value at $k = 0$ necessitating the use of L'Hospital's rule. Without any precautions, line 3 of the code given above would produce a division by zero error for $k = 0$. To avoid this, we add a very small positive value to all indices. This is done in line 2 of the code with the statement

```
4    k = k+eps;
```

MATLAB has a built-in variable `eps` for this purpose. It is the smallest positive floating-point number MATLAB can represent, and therefore its effect on the end result should be negligible. An alternative and perhaps tidier method of avoiding the division-by-zero error is to use the sinc function defined as

$$\operatorname{sinc}(\alpha) \triangleq \frac{\sin(\pi\alpha)}{\pi\alpha}$$

which allows the result of Eqn. (4.73) to be written in the form

$$c_k = \tfrac{1}{3} \operatorname{sinc}(k/3) \tag{4.388}$$

MATLAB function `sinc(..)` computes the value at $k = 0$ correctly without requiring any special precautions on our part. A modified script file that uses the function `sinc(..)` and produces the same graph is given below:

```
1    % Script: matex_4_3b.m
2    %
3    k = [-25:25];          % Create a vector of index values.
4    ck = 1/3*sinc(k/3);    % Compute the EFS coefficients.
5    % Graph the line spectrum.
6    clf;
7    stem(k,ck);
8    axis([-25.5,25.5,-0.1,0.4]);
9    xlabel('Index k');
10   ylabel('Coefficient');
```

If the line spectrum is desired in polar form with magnitude and phase, lines 7 through 10 of the script listed above may be replaced with the following:

```
8    subplot(2,1,1);
9    stem(k,abs(ck));
10   axis([-25.5,25.5,-0.1,0.4]);
```

```
11    xlabel('Index');
12    ylabel('Magnitude');
13    subplot(2,1,2);
14    stem(k,angle(ck));
15    axis([-25.5,25.5,-4,4]);
16    xlabel('Index');
17    ylabel('Phase (rad)');
```

Magnitude and phase of the EFS coefficients are graphed through the use of MATLAB functions `abs(..)` and `angle(..)` respectively. The resulting graph is shown in Fig. 4.99.

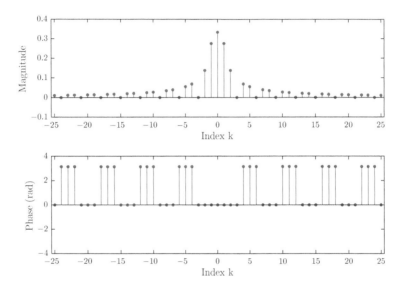

Figure 4.99 – The line spectrum in polar form for MATLAB Exercise 4.3.

Software resources:

`matex_4_3a.m`

`matex_4_3b.m`

`matex_4_3c.m`

MATLAB Exercise 4.4: Line spectrum for Example 4.6

MATLAB listing given below can be used for computing and displaying the EFS coefficients for the signal $\tilde{x}(t)$ of Example 4.6 for $k = -25, \ldots, 25$. Eqns. (4.74) and (4.75) are graphed as functions of the integer index k.

```
1     % Script: matex_4_4a.m
2     %
3     k = [-25:25];        % Create a vector of index values.
4     k = k+eps;           % Avoid division by zero.
5     cReal = 1./(2*pi*k).*sin(2*pi*k/3);
6     cImag = 1./(2*pi*k).*(cos(2*pi*k/3)-1);
7     cMag = abs(cReal+j*cImag);
8     cPhs = angle(cReal+j*cImag);
9     % Graph the line spectrum.
10    clf;
11    subplot(2,1,1);
12    stem(k,cMag);
13    axis([-25.5,25.5,-0.1,0.4]);
```

```
14    title('|c_{k}|');
15    xlabel('Index k');
16    ylabel('Magnitude');
17    subplot(2,1,2);
18    stem(k,cPhs);
19    axis([-25.5,25.5,-4,4]);
20    title('\angle c_{k}');
21    xlabel('Index k');
22    ylabel('Phase (rad)');
```

A finite-harmonic approximation to the signal $\tilde{x}(t)$ can be computed and graphed using the script below:

```
1     % Script: matex_4_4b.m
2     %
3     f0 = 1/3;              % Fundamental frequency.
4     omega0 = 2*pi*f0;
5     t = [-3:0.01:6];       % Create a vector of time instants.
6     x=zeros(size(t));      % Start with a vector of all zeros.
7     for k=-25:25
8         k = k+eps;         % Avoid division by zero.
9         cReal = 1/(2*pi*k)*sin(k*omega0);
10        cImag = 1/(2*pi*k)*(cos(k*omega0)-1);
11        x = x+(cReal+j*cImag)*exp(j*k*omega0*t);
12    end;
13    % Graph the signal.
14    clf;
15    plot(t,real(x)); grid;
16    title('Approximation to x(t)');
17    xlabel('Time (sec)');
18    ylabel('Amplitude');
```

On line 15 of the code we use the `real(..)` function on the vector "x" to remove any residual imaginary part that may be introduced as a result of round-off error.
Software resources:
matex_4_4a.m
matex_4_4b.m

MATLAB Exercise 4.5: Graphing system function for RC circuit

MATLAB code listing given below can be used for computing and graphing the system function found in Eqn. (4.351) of Example 4.43 in the frequency range $-1\,\mathrm{Hz} \le f \le 1\,\mathrm{Hz}$.

```
1     % File: matex_4_5.m
2     %
3     R = 1e6;
4     C = 1e-6;
5     fc = 1/(2*pi*R*C);     % The critical frequency
6     f = [-500:500]/500;    % Vector of frequency values
7     Hf = 1./(1+j*f/fc);    % Evaluate system function
8     % Graph the magnitude and the phase
9     clf;
10    subplot(2,1,1);
11    plot(f,abs(Hf)); grid;
12    axis([-1,1,-0.2,1.2]);
13    ylabel('Magnitude');
14    xlabel('Frequency (Hz)');
15    subplot(2,1,2);
16    plot(f,angle(Hf)); grid;
17    ylabel('Phase (rad)');
18    xlabel('Frequency (Hz)');
```

Note that line 7 of the code is a direct implementation of Eqn. (4.351) with f and f_c in Hz used instead of radian frequencies ω and ω_c.

Software resources:

matex_4_5.m

Problems

4.1. Consider the periodic square-wave signal $\tilde{x}(t)$ shown in Fig. 4.1 on page 267.

  a. Set up the cost function for approximating this signal using three terms in the form

$$\tilde{x}^{(3)}(t) = b_1 \sin(\omega_0 t) + b_2 \sin(2\omega_0 t) + b_3 \sin(3\omega_0 t)$$

  b. Show that the optimum values of the coefficients are

$$b_1 = \frac{4A}{\pi}, \quad b_2 = 0, \quad \text{and} \quad b_3 = \frac{4A}{3\pi}$$

4.2. Consider the pulse train shown in Fig. P.4.2.

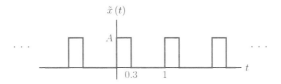

Figure P. 4.2

  a. Determine the fundamental period T_0 and the fundamental frequency ω_0 for the signal.
  b. Using the technique described in Section 4.2 find an approximation to $\tilde{x}(t)$ in the form

$$\tilde{x}^{(1)}(t) \approx a_0 + a_1 \cos(\omega_0 t) + b_1 \sin(\omega_0 t)$$

  Determine the optimum coefficients a_0, a_1 and b_1.

4.3. Consider again the pulse train shown in Fig. P.4.2. Using the technique described in Section 4.2 find an approximation to $\tilde{x}(t)$ in the form

$$\tilde{x}^{(2)}(t) \approx a_0 + a_1 \cos(\omega_0 t) + b_1 \sin(\omega_0 t) + a_2 \cos(2\omega_0 t) + b_2 \sin(2\omega_0 t)$$

4.4. Consider again the periodic square-wave signal $\tilde{x}(t)$ shown in Fig. 4.1. The approximation using M trigonometric functions would be in the form

$$\tilde{x}^{(M)}(t) = b_1 \sin(\omega_0 t) + b_2 \sin(2\omega_0 t) + \ldots + b_M \sin(M\omega_0 t) = \sum_{k=1}^{M} b_k \sin(k\omega_0 t)$$

Show that the approximation error is equal to $+A$ at discontinuities of the signal $\tilde{x}(t)$ independent of the number of terms M, that is,

$$\tilde{\varepsilon}_M\left(n\frac{T_0}{2}\right) = \pm A, \quad n: \text{Integer}$$

Hint: It should not be necessary to determine the optimum coefficient values $b_1, b_2, \ldots, b_M$ for this problem.

4.5. Consider the pulse train $\tilde{x}(t)$ shown in Fig. P.4.5.

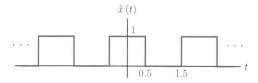

$$\tilde{x}(t)$$

Figure P. 4.5

a. Determine the fundamental period T_0 and the fundamental frequency ω_0 for the signal.
b. Using the approach followed in Section 4.2 determine the coefficients of the approximation

$$\tilde{x}^{(2)}(t) = a_0 + a_1 \cos(\omega_0 t) + a_2 \cos(2\omega_0 t)$$

to the signal $\tilde{x}(t)$ that results in the minimum mean-squared error.

4.6. Refer to the TFS representation of a periodic signal $\tilde{x}(t)$ given by Eqn. (4.26). Using the orthogonality properties of the basis functions show that the coefficients of the sine terms are computed as

$$b_k = \frac{2}{T_0} \int_{t_0}^{t_0 + T_0} \tilde{x}(t) \sin(k\omega_0 t) \, dt, \quad \text{for} \quad k = 1, \ldots, \infty$$

where T_0 is the fundamental period, $\omega_0 = 2\pi/T_0$ is the fundamental frequency in rad/s and t_0 is an arbitrary time instant.

4.7. Determine the TFS coefficients for the periodic signal shown in Fig. P.4.7.

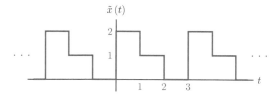

Figure P. 4.7

4.8. Determine the TFS coefficients for the periodic signal shown in Fig. P. 4.8. One period of the signal is $\tilde{x}(t) = e^{-2t}$ for $0 < t < 1$ s.

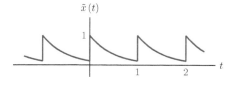

Figure P. 4.8

4.9. Consider the periodic signal $\tilde{x}(t)$ shown in Fig. P.4.7.

a. Determine the EFS coefficients from the TFS coefficients obtained in Problem 4.7.
b. Determine the EFS coefficients by direct application of the analysis equation given by Eqn. (4.72).
c. Sketch the line spectrum (magnitude and phase).

4.10. Consider the periodic signal $\tilde{x}(t)$ shown in Fig. P.4.8.

a. Determine the EFS coefficients from the TFS coefficients obtained in Problem 4.8.
b. Determine the EFS coefficients by direct application of the analysis equation given by Eqn. (4.72).
c. Sketch the line spectrum (magnitude and phase).

4.11. Consider the periodic sawtooth signal $\tilde{g}(t)$ shown in Fig. P.4.11.

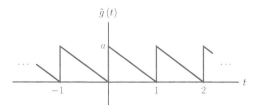

Figure P. 4.11

a. Determine EFS coefficients of this signal using the analysis equation in Eqn. (4.72).
b. Sketch the line spectrum in terms of magnitude and phase.
c. Determine the TFS coefficients of $\tilde{g}(t)$ from the EFS coefficients using the conversion formulas given by Eqns. (4.58) and (4.59).

4.12. Let the periodic signal $\tilde{x}(t)$ have the EFS coefficients c_k. A new periodic signal $\tilde{g}(t)$ is defined as a time reversed version of $\tilde{x}(t)$, that is,

$$\tilde{g}(t) = \tilde{x}(-t)$$

a. If $\tilde{g}(t)$ has the EFS coefficients d_k, find the relationship between the two sets of coefficients c_k and d_k.
b. Consider the signal $\tilde{x}(t)$ shown in Fig. 4.17 on page 290. Its EFS coefficients of were determined in Example 4.8 and given by Eqns. (4.81) and (4.82). Obtain the EFS coefficients of $\dot{g}(t)$ shown in Fig. P.4.11 from the EFS coefficients of $\tilde{x}(t)$ using the results of part (a).

4.13. The signal $\tilde{x}(t)$ shown in Fig. 4.17 and the signal $\tilde{g}(t)$ shown in Fig. P4.11 add up to a constant:

$$\tilde{x}(t) + \tilde{g}(t) = a$$

Explain how this property can be used for finding the EFS coefficients of $\tilde{g}(t)$ from the EFS coefficients of $\tilde{x}(t)$ that were determined in Example 4.8 and given by Eqns. (4.81) and (4.82).

4.14. Determine the EFS coefficients of the full-wave rectified signal shown in Fig. P.4.14.

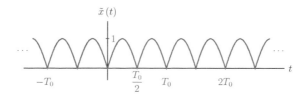

Figure P. 4.14

4.15. Prove the time shifting property of the exponential Fourier series given by Eqn. (4.113). Specifically show that, if the periodic signal $\tilde{x}(t)$ has the EFS coefficients c_k, then the signal $\tilde{x}(t - \tau)$ has the coefficients

$$d_k = c_k \, e^{-jk\omega_0 \tau}$$

4.16. Refer to the half-wave rectified sinusoid shown in Fig. 4.21 on page 293. Its EFS coefficients were determined in Example 4.10. Explain how the EFS coefficients of the full-wave rectified sinusoid in Fig. P.4.14 could be obtained from the results of Example 4.10 through the use of the time shifting property.

4.17. Refer to the half-wave rectified sinusoid shown in Fig. 4.21. Its EFS coefficients were determined in Example 4.10. Using the conversion formulas given by Eqns. (4.92) and (4.93) find the compact Fourier series (CFS) representation of the signal.

4.18. Find the Fourier transform of each of the pulse signals given below:

   a. $x(t) = 3\,\Pi(t)$
   b. $x(t) = 3\,\Pi(t - 0.5)$

   c. $x(t) = 2\,\Pi\left(\dfrac{t}{4}\right)$

   d. $x(t) = 2\,\Pi\left(\dfrac{t - 3}{2}\right)$

4.19. Starting with the Fourier transform integral given by Eqn. (4.129) find the Fourier transform of the signal

$$\Pi(t - 0.5) - \Pi(t - 1.5)$$

shown in Fig. P.4.19.

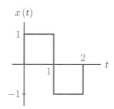

Figure P. 4.19

4.20. Starting with the Fourier transform integral given by Eqn. (4.129) find the Fourier transform of the signal

$$e^{-at}\,u(t) - e^{at}\,u(-t)$$

shown in Fig. P.4.20. Assume $a > 0$.

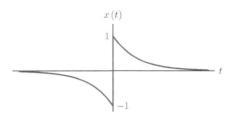

Figure P. 4.20

4.21. Refer to the signal shown in Fig. P.4.19. Find its Fourier transform by starting with the transform of the unit pulse and using linearity and time shifting properties.

4.22. Refer to the signal shown in Fig. P.4.20. Find its Fourier transform by starting with the transform of the right-sided exponential signal and using linearity and time scaling properties.

4.23. The Fourier transform of the triangular pulse with peak amplitude A and two corners at $\pm\tau$ was found in Example 4.17 as

$$A \Lambda \left(\frac{t}{\tau}\right) \xleftrightarrow{\mathcal{F}} A\tau \operatorname{sinc}^2 \left(\frac{\omega\tau}{2\pi}\right)$$

Using this result along with linearity and time shifting properties of the Fourier transform, find the transform of the signal shown in Fig. P.4.23.

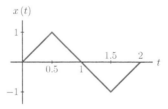

Figure P. 4.23

4.24. The transform pair

$$e^{-a|t|} \xleftrightarrow{\mathcal{F}} \frac{2a}{a^2 + \omega^2}$$

was obtained in Example 4.16. Using this pair along with the duality property, find the Fourier transform of the signal

$$x(t) = \frac{2}{1 + 4t^2}$$

4.25. Prove that, for a real-valued signal with odd symmetry, the Fourier transform is purely imaginary.
Hint: Start with the conditions $x^*(t) = x(t)$ and $x(-t) = -x(t)$. Show that $\operatorname{Re}\{X(\omega)\} = 0$.

4.26.

  a. Find the Fourier transform $X(\omega)$ of the pulse signal

$$x(t) = \Pi\left(\frac{t-1}{2}\right)$$

b. Express $x(t)$ as the sum of its even and odd components, that is,

$$x(t) = x_e(t) + x_o(t)$$

Determine and sketch each component.

c. Find the Fourier transforms $X_e(\omega)$ and $X_o(\omega)$ of the even and odd components respectively. Show that

$$X_e(\omega) = \text{Re}\{X(\omega)\} \qquad \text{and} \qquad X_o(\omega) = j\,\text{Im}\{X(\omega)\}$$

4.27. Compute and sketch the Fourier transforms of the modulated pulse signals given below:

a. $x(t) = \cos(10\pi t)\,\Pi(t)$

b. $x(t) = \cos(10\pi t)\,\Pi\left(\dfrac{t}{2}\right)$

c. $x(t) = \cos(10\pi t)\,\Pi(2t)$

d. $x(t) = \cos(10\pi t)\,\Pi(4t)$

4.28.

a. Starting with the inverse Fourier transform integral given by Eqn. (4.129) prove the second form of the differentiation-in-time property stated in Eqn. (4.249) for $n = 1$.

b. Show that, if Eqn. (4.249) holds true for $n = k$ where k is any positive integer, then it must also hold true for $n = k + 1$.

4.29.

a. Starting with the Fourier transform integral given by Eqn. (4.130) prove the second form of the differentiation-in-frequency property stated in Eqn. (4.265) for $n = 1$.

b. Show that, if Eqn. (4.265) holds true for $n = k$ where k is any positive integer, then it must also hold true for $n = k + 1$.

4.30. Taking Eqn. (4.249) as a starting point and employing the duality property of the Fourier transform, prove the second form of the differentiation-in-frequency property stated in Eqn. (4.265) for $n = 1$.

4.31. Using the differentiation-in-time property of the Fourier transform, determine the transform of the signal shown in Fig. P.4.31.

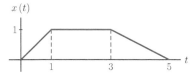

Figure P. 4.31

4.32. The Fourier transform of the trapezoidal signal $x(t)$ shown in Fig. 4.70 on page 355 was found in Example 4.32 to be

$$X(f) = 2\lambda\,\text{sinc}(f\tau)\,\text{sinc}(2f\lambda)$$

Find the inverse transform of the spectrum shown in Fig. 4.74 on page 360 by using the result above in conjunction with the duality property.

4.33. Determine the Fourier transform of the signal

$$x\left(t\right) = \sin\left(\pi t\right) \Pi\left(t - \frac{1}{2}\right) = \begin{cases} \sin\left(\pi t\right), & 0 \le t \le 1 \\ 0, & \text{otherwise} \end{cases}$$

a. Using the modulation property of the Fourier transform
b. Using the multiplication property of the Fourier transform

4.34. Consider the periodic pulse train shown in Fig. P.4.34.

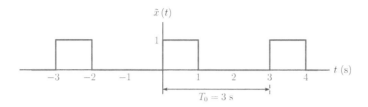

Figure P. 4.34

Recall that the EFS coefficients of this signal were found in Example 4.6. Express the Fourier transform of $\tilde{x}\left(t\right)$ using impulse functions.

4.35. Consider the pulse train with duty cycle d shown in Fig. 4.15 on page 289. Its EFS coefficients were determined in Example 4.7 to be

$$c_k = d \operatorname{sinc}\left(kd\right)$$

a. Working in the time domain, compute the power of the pulse train as a function of the duty cycle d.
b. Sketch the power spectral density based on the EFS coefficients.
c. Let $d = 0.5$. Suppose this signal is processed through a lowpass system that only retains the first m harmonics and eliminates the others. How many harmonics should be retained if we want to preserve at least 90 percent of the signal power?
d. How many harmonics should be retained to preserve at least 95 percent of the signal power?
e. How many harmonics should be retained to preserve at least 99 percent of the signal power?

4.36. Repeat parts (c)–(e) of Problem 4.35 using $d = 0.2$. Does it take fewer or more harmonics to preserve the same percentage of power when the duty cycle is reduced?

4.37. Use Parseval's theorem to prove that

$$\int_{-\infty}^{\infty} \left| \operatorname{sinc}\left(f\right)\right|^2 \, df = 1$$

4.38. Determine and sketch the power spectral density of the following signals:

a. $x\left(t\right) = 3\cos\left(20\pi t\right)$
b. $x\left(t\right) = 2\cos\left(20\pi t\right) + 3\cos\left(30\pi t\right)$
c. $x\left(t\right) = 5\cos\left(200\pi t\right) + 5\cos\left(200\pi t\right)\cos\left(30\pi t\right)$

MATLAB Problems

4.39. Refer to Problem 4.1 in which a periodic square-wave signal was approximated using three terms. Let the parameters of the square wave be $A = 1$ and $T_0 = 1$. Using MATLAB, compute and graph the three-term approximation superimposed with the original signal, that is, graph $\tilde{x}(t)$ and $\tilde{x}^{(3)}(t)$ on the same coordinate system. Display the signals in the time interval $-1.5 < t < 2.5$ s.

4.40. Refer to Problem 4.2 in which a periodic pulse train was approximated using using both cosine and sine functions. Using MATLAB, compute and graph the approximation $\tilde{x}^{(1)}(t)$ superimposed with the original signal $\tilde{x}(t)$ and on the same coordinate system. Show the time interval $-1.5 < t < 2.5$ s.

4.41. Compute and graph finite-harmonic approximations to the signal $\tilde{x}(t)$ shown in Fig. P.4.7 using 3, 4, and 5 harmonics. Also graph the approximation error for each case.

4.42. Compute and graph finite-harmonic approximations to the signal $\tilde{x}(t)$ shown in Fig. P.4.8 using 3, 4, and 5 harmonics. Also graph the approximation error for each case.

4.43. Consider the periodic pulse train shown in Fig. 4.15 on page 289. Its EFS coefficients were determined in Example 4.7 and given in Eqn. (4.77). Write a script to compute and graph the EFS line spectrum for duty cycle values $d = 0.4$, 0.6, 0.8 and 1. Comment on the changes in the spectrum as d is increased.

4.44. Refer to Problem 4.17 in which the CFS representation of a half-wave rectified sinusoidal signal was determined. Write a script to compute and graph finite-harmonic approximations to the signal using $m = 2$, 4 and 6 harmonics.

4.45. Refer to Problem 4.27. Write a script to compute and graph the transform of the modulated pulse

$$x(t) = \cos(2\pi f_0 t) \, \Pi\left(\frac{t}{\tau}\right)$$

In your script, define an anonymous function that takes two arguments, namely "f" and "tau", and returns the spectrum of a pulse with width τ centered around the time origin. Use the result with modulation property of the Fourier transform to compute the transform sought. Use the script to verify the results obtained in Problem 4.27.

4.46. Refer to Problem 4.35.

a. Write a script to compute and graph the finite-harmonic approximation to the signal $\tilde{x}(t)$ using the first m harmonics. Leave m as a parameter that can be adjusted.

b. Using the values of m found in parts (c), (d) and (e) of Problem 4.35 compute and graph the finite-harmonic approximations to the signal. Comment on the correlation between the percentage of power preserved and the quality of the finite-harmonic approximation.

MATLAB Projects

4.47. Consider the RC circuit shown in Fig. 4.87 on page 382. Its system function was determined in Example 4.43 to be

$$H(\omega) = \frac{1}{1 + j(\omega/\omega_c)}$$

Assume that the element values of the circuit are adjusted to obtain a critical frequency of $f_c = 80$ Hz or, equivalently, $\omega_c = 160\pi$ rad/s. Let the input signal to this circuit be a periodic pulse train with a period of $T_0 = 20$ milliseconds as shown in Fig. P.4.47.

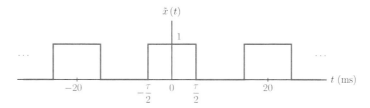

Figure P. 4.47

a. Determine the EFS coefficients of the signal $\tilde{x}(t)$ as a function of the duty cycle d. Write a script to graph the EFS spectrum of the signal for duty cycle values $d = 0.5, 0.3, 0.1$.

b. Let the pulse width be $\tau = 10$ ms corresponding to a duty cycle of $d = 0.5$. The signal $\tilde{x}(t)$ is used as input to the RC circuit. Write a script to compute the EFS coefficients of the output signal $\tilde{y}(t)$. Also graph the EFS spectrum for the output signal.

c. Write a script to construct and graph an approximation to the output signal $\tilde{y}(t)$ using the first 30 harmonics in the EFS spectrum.

d. An alternative to approximating the output of the RC circuit in response to a periodic signal is to determine its response to one isolated period, and then extend the result periodically. Let one period of the input signal be $f(t)$ so that $\tilde{x}(t)$ can be expressed as

$$\tilde{x}(t) = \sum_{r=-\infty}^{\infty} f(t - rT_0) = \sum_{r=-\infty}^{\infty} f(t - 0.02\,r)$$

For the input signal we are considering, $f(t)$ would be

$$f(t) = \Pi\left(\frac{t}{\tau}\right)$$

If the response of the RC circuit to one period $f(t)$ is

$$g(t) = \mathrm{Sys}\,\{f(t)\}$$

then the response to the periodic signal $\tilde{x}(t)$ is

$$\tilde{y}(t) = \sum_{r=-\infty}^{\infty} g(t - rT_0) = \sum_{r=-\infty}^{\infty} g(t - 0.02\,r)$$

For $d = 0.5$ manually determine $g(t)$, the response to a single pulse with $\tau = 10$ ms. Write a script to compute the response using this approach and graph it. In your script use an anonymous function to compute the values of $g(t)$. Approximate the summation for $\tilde{y}(t)$ using the terms for $r = -20, \ldots, 20$. Compare the result obtained through this approach with the one obtained in part (c). What are the differences, and how would you explain them?

e. Repeat parts (b) and (d) for duty cycle values $d = 0.3$ and 0.1.

4.48.   Problem 4.47 should illustrate that the signal at the output of the RC circuit is a distorted version of the input signal. It should also illustrate that the severity of the distortion is dependent on the pulse-width τ and consequently the duty cycle d.

The amount of distortion is also dependent on the shape of the input signal. In digital communication systems that rely on pulses for transmitting information, the fidelity of the output signal is important; the received pulses should be as similar to the transmitted pulses as possible. If alternative pulse shapes can provide better fidelity, they may be preferred over rectangular pulses.

In this problem we will explore the use of an alternative pulse train. Consider the signal shown in Fig. P.4.48 to be used as input in place of the signal of Problem 4.47 with $d = 0.5$.

Figure P. 4.48

A general expression for the EFS coefficients for the signal was obtained in Example 4.10. Repeat parts (a) through (c) of Problem 4.47 using this alternative signal. Comment on the results, especially in comparison to those obtained in Problem 4.47.

Chapter 5

Fourier Analysis for Discrete-Time Signals and Systems

Chapter Objectives

- Learn the use of discrete-time Fourier series for representing periodic signals using orthogonal basis functions.

- Learn the discrete-time Fourier transform (DTFT) for non-periodic signals as an extension of discrete-time Fourier series for periodic signals.

- Study properties of the DTFT. Understand energy and power spectral density concepts.

- Explore frequency-domain characteristics of DTLTI systems. Understand the system function concept.

- Learn the use of frequency-domain analysis methods for solving signal-system interaction problems with periodic and non-periodic input signals.

- Understand the fundamentals of the discrete Fourier transform (DFT) and fast Fourier transform (FFT). Learn how to compute linear convolution using the DFT.

5.1   Introduction

In Chapters 1 and 3 we have developed techniques for analyzing discrete-time signals and systems from a time-domain perspective. A discrete-time signal can be modeled as a function of the sample index. A DTLTI system can be represented by means of a constant-coefficient linear difference equation, or alternatively by means of an impulse response.

The output signal of a DTLTI system can be determined by solving the corresponding difference equation or by using the convolution operation.

In this chapter frequency-domain analysis methods are developed for discrete-time signals and systems. Section 5.2 focuses on analyzing periodic discrete-time signals in terms of their frequency content. Discrete-time Fourier series (DTFS) is presented as the counterpart of the exponential Fourier series (EFS) studied in Chapter 4 for continuous-time signals. Frequency-domain analysis methods for non-periodic discrete-time signals are presented in Section 5.3 through the use of the discrete-time Fourier transform (DTFT). Energy and power spectral density concepts for discrete-time signals are introduced in Section 5.4. Sections 5.5, 5.6 and 5.7 cover the application of frequency-domain analysis methods to the analysis of DTLTI systems. Section 5.8 introduces the discrete Fourier transform (DFT) and the fast Fourier transform (FFT).

5.2   Analysis of Periodic Discrete-Time Signals

In Section 4.2 of Chapter 4 we have focused on expressing continuous-time periodic signals as weighted sums of sinusoidal or complex exponential basis functions. It is also possible to express a discrete-time periodic signal in a similar manner, as a linear combination of discrete-time periodic basis functions. In this section we will explore one such periodic expansion referred to as the *discrete-time Fourier series (DTFS)*. In the process we will discover interesting similarities and differences between DTFS and its continuous-time counterpart, the exponential Fourier series (EFS) that was discussed in Section 4.2. One fundamental difference is regarding the number of series terms needed. We observed in Chapter 4 that a continuous-time periodic signal may have an infinite range of frequencies, and therefore may require an infinite number of harmonically related basis functions. In contrast, discrete-time periodic signals contain a finite range of angular frequencies, and will therefore require a finite number of harmonically related basis functions. We will see in later parts of this section that a discrete-time signal with a period of N samples will require at most N basis functions.

5.2.1   Discrete-Time Fourier Series (DTFS)

Consider a discrete-time signal $\tilde{x}[n]$ periodic with a period of N samples, that is, $\tilde{x}[n]=\tilde{x}[n+N]$ for all n. As in Chapter 4 periodic signals will be distinguished through the use of the tilde ($\tilde{\ }$) character over the name of the signal. We would like to explore the possibility of writing $\tilde{x}[n]$ as a linear combination of complex exponential basis functions in the form

$$\phi_k[n] = e^{j\Omega_k n} \tag{5.1}$$

using a series expansion

$$\tilde{x}[n] = \sum_k \tilde{c}_k\, \phi_k[n] = \sum_k \tilde{c}_k\, e^{j\Omega_k n} \tag{5.2}$$

Two important questions need to be answered:

  1. How should the angular frequencies Ω_k be chosen?
  2. How many basis functions are needed? In other words, what should be the limits of the summation index k in Eqn. (5.2)?

Intuitively, since the period of $\tilde{x}[n]$ is N samples long, the basis functions used in constructing the signal must also be periodic with N. Therefore we require

$$\phi_k[n + N] = \phi_k[n] \tag{5.3}$$

for all n. Substituting Eqn. (5.1) into Eqn. (5.3) leads to

$$e^{j\Omega_k(n+N)} = e^{j\Omega_k n} \tag{5.4}$$

For Eqn. (5.4) to be satisfied we need $e^{j\Omega_k N} = 1$, and consequently $\Omega_k N = 2\pi k$. The angular frequency of the basis function $\phi_k[n]$ must be

$$\Omega_k = \frac{2\pi k}{N} \tag{5.5}$$

leading to the set of basis functions

$$\phi_k[n] = e^{j(2\pi/N)kn} \tag{5.6}$$

Using $\phi_k[n]$ found, the series expansion of the signal $\tilde{x}[n]$ is in the form

$$\tilde{x}[n] = \sum_k \tilde{c}_k \, e^{j(2\pi/N)kn} \tag{5.7}$$

To address the second question, it can easily be shown that only the first N basis functions $\phi_0[n], \phi_1[n], \ldots, \phi_{N-1}[n]$ are unique; all other basis functions, i.e., the ones obtained for $k < 0$ or $k \geq N$, are duplicates of the basis functions in this set. To see why this is the case, let us write $\phi_{k+N}[n]$:

$$\phi_{k+N}[n] = e^{j(2\pi/N)(k+N)n} \tag{5.8}$$

Factoring Eqn. (5.8) into two exponential terms and realizing that $e^{j2\pi n} = 1$ for all integers n we obtain

$$\phi_{k+N}[n] = e^{j(2\pi/N)kn} \, e^{j2\pi n} = e^{j(2\pi/N)kn} = \phi_k[n] \tag{5.9}$$

Since $\phi_{k+N}[n]$ is equal to $\phi_k[n]$, we only need to include N terms in the summation of Eqn. (5.7).

The signal $\tilde{x}[n]$ can be constructed as

$$\tilde{x}[n] = \sum_{k=0}^{N-1} \tilde{c}_k \, e^{j(2\pi/N)kn} \tag{5.10}$$

As a specific example, if the period of the signal $\tilde{x}[n]$ is $N = 5$, then the only basis functions that are unique would be

$$\phi_k[n] \quad \text{for} \quad k = 0, 1, 2, 3, 4$$

Increasing the summation index k beyond $k = 4$ would not create any additional unique terms since $\phi_5[n] = \phi_0[n]$, $\phi_6[n] = \phi_1[n]$, and so on.

Eqn. (5.10) is referred to as the *discrete-time Fourier series (DTFS)* expansion of the periodic signal $\tilde{x}[n]$. The coefficients $\tilde{c}_k$ used in the summation of Eqn. (5.10) are the DTFS coefficients of the signal $\tilde{x}[n]$.

Example 5.1: **DTFS for a discrete-time sinusoidal signal**

Determine the DTFS representation of the signal $\tilde{x}[n] = \cos\left(\sqrt{2}\pi n\right)$.

Solution: The angular frequency of the signal is

$$\Omega_0 = \sqrt{2}\pi \ \ \text{rad}$$

and it corresponds to the normalized frequency

$$F_0 = \frac{\Omega_0}{2\pi} = \frac{1}{\sqrt{2}}$$

Since normalized frequency F_0 is an irrational number, the signal specified is not periodic (refer to the discussion in Section 1.4.4 of Chapter 1). Therefore it cannot be represented in series form using periodic basis functions. It can still be analyzed in the frequency domain, however, using the *discrete-time Fourier transform (DTFT)* which will be explored in Section 5.3.

Example 5.2: **DTFS for a discrete-time sinusoidal signal revisited**

Determine the DTFS representation of the signal $\tilde{x}[n] = \cos(0.2\pi n)$.

Solution: The angular frequency of the signal is

$$\Omega_0 = 0.2\pi \ \ \text{rad}$$

and the corresponding normalized frequency is

$$F_0 = \frac{\Omega_0}{2\pi} = \frac{1}{10}$$

Based on the normalized frequency, the signal is periodic with a period of $N = 1/F_0 = 10$ samples. A general formula for obtaining the DTFS coefficients will be derived later in this section. For the purpose of this example, however, we will take a shortcut afforded by the sinusoidal nature of the signal $\tilde{x}[n]$, and express it using Euler's formula:

$$\begin{aligned} \tilde{x}[n] =& \frac{1}{2}\, e^{j0.2\pi n} + \frac{1}{2}\, e^{-j0.2\pi n} \\ =& \frac{1}{2}\, e^{j(2\pi/10)n} + \frac{1}{2}\, e^{-j(2\pi/10)n} \end{aligned} \tag{5.11}$$

The two complex exponential terms in Eqn. (5.11) correspond to $\phi_1[n]$ and $\phi_{-1}[n]$, therefore their coefficients must be $\tilde{c}_1$ and $\tilde{c}_{-1}$, respectively. As a result we have

$$\tilde{c}_1 = \frac{1}{2}, \quad \text{and} \quad \tilde{c}_{-1} = \frac{1}{2} \tag{5.12}$$

As discussed in the previous section we would like to see the series coefficients $\tilde{c}_k$ in the index range $k = 0, \ldots, N-1$ where N is the period. In this case we need to obtain $\tilde{c}_k$ for $k = 0, \ldots, 9$. The basis functions have the property

$$\phi_k[n] = \phi_{k+N}[n]$$

The term $\phi_{-1}[n]$ in Eqn. (5.11) can be written as

$$\begin{aligned} \phi_{-1}[n] =& e^{-j(2\pi/10)n} \\ =& e^{-j(2\pi/10)n}\, e^{j2\pi n} \\ =& e^{j(18\pi/10)n} = \phi_9[n] \end{aligned} \tag{5.13}$$

Eqn. (5.11) becomes

$$\tilde{x}[n] = \cos(0.2\pi n)$$
$$= \frac{1}{2} e^{j(2\pi/10)n} + \frac{1}{2} e^{j(18\pi/10)n}$$
$$= \tilde{c}_1 e^{j(2\pi/10)n} + \tilde{c}_9 e^{j(18\pi/10)n}$$

DTFS coefficients are

$$\tilde{c}_k = \begin{cases} \dfrac{1}{2}, & k = 1 \text{ or } k = 9 \\ 0, & \text{otherwise} \end{cases}$$

The signal $\tilde{x}[n]$ and its DTFS spectrum $\tilde{c}_k$ are shown in Fig. 5.1.

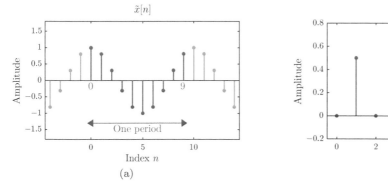

Figure 5.1 – (a) The signal $\tilde{x}[n]$ for Example 5.2, and (b) its DTFS spectrum.

Software resources:
ex_5_2.m

Example 5.3: **DTFS for a multi-tone signal**

Determine the DTFS coefficients for the signal

$$\tilde{x}[n] = 1 + \cos(0.2\pi n) + 2\sin(0.3\pi n)$$

Solution: The two angular frequencies $\Omega_1 = 0.2\pi$ and $\Omega_2 = 0.3\pi$ radians correspond to normalized frequencies $F_1 = 0.1$ and $F_2 = 0.15$ respectively. The normalized fundamental frequency of $\tilde{x}[n]$ is $F_0 = 0.05$, and it corresponds to a period of $N = 20$ samples. Using this value of N, the angular frequencies of the two sinusoidal terms of $\tilde{x}[n]$ are

$$\Omega_1 = 2\Omega_0 = \frac{2\pi}{20}(2)$$

and

$$\Omega_2 = 3\Omega_0 = \frac{2\pi}{20}(3)$$

We will use Euler's formula to write $\tilde{x}[n]$ in the form

$$\tilde{x}[n] = 1 + \frac{1}{2} e^{j(2\pi/20)2n} + \frac{1}{2} e^{-j(2\pi/20)2n} + \frac{1}{j} e^{j(2\pi/20)3n} - \frac{1}{j} e^{-j(2\pi/20)3n}$$

$$= 1 + \frac{1}{2} \phi_2[n] + \frac{1}{2} \phi_{-2}[n] + \frac{1}{j} \phi_3[n] - \frac{1}{j} \phi_{-3}[n]$$

$$= 1 + \frac{1}{2} \phi_2[n] + \frac{1}{2} \phi_{-2}[n] + e^{-j\pi/2} \phi_3[n] - e^{j\pi/2} \phi_{-3}[n]$$

The DTFS coefficients are

$$\tilde{c}_0 = 1$$

$$\tilde{c}_2 = \frac{1}{2}, \qquad \tilde{c}_{-2} = \frac{1}{2}$$

$$\tilde{c}_3 = e^{-j\pi/2}, \qquad \tilde{c}_{-3} = e^{j\pi/2}$$

We know from the periodicity of the DTFS representation that

$$\tilde{c}_{-2} = \tilde{c}_{18} \qquad \text{and} \qquad \tilde{c}_{-3} = \tilde{c}_{17} \tag{5.14}$$

The signal $\tilde{x}[n]$ and its DTFS spectrum $\tilde{c}_k$ are shown in Fig. 5.2.

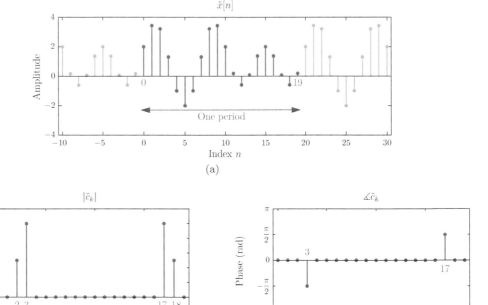

Figure 5.2 – (a) The signal $\tilde{x}[n]$ for Example 5.3, (b) magnitude of the DTFS spectrum, and (c) phase of the DTFS spectrum.

Software resources:
ex_5_3.m

Finding DTFS coefficients

In Examples 5.2 and 5.3 the DTFS coefficients for discrete-time sinusoidal signals were easily determined since the use of Euler's formula gave us the ability to express those signals in a form very similar to the DTFS expansion given by Eqn. (5.7). In order to determine the DTFS coefficients for a general discrete-time signal we will take advantage of the orthogonality of the basis function set $\{\phi_k[n], \; k = 0, \ldots, N-1\}$. It can be shown that (see Appendix D)

$$\sum_{n=0}^{N-1} e^{j(2\pi/N)(m-k)n} = \begin{cases} N, & (m-k) = rN, \; r \text{ integer} \\ 0, & \text{otherwise} \end{cases} \tag{5.15}$$

To derive the expression for the DTFS coefficients, let us first write Eqn. (5.10) using m instead of k as the summation index:

$$\tilde{x}[n] = \sum_{m=0}^{N-1} \tilde{c}_m \, e^{j(2\pi/N)mn} \tag{5.16}$$

Multiplication of both sides of Eqn. (5.7) by $e^{-j(2\pi/N)kn}$ leads to

$$\tilde{x}[n] \, e^{-j(2\pi/N)kn} = \sum_{m=0}^{N-1} \tilde{c}_m \, e^{j(2\pi/N)mn} \, e^{-j(2\pi/N)kn} \tag{5.17}$$

Summing the terms on both sides of Eqn. (5.17) for $n = 0, \ldots, N-1$, rearranging the summations, and using the orthogonality property yields

$$\sum_{n=0}^{N-1} \tilde{x}[n] \, e^{-j(2\pi/N)kn} = \sum_{n=0}^{N-1} \sum_{m=0}^{N-1} \tilde{c}_m \, e^{j(2\pi/N)mn} \, e^{-j(2\pi/N)kn}$$

$$= \sum_{m=0}^{N-1} \tilde{c}_m \sum_{n=0}^{N-1} e^{j(2\pi/N)(m-k)n}$$

$$= \tilde{c}_k N \tag{5.18}$$

The DTFS coefficients are computed from Eqn. (5.18) as

$$\tilde{c}_k = \frac{1}{N} \sum_{n=0}^{N-1} \tilde{x}[n] \, e^{-j(2\pi/N)kn} \, , \qquad k = 0, \ldots, N-1 \tag{5.19}$$

In Eqn. (5.19) the DTFS coefficients are computed for the index range $k = 0, \ldots, N-1$ since those are the only coefficient indices that are needed in the DTFS expansion in Eqn. (5.10). If we were to use Eqn. (5.19) outside the specified index range we would discover that

$$\tilde{c}_{k+rN} = \tilde{c}_k \tag{5.20}$$

for all integers r. The DTFS coefficients evaluated outside the range $k = 0, \ldots, N-1$ exhibit periodic behavior with period N. This was evident in Examples 5.2 and 5.3 as well. The development so far can be summarized as follows:

Discrete-Time Fourier Series (DTFS):

1. Synthesis equation:

$$\tilde{x}[n] = \sum_{k=0}^{N-1} \tilde{c}_k \, e^{j(2\pi/N)kn} \, , \qquad \text{all } n \tag{5.21}$$

2. Analysis equation:

$$\tilde{c}_k = \frac{1}{N} \sum_{n=0}^{N-1} \tilde{x}[n] \, e^{-j(2\pi/N)kn} \tag{5.22}$$

Note that the coefficients $\tilde{c}_k$ are computed for all indices k in Eqn. (5.22), however, only the ones in the range $k = 0, \ldots, N - 1$ are needed in constructing the signal $\tilde{x}[n]$ in Eqn. (5.21). Due to the periodic nature of the DTFS coefficients $\tilde{c}_k$, the summation in the synthesis equation can be started at any arbitrary index, provided that the summation includes exactly N terms. In other words, Eqn. (5.21) can be written in the alternative form

$$\tilde{x}[n] = \sum_{k=N_0}^{N_0+N-1} \tilde{c}_k \, e^{j(2\pi/N)kn} \,, \qquad \text{all } n \tag{5.23}$$

Example 5.4: **Finding DTFS representation**

Consider the periodic signal $\tilde{x}[n]$ defined as

$$\tilde{x}[n] = n \quad \text{for } n = 0, 1, 2, 3, 4 \quad \text{and} \quad \tilde{x}[n + 5] = \tilde{x}[n]$$

and shown in Fig. 5.3. Determine the DTFS coefficients for $\tilde{x}[n]$. Afterwards, verify the synthesis equation in Eqn. (5.21).

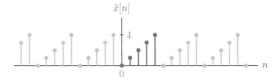

Figure 5.3 – The signal $\tilde{x}[n]$ for Example 5.4.

Solution: Using the analysis equation given by Eqn. (5.22) the DTFS coefficients are

$$\begin{aligned}
\tilde{c}_k &= \frac{1}{5} \sum_{n=0}^{4} \tilde{x}[n] \, e^{-j(2\pi/5)kn} \\
&= \frac{1}{5} e^{-j2\pi k/5} + \frac{2}{5} e^{-j4\pi k/5} + \frac{3}{5} e^{-j6\pi k/5} + \frac{4}{5} e^{-j8\pi k/5}
\end{aligned} \tag{5.24}$$

Evaluating Eqn. (5.24) for $k = 0, \ldots, 4$ we get

$$\begin{aligned}
\tilde{c}_0 &= 2 \\
\tilde{c}_1 &= -0.5 + j\,0.6882 \\
\tilde{c}_2 &= -0.5 + j\,0.1625 \\
\tilde{c}_3 &= -0.5 - j\,0.1625 \\
\tilde{c}_4 &= -0.5 - j\,0.6882
\end{aligned}$$

The signal $\tilde{x}[n]$ can be constructed from DTFS coefficients as

$$\begin{aligned}
\tilde{x}[n] &= \tilde{c}_0 + \tilde{c}_1 \, e^{j2\pi n/5} + \tilde{c}_2 \, e^{j4\pi n/5} + \tilde{c}_3 \, e^{j6\pi n/5} + \tilde{c}_4 \, c^{j8\pi n/5} \\
&= 2 + (-0.5 + j\,0.6882) \, e^{j2\pi n/5} + (-0.5 + j\,0.1625) \, e^{j4\pi n/5} \\
&\quad + (-0.5 - j\,0.1625) \, e^{j6\pi n/5} + (-0.5 - j\,0.6882) \, e^{j8\pi n/5}
\end{aligned}$$

Software resources:

`ex_5_4a.m`

`ex_5_4b.m`

Example 5.5: **DTFS for discrete-time pulse train**

Consider the periodic pulse train $\tilde{x}[n]$ defined by

$$\tilde{x}[n] = \begin{cases} 1, & -L \leq n \leq L \\ 0, & L < n < N - L \end{cases} \quad \text{and} \quad \tilde{x}[n + N] = \tilde{x}[n]$$

where $N > 2L + 1$ as shown in Fig. 5.4. Determine the DTFS coefficients of the signal $\tilde{x}[n]$ in terms of L and N.

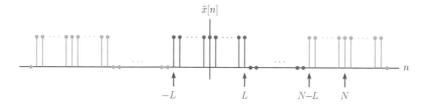

Figure 5.4 – The signal $\tilde{x}[n]$ for Example 5.5.

Solution: Using Eqn. (5.22) the DTFS coefficients are

$$\tilde{c}_k = \frac{1}{N} \sum_{n=-L}^{L} (1) \, e^{-j(2\pi/N)kn}$$

The closed form expression for a finite-length geometric series is (see Appendix C for derivation)

$$\text{Eqn. (C.13):} \qquad \sum_{n=L_1}^{L_2} a^n = \frac{a^{L_1} - a^{L_2+1}}{1 - a}$$

Using Eqn. (C.13) with $a = e^{-j(2\pi/N)k}$, $L_1 = -L$ and $L_2 = L$, the closed form expression for $\tilde{c}_k$ is

$$\tilde{c}_k = \frac{1}{N} \frac{e^{j(2\pi/N)Lk} - e^{-j(2\pi/N)(L+1)k}}{1 - e^{-j(2\pi/N)k}} \tag{5.25}$$

In order to get symmetric complex exponentials in Eqn. (5.25) we will multiply both the numerator and the denominator of the fraction on the right side of the equal sign with $e^{j\pi k/N}$ resulting in

$$\tilde{c}_k = \frac{1}{N} \frac{e^{j(2\pi/N)(L+1/2)k} - e^{-j(2\pi/N)(L+1/2)k}}{e^{j(2\pi/N)(k/2)} - e^{-j(2\pi/N)(k/2)}} \tag{5.26}$$

which, using Euler's formula, can be simplified to

$$\tilde{c}_k = \frac{\sin\left(\dfrac{\pi k}{N}(2L+1)\right)}{N \sin\left(\dfrac{\pi k}{N}\right)}, \quad k = 0, \ldots, N-1$$

The coefficient $\tilde{c}_0$ needs special attention. Using L'Hospital's rule we obtain

$$\tilde{c}_0 = \frac{2L+1}{N}$$

The DTFS representation of the signal $\tilde{x}[n]$ is

$$\tilde{x}[n] = \sum_{k=0}^{N-1} \left[\frac{\sin\left(\dfrac{\pi k}{N}\,(2L+1)\right)}{N\sin\left(\dfrac{\pi k}{N}\right)} \right] e^{j(2\pi/N)kn}$$

The DTFS coefficients are graphed in Fig. 5.5 for $N=40$ and $L=3,5,$ and 7.

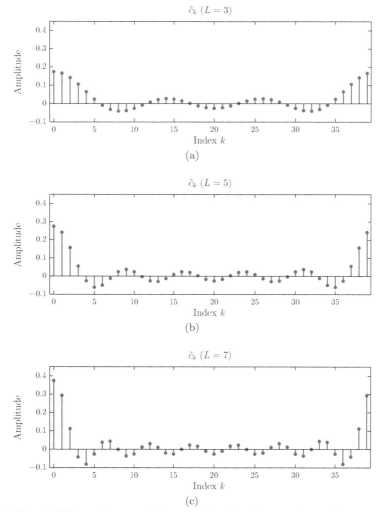

Figure 5.5 – The DTFS coefficients of the signal $\tilde{x}[n]$ of Example 5.5 for $N=40$ and (a) $L=3$, (b) $L=5$, and (c) $L=7$.

Software resources:
ex_5_5.m

Interactive Demo: dtfs_demo1

The demo program in "dtfs_demo1.m" provides a graphical user interface for computing the DTFS representation of the periodic discrete-time pulse train of Example 5.5. The period is fixed at $N = 40$ samples. The parameter L can be varied from 0 to 19. The signal $\tilde{x}[n]$ and its DTFS coefficients $\tilde{c}_k$ are displayed.

1. Set $L = 0$. Observe the coefficients and comment.
2. Increment L to $2, 3, 4, \ldots$ and observe the changes to DTFS coefficients. Pay attention to the outline (or the envelope) of the coefficients. Can you identify a pattern that emerges as L is incremented.

Software resources:
dtfs_demo1.m

Software resources: See MATLAB Exercises 5.1 and 5.2.

5.2.2   Properties of the DTFS

In this section we will summarize a few important properties of the DTFS representation of a periodic signal. To keep the notation compact, we will denote the relationship between the periodic signal $\tilde{x}[n]$ and its DTFS coefficients $\tilde{c}_k$ as

$$\tilde{x}[n] \overset{\text{DTFS}}{\longleftrightarrow} \tilde{c}_k \tag{5.27}$$

Periodicity
DTFS coefficients are periodic with the same period N as the signal $\tilde{x}[n]$.

Given
$$\tilde{x}[n] = \tilde{x}[n + rN] \,, \quad \text{all integer } r$$

it can be shown that

$$\tilde{c}_k = \tilde{c}_{k+rN} \,, \quad \text{all integer } r \tag{5.28}$$

Periodicity of DTFS coefficients follows easily from the analysis equation given by Eqn. (5.22).

Linearity

Let $\tilde{x}_1[n]$ and $\tilde{x}_2[n]$ be two signals, both periodic with the same period, and with DTFS representations
$$\tilde{x}_1[n] \overset{\text{DTFS}}{\longleftrightarrow} \tilde{c}_k \quad \text{and} \quad \tilde{x}_2[n] \overset{\text{DTFS}}{\longleftrightarrow} \tilde{d}_k$$

It can be shown that
$$\alpha_1 \, \tilde{x}_1[n] + \alpha_2 \, \tilde{x}_2[n] \overset{\text{DTFS}}{\longleftrightarrow} \alpha_1 \, \tilde{c}_k + \alpha_2 \, \tilde{d}_k \tag{5.29}$$

for any two arbitrary constants α_1 and α_2.

Linearity property is easily proven starting with the synthesis equation in Eqn. (5.21).

Time shifting

Given that
$$\tilde{x}[n] \overset{\text{DTFS}}{\longleftrightarrow} \tilde{c}_k$$

it can be shown that
$$\tilde{x}[n-m] \overset{\text{DTFS}}{\longleftrightarrow} e^{-j(2\pi/N)km} \tilde{c}_k \qquad (5.30)$$

Time shifting the signal $\tilde{x}[n]$ causes the DTFS coefficients to be multiplied by a complex exponential function.

Consistency check: Let the signal be time shifted by exactly one period, that is, $m = N$. We know that $\tilde{x}[n - N] = \tilde{x}[n]$ due to the periodicity of $\tilde{x}[n]$. The exponential function on the right side of Eqn. (5.30) would be $e^{-j(2\pi/N)kN} = 1$, and the DTFS coefficients remain unchanged, as expected.

Symmetry of DTFS coefficients

Conjugate symmetry and conjugate antisymmetry properties were defined for discrete-time signals in Section 1.4.6 of Chapter 1. Same definitions apply to DTFS coefficients as well. A *conjugate symmetric* set of coefficients satisfy

$$\tilde{c}_k^* = \tilde{c}_{-k} \qquad (5.31)$$

for all k. Similarly, the coefficients form a *conjugate antisymmetric* set if they satisfy

$$\tilde{c}_k^* = -\tilde{c}_{-k} \qquad (5.32)$$

for all k. For a signal $\tilde{x}[n]$ which is periodic with N samples it is customary to use the DTFS coefficients in the index range $k = 0, \ldots, N - 1$. The definitions in Eqns. (5.31) and (5.32) can be adjusted in terms of their indices using the periodicity of the DTFS coefficients. Since $\tilde{c}_{-k} = \tilde{c}_{N-k}$, a conjugate symmetric set of DTFS coefficients have the property

$$\tilde{c}_k^* = \tilde{c}_{N-k}, \qquad k = 0, \ldots, N - 1 \qquad (5.33)$$

Similarly, a conjugate antisymmetric set of DTFS coefficients have the property

$$\tilde{c}_k^* = -\tilde{c}_{N-k}, \qquad k = 0, \ldots, N - 1 \qquad (5.34)$$

If the signal $\tilde{x}[n]$ is real-valued, it can be shown that its DTFS coefficients form a conjugate symmetric set. Conversely, if the signal $\tilde{x}[n]$ is purely imaginary, its DTFS coefficients form a conjugate antisymmetric set.

Symmetry of DTFS coefficients:

$$\tilde{x}[n]: \text{Real}, \quad \text{Im}\{\tilde{x}[n]\} = 0 \quad \text{implies that} \quad \tilde{c}_k^* = \tilde{c}_{N-k} \qquad (5.35)$$

$$\tilde{x}[n]: \text{Imag}, \quad \text{Re}\{\tilde{x}[n]\} = 0 \quad \text{implies that} \quad \tilde{c}_k^* = -\tilde{c}_{N-k} \qquad (5.36)$$

Polar form of DTFS coefficients

DTFS coefficients can be written in polar form as

$$\tilde{c}_k = |\tilde{c}_k|\, e^{j\tilde{\theta}_k} \qquad (5.37)$$

If the set $\{\tilde{c}_k\}$ is conjugate symmetric, the relationship in Eqn. (5.33) leads to

$$|\tilde{c}_k| \, e^{-j\tilde{\theta}_k} = |\tilde{c}_{N-k}| \, e^{j\tilde{\theta}_{N-k}} \tag{5.38}$$

using the polar form of the coefficients. The consequences of Eqn. (5.38) are obtained by equating the magnitudes and the phases on both sides.

Conjugate symmetric coefficients: $\quad \tilde{c}_k^* = \tilde{c}_{N-k}$

$$\text{Magnitude:} \quad |\tilde{c}_k| = |\tilde{c}_{N-k}| = |\tilde{c}_{-k}| \tag{5.39}$$

$$\text{Phase:} \quad \tilde{\theta}_k = -\tilde{\theta}_{N-k} = -\tilde{\theta}_{-k} \tag{5.40}$$

It was established in Eqn. (5.35) that the DTFS coefficients of a real-valued $\tilde{x}[n]$ are conjugate symmetric. Based on the results in Eqns. (5.39) and (5.40) the magnitude spectrum is an even function of k, and the phase spectrum is an odd function of k.

Similarly, if the set $\{\tilde{c}_k\}$ is conjugate antisymmetric, the relationship in Eqn. (5.34) reflects on polar form of $\tilde{c}_k$ as

$$|\tilde{c}_k| \, e^{-j\tilde{\theta}_k} = -|\tilde{c}_{N-k}| \, e^{j\tilde{\theta}_{N-k}} \tag{5.41}$$

The negative sign on the right side of Eqn. (5.41) needs to be incorporated into the phase since we could not write $|\tilde{c}_k| = -|\tilde{c}_{N-k}|$ (recall that magnitude must to be non-negative). Using $e^{\mp j\pi} = -1$, Eqn. (5.41) becomes

$$|\tilde{c}_k| \, e^{-j\tilde{\theta}_k} = |\tilde{c}_{N-k}| \, e^{j\tilde{\theta}_{N-k}} \, e^{\mp j\pi}$$

$$= |\tilde{c}_{N-k}| \, e^{j(\tilde{\theta}_{N-k} \mp \pi)} \tag{5.42}$$

The consequences of Eqn. (5.42) are summarized below.

Conjugate antisymmetric coefficients: $\quad \tilde{c}_k^* = -\tilde{c}_{N-k}$

$$\text{Magnitude:} \quad |\tilde{c}_k| = |\tilde{c}_{N-k}| = |\tilde{c}_{-k}| \tag{5.43}$$

$$\text{Phase:} \quad \tilde{\theta}_k = -\tilde{\theta}_{N-k} \pm \pi = -\tilde{\theta}_{-k} \pm \pi \tag{5.44}$$

A purely imaginary signal $\tilde{x}[n]$ leads to a set of DTFS coefficients with conjugate antisymmetry. The corresponding magnitude spectrum is an even function of k as suggested by Eqn. (5.43). The phase spectrum is neither even nor odd.

Example 5.6: **Symmetry of DTFS coefficients**

Recall the real-valued periodic signal $\tilde{x}[n]$ of Example 5.4 shown in Fig. 5.3. Its DTFS coefficients were found as

$$\tilde{c}_0 = 2$$
$$\tilde{c}_1 = -0.5 + j\,0.6882$$
$$\tilde{c}_2 = -0.5 + j\,0.1625$$
$$\tilde{c}_3 = -0.5 - j\,0.1625$$
$$\tilde{c}_4 = -0.5 - j\,0.6882$$

It can easily be verified that coefficients form a conjugate symmetric set. With $N = 5$ we have

$$
\begin{aligned}
k &= 1 \quad \implies \quad \tilde{c}_1^* = \tilde{c}_4 \\
k &= 2 \quad \implies \quad \tilde{c}_2^* = \tilde{c}_3
\end{aligned}
$$

DTFS spectra of even and odd signals

If the real-valued signal $\tilde{x}[n]$ is an even function of index n, the resulting DTFS spectrum $\tilde{c}_k$ is real-valued for all k.

$$
\tilde{x}[n] = \tilde{x}[n] \,, \text{all } n \quad \text{implies that} \quad \operatorname{Im}\{\tilde{c}_k\} = 0 \,, \text{all } k \tag{5.45}
$$

Conversely it can also be proven that, if the real-valued signal $\tilde{x}[n]$ has odd-symmetry, the resulting DTFS spectrum is purely imaginary.

$$
\tilde{x}[n] = -\tilde{x}[n] \,, \text{all } n \quad \text{implies that} \quad \operatorname{Re}\{\tilde{c}_k\} = 0 \,, \text{all } k \tag{5.46}
$$

Example 5.7: **DTFS symmetry for periodic waveform**

Explore the symmetry properties of the periodic waveform $\tilde{x}[n]$ shown in Fig. 5.6. One period of $\tilde{x}[n]$ has the sample amplitudes

$$
\tilde{x}[n] = \{ \ldots, \; \underset{\substack{\uparrow \\ n=0}}{0} \;, \tfrac{1}{2}, 1, \tfrac{3}{4}, 0, -\tfrac{3}{4}, -1, -\tfrac{1}{2}, \ldots \}
$$

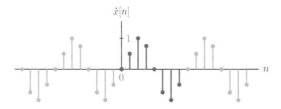

Figure 5.6 – The signal $\tilde{x}[n]$ for Example 5.7.

Solution: The DTFS coefficients for $\tilde{x}[n]$ are computed as

$$
\tilde{c}_k = \sum_{n=0}^{7} \tilde{x}[n] \, e^{-j(2\pi/8)kn}
$$

and are listed for $k = 0, \ldots, 8$ in Table 5.1 along with magnitudes and phase values for each. Symmetry properties can be easily observed. The signal is real-valued, therefore the DTFS spectrum is conjugate symmetric:

$$
\begin{aligned}
k &= 1 \quad \implies \quad \tilde{c}_1^* = \tilde{c}_7 \\
k &= 2 \quad \implies \quad \tilde{c}_2^* = \tilde{c}_6 \\
k &= 2 \quad \implies \quad \tilde{c}_3^* = \tilde{c}_5
\end{aligned}
$$

k	$\tilde{c}_k$	$|\tilde{c}_k|$	$\tilde{\theta}_k$
0	0.0	0.0	N/A
1	j0.4710	0.4710	$\pi/2$
2	$-$j0.0625	0.0625	$-\pi/2$
3	$-$j0.0290	0.0290	$-\pi/2$
4	0.0	0.0	N/A
5	j0.0290	0.0290	$\pi/2$
6	j0.0625	0.0625	$\pi/2$
7	$-$j0.4710	0.4710	$-\pi/2$

Table 5.1 – DTFS coefficients for the pulse train of Example 5.7.

Furthermore, the odd symmetry of $\tilde{x}[n]$ causes coefficients to be purely imaginary:

$$\operatorname{Re}\{\tilde{c}_k\} = 0 , \qquad k = 0,\ldots,7$$

In terms of the magnitude values we have

$$k = 1 \quad \Longrightarrow \quad |\tilde{c}_1| = |\tilde{c}_7|$$
$$k = 2 \quad \Longrightarrow \quad |\tilde{c}_2| = |\tilde{c}_6|$$
$$k = 2 \quad \Longrightarrow \quad |\tilde{c}_3| = |\tilde{c}_5|$$

For the phase angles the following relationships hold:

$$k = 1 \quad \Longrightarrow \quad \tilde{\theta}_1 = -\tilde{\theta}_7$$
$$k = 2 \quad \Longrightarrow \quad \tilde{\theta}_2 = -\tilde{\theta}_6$$
$$k = 2 \quad \Longrightarrow \quad \tilde{\theta}_3 = -\tilde{\theta}_5$$

The phase values for $\tilde{\theta}_0$ and $\tilde{\theta}_4$ are insignificant since the corresponding magnitude values $|\tilde{c}_0|$ and $|\tilde{c}_4|$ are equal to zero.
Software resources:
ex_5_7.m

Periodic convolution

Consider the convolution of two discrete-time signals defined by Eqn. (3.128) and repeated here for convenience:

$$\text{Eqn. (3.128):} \qquad y[n] = x[n] * h[n] = \sum_{k=-\infty}^{\infty} x[k]\,h[n-k]$$

This summation would obviously fail to converge if both signals $x[n]$ and $h[n]$ happen to be periodic with periods of N. For such a case, a *periodic convolution* operator can be defined as

$$\tilde{y}[n] = \tilde{x}[n] \otimes \tilde{h}[n] = \sum_{k=0}^{N-1} \tilde{x}[k]\,\tilde{h}[n-k] , \quad \text{all } n \tag{5.47}$$

Eqn. (5.47) is essentially an adaptation of the convolution sum to periodic signals where the limits of the summation are modified to cover only one period (we are assuming that both

$\tilde{x}[n]$ and $\tilde{h}[n]$ have the same period N). It can be shown that (see Problem 5.7) the periodic convolution of two signals $\tilde{x}[n]$ and $\tilde{h}[n]$ that are both periodic with N is also periodic with the same period.

Example 5.8: **Periodic convolution**

Two signals $\tilde{x}[n]$ and $\tilde{h}[n]$, each periodic with $N = 5$ samples, are shown in Fig. 5.7(a) and (b). Determine the periodic convolution

$$\tilde{y}[n] = \tilde{x}[n] \otimes \tilde{h}[n] \tag{5.48}$$

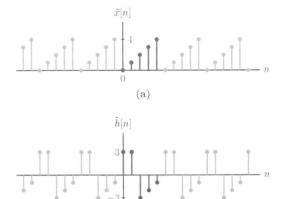

(a)

(b)

Figure 5.7 – Signals $\tilde{x}[n]$ and $\tilde{h}[n]$ for Example 5.4.

Solution: Sample amplitudes for one period are

$$\tilde{x}[n] = \{ \ \dots, \ 0, \ 1, \ 2, \ 3, \ 4, \ \dots \ \}$$
$$\underset{n=0}{\uparrow}$$

and

$$\tilde{h}[n] = \{ \ \dots, \ 3, \ 3, \ -3, \ -2, \ -1, \ \dots \ \}$$
$$\underset{n=0}{\uparrow}$$

The periodic convolution is given by

$$\tilde{y}[n] = \sum_{k=0}^{4} \tilde{x}[k] \, \tilde{h}[n-k]$$

To start, let $n = 0$. The terms $\tilde{x}[k]$ and $\tilde{h}[0-k]$ are shown in Fig. 5.8. The main period of each signal is indicated with sample amplitudes colored blue. The shaded area contains the terms included in the summation for periodic convolution. The sample $\tilde{y}[0]$ is computed as

$$\tilde{y}[0] = \tilde{x}[0] \, \tilde{h}[0] + \tilde{x}[1] \, \tilde{h}[4] + \tilde{x}[2] \, \tilde{h}[3] + \tilde{x}[3] \, \tilde{h}[2] + \tilde{x}[4] \, \tilde{h}[1]$$
$$= (0)\,(3) + (1)\,(-1) + (2)\,(-3) + (3)\,(-3) + (4)\,(3) = -2$$

Next we will set $n = 1$. The terms $\tilde{x}[k]$ and $\tilde{h}[1-k]$ are shown in Fig. 5.9.

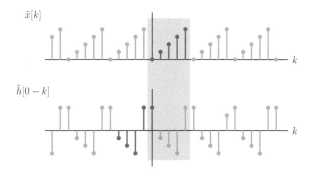

Figure 5.8 – The terms $\tilde{x}[k]$ and $\tilde{h}[0-k]$ in the convolution sum.

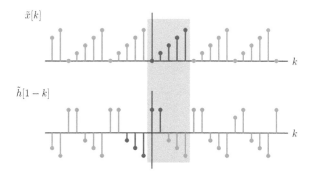

Figure 5.9 – The terms $\tilde{x}[k]$ and $\tilde{h}[1-k]$ in the convolution sum.

The sample $\tilde{y}[1]$ is computed as

$$\begin{aligned}
\tilde{y}[1] =& \tilde{x}[0]\,\tilde{h}[1] + \tilde{x}[1]\,\tilde{h}[0] + \tilde{x}[2]\,\tilde{h}[4] + \tilde{x}[3]\,\tilde{h}[3] + \tilde{x}[4]\,\tilde{h}[2] \\
=& (0)\,(3) + (1)\,(3) + (2)\,(-1) + (3)\,(-2) + (4)\,(-3) = -17
\end{aligned}$$

Finally, for $n = 2$ the terms involved in the summation are shown in Fig. 5.10.

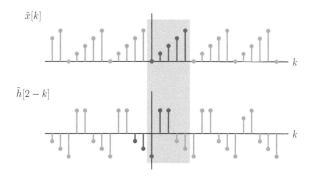

Figure 5.10 – The terms $\tilde{x}[k]$ and $\tilde{h}[2-k]$ in the convolution sum.

The sample $\tilde{y}[2]$ is computed as

$$\begin{aligned}
\tilde{y}[2] =& \tilde{x}[0]\,\tilde{h}[2] + \tilde{x}[1]\,\tilde{h}[1] + \tilde{x}[2]\,\tilde{h}[0] + \tilde{x}[3]\,\tilde{h}[4] + \tilde{x}[4]\,\tilde{h}[3] \\
=& (0)\,(-3) + (1)\,(3) + (2)\,(3) + (3)\,(-1) + (4)\,(-2) = -2
\end{aligned}$$

Continuing in this fashion, it can be shown that $\tilde{y}[3] = 8$ and $\tilde{y}[4] = 13$. Thus, one period of the signal $\tilde{y}[n]$ is

$$\tilde{y}[n] = \{ \ \ldots, -2, -17, -2, 8, 13, \ldots \ \}$$
$$\underset{n=0}{\uparrow}$$

The signal $\tilde{y}[n]$ is shown in Fig. 5.11.

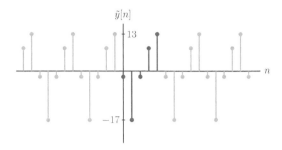

Figure 5.11 – Periodic convolution result $\tilde{y}[n]$ for Example 5.8.

The periodic convolution property of the discrete-time Fourier series can be stated as follows:

Let $\tilde{x}[n]$ and $\tilde{h}[n]$ be two signals both periodic with the same period, and with DTFS representations

$$\tilde{x}[n] \overset{\text{DTFS}}{\longleftrightarrow} \tilde{c}_k \qquad \text{and} \qquad \tilde{h}[n] \overset{\text{DTFS}}{\longleftrightarrow} \tilde{d}_k$$

It can be shown that

$$\tilde{x}[n] \otimes \tilde{h}[n] \overset{\text{DTFS}}{\longleftrightarrow} N\,\tilde{c}_k\,\tilde{d}_k \tag{5.49}$$

The DTFS coefficients of the periodic convolution result are equal to N times the product of the DTFS coefficients of the two signals.

Let $\tilde{y}[n] = \tilde{x}[n] \otimes \tilde{h}[n]$, and let the DTFS coefficients of $\tilde{y}[n]$ be $\tilde{e}_k$. Using the DTFS analysis equation, we can write

$$\tilde{e}_k = \frac{1}{N} \sum_{n=0}^{N-1} \tilde{y}[n]\, e^{-j(2\pi/N)kn} \tag{5.50}$$

Substituting

$$\tilde{y}[n] = \sum_{m=0}^{N-1} \tilde{x}[m]\, \tilde{h}[n-m] \tag{5.51}$$

into Eqn. (5.50) we obtain

$$\tilde{e}_k = \frac{1}{N} \sum_{n=0}^{N-1} \left[\sum_{m=0}^{N-1} \tilde{x}[m]\, \tilde{h}[n-m] \right] e^{-j(2\pi/N)kn} \tag{5.52}$$

Changing the order of the two summations and rearranging terms leads to

$$\tilde{e}_k = \sum_{m=0}^{N-1} \tilde{x}[m] \left[\frac{1}{N} \sum_{n=0}^{N-1} \tilde{h}[n-m]\, e^{-j(2\pi/N)kn} \right] \tag{5.53}$$

In Eqn. (5.53) the term in square brackets represents the DTFS coefficients for the time shifted periodic signal $\tilde{h}[n-m]$, and is evaluated as

$$\frac{1}{N} \sum_{n=0}^{N-1} \tilde{h}[n-m] \, e^{-j(2\pi/N)kn} = e^{-j(2\pi/N)km} \, \tilde{d}_k \tag{5.54}$$

Using this result Eqn. (5.53) becomes

$$\tilde{e}_k = \tilde{d}_k \sum_{m=0}^{N-1} \tilde{x}[m] \, e^{-j(2\pi/N)km} = N \, \tilde{c}_k \, \tilde{d}_k \tag{5.55}$$

completing the proof.

Example 5.9: **Periodic convolution**

Refer to the signals $\tilde{x}[n]$, $\tilde{h}[n]$ and $\tilde{y}[n]$ of Example 5.8. The DTFS coefficients of $\tilde{x}[n]$ were determined in Example 5.4. Find the DTFS coefficients of $\tilde{h}[n]$ and $\tilde{y}[n]$. Afterwards verify that the convolution property given by Eqn. (5.49) holds.

Solution: Let $\tilde{c}_k$, $\tilde{d}_k$ and $\tilde{e}_k$ represent the DTFS coefficients of $\tilde{x}[n]$, $\tilde{h}[n]$ and $\tilde{y}[n]$ respectively. Table 5.2 lists the DTFS coefficients for the three signals. It can easily be verified that

$$\tilde{e}_k = 5 \, \tilde{c}_k \, \tilde{d}_k \,, \qquad k = 0, \dots, 4$$

k	$\tilde{c}_k$	$\tilde{d}_k$	$\tilde{e}_k$
0	$2.0000 + j\,0.0000$	$0.0000 + j\,0.0000$	$0.0000 + j\,0.0000$
1	$-0.5000 + j\,0.6882$	$1.5326 - j\,0.6433$	$-1.6180 + j\,6.8819$
2	$-0.5000 + j\,0.1625$	$-0.0326 - j\,0.6604$	$0.6180 + j\,1.6246$
3	$-0.5000 - j\,0.1625$	$-0.0326 + j\,0.6604$	$0.6180 - j\,1.6246$
4	$-0.5000 - j\,0.6882$	$1.5326 + j\,0.6433$	$-1.6180 - j\,6.8819$

Table 5.2 – DTFS coefficients for Example 5.9.

Software resources:
ex_5_9.m

Software resources:	See MATLAB Exercise 5.3.

5.3 Analysis of Non-Periodic Discrete-Time Signals

In the previous section we have focused on representing periodic discrete-time signals using complex exponential basis functions. The end result was the discrete-time Fourier series (DTFS) that allowed a signal periodic with a period of N samples to be constructed using N harmonically related exponential basis functions. In this section we extend the DTFS concept for use in non-periodic signals.

5.3.1 Discrete-time Fourier transform (DTFT)

In the derivation of the Fourier transform for non-periodic discrete-time signals we will take an approach similar to that employed in Chapter 4. Recall that in Section 4.3.1 a non-periodic continuous-time signal was viewed as a limit case of a periodic continuous-time signal, and the Fourier transform was derived from the exponential Fourier series. The resulting development is not a mathematically rigorous derivation of the Fourier transform, but it is intuitive. Let us begin by considering a non-periodic discrete-time signal $x[n]$ as shown in Fig. 5.12.

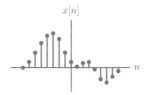

Figure 5.12 – A non-periodic signal $x[n]$.

Initially we will assume that $x[n]$ is finite-length with its significant samples confined into the range $-M \le n \le M$ of the index, that is, $x[n] = 0$ for $n < -M$ and for $n > M$. A periodic extension $\tilde{x}[n]$ can be constructed by taking $x[n]$ as one period in $-M \le n \le M$, and repeating it at intervals of $2M + 1$ samples.

$$\tilde{x}[n] = \sum_{k=-\infty}^{\infty} x[n + k(2M + 1)] \tag{5.56}$$

This is illustrated in Fig. 5.13.

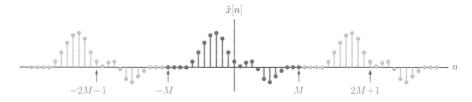

Figure 5.13 – Periodic extension $\tilde{x}[n]$ of the signal $x[n]$.

The periodic extension $\tilde{x}[n]$ can be expressed in terms of its DTFS coefficients. Using Eqn. (5.21) with $N = 2M + 1$ we obtain

$$\tilde{x}[n] = \sum_{k=0}^{2M} \tilde{c}_k \, e^{j\left(2\pi/(2M+1)\right)kn} = \sum_{k=-M}^{M} \tilde{c}_k \, e^{j\left(2\pi/(2M+1)\right)kn} \tag{5.57}$$

The coefficients are computed through the use of Eqn. (5.22) as

$$\tilde{c}_k = \frac{1}{2M + 1} \sum_{n=-M}^{M} \tilde{x}[n] \, e^{-j\left(2\pi/(2M+1)\right)kn} \, , \quad k = -M, \dots, M \tag{5.58}$$

Fundamental angular frequency is

$$\Omega_0 = \frac{2\pi}{2M + 1} \quad \text{radians.}$$

The k-th DTFS coefficient is associated with the angular frequency $\Omega_k = k\Omega_0 = 2\pi k/\left(2M+1\right)$. The set of DTFS coefficients span the range of discrete frequencies

$$\Omega_k: \quad \frac{-2M}{2M+1}\pi\,,\,\ldots\,,0,\,\ldots\,,\frac{2M}{2M+1}\pi \tag{5.59}$$

It is worth noting that the set of coefficients in Eqn. (5.59) are roughly in the interval $(-\pi, \pi)$, just slightly short of either end of the interval. Realizing that $\tilde{x}[n] = x[n]$ within the range $-M \leq n \leq M$, Eqn. (5.58) can be written using $x[n]$ instead of $\tilde{x}[n]$ to yield

$$\tilde{c}_k = \frac{1}{2M+1}\sum_{n=-M}^{M} x[n]\,e^{-j\left(2\pi/(2M+1)\right)kn}\,,\quad k = -M,\ldots,M \tag{5.60}$$

If we were to stretch out the period of the signal by increasing the value of M, then $\tilde{x}[n]$ would start to resemble $x[n]$ more and more. Other effects of increasing M would be an increase in the coefficient count and a decrease in the magnitudes of the coefficients $\tilde{c}_k$ due to the $1/\left(2M+1\right)$ factor in front of the summation. Let us multiply both sides of Eqn. (5.60) by $2M+1$ to obtain

$$\left(2M+1\right)\tilde{c}_k = \sum_{n=-M}^{M} x[n]\,e^{-j\left(2\pi/(2M+1)\right)kn} \tag{5.61}$$

As M becomes very large, the fundamental angular frequency Ω_0 becomes very small, and the spectral lines get closer to each other in the frequency domain, resembling a continuous transform.

$$M \to \infty \quad \text{implies that} \quad \Omega_0 = \frac{2\pi}{2M+1} \to \Delta\Omega \quad \text{and} \quad k\,\Delta\Omega \to \Omega \tag{5.62}$$

In Eqn. (5.62) we have switched the notation from Ω_0 to $\Delta\Omega$ due to the infinitesimal nature of the fundamental angular frequency. In the limit we have

$$\lim_{M\to\infty} [\tilde{x}[n]] = x[n] \tag{5.63}$$

in the time domain. Using the substitutions $2\pi k/\left(2M+1\right) \to \Omega$ and $\left(2M+1\right)\tilde{c}_k \to X\left(\Omega\right)$ Eqn. (5.61) becomes

$$X\left(\Omega\right) = \sum_{n=-\infty}^{\infty} x[n]\,e^{-j\Omega n} \tag{5.64}$$

The result in Eqn. (5.64) is referred to as the *discrete-time Fourier transform (DTFT)* of the signal $x[n]$. In deriving this result we assumed a finite-length signal $x[n]$, the samples of which are confined into the range $-M \leq n \leq M$, and then took the limit as $M \to \infty$. Would Eqn. (5.64) still be valid for an infinite-length $x[n]$? The answer is yes, provided that the summation in Eqn. (5.64) converges.

Next we will try to develop some insight about the meaning of the transform $X\left(\Omega\right)$. Let us apply the limit operation to the periodic extension signal $\tilde{x}[n]$ defined in Eqn. (5.57).

$$x[n] = \lim_{M\to\infty} [\tilde{x}[n]] = \lim_{M\to\infty}\left[\sum_{k=-M}^{M} \tilde{c}_k\,e^{j\left(2\pi/(2M+1)\right)kn}\right] \tag{5.65}$$

For large M we have from Eqn. (5.62)

$$\frac{(2M+1)\,\Delta\Omega}{2\pi} \to 1 \tag{5.66}$$

Using this result in Eqn. (5.65) leads to

$$x[n] = \lim_{M\to\infty} \left[\sum_{k=-M}^{M} (2M+1)\,\tilde{c}_k\, e^{jk\Delta\Omega\,n}\, \frac{\Delta\Omega}{2\pi} \right] \tag{5.67}$$

In the limit we have

$$(2M+1)\,\tilde{c}_k \to X\,(\Omega)\;,\quad k\,\Delta\Omega \to \Omega \quad \text{and} \quad \Delta\Omega \to d\Omega \tag{5.68}$$

Furthermore, the summation turns into an integral to yield

$$x[n] = \frac{1}{2\pi} \int_{-\pi}^{\pi} X\,(\Omega)\, e^{j\Omega n}\, d\Omega \tag{5.69}$$

This result explains how the transform $X\,(\Omega)$ can be used for constructing the signal $x[n]$. We can interpret the integral in Eqn. (5.69) as a continuous sum of complex exponentials at harmonic frequencies that are infinitesimally close to each other.

In summary, we have derived a transform relationship between $x[n]$ and $X\,(\Omega)$ through the following equations:

Discrete-Time Fourier Transform (DTFT):

  1. Synthesis equation:

$$x[n] = \frac{1}{2\pi} \int_{-\pi}^{\pi} X\,(\Omega)\, e^{j\Omega n}\, d\Omega \tag{5.70}$$

  2. Analysis equation:

$$X\,(\Omega) = \sum_{n=-\infty}^{\infty} x[n]\, e^{-j\Omega n} \tag{5.71}$$

Often we will use the Fourier transform operator $\mathcal{F}$ and its inverse $\mathcal{F}^{-1}$ in a shorthand notation as

$$X\,(\Omega) = \mathcal{F}\,\{x[n]\} \tag{5.72}$$

for the *forward transform*, and

$$x[n] = \mathcal{F}^{-1}\,\{X\,(\Omega)\} \tag{5.73}$$

for the *inverse transform*. Sometimes we will use an even more compact notation to express the relationship between $x[n]$ and $X\,(\Omega)$ by

$$x[n] \overset{\mathcal{F}}{\longleftrightarrow} X\,(\Omega) \tag{5.74}$$

In general, the Fourier transform, as computed by Eqn. (5.71), is a complex function of Ω. It can be written in Cartesian form as

$$X\,(\Omega) = X_r\,(\Omega) + jX_i\,(\Omega) \tag{5.75}$$

or in polar form as

$$X\,(\Omega) = |X\,(\Omega)|\, e^{j\angle X(\Omega)} \tag{5.76}$$

5.3.2 Developing further insight

In this section we will build on the idea of obtaining the DTFT as the limit case of DTFS coefficients when the signal period is made very large. Consider a discrete-time pulse with 7 unit-amplitude samples as shown in Fig. 5.14.

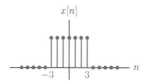

Figure 5.14 – Discrete-time rectangular pulse.

The analytical definition of $x[n]$ is

$$x[n] = \begin{cases} 1 \,, & n = -3, \ldots, 3 \\ 0 \,, & \text{otherwise} \end{cases} \tag{5.77}$$

Let the signal $\tilde{x}[n]$ be defined as the periodic extension of $x[n]$ with a period of $2M + 1$ samples so that

$$\tilde{x}[n] = \sum_{r=-\infty}^{\infty} x[n + r\,(2M + 1)] \tag{5.78}$$

as shown in Fig. 5.15.

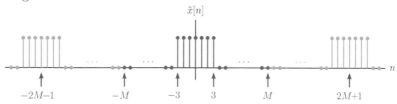

Figure 5.15 – Periodic extension of discrete-time pulse into a pulse train.

One period of $\tilde{x}[n]$ extends from $n = -M$ to $n = M$ for a total of $2M + 1$ samples. The general expression for DTFS coefficients can be found by adapting the result obtained in Example 5.5 to the signal $\tilde{x}[n]$ with $L = 3$ and $N = 2M + 1$:

$$\tilde{c}_k = \frac{\sin\,(7k\Omega_0/2)}{(2M + 1)\,\sin\,(k\Omega_0/2)} \tag{5.79}$$

The parameter Ω_0 is the fundamental angular frequency given by

$$\Omega_0 = \frac{2\pi}{2M + 1} \quad \text{rad.} \tag{5.80}$$

Let us multiply both sides of Eqn. (5.79) by $(2M + 1)$ and write the scaled DTFS coefficients as

$$(2M + 1)\,\tilde{c}_k = \frac{\sin\,(7k\Omega_0/2)}{\sin\,(k\Omega_0/2)} \tag{5.81}$$

Let $M = 8$ corresponding to a period length of $2M + 1 = 17$. The scaled DTFS coefficients are shown in Fig. 5.16(a) for the index range $k = -8, \ldots, 8$. In addition, the outline (or the envelope) of the scaled DTFS coefficients is also shown.

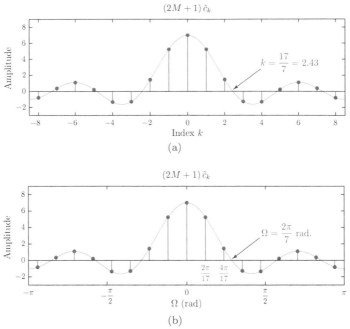

Figure 5.16 – (a) The scaled DTFS spectrum $(2M + 1)\,\tilde{c}_k$ for $M = 8$ for the signal $\tilde{x}[n]$, (b) the scaled DTFS spectrum as a function of angular frequency Ω.

In Fig. 5.16(a) the leftmost coefficient has the index $k = -8$, and is associated with the angular frequency $-8\Omega_0 = -16\pi/17$. Similarly, the rightmost coefficient is at $k = 8$, and is associated with the angular frequency $8\Omega_0 = 16\pi/17$. Fig. 5.16(b) shows the same coefficients and envelope as functions of the angular frequency Ω instead of the integer index k.

It is interesting to check the locations for the zero crossings of the envelope. The first positive zero crossing of the envelope occurs at the index value

$$\frac{7k\Omega_0}{2} = \pi \quad \Rightarrow \quad k = \frac{2\pi}{7\Omega_0} = \frac{2M + 1}{7} \tag{5.82}$$

which may or may not be an integer. For $M = 8$ the first positive zero crossing is at index value $k = 17/7 = 2.43$ as shown in Fig. 5.16(a). This corresponds to the angular frequency $(17/7)\,\Omega_0 = 2\pi/7$ radians, independent of the value of M.

If we increase the period M, the following changes occur in the scaled DTFS spectrum:

1. The number of DTFS coefficients increases since the total number of unique DTFS coefficients is the same as the period length of $\tilde{x}[n]$ which, in this case, is $2M + 1$.
2. The fundamental angular frequency decreases since it is inversely proportional to the period of $\tilde{x}[n]$. The spectral lines in Fig. 5.16(b) move in closer to each other.
3. The leftmost and the rightmost spectral lines get closer to $\pm\pi$ since they are $\pm M\Omega_0 = \pm 2M\pi/\,(2M + 1)$.
4. As $M \to \infty$ the fundamental angular frequency becomes infinitesimally small, and the spectral lines come together to form a continuous transform.

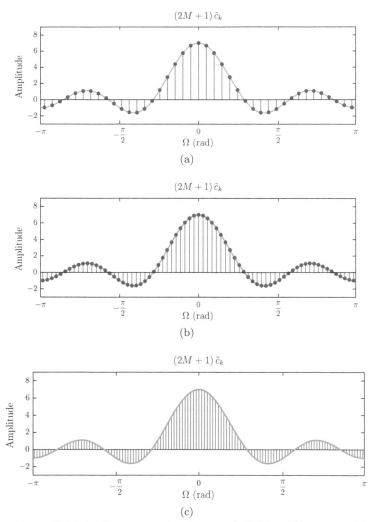

Figure 5.17 – The scaled DTFS spectrum for the signal $\tilde{x}[n]$ for (a) $M = 20$, (b) for $M = 35$, and (c) for $M = 60$.

We can conclude from the foregoing development that, as the period becomes infinitely large, spectral lines of the DTFS representation of $\tilde{x}[n]$ converge to a continuous function of Ω to form the DTFT of $x[n]$. Taking the limit of Eqn. (5.81) we get

$$X(\Omega) = \lim_{M \to \infty} \left[(2M+1)\,\tilde{c}_k \right] = \frac{\sin(7\Omega/2)}{\sin(\Omega/2)} \qquad (5.83)$$

We will also obtain this result in Example 5.12 through direct application of the DTFT equation.

Interactive Demo: `dtft_demo1`

The demo program in "`dtft_demo1.m`" provides a graphical user interface for experimenting with the development in Section 5.3.2. Refer to Figs. 5.14 through 5.17, Eqns. (5.77) through (5.81). The discrete-time pulse train has a period of $2M+1$. In each period $2L+1$ contiguous samples have unit amplitude. (Keep in mind that $M > L$.) Parameters L and M can be adjusted and the resulting scaled DTFS spectrum can be observed. As the period $2M+1$ is increased the DTFS coefficients move in closer due to the fundamental angular frequency Ω_0 becoming smaller. Consequently, the DTFS spectrum of the periodic pulse train approaches the DTFT spectrum of the non-periodic discrete-time pulse with $2L+1$ samples.

1. Set $L = 3$ and $M = 8$. This should duplicate Fig. 5.16. Observe the scaled DTFS coefficients. Pay attention to the envelope of the DTFS coefficients.
2. While keeping $L = 3$ increment M and observe the changes in the scaled DTFS coefficients. Pay attention to the fundamental angular frequency Ω_0 change, causing the coefficients to move in closer together. Observe that the envelope does not change as M is increased.

Software resources:
dtft_demo1.m

5.3.3 Existence of the DTFT

A mathematically thorough treatment of the conditions for the existence of the DTFT is beyond the scope of this text. It will suffice to say, however, that the question of existence is a simple one for the types of signals we encounter in engineering practice. A sufficient condition for the convergence of Eqn. (5.71) is that the signal $x[n]$ be absolute summable, that is,

$$\sum_{n=-\infty}^{\infty} |x[n]| < \infty \tag{5.84}$$

Alternatively, it is also sufficient for the signal $x[n]$ to be square-summable:

$$\sum_{n=-\infty}^{\infty} |x[n]|^2 < \infty \tag{5.85}$$

In addition, we will see in the next section that some signals that do not satisfy either condition may still have a DTFT if we are willing to resort to the use of singularity functions in the transform.

5.3.4 DTFT of some signals

In this section we present examples of determining the DTFT for a variety of discrete-time signals.

Example 5.10: **DTFT of right-sided exponential signal**

Determine the DTFT of the signal $x[n] = \alpha^n \, u[n]$ as shown in Fig. 5.18. Assume $|\alpha| < 1$.

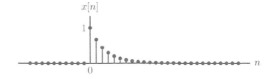

Figure 5.18 – The signal $x[n]$ for Example 5.10.

Solution: The use of the DTFT analysis equation given by Eqn. (5.71) yields

$$X\left(\Omega\right) = \sum_{n=-\infty}^{\infty} \alpha^n\, u[n]\, e^{-j\Omega n}$$

The factor $u[n]$ causes terms of the summation for $n < 0$ to equal zero. Consequently, we can start the summation at $n = 0$ and drop the term $u[n]$ to write

$$X\left(\Omega\right) = \sum_{n=0}^{\infty} \alpha^n\, e^{-j\Omega n} = \frac{1}{1 - \alpha\, e^{-j\Omega}} \qquad (5.86)$$

provided that $|\alpha| < 1$. In obtaining the result in Eqn. (5.86) we have used the closed form of the sum of infinite-length geometric series (see Appendix C). The magnitude of the transform is

$$|X\left(\Omega\right)| = \frac{1}{|1 - \alpha\, e^{-j\Omega}|} = \frac{1}{\sqrt{1 + \alpha^2 - 2\alpha\, \cos\left(\Omega\right)}}$$

The phase of the transform is found as the difference of the phases of numerator and denominator of the result in Eqn. (5.86):

$$\angle X\left(\Omega\right) = 0 - \angle \left(1 - \alpha\, e^{-j\Omega}\right) = -\tan^{-1}\left[\frac{\alpha \sin\left(\Omega\right)}{1 - \alpha\, \cos\left(\Omega\right)}\right]$$

The magnitude and the phase of the transform are shown in Fig. 5.19(a) and (b) for the case $\alpha = 0.4$.

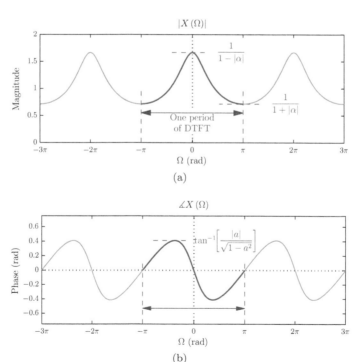

Figure 5.19 – The DTFT of the signal $x[n]$ for Example 5.10 for $\alpha = 0.4$: (a) the magnitude, (b) the phase.

Software resources:
ex_5_10.m

Interactive Demo: dtft_demo2

The demo program in "dtft_demo2.m" is based on Example 5.10. The signal $x[n]$ is graphed along with the magnitude and the phase of its DTFT $X(\Omega)$. The parameter α may be varied, and its effects on the spectrum may be observed.
Software resources:
dtft_demo2.m

Example 5.11: **DTFT of unit-impulse signal**

Determine the DTFT of the unit-impulse signal $x[n] = \delta[n]$.

Solution: Direct application of Eqn. (5.71) yields

$$\mathcal{F}\{\delta[n]\} = \sum_{n=-\infty}^{\infty} \delta[n]\, e^{-j\Omega n} \tag{5.87}$$

Using the sifting property of the impulse function, Eqn. (5.87) reduces to

$$\mathcal{F}\{\delta[n]\} = 1\,, \quad \text{all } \Omega \tag{5.88}$$

Example 5.12: **DTFT for discrete-time pulse**

Determine the DTFT of the discrete-time pulse signal $x[n]$ given by

$$x[n] = \begin{cases} 1\,, & -L \le n \le L \\ 0\,, & \text{otherwise} \end{cases}$$

Solution: Using Eqn. (5.71) the transform is

$$X(\Omega) = \sum_{n=-L}^{L} (1)\, e^{-j\Omega n}$$

The closed form expression for a finite-length geometric series is (see Appendix C for derivation)

$$\text{Eqn. (C.13):} \qquad \sum_{n=L_1}^{L_2} a^n = \frac{a^{L_1} - a^{L_2+1}}{1-a}$$

Using Eqn. (C.13) with $a = e^{-j\Omega}$, $L_1 = -L$ and $L_2 = L$, the closed form expression for $X(\Omega)$ is

$$X(\Omega) = \frac{e^{j\Omega L} - e^{-j\Omega(L+1)}}{1 - e^{-j\Omega}} \tag{5.89}$$

In order to get symmetric complex exponentials in Eqn. (5.89) we will multiply both the numerator and the denominator of the fraction on the right side of the equal sign with $e^{j\Omega/2}$. The result is

$$X(\Omega) = \frac{e^{j\Omega(L+1/2)} - e^{-j\Omega(L+1/2)}}{e^{j\Omega/2} - e^{-j\Omega/2}}$$

which, using Euler's formula, can be simplified to

$$X\left(\Omega\right) = \frac{\sin\left(\dfrac{\Omega}{2}\left(2L+1\right)\right)}{\sin\left(\dfrac{\Omega}{2}\right)}$$

The value of the transform at $\Omega = 0$ must be resolved through the use of L'Hospital's rule:

$$X\left(0\right) = 2L + 1$$

The transform $X\left(\Omega\right)$ is graphed in Fig. 5.20 for $L = 3, 4, 5$.

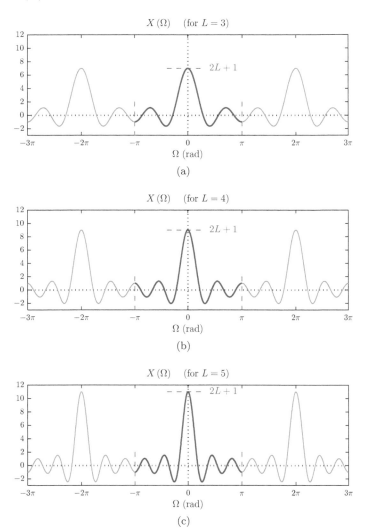

Figure 5.20 – The transform of the pulse signal $x[n]$ of Example 5.12 for (a) $L = 3$, (b) $L = 4$, and (c) $L = 5$.

Software resources:
ex_5_12.m

Interactive Demo: dtft_demo3

The demo program in "dtft_demo3.m" is based on computing the DTFT of a discrete-time pulse explored in Example 5.12. The pulse with $2L + 1$ unit-amplitude samples centered around $n = 0$ leads to the DTFT shown in Fig. 5.20. The demo program allows experimentation with that spectrum as L is varied.

Software resources:
dtft_demo3.m

Example 5.13: Inverse DTFT of rectangular spectrum

Determine the inverse DTFT of the transform $X(\Omega)$ defined in the angular frequency range $-\pi < \Omega < \pi$ by

$$X(\Omega) = \begin{cases} 1, & -\Omega_c < \Omega < \Omega_c \\ 0, & \text{otherwise} \end{cases}$$

Solution: To be a valid transform, $X(\Omega)$ must be 2π-periodic, therefore we need $X(\Omega) = X(\Omega + 2\pi)$. The resulting transform is shown in Fig. 5.21.

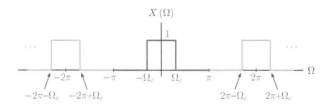

Figure 5.21 – The transform $X(\Omega)$ for Example 5.13.

The inverse $x[n]$ is found by application of the DTFT synthesis equation given by Eqn. (5.70).

$$x[n] = \frac{1}{2\pi} \int_{-\Omega_c}^{\Omega_c} (1)\, e^{j\Omega n}\, d\Omega = \frac{\sin(\Omega_c n)}{\pi n} \tag{5.90}$$

It should be noted that the expression found in Eqn. (5.90) for the signal $x[n]$ is for all n; therefore, $x[n]$ is non-causal. For convenience let us use the normalized frequency F_c related to the angular frequency Ω_c by

$$\Omega_c = 2\pi F_c$$

and the signal $x[n]$ becomes

$$x[n] = 2F_c \operatorname{sinc}(2F_c n)$$

The result is shown in Fig. 5.22 for the case $F_c = 1/9$. Zero crossings of the sinc function in Eqn. (5.91) are spaced $1/(2F_c)$ apart which has the non-integer value of 4.5 in this case. Therefore, the envelope of the signal $x[n]$ crosses the axis at the midpoint of the samples for $n = 4$ and $n = 5$.

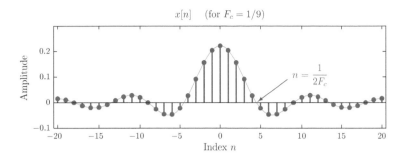

Figure 5.22 – The signal $x[n]$ for Example 5.13.

Software resources:
ex_5_13.m

Interactive Demo: dtft_demo4

The demo program in "dtft_demo4.m" is based on Example 5.13 where the inverse DTFT of the rectangular spectrum shown in Fig. 5.21 was determined using the DTFT synthesis equation. The demo program displays graphs of the spectrum and the corresponding time-domain signal. The normalized cutoff frequency parameter F_c of the spectrum may be varied through the use of the slider control, and its effects on the signal $x[n]$ may be observed. Recall that $\Omega_c = 2\pi F_c$.
Software resources:
dtft_demo4.m

Example 5.14: **Inverse DTFT of the unit-impulse function**

Find the signal the DTFT of which is $X(\Omega) = \delta(\Omega)$ in the range $-\pi < \Omega < \pi$.

Solution: To be a valid DTFT, the transform $X(\Omega)$ must be 2π-periodic. Therefore, the full expression for $X(\Omega)$ must be

$$X(\Omega) = \sum_{m=-\infty}^{\infty} \delta(\Omega - 2\pi m)$$

as shown in Fig. 5.23.

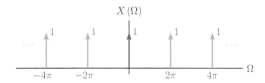

Figure 5.23 – The transform $X(\Omega)$ for Example 5.14.

The inverse transform is

$$x[n] = \frac{1}{2\pi} \int_{-\pi}^{\pi} \delta(\Omega) \, e^{j\Omega n} \, d\Omega$$

Using the sifting property of the impulse function we get

$$x[n] = \frac{1}{2\pi} , \quad \text{all } n$$

Thus we have the DTFT pair

$$\frac{1}{2\pi} \overset{\mathcal{F}}{\longleftrightarrow} \sum_{m=-\infty}^{\infty} \delta\left(\Omega - 2\pi m\right) \tag{5.91}$$

An important observation is in order: A signal that has constant amplitude for all index values is a power signal; it is neither absolute summable nor square summable. Strictly speaking, its DTFT does not converge. The transform relationship found in Eqn. (5.91) is a compromise made possible by our willingness to allow the use of singularity functions in $X(\Omega)$. Nevertheless, it is a useful relationship since it can be used in solving problems in the frequency domain.

Multiplying both sides of the relationship in Eqn. (5.91) by 2π results in

$$\mathcal{F}\{1\} = 2\pi \sum_{m=-\infty}^{\infty} \delta\left(\Omega - 2\pi m\right) \tag{5.92}$$

which is illustrated in Fig. 5.24. This relationship is fundamental, and will be explored further in the next example.

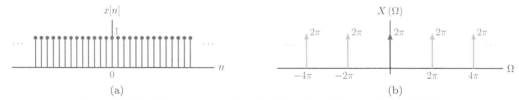

(a)   (b)

Figure 5.24 – The constant signal $x[n] = 1$ and its transform $X(\Omega)$.

Example 5.15: **Inverse DTFT of impulse function revisited**

Consider again the DTFT transform pair found in Eqn. (5.92) of Example 5.14. The unit-amplitude signal $x[n] = 1$ can be thought of as the limit of the rectangular pulse signal explored in Example 5.12. Let

$$x[n] = 1 , \text{ all } n \quad \text{and} \quad w[n] = \begin{cases} 1 , & -L \leq n \leq L \\ 0 , & \text{otherwise} \end{cases}$$

so that

$$x[n] = \lim_{L \to \infty} [w[n]]$$

Adapting from Example 5.12, the transform of $w[n]$ is

$$W(\Omega) = \frac{\sin\left(\dfrac{\Omega}{2}(2L+1)\right)}{\sin\left(\dfrac{\Omega}{2}\right)}$$

and the transform of $x[n]$ was found in Example 5.14 to be

$$X\left(\Omega\right) = 2\pi \sum_{m=-\infty}^{\infty} \delta\left(\Omega - 2\pi m\right)$$

Intuitively we would expect $W\left(\Omega\right)$ to resemble $X\left(\Omega\right)$ more and more closely as L is increased. Fig. 5.25 shows the transform $W\left(\Omega\right)$ for $L = 10, 20$, and 50. Observe the transition from $W\left(\Omega\right)$ to $X\left(\Omega\right)$ as L is increased.

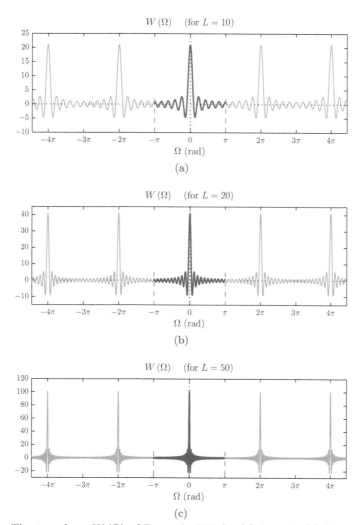

Figure 5.25 – The transform $W\left(\Omega\right)$ of Example 5.15 for (a) $L = 10$, (b) $L = 20$, (c) $L = 50$.

Software resources:
ex_5_15.m

Interactive Demo: dtft_demo5

The demo program in "dtft_demo5.m" is based on Examples 5.14 and 5.15. The signal $x[n]$ is a discrete-time pulse with $2L + 1$ unit-amplitude samples centered around $n = 0$.

The parameter L may be varied. As L is increased, the signal $x[n]$ becomes more and more similar to the constant amplitude signal of Example 5.14, and its spectrum approaches the spectrum shown in Fig. 5.24(b).

Software resources:

dtft_demo5.m

5.3.5   Properties of the DTFT

In this section we will explore some of the fundamental properties of the DTFT. As in the case of the Fourier transform for continuous-time signals, careful use of DTFT properties simplifies the solution of many types of problems.

Periodicity

The DTFT is periodic:

$$X\left(\Omega + 2\pi r\right) = X\left(\Omega\right) \quad \text{for all integers } r \tag{5.93}$$

Periodicity of the DTFT is a direct consequence of the analysis equation given by Eqn. (5.71).

Linearity

DTFT is a linear transform.

For two transform pairs

$$x_1[n] \overset{\mathcal{F}}{\longleftrightarrow} X_1\left(\Omega\right)$$

$$x_2[n] \overset{\mathcal{F}}{\longleftrightarrow} X_2\left(\Omega\right)$$

and two arbitrary constants α_1 and α_2 it can be shown that

$$\alpha_1\, x_1[n] + \alpha_2\, x_2[n] \overset{\mathcal{F}}{\longleftrightarrow} \alpha_1\, X_1\left(\Omega\right) + \alpha_2\, X_2\left(\Omega\right) \tag{5.94}$$

Proof: Use of the forward transform given by Eqn. (5.71) with the signal $\left(\alpha_1\, x_1[n] + \alpha_2\, x_2[n]\right)$ yields

$$\mathcal{F}\left\{\alpha_1\, x_1[n] + \alpha_2\, x_2[n]\right\} = \sum_{n=-\infty}^{\infty} \left(\alpha_1\, x_1[n] + \alpha_2\, x_2[n]\right) e^{-j\Omega n}$$

$$= \sum_{n=-\infty}^{\infty} \alpha_1\, x_1[n]\, e^{-j\Omega n} + \sum_{n=-\infty}^{\infty} \alpha_2\, x_2[n]\, e^{-j\Omega n}$$

$$- \alpha_1 \sum_{n=-\infty}^{\infty} x_1[n]\, e^{-j\Omega n} + \alpha_2 \sum_{n=-\infty}^{\infty} x_2[n]\, e^{-j\Omega n}$$

$$= \alpha_1\, \mathcal{F}\left\{x_1[n]\right\} + \alpha_2\, \mathcal{F}\left\{x_2[n]\right\} \tag{5.95}$$

Time shifting

For a transform pair

$$x[n] \overset{\mathcal{F}}{\longleftrightarrow} X\left(\Omega\right)$$

it can be shown that

$$x[n-m] \overset{\mathcal{F}}{\longleftrightarrow} X\left(\Omega\right) e^{-j\Omega m} \tag{5.96}$$

The consequence of shifting, or delaying, a signal in time is multiplication of its DTFT by a complex exponential function of angular frequency.

Proof: Applying the forward transform in Eqn. (5.71) to the signal $x[n-m]$ we obtain

$$\mathcal{F}\left\{x[n-m]\right\} = \sum_{n=-\infty}^{\infty} x[n-m]\, e^{-j\Omega n} \tag{5.97}$$

Let us apply the variable change $k = n - m$ to the summation on the right side of Eqn. (5.97) so that

$$\mathcal{F}\left\{x[n-m]\right\} = \sum_{k=-\infty}^{\infty} x(k)\, e^{-j\Omega(k+m)} \tag{5.98}$$

The exponential function in the summation of Eqn. (5.98) can be written as the product of two exponential functions to obtain

$$\mathcal{F}\left\{x[n-m]\right\} = \sum_{k=-\infty}^{\infty} x[k]\, e^{-j\Omega k}\, e^{-j\Omega m}$$

$$= e^{-j\Omega m} \sum_{k=-\infty}^{\infty} x[k]\, e^{-j\Omega k}$$

$$= e^{-j\Omega m}\, X\left(\Omega\right) \tag{5.99}$$

Example 5.16: **DTFT of a time-shifted signal**

Determine the DTFT of the signal $x[n] = e^{-\alpha(n-1)}\, u[n-1]$ as shown in Fig. 5.26. (Assume $|\alpha| < 1$).

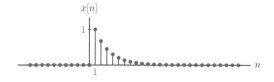

Figure 5.26 – The signal $x[n]$ for Example 5.16.

Solution: The transform of a right-sided exponential signal was determined in Example 5.10.

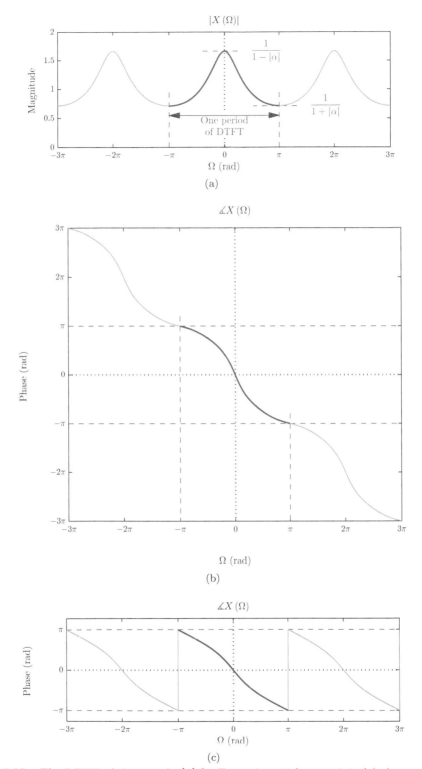

Figure 5.27 – The DTFT of the signal $x[n]$ for Example 5.16 for $\alpha = 0.4$: (a) the magnitude, (b) the phase as computed by Eqn. (5.100), (c) the phase shown in customary form.

As a result we have the transform pair

$$e^{-\alpha n}\, u[n] \xleftrightarrow{\mathcal{F}} \frac{1}{1 - \alpha\, e^{-j\Omega}}$$

Applying the time shifting property of the DTFT given by Eqn. (5.96) with $m = 1$ leads to the result

$$X(\Omega) = \mathcal{F}\left\{ e^{-\alpha(n-1)}\, u[n-1] \right\} = \frac{e^{-j\Omega}}{1 - \alpha\, e^{-j\Omega}}$$

Since $\left| e^{-j\Omega} \right| = 1$, the magnitude of $X(\Omega)$ is the same as that obtained in Example 5.10. Time shifting a signal does not change the magnitude of its transform.

$$|X(\Omega)| = \frac{1}{\sqrt{1 + \alpha^2 - 2\alpha\, \cos(\Omega)}}$$

The magnitude $|X(\Omega)|$ is shown in Fig. 5.27(a) for $\alpha = 0.4$. The phase of $X(\Omega)$ is found by first determining the phase angles of the numerator and the denominator, and then subtracting the latter from the former. In Example 5.10 the numerator of the transform was a constant equal to unity; in this case it is a complex exponential. Therefore

$$\angle X(\Omega) = \angle \left(e^{-j\Omega} \right) - \angle \left(1 - \alpha\, e^{-j\Omega} \right)$$

$$= -\Omega - \tan^{-1}\left[\frac{\alpha \sin(\Omega)}{1 - \alpha \cos(\Omega)} \right] \tag{5.100}$$

Thus, the phase of the transform differs from that obtained in Example 5.10 by $-\Omega$, a ramp with a slope of -1. The phase, as computed by Eqn. (5.100), is shown in Fig. 5.27(b). In graphing the phase of the transform it is customary to fit phase values to in the range $(-\pi, \pi)$ radians. Fig. 5.27(c) depicts the phase of the transform in the more traditional form.

Software resources:
ex_5_16.m

Interactive Demo: dtft_demo6

The demo program in "dtft_demo6.m" is based on Example 5.16. The signal

$$x[n] = e^{-\alpha(n-m)}\, u[n-m]$$

is shown along with the magnitude and the phase of its transform $X(\Omega)$. The time delay m can be varied, and its effect on the spectrum can be observed.

1. Observe that changing the amount of delay does not affect the magnitude of the spectrum.
2. Pay attention to the phase characteristic. In Example 5.16 we have used $m = 1$ and obtained the expression given by Eqn. (5.100) for the phase. When the signal is delayed by m samples the corresponding phase is

$$-m\Omega - \tan^{-1}\left[\frac{\alpha \sin(\Omega)}{1 - \alpha \cos(\Omega)} \right]$$

Software resources:
dtft_demo6.m

Time reversal

For a transform pair
$$x[n] \overset{\mathcal{F}}{\longleftrightarrow} X(\Omega)$$

it can be shown that
$$x[-n] \overset{\mathcal{F}}{\longleftrightarrow} X(-\Omega) \qquad (5.101)$$

Time reversal of the signal causes angular frequency reversal of the transform. This property will be useful when we consider symmetry properties of the DTFT.

Proof: Direct application of the forward transform in Eqn. (5.71) to the signal $x[-n]$ yields

$$\mathcal{F}\{x[-n]\} = \sum_{n=-\infty}^{\infty} x[-n] e^{-j\Omega n} \qquad (5.102)$$

Let us apply the variable change $k = -n$ to the summation on the right side of Eqn. (5.102) to obtain

$$\mathcal{F}\{x[-n]\} = \sum_{k=\infty}^{-\infty} x[k] e^{j\Omega k} \qquad (5.103)$$

which can be written as

$$\mathcal{F}\{x[-n]\} = \sum_{k=-\infty}^{\infty} x[k] e^{-j(-\Omega)k} = X(-\Omega) \qquad (5.104)$$

Example 5.17: **DTFT of two-sided exponential signal**

Determine the DTFT of the signal $x[n] = \alpha^{|n|}$ with $|\alpha| < 1$.

Solution: The signal $x[n]$ can be written as

$$x[n] = \begin{cases} \alpha^n, & n \geq 0 \\ \alpha^{-n}, & n < 0 \end{cases}$$

It can also be expressed as
$$x[n] = x_1[n] + x_2[n]$$

where $x_1[n]$ is a causal signal and $x_2[n]$ is an anti-causal signal defined as

$$x_1[n] = \alpha^n\, u[n] \quad \text{and} \quad x_2[n] = \alpha^{-n}\, u[-n - 1]$$

This is illustrated in Fig. 5.28.

Based on the linearity property of the DTFT, the transform of $x[n]$ is the sum of the transforms of $x_1[n]$ and $x_2[n]$.

$$X(\Omega) = X_1(\Omega) + X_2(\Omega)$$

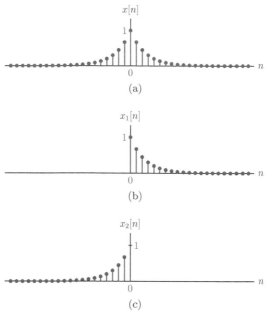

Figure 5.28 – (a) Two-sided exponential signal of Example 5.17, (b) its causal component, and (c) its anti-causal component.

For the transform of $x_1 = \alpha^n \, u[n]$ we will make use of the result obtained in Example 5.10.

$$X_1 \left(\Omega \right) = \frac{1}{1 - \alpha \, e^{-j\Omega}}$$

Time shifting and time reversal properties of the DTFT will be used for obtaining the transform of $x_2[n]$. Let a new signal $g[n]$ be defined as a scaled and time-shifted version of $x_1[n]$:

$$g[n] = \alpha \, x_1[n-1] = \alpha \left(\alpha^{n-1} \, u[n-1] \right) = \alpha^n \, u[n-1]$$

Using the time shifting property, the transform of $g[n]$ is

$$G \left(\Omega \right) = \alpha \, X \left(\Omega \right) e^{-j\Omega} = \frac{\alpha \, e^{-j\Omega}}{1 - \alpha \, e^{-j\Omega}}$$

The signal $x_2[n]$ is a time reversed version of $g[n]$, that is,

$$x_2[n] = g[-n]$$

Applying the time reversal property, the transform of $x_2[n]$ is found as

$$X_2 \left(\Omega \right) = G \left(-\Omega \right) = \frac{\alpha \, e^{j\Omega}}{1 - \alpha \, e^{j\Omega}}$$

Finally, $X \left(\Omega \right)$ is found by adding the two transforms:

$$X \left(\Omega \right) = \frac{1}{1 - \alpha \, e^{-j\Omega}} + \frac{\alpha \, e^{j\Omega}}{1 - \alpha \, e^{j\Omega}}$$

$$= \frac{1 - \alpha^2}{1 - 2\alpha \, \cos \left(\Omega \right) + \alpha^2} \tag{5.105}$$

The transform $X(\Omega)$ obtained in Eqn. (5.105) is real-valued, and is shown in Fig. 5.29 for $\alpha = 0.4$.

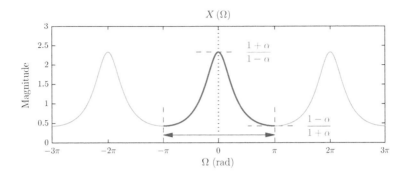

$X(\Omega)$

Figure 5.29 – The DTFT of the signal $x[n]$ of Example 5.17 for $a = 0.4$.

Software resources:
ex_5_17.m

Interactive Demo: dtft_demo7

The demo program in "dtft_demo7.m" is based on Example 5.17. It graphs the signal $x[n]$ of Example 5.17 and the corresponding spectrum which is purely real due to the even symmetry of the signal. Parameter α can be varied in order to observe its effect on the spectrum.
Software resources:
dtft_demo7.m

Conjugation property

For a transform pair

$$x[n] \xleftrightarrow{\mathcal{F}} X(\Omega)$$

it can be shown that

$$x^*[n] \xleftrightarrow{\mathcal{F}} X^*(-\Omega) \tag{5.106}$$

Conjugation of the signal causes both conjugation and angular frequency reversal of the transform. This property will also be useful when we consider symmetry properties of the DTFT.

Proof: Using the signal $x^*[n]$ in the forward transform equation results in

$$\mathcal{F}\{x^*[n]\} = \sum_{n=-\infty}^{\infty} x^*[n] e^{-j\Omega n} \tag{5.107}$$

Writing the right side of Eqn. (5.107) in a slightly modified form we obtain

$$\mathcal{F}\left\{ x^*[n] \right\} = \left[\sum_{n=\infty}^{-\infty} x[n]\, e^{j\Omega n} \right]^* = X^*\left(-\Omega \right) \tag{5.108}$$

Symmetry of the DTFT

If the signal $x[n]$ is real-valued, it can be shown that its DTFT is conjugate symmetric. Conversely, if the signal $x[n]$ is purely imaginary, its transform is conjugate antisymmetric.

Symmetry of the DTFT:

$$x[n]:\ \text{Real},\ \ \text{Im}\left\{ x[n] \right\} = 0 \quad \text{implies that} \quad X^*\left(\Omega \right) = X\left(-\Omega \right) \tag{5.109}$$

$$x[n]:\ \text{Imag},\ \ \text{Re}\left\{ x[n] \right\} = 0 \quad \text{implies that} \quad X^*\left(\Omega \right) = -\,X\left(-\Omega \right) \tag{5.110}$$

Proof:

1. **Real $x\left(t \right)$:**
   Any real-valued signal is equal to its own conjugate, therefore we have

$$x^*[n] = x[n] \tag{5.111}$$

   Taking the transform of each side of Eqn. (5.111) and using the conjugation property stated in Eqn. (5.106) we obtain

$$X^*\left(-\Omega \right) = X\left(\Omega \right) \tag{5.112}$$

   which is equivalent to Eqn. (5.109).

2. **Imaginary $x\left(t \right)$:**
   A purely imaginary signal is equal to the negative of its conjugate, i.e.,

$$x^*[n] = -x[n] \tag{5.113}$$

   Taking the transform of each side of Eqn. (5.113) and using the conjugation property given by Eqn. (5.106) yields

$$X^*\left(-\Omega \right) = -X\left(\Omega \right) \tag{5.114}$$

   This is equivalent to Eqn. (5.110). Therefore the transform is conjugate antisymmetric.

Cartesian and polar forms of the transform

A complex transform $X\left(\Omega \right)$ can be written in polar form as

$$X\left(\Omega \right) = \left| X\left(\Omega \right) \right|\, e^{j\Theta\left(\Omega \right)} \tag{5.115}$$

and in Cartesian form as

$$X\left(\Omega \right) = X_r\left(\Omega \right) + jX_i\left(\Omega \right) \tag{5.116}$$

Case 1: If $X(\Omega)$ is conjugate symmetric, the relationship in Eqn. (5.109) can be written as

$$|X(-\Omega)|\, e^{j\Theta(-\Omega)} = |X(\Omega)|\, e^{-j\Theta(\Omega)} \tag{5.117}$$

using the polar form of the transform, and

$$X_r(-\Omega) + jX_i(-\Omega) = X_r(\Omega) - jX_i(\Omega) \tag{5.118}$$

using its Cartesian form. The consequences of Eqns. (5.117) (5.118) can be obtained by equating the magnitudes and the phases on both sides of Eqn. (5.117) and by equating real and imaginary part on both sides of Eqn. (5.118). The results are summarized below:

Conjugate symmetric transform: $\quad X(-\Omega) = X^*(\Omega)$

Magnitude:	$|X(-\Omega)| = |X(\Omega)|$	(5.119)
Phase:	$\Theta(-\Omega) = -\Theta(\Omega)$	(5.120)
Real part:	$X_r(-\Omega) = X_r(\Omega)$	(5.121)
Imag. part:	$X_i(-\Omega) = -X_i(\Omega)$	(5.122)

The transform of a real-valued $x[n]$ is conjugate symmetric. For such a transform, the magnitude is an even function of Ω, and the phase is an odd function. Furthermore, the real part of the transform has even symmetry, and its imaginary part has odd symmetry.

Case 2: If $X(\Omega)$ is conjugate antisymmetric, the polar form of $X(\Omega)$ has the property

$$|X(-\Omega)|\, e^{j\Theta(-\Omega)} = -|X(\Omega)|\, e^{-j\Theta(\Omega)} \tag{5.123}$$

The negative sign on the right side of Eqn. (5.123) needs to be incorporated into the phase since we could not write $|X(-\Omega)| = -|X(\Omega)|$ (recall that magnitude needs to be a non-negative function for all Ω). Using $e^{\mp j\pi} = -1$, Eqn. (5.123) can be written as

$$|X(-\Omega)|\, e^{j\Theta(-\Omega)} = |X(\Omega)|\, e^{-j\Theta(\Omega)}\, e^{\mp j\pi}$$
$$= |X(\Omega)|\, e^{-j[\Theta(\Omega)\mp\pi]} \tag{5.124}$$

Conjugate antisymmetry property of the transform can also be expressed in Cartesian form as

$$X_r(-\Omega) + jX_i(-\Omega) = -X_r(\Omega) + jX_i(\Omega) \tag{5.125}$$

The consequences of Eqns. (5.124) and (5.125) are given below.

Conjugate antisymmetric transform: $\quad X(-\Omega) = -X^*(\Omega)$

Magnitude:	$|X(-\Omega)| = |X(\Omega)|$	(5.126)
Phase:	$\Theta(-\Omega) = -\Theta(\Omega) \mp \pi$	(5.127)
Real part:	$X_r(-\Omega) = -X_r(\Omega)$	(5.128)
Imag. part:	$X_i(-\Omega) = X_i(\Omega)$	(5.129)

We know that a purely-imaginary signal leads to a DTFT that is conjugate antisymmetric. For such a transform the magnitude is still an even function of Ω as suggested by Eqn. (5.126). The phase is neither even nor odd. The real part is an odd function of Ω, and the imaginary part is an even function.

Transforms of even and odd signals

If the real-valued signal $x[n]$ is an even function of time, the resulting transform $X(\Omega)$ is real-valued for all Ω.

$$x[-n] = x[n] \text{, all } n \quad \text{implies that} \quad \text{Im}\{X(\Omega)\} = 0 \text{, all } \Omega \qquad (5.130)$$

Proof: Using the time reversal property of the DTFT, Eqn. (5.130) implies that

$$X(-\Omega) = X(\Omega) \qquad (5.131)$$

Furthermore, since $x[n]$ is real-valued, the transform is conjugate symmetric, therefore

$$X^*(\Omega) = X(-\Omega) \qquad (5.132)$$

Combining Eqns. (5.131) and (5.132) we reach the conclusion

$$X^*(\Omega) = X(\Omega) \qquad (5.133)$$

Therefore, $X(\Omega)$ must be real.

Conversely it can also be proven that, if the real-valued signal $x[n]$ has odd-symmetry, the resulting transform is purely imaginary. This can be stated mathematically as follows:

$$x[-n] = -x[n] \text{, all } n \quad \text{implies that} \quad \text{Re}\{X(\Omega)\} = 0 \text{, all } \Omega \qquad (5.134)$$

Conjugating a purely imaginary transform is equivalent to negating it, so an alternative method of expressing the relationship in Eqn. (5.134) is

$$X^*(\Omega) = -X(\Omega) \qquad (5.135)$$

Proof of Eqn. (5.135) is similar to the procedure used above for proving Eqn. (5.130) (see Problem 5.14 at the end of this chapter).

Frequency shifting

For a transform pair

$$x[n] \xleftrightarrow{\mathcal{F}} X(\Omega)$$

it can be shown that

$$x[n]\, e^{j\Omega_0 n} \xleftrightarrow{\mathcal{F}} X(\Omega - \Omega_0) \qquad (5.136)$$

Proof: Applying the DTFT definition given by Eqn. (5.71) to the signal $x[n]\, e^{j\Omega_0 n}$ we obtain

$$
\begin{aligned}
\mathcal{F}\left\{ x[n]\, e^{j\Omega_0 n} \right\} &= \sum_{n=-\infty}^{\infty} x[n]\, e^{j\Omega_0 n}\, e^{-j\Omega n} \\
&= \sum_{n=-\infty}^{\infty} x[n]\, e^{-j(\Omega - \Omega_0)n} \\
&= X\left(\Omega - \Omega_0\right)
\end{aligned}
\tag{5.137}
$$

Modulation property

For a transform pair

$$
x[n] \xleftrightarrow{\ \mathcal{F}\ } X\left(\Omega\right)
$$

it can be shown that

$$
x[n]\, \cos\left(\Omega_0 n\right) \xleftrightarrow{\ \mathcal{F}\ } \frac{1}{2}\left[X\left(\Omega - \Omega_0\right) + X\left(\Omega + \Omega_0\right) \right]
\tag{5.138}
$$

and

$$
x[n]\, \sin\left(\Omega_0 n\right) \xleftrightarrow{\ \mathcal{F}\ } \frac{1}{2}\left[X\left(\Omega - \Omega_0\right) e^{-j\pi/2} + X\left(\Omega + \Omega_0\right) e^{j\pi/2} \right]
\tag{5.139}
$$

Modulation property is an interesting consequence of the frequency shifting property combined with the linearity of the Fourier transform. Multiplication of a signal by a cosine waveform causes its spectrum to be shifted in both directions by the angular frequency of the cosine waveform, and to be scaled by $\frac{1}{2}$. Multiplication of the signal by a sine waveform causes a similar effect with an added phase shift of $-\pi/2$ radians for positive frequencies and $\pi/2$ radians for negative frequencies.

Proof: Using Euler's formula, the left side of the relationship in Eqn. (5.138) can be written as

$$
x[n]\, \cos\left(\Omega_0 n\right) = \frac{1}{2}\, x[n]\, e^{\Omega_0 n} + \frac{1}{2}\, x[n]\, e^{-\Omega_0 n}
\tag{5.140}
$$

The desired proof is obtained by applying the frequency shifting theorem to the terms on the right side of Eqn. (5.140). Using Eqn. (5.136) we obtain

$$
x[n]\, e^{j\Omega_0 n} \xleftrightarrow{\ \mathcal{F}\ } X\left(\Omega - \Omega_0\right)
\tag{5.141}
$$

and

$$
x[n]\, e^{-j\Omega_0 n} \xleftrightarrow{\ \mathcal{F}\ } X\left(\Omega + \Omega_0\right)
\tag{5.142}
$$

which could be used together in Eqn. (5.140) to arrive at the result in Eqn. (5.138).

The proof of Eqn. (5.139) is similar, but requires one additional step. Again using Euler's formula, let us write the left side of Eqn. (5.139) as

$$
x[n]\, \sin\left(\Omega_0 n\right) = \frac{1}{2j}\, x[n]\, e^{j\Omega_0 n} - \frac{1}{2j}\, x[n]\, e^{-j\Omega_0 n}
\tag{5.143}
$$

Realizing that

$$\frac{1}{j} = -j = e^{-j\pi/2} \quad \text{and} \quad \frac{-1}{j} = j = e^{j\pi/2}$$

Eqn. (5.143) can be rewritten as

$$x[n]\sin(\Omega_0 n) = \frac{1}{2}x[n]\,e^{j\Omega_0 n}\,e^{-j\pi/2} + \frac{1}{2}x[n]\,e^{-j\Omega_0 n}\,e^{j\pi/2} \tag{5.144}$$

The proof of Eqn. (5.139) can be completed by using Eqns. (5.141) and (5.142) on the right side of Eqn. (5.144).

Differentiation in the frequency domain

For a transform pair

$$x[n] \xleftrightarrow{\mathcal{F}} X(\Omega)$$

it can be shown that

$$n\,x[n] \xleftrightarrow{\mathcal{F}} j\,\frac{dX(\Omega)}{d\Omega} \tag{5.145}$$

and

$$n^m\,x[n] \xleftrightarrow{\mathcal{F}} j^m\,\frac{d^m X(\Omega)}{d\Omega^m} \tag{5.146}$$

Proof: Differentiating both sides of Eqn. (5.71) with respect to Ω yields

$$\frac{dX(\Omega)}{d\Omega} = \frac{d}{d\Omega}\left[\sum_{n=-\infty}^{\infty} x[n]\,e^{-j\Omega n}\right]$$

$$= \sum_{n=-\infty}^{\infty} -jn\,x[n]\,e^{-j\Omega n} \tag{5.147}$$

The summation on the right side of Eqn. (5.147) is the DTFT of the signal $-jn\,x[n]$. Multiplying both sides of Eqn. (5.147) by j results in the transform pair in Eqn. (5.145). Eqn. (5.146) is proven by repeated use of Eqn. (5.145).

Example 5.18: **Use of differentiation in frequency property**

Determine the DTFT of the signal $x[n] = n\,e^{-\alpha n}\,u[n]$ shown in Fig. 5.30. Assume $|\alpha| < 1$.

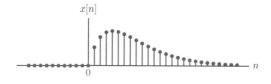

Figure 5.30 – The signal $x[n]$ for Example 5.18.

Solution: In Example 5.10 we have established the following transform pair:

$$e^{-\alpha n}\, u[n] \quad\overset{\mathcal{F}}{\longleftrightarrow}\quad \frac{1}{1-\alpha\, e^{-j\Omega}}$$

Applying the differentiation in frequency property of the DTFT leads to

$$n\, e^{-\alpha n}\, u[n] \quad\overset{\mathcal{F}}{\longleftrightarrow}\quad j\,\frac{d}{d\Omega}\left[\frac{1}{1-\alpha\, e^{-j\Omega}}\right]$$

Differentiation is carried out easily:

$$\frac{d}{d\Omega}\left[\frac{1}{1-\alpha\, e^{-j\Omega}}\right] = \frac{d}{d\Omega}\left[\left(1-\alpha\, e^{-j\Omega}\right)^{-1}\right]$$
$$= -\left(1-\alpha\, e^{-j\Omega}\right)^{-2}\left(j\alpha\, e^{-j\Omega}\right)$$

and the transform we seek is

$$X\left(\Omega\right) = \frac{\alpha\, e^{-j\Omega}}{\left(1-\alpha\, e^{-j\Omega}\right)^{2}}$$

The magnitude and the phase of the transform $X\left(\Omega\right)$ are shown in Fig. 5.31(a) and (b) for the case $\alpha = 0.4$.

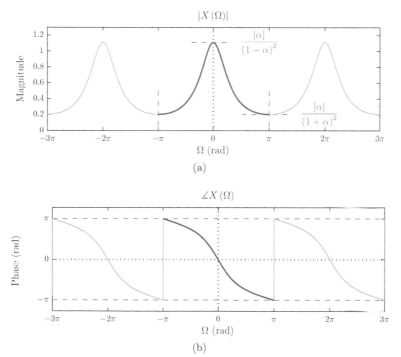

(a)

(b)

Figure 5.31 – The transform $X\left(\Omega\right)$ found in Example 5.18: (a) magnitude, (b) phase.

Software resources:
ex_5_18.m

Convolution property

For two transform pairs

$$x_1[n] \xleftrightarrow{\mathcal{F}} X_1(\Omega) \quad \text{and} \quad x_2[n] \xleftrightarrow{\mathcal{F}} X_2(\Omega)$$

it can be shown that

$$x_1[n] * x_2[n] \xleftrightarrow{\mathcal{F}} X_1(\Omega) X_2(\Omega) \tag{5.148}$$

Convolving two signals in the time domain corresponds to multiplying the corresponding transforms in the frequency domain.

Proof: The convolution of signals $x_1[n]$ and $x_2[n]$ is given by

$$x_1[n] * x_2[n] = \sum_{k=-\infty}^{\infty} x_1[k] \, x_2[n-k] \tag{5.149}$$

Using Eqn. (5.149) in the DTFT definition yields

$$\mathcal{F}\{x_1[n] * x_2[n]\} = \sum_{n=-\infty}^{\infty} \left[\sum_{k=-\infty}^{\infty} x_1[k] \, x_2[n-k] \right] e^{-j\Omega n} \tag{5.150}$$

Changing the order of the two summations in Eqn. (5.150) and rearranging terms we obtain

$$\mathcal{F}\{x_1[n] * x_2[n]\} = \sum_{k=-\infty}^{\infty} x_1[k] \left[\sum_{n=-\infty}^{\infty} x_2[n-k] \, e^{-j\Omega n} \right] \tag{5.151}$$

In Eqn. (5.151) the expression in square brackets should be recognized as the transform of time-shifted signal $x_2[n-k]$, and is equal to $X_2(\Omega) \, e^{-j\Omega k}$. Using this result in Eqn. (5.151) leads to

$$\mathcal{F}\{x_1[n] * x_2[n]\} = \sum_{k=-\infty}^{\infty} x_1[k] \left[X_2(\Omega) \, e^{-j\Omega k} \right]$$

$$= X_2(\Omega) \sum_{k=-\infty}^{\infty} x_1[k] \, e^{-j\Omega k}$$

$$= X_1(\Omega) X_2(\Omega) \tag{5.152}$$

Example 5.19: Convolution using the DTFT

Two signals are given as

$$h[n] = \left(\frac{2}{3}\right)^n u[n] \quad \text{and} \quad x[n] = \left(\frac{3}{4}\right)^n u[n]$$

Determine the convolution $y[n] = h[n] * x[n]$ of these two signals using the DTFT.

Solution: Transforms of $H(\Omega)$ and $X(\Omega)$ can easily be found using the result obtained in Example 5.10:

$$H(\Omega) = \frac{1}{1 - \frac{2}{3} e^{-j\Omega}}, \qquad X(\Omega) = \frac{1}{1 - \frac{3}{4} e^{-j\Omega}}$$

Using the convolution property, the transform of $y[n]$ is

$$
\begin{aligned}
Y(\Omega) &= H(\Omega) X(\Omega) \\
&= \frac{1}{\left(1 - \frac{2}{3} e^{-j\Omega}\right)\left(1 - \frac{3}{4} e^{-j\Omega}\right)}
\end{aligned}
\tag{5.153}
$$

The transform $Y(\Omega)$ found in Eqn. (5.153) can be written in the form

$$Y(\Omega) = \frac{-8}{1 - \frac{2}{3} e^{-j\Omega}} + \frac{9}{1 - \frac{3}{4} e^{-j\Omega}}$$

The convolution result $y[n]$ is the inverse DTFT of $Y(\Omega)$ which can be obtained using the linearity property of the transform:

$$y[n] = -8 \left(\frac{2}{3}\right)^n u[n] + 9 \left(\frac{3}{4}\right)^n u[n]$$

Solution of problems of this type will be more practical through the use of z-transform in Chapter 8.

Software resources:
ex_5_19.m

Multiplication of two signals

For two transform pairs

$$x_1[n] \xleftrightarrow{\mathcal{F}} X_1(\Omega) \quad \text{and} \quad x_2[n] \xleftrightarrow{\mathcal{F}} X_2(\Omega) \tag{5.154}$$

it can be shown that

$$x_1[n] x_2[n] \xleftrightarrow{\mathcal{F}} \frac{1}{2\pi} \int_{-\pi}^{\pi} X_1(\lambda) X_2(\Omega - \lambda) \, d\lambda \tag{5.155}$$

The DTFT of the product of two signals $x_1[n]$ and $x_2[n]$ is equal the 2π-periodic convolution of the individual transforms $X_1(\Omega)$ and $X_2(\Omega)$.

Proof: Applying the DTFT definition to the product $x_1[n] x_2[n]$ leads to

$$\mathcal{F}\{x_1[n] x_2[n]\} = \sum_{n=-\infty}^{\infty} x_1[n] x_2[n] e^{-j\Omega n} \tag{5.156}$$

Using the DTFT synthesis equation given by Eqn. (5.70) we get

$$x_1[n] = \frac{1}{2\pi} \int_{-\pi}^{\pi} X_1(\lambda) e^{j\lambda n} \, d\lambda \tag{5.157}$$

Substituting Eqn. (5.157) into Eqn. (5.156)

$$\mathcal{F}\{x_1[n]\, x_2[n]\} = \sum_{n=-\infty}^{\infty} \left[\frac{1}{2\pi} \int_{-\pi}^{\pi} X_1(\lambda)\, e^{j\lambda n}\, d\lambda \right] x_2[n]\, e^{-j\Omega n} \qquad (5.158)$$

Interchanging the order of summation and integration, and rearranging terms yields

$$\mathcal{F}\{x_1[n]\, x_2[n]\} = \frac{1}{2\pi} \int_{-\pi}^{\pi} X_1(\lambda) \left[\sum_{n=-\infty}^{\infty} x_2[n]\, e^{-j(\Omega-\lambda)n} \right] d\lambda$$

$$= \frac{1}{2\pi} \int_{-\pi}^{\pi} X_1(\lambda)\, X_2(\Omega-\lambda)\, d\lambda \qquad (5.159)$$

Table 5.3 contains a summary of key properties of the DTFT. Table 5.4 lists some of the fundamental DTFT pairs.

Theorem	Signal	Transform	
Linearity	$\alpha\, x_1[n] + \beta\, x_2[n]$	$\alpha\, X_1(\Omega) + \beta\, X_2(\Omega)$	
Periodicity	$x[n]$	$X(\Omega) = X(\Omega + 2\pi r)$ for all integers r	
Conjugate symmetry	$x[n]$ real	$X^*(\Omega) = X(-\Omega)$	
		Magnitude:	$|X(-\Omega)| = |X(\Omega)|$
		Phase:	$\Theta(-\Omega) = -\Theta(\Omega)$
		Real part:	$X_r(-\Omega) = X_r(\Omega)$
		Imaginary part:	$X_i(-\Omega) = -X_i(\Omega)$
Conjugate antisymmetry	$x[n]$ imaginary	$X^*(\Omega) = -X(-\Omega)$	
		Magnitude:	$|X(-\Omega)| = |X(\Omega)|$
		Phase:	$\Theta(-\Omega) = -\Theta(\Omega) \mp \pi$
		Real part:	$X_r(-\Omega) = -X_r(\Omega)$
		Imaginary part:	$X_i(-\Omega) = X_i(\Omega)$
Even signal	$x[n] = x[-n]$	$\mathrm{Im}\{X(\Omega)\} = 0$	
Odd signal	$x[n] = -x[-n]$	$\mathrm{Re}\{X(\Omega)\} = 0$	
Time shifting	$x[n-m]$	$X(\Omega)\, e^{-j\Omega m}$	
Time reversal	$x[-n]$	$X(-\Omega)$	
Conjugation	$x^*[n]$	$X^*(-\Omega)$	
Frequency shifting	$x[n]\, e^{j\Omega_0 n}$	$X(\Omega - \Omega_0)$	
Modulation	$x[n]\cos(\Omega_0 n)$	$\frac{1}{2}\left[X(\Omega - \Omega_0) + X(\Omega + \Omega_0) \right]$	
	$x[n]\sin(\Omega_0 n)$	$\frac{1}{2}\left[X(\Omega - \Omega_0)\, e^{-j\pi/2} - X(\Omega + \Omega_0)\, e^{j\pi/2} \right]$	
Differentiation in frequency	$n^m\, x[n]$	$j^m \dfrac{d^m}{d\Omega^m}\left[X(\Omega) \right]$	
Convolution	$x_1[n] * x_2[n]$	$X_1(\Omega)\, X_2(\Omega)$	
Multiplication	$x_1[n]\, x_2[n]$	$\dfrac{1}{2\pi} \displaystyle\int_{-\pi}^{\pi} X_1(\lambda)\, X_2(\Omega - \lambda)\, d\lambda$	
Parseval's theorem	$\displaystyle\sum_{n=-\infty}^{\infty} |x[n]|^2 = \frac{1}{2\pi} \int_{-\pi}^{\pi} |X(\Omega)|^2\, d\Omega$		

Table 5.3 – DTFT properties.

Name	Signal	Transform
Discrete-time pulse	$x[n] = \begin{cases} 1, & |n| \leq L \\ 0 & \text{otherwise} \end{cases}$	$X(\Omega) = \dfrac{\sin\left(\dfrac{(2L+1)\,\Omega}{2}\right)}{\sin\left(\dfrac{\Omega}{2}\right)}$
Unit-impulse signal	$x[n] = \delta[n]$	$X(\Omega) = 1$
Constant-amplitude signal	$x[n] = 1$, all n	$X(\Omega) = 2\pi \displaystyle\sum_{m=-\infty}^{\infty} \delta(\Omega - 2\pi m)$
Sinc function	$x[n] = \dfrac{\Omega_c}{\pi}\,\text{sinc}\left(\dfrac{\Omega_c n}{\pi}\right)$	$X(\Omega) = \begin{cases} 1, & |\Omega| < \Omega_c \\ 0, & \text{otherwise} \end{cases}$
Right-sided exponential	$x[n] = \alpha^n\, u[n]$, $|\alpha| < 1$	$X(\Omega) = \dfrac{1}{1 - \alpha\,e^{-j\Omega}}$
Complex exponential	$x[n] = e^{j\Omega_0 n}$	$X(\Omega) = 2\pi \displaystyle\sum_{m=-\infty}^{\infty} \delta(\Omega - \Omega_0 - 2\pi m)$

Table 5.4 – Some DTFT transform pairs.

5.3.6   Applying DTFT to periodic signals

In previous sections of this chapter we have distinguished between two types of discrete-time signals: periodic and non-periodic. For periodic discrete-time signals we have the DTFS as an analysis and problem solving tool; for non-periodic discrete-time signals the DTFT serves a similar purpose. This arrangement will serve us adequately in cases where we work with one type of signal or the other. There may be times, however, when we need to mix periodic and non-periodic signals within one system. For example, in amplitude modulation, a non-periodic signal may be multiplied with a periodic signal, and we may need to analyze the resulting product in the frequency domain. Alternately, a periodic signal may be used as input to a system the impulse response of which is non-periodic. In these types of scenarios it would be convenient to find a way to use the DTFT for periodic signals as well. We have seen in Example 5.14 that a DTFT can be found for a constant-amplitude signal that does not satisfy the existence conditions, as long as we are willing to accept singularity functions in the transform. The next two examples will expand on this idea. Afterwards we will develop a technique for converting the DTFS of any periodic discrete-time signal to a DTFT.

Example 5.20: **DTFT of complex exponential signal**

Determine the transform of the complex exponential signal $x[n] = e^{j\Omega_0 n}$ with $-\pi < \Omega_0 < \pi$.

Solution: The transform of the constant unit-amplitude signal was found in Example 5.14 to be

$$\mathcal{F}\{1\} = 2\pi \sum_{m=-\infty}^{\infty} \delta(\Omega - 2\pi m)$$

Using this result along with the frequency shifting property of the DTFT given by Eqn. (5.136), the transform of the complex exponential signal is

$$\mathcal{F}\left\{e^{j\Omega_0 n}\right\} = 2\pi \sum_{m=-\infty}^{\infty} \delta(\Omega - \Omega_0 - 2\pi m) \tag{5.160}$$

This is illustrated in Fig. 5.32.

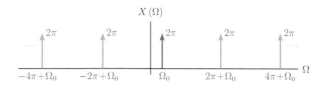

Figure 5.32 – The transform of complex exponential signal $x[n] = e^{j\Omega_0 n}$.

Example 5.21: **DTFT of sinusoidal signal**

Determine the transform of the sinusoidal signal $x[n] = \cos{(\Omega_0 n)}$ with $-\pi < \Omega_0 < \pi$.

Solution: The transform of the constant unit-amplitude signal was found in Example 5.14 to be

$$\mathcal{F}\{1\} = 2\pi \sum_{m=-\infty}^{\infty} \delta\left(\Omega - 2\pi m\right)$$

Using this result along with the modulation property of the DTFT given by Eqn. (5.138), the transform of the sinusoidal signal is

$$\mathcal{F}\{\cos{(\Omega_0 n)}\} = \pi \sum_{m=-\infty}^{\infty} \delta\left(\Omega - \Omega_0 - 2\pi m\right) + \pi \sum_{m=-\infty}^{\infty} \delta\left(\Omega + \Omega_0 - 2\pi m\right) \tag{5.161}$$

Let $\bar{X}(\Omega)$ represent the part of the transform in the range $-\pi < \Omega < \pi$.

$$\bar{X}(\Omega) = \pi\,\delta\left(\Omega - \Omega_0\right) + \pi\,\delta\left(\Omega + \Omega_0\right) \tag{5.162}$$

The DTFT can now be expressed as

$$X(\Omega) = \sum_{m=-\infty}^{\infty} \bar{X}\left(\Omega - 2\pi m\right) \tag{5.163}$$

This is illustrated in Fig. 5.33.

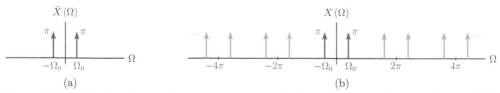

Figure 5.33 – The transform of complex exponential signal $x[n] = \cos{(\Omega_0 n)}$: (a) the middle part $\bar{X}(\Omega)$, (b) the complete transform $X(\Omega)$.

In Examples 5.20 and 5.21 we were able to obtain the DTFT of two periodic signals, namely a complex exponential signal and a sinusoidal signal. The idea can be generalized to apply to any periodic discrete-time signal. The DTFS synthesis equation for a periodic signal $\tilde{x}[n]$ was given by Eqn. (5.21) which is repeated here for convenience:

$$\text{Eqn. (5.21):} \qquad \tilde{x}[n] = \sum_{k=0}^{N-1} \tilde{c}_k\, e^{j(2\pi/N)kn}\,, \qquad \text{all } n$$

If we were to attempt to find the DTFT of the signal $\tilde{x}[n]$ by direct application of the DTFT analysis equation given by Eqn. (5.71) we would need to evaluate

$$X(\Omega) = \sum_{n=-\infty}^{\infty} \tilde{x}[n] e^{-j\Omega n} \tag{5.164}$$

Substituting Eqn. (5.21) into Eqn. (5.164) leads to

$$X(\Omega) = \sum_{n=-\infty}^{\infty} \left[\sum_{k=0}^{N-1} \tilde{c}_k e^{j(2\pi/N)kn} \right] e^{-j\Omega n} \tag{5.165}$$

Interchanging the order of the two summations in Eqn. (5.165) and rearranging terms we obtain

$$X(\Omega) = \sum_{k=0}^{N-1} \tilde{c}_k \left[\sum_{n=-\infty}^{\infty} e^{j(2\pi/N)kn} e^{-j\Omega n} \right] \tag{5.166}$$

In Eqn. (5.166) the expression in square brackets is the DTFT of the signal $e^{j(2\pi/N)kn}$. Using the result obtained in Example 5.20, and remembering that $2\pi/N = \Omega_0$ is the fundamental angular frequency for the periodic signal $\tilde{x}[n]$, we get

$$\sum_{n=-\infty}^{\infty} e^{j(2\pi/N)kn} e^{-j\Omega n} = 2\pi \sum_{m=-\infty}^{\infty} \delta\left(\Omega - \frac{2\pi k}{N} - 2\pi m\right)$$

$$= 2\pi \sum_{m=-\infty}^{\infty} \delta\left(\Omega - k\Omega_0 - 2\pi m\right) \tag{5.167}$$

and

$$X(\Omega) = 2\pi \sum_{k=0}^{N-1} \tilde{c}_k \sum_{m=-\infty}^{\infty} \delta\left(\Omega - k\Omega_0 - 2\pi m\right) \tag{5.168}$$

The part of the transform in the range $0 < \Omega < 2\pi$ is found by setting $m = 0$ in Eqn. (5.168):

$$\bar{X}(\Omega) = 2\pi \sum_{k=0}^{N-1} \tilde{c}_k \, \delta\left(\Omega - k\Omega_0\right) \tag{5.169}$$

Thus, $\bar{X}(\Omega)$ for a periodic discrete-time signal is obtained by converting each DTFS coefficient $\tilde{c}_k$ to an impulse with area equal to $2\pi\tilde{c}_k$ and placing it at angular frequency $\Omega = k\Omega_0$. The DTFT for the signal is then obtained as

$$X(\Omega) = \sum_{m=-\infty}^{\infty} \bar{X}(\Omega - 2\pi m) \tag{5.170}$$

This process is illustrated in Fig. 5.34.

Note: In Eqn. (5.162) of Example 5.21 we have used $\bar{X}(\Omega)$ to represent the part of the transform in the range $-\pi < \Omega < \pi$. On the other hand, in Eqn. (5.169) above, $\bar{X}(\Omega)$ was used as the part of the transform in the range $0 < \Omega < 2\pi$. This should not cause any confusion. In general, $\bar{X}(\Omega)$ represents one period of the 2π-periodic transform, and the starting value of Ω is not important.

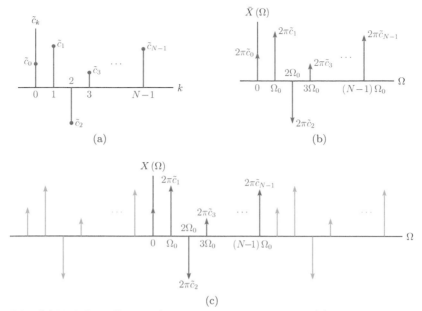

(a)

(b)

(c)

Figure 5.34 – (a) DTFS coefficients for a signal periodic with N, (b) DTFT for $-\pi < \Omega < \pi$, (c) complete DTFT.

Example 5.22: **DTFT of the periodic signal of Example 5.4**

Determine the DTFT of the signal $x[n]$ used in Example 5.4 and shown in Fig. 5.3.

Solution: We will first write the transform in the interval $0 < \Omega < 2\pi$.

$$\bar{X}(\Omega) = 2\pi \left[\tilde{c}_0\,\delta(\Omega) + \tilde{c}_1\,\delta(\Omega-\Omega_0) + \tilde{c}_2\,\delta(\Omega-2\Omega_0) + \tilde{c}_3\,\delta(\Omega-3\Omega_0) + \tilde{c}_4\,\delta(\Omega-4\Omega_0) \right]$$

The period of the signal $\tilde{x}[n]$ is $N = 5$, therefore the fundamental angular frequency is $\Omega_0 = 2\pi/5$. Using the values of DTFS coefficients found in Example 5.4 we obtain

$$\begin{aligned}
\bar{X}(\Omega) &= 12.566\,\delta(\Omega) + (-3.142 - j4.324)\,\delta(\Omega-2\pi/5) \\
&\quad + (-3.142 - j1.021)\,\delta(\Omega-4\pi/5) \\
&\quad + (-3.142 + j1.021)\,\delta(\Omega-6\pi/5) \\
&\quad + (-3.142 + j4.324)\,\delta(\Omega-8\pi/5)
\end{aligned} \tag{5.171}$$

The complete transform $X(\Omega)$ is found by periodically extending $\bar{X}(\Omega)$.

$$\begin{aligned}
X(\Omega) = \sum_{m=-\infty}^{\infty} &\Big[12.566\,\delta(\Omega-2\pi m) + (-3.142 - j4.324)\,\delta(\Omega-2\pi/5-2\pi m) \\
&+ (-3.142 - j1.021)\,\delta(\Omega-4\pi/5-2\pi m) \\
&+ (-3.142 + j1.021)\,\delta(\Omega-6\pi/5-2\pi m) \\
&+ (-3.142 + j4.324)\,\delta(\Omega-8\pi/5-2\pi m) \Big]
\end{aligned} \tag{5.172}$$

5.4   Energy and Power in the Frequency Domain

Parseval's theorem and its application to the development of energy and power spectral density concepts were discussed for continuous-time signals in Section 4.4 of Chapter 4. In this section we will discuss Parseval's theorem for periodic and non-periodic discrete-time signals, and introduce the corresponding energy and power spectral density concepts.

5.4.1   Parseval's theorem

For a periodic power signal $\tilde{x}[n]$ with period N and DTFS coefficients $\{\tilde{c}_k,\ k = 0, \ldots, N-1\}$ it can be shown that

$$\frac{1}{N} \sum_{n=0}^{N-1} |\tilde{x}[n]|^2 = \sum_{k=0}^{N-1} |\tilde{c}_k|^2 \tag{5.173}$$

For a non-periodic energy signal $x[n]$ with DTFT $X(\Omega)$, the following holds true:

$$\sum_{n=-\infty}^{\infty} |\tilde{x}[n]|^2 = \frac{1}{2\pi} \int_{-\pi}^{\pi} |X(\Omega)|^2 \, d\Omega \tag{5.174}$$

The left side of Eqn. (5.173) represents the normalized average power in a periodic signal which was derived in Chapter 1 Eqn. (1.187). The left side of Eqn. (5.174) represents the normalized signal energy as derived in Eqn. (1.180). The relationships given by Eqns. (5.173) and (5.174) relate signal energy or signal power to the frequency-domain representation of the signal. They are two forms of *Parseval's theorem*.

Proofs: First we will prove Eqn. (5.173). DTFS representation of a periodic discrete-time signal with period N was given by Eqn. (5.21), and is repeated here for convenience:

$$\text{Eqn. (5.21):} \qquad \tilde{x}[n] = \sum_{k=0}^{N-1} \tilde{c}_k \, e^{j(2\pi/N)kn}, \qquad \text{all } n$$

Using $|x[n]|^2 = x[n]\, x^*[n]$, the left side of Eqn. (5.173) can be written as

$$\frac{1}{N} \sum_{n=0}^{N-1} |\tilde{x}[n]|^2 = \frac{1}{N} \sum_{n=0}^{N-1} \left[\sum_{k=0}^{N-1} \tilde{c}_k \, e^{j(2\pi/N)kn} \right] \left[\sum_{m=0}^{N-1} \tilde{c}_m \, e^{j(2\pi/N)mn} \right]^*$$

$$= \frac{1}{N} \sum_{n=0}^{N-1} \left[\sum_{k=0}^{N-1} \tilde{c}_k \, e^{j(2\pi/N)kn} \right] \left[\sum_{m=0}^{N-1} \tilde{c}_m^* \, e^{-j(2\pi/N)mn} \right] \tag{5.175}$$

Rearranging the order of the summations in Eqn. (5.175) we get

$$\frac{1}{N} \sum_{n=0}^{N-1} |\tilde{x}[n]|^2 = \frac{1}{N} \sum_{k=0}^{N-1} \tilde{c}_k \left[\sum_{m=0}^{N-1} \tilde{c}_m^* \left[\sum_{n=0}^{N-1} e^{j(2\pi/N)(k-m)n} \right] \right] \tag{5.176}$$

Using orthogonality of the basis function set $\left\{e^{j(2\pi/N)kn}\ ,\ k=0,\ldots,N-1\right\}$ it can be shown that (see Appendix D)

$$\sum_{n=0}^{N-1} e^{j(2\pi/N)(k-m)n} = \left\{ \begin{array}{ll} N\,, & k-m=0,\mp N,\mp 2N,\ldots \\ 0\,, & \text{otherwise} \end{array} \right. \tag{5.177}$$

Using Eqn. (5.177) in Eqn. (5.176) leads to the desired result:

$$\frac{1}{N} \sum_{n=0}^{N-1} |\tilde{x}[n]|^2 = \sum_{k=0}^{N-1} \tilde{c}_k \tilde{c}_k^* = \sum_{k=0}^{N-1} |\tilde{c}_k|^2 \tag{5.178}$$

The proof for Eqn. (5.174) is similar. The DTFT synthesis equation was given by Eqn. (5.70) and is repeated here for convenience.

$$\text{Eqn. (5.70):} \qquad x[n] = \frac{1}{2\pi} \int_{-\pi}^{\pi} X\left(\Omega\right) e^{j\Omega n}\, d\Omega$$

The left side of Eqn. (5.174) can be written as

$$\sum_{n=-\infty}^{\infty} |\tilde{x}[n]|^2 = \sum_{n=-\infty}^{\infty} x[n] \left[\frac{1}{2\pi} \int_{-\pi}^{\pi} X\left(\Omega\right) e^{j\Omega n}\, d\Omega \right]^*$$

$$= \sum_{n=-\infty}^{\infty} x[n] \left[\frac{1}{2\pi} \int_{-\pi}^{\pi} X^*\left(\Omega\right) e^{-j\Omega n}\, d\Omega \right] \tag{5.179}$$

Interchanging the order of the integral and the summation in Eqn. (5.179) and rearranging terms yields

$$\sum_{n=-\infty}^{\infty} |\tilde{x}[n]|^2 = \frac{1}{2\pi} \int_{-\pi}^{\pi} X^*\left(\Omega\right) \left[\sum_{n=-\infty}^{\infty} x[n]\, e^{-j\Omega n} \right] d\Omega$$

$$= \frac{1}{2\pi} \int_{-\pi}^{\pi} |X\left(\Omega\right)|^2\, d\Omega \tag{5.180}$$

5.4.2 Energy and power spectral density

The two statements of Parseval's theorem given by Eqns. (5.173) and (5.174) lead us to the following conclusions:

1. In Eqn. (5.173) the left side corresponds to the normalized average power of the periodic signal $\tilde{x}[n]$, and therefore the summation on the right side must also represent normalized average power. The term $|\tilde{c}_k|^2$ corresponds to the power of the signal component at angular frequency $\Omega = k\Omega_0$. (Remember that $\Omega_0 = 2\pi/N$.) Let us construct a new function $S_x\left(\Omega\right)$ as follows:

$$S_x\left(\Omega\right) = \sum_{k=-\infty}^{\infty} 2\pi\, |\tilde{c}_k|^2\, \delta\left(\Omega - k\Omega_0\right) \tag{5.181}$$

The function $S_x(\Omega)$ consists of impulses placed at angular frequencies $k\Omega_0$ as illustrated in Fig. 5.35. Since the DTFS coefficients are periodic with period N, the function $S_x(\Omega)$ is 2π-periodic.

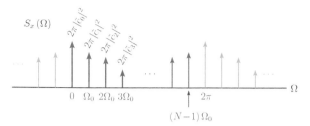

Figure 5.35 – The function $S_x(\Omega)$ constructed using the DTFS coefficents of the signal $\tilde{x}[n]$.

Integrating $S_x(\Omega)$ over an interval of 2π radians leads to

$$\int_0^{2\pi} S_x(\Omega)\, d\Omega = \int_0^{2\pi} \left[\sum_{k=-\infty}^{\infty} 2\pi\, |\tilde{c}_k|^2\, \delta(\Omega - k\Omega_0) \right] d\Omega \qquad (5.182)$$

Interchanging the order of integration and summation in Eqn. (5.182) and rearranging terms we obtain

$$\int_0^{2\pi} S_x(\Omega)\, d\Omega = 2\pi \sum_{k=-\infty}^{\infty} |\tilde{c}_k|^2 \left[\int_0^{2\pi} \delta(\Omega - k\Omega_0)\, d\Omega \right] \qquad (5.183)$$

Recall that $\Omega_0 = 2\pi/N$. Using the sifting property of the impulse function, the integral between square brackets in Eqn. (5.183) is evaluated as

$$\int_0^{2\pi} \delta(\Omega - k\Omega_0)\, d\Omega = \begin{cases} 1\,, & k = 0, \ldots, N-1 \\ 0\,, & \text{otherwise} \end{cases} \qquad (5.184)$$

and Eqn. (5.183) becomes

$$\int_0^{2\pi} S_x(\Omega)\, d\Omega = 2\pi \sum_{k=0}^{N-1} |\tilde{c}_k|^2 \qquad (5.185)$$

The normalized average power of the signal $\tilde{x}[n]$ is therefore found as

$$P_x = \sum_{k=0}^{N-1} |\tilde{c}_k|^2 = \frac{1}{2\pi} \int_0^{2\pi} S_x(\Omega)\, d\Omega \qquad (5.186)$$

Consequently, the function $S_x(\Omega)$ is the *power spectral density* of the signal $\tilde{x}[n]$. Since $S_x(\Omega)$ is 2π-periodic, the integral in Eqn. (5.186) can be started at any value of Ω as long as it covers a span of 2π radians. It is usually more convenient to write the integral in Eqn. (5.186) to start at $-\pi$.

$$P_x = \frac{1}{2\pi} \int_{-\pi}^{\pi} S_x(\Omega)\, d\Omega \qquad (5.187)$$

As an interesting by-product of Eqn. (5.186), the normalized average power of $\tilde{x}[n]$ that is within a specific angular frequency range can be determined by integrating

$S_x(\Omega)$ over that range. For example, the power contained at angular frequencies in the range $(-\Omega_0, \Omega_0)$ is

$$P_x \text{ in } (-\Omega_0, \Omega_0) = \frac{1}{2\pi} \int_{-\Omega_0}^{\Omega_0} S_x(\Omega) \, d\Omega \tag{5.188}$$

2. In Eqn. (5.174) the left side is the normalized signal energy for the signal $x[n]$, and the right side must be the same. The integrand $|X(\Omega)|^2$ is therefore the *energy spectral density* of the signal $x[n]$. Let the function $G_x(\Omega)$ be defined as

$$G_x(\Omega) = |X(\Omega)|^2 \tag{5.189}$$

Substituting Eqn. (5.189) into Eqn. (5.174), the normalized energy in the signal $x[n]$ can be expressed as

$$E_x = \sum_{n=-\infty}^{\infty} |\tilde{x}[n]|^2 = \frac{1}{2\pi} \int_{-\pi}^{\pi} G(\Omega) \, d\Omega \tag{5.190}$$

The energy contained at angular frequencies in the range $(-\Omega_0, \Omega_0)$ is found by integrating $G_x(\Omega)$ in the frequency range of interest:

$$E_x \text{ in } (-\Omega_0, \Omega_0) = \frac{1}{2\pi} \int_{-\Omega_0}^{\Omega_0} G_x(\Omega) \, d\Omega \tag{5.191}$$

3. In Eqn. (5.174) we have expressed Parseval's theorem for an energy signal, and used it to lead to the derivation of the energy spectral density in Eqn. (5.189). We know that some non-periodic signals are power signals, therefore their energy cannot be computed. The example of one such signal is the unit-step function. The power of a non-periodic signal was defined in Eqn. (1.188) in Chapter 1 which will be repeated here:

$$\text{Eqn. (1.188):} \qquad P_x = \lim_{M \to \infty} \left[\frac{1}{2M+1} \sum_{n=-M}^{M} |x[n]|^2 \right]$$

In order to write the counterpart of Eqn. (5.174) for a power signal, we will first define a *truncated* version of $x[n]$ as

$$x_T[n] = \begin{cases} x[n], & -M \leq n \leq M \\ 0, & \text{otherwise} \end{cases} \tag{5.192}$$

Let $X_T(\Omega)$ be the DTFT of the truncated signal $x_T[n]$:

$$X_T(\Omega) = \mathcal{F}\{x_T[n]\} = \sum_{n=-M}^{M} x_T[n] \, e^{-j\Omega n} \tag{5.193}$$

Now Eqn. (5.174) can be written in terms of the truncated signal and its transform as

$$\sum_{n=-M}^{M} |x_T[n]|^2 = \frac{1}{2\pi} \int_{-\pi}^{\pi} |X_T(\Omega)|^2 \, d\Omega \tag{5.194}$$

Scaling both sides of Eqn. (5.194) by $(2M+1)$ and taking the limit as M becomes infinitely large, we obtain

$$\lim_{M \to \infty} \left[\frac{1}{2M+1} \sum_{n=-M}^{M} |x_T[n]|^2 \right] = \lim_{M \to \infty} \left[\frac{1}{2\pi(2M+1)} \int_{-\pi}^{\pi} |X_T(\Omega)|^2 \, d\Omega \right] \tag{5.195}$$

The left side of Eqn. (5.195) is the average normalized power in a non-periodic signal as we have established in Eqn. (1.188). Therefore, the power spectral density of $x[n]$ is

$$S_x(\Omega) = \lim_{M \to \infty} \left[\frac{1}{2M+1} |X_T(\Omega)|^2 \right] \tag{5.196}$$

Example 5.23: **Normalized average power for waveform of Example 5.7**

Consider again the periodic waveform used in Example 5.7 and shown in Fig. 5.6. Using the DTFS coefficients found in Example 5.7, verify Parseval's theorem. Also determine the percentage of signal power in the angular frequency range $-\pi/3 < \Omega_0 < \pi/3$.

Solution: The average power computed from the signal is

$$
\begin{aligned}
P_x &= \frac{1}{8} \sum_{n=0}^{7} |\tilde{x}[n]|^2 \\
&= \frac{1}{8} \left[0 + \left(\frac{1}{2}\right)^2 + (1)^2 + \left(\frac{3}{4}\right)^2 + 0 + \left(-\frac{3}{4}\right)^2 + (-1)^2 + \left(-\frac{1}{2}\right)^2 \right] \\
&= 0.4531
\end{aligned}
$$

Using the DTFS coefficients in Table 5.1 yields

$$
\begin{aligned}
\sum_{k=0}^{7} |\tilde{c}_k|^2 &= 0 + (0.4710)^2 + (0.0625)^2 + (0.0290)^2 + 0 + (0.0290)^2 + (0.0625)^2 + (0.4710)^2 \\
&= 0.4531
\end{aligned}
$$

As expected, the value found from DTFS coefficients matches that found from the signal. One period of the power spectral density $S_x(\Omega)$ is

$$
\begin{aligned}
S_x(\Omega) =\ &0.0017\,\pi\,\delta\,(\Omega + 3\pi/4) + 0.0078\,\pi\,\delta\,(\Omega + \pi/2) + 0.4436\,\pi\,\delta\,(\Omega + \pi/4) \\
&+ 0.4436\,\pi\,\delta\,(\Omega - \pi/4) + 0.0078\,\pi\,\delta\,(\Omega - \pi/2) + 0.0017\,\pi\,\delta\,(\Omega - 3\pi/4)
\end{aligned}
$$

The power in the angular frequency range of interest is

$$
\begin{aligned}
P_x \text{ in } (-\pi/3, \pi/3) &= \frac{1}{2\pi} \int_{-\pi/3}^{\pi/3} S_x(\Omega)\, d\Omega \\
&= \frac{1}{2\pi} (0.4436\,\pi + 0.4436\,\pi) \\
&= 0.4436
\end{aligned}
$$

The ratio of the power in $-\pi/3 < \Omega_0 < \pi/3$ to the total signal power is

$$\frac{P_x \text{ in } (-\pi/3,\ \pi/3)}{P_x} = \frac{0.4436}{0.4531} = 0.979$$

It appears that 97.9 percent of the power of the signal is in the frequency range $-\pi/3 < \Omega_0 < \pi/3$.

Software resources:

ex_5_23.m

Example 5.24: **Power spectral density of a discrete-time sinusoid**

Find the power spectral density of the signal

$$\tilde{x}[n] = 3 \cos (0.2\pi n)$$

Solution: Using Euler's formula, the signal can be written as

$$\tilde{x}[n] = \frac{3}{2} e^{-j0.2\pi n} + \frac{3}{2} e^{j0.2\pi n}$$

The normalized frequency of the sinusoidal signal is $F_0 = 0.1$ corresponding to a period length of $N = 10$ samples. Non-zero DTFS coefficients for $k = 0, \ldots, 9$ are

$$\tilde{c}_1 = \tilde{c}_9 = \frac{3}{2}$$

Using Eqn. (5.181), the power spectral density is

$$S_x (\Omega) = \sum_{r=-\infty}^{\infty} \left[\frac{9\pi}{2} \delta (\Omega - 0.2\pi - 2\pi r) + \frac{9\pi}{2} \delta (\Omega - 1.8\pi - 2\pi r) \right]$$

which is shown in Fig. 5.36.

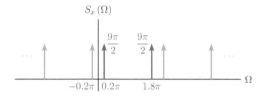

Figure 5.36 – Power spectral density for Example 5.24.

The power in the sinusoidal signal is computed from the power spectral density as

$$P_x = \frac{1}{2\pi} \int_{-\pi}^{\pi} S_x (\Omega) \, d\Omega = \frac{1}{2\pi} \left(\frac{9\pi}{2} + \frac{9\pi}{2} \right) = \frac{9}{2}$$

Example 5.25: **Energy spectral density of a discrete-time pulse**

Determine the energy spectral density of the rectangular pulse

$$x[n] = \begin{cases} 1, & n = -5, \ldots, 5 \\ 0, & \text{otherwise} \end{cases}$$

Also compute the energy of the signal in the frequency interval $-\pi/10 < \Omega < \pi/10$.

Solution: Using the general result obtained in Example 5.12 with $L = 5$, the DTFT of the signal $x[n]$ is

$$X (\Omega) = \frac{\sin \left(\dfrac{11\,\Omega}{2} \right)}{\sin \left(\dfrac{\Omega}{2} \right)}$$

The energy spectral density for $x[n]$ is found as

$$G_x\left(\Omega\right) = \left|X\left(\Omega\right)\right|^2 = \frac{\sin^2\left(\dfrac{11\,\Omega}{2}\right)}{\sin^2\left(\dfrac{\Omega}{2}\right)}$$

which is shown in Fig. 5.37.

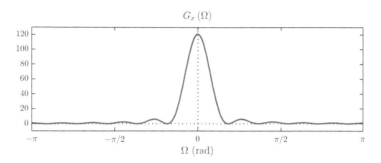

Figure 5.37 – Power spectral density for Example 5.25.

The energy of the signal within the frequency interval $-\pi/10 < \Omega < \pi/10$ is computed as

$$E_x \ \text{in} \ \left(\pi/10,\,\pi/10\right) = \frac{1}{2\pi}\int_{-\pi/10}^{\pi/10} G_x\left(\Omega\right)\,d\Omega$$

which is proportional to the shaded area under $G\left(\Omega\right)$ in Fig. 5.38.

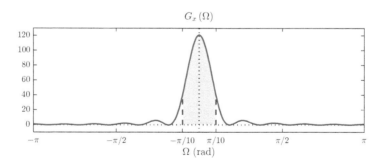

Figure 5.38 – The area under $G_x\left(\Omega\right)$ for $-\pi/10 < \Omega/\pi/10$.

Direct evaluation of the integral in Eqn. (5.197) is difficult; however, the result can be obtained by numerical approximation of the integral, and is

$$E_x \ \text{in} \ \left(\pi/10,\,\pi/10\right) \approx 8.9309$$

Software resources:
ex_5_25.m

Energy or power in a frequency range

Signal power or signal energy that is within a finite range of frequencies is found through the use of Eqns. (5.188) and (5.191) respectively. An interesting interpretation of the result in Eqn. (5.188) is that, for the case of a power signal, the power of $x[n]$ in the frequency range $-\Omega_0 < \Omega < \Omega_0$ is the same as the power of the output signal of a system with system function

$$H\left(\Omega\right) = \begin{cases} 1\,, & |\Omega| < \Omega_0 \\ 0\,, & \Omega_0 < |\Omega| < \pi \end{cases} \tag{5.197}$$

driven by the signal $x[n]$. Similarly, for the case of an energy signal, the energy of $x[n]$ in the frequency range $-\Omega_0 < \Omega < \Omega_0$ is the same as the energy of the output signal of a system with the system function in Eqn. (5.197) driven by the signal $x[n]$. These relationships are illustrated in Fig. 5.39(a) and (b).

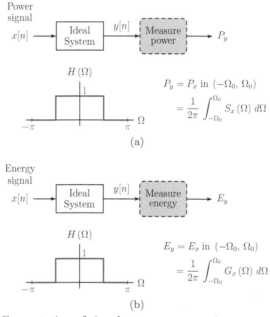

Figure 5.39 – Computation of signal power or energy in a range of frequencies.

5.4.3   Autocorrelation

The energy spectral density $G_x\left(\Omega\right)$ or the power spectral density $S_x\left(\Omega\right)$ for a signal can be computed through direct application of the corresponding equations derived in Section 5.4.2; namely Eqn. (5.189) for an energy signal, and either Eqn. (5.181) or Eqn. (5.196) for a power signal, depending on its type.   In some circumstances it is also possible to compute either function from the knowledge of the *autocorrelation function* which will be defined in this section. Let $x[n]$ be a real-valued signal.

For an energy signal $x[n]$ the *autocorrelation function* is defined as

$$r_{xx}[m] = \sum_{n=-\infty}^{\infty} x[n]\,x[n+m] \tag{5.198}$$

For a periodic power signal $\tilde{x}[n]$ with period N, the corresponding definition of the auto-correlation function is

$$\tilde{r}_{xx}[m] = \langle \tilde{x}[n]\, \tilde{x}[n+m] \rangle$$

$$= \frac{1}{N} \sum_{n=0}^{N-1} \tilde{x}[n]\, \tilde{x}[n+m] \qquad (5.199)$$

The triangle brackets indicate time average. The autocorrelation function for a periodic signal is also periodic as signified by the tilde ($\tilde{\ }$) character used over the symbol r_{xx} in Eqn. (5.199). Finally, for a non-periodic power signal, the corresponding definition is

$$r_{xx}[m] = \langle x[n]\, x[n+m] \rangle$$

$$= \lim_{M\to\infty} \left[\frac{1}{2M+1} \sum_{n=-M}^{M} x[n]\, x[n+m] \right] \qquad (5.200)$$

Even though we refer to $r_{xx}[m]$ as a *function*, we will often treat it as if it is a discrete-time signal, albeit one that uses a different index, m, from the signal $x[n]$ for which it is computed. The index m simply corresponds to the time shift between the two copies of $x[n]$ used in the definitions of Eqns. (5.198) through (5.200).

It can be shown that, for an energy signal, the energy spectral density is the DTFT of the autocorrelation function, that is,

$$\mathcal{F}\{r_{xx}[m]\} = G_x\,(\Omega) \qquad (5.201)$$

Proof: Let us begin by applying the DTFT definition to the autocorrelation function $r_{xx}[m]$ treated as a discrete-time signal indexed by m:

$$\mathcal{F}\{r_{xx}[m]\} = \sum_{m=-\infty}^{\infty} \left[\sum_{n=-\infty}^{\infty} x[n]\, x[n+m] \right] e^{-j\Omega m} \qquad (5.202)$$

The two summations in Eqn. (5.202) can be rearranged to yield

$$\mathcal{F}\{r_{xx}[m]\} = \sum_{n=-\infty}^{\infty} x[n] \left[\sum_{m=-\infty}^{\infty} x[n+m]\, e^{-j\Omega m} \right] \qquad (5.203)$$

Realizing that the inner summation in Eqn. (5.203) is equal to

$$\sum_{m=-\infty}^{\infty} x[n+m]\, e^{-j\Omega m} = \mathcal{F}\{x[m+n]\} = e^{j\Omega n}\, X\,(\Omega) \qquad (5.204)$$

Eqn. (5.203) becomes

$$\mathcal{F}\{r_{xx}[m]\} = X\,(\Omega) \sum_{n=-\infty}^{\infty} x[n]\, e^{j\Omega n} \qquad (5.205)$$

and since $x[n]$ is real, we obtain

$$\mathcal{F}\{r_{xx}[m]\} = X(\Omega) \, X^*(\Omega) = |X(\Omega)|^2 \tag{5.206}$$

which completes the proof.

The definition of the autocorrelation function for a periodic signal $\tilde{x}[n]$ can be used in a similar manner in determining the power spectral density of the signal.

Let the signal $\tilde{x}[n]$ and the autocorrelation function $\tilde{r}_{xx}[m]$, both periodic with period N, have the DTFS representations given by

$$\tilde{x}[n] = \sum_{k=0}^{N-1} \tilde{c}_k \, e^{j(2\pi/N)kn} \tag{5.207}$$

and

$$\tilde{r}_{xx}[m] = \sum_{k=0}^{N-1} \tilde{d}_k \, e^{j(2\pi/N)km} \tag{5.208}$$

respectively. It can be shown that the DTFS coefficients $\{\tilde{c}_k\}$ and $\left\{\tilde{d}_k\right\}$ are related by

$$\tilde{d}_k = |\tilde{c}_k|^2 \tag{5.209}$$

and the power spectral density is the DTFT of the autocorrelation function, that is,

$$\tilde{S}_x(\Omega) = \mathcal{F}\{\tilde{r}_{xx}[m]\} \tag{5.210}$$

Proof: Using the DTFS analysis equation given by Eqn. (5.22) with the definition of the autocorrelation function in Eqn. (5.199) leads to

$$\tilde{d}_k = \frac{1}{N} \sum_{m=0}^{N-1} \tilde{r}_{xx}[m] \, e^{-j(2\pi/N)km}$$

$$= \frac{1}{N} \sum_{m=0}^{N-1} \left[\frac{1}{N} \sum_{n=0}^{N-1} \tilde{x}[n] \, \tilde{x}[n+m] \right] e^{-j(2\pi/N)km} \tag{5.211}$$

Rearranging the order of the two summations Eqn. (5.211) can be written as

$$\tilde{d}_k = \frac{1}{N} \sum_{n=0}^{N-1} \tilde{x}[n] \left[\frac{1}{N} \sum_{m=0}^{N-1} \tilde{x}[n+m] \, e^{-j(2\pi/N)km} \right] \tag{5.212}$$

Using the time shifting property of the DTFS given by Eqn. (5.30) the expression in square brackets in Eqn. (5.212) is

$$\frac{1}{N} \sum_{m=0}^{N-1} \tilde{x}[n+m] \, e^{-j(2\pi/N)km} = e^{j(2\pi/N)kn} \, \tilde{c}_k \tag{5.213}$$

Substituting Eqn. (5.213) into Eqn. (5.212) yields

$$
\begin{aligned}
\tilde{d}_k &= \frac{1}{N} \sum_{n=0}^{N-1} \tilde{x}[n]\, e^{j(2\pi/N)kn}\, \tilde{c}_k \\
&= \tilde{c}_k \frac{1}{N} \sum_{n=0}^{N-1} \tilde{x}[n]\, e^{j(2\pi/N)kn} \\
&= \tilde{c}_k\, \tilde{c}_k^{*} = \left| \tilde{c}_k \right|^2
\end{aligned}
\tag{5.214}
$$

Using the technique developed in Section 5.3.6 the DTFT of $\tilde{r}_{xx}[m]$ is

$$
\mathcal{F}\{\tilde{r}_{xx}[m]\} = 2\pi \sum_{k=-\infty}^{\infty} \tilde{d}_k\, \delta\left(\Omega - k\Omega_0\right)
\tag{5.215}
$$

Using Eqn. (5.214) in Eqn. (5.215) results in

$$
\mathcal{F}\{\tilde{r}_{xx}[m]\} = 2\pi \sum_{k=-\infty}^{\infty} \left| \tilde{c}_k \right|^2 \delta\left(\Omega - k\Omega_0\right)
\tag{5.216}
$$

which is identical to the expression given by Eqn. (5.181) for the power spectral density of a periodic signal.

A relationship similar to the one expressed by Eqn. (5.210) applies to random processes that are *wide-sense stationary*, and is known as the Wiener-Khinchin theorem. It is one of the fundamental theorems of random signal processing.

Example 5.26: **Power spectral density of a discrete-time sinusoid revisited**

Consider again the discrete-time sinusoidal signal

$$
\tilde{x}[n] = 3 \cos\left(0.2\pi n\right)
$$

the power spectral density of which was determined in Example 5.24. Determine the autocorrelation function $\tilde{r}_{xx}[m]$ for this signal, and then find the power spectral density $S_x\left(\Omega\right)$ from the autocorrelation function.

Solution: The period of $\tilde{x}[n]$ is $N = 10$ samples. Using the definition of the autocorrelation function for a periodic signal

$$
\begin{aligned}
\tilde{r}_{xx}[m] &= \frac{1}{10} \sum_{n=0}^{9} \tilde{x}[n]\, \tilde{x}[n+m] \\
&= \frac{1}{10} \sum_{n=0}^{9} \left[3 \cos\left(0.2\pi n\right)\right] \left[3 \cos\left(0.2\pi \left[n+m\right]\right)\right]
\end{aligned}
\tag{5.217}
$$

Through the use of the appropriate trigonometric identity, the result in Eqn. (5.217) is simplified to

$$\tilde{r}_{xx}[m] = \frac{1}{10} \sum_{n=0}^{9} \left[\frac{9}{2} \cos\left(0.4\pi n + 0.2\pi m\right) + \frac{9}{2} \cos\left(0.2\pi m\right) \right]$$

$$= \frac{9}{20} \sum_{n=0}^{9} \cos\left(0.4\pi n + 0.2\pi m\right) + \frac{9}{20} \sum_{n=0}^{9} \cos\left(0.2\pi m\right) \qquad (5.218)$$

In Eqn. (5.218) the first summation is equal to zero since its term $\cos\left(0.4\pi n + 0.2\pi m\right)$ is periodic with a period of five samples, and the summation is over two full periods. The second summation yields

$$\tilde{r}_{xx}[m] = \frac{9}{20} \sum_{n=0}^{9} \cos\left(0.2\pi m\right)$$

$$= \frac{9}{20} \cos\left(0.2\pi m\right) \sum_{n=0}^{9} (1)$$

$$= \frac{9}{2} \cos\left(0.2\pi m\right)$$

for the autocorrelation function. The power spectral density is found as the DTFT of the autocorrelation function. Application of the DTFT was discussed in Section 5.3.6. Using the technique developed in that section (specifically see Example 5.21) the transform is found as

$$S_x\left(\Omega\right) = \mathcal{F}\left\{\tilde{r}_{xx}[m]\right\} = \mathcal{F}\left\{ \frac{9}{2} \cos\left(0.2\pi m\right) \right\}$$

$$= \sum_{r=-\infty}^{\infty} \left[\frac{9\pi}{2} \delta\left(\Omega + 0.2\pi - 2\pi r\right) + \frac{9\pi}{2} \delta\left(\Omega - 0.2\pi - 2\pi r\right) \right]$$

which matches the answer found earlier in Example 5.24.
Software resources:
ex_5_26.m

Properties of the autocorrelation function

The autocorrelation function as defined by Eqns. (5.198), (5.199) and (5.200) has a number of important properties that will be summarized here:

1. $r_{xx}[0] \geq |r_{xx}[m]|$ for all m.
   To see why this is the case, we will consider the non-negative function $(x[n] \mp x[n+m])^2$. The time average of this function must also be non-negative, therefore

   $$\left\langle \left(x[n] \mp x[n+m]\right)^2 \right\rangle \geq 0$$

   or equivalently
   $$\left\langle x^2[n] \right\rangle \mp 2\left\langle x[n]\, x[n+m] \right\rangle + \left\langle x^2[n+m] \right\rangle \geq 0$$

   which implies that
   $$r_{xx}[0] \mp 2\, r_{xx}[m] + r_{xx}[0] \geq 0$$

   which is the same as property 1.

2. $r_{xx}[-m] = r_{xx}[m]$ for all m, that is, the autocorrelation function has even symmetry. Recall that the autocorrelation function is the inverse Fourier transform of either the energy spectral density or the power spectral density. Since $G_x(\Omega)$ and $S_x(\Omega)$ are purely real, $r_{xx}[m]$ must be an even function of m.

3. If the signal $\tilde{x}[n]$ is periodic with period N, then its autocorrelation function $\tilde{r}_{xx}[m]$ is also periodic with the same period. This property easily follows from the time-average based definition of the autocorrelation function given by Eqn. (5.199).

5.5 System Function Concept

In time-domain analysis of systems in Chapter 3 two distinct description forms were used for DTLTI systems:

1. A linear constant-coefficient difference equation that describes the relationship between the input and the output signals

2. An impulse response which can be used with the convolution operation for determining the response of the system to an arbitrary input signal

In this section, the concept of *system function* will be introduced as the third method for describing the characteristics of a system.

The system function is simply the DTFT of the impulse response:

$$H(\Omega) = \mathcal{F}\{h[n]\} = \sum_{n=-\infty}^{\infty} h[n] e^{-j\Omega n} \tag{5.219}$$

Recall that the impulse response is only meaningful for a system that is both linear and time-invariant (since the convolution operator could not be used otherwise). It follows that the system function concept is valid for linear and time-invariant systems only. In general, $H(\Omega)$ is a complex function of Ω, and can be written in polar form as

$$H(\Omega) = |H(\Omega)| \, e^{j\Theta(\Omega)} \tag{5.220}$$

Obtaining the system function from the difference equation

In finding a system function for a DTLTI system described by means of a difference equation, two properties of the DTFT will be useful: the convolution property and the time shifting property.

1. Since the output signal is computed as the convolution of the impulse response and the input signal, that is, $y[n] = h[n] * x[n]$, the corresponding relationship in the frequency domain is $Y(\Omega) = H(\Omega) X(\Omega)$. Consequently, the system function is equal to the ratio of the output transform to the input transform:

$$H(\Omega) = \frac{Y(\Omega)}{X(\Omega)} \tag{5.221}$$

2. Using the time shifting property, transforms of the individual terms in the difference equation are found as

$$y[n-m] \overset{\mathcal{F}}{\longleftrightarrow} e^{-j\Omega m} Y(\Omega) \qquad m = 0, 1, \ldots \tag{5.222}$$

and

$$x[n - m] \overset{\mathcal{F}}{\longleftrightarrow} e^{-j\Omega m} X(\Omega) \qquad m = 0, 1, \ldots \tag{5.223}$$

The system function is obtained from the difference equation by first transforming both sides of the difference equation through the use of Eqns. (5.222) and (5.223), and then using Eqn. (5.221).

Example 5.27: **Finding the system function from the difference equation**

Determine the system function for a DTLTI system described by the difference equation

$$y[n] - 0.9\, y[n - 1] + 0.36\, y[n - 2] = x[n] - 0.2\, x[n - 1]$$

Solution: Taking the DTFT of both sides of the difference equation we obtain

$$Y(\Omega) - 0.9\, e^{-j\Omega}\, Y(\Omega) + 0.36\, e^{-j2\Omega}\, Y(\Omega) = X(\Omega) - 0.2\, e^{-j\Omega}\, X(\Omega)$$

which can be written as

$$\left[1 - 0.9\, e^{-j\Omega} + 0.36\, e^{-j2\Omega}\right] Y(\Omega) = \left[1 - 0.2\, e^{-j\Omega}\right] X(\Omega)$$

The system function is found through the use of Eqn. (5.221)

$$H(\Omega) = \frac{Y(\Omega)}{X(\Omega)} = \frac{1 - 0.2\, e^{-j\Omega}}{1 - 0.9\, e^{-j\Omega} + 0.36\, e^{-j2\Omega}}$$

The magnitude and the phase of the system function are shown in Fig. 5.40.

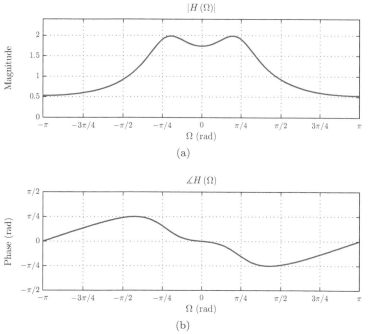

Figure 5.40 – The system function for Example 5.27: (a) magnitude, (b) phase.

Software resources:
ex_5_27.m

Example 5.28: **System function for length-N moving average filter**

Recall the length-N moving average filter with the difference equation

$$y[n] = \frac{1}{N} \sum_{k=0}^{N-1} x[n-k]$$

Determine the system function. Graph its magnitude and phase as functions of Ω.

Solution: Taking the DTFT of both sides of the difference equation we obtain

$$Y(\Omega) = \frac{1}{N} \sum_{k=0}^{N-1} e^{-j\Omega k} X(\Omega)$$

from which the system function is found as

$$H(\Omega) = \frac{Y(\Omega)}{X(\Omega)} = \frac{1}{N} \sum_{k=0}^{N-1} e^{-j\Omega k} \tag{5.224}$$

The expression in Eqn. (5.224) can be put into closed form as

$$H(\Omega) = \frac{1 - e^{-j\Omega N}}{1 - e^{-j\Omega}}$$

As the first step in expressing $H(\Omega)$ in polar complex form, we will factor out $e^{-j\Omega N/2}$ from the numerator and $e^{-j\Omega/2}$ from the denominator to obtain

$$H(\Omega) = \frac{e^{-j\Omega N/2} \left(e^{j\Omega N/2} - e^{-j\Omega N/2} \right)}{e^{-j\Omega/2} \left(e^{j\Omega/2} - e^{-j\Omega/2} \right)}$$

$$= \frac{\sin(\Omega N/2)}{\sin(\Omega/2)} e^{j\Omega (N-1)/2}$$

The magnitude and the phase of the system function are shown in Fig. 5.41 for $N = 4$.

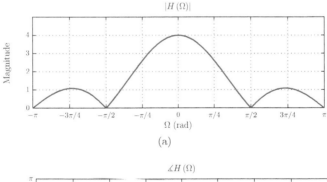

(a)

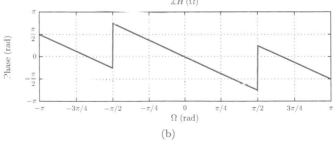

(b)

Figure 5.41 – The system function for Example 5.28: (a) magnitude, (b) phase.

Software resources:
ex_5_28.m

Interactive Demo: sf_demo5.m

The demo program in "sf_demo5.m" is based on Example 5.28. It computes and graphs the magnitude and the phase of the system function $H(\Omega)$ for the length-N moving average filter. The filter length N may be varied.

1. Observe that zero crossings (dips) of the magnitude spectrum divide the angular frequency interval $0 \leq \Omega \leq 2\pi$ into N equal segments.
2. Pay attention to the phase of the system function. Its sloped sections all have the same slope, indicating that the phase response is linear. How does the slope of the phase response relate to the filter length N?

Software resources:
sf_demo5.m

5.6   DTLTI Systems with Periodic Input Signals

In earlier sections of this chapter we have explored the use of the discrete-time Fourier series (DTFS) for representing periodic discrete-time signals. It was shown that a periodic discrete-time signal can be expressed using complex exponential basis functions in the form

$$\text{Eqn. (5.21):} \qquad \tilde{x}[n] = \sum_{k=0}^{N-1} \tilde{c}_k \, e^{j(2\pi/N)kn} \quad \text{for all } n$$

If a periodic signal is used as input to a DTLTI system, the use of the superposition property allows the response of the system to be determined as a linear combination of its responses to individual basis functions

$$\phi_k[n] = e^{j(2\pi/N)kn}$$

5.6.1   Response of a DTLTI system to complex exponential signal

Consider a DTLTI system with impulse response $h[n]$, driven by a complex exponential input signal

$$\tilde{x}[n] = e^{j\Omega_0 n} \tag{5.225}$$

The response $y[n]$ of the system is found through the use of the convolution relationship that was derived in Section 3.7.2 of Chapter 3.

$$y[h] = h[n] * \tilde{x}[n]$$
$$= \sum_{k=-\infty}^{\infty} h[k] \, \tilde{x}[n-k] \tag{5.226}$$

Using the signal $\tilde{x}[n]$ given by Eqn. (5.225) in Eqn. (5.226) we get

$$y[n] = \sum_{k=-\infty}^{\infty} h[k] \, e^{j\Omega_0(n-k)} \tag{5.227}$$

or, equivalently

$$y[n] = e^{j\Omega_0 n} \sum_{k=-\infty}^{\infty} h[k] \, e^{-j\Omega_0 n} \tag{5.228}$$

The summation in Eqn. (5.228) should be recognized as the system function evaluated at the specific angular frequency $\Omega = \Omega_0$. Therefore

$$\tilde{y}[n] = e^{j\Omega_0 n} \, H(\Omega_0) \tag{5.229}$$

We have used the tilde ($\tilde{\ }$) character over the name of the output signal in realization of the fact that it is also periodic. The development in Eqns. (5.227) through (5.229) is based on the inherent assumption that the Fourier transform of $h[n]$ exists. This in turn requires the corresponding DTLTI system to be stable. Any natural response the system may have exhibited at one point would have disappeared a long time ago. Consequently, the response found in Eqn. (5.229) is the *steady-state response* of the system.

For a DTLTI system driven by a complex exponential input signal, we have the following important relationship:

Response to complex exponential input:

$$\tilde{y}[n] = \text{Sys}\left\{e^{j\Omega_0 n}\right\} = e^{j\Omega_0 n} \, H(\Omega_0)$$
$$= |H(\Omega_0)| \, e^{j[\Omega_0 t + \Theta(\Omega_0)]} \tag{5.230}$$

1. The response of a DTLTI system to a complex exponential input signal is a complex exponential output signal with the same angular frequency Ω_0.
2. The effect of the system on the complex exponential input signal is to

   (a) Scale its amplitude by an amount equal to the magnitude of the system function at $\Omega = \Omega_0$
   (b) Shift its phase by an amount equal to the phase of the system function at $\Omega = \Omega_0$

5.6.2   Response of a DTLTI system to sinusoidal signal

Let the input signal to the DTLTI system under consideration be a sinusoidal signal in the form

$$\tilde{x}[n] = \cos(\Omega_0 n) \tag{5.231}$$

The response of the system in this case will be determined by making use of the results of the previous section.

We will use Euler's formula to write the input signal using two complex exponential functions as

$$\tilde{x}[n] = \cos\left(\Omega_0 n\right) = \frac{1}{2} e^{j\Omega_0 n} + \frac{1}{2} e^{-j\Omega_0 n} \tag{5.232}$$

This representation of $\tilde{x}[n]$ allows the results of the previous section to be used. The output signal can be written using superposition:

$$
\begin{aligned}
\tilde{y}[n] &= \frac{1}{2} \operatorname{Sys}\left\{e^{j\Omega_0 n}\right\} + \frac{1}{2} \operatorname{Sys}\left\{e^{-j\Omega_0 n}\right\} \\
&= \frac{1}{2} e^{j\Omega_0 n} H\left(\Omega_0\right) + \frac{1}{2} e^{-j\Omega_0 n} H\left(-\Omega_0\right) \\
&= \frac{1}{2} e^{j\Omega_0 n} \left|H\left(\Omega_0\right)\right| e^{j\Theta(\Omega_0)} + \frac{1}{2} e^{-j\Omega_0 n} \left|H\left(-\Omega_0\right)\right| e^{j\Theta(-\Omega_0)}
\end{aligned} \tag{5.233}
$$

If the impulse response $h[n]$ is real-valued, the result in Eqn. (5.233) can be further simplified. Recall from Section 5.3.5 that, for real-valued $h[n]$, the transform $H(\Omega)$ is conjugate symmetric, resulting in

$$\left|H\left(-\Omega_0\right)\right| = \left|H\left(\Omega_0\right)\right| \quad \text{and} \quad \Theta\left(-\Omega_0\right) = -\Theta\left(\Omega_0\right) \tag{5.234}$$

Using these relationships, Eqn. (5.233) becomes

$$
\begin{aligned}
\tilde{y}[n] &= \frac{1}{2} \left|H\left(\Omega_0\right)\right| e^{j[\Omega_0 t + \Theta(\Omega_0)]} + \frac{1}{2} \left|H\left(\Omega_0\right)\right| e^{-j[\Omega_0 t + \Theta(\Omega_0)]} \\
&= \left|H\left(\Omega_0\right)\right| \cos\left(\omega_0 t + \Theta\left(\Omega_0\right)\right)
\end{aligned} \tag{5.235}
$$

For a DTLTI system driven by a cosine input signal, we have the following important relationship:

Response to cosine input:

$$\tilde{y}[n] = \operatorname{Sys}\left\{\cos\left(\Omega_0 n\right)\right\} = \left|H\left(\Omega_0\right)\right| \cos\left(\omega_0 n + \Theta\left(\Omega_0\right)\right) \tag{5.236}$$

1. When a DTLTI system is driven by single-tone input signal at angular frequency Ω_0, its output signal is also a single-tone signal at the same angular frequency.
2. The effect of the system on the input signal is to
   (a) Scale its amplitude by an amount equal to the magnitude of the system function at $\Omega = \Omega_0$
   (b) Shift its phase by an amount equal to the phase of the system function at $\Omega = \Omega_0$

Example 5.29: **Steady-state response of DTLTI system to sinusoidal input**

Consider a DTLTI system characterized by the difference equation

$$y[n] - 0.9y[n-1] + 0.36y[n-2] = x[n] - 0.2x[n-1]$$

The system function for this system was determined in Example 5.27 to be

$$H\left(\Omega\right) = \frac{Y\left(\Omega\right)}{X\left(\Omega\right)} = \frac{1 - 0.2\,e^{-j\Omega}}{1 - 0.9\,e^{-j\Omega} + 0.36\,e^{-j2\Omega}}$$

Find the response of the system to the sinusoidal input signal

$$\tilde{x}[n] = 5\,\cos\left(\frac{\pi n}{5}\right)$$

Solution: Evaluating the system function at the angular frequency $\Omega_0 = \pi/5$ yields

$$H\left(\pi/5\right) = \frac{1 - 0.2\,e^{-j\pi/5}}{1 - 0.9\,e^{-j\pi/5} + 0.36\,e^{-j2\pi/5}} = 1.8890 - j0.6133$$

which can be written in polar form as

$$H\left(\pi/5\right) = \left|H\left(\pi/5\right)\right|\,e^{j\Theta(\pi/5)} = 1.9861\,e^{-j0.3139}$$

The magnitude and the phase of the system function are shown in Fig. 5.42. The values of magnitude and phase at the angular frequency of interest are marked on the graphs.

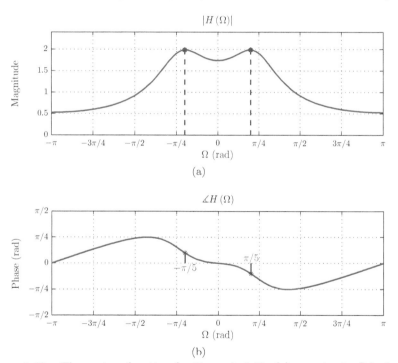

Figure 5.42 – The system function for Example 5.29: (a) magnitude, (b) phase.

The steady-state response of the system to the specified input signal $\tilde{x}[n]$ is

$$
\begin{aligned}
\tilde{y}[n] &= 5\,\left|H\left(\pi/5\right)\right|\,\cos\left(\frac{\pi n}{5} + \Theta\left(\pi/5\right)\right) \\
&= 9.9305\,\cos\left(\frac{\pi n}{5} - 0.3139\right)
\end{aligned}
$$

Software resources:
ex_5_29.m

> Software resources: See MATLAB Exercise 5.4.

Interactive Demo: `sf_demo6.m`

The demo program "`sf_demo6.m`" is based on Example 5.29. It computes and graphs the steady-state response of the system under consideration to the sinusoidal input signal

$$x[n] = 5 \cos\left(\Omega_0 n\right) = 5 \cos\left(2\pi F_0 n\right) =$$

The normalized frequency F_0 can be varied from $F_0 = 0.01$ to $F_0 = 0.49$ using a slider control. As F_0 is varied, the display is updated

1. To show the changes to the steady-state response
2. To show the critical magnitude and phase and values on the graphs of the system function

Software resources:
`sf_demo6.m`

5.6.3   Response of a DTLTI system to periodic input signal

Using the development in the previous sections, we are now ready to consider the use of a general periodic signal $\tilde{x}[n]$ as input to a DTLTI system. Using the DTFS representation of the signal, the response of the system is

$$\text{Sys}\left\{\tilde{x}[n]\right\} = \text{Sys}\left\{\sum_{k=0}^{N-1} \tilde{c}_k\, e^{j(2\pi/N)kn}\right\} \qquad (5.237)$$

Let us use the linearity of the system to write the response as

$$\text{Sys}\left\{\tilde{x}[n]\right\} = \sum_{k=0}^{N-1} \text{Sys}\left\{\tilde{c}_k\, e^{j(2\pi/N)kn}\right\} = \sum_{k=0}^{N-1} \tilde{c}_k\, \text{Sys}\left\{e^{j(2\pi/N)kn}\right\} \qquad (5.238)$$

Based on Eqn. (5.230) the response of the system to an exponential basis function at frequency $\Omega = 2\pi k/N$ is given by

$$\text{Sys}\left\{e^{j(2\pi/N)kn}\right\} = e^{j(2\pi/N)kn}\, H\left(\frac{2\pi k}{N}\right) \qquad (5.239)$$

Using Eqn. (5.239) in Eqn. (5.238), the response of the system to $\tilde{x}[n]$ is found as

$$\text{Sys}\left\{\tilde{x}[n]\right\} = \sum_{k=0}^{N-1} \tilde{c}_k\, H\left(\frac{2\pi k}{N}\right) e^{j(2\pi/N)kn} \qquad (5.240)$$

Two important observations should be made based on Eqn. (5.240):

1. For a DTLTI system driven by an input signal $\tilde{x}[n]$ with period N, the output signal is also periodic with the same period.

2. If the DTFS coefficients of the input signal are $\{\tilde{c}_k \; ; \; k = 1, \ldots, N-1\}$ then the DTFS coefficients of the output signal are $\left\{\tilde{d}_k \; ; \; k = 1, \ldots, N-1\right\}$ such that

$$\tilde{d}_k = \tilde{c}_k \, H\left(\frac{2\pi k}{N}\right), \quad k = 0, \ldots, N-1 \tag{5.241}$$

Example 5.30: Response of DTLTI system to discrete-time sawtooth signal

Let the discrete-time sawtooth signal used in Example 5.4 and shown again in Fig. 5.43 be applied to the system with system function

$$H\left(\Omega\right) = \frac{1 - 0.2\,e^{-j\Omega}}{1 - 0.9\,e^{-j\Omega} + 0.36\,e^{-j2\Omega}}$$

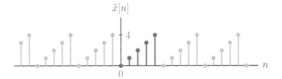

Figure 5.43 – The signal $\tilde{x}[n]$ for Example 5.30.

Find the steady-state response of the system.

Solution: The DTFS coefficients for $\tilde{x}[n]$ were determined in Example 5.4, and are repeated below:

$$\begin{aligned}
\tilde{c}_0 &= 2 \\
\tilde{c}_1 &= -0.5 + j\,0.6882 \\
\tilde{c}_2 &= -0.5 + j\,0.1625 \\
\tilde{c}_3 &= -0.5 - j\,0.1625 \\
\tilde{c}_4 &= -0.5 - j\,0.6882
\end{aligned}$$

Evaluating the system function at angular frequencies

$$\Omega_k = \frac{2\pi k}{5}, \quad k = 0, 1, 2, 3, 4$$

we obtain

$$\begin{aligned}
H\left(0\right) &= 1.7391 \\
H\left(\frac{2\pi}{5}\right) &= 0.8767 - j\,0.8701 \\
H\left(\frac{4\pi}{5}\right) &= 0.5406 - j\,0.1922 \\
H\left(\frac{6\pi}{5}\right) &= 0.5406 + j\,0.1922 \\
H\left(\frac{8\pi}{5}\right) &= 0.8767 + j\,0.8701
\end{aligned}$$

The DTFS coefficients for the output signal are found using Eqn. (5.241):

$$\tilde{d}_0 = \tilde{c}_0\, H(0) = (2)(1.7391) = 3.4783$$

$$\tilde{d}_1 = \tilde{c}_1\, H\left(\frac{2\pi}{5}\right) = (-0.5000 + j\,0.6882)(0.8767 - j\,0.8701) = 0.1604 + j\,1.0384$$

$$\tilde{d}_2 = \tilde{c}_2\, H\left(\frac{4\pi}{5}\right) = (-0.5000 + j\,0.1625)(0.5406 - j\,0.1922) = -0.2391 + j\,0.1839$$

$$\tilde{d}_3 = \tilde{c}_3\, H\left(\frac{6\pi}{5}\right) = (-0.5000 - j\,0.1625)(0.5406 + j\,0.1922) = -0.2391 - j\,0.1839$$

$$\tilde{d}_4 = \tilde{c}_4\, H\left(\frac{8\pi}{5}\right) = (-0.5000 - j\,0.6882)(0.8767 + j\,0.8701) = 0.1604 - j\,1.0384$$

The output signal $y[n]$ can now be constructed using the DTFS coefficients $\{\tilde{d}_k\,;\ k = 0, 1, 2, 3, 4\}$:

$$\tilde{y}[n] = \sum_{k=0}^{4} \tilde{d}_k\, e^{j(2\pi/N)kn}$$

$$= 3.4783 + (0.1604 + j\,1.0384)\, e^{j2\pi n/5} + (-0.2391 + j\,0.1839)\, e^{j4\pi n/5}$$

$$+ (-0.2391 - j\,0.1839)\, e^{j6\pi n/5} + (0.1604 - j\,1.0384)\, e^{j8\pi n/5}$$

and is shown in Fig. 5.44.

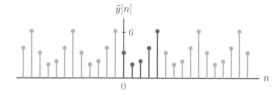

Figure 5.44 – The output signal $\tilde{y}[n]$ for Example 5.30.

Software resources:
ex_5_30.m

5.7 DTLTI Systems with Non-Periodic Input Signals

Let us consider the case of using a non-periodic signal $x[n]$ as input to a DTLTI system. It was established in Section 3.7.2 of Chapter 3 that the output of a DTLTI system is equal to the convolution of the input signal with the impulse response, that is

$$y[n] = h[n] * x[n]$$

$$= \sum_{k=-\infty}^{\infty} h[k]\, x[n-k] \tag{5.242}$$

Let us assume that

1. The system is stable ensuring that $H(\Omega)$ converges.
2. The DTFT of the input signal also converges.

We have seen in Section 5.3.5 of this chapter that the DTFT of the convolution of two signals is equal to the product of individual transforms:

$$Y(\Omega) = H(\Omega) X(\Omega) \qquad (5.243)$$

The output transform is the product of the input transform and the system function.

Writing each transform involved in Eqn. (5.243) in polar form using its magnitude and phase we obtain the corresponding relationships:

$$|Y(\Omega)| = |H(\Omega)| \, |X(\Omega)| \qquad (5.244)$$

$$\angle Y(\Omega) = \angle X(\Omega) + \Theta(\Omega) \qquad (5.245)$$

1. The magnitude of the output spectrum is equal to the product of the magnitudes of the input spectrum and the system function.
2. The phase of the output spectrum is found by adding the phase characteristics of the input spectrum and the system function.

5.8   Discrete Fourier Transform

In previous sections of this chapter we have studied the discrete-time Fourier series (DTFS) for periodic discrete-time signals, and the discrete-time Fourier transform (DTFT) for non-periodic discrete-time signals. The result of DTFT analysis of a discrete-time signal $x[n]$ is a transform $X(\Omega)$ which, if it exists, is a 2π-periodic function of the continuous variable Ω. Storing the DTFT of a signal on a digital computer is impractical because of the continuous nature of Ω. On the other hand, the DTFS representation of a signal $\tilde{x}[n]$ that is periodic with N samples is a set of coefficients $\tilde{c}_k$ that is also periodic with N. While this combination would certainly be suitable for computer implementation and storage, it is only for periodic signals. We often deal with signals that are not necessarily periodic.

In the analysis of non-periodic discrete-time signals, sometimes it is desirable to have a transform that is also discrete. This can be accomplished through the use of the *discrete Fourier transform (DFT)* provided that the signal under consideration is finite-length.

Consider a signal $x[n]$ the meaningful samples of which are limited to the index range $n = 0, \ldots, N-1$, that is,

$$x[n] = 0 \quad \text{for } n < 0 \text{ or } n \geq N$$

We will refer to $x[n]$ as a length-N signal since it has N non-trivial samples. An easy method of representing $x[n]$ with a transform that is also length-N would be as follows:

1. Consider $x[n]$ as one period of a periodic signal $\tilde{x}[n]$ defined as

$$\tilde{x}[n] = \sum_{m=-\infty}^{\infty} x[n - mN] \qquad (5.246)$$

We will refer to $\tilde{x}[n]$ as the *periodic extension* of $x[n]$.

2. Determine the DTFS coefficients $\tilde{c}_k$ for the periodic extension $\tilde{x}[n]$ obtained from $x[n]$ through Eqn. (5.246):

$$\tilde{x}[n] \overset{\text{DTFS}}{\longleftrightarrow} \tilde{c}_k \qquad (5.247)$$

3. DTFS coefficients of $\tilde{x}[n]$ form a set that is also periodic with N. Let us extract just one period $\{c_k \; ; \; k = 0, \ldots, N-1\}$ from the DTFS coefficients $\{\tilde{c}_k \; ; \; \text{all } k\}$:

$$c_k = \begin{cases} \tilde{c}_k, & k = 0, \ldots, N-1 \\ 0, & \text{otherwise} \end{cases} \qquad (5.248)$$

This gives us the ability to represent the signal $x[n]$ with the set of coefficients c_k for $k = 0, \ldots, N-1$. The coefficients can be obtained from the signal using the three steps outlined above. Conversely, the signal can be reconstructed from the coefficients by simply reversing the order of the steps.

The discrete Fourier transform (DFT) will be defined by slightly modifying the idea presented above. The forward transform is

$$X[k] = \sum_{n=0}^{N-1} x[n] \, e^{-j(2\pi/N)kn}, \qquad k = 0, \ldots, N-1 \qquad (5.249)$$

The length-N signal $x[n]$ leads to the length-N transform $X[k]$. Compare Eqn. (5.249) with DTFS analysis equation given by Eqn. (5.22). The only difference is the scale factor $1/N$:

$$X[k] = N \, c_k, \qquad k = 0, \ldots, N-1 \qquad (5.250)$$

It is also possible to obtain the signal $x[n]$ from the transform $X[k]$ using the inverse DFT relationship

$$x[n] = \frac{1}{N} \sum_{k=0}^{N-1} X[k] \, e^{j(2\pi/N)kn}, \qquad n = 0, \ldots, N-1 \qquad (5.251)$$

The notation can be made a bit more compact by defining w_N as

$$w_N = e^{-j(2\pi/N)} \qquad (5.252)$$

Using w_N the forward DFT equation becomes

$$X[k] = \sum_{n=0}^{N-1} x[n] \, w_N^{kn}, \qquad k = 0, \ldots, N-1 \qquad (5.253)$$

and the inverse DFT is found as

$$x[n] = \frac{1}{N} \sum_{k=0}^{N-1} X[k] \, w_N^{-kn}, \qquad n = 0, \ldots, N-1 \qquad (5.254)$$

Notationally the DFT relationship between a signal and its transform can be represented as

$$x[n] \overset{\text{DFT}}{\longleftrightarrow} X[k]$$

or as

$$X[k] = \text{DFT}\,\{x[n]\}$$
$$x[n] = \text{DFT}^{-1}\,\{X[k]\}$$

The analysis and the synthesis equations for the DFT can be summarized as follows:

Discrete Fourier transform (DFT)

1. Analysis equation (Forward transform):

$$X[k] = \sum_{n=0}^{N-1} x[n]\, e^{-j(2\pi/N)kn}\,, \qquad k = 0, \ldots, N-1 \tag{5.255}$$

2. Synthesis equation (Inverse transform):

$$x[n] = \frac{1}{N} \sum_{k=0}^{N-1} X[k]\, e^{j(2\pi/N)kn}, \qquad n = 0, \ldots, N-1 \tag{5.256}$$

The DFT is a very popular tool in a wide variety of engineering applications for a number of reasons:

1. The signal $x[n]$ and its DFT $X[k]$ each have N samples, making the discrete Fourier transform practical for computer implementation. N samples of the signal can be replaced in memory with N samples of the transform without losing any information. It does not matter whether we store the signal or the transform since one can always be obtained from the other.
2. Fast and efficient algorithms, known as *fast Fourier transforms (FFTs)*, are available for the computation of the DFT.
3. DFT can be used for approximating other forms of Fourier series and transforms for both continuous-time and discrete-time systems. It can also be used for fast convolution in filtering applications.
4. Dedicated processors are available for fast and efficient computation of the DFT with minimal or no programming needed.

Example 5.31: **DFT of simple signal**

Determine the DFT of the signal

$$x[n] = \{\, 1,\ -1, 2 \,\}$$
$$\uparrow \atop n=0$$

Solution: The discrete Fourier transform is

$$X[k] = e^{-j(2\pi/3)k(0)} - e^{-j(2\pi/3)k(1)} + 2\, e^{-j(2\pi/3)k(2)}$$
$$= 1 - e^{-j2\pi k/3} + 2\, e^{-j4\pi k/3}$$

We will evaluate this result for $k = 1, 2, 3$:

$$X[0] = 1 - 1 + 2 = 2$$
$$X[1] = 1 - e^{-j2\pi/3} + 2\, e^{-j4\pi/3} = 0.5 + j\, 2.5981$$
$$X[2] = 1 - e^{-j4\pi k/3} + 2\, e^{-j8\pi k/3} = 0.5 - j\, 2.5981$$

Example 5.32: **DFT of discrete-time pulse**

Determine the DFT of the discrete-time pulse signal

$$
\begin{aligned}
x[n] &= u[n] - u[n-10] \\
&= \{\ 1,\ 1,\ 1,\ 1,\ 1,\ 1,\ 1,\ 1,\ 1,\ 1,\ 1\ \} \\
&\quad\ \ \underset{n=0}{\uparrow}
\end{aligned}
$$

Solution: The discrete Fourier transform is

$$
X[k] = \sum_{n=0}^{9} e^{-j(2\pi/10)kn}\,, \qquad k = 0, \ldots, 9
$$

which can be put into closed form using the finite-length geometric series formula (see Appendix C)

$$
X[k] = \frac{1 - e^{-j2\pi k}}{1 - e^{-j2\pi k/10}} = \begin{cases} 10\,, & n = 0 \\ 0\,, & k = 1, \ldots, 9 \end{cases} \tag{5.257}
$$

Note that L'Hospital's rule was used for determining the value $X[0]$. The signal $x[n]$ and its DFT are shown in Fig. 5.45.

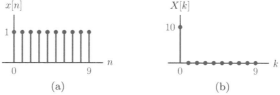

(a)   (b)

Figure 5.45 – The signal $x[n]$ and the transform $X[k]$ for Example 5.32.

5.8.1   Relationship of the DFT to the DTFT

Consider again a length-N signal $x[n]$. The DTFT of such a signal, defined by Eqn. (5.71), can be written as

$$
X(\Omega) = \sum_{n=0}^{N-1} x[n]\, e^{-j\Omega n} \tag{5.258}
$$

where the summation limits have been adjusted to account for the fact that the only significant samples of $x[n]$ are in the interval $n = 0, \ldots, N-1$. A comparison of Eqn. (5.258) with Eqn. (5.71) reveals the simple relationship between the DTFT and the DFT:

The DFT of a length-N signal is equal to its DTFT evaluated at a set of N angular frequencies equally spaced in the interval $[0, 2\pi)$. Let an indexed set of angular frequencies be defined as

$$
\Omega_k = \frac{2\pi k}{N}, \quad k = 0, \ldots, N-1
$$

The DFT of the signal is written as

$$X[k] = X\left(\Omega_k\right) = \sum_{n=0}^{N-1} x[n]\, e^{-j(2\pi k/N)n} \tag{5.259}$$

It is obvious from Eqn. (5.259) that, for a length-N signal, the DFT is very similar to the DTFT with one fundamental difference: In the DTFT, the transform is computed at every value of Ω in the range $0 \leq \Omega < 2\pi$. In the DFT, however, the same is computed only at frequencies that are integer multiples of $2\pi/N$. In a way, looking at the DFT is similar to looking at the DTFT placed behind a picket fence with N equally spaced openings. This is referred to as the *picket-fence effect*. We will elaborate on this relationship further in the next example.

Example 5.33: **DFT of a discrete-time pulse revisited**

Consider again the discrete-time pulse that was used in Example 5.32.

$$x(n) = u[n] - u[n - 10]$$

The DFT of this pulse was determined in Example 5.32 and shown graphically in Fig. 5.45(b). The DTFT of $x[n]$ is

$$X\left(\Omega\right) = \sum_{n=0}^{9} e^{-j\Omega n}$$

and it can easily be put into a closed form as

$$X\left(\Omega\right) = \frac{\sin\left(5\,\Omega\right)}{\sin\left(0.5\,\Omega\right)}\, e^{-j\,4.5\Omega}$$

Recall from earlier discussion (see Eqn. (5.259)) that the transform sample with index k corresponds to the angular frequency $\Omega_k = 2\pi k/N$. If we want to graph the DTFT and the DFT on the same frequency axis, we need to place the DFT samples using an angular frequency spacing of $2\pi/N$ radians. Fig. 5.46 shows both the DTFT and the DFT of the signal $x[n]$, and reveals why we obtained such a trivial looking result in Eqn. (5.257) for the DFT.

For $k = 1, \ldots, 9$, the locations of the transform samples in $X[k]$ coincide with the zero crossings of the DTFT. This is the so-called *picket-fence effect*. It is as though we are looking at the DTFT of the signal $x[n]$ through a picket fence that has narrow openings spaced $2\pi/N$ radians apart. In this case we see mostly the zero crossings of the DTFT and miss the detail between them.

The DFT as given by Eqn. (5.257) is still a complete transform, and the signal $x[n]$ can be obtained from it using the inverse DFT equation given by Eqn. (5.256).

Software resources:
ex_5_33.m

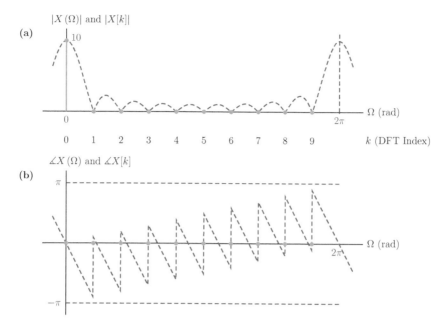

Figure 5.46 – Relationship between the DFT and the DTFT of the signal $x[n]$ used in Example 5.33: (a) magnitudes, and (b) phases of the two transforms.

5.8.2 Zero padding

Even though the DFT computed in Examples 5.32 and 5.33 is a complete and accurate representation of signal $x[n]$, from a visual perspective it does not show much. Sometimes we would like more visual detail than provided by the DFT result. Continuing with the picket-fence analogy, we may want to observe the DTFT through a more dense picket fence with more openings in a 2π-radian range of the angular frequency. This can be accomplished by *zero-padding* the original signal, that is, by extending it with zero-amplitude samples before computing the DFT.

Consider again the length-N signal $x[n]$. Let us define a length-$(N+M)$ signal $q[n]$ as follows:

$$q[n] = \begin{cases} x[n] , & n = 0, \ldots, N-1 \\ 0 , & n = N, \ldots, N+M-1 \end{cases} \tag{5.260}$$

The DFT of the newly defined signal $q[n]$ is

$$Q[k] = \sum_{n=0}^{N+M-1} q[n] \, e^{-j2\pi kn/(N+M)} \tag{5.261}$$

$$= \sum_{n=0}^{N-1} x[n] \, e^{-j2\pi kn/(N+M)} \tag{5.262}$$

Comparing Eqn. (5.261) with Eqn. (5.258) we conclude that

$$Q[k] = X(\Omega)|_{\Omega=2\pi k/(N+M)} \tag{5.263}$$

Thus, $Q[k]$ corresponds to observing the DTFT of $x[n]$ through a picket fence with openings spaced $2\pi/(N+M)$ radians apart as opposed to $2\pi/N$ radians apart. The number of zeros to be appended to the end of the original signal can be chosen to obtain any desired angular frequency spacing.

Example 5.34: **Zero padding the discrete-time pulse**

Consider again the length-10 discrete-time pulse used in Example 5.33. Create a new signal $q[n]$ by zero-padding it to 20 samples, and compare the 20-point DFT of $q[n]$ to the DTFT of $x[n]$.

Solution: The new signal $q[n]$ is

$$q[n] = \begin{cases} 1, & n = 0, \ldots, 9 \\ 0, & n = 10, \ldots, 19 \end{cases}$$

The 20-point DFT of $q[n]$ is

$$Q[k] = \sum_{n=0}^{19} q[n]\, e^{-j2\pi nk/20} \tag{5.264}$$

Since $q[n] = 1$ for $n = 0, \ldots, 9$ and $q[n] = 0$ for $n = 10, \ldots, 19$, Eqn. (5.264) can be written as

$$Q[k] = \sum_{n=0}^{9} e^{-j2\pi nk/20} = \frac{1 - e^{-j\pi k}}{1 - e^{-j2\pi k/20}}, \qquad k = 0, \ldots, 19$$

$Q[k]$ is graphed in Fig. 5.47 along with the DTFT $X(\Omega)$. Compare the figure to Fig. 5.33. Notice how 10 new transform samples appear between the 10 transform samples that were there previously. This is equivalent to obtaining new points between existing ones through some form of interpolation. After zero-padding, the angular frequency spacing between the transform samples is $\Omega_k = 2\pi/20$.

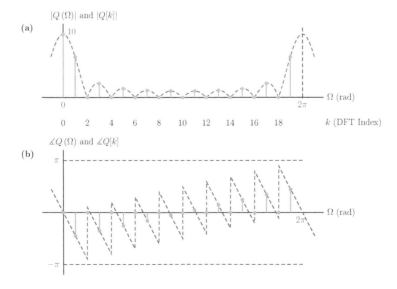

Figure 5.47 – Relationship between the DFT and the DTFT of the zero-padded signal $q[n]$ used in Example 5.34: (a) magnitudes, and (b) phases of the two transforms.

Software resources:
ex_5_34.m

Interactive Demo: pf_demo.m

The demo program pf_demo.m is based on Examples 5.33 and 5.34 as well as Figs. 5.46 and 5.47. The DFT and the DTFT of the 10-sample discrete-time pulse signal are graphed on the same coordinate system to show the picket-fence effect of the DFT. The input signal may be padded with a user-specified number (between 0 and 30) of zero-amplitude samples before the transform is computed. Accordingly, the case in Fig. 5.46 may be duplicated with no additional zero-amplitude samples whereas padding the signal with 10 zero-amplitude samples leads to the situation in Fig. 5.47. The horizontal axis variable for the graphs can be one of three choices: It can display the angular frequency Ω in the range from 0 to 2π, the normalized frequency F in the range from 0 to 1, or the DFT index k in the range from 0 to $N - 1$ where N is the size of the DFT. Recall that the relationships between these parameters are $\Omega = 2\pi F$, $\Omega_k = 2\pi k/N$, and $F_k = k/N$.

- Let L be the number of zero-amplitude data samples with which the 10 sample pulse signal is padded. Compare the graphs for $L = 0$ and $L = 10$. In going from $L = 0$ to $L = 10$, observe how the additional DFT samples for $L = 10$ interpolate between the existing DFT samples.

- Compare the graphs for $L = 0$ and $L = 20$. In going from $L = 0$ to $L = 20$, observe that now two additional DFT samples are inserted between existing DFT samples.

- Change the horizontal axis to display the DFT index k. Pay attention to the locations of DFT samples in terms of the angular frequency Ω. Explain why the last DFT sample does not coincide with the second peak of the magnitude function, but rather appears slightly to the left of it.

Software resources:
pf_demo.m

Software resources: See MATLAB Exercises 5.5 and 5.6.

5.8.3 Properties of the DFT

Important properties of the DFT will be summarized in this section. It will become apparent in that process that the properties of the DFT are similar to those of DTFS and DTFT with one significant difference: Any shifts in the time domain or the transform domain are *circular shifts* rather than linear shifts. Also, any time reversals used in conjunction with the DFT are *circular time reversals* rather than linear ones. Therefore, the concepts of circular shift and circular time reversal will be introduced here in preparation for the discussion of DFT properties.

In the derivation leading to forward and inverse DFT relationships in Eqns. (5.255) and (5.256) for a length-N signal $x[n]$ we have relied on the DTFS representation of the periodic extension signal $\tilde{x}[n]$. Let us consider the following scenario:

1. Obtain periodic extension $\tilde{x}[n]$ from $x[n]$ using Eqn. (5.246).

2. Apply a time shift to $\tilde{x}[n]$ to obtain $\tilde{x}[n-m]$. The amount of the time shift may be positive or negative.

3. Obtain an length-N signal $g[n]$ by extracting the main period of $\tilde{x}[n-m]$.

$$g[n] = \begin{cases} \tilde{x}[n-m], & n = 0, \ldots, N-1 \\ 0, & \text{otherwise} \end{cases} \tag{5.265}$$

The resulting signal $g[n]$ is a *circularly shifted* version of $x[n]$, that is

$$g[n] = x[n-m]_{\text{mod } N} \tag{5.266}$$

The term

$$x[n-m]_{\text{mod } N} \tag{5.267}$$

on the right side of Eqn. (5.266) uses *modulo indexing*. The signal $x[n]$ has meaningful samples only for $n = 0, \ldots, N-1$. The index $n-m$ for a particular set of n and m values may or may not be in this range. Modulo N value of the index is found by adding integer multiples of N to the index until the result is within the range $n = 0, \ldots, N-1$. A few examples are given below:

$$\begin{array}{lll} x[-3]_{\text{mod } 8} = x[5] & x[12]_{\text{mod } 10} = x[2] & x[-7]_{\text{mod } 25} = x[18] \\ x[16]_{\text{mod } 16} = x[0] & x[-3]_{\text{mod } 4} = x[1] & x[95]_{\text{mod } 38} = x[19] \end{array}$$

The process that led to Eqn. (5.266) is illustrated in Figs. 5.48 and 5.49 for an example length-8 signal.

For the example we are considering in Figs. 5.48 and 5.49 imagine a picture frame that fits samples $0, \ldots, 7$. Right shifting the signal by two samples causes two samples to leave the frame from the right edge and re-enter from the left edge, as shown in Fig. 5.48. Left shifting the signal has the opposite effect. Samples leave the frame from the left edge and re-enter from the right edge, as shown in Fig. 5.49. Fig. 5.50 further illustrates the concept of circular shifting.

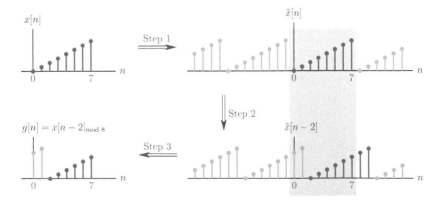

Figure 5.48 – Obtaining a circular shift to the right by two samples. In step 1 a periodic extension $\tilde{x}[n]$ is formed. In step 2 the periodic extension is time shifted to obtain $\tilde{x}[n-2]$. In step 3 the main period is extracted to obtain $g[n]$.

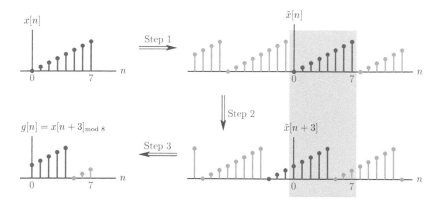

Figure 5.49 – Obtaining a circular shift to the left by three samples.

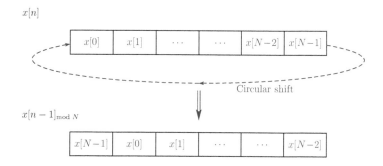

Figure 5.50 – Circular shifting a length-N signal.

For the time reversal operation consider the following steps:

1. Obtain periodic extension $\tilde{x}[n]$ from $x[n]$ using Eqn. (5.246).
2. Apply the time reversal operation to $\tilde{x}[n]$ to obtain $\tilde{x}[-n]$.
3. Obtain an length-N signal $g[n]$ by extracting the main period of $\tilde{x}[-n]$.

$$g[n] = \begin{cases} \tilde{x}[-n] , & n = 0, \ldots, N-1 \\ 0 , & \text{otherwise} \end{cases} \tag{5.268}$$

The resulting signal $g[n]$ is a *circularly time reversed* version of $x[n]$, that is

$$g[n] = x[-n]_{\text{mod } N} \tag{5.269}$$

The process that led to Eqn. (5.269) is illustrated in Fig. 5.51 for an example length-8 signal.

Software resources: See MATLAB Exercise 5.7.

For DFT-related operations the definitions of conjugate symmetry properties also need to be adjusted so that they utilize circular time reversals. A length-8 signal $x[n]$ is *circularly conjugate symmetric* if it satisfies

$$x^*[n] = x[-n]_{\text{mod } N} \tag{5.270}$$

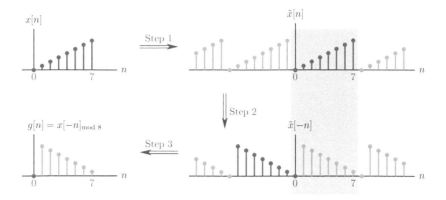

Figure 5.51 – Circular time reversal of a length-8 signal.

or *circularly conjugate antisymmetric* if it satisfies

$$x^*[n] = -x[-n]_{\text{mod } N} \tag{5.271}$$

A signal that satisfies neither Eqn. (5.270) nor Eqn. (5.271) can still be decomposed into two components such that one is circularly conjugate symmetric and the other is circularly conjugate antisymmetric. The conjugate symmetric component is computed as

$$x_E[n] = \frac{x[n] + x^*[-n]_{\text{mod } N}}{2} \tag{5.272}$$

and the conjugate antisymmetric component is computed as

$$x_O[n] = \frac{x[n] - x^*[-n]_{\text{mod } N}}{2} \tag{5.273}$$

respectively, so that

$$x[n] = x_E[n] + x_O[n] \tag{5.274}$$

Software resources: See MATLAB Exercise 5.8.

We are now ready to explore the properties of the DFT. All properties listed in this section assume length-N signals and transforms.

Linearity

Let $x_1[n]$ and $x_2[n]$ be two length-N signals with discrete Fourier transforms

$$x_1[n] \overset{\text{DFT}}{\longleftrightarrow} X_1[k] \qquad \text{and} \qquad x_2[n] \overset{\text{DFT}}{\longleftrightarrow} X_2[k]$$

It can be shown that

$$\alpha_1\, x_1[n] + \alpha_2\, x_2[n] \overset{\text{DFT}}{\longleftrightarrow} \alpha_1\, X_1[k] + \alpha_2\, X_2[k] \tag{5.275}$$

for any two arbitrary constants α_1 and α_2.

Linearity property is easily proven starting with the forward DFT equation in Eqn. (5.255).

Time shifting

Given a transform pair

$$x[n] \overset{\text{DFT}}{\longleftrightarrow} X[k]$$

it can be shown that

$$x[n-m]_{\text{mod }N} \overset{\text{DFT}}{\longleftrightarrow} e^{-j(2\pi/N)km} X[k] \qquad (5.276)$$

Circular shifting of the signal $x[n]$ causes the DFT to be multiplied by a complex exponential function.

Consistency check: Let the signal be circularly shifted by exactly one period, that is, $m = N$. We know that

$$x[n-N]_{\text{mod }N} = x[n]$$

In this case the exponential function on the right side of Eqn. (5.276) would be $e^{-j(2\pi/N)kN} = 1$, and the transform remains unchanged, as expected.

Example 5.35: **Gaining insight into the time shifting property of DFT**

Let a signal $x[n]$ be given by

$$x[n] = \{\ a,\ b,\ c,\ d\ \}$$
$$\underset{n=0}{\uparrow}$$

where a, b, c, d represent arbitrary signal amplitudes. Write $X[k]$, the DFT of $x[n]$, in terms of the parameters a, b, c, d. Afterwards construct the transform

$$G[k] = e^{-2\pi k/4} X[k]$$

and determine the signal $g[n]$ to which it corresponds.

Solution: The DFT of $x[n]$ is

$$X[k] = a + b\,e^{-j\pi k/2} + c\,e^{-j\pi k} + d\,e^{-j3\pi k/2} \qquad (5.277)$$

The transform $G[k]$ is obtained as

$$G[k] = e^{-2\pi k/4} X[k] = a\,e^{-j\pi k/2} + b\,e^{-j\pi k} + c\,e^{-j3\pi k/2} + d\,e^{-j2\pi k}$$

Realizing that $e^{-j2\pi k} = 1$ for any integer value of k, Eqn. (5.278) can be written as

$$G[k] = d + a\,e^{-j\pi k/2} + b\,e^{-j\pi k} + c\,e^{-j3\pi k/2} \qquad (5.278)$$

Comparing Eqn. (5.278) with Eqn. (5.277) we conclude that $G[k]$ is the DFT of the signal

$$g[n] = \{\ d,\ a,\ b,\ c\ \}$$
$$\underset{n=0}{\uparrow}$$

which is a circularly shifted version of $x[n]$, that is,

$$g[n] = x[n-1]_{\text{mod }4}$$

Time reversal

For a transform pair

$$x[n] \overset{\text{DFT}}{\longleftrightarrow} X[k]$$

it can be shown that

$$x[-n]_{\text{mod } N} \overset{\text{DFT}}{\longleftrightarrow} X[-k]_{\text{mod } N} \qquad (5.279)$$

Example 5.36: **Gaining insight into the time reversal property of DFT**

Consider again the signal $x[n]$ used in Example 5.35:

$$x[n] = \{\ a,\ b,\ c,\ d\ \}$$
$$\underset{n=0}{\uparrow}$$

The transform $X[k] = \text{DFT}\{x[n]\}$ was derived in Eqn. (5.277). Construct the transform

$$G[k] = X[-k]_{\text{mod } 4}$$

and determine the signal $g[n]$ to which it corresponds.

Solution: Writing $X[k]$ for each value of the index k we get

$$\begin{aligned}
X[0] &= a + b + c + d \\
X[1] &= (a - c) - j(b - d) \\
X[2] &= a - b + c - d \\
X[3] &= (a - c) + j(b - d)
\end{aligned}$$

The transform $G[k]$ is a circularly time reversed version of $X[k]$. Its samples are

$$\begin{aligned}
G[0] &= X[0] = a + b + c + d \\
G[1] &= X[3] = (a - c) - j(d - b) \\
G[2] &= X[2] = a - b + c - d \\
G[3] &= X[1] = (a - c) + j(d - b)
\end{aligned}$$

Comparing $G[k]$ with $X[k]$ we conclude that the expressions for $G[0]$ through $G[3]$ can be obtained from those for $X[0]$ through $X[3]$ by simply swapping the roles of the parameters b and d. Consequently the signal $g[n]$ is

$$g[n] = \{\ a,\ d,\ c,\ b\ \}$$
$$\underset{n=0}{\uparrow}$$

which is a circularly reversed version of $x[n]$, that is,

$$g[n] = x[-n]_{\text{mod } 4}$$

Conjugation property

For a transform pair

$$x[n] \overset{\text{DFT}}{\longleftrightarrow} X[k]$$

it can be shown that

$$x^*[n] \overset{\text{DFT}}{\longleftrightarrow} X^*[-k]_{\text{mod } N} \qquad (5.280)$$

Symmetry of the DFT

If the signal $x[n]$ is real-valued, it can be shown that its DFT is circularly conjugate symmetric. Conversely, if the signal $x[n]$ is purely imaginary, its transform is circularly conjugate antisymmetric. When we discuss symmetry properties in the context of the DFT we will always imply circular symmetry. If the signal $x[n]$ is conjugate symmetric, its DFT is purely real. In contrast, the DFT of a conjugate antisymmetric signal is purely imaginary.

Symmetry properties of the DFT:

$$x[n]: \text{Real}, \ \text{Im}\{x[n]\} = 0 \ \Longrightarrow \ X^*[k] = X[-k]_{\text{mod } N} \tag{5.281}$$

$$x[n]: \text{Imag}, \ \text{Re}\{x[n]\} = 0 \ \Longrightarrow \ X^*[k] = -X[-k]_{\text{mod } N} \tag{5.282}$$

$$x^*[n] = x[-n]_{\text{mod } N} \ \Longrightarrow \ X[k] : \text{Real} \tag{5.283}$$

$$x^*[n] = -x[-n]_{\text{mod } N} \ \Longrightarrow \ X[k] : \text{Imag} \tag{5.284}$$

Consider a length-N signal $x[n]$ that is complex-valued. In Cartesian complex form $x[n]$ can be written as

$$x[n] = x_r[n] + jx_i[n]$$

Let the discrete Fourier transform $X[k]$ of the signal $x[n]$ be written in terms of its conjugate symmetric and conjugate antisymmetric components as

$$X[k] = X_E[k] + X_O[k]$$

The transform relationship between $x[n]$ and $X[k]$ is

$$x_r[n] + jx_i[n] \xleftrightarrow{\text{DFT}} X_E[k] + X_O[k]$$

We know from Eqns. (5.281) and (5.282) that the DFT of a real signal must be conjugate symmetric, and the DFT of a purely imaginary signal must be conjugate antisymmetric. Therefore it follows that the following must be valid transform pairs:

$$x_r[n] \xleftrightarrow{\text{DFT}} X_E[k] \tag{5.285}$$

$$jx_i[n] \xleftrightarrow{\text{DFT}} X_O[k] \tag{5.286}$$

A similar argument can be made by writing the signal $x[n]$ as the sum of a conjugate symmetric signal and a conjugate antisymmetric signal

$$x[n] = x_E[n] + x_O[n]$$

and writing the transform $X[k]$ in Cartesian complex form

$$X[k] = X_r[k] + jX_i[k]$$

The transform relationship between the two is

$$x_E[n] + x_O[n] \xleftrightarrow{\text{DFT}} X_r[k] + jX_i[k]$$

which leads to the following transform pairs:

$$x_E[n] \overset{\text{DFT}}{\longleftrightarrow} X_r[k] \tag{5.287}$$

$$x_O[n] \overset{\text{DFT}}{\longleftrightarrow} jX_i[k] \tag{5.288}$$

Example 5.37: **Using symmetry properties of the DFT**

The DFT of a length-4 signal $x[n]$ is given by

$$X[k] = \{\, (2 + j3), (1 + j5), (-2 + j4), (-1 - j3)\,\}$$
$$\underset{k=0}{\uparrow}$$

Without computung $x[n]$ first, determine the DFT of $x_r[n]$, the real part of $x[n]$.

Solution: We know from the symmetry properties of the DFT that the transform of the real part of $x[n]$ is the conjugate symmetric part of $X[k]$:

$$\text{DFT}\{x_r[n]\} = X_E[k] = \frac{X[k] + X^*[-k]_{\text{mod } N}}{2}$$

The complex conjugate of the time reversed transform is

$$X^*[-k]_{\text{mod } 4} = \{\, (2 - j3), (-1 + j3), (-2 - j4), (1 - j5)\,\}$$
$$\underset{k=0}{\uparrow}$$

The conjugate symmetric component of $X[k]$ is

$$X_E[k] = \{\, 2, \, j4, \, -2, \, -j4\,\}$$
$$\underset{k=0}{\uparrow}$$

The real part of $x[n]$ can be found as the inverse transform of $X_E[k]$:

$$x_r[n] = \{\, 0, \, -1, 0, 3\,\}$$
$$\underset{n=0}{\uparrow}$$

Example 5.38: **Using symmetry properties of the DFT to increase efficiency**

Consider the real-valued signals $g[n]$ and $h[n]$ specified as

$$g[n] = \{\, 11, \, -2, 7, 9\,\}$$
$$\underset{n=0}{\uparrow}$$

$$h[n] = \{\, 6, \, 14, \, -13, 8\,\}$$
$$\underset{n=0}{\uparrow}$$

Devise a method of obtaining the DFTs of $g[h]$ and $h[n]$ by computing only one 4-point DFT and utilizing symmetry properties.

Solution: Let us construct a complex signal $x[n]$ as

$$x[n] = g[n] + jh[n]$$
$$= \{\, (11 + j6), (-2 + j14), (7 - j13), (9 + j8)\,\}$$
$$\underset{n=0}{\uparrow}$$

The 4-point XFT of $x[n]$ is

$$X[k] = \{\ (25 + j15),\ (10 + j30),\ (11 - j29),\ (-2 + j8)\ \}$$
$$\underset{k=0}{\uparrow}$$

The conjugate symmetric component of $X[k]$ is

$$X_E[k] = \frac{X[k] + X^*[-k]_{\text{mod } 4}}{2} = \{\ \underset{\underset{k=0}{\uparrow}}{25},\ (4 + j11),\ 11,\ (4 - j11)\ \}$$

and its conjugate antisymmetric component is

$$X_O[k] = \frac{X[k] - X^*[-k]_{\text{mod } 4}}{2} = \{\ \underset{\underset{k=0}{\uparrow}}{j15},\ (6 + j19),\ -j29,\ (-6 + j19)\ \}$$

Based on the symmetry properties of the DFT we have DFT $\{g[n]\} = X_E[k]$ and DFT $\{jh[n]\}$ $= X_O[k]$. Therefore

$$G[k] = X_E[k] = \{\ \underset{\underset{k=0}{\uparrow}}{25},\ (4 + j11),\ 11,\ (4 - j11)\ \}$$

and

$$H[k] = -jX_O[k] = \{\ \underset{\underset{k=0}{\uparrow}}{15},\ (19 - j6),\ -29,\ (19 + j6)\ \}$$

It can easily be verified that $G[k]$ and $H[k]$ found above are indeed the DFTs of the two signals $g[n]$ and $h[n]$.

Software resources:

ex_5_38.m

Software resources: See MATLAB Exercise 5.9.

Frequency shifting

For a transform pair

$$x[n] \overset{\text{DFT}}{\longleftrightarrow} X[k]$$

it can be shown that

$$x[n]\, e^{j(2\pi/N)mn} \overset{\text{DFT}}{\longleftrightarrow} X[k - m]_{\text{mod } N} \qquad (5.289)$$

Circular convolution

Periodic convolution of two periodic signals $\tilde{x}[n]$ and $\tilde{h}[n]$ was defined in Section 5.2.2, Eqn. (5.47). In this section we will define *circular convolution* for length-N signals in the context of the discrete Fourier transform. Let $x[n]$ and $h[n]$ be length-N signals. Consider the following set of steps:

1. Obtain periodic signals $\tilde{x}[n]$ and $\tilde{h}[n]$ as periodic extensions of $x[n]$ and $h[n]$:

$$\tilde{x}[n] = \sum_{m=-\infty}^{\infty} x[n+mN]$$

$$\tilde{h}[n] = \sum_{m=-\infty}^{\infty} h[n+mN]$$

2. Compute $\tilde{y}[n]$ as the periodic convolution of $\tilde{x}[n]$ and $\tilde{h}[n]$.

$$\tilde{y}[n] = \tilde{x}[n] \otimes \tilde{h}[n] = \sum_{k=0}^{N-1} \tilde{x}[k]\,\tilde{h}[n-k]$$

3. Let $y[n]$ be the length-N signal that is equal to one period of $\tilde{y}[n]$:

$$y[n] = \tilde{y}[n]\,, \quad \text{for } n = 0,\dots,N-1$$

The signal $y[n]$ is the *circular convolution* of $x[n]$ and $h[n]$. It can be expressed in compact form as

$$y[n] = x[n] \otimes h[n] = \sum_{k=0}^{N-1} x[k]\,h[n-k]_{\mathrm{mod}\ N}\,, \qquad n = 0,\dots,N-1 \qquad (5.290)$$

Example 5.39: **Circular convolution of two signals**

Determine the circular convolution of the length-5 signals

$$x[n] = \{\ 1,\,3,\,2,\,-4,\,6\ \}$$
$$\underset{n=0}{\uparrow}$$

and

$$h[n] = \{\ 5,\,4,\,3,\,2,\,1\ \}$$
$$\underset{n=0}{\uparrow}$$

using the definition of circular convolution given by Eqn. (5.290).

Solution: Adapting Eqn. (5.290) to length-5 signals we have

$$y[n] = x[n] \otimes h[n] = \sum_{k=0}^{4} x[k]\,h[n-k]_{\mathrm{mod}\ 5}\,, \qquad n = 0,1,2,3,4$$

Fig. 5.52 illustrates the steps involved in computing the circular convolution. The result is

$$y[n] = \{\ 24, 31,\ 33,\ 5,\ 27\ \}$$
$$\underset{n=0}{\uparrow}$$

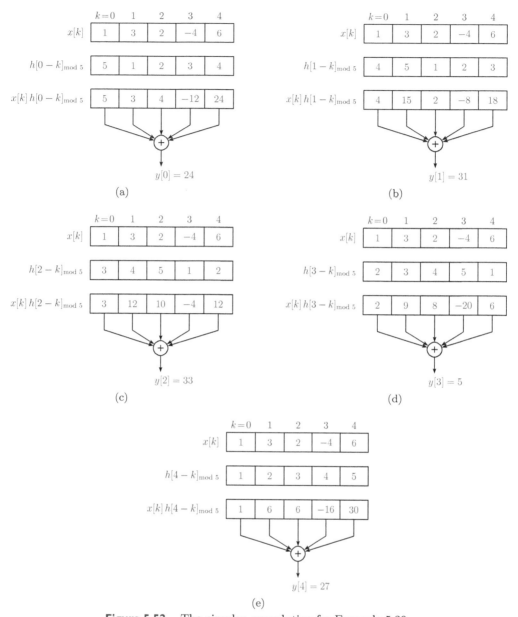

Figure 5.52 – The circular convolution for Example 5.39.

The circular convolution property of the discrete Fourier transform can be stated as follows:

Let $x[n]$ and $h[n]$ be two length-N signals with discrete-Fourier transforms

$$x[n] \overset{\text{DFT}}{\longleftrightarrow} X[k] \quad \text{and} \quad h[n] \overset{\text{DFT}}{\longleftrightarrow} H[k]$$

It can be shown that

$$x[n] \otimes h[n] \overset{\text{DFT}}{\longleftrightarrow} X[k]\,H[k] \tag{5.291}$$

The DFT of the circular convolution of two signals $x[n]$ and $h[n]$ is equal to the product of individual DFTs $X[k]$ and $H[k]$.

This is a very significant result in the use of the DFT for signal processing applications. The proof is straightforward using the periodic convolution property of the discrete-time Fourier series (DTFS), and will be given here.

Let $\tilde{x}[n]$ and $\tilde{h}[n]$ be periodic extensions of the length-N signals $x[n]$ and $h[n]$. Furthermore, let $\tilde{c}_k$ and $\tilde{d}_k$ be the DTFS coefficients for $\tilde{x}[n]$ and $\tilde{h}[n]$ respectively:

$$\tilde{x}[n] \overset{\text{DTFS}}{\longleftrightarrow} \tilde{c}_k \qquad \text{and} \qquad \tilde{h}[n] \overset{\text{DTFS}}{\longleftrightarrow} \tilde{d}_k$$

The periodic convolution of $\tilde{x}[n]$ and $\tilde{h}[n]$ is

$$\tilde{y}[n] = \tilde{x}[n] \otimes \tilde{h}[n] = \sum_{m=0}^{N-1} \tilde{x}[m]\, \tilde{h}[n-m]$$

and the DTFS coefficients of $\tilde{y}[n]$ are

$$\tilde{e}_k = \frac{1}{N} \sum_{n=0}^{N-1} \tilde{y}[n]\, e^{-j(2\pi/N)kn}$$

We know from Eqn. (5.49) that

$$\tilde{e}_k = N\, \tilde{c}_k\, \tilde{d}_k \tag{5.292}$$

Recall the relationship between the DTFS and the DFT given by Eqn. (5.250). The DFTs of length-N signals $x[n]$, $h[n]$ and $y[n]$ are related to the DTFS coefficients of the periodic extensions by

$$X[k] = N\, \tilde{c}_k , \quad k = 0, \ldots, N-1 \tag{5.293}$$

$$H[k] = N\, \tilde{d}_k , \quad k = 0, \ldots, N-1 \tag{5.294}$$

$$Y[k] = N\, \tilde{e}_k , \quad k = 0, \ldots, N-1 \tag{5.295}$$

Using Eqns. (5.293), (5.294) and (5.295) in Eqn. (5.292) we obtain the desired result:

$$Y[k] = X[k]\, H[k] \tag{5.296}$$

Example 5.40: **Circular convolution through DFT**

Consider again the length-5 signals $x[n]$ and $h[n]$ of Example 5.39. The circular convolution

$$y[n] = x[n] \otimes h[n]$$

was determined in Example 5.39 in the time-domain. Verify the circular convolution property of the DFT using these signals.

Solution: Table 5.5 lists the DFT for the three signals. It can easily be verified that

$$Y[k] = X[k] H[k]$$

k	$X[k]$	$H[k]$	$Y[k]$
0	$8.0000+j\,0.0000$	$15.0000+j\,0.0000$	$120.0000+j\,0.0000$
1	$5.3992+j\,0.6735$	$2.5000+j\,3.4410$	$11.1803+j\,20.2622$
2	$-6.8992-j\,7.4697$	$2.5000+j\,0.8123$	$-11.1803-j\,24.2784$
3	$-6.8992+j\,7.4697$	$2.5000-j\,0.8123$	$-11.1803+j\,24.2784$
4	$5.3992-j\,0.6735$	$2.5000-j\,3.4410$	$11.1803-j\,20.2622$

Table 5.5 – DFTs of the three signals in Example 5.40.

Software resources:
ex_5_40.m

If the circular convolution of two length-N signals is desired, the convolution property of the DFT provides an easy and practical method of computing it.

Obtaining circular convolution $y[n] = x[n] \otimes h[n]$:

1. Compute the DFTs

$$X[k] = \text{DFT}\{x[n]\}, \quad \text{and} \quad H[k] = \text{DFT}\{h[n]\}$$

2. Multiply the two DFTs to obtain $Y[k]$.

$$Y[k] = X[k] H[k]$$

3. Compute $y[n]$ through inverse DFT:

$$y[n] = \text{DFT}^{-1}\{Y[k]\}$$

In most applications of signal processing, however, we are interested in computing the *linear convolution* of two signals rather than their circular convolution. The output signal of a DTLTI system is equal to the linear convolution of its impulse response with the input signal. The ability to use the DFT as a tool for the computation of linear convolution is very important due to the availability of fast and efficient algorithms for computing the DFT. Therefore, the following two questions need to be answered:

1. How is the circular convolution of two length-N signals related to their linear convolution?
2. What can be done to ensure that the circular convolution result obtained using the DFT method matches the linear convolution result?

The next example will address the first question.

Example 5.41: **Linear vs. circular convolution**

Consider again the length-5 signals $x[n]$ and $h[n]$ of Example 5.39.

$$x[n] = \{\ 1,\ 3,\ 2,\ -4,\ 6\ \}$$
$$\underset{n=0}{\uparrow}$$

$$h[n] = \{\ 5,\ 4,\ 3,\ 2,\ 1\ \}$$
$$\underset{n=0}{\uparrow}$$

The circular convolution of these two signals was determined in Example 5.39 as

$$y[n] = x[n] \otimes h[n] = \{\ 24,\ 31,\ 33,\ 5,\ 27\ \} \tag{5.297}$$
$$\underset{n=0}{\uparrow}$$

The linear convolution of $x[n]$ and $h[n]$, computed using the convolution sum

$$y_\ell[n] = \sum_{k=-\infty}^{\infty} x[k]\, h[n-k]$$

is found as

$$y_\ell[n] = \{\ 5,\ 19,\ 25,\ -1,\ 27,\ 19,\ 12,\ 8,\ 6\ \} \tag{5.298}$$
$$\underset{n=0}{\uparrow}$$

Note that in Eqn. (5.298) we have used the notation $y_\ell[n]$ for linear convolution to differentiate it from the circular convolution result of Eqn. (5.297). The most obvious difference between the two results $y[n]$ and $y_\ell[n]$ is the length of each. The circular convolution result is 5 samples long, however the linear convolution result is 9 samples long. This is the first step toward explaining why the DFT method does not produce the linear convolution result. $X[k]$ and $H[k]$ are length-5 transforms, and the inverse DFT of their product yields a length-5 result for $y[n]$.

How does $y_\ell[n]$ relate to $y[n]$? Imagine filling out a form that has 5 boxes for entering values, yet we have 9 values we must enter. We start from with the leftmost box and enter the first 5 of 9 values. At this point each box has a value in it, and we still have 4 more values not entered into any box. Suppose we decide to go back to the leftmost box, and start entering additional values into each box as needed. This is illustrated in Fig. 5.53 using the samples of the linear convolution result to fill the boxes.

$n=0$	1	2	3	4
$y_\ell[0]$	$y_\ell[1]$	$y_\ell[2]$	$y_\ell[3]$	$y_\ell[4]$
$y_\ell[5]$	$y_\ell[6]$	$y_\ell[7]$	$y_\ell[8]$	

Totals: $y[0]$ $y[1]$ $y[2]$ $y[3]$ $y[4]$

$n=0$	1	2	3	4
5	19	25	-1	27
19	12	8	6	

Totals: 24   31   33   5   27

Figure 5.53 – Relationship between linear and circular convolution in Example 5.41.

Each sample of the circular convolution result is equal to the sum of values in the corresponding box. For example, $y[0] = y_\ell[0] + y_\ell[5]$ and $y[1] = y_\ell[1] + y_\ell[6]$. If a box has a single value, the circular convolution result is identical to the linear convolution result for the corresponding n. For boxes that have multiple entries, the circular convolution result is a corrupted version of the linear convolution result.

Software resources:

ex_5_41.m

Generalizing the results of Example 5.41 the circular convolution of two signals can be expressed in terms of their linear convolution as

$$y[n] = \sum_{m=-\infty}^{\infty} y_\ell[n + mN] \qquad (5.299)$$

If the circular convolution result is desired to be identical to the linear convolution result, the length of the circular convolution result must be sufficient to accommodate the number of samples expected from linear convolution. Using the analogy employed in Example 5.41, namely filling out a form with the results, there must be enough "boxes" to accommodate all samples of $y_\ell[k]$ without any overlaps. One method of achieving this is through zero-padding $x[n]$ and $h[n]$ before the computation of the DFT.

Computing linear convolution using the DFT:

Given two finite length signals with N_x and N_h samples respectively

$$x[n]\,,\quad n=0,\dots,N_x-1 \qquad\text{and}\qquad h[n]\,,\quad n=0,\dots,N_h-1$$

the linear convolution $y_\ell[n] = x[n] * h[n]$ can be computed as follows:

1. Anticipating the length of the linear convolution result to be $N_y = N_x + N_h - 1$, extend the length of each signal to N_y through zero padding:

$$x_p[n] = \begin{cases} x[n]\,, & n=0,\dots,N_x-1 \\ 0\,, & n=N_x,\dots,N_y-1 \end{cases}$$

$$h_p[n] = \begin{cases} h[n]\,, & n=0,\dots,N_h-1 \\ 0\,, & n=N_h,\dots,N_y-1 \end{cases}$$

2. Compute the DFTs of the zero-padded signals $x_p[n]$ and $h_p[n]$.

$$X_p[k] = \text{DFT}\{x_p[n]\}\,,\quad\text{and}\quad H_p[k] = \text{DFT}\{h_p[n]\}$$

3. Multiply the two DFTs to obtain $Y_p[k]$.

$$Y_p[k] = X_p[k]\,H_p[k]$$

4. Compute $y_p[n]$ through inverse DFT:

$$y_p[n] = \text{DFT}^{-1}\{Y_p[k]\}$$

The result $y_p[n]$ is the same as the linear convolution of the signals $x[n]$ and $y[n]$.

$$y_p[n] = y_\ell[n]\quad\text{for}\quad n=0,\dots,N_y-1$$

Software resources: See MATLAB Exercises 5.10 and 5.11.

5.8.4   Using the DFT to approximate the EFS coefficients

EFS representation of a continuous-time signal $x_a(t)$ periodic with a period T_0 was given in Eqn. (4.72) in Chapter 4 which is repeated below:

$$\text{Eqn. (4.72):} \qquad c_k = \frac{1}{T_0} \int_{t_0}^{t_0+T_0} x_a(t)\, e^{-jk\omega_0 t}\, dt$$

One method of approximating the coefficients c_k on a computer is by approximating the integral in Eqn. (4.72) using the rectangular approximation method. More sophisticated methods for approximating integrals exist, and are explained in detail in a number of excellent texts on numerical analysis. The rectangular approximation method is quite simple, and will be sufficient for our purposes.

Suppose we would like to approximate the following integral:

$$G = \int_0^{T_0} g(t)\, dt \tag{5.300}$$

Since integrating the function $g(t)$ amounts to computing the area under the function, a simple approximation is

$$G \approx \sum_{n=0}^{N-1} g(nT)\, T \tag{5.301}$$

We have used the sampling interval $T = T_0/N$ and assumed that N is sufficiently large for the approximation to be a good one. The area under the function $g(t)$ is approximated using the areas of successive rectangles formed by the samples $g(nT)$. This is depicted graphically in Fig. 5.54.

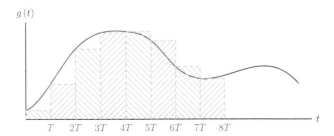

Figure 5.54 – Rectangular approximation to an integral.

For the purpose of approximating Eqn. (4.72) let $g(t)$ be chosen as

$$g(t) = \frac{1}{T_0}\, x_a(t)\, e^{-jk\omega_0 t} \tag{5.302}$$

so that the integral in Eqn. (4.72) is approximated as

$$c_k \approx \frac{1}{T_0} \sum_{n=0}^{N-1} x_a(nT)\, e^{-jk\omega_0 nT}\, T \tag{5.303}$$

Recalling that $T_0 = NT$ and $\omega_0 = 2\pi/T_0$, and using the discrete-time signal $x[n] = x_a(nT)$, we have

$$c_k \approx \frac{1}{N} \sum_{n=0}^{N-1} x[n]\, e^{-j(2\pi/N)kn} = \frac{1}{N} X[k] \qquad (5.304)$$

The EFS coefficients of a periodic signal can be approximated by sampling the signal at N equally spaced time instants over one period, computing the DFT of the resulting discrete-time signal $x[n]$, and scaling the DFT result by $1/N$. Some caveats are in order:

1. The number of samples over one period, N, should be sufficiently large. If we want to be more rigorous, the conditions of the Nyquist sampling theorem must be observed, and the sampling rate $f_s = 1/T$ should be at least twice the highest frequency in the periodic signal $x(t)$.
2. Only the first half of the DFT values can be used as approximations for positive indexed coefficients, that is

$$c_k \approx \frac{1}{N} X[k], \quad k = 0, \ldots, \frac{N}{2} - 1 \qquad (5.305)$$

   with negative indexed coefficients obtained from the second half of the DFT by

$$c_{-k} \approx \frac{1}{N} X[N-k], \quad k = 1, \ldots, \frac{N}{2} \qquad (5.306)$$

3. If the conditions of the Nyquist sampling theorem cannot be satisfied in a strict sense (such as when we try to approximate the Fourier series of a pulse train which is not limited in terms of the highest frequency it contains), only the first few terms of each set in Eqns. (5.305)and (5.306) should be used to keep the quality of the approximation acceptable.

Software resources: See MATLAB Exercises 5.12 and 5.13.

5.8.5   Using the DFT to approximate the continuous Fourier transform

It is also possible to use the DFT for approximating the Fourier transform of a non-periodic continuous-time signal. For a signal $x_a(t)$ the Fourier transform was defined by Eqn. (4.127) in Chapter 4 which is repeated here:

$$\text{Eqn. (4.127):} \quad X_a(\omega) = \int_{-\infty}^{\infty} x_a(t)\, e^{-j\omega t}\, dt \qquad (5.307)$$

Suppose $x_a(t)$ is a finite-length signal that is zero outside the interval $0 \le t \le t_1$. The integral in Eqn. (4.127) can be written as

$$X_a(\omega) = \int_0^{t_1} x_a(t)\, e^{-j\omega t}\, dt \qquad (5.308)$$

The integral can be evaluated in an approximate sense using the rectangular approximation technique outlined in Eqn. (5.301) with

$$g(t) = x_a(t)\, e^{-j\omega t} \qquad (5.309)$$

The function $g(t)$ needs to be sampled at N equally spaced time instants in the interval $0 \leq t \leq t_1$, and thus $NT = t_1$. Rectangular rule approximation to the integral in Eqn. (5.308) is

$$X_a(\omega) \approx \sum_{n=0}^{N-1} x_a(nT) e^{-j\omega nT} T \qquad (5.310)$$

Using the discrete-time signal $x[n] = x_a(nT)$ and evaluating Eqn. (5.310) at a discrete set of frequencies

$$\omega_k = \frac{k\omega_s}{N} = \frac{2\pi k}{NT}, \qquad k = 0, \ldots, N-1$$

where ω_s is the sampling rate in rad/s, we obtain

$$X_a(\omega_k) \approx T \sum_{n=0}^{N-1} x[n] e^{-j(2\pi/N)kn} = T X[k] \qquad (5.311)$$

The Fourier transform of a continuous-time signal can be approximated by sampling the signal at N equally spaced time instants, computing the DFT of the resulting discrete-time signal $x[n]$, and scaling the DFT result by the sampling interval T. It is also possible to obtain the approximation for $X_a(\omega)$ at a more closely spaced set of frequencies than $\omega_k = k\omega_s/N$ by zero-padding $x[n]$ prior to computing the DFT. Let

$$\bar{\omega}_k = \frac{k\omega_s}{N+M} = \frac{2\pi k}{(N+M)T}, \qquad k = 0, \ldots, N+M-1$$

where M is an integer. Using $\bar{\omega}_k$ in Eqn. (5.311) leads to

$$X_a(\bar{\omega}_k) \approx T \sum_{n=0}^{N-1} x[n] e^{-j2\pi kn(N+M)} \qquad (5.312)$$

The summation on the right side of Eqn. (5.312) is the DFT of the signal $x[n]$ zero-padded with M additional samples.

As noted in the discussion of the previous section, the conditions of the Nyquist sampling theorem apply to this case as well. Ideally we would have liked the sampling rate ω_s to be at least twice the highest frequency of the signal the transform of which is being approximated. On the other hand, it can be shown that a time-limited signal contains an infinite range of frequencies, and strict adherence to the Nyquist sampling theorem is not even possible. Approximation errors are due to the aliasing that occurs in sampling a time-limited signal. For the approximation to be acceptable, we need to ensure that the effect of aliasing is negligible.

Based on the Nyquist sampling theorem, only the first half of the DFT samples in Eqns. (5.311) and (5.312) should be used for approximating the transform $X_a(\omega)$ at positive frequencies in the range $0 \leq \omega < \omega_s/2$. The second half of the DFT samples represent an approximation to the transform $X_a(\omega)$ in the negative frequency range $-\omega_s/2 \leq \omega < 0$.

Software resources: See MATLAB Exercise 5.14.

Interactive Demo: `dft_demo.m`

The demo program `dft_demo.m` illustrates the use of DFT for approximating the Fourier transform of a continuous-time function. It is based on Eqn. (5.311), MATLAB Exercise 5.14 and Fig. 5.57.

The continuous-time function $x_a(t)$ used in MATLAB Exercise 5.14 is graphed on the left side of the screen along with its sampled form $x[n]$. The parameter N represents the number of samples in the time interval $0 \leq t \leq 1$ s, and the parameter M represents the number of padded zero-amplitude samples, paralleling the development in Eqns. (5.307) through (5.312). Magnitude and phase of the actual spectrum $X_a(f)$ as well as the DFT based approximation are shown on the right side.

- The case in Fig. 5.57 may be duplicated with the choices of $N = 8$ and $M = 0$. We know that $NT = 1$ s, so the sampling interval is $T = 1/N = 0.125$ s, and the sampling rate is $f_s = 8$ Hz. The approximated samples appear from $f_1 = -4$ Hz to $f_2 = 3$ Hz with a frequency increment of $\Delta f = 1$ Hz.

- While keeping N unchanged, set $M = 8$ which causes 8 zero-amplitude samples to be appended to the right of the existing samples before the DFT computation. Now there are 16 estimated samples of the transform $X_a(f)$ starting at $f_1 = -4$ Hz with an increment of $\Delta f = 0.5$ s.

- Set $N = 16$ and $M = 0$. The sampling rate is $f_s = 16$ Hz now, and therefore the estimated samples start at $f_1 = -8$ Hz and go up to $f_2 = 7$ Hz with an increment of $\Delta f = 1$ Hz.

- Understanding these relationships is key to understanding effective use of the DFT as a tool for analyzing continuous-time signals.

Software resources:
dft_demo.m

5.9 Further Reading

[1] R.J. Beerends. *Fourier and Laplace Transforms*. Cambridge University Press, 2003.

[2] A. Boggess and F.J. Narcowich. *A First Course in Wavelets with Fourier Analysis*. Wiley, 2011.

[3] S.C. Chapra. *Applied Numerical Methods with MATLAB for Engineers and Scientists*. McGraw-Hill, 2008.

[4] S.C. Chapra and R.P. Canale. *Numerical Methods for Engineers*. McGraw-Hill, 2010.

[5] D.G. Manolakis and V.K. Ingle. *Applied Digital Signal Processing: Theory and Practice*. Cambridge University Press, 2011.

[6] A.V. Oppenheim and R.W. Schafer. *Discrete-Time Signal Processing*. Prentice Hall, 2010.

[7] L. Tan and J. Jiang. *Digital Signal Processing: Fundamentals and Applications*. Elsevier Science, 2013.

[8] J.S. Walker. *Fast Fourier transforms*. Studies in Advanced Mathematics. Taylor & Francis, 1996.

MATLAB Exercises

MATLAB Exercise 5.1: Developing functions to implement DTFS analysis and synthesis

In this exercise we will develop two MATLAB functions to implement DTFS analysis and synthesis equations. Function `ss_dtfs(..)` given below computes the DTFS coefficients for the periodic signal $\tilde{x}[n]$. The vector "x" holds one period of the signal $\tilde{x}[n]$ for $n = 0, \ldots, N - 1$. The vector "idx" holds the values of the index k for which the DTFS coefficients $\tilde{c}_k$ are to be computed. The coefficients are returned in the vector "c"

```
1   function c = ss_dtfs(x,idx)
2     c = zeros(size(idx)); % Create all-zero vector.
3     N = length(x);        % Period of the signal.
4     for kk = 1:length(idx),
5       k = idx(kk);
6       tmp = 0;
7       for nn = 1:length(x),
8         n = nn-1;          % MATLAB indices start with 1.
9         tmp = tmp+x(nn)*exp(-j*2*pi/N*k*n);
10      end;
11      c(kk) = tmp/N;
12    end;
13  end
```

The inner loop between lines 6 and 11 implements the summation in Eqn. (5.22) for one specific value of k. The outer loop between lines 4 and 12 causes this computation to be performed for all values of k in the vector "idx".

Function `ss_invdtfs(..)` implements the DTFS synthesis equation. The vector "c" holds one period of the DTFS coefficients $\tilde{c}_k$ for $k = 0, \ldots, N - 1$. The vector "idx" holds the values of the index n for which the signal samples $\tilde{x}[n]$ are to be computed. The synthesized signal $\tilde{x}[n]$ is returned in the vector "x"

```
1   function x = ss_invdtfs(c,idx)
2     x = zeros(size(idx)); % Create all-zero vector.
3     N = length(c);        % Period of the coefficient set.
4     for nn = 1:length(idx),
5       n = idx(nn);
6       tmp = 0;
7       for kk = 1:length(c),
8         k = kk-1;          % MATLAB indices start with 1.
9         tmp = tmp+c(kk)*exp(j*2*pi/N*k*n);
10      end;
11      x(nn) = tmp;
12    end;
13  end
```

Note the similarity in the structures of the two functions. This is due to the similarity of DTFS analysis and synthesis equations.

The functions `ss_dtfs(..)` and `ss_invdtfs(..)` are not meant to be computationally efficient or fast. Rather, they are designed to correlate directly with DTFS analysis and synthesis equations (5.22) and (5.21) respectively. A more efficient method of obtaining the same results will be discussed later in this chapter.

Software resources:

ss_dtfs.m

ss_invdtfs.m

MATLAB Exercise 5.2: Testing DTFS functions

In this exercise we will test the two functions `ss_dtfs(..)` and `ss_invdtfs(..)` that were developed in MATLAB Exercise 5.1.

Consider the signal $\tilde{x}[n]$ used in Example 5.4 and shown in Fig. 5.3. It is defined by

$$\tilde{x}[n] = n \, , \quad \text{for } n = 0, \ldots, 4 \, ; \quad \text{and} \quad \tilde{x}[n+5] = \tilde{x}[n]$$

Its DTFS coefficients can be computed using the function `ss_dtfs(..)` as

```
>>   x = [0,1,2,3,4]
>>   c = ss_dtfs(x,[0:4])
```

The signal can be reconstructed from its DTFS coefficients using the function `ss_invdtfs(..)` as

```
>>   x = ss_invdtfs(c,[-12:15])
>>   stem([-12:15],real(x))
```

The stem graph produced matches Fig. 5.3. The use of the function `real(..)` is necessary to remove very small imaginary parts due to round-off error.

Next, consider the signal of Example 5.5 which is a discrete-time pulse train. We will assume a period of 40 samples and duplicate the DTFS spectra in Fig. 5.5(a), (b), and (c). Let the signal $\tilde{x}_a[n]$ have $L = 3$. Its DTFS coefficients are computed and graphed as follows:

```
>>   xa=[ones(1,4),zeros(1,33),ones(1,3)]
>>   ca = ss_dtfs(xa,[0:39])
>>   stem([0:39],real(ca))
```

Note that one period of the signal must be specified using the index range $n = 0, \ldots, 39$. The DTFS coefficients for the signal $\tilde{x}_b[n]$ with $L = 5$ are computed and graphed with the following lines:

```
>>   xb=[ones(1,6),zeros(1,29),ones(1,5)]
>>   cb = ss_dtfs(xb,[0:39])
>>   stem([0:39],real(cb))
```

Finally for $\tilde{x}_c[n]$ with $L = 7$ we have

```
>>   xc=[ones(1,8),zeros(1,25),ones(1,7)]
>>   cc = ss_dtfs(xc,[0:39])
>>   stem([0:39],real(cc))
```

Software resources:
`matex_5_2a.m`
`matex_5_2b.m`

MATLAB Exercise 5.3: Developing and testing a function to implement periodic convolution

In this exercise we will develop a MATLAB function to implement the periodic convolution operation defined by Eqn. (5.47). The function `ss_pconv(..)` given below computes the periodic convolution of two length-N signals $\tilde{x}[n]$ and $\tilde{h}[n]$. The vector "x" holds one period of the signal $\tilde{x}[n]$ for $n = 0, \ldots, N-1$. Similarly, the vector "h" holds one period of the signal $\tilde{h}[n]$ for $n = 0, \ldots, N-1$. One period of the periodic convolution result $\tilde{y}[n]$ is returned in vector "y".

```
 1    function y = ss_pconv (x,h)
 2      N = length(x);          % Period for all three signals.
 3      y = zeros(size(x));     % Create all-zero vector.
 4      for n = 0:N-1,
 5        tmp = 0;
 6        for k = 0:N-1,
 7          tmp = tmp+ss_per(x,k)*ss_per(h,n-k);
 8        end;
 9        nn = n+1;
10        y(nn) = tmp;
11      end;
12    end
```

Line 7 of the code is a direct implementation of Eqn. (5.47). It utilizes the function ss_per(..), which was developed in MATLAB Exercise 1.7 for periodically extending a discrete-time signal.

The function ss_pconv(..) can easily be tested with the signals used in Example 5.8. Recall that the two signals were

$$\tilde{x}[n] = \{ \ \ldots, \ 0, \ 1, \ 2, \ 3, \ 4, \ \ldots \ \}$$
$$\underset{n=0}{\uparrow}$$

and

$$\tilde{h}[n] = \{ \ \ldots, \ 3, \ 3, \ -3, \ -2, \ -1, \ \ldots \ \}$$
$$\underset{n=0}{\uparrow}$$

each with a period of $N = 5$. The circular convolution result is obtained as follows:

```
>>   x = [0,1,2,3,4]
>>   h = [3,3,-3,-2,-1]
>>   y = ss_pconv (x,h)
```

Software resources:
matex_5_3.m
ss_pconv.m

MATLAB Exercise 5.4: Steady-state response of DTLTI system to sinusoidal input

Consider the DTLTI system of Example 5.29 described by the difference equation

$$y[n] - 0.9y[n-1] + 0.36y[n-2] = x[n] - 0.2x[n-1]$$

The steady-state response of the system to the input signal

$$\tilde{x}[n] = 5 \cos \left(\frac{\pi n}{5} \right)$$

was found to be

$$\tilde{y}[n] = 9.9305 \cos \left(\frac{\pi n}{5} - 0.3139 \right)$$

In this exercise we will obtain the response of the system to a sinusoidal signal turned on at $n = 0$, that is,

$$x[n] = 5 \cos \left(\frac{\pi n}{5} \right) u[n]$$

by iteratively solving the difference equation, and compare it to the steady-state response found in Example 5.29. MATLAB function filter(..) will be used in iteratively solving the difference equation. The script listed below computes both responses and compares them.

```
1   % Script: matex_5_4.m
2   %
3   n = [-10:30];
4   ytilde = 9.9305*cos(pi*n/5-0.3139);     % Steady-state
5   x = 5*cos(pi*n/5).*(n>=0);
6   y = filter([1,-0.2],[1,-0.9,0.36],x);   % Solve diff. eqn.
7   p1 = stem(n-0.125,ytilde,'b');
8   hold on
9   p2 = stem(n+0.125,y,'r');
10  hold off
11  axis([-11,31,-12,12]);
12  xlabel('n');
```

The lines 7 and 9 of the script create two stem plots that are overlaid. In order to observe the two discrete-time signals comparatively, two different colors are used. In addition, horizontal positions of the stems are offset slightly from their integer values, to the left for $\tilde{y}[n]$ and to the right for $y[n]$. The graph produced by the script is shown in Fig. 5.55. The steady-state response is shown in blue and the response to the sinusoidal signal turned on at $n = 0$ is shown in red. Notice how the two responses become the same after the transient dies out after about $n = 10$.

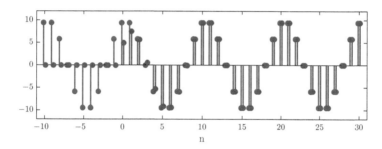

Figure 5.55 – The graph obtained in MATLAB Exercise 5.4.

Software resources:
matex_5_4.m

MATLAB Exercise 5.5: Exploring the relationship between the DFT and the DTFT

The DFT and the DTFT of the length-10 discrete-time pulse were computed in Example 5.33. In this exercise we will duplicate the results of that example in MATLAB. The script listed below computes the 10-point DFT and graphs it on the same coordinate system with the DTFT of the signal.

```
1   % Script matex_5_5a.m
2   %
3   xn = ones(1,10);           % Length-10 pulse signal.
4   Xk = fft(xn);              % X[k] = DFT of x[n].
5   % Create a vector 'Omega' and compute the DTFT.
6   Omega = [-0.1:0.01:1.1]*2*pi+eps;
7   XDTFT = sin(5*Omega)./sin(0.5*Omega).*exp(-j*4.5*Omega);
8   % Compute frequencies that correspond to DFT samples.
9   k = [0:9];
10  Omega_k = 2*pi*k/10;
11  % Graph the DTFT and the DFT on the same coordinate system.
12  clf;
13  subplot(211);
```

```
14   plot(Omega,abs(XDTFT),'-',Omega_k,abs(Xk),'ro'); grid;
15   axis([-0.2*pi,2.2*pi,-1,11]);
16   xlabel('\Omega (rad)');
17   ylabel('Magnitude')
18   subplot(212)
19   plot(Omega,angle(XDTFT),'-',Omega_k,angle(Xk),'ro'); grid;
20   axis([-0.2*pi,2.2*pi,-pi,pi]);
21   xlabel('\Omega (rad)');
22   ylabel('Phase (rad)');
```

A slightly modified script is given below. The signal is padded with 10 additional zero-amplitude samples to extend its length to 20. This causes new transform samples to be displayed between the existing transform samples.

```
1    % Script matex_5_5b.m
2    %
3    xn = ones(1,10);         % Length-10 pulse signal.
4    Xk = fft(xn,20);         % X[k] = DFT of x[n] zero padded to 20.
5    % Create a vector 'Omega' and compute the DTFT.
6    Omega = [-0.1:0.01:1.1]*2*pi+eps;
7    XDTFT = sin(5*Omega)./sin(0.5*Omega).*exp(-j*4.5*Omega);
8    % Compute frequencies that correspond to DFT samples.
9    k = [0:19];
10   Omega_k = 2*pi*k/20;
11   % Graph the DTFT and the DFT on the same coordinate system.
12   clf;
13   subplot(211);
14   plot(Omega,abs(XDTFT),'-',Omega_k,abs(Xk),'ro'); grid;
15   axis([-0.2*pi,2.2*pi,-1,11]);
16   xlabel('\Omega (rad)');
17   ylabel('Magnitude')
18   subplot(212)
19   plot(Omega,angle(XDTFT),'-',Omega_k,angle(Xk),'ro'); grid;
20   axis([-0.2*pi,2.2*pi,-pi,pi]);
21   xlabel('\Omega (rad)');
22   ylabel('Phase (rad)');
```

Software resources:

matex_5_5a.m

matex_5_5b.m

MATLAB Exercise 5.6: Using the DFT to approximate the DTFT

Recall from the discussion in Section 5.8.2 that zero padding a length-N signal by M additional zero-amplitude samples before the computation of the DFT results in a frequency spacing of

$$\Omega_k = \frac{2\pi}{N+M} \text{ radians}$$

By choosing M large, the angular frequency spacing can be made as small as desired. This allows us to use the DFT as a tool for approximating the DTFT. If sufficient DFT samples are available, they can be graphed in the form of a continuous function of Ω to mimic the appearance of the DTFT graph. The script listed below graphs the approximate DTFT of the length 10 discrete-time pulse

$$x[n] = u[n] - u[n-10]$$

The signal is padded with 490 zero-amplitude samples to extend its length to 500. The DFT result is obtained for $k = 0, \ldots, 499$. The DFT sample for index k corresponds to the angular frequency $\Omega_k = 2\pi k/500$. The last DFT sample is at the angular frequency $\Omega_{499} = 998\pi/500$ which is just slightly less than 2π radians.

```
1    % Script: matex_5_6.m
2    %
3    x = ones(1,10);      % Generate x[n].
4    Xk = fft(x,500);     % Compute 500-point DFT.
5    k = [0:499];         % DFT indices
6    Omega_k = 2*pi*k/500;   % Angular frequencies.
7    % Graph the magnitude and the phase of the DTFT.
8    clf;
9    subplot(211);
10   plot(Omega_k,abs(Xk)); grid;
11   title('|X(\Omega)|');
12   xlabel('\Omega (rad)');
13   ylabel('Magnitude');
14   axis([0,2*pi,-1,11]);
15   subplot(212);
16   plot(Omega_k,angle(Xk)); grid;
17   title('\angle X(\Omega)');
18   xlabel('\Omega (rad)');
19   ylabel('Phase (rad)');
20   axis([0,2*pi,-pi,pi]);
```

Software resources:
matex_5_6.m

MATLAB Exercise 5.7: Developing functions for circular time shifting and time reversal

Circular versions of time shifting and time reversal operations can be implemented in MAT-LAB using the steps presented in Section 5.8.3. Recall that the idea behind circularly shifting a length-N signal was to first create its periodic extension with period N, then to apply time shifting to the periodic extension, and finally to extract the main period for $n = 0, \ldots, N-1$. The function ss_per(..) which was developed in MATLAB Exercise 1.7 for periodically extending a discrete-time signal will be used in constructing the functions of this exercise.

The function ss_cshift(..) listed below circularly shifts a signal by the specified number of samples. The vector "x" holds the samples of the signal $x[n]$ for $n = 0, \ldots, N-1$. The second argument "m" is the number of samples by which the signal is to be circularly shifted. The vector "g" returned holds samples of the circularly shifted signal for $n = 0, \ldots, N-1$.

```
1    function g = ss_cshift(x,m)
2      N = length(x);   % Length of x[n].
3      n = [0:N-1];     % Vector of indices.
4      g = ss_per(x,n-m);
5    end
```

The function ss_crev(..) listed below circularly reverses a signal. Again, the vector "x" holds the samples of the signal $x[n]$ for $n = 0, \ldots, N-1$.

```
1    function g = ss_crev(x)
2      N = length(x);   % Length of x[n].
3      n = [0:N-1];     % Vector of indices.
4      g = ss_per(x,-n);
5    end
```

Let us create a length-10 signal to test the functions.

```
>>  n = [0:9];
>>  x = [0,2,3,4,4.5,3,1,-1,2,1];
```

The signal $g[n] = x[n-3]_{\mathrm{mod}\ 10}$ is obtained and graphed by typing

```
>>  g = ss_cshift(x,3);
>>  stem(n,g);
```

A circular shift to the left can be obtained as well. To compute $g[n] = x[n+2]_{\mathrm{mod}\ 10}$ type the following:

```
>>  g = ss_cshift(x,-2);
```

A circularly time reversed version of $x[n]$ is obtained by

```
>>  g = ss_crev(x);
```

Software resources:
matex_5_7.m
ss_cshift.m
ss_crev.m

MATLAB Exercise 5.8: **Computing conjugate symmetric and antisymmetric components**

Any complex length-N signal or transform can be written as the sum of two components, one of which is circularly conjugate symmetric and the other circularly conjugate antisymmetric. The two components are computed using Eqns. (5.272) and (5.273) which will be repeated here:

$$\text{Eqn. (5.272):} \qquad x_E[n] = \frac{x[n] + x^*[-n]_{\mathrm{mod}\ N}}{2}$$

$$\text{Eqn. (5.273):} \qquad x_O[n] = \frac{x[n] - x^*[-n]_{\mathrm{mod}\ N}}{2}$$

The terms needed in the two equations above can easily be computed using the function ss_crev(..) developed in MATLAB Exercise 5.7. Given a vector "x" that holds samples of $x[n]$ for $n = 0, \ldots, N-1$, the signals $x_E[n]$ and $x_O[n]$ are computed as follows:

```
>>  xE = 0.5*(x+conj(ss_crev(x)))
>>  xO = 0.5*(x-conj(ss_crev(x)))
```

Software resources:
matex_5_8.m

MATLAB Exercise 5.9: **Using the symmetry properties of the DFT**

Consider Example 5.38 where two real signals $g[n]$ and $h[n]$ were combined to construct a complex signal $x[n]$. The DFT of the complex signal was computed and separated into its circularly conjugate symmetric and antisymmetric components, and the individual transforms of $g[n]$ and $h[n]$ were extracted.

If the samples of the two signals are placed into MATLAB vectors "g" and "h", the following script can be used to obtain the transforms $G[k]$ and $H[k]$.

```
1    % Script matex_5_9.m
2    %
3    x = g+j*h;              % Construct complex signal.
4    Xk = fft(x);            % Compute DFT.
5    % Compute the two components of the DFT.
6    XE = 0.5*(Xk+conj(ss_crev(Xk)));
7    XO = 0.5*(Xk-conj(ss_crev(Xk)));
8    % Extract DFTs of the two signals.
9    Gk = XE;
10   Hk = -j*XO;
11   disp(ifft(Gk))         % Should be the same as vector 'g'.
12   disp(ifft(Hk))         % Should be the same as vector 'h'.
```

Software resources:
matex_5_9.m

MATLAB Exercise 5.10: Circular and linear convolution using the DFT

Consider the two length-5 signals $x[n]$ and $h[n]$ used in Example 5.41. The script listed below computes the circular convolution of the two signals using the DFT method:

```
1    % Script matex_5_10a.m
2    %
3    x = [1,3,2,-4,6];   % Signal x[n]
4    h = [5,4,3,2,1];    % Signal h[n]
5    Xk = fft(x);        % DFT of x[n]
6    Hk = fft(h);        % DFT of y[n]
7    Yk = Xk.*Hk;        % Eqn.(5.296)
8    y = ifft(Yk);
```

If linear convolution of the two signals is desired, the signals must be extended to at least nine samples each prior to computing the DFTs. The script given below computes the linear convolution of the two signals using the DFT method.

```
1    % Script: mex_5_10b.m
2    %
3    x = [1,3,2,-4,6];   % Signal x[n]
4    h = [5,4,3,2,1];    % Signal h[n]
5    xp = [x,zeros(1,4)];     % Extend by zero padding.
6    hp = [h,zeros(1,4)];     % Extend by zero padding.
7    Xpk = fft(xp);      % DFT of xp[n]
8    Hpk = fft(hp);      % DFT of hp[n]
9    Ypk = Xpk.*Hpk;     % Eqn.(5.296)
10   yp = ifft(Ypk);
```

The script "matex_5_10b.m" can be further simplified by using an alternative syntax for the function fft(..). The statement

```
>>   fft(x,9)
```

computes the DFT of the signal in vector "x" after internally extending the signal to nine samples by zero padding. Consider the modified script listed below:

```
1    % Script: matex_5_10c.m
2    %
3    x = [1,3,2,-4,6];   % Signal x[n]
4    h = [5,4,3,2,1];    % Signal h[n]
5    Xpk = fft(x,9);     % 9-point DFT of x[n]
6    Hpk = fft(h,9);     % 9-point DFT of h[n]
7    Ypk = Xpk.*Hpk;     % Eqn.(5.296)
8    yp = ifft(Ypk);
```

Software resources:
matex_5_10a.m
matex_5_10b.m
matex_5_10c.m

MATLAB Exercise 5.11: Developing a convolution function using the DFT

In this exercise we will develop a function for convolving two signals. Even though the built-in MATLAB function conv(..) accomplishes this task very well, it is still instructive to develop our own function that uses the DFT. As discussed in Section 5.8.3, linear convolution can be obtained through the use of the DFT as long as the length of the transform is adjusted carefully. The function ss_conv2(..) given below computes the linear convolution of two finite-length signals $x[n]$ and $h[n]$ the samples of which are stored in vectors "x" and "h".

```
1  function y = ss_conv2(x,h)
2    N1 = length(x);   % Length of signal 'x'
3    N2 = length(h);   % Length of signal 'h'
4    N = N1+N2-1;      % Length of linear convolution result.
5    Xk = fft(x,N);    % DFT of x[n] zero padded to N.
6    Hk = fft(h,N);    % DFT of h[n] zero padded to N.
7    Yk = Xk.*Hk;
8    y = ifft(Yk,N);
9  end
```

Software resources:
ss_conv2.m

MATLAB Exercise 5.12: Exponential Fourier series approximation using the DFT

In this example, we will develop a MATLAB function to approximate the EFS coefficients of a periodic signal. The strategy used will parallel the development in Eqns. (5.300) through (5.306). We have concluded in Section 5.8.4 that the DFT can be used for approximating the EFS coefficients through the use of Eqns. (5.304) and (5.305) provided that the largest coefficient index k that is used in approximation is sufficiently small compared to the DFT size N.

The code for the function ss_efsapprox(..) is given below.

```
1  function c = ss_efsapprox(x,k)
2    Nx = length(x);   % Size of vector 'x'
3    Nk = length(k);   % Size of vector 'k'
4    % Create a return vector same size as 'k'.
5    c = zeros(1,Nk);
6    Xk = fft(x)/Nx;   % Eqn. (5.304)
7    % Copy the coefficients requested.
8    for i = 1:Nk,
9      kk = k(i);
10     if (kk >= 0),
11       c(i) = Xk(kk+1);
12     else
13       c(i) = Xk(Nx+1+kk);
14     end;
15   end;
16  end
```

The input argument x is a vector holding samples of one period of the periodic signal. The argument k is the index (or a set of indices) at which we wish to approximate the exponential

Fourier series coefficients, so it can be specified as either a scalar index or a vector of index values. The return value c is either a scalar or a vector of coefficients. Its dimensions are compatible with those of k.

Suppose that 500 samples representing one period of a periodic signal $\tilde{x}(t)$ have been placed into the vector x. Using the statement

```
>>  c = ss_efsapprox(x,2)
```

causes an approximate value for the coefficient c_2 to be computed and returned. On the other hand, issuing the statement

```
>>  c = ss_efsapprox(x,[-2,-1,0,1,2])
```

results in the approximate values of coefficients $\{c_{-2}, c_{-1}, c_0, c_1, c_2,\}$ to be computed and returned as a vector.

To keep the function listing simple, no error checking code has been included. Therefore, it is the user's responsibility to call the function with an appropriate set of arguments. Specifically, the index values in k must not cause an out-of-bounds error with the dimensions of the vector x. Furthermore, for good quality approximate values to be obtained, indices used should satisfy $|k| << N$.

Software resources:

ss_efsapprox.m

MATLAB Exercise 5.13: **Testing the EFS approximation function**

In this exercise we will test the function ss_efsapprox(..) that was developed in MATLAB Exercise 5.12. Consider the periodic pulse train shown in Fig. 5.56.

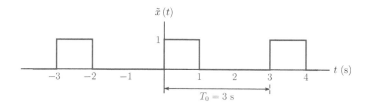

Figure 5.56 – The periodic pulse train for MATLAB Exercise 5.13.

One period is defined as

$$\tilde{x}(t) = \begin{cases} 1, & 0 \le t < 1 \\ 0, & 1 \le t < 3 \end{cases}$$

We will compute 1024 samples of this signal within one period. The test code for approximating c_k for $k = 0, \ldots, 10$ is given below:

```
1   % Script: matex_5_13.m
2   %
3   t = [0:1023]/1024*3;    % 1024 samples in one period.
4   x = (t<=1);             % x(t)=1 if t<=1
5   % Compute and print approximate EFS coefficients for
6   % k=0,...,10.
7   for k = 0:10,
8     coeff = ss_efsapprox(x,k);
9     str = sprintf('k=%3d, magnitude=%0.5f, phase=%0.5f',...
10      k,abs(coeff),angle(coeff));
11    disp(str);
12  end;
```

One point needs to be clarified: In line 3 of the code we generate a vector "t" of time samples spanning one period of $T = 3$ seconds. The first element of the time vector is equal to 0. We might intuitively think that the last element of this vector should be set equal to 3, the length of the period, yet it is equal to $1023(3)/(1024)$ which is slightly less than 3. The reason for this subtle difference becomes obvious when we look at Fig. 5.54. The width of each rectangle used in the numerical approximation of the integral is $T = 3/1024$. The first rectangle starts at $t = 0$ and extends to $t = T = 3/1024$. The second rectangle rectangle starts at $t = T = 3/1024$ and extends to $t = 2T = 6/1024$. Continuing to reason in this fashion, the left edge of the last rectangle of the period is at $t = 1023T = 1023(3)/(1024)$, and the right edge of it is at $t = 3$. The vector "t" holds the left edge of each rectangle in the approximation, and therefore its last value is just shy of the length of the period by one rectangle width.

Actual versus approximate magnitude and phase values of exponential Fourier series coefficients of the signal $\tilde{x}(t)$ are listed in Table 5.6. The same table also includes percent error calculations between actual and approximated values. Note that for index values at which the magnitude of a coefficient is zero or near-zero, phase calculations are meaningless, and percent error values are skipped.

k	Magnitude			Phase (rad)		
	Actual	Approx.	% Error	Actual	Approx.	% Error
0	0.3333	0.3340	0.20	0.00	0.000	0.000
1	0.2757	0.2760	0.12	-1.047	-1.046	-0.098
2	0.1378	0.1375	-0.24	-2.094	-2.092	-0.098
3	0.0000	0.0007	0.00	3.142	0.003	
4	0.0689	0.0692	0.47	-1.047	-1.043	-0.391
5	0.0551	0.0548	-0.59	-2.094	-2.089	-0.244
6	0.0000	0.0007	0.00	3.142	0.006	
7	0.0394	0.0397	0.82	-1.047	-1.040	-0.683
8	0.0345	0.0341	-0.94	-2.094	-2.086	-0.390
9	0.0000	0.0007	0.00	3.142	0.009	
10	0.0276	0.0279	1.17	-1.047	-1.037	-0.976

Table 5.6 – Exact vs. approximate exponential Fourier series coefficients for the test case in MATLAB Example 5.13.

MATLAB Exercise 5.14: Fourier transform approximation using DFT

The continuous-time signal

$$x_a(t) = \sin(\pi t)\, \Pi\left(t - \frac{1}{2}\right)$$

$$= \begin{cases} \sin(\pi t), & 0 \le t \le 1 \\ 0, & \text{otherwise} \end{cases}$$

has the Fourier transform (see Problem 4.33)

$$X_a(f) = \frac{1}{2}\left[\text{sinc}\left(f + \frac{1}{2}\right) + \text{sinc}\left(f - \frac{1}{2}\right)\right] e^{-j\pi f}$$

In this example we will develop the MATLAB code to approximate this transform using the DFT as outlined by Eqns. (5.311) and (5.312). The MATLAB script given below computes and graphs both the actual and the approximated transforms for the signal $x_a(t)$.

1. The actual transform is computed in the frequency range $-10 \text{ Hz} \leq f \leq 10 \text{ Hz}$.
2. For the approximate solution, $N = 16$ samples of the signal $x_a(t)$ are taken in the time interval $0 \ll t \leq 1$ s. This corresponds to a sampling rate of $f_s = 16$ Hz, and to a sampling interval of $T = 0.0625$ s.
3. Initially no zero padding is performed, therefore $M = 0$.
4. Approximations to the continuous Fourier transform are obtained at frequencies that are spaced $f_s/N = 1$ Hz apart.
5. The DFT result has 16 samples as approximations to the continuous Fourier transform at frequencies $f_k = kf_s/N = k$ Hz. Only the first 8 DFT samples are usable for positive frequencies from 0 to 7 Hz. This is due to Nyquist sampling theorem. The second half of the transform represents negative frequencies from -8 Hz to -1 Hz, and must be placed in front of the first half. The function fftshift(..) used in line 21 accomplishes that.

```
1   % Script: matex_5_14.m
2   %
3   f = [-10:0.01:10];        % Create a vector of frequencies
4   % Compute the actual transform.
5   X_actual = 0.5*(sinc(f+0.5)+sinc(f-0.5)).*exp(-j*pi*f);
6   % Set parameters for the approximate transform
7   t1 = 1;                   % Upper limit of the time range
8   N = 16;                   % Number of samples
9   M = 0;                    % Number of samples for zero-padding
10  T = t1/N;                 % Sampling interval
11  fs = 1/T;                 % Sampling rate.
12  n = [0:N-1];              % Index n for the sampled signal x[n]
13  k = [0:N+M-1];            % Index k for the DFT X[k]
14  time = n*T;               % Sampling instants
15  % Sample the signal and compute the DFT.
16  xn = sin(pi*time);
17  Xk = T*fft(xn,N+M);
18  fk = k*fs/(N+M);
19  % Use fftshift() function on the DFT result to bring the zero-
20  % frequency to the middle. Also adjust vector fk for it.
21  Xk = fftshift(Xk);
22  fk = fk-0.5*fs;
23  % Graph the results.
24  clf;
25  subplot(2,1,1);
26  plot(f,abs(X_actual),'-',fk,abs(Xk),'r*'); grid;
27  title('Magnitude of actual and approximate transforms');
28  xlabel('f (Hz)');
29  ylabel('Magnitude');
30  subplot(2,1,2);
31  plot(f,angle(X_actual),'-',fk,angle(Xk),'r*'); grid;
32  title('Phase of actual and approximate transforms');
33  xlabel('f (Hz)');
34  ylabel('Phase (rad)');
```

The MATLAB graph produced by the script is shown in Fig. 5.57. The zero padding parameter M can be used to control the frequency spacing. For example, setting $M = 16$ causes the continuous Fourier transform to be estimated at intervals of 0.5 Hz starting at -8 Hz and ending at 7.5 Hz.

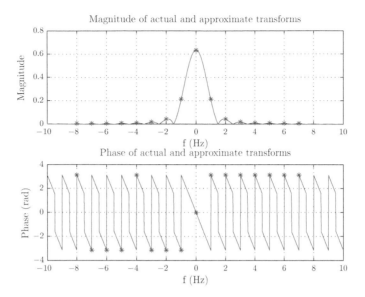

Figure 5.57 – The graph obtained in MATLAB Exercise 5.14. Magnitude and phase of the actual spectrum shown are in blue, and approximated values are shown with red asterisks.

Problems

5.1.   Determine the DTFS representation of the signal $\tilde{x}[n] = \cos(0.3\pi n)$. Sketch the DTFS spectrum.

5.2.   Determine the DTFS representation of the signal $\tilde{x}[n] = 1 + \cos(0.24\pi n) + 3\sin(0.56\pi n)$. Sketch the DTFS spectrum.

5.3.   Determine the DTFS coefficients for each of the periodic signals given in Fig. P.5.3.

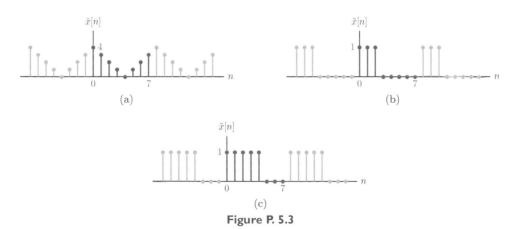

Figure P. 5.3

5.4.   Consider the periodic signal of Example 5.4. Let $\tilde{g}[n]$ be one sample delayed version of it, that is,

$$\tilde{g}[n] = \tilde{x}[n-1]$$

as shown in Fig. P.5.4.

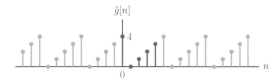

Figure P. 5.4

a. Determine the DTFS coefficients of $\tilde{g}[n]$ directly from the DTFS analysis equation.
b. Determine the DTFS coefficients of $\tilde{g}[n]$ by applying the time shifting property to the coefficients of $\tilde{x}[n]$ that were determined in Example 5.4.

5.5. Verify that the DTFS coefficients found for the signals in Problem 5.3 satisfy the conjugate symmetry properties outlined in Section 5.2.2. Since the signals are real, DTFS spectra should be conjugate symmetric.

5.6. Consider the periodic signal shown in Fig. P.5.4.

a. Find even and odd components of $\tilde{g}[n]$ so that

$$\tilde{g}[n] = \tilde{g}_e[n] + \tilde{g}_o[n]$$

b. Determine the DTFS coefficients of $\tilde{g}_e[n]$ and $\tilde{g}_o[n]$. Verify the symmetry properties outlined in Section 5.2.2. The spectrum of the even component should be real, and the spectrum of the odd component should be purely imaginary.

5.7. Let $\tilde{x}[n]$ and $\tilde{h}[n]$ be both periodic with period N. Using the definition of periodic convolution given by Eqn. (5.49), show that the signal

$$\tilde{y}[n] = \tilde{x}[n] \otimes \tilde{h}[n]$$

is also periodic with period N.

5.8. Consider the two periodic signals shown in Fig. P.5.8.

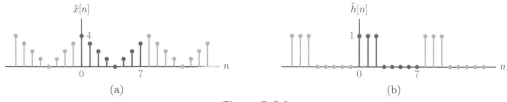

Figure P. 5.8

a. Compute the periodic convolution

$$\tilde{y}[n] = \tilde{x}[n] \otimes \tilde{h}[n]$$

of the two signals.
b. Determine the DTFS coefficients of signals $\tilde{x}[n]$, $\tilde{h}[n]$ and $\tilde{y}[n]$. Verify that the periodic convolution property stated by Eqn. (5.49) holds.

5.9. Repeat Problem 5.8 for the two signals shown in Fig. P.5.9.

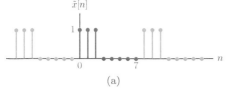

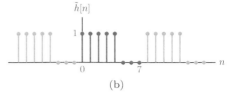

(a)   (b)

Figure P. 5.9

5.10. Find the DTFT of each signal given below. For each, sketch the magnitude and the phase of the transform.

a. $x[n] = \delta[n] + \delta[n - 1]$
b. $x[n] = \delta[n] - \delta[n - 1]$

c. $x[n] = \begin{cases} 1, & n = 0, 1, 2, 3 \\ 0, & \text{otherwise} \end{cases}$

d. $x[n] = \begin{cases} 1, & n = -2, \ldots, 2 \\ 0, & \text{otherwise} \end{cases}$

e. $x[n] = (0.7)^n \, u[n]$
f. $x[n] = (0.7)^n \, \cos(0.2\pi n) \, u[n]$

5.11. Find the inverse DTFT of each transform specified below for $-\pi \leq \Omega < \pi$.

a. $X(\Omega) = \begin{cases} 1, & |\Omega| < 0.2\pi \\ 0, & \text{otherwise} \end{cases}$

b. $X(\Omega) = \begin{cases} 1, & |\Omega| < 0.4\pi \\ 0, & \text{otherwise} \end{cases}$

c. $X(\Omega) = \begin{cases} 0, & |\Omega| < 0.2\pi \\ 1, & \text{otherwise} \end{cases}$

d. $X(\Omega) = \begin{cases} 1, & 0.1\pi < |\Omega| < 0.2\pi \\ 0, & \text{otherwise} \end{cases}$

5.12. Use linearity and time shifting properties of the DTFT to find the transform of each signal given below.

a. $x[n] = (0.5)^n \, u[n - 2]$
b. $x[n] = (0.8)^n \, (u[n] - u[n - 10])$
c. $x[n] = (0.8)^n \, (u[n + 5] - u[n - 5])$

5.13. Use linearity and time reversal properties of the DTFT to find the transform of the signals listed below:

a. $x[n] = (2)^n \, u[-n - 1]$
b. $x[n] = (1.25)^n \, u[-n]$
c. $x[n] = (0.8)^{|n|}$
d. $x[n] = (0.8)^{|n|} \, (u[n + 5] + u[n - 5])$

5.14. Prove that the DTFT of a signal with odd symmetry is purely imaginary. In mathematical terms, if

$$x[-n] = -x[n]$$

then
$$X^* \left(\Omega \right) = -X \left(\Omega \right)$$

5.15. Signals listed below have even symmetry. Determine the DTFT of each, and graph the magnitude of the transform.

a. $\quad x[n] = \begin{cases} 1, & n = 0 \\ 1/2, & n = \pm 1 \\ 0, & \text{otherwise} \end{cases}$

b. $\quad x[n] = \begin{cases} 5 - |n|, & |n| \le 4 \\ 0, & \text{otherwise} \end{cases}$

c. $\quad x[n] = \begin{cases} \cos\left(0.2\pi n\right), & n = -4, \dots, 4 \\ 0, & \text{otherwise} \end{cases}$

5.16. Signals listed below have odd symmetry. For each signal determine the DTFT. Graph the magnitude and the phase of the transform.

a. $\quad x[n] = \begin{cases} -1/4, & n = -2 \\ -1/2, & n = -1 \\ 1/2, & n = 1 \\ 1/4, & n = 2 \\ 0, & \text{otherwise} \end{cases}$

b. $\quad x[n] = \begin{cases} n, & n = -5, \dots, 5 \\ 0, & \text{otherwise} \end{cases}$

c. $\quad x[n] = \begin{cases} \sin\left(0.2\pi n\right), & n = -4, \dots, 4 \\ 0, & \text{otherwise} \end{cases}$

5.17. Determine the transforms of the signals listed below using the modulation property of the DTFT. Sketch the magnitude and the phase of the transform for each.

a. $\quad x[n] = (0.7)^n \cos\left(0.2\pi n\right) u[n]$
b. $\quad x[n] = (0.7)^n \sin\left(0.2\pi n\right) u[n]$
c. $\quad x[n] = \cos\left(\pi n/5\right) \left(u[n] - u[n-10]\right)$

5.18. Use the differentiation in frequency property of the DTFT to find the transforms of the signals listed below.

a. $\quad x[n] = n \left(0.7\right)^n u[n]$
b. $\quad x[n] = n \left(n+1\right) \left(0.7\right)^n u[n]$
c. $\quad x[n] = n \left(0.7\right)^n \left(u[n] - u[n-10]\right)$

5.19. Two signals $x[n]$ and $h[n]$ are given by

$$x[n] = u[n] - u[n-10] \quad \text{and} \quad h[n] = (0.8)^n u[n]$$

a. Determine the DTFT for each signal.
b. Let $y[n]$ be the convolution of these two signals, that is, $y[n] = x[n] * h[n]$. Compute $y[n]$ by direct application of the convolution sum.
c. Determine the DTFT of $y[n]$ by direct application of the DTFT analysis equation. Verify that it is equal to the product of the individual transforms of $x[n]$ and $h[n]$:

$$Y \left(\Omega \right) = X \left(\Omega \right) H \left(\Omega \right)$$

5.20. Determine and sketch the DTFTs of the following periodic signals.

- **a.** $x[n] = e^{j0.3\pi n}$
- **b.** $x[n] = e^{j0.2\pi n} + 3\,e^{j0.4\pi n}$
- **c.** $x[n] = \cos(\pi n/5)$
- **d.** $x[n] = 2\cos(\pi n/5) + 3\cos(2\pi n/5)$

5.21. Determine and sketch the power spectral density for each signal shown in Fig. P.5.3.

5.22. Consider the periodic signal

$$\tilde{x}[n] = \begin{cases} 1\,, & -L \leq n \leq L \\ 0\,, & L < n < N - L \end{cases} \qquad \text{and} \qquad \tilde{x}[n+N] = \tilde{x}[n]$$

Recall that the DTFS representation of this signal was found in Example 5.5.

- **a.** Find the power spectral density $S_x(\Omega)$.
- **b.** Let $N = 40$ and $L = 3$. Determine what percentage of signal power is preserved if only three harmonics are kept and the others are discarded.
- **c.** Repeat part (b) with $N = 40$ and $L = 6$.

5.23. A DTLTI system is characterized by the difference equation

$$y[n] + y[n-1] + 0.89\,y[n-2] = x[n] + 2\,x[n-1]$$

Determine the steady-state response of the system to the following input signals:

- **a.** $x[n] = e^{j0.2\pi n}$
- **b.** $x[n] = \cos(0.2\pi n)$
- **c.** $x[n] = 2\sin(0.3\pi n)$
- **d.** $x[n] = 3\cos(0.1\pi n) - 5\sin(0.2\pi n)$

Hint: First find the system function $H(\Omega)$.

5.24. Determine the steady-state response of a length-4 moving average filter to the following signals.

- **a.** $x[n] = e^{j0.2\pi n}$
- **b.** $x[n] = \cos(0.2\pi n)$
- **c.** $x[n] = 2\sin(0.3\pi n)$
- **d.** $x[n] = 3\cos(0.1\pi n) - 5\sin(0.2\pi n)$

5.25. A DTLTI system is characterized by the difference equation

$$y[n] + y[n-1] + 0.89\,y[n-2] = x[n] + 2\,x[n-1]$$

Using the technique outlined in Section 5.6.3 find the steady-state response of the system to each of the periodic signals shown in Fig. P.5.3.

5.26. Compute the DFTs of the signals given below.

- **a.** $x[n] = \{\,1, 1, 1\,\}$
  $$\underset{n=0}{\uparrow}$$
- **b.** $x[n] = \{\,1, 1, 1, 0, 0\,\}$
  $$\underset{n=0}{\uparrow}$$

c. $x[n] = \{\ 1, 1, 1, 0, 0, 0, 0\ \}$
           $\uparrow$
           $n=0$

5.27.   Compute the DFTs of the signals given below. Simplify the results and show that each transform is purely real.

a. $x[n] = \{\ 1, 1, 0, 0, 0, 0, 0, 1\ \}$
           $\uparrow$
           $n=0$

b. $x[n] = \{\ 1, 1, 1, 0, 0, 0, 1, 1\}$
           $\uparrow$
           $n=0$

c. $x[n] = \{\ 1, 1, 1, 1, 0, 1, 1, 1\ \}$
           $\uparrow$
           $n=0$

5.28.   It is possible to express the forward DFT in matrix form using the compact form of the DFT definition given by Eqn. (5.253). Given the signal $x[n]\ ;n = 0,\ldots,N-1$ and its DFT $X[k]\ ;k = 0,\ldots,N-1$, let the vectors $\mathbf{x}$ and $\mathbf{X}$ be defined as

$$\mathbf{x} = \begin{bmatrix} x[0] \\ x[1] \\ \vdots \\ x[N-1] \end{bmatrix}, \qquad \mathbf{X} = \begin{bmatrix} X[0] \\ X[1] \\ \vdots \\ X[N-1] \end{bmatrix}$$

Show that the discrete Fourier transform can be computed as

$$\mathbf{X} = \mathbf{W}\,\mathbf{x}$$

Construct the $N \times N$ coefficient matrix $\mathbf{W}$.

5.29.   For each finite-length signal listed below, find the specified circularly shifted version.

a. $x[n] = \{\ 4, 3, 2, 1\ \}$ find $x[n-2]_{\mathrm{mod}\ 4}$
           $\uparrow$
           $n=0$

b. $x[n] = \{\ 1, 1, 1, 0, 0\ \}$ find $x[n-4]_{\mathrm{mod}\ 5}$
           $\uparrow$
           $n=0$

c. $x[n] = \{\ 1, 4, 2, 3, 1, -2, -3, 1\ \}$ find $x[-n]_{\mathrm{mod}\ 8}$
           $\uparrow$
           $n=0$

d. $x[n] = \{\ 1, 4, 2, 3, 1, -2, -3, 1\ \}$ find $x[-n+2]_{\mathrm{mod}\ 8}$
           $\uparrow$
           $n=0$

5.30.   Refer to Problem 5.29. Find the circularly conjugate symmetric and antisymmetric components $x_E[n]$ and $x_O[n]$ for each signal listed.

5.31.   Consider the finite-length signal

$$x[n] = \{\ 1, 1, 1, 0, 0\ \}$$
$$\uparrow$$
$$n=0$$

a.   Compute the 5-point DFT $X[k]$ for $k = 0,\ldots,4$.
b.   Multiply the DFT found in part (a) with $e^{-j2\pi k/5}$ to obtain the product

$$R[k] = e^{-j2\pi k/5}\,X[k] \qquad \text{for}\ \ k = 0,\ldots,4$$

   c. Compute the signal $r[n]$ as the inverse DFT of $R[k]$ and compare it to the original signal $x[n]$.

   d. Repeat parts (b) and (c) with

$$S[k] = e^{-j4\pi k/5}\, X[k] \qquad \text{for } k = 0, \ldots, 4$$

   e. Provide justification for the answers found using the properties of the DFT.

5.32. Consider the finite-length signal

$$x[n] = \{\ 5,\, 4,\, 3,\, 2,\, 1\ \}$$
$$\uparrow$$
$$n=0$$

   a. Compute the 5-point DFT $X[k]$ for $k = 0, \ldots, 4$.

   b. Reverse the DFT found in part (a) to obtain a new transform $R[k]$ as

$$R[k] = X[-k]_{\text{mod } 5} \qquad \text{for } k = 0, \ldots, 4$$

   c. Compute the signal $r[n]$ as the inverse DFT of $R[k]$ and compare it to the original signal $x[n]$. Justify the answer found using the properties of the DFT.

5.33. The DFT of a length-6 signal $x[n]$ is given by

$$X[k] = \{\ (2 + j3),\, (1 + j5),\, (-2 + j4),\, (-1 - j3),\, (2),\, (3 + j1)\ \}$$
$$\uparrow$$
$$k=0$$

Without computing $x[n]$ first, determine

   a. The DFT of $x_r[n]$, the real part of $x[n]$

   b. The DFT of $x_i[n]$, the imaginary part of $x[n]$

5.34. Two signals $x[n]$ and $h[n]$ are given by

$$x[n] = \{\ 2,\, -3,\, 4,\, 1,\, 6\ \}$$
$$\uparrow$$
$$n=0$$
$$h[n] = \{\ 1,\, 1,\, 1,\, 0,\, 0\ \}$$
$$\uparrow$$
$$n=0$$

   a. Compute the circular convolution $y[n] = x[n] \otimes h[n]$ through direct application of the circular convolution sum.

   b. Compute the 5-point transforms $X[k]$ and $H[k]$.

   c. Compute $Y[k] = X[k]\,H[k]$, and the obtain $y[n]$ as the inverse DFT of $Y[k]$. Verify that the same result is obtained as in part (a).

5.35. Refer to the signals $x[n]$ and $h[n]$ in Problem 5.34. Let $X[k]$ and $H[k]$ represent 7-point DFTs of these signals, and let $Y[k]$ be their product, that is, $Y[k] = X[k]\,H[k]$ for $k = 0, \ldots, 6$. The signal $y[n]$ is the inverse DFT of $Y[k]$. Determine $y[n]$ without actually computing any DFTs.

5.36.

a. The DFT of a length-N signal is equal to its DTFT sampled at N equally spaced angular frequencies. In this problem we will explore the possibility of sampling a DTFT at fewer than N frequency values. Let $X(\Omega)$ be the DTFT of the length-12 signal

$$x[n] = u[n] - u[n-12]$$

Obtain $S[k]$ by sampling the transform $X(\Omega)$ at 10 equally spaced frequencies, that is,

$$S[k] = X\left(\frac{2\pi k}{10}\right), \qquad k = 0, \ldots, 9$$

Determine $s[n]$, the inverse DFT of $S[k]$.

b. Consider the general case where $x[n]$ may be an infinitely long signal, and its DTFT is

$$X(\Omega) = \sum_{n=-\infty}^{\infty} x[n] e^{-j\Omega n}$$

$S[k]$ is obtained by sampling $X(\Omega)$ at N equally spaced angular frequencies:

$$S[k] = X\left(\frac{2\pi k}{N}\right), \qquad k = 0, \ldots, N-1$$

$s[n]$ is the length-N inverse DFT of $S[k]$. Show that $s[n]$ is related to $x[n]$ by

$$s[n] = \sum_{r=-\infty}^{\infty} x[n+rN], \qquad n = 0, \ldots, N-1$$

MATLAB Problems

5.37. Consider the periodic signals $\tilde{x}[n]$ and $\tilde{g}[n]$ shown in Fig. P.5.37.

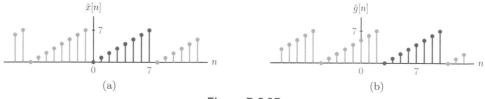

(a)   (b)

Figure P. 5.37

Write a MATLAB script to do the following:

a. Compute the DTFS coefficients of each signal using the function `ss_dtfs(..)` developed in MATLAB Exercise 5.1.
b. Compute the DTFS coefficients of $\tilde{g}[n]$ by multiplying the DTFS coefficients of $\tilde{x}[n]$ with the proper exponential sequence as dictated by the time shifting property.
c. Compare the results found in parts (a) and (b).

5.38. Consider again the periodic signals used in Problem 5.37 and shown in Fig. P.5.37. Both signals are real-valued. Consequently, their DTFS coefficients must satisfy the conjugate symmetry property as stated by Eqn. (5.35). Verify this with the values obtained in Problem 5.37.

5.39.   Refer to the periodic pulse train used in Example 5.5. Write a MATLAB script to compute and graph its DTFS spectrum. Use the script to obtain graphs for the following parameter configurations:

a. $N = 30$, $L = 5$
b. $N = 30$, $L = 8$
c. $N = 40$, $L = 10$
d. $N = 40$, $L = 15$

5.40. Use the MATLAB script developed in Problem 5.39 to produce graphs that duplicate the three cases shown in Fig. 5.17 parts (a), (b) and (c).

5.41. Refer to Problem 5.22.

a.   Let $N = 40$ and $L = 3$. Develop a script to compute and graph a finite-harmonic approximation to $x[n]$ using three harmonics.
b.   Repeat part (a) using a signal with $N = 40$ and $L = 6$.

5.42.   Refer to Problem 5.24. Write a script to compute the response of the length-4 moving average filter to each of the the signals specified in the problem. Use the function `ss_movavg4(..)` that was developed in MATLAB Exercise 3.1 in Chapter 3. For each input signal compute the response for $n = 0, \ldots, 49$. Compare the steady-state part of the response (in this case it will be after the first three samples) to the theoretical result obtained in Problem 5.24.

5.43. Refer to Problem 5.25.

a.   Develop a script to iteratively solve the difference equation given. The system should be assumed initially relaxed, that is, $y[-1] = y[-2] = 0$.
b.   Find the response of the system to each of the input signals in Fig. P.5.3 for $n = 0, \ldots, 49$.
c.   In each of the responses, disregard the samples for $n = 0, \ldots, 24$, and compare samples $n = 25, \ldots, 49$ to the theoretical steady-state response computed in Problem 5.25.

Hint: You may wish to use the function `ss_per(..)` that was developed in MATLAB Exercise 1.7 for creating the periodic signals needed.

5.44.   Refer to Problem 5.28 in which the matrix form of the forward DFT equation was explored.

a.   Develop a MATLAB function `ss_dftmat(..)` to construct the coefficient matrix **W** needed in computing the DFT. Its syntax should be

```
W = ss_dftmat(N)
```

where N is the size of the DFT to be computed.
b.   Use the function `ss_dftmat(..)` to compute the 10-point DFT of the discrete-time pulse signal

$$x[n] = u[n] - u[n - 10]$$

using the matrix method. Compare the result to that obtained in Example 5.33.
c.   Use the function `ss_dftmat(..)` to compute the 20-point DFT of the discrete-time pulse signal through zero padding the vector **x**. Compare the result to that obtained in Example 5.34.

5.45. Write a script to approximate the EFS coefficients of the half-wave rectified sinusoidal signal $\tilde{x}(t)$ shown in Fig. P.5.45.

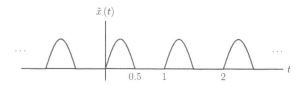

$\tilde{x}(t)$

0.5 1 2 t

Figure P. 5.45

You may wish to use the function `ss_efsapprox(..)` developed in MATLAB Exercise 5.12. Sample the signal with a step size of $T = 0.01$ s to obtain 100 samples in one period. Obtain approximate values for the coefficients c_k for $k = -15, \ldots, 15$ and compare to the correct values obtained in Chapter 4, Example 4.10.

5.46. Refer to Problem 5.36 in which the issue of sampling a DTFT was explored. Develop a MATLAB script to perform the following steps:

a. Define an anonymous function to return the DTFT $X(\Omega)$ of the signal

$$x[n] = \begin{cases} n, & n = 0, \ldots, 11 \\ 0, & \text{otherwise} \end{cases}$$

b. Obtain $S[k]$ by sampling the transform $X(\Omega)$ at 10 equally spaced frequencies, that is,

$$S[k] = X\left(\frac{2\pi k}{10}\right), \qquad k = 0, \ldots, 9$$

Compute $s[n] = \mathrm{DFT}^{-1}\{S[k]\}$, and create a stem plot.

c. Repeat part (b) by evaluating the DTFT at 8 equally-spaced angular frequencies

$$\Omega_k = \frac{2\pi k}{8}, \qquad k = 0, \ldots, 7$$

MATLAB Projects

5.47. This problem is about increasing the efficiency of DFT computation by recognizing the symmetry properties of the DFT and using them to our advantage. Recall from the symmetry properties discussed in Section 5.8.3 that for a signal expressed in Cartesian complex form

$$x[n] = x_r[n] + jx_i[n]$$

and its transform expressed using circularly conjugate symmetric and antisymmetric components as

$$X[k] = X_E[k] + X_O[k]$$

the following relationships exist:

$$x_r[n] \overset{\mathrm{DFT}}{\longleftrightarrow} X_E[k]$$
$$jx_i[n] \overset{\mathrm{DFT}}{\longleftrightarrow} X_O[k]$$

Develop a MATLAB script to compute the DFTs of two real-valued length-N signals $r[n]$ and $s[n]$ with one call to the function $\texttt{fft(..)}$ using the following approach:

a. Construct a complex signal $x[n] = r[n] + j\,s[n]$.

b. Compute the DFT of the complex signal $x[n]$ using the function $\texttt{fft(..)}$.

c. Find the DFTs of the signals $r[n]$ and $s[n]$ by splitting the transform $X[k]$ into its two components $X_E[k]$ and $X_O[k]$ and making the necessary adjustments. You may wish to use the function $\texttt{ss\_crev(..)}$ developed in MATLAB Exercise 5.8.

Test your script using two 128-point signals with random amplitudes. (Use MATLAB function $\texttt{rand(..)}$ to generate them.) Compare the transforms for $R[k]$ and $S[k]$ to results obtained by direct application of the function $\texttt{fft(..)}$ to signals $r[n]$ and $s[n]$.

5.48. One of the popular uses of the DFT is for analyzing the frequency spectrum of a continuous-time signal that has been sampled. The theory of sampling is covered in detail in Chapter 6. In this problem we will focus on computing and graphing the approximate spectrum of a recorded audio signal using the technique discussed in Section 5.8.5.

Provided that a computer with a microphone and a sound processor is available, the following MATLAB code allows 3 seconds of audio signal to be recorded, and a vector "$\texttt{x}$" to be created with the recording:

```
hRec = audiorecorder;
disp('Press a key to start recording');
pause;
recordblocking(hRec, 3);
disp('Finished recording');
x = getaudiodata(hRec);
```

By default the analog signal $x_a(t)$ captured by the microphone and the sound device is sampled at the rate of 8000 times per second, corresponding to $T = 125\ \mu\text{s}$. For a 3-second recording the vector "$\texttt{x}$" contains 24000 samples that represent

$$x[n] = x_a\left(\frac{n}{8000}\right), \qquad n = 0, \ldots, 23999$$

Develop a MATLAB script to perform the following steps to display the approximate spectrum of a speech signal $x_a(t)$:

a. Extract 1024 samples of the vector $x[n]$ into a new vector.

$$r[n] = x[n + 8000], \qquad n = 0, \ldots, 1023$$

We skip the first 8000 samples so that we do not get a blank period before the person begins to speak.

b. Compute the DFT $R[k]$ of the signal $r[n]$. Scale it as outlined in Section 5.8.5 so that it can be used for approximating the continuous Fourier transform of $x_a(t)$. Also create an approximate vector "$\texttt{f}$" of frequencies in Hz for use in graphing the approximate spectrum $X_a(f)$.

c. Compute and graph the decibel (dB) magnitude of the approximate spectrum as a function of frequency.

d. Record your own voice and use it to test the script.

Chapter 6

Sampling and Reconstruction

Chapter Objectives

- Understand the concept of sampling for converting a continuous-time signal to a discrete-time signal.

- Learn how sampling affects the frequency-domain characteristics of the signal, and what precautions must be taken to ensure that the signal obtained through sampling is an accurate representation of the original.

- Consider the issue of reconstructing an analog signal from its sampled version. Understand various interpolation methods used and the spectral relationships involved in reconstruction.

- Discuss methods for changing the sampling rate of a discrete-time signal.

6.1 Introduction

The term *sampling* refers to the act of periodically measuring the amplitude of a continuous-time signal and constructing a discrete-time signal with the measurements. If certain conditions are satisfied, a continuous-time signal can be completely represented by measurements (samples) taken from it at uniform intervals. This allows us to store and manipulate continuous-time signals on a digital computer.

Consider, for example, the problem of keeping track of temperature variations in a classroom. The temperature of the room can be measured at any time instant, and can therefore be modeled as a continuous-time signal $x_a(t)$. Alternatively, we may choose to check the room temperature once every 10 minutes and construct a table similar to Table 6.1.

If we choose to index the temperature values with integers as shown in the third row of Table 6.1 then we could view the result as a discrete time signal $x[n]$ in the form

$$x[n] = \{\, 22.4,\ 22.5,\ 22.8,\ 21.6,\ 21.7,\ 21.7,\ 21.9,\ 22.2,\ \dots \,\} \tag{6.1}$$
$$\underset{n=0}{\uparrow}$$

Time	8:30	8:40	8:50	9:00	9:10	9:20	9:30	9.40
Temp. (°C)	22.4	22.5	22.8	21.6	21.7	21.7	21.9	22.2
Index n	0	1	2	3	4	5	6	7

Table 6.1 – Sampling the temperature signal.

Thus the act of sampling allows us to obtain a discrete-time signal $x[n]$ from the continuous-time signal $x_a(t)$. While any signal can be sampled with any time interval between consecutive samples, there are certain questions that need to be addressed before we can be confident that $x[n]$ provides an accurate representation of $x_a(t)$. We may question, for example, the decision to wait for 10 minutes between temperature measurements. Do measurements taken 10 minutes apart provide enough information about the variations in temperature? Are we confident that no significant variations occur between consecutive measurements? If that is the case, then could we have waited for 15 minutes between measurements instead of 10 minutes?

Generalizing the temperature example used above, the relationship between the continuous-time signal $x_a(t)$ and its discrete-time counterpart $x[n]$ is

$$x[n] = x_a(t)\Big|_{t=nT_s} = x_a(nT_s) \tag{6.2}$$

where T_s is the *sampling interval*, that is, the time interval between consecutive samples. It is also referred to as the *sampling period*. The reciprocal of the sampling interval is called the *sampling rate* or the *sampling frequency*:

$$f_s = \frac{1}{T_s} \tag{6.3}$$

The relationship between a continuous-time signal and its discrete-time version is illustrated in Fig. 6.1.

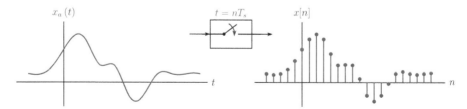

Figure 6.1 – Graphical representation of sampling relationship.

The claim that it may be possible to represent a continuous-time signal without any loss of information by a discrete set of amplitude values measured at uniformly spaced time intervals may be a bit counter-intuitive at first. How is it possible that we do not lose any information by merely measuring the signal at a discrete set of time instants and ignoring what takes place between those measurements? This question is perhaps best answered by posing another question: Does the behavior of the signal between measurement instants constitute worthwhile information, or is it just redundant behavior that could be completely predicted from the set of measurements? If it is the latter, then we will see that the measurements (samples) taken at intervals of T_s will be sufficient to reconstruct the continuous-time signal $x_a(t)$.

Sampling forms the basis of digital signals we encounter everyday in our lives. For example, an audio signal played back from a compact disc is a signal that has been captured and recorded at discrete time instants. When we look at the amplitude values stored on the disc, we only see values taken at equally spaced time instants (at a rate of 44,100 times per second) with missing amplitude values between these instants. This is perfectly fine since all the information contained in the original audio signal in the studio can be accounted for in these samples. An image captured by a digital camera is stored in the form of a dense rectangular grid of colored dots (known as pixels). When printed and viewed from an appropriate distance, we cannot tell the individual pixels apart. Similarly, a movie stored on a video cassette or a disc is stored in the form of consecutive snapshots, taken at equal time intervals. If enough snapshots are taken from the scene and are played back in sequence with the right timing, we perceive motion.

We begin by considering the sampling of continuous-time signals in Section 6.2. The idea of impulse sampling and its implications on the frequency spectrum are studied. Nyquist sampling criterion is introduced. Conversion of the impulse-sampled signal to a discrete-time signal is discussed along with the effect of the conversion on the frequency spectrum. Practical issues in sampling applications are also briefly discussed. The issue of reconstructing a continuous-time signal from its sampled version is the topic of Section 6.3. Section 6.4 covers the topic of changing the sampling rate of a signal that has already been sampled.

6.2   Sampling of a Continuous-Time Signal

Consider a periodic impulse train $p(t)$ with period T_s:

$$\tilde{p}(t) = \sum_{n=-\infty}^{\infty} \delta(t - nT_s) \tag{6.4}$$

Multiplication of any signal $x(t)$ with this impulse train $\tilde{p}(t)$ would result in amplitude information for $x(t)$ being retained only at integer multiples of the period T_s. Let the signal $x_s(t)$ be defined as the product of the original signal and the impulse train, i.e.,

$$\begin{aligned} x_s(t) &= x_a(t)\, \tilde{p}(t) \\ &= x_a(t) \sum_{n=-\infty}^{\infty} \delta(t - nT_s) \\ &= \sum_{n=-\infty}^{\infty} x_a(nT_s)\, \delta(t - nT_s) \end{aligned} \tag{6.5}$$

We will refer to the signal $x_s(t)$ as the *impulse-sampled* version of $x(t)$. Fig. 6.2 illustrates the relationship between the signals involved in impulse sampling.

It is important to understand that the impulse-sampled signal $x_s(t)$ is still a continuous-time signal. The subject of converting $x_s(t)$ to a discrete-time signal will be discussed in Section 6.2.2.

At this point, we need to pose a critical question: How dense must the impulse train $\tilde{p}(t)$ be so that the impulse-sampled signal $x_s(t)$ is an accurate and complete representation of the original signal $x_a(t)$? In other words, what are the restrictions on the sampling interval T_s or, equivalently, the sampling rate f_s? In order to answer this question, we need to

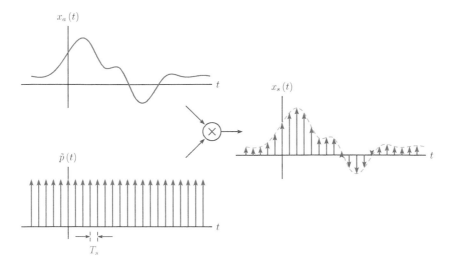

Figure 6.2 – Impulse-sampling a signal: (a) continuous-time signal $x(t)$, (b) the pulse train $p(t)$, (c) impulse-sampled signal $x_s(t)$.

develop some insight into how the frequency spectrum of the impulse-sampled signal $x_s(t)$ relates to the spectrum of the original signal $x_a(t)$.

Let us focus on the periodic impulse train $\tilde{p}(t)$ which is shown in detail in Fig. 6.3.

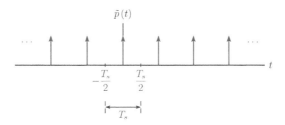

Figure 6.3 – Periodic impulse train $\tilde{p}(t)$.

As discussed in Section 4.2.3 of Chapter 4, $\tilde{p}(t)$ can be represented in an exponential Fourier series expansion in the form

$$\tilde{p}(t) = \sum_{k=-\infty}^{\infty} c_k\, e^{jk\omega_s t} \tag{6.6}$$

where ω_s is both the sampling rate in rad/s and the fundamental frequency of the impulse train. It is computed as $\omega_s = 2\pi f_s = 2\pi/T_s$. The EFS coefficients for $\tilde{p}(t)$ are found as

$$c_k = \frac{1}{T_s} \int_{-T_s/2}^{T_s/2} \tilde{p}(t)\, e^{-jk\omega_s t}\, dt$$

$$= \frac{1}{T_s} \int_{-T_s/2}^{T_s/2} \delta(t)\, e^{-jk\omega_s t}\, dt = \frac{1}{T_s}, \qquad \text{all } k \tag{6.7}$$

Substituting the EFS coefficients found in Eqn. (6.7) into Eqn. (6.6), the impulse train $\tilde{p}(t)$ becomes

$$\tilde{p}(t) = \frac{1}{T_s} \sum_{k=-\infty}^{\infty} e^{jk\omega_s t} \tag{6.8}$$

Finally, using Eqn. (6.8) in Eqn. (6.5) we get

$$x_s(t) = \frac{1}{T_s} \sum_{k=-\infty}^{\infty} x_a(t) e^{jk\omega_s t} \tag{6.9}$$

for the sampled signal $x_s(t)$. In order to determine the frequency spectrum of the impulse sampled signal $x_s(t)$ let us take the Fourier transform of both sides of Eqn. (6.9).

$$\mathcal{F}\{x_s(t)\} = \mathcal{F}\left\{ \frac{1}{T_s} \sum_{k=-\infty}^{\infty} x_a(t) e^{jk\omega_s t} \right\}$$

$$= \frac{1}{T_s} \sum_{k=-\infty}^{\infty} \mathcal{F}\{x_a(t) e^{jk\omega_s t}\} \tag{6.10}$$

Linearity property of the Fourier transform was used in obtaining the result in Eqn. (6.10). Furthermore, using the frequency shifting property of the Fourier transform, the term inside the summation becomes

$$\mathcal{F}\{x_a(t) e^{jk\omega_s t}\} = X_a(\omega - k\omega_s) \tag{6.11}$$

The frequency-domain relationship between the signal $x_a(t)$ and its impulse-sampled version $x_s(t)$ follows from Eqns. (6.10) and (6.11).

The Fourier transform of the impulse-sampled signal is related to the Fourier transform of the original signal by

$$X_s(\omega) = \frac{1}{T_s} \sum_{k=-\infty}^{\infty} X_a(\omega - k\omega_s) \tag{6.12}$$

This relationship can also be written using frequencies in Hertz as

$$X_s(f) = \frac{1}{T_s} \sum_{k=-\infty}^{\infty} X_a(f - kf_s) \tag{6.13}$$

This is a very significant result. The spectrum of the impulse-sampled signal is obtained by adding frequency-shifted versions of the spectrum of the original signal, and then scaling the sum by $1/T_s$. The terms of the summation in Eqn. (6.12) are shifted by all integer multiples of the sampling rate ω_s. Fig. 6.4 illustrates this.

For the impulse-sampled signal to be an accurate and complete representation of the original signal, $x_a(t)$ should be recoverable from $x_s(t)$. This in turn requires that the frequency spectrum $X_a(\omega)$ be recoverable from the frequency spectrum $X_s(\omega)$. In Fig. 6.4 the example spectrum $X_a(\omega)$ used for the original signal is bandlimited to the frequency

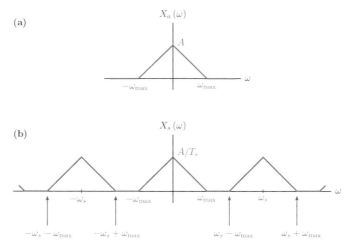

Figure 6.4 – Effects of impulse-sampling on the frequency spectrum: (a) the example spectrum $X_a(\omega)$ of the original signal $x_a(t)$, (b) the spectrum $X_s(\omega)$ of the impulse-sampled signal $x_s(t)$.

range $|\omega| \leq \omega_{\max}$. Sampling rate ω_s is chosen such that the repetitions of $X_a(\omega)$ do not overlap with each other in the construction of $X_s(\omega)$. As a result, the shape of the original spectrum $X_a(\omega)$ is preserved within the sampled spectrum $X_s(\omega)$. This ensures that $x_a(t)$ is recoverable from $x_s(t)$.

Alternatively, consider the scenario illustrated by Fig. 6.5 where the sampling rate chosen causes overlaps to occur between the repetitions of the spectrum. In this case $X_a(\omega)$ cannot be recovered from $X_s(\omega)$. Consequently, the original signal $x_a(t)$ cannot be recovered from its sampled version. Under this scenario, replacing the signal with its sampled version represents an irrecoverable loss of information.

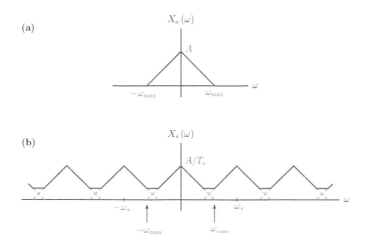

Figure 6.5 – Effects of impulse sampling on the frequency spectrum when the sampling rate chosen is too low: (a) the example spectrum $X_a(\omega)$ of the original signal $x_a(t)$, (b) the spectrum $X_s(\omega)$ of the impulse-sampled signal $x_s(t)$.

Interactive Demo: `smp_demo1`

The demo program "`smp_demo1.m`" illustrates the process of obtaining the spectrum $X_s(\omega)$ from the original spectrum $X_a(\omega)$ based on Eqn. (6.12) and Figs. 6.4 and 6.5. Sampling rate f_s and the bandwidth f_{max} of the signal to be sampled can be varied using slider controls. (In the preceding development we have used radian frequencies ω_s and ω_{max}. They are related to related to frequencies in Hertz used by the demo program through $\omega_s = 2\pi f_s$ and $\omega_{max} = 2\pi f_{max}$.)

Spectra $X_a(f)$ and $T_s X_s(f)$ are computed and graphed. Note that, in Fig. 6.4(b), the peak magnitude of $X_s(f)$ is proportional to $f_s = 1/T_s$. Same can be observed from the $1/T_s$ factor in Eqn. (6.12). As a result, graphing $X_s(f)$ directly would have required a graph window tall enough to accommodate the necessary magnitude changes as the sampling rate $f_s = 1/T_s$ is varied, and still show sufficient detail. Instead, we opt to graph

$$T_s X_s(f) = \sum_{k=-\infty}^{\infty} X_a(f - kf_s)$$

to avoid the need to deal with scaling issues.

Individual terms $X_a(f - kf_s)$ in Eqn. (6.12) are also shown, although they may be under the sum $T_s X_s(f)$, and thus invisible when the spectral sections do not overlap. When there is an overlap of spectral sections as in Fig. 6.5(b), part of each individual term becomes visible in red dashed lines. They may also be made visible by unchecking the "Show sum" box.

When spectral sections overlap, the word "Aliasing" is displayed, indicating that the spectrum is being corrupted through the sampling process.

Software resources:
`smp_demo1.m`

Example 6.1: Impulse sampling a right-sided exponential

Consider a right-sided exponential signal

$$x_a(t) = e^{-100t} u(t)$$

This signal is to be impulse sampled. Determine and graph the spectrum of the impulse-sampled signal $x_s(t)$ for sampling rates $f_s = 200$ Hz, $f_s = 400$ Hz and $f_s = 600$ Hz.

Solution: Using the techniques developed in Chapter 4, the frequency spectrum of the signal $x_a(t)$ is

$$X_a(f) = \frac{1}{100 + j2\pi f}$$

which is graphed in Fig. 6.6(a).

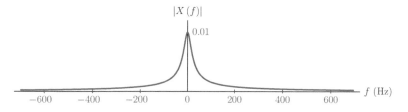

Figure 6.6 – Frequency spectrum of the signal $x_a(t)$ for Example 6.1.

Impulse-sampling $x_a\left(t\right)$ at a sampling rate of $f_s = /1/T_s$ yields the signal

$$x_s\left(t\right) = \sum_{n=-\infty}^{\infty} e^{-100nT_s}\, u\left(nT_s\right) \delta\left(t - nT_s\right)$$

$$= \sum_{n=0}^{\infty} e^{-100nT_s}\, \delta\left(t - nT_s\right)$$

The frequency spectrum of this impulse sampled signal is

$$X_s\left(f\right) = \int_{-\infty}^{\infty} \left[\sum_{n=0}^{\infty} e^{-100nT_s} \delta\left(t - nT_s\right)\right] e^{-j2\pi ft}\, dt$$

$$= \sum_{n=0}^{\infty} e^{-100nT_s} \left[\int_{-\infty}^{\infty} \delta\left(t - nT_s\right) e^{-j2\pi ft}\, dt\right]$$

$$= \sum_{n=0}^{\infty} e^{-100nT_s}\, e^{-j2\pi fnT_s}$$

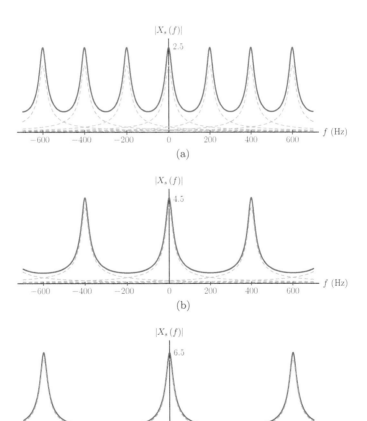

Figure 6.7 – The spectrum of impulse-sampled signal for Example 6.1 (a) for $f_s = 200$ Hz, (b) for $f_s = 400$ Hz, (c) for $f_s = 600$ Hz.

which can be put into a closed form through the use of the geometric series formula (see Appendix C) as

$$X_s \left(f \right) = \frac{1}{1 - e^{-100T_s} \, e^{-j2\pi fT_s}}$$

The resulting spectrum is shown in Fig. 6.7 for $f_s = 200$ Hz, $f_s = 400$ Hz and $f_s = 600$ Hz respectively.

Note that the overlap of spectral segments $(1/T_s) \, X_a \left(f - kf_s \right)$ causes the shape of the spectrum $X_s \left(f \right)$ to be different from the shape of $X \left(f \right)$ since the right-sided exponential is not band-limited. This distortion of the spectrum is present in all three cases, but seems to be more pronounced for $f_s = 200$ Hz than it is for the other two choices.

Software resources:

ex_6_1.m

Interactive Demo: smp_demo2

The demo program "smp_demo2.m" is based on Example 6.2. The signal $x_a \left(t \right) = \exp \left(-100t \right)$ $\cdot u \left(t \right)$ and its impulse sampled version $x_s \left(t \right)$ are shown at the top. The bottom graph displays the magnitude of the spectrum

$$T_s \, X_s \left(f \right) = \sum_{k=-\infty}^{\infty} X \left(f - kf_s \right)$$

as well as the magnitudes of contributing terms $X \left(f - kf_s \right)$. Our logic in graphing the magnitude of $T_s \, X_s \left(t \right)$ instead of $X_s \left(t \right)$ is the same as in the previous interactive demo program "smp_demo1.m". Sampling rate f_s may be adjusted using a slider control. Time and frequency domain plots are simultaneously updated to show the effect of sampling rate adjustment.

Software resources:

smp_demo2.m

Software resources: See MATLAB Exercise 6.1.

6.2.1   Nyquist sampling criterion

As illustrated in Example 6.1, if the range of frequencies in the signal $x_a \left(t \right)$ is not limited, then the periodic repetition of spectral components dictated by Eqn. (6.12) creates over-lapped regions. This effect is known as *aliasing*, and it results in the shape of the spectrum $X_s \left(f \right)$ being different than the original spectrum $X_a \left(f \right)$. Once the spectrum is aliased, the original signal is no longer recoverable from its sampled version. Aliasing could also occur when sampling signals that contain a finite range of frequencies if the sampling rate is not chosen carefully.

Let $x_a \left(t \right)$ be a signal the spectrum $X_a \left(f \right)$ of which is band-limited to $f_{\max}$, meaning it exhibits no frequency content for $|f| > f_{\max}$. If $x_a \left(t \right)$ is impulse sampled to obtain the signal $x_s \left(t \right)$, the frequency spectrum of the resulting signal is given by Eqn. (6.12). If we want to be able to recover the signal from its impulse-sampled version, then the spectrum $X_a \left(f \right)$ must also be recoverable from the spectrum $X_s \left(f \right)$. This in turn requires that no overlaps occur between periodic repetitions of spectral segments. Refer to Fig. 6.8.

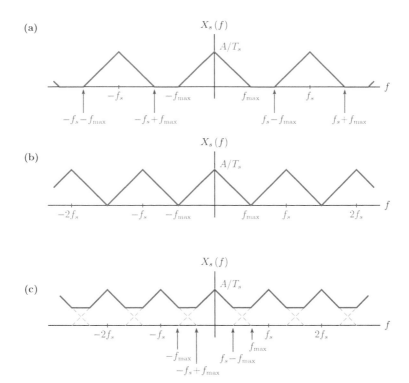

Figure 6.8 – Determination of an appropriate sampling rate: (a) the spectrum of a properly impulse-sampled signal, (b) the spectrum of the signal with critical impulse-sampling, (c) the spectrum of the signal after improper impulse-sampling.

In order to keep the left edge of the spectral segment centered at $f = f_s$ from interfering with the right edge of the spectral segment centered at $f = 0$, we need

$$f_s - f_{\max} \geq f_{\max} \tag{6.14}$$

and therefore

$$f_s \geq 2f_{\max} \tag{6.15}$$

For the impulse-sampled signal to form an accurate representation of the original signal, the sampling rate must be at least twice the highest frequency in the spectrum of the original signal. This is known as the *Nyquist sampling criterion*. It was named after Harry Nyquist (1889-1976) who first introduced the idea in his work on telegraph transmission. Later it was formally proven by his colleague Claude Shannon (1916-2001) in his work that formed the foundations of *information theory*.

In practice, the condition in Eqn. (6.15) is usually met with inequality, and with sufficient margin between the two terms to allow for the imperfections of practical samplers and reconstruction systems. In practical implementations of samplers, the sampling rate f_s is typically fixed by the constraints of the hardware used. On the other hand, the highest frequency of the actual signal to be sampled is not always known a priori. One example of this is the sampling of speech signals where the highest frequency in the signal depends on the speaker, and may vary. In order to ensure that the Nyquist sampling criterion in Eqn.

(6.15) is met regardless, the signal is processed through an *anti-aliasing filter* before it is sampled, effectively removing all frequencies that are greater than half the sampling rate. This is illustrated in Fig. 6.9.

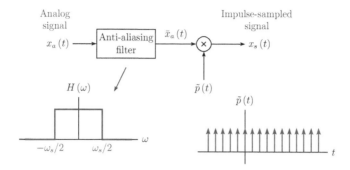

Figure 6.9 – Use of an antialiasing filter to ensure compliance with the requirements of Nyquist sampling criterion.

6.2.2   DTFT of sampled signal

The relationship between the Fourier transforms of the continuous-time signal and its impulse-sampled version is given by Eqns. (6.12) and (6.13). As discussed in Section 6.1, the purpose of sampling is to ultimately create a discrete-time signal $x[n]$ from a continuous-time signal $x_a(t)$. The discrete-time signal can then be converted to a digital signal suitable for storage and manipulation on digital computers.

Let $x[n]$ be defined in terms of $x_a(t)$ as

$$x[n] = x_a(nT_s) \tag{6.16}$$

The Fourier transform of $x_a(t)$ is defined by

$$X_a(\omega) = \int_{-\infty}^{\infty} x_a(t)\, e^{-j\omega t}\, dt \tag{6.17}$$

Similarly, the DTFT of the discrete-time signal $x[n]$ is

$$X(\Omega) = \sum_{n=-\infty}^{\infty} x[n]\, e^{-j\Omega n} \tag{6.18}$$

We would like to understand the relationship between the two transforms in Eqns. (6.17) and (6.18).

The impulse-sampled signal, given by Eqn. (6.5) can be written as

$$x_s(t) = \sum_{n=-\infty}^{\infty} x_a(nT_s)\, \delta(t - nT_s) \tag{6.19}$$

making use of the *sampling property* of the impulse function (see Eqn. (1.26) in Section

1.3.2 of Chapter 1). The Fourier transform of the impulse-sampled signal is

$$X_s(\omega) = \int_{-\infty}^{\infty} x_s(t) \, e^{-j\omega t} \, dt$$

$$= \int_{-\infty}^{\infty} \left[\sum_{n=-\infty}^{\infty} x_a(nT_s) \, \delta(t - nT_s) \right] e^{-j\omega t} \, dt \tag{6.20}$$

Interchanging the order of integration and summation, and rearranging terms, Eqn. (6.20) can be written as

$$X_s(\omega) = \sum_{n=-\infty}^{\infty} x_a(nT_s) \left[\int_{-\infty}^{\infty} \delta(t - nT_s) \, e^{-j\omega t} \, dt \right] \tag{6.21}$$

Using the sifting property of the impulse function (see Eqn. (1.27) in Section 1.3.2 of Chapter 1), Eqn. (6.21) becomes

$$X_s(\omega) = \sum_{n=-\infty}^{\infty} x_a(nT_s) \, e^{-j\omega n T_s} \tag{6.22}$$

Compare Eqn. (6.22) to Eqn. (6.18). The two equations would become identical with the adjustment

$$\Omega = \omega \, T_s \tag{6.23}$$

This leads us to the conclusion

$$X(\Omega) = X_s\left(\frac{\Omega}{T_s}\right) \tag{6.24}$$

Using Eqn. (6.24) with Eqn. (6.12) yields the relationship between the spectrum of the original continuous-time signal and the DTFT of the discrete-time signal obtained by sampling it:

$$X(\Omega) = \frac{1}{T_s} \sum_{k=-\infty}^{\infty} X_a\left(\frac{\Omega - 2\pi k}{T_s}\right) \tag{6.25}$$

This relationship is illustrated in Fig. 6.10. It is evident from Fig. 6.10 that, in order to avoid overlaps between repetitions of the segments of the spectrum, we need

$$\omega_{\max} T_s \leq \pi \quad \implies \quad \omega_{\max} \leq \frac{\pi}{T_s} \quad \implies \quad f_{\max} \leq \frac{f_s}{2} \tag{6.26}$$

consistent with the earlier conclusions.

Software resources: See MATLAB Exercise 6.2.

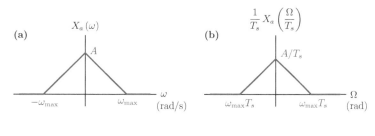

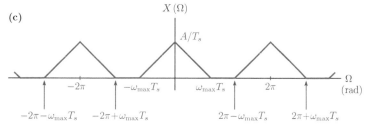

Figure 6.10 – Relationships between the frequency spectra of the original signal and the discrete-time signal obtained from it: (a) the example spectrum $X_a(\omega)$ of the original signal $x_a(t)$, (b) the term in the summation of Eqn. (6.25) for $k = 0$, (c) the spectrum $X(\Omega)$ of the sampled signal $x[n]$.

6.2.3 Sampling of sinusoidal signals

In this section we consider the problem of obtaining a discrete-time sinusoidal signal by sampling a continuous-time sinusoidal signal. Let $x_a(t)$ be defined as

$$x_a(t) = \cos(2\pi f_0 t) \tag{6.27}$$

and let $x[n]$ be obtained by sampling $x_a(t)$ as

$$x[n] = x_a(nT_s) = \cos(2\pi f_0 n T_s) \tag{6.28}$$

Using the normalized frequency $F_0 = f_0 T_s = f_0/f_s$ the signal $x[n]$ becomes

$$x[n] = \cos(2\pi F_0 n) \tag{6.29}$$

The DTFT of a discrete-time sinusoidal signal was derived in Section 5.3.6 of Chapter 5, and can be applied to $x[n]$ to yield

$$X(\Omega) = \sum_{k=-\infty}^{\infty} [\pi\,\delta(\Omega + 2\pi F_0 - 2\pi k) + \pi\,\delta(\Omega - 2\pi F_0 - 2\pi k)] \tag{6.30}$$

For the continuous-time signal $x_a(t)$ to be recoverable from $x[n]$, the sampling rate must be chosen properly. In terms of the normalized frequency F_0 we need $|F_0| < 0.5$. Fig. 6.11 illustrates the spectrum of $x[n]$ for proper and improper choices of the sampling rate and the corresponding normalized frequency.

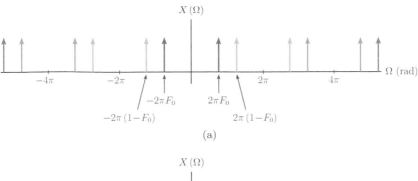

(a)

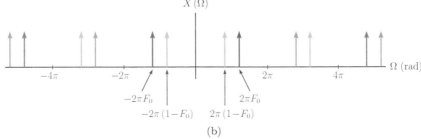

(b)

Figure 6.11 – The spectrum of sampled sinusoidal signal with (a) proper sampling rate, (b) improper sampling rate.

Example 6.2: **Sampling a sinusoidal signal**

The sinusoidal signals

$$x_{1a}(t) = \cos(12\pi t)$$
$$x_{2a}(t) = \cos(20\pi t)$$
$$x_{3a}(t) = \cos(44\pi t)$$

are sampled using the sampling rate $f_s = 16$ Hz and $T_s = 1/f_s = 0.0625$ seconds to obtain the discrete-time signals

$$x_1[n] = x_{1a}(0.0625\,n)$$
$$x_2[n] = x_{2a}(0.0625\,n)$$
$$x_3[n] = x_{3a}(0.0625\,n)$$

Show that the three discrete-time signals are identical, that is,

$$x_1[n] = x_2[n] = x_3[n], \qquad \text{all } n$$

Solution: Using the specified value of T_s the signals can be written as

$$x_1[n] = \cos(0.75\pi n)$$
$$x_2[n] = \cos(1.25\pi n)$$
$$x_3[n] = \cos(2.75\pi n)$$

Incrementing the phase of a sinusoidal function by any integer multiple of 2π radians does not affect the result. Therefore, $x_2[n]$ can be written as

$$x_2[n] = \cos(1.25\pi n - 2\pi n) = \cos(-0.75\pi n)$$

Since cosine is an even function, we have

$$x_2[n] = \cos\left(0.75\pi n\right) = x_1[n]$$

Similarly, $x_3[n]$ can be written as

$$x_3[n] = \cos\left(2.75\pi n - 2\pi n\right) = \cos\left(0.75\pi n\right) = x_1[n]$$

Thus, three different continuous-time signals correspond to the same discrete-time signal. Fig. 6.12 shows the three signals $x_{1a}\left(t\right)$, $x_{2a}\left(t\right)$, $x_{3a}\left(t\right)$ and their values at the sampling instants.

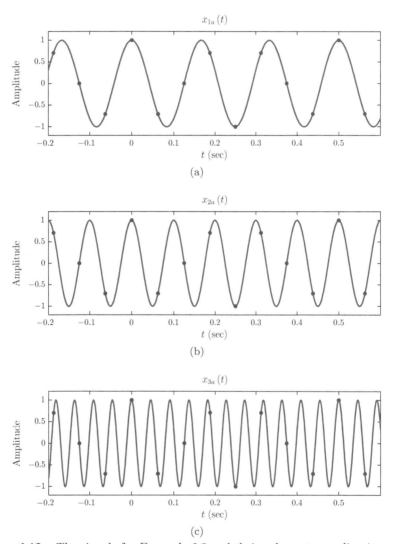

Figure 6.12 – The signals for Example 6.2 and their values at sampling instants.

An important issue in sampling is the *reconstruction* of the original signal from its sampled version. Given the discrete-time signal $x[n]$, how do we determine the continuous-time signal it represents? In this particular case we have at least three possible candidates, $x_{1a}\left(t\right)$, $x_{2a}\left(t\right)$, and $x_{3a}\left(t\right)$, from which $x[n]$ could have been obtained by sampling. Other possible answers could also be found. In fact it can be shown that an infinite number of different sinusoidal signals can be passed through the points shown with red dots in Fig. 6.12. Which one is the right answer?

Let us determine the actual and the normalized frequencies for the three signals. For $x_{1a}(t)$ we have

$$f_1 = 6 \text{ Hz}, \qquad F_1 = \frac{6}{16} = 0.375$$

Similarly for $x_{2a}(t)$

$$f_2 = 10 \text{ Hz}, \qquad F_2 = \frac{10}{16} = 0.625$$

and for $x_{3a}(t)$

$$f_3 = 22 \text{ Hz}, \qquad F_3 = \frac{22}{16} = 1.375$$

Of the three normalized frequencies only F_1 satisfies the condition $|F| \leq 0.5$, and the other two violate it. In terms of the Nyquist sampling theorem, the signal $x_{1a}(t)$ is sampled using a proper sampling rate, that is, $f_s > 2f_1$. The other two signals are sampled improperly since $f_s < 2f_2$ and $f_s < 2f_3$. Therefore, in the reconstruction process, we would pick $x_{1a}(t)$ based on the assumption that $x[n]$ is a properly sampled signal. Fig. 6.13 shows the signal $x_{2a}(t)$ being improperly sampled with a sampling rate of $f_s = 16$ Hz to obtain $x[n]$. In reconstructing the continuous-time signal from its sampled version, the signal $x_{1a}(t)$ is incorrectly identified as the signal that led to $x[n]$. This is referred to as *aliasing*. In this sampling scheme, $x_{1a}(t)$ is an *alias* for $x_{2a}(t)$.

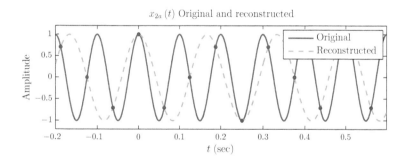

Figure 6.13 – Sampling a 10 Hz sinusoidal signal with sampling rate $f_s = 16$ Hz and attempting to reconstruct it from its sampled version.

Software resources:
ex_6_2a.m
ex_6_2b.m

Example 6.3: **Spectral relationships in sampling a sinusoidal signal**

Refer to the signals in Example 6.2. Sketch the frequency spectrum for each, and then use it in obtaining the DTFT spectrum of the sampled signal.

Solution: The Fourier transform of $x_{1a}(t)$ is

$$X_{1a}(\omega) = \pi \, \delta(\omega + 12\pi) + \pi \, \delta(\omega - 12\pi)$$

and is shown in Fig. 6.14(a). Referring to Eqn. (6.25) the term for $k = 0$ is

$$X_{1a}\left(\frac{\Omega}{T_s}\right) = \pi\,\delta\left(\frac{\Omega + 0.75\pi}{0.0625}\right) + \pi\,\delta\left(\frac{\Omega - 0.75\pi}{0.0625}\right)$$

shown in Fig. 6.14(b). The spectrum of the sampled signal $x_1[n]$ is obtained as

$$X_1(\Omega) = \frac{1}{T_s}\sum_{k=-\infty}^{\infty} X_{1a}\left(\frac{\Omega - 2\pi k}{T_s}\right)$$

$$= 16\sum_{k=-\infty}^{\infty}\left[\pi\,\delta\left(\frac{\Omega + 0.75\pi - 2\pi k}{0.0625}\right) + \pi\,\delta\left(\frac{\Omega - 0.75\pi - 2\pi k}{0.0625}\right)\right]$$

which is shown in Fig. 6.14(c). Each impulse in $X_1(\Omega)$ has an area of 16π. The term for $k = 0$ is shown in blue whereas the terms for $k = \mp1$, $k = \mp2$, $k = \mp3$ are shown in green, orange and brown respectively. In the reconstruction process, the assumption that $x_1[n]$ is a properly sampled signal would lead us to correctly picking the two blue colored impulses at $\Omega = \pm0.75\pi$ and ignoring the others.

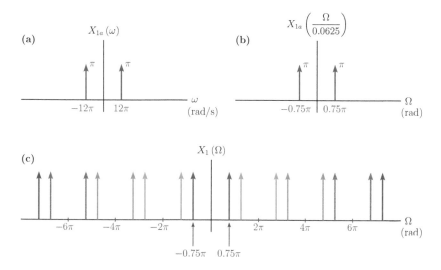

Figure 6.14 – Obtaining the spectrum $X_1(\Omega)$ from the spectrum $X_{1a}(\omega)$ for Example 6.3.

Similar analyses can be carried out for the spectra of the other two continuous-time signals $x_{2a}(t)$ and $x_{3a}(t)$ as well as the discrete-time signals that result from sampling them. Spectral relationships for these cases are shown in Figs. 6.15 and 6.16. Even though $X_1(\Omega) = X_2(\Omega) = X_3(\Omega)$ as expected, notice the differences in the contributions of the $k = 0$ term of Eqn. (6.25) and the others in obtaining the DTFT spectrum. In reconstructing the continuous-time signals from $x_2[n]$ and $x_3[n]$ we would also work with the assumption that the signals have been sampled properly, and incorrectly pick the green colored impulses at $\Omega = \pm0.75\pi$.

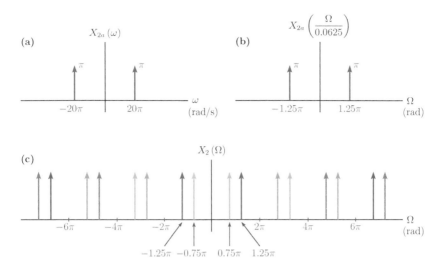

Figure 6.15 – Obtaining the spectrum $X_2(\Omega)$ from the spectrum $X_{2a}(\omega)$ for Example 6.3.

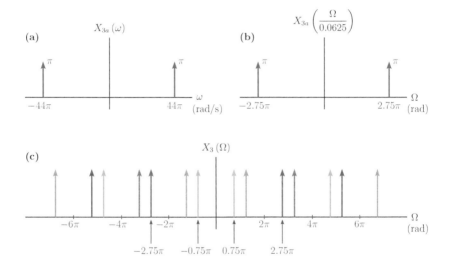

Figure 6.16 – Obtaining the spectrum $X_3(\Omega)$ from the spectrum $X_{3a}(\omega)$ for Example 6.3.

Interactive Demo: smp_demo3

The demo program "smp_demo3.m" is based on Examples 6.2 and 6.3. It allows experimentation with sampling the sinusoidal signal $x_a(t) = \cos(2\pi f_a t)$ to obtain a discrete-time signal $x[n]$. The two signals and their frequency spectra are displayed. The signal $\hat{x}_a(t)$ reconstructed from $x[n]$ may also be displayed, if desired. The signal frequency f_a and the sampling rate f_s may be varied through the use of slider controls.

1. Duplicate the results of Example 6.3 by setting the signal frequency to $f_a = 6$ Hz and sampling rate to $f_s = 16$ Hz. Slowly increase the signal frequency while observing the changes in the spectra of $x_a(t)$ and $x[n]$.

2. Set the signal frequency to $f_a = 10$ Hz and sampling rate to $f_s = 16$ Hz. Observe the aliasing effect. Afterwards slowly increase the sampling rate until the aliasing disappears, that is, the reconstructed signal is the same as the original.

Software resources:
smp_demo3.m

Software resources: See MATLAB Exercise 6.3.

6.2.4   Practical issues in sampling

In previous sections the issue of sampling a continuous-time signal through multiplication by an impulse train was discussed. A practical consideration in the design of samplers is that we do not have ideal impulse trains, and must therefore approximate them with pulse trains. Two important questions that arise in this context are:

1. What would happen to the spectrum if we used a pulse train instead of an impulse train in Eqn. (6.25)?

2. How would the use of pulses affect the methods used in recovering the original signal from its sampled version?

When pulses are used instead of impulses, there are two variations of the sampling operation that can be used, namely *natural sampling* and *zero-order hold sampling*. The former is easier to generate electronically while the latter lends itself better to digital coding through techniques known as *pulse-code modulation* and *delta modulation*. We will review each sampling technique briefly.

Natural sampling

Instead of using the periodic impulse train of Eqn. (6.5), let the multiplying signal $\tilde{p}(t)$ be defined as a periodic pulse train with a duty cycle of d:

$$\tilde{p}(t) = \sum_{n=-\infty}^{\infty} \Pi\left(\frac{t - nT_s}{dT_s}\right) \tag{6.31}$$

where $\Pi(t)$ represents a unit pulse, that is, a pulse with unit amplitude and unit width centered around the time origin $t = 0$. The period of the pulse train is T_s, the same as the sampling interval. The width of each pulse is dT_s as shown in Fig. 6.17.

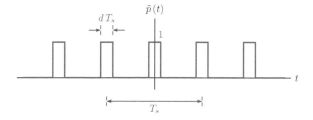

Figure 6.17 – Periodic pulse train for natural sampling.

Multiplication of the signal $x_a(t)$ with $\tilde{p}(t)$ yields a *natural sampled* version of the signal $x_a(t)$:

$$\begin{aligned}\bar{x}_s(t) &= x_a(t)\,\tilde{p}(t)\\ &= x_a(t)\sum_{n=-\infty}^{\infty}\Pi\left(\frac{t-nT_s}{dT_s}\right)\end{aligned} \tag{6.32}$$

This is illustrated in Fig. 6.18.

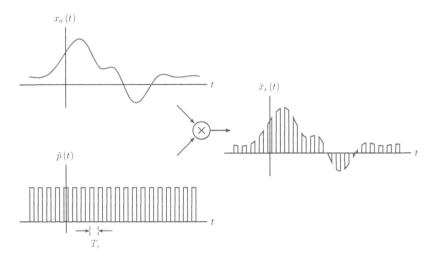

Figure 6.18 – Illustration of natural sampling.

An alternative way to visualize the natural sampling operation is to view the naturally sampled signal as the output of an electronic switch which is controlled by the pulse train $\tilde{p}(t)$. This implementation is shown in Fig. 6.19. The switch is closed when the pulse is present, and is opened when the pulse is absent.

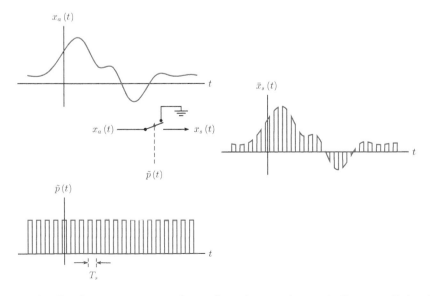

Figure 6.19 – Implementing a natural sampler using an electronically controlled switch.

In order to derive the relationship between frequency spectra of the signal $x_a(t)$ and its naturally sampled version $\bar{x}_s(t)$, we will make use of the exponential Fourier series representation of $\tilde{p}(t)$.

The EFS coefficients for a pulse train with duty cycle d were found in Example 4.7 of Chapter 4 as

$$c_k = d \operatorname{sinc}(kd)$$

Therefore the EFS representation of $\tilde{p}(t)$ is

$$\tilde{p}(t) = \sum_{k=-\infty}^{\infty} c_k\, e^{jk\omega_s t}$$

$$= \sum_{k=-\infty}^{\infty} d \operatorname{sinc}(kd)\, e^{jk\omega_s t} \tag{6.33}$$

Fundamental frequency is the same as the sampling rate $\omega_s = 2\pi/T_s$. Using Eqn. (6.33) in Eqn. (6.32) the naturally sampled signal is

$$\bar{x}_s(t) = x_a(t) \sum_{k=-\infty}^{\infty} d \operatorname{sinc}(kd)\, e^{jk\omega_s t} \tag{6.34}$$

The Fourier transform of the natural sampled signal is

$$\bar{X}_s(\omega) = \mathcal{F}\{\bar{x}_s(t)\} = \int_{-\infty}^{\infty} \bar{x}_s(t)\, e^{-j\omega t}\, dt$$

Using Eqn. (6.34) in Eqn. (6.35) yields

$$\bar{X}_s(\omega) = \int_{-\infty}^{\infty} \left[x_a(t) \sum_{k=-\infty}^{\infty} d \operatorname{sinc}(kd)\, e^{jk\omega_s t} \right] e^{-j\omega t}\, dt \tag{6.35}$$

Interchanging the order of integration and summation and rearranging terms we obtain

$$\bar{X}_s(\omega) = d \sum_{k=-\infty}^{\infty} \operatorname{sinc}(kd) \left[\int_{-\infty}^{\infty} x_a(t)\, e^{-j(\omega - k\omega_s)t}\, dt \right] \tag{6.36}$$

The expression in square brackets is the Fourier transform of $x_a(t)$ evaluated for $\omega - k\omega_s$, that is,

$$\int_{-\infty}^{\infty} x_a(t)\, e^{-j(\omega - k\omega_s)t}\, dt = X_a(\omega - k\omega_s) \tag{6.37}$$

Substituting Eqn. (6.37) into Eqn. (6.36) yields the desired result.

Spectrum of the signal obtained through natural sampling:

$$\bar{X}_s(\omega) = d \sum_{k=-\infty}^{\infty} \operatorname{sinc}(kd)\, X_a(\omega - k\omega_s) \tag{6.38}$$

The effect of natural sampling on the frequency spectrum is shown in Fig. 6.20.

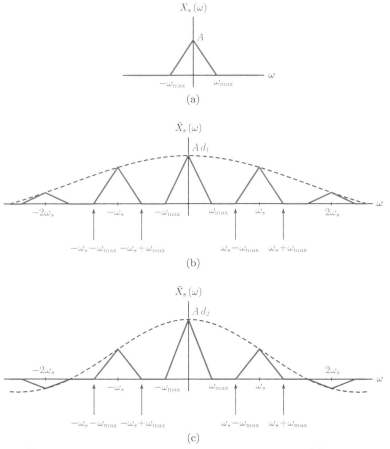

Figure 6.20 – Effects of natural sampling on the frequency spectrum: (a) the example spectrum $X_a(\omega)$ of the original signal $x_a(t)$, (b) the spectrum $\bar{X}_s(\omega)$ of the naturally sampled signal $\bar{x}_s(t)$ obtained using a pulse train with duty cycle d_1, (c) the spectrum obtained using a pulse train with duty cycle $d_2 > d_1$.

It is interesting to compare the spectrum $\bar{X}_s(\omega)$ obtained in Eqn. (6.38) to the spectrum $X_s(\omega)$ given by Eqn. (6.12) for the impulse-sampled signal:

1. The spectrum of the impulse-sampled signal has a scale factor of $1/T_s$ in front of the summation while the spectrum of the naturally sampled signal has a scale factor of d. This is not a fundamental difference. Recall that, in the multiplying signal $p(t)$ of the impulse-sampled signal, each impulse has unit area. On the other hand, in the pulse train $\tilde{p}(t)$ used for natural sampling, each pulse has a width of dT_s and a unit amplitude, corresponding to an area of dT_s. If we were to scale the pulse train $\tilde{p}(t)$ so that each of its pulses has unit area under it, then the amplitude scale factor would have to be $1/dT_s$. Using $(1/dT_s)\,\tilde{p}(t)$ in natural sampling would cause the spectrum $\bar{X}_s(\omega)$ in Eqn. (6.38) to be scaled by $1/dT_s$ as well, matching the scaling of $X_s(\omega)$.

2. As in the case of impulse sampling, frequency-shifted versions of the spectrum $X_a(\omega)$ are added together to construct the spectrum of the sampled signal. A key difference is that each term $X_a(\omega - k\omega_s)$ is scaled by $\mathrm{sinc}(kd)$ as illustrated in Fig. 6.20.

Zero-order hold sampling

In natural sampling the tops of the pulses are not flat, but are rather shaped by the signal $x_a(t)$. This behavior is not always desired, especially when the sampling operation is to be followed by conversion of each pulse to digital format. An alternative is to hold the amplitude of each pulse constant, equal to the value of the signal at the left edge of the pulse. This is referred to as *zero-order hold sampling* or *flat-top sampling*, and is illustrated in Fig. 6.21.

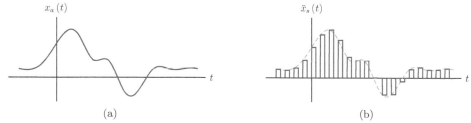

(a)   (b)

Figure 6.21 – Illustration of zero-order hold sampling.

Conceptually the signal $\bar{x}_s(t)$ can be modeled as the convolution of the impulse sampled signal $x_s(t)$ and a rectangular pulse with unit amplitude and a duration of dT_s as shown in Fig. 6.22.

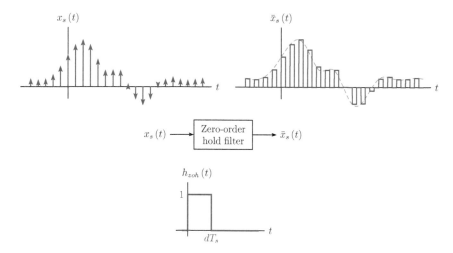

Figure 6.22 – Modeling sampling operation with flat-top pulses.

The impulse response of the zero-order hold filter in Fig. 6.22 is

$$h_{zoh}(t) = \Pi\left(\frac{t - 0.5\,dT_s}{dT_s}\right) = u(t) - u(t - dT_s) \tag{6.39}$$

Zero-order hold sampled signal $\bar{x}_s(t)$ can be written as

$$\bar{x}_s(t) = h_{zoh}(t) * x_s(t) \tag{6.40}$$

where $x_s(t)$ represents the impulse sampled signal given by Eqn. (6.5). The frequency spectrum of the zero-order hold sampled signal is found as

$$\bar{X}_s(\omega) = H_{zoh}(\omega)\, X_s(\omega) \tag{6.41}$$

The system function for the zero-order hold filter is

$$H_{zoh}\left(\omega\right) = dT_s \, \text{sinc}\left(\frac{\omega dT_s}{2\pi}\right) e^{-j\omega dT_s/2} \tag{6.42}$$

The spectrum of the zero-order hold sampled signal is found by using Eqns. (6.12) and (6.42) in Eqn. (6.41).

Spectrum of the signal obtained through zero-order hold sampling:

$$\bar{X}_s\left(\omega\right) = d \, \text{sinc}\left(\frac{\omega dT_s}{2\pi}\right) e^{-j\omega dT_s/2} \sum_{k=-\infty}^{\infty} X_a\left(\omega - k\omega_s\right) \tag{6.43}$$

Software resources: See MATLAB Exercises 6.4, 6.5 and 6.6.

6.3   Reconstruction of a Signal from Its Sampled Version

In previous sections of this chapter we explored the issue of sampling an analog signal to obtain a discrete-time signal. Ideally, a discrete-time signal is obtained by multiplying the analog signal with a periodic impulse train. Approximations to the ideal scenario can be obtained through the use of a pulse train instead of an impulse train.

Often the purpose of sampling an analog signal is to store, process and/or transmit it digitally, and to later convert it back to analog format. To that end, one question still remains: How can the original analog signal be reconstructed from its sampled version? Given the discrete-time signal $x[n]$ or the impulse-sampled signal $x_s\left(t\right)$, how can we obtain a signal identical, or at least reasonably similar, to $x_a\left(t\right)$? Obviously we need a way to "fill the gaps" between the impulses of the signal $x_s\left(t\right)$ in some meaningful way. In more technical terms, signal amplitudes between sampling instants need to be computed by some form of interpolation.

Let us first consider the possibility of obtaining a signal similar to $x_a\left(t\right)$ using rather simple methods. One such method would be to start with the impulse sampled signal $x_s\left(t\right)$ given by Eqn. (6.5) and repeated here

$$\text{Eqn. (6.5):} \qquad x_s\left(t\right) = \sum_{n=-\infty}^{\infty} x_a\left(nT_s\right) \delta\left(t - nT_s\right)$$

and to hold the amplitude of the signal equal to the value of each sample for the duration of T_s immediately following each sample. The result is a "staircase" type of approximation to the original signal. Interpolation is performed using horizontal lines, or polynomials of order zero, between sampling instants. This is referred to as *zero-order hold* interpolation, and is illustrated in Fig. 6.23(a),(b).

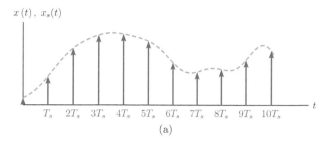

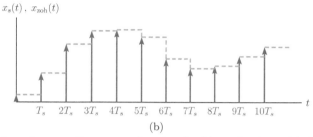

Figure 6.23 – (a) Impulse sampling an analog signal $x(t)$ to obtain $x_s(t)$, (b) reconstruction using zero-order hold interpolation.

Zero-order hold interpolation can be achieved by processing the impulse sampled signal $x_s(t)$ through *zero-order hold reconstruction filter*, a linear system the impulse response of which is a rectangle with unit amplitude and a duration of T_s.

$$h_{\text{zoh}}(t) = \Pi\left(\frac{t - T_s/2}{T_s}\right) \tag{6.44}$$

This is illustrated in Fig. 6.24. The linear system that performs the interpolation is called a *reconstruction filter*.

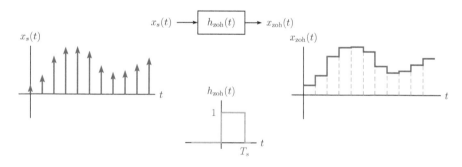

Figure 6.24 – Zero-order hold interpolation using an interpolation filter.

Notice the similarity between Eqn. (6.44) for zero-order hold interpolation and Eqn. (6.39) derived in the discussion of zero-order hold sampling. The two become the same if the duty cycle is set equal to $d = 1$. Therefore, the spectral relationship between the analog signal and its naturally sampled version given by Eqn. (6.43) can be used for obtaining the relationship between the analog signal and the signal reconstructed from samples using zero-order hold:

$$X_{zoh}(\omega) = \text{sinc}\left(\frac{\omega T_s}{2\pi}\right) e^{-j\omega T_s/2} \sum_{k=-\infty}^{\infty} X_a(\omega - k\omega_s) \tag{6.45}$$

As an alternative to zero-order hold, the gaps between the sampling instants can be filled by linear interpolation, that is, by connecting the tips of the samples with straight lines as shown in Fig. 6.25. This is also known as *first-order hold* interpolation since the straight line segments used in the interpolation correspond to first order polynomials.

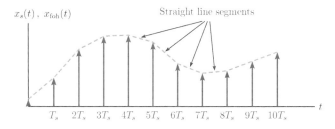

Figure 6.25 – Reconstruction using first-order hold interpolation.

First-order hold interpolation can also be implemented using a *first-order hold reconstruction filter*. The impulse response of such a filter is a triangle in the form

$$h_{\text{foh}}(t) = \begin{cases} 1 + t/T_s, & -T_s < t < 0 \\ 1 - t/T_s, & 0 < t < T_s \\ 0, & |t| \geq T_s \end{cases} \tag{6.46}$$

as illustrated in Fig. 6.26.

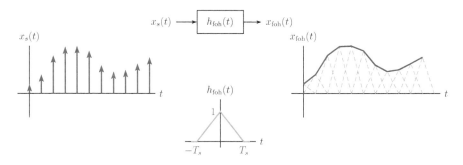

Figure 6.26 – First-order hold interpolation using an interpolation filter.

The impulse response $h_{\text{foh}}(t)$ of the first-order hold interpolation filter is non-causal since it starts at $t = -T_s$. If a practically realizable interpolator is desired, it would be a simple matter to achieve causality by using a delayed version of $h_{\text{foh}}(t)$ for the impulse response:

$$\bar{h}_{\text{foh}}(t) = h_{\text{foh}}(t - T_s) \tag{6.47}$$

In this case, the reconstructed signal would naturally lag behind the sampled signal by T_s.

It is insightful to derive the frequency spectra of the reconstructed signals obtained through zero-order hold and first-order hold interpolation. For convenience we will use f rather than ω in this derivation. The system function for the zero-order hold filter is

$$H_{\text{zoh}}(f) = T_s \, \text{sinc}(fT_s) \, e^{-j\pi T_s} \tag{6.48}$$

and the spectrum of the analog signal constructed using the zero-order hold filter is

$$X_{\text{zoh}}(f) = H_{\text{zoh}}(f) \, X_s(f) \tag{6.49}$$

Similarly, the system function for first-order hold filter is

$$H_{\text{foh}}(f) = T_s \operatorname{sinc}^2(fT_s) \tag{6.50}$$

and the spectrum of the analog signal constructed using the first-order hold filter is

$$X_{\text{foh}}(f) = H_{\text{foh}}(f) X_s(f) \tag{6.51}$$

Fig. 6.27 illustrates the process of obtaining the magnitude spectrum of the output of the zero-order hold interpolation filter for the sample input spectrum used earlier in Section 6.2. Fig. 6.28 illustrates the same concept for the first-order hold interpolation filter.

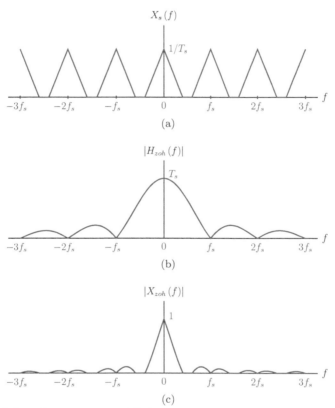

Figure 6.27 – (a) Sample spectrum $X_s(f)$ for an impulse-sampled signal, (b) magnitude spectrum $|H_{\text{zoh}}(f)|$ for the zero-order hold interpolation filter, (c) magnitude spectrum $|X_{\text{zoh}}(f)|$ for the signal reconstructed using zero-order hold interpolation.

Interestingly, both zero-order hold and first-order hold filters exhibit lowpass characteristics. A comparison of Figs. 6.27(c) and 6.28(c) reveals that the first-order hold interpolation filter does a better job in isolating the main section of the signal spectrum around $f = 0$ and suppressing spectral repetitions in $X_s(f)$ compared to the zero-order hold interpolation filter. A comparison of the time-domain signals obtained through zero-order hold and first-order hold interpolation in Figs. 6.23(b) and 6.25 supports this conclusion as well. The reconstructed signal $x_{\text{foh}}(t)$ is closer to the original signal $x_a(t)$ than $x_{\text{zoh}}(t)$ is.

The fact that both interpolation filters have lowpass characteristics warrants further exploration. The Nyquist sampling theorem states that a properly sampled signal can be recovered perfectly from its sampled version. What kind of interpolation is needed for

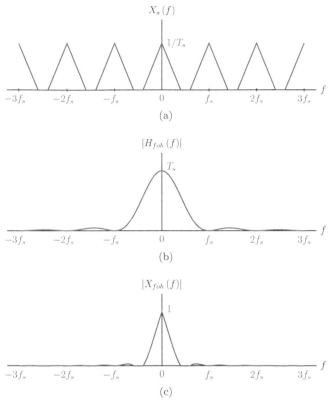

Figure 6.28 – (a) Sample spectrum $X_s(f)$ for an impulse-sampled signal, (b) magnitude spectrum $|H_{\text{foh}}(f)|$ for the first-order hold interpolation filter, (c) magnitude spectrum $|X_{\text{foh}}(f)|$ for the signal reconstructed using first-order hold interpolation.

perfect reconstruction of the analog signal from its impulse-sampled version? The answer must be found through the frequency spectrum of the sampled signal. Recall that the relationship between $X_s(f)$, the spectrum of the impulse-sampled signal, and $X_a(f)$, the frequency spectrum of the original signal, was found in Eqn. (6.13) to be

$$\text{Eqn. (6.13):} \quad X_s(f) = \frac{1}{T_s} \sum_{k=-\infty}^{\infty} X_a(f - kf_s)$$

As long as the choice of the sampling rate satisfies the Nyquist sampling criterion, the spectrum of the impulse-sampled signal is simply a sum of frequency shifted versions of the original spectrum, shifted by every integer multiple of the sampling rate. An ideal lowpass filter that extracts the term for $k = 0$ from the summation in Eqn. (6.13) and suppresses all other terms for $k = \pm 1, \ldots, \pm\infty$ would recover the original spectrum $X_a(f)$, and therefore the original signal $x_a(t)$. Since the highest frequency in a properly sampled signal would be equal to or less than half the sampling rate, an ideal *lowpass filter* with cutoff frequency set equal to $f_s/2$ is needed. The system function for such a reconstruction filter is

$$H_r(f) = T_s \, \Pi\left(\frac{f}{f_s}\right) \tag{6.52}$$

where we have also included a magnitude scaling by a factor of T_s within the system function of the lowpass filter in order to compensate for the $1/T_s$ term in Eqn. (6.13).

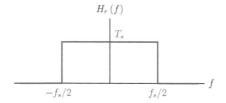

Figure 6.29 – Ideal lowpass reconstruction filter.

The frequency spectrum of the output of the filter defined by Eqn. (6.52) is

$$X_r\left(f\right) = H_r\left(f\right) X_s\left(f\right)$$

$$= T_s \, \Pi\left(\frac{f}{f_s}\right) \frac{1}{T_s} \sum_{k=-\infty}^{\infty} X_a\left(f - kf_s\right) = X_a\left(f\right) \tag{6.53}$$

This is illustrated in Fig. 6.30.

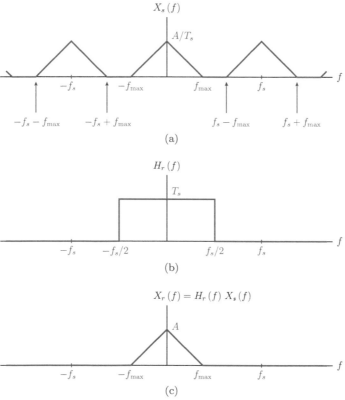

(a)

(b)

(c)

Figure 6.30 – Reconstruction using an ideal lowpass reconstruction filter: (a) sample spectrum $X_s\left(f\right)$ for an impulse-sampled signal, (b) system function $H_r\left(f\right)$ for the ideal lowpass filter with cutoff frequency $f_s/2$, (c) spectrum $X_r\left(f\right)$ for the signal at the output of the ideal lowpass filter.

Since $X_r\left(f\right) = X_a\left(f\right)$ we have $x_r\left(t\right) = x_a\left(t\right)$. It is also interesting to determine what type of interpolation is implied in the time domain by the ideal lowpass filter of Eqn. (6.52).

The impulse response of the filter is

$$h_r\left(t\right) = \mathrm{sinc}\left(t f_s\right) = \mathrm{sinc}\left(\frac{t}{T_s}\right) \tag{6.54}$$

which, due to the sinc function, has equally-spaced zero crossings that coincide with the sampling instants. The signal at the output of the ideal lowpass filter is obtained by convolving $h_r\left(t\right)$ with the impulse-sampled signal given by Eqn. (6.5):

$$x_r\left(t\right) = \sum_{n=-\infty}^{\infty} x_a\left(nT_s\right) \mathrm{sinc}\left(\frac{t - nT_s}{T_s}\right) \tag{6.55}$$

The nature of interpolation performed by the ideal lowpass reconstruction filter is evident from Eqn. (6.55). Let us consider the output of the filter at one of the sampling instants, say $t = kT_s$:

$$
\begin{aligned}
x_r\left(kT_s\right) &= \sum_{n=-\infty}^{\infty} x_a\left(nT_s\right) \mathrm{sinc}\left(\frac{kT_s - nT_s}{T_s}\right) \\
&= \sum_{n=-\infty}^{\infty} x_a\left(nT_s\right) \mathrm{sinc}\left(k - n\right) \tag{6.56}
\end{aligned}
$$

We also know that

$$\mathrm{sinc}\left(k - n\right) = \left\{ \begin{array}{ll} 1\,, & n = k \\ 0\,, & n \neq k \end{array} \right.$$

and Eqn. (6.56) becomes $x_r\left(kT_s\right) = x_a\left(kT_s\right)$.

1. The output $x_r\left(t\right)$ of the ideal lowpass reconstruction filter is equal to the sampled signal at each sampling instant.
2. Between sampling instants, $x_r\left(t\right)$ is obtained by interpolation through the use of sinc functions. This is referred to as *bandlimited interpolation* and is illustrated in Fig. 6.31.

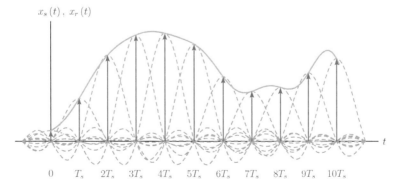

Figure 6.31 – Reconstruction of a signal from its sampled version through bandlimited interpolation.

Up to this point we have discussed three possible methods of reconstruction by interpolating between the amplitudes of the sampled signal, namely zero-order hold, first-order hold, and bandlimited interpolation. All three methods result in reconstructed signals that have the correct amplitude values at the sampling instants, and interpolated amplitude values between them. Some interesting questions might be raised at this point:

1. What makes the signal obtained by bandlimited interpolation more accurate than the other two?
2. In a practical situation we would only have the samples $x_a(nT_s)$ and would not know what the original signal $x_a(t)$ looked like between sampling instants. What if $x_a(t)$ had been identical to the zero-order hold result $x_{\text{zoh}}(t)$ or the first-order hold result $x_{\text{foh}}(t)$ before sampling?

The answer to both questions lies in the fact that, the signal obtained by bandlimited interpolation is the only signal among the three that is limited to a bandwidth of $f_s/2$. The bandwidth of each of the other two signals is greater than $f_s/2$, therefore, neither of them could have been the signal that produced a properly sampled $x_s(t)$. From a practical perspective, however, both zero-order hold and first-order hold interpolation techniques are occasionally utilized in cases where exact or very accurate interpolation is not needed and simple approximate reconstruction may suffice.

Software resources: See MATLAB Exercises 6.7 and 6.8.

6.4   Resampling Discrete-Time Signals

Sometimes we have the need to change the sampling rate of a discrete-time signal. Subsystems of a large scale system may operate at different sampling rates, and the ability to convert from one sampling rate to another may be necessary to get the subsystems to work together.

Consider a signal $x_1[n]$ that may have been obtained from an analog signal $x_a(t)$ by means of sampling with a sampling rate $f_{s1} = 1/T_1$.

$$x_1[n] = x_a(nT_1) \tag{6.57}$$

Suppose an alternative version of the signal is needed, one that corresponds to sampling $x_a(t)$ with a different sampling rate $f_{s2} = 1/T_2$.

$$x_2[n] = x_a(nT_2) \tag{6.58}$$

The question is: *How can $x_2[n]$ be obtained from $x_1[n]$?*

If $x_1[n]$ is a properly sampled signal, that is, if the conditions of the Nyquist sampling theorem have been satisfied, the analog signal $x_a(t)$ may be reconstructed from it and then resampled at the new rate to obtain $x_2[n]$. This approach may not always be desirable or practical. Realizable reconstruction filters are far from the ideal filters called for in perfect reconstruction of the analog signal $x_a(t)$, and a loss of signal quality would occur in the conversion. We prefer to obtain $x_2[n]$ from $x_1[n]$ using discrete-time processing methods without the need to convert $x_1[n]$ to an intermediate analog signal.

6.4.1   Reducing the sampling rate by an integer factor

Reduction of sampling rate by an integer factor D is easily accomplished by defining the signal $x_d[n]$ as

$$x_d[n] = x[nD] \tag{6.59}$$

This operation is known as *downsampling*. The parameter D is the *downsampling rate*. Graphical representation of a downsampler is shown in Fig. 6.32.

Figure 6.32 – Graphical representation of a downsampler.

Downsampling operation was briefly discussed in Section 1.4.1 of Chapter 1 in the context of time scaling for discrete-time signals. Figs. 6.33, 6.34 and 6.35 illustrate the downsampling of a signal using downsampling rates of $D = 2$ and $D = 3$.

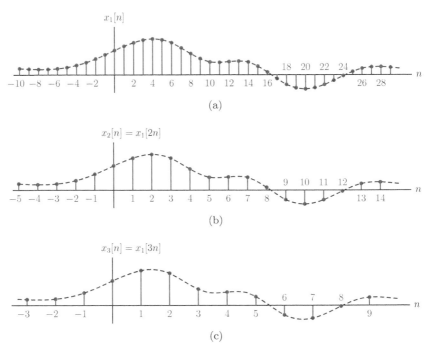

Figure 6.33 – Downsampling a signal $x_1[n]$: (a) original signal $x_1[n]$, (b) $x_2[n] = x_1[2n]$, (c) $x_3[n] = x_1[3n]$.

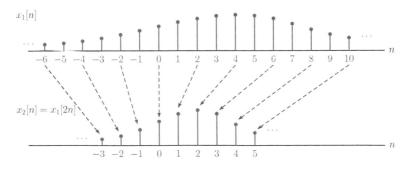

Figure 6.34 – Downsampling by a factor of $D = 2$.

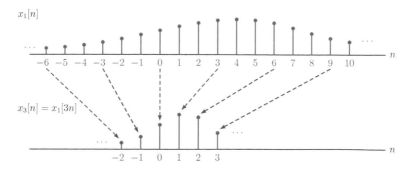

Figure 6.35 – Downsampling by a factor of $D = 3$.

Let us consider the general downsampling relationship in Eqn. (6.59). The signal $x_d[n]$ retains one sample out of each set of D samples of the original signal. For each sample retained $(D - 1)$ samples are discarded. The natural question that must be raised is: *Are we losing information by discarding samples, or were those discarded samples just redundant or unnecessary?* In order to answer this question we need to focus on the frequency spectra of the signals involved.

Assume that the original signal was obtained by sampling an analog signal $x_a(t)$ with a sampling rate f_s so that

$$x[n] = x_a\left(\frac{n}{f_s}\right) \tag{6.60}$$

For sampling to be appropriate, the highest frequency of the signal $x_a(s)$ must not exceed $f_s/2$. The downsampled signal $x_d[n]$ may be obtained by sampling $x_a(t)$ with a sampling rate f_s/D:

$$x_d[n] = x_a\left(\frac{nD}{f_s}\right) \tag{6.61}$$

For $x_d[n]$ to represent an appropriately sampled signal the highest frequency in $x_a(t)$ must not exceed $f_s/2D$. This is the more restricting of the two conditions on the bandwidth of $x_a(n)$ as illustrated in Fig. 6.36.

The relationship between $X(\Omega)$ and $X_d(\Omega)$ could also be derived without resorting to the analog signal $x_a(t)$.

The spectrum of $x_d[n]$ is computed as

$$X_d(\Omega) = \sum_{n=-\infty}^{\infty} x_d[n] e^{-j\Omega n} \tag{6.62}$$

Substituting Eqn. (6.59) into Eqn. (6.62) yields

$$X_d(\Omega) = \sum_{n=-\infty}^{\infty} x[nD] e^{-j\Omega n} \tag{6.63}$$

Using the variable change $m = nD$, Eqn. (6.63) becomes

$$X_d(\Omega) = \sum_{\substack{m=-\infty \\ m=nD}}^{\infty} x[m] e^{-j\Omega m/D} \tag{6.64}$$

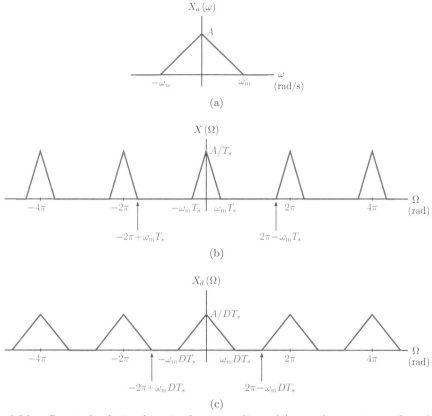

Figure 6.36 – Spectral relationships in downsampling: (a) sample spectrum of analog signal $x_a(t)$, (b) corresponding spectrum for $x[n] = x_a(n/f_s)$, (c) corresponding spectrum for $x_d[n] = x_a(nD/f_s)$.

The restriction on the index of the summation in Eqn. (6.64) makes it difficult for us to put the summation into a closed form. If a signal $w[m]$ with the definition

$$w[m] = \begin{cases} 1, & m = nD, \quad n : \text{integer} \\ 0, & \text{otherwise} \end{cases} \tag{6.65}$$

is available, it can be used in Eqn. (6.65) to obtain

$$X_d(\Omega) = \sum_{m=-\infty}^{\infty} x[m]\, w[m] e^{-j\Omega m/D} \tag{6.66}$$

with no restrictions on the summation index m. Since $w[m]$ is equal to zero for index values that are not integer multiples of D, the restriction that m be an integer multiple of D may be safely removed from the summation. It can be shown (see Problem 6.16 at the end of this chapter) that the signal $w[m]$ can be expressed as

$$w[m] = \frac{1}{D} \sum_{k=0}^{D-1} e^{j2\pi mk/D} \tag{6.67}$$

Substituting Eqn. (6.67) into Eqn. (6.66) leads to

$$X_d\left(\Omega\right) = \sum_{m=-\infty}^{\infty} \left[\frac{1}{D}\sum_{k=0}^{D-1} e^{j2\pi mk/D}\right] e^{-j\Omega m/D}$$

$$= \frac{1}{D}\sum_{k=0}^{D-1}\left[\sum_{m=-\infty}^{\infty} x[m]\, e^{-j\left(\Omega-2\pi k\right)m/D}\right] \tag{6.68}$$

Using Eqn. (6.68) the spectrum of the downsampled signal $x_d[n]$ is related to the spectrum of the original signal $x[n]$ by the following:

$$X_d\left(\Omega\right) = \frac{1}{D}\sum_{k=0}^{D-1} X\left(\frac{\Omega - 2\pi k}{D}\right) \tag{6.69}$$

This result along with a careful comparison of Fig. 6.36(b) and (c) reveals that the downsampling operation could lead to aliasing through overlapping of spectral segments if care is not exercised. Spectral overlap occurs if the highest angular frequency in $x[n]$ exceeds π/D. It is therefore customary to process the signal $x[n]$ through a lowpass anti-aliasing filter before it is downsampled. The combination of the anti-aliasing filter and the downsampler is referred to as a *decimator*.

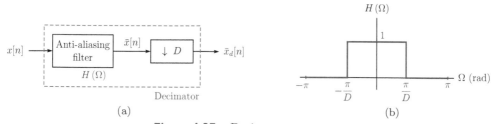

Figure 6.37 – Decimator structure.

6.4.2 Increasing the sampling rate by an integer factor

In contrast with downsampling to reduce the sampling rate, the *upsampling* operation is used as the first step in increasing the sampling rate of a discrete-time signal. Upsampled version of a signal $x[n]$ is defined as

$$x_u[n] = \begin{cases} x[n/L]\,, & n = kL\,, \quad k : \text{integer} \\ 0\,, & \text{otherwise} \end{cases} \tag{6.70}$$

where L is the upsampling rate. Graphical representation of an upsampler is shown in Fig. 6.38.

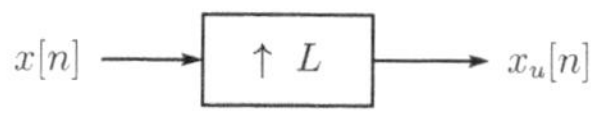

$$x[n] \longrightarrow \boxed{\uparrow L} \longrightarrow x_u[n]$$

Figure 6.38 – Graphical representation of an upsampler.

Upsampling operation was also briefly discussed in Section 1.4.1 of Chapter 1 in the context of time scaling for discrete-time signals. Fig. 6.39, illustrates the upsampling of a signal using $L = 2$.

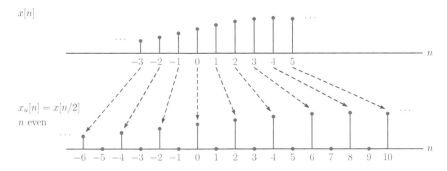

Figure 6.39 – Upsampling by a factor of $L = 2$.

Upsampling operation produces the additional samples needed for increasing the sampling rate by an integer factor; however, the new samples inserted into the signal all have zero amplitudes. Further processing is needed to change the zero-amplitude samples to more meaningful values. In order to understand what type of processing is necessary, we will again use spectral relationships. The DTFT of $x_u[n]$ is

$$X_u\left(\Omega\right) = \sum_{n=-\infty}^{\infty} x_u[n]\,e^{-j\Omega n} = \sum_{\substack{n=-\infty \\ n=mL}}^{\infty} x[n/L]\,e^{-j\Omega n} \tag{6.71}$$

Using the variable change $n = mL$, Eqn. (6.71) becomes

$$X_u\left(\Omega\right) = \sum_{m=-\infty}^{\infty} x[m]\,e^{-j\Omega L m} = X\left(\Omega L\right) \tag{6.72}$$

This result is illustrated in Fig. 6.40.

A lowpass *interpolation filter* is needed to make the zero-amplitude samples "blend-in" with the rest of the signal. The combination of an upsampler and a lowpass interpolation filter is referred to as an *interpolator*.

Ideally, the interpolation filter should remove the extraneous spectral segments in $X_u\left(\Omega\right)$ as shown in Fig. 6.42.

Thus the ideal interpolation filter is a discrete-time lowpass filter with an angular cutoff frequency of $\Omega_c = \pi/L$ and a gain of L as shown in Fig. 6.43(a). The role of the interpolation filter is similar to that of the ideal reconstruction filter discussed in Section 6.3. The impulse response of the filter is (see Problem 6.17 at the end of this chapter)

$$h_r[n] = \text{sinc}\left(n/L\right) \tag{6.73}$$

and is shown in Fig. 6.43(b) for the sample case of $L = 5$. Notice how every 5-th sample has zero amplitude so that convolution of $x_u[n]$ with this filter causes interpolation between the original samples of the signal without changing the values of the original samples.

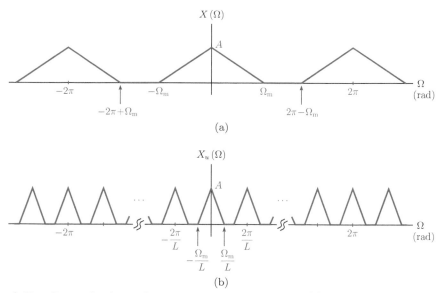

Figure 6.40 – Spectral relationships in upsampling a signal: (a) sample spectrum of signal $x[n]$, (b) corresponding spectrum for $x_u[n] = x[n/L]$.

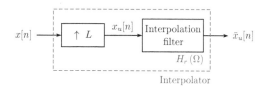

Figure 6.41 – Practical interpolator.

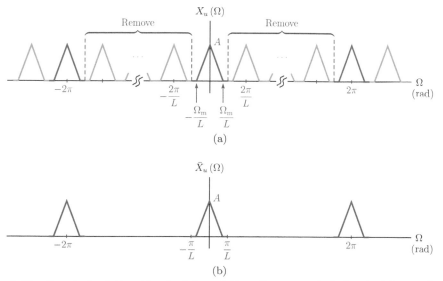

Figure 6.42 – Spectral relationships in interpolating the upsampled signal: (a) sample spectrum of signal $x_u[n]$, (b) corresponding spectrum for $\bar{x}_u[n]$ at filter output.

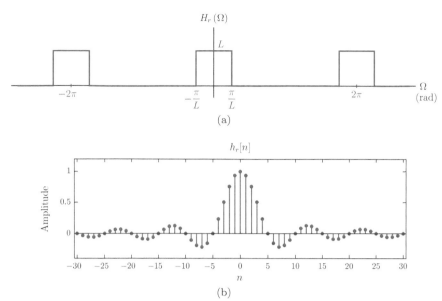

Figure 6.43 – (a) Spectrum of ideal interpolation filter, (b) impulse response of ideal interpolation filter.

In practical situations simpler interpolation filters may be used as well. A zero-order hold interpolation filter has the impulse response

$$h_{zoh}[n] = u[n] - u[n - L] \tag{6.74}$$

whereas the impulse response of a first-order hold interpolation filter is

$$h_r[n] = \begin{cases} 1 - \dfrac{|n|}{L}, & n = -L, \ldots, L \\ 0, & \text{otherwise} \end{cases} \tag{6.75}$$

6.5   Further Reading

[1] U. Graf. *Applied Laplace Transforms and Z-Transforms for Scientists and Engineers: A Computational Approach Using a Mathematica Package.* Birkhäuser, 2004.

[2] D.G. Manolakis and V.K. Ingle. *Applied Digital Signal Processing: Theory and Practice.* Cambridge University Press, 2011.

[3] A.V. Oppenheim and R.W. Schafer. *Discrete-Time Signal Processing.* Prentice Hall, 2010.

[4] R.A. Schilling, R.J. Schilling, and P.D. Sandra L. Harris. *Fundamentals of Digital Signal Processing Using MATLAB.* Cengage Learning, 2010.

[5] L. Tan and J. Jiang. *Digital Signal Processing: Fundamentals and Applications.* Elsevier Science, 2013.

MATLAB Exercises

MATLAB Exercise 6.1: Spectral relations in impulse sampling

Consider the continuous-time signal

$$x_a(t) = e^{-|t|}$$

Its Fourier transform is (see Example 4.16 in Chapter 4)

$$X_a(f) = \frac{2}{1 + 4\pi^2 f^2}$$

Compute and graph the spectrum of $x_a(t)$. If the signal is impulse-sampled using a sampling rate of $f_s = 1$ Hz to obtain the signal $x_s(t)$, compute and graph the spectrum of the impulse-sampled signal. Afterwards repeat with $f_s = 2$ Hz.

Solution: The script listed below utilizes an anonymous function to define the transform $X_a(f)$. It then uses Eqn. (6.13) to compute and graph $X_s(f)$ superimposed with the contributing terms.

```
1   % Script: matex_6_1a.m
2   %
3   Xa = @(f) 2./(1+4*pi*pi*f.*f);     % Original spectrum
4   f = [-3:0.01:3];
5   fs = 1;           % Sampling rate
6   Ts = 1/fs;        % Sampling interval
7   % Approximate spectrum of impulse-sampled signal
8   Xs = zeros(size(Xa(f)));
9   for k=-5:5,
10     Xs = Xs+fs*Xa(f-k*fs);
11  end;
12  % Graph the original spectrum
13  clf;
14  subplot(2,1,1);
15  plot(f,Xa(f)); grid;
16  axis([-3,3,-0.5,2.5]);
17  title('X_{a}(f)');
18  % Graph spectrum of impulse-sampled signal
19  subplot(2,1,2);
20  plot(f,Xs); grid;
21  axis([-3,3,-0.5,2.5]);
22  hold on;
23  for k=-5:5,
24     tmp = plot(f,fs*Xa(f-k*fs),'g:');
25  end;
26  hold off;
27  title('X_{s}(f)');
28  xlabel('f (Hz)');
```

Lines 8 through 11 of the code approximate the spectrum $X_s(f)$ using terms for $k = -5, \ldots, 5$ of the infinite summation in Eqn. (6.13). The graphs produced are shown in Fig. 6.44.

Different sampling rates may be tried easily by changing the value of the variable "fs" in line 5. The spectrum $X_s(f)$ for $f_s = 2$ Hz is shown in Fig. 6.45.

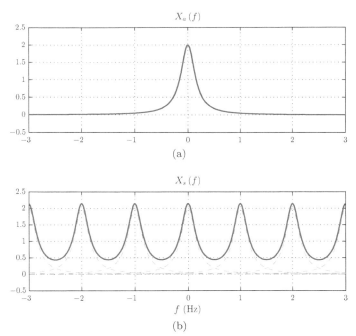

$X_a(f)$

$X_s(f)$

f (Hz)

(a)

(b)

Figure 6.44 – MATLAB graphs for the script `matex_6_1a.m`.

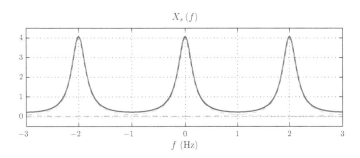

$X_s(f)$

f (Hz)

Figure 6.45 – The spectrum $X_s(f)$ for $f_s = 2$ Hz.

Software resources:
`matex_6_1a.m`
`matex_6_1b.m`

MATLAB Exercise 6.2: **DTFT of discrete-time signal obtained through sampling**

The two-sided exponential signal

$$x_a(t) = e^{-|t|}$$

which was also used in MATLAB Exercise 6.1 is sampled using a sampling rate of $f_s = 1$ Hz to obtain a discrete-time signal $x[n]$. Compute and graph $X(\Omega)$, the DTFT spectrum of $x[n]$ for the angular frequency range $-\pi \leq \Omega \pi$. Afterwards repeat with $f_s = 2$ Hz.

Solution: The script listed below graphs the signal $x_a(t)$ and its sampled version $x[n]$.

```
1   % Script: matex_6_2a.m
2   %
```

```
3    t = [-5:0.01:5];
4    xa = @(t) exp(-abs(t));
5    fs = 2;
6    Ts = 1/fs;
7    n = [-15:15];
8    xn = xa(n*Ts);
9    clf;
10   subplot(2,1,1);
11   plot(t,xa(t)); grid;
12   title('Signal x_{a}(t)');
13   subplot(2,1,2)
14   stem(n,xn);
15   title('Signal x[n]');
```

Recall that the spectrum of the analog signal is

$$X_a(\omega) = \frac{2}{1+\omega^2}$$

The DTFT spectrum $X(\Omega)$ is computed using Eqn. (6.25). The script is listed below:

```
1    % Script matex_6_2b.m
2    %
3    Xa = @(omg) 2./(1+omg.*omg);
4    fs = 1;
5    Ts = 1/fs;
6    Omg = [-1:0.001:1]*pi;
7    XDTFT = zeros(size(Xa(Omg/Ts)));
8    for k=-5:5,
9      XDTFT = XDTFT+fs*Xa((Omg-2*pi*k)/Ts);
10   end;
11   plot(Omg,XDTFT); grid;
12   axis([-pi,pi,-0.5,2.5]);
13   title('X(\Omega)');
14   xlabel('\Omega (rad)');
```

The graph produced is shown in Fig. 6.46.

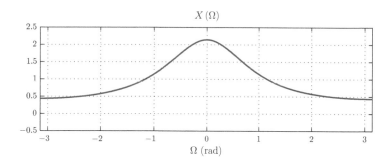

Figure 6.46 – The spectrum $X(\Omega)$ for $f_s = 1$ Hz.

The sampling rate $f_s = 2$ Hz can be obtained by changing the value of the variable "**fs**" in line 4.

Software resources:

matex_6_2a.m

matex_6_2b.m

MATLAB Exercise 6.3: Sampling a sinusoidal signal

In Example 6.2 three sinusoidal signals

$$x_{1a}(t) = \cos(12\pi t)$$
$$x_{2a}(t) = \cos(20\pi t)$$
$$x_{3a}(t) = \cos(44\pi t)$$

were each sampled using the sampling rate $f_s = 16$ Hz to obtain three discrete-time signals. It was shown that the resulting signals $x_1[n]$, $x_2[n]$ and $x_3[n]$ were identical. In this exercise we will verify this result using MATLAB. The script listed below computes and displays the first few samples of each discrete-time signal:

```
1   % Script matex_6_3a.m
2   %
3   x1a = @(t) cos(12*pi*t);
4   x2a = @(t) cos(20*pi*t);
5   x3a = @(t) cos(44*pi*t);
6   t = [0:0.001:0.5];
7   fs = 16;
8   Ts = 1/fs;
9   n = [0:5];
10  x1n = x1a(n*Ts)
11  x2n = x2a(n*Ts)
12  x3n = x3a(n*Ts)
```

The script listed below computes and graphs the three signals and the discrete-time samples obtained by sampling them.

```
1   % Script: matex_6_3b.m
2   %
3   x1a = @(t) cos(12*pi*t);
4   x2a = @(t) cos(20*pi*t);
5   x3a = @(t) cos(44*pi*t);
6   t = [0:0.001:0.5];
7   fs = 16;
8   Ts = 1/fs;
9   n = [0:20];
10  x1n = x1a(n*Ts);
11  plot(t,x1a(t),t,x2a(t),t,x3a(t));
12  hold on;
13  plot(n*Ts,x1n,'ro');
14  hold off;
15  axis([0,0.5,-1.2,1.2]);
```

The graph produced is shown in Fig. 6.47.

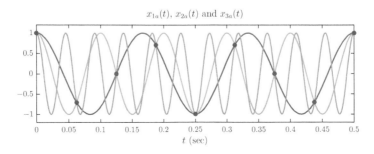

Figure 6.47 – Three sinusoidal signals that produce the same samples.

Software resources:
matex_6_3a.m
matex_6_3b.m

MATLAB Exercise 6.4: Natural sampling

The two-sided exponential signal

$$x_a(t) = e^{-|t|}$$

is sampled using a natural sampler with a sampling rate of $f_s = 4$ Hz and a duty cycle of $d = 0.6$. Compute and graph $X_s(f)$ in the frequency interval $-12 \leq f \leq 12$ Hz.

Solution: The spectrum given by Eqn. (6.38) may be written using f instead of ω as

$$\bar{X}_s(f) = d \sum_{k=-\infty}^{\infty} \text{sinc}(kd) \, X_a(f - kf_s)$$

The script to compute and graph $\bar{X}_s(f)$ is listed below. It is obtained by modifying the script "matex_6_1a.m" developed in MATLAB Exercise 6.1. The sinc envelope is also shown.

```
1   % Script: matex_6_4.m
2   %
3   Xa = @(f) 2./(1+4*pi*pi*f.*f);
4   f = [-12:0.01:12];
5   fs = 4;          % Sampling rate.
6   Ts = 1/fs;       % sampling interval.
7   d = 0.6;         % Duty cycle
8   Xs = zeros(size(Xa(f)));
9   for k=-5:5,
10      Xs = Xs+d*sinc(k*d)*Xa(f-k*fs);
11  end;
12  plot(f,Xs,'b-',f,2*d*sinc(f*d/fs),'r--'); grid;
13  axis([-12,12,-0.5,1.5]);
```

The spectrum $\bar{X}_s(f)$ is shown in Fig. 6.49.

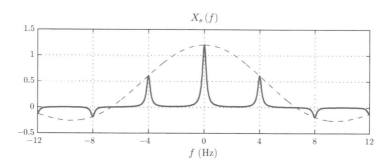

Figure 6.48 – The spectrum $X_s(f)$ for MATLAB Exercise 6.4.

Software resources:
matex_6_4.m

MATLAB Exercise 6.5: Zero-order hold sampling

The two-sided exponential signal

$$x_a(t) = e^{-|t|}$$

is sampled using a zero-order hold sampler with a sampling rate of $f_s = 3$ Hz and a duty cycle of $d = 0.3$. Compute and graph $|X_s(f)|$ in the frequency interval $-12 \le f \le 12$ Hz.

Solution: The spectrum given by Eqn. (6.43) may be written using f instead of ω as

$$\bar{X}_s(f) = d \operatorname{sinc}(fdT_s)\, e^{-j\pi fdT_s} \sum_{k=-\infty}^{\infty} X_a(f - kf_s)$$

The script to compute and graph $\left|\bar{X}_s(f)\right|$ is listed below. It is obtained by modifying the script "matex_6_1a.m.m" developed in MATLAB Exercise 6.1.

```
1   % Script: matex_6_5.m
2   %
3   Xa = @(f) 2./(1+4*pi*pi*f.*f);
4   f = [-12:0.01:12];
5   fs = 3;          % Sampling rate
6   Ts = 1/fs;       % Sampling interval
7   d = 0.3;         % Duty cycle
8   Xs = zeros(size(Xa(f)));
9   for k=-5:5,
10    Xs = Xs+fs*Xa(f-k*fs);
11  end;
12  Xs = d*Ts*sinc(f*d*Ts).*exp(-j*pi*f*d*Ts).*Xs;
13  plot(f,abs(Xs)); grid;
14  axis([-12,12,-0.1,0.8]);
15  title('|X_s(f)|');
16  xlabel('f (Hz)');
```

The magnitude spectrum $\left|\bar{X}_s(f)\right|$ is shown in Fig. 6.49.

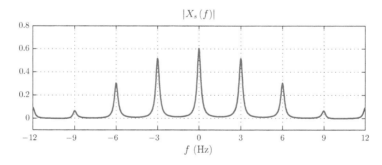

Figure 6.49 – The magnitude spectrum $|X_s(f)|$ for MATLAB Exercise 6.5.

Software resources:
matex_6_5.m

MATLAB Exercise 6.6: Graphing signals for natural and zero-order hold sampling

In this exercise we will develop and test two functions ss_natsamp(..) and ss_zohsamp(..) for obtaining graphical representation of signals samples using natural sampling and zero-order sampling respectively. The function ss_natsamp(..) evaluates a naturally sampled signal at a specified set of time instants.

```
1    function xnat = ss_natsamp(xa,Ts,d,t)
2      t1 = (mod(t,Ts)<=d*Ts);
3      xnat = xa(t).*t1;
4    end
```

The function `ss_zohsamp(..)` evaluates and returns a zero-order hold sampled version of the signal.

```
1    function xzoh = ss_zohsamp(xa,Ts,d,t)
2      t1 = (mod(t,Ts)<=d*Ts);
3      xzoh = xa(t).*t1;
4      flg = 0;
5      for i=1:length(t),
6        if not(t1(i)),
7          flg = 0;
8        elseif (t1(i) & (flg==0)),
9          flg = 1;
10         value = xzoh(i);
11       end;
12       if (flg == 1),
13         xzoh(i) = value;
14       end;
15     end;
16   end
```

For both functions the input arguments are as follows:

xa: Name of an anonymous function that can be used for evaluating the analog signal $x_a(t)$ at any specified time instant.

Ts: The sampling interval in seconds.

d: The duty cycle. Should be $0 < d \leq 1$.

t: Vector of time instants at which the sampled signal should be evaluated. For a detailed graph, choose the time increment for the values in vector "t" to be significantly smaller than T_s.

The function `ss_natsamp(..)` can be tested with the double sided exponential signal using the following statements:

```
>>   x = @(t) exp(-abs(t));
>>   t = [-4:0.001:4];
>>   xnat = ss_natsamp(x,0.2,0.5,t);
>>   plot(t,xnat);
```

The function `ss_zohsamp(..)` can be tested with the following:

```
>>   xzoh = ss_zohsamp(x,0.2,0.5,t);
>>   plot(t,xzoh);
```

Software resources:
ss_natsamp.m
ss_zohsamp.m
matex_6_6a.m
matex_6_6b.m

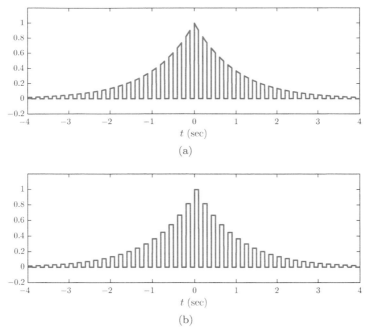

Figure 6.50 – Graphs for MATLAB Exercise 6.6: (a) natural sampling, (b) zero-order hold sampling.

MATLAB Exercise 6.7: Reconstruction of right-sided exponential

Recall that in Example 6.1 we considered impulse-sampling the right-sided exponential signal

$$x_a(t) = e^{-100t} u(t)$$

The spectrum $X(f)$ shown in Fig. 6.6 indicates that the right-sided exponential signal is not bandlimited, and therefore there is no sampling rate that would satisfy the requirements of the Nyquist sampling theorem. As a result, aliasing will be present in the spectrum regardless of the sampling rate used. In Example 6.1 three sampling rates, $f_s = 200$ Hz, $f_s = 400$ Hz, and $f_s = 600$ Hz, were used. The aliasing effect is most noticeable for $f_s = 200$ Hz, and less so for $f_s = 600$ Hz.

In a practical application of sampling, we would have processed the signal $x_a(t)$ through an anti-aliasing filter prior to sampling it, as shown in Fig. 6.9. However, in this exercise we will omit the anti-aliasing filter, and attempt to reconstruct the signal from its sampled version using the three techniques we have discussed. The script given below produces a graph of the impulse-sampled signal and the zero-order hold approximation to the analog signal $x_a(t)$.

```
1    % Script: matex_6_7a.m
2    %
3    fs = 200;      % Sampling rate
4    Ts = 1/fs;     % Sampling interval
5    % Set index limits "n1" and "n2" to cover the time interval
6    % from -25 ms to +75 ms
7    n1 = -fs/40;
8    n2 = -3*n1;
9    n = [n1:n2];
10   t = n*Ts;      % Vector of time instants
```

```
11   xs = exp(-100*t).*(n>=0);      % Samples of the signal
12   clf;
13   stem(t,xs,'^'); grid;
14   hold on;
15   stairs(t,xs,'r-');
16   hold off;
17   axis([-0.030,0.080,-0.2,1.1]);
18   title('Reconstruction using zero-order hold');
19   xlabel('t (sec)');
20   ylabel('Amplitude');
21   text(0.015,0.7,sprintf('Sampling rate = %.3g Hz',fs));
```

The sampling rate can be modified by editing line 3 of the code. The graph generated by this function is shown in Fig. 6.51 for sampling rates 200 Hz and 400 Hz.

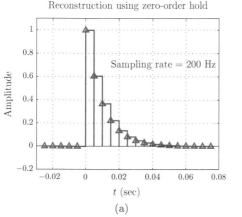

(a)

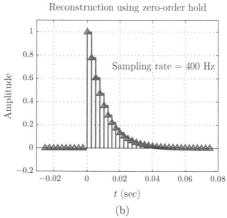

(b)

Figure 6.51 – Impulse-sampled right-sided exponential signal and zero-order hold reconstruction for (a) $f_s = 200$ Hz, and (b) $f_s = 400$ Hz.

Modifying this script to produce first-order hold interpolation is almost trivial. The modified script is given below.

```
1    % Script: matex_6_7b.m
2    %
3    fs = 200;      % Sampling rate
4    Ts = 1/fs;     % Sampling interval
5    % Set index limits "n1" and "n2" to cover the time interval
6    % from -25 ms to +75 ms
7    n1 = -fs/40;
8    n2 = -3*n1;
9    n = [n1:n2];
10   t = n*Ts;      % Vector of time instants
11   xs = exp(-100*t).*(n>=0);      % Samples of the signal
12   clf;
13   stem(t,xs,'^'); grid;
14   hold on;
15   plot(t,xs,'r-');
16   hold off;
17   axis([-0.030,0.080,-0.2,1.1]);
18   title('Reconstruction using first-order hold');
19   xlabel('t (sec)');
20   ylabel('Amplitude');
21   text(0.015,0.7,sprintf('Sampling rate = %.3g Hz',fs));
```

The only functional change is in line 15 where we use the function plot(..) instead of the function stairs(..). The graph generated by this modified script is shown in Fig. 6.52 for sampling rates 200 Hz and 400 Hz.

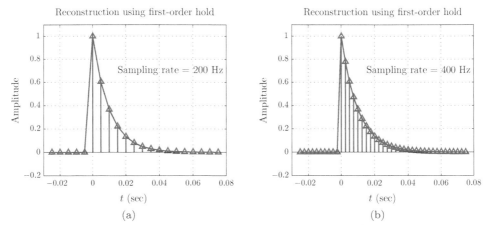

Figure 6.52 – Impulse-sampled right-sided exponential signal and first-order hold reconstruction for (a) $f_s = 200$ Hz, and (b) $f_s = 400$ Hz.

Reconstruction through bandlimited interpolation requires a bit more work. The script for this purpose is given below. Note that we have added a new section between lines 12 and 19 to compute the shifted sinc functions called for in Eqn. (6.55).

```
1   % Script: matex_6_7c.m
2   %
3   fs = 200;      % Sampling rate
4   Ts = 1/fs;     % Sampling interval
5   % Set index limits "n1" and "n2" to cover the time interval
6   % from -25 ms to +75 ms
7   n1 = -fs/40;
8   n2 = -3*n1;
9   n = [n1:n2];
10  t = n*Ts;      % Vector of time instants
11  xs = exp(-100*t).*(n>=0);    % Samples of the signal
12  % Generate a new, more dense, set of time values for the
13  % sinc interpolating functions
14  t2 = [-0.025:0.0001:0.1];
15  xr = zeros(size(t2));
16  for n=n1:n2,
17     nn = n-n1+1;  % Because MATLAB indices start at 1
18     xr = xr+xs(nn)*sinc((t2-n*Ts)/Ts);
19  end;
20  clf;
21  stem(t,xs,'^'); grid;
22  hold on;
23  plot(t2,xr,'r-');
24  hold off;
25  axis([-0.030,0.08,-0.2,1.1]);
26  title('Reconstruction using bandlimited interpolation');
27  xlabel('t (sec)');
28  ylabel('Amplitude');
29  text(0.015,0.7,sprintf('Sampling rate = %.3g Hz',fs));
```

The graph generated by this script is shown in Fig. 6.53 for sampling rates 200 Hz and 400 Hz.

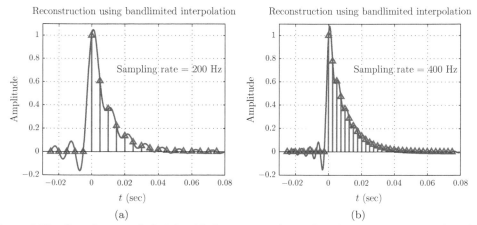

Figure 6.53 – Impulse-sampled right-sided exponential signal and its reconstruction based on bandlimited interpolation for (a) $f_s = 200$ Hz, and (b) $f_s = 400$ Hz.

Software resources:
matex_6_7a.m
matex_6_7b.m
matex_6_7c.m

MATLAB Exercise 6.8: Frequency spectrum of reconstructed signal

In this exercise we will compute and graph the frequency spectra for the reconstructed signals $x_{zoh}(t)$, $x_{foh}(t)$ and $x_r(t)$ obtained in MATLAB Exercise 6.7. Recall that the frequency spectra for the original right-sided exponential signal and its impulse-sampled version were found in Example 6.1. System functions for zero-order hold and first-order hold interpolation filters are given by Eqns. (6.48) and (6.50) respectively. Spectra for reconstructed signals are found through Eqns. (6.49) and (6.51). In the case of bandlimited interpolation, the spectrum of the reconstructed signal can be found by simply truncating the spectrum $X_s(f)$ to retain only the part of it in the frequency range $-f_s/2 \le f \le f_s/2$. The script listed below computes and graphs each spectrum along with the original spectrum $X_a(f)$ for comparison. The sampling rate used in each case is $f_s = 200$ Hz, and may be modified by editing line 3 of the code.

```
1    % Script: matex_6_8.m
2    %
3    fs = 200;     % Sampling rate
4    Ts = 1/fs;    % Sampling interval
5    f = [-700:0.5:700];     % Vector of frequencies
6    Xa = 1./(100+j*2*pi*f); % Original spectrum
7    % Compute the spectrum of the impulse-sampled signal.
8    Xs = 1./(1-exp(-100*Ts)*exp(-j*2*pi*f*Ts));
9    % Compute system functions of reconstruction filters.
10   Hzoh = Ts*sinc(f*Ts).*exp(-j*pi*Ts);
11   Hfoh = Ts*sinc(f*Ts).*sinc(f*Ts);
12   Hr  = Ts*((f>=-0.5*fs)&(f<=0.5*fs));
13   % Compute spectra of reconstructed signals.
14   Xzoh = Xs.*Hzoh;  % Eqn. (6.49)
```

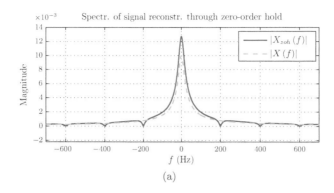

(a)

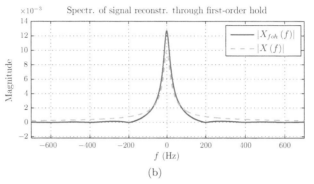

(b)

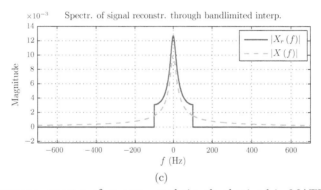

(c)

Figure 6.54 – Frequency spectra of reconstructed signals obtained in MATLAB Exercise 6.8 through (a) zero-order hold, (b) first-order hold, (c) bandlimited interpolation.

```
15   Xfoh = Xs.*Hfoh;    % Eqn. (6.51)
16   Xr = Xs.*Hr;        % Eqn. (6.53)
17   % Graph the results.
18   clf;
19   subplot(3,1,1);
20   plot(f,abs(Xzoh),'-',f,abs(Xa),'--'); grid;
21   title('Spectr. of signal reconstr. through zero-order hold');
22   xlabel('f (Hz)');
23   ylabel('Magnitude');
24   legend('|X_{zoh}(f)|','|X(f)|');
25   subplot(3,1,2);
26   plot(f,abs(Xfoh),'-',f,abs(Xa),'--'); grid;
27   title('Spectr. of signal reconstr. through first-order hold');
28   xlabel('f (Hz)');
29   ylabel('Magnitude');
```

```
30    legend('|X_{foh}(f)|','|X(f)|');
31    subplot(3,1,3);
32    plot(f,abs(Xr),'-',f,abs(Xa),'--'); grid;
33    title('Spectr. of signal reconstr. through bandlimited interp.');
34    xlabel('f (Hz)');
35    ylabel('Magnitude');
36    legend('|X_{r}(f)|','|X(f)|');
```

The graphs generated by the script are shown in Fig. 6.54. Observe the effect of aliasing on each spectrum.

Software resources:

matex_6_8.m

MATLAB Exercise 6.9: Resampling discrete-time signals

Consider the system shown in Fig. 6.55.

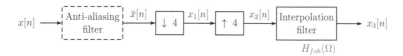

Figure 6.55 – The system for MATLAB Exercise 6.9.

The input signal is

$$x[n] = \cos(0.1\pi n) + 0.7\sin(0.2\pi n) + \cos(0.4\pi n)$$

Simulate the system first without the anti-aliasing filter and then with the anti-aliasing filter. Use first-order hold interpolation after the upsampler.

Solution: The interpolation filter for first-order hold interpolation is a discrete-time triangle with a peak amplitude of 1 and the two corners at $n = \mp L = \mp 4$. Its impulse response is

$$h_r[n] = \{\, 0,\ 0.25,\ 0.5,\ 0.75,\ \underset{\underset{n=0}{\uparrow}}{1.0},\ 0.75,\ 0.5,\ 0.25,\ 0 \,\} \tag{6.76}$$

The script listed below implements this system without the anti-aliasing filter.

```
1     % Script: matex_6_9a.m
2     %
3     n = [0:99];
4     x = cos(0.1*pi*n)+0.7*sin(0.2*pi*n)+cos(0.4*pi*n);
5     x1 = downsample(x,4);
6     x2 = upsample(xd,4);
7     hr = [0,0.25,0.5,0.75,1,0.75,0.5,0.25,0];    % FOH filter
8     x3 = conv(x2,hr);
9     x3 = x3(5:104);    % Compensate for 4 samples of delay
10    n = [0:99];
11    stem(n,x);
12    hold on;
13    plot(n,x3,'r');
14    hold off;
15    xlabel('n');
```

The vector "**x3**" obtained through convolution in line 8 has 108 samples. Line 9 of the code discards the first and the last 4 elements of "**x3**". The first 4 samples discarded correspond to sample indices $n = -4, \ldots, -1$ since the impulse response $h_r[n]$ starts at index $n = -4$. The last 4 samples discarded are the tail end of the convolution result. We are left with 100 samples of the signal $x_3[n]$ suitable for direct comparison with the 100-sample input signal $x[n]$. Input and output signals are shown in Fig. 6.56. For display purposes the output signal is shown in red with tips of samples connected by straight lines. The effect of aliasing due to the missing anti-aliasing filter is evident.

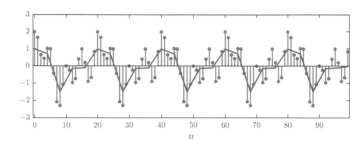

Figure 6.56 – Input and output signals of the system in Fig. 6.55 without anti-aliasing filter.

For this particular signal the implementation of an anti-aliasing filter is almost trivial. Recall from the discussion in Section 6.4.1 that the signal to be downsampled must not have any frequencies greater than $\Omega_c = \pi/D = \pi/4$. An ideal anti-aliasing filter would simply remove the third term in $x[n]$, the term that has an angular frequency of 0.4π. In the script code we simply modify line 4 to that effect:

```
1   % Script: matex_6_9b.m
2   %
3   n = [0:99];
4   x = cos(0.1*pi*n)+0.7*sin(0.2*pi*n);
5   x1 = downsample(x,4);
6   x2 = upsample(xd,4);
7   hr = [0,0.25,0.5,0.75,1,0.75,0.5,0.25,0];   % FOH filter
8   x3 = conv(x2,hr);
9   x3 = x3(5:104);   % Compensate for 4 samples of delay
10  n = [0:99];
11  stem(n,x);
12  hold on;
13  plot(n,x3,'r');
14  hold off;
15  xlabel('n');
```

Input and output signals for this case are shown in Fig. 6.57.

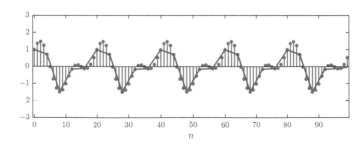

Figure 6.57 – Input and output signals of the system in Fig. 6.55 with anti-aliasing filter.

Software resources:
matex_6_9a.m
matex_6_9b.m

Problems

6.1. Consider the triangular waveform shown in Fig. P.6.1.

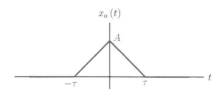

Figure P. 6.1

Its Fourier transform is (see Example 4.17 in Chapter 4)

$$X_a(f) = A\tau \operatorname{sinc}^2(f\tau)$$

Let $A = 1$ and $\tau = 1$ s. The signal $x_a(t)$ is impulse-sampled using a sampling rate of $f_s = 5$ Hz.

   a. Sketch the impulse-sampled signal $x_s(t)$.
   b. Find an expression for $X_s(f)$.
   a. Sketch $X_s(f)$ for $-10 \leq f \leq 10$ Hz.

6.2. An analog signal $x_a(t)$ has the Fourier transform shown in Fig. P.6.2.

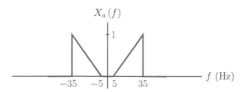

Figure P. 6.2

The signal is impulse sampled using a sampling rate of $f_s = 100$ Hz to obtain the signal $x_s(t)$. Sketch the spectrum $X_s(\omega)$.

6.3. The analog signal $x_a(t)$ with the Fourier transform shown in Fig. P.6.2 is sampled using a sampling rate of $f_s = 100$ Hz to obtain a discrete-time signal $x[n]$. Sketch the spectrum $X(\Omega)$.

6.4. If the analog signal $x_a(t)$ with the Fourier transform shown in Fig. P.6.2 is sampled using a sampling rate that is 10 percent less than the minimum requirement, sketch the spectrum $X(\Omega)$.

6.5. Consider the triangular waveform $x_a(t)$ shown in Fig. P.6.1. Let $A = 1$ and $\tau = 1$ s. This signal is sampled with a sampling rate of $f_s = 12$ Hz to obtain a discrete-time signal $x[n]$.

a. Sketch the signal $x[n]$.
b. Determine and sketch the DTFT spectrum of $x[n]$.

6.6. The signal

$$x_a(t) = \begin{cases} \sin(\pi t), & 0 \le t \le 1 \\ 0, & \text{otherwise} \end{cases}$$

is sampled with a sampling rate of $f_s = 15$ Hz to obtain a discrete-time signal $x[n]$. Determine and sketch the DTFT spectrum of $x[n]$.

6.7. Indicate which of the following signals can be sampled without any loss of information? For signals that can be sampled properly, determine the minimum sampling rate that can be used.

a. $x_a(t) = u(t) - u(t-3)$
b. $x_a(t) = e^{-2t} u(t)$
c. $x_a(t) = \cos(100\pi t) + 2\sin(150\pi t)$
d. $x_a(t) = \cos(100\pi t) + 2\sin(150\pi t)\sin(200\pi t)$
e. $x_a(t) = e^{-t}\cos(100\pi t)$

6.8.   A sinusoidal signal $x_a(t) = \sin(2\pi f_a t)$ with a frequency of $f_a = 1$ kHz is sampled using a sampling rate of $f_s = 2.4$ kHz to obtain a discrete-time signal $x[n]$.

a. Manually sketch the signal $x_a(t)$ for the time interval $0 < t < 5$ ms.
b. Show the sample amplitudes of the discrete-time signal $x[n]$ on the sketch of $x_a(t)$.
c. Find three alternative frequencies for the analog signal that result in the same discrete-time signal $x[n]$ when sampled with the sampling rate $f_s = 2.4$ kHz.

6.9. Refer to Problem 6.8.

a. Sketch the frequency spectrum of the analog signal for the original sinusoid and each of the three alternative frequencies.
b. For each of the signals and corresponding spectra in part (a), determine the DTFT spectrum of the discrete-time signal that results from sampling the analog signal with a sampling rate of $f_s = 2.4$ kHz.

6.10.   The analog sinusoidal signal $x_a(t) = \sin(500\pi t)$ is sampled to obtain a discrete-time signal $x[n] = \sin(0.4\pi n)$.

a. Assuming that the signal is sampled properly, determine the sampling interval T and the sampling rate f_s.
b. Find two other sampling rates that would produce the same discrete-time signal $x[n]$.
c. Using the sampling rate found in part (a), find two other analog signals that could be sampled to produce the same discrete-time signal $x[n]$.

6.11.   The sinusoidal signal

$$x_a(t) = 3\cos(100\pi t) + 5\sin(250\pi t)$$

is sampled at the rate of 100 times per second to obtain a discrete-time signal $x[n]$.

a. Sketch the spectrum $X(\Omega)$ of the signal $x[n]$.
b. If we incorrectly assume that $x[n]$ is the result of properly sampling an analog signal and try to reconstruct that analog signal, what signal would we get?

6.12. The analog signal
$$x_a\left(t\right) = \cos\left(100\pi t\right) + \cos\left(120\pi t\right)$$
is sampled using natural sampling as shown in Fig. 6.18. The sampling rate used is $f_s = 400$ Hz, and the width of each pulse is $\tau = 0.5$ ms.

   a. Write an analytical expression for the Fourier transform $X_a\left(\omega\right)$ and sketch it.
   b. Find an analytical expression for $\bar{X}_s\left(\omega\right)$ the Fourier transform of the naturally-sampled signal $\bar{x}_s\left(t\right)$.
   c. Sketch the transform $\bar{X}_s\left(\omega\right)$.

6.13. Repeat Problem 6.12 if zero-order hold sampling is used instead of natural sampling.

6.14. The signal $x_a\left(t\right) = \cos\left(150\pi n\right)$ is impulse-sampled with a sampling rate of $f_s = 200$ Hz and applied to a zero-order hold reconstruction filter as shown in Fig. P.6.14.

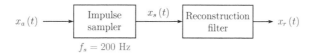

$$f_s = 200 \text{ Hz}$$

Figure P. 6.14

Sketch the signal at the output of the reconstruction filter.

6.15. Repeat Problem 6.14 if the reconstruction filter is a delayed first-order hold filter with the impulse response shown in Fig. P.6.15.

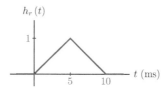

Figure P. 6.15

6.16. Show that the signal $w[m]$ defined as

$$w[m] = \frac{1}{D}\sum_{k=0}^{D-1} e^{j2\pi mk/D}$$

and used in the derivation of the spectrum of a downsampled signal in Section 6.4 satisfies the condition
$$w[m] = \begin{cases} 1, & m = nD, \quad n : \text{integer} \\ 0, & \text{otherwise} \end{cases}$$

6.17. Refer to the ideal interpolation filter spectrum $H_r\left(\Omega\right)$ shown in Fig. 6.43(a). Show that the impulse response of the ideal interpolation filter is

$$h_r[n] = \text{sinc}\left(n/L\right)$$

6.18. Indicate which of the following discrete-time signals can be downsampled without any loss of information? For signals that can be downsampled properly, determine the maximum downsampling rate D that can be used.

   a. $x[n] = u[n] - u[n - 25]$

   b. $x[n] = n\,u[n] - (n - 10)\,u[n - 10]\,,\qquad n = 0, \ldots, 19$

   c. $x[n] = \cos\left(\dfrac{\pi n}{3}\right)$

   d. $x[n] = \cos\left(\dfrac{\pi n}{10}\right) + 3\sin\left(\dfrac{2\pi n}{7}\right)$

6.19. The signal $x[n] = \sin(0.1\pi n)$ is applied to the system shown in Fig. P.6.19.

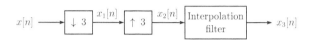

Figure P. 6.19

The interpolation filter is a zero-order hold filter with impulse response

$$h_{zoh}[n] = u[n] - u[n - 3]$$

  a. Find the spectrum of the zero-order hold filter.

  b. Determine and sketch the frequency spectra $X(\Omega)$, $X_1(\Omega)$, $X_2(\Omega)$ and $X_3(\Omega)$.

6.20. Rework Problem 6.19 if the interpolation filter is changed to a first-order hold filter with impulse response

$$h_{foh}[n] = \begin{cases} 1 - \dfrac{|n|}{3}\,, & n = -3, \ldots, 3 \\ 0\,, & \text{otherwise} \end{cases}$$

MATLAB Problems

6.21. Refer to the problem described in MATLAB Exercise 6.1.

  a. Set the sampling rate as $f_s = 3$ Hz. Compute and graph the spectrum $X_s(f)$ in the interval $-7 \le f \le 7$ Hz.

  b. Repeat part (a) with the sampling rate set equal to $f_s = 4$ Hz.

  c. Comment on the amount of aliasing for each case.

6.22. Refer to Problem 6.1. Write a script to

  a. Compute and graph the spectrum $X_a(f)$.

  b. Compute and graph the spectrum $X_s(s)$ for sampling rates $f_s - 5\,, 3\,, 2$ Hz.

Use the frequency range for $-10 \le f \le 10$ Hz. for all graphs.

6.23. Write a script to compute and graph the DTFT spectrum of the discrete-time signal obtained in Problem 6.5.

6.24. Write a script to compute and graph the DTFT spectrum of the discrete-time signal obtained in Problem 6.6.

6.25. Refer to Problem 6.8.

a. Write a script to graph the signal $x_a(t)$ and the samples of the signal $x[n]$ simultaneously for the time interval $0 < t < 5$ ms.

b. Three alternative frequencies were found that produce the same discrete-time signal. Modify the script in part (a) so that these alternative analog signals are also computed and graphed simultaneously with the original graph.

6.26. Refer to the function `ss_zohsamp(..)` that was developed in MATLAB Exercise 6.6.

a. Explain how the function works. Pay special attention to the conditional statements between lines 6 and 14 of the code.

b. The function does not work properly when the duty cycle is $d = 1$. Modify it so that it also works when $d = 1$. Assume that the sampling rate T_s is always an integer multiple of the time increment used in vector "t" and explore the use of MATLAB function `kron(..)` to implement the zero-order hold effect.

6.27. The two-sided exponential signal

$$x_a(t) = e^{-|t|}$$

is sampled using a zero-order hold sampler using a sampling rate of $f_s = 5$ Hz and a duty-cycle of $d = 0.9$.

a. Generate samples of the signal $x_{zoh}(t)$ for $-4 < t < 4$ s using the function `ss_zohsamp(..)`. Use a time vector with increments of 1 ms.

b. We will explore the possibility of obtaining a smoothly reconstructed signal from the zero-order held signal by filtering $x_{zoh}(t)$. A system with system function

$$H(\omega) = \frac{a}{j\omega + a}$$

may be simulated with statements

```
>>  sys = tf([a],[1,a]);
>>  y = lsim(sys,xzoh,t);
```

Compute and graph the filter output with parameter values $a = 1, 2, 3$ and comment on the results.

6.28. Write a script to simulate the system shown in Fig. P.6.19 with the input signal $x[n] = \sin(0.1\pi n)$ and using a zero-order hold interpolation filter. Generate 200 samples of the signal $x[n]$. Compute and graph the signals $x_1[n]$, $x_2[n]$ and $x_3[n]$.

6.29. Write a script to repeat the simulation in Problem 6.28 using a first-order hold interpolation filter instead of the zero-order hold filter.

MATLAB Projects

6.30. In this project we will explore the concept of aliasing especially in the way it exhibits itself in an audio waveform. MATLAB has a built-in sound file named "`handel`" which contains a recording of Handel's Hallelujah Chorus. It was recorded with a sampling rate of $f_s = 8192$ Hz. Use the statement

```
>>  load handel
```

which loads two new variables named "`Fs`" and "`y`" into the workspace. The scalar variable "`Fs`" holds the sampling rate, and the vector "`y`" holds 73113 samples of the recording that corresponds to about 9 seconds of audio. Once loaded, the audio waveform may be played back with the statement

```
>>  sound(y,Fs)
```

   a. Graph the audio signal as a function of time. Create a continuous time variable that is correctly scaled for this purpose.

   b. Compute and graph the frequency spectrum of the signal using the function `fft(..)`. Create an appropriate frequency variable to display frequencies in Hertz, and use it in graphing the spectrum of the signal. Refer to the discussion in Section 5.8.5 of Chapter 5 for using the DFT to approximate the continuous Fourier transform.

   c. Downsample the audio signal in vector "`y`" using a downsampling rate of $D = 2$. Do not use an anti-aliasing filter for this part. Play back the resulting signal using the function `sound(..)`. Be careful to adjust the sampling rate to reflect the act of downsampling or it will play too fast.

   d. Repeat part (c) this time using an anti-aliasing filter. A simple Chebyshev type-I lowpass filter may be designed with the following statement (see Chapter 10 for details):

```
>>  [num,den] = cheby1(5,1,0.45)
```

Process the audio signal through the anti-aliasing filter using the statement

```
>>  yfilt = filter(num,den,y);
```

The vector "`yfilt`" represents the signal at the output of the anti-aliasing filter. Downsample this output signal using $D = 2$ and listen to it. How does it compare to the sound obtained in part (c)? How would you explain the difference between the two sounds?

   e. Repeat parts (c) and (d) using a downsampling rate of $D = 4$. The anti-aliasing filter for this case should be obtained by

```
>>  [num,den] = cheby1(5,1,0.23)
```

6.31. On a computer with a microphone and a sound processor, the following MATLAB code allows 2 seconds of audio signal to be recorded, and a vector "`x`" to be created with the recording:

```
hRec = audiorecorder;
disp('Press a key to start recording');
pause;
recordblocking(hRec, 2);
disp('Finished recording');
x = getaudiodata(hRec);
```

By default the analog signal $x_a(t)$ captured by the microphone and the sound device is sampled at the rate of 8000 times per second, corresponding to $T = 125$ μs. For a 2-second recording the vector "`x`" contains 16000 samples that represent

$$x[n] = x_a\left(\frac{n}{8000}\right), \qquad n = 0,\ldots,15999$$

Develop a MATLAB script to perform the following steps:

a. Extract 8000 samples of the vector $x[n]$ into a new vector.

$$r[n] = x[n + 8000] , \qquad n = 0, \ldots, 7999$$

We skip the first 8000 samples so that we do not get a blank period before the person begins to speak. This should create a vector with 8000 elements representing one full second of speech.

b. Convert the sampling rate of the one-second speech waveform to 12 kHz. Use a combination of downsampling and upsampling along with any necessary filters to achieve this. Justify your choice of which operation should be carried out first.

c. Play back the resulting 12000-sample waveform using the function `sound(..)`.

d. Use the function `fft(..)` to compute and graph the frequency spectra of the signals involved at each step of sampling rate conversion.

Chapter 7

Laplace Transform for Continuous-Time Signals and Systems

Chapter Objectives

- Learn the Laplace transform as a more generalized version of the Fourier transform studied in Chapter 4.

- Understand the convergence characteristics of the Laplace transform and the concept of region of convergence.

- Explore the properties of the Laplace transform.

- Understand the use of the Laplace transform for modeling CTLTI systems. Learn the s-domain system function and its use for solving signal-system interaction problems.

- Learn techniques for obtaining simulation diagrams for CTLTI systems based on the s-domain system function.

- Learn the use of the unilateral Laplace transform for solving differential equations with specified initial conditions.

7.1 Introduction

In earlier chapters we have studied Fourier analysis techniques for understanding characteristics of continuous-time signals and systems. Chapter 4 focused on representing continuous-time signals in the frequency domain through the use of Fourier series and Fourier transform. We have also adapted frequency domain analysis techniques for use with linear and time-invariant systems in the form of a system function.

Some limitations exist in the use of Fourier series and Fourier transform for analyzing signals and systems. A signal must be absolute integrable to have a representation based on Fourier series or Fourier transform. Consider, for example, a ramp signal in the form $x(t) = t\,u(t)$. Such a signal cannot be analyzed using the Fourier transform since it is not absolute integrable. Similarly, a system function based on the Fourier transform is only available for systems for which the impulse response is absolute integrable. Consequently, Fourier transform techniques are usable only with stable systems.

Laplace transform, named after the French mathematician Pierre Simon Laplace (1749-1827), overcomes these limitations. It can be seen as an extension, or a generalization, of the Fourier transform. In contrast, the Fourier transform is a special case, or a limited view, of the Laplace transform. A signal that does not have a Fourier transform may still have a Laplace transform for a certain range of the transform variable. Similarly, an unstable system that does not have a system function based on the Fourier transform may still have a system function based on the Laplace transform.

The Fourier transform of a signal, if it exists, can be obtained from its Laplace transform while the reverse is not generally true. In addition to analysis of signals and systems, block diagram and signal flow graph structures for simulating continuous-time systems can be developed using Laplace transform techniques. The unilateral variant of the Laplace transform can be used for solving differential equations subject to specified initial conditions.

We begin with the basic definition of the Laplace transform and its application to some simple signals. The significance of the issue of convergence of the Laplace transform will become apparent throughout this discussion. Section 7.2 is dedicated to the convergence properties of the Laplace transform, and to the concept of region of convergence. We cover the fundamental properties of the Laplace transform in Section 7.3. These properties will prove useful in working with the Laplace transform for analyzing signals and systems, for understanding system characteristics, and for working with signal-system interaction problems. Proofs of significant Laplace transform properties are given not just for the sake of providing proofs, but also to provide further insight and experience on working with transforms. Techniques for computing the inverse Laplace transform are presented in Section 7.4. Use of the Laplace transform for the analysis of linear and time-invariant systems is discussed in Section 7.5, and the s-domain system function concept is introduced. Derivations of block diagram structures for the simulation of continuous-time systems based on the s-domain system function are covered in Section 7.6. In Section 7.7 the unilateral variant of the Laplace transform is introduced, and it is shown that it is useful in solving differential equations with specified initial conditions.

The Laplace transform of a continuous-time signal $x(t)$ is defined as

$$X(s) = \int_{-\infty}^{\infty} x(t)\, e^{-st}\, dt \tag{7.1}$$

where s, the independent variable of the transform, is a complex variable.

Notationally, the relationship between the signal $x(t)$ and the transform $X(s)$ can be expressed in the form

$$X(s) = \mathcal{L}\{x(t)\} \tag{7.2}$$

or in the form

$$x(t) \overset{\mathcal{L}}{\longleftrightarrow} X(s) \tag{7.3}$$

Since the parameter s is a complex variable, we will represent it graphically as a point in the complex s-plane. It is customary to draw the s-plane with σ as its real axis and ω as its imaginary axis. Based on this convention the complex variable s is written in Cartesian form as

$$s = \sigma + j\omega \tag{7.4}$$

A specific value $s_1 = \sigma_1 + j\omega_1$ of the variable s can be shown as a point in the s-plane as illustrated in Fig. 7.1 with σ_1 equal to the horizontal displacement of the point from the origin and ω_1 as the vertical displacement from the origin. If the Laplace transform converges at the point $s = s_1$, its value can be computed by evaluating the integral in Eqn. (7.1) at that point.

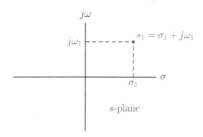

Figure 7.1 – The point $s_1 = \sigma_1 + j\omega_1$ in the complex s-plane.

Using the Cartesian form of the variable s given by Eqn. (7.4), the transform of the signal $x(t)$ becomes

$$X(s)\Big|_{s=\sigma+j\omega} = \int_{-\infty}^{\infty} x(t)\, e^{-(\sigma+j\omega)t}\, dt \tag{7.5}$$

At this point it would be interesting to let the value of σ, the real part of s, be fixed at $\sigma = \sigma_1$. Thus, we have $s = \sigma_1 + j\omega$. If the imaginary part ω is allowed to vary in the range $-\infty < \omega < \infty$, the trajectory of the points s in the complex plane would be a vertical line with a horizontal displacement of σ_1 from the origin, that is, a vertical line that passes through the point $s = \sigma_1 + j0$. This is illustrated in Fig. 7.2.

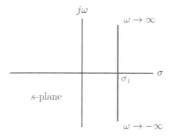

Figure 7.2 – Trajectory of the points $s = \sigma_1 + j\omega$ in the s-plane.

If the transform in Eqn. (7.5) is evaluated at points on the vertical line represented by the trajectory $s = \sigma_1 + j\omega$, we get

$$X(\sigma_1, \omega) = X(s)\Big|_{s=\sigma_1+j\omega} = \int_{-\infty}^{\infty} x(t)\, e^{-(\sigma_1+j\omega)t}\, dt$$

$$= \int_{-\infty}^{\infty} x(t)\, e^{-\sigma_1 t} e^{-j\omega t}\, dt \tag{7.6}$$

Comparison of the result in Eqn. (7.6) with the definition of the Fourier transform in Eqn. (4.127) in Chapter 4 suggests that the Laplace transform $X(s)$ evaluated on the trajectory $s = \sigma_1 + j\omega$ is identical to the Fourier transform of the signal $[x(t) e^{-\sigma_1 t}]$.

$$X(s)\Big|_{s=\sigma_1+j\omega} = \mathcal{F}\left\{x(t) e^{-\sigma_1 t}\right\} \tag{7.7}$$

This important conclusion can be summarized as follows:

In the s-plane consider a vertical line passing through the point $s = \sigma_1$ for a fixed value of σ_1. The Laplace transform of a signal $x(t)$ evaluated on this line as the parameter ω is varied from $\omega = -\infty$ to $\omega = \infty$ is the same as the Fourier transform of the signal $x(t) e^{-\sigma_1 t}$.

If we choose $\sigma_1 = 0$ then the vertical line in Fig. 7.2 coincides with the $j\omega$ axis of the s-plane, and the Laplace transform evaluated on that trajectory becomes

$$X(s)\Big|_{s=0+j\omega} = \int_{-\infty}^{\infty} x(t) e^{-j\omega t}\, dt = \mathcal{F}\{x(t)\}$$

Consider the Laplace transform of a signal $x(t)$ evaluated at all points on the $j\omega$ axis of the s-plane as ω is varied from $\omega = -\infty$ to $\omega = \infty$. The result is the same as the Fourier transform $X(\omega)$ of the signal.

An easy way to visualize the relationship between the Laplace transform and the Fourier transform is the following: Imagine that the Laplace transform evaluated at every point in the s-plane results in a three-dimensional surface. The transform $X(s)$ is complex-valued in general; therefore, it may be difficult to visualize it as a surface unless we are willing to accept the notion of a complex-valued surface. Alternatively, we may split the complex function $X(s)$ into its magnitude and phase, each defining a surface. In any case, if we were to take the surface represented by the Laplace transform and cut it the through the length of the $j\omega$ axis, the profile of the cross-section would be the same as the Fourier transform of the signal. This relationship is illustrated graphically in Fig. 7.3.

Consider a particular transform in the form

$$X(s) = \frac{s + 0.5}{(s + 0.5)^2 + 4\pi^2} \tag{7.8}$$

which is complex-valued. We will choose to graph only its magnitude $|X(s)|$ as a three-dimensional surface which is shown in Fig. 7.3(a). Also shown in the figure is the set of values of $|X(s)|$ computed at points on the $j\omega$ axis of the s-plane. In Fig. 7.3(b) the magnitude of the Fourier transform $X(\omega)$ of the same signal is graphed as a function of the radian frequency ω. Notice how the $j\omega$ axis of the s-plane becomes the frequency axis for the Fourier transform.

For the same transform under consideration, Fig. 7.4 shows the values on the trajectory $s = 0.5 + j\omega$. Recall that the values of $|X(s)|$ on this trajectory are the same as the magnitude of the Fourier transform of the modified signal $[x(t) e^{-0.5t}]$.

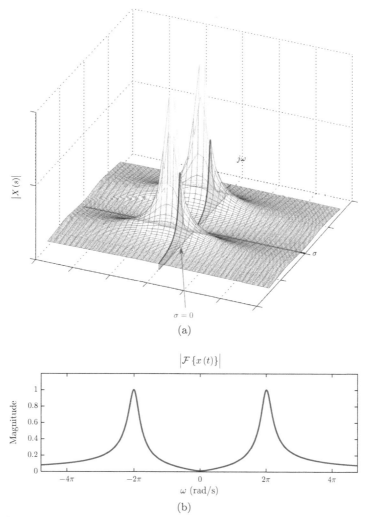

(a)

(b)

Figure 7.3 – (a) The magnitude $|X(s)|$ shown as a surface plot along with the the magnitude computed on the $j\omega$ axis of the s-plane, (b) the magnitude of the Fourier transform as a function of ω.

The Laplace transform as defined by Eqn. (7.1) is referred to as the *bilateral* (two-sided) Laplace transform. A variant of the Laplace transform known as the *unilateral* (one-sided) Laplace transform will be introduced in Section 7.7 as an alternative analysis tool. In this text, when we refer to Laplace transform without the qualifier word "bilateral" or "unilateral", we will always imply the more general bilateral Laplace transform as defined in Eqn. (7.1).

Interactive Demo: `lap_demo1.m`

The demo program "`lap_demo1.m`" is based on the Laplace transform given by Eqn. (7.8). The magnitude of the transform in question is shown in Fig. 7.3(a). The demo program computes this magnitude and graphs it as a three-dimensional mesh. It can be rotated freely for viewing from any angle by using the rotation tool in the toolbar. The magnitude of the

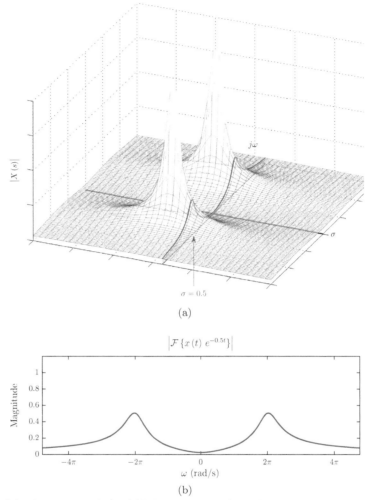

(a)

(b)

Figure 7.4 – (a) The magnitude $|X(s)|$ shown as a surface plot along with the the magnitude computed on the trajectory $s = 0.5 + j\omega$ on the s-plane, (b) the magnitude of the Fourier transform of $x(t) e-0.5t$.

transform is also evaluated on the vertical line $s = \sigma_1 + j\omega$ in the s-plane, and displayed as a two-dimensional graph. The value σ_1 may be varied through the use of a slider control.
Software resources:
`lap_demo1.m`

Software resources:	See MATLAB Exercises 7.1 and 7.2.

Example 7.1: **Laplace transform of the unit impulse**

Find the Laplace transform of the unit-impulse signal

$$x(t) = \delta(t)$$

Solution: Applying the definition of the Laplace transform given by Eqn. (7.1) we get

$$X(s) = \mathcal{L}\{\delta(t)\} = \int_{-\infty}^{\infty} \delta(t) e^{-st} dt = 1 \tag{7.9}$$

using the sifting property of the unit-impulse function. Therefore, the Laplace transform of the unit-impulse signal is constant and equal to unity. Since it does not depend on the value of s, it converges at every point in the s-plane with no exceptions.

Example 7.2: **Laplace Transform of a time-shifted unit impulse**

Find the Laplace transform of the time-shifted unit-impulse signal

$$x(t) = \delta(t - \tau)$$

Solution: Applying the Laplace transform definition we obtain

$$X(s) = \int_{-\infty}^{\infty} \delta(t - \tau) e^{-st} dt = e^{-s\tau} \tag{7.10}$$

Again we have used the sifting property of the unit-impulse function. If $s = \sigma + j\omega$, then Eqn. (7.10) becomes

$$X(s)\Big|_{s=\sigma+j\omega} = e^{-\sigma\tau} e^{-j\omega\tau}$$

The transform obtained in Eqn. (7.10) converges as long as $\sigma = \mathrm{Re}\{s\} > -\infty$.

Example 7.3: **Laplace transform of the unit-step signal**

Find the Laplace transform of the unit-step signal

$$x(t) = u(t)$$

Solution: Substituting $x(t) = u(t)$ into the Laplace transform definition we obtain

$$X(s) = \int_{-\infty}^{\infty} u(t) e^{-st} dt$$

Since $u(t) = 1$ for $t > 0$ and $u(t) = 0$ for $t < 0$, the lower limit of the integral can be changed to $t = 0$ and the unit-step term can be dropped to yield

$$X(s) = \int_{0}^{\infty} e^{-st} dt = \left. -\frac{1}{s} e^{-st} \right|_{0}^{\infty} \tag{7.11}$$

To evaluate the integral of Eqn. (7.11) for the specified limits we will use the Cartesian form of the complex variable s. Substituting $s = \sigma + j\omega$

$$X(s)\Big|_{s=\sigma+j\omega} = \left. -\frac{1}{\sigma + j\omega} e^{-(\sigma+j\omega)t} \right|_{0}^{\infty}$$

It is obvious that, for the exponential term $e^{-\sigma t}$ to converge as $t \to \infty$, we need $\sigma > 0$:

$$\text{If } \sigma > 0 \implies \left. e^{-\sigma t} \right|_{t \to \infty} = 0$$

The transform becomes

$$X(s) = -\frac{1}{\sigma + j\omega}[0 - 1] = \frac{1}{\sigma + j\omega} = \frac{1}{s}, \qquad \text{Re}\{s\} > 0 \qquad (7.12)$$

The transform expression found in Eqn. (7.12) is valid only for points in the right half of the s-plane. This region is shown shaded in Fig. 7.5. Note that the transform does not converge at points on the $j\omega$ axis. It converges at any point to the right of the $j\omega$ axis regardless of how close to the axis it might be.

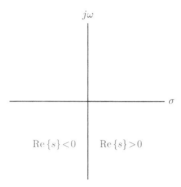

Figure 7.5 – The region in which the Laplace-transform of $x(t) = u(t)$ converges.

Example 7.4: **Laplace transform of a time-shifted unit-step signal**

Find the Laplace transform of the time-shifted unit-step signal

$$x(t) = u(t - \tau)$$

Solution: Using $x(t) = u(t - \tau)$ in the Laplace transform definition we get

$$X(s) = \int_{-\infty}^{\infty} u(t - \tau) e^{-st} \, dt$$

Since $u(t - \tau) = 1$ for $t > \tau$ and $u(t - \tau) = 0$ for $t < \tau$, the lower limit of the integral can be changed to $t = \tau$ and the unit-step function can be dropped without affecting the result.

$$X(s) = \int_{\tau}^{\infty} e^{-st} \, dt = \left. -\frac{1}{s} e^{-st} \right|_{\tau}^{\infty}$$

As in Example 7.3, the integral can be evaluated only for $\sigma = \text{Re}\{s\} > 0$, and its value is

$$X(s) = -\frac{1}{s}\left[0 - e^{-s\tau}\right] = \frac{e^{-s\tau}}{s}, \qquad \text{Re}\{s\} > 0 \qquad (7.13)$$

Convergence conditions for the the transform of $u(t - \tau)$ are the same as those found in Example 7.3 for the transform of $u(t)$.

We observe from the last two examples that, when we find the Laplace transform $X(s)$ of a signal, we also need to specify the region in which the transform is valid. The collection of all points in the s-plane for which the Laplace transform converges is called the *region of convergence (ROC)*.

Recall that in Eqn. (7.7) we have represented the Laplace transform of a signal $x(t)$ as equivalent to the Fourier transform of the modified signal $[x(t)e^{-\sigma t}]$. Consequently, the conditions for the convergence of the Laplace transform of $x(t)$ are identical to the conditions for the convergence of the Fourier transform of $[x(t)e^{-\sigma t}]$. Using Dirichlet conditions for the latter, we need the signal $[x(t)e^{-\sigma t}]$ to be absolute integrable for the transform to exist, that is,

$$\int_{-\infty}^{\infty} \left| x(t)e^{-\sigma t} \right| dt < \infty \tag{7.14}$$

The convergence condition stated in Eqn. (7.14) highlights the versatility of the Laplace transform over the Fourier transform. The Fourier transform of a signal $x(t)$ exists is the signal is absolute integrable, and does not exist otherwise. Therefore, the existence of the Fourier transform is a binary question; the answer to it is either yes or no. In contrast, if the signal $x(t)$ is not absolute integrable, we may still be able to find values of σ for which the modified signal $[x(t)e^{-\sigma t}]$ is absolute integrable. Therefore, the Laplace transform of the signal $x(t)$ may exist for some values of σ. The question of existence for the Laplace transform is not a binary one; it is a question of which values of σ allow the transform to converge.

The next two examples will further highlight the significance of the region of convergence for the Laplace transform. The fundamental characteristics of the region of convergence will be discussed in detail in Section 7.2.

Example 7.5: **Laplace transform of a causal exponential signal**

Find the Laplace transform of the signal

$$x(t) = e^{at}u(t)$$

where a is any real or complex constant.

Solution: The signal $x(t)$ is causal since $x(t) = 0$ for $t < 0$. Applying the Laplace transform definition to $x(t)$ results in

$$X(s) = \int_{-\infty}^{\infty} e^{at}u(t)e^{-st}\,dt$$

Changing the lower limit of the integral to $t = 0$ and dropping the factor $u(t)$ we get

$$X(s) = \int_{0}^{\infty} e^{at}e^{-st}\,dt = \int_{0}^{\infty} e^{(a-s)t}\,dt = \frac{1}{a-s}\,e^{(a-s)t}\Big|_{0}^{\infty} \tag{7.15}$$

The parameter a may be real or complex-valued. Let us assume that it is in the form $a = a_r + ja_i$ to keep the results general. For the purpose of evaluating the integral in Eqn. (7.15) we will substitute $s = \sigma + j\omega$ and write the transform as

$$X(s) = \frac{1}{a_r + ja_i - \sigma - j\omega}\,e^{(a_r + ja_i - \sigma - j\omega)t}\Big|_{0}^{\infty}$$

$$= \frac{1}{a_r + ja_i - \sigma - j\omega}\,e^{(a_r - \sigma)t}\,e^{j(a_i - \omega)t}\Big|_{0}^{\infty} \tag{7.16}$$

For the result of Eqn. (7.16) to converge at the upper limit, we need

$$a_r - \sigma < 0 \quad \Longrightarrow \quad \sigma > a_r$$

or equivalently

$$\mathrm{Re}\left\{s\right\} > \mathrm{Re}\left\{a\right\} \tag{7.17}$$

which establishes the ROC for the transform. With the condition in Eqn. (7.17) satisfied, the exponential term in Eqn. (7.16) becomes zero as $t \to \infty$, and we obtain

$$X\left(s\right) = \frac{1}{a-s}\left[0-1\right] = \frac{1}{s-a}, \qquad \mathrm{Re}\left\{s\right\} > \mathrm{Re}\left\{a\right\} \tag{7.18}$$

The ROC is the region to the right of the vertical line that goes through the point $s = a_r + ja_i$ as shown in Fig. 7.6.

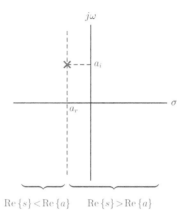

Figure 7.6 – The ROC for the transform $X\left(s\right)$ of Example 7.5.

It is interesting to look at the shape of the signal $x\left(t\right)$ in conjunction with the ROC.

1. Fig. 7.7 shows the signal for the case where a is real-valued. The signal $x\left(t\right)$ is a decaying exponential for $a < 0$, and a growing exponential for $a > 0$.
2. Real and imaginary parts of the signal $x\left(t\right)$ are shown in Figs. 7.8 and 7.9 for the case of parameter a being complex-valued. Fig. 7.8 illustrates the possibility of $\mathrm{Re}\left\{a\right\} < 0$. Real and imaginary parts of $x\left(t\right)$ exhibit oscillatory behavior with exponential damping. In contrast, if $\mathrm{Re}\left\{a\right\} > 0$, both real and imaginary parts of $x\left(t\right)$ grow exponentially as shown in Fig. 7.9.

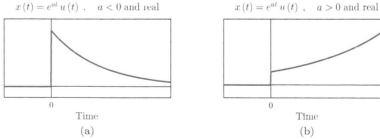

Figure 7.7 – The signal $x\left(t\right)$ of Example 7.5 for real-valued parameter a and (a) $a < 0$ and (b) $a > 0$.

$$x(t) = e^{at} u(t), \quad \text{Re}\{a\} < 0$$

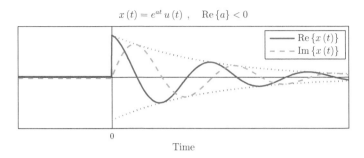

Figure 7.8 – The signal $x(t)$ of Example 7.5 for $\text{Re}\{a\} < 0$.

$$x(t) = e^{at} u(t), \quad \text{Re}\{a\} > 0$$

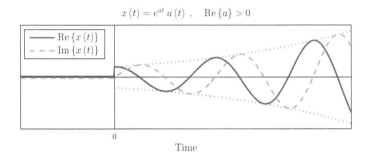

Figure 7.9 – The signal $x(t)$ of Example 7.5 for $\text{Re}\{a\} > 0$.

Since the ROC is to the right of a vertical line, for the case $\text{Re}\{a\} < 0$ the $j\omega$ axis is part of the ROC whereas, for the case $\text{Re}\{a\} > 0$, it is not.

Recall from the previous discussion that the Fourier transform of the signal $x(t)$ is equal to the Laplace transform evaluated on the $j\omega$ axis of the s-plane. Consequently, the Fourier transform of the signal exists only if the region of convergence includes the $j\omega$ axis. We conclude that, for the Fourier transform of the signal $x(t)$ to exist, we need $\text{Re}\{a\} < 0$. Signals in Figs. 7.7(a) and 7.8 have Fourier transforms whereas signals in Figs. 7.7(b) and 7.9 do not. This is consistent with the existence conditions discussed in Chapter 4.

Software resources:

ex_7_5a.m

ex_7_5b.m

Example 7.6: **Laplace transform of an anti-causal exponential signal**

Find the Laplace transform of the signal

$$x(t) = -e^{at} u(-t)$$

where a is any real or complex constant.

Solution: In this case the signal $x(t)$ is anti-causal since it is equal to zero for $t > 0$. Substituting $x(t)$ into the Laplace transform definition we get

$$X(s) = \int_{-\infty}^{\infty} -e^{at} u(-t) e^{-st} dt$$

Since $u(-t) = 1$ for $t < 0$ and $u(-t) = 0$ for $t > 0$, changing the upper limit of the integral to $t = 0$ and dropping the factor $u(-t)$ would have no effect on the result. Therefore

$$X(s) = \int_{-\infty}^{0} -e^{at} e^{-st} \, dt = \int_{-\infty}^{0} -e^{(a-s)t} \, dt = \left. -\frac{1}{a-s} e^{(a-s)t} \right|_{-\infty}^{0} \tag{7.19}$$

The parameter a may be real or complex valued. We will assume that it is in the general form $a = a_r + ja_i$. For the purpose of evaluating the integral in Eqn. (7.19) let us substitute $s = \sigma + j\omega$ and write the transform as

$$X(s) = -\frac{1}{a_r + ja_i - \sigma - j\omega} \left. e^{(a_r + ja_i - \sigma - j\omega)t} \right|_{-\infty}^{0}$$

$$= -\frac{1}{a_r + ja_i - \sigma - j\omega} \left. e^{(a_r - \sigma)t} e^{j(a_i - \omega)t} \right|_{-\infty}^{0} \tag{7.20}$$

In contrast with Example 7.5, the critical end of the integral with respect to convergence is its lower limit. To evaluate the result of Eqn. (7.20) at the lower limit, we need

$$a_r - \sigma > 0 \quad \Longrightarrow \quad \sigma < a_r$$

or equivalently

$$\mathrm{Re}\,\{s\} < \mathrm{Re}\,\{a\} \tag{7.21}$$

which establishes the ROC for the transform. With the condition in Eqn. (7.21) satisfied, the exponential term in Eqn. (7.20) becomes zero as $t \to -\infty$, and we obtain

$$X(s) = -\frac{1}{a-s} [1-0] = \frac{1}{s-a}, \qquad \mathrm{Re}\,\{s\} < \mathrm{Re}\,\{a\}$$

The ROC is the region to the left of the vertical line that goes through the point $s = a_r + ja_i$ as shown in Fig. 7.10.

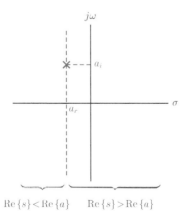

Figure 7.10 – The ROC for the transform $X(s)$ of Example 7.6.

As in Example 7.5 we will look at the shape of the signal $x(t)$ in conjunction with the ROC.

1. Fig. 7.11 shows the signal for the case where a is real-valued. For $a < 0$, the signal grows exponentially as $t \to -\infty$. In contrast, for $a > 0$ the signal decays exponentially.

2. For complex a, real and imaginary parts of $x(t)$ exhibit oscillatory behavior. For $\mathrm{Re}\{a\} < 0$ the signal grows exponentially as shown in Fig. 7.12. In contrast, oscillations are damped exponentially for $\mathrm{Re}\{a\} > 0$. This is shown in Fig. 7.13.

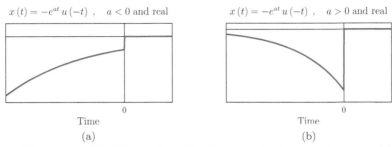

$x(t) = -e^{at}u(-t)$, $a < 0$ and real $x(t) = -e^{at}u(-t)$, $a > 0$ and real

Time   Time

(a)   (b)

Figure 7.11 – The signal $x(t)$ of Example 7.6 for the real-valued parameter a and (a) $a < 0$ and (b) $a > 0$.

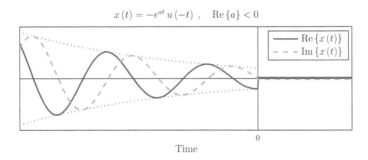

$x(t) = -e^{at}u(-t)$, $\mathrm{Re}\{a\} < 0$

Time

Figure 7.12 – The signal $x(t)$ of Example 7.6 for $\mathrm{Re}\{a\} < 0$.

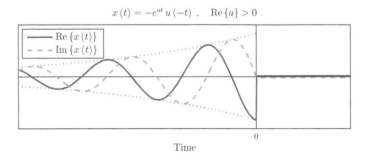

$x(t) = -e^{at}u(-t)$, $\mathrm{Re}\{a\} > 0$

Time

Figure 7.13 – The signal $x(t)$ of Example 7.6 for $\mathrm{Re}\{a\} > 0$.

For the case $\mathrm{Re}\{a\} > 0$ the $j\omega$ axis of the s-plane is part of the ROC. For the case $\mathrm{Re}\{a\} < 0$, however, it is not. Thus, the Fourier transform of the signal $x(t)$ exists only when $\mathrm{Re}\{a\} > 0$.

Software resources:
ex_7_6a.m
ex_7_6b.m

Examples 7.5 and 7.6 demonstrate a fundamental concept for the Laplace transform: It is possible for two different signals to have the same transform expression for $X(s)$. In both of the examples above we have found the transform

$$X(s) = \frac{1}{s-a}$$

What separates the two results apart is the ROC. In order for us to uniquely identify which signal among the two led to a particular transform, the ROC must be specified along with the transform. The following two transform pairs will be fundamental in determining to which signal a given transform corresponds:

$$e^{at}u(t) \overset{\mathcal{L}}{\longleftrightarrow} \frac{1}{s-a}, \qquad \text{ROC: } \mathrm{Re}\{s\} > \mathrm{Re}\{a\} \qquad (7.22)$$

$$-e^{at}u(-t) \overset{\mathcal{L}}{\longleftrightarrow} \frac{1}{s-a}, \qquad \text{ROC: } \mathrm{Re}\{s\} < \mathrm{Re}\{a\} \qquad (7.23)$$

Sometimes we need to solve the inverse problem of finding the signal $x(t)$ with a given Laplace transform $X(s)$. Given a transform

$$X(s) = \frac{1}{s-a}$$

we need to know the ROC to determine if the signal $x(t)$ is the one in Eqn. (7.22) or the one in Eqn. (7.23). This will be very important when we work with inverse Laplace transform in Section 7.4 later in this chapter. In order to avoid ambiguity, we will adopt the convention that the ROC is an integral part of the Laplace transform. It must be specified explicitly, or implied by means of another property of the signal or the system under consideration every time a transform is given.

In each of the Examples 7.3, 7.5 and 7.6 we have obtained the Laplace transform of the specified signal in the form of a rational function of s, that is, a ratio of two polynomials in s. In the general case, a rational transform $X(s)$ is expressed in the form

$$X(s) = K \frac{B(s)}{A(s)} \qquad (7.24)$$

where the numerator $B(s)$ and the denominator $A(s)$ are polynomials of s. Let the numerator polynomial be written in factored form as

$$B(s) = (s - z_1)(s - z_2) \dots (s - z_M) \qquad (7.25)$$

where $z_1, z_2, \dots, z_M$ are its roots. Similarly, let $p_1, p_2, \dots, p_N$ be the roots of the denominator polynomial so that

$$A(s) = (s - p_1)(s - p_2) \dots (s - p_N) \qquad (7.26)$$

The transform can be written as

$$X(s) = K \frac{(s - z_1)(s - z_2) \dots (s - z_M)}{(s - p_1)(s - p_2) \dots (s - p_N)} \qquad (7.27)$$

In Eqns. (7.26) and (7.27) the parameters M and N are the numerator order and the denominator order respectively. The larger of M and N is the order of the transform $X(s)$. The roots of the numerator polynomial are referred to as the *zeros* of $X(s)$. In contrast, the roots of the denominator polynomial are the *poles* of $X(s)$. The transform does not converge at a pole; therefore, the ROC cannot contain any poles. In the next section we will see that the poles of the transform also determine the boundaries of the ROC.

Software resources: See MATLAB Exercise 7.3.

Example 7.7: **Laplace transform of a pulse signal**

Determine the Laplace transform of the pulse signal

$$x(t) = \Pi\left(\frac{t - \tau/2}{\tau}\right)$$

which is shown in Fig. 7.14.

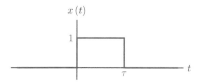

Figure 7.14 – The pulse signal $x(t)$ for Example 7.7.

Solution: The Laplace transform of the signal $x(t)$ is computed as

$$X(s) = \int_0^\tau (1) e^{-st}\, dt = \left.\frac{e^{-st}}{-s}\right|_0^\tau = \frac{1 - e^{-s\tau}}{s}$$

At a first glance we may be tempted to think that the transform $X(s)$ might not converge at $s = 0$ since the denominator of $X(s)$ becomes equal to zero at $s = 0$. We must realize, however, that the numerator of $X(s)$ is also equal to zero at $s = 0$. Using L'Hospital's rule on $X(s)$ we obtain its value at $s = 0$ as

$$\left. X(s) \right|_{s=0} = \left.\frac{\tau e^{-s\tau}}{1}\right|_{s=0} = \tau$$

confirming the fact that $X(s)$ does indeed converge at $s = 0$. As a matter of fact, $X(s)$ converges everywhere on the s-plane with the exception of $s = -\infty \pm j\omega$. In order to observe this, let us write the Taylor series representation of the exponential term $e^{-s\tau}$:

$$e^{-s\tau} = 1 - s\tau + \frac{s^2\tau^2}{2!} - \frac{s^3\tau^3}{3!} + \frac{s^4\tau^4}{4!} - \frac{s^5\tau^5}{5!} + \dots \tag{7.28}$$

The numerator of $X(s)$ can be written as

$$1 - e^{-s\tau} = s\tau - \frac{s^2\tau^2}{2!} + \frac{s^3\tau^3}{3!} - \frac{s^4\tau^4}{4!} + \frac{s^5\tau^5}{5!} - \dots$$

and the transform $X(s)$ can be written in series form

$$X(s) = \frac{1 - e^{-s\tau}}{s} = \tau - \frac{s\tau^2}{2!} + \frac{s^2\tau^3}{3!} - \frac{s^3\tau^4}{4!} + \frac{s^4\tau^5}{5!} - \cdots$$

Thus, the transform $X(s)$ has an infinite number of zeros and no finite poles. Its ROC is

$$\mathrm{Re}\{s\} > -\infty$$

It will be interesting to determine the zeros of the transform. The roots of the numerator are found by solving the equation

$$1 - e^{-s\tau} = 0 , \quad \Rightarrow e^{-s\tau} = 1 \tag{7.29}$$

The value $s = 0$ is a solution for Eqn. (7.29), however, there must be other solutions as well. Realizing that $e^{j2\pi k} = 1$ for all integer values of k, Eqn. (7.29) can be written as

$$e^{-s\tau} = e^{-j2\pi k}$$

which leads to the set of solutions

$$s = \frac{j2\pi k}{\tau} , \quad k \text{ integer}$$

The numerator of $X(s)$ has an infinite number of roots, all on the $j\omega$-axis of the s-plane. The roots are uniformly spaced, and occur at integer multiples of $j2\pi/\tau$. The denominator polynomial of $X(s)$ is just equal to s, and has only one root at $s = 0$. This single root of the denominator cancels the root of the numerator at $s = 0$, so that there is neither a zero nor a pole at the origin of the s-plane. We are left with zeros at

$$z_k = \frac{j2\pi k}{\tau} , \quad k \text{ integer, and } k \neq 0$$

Numerator and denominator roots are shown in Fig. 7.15. The pole-zero diagram for $X(s)$ is shown in Fig. 7.16.

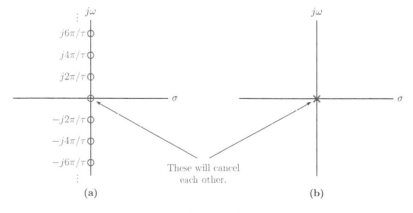

Figure 7.15 – (a) Roots of the numerator $(1 - e^{-s\tau})$, and (b) root of the denominator polynomial s.

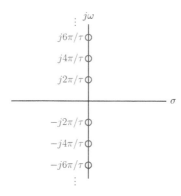

Figure 7.16 – Pole-zero diagram for $X(s)$.

To get a sense of what the transform looks like in the s-plane, we will compute and graph the magnitude $|X(s)|$ of the transform as a three-dimensional surface (recall that the transform $X(s)$ is complex-valued, so we can only graph part of it as a surface, that is, we can graph magnitude, phase, real part or imaginary part). Let $s = \sigma + j\omega$, and write the transform as

$$X(s)\Big|_{s=\sigma+j\omega} = \frac{1 - e^{-\sigma\tau}e^{-j\omega\tau}}{\sigma + j\omega} \tag{7.30}$$

The squared magnitude of a complex function is computed by multiplying it with its own complex conjugate; therefore

$$
\begin{aligned}
|X(s)|^2 &= X(s)\,X^*(s) \\[2mm]
&= \left(\frac{1 - e^{-\sigma\tau}e^{-j\omega\tau}}{\sigma + j\omega}\right)\left(\frac{1 - e^{-\sigma\tau}e^{-j\omega\tau}}{\sigma + j\omega}\right)^* \\[2mm]
&= \left(\frac{1 - e^{-\sigma\tau}e^{-j\omega\tau}}{\sigma + j\omega}\right)\left(\frac{1 - e^{-\sigma\tau}e^{j\omega\tau}}{\sigma - j\omega}\right) \\[2mm]
&= \frac{1 - 2e^{-\sigma\tau}\cos(\omega\tau) + e^{-2\sigma\tau}}{\sigma^2 + \omega^2}
\end{aligned} \tag{7.31}
$$

and the magnitude of the transform is

$$|X(s)| = \left[\frac{1 - 2e^{-\sigma\tau}\cos(\omega\tau) + e^{-2\sigma\tau}}{\sigma^2 + \omega^2}\right]^{1/2}$$

which is shown in Fig. 7.17(a) as a surface graph. Note how the zeros equally spaced on the $j\omega$-axis cause the magnitude surface to dip down. In addition to the surface graph, magnitude values computed at points on the $j\omega$-axis of the s-plane are marked on the surface in Fig. 7.17(a). For comparison, the magnitude of the Fourier transform of the signal $x(t)$ computed for the range of angular frequency values $-10\pi/\tau < \omega < 10\pi/\tau$ is shown in Fig. 7.17(b). It should be compared to the values marked on the surface graph that correspond to the magnitude of the Laplace transform evaluated on the $j\omega$-axis.

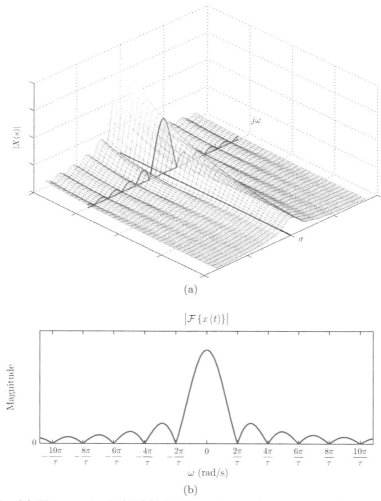

(a)

(b)

Figure 7.17 – (a) The magnitude $|X(s)|$ for Example 7.7 shown as a surface plot along with the $j\omega$-axis of the s-plane and the magnitude of the transform computed on that axis, (b) the magnitude of the Fourier transform of $x(t)$ as a function of ω.

Software resources:
ex_7_7a.m
ex_7_7b.m

Interactive Demo: lap_demo2.m

The demo program "lap_demo2.m" is based on the Laplace transform of the pulse signal analyzed in Example 7.7. The magnitude of the transform in question was shown in Fig. 7.17(a). The demo program computes this magnitude and graphs it as a three-dimensional mesh. It can be rotated freely for viewing from any angle by using the rotation tool in the toolbar. The magnitude of the transform is also evaluated on the vertical line $s = \sigma_1 + j\omega$ in the s-plane, and displayed as a two-dimensional graph. This corresponds to the Fourier transform of the modified signal $x(t)\,e^{-\sigma_1 t}$. The value σ_1 and the pulse width τ may be varied through the use of slider controls.

Software resources:
lap_demo2.m

Example 7.8: Laplace Transform of complex exponential signal

Find the Laplace transform of the signal

$$x\left(t\right) = e^{j\omega_0 t} u\left(t\right)$$

The parameter ω_0 is real and positive, and is in radians.

Solution: Applying the Laplace transform definition directly to the signal $x\left(t\right)$ leads to

$$X\left(s\right) = \int_{-\infty}^{\infty} e^{j\omega_0 t} u\left(t\right) e^{-st} \, dt$$

Since $x\left(t\right)$ is causal, the lower limit of the summation can be set to $t = 0$, and the $u\left(t\right)$ term can be dropped to yield

$$X\left(s\right) = \int_{0}^{\infty} e^{j\omega_0 t} e^{-st} \, dt = \int_{0}^{\infty} e^{(j\omega_0 - s)t} \, dt = \left. \frac{e^{(j\omega_0 - s)t}}{j\omega_0 - s} \right|_{0}^{\infty} \tag{7.32}$$

The upper limit is critical in Eqn. (7.32). To be able to evaluate the expression at $s \to \infty$ we need

$$\mathrm{Re}\left\{j\omega_0 - s\right\} < 0$$

and therefore

$$\mathrm{Re}\left\{s\right\} > 0$$

Provided that this condition is satisfied, the transform in Eqn. (7.32) is computed as

$$X\left(s\right) = \frac{1}{j\omega_0 - s}\left[0 - 1\right] = \frac{1}{s - j\omega_0}, \qquad \mathrm{ROC:}\ \mathrm{Re}\left\{s\right\} > 0$$

The transform $X\left(s\right)$ has one pole at $s = j\omega_0$, and its ROC is the region to the right of the vertical line that passes through its pole. This is illustrated in Fig. 7.18.

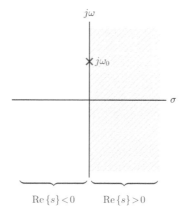

Figure 7.18 – Region of convergence for the transform computed in Example 7.8.

7.2   Characteristics of the Region of Convergence

Through the examples we have worked on so far, we have observed that the Laplace transform of a continuous-time signal always needs to be considered in conjunction with its region of convergence, that is, the region in the s-plane in which the transform converges. In this section we will summarize and justify the fundamental characteristics of the region of convergence.

1. For a finite-duration signal the ROC is the entire s-plane provided that the signal is absolute integrable. The extreme points such as $\mathrm{Re}\,\{s\} \to \pm\infty$ need to be checked separately.

Let the signal $x\,(t)$ be equal to zero outside the interval $t_1 < t < t_2$. The Laplace transform is

$$X\,(s) = \int_{t_1}^{t_2} x\,(t)\,e^{-st}\,dt \tag{7.33}$$

If the signal $x\,(t)$ is absolute integrable, then the transform given by Eqn. (7.33) converges on the vertical line $s = j\omega$. Since the limits of the integral are finite, it also converges for all other values of s.

2. For a general signal the ROC is in the form of a vertical strip. It is either to the right of a vertical line, to the left of a vertical line, or between two vertical lines as shown in Fig. 7.19.

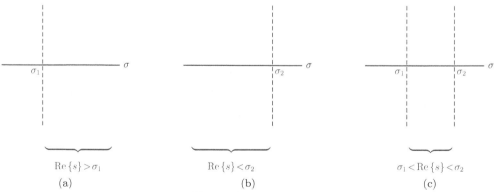

$$\mathrm{Re}\,\{s\} > \sigma_1 \qquad\qquad \mathrm{Re}\,\{s\} < \sigma_2 \qquad\qquad \sigma_1 < \mathrm{Re}\,\{s\} < \sigma_2$$

(a)   (b)   (c)

Figure 7.19 – Shape of the region of convergence.

This property is easy to justify. We have established in Eqn. (7.7) of the previous section that, for $s = \sigma + j\omega$, the values of the Laplace transform of the signal $x\,(t)$ are identical to the values of the Fourier transform of the signal $x\,(t)\,e^{-\sigma t}$. Therefore, the convergence of the Laplace transform for $s = \sigma + j\omega$ is equivalent to the convergence of the Fourier transform of the signal $x\,(t)\,e^{-\sigma t}$, and it requires that

$$\int_{-\infty}^{\infty} \left|x\,(t)\,e^{-\sigma t}\right|\,dt < \infty \tag{7.34}$$

Thus, the ROC depends only on σ and not on ω, explaining why the region is in the form of a vertical strip. Following are the possibilities for the ROC:

$$\operatorname{Re}\{s\} > \sigma_1 \;:\; \text{To the right of a vertical line}$$

$$\operatorname{Re}\{s\} < \sigma_2 \;:\; \text{To the left of a vertical line}$$

$$\sigma_1 < \operatorname{Re}\{s\} < \sigma_2 \;:\; \text{Between two vertical lines}$$

3. For a rational transform $X(s)$ the ROC cannot contain any poles.

By definition, poles of $X(s)$ are values of s that make the transform infinitely large. For rational Laplace transforms, poles are the roots of the denominator polynomial. Since the transform does not converge at a pole, the ROC must naturally exclude all poles of the transform.

4. The ROC for the Laplace transform of a causal signal is the region to the right of a vertical line, and is expressed as

$$\operatorname{Re}\{s\} > \sigma_1$$

Since a causal signal equals zero for all $t < 0$, its Laplace transform can be written as

$$X(s) = \int_0^\infty x(t)\, e^{-st}\, dt$$

Using $s = \sigma + j\omega$ and remembering that the convergence of the Laplace transform is equivalent to the signal $x(t)\, e^{-\sigma t}$ being absolute integrable leads to the convergence condition

$$\int_0^\infty \left| x(t)\, e^{-\sigma t} \right| dt < \infty \tag{7.35}$$

If we can find a value of σ for which Eqn. (7.35) is satisfied, then any larger value of σ will satisfy Eqn. (7.35) as well. All we need to do is find the value $\sigma = \sigma_1$ for the boundary, and the ROC will be the region to the right of the vertical line that passes through the point $s = \sigma_1$.

5. The ROC for the Laplace transform of an anti-causal signal is the region to the left of a vertical line, and is expressed as

$$\operatorname{Re}\{s\} < \sigma_2$$

The justification of this property will be similar to that of the previous one. Since an anti-causal signal is equal to zero for all $t > 0$, its Laplace transform can be written as

$$X(s) = \int_{-\infty}^0 x(t)\, e^{-st}\, dt$$

Using $s = \sigma + j\omega$ the condition for the convergence of the Laplace transform can be expressed through the equivalent condition for the absolute integrability of the signal $x(t)\, e^{-\sigma t}$ as

$$\int_{-\infty}^0 \left| x(t)\, e^{-\sigma t} \right| dt < \infty \tag{7.36}$$

If we can find a value of σ for which Eqn. (7.36) is satisfied, then any smaller value of σ will satisfy Eqn. (7.36) as well. Once a value $\sigma = \sigma_2$ is found for the boundary, the ROC will be the region to the left of the vertical line that passes through the point $s = \sigma_2$.

6. The ROC for the Laplace transform of a signal that is neither causal nor anti-causal is a strip between two vertical lines, and can be expressed as

$$\sigma_1 < \text{Re}\left\{s\right\} < \sigma_2$$

Any signal $x(t)$ can be written as the sum of a causal signal and an anti-causal signal. Let the two signals $x_R(t)$ and $x_L(t)$ be defined in terms of the signal $x(t)$ as

$$x_R(t) = x(t)\, u(t) = \begin{cases} x(t), & t > 0 \\ 0, & t < 0 \end{cases}$$

and

$$x_L(t) = x(t)\, u(-t) = \begin{cases} x(t), & t < 0 \\ 0, & t > 0 \end{cases}$$

so that $x_R(t)$ is causal, $x_L(t)$ is anti-causal, and

$$x(t) = x_R(t) + x_L(t)$$

Let the Laplace transforms of these two signals be

$$X_R(s) = \mathcal{L}\left\{x_R(t)\right\}, \qquad \text{ROC: } \text{Re}\left\{s\right\} > \sigma_1$$
$$X_L(s) = \mathcal{L}\left\{x_L(t)\right\}, \qquad \text{ROC: } \text{Re}\left\{s\right\} < \sigma_2$$

The Laplace transform of the signal $x(t)$ is

$$X(s) = X_R(s) + X_L(s) \tag{7.37}$$

The ROC for $X(s)$ is at least the overlap of the two regions, that is,

$$\sigma_1 < \text{Re}\left\{s\right\} < \sigma_2 \tag{7.38}$$

provided that $\sigma_2 > \sigma_1$ (otherwise there may be no overlap, and the Laplace transform may not exist at any point in the s-plane).

In some cases the ROC may actually be larger than the overlap in Eqn. (7.38) if the addition of the two transforms in Eqn. (7.37) results in the cancellation of a pole that sets one of the boundaries.

Example 7.9: **Laplace transform of two-sided exponential signal**

Find the Laplace transform of the signal

$$x(t) = e^{-|t|}$$

Solution: The specified signal exists for all t, and can be written as

$$x(t) = e^{-t}\, u(t) + e^{t}\, u(-t) = \begin{cases} e^{-t}, & t > 0 \\ e^{t}, & t < 0 \end{cases}$$

Thus, $x(t)$ is the sum of a causal signal $x_R(t)$ and an anti-causal signal $x_L(t)$:

$$x(t) = x_R(t) + x_L(t)$$

Two components of $x(t)$ are

$$x_R(t) = e^{-t}u(t)$$
$$x_L(t) = e^t u(-t)$$

When we discuss the properties of the Laplace transform later in Section 7.3 we will show that the Laplace transform is linear. Consequently, the transform of the sum of two signals is equal to the sum of their respective transforms, and $X(s)$ can be written as

$$\begin{aligned} X(s) &= X_R(s) + X_L(s) \\ &= \mathcal{L}\left\{e^{-t}u(t)\right\} + \mathcal{L}\left\{e^t u(-t)\right\} \end{aligned} \tag{7.39}$$

The individual transforms that make up $X(s)$ in Eqn. (7.39) can be determined by adapting the results obtained earlier in Eqns. (7.22) and (7.23) as:

$$\mathcal{L}\left\{e^{-t}u(t)\right\} = \frac{1}{s+1}, \qquad \text{ROC:} \quad \text{Re}\{s\} > -1 \tag{7.40}$$

$$\mathcal{L}\left\{e^t u(-t)\right\} = -\frac{1}{s-1}, \qquad \text{ROC:} \quad \text{Re}\{s\} < 1 \tag{7.41}$$

The ROC for $X_R(s)$ and $X_L(s)$ are shown in Fig. 7.20.

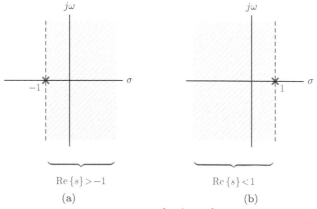

Re$\{s\} > -1$ Re$\{s\} < 1$
(a)   (b)

Figure 7.20 – The ROC for (a) $X_R(s) = \mathcal{L}\left\{e^{-t}u(t)\right\}$, and (b) $X_L(s) = \mathcal{L}\left\{e^t u(-t)\right\}$.

The transform $X(s)$ can be computed as

$$X(s) = \frac{1}{s+1} - \frac{1}{s-1} = \frac{-2}{s^2-1}$$

$X(s)$ has a two poles at $s = \pm 1$. For this transform to converge both $X_R(s)$ and $X_L(s)$ must converge. Therefore, the ROC for $X(s)$ is the overlap of the ROCs of $X_R(s)$ and $X_L(s)$

$$\text{ROC:} \quad -1 < \text{Re}\{s\} < 1$$

and is shown in Fig. 7.21.

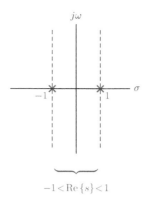

Figure 7.21 – The ROC for the transform $X(s)$.

Software resources:
ex_7_9.m

7.3 Properties of the Laplace Transform

The important properties of the Laplace transform will be summarized in this section, and the proof of each will be given. The use of these properties greatly simplifies the application of the Laplace transform to problems involving the analysis of continuous-time signals and systems.

7.3.1 Linearity

Laplace transform is linear. For any two signals $x_1(t)$ and $x_2(t)$ with their respective transforms

$$x_1(t) \xleftrightarrow{\mathcal{L}} X_1(s)$$

$$x_2(t) \xleftrightarrow{\mathcal{L}} X_2(s)$$

and any two constants α_1 and α_2, it can be shown that the following relationship holds:

Linearity of the Laplace transform:

$$\alpha_1 x_1(t) + \alpha_2 x_2(t) \xleftrightarrow{\mathcal{L}} \alpha_1 X_1(s) + \alpha_2 X_2(s) \tag{7.42}$$

Proof: The proof is straightforward using the Laplace transform definition given by Eqn. (7.1). Substituting $x(t) = \alpha_1 x_1(t) + \alpha_2 x_2(t)$ into Eqn. (7.1) we get

$$\mathcal{L}\{\alpha_1 x_1(t) + \alpha_2 x_2(t)\} = \int_{-\infty}^{\infty} \left[\alpha_1 x_1(t) + \alpha_2 x_2(t)\right] e^{-st}\, dt \tag{7.43}$$

The integral in Eqn. (7.43) can be separated into two integrals as

$$\mathcal{L}\left\{\alpha_1\, x_1\left(t\right)+\alpha_2\, x_2\left(t\right)\right\} = \int_{-\infty}^{\infty} \alpha_1\, x_1\left(t\right)\, e^{-st}\, dt + \int_{-\infty}^{\infty} \alpha_2\, x_2\left(t\right)\, e^{-st}\, dt$$

$$=\alpha_1 \int_{-\infty}^{\infty} x_1\left(t\right)\, e^{-st}\, dt + \alpha_2 \int_{-\infty}^{\infty} x_2\left(t\right)\, e^{-st}\, dt$$

$$=\alpha_1\, \mathcal{L}\left\{x_1\left(t\right)\right\} + \alpha_2\, \mathcal{L}\left\{x_2\left(t\right)\right\} \qquad (7.44)$$

The Laplace transform of a weighted sum of two signals is equal to the same weighted sum of their respective transforms $X_1\left(s\right)$ and $X_2\left(s\right)$. The ROC for the resulting transform is at least the overlap of the two individual transforms, if such an overlap exists. The ROC may be greater than the overlap of the two regions if the addition of the two transforms results in the cancellation of a pole.

The linearity property proven above for two signals can be generalized to any arbitrary number of signals. The Laplace transform of a weighted sum of any number of signals is equal to the same weighted sum of their respective transforms.

Example 7.10: **Using the linearity property of the Laplace transform**

Determine the Laplace transform of the signal

$$x\left(t\right) = 2e^{-t}u\left(t\right) + 5e^{-2t}u\left(t\right)$$

Solution: A general expression for the Laplace transform of the causal exponential signal was found in Example 7.5 as

$$\mathcal{L}\left\{e^{at}\, u\left(t\right)\right\} = \frac{1}{s-a}, \qquad \text{ROC:} \quad \text{Re}\left\{s\right\} > \text{Re}\left\{a\right\}$$

Applying this result to the exponential terms in $x\left(t\right)$ we get

$$\mathcal{L}\left\{e^{-t}\, u\left(t\right)\right\} = \frac{1}{s+1}, \qquad \text{ROC:} \quad \text{Re}\left\{s\right\} > -1 \qquad (7.45)$$

and

$$\mathcal{L}\left\{e^{-2t}\, u\left(t\right)\right\} = \frac{1}{s+2}, \qquad \text{ROC:} \quad \text{Re}\left\{s\right\} > -2 \qquad (7.46)$$

Combining the results in Eqns. (7.45) and (7.46) and using the linearity property we arrive at the desired result:

$$X\left(s\right) = 2\left(\frac{1}{s+1}\right) + 5\left(\frac{1}{s+2}\right) = \frac{7s+9}{s^2+3s+2}$$

The ROC for $X\left(s\right)$ is the overlap of the two regions in Eqns. (7.45) and (7.46), namely

$$\text{Re}\left\{s\right\} > -1$$

Fig. 7.22 illustrates the poles and the ROCs of the individual terms $X_1\left(s\right)$ and $X_2\left(s\right)$ as well as the ROC of the transform $X\left(s\right)$ obtained as the overlap of the two.

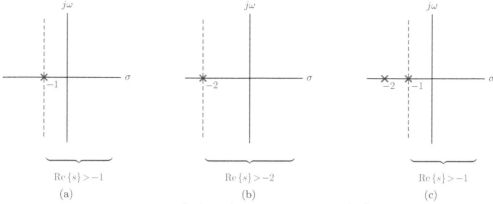

Figure 7.22 – (a) The ROC for $\mathcal{L}\left\{e^{-t}\,u\,(t)\right\}$, (b) the ROC for $\mathcal{L}\left\{e^{-2t}\,u\,(t)\right\}$, (c) the ROC for $X\,(s)$.

Example 7.11: **Laplace transform of a cosine signal**

Find the Laplace transform of the signal

$$x\,(t) = \cos\,(\omega_0 t)\,u\,(t)$$

Solution: Using Euler's formula for the $\cos\,(\omega_0 t)$ term, the signal $x\,(t)$ can be written as

$$\cos\,(\omega_0 t)\,u\,(t) = \frac{1}{2}\,e^{j\omega_0 t}\,u\,(t) + \frac{1}{2}\,e^{-j\omega_0 t}\,u\,(t)$$

and its Laplace transform is

$$\mathcal{L}\left\{\cos\,(\omega_0 t)\,u\,(t)\right\} = \mathcal{L}\left\{\frac{1}{2}\,e^{j\omega_0 t}\,u\,(t) + \frac{1}{2}\,e^{-j\omega_0 t}\,u\,(t)\right\} \tag{7.47}$$

Using the linearity property of the Laplace transform Eqn. (7.47) can be written as

$$\mathcal{L}\left\{\cos\,(\omega_0 t)\,u\,(t)\right\} = \frac{1}{2}\,\mathcal{L}\left\{e^{j\omega_0 t}\,u\,(t)\right\} + \frac{1}{2}\,\mathcal{L}\left\{e^{-j\omega_0 t}\,u\,(t)\right\}$$

$$= \frac{1/2}{s - j\omega_0} + \frac{1/2}{s + j\omega_0} \tag{7.48}$$

Combining the terms of Eqn. (7.48) under a common denominator we have

$$X\,(s) = \frac{s}{s^2 + \omega_0^2} \tag{7.49}$$

The transform $X\,(s)$ has a zero at $s = 0$ and a pair of complex conjugate poles at $s = \pm j\omega_0$. The ROC is the region to the right of the $j\omega$-axis of the s-plane, as illustrated in Fig. 7.23. The $j\omega$-axis itself is not included in the ROC since there are poles on it.

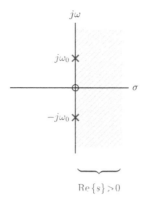

Figure 7.23 – The ROC for the transform in Eqn. (7.49).

Example 7.12: **Laplace transform of a sine signal**

Find the Laplace transform of the signal

$$x\left(t\right) = \sin\left(\omega_0 t\right) u\left(t\right)$$

Solution: This problem is quite similar to that in Example 7.11. Using Euler's formula for the signal $x\left(t\right)$ we get

$$\sin\left(\omega_0 t\right) u\left(t\right) = \frac{1}{2j} e^{j\omega_0 t} u\left(t\right) - \frac{1}{2j} e^{-j\omega_0 t} u\left(t\right)$$

and its Laplace transform is

$$\mathcal{L}\left\{\sin\left(\omega_0 t\right) u\left(t\right)\right\} = \mathcal{L}\left\{\frac{1}{2j} e^{j\omega_0 t} u\left(t\right) - \frac{1}{2j} e^{-j\omega_0 t} u\left(t\right)\right\}$$

$$= \frac{1}{2j} \mathcal{L}\left\{e^{j\omega_0 t} u\left(t\right)\right\} - \frac{1}{2j} \mathcal{L}\left\{e^{-j\omega_0 t} u\left(t\right)\right\}$$

$$= \frac{1}{2j} \left(\frac{1}{s - j\omega_0}\right) - \frac{1}{2j} \left(\frac{1}{s + j\omega_0}\right) \tag{7.50}$$

Combining the terms of Eqn. (7.50) under a common denominator we have

$$X\left(s\right) = \frac{\omega_0}{s^2 + \omega_0^2} \tag{7.51}$$

The transform $X\left(s\right)$ has a pair of complex conjugate poles at $s = \pm j\omega_0$. The ROC of the transform is $\mathrm{Re}\left\{s\right\} > 0$ as in the previous example.

7.3.2   Time shifting

Given the transform pair

$$x\left(t\right) \overset{\mathcal{L}}{\longleftrightarrow} X\left(s\right)$$

the following is also a valid transform pair:

$$x\left(t - \tau\right) \overset{\mathcal{L}}{\longleftrightarrow} e^{-s\tau} X\left(s\right) \tag{7.52}$$

Time shifting the signal $x(t)$ in the time domain by τ corresponds to multiplication of the transform $X(s)$ by $e^{-s\tau}$.

Proof: The Laplace transform of $x(t-\tau)$ is

$$\mathcal{L}\{x(t-\tau)\} = \int_{-\infty}^{\infty} x(t-\tau)\, e^{-st}\, dt \tag{7.53}$$

Let us define a new variable $\lambda = t - \tau$, and write the integral in Eqn. (7.52) in terms of this new variable:

$$\mathcal{L}\{x(t-\tau)\} = \int_{-\infty}^{\infty} x(\lambda)\, e^{-s(\lambda+\tau)}\, d\lambda$$

$$= e^{-s\tau} \int_{-\infty}^{\infty} x(\lambda)\, e^{-s\lambda}\, d\lambda$$

$$= e^{-s\tau} X(s) \tag{7.54}$$

The ROC for the resulting transform $e^{-s\tau} X(s)$ is the generally same as that of $X(s)$.[1]

Example 7.13: **Laplace transform of a pulse signal revisited**

The Laplace transform of the pulse signal

$$x(t) = \Pi\left(\frac{t-\tau/2}{\tau}\right)$$

was determined earlier in Example 7.7. Find the same transform through the use of linearity and time-shifting properties.

Solution: The signal $x(t)$ can be expressed as the difference of a unit-step signal and a time-shifted unit-step signal in the form

$$x(t) = u(t) - u(t-\tau)$$

Using the linearity of the Laplace transform, $X(s)$ can be expressed as

$$X(s) = \mathcal{L}\{u(t) - u(t-\tau)\}$$

$$= \mathcal{L}\{u(t)\} - \mathcal{L}\{u(t-\tau)\}$$

The Laplace transform of the unit-step function was found in Example 7.3 to be

$$\mathcal{L}\{u(t)\} = \frac{1}{s}, \qquad \text{ROC:}\quad \text{Re}\{s\} > 0 \tag{7.55}$$

Using the time-shifting property of the Laplace transform, we find the transform of the time-shifted unit-step signal as

$$\mathcal{L}\{u(t-\tau)\} = \frac{e^{-s\tau}}{s}, \qquad \text{ROC:}\quad \text{Re}\{s\} > 0 \tag{7.56}$$

---

[1] Care must be taken in certain circumstances. If the time shift makes a causal signal non-causal then the points $s = \infty + j\omega$ would need to be excluded from the ROC. Similarly, if an anti-causal signal loses its anti-causal property as the result of a shift, then the points $s = -\infty + j\omega$ need to be excluded.

Subtracting the transform in Eqn. (7.56) from the transform in Eqn. (7.55) yields

$$X(s) = \frac{1 - e^{-s\tau}}{s}$$

which matches the earlier result found in Example 7.7. Since $x(t)$ is a finite-duration signal, the ROC is the entire s-plane with the exception of $s = -\infty$. This is one example of the possibility mentioned earlier regarding the boundaries. The ROC for $X(s)$ is larger than the overlap of the two regions given by Eqns. (7.55) and (7.56). The individual transforms $\mathcal{L}\{u(t)\}$ and $\mathcal{L}\{u(t-\tau)\}$ each have a pole at $s = 0$, and therefore the individual ROCs are to the right of the $j\omega$-axis. When the two terms are added to construct $X(s)$, however, the pole at $s = 0$ is canceled, resulting in the ROC for $X(s)$ being larger than the overlap of the individual regions.

Example 7.14: Laplace transform of a truncated sine function

Determine the Laplace transform of the signal

$$x(t) = \begin{cases} \sin(\pi t), & 0 < t < 1 \\ 0, & \text{otherwise} \end{cases}$$

which is graphed in Fig. 7.24.

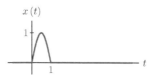

Figure 7.24 – The signal $x(t)$ for Example 7.14.

Solution: We will solve this problem first by direct application of the Laplace transform definition, and then through the use of linearity and shifting properties of the Laplace transform. Applying the definition of the Laplace transform we get

$$X(s) = \int_0^1 \sin(\pi t) e^{-st} \, dt \qquad (7.57)$$

Writing $\sin(\pi t)$ using Euler's formula leads to

$$\sin(\pi t) = \frac{1}{2j} e^{j\pi t} - \frac{1}{2j} e^{-j\pi t}$$

and substituting this result into Eqn. (7.57) yields

$$X(s) = \frac{1}{2j} \int_0^1 e^{j\pi t} e^{-st} \, dt - \frac{1}{2j} \int_0^1 e^{-j\pi t} e^{-st} \, dt$$

which can be evaluated as

$$X(s) = \frac{1}{2j} \left. \frac{e^{(j\pi - s)t}}{(j\pi - s)} \right|_0^1 + \frac{1}{2j} \left. \frac{e^{-(j\pi + s)t}}{(j\pi + s)} \right|_0^1 = \frac{\pi(1 + e^{-s})}{s^2 + \pi^2}$$

In Eqn. (7.58) we have used the fact that $e^{\pm j\pi} = 1$. Since the signal $x(t)$ is of finite duration, the ROC of the transform is the entire s-plane with just the exception of points where $\mathrm{Re}\{s\} \to -\infty$.

The solution can be simplified by clever use of the properties of the Laplace transform. The first step is to recognize that the signal $x(t)$ can be expressed as the sum of a causal sinusoidal signal and its time-shifted version as shown in Fig. 7.25.

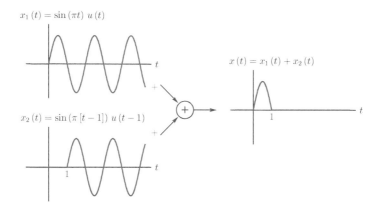

Figure 7.25 – Expressing a truncated sinusoidal signal as the sum of two signals.

Mathematically we have

$$x(t) = \sin(\pi t)\, u(t) + \sin(\pi[t-1])\, u(t-1) \tag{7.58}$$

Recall that the Laplace transform of a causal sinusoidal signal was found in Example 7.12 as

$$\mathcal{L}\{\sin(\omega_0 t)\, u(t)\} = \frac{\omega_0}{s^2 + \omega_0^2}, \qquad \text{ROC:} \quad \mathrm{Re}\{s\} > 0$$

Using this result with $\omega_0 = \pi$ along with the linearity of the Laplace transform and the time-shifting property, we obtain the transform as

$$X(s) = \mathcal{L}\{\sin(\pi t)\, u(t)\} + \mathcal{L}\{\sin(\pi[t-1])\, u(t-1)\}$$

$$= \frac{\pi}{s^2 + \pi^2} + e^{-s}\,\frac{\pi}{s^2 + \pi^2}$$

$$= \frac{\pi(1 + e^{-s})}{s^2 + \pi^2} \tag{7.59}$$

which matches the result found earlier.
Software resources:
ex_7_14.m

7.3.3 Shifting in the s-domain

Given the transform pair

$$x(t) \xleftrightarrow{\ \mathcal{L}\ } X(s)$$

the following is also a valid transform pair:

$$x(t)\, e^{s_0 t} \overset{\mathcal{L}}{\longleftrightarrow} X(s - s_0) \tag{7.60}$$

Proof: Through direct application of the Laplace transform definition given by Eqn. (7.1) we get

$$\mathcal{L}\left\{x(t)\, e^{s_0 t}\right\} = \int_{-\infty}^{\infty} x(t)\, e^{s_0 t}\, e^{-st}\, dt$$

$$= \int_{-\infty}^{\infty} x(t)\, e^{-(s-s_0)t}\, dt$$

$$= X(s - s_0) \tag{7.61}$$

Let the ROC for original transform $X(s)$ be

$$\sigma_1 < \text{Re}\{s\} < \sigma_2$$

For $X(s - s_0)$ to converge, we need

$$\sigma_1 < \text{Re}\{s - s_0\} < \sigma_2$$

Therefore, the ROC for $X(s - s_0)$ is

$$\sigma_1 + \text{Re}\{s_0\} < \text{Re}\{s\} < \sigma_2 + \text{Re}\{s_0\}$$

The ROC for $X(s - s_0)$ is a shifted version of the ROC for $X(s)$, shifted horizontally by an amount equal to the real part of the parameter s_0. This is illustrated in Fig. 7.26.

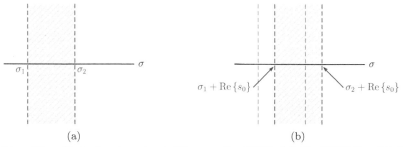

Figure 7.26 – The effect of s-domain shifting on the ROC: (a) the ROC for $X(s)$, (b) the ROC for $X(s - s_0)$.

Example 7.15: Laplace transform of exponentially damped sinusoidal signal

Determine the Laplace transform of the signal

$$x(t) = e^{-2t} \cos(3t)\, u(t)$$

Solution: The transform can be found easily by the use of the s-domain shifting property. Let the signal $x_1(t)$ be defined as

$$x_1(t) = \cos(3t)\,u(t)$$

so that

$$x(t) = e^{-2t}\,x_1(t)$$

The Laplace transform of a causal cosine signal was derived in Example 7.11. Using the result found in that example with $\omega_0 = 3$ yields

$$X_1(s) = \mathcal{L}\{\cos(3t)\,u(t)\} = \frac{s}{s^2+9}\,, \qquad \text{ROC:} \quad \text{Re}\{s\} > 0$$

We are now ready to apply the s-domain shifting property.

$$X(s) = X_1(s+2) = \left.\frac{s}{s^2+9}\right|_{s \to s+2} = \frac{s+2}{(s+2)^2+9}$$

The resulting transform has a zero at $s = -2$ and a pair of complex conjugate poles at $s = -2 \pm j3$ as shown in Fig. 7.27. The boundary for the ROC is set by the two poles, therefore the ROC is

$$\text{ROC:} \quad \text{Re}\{s\} > -2$$

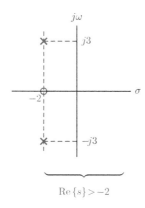

Figure 7.27 – The pole-zero diagram and the ROC for the transform in Example 7.15.

7.3.4 Scaling in time and s-domains

Given the transform pair

$$x(t) \overset{\mathcal{L}}{\longleftrightarrow} X(s)$$

and a real-valued parameter a, the following is also a valid transform pair:

$$x(at) \overset{\mathcal{L}}{\longleftrightarrow} \frac{1}{|a|} X\left(\frac{s}{a}\right) \tag{7.62}$$

Proof: Once again, we will begin by applying the Laplace transform definition given by Eqn. (7.1) to obtain

$$\mathcal{L}\left\{x\left(at\right)\right\} = \int_{-\infty}^{\infty} x\left(at\right) e^{-st}\, dt \tag{7.63}$$

If $a > 0$, employing a variable change $\lambda = at$ and its consequence $d\lambda = a\, dt$ on the integral of Eqn. (7.63) leads to

$$\mathcal{L}\left\{x\left(at\right)\right\} = \int_{-\infty}^{\infty} x\left(\lambda\right) e^{-s\lambda/a}\, \frac{d\lambda}{a}$$

$$= \frac{1}{a} \int_{-\infty}^{\infty} x\left(\lambda\right) e^{-(s/a)\lambda}\, d\lambda$$

$$= \frac{1}{a} X\left(\frac{s}{a}\right) \tag{7.64}$$

On the other hand, if $a < 0$, the same variable change leads to

$$\mathcal{L}\left\{x\left(at\right)\right\} = \int_{\infty}^{-\infty} x\left(\lambda\right) e^{-s\lambda/a}\, \frac{d\lambda}{a}$$

$$= -\frac{1}{a} \int_{-\infty}^{\infty} x\left(\lambda\right) e^{-(s/a)\lambda}\, d\lambda$$

$$= -\frac{1}{a} X\left(\frac{s}{a}\right) \tag{7.65}$$

The reason for the negative sign in Eqn. (7.65) is that, for $a < 0$, the integration limits for $t \to \pm\infty$ translate to limits $\lambda \to \mp\infty$. In order to account for both possibilities of $a > 0$ and $a < 0$, the results in Eqns. (7.64) and (7.65) may be combined to yield

$$\mathcal{L}\left\{x\left(at\right)\right\} = \frac{1}{|a|} \int_{-\infty}^{\infty} x\left(\lambda\right) e^{-(s/a)\lambda}\, d\lambda$$

$$= \frac{1}{|a|} X\left(\frac{s}{a}\right) \tag{7.66}$$

The use of the absolute value on the scale factor eliminates the need for the negative sign when $a < 0$. This completes the proof of the scaling property given by Eqn. (7.62).

The ROC of the result still needs to be related to the ROC for the original transform $X\left(s\right)$. Let the latter be

$$\sigma_1 < \text{Re}\left\{s\right\} < \sigma_2$$

For the term $X\left(s/a\right)$ in Eqn. (7.66) to converge, we need

$$\sigma_1 < \text{Re}\left\{\frac{s}{a}\right\} < \sigma_2 \tag{7.67}$$

Since the parameter a is real-valued, the ROC in Eqn. (7.67) can be written as

$$\sigma_1 < \frac{\text{Re}\left\{s\right\}}{a} < \sigma_2 \tag{7.68}$$

Depending on the sign of the parameter a, two possibilities need to be considered for the ROC of $\mathcal{L}\{x(at)\}$:

a.   If $a > 0$: $a\,\sigma_1 < \mathrm{Re}\{s\} < a\,\sigma_2$
b.   If $a < 0$: $a\,\sigma_2 < \mathrm{Re}\{s\} < a\,\sigma_1$

Example 7.16: Using the scaling property of the Laplace transform

Use the scaling property to find the Laplace transform of the signal

$$x(t) = e^{2t}\,u(-t)$$

from the knowledge of the transform pair

$$g(t) = e^{-2t}\,u(t) \qquad \Longrightarrow \qquad G(s) = \frac{1}{s+2}, \qquad \text{ROC:} \quad \mathrm{Re}\{s\} > -2$$

Solution: It is obvious from a comparison of the signals $x(t)$ and $g(t)$ that

$$x(t) = g(-t)$$

so that the scaling property given by Eqn. (7.62) can be used with the scale factor $a = -1$ to yield

$$X(s) = \frac{1}{|-1|}G\left(\frac{s}{-1}\right) = G(-s) = \frac{1}{-s+2}$$

and the ROC is

$$\mathrm{Re}\{-s\} > -2 \quad \Rightarrow \quad \mathrm{Re}\{s\} < 2$$

The result found is consistent with Eqn. (7.23).

7.3.5   Differentiation in the time domain

Given the transform pair

$$x(t) \overset{\mathcal{L}}{\longleftrightarrow} X(s)$$

the following is also a valid transform pair:

$$\frac{dx(t)}{dt} \overset{\mathcal{L}}{\longleftrightarrow} s\,X(s) \tag{7.69}$$

Proof: Using the Laplace transform definition in Eqn. (7.1), the transform of $dx(t)/dt$ is

$$\mathcal{L}\left\{\frac{dx(t)}{dt}\right\} = \int_{-\infty}^{\infty} \frac{dx(t)}{dt} e^{-st}\,dt$$

$$= \int_{-\infty}^{\infty} e^{-st}\,dx(t) \tag{7.70}$$

Integrating Eqn. (7.70) by parts yields

$$\mathcal{L}\left\{\frac{dx\,(t)}{dt}\right\} = x\,(t)\,e^{-st}\Big|_{-\infty}^{\infty} + s\int_{-\infty}^{\infty} x\,(t)\,e^{-st}\,dt \tag{7.71}$$

The term $x\,(t)\,e^{-st}$ must evaluate to zero for $t \to \pm\infty$ for the transform $X\,(s)$ to exist. Therefore, Eqn. (7.71) reduces to

$$\mathcal{L}\left\{\frac{dx\,(t)}{dt}\right\} = s\int_{-\infty}^{\infty} x\,(t)\,e^{-st}\,dt = s\,X\,(s) \tag{7.72}$$

completing the proof.

The ROC for the transform of $dx\,(t)\,/dt$ is at least equal to the ROC of the original transform. If the original transform $X\,(s)$ has a single pole at $s = 0$ that sets the boundary of its ROC, then the cancellation of that pole due to multiplication by s causes the ROC of the new transform $s\,X\,(s)$ to be larger.

Example 7.17: **Using the time-domain differentiation property of the Laplace transform**

Use the time-domain differentiation property to find the Laplace transform of the signal

$$x\,(t) = \cos\,(3t)\,u\,(t)$$

from the knowledge of the transform pair

$$g\,(t) = \sin\,(3t)\,u\,(t) \qquad \Longrightarrow \qquad G\,(s) = \frac{3}{s^2 + 9}, \qquad \text{ROC:} \quad \text{Re}\,\{s\} > 0$$

Solution: Let us differentiate the signal $g\,(t)$:

$$\frac{dg\,(t)}{dt} = 3\,\cos\,(3t)\,u\,(t) \tag{7.73}$$

Based on the result in Eqn. (7.73), the signal $x\,(t)$ can be written as

$$x\,(t) = \left(\frac{1}{3}\right)\frac{dg\,(t)}{dt}$$

and its Laplace transform is

$$X\,(s) = \left(\frac{1}{3}\right)s\,G\,(s) = \frac{s}{s^2 + 9}$$

as expected. The ROC of $X\,(s)$ is

$$\text{Re}\,\{s\} > 0$$

7.3.6 Differentiation in the s-domain

Given the transform pair

$$x\,(t) \xleftrightarrow{\;\mathcal{L}\;} X\,(s)$$

the following is also a valid transform pair:

$$t\,x\,(t) \xleftrightarrow{\;\mathcal{L}\;} -\frac{dX\,(s)}{ds} \tag{7.74}$$

Proof: The proof is straightforward by differentiating both sides of the Laplace transform definition given by Eqn. (7.1):

$$\frac{dX(s)}{ds} = \frac{d}{ds}\left[\int_{-\infty}^{\infty} x(t)\,e^{-st}\,dt\right] \tag{7.75}$$

Interchanging the order of integration and differentiation on the right side of Eqn. (7.75) leads to

$$\frac{dX(s)}{ds} = \int_{-\infty}^{\infty} \frac{d}{ds}\left[x(t)\,e^{-st}\right]\,dt$$

$$= -\int_{-\infty}^{\infty} t\,x(t)\,e^{-st}\,dt$$

$$= -\mathcal{L}\{t\,x(t)\} \tag{7.76}$$

Eqn. (7.74) follows from Eqn. ((7.76).

The ROC for the transform of $t\,x(t)$ is the same as the ROC for the original transform $X(s)$.

Example 7.18: **Using the s-domain differentiation property of the Laplace transform**

Determine the Laplace transform of the unit-ramp signal

$$r(t) = t\,u(t)$$

Solution: The Laplace transform of the unit-step function is

$$\mathcal{L}\{u(t)\} = \frac{1}{s}, \qquad \text{ROC:} \quad \text{Re}\{s\} > 0$$

Using the s-domain differentiation property, the Laplace transform of the unit-ramp signal is

$$R(s) = \mathcal{L}\{t\,u(t)\} = -\frac{d}{ds}\left[\frac{1}{s}\right] = \frac{1}{s^2} \tag{7.77}$$

and the ROC of the transform is

$$\text{Re}\{s\} > 0$$

Example 7.19: **Transform of a signal using multiple ramp functions**

Determine the Laplace transform of the signal $x(t)$ shown in Fig. 7.28.

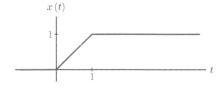

Figure 7.28 – The signal $x(t)$ for Example 7.19.

Solution: The signal $x(t)$ can be expressed in terms of a unit-ramp function and its delayed version as

$$x(t) = r(t) - r(t-1)$$
$$= t\, u(t) - (t-1)\, u(t-1)$$

The transform of the unit-ramp function was determined in Example 7.18. Using the result in Eqn. (7.77) along with the time-shifting property we obtain

$$X(s) = \frac{1}{s^2} - e^{-s}\frac{1}{s^2} = \frac{1 - e^{-s}}{s^2}$$

with the ROC

$$\operatorname{Re}\{s\} > 0$$

Example 7.20: **Transform of an exponentially damped ramp function**

Determine the Laplace transform of the signal

$$x(t) = t\, e^{-2t}\, u(t)$$

which is shown in Fig. 7.29.

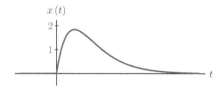

Figure 7.29 – The signal $x(t)$ for Example 7.20.

Solution: The transform of the causal exponential signal $e^{-2t}u(t)$ is

$$\mathcal{L}\left\{e^{-2t}\, u(t)\right\} = \frac{1}{s+2}, \qquad \text{ROC:} \quad \operatorname{Re}\{s\} > -2$$

Using the s-domain differentiation property

$$X(s) = \mathcal{L}\left\{t\, e^{-2t}\, u(t)\right\} = -\frac{d}{ds}\left[\frac{1}{s+2}\right] = \frac{1}{(s+2)^2}$$

with the ROC

$$\operatorname{Re}\{s\} > -2$$

7.3.7   Convolution property

For any two signals $x_1(t)$ and $x_2(t)$ with their respective transforms

$$x_1(t) \; \xleftrightarrow{\mathcal{L}} \; X_1(s)$$

and

$$x_2(t) \; \xleftrightarrow{\mathcal{L}} \; X_2(s)$$

it can be shown that the following transform relationship holds:

Convolution property of the Laplace transform:

$$x_1(t) * x_2(t) \xleftrightarrow{\mathcal{L}} X_1(s)\, X_2(s) \tag{7.78}$$

Convolution property of the Laplace transform is fundamental in its application to CTLTI systems.

Proof: The proof will be carried out by using the convolution result inside the Laplace transform definition. The convolution of two continuous-time signals $x_1(t)$ and $x_2(t)$ is given by

$$x_1(t) * x_2(t) = \int_{-\infty}^{\infty} x_1(\lambda)\, x_2(t - \lambda)\, d\lambda \tag{7.79}$$

Substituting Eqn. (7.79) into the Laplace transform definition leads to

$$\mathcal{L}\{x_1(t) * x_2(t)\} = \mathcal{L}\left\{ \int_{-\infty}^{\infty} x_1(\lambda)\, x_2(t - \lambda)\, d\lambda \right\}$$

$$= \int_{-\infty}^{\infty} \left[\int_{-\infty}^{\infty} x_1(\lambda)\, x_2(t - \lambda)\, d\lambda \right] e^{-st}\, dt \tag{7.80}$$

Interchanging the order of the two integrals, Eqn. (7.80) can be written as

$$\mathcal{L}\{x_1(t) * x_2(t)\} = \int_{-\infty}^{\infty} \left[\int_{-\infty}^{\infty} x_1(\lambda)\, x_2(t - \lambda)\, e^{-st}\, dt \right] d\lambda$$

$$= \int_{-\infty}^{\infty} x_1(\lambda) \left[\int_{-\infty}^{\infty} x_2(t - \lambda)\, e^{-st}\, dt \right] d\lambda \tag{7.81}$$

We will focus our attention on the inner integral in Eqn. (7.81). Using the time-shifting property it follows that

$$\int_{-\infty}^{\infty} x_2(t - \lambda)\, e^{-st}\, dt = \mathcal{L}\{x_2(t - \lambda)\} = e^{-s\lambda} X_2(s) \tag{7.82}$$

Substituting Eqn. (7.82) into Eqn. (7.81) we obtain

$$\mathcal{L}\{x_1(t) * x_2(t)\} = \int_{-\infty}^{\infty} x_1(\lambda)\, e^{-s\lambda} X_2(s)\, d\lambda$$

$$= X_2(s) \int_{-\infty}^{\infty} x_1(\lambda)\, e^{-s\lambda}\, d\lambda$$

$$= X_2(s)\, X_1(s) \tag{7.83}$$

As before, the ROC for the resulting transform is at least the overlap of the ROCs of two individual transforms, if such an overlap exists. It may be greater than the overlap of the two ROCs if the multiplication of $X_1(s)$ and $X_2(s)$ results in the cancellation of a pole that determines the boundary of one of the ROCs.

Convolution property of the Laplace transform is very useful in the sense that it provides an alternative to computing the convolution of two signals directly in the time-domain. Instead, the convolution result $x(t) = x_1(t) * x_2(t)$ can be obtained using the procedure outlined below:

Finding convolution result through the Laplace transform:
To compute $x(t) = x_1(t) * x_2(t)$:

1. Find the Laplace transforms of the two signals.

$$X_1(s) = \mathcal{L}\{x_1(t)\}$$
$$X_2(s) = \mathcal{L}\{x_2(t)\}$$

2. Multiply the two transforms to obtain $X(s)$.

$$X(s) = X_1(s)\, X_2(s)$$

3. Compute $x(t)$ as the inverse Laplace transform of $X(s)$.

$$x(t) = \mathcal{L}^{-1}\{X(s)\}$$

Example 7.21: Using the convolution property of the Laplace transform

Consider two signals $x_1(t)$ and $x_2(t)$ given by

$$x_1(t) = e^{-t}u(t)$$

and

$$x_2(t) = \delta(t) - e^{-2t}u(t)$$

Determine

$$x(t) = x_1(t) * x_2(t)$$

using Laplace transform techniques.

Solution: Individual transforms of the signals $x_1(t)$ and $x_2(t)$ are

$$X_1(s) = \frac{1}{s+1}, \qquad \text{ROC:} \ \ \text{Re}\{s\} > -1 \qquad\qquad (7.84)$$

and

$$X_2(s) = 1 - \frac{1}{s+2} = \frac{s+1}{s+2}, \qquad \text{ROC:} \ \ \text{Re}\{s\} > -2 \qquad\qquad (7.85)$$

Applying the convolution property, the transform of $x(t)$ is the product of the two transforms:

$$X(s) = X_1(s)\, X_2(s) = \left(\frac{1}{s+1}\right)\left(\frac{s+1}{s+2}\right) = \frac{1}{s+2}$$

The overlap of the two ROCs in Eqns. (7.84) and (7.85) would be $\text{Re}\{s\} > -1$; however, the pole at $s = -1$ is cancelled in the process of multiplying the two transforms. Consequently, the ROC for $X(s)$ is

$$\text{Re}\{s\} > -2$$

as shown in Fig. 7.30.

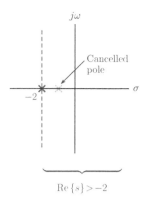

Figure 7.30 – The pole-zero diagram and the ROC for the transform in Example 7.21.

The signal $x(t)$ is the inverse Laplace transform of $X(s)$:

$$x(t) = \mathcal{L}^{-1}\left\{\frac{1}{s+2}\right\} = e^{-2t}\,u(t)$$

7.3.8 Integration property

Given the transform pair

$$x(t) \overset{\mathcal{L}}{\longleftrightarrow} X(s)$$

the following is also a valid transform pair:

$$\int_{-\infty}^{t} x(\lambda)\,d\lambda \overset{\mathcal{L}}{\longleftrightarrow} \frac{1}{s}X(s) \qquad (7.86)$$

Proof: Let a new signal be defined from $x(t)$ as

$$w(t) = \int_{-\infty}^{t} x(t)\,d\lambda \qquad (7.87)$$

Using the definition of the Laplace transform in conjunction with the signal $w(t)$ we get

$$W(s) = \int_{-\infty}^{\infty} w(t)\,e^{-st}\,dt$$

$$= \int_{-\infty}^{\infty}\left[\int_{-\infty}^{t} x(\lambda)\,d\lambda\right]e^{-st}\,dt$$

which would be difficult to evaluate directly. Instead, we will make use of other properties of the Laplace transform discussed earlier. Differentiating both sides of Eqn. (7.87) we get

$$\frac{dw(t)}{dt} = x(t) \qquad (7.88)$$

Using the time-domain differentiation property of the Laplace transform, Eqn. (7.88) leads to the relationship

$$s\, W\,(s) = X\,(s)$$

from which the desired result is obtained:

$$W\,(s) = \frac{1}{s} X\,(s) \qquad (7.89)$$

An alternative method of proving the integration property will be presented to provide further insight. Let us write the convolution of $x\,(t)$ with the unit-step function:

$$x\,(t) * u\,(t) = \int_{-\infty}^{\infty} x\,(\lambda)\, u\,(t - \lambda)\, d\lambda \qquad (7.90)$$

We know that

$$u\,(t - \lambda) = \begin{cases} 1, & t - \lambda > 0 \Rightarrow \lambda < t \\ 0, & t - \lambda < 0 \Rightarrow \lambda > t \end{cases}$$

Therefore, we can set the upper limit of the integral in Eqn. (7.90) to $\lambda = t$ and drop the unit-step function without changing the integration result. Doing so leads to

$$x\,(t) * u\,(t) = \int_{-\infty}^{t} x\,(\lambda)\, d\lambda = w\,(t) \qquad (7.91)$$

Since $w\,(t)$ is equal to the convolution of $x\,(t)$ and the unit-step function, its Laplace transform must be equal to the product of the respective transforms:

$$W\,(s) = \mathcal{L}\,\{u\,(t)\}\, \mathcal{L}\,\{x\,(t)\} = \frac{1}{s} X\,(s) \qquad (7.92)$$

The ROC may need to be adjusted. Let the ROC of the original transform $X\,(s)$ be

$$\sigma_1 < \text{Re}\,\{s\} < \sigma_2$$

The ROC for the Laplace transform of the unit-step function is

$$\text{Re}\,\{s\} > 0$$

Therefore, the ROC of the transform $W\,(s)$ must be at least the overlap of these regions. It may be larger than the overlap if $X\,(s)$ has a zero at $s = 0$ to counter the pole at $s = 0$ introduced by the transform of the unit-step function.

7.4 Inverse Laplace Transform

Consider a transform pair

$$x\,(t) \overset{\mathcal{L}}{\longleftrightarrow} X\,(s)$$

If $X\,(s)$ is given, the signal $x\,(t)$ can be found using the inverse Laplace transform

$$x\,(t) = \mathcal{L}^{-1}\,\{X\,(s)\} = \frac{1}{2\pi j} \int_{C} X\,(s)\, e^{st}\, ds \qquad (7.93)$$

The contour C is a vertical line within the ROC of the transform as shown in Fig. 7.31.

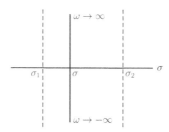

Figure 7.31 – Integration path C in the s-plane.

Even though the contour integral in Eqn. (7.93) will not be used in this text for the actual computation of the inverse Laplace transform, it will be explored a bit further to provide insight.

Consider the Laplace transform $X(s)$ evaluated at $s = \sigma + j\omega$, first given by Eqn. (7.5) and repeated below.

$$\text{Eqn. (7.5):} \quad X(s)\Big|_{s=\sigma+j\omega} = \int_{-\infty}^{\infty} x(t)\, e^{-(\sigma+j\omega)t}\, dt$$

$$= \int_{-\infty}^{\infty} x(t)\, e^{-\sigma t}\, e^{-j\omega t}\, dt$$

The integral on the right side represents the Fourier transform of the modified signal $x(t)\, e^{-\sigma t}$ provided that it exists. Let us assume that the point $s = \sigma + j\omega$ is in the ROC of $X(s)$ to satisfy the existence condition. The signal $x(t)\, e^{-\sigma t}$ can be found using the inverse Fourier transform:

$$x(t)\, e^{-\sigma t} = \frac{1}{2\pi} \int_{-\infty}^{\infty} X(s)\Big|_{s=\sigma+j\omega} e^{j\omega t}\, d\omega \tag{7.94}$$

Multiplying both sides of Eqn. (7.94) by $e^{\sigma t}$ yields

$$x(t) = \frac{1}{2\pi} \int_{-\infty}^{\infty} X(s)\Big|_{s=\sigma+j\omega} e^{\sigma t}\, e^{j\omega t}\, d\omega$$

$$= \frac{1}{2\pi} \int_{-\infty}^{\infty} X(s)\Big|_{s=\sigma+j\omega} e^{(\sigma+j\omega)t}\, d\omega \tag{7.95}$$

Substituting

$$s = \sigma + j\omega \quad \Longrightarrow \quad ds = j\, d\omega$$

into Eqn. (7.95) we obtain

$$x(t) = \frac{1}{2\pi j} \int_{\sigma-j\infty}^{\sigma+j\infty} X(s)\, e^{st}\, ds \tag{7.96}$$

which is the contour integral given by Eqn. (7.93). The integral needs to be carried out at points on a vertical line in the ROC of the transform.

For a rational function $X(s)$ it is usually easier to compute the inverse Laplace transform through the use of partial fraction expansion (PFE).

7.4.1 Partial fraction expansion with simple poles

Consider a rational transform in the form

$$X(s) = \frac{B(s)}{(s-p_1)(s-p_2)\ldots(s-p_N)} \tag{7.97}$$

where the poles $p_1, p_2, \ldots, p_N$ are distinct. Furthermore, let the order of the numerator polynomial $B(s)$ be less than N, the order of the denominator polynomial. The transform $X(s)$ can be expanded into partial fractions in the form

$$X(s) = \frac{k_1}{s-p_1} + \frac{k_2}{s-p_2} + \ldots + \frac{k_N}{s-p_N} \tag{7.98}$$

The coefficients $k_1, k_2, \ldots, k_N$ are called the *residues* of the partial fraction expansion. They can be computed by (see Appendix E)

$$k_i = (s-p_i) X(s)\Big|_{s=p_i}, \qquad i = 1, 2, \ldots, N \tag{7.99}$$

Once the residues are determined, the inverse transform of each term in the partial fraction expansion is determined using either Eqn. (7.22) or Eqn. (7.23), depending on the placement of each pole relative to the ROC.

$$\mathcal{L}^{-1}\left\{ \frac{1}{s-p_i} \right\} = \begin{cases} e^{p_i t}\, u(t)\,, & \text{if the ROC is to the right of } p_i \\ -e^{p_i t}\, u(-t)\,, & \text{if the ROC is to the left of } p_i \end{cases} \tag{7.100}$$

This will be illustrated in the next several examples.

Software resources: See MATLAB Exercise 7.4.

Example 7.22: **Inverse Laplace transform using PFE**

A causal signal $x(t)$ has the Laplace transform

$$X(s) = \frac{s+1}{s(s+2)}$$

Determine $x(t)$ using partial fraction expansion.

Solution: Since $x(t)$ is specified to be causal, the ROC of the transform must be to the right of a vertical line. $X(s)$ has poles at $s = -2$ and $s = 0$, therefore the ROC is

$$\text{Re}\{s\} > 0$$

Partial fraction expansion of $X(s)$ is in the form

$$X(s) = \frac{k_1}{s} + \frac{k_2}{s+2}$$

Residues are found by the application of residue formulas:

$$k_1 = s\, X\,(s)\Big|_{s=0} = \frac{s+1}{s+2}\Big|_{s=0} = \frac{1}{2}$$

$$k_2 = (s+2)\, X\,(s)\Big|_{s=-2} = \frac{s+1}{s}\Big|_{s=-2} = \frac{1}{2}$$

Using the values found, $X\,(s)$ is

$$X\,(s) = \frac{1/2}{s} + \frac{1/2}{s+2} \tag{7.101}$$

Both terms of $X\,(s)$ in Eqn. (7.101) correspond to causal terms in the time domain. Use of the transform pair given by Eqn. (7.22) for both terms yields

$$x\,(t) = \frac{1}{2}\, u\,(t) + \frac{1}{2}\, e^{-2t}\, u\,(t)$$

Software resources:
ex_7_22a.m
ex_7_22b.m

Example 7.23: **Using PFE with complex poles**

The Laplace transform of a signal $x\,(t)$ is

$$X\,(s) = \frac{s+1}{s\,(s^2+9)}$$

with the ROC specified as

$$\mathrm{Re}\,\{s\} > 0$$

Determine $x\,(t)$.

Solution: The transform $X\,(s)$ can be written in factored form as

$$X\,(s) = \frac{s+1}{s\,(s+j3)\,(s-j3)}$$

and expanded into partial fractions as

$$X\,(s) = \frac{k_1}{s} + \frac{k_2}{s+j3} + \frac{k_3}{s-j3} \tag{7.102}$$

The residues are determined using the residue formulas. The residues associated with complex poles at $\pm j3$ will be complex-valued, however, the method used for computing them is the same.

$$k_1 - s\, X\,(s)\Big|_{s=0} = \frac{s+1}{(s+j3)\,(s-j3)}\Big|_{s=0} = \frac{1}{9}$$

$$k_2 = (s+j3)\, X\,(s)\Big|_{s=-j3} = \frac{s+1}{s\,(s-j3)}\Big|_{s=-j3} = -\frac{1}{18} + j\frac{1}{6}$$

The residue k_3 can be computed using the same method, however, there is a shortcut available. Since the transform $X\,(s)$ is a rational function with real coefficients, residues of complex conjugate poles must be complex conjugates of each other. Therefore

$$k_3 = k_2^* = -\frac{1}{18} - j\frac{1}{6}$$

Based on the specified ROC, the signal $x(t)$ is causal. Using Eqn. (7.22) for inverting each term of the partial fraction expansion we get

$$x(t) = k_1 u(t) + k_2 e^{-j3t} u(t) + k_3 e^{j3t} u(t)$$

$$= \frac{1}{9} u(t) + \left(-\frac{1}{18} + j\frac{1}{6} \right) e^{-j3t} u(t) + \left(-\frac{1}{18} - j\frac{1}{6} \right) e^{j3t} u(t)$$

which can be simplified as

$$x(t) = \frac{1}{9} u(t) - \frac{1}{18} \left[e^{-j3t} + e^{j3t} \right] u(t) + j\frac{1}{6} \left[e^{-j3t} - e^{j3t} \right] u(t)$$

$$= \frac{1}{9} u(t) - \frac{1}{9} \cos(3t) u(t) + \frac{1}{3} \sin(3t) u(t)$$

An alternative approach would be to combine the two partial fractions with complex poles into a second-order term and write Eqn. (7.102) as

$$X(s) = \frac{1/9}{s} + \frac{-\frac{1}{9}s + 1}{s^2 + 9}$$

Afterwards the transform pairs

$$\cos(3t) u(t) \overset{\mathcal{L}}{\longleftrightarrow} \frac{s}{s^2 + 9} \quad \text{and} \quad \sin(3t) u(t) \overset{\mathcal{L}}{\longleftrightarrow} \frac{3}{s^2 + 9}$$

can be used for arriving at the same result.
Software resources:
ex_7_23a.m
ex_7_23b.m

Example 7.24: **Using PFE in conjunction with the ROC**

The Laplace transform of a signal $x(t)$ is

$$X(s) = \frac{5(s-1)}{(s+1)(s+2)(s-2)(s-3)}$$

with the ROC specified as

$$-1 < \mathrm{Re}\{s\} < 2$$

Determine $x(t)$.

Solution: We will first find the partial fraction expansion of $X(s)$. The ROC will be taken into consideration in the next step. Partial fraction expansion of $X(s)$ is in the form

$$X(s) = \frac{k_1}{s+1} + \frac{k_2}{s+2} + \frac{k_3}{s-2} + \frac{k_4}{s-3}$$

The residues are

$$k_1 = \left. (s+1) X(s) \right|_{s=-1} = \left. \frac{5(s-1)}{(s+2)(s-2)(s-3)} \right|_{s=-1} = -0.8333$$

$$k_2 = \left. (s+2) X(s) \right|_{s=-2} = \left. \frac{5(s-1)}{(s+1)(s-2)(s-3)} \right|_{s=-2} = 0.75$$

$$k_3 = (s-2) X(s)\Big|_{s=2} = \frac{5(s-1)}{(s+1)(s+2)(s-3)}\Big|_{s=2} = -0.4167$$

$$k_4 = (s-3) X(s)\Big|_{s=} = \frac{5(s-1)}{(s+1)(s+2)(s-2)}\Big|_{s=3} = 0.5$$

The completed form of the partial fraction expansion is

$$X(s) = -\frac{0.8333}{s+1} + \frac{0.75}{s+2} - \frac{0.4167}{s-2} + \frac{0.5}{s-3}$$

Next we need to pay attention to the ROC which is a vertical strip in the s-plane between the poles at $s = -1$ and $s = 2$ as shown in Fig. 7.32.

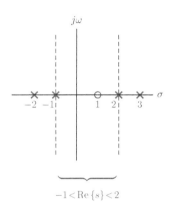

Figure 7.32 – The pole-zero plot and the ROC for $X(s)$ in Example 7.24.

The shape of the ROC indicates a two-sided signal. Accordingly, we will write the transform as the sum of two transforms belonging to causal and anti-causal components of $x(t)$:

$$X(s) = X_R(s) + X_L(s)$$

The poles at $s = -1$ and $s = -2$ are associated with $x_R(t)$, the causal component of $x(t)$. Therefore $X_R(s)$ can be written as

$$X_R(s) = -\frac{0.8333}{s+1} + \frac{0.75}{s+2} = \frac{-0.0833(s+11)}{(s+1)(s+2)}, \qquad \text{ROC:} \quad \text{Re}\{s\} > -1$$

Conversely the poles at $s = 2$ and $s = 3$ are associated with $x_L(t)$, the anti-causal component of $x(t)$.

$$X_L(s) = -\frac{0.4167}{s-2} + \frac{0.5}{s-3} = \frac{0.0833(s+3)}{(s-2)(s-3)};, \qquad \text{ROC:} \quad \text{Re}\{s\} < 2$$

The individual ROCs of the transforms $X_R(s)$ and $X_L(s)$ are shown in Fig. 7.33(a) and (b).

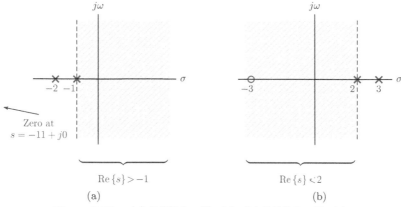

Figure 7.33 – (a) ROC for $X_R(s)$, (b) ROC for $X_L(s)$.

The signal $x_R(t)$ is found by using Eqn. (7.22) on the terms of $X_R(s)$:

$$x_R(t) = -0.8333\, e^{-t}\, u(t) + 0.75\, e^{-2t}\, u(t)$$

The signal $x_L(t)$ is found by using Eqn. (7.23) on the terms of $X_L(s)$:

$$x_L(t) = 0.4167\, e^{2t}\, u(-t) - 0.5\, e^{3t}\, u(-t)$$

The signal $x(t)$ is the sum of these two components:

$$
\begin{aligned}
x(t) =\, & x_R(t) + x_L(t) \\
=\, & -0.8333\, e^{-t}\, u(t) + 0.75\, e^{-2t}\, u(t) + 0.4167\, e^{2t}\, u(-t) - 0.5\, e^{3t}\, u(-t)
\end{aligned}
$$

Software resources:
ex_7_24a.m
ex_7_24b.m

If the order of the numerator polynomial $B(s)$ in Eqn. (7.97) is equal to or greater than the order of the denominator polynomial, special precautions need to be taken before partial fraction expansion can be used. In this case $X(s)$ must be written as

$$X(s) = C(s) + \frac{\bar{B}(s)}{(s - p_1)(s - p_2)\ldots(s - p_N)} \tag{7.103}$$

where $C(s)$ is a polynomial of S, and the order of the new numerator polynomial $\bar{B}(s)$ is $N - 1$.

Example 7.25: **Using PFE when numerator order is not less than denominator order**

The Laplace transform of a causal signal is

$$X(s) = \frac{s(s+1)}{(s+2)(s+3)}$$

Determine $x(t)$.

Solution: Since the numerator order is equal to the denominator order, $X(s)$ cannot be expanded into partial fractions directly. Let us first multiply numerator and denominator factors to obtain

$$X(s) = \frac{s^2 + s}{s^2 + 5s + 6}$$

which can be written as

$$X(s) = 1 + \frac{-4s - 6}{s^2 + 5s + 6}$$

Let $X_1(s)$ be defined as

$$X_1(s) = \frac{-4s - 6}{s^2 + 5s + 6} = \frac{-4s - 6}{(s+2)(s+3)}$$

so that $X(s) = 1 + X_1(s)$. Expanding $X_1(s)$ into partial fractions yields

$$X_1(s) = \frac{2}{s+2} - \frac{6}{s+3}$$

Consequently we have

$$x_1(t) = 2\,e^{-2t}\,u(t) - 6\,e^{-3t}\,u(t)$$

and

$$\begin{aligned} x(t) &= \delta(t) + x_1(t) \\ &= \delta(t) + 2\,e^{-2t}\,u(t) - 6\,e^{-3t}\,u(t) \end{aligned} \tag{7.104}$$

Software resources:
ex_7_25a.m
ex_7_25b.m

7.4.2   Partial fraction expansion with multiple poles

If the transform $X(s)$ has repeated roots, residue calculations become a bit more complicated. Consider a rational transform in the form

$$X(s) = \frac{B(s)}{(s - p_1)^r (s - p_2) \ldots (s - p_N)} \tag{7.105}$$

The pole of multiplicity r at $s = p_1$ requires r terms in the partial fraction expansion:

$$X(s) = \frac{k_{1,1}}{s - p_1} + \frac{k_{1,2}}{(s - p_1)^2} + \ldots + \frac{k_{1,r}}{(s - p_1)^r} + \frac{k_2}{s - p_2} + \ldots + \frac{k_N}{s - p_N} \tag{7.106}$$

The residues of single poles at $p_2, \ldots, p_N$ are still computed using the residue formulas discussed earlier in the previous section. The residue $k_{1,r}$ is also easy to compute:

$$k_{1,r} = \left. (s - p_1)^r\, X(s) \right|_{s=p_1} \tag{7.107}$$

The residue $k_{1,r-1}$ for the partial fraction with $(s - p_1)^{r-1}$ in its denominator requires more work, and is computed as

$$k_{1,r-1} = \left. \frac{1}{1!} \frac{d}{ds} \left[(s - p_1)^r\, X(s) \right] \right|_{s=p_1} \tag{7.108}$$

The residue $k_{1,r-2}$ is

$$k_{1,r-2} = \left. \frac{1}{2!} \frac{d^2}{ds^2} \left[(s - p_1)^r\, X(s) \right] \right|_{s=p_1} \tag{7.109}$$

and so on. See Appendix E for details.

Example 7.26: **Multiple-order poles**

A causal signal $x(t)$ has the Laplace transform

$$X(s) = \frac{s-1}{(s+1)^3 (s+2)}$$

Determine $x(t)$ using partial fraction expansion.

Solution: The partial fraction expansion for $X(s)$ is in the form

$$X(s) = \frac{k_{1,1}}{s+1} + \frac{k_{1,2}}{(s+1)^2} + \frac{k_{1,3}}{(s+1)^3} + \frac{k_2}{s+2}$$

The residue k_2 for the single pole at $s = 2$ is easily determined using the residue formula:

$$k_2 = \left. (s+2) X(s) \right|_{s=-2} = \left. \frac{s-1}{(s+1)^3} \right|_{s=-2} = 3$$

The residues of the third-order pole at $s = -1$ are found using Eqns. (7.107) through (7.109).

$$k_{1,3} = \left. (s+1)^3 X(s) \right|_{s=-1} = \left. \frac{s-1}{s+2} \right|_{s=-1} = -2$$

$$k_{1,2} = \left. \frac{d}{ds} \left[(s+1)^3 X(s) \right] \right|_{s=-1} = \left. \frac{3}{(s+2)^2} \right|_{s=-1} = 3$$

$$k_{1,1} = \left. \frac{1}{2} \frac{d^2}{ds^2} \left[(s+1)^3 X(s) \right] \right|_{s=-1} = \left. \frac{-3}{(s+2)^3} \right|_{s=-1} = -3$$

and the partial fraction expansion of $X(s)$ is

$$X(s) = -\frac{3}{s+1} + \frac{3}{(s+1)^2} - \frac{2}{(s+1)^3} + \frac{3}{s+2}$$

Using the transform pairs

$$e^{-t} u(t) \xleftrightarrow{\mathcal{L}} \frac{1}{s+1}$$

$$t e^{-t} u(t) \xleftrightarrow{\mathcal{L}} -\frac{d}{ds} \left[\frac{1}{s+1} \right] = \frac{1}{(s+1)^2}$$

and

$$t^2 e^{-t} u(t) \xleftrightarrow{\mathcal{L}} -\frac{d}{ds} \left[\frac{1}{(s+1)^2} \right] = \frac{2}{(s+1)^3}$$

to invert the terms of the partial fraction expansion, we arrive at the solution

$$x(t) = -3 e^{-t} u(t) + 3t e^{-t} u(t) - t^2 e^{-t} u(t) + 3 e^{-2t} u(t)$$

Software resources:
ex_7_26.m

Software resources: See MATLAB Exercises 7.4 and 7.5.

7.5   Using the Laplace Transform with CTLTI Systems

We have shown in Chapter 2 that the output signal of a CTLTI system can be computed from its input signal and its impulse response through the use of the convolution integral. If the impulse response of a CTLTI system is $h(t)$, and if the signal $x(t)$ is applied to the system as input, the output signal $y(t)$ is found as

$$y(t) = x(t) * h(t) = \int_{-\infty}^{\infty} x(\lambda) h(t - \lambda) \, d\lambda$$

Based on the convolution property of the Laplace transform introduced in Eqn. (7.78) in Section 7.3.7, the transform of the convolution of two signals is equal to the product of their individual transforms. Therefore, the transform of the output signal is

$$Y(s) = X(s) H(s) \tag{7.110}$$

As an alternative to computing the output signal by direct application of the convolution integral, we could

1. Find the Laplace transforms of the input signal $x(t)$ and the impulse response $h(t)$.
2. Multiply the two transforms to obtain the transform $Y(s)$.
3. Determine the output signal $y(t)$ from $Y(s)$ by means of the inverse Laplace transform.

Solving Eqn. (7.110) for $H(s)$ we get

$$H(s) = \frac{Y(s)}{X(s)} \tag{7.111}$$

The function $H(s)$ is the *s-domain system function* of the system under consideration.

If the input and the output signals $x(t)$ and $y(t)$ are specified, the impulse response of the CTLTI system can be found by

1. Finding the Laplace transform of each.
2. Finding the transform $H(s)$ as the ratio of the two, as given by Eqn. (7.111).
3. Using the inverse Laplace transform operation on $H(s)$.

We already know that a CTLTI system can be completely and uniquely described by means of its impulse response $h(t)$. Since the system function $H(s)$ is just the Laplace transform of the impulse response $h(t)$, it also represents a complete description of the CTLTI system.

7.5.1   Relating the system function to the differential equation

As discussed in Section 2.4 of Chapter 2, the input-output relationship of a CTLTI system can be modeled by a constant-coefficient linear differential equation given in the standard form

$$\sum_{k=0}^{N} a_k \frac{d^k y(t)}{dt^k} = \sum_{k=0}^{M} b_k \frac{d^k x(t)}{dt^k} \tag{7.112}$$

The order of the CTLTI system is the larger of M and N. Up to this point we have considered three different forms of modeling for a CTLTI system, namely the differential equation, the impulse response and the system function.   (A fourth method, state-space modeling, will be introduced in Chapter 9.) It must be possible to obtain any of the three models from the knowledge of any other.   In this section we will focus on the problem of determining the system function from the differential equation. If we take the Laplace transform of both sides of Eqn. (7.112) the equality would still be valid:

$$\mathcal{L}\left\{\sum_{k=0}^{N} a_k \frac{d^k y(t)}{dt^k}\right\} = \mathcal{L}\left\{\sum_{k=0}^{M} b_k \frac{d^k x(t)}{dt^k}\right\} \tag{7.113}$$

Laplace transform is linear; therefore, the transform of a summation is equal to the sum of individual terms, allowing us to write Eqn. (7.113) as

$$\sum_{k=0}^{N} \mathcal{L}\left\{a_k \frac{d^k y(t)}{dt^k}\right\} = \sum_{k=0}^{M} \mathcal{L}\left\{b_k \frac{d^k x(t)}{dt^k}\right\} \tag{7.114}$$

and subsequently as

$$\sum_{k=0}^{N} a_k \,\mathcal{L}\left\{\frac{d^k y(t)}{dt^k}\right\} = \sum_{k=0}^{M} b_k \,\mathcal{L}\left\{\frac{d^k x(t)}{dt^k}\right\} \tag{7.115}$$

Using the time-domain differentiation property of the Laplace transform, given by Eqn. (7.69), in Eqn. (7.115) yields

$$\sum_{k=0}^{N} a_k \, s^k \, Y(s) = \sum_{k=0}^{M} b_k \, s^k \, X(s) \tag{7.116}$$

The transforms $X(s)$ and $Y(s)$ do not depend on the summation indices k, and can therefore be factored out of the summations in Eqn. (7.116) resulting in

$$Y(s) \sum_{k=0}^{N} a_k \, s^k = X(s) \sum_{k=0}^{M} b_k \, s^k \tag{7.117}$$

The system function can now be obtained from Eqn. (7.117) as

$$H(s) = \frac{Y(s)}{X(s)} = \frac{\displaystyle\sum_{k=0}^{M} b_k s^k}{\displaystyle\sum_{k=0}^{N} a_k s^k} \tag{7.118}$$

Finding the system function from the differential equation:

1. Separate the terms of the differential equation so that $y(t)$ and its derivatives are on the left of the equal sign, and $x(t)$ and its derivatives are on the right of the equal sign, as in Eqn. (7.112).
2. Take the Laplace transform of each side of the differential equation, and use the time-differentiation property of the Laplace transform as in Eqn. (7.116).
3. Determine the system function as the ratio of $Y(s)$ to $X(s)$ as in Eqn. (7.118).
4. If the impulse response is needed, it can now be determined as the inverse Laplace transform of $H(s)$.

System function, linearity, and initial conditions

Two important observations will be made at this point:

1. The development leading up to the s-domain system function in Eqn. (7.111) has relied heavily on the convolution operation and the convolution property of the Laplace transform. We know from Chapter 2 that the convolution operation is only applicable to problems involving linear and time-invariant systems. Therefore it follows that the system function concept is meaningful only for systems that are both linear and time-invariant. This notion was introduced in earlier discussions involving system functions as well.
2. Furthermore, it was justified in Section 2.4 of Chapter 2 that a constant-coefficient differential equation corresponds to a linear and time-invariant system only if all initial conditions are set equal to zero.

We conclude that, in determining the system function from the differential equation, all initial conditions must be assumed to be zero.

If we need to use Laplace transform-based techniques to solve a differential equation subject to non-zero initial conditions, that can be done through the use of the unilateral Laplace transform, but not through the use of the system function. The unilateral Laplace transform and its use for solving differential equations will be discussed in Section 7.7.

Example 7.27: **Finding the system function from the differential equation**

A CTLTI system is defined by means of the differential equation

$$\frac{d^3 y(t)}{dt^3} + 5\frac{d^2 y(t)}{dt^2} + 17\frac{dy(t)}{dt} + 13\,y(t) = \frac{d^2 x(t)}{dt^2} + x(t)$$

Find the system function $H(s)$ for this system.

Solution: We will assume that all initial conditions are equal to zero, and take the Laplace transform of each side of the differential equation to obtain

$$s^3\,Y(s) + 5s^2\,Y(s) + 17s\,Y(s) + 13\,Y(s) = s^2\,X(s) + X(s) \qquad (7.119)$$

The system function can be obtained from Eqn. (7.119) as

$$H(s) = \frac{Y(s)}{X(s)} = \frac{s^2 + 1}{s^3 + 5s^2 + 17s + 13}$$

Characteristic polynomial vs. the denominator of the system function

Another important observation will be made based on the result obtained in Example 7.27: The characteristic equation for the system considered in Example 7.27 is

$$s^3 + 5s^2 + 17s + 13 = 0$$

and the solutions of the characteristic equation are the modes of the system as defined in Section 2.5.3 of Chapter 2. When we find the system function $H(s)$ from the differential equation we see that its denominator polynomial is identical to the characteristic polynomial. The roots of the denominator polynomial are the poles of the system function in the s-domain, and consequently, they are identical to the modes of the differential equation of the system.

Recall that in Section 2.5.3 we have reached some conclusions about the relationship between the modes of the differential equation and the natural response of the corresponding system. The same conclusions would apply to the poles of the system function. Specifically, if all poles of the system are real-valued and distinct, then the transient response of the system is in the form

$$y[n] = \sum_{k=1}^{N} c_k e^{p_k t}$$

Complex poles appear in conjugate pairs provided that all denominator coefficients of the system function are real-valued. A pair of complex conjugate poles

$$p_{1a} = \sigma_1 + j\omega_1 , \quad \text{and} \quad p_{1b} = \sigma_1 - j\omega_1$$

yields a response of the type

$$y_1(t) = d_1 \, e^{\sigma_1 t} \, \cos(\omega_1 t + \theta_1)$$

Finally, a pole of multiplicity m at $s = p_1$ leads to a response in the form

$$y_1(t) = c_{11} \, e^{p_1 t} + c_{12} \, t \, e^{p_1 t} + \ldots + c_{1m} \, t^{m-1} \, e^{p_1 t} + \text{other terms}$$

regardless of whether p_1 is real or complex-valued. Justifications for these relationships were given in Section 2.5.3 of Chapter 2 through the use of the modes of the differential equation, and will not be repeated here.

Sometimes we need to reverse the problem represented in Example 7.27 and find the differential equation from the knowledge of the system function. The next three examples will demonstrate this.

Software resources: See MATLAB Exercise 7.5.

Example 7.28: **Finding the differential equation from the system function**

A causal CTLTI system is defined by the system function

$$H(s) = \frac{2s + 5}{s^2 + 5s + 6}$$

Find a differential equation for this system.

Solution: The system function is the ratio of the output transform to the input transform, that is,

$$H(s) = \frac{Y(s)}{X(s)}$$

Therefore we can write

$$(s^2 + 5s + 6) \, Y(s) = (2s + 5) \, X(s) \qquad\qquad (7.120)$$

The differential equation follows from Eqn. (7.120) as

$$\frac{d^2 y(t)}{dt^2} + 5 \frac{dy(t)}{dt} + 6 \, y(t) = 2 \frac{dx(t)}{dt} + 5 \, x(t)$$

Example 7.29: **Finding the differential equation from input and output signals**

The unit-step response of a CTLTI system is

$$y(t) = \mathrm{Sys} \{ u(t) \} = \left(2 - 4e^{-t} + 2e^{-2t} \right) u(t)$$

Find a differential equation for this system.

Solution: Since the input signal is a unit-step function, that is, $x(t) = u(t)$, its Laplace transform is

$$X(s) = \frac{1}{s}$$

The Laplace transform of the output signal is found as

$$Y(s) = \frac{2}{s} - \frac{4}{s+1} + \frac{2}{s+2} = \frac{4}{s(s+1)(s+2)}$$

The system function can be obtained as the ratio of the output transform to the input transform:

$$H(s) = \frac{Y(s)}{X(s)} = \frac{4}{(s+1)(s+2)} = \frac{4}{s^2 + 3s + 2}$$

The differential equation follows from the system function as

$$\frac{d^2 y(t)}{dt^2} + 3 \frac{dy(t)}{dt} + 2 \, y(t) = 4 \, x(t)$$

Software resources:
ex_7_29.m

Example 7.30: **Finding the impulse response from input and output signals**

Determine the impulse response of the CTLTI system system the unit-step response of which was given in Example 7.29.

Solution: The system function was found in Example 7.29 to be

$$H(s) = \frac{4}{s^2 + 3s + 2}$$

which can be expanded into partial fractions to yield

$$H(s) = \frac{4}{s+1} - \frac{4}{s+2}$$

The impulse response is

$$h(t) = \left(4e^{-t} - 4e^{-2t}\right) u(t)$$

It can be easily verified that the unit-step response specified in Example 7.29 for this system is indeed the convolution of $h(t)$ found above with the unit-step function.

Software resources:

ex_7_30.m

7.5.2 Response of a CTLTI system to a complex exponential signal

In this section we will consider the response of a CTLTI system to a complex exponential input signal. This will help us gain further insight into the system function concept.

Let a CTLTI system with impulse response $h(t)$ be driven by a complex exponential input signal in the form

$$x(t) = e^{s_0 t}$$

where s_0 represents a point in the s-plane within the ROC of the system function. The output signal can be determined through the use of the convolution integral

$$y(t) = h(t) * x(t) = \int_{-\infty}^{\infty} h(\lambda) \, x(t - \lambda) \, d\lambda \tag{7.121}$$

Substituting $x(t - \lambda) = e^{s_0(t-\lambda)}$ into Eqn. (7.121) and simplifying the resulting integral we obtain

$$y(t) = \int_{-\infty}^{\infty} h(\lambda) \, e^{s_0(t-\lambda)} \, d\lambda$$

$$= e^{s_0 t} \underbrace{\int_{-\infty}^{\infty} h(\lambda) \, e^{-s_0 \lambda} \, d\lambda}_{H(s_0)} \tag{7.122}$$

The integral in Eqn. (7.122) should be recognized as the system function $H(s)$ evaluated at the point $s = s_0$ in the s-plane. Therefore, we reach the following important conclusion:

The response of the CTLTI system to the input signal $x(t) = e^{s_0 t}$ is

$$y(t) = \text{Sys}\left\{e^{s_0 t}\right\} = e^{s_0 t} H(s_0) \tag{7.123}$$

The CTLTI system responds to the complex exponential signal $x(t) = e^{s_0 t}$ by scaling it with the (generally complex) value of the system function at the point $s = s_0$.

In Eqn. (7.123) the complex exponential input signal is assumed to have been in existence forever; it is not turned on at a specific time instant. Therefore, the response found is the *steady-state response* of the system.

Example 7.31: **Response to a complex exponential signal**

A CTLTI system with the system function

$$H(s) = \frac{s + 6}{s^2 + 5s + 6}$$

is driven by the complex exponential input signal

$$x\left(t\right) = e^{(-0.4+j4)t}$$

Determine the steady-state response of the system.

Solution: The input signal is complex-valued, and can be written in Cartesian form using Euler's formula:

$$x\left(t\right) = e^{-0.4t}\left[\cos\left(4t\right) + j\sin\left(4t\right)\right] \tag{7.124}$$

Real and imaginary parts of $x\left(t\right)$ are shown in Fig. 7.34.

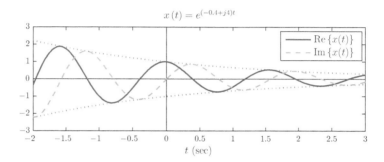

Figure 7.34 – The signal $x\left(t\right)$ for Example 7.31.

The value of the system function at $s = s_0 = -0.4 + j4$ is

$$H\left(-0.4 + j4\right) = \frac{\left(-0.4 + j4\right) + 6}{\left(-0.4 + j4\right)^2 + 5\left(-0.4 + j4\right) + 6}$$

$$= 0.0021 - j\,0.3348$$

$$= 0.3348\,e^{-j1.5645}$$

The output signal is found using Eqn. (7.123) as

$$y\left(t\right) = e^{(-0.4+j4)t}\left(0.3348\,e^{-j1.5645}\right)$$

$$= 0.3348\,e^{-0.4t}\,e^{j(4t-1.5645)}$$

or, in Cartesian form as

$$y\left(t\right) = 0.3348\,e^{-0.4t}\cos\left(4t - 1.5645\right) + j\,0.3348\,e^{-0.4t}\sin\left(4t - 1.5645\right) \tag{7.125}$$

Real and imaginary parts of the output signal $y\left(t\right)$ are shown in Fig. 7.35.

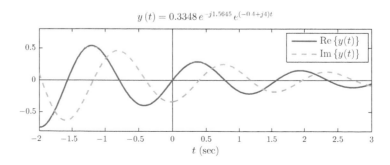

Figure 7.35 – The signal $y\left(t\right)$ for Example 7.31.

Software resources:
ex_7_31.m

7.5.3   Response of a CTLTI system to an exponentially damped sinusoid

Next we will consider an input signal in the form of an exponentially damped sinusoid

$$x(t) = e^{\sigma_0 t} \cos(\omega_0 t) \tag{7.126}$$

In order to find the response of a CTLTI system to this signal, let us write $x(t)$ using Euler's formula:

$$x(t) = \tfrac{1}{2} e^{\sigma_0 t} e^{j\omega_0 t} + \tfrac{1}{2} e^{\sigma_0 t} e^{-j\omega_0 t} \tag{7.127}$$

Let the parameter s_0 be defined as

$$s_0 = \sigma_0 + j\omega_0$$

so that Eqn. (7.127) becomes

$$x(t) = \tfrac{1}{2} e^{s_0 t} + \tfrac{1}{2} e^{s_0^* t} \tag{7.128}$$

Using the linearity of the system, its response to $x(t)$ can be written as

$$y(t) = \text{Sys} \left\{ \tfrac{1}{2} e^{s_0 t} + \tfrac{1}{2} e^{s_0^* t} \right\}$$

$$= \tfrac{1}{2} \, \text{Sys} \left\{ e^{s_0 t} \right\} + \tfrac{1}{2} \, \text{Sys} \left\{ e^{s_0^* t} \right\} \tag{7.129}$$

We already know that the response of the system to the term $e^{s_0 t}$ is

$$\text{Sys} \left\{ e^{s_0 t} \right\} = e^{s_0 t} H(s_0) \tag{7.130}$$

The response to the term $e^{s_0^* t}$ is found similarly:

$$\text{Sys} \left\{ e^{s_0^* t} \right\} = e^{s_0^* t} H(s_0^*) \tag{7.131}$$

Using Eqns. (7.130) and (7.131) in Eqn. (7.129) the output signal is

$$y(t) = \tfrac{1}{2} e^{s_0 t} H(s_0) + \tfrac{1}{2} e^{s_0^* t} H(s_0^*) \tag{7.132}$$

It is possible to further simplify the result obtained in Eqn. (7.132). Let the value of the system function evaluated at the point $s = s_0$ be written in polar complex form as

$$H(s_0) = H_0 \, e^{j\theta_0} \tag{7.133}$$

where H_0 and θ_0 represent the magnitude and the phase of the system function at the point $s = s_0$ respectively:

$$H_0 = |H(s_0)| \tag{7.134}$$

and

$$\theta_0 = \angle H(s_0) \tag{7.135}$$

For a real-valued impulse response it can be shown (see Problem 7.27 at the end of this chapter) that the value of the system function at the point $s = s_0^*$ is the complex conjugate of its value at the point $s = s_0$, that is,

$$H\left(s_0^*\right) = \left[H\left(s_0\right)\right]^* = H_0\, e^{-j\theta_0} \tag{7.136}$$

Using Eqns. (7.133) and (7.136) in Eqn. (7.132), the output signal $y\left(t\right)$ becomes

$$\begin{aligned}
y\left(t\right) &= \tfrac{1}{2}\, e^{s_0 t}\, H_0\, e^{j\theta_0} + \tfrac{1}{2}\, e^{s_0^* t}\, H_0\, e^{-j\theta_0}\\[4pt]
&= \tfrac{1}{2}\left(e^{\sigma_0 t}\, e^{j\omega_0 t}\right) H_0\, e^{j\theta_0} + \tfrac{1}{2}\left(e^{\sigma_0 t}\, e^{-j\omega_0 t}\right) H_0\, e^{-j\theta_0}\\[4pt]
&= \tfrac{1}{2}\, H_0\, e^{\sigma_0 t}\left[e^{j\left(\omega_0 t + \theta_0\right)} + e^{-j\left(\omega_0 t + \theta_0\right)}\right]\\[4pt]
&= H_0\, e^{\sigma_0 t}\, \cos\left(\omega_0 t + \theta_0\right)
\end{aligned} \tag{7.137}$$

The derivation outlined in Eqns. (7.127) through (7.137) can be summarized as shown in Fig. 7.36.

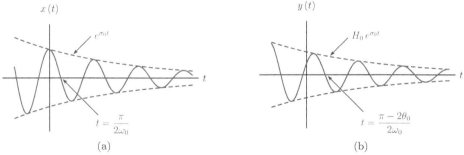

Figure 7.36 – The response of a CTLTI system to an exponentially damped sinusoid: (a) input signal, (b) output signal.

Comparison of the input signal in Eqn. (7.126) and the output signal Eqn. (7.137) reveals the following:

1. The amplitude of the signal is multiplied by the magnitude of the system function evaluated at the point $s = s_0 = \sigma_0 + j\omega_0$.
2. The phase of the cosine function is incremented by an amount equal to the phase of the system function evaluated at the point $s = s_0 = \sigma_0 + j\omega_0$.

Example 7.32: **Response to an exponentially damped sinusoid**

Consider again the system function used in Example 7.31. Determine the response of the system to the input signal

$$x\left(t\right) = e^{-0.3t}\, \cos\left(2t\right)$$

Solution: Let s_0 be defined as

$$s_0 = -0.3 + j\,2$$

The system function evaluated at $s = s_0$ is

$$H\left(-0.3 + j\,2\right) = \frac{\left(-0.3 + j2\right) + 6}{\left(-0.3 + j2\right)^2 + 5\left(-0.3 + j2\right) + 6} = 0.2695 - j0.6297$$

or in polar form

$$H\left(-0.3 + j\,2\right) = 0.6849\,e^{-j1.1664}$$

The output signal is found using Eqn. (7.137):

$$y\left(t\right) = 0.6849\,e^{-0.3t}\,\cos\left(2t - 1.1664\right)$$

The input and the output signals are shown in Fig. 7.37.

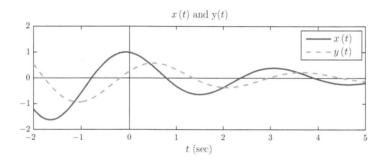

Figure 7.37 – Input and output signals for Example 7.32.

Software resources:
ex_7_32.m

7.5.4   Pole-zero plot for a system function

A rational system function $H\left(s\right)$ can be expressed in *pole-zero form* as

$$H\left(s\right) = K\,\frac{\left(s - z_1\right)\left(s - z_2\right)\ldots\left(s - z_M\right)}{\left(s - p_1\right)\left(s - p_2\right)\ldots\left(s - p_N\right)} \tag{7.138}$$

The parameters M and N are the numerator order and the denominator order respectively. The larger of M and N is the order of the system. The roots $z_1, \ldots, z_M$ of the numerator polynomial are referred to as the *zeros* of the system function. In contrast, the roots of the denominator polynomial are the *poles* of the system function. A *pole-zero plot* for a system function is obtained by marking the poles and the zeros of the system function on the s-plane. It is customary to use "$\times$" and "$\circ$" to mark each pole and each zero respectively.

Example 7.33: **Pole-zero plot for system function**

Construct a pole-zero plot for a CTLTI system with system function

$$H\left(s\right) = \frac{s^2 + 1}{s^3 + 5s^2 + 17s + 13}$$

Solution: The zeros of the system function are the roots of the numerator polynomial, and are found by solving the equation

$$s^2 + 1 = 0$$

which yields $z_1 = j$ and $z_2 = -j$. The poles of the system function are the roots of the denominator polynomial or, equivalently, the solutions of the equation

$$s^3 + 5s^2 + 17s + 13 = 0$$

The three poles are at $p_1 = -1$, $p_2 = -2 + j3$ and $p_3 = -2 - j3$. The pole-zero diagram for the system function is shown in Fig. 7.38.

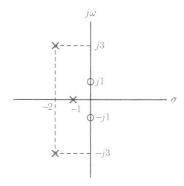

Figure 7.38 – The pole-zero plot for $H(s)$ of Example 7.33.

Software resources:
ex_7_33.m

Example 7.34: **ROC and the pole-zero plot**

Assume that the system function used in Example 7.33 represents a causal system. Indicate the ROC on the pole-zero plot.

Solution: Following is the knowledge we possess:

1. For a causal system, the ROC of the system function must be the area to the right of a vertical line.
2. The ROC must be bound by one or more poles.
3. There may be no poles within the ROC.

Consequently, the ROC of the system function must be

$$\text{ROC:} \qquad \text{Re}\,\{s\} > -1$$

which is shown in Fig. 7.39.

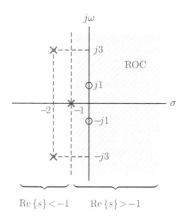

Figure 7.39 – The ROC superimposed on pole-zero plot for Example 7.34.

7.5.5 Graphical interpretation of the pole-zero plot

In this section we will explore the significance of the geometric placement of poles and zeros of the system function. The pole-zero plot, introduced in Section 7.5.4, can be used for understanding the behavior characteristics of a CTLTI system, especially in terms of the magnitude and the phase of the system function. The relationship between the locations of the poles and the zeros of the system function and the frequency-domain behavior of the system is significant.

Let us begin by considering the pole-zero form of the system function given by Eqn. (7.138). Assuming the system is stable, the Fourier transform-based system function $H(\omega)$ exists, and can be found by evaluating $H(s)$ for $s = j\omega$:

$$H(\omega) = H(s)\Big|_{s=j\omega} = K \frac{(j\omega - z_1)(j\omega - z_2)\dots(j\omega - z_M)}{(j\omega - p_1)(j\omega - p_2)\dots(j\omega - p_N)} \tag{7.139}$$

The numerator of Eqn. (7.139) has M factors each in the form

$$B_i(\omega) = (j\omega - z_i), \qquad i = 1,\dots,M \tag{7.140}$$

Similarly, the denominator of Eqn. (7.139) consists of factors in the form

$$A_i(\omega) = (j\omega - p_i), \qquad i = 1,\dots,N \tag{7.141}$$

In a sense, the function $B_i(\omega)$ describes the contribution of the zero at $s = z_i$ to the frequency response of the system. Similarly, the function $A_i(\omega)$ describes the contribution of the pole at $s = p_i$. Using the definitions in Eqns. (7.140) and (7.141), $H(\omega)$ becomes

$$H(\omega) = K \frac{B_1(\omega) B_2(\omega)\dots B_M(\omega)}{A_1(\omega) A_2(\omega)\dots A_N(\omega)} \tag{7.142}$$

If we wanted to compute the frequency response of the system at a specific frequency $\omega = \omega_0$ we would get

$$H(\omega_0) = K \frac{B_1(\omega_0) B_2(\omega_0)\dots B_M(\omega_0)}{A_1(\omega_0) A_2(\omega_0)\dots A_N(\omega_0)} \tag{7.143}$$

The magnitude of the frequency response at $\omega = \omega_0$ is found by computing the magnitude of each complex function $B_i(\omega_0)$ and $A_i(\omega_0)$, and forming the ratio

$$|H(\omega_0)| = K \frac{|B_1(\omega_0)| \cdot |B_2(\omega_0)| \dots |B_M(\omega_0)|}{|A_1(\omega_0)| \cdot |A_2(\omega_0)| \dots |A_N(\omega_0)|} \tag{7.144}$$

The phase of the frequency response is computed as the algebraic sum of the phases of numerator and denominator factors:

$$\angle H(\omega_0) = \angle B_1(\omega_0) + \angle B_2(\omega_0) + \dots + \angle B_M(\omega_0) - \angle A_1(\omega_0) - \angle A_2(\omega_0) - \dots - \angle A_N(\omega_0) \tag{7.145}$$

We would like to gain a graphical understanding of the computations in Eqns. (7.144) and (7.145). Let us focus on one of the numerator terms, $B_i(\omega_0)$:

$$B_i(\omega_0) = j\omega_0 - z_i \tag{7.146}$$

Clearly, $B(\omega_0)$ is a complex quantity. Its two terms z_i and $j\omega_0$ are shown in Fig. 7.40(a) as points in the s-plane. It is also possible to represent the terms in Eqn. (7.146) with vectors. Treating each term in Eqn. (7.146) as a vector, we obtain the corresponding vector expression

$$\overrightarrow{B_i}(\omega_0) = \overrightarrow{j\omega_0} - \overrightarrow{z_i} \tag{7.147}$$

which can be written in alternative form

$$\overrightarrow{z_i} + \overrightarrow{B_i}(\omega_0) = \overrightarrow{j\omega_0} \tag{7.148}$$

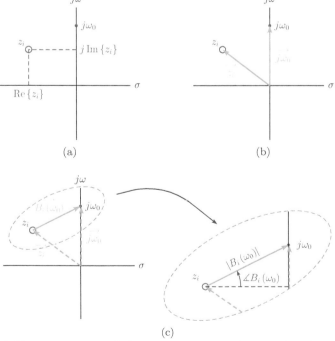

(a)   (b)

(c)

Figure 7.40 – (a) The zero $s = z_i$ and the point $s = j\omega_0$ in the complex plane, (b) the vectors for z_i and $j\omega_0$ in the complex plane, (c) the relationship expressed by Eqn. (7.148).

We will draw the vector $\overrightarrow{z_i}$ starting at the origin and ending at the point $s = z_i$. Similarly the vector $\overrightarrow{j\omega_0}$ will be drawn starting at the origin and ending at the point $s = j\omega_0$. These two vectors are shown in Fig. 7.40(b). The relationship expressed by Eqn. (7.148) is illustrated in Fig. 7.40(c). The following important conclusions can be drawn from Fig. 7.40(c):

1. The vector $\overrightarrow{B_i}(\omega_0)$ is drawn starting at the zero at $s = z_i$ and ending at the point $s = j\omega_0$ on the $j\omega$-axis.

2. The magnitude $|B_i(\omega_0)|$ is equal to the norm of the vector $\overrightarrow{B_i}(\omega_0)$.

3. The phase $\angle B_i(\omega_0)$ is equal to the angle that the vector $\overrightarrow{B_i}(\omega_0)$ makes with the positive real axis, measured counterclockwise.

If Fig. 7.40(c) is drawn to scale, the norm and the angle of the vector $\overrightarrow{B_i}(\omega_0)$ could simply be measured to determine the contributions of the zero at $s = z_i$ to the magnitude and the phase expressions for $|H(\omega_0)|$ and $\angle H(\omega_0)$ in Eqns. (7.144) and (7.145).

What if we need to determine the contributions of the zero at $s = z_i$ to the magnitude $|H(\omega)|$ and phase $\angle H(\omega)$ not just for a specific frequency ω_0 but for all frequencies ω? Imagine the vector $\overrightarrow{B_i}(\omega)$ to be a piece of rubber band, the tail end of which is permanently attached to the point $s = z_i$. Assume that, as the frequency ω varies, the tip of the rubber band vector moves on the $j\omega$-axis. The length of the rubber band and its angle with the positive real axis vary as the tip is moved. The contributions of the zero at $s = z_i$ to the magnitude and the phase of the frequency response $H(\omega)$ can be graphed by tracking these variations as shown in Fig. 7.41.

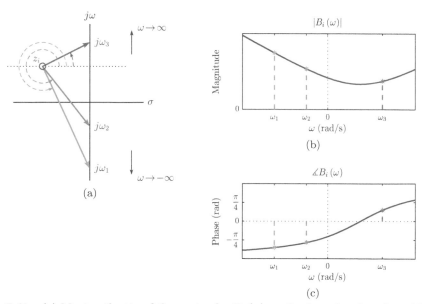

Figure 7.41 – (a) Moving the tip of the vector for $B_i(\omega)$ on the $j\omega$-axis using the rubber band analogy, (b) contribution of the zero at $s = z_i$ to the magnitude of the frequency response, (c) contribution of the zero at $s = z_i$ to the phase of the frequency response.

Similar analysis can be carried out for determining the contribution of a pole p_i to magnitude and phase characteristics of the system. Fig. 7.42 illustrates the effect of a pole on the frequency response.

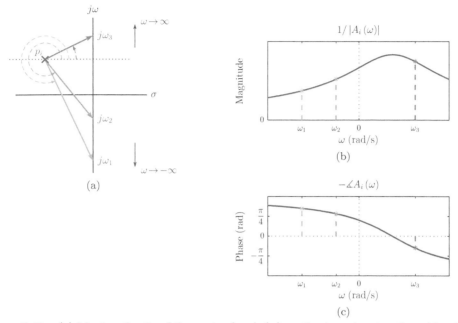

Figure 7.42 – (a) Moving the tip of the vector for $A_i(\omega)$ on the $j\omega$-axis using the rubber band analogy, (b) contribution of the pole at $s = p_i$ to the magnitude of the frequency response, (c) contribution of the pole at $s = p_i$ to the phase of the frequency response.

For a system with M zeros and N poles, magnitude and phase of the system function at $\omega = \omega_0$ are found by taking into account the contributions of each zero and pole. The next example will illustrate this.

Example 7.35: **Frequency response from pole-zero plot**

A CTLTI system is described by the system function

$$H(s) = \frac{s^2 + s - 2}{s^2 + 2s + 5}$$

Construct a pole-zero plot and use it to determine the magnitude and the phase of the frequency response of the system at the frequency $\omega_0 = 1.5$ rad/s.

Solution: The system has two zeros and two poles. The zeros of the system function are at

$$z_1 = 1, \quad z_2 = -2$$

and the poles are at

$$p_1 = -1 + j2, \quad p_2 = -1 - j2$$

Vector forms of the contribution of each zero and pole to the system function at $\omega = \omega_0$

can be written using Eqns. (7.140) and (7.141) as

$$\overrightarrow{B_1}(\omega_0) = \overrightarrow{j\omega_0} - \overrightarrow{z_1} = (-1 + j1.5)$$

$$\overrightarrow{B_2}(\omega_0) = \overrightarrow{j\omega_0} - \overrightarrow{z_2} = (2 + j1.5)$$

$$\overrightarrow{A_1}(\omega_0) = \overrightarrow{j\omega_0} - \overrightarrow{p_1} = (1 - j0.5)$$

$$\overrightarrow{A_2}(\omega_0) = \overrightarrow{j\omega_0} - \overrightarrow{p_2} = (1 + j3.5)$$

The vectors $\overrightarrow{B_1}(\omega_0)$, $\overrightarrow{B_2}(\omega_0)$, $\overrightarrow{A_1}(\omega_0)$, and $\overrightarrow{A_2}(\omega_0)$ are shown in Fig. 7.43.

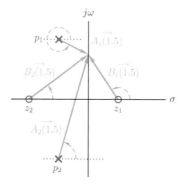

Figure 7.43 – Graphical computation of $H(\omega_0)$ for Example 7.35.

The norm and the phase of each vector are

$$\left|\overrightarrow{B_1}(\omega_0)\right| = 1.8028\,, \qquad \angle\overrightarrow{B_1}(\omega_0) = 2.1588 \text{ rad}$$

$$\left|\overrightarrow{B_2}(\omega_0)\right| = 2.5000\,, \qquad \angle\overrightarrow{B_2}(\omega_0) = 0.6435 \text{ rad}$$

$$\left|\overrightarrow{A_1}(\omega_0)\right| = 1.1180\,, \qquad \angle\overrightarrow{A_1}(\omega_0) = -0.4636 \text{ rad}$$

$$\left|\overrightarrow{A_2}(\omega_0)\right| = 3.6401\,, \qquad \angle\overrightarrow{A_2}(\omega_0) = 1.2925 \text{ rad}$$

Based on the values measured, the magnitude of the frequency response at ω_0 is computed as

$$|H(\omega_0)| = \frac{|B_1(\omega_0)| \cdot |B_2(\omega_0)|}{|A_1(\omega_0)| \cdot |A_2(\omega_0)|} = \frac{(1.8028)(2.5000)}{(1.1180)(3.6401)} = 1.1074 \qquad (7.149)$$

The phase of the frequency response is computed as

$$\angle H(\omega_0) = \angle B_1(\omega_0) + \angle B_2(\omega_0) - \angle A_1(\omega_0) - \angle A_2(\omega_0)$$
$$= 2.1588 + 0.6435 - (-0.4636) - 1.2925$$
$$= 1.9735 \text{ rad}$$

Complete magnitude and phase characteristics for the system can be obtained by repeating this process for all values of ω that are of interest. Fig. 7.44 shows the magnitude and the phase of the system. The values at $\omega_0 = 1.5$ rad/s are marked on magnitude and phase plots.

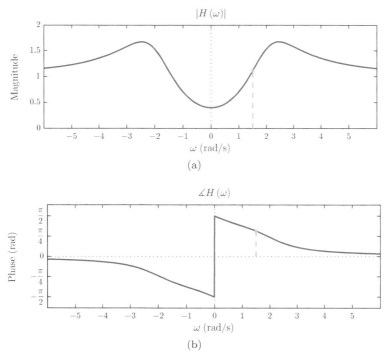

Figure 7.44 – The frequency response of the system in Example 7.35: (a) magnitude, (b) phase.

Software resources:
ex_7_35.m

Software resources:	See MATLAB Exercises 7.6 and 7.7.

Interactive Demo: pz_demo1.m

The pole-zero explorer demo program "pz_demo1.m" allows experimentation with the placement of poles and zeros of the system function. Before using it two vectors should be created in MATLAB workspace: one containing the poles of the system and one containing its zeros. In the pole-zero explorer user interface, the "import" button is used for importing these vectors. Pole-zero layout in the s-plane is displayed along with the magnitude and the phase of the system function. The vectors from each zero and each pole to a point on the $j\omega$-axis may optionally be displayed. Individual poles and zeros may be moved, and the effects on magnitude and phase may be observed. Complex conjugate poles and zeros move together to keep the conjugate relationship.

As an example, the system in Example 7.32 may be duplicated by creating and importing the two vectors

$$\texttt{zrs} = [1,\ -2]$$

$$\texttt{pls} = [-1 + j2,\ -1 - j2]$$

Software resources:
pz_demo1.m

7.5.6   System function and causality

Causality in linear and time-invariant systems was discussed in Section 2.8 of Chapter 2. For a CTLTI system to be causal, its impulse response $h(t)$ needs to be equal to zero for $t < 0$. Thus, by changing the lower limit of the integral to $t = 0$ in the definition of the Laplace transform, the s-domain system function for a causal CTLTI system can be written as

$$H(s) = \int_{-\infty}^{\infty} h(t)\, e^{-st}\, dt = \int_{0}^{\infty} h(t)\, e^{-st}\, dt \tag{7.150}$$

As we have discussed in Section 7.2 of this chapter, the ROC for the system function of a causal system is to the right of a vertical line in the s-plane. As a consequence, the system function must also converge at $\mathrm{Re}\{s\} \to \infty$. Consider a system function in the form

$$H(z) = \frac{B(s)}{A(s)} = \frac{b_M s^M + b_{M-1} s^{M-1} + \ldots + b_1 s + b_0}{a_N s^N + a_{N-1} s^{N-1} + \ldots + a_1 s + a_0}$$

For the system described by $H(s)$ to be causal we need

$$\lim_{s \to \infty} [H(s)] = \lim_{s \to \infty} \left[\frac{b_M}{a_N} s^{M-N} \right] < \infty \tag{7.151}$$

which requires that $M - N \leq 0$ and consequently $M \leq N$. Thus we arrive at an important conclusion:

Causality condition:
    In the s-domain system function of a causal CTLTI system the order of the numerator must not be greater than the order of the denominator.

Note that this condition is necessary for a system to be causal, but it is not sufficient. It is also possible for a non-causal system to have a system function with $M \leq N$.

7.5.7   System function and stability

In Section 2.9 of Chapter 2 we have concluded that for a CTLTI system to be stable its impulse response must be absolute integrable, that is,

$$\int_{-\infty}^{\infty} |h(t)| < \infty$$

Furthermore, we have established in Section 4.3.2 of Chapter 4 that the Fourier transform of a signal exists if the signal is absolute integrable. But the Fourier transform of the impulse response is equal to the s-domain system function evaluated on the $j\omega$-axis of the s-plane, that is,

$$H\left(\omega\right) = H\left(s\right)\Big|_{s=j\omega}$$

provided that the $j\omega$-axis of the s-plane is within the ROC.

Stability condition:
Therefore, it follows that, for a CTLTI system to be stable, the ROC of its s-domain system function must include the $j\omega$-axis.

What are the corresponding conditions that must be imposed on the locations of poles and zeros for stability? We will answer this question by taking three-separate cases into account:

1. **Causal system:**
   The ROC for the system function of a causal system is to the right of a vertical line in the s-plane, and is expressed in the form

   $$\mathrm{Re}\left\{s\right\} > \sigma_1$$

   For the ROC to include the $j\omega$-axis we need $\sigma_1 < 0$. Since the ROC cannot have any poles in it, all the poles of the system function must be on or to the left of the vertical line $\sigma = \sigma_1$.

   For a causal system to be stable, the system function must not have any poles on the $j\omega$-axis or in the right half s-plane.

2. **Anti-causal system:**
   If the system is anti-causal, its impulse response is equal to zero for $t \geq 0$. The ROC for the system function is to the left of a vertical line in the s-plane, and is expressed in the form
   $$\mathrm{Re}\left\{s\right\} < \sigma_2$$

   For the ROC to include the $j\omega$-axis we need $\sigma_2 > 0$. All the poles of the system function must reside on or to the right of the vertical line $\sigma = \sigma_2$.

   For an anti-causal system to be stable, the system function must not have any poles on the $j\omega$-axis or in the left half s-plane.

3. **Neither causal nor anti-causal system:**
   In this case the ROC for the system function, if it exists, is the region between two vertical lines at $\sigma = \sigma_1$ and $\sigma = \sigma_2$, and is expressed in the form

   $$\sigma_1 < \mathrm{Re}\left\{s\right\} < \sigma_2$$

   For stability we need $\sigma_1 < 0$ and $\sigma_2 > 0$. The poles of the system function may be either

   a. On or to the left of the vertical line $\sigma = \sigma_1$
   b. On or to the right of the vertical line $\sigma = \sigma_2$

Fig. 7.45 illustrates the three possibilities discussed above.

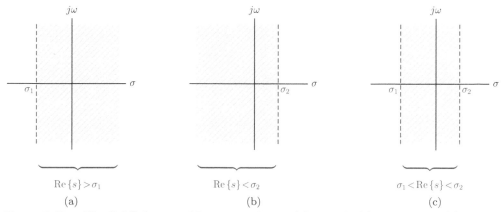

$\mathrm{Re}\{s\} > \sigma_1$ $\mathrm{Re}\{s\} < \sigma_2$ $\sigma_1 < \mathrm{Re}\{s\} < \sigma_2$

(a)   (b)   (c)

Figure 7.45 – The ROC for a stable system that is (a) causal, (b) anti-causal, (c) neither causal nor anti-causal.

Example 7.36: **Impulse response of a stable system**

A stable system is characterized by the system function

$$H(s) = \frac{15\,s\,(s+1)}{(s+3)\,(s-1)\,(s-2)}$$

Determine the ROC of the system function. Afterwards find the impulse response of the system.

Solution: The ROC for the system function is not directly stated; however, we are given enough information to deduce it. The three poles of the system function are at $s = -3, 1, 2$. Since the system is known to be stable, its ROC must include the $j\omega$-axis of the s-plane. The only possible choice is

$$-3 < \mathrm{Re}\{s\} < 1$$

as shown in Fig. 7.46(a). Partial fraction expansion of $H(s)$ is

$$H(s) = \frac{4.5}{s+3} - \frac{7.5}{s-1} + \frac{18}{s-2}$$

Based on the ROC determined above, the first term in the partial fraction expansion corresponds to a causal signal, and the other two terms correspond to anti-causal signals. The impulse response of the system is

$$h(t) = 4.5\,e^{-3t}\,u(t) + 7.5\,e^{t}\,u(-t) - 18\,e^{2t}\,u(-t) \tag{7.152}$$

which is shown in Fig. 7.46(b). We observe that $h(t)$ tends to zero as t is increased in both directions, consistent with the fact that $h(t)$ must be absolute summable for a stable system.

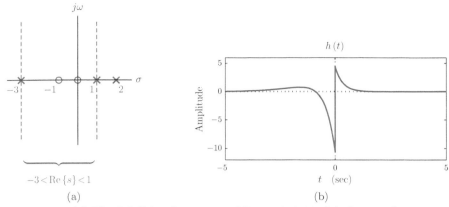

Figure 7.46 – (a) The ROC for the system of Example 7.36, (b) the impulse response.

Software resources:
ex_7_36a.m
ex_7_36b.m

Example 7.37: **Stability of a system described by a differential equation**

A causal CTLTI system is characterized by the differential equation

$$\frac{d^2 y(t)}{dt^2} + 2\frac{dy(t)}{dt} + 2\,y(t) = \frac{dx(t)}{dt} + x(t)$$

Determine if the system is stable.

Solution: The ROC for the system function is not directly stated. On the other hand, we are told that the system is causal. This bit of information should allow us to determine the ROC. Taking the Laplace transform of both sides of the differential equation and using the time differentiation property we have

$$s^2\,Y(s) + 2s\,Y(s) + 2\,Y(s) = s\,X(s) + X(s) \tag{7.153}$$

The system function is found by forming the ratio of the output and the input transforms.

$$H(s) = \frac{Y(s)}{X(s)} = \frac{s+1}{s^2 + 2s + 2}$$

The system has a zero at $s = -1$ and a pair of complex conjugate poles at $s = -1 \pm j1$. Since it is causal, the ROC must be to the right of a vertical line going through $\sigma = -1$, that is

$$\text{ROC:} \quad \text{Re}\,\{s\} > \quad 1 \tag{7.154}$$

Since the ROC includes the $j\omega$-axis of the s-plane, the system is stable. The impulse response can be determined easily by writing the $H(s)$ in the form

$$H(s) = \frac{s+1}{(s+1)^2 + 1} = \left.\frac{s}{s^2+1}\right|_{s\to s+1}$$

Using s-domain shifting property of the Laplace transform it follows that

$$h(t) = e^{-t}\cos(t)\,u(t)$$

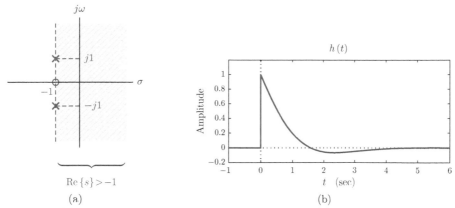

Figure 7.47 – (a) Pole-zero diagram and ROC for the system of Example 7.37, (b) impulse response $h(t)$.

Software resources:
ex_7_37a.m
ex_7_37b.m

7.5.8   Allpass systems

A system the magnitude characteristic of which is constant across all frequencies is called an *allpass system*. For a system to be considered an allpass system we need

$$|H(\omega)| = C \text{ (constant)} \qquad \text{for all } \omega \tag{7.155}$$

Consider a first-order CTLTI system with a pole at $p_1 = -\sigma_1 + j0$ and a zero at $z_1 = \sigma_1 + j0$ so that the system function is

$$H(s) = \frac{s - z_1}{s - p_1} = \frac{s - \sigma_1}{s + \sigma_1} \tag{7.156}$$

For the system to be causal and stable the parameter σ_1 needs to be positive so that the resulting pole is in the left half s-plane. The magnitude and the phase of the system function can be expressed using the conventions established in Section 7.5.5. The frequency response $H(\omega)$ is found by substituting $s = j\omega$:

$$H(\omega) = \frac{B_1(\omega)}{A_1(\omega)} = \frac{j\omega - \sigma_1}{j\omega + \sigma_1} \tag{7.157}$$

The magnitude of the frequency response is

$$|H(\omega)| = \frac{|B_1(\omega)|}{|A_1(\omega)|} = \frac{\sqrt{\omega^2 + \sigma_1^2}}{\sqrt{\omega^2 + \sigma_1^2}} = 1 \tag{7.158}$$

The phases of numerator and denominator terms are

$$\angle B_1(\omega) = \pi - \tan^{-1}\left(\frac{\omega}{\sigma_1}\right), \qquad \text{and} \qquad \angle A_1(\omega) = -\tan^{-1}\left(\frac{\omega}{\sigma_1}\right) \tag{7.159}$$

and the phase of the frequency response is

$$\angle H\left(\omega\right) = \angle B_1\left(\omega\right) - \angle A_1\left(\omega\right) = \pi - 2\tan^{-1}\left(\frac{\omega}{\sigma_1}\right) \tag{7.160}$$

The pole-zero diagram and the vector representation of the system function are shown in Fig. 7.48(a) along with the vectors

$$\overrightarrow{B_1}\left(\omega_0\right) = \overrightarrow{j\omega_0} - \overrightarrow{\sigma_1} \qquad \text{and} \qquad \overrightarrow{A_1}\left(\omega_0\right) = \overrightarrow{j\omega_0} + \overrightarrow{\sigma_1} \tag{7.161}$$

Fig. 7.48(b) shows the phase characteristics for $\sigma_1 = 1$, $\sigma_1 = 2$ and $\sigma_1 = 3$.

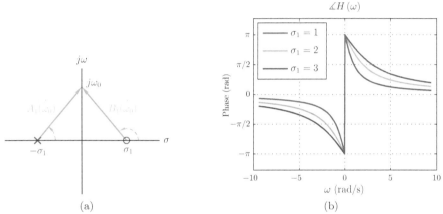

Figure 7.48 – (a) Pole-zero diagram for first-order allpass system in Eqn. (7.156), (b) phase response of the first-order allpass system.

Since the shape of the phase response can be controlled by the choice of parameter σ_1 while keeping the magnitude response constant, an allpass system is also referred to as a *phase-shifter*. Increased versatility in controlling the phase response can be obtained by choosing the pole and the zero to be complex-valued. Consider a first-order system with a pole at $p_1 = -\sigma_1 + j\omega_1$ and a zero at $z_1 = \sigma_1 + j\omega_1$. Again we choose $\sigma_1 > 0$ to obtain a system that is both causal and stable. The corresponding system function is

$$H\left(s\right) = \frac{s - z_1}{s - p_1} = \frac{s - \sigma_1 - j\omega_1}{s + \sigma_1 - j\omega_1} \tag{7.162}$$

The frequency response of the system, found by substituting $s = j\omega$ into the system function, is

$$H\left(\omega\right) = \frac{j\omega - z_1}{j\omega - p_1} = \frac{j\omega - \sigma_1 - j\omega_1}{j\omega + \sigma_1 - j\omega_1} \tag{7.163}$$

It is a trivial matter to show that the magnitude of $H\left(\omega\right)$ is still equal to unity. Its phase is

$$\angle H\left(\omega\right) = \angle\left(j\omega - \sigma_1 - j\omega_1\right) - \angle\left(j\omega + \sigma_1 - j\omega_1\right) = \pi - 2\tan^{-1}\left(\frac{\omega - \omega_1}{\sigma_1}\right) \tag{7.164}$$

The pole-zero diagram and the vector representation of the system function are shown in Fig. 7.49(a) along with the vectors

$$\overrightarrow{B_1}\left(\omega_0\right) = \overrightarrow{j\omega_0} - \overrightarrow{z_1} \qquad \text{and} \qquad \overrightarrow{A_1}\left(\omega_0\right) = \overrightarrow{j\omega_0} + \overrightarrow{p_1} \tag{7.165}$$

The phase response is shown in Fig. 7.49(b) for $\omega_1 = 1.5$ rad/s and for three different values of σ_1, namely $\sigma_1 = 1$, $\sigma_1 = 2$ and $\sigma_1 = 3$.

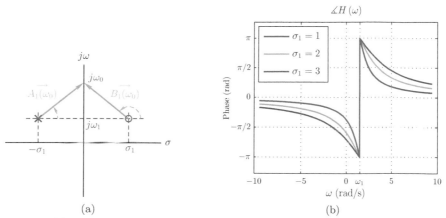

(a)   (b)

Figure 7.49 – (a) Pole-zero diagram for first-order allpass system in Eqn. (7.162), (b) phase response of the first-order allpass system.

Naturally, the system function in Eqn. (7.162) has complex coefficients. If an allpass system with real coefficients is desired, complex poles and zeros must occur in conjugate pairs. A second-order system with zeros at $z_{1,2} = \sigma_1 \pm j\omega_1$ and poles at $p_{1,2} = -\sigma_1 \pm j\omega_1$ has allpass characteristics. As before, $\sigma_1 > 0$ for a causal and stable system. The system function is

$$H(s) = \frac{(s - \sigma_1 - j\omega_1)(s - \sigma_1 + j\omega_1)}{(s + \sigma_1 - j\omega_1)(s + \sigma_1 + j\omega_1)} = \frac{(s - \sigma_1)^2 + \omega_1^2}{(s + \sigma_1)^2 + \omega_1^2}$$

The pole-zero diagram for a second-order allpass system is shown in Fig. 7.50(a). The phase response is shown in Fig. 7.50(b) for $\omega_1 = 1.5$ rad/s and for three different values of σ_1.

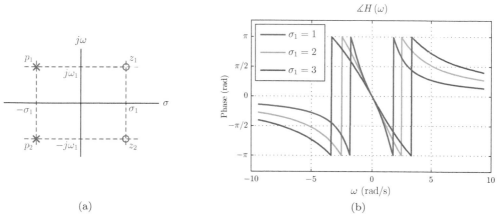

(a)   (b)

Figure 7.50 – (a) Pole-zero diagram for second-order allpass system in Eqn. (7.166), (b) phase response of the second-order allpass system.

7.5.9   Inverse systems

The inverse of a system is another system which, when connected in cascade with the original system, forms an identity system. This relationship is depicted in Fig. 7.51.

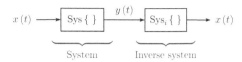

Figure 7.51 – A system and its inverse connected in cascade.

The output signal of the original system is

$$y(t) = \text{Sys}\{x(t)\} \tag{7.166}$$

We require the inverse system to recover the original input signal $x(t)$ from the output signal $y(t)$, therefore

$$x(t) = \text{Sys}_i\{y(t)\} \tag{7.167}$$

Combining Eqns. (7.166) and (7.166) yields

$$\text{Sys}_i\Big\{\text{Sys}\{x(t)\}\Big\} = x(t) \tag{7.168}$$

Let the original system and its inverse be both CTLTI systems with impulse responses $h(t)$ and $h_i(t)$ respectively as shown in Fig. 7.52. For the output signal of the inverse system to be identical to the input signal of the original system, the impulse response of the cascade combination must be equal to $\delta(t)$, that is,

$$h_{eq}(t) = h(t) * h_i(t) = \delta(t) \tag{7.169}$$

or, using the convolution integral

$$h_{eq}(t) = \int_{-\infty}^{\infty} h(\lambda)\, h_i(t - \lambda)\, d\lambda = \delta(t) \tag{7.170}$$

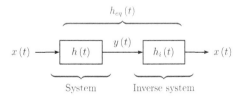

Figure 7.52 – A CTLTI system and its inverse connected in cascade.

The corresponding relationship between the system functions of the original system and the inverse system is found by taking the Laplace transform of Eqn. (7.169):

$$H_{eq}(s) = H(s)\, H_i(s) = 1 \tag{7.171}$$

Consequently, the system function of the inverse system is

$$H_i(s) = \frac{1}{H(s)} \tag{7.172}$$

Two important characteristics of the inverse system are causality and stability. We will first focus on causality. Consider the system $H(s)$ function in the form

$$H(s) = \frac{B(s)}{A(s)} = \frac{b_M s^M + b_{M-1} s^{M-1} + \ldots + b_1 s + b_0}{a_N s^N + a_{N-1} s^{N-1} + \ldots + a_1 s + a_0} \tag{7.173}$$

The system function for the inverse system is

$$H_i(s) = \frac{A(s)}{B(s)} = \frac{a_N s^N + a_{N-1} s^{N-1} + \ldots + a_1 s + a_0}{b_M s^M + b_{M-1} s^{M-1} + \ldots + b_1 s + b_0}$$

If the original system with system function $H(s)$ is causal then $M < N$ as we have established in Section 7.5.6. By the same token, causality of the inverse system with system function $H_i(s)$ requires $N \leq M$. Hence we need $N = M$ if both the original system and its inverse are required to be causal.

To analyze the stability of the inverse system we will find it more convenient to write the system function $H(s)$ in pole-zero form. Using $M = N$ we have

$$H(s) = \frac{b_N (s - z_1)(s - z_2) \ldots (s - z_N)}{a_N (s - p_1)(s - p_2) \ldots (s - p_N)} \tag{7.174}$$

If the original system is both causal and stable, all its poles must be in the left half s-plane (see Section 7.5.7), therefore

$$\operatorname{Re}\{p_k\} < 0, \qquad k = 1, \ldots, N \tag{7.175}$$

The system function of the inverse system, written in pole-zero form, is

$$H_i(s) = \frac{a_N (s - p_1)(s - p_2) \ldots (s - p_N)}{b_N (s - z_1)(s - z_2) \ldots (s - z_N)} \tag{7.176}$$

For the inverse system to be stable, its poles must also lie in the left half s-plane. The poles of the inverse system are the zeros of the original system. Therefore, for the inverse system to be stable, both zeros and poles of the original system must be in the left half s-plane. In addition to Eqn. (7.175) we also need

$$\operatorname{Re}\{z_k\} < 0, \qquad k = 1, \ldots, N \tag{7.177}$$

> A causal CTLTI system that has all of its zeros and poles in the left half s-plane is referred to as a *minimum-phase system*. A minimum-phase system and its inverse are both causal and stable.

Example 7.38: **Inverse of a system described by a differential equation**

A causal CTLTI system is described by a differential equation

$$\frac{dy(t)}{dt} + 2y(t) = \frac{dx(t)}{dt} + x(t)$$

Determine if a causal and stable inverse can be found for this system. If yes, find a differential equation for the inverse system.

Solution: Taking the Laplace transform of both sides of the differential equation we get

$$(s+2) \, Y \, (s) = (s+1) \, X \, (s)$$

The system function for the original system is

$$H \, (s) = \frac{s+1}{s+2}$$

The system function for the inverse system is found as the reciprocal of $H \, (s)$:

$$H_i \, (s) = \frac{1}{H \, (s)} = \frac{s+2}{s+1}$$

The inverse system is also causal and stable. It leads to the differential equation

$$\frac{dy \, (t)}{dt} + y \, (t) = \frac{dx \, (t)}{dt} + 2 \, x \, (t)$$

7.5.10   Bode plots

Bode plots of the frequency response are used in the analysis and design of feedback control systems. A Bode plot consists of the dB magnitude $20 \log_{10} |H \, (\omega)|$ and the phase $\angle H \, (\omega)$, each graphed as a function of $\log_{10} (\omega)$. Because of the use of the logarithm, individual contributions of the zeros and the poles of $H \, (s)$ to the magnitude of the frequency response are additive rather than multiplicative. Also, since $\log_{10} (\omega)$ is used for the horizontal axis, only positive values of ω are of interest.

Consider again the pole-zero form of the system function $H \, (s)$ given by Eqn. (7.138). A slightly modified form of it will be more convenient for use in deriving the Bode plot. Let us scale each factor in the numerator and the denominator, and write Eqn. (7.138) as

$$H \, (s) = K_1 \frac{(1 - s/z_1) \, (1 - s/z_2) \, \dots (1 - s/z_M)}{(1 - s/p_1) \, (1 - s/p_2) \, \dots (1 - s/p_N)} \qquad (7.178)$$

The new gain factor K_1 is chosen to compensate for all the scale factors used in individual terms. The dB magnitude of $H \, (\omega)$ is obtained as

$$20 \log_{10} |H \, (\omega)| = 20 \, log_{10} |K_1|$$

$$+ 20 \log_{10} |1 - j\omega/z_1| + 20 \log_{10} |1 - j\omega/z_2| + \dots + 20 \log_{10} |1 - j\omega/z_M|$$

$$- 20 \log_{10} |1 - j\omega/p_1| - 20 \log_{10} |1 - j\omega/p_2| - \dots - 20 \log_{10} |1 - j\omega/p_N|$$
$$(7.179)$$

and the phase is

$$\angle H \, (\omega) = \angle K_1 + \angle (1 - j\omega/z_1) + \angle (1 - j\omega/z_2) + \dots + \angle (1 - j\omega/z_M)$$

$$- \angle (1 - j\omega/p_1) - \angle (1 - j\omega/p_2) - \dots - \angle (1 - j\omega/p_N) \qquad (7.180)$$

To facilitate the analysis of individual contributions from zeros and poles of the system function, let us write $H \, (s)$ as a cascade combination of $M + N$ subsystems such that

$$H \, (s) = K_1 \, H_1 \, (s) \, H_2 \, (s) \, \dots H_M \, (s) \, H_{M+1} \, (s) \, H_{M+2} \, (s) \, \dots H_{M+N} \, (s) \qquad (7.181)$$

with

$$H_i(s) = 1 - s/z_i, \qquad i = 1, \ldots, M \tag{7.182}$$

and

$$H_{M+i}(s) = \frac{1}{1 - s/p_i}, \qquad i = 1, \ldots, N \tag{7.183}$$

The Bode plot for magnitude and phase can be constructed by computing the contribution of each term in Eqns. (7.182) and (7.183). We will consider four different cases.

Zero at the origin

Let $H_k(s) = s$. Corresponding dB magnitude and phase are computed as

$$20 \log_{10} |H_k(\omega)| = 20 \log_{10} |j\omega| = 20 \log_{10}(\omega) \tag{7.184}$$

and

$$\angle H_k(\omega) = \angle j\omega = 90° \tag{7.185}$$

The magnitude characteristic, when graphed as a function of $\log_{10}(\omega)$, is a straight line that goes through 0 dB at $\omega = 1$. Its slope is 20 dB per decade.[2] This is illustrated in Fig. 7.53(a).

Pole at the origin

Let $H_k(s) = 1/s$. Corresponding dB magnitude and phase are computed as

$$20 \log_{10} |H_k(\omega)| = 20 \log_{10} \left| \frac{1}{j\omega} \right| = -20 \log_{10}(\omega) \tag{7.186}$$

and

$$\angle H_k(\omega) = \angle \left(\frac{1}{j\omega} \right) = -90° \tag{7.187}$$

The magnitude characteristic, when graphed as a function of $\log_{10}(\omega)$, is a straight line that goes through 0 dB at $\omega = 1$. Its slope is -20 dB per decade. This is illustrated in Fig. 7.53(b).

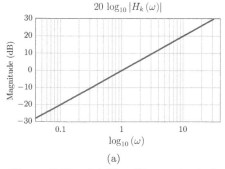
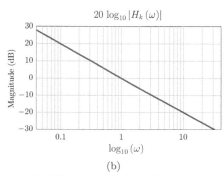

Figure 7.53 – (a) The dB magnitude for $H_k(s) = s$, (b) dB magnitude for $H_k(s) = 1/s$.

---

2 One decade corresponds to a tenfold change in ω. If $\omega_2 = 10\omega_1$, then $\log_{10}(\omega_2) = \log_{10}(\omega_1) + 1$.

Single real zero

Let z_k be a real-valued zero of the system. Substituting $s = j\omega$ into Eqn. (7.182) we obtain

$$H_k(\omega) = H_k(s)\Big|_{s=j\omega} = 1 - j\omega/z_k \tag{7.188}$$

Corresponding dB magnitude is

$$20 \log_{10}|H_k(\omega)| = 20 \log_{10}\sqrt{1 + \left(\frac{\omega}{z_k}\right)^2} = 10 \log_{10}\left[1 + \left(\frac{\omega}{z_k}\right)^2\right] \tag{7.189}$$

and the corresponding phase characteristic is

$$\angle H_k(\omega) = \angle\left(1 - \frac{j\omega}{z_k}\right) = \tan^{-1}\left(-\frac{\omega}{z_k}\right) = -\tan^{-1}\left(\frac{\omega}{z_k}\right) \tag{7.190}$$

Keep in mind that we need to graph $20 \log_{10}|H_k(\omega)|$ and $\angle H_k(\omega)$ as functions of $\log_{10}(\omega)$, and we are only interested in positive values of ω.

1. If $\omega \ll |z_k|$ we have

$$\left(\frac{\omega}{z_k}\right)^2 \ll 1 \tag{7.191}$$

The magnitude in Eqn. (7.189) can be approximated as

$$20 \log_{10}|H_k(\omega)| \approx 10 \log_{10}(1) = 0 \tag{7.192}$$

The phase angle can be approximated as

$$\angle H_k(\omega) \approx -\tan^{-1}(0) = 0 \tag{7.193}$$

2. At the opposite extreme, for $\omega \gg |z_k|$ we have

$$\left(\frac{\omega}{z_k}\right)^2 \gg 1 \tag{7.194}$$

The magnitude in Eqn. (7.189) can be approximated as

$$20 \log_{10}|H_k(\omega)| \approx 10 \log_{10}\left(\frac{\omega}{z_k}\right)^2 = 20 \log_{10}\left(\frac{\omega}{|z_k|}\right) \tag{7.195}$$

which can also be written in the form

$$20 \log_{10}|H_k(\omega)| \approx 20 \log_{10}(\omega) - 20 \log_{10}|z_k| \tag{7.196}$$

In this case the phase angle depends on the sign of z_k:

$$\angle H_k(\omega) \approx \begin{cases} 90°, & z_k < 0 \\ -90°, & z_k > 0 \end{cases} \tag{7.197}$$

3. At the frequency $\omega = |z_k|$ the magnitude is

$$20 \log_{10}|H_k(\omega)| = 10 \log_{10}(2) \approx 3 \text{ dB} \tag{7.198}$$

The phase angle depends on the sign of z_k:

$$\angle H_k(\omega) \approx \begin{cases} 45°, & z_k < 0 \\ -45°, & z_k > 0 \end{cases} \tag{7.199}$$

The conclusions obtained above can be summarized as follows:

For the term $H_k(s) = 1 - s/z_k$ with real-valued z_k:

1. **Magnitude:** For $\omega << |z_k|$ the magnitude characteristic is asymptotic to 0 dB. For $\omega >> |z_k|$ it becomes asymptotic to a straight line with a slope of 20 dB per decade, which intersects the horizontal axis at $\omega = |z_k|$. At the frequency $\omega = |z_k|$ the actual magnitude is approximately equal to 3 dB; therefore, the characteristic passes 3 dB above the intersection of the two asymptotes.

2. **Phase:** For $\omega << |z_k|$ the phase is asymptotic to 0 degrees. For $\omega >> |z_k|$ the phase is 90 degrees for $z_k < 0$ and -90 degrees for $z_k > 0$. At $\omega = z_k$ the phase is 45 degrees for $z_k < 0$ and -45 degrees for $z_k > 0$.

This is illustrated in Figs. 7.54 and 7.55.

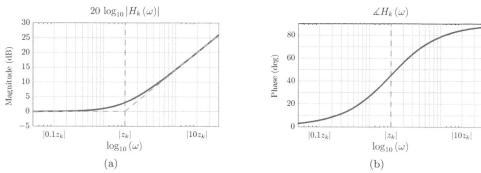

Figure 7.54 – Bode plots for $H_k(s) = (1 - s/z_k)$ with $z_k < 0$: (a) magnitude, (b) phase.

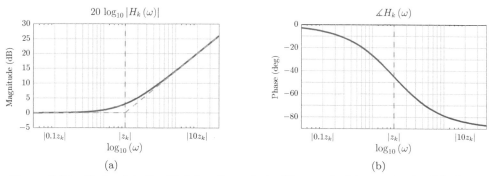

Figure 7.55 – Bode plots for $H_k(s) = (1 - s/z_k)$ with $z_k > 0$: (a) magnitude, (b) phase.

Single real pole

Derivation of the Bode plot for a single real pole is similar. Let p_k be a real-valued pole of the system. Substituting $s = j\omega$ into Eqn. (7.183) yields

$$H_k(\omega) = \left.\frac{1}{1 - s/p_k}\right|_{s=j\omega} = \frac{1}{1 - j\omega/p_k} \tag{7.200}$$

Corresponding dB magnitude is

$$20 \log_{10} |H_k(\omega)| = 20 \log_{10} \frac{1}{\sqrt{1 + \left(\dfrac{\omega}{p_k}\right)^2}} = -10 \log_{10} \left[1 + \left(\frac{\omega}{p_k}\right)^2\right] \qquad (7.201)$$

and the corresponding phase characteristic is

$$\angle H_k(\omega) = \angle \left[\frac{1}{1 - j\omega/p_k}\right] = -\tan^{-1}\left(-\frac{\omega}{p_k}\right) = \tan^{-1}\left(\frac{\omega}{p_k}\right) \qquad (7.202)$$

1. If $\omega << |p_k|$ we have

$$\left(\frac{\omega}{p_k}\right)^2 << 1 \qquad (7.203)$$

The magnitude in Eqn. (7.201) can be approximated as

$$20 \log_{10} |H_k(\omega)| \approx -10 \log_{10}(1) = 0 \qquad (7.204)$$

The phase angle can be approximated as

$$\angle H_k(\omega) \approx \tan^{-1}(0) = 0 \qquad (7.205)$$

2. At the opposite extreme, for $\omega >> |p_k|$ we have

$$\left(\frac{\omega}{p_k}\right)^2 >> 1 \qquad (7.206)$$

The magnitude in Eqn. (7.201) can be approximated as

$$20 \log_{10} |H_k(\omega)| \approx -10 \log_{10}\left(\frac{\omega}{p_k}\right)^2 = -20 \log_{10}\left(\frac{\omega}{|p_k|}\right) \qquad (7.207)$$

which can also be written in the form

$$20 \log_{10} |H_k(\omega)| \approx -20 \log_{10}(\omega) + 20 \log_{10} |p_k| \qquad (7.208)$$

In this case the phase angle depends on the sign of p_k:

$$\angle H_k(\omega) \approx \begin{cases} -90°, & p_k < 0 \\ 90°, & p_k > 0 \end{cases} \qquad (7.209)$$

3. At the frequency $\omega = |p_k|$ the magnitude is

$$20 \log_{10} |H_k(\omega)| = -10 \log_{10}(2) \approx -3 \text{ dB} \qquad (7.210)$$

The phase angle depends on the sign of p_k:

$$\angle H_k(\omega) \approx \begin{cases} -45°, & p_k < 0 \\ 45°, & p_k > 0 \end{cases} \qquad (7.211)$$

The conclusions obtained above can be summarized as follows:

For the term $H_k\left(s\right)=1/\left(1-s/p_k\right)$ with real-valued p_k:

1. **Magnitude:** For $\omega << |p_k|$ the magnitude characteristic is asymptotic to 0 dB. For $\omega >> |p_k|$ it becomes asymptotic to a straight line with a slope of -20 dB per decade, which intersects the 0-dB axis at $\omega = |p_k|$. At the frequency $\omega = |p_k|$ the actual magnitude is approximately equal to -3 dB, therefore the characteristic passes 3 dB below the intersection of the two asymptotes.

2. **Phase:** For $\omega << |p_k|$ the phase is asymptotic to 0 degrees. For $\omega >> |p_k|$ the phase is -90 degrees for $p_k < 0$ and 90 degrees for $p_k > 0$. At $\omega = p_k$ the phase is -45 degrees for $p_k < 0$ and 45 degrees for $p_k > 0$.

This is illustrated in Figs. 7.56 and 7.57.

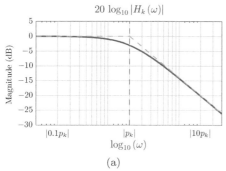

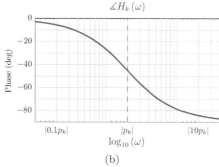

(a)   (b)

Figure 7.56 – Bode plots for $H_k\left(s\right)=1/\left(1-s/p_k\right)$ with $p_k < 0$: (a) magnitude, (b) phase.

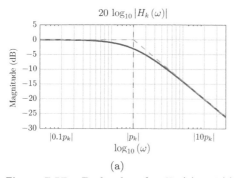

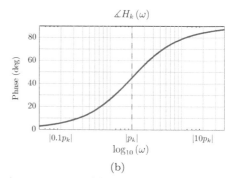

(a)   (b)

Figure 7.57 – Bode plots for $H_k\left(s\right)=1/\left(1-s/p_k\right)$ with $p_k > 0$: (a) magnitude, (b) phase.

Example 7.39: **Constructing a Bode plot**

A CTLTI system is described by the system function

$$H\left(s\right)=\frac{s\left(1+s/300\right)}{\left(1+s/5\right)\left(1+s/40\right)}$$

Plot the Bode magnitude and the phase characteristics of the system.

Solution: Let us express the system function in the form of four subsystems connected in cascade:

$$H\left(s\right)=H_1\left(s\right)H_2\left(s\right)H_3\left(s\right)H_4\left(s\right)$$

The subsystems are

$$H_1\left(s\right) = s\,,\qquad H_2\left(s\right) = 1 + s/300\,,\qquad H_3\left(s\right) = \frac{1}{1 + s/5}\,,\qquad H_4\left(s\right) = \frac{1}{1 + s/40}$$

Magnitude and phase contributions of the four subsystems are shown in Fig. 7.58.

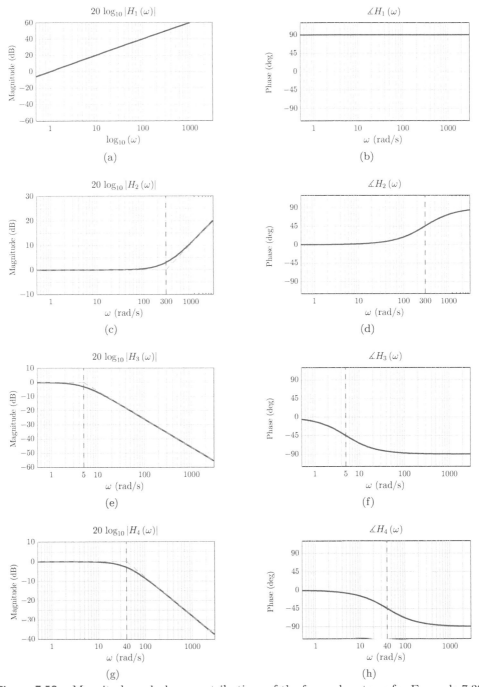

Figure 7.58 – Magnitude and phase contributions of the four subsystems for Example 7.39.

Bode plot for the magnitude of the system function is obtained by adding the magnitude contributions in Fig. 7.58(a), (c), (e) and (g). Asymptote lines of the magnitude characteristics can also be added together. Four sections of asymptotes are shown in Fig. 7.59(a) numbered 1 through 4.

1. Section 1 covers frequencies up to $\omega = 5$ rad/s, and has a slope of 20 dB per decade which is the sum of slopes of all four contributing asymptote lines up to that point. At $\omega = 5$ rad/s, its value is 13.98 dB (see Problem 7.43 at the end of this chapter).
2. Section 2 covers frequencies from $\omega = 5$ rad/s to $\omega = 40$ rad/s, and has a slope of zero (20 dB per decade from H_1 and -20 dB per decade from $H3$.)
3. Section 3 covers frequencies from $\omega = 40$ rad/s to $\omega = 300$ rad/s, and has a slope of -20 dB per decade (20 dB per decade from H_1; -20 dB per decade each from $H3$ and H_4). At the endpoint $\omega = 300$ rad/s, its value is -3.52 dB (see Problem 7.43 at the end of this chapter).
4. Section 4 covers frequencies greater than $\omega = 300$ rad/s, and has a slope of zero (20 dB per decade each from H_1 and H_2; -20 dB per decade each from $H3$ and H_4).

The actual Bode magnitude plot is also shown in Fig. 7.59. Notice how it passes approximately 3 dB above or below each corner point depending on whether it belongs to a zero or a pole.

Bode plot for the phase of the system function is obtained by adding the phase contributions in Fig. 7.58(b), (d), (f) and (h), and is shown in Fig. 7.59(b).

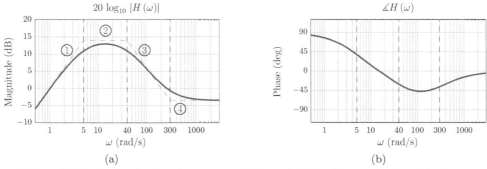

Figure 7.59 – Bode plot for the system of Example 7.39: (a) magnitude, (b) phase.

Software resources:
ex_7_39a.m
ex_7_39b.m

Conjugate pair of poles

Consider a causal and stable second-order system with a pair of complex conjugate poles, that is, $p_2 = p_1^*$. The system function is in the form

$$H(s) = \frac{1}{(1 - s/p_1)(1 - s/p_1^*)} \tag{7.212}$$

$H(s)$ can be written in a slightly different form by multiplying both the numerator and the denominator of Eqn. (7.212) with the product $(p_1 p_1^*)$:

$$H(s) = \frac{|p_1|^2}{(s - p_1)(s - p_1^*)} \tag{7.213}$$

Let us put $H(s)$ into the standard form

$$H(s) = \frac{\omega_0^2}{s^2 + 2\zeta\omega_0 s + \omega_0^2} \tag{7.214}$$

with

$$\omega_0^2 = |p_1|^2, \quad \text{and} \quad 2\zeta\omega_0 = -2\,\mathrm{Re}\{p_1\} \tag{7.215}$$

It follows from Eqn. (7.215) that

$$\zeta = -\frac{\mathrm{Re}\{p_1\}}{|p_1|} \tag{7.216}$$

Since the system is causal and stable, $\mathrm{Re}\{p_1\} < 0$. Consequently, when the poles of the system form a complex conjugate pair, we have $0 < \zeta < 1$.

Analysis of the second-order system

The form of the second-order system function given by Eqn. (7.214) can be used for a system with either two real poles or a complex conjugate pair. Let us write $H(s)$ using two poles p_1 and p_2:

$$H(s) = \frac{1}{(1 - s/p_1)(1 - s/p_2)} = \frac{p_1\,p_2}{(s - p_1)(s - p_2)} \tag{7.217}$$

Equating $H(s)$ in Eqn. (7.217) with the form given in Eqn. (7.214) we obtain the relationships

$$\omega_0^2 = p_1\,p_2, \quad \text{and} \quad 2\zeta\omega_0 = -\,\mathrm{Re}\{p_1\} - \mathrm{Re}\{p_2\} \tag{7.218}$$

The parameter $\omega_0 = \sqrt{p_1\,p_2}$ is called the *natural undamped frequency* of the system. The parameter ζ is called the *damping ratio*, and is computed as

$$\zeta = \frac{-\,\mathrm{Re}\{p_1\} - \mathrm{Re}\{p_2\}}{2\sqrt{p_1\,p_2}} \tag{7.219}$$

The poles of the system function can be related to the parameters ω_0 and ζ as

$$p_{1,2} = -\zeta\omega_0 \pm \omega_0\sqrt{\zeta^2 - 1} \tag{7.220}$$

Since the system is causal and stable, both poles are in the left half of the s-plane. As a result the numerator of Eqn. (7.219) is positive, and therefore the damping ratio ζ must be positive. Locations of the poles depend on the value of ζ. We will observe three distinct possibilities:

$\zeta > 1$: The poles p_1 and p_2 are real-valued and distinct. The system is said to be *overdamped*.

$\zeta = 1$: The expression in square root in Eqn. (7.220) equals zero. The two poles of the system function are both at $p_1 = p_2 = -\zeta\omega_0$. In this case the system is said to be *critically damped*.

$\zeta < 1$: The expression in square root in Eqn. (7.220) is negative. The two poles of the system function are a complex conjugate pair:

$$p_{1,2} = -\zeta\omega_0 \pm j\omega_0\sqrt{1 - \zeta^2} = -\zeta\omega_0 \pm j\omega_d$$

In this case the system is said to be *underdamped*.

Fig. 7.60 illustrates the placement of the two poles based on the values of the parameters ζ and ω_0.

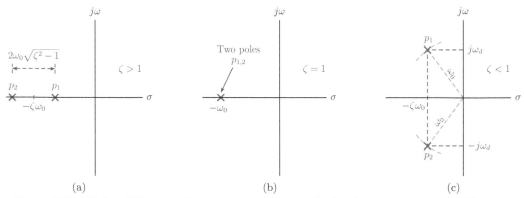

Figure 7.60 – Poles of the second-order system function for (a) $\zeta > 1$, overdamped, (b) $\zeta = 1$, critically damped, (c) $\zeta < 1$, underdamped.

It is also interesting to look at the movement of the two poles as the parameter ζ is changed. Fig. 7.61 illustrates the trajectories of p_1 and p_2.

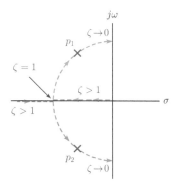

Figure 7.61 – Trajectories of the two poles p_1 and p_2 as ζ is varied.

Let us determine the magnitude and phase responses of the system. Substituting $s = j\omega$ into Eqn. (7.214) yields

$$H(\omega) = \frac{\omega_0^2}{(j\omega)^2 + 2\zeta\omega_0 \, j\omega + \omega_0^2} = \frac{1}{1 - \left(\dfrac{\omega}{\omega_0}\right)^2 + j2\zeta\left(\dfrac{\omega}{\omega_0}\right)} \tag{7.221}$$

The dB magnitude of the system function is

$$20 \log_{10}|H(\omega)| = -10 \log_{10}\left\{\left[1 - \left(\frac{\omega}{\omega_0}\right)^2\right]^2 + \left[2\zeta\left(\frac{\omega}{\omega_0}\right)\right]^2\right\} \tag{7.222}$$

and the phase characteristic is

$$\angle H\left(\omega\right) = -\tan^{-1}\left[\frac{2\zeta\left(\dfrac{\omega}{\omega_0}\right)}{1-\left(\dfrac{\omega}{\omega_0}\right)^2}\right] \tag{7.223}$$

If $\omega << \omega_0$ then we have

$$20\log_{10}\left|H\left(\omega\right)\right| \approx 0\,, \quad \text{and} \quad \angle H\left(\omega\right) \approx 0$$

If $\omega >> \omega_0$, the dB magnitude becomes

$$20\log_{10}\left|H\left(\omega\right)\right| \approx -10\log_{10}\left(\frac{\omega}{\omega_0}\right)^4 = -40\log_{10}\left(\omega\right)+40\log_{10}\left(\omega\right)$$

and the phase becomes

$$\angle H\left(\omega\right) \approx -\tan^{-1}\left[2\zeta\left(\frac{\omega_0}{\omega}\right)\right] = -180°$$

It will also be interesting to check the magnitude and the phase at the corner frequency. Substituting $\omega = \omega_0$ into Eqn. (7.221) we obtain

$$H\left(\omega_0\right) = \frac{1}{j2\zeta}$$

In this case the phase angle is $\angle H\left(\omega\right) = -90$ degrees. Before we compute the dB magnitude at the corner frequency we will define the *quality factor* as

$$Q = \frac{1}{2\zeta} \tag{7.224}$$

so that $H\left(\omega_0\right) = -jQ$. The dB magnitude at $\omega = \omega_0$ is

$$20\log_{10}\left|H\left(\omega_0\right)\right| = 20\log_{10}Q$$

The conclusions obtained above can be summarized as follows:

For the second-order system function of Eqn. (7.214):

1. **Magnitude:** For $\omega << \omega_0$ the magnitude characteristic is asymptotic to 0 dB. For $\omega >> \omega_0$ it becomes asymptotic to a straight line with a slope of -40 dB per decade, which intersects the 0-dB axis at ω_0. At the corner frequency $\omega = \omega_0$ the actual magnitude is $20\log_{10}Q = -20\log_{10}\left(2\zeta\right)$.
2. **Phase:** For $\omega << \omega_0$ the phase is asymptotic to 0 degrees. For $\omega >> \omega_0$ it becomes asymptotic to -180 degrees. At the corner frequency $\omega = \omega_0$ the phase is -90 degrees.

The asymptotic behavior of the dB magnitude characteristic is illustrated in Fig. 7.62 for $\omega_0 = 3$ rad/s and for two different values of ζ, namely $\zeta = 0.1$ and $\zeta = 1.2$.

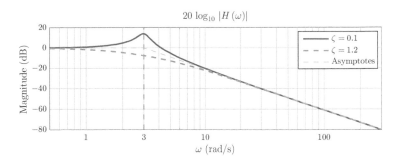

Figure 7.62 – The dB magnitude for the second-order system of Eqn. (7.214) for $\zeta = 0.1$ and $\zeta = 1.2$.

The definitions of overdamped, critically damped and underdamped systems can be related to the new parameter Q as follows:

$$\text{Overdamped:} \quad \zeta > 1 \implies Q < 0.5$$
$$\text{Critically damped:} \quad \zeta = 1 \implies Q = 0.5$$
$$\text{Underdamped:} \quad \zeta < 1 \implies Q > 0.5$$

Bode plots for dB magnitude and phase of the second-order system are shown in Fig. 7.63 for several values of Q.

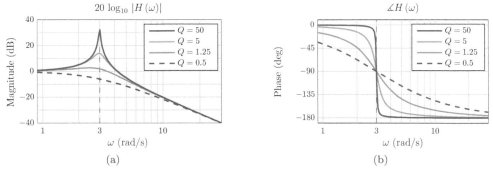

(a)   (b)

Figure 7.63 – Decibel magnitude and phase characteristics of the second-order system of Eqn. (7.214) for $\omega_0 = 3$ rad/s and several values of Q.

The responses of the second-order system to unit-impulse and unit-step input signals are also of interest. Let us use partial fraction expansion to find the impulse response $h(t)$. For the case of two distinct poles (real or complex) the system function $H(s)$ in partial fraction form is

$$H(s) = \frac{k_1}{s - p_1} + \frac{k2}{s - p_2} \tag{7.225}$$

The residues in Eqn. (7.225) are found as

$$k_1 = \frac{p_1 \, p_2}{p_1 - p_2} = \frac{\omega_0}{2\sqrt{\zeta^2 - 1}}, \quad \text{and} \quad k_2 = \frac{p_1 \, p_2}{p_2 - p_1} = -\frac{\omega_0}{2\sqrt{\zeta^2 - 1}}$$

and the impulse response is

$$
\begin{aligned}
h\left(t\right) &= k_1\, e^{p_1 t} + k_2\, e^{p_1 t} \\
&= \frac{\omega_0}{2\sqrt{\zeta^2 - 1}}\, e^{-\zeta \omega_0 t}\left[e^{\left(\omega_0\sqrt{\zeta^2 - 1}\right)t} - e^{-\left(\omega_0\sqrt{\zeta^2 - 1}\right)t}\right] u\left(t\right)
\end{aligned}
\tag{7.226}
$$

If $\zeta < 1$, the result in Eqn. (7.226) becomes

$$
h\left(t\right) = \frac{\omega_0}{\sqrt{1 - \zeta^2}}\, e^{-\zeta \omega_0 t}\sin\left(\omega_0\sqrt{1 - \zeta^2}t\right)u\left(t\right) = \frac{\omega_0}{\sqrt{1 - \zeta^2}}\sin\left(\omega_d t\right)u\left(t\right)
\tag{7.227}
$$

If $\zeta = 1$ then the expression in Eqn. (7.226) or the one in Eqn. (7.227) will not work since $p_1 = p_2$, and the partial fraction expansion in Eqn. (7.225) is not valid in this case. For $\zeta = 1$ we have

$$
H\left(s\right) = \frac{\omega_0}{\left(s + \omega_0\right)^2}
$$

Using the s-domain differentiation property of the Laplace transform it can be shown that

$$
h\left(t\right) = \omega_0^2\, t\, e^{-\omega_0 t}\, u\left(t\right)
\tag{7.228}
$$

The unit-step response can be found by convolving the impulse response found with the unit-step input signal:

$$
\mathrm{Sys}\left\{u\left(t\right)\right\} = h\left(t\right) * u\left(t\right) = \int_0^t h\left(\lambda\right)d\lambda, \qquad t \geq 0
\tag{7.229}
$$

For $\zeta \neq 1$ the use of Eqn. (7.226) in Eqn. (7.229) yields

$$
\begin{aligned}
\mathrm{Sys}\left\{u\left(t\right)\right\} &= 1 + \frac{1}{p_1 - p_2}\left[p_2\, e^{p_1 t} - p_1\, e^{p_2 t}\right]u\left(t\right) \\
&= 1 + \frac{e^{-\omega_0 \zeta t}}{2\omega_0\sqrt{1 - \zeta^2}}\left[\left(-\omega_0\zeta - \omega_0\sqrt{1 - \zeta^2}\right)e^{\left(\omega_0\sqrt{1 - \zeta^2}\right)t}\right. \\
&\qquad\qquad\qquad \left. + \left(\omega_0\zeta - \omega_0\sqrt{1 - \zeta^2}\right)e^{\left(-\omega_0\sqrt{1 - \zeta^2}\right)t}\right]u\left(t\right)
\end{aligned}
\tag{7.230}
$$

If $\zeta = 1$, the unit-step response is found by integrating the result Eqn. (7.228):

$$
\mathrm{Sys}\left\{u\left(t\right)\right\} = \left[1 - e^{-\omega_0 t} - \omega_0\, t\, e^{-\omega_0 t}\right]u\left(t\right)
\tag{7.231}
$$

Impulse response of the system is shown in Fig. 7.64(a) for several values of the damping ratio ζ. Fig. 7.64(b) shows the unit-step response of the system for the same set of values for ζ.

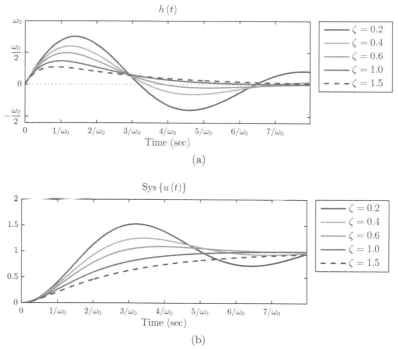

Figure 7.64 – (a) Impulse responses of second-order systems with varying damping ratios, (b) unit-step responses of second-order systems with varying damping ratios.

7.6 Simulation Structures for CTLTI Systems

Development of simulation structures for continuous-time systems was discussed briefly in Section 2.6 of Chapter 2 in the context of obtaining a block diagram from a differential equation. Our discussion in this section will parallel that of Section 2.6, and will utilize the s-domain system function as the starting point.

7.6.1 Direct-form implementation

The method of obtaining a block diagram from an s-domain system function will be derived using a third-order system, but its generalization to higher-order system functions is quite straightforward. Consider a CTLTI system described by a system function $H(s)$.

$$H(s) = \frac{Y(s)}{X(s)} = \frac{b_2 s^2 + b_1 s + b_0}{s^3 + a_2 s^2 + a_1 s + a_0} \tag{7.232}$$

$X(s)$ and $Y(s)$ are the Laplace transforms of the input and the output signals respectively. Let us use an intermediate function $W(s)$ and express the system function as

$$H(s) = \frac{Y(s)}{W(s)} \frac{W(s)}{X(s)} = \frac{b_2 s^{-1} + b_1 s^{-2} + b_0 s^{-3}}{1 + a_2 s^{-1} + a_1 s^{-2} + a_0 s^{-3}} \tag{7.233}$$

where we have also multiplied both the numerator and the denominator of the system function with s^{-3} to ensure that no positive powers of s appear. The relationships described by Eqns. (7.232) and (7.233) are illustrated in Fig. 7.65.

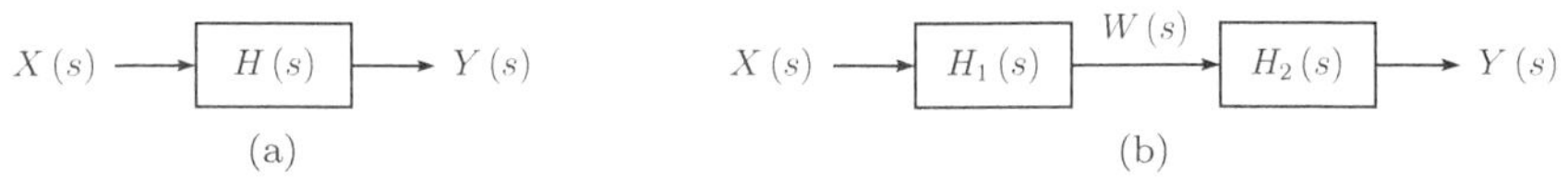

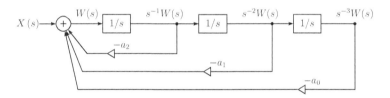

Figure 7.65 – (a) CTLTI system with system function $H(s)$, (b) cascade form using an intermediate function $W(s)$.

If we choose to associate $W(s)/X(s)$ with the denominator of Eqn. (7.233), that is,

$$H_1(s) = \frac{W(s)}{X(s)} = \frac{1}{1 + a_2 s^{-1} + a_1 s^{-2} + a_0 s^{-3}} \tag{7.234}$$

then we have

$$H_2(s) = \frac{Y(s)}{W(s)} = b_2 s^{-1} + b_1 s^{-2} + b_0 s^{-3} \tag{7.235}$$

to satisfy Eqn. (7.233). Rearranging the terms in Eqn. (7.234) yields

$$W(s) = X(s) - a_2 s^{-1} W(s) - a_1 s^{-2} W(s) - a_0 s^{-3} W(s) \tag{7.236}$$

The relationship in Eqn. (7.236) can easily be translated to a simulation diagram as shown in Fig. 7.66.

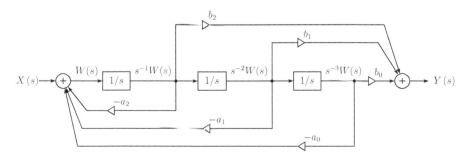

Figure 7.66 – Block diagram implementation of Eqn. (7.236).

Solving Eqn. (7.235) for the output transform $Y(s)$ yields

$$Y(s) = b_2 s^{-1} W(s) + b_1 s^{-2} W(s) + b_0 s^{-3} W(s) \tag{7.237}$$

The terms $s^{-1}W(s)$, $s^{-2}W(s)$ and $s^{-3}W(s)$ are already available in the simulation diagram of Fig. 7.66. Utilizing them to implement the relationship described by Eqn. (7.237), the diagram can be completed as shown in Fig. 7.67.

Figure 7.67 – Completed block diagram for simulating the system function $H(s)$ of Eqn. (7.232).

Using the integration property of the Laplace transform stated by Eqn. (7.86), multiplication by s^{-1} in the transform domain corresponds to integration of the signal in the time domain. Consequently, the blocks with system function $1/s$ in the diagram of Fig. 7.67 represent integrators.

$$x(t) \longrightarrow \boxed{\int dt} \longrightarrow y(t) \qquad\qquad X(s) \longrightarrow \boxed{1/s} \longrightarrow Y(s)$$

$$\text{(a)} \qquad\qquad\qquad\qquad\qquad \text{(b)}$$

Figure 7.68 – Integrator component (a) in the time domain, (b) in the transform domain.

It should be noted that the diagram in Fig. 7.67 could easily have been obtained directly from the system function in Eqn. (7.233) by inspection, using the following set of rules:

1. Begin by ordering terms of numerator and denominator polynomials from highest to lowest order of s.
2. Ensure that the leading coefficient in the denominator, that is, the coefficient of the highest order term, is equal to unity. (If it is not equal to unity, simply scale all coefficients to satisfy this rule.)
3. Set gain factors of feed-forward branches equal to the numerator coefficients.
4. Set gain factors of feedback branches equal to the negatives of the denominator coefficients.
5. Be careful to account for any missing powers of s in either polynomial, and treat them as terms with their coefficients equal to zero.

Example 7.40: **Obtaining a block diagram from system function**

A CTLTI system is described through the system function

$$H(s) = \frac{2s^3 - 26s + 24}{s^4 + 7s^3 + 21s^2 + 37s + 30}$$

Draw a block diagram for simulating this system.

Solution: Using the technique outlined above, the block diagram shown in Fig. 7.69 is obtained.

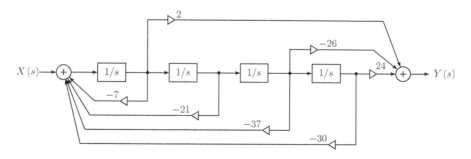

Figure 7.69 – Block diagram for Example 7.40.

7.6.2   Cascade and parallel forms

Instead of simulating a system with the direct-form block diagram discussed in the previous section, it is also possible to express the system function as either the product or the sum of lower order sections, and base the block diagram on cascade or parallel combination smaller diagrams. Consider a system function of order M that can be expressed in the form

$$
\begin{aligned}
H\left(s\right) &= H_1\left(s\right) H_2\left(s\right) \ldots H_M\left(s\right) \\
&= \frac{W_1\left(s\right)}{X\left(s\right)} \frac{W_2\left(s\right)}{W_1\left(s\right)} \cdots \frac{Y\left(s\right)}{W_{M-1}\left(s\right)}
\end{aligned}
\tag{7.238}
$$

One method of simulating this system would be to build a diagram for each of the subsections $H_i\left(s\right)$ using the direct-form approach discussed previously, and then to connect those sections in cascade as shown in Fig. 7.70.

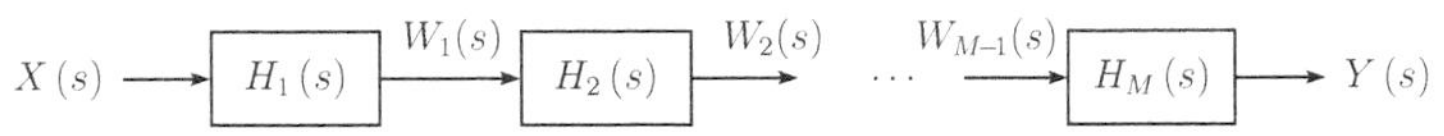

Figure 7.70 – Cascade implementation of $H\left(s\right)$.

An easy method of sectioning a system function in the style of Eqn. (7.238) would be to determine the poles and the zeros of the system function, and to use them for factoring numerator and denominator polynomials. Afterwards, each section may be constructed by using one of the poles. Each zero is incorporated into one of the sections, and some sections may have constant numerators. If some poles and zeros are complex-valued, we may choose to keep conjugate pairs together in second-order sections to avoid the need for complex gain factors in the diagram. The next example will illustrate this process.

Example 7.41: **Cascade form block diagram**

Develop a cascade form block diagram for simulating the system used in Example 7.40.

Solution: The system function specified in Example 7.40 can be factored into the form

$$
H\left(s\right) = \frac{2\left(s+4\right)\left(s-3\right)\left(s-1\right)}{\left(s+1-j2\right)\left(s+1+j2\right)\left(s+3\right)\left(s+2\right)}
$$

The roots can be found easily using MATLAB (see MATLAB Exercise 7.3). Let us write $H\left(s\right)$ as

$$
H\left(s\right) = H_1\left(s\right) H_2\left(s\right) H_3\left(s\right)
$$

by choosing

$$
H_1\left(s\right) = \frac{2\left(s+4\right)}{\left(s+1-j2\right)\left(s+1+j2\right)} = \frac{2s+8}{s^2+2s+5}, \qquad H_2\left(s\right) = \frac{s-3}{s+3}, \qquad H_3\left(s\right) = \frac{s-1}{s+2}
$$

The cascade form simulation diagram is shown in Fig. 7.71.

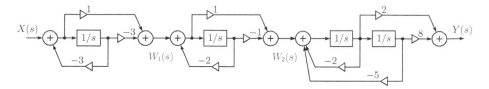

Figure 7.71 – Cascade form block diagram for Example 7.41.

It is also possible to consolidate the neighboring adders although this would cause the intermediate signals $W_1(s)$ and $W_2(s)$ to be lost. The resulting diagram is shown in Fig. 7.72.

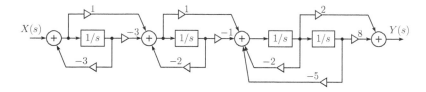

Figure 7.72 – Further simplified cascade form block diagram for Example 7.41.

An alternative to the cascade form simulation diagram is a parallel form diagram which is based on writing the system function as a sum of lower-order functions:

$$
\begin{aligned}
H(s) &= \bar{H}_1(s) + \bar{H}_2(s) + \ldots + \bar{H}_M(s) \\
&= \frac{\bar{W}_1(s)}{X(s)} + \frac{\bar{W}_2(s)}{X(s)} + \ldots \frac{\bar{W}_M(s)}{X(s)}
\end{aligned}
\tag{7.239}
$$

A simulation diagram can be constructed by implementing each term in Eqn. (7.239) using the direct-form approach, and then connecting the resulting subsystems in a parallel configuration as shown in Fig. 7.73.

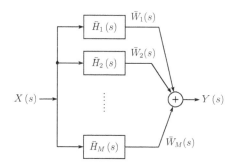

Figure 7.73 – Parallel implementation of $H(s)$.

A rational system function $H(s)$ can be sectioned in the form of Eqn. (7.239) using partial fraction expansion. If some poles and zeros are complex-valued, we may choose to keep conjugate pairs together in second-order sections to avoid the need for complex gain factors in the diagram. This process will be illustrated in the next example.

Example 7.42: Parallel form block diagram

Develop a parallel form block diagram for simulating the system used in Example 7.40.

Solution: The system function specified in Example 7.40 can be expanded into partial fractions as

$$H\left(s\right) = \frac{-2+j3}{s+1-j2} + \frac{-2-j3}{s+1+j2} + \frac{12}{s+2} + \frac{-6}{s+3}$$

Since the first two terms have complex poles and complex conjugate residues, we will combine them back into a second-order section to avoid the need for complex gain factors in the diagram. This results in

$$H\left(s\right) = \bar{H}_1\left(s\right) + \bar{H}_2\left(s\right) + \bar{H}_3\left(s\right)$$

with

$$\bar{H}_1\left(s\right) = \frac{-2+j3}{s+1-j2} + \frac{-2-j3}{s+1+j2} = \frac{-4s-16}{s^2+2s+5}, \qquad \bar{H}_2\left(s\right) = \frac{12}{s+2}, \qquad \bar{H}_3\left(s\right) = \frac{-6}{s+3}$$

The parallel form simulation diagram is shown in Fig. 7.74.

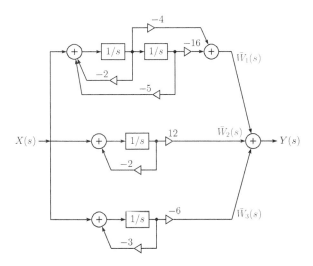

Figure 7.74   Parallel form block diagram for Example 7.42.

7.7   Unilateral Laplace Transform

It was mentioned in earlier discussion that the Laplace transform as defined by Eqn. (7.1) is sometimes referred to as the *bilateral Laplace transform*.

An alternative version of the Laplace transform, known as the *unilateral Laplace transform*, is defined by

$$X_u(s) = \mathcal{L}_u\{x(t)\}$$

$$= \int_{0+}^{\infty} x(t)\, e^{-st}\, dt \tag{7.240}$$

We use the subscript "u" to distinguish the unilateral Laplace transform from its bilateral counterpart. Comparing the definition in Eqn. (7.240) with the definition of the bilateral transform given by Eqn. (7.1) it is clear that the only difference is the lower limit of the integral. In fact, if the signal $x(t)$ is causal, that is, if $x(t) = 0$ for $t < 0$, then both definitions of the Laplace transform produce the same result.

$$X_u(s) = X(s) \quad \text{if} \quad x(t)\colon \text{causal} \tag{7.241}$$

In many engineering applications we work with causal signals, and may not need to pay attention to which Laplace transform definition we use. On the other hand, for a non-causal signal, the two definitions produce different results. Consider a signal $x(t)$ that is non-causal. The unilateral Laplace transform of $x(t)$ is

$$\mathcal{L}_u\{x(t)\} = \int_{0+}^{\infty} x(t)\, e^{-st}\, dt$$

$$= \int_{-\infty}^{\infty} x(t)\, u(t)\, e^{-st}\, dt = \mathcal{L}\{x(t)\, u(t)\} \tag{7.242}$$

Thus, the unilateral Laplace transform of a signal $x(t)$ is the same as the bilateral Laplace transform of the signal $[x(t)\, u(t)]$.

Because of the way $X_u(s)$ is defined, its region of convergence is always to the right of a vertical line in the s-plane, and does not have to be explicitly stated. The ambiguity that we have observed with the bilateral Laplace transform in Eqns. (7.22) and (7.23) does not exist with the unilateral transform.

Most of the properties discussed in Section 7.3 for the bilateral Laplace transform apply to the unilateral Laplace transform as well. A few of the properties need to be modified, and a few new ones need to be introduced. These will be discussed briefly.

7.7.1　Time shifting

The use of the time shifting property with the unilateral Laplace transform requires special care. Recall the time shifting property derived in Eqn. (7.52) and repeated here:

$$\text{Eqn. (7.52):} \qquad \mathcal{L}\{x(t-\tau)\} = e^{-s\tau}\, \mathcal{L}\{x(t)\}$$

Using Eqn. (7.242), the unilateral Laplace transform of $x(t-\tau)$ is

$$\mathcal{L}_u\{x(t-\tau)\} = \mathcal{L}\{x(t-\tau)\, u(t)\} \tag{7.243}$$

Correspondingly, the time shifting property will work for the unilateral Laplace transform only if the shift by τ does not cause any signal components to move from the negative time territory to positive time territory or vice versa. Mathematically we have the following:

Given the transform pair

$$x\left(t\right) \overset{\mathcal{L}_u}{\longleftrightarrow} X_u\left(s\right)$$

it can be shown that

$$x\left(t-\tau\right) \overset{\mathcal{L}_u}{\longleftrightarrow} e^{-s\tau} X_u\left(s\right) \tag{7.244}$$

provided that

$$x\left(t-\tau\right) u\left(t\right) = x\left(t-\tau\right) u\left(t-\tau\right), \quad \text{all } t$$

Fig. 7.75 depicts a signal $x\left(t\right)$ and three of its shifted versions. It can be shown that the time shifting property in Eqn. (7.244) holds for $x_1\left(t\right)$ and $x_2\left(t\right)$, but not for $x_3\left(t\right)$.

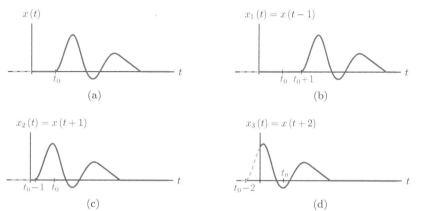

Figure 7.75 – (a) The signal $x\left(t\right)$, (b) the signal $x_1\left(t\right) = x\left(t-1\right)$, (c) the signal $x_2\left(t\right) = x\left(t+1\right)$, (d) the signal $x_3\left(t\right) = x\left(t+2\right)$.

In contrast, the time shifting property does not hold for any amount of shift of the signal $x\left(t\right)$ in Fig. 7.76 due to signal components that cross from left of the vertical axis to the right or vice versa.

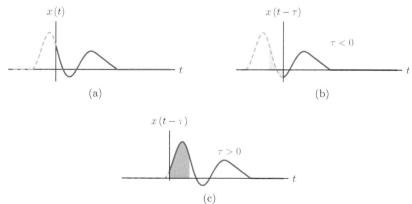

Figure 7.76 – (a) The signal $x\left(t\right)$, (b) the signal $x_1\left(t\right) = x\left(t-1\right)$, (c) the signal $x_2\left(t\right) = x\left(t+1\right)$.

7.7.2 Differentiation in time

Given the transform pair

$$x(t) \xleftrightarrow{\mathcal{L}_u} X_u(s)$$

the following is also a valid transform pair:

$$\frac{dx(t)}{dt} \xleftrightarrow{\mathcal{L}_u} s\,X_u(s) - x(0^+) \qquad (7.245)$$

This property will be very important in using the unilateral Laplace transform for solving differential equations with initial conditions. The proof is similar to the proof of the corresponding property for the bilateral Laplace transform, and will be given here.

Proof: Using the Laplace transform definition in Eqn. (7.1), the transform of $dx(t)/dt$ is

$$\mathcal{L}_u\left\{\frac{dx(t)}{dt}\right\} = \int_{0^+}^{\infty} \frac{dx(t)}{dt} e^{-st}\, dt$$

$$= \int_{0^+}^{\infty} e^{-st}\, d\,[x(t)] \qquad (7.246)$$

Integrating Eqn. (7.246) by parts yields

$$\mathcal{L}_u\left\{\frac{dx(t)}{dt}\right\} = x(t)\,e^{-st}\Big|_{0^+}^{\infty} + s \int_{0^+}^{\infty} x(t)\,e^{-st}\, dt \qquad (7.247)$$

The term $x(t)\,e^{-st}$ must evaluate to zero for $t \to \infty$ for the transform $X_u(s)$ to exist. Therefore, Eqn. (7.247) reduces to

$$\mathcal{L}_u\left\{\frac{dx(t)}{dt}\right\} = -x(0^+) + s \int_{0^+}^{\infty} x(t)\,e^{-st}\, dt = -x(0^+) + s\,X_u(s) \qquad (7.248)$$

to complete the proof.

The unilateral Laplace transforms of higher-order derivatives can be found through repeated use of Eqn. (7.245). For example,

$$\mathcal{L}_u\left\{\frac{dx^2(t)}{dt^2}\right\} = \mathcal{L}_u\left\{\frac{d}{dt}\left(\frac{dx(t)}{dt}\right)\right\}$$

$$= s\,\mathcal{L}_u\left\{\frac{dx(t)}{st}\right\} - \frac{dx(t)}{dt}\Big|_{t=0^+} \qquad (7.249)$$

Using Eqn. (7.245) in Eqn. (7.249) yields

$$\mathcal{L}_u\left\{\frac{dx^2(t)}{dt^2}\right\} = s\,[s\,X_u(s) - x(0^+)] - \frac{dx(t)}{dt}\Big|_{t=0^+}$$

$$= s^2\,X_u(s) - s\,x(0^+) - \frac{dx(t)}{dt}\Big|_{t=0^+} \qquad (7.250)$$

Similarly it can be shown that

$$\mathcal{L}_u \left\{ \frac{dx^3(t)}{dt^3} \right\} = s^3 X_u(s) - s^2 x(0^+) - s \left. \frac{dx(t)}{dt} \right|_{t=0^+} - \left. \frac{d^2 x(t)}{dt^2} \right|_{t=0^+} \tag{7.251}$$

and for the general case

$$\mathcal{L}_u \left\{ \frac{dx^n(t)}{dt^n} \right\} = s^n X_u(s) - s^{n-1} x(0^+) - s^{n-2} \left. \frac{dx(t)}{dt} \right|_{t=0^+} - \ldots - \left. \frac{d^{n-1} x(t)}{dt^{n-1}} \right|_{t=0^+} \tag{7.252}$$

As mentioned before, the primary utility of the unilateral Laplace transform is in solving differential equations with specified initial conditions. The next couple of examples will illustrate this.

Example 7.43: **Using Laplace transform to solve a differential equation**

Consider the circuit shown in Fig. 7.77(a) driven by the pulse signal shown in Fig. 7.77(b). Determine the output signal $y(t)$ for $t > 0$ subject to the initial condition $y(0^+) = -2$ V.

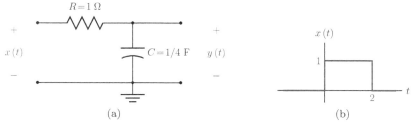

(a)   (b)

Figure 7.77 – (a) The RC circuit for Example 7.43, and (b) its input signal $x(t)$.

Solution: The system can be modeled with the following differential equation (see Example 2.9 in Chapter 2 for derivation):

$$\frac{dy(t)}{dt} + 4 y(t) = 4 x(t)$$

Taking the unilateral Laplace transform of each side of the differential equation we obtain

$$s Y_u(s) - y(0^+) + 4 Y_u(s) = 4 X_u(s) \tag{7.253}$$

The input signal $x(t)$ can be written as

$$x(t) = u(t) - u(t - 2)$$

and its unilateral Laplace transform is

$$X_u(s) = \left(1 - e^{-2s}\right) \frac{1}{s} \tag{7.254}$$

Using Eqn. (7.254) in Eqn. (7.253) and substituting the initial value $y(0^+)$ leads to

$$Y_u(s) = \frac{-2}{s+4} + \left(1 - e^{-2s}\right) \frac{4}{s(s+4)}$$

which can be written in the form

$$Y_u(s) = \frac{-2s+4}{s(s+4)} - e^{-2s} \frac{4}{s(s+4)}$$

Let

$$Y_{1u}(s) = \frac{-2s+4}{s(s+4)}, \quad \text{and} \quad Y_{2u}(s) = \frac{4}{s(s+4)}$$

so that

$$Y_u(s) = Y_{1u}(s) - e^{-2s}\, Y_{2u}(s)$$

and consequently

$$y(t) = y_1(t) - y_2(t-2)$$

Using partial fraction expansion, we can write

$$Y_{1u}(s) = \frac{1}{s} - \frac{3}{s+4} \qquad \Longrightarrow \qquad y_1(t) = \left(1 - 3\,e^{-4t}\right) u(t) \qquad (7.255)$$

$$Y_{2u}(s) = \frac{1}{s} - \frac{1}{s+4} \qquad \Longrightarrow \qquad y_2(t) = \left(1 - e^{-4t}\right) u(t) \qquad (7.256)$$

The output signal is

$$y(t) = \left(1 - 3\,e^{-4t}\right) u(t) - \left(1 - e^{-4(t-2)}\right) u(t-2)$$

and is shown in Fig. 7.78.

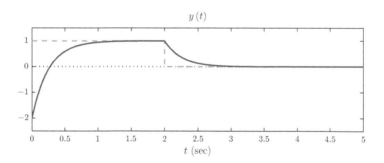

Figure 7.78 – The output signal $y(t)$ for Example 7.43.

Software resources: See MATLAB Exercise 7.10.

Example 7.44: **Solving second-order differential equation using Laplace transform**

Find the solution of the second-order homogeneous differential equation

$$\frac{d^2y(t)}{dt^2} + 9\,y(t) = 0$$

subject to initial conditions

$$y(0) = 1, \quad \text{and} \quad \left.\frac{dy(t)}{dt}\right|_{t=0^+} = 1$$

Solution: Taking the unilateral Laplace transform of the differential equation leads to

$$s^2 Y_u(s) - s\, y\left(0^+\right) - \left.\frac{dy(t)}{dt}\right|_{t=0^+} + 9\, Y_u(s) = 0$$

Substituting initial conditions we obtain

$$s^2 Y_u(s) - s - 1 + 9\, Y_u(s) = 0$$

which can be solved for $Y_u(s)$ to yield

$$Y_u(s) = \frac{s+1}{s^2 + 9}$$

The transform $Y_u(s)$ can be written in partial fraction form as

$$Y_u(s) = \frac{k_1}{s + j3} + \frac{k_2}{s - j3}$$

with residues

$$k_1 = \frac{1}{2} + j\frac{1}{6}, \quad \text{and} \quad k_2 = \frac{1}{2} - j\frac{1}{6}$$

Therefore the solution is

$$y(t) = \left(\frac{1}{2} + j\frac{1}{6}\right) e^{-j3t} + \left(\frac{1}{2} - j\frac{1}{6}\right) e^{j3t}$$

$$= \cos(3t) + \frac{1}{3}\sin(3t), \qquad t \geq 0$$

The solution $y(t)$ is shown in Fig. 7.79. It can easily be verified that $y(t)$ satisfies the differential equation and the specified initial conditions.

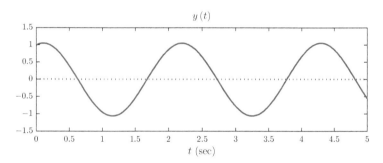

Figure 7.79 – The solution $y(t)$ for Example 7.44.

7.7.3 Initial and final value theorems

The initial value theorem is stated as follows:

Given the transform pair

$$x(t) \overset{\mathcal{L}_u}{\longleftrightarrow} X_u(s)$$

we have

$$\lim_{t\to 0^+} x(t) = \lim_{s\to\infty} \left[s\, X_u(s)\right] \tag{7.257}$$

provided that the limit exists.

The initial value theorem does not apply to rational transforms in which the order of the numerator is equal to or greater than that of the denominator, since the limit in Eqn. (7.257) would not exist in that case.

The final value theorem is stated as follows:

Given the transform pair

$$x\left(t\right) \overset{\mathcal{L}_u}{\longleftrightarrow} X_u\left(s\right)$$

we have

$$\lim_{t \to \infty} x\left(t\right) = \lim_{s \to 0}\left[s\, X_u\left(s\right)\right] \tag{7.258}$$

The final value theorem does not apply to a transform $X_u\left(t\right)$ if the corresponding signal $x\left(t\right)$ is unbounded, or if it has undamped oscillations. An unbounded signal indicates the presence of poles in the right half s-plane. Undamped oscillations are associated with poles on the $j\omega$ axis of the s-plane other than at the origin $s = 0$. Consequently, the final value theorem should not be used with transforms that have poles on the $j\omega$ axis or in the right half s-plane. A single pole at the origin $s = 0$ is permissible.

Example 7.45: **Application of initial and final value theorems**

Consider the transform

$$X_u\left(s\right) = \frac{s}{s^2 + 4}$$

Can initial and final values be determined using initial and final value theorems?

Solution: Since the numerator order is lower than the denominator order, the initial value theorem can be used.

$$x\left(0^+\right) = \lim_{s \to \infty}\left[s\left(\frac{s}{s^2 + 4}\right)\right] = \lim_{s \to \infty}\left[\frac{s^2}{s^2}\right] = 1$$

The final value theorem does not apply since $X_u\left(s\right)$ has poles on the $j\omega$ axis at $s = \pm j2$. We may also recognize that $X_u\left(s\right)$ is the Laplace transform of the signal

$$x\left(t\right) = \cos\left(2t\right) u\left(t\right)$$

which does not have a final value due to the undamped oscillating term.

Example 7.46: **Further exploring initial and final value theorems**

Determine initial and final values of the signal with the Laplace transform

$$X_u\left(s\right) = \frac{2}{s\left(s^2 + 4s + 13\right)}$$

Solution: The numerator order is lower than the denominator order; therefore, the initial value theorem can be used.

$$x\left(0^+\right) = \lim_{s \to \infty}\left[\frac{2}{s^2 + 4s + 13}\right] = 0$$

The transform $X_u\left(s\right)$ has a pair of complex conjugate poles at $s = -2 \pm j3$ and a single pole at $s = 0$. Complex conjugate poles are in the left half s-plane. The single pole at the origin does not cause any problems in the use of the final value theorem since it is canceled by the factor s in Eqn. (7.258). The final value of $x(t)$ is

$$\lim_{t \to \infty} [x(t)] = \lim_{s \to 0} \left[\frac{2}{s^2 + 4s + 13} \right] = \frac{2}{13}$$

7.8   Further Reading

[1] J. Bak and D.J. Newman. *Complex Analysis*. Undergraduate Texts in Mathematics. Springer, 2010.

[2] R.J. Beerends. *Fourier and Laplace Transforms*. Cambridge University Press, 2003.

[3] W.R.L. Page. *Complex Variables and the Laplace Transform for Engineers*. Dover Books on Electrical Engineering Series. Dover Publications, 1980.

[4] J.L. Schiff. *The Laplace Transform: Theory and Applications*. Springer Undergraduate Texts in Mathematics and Technology. Springer, 1999.

[5] J.L. Taylor. *Complex Variables*. Pure and Applied Undergraduate Texts. American Mathematical Society, 2011.

MATLAB Exercises

MATLAB Exercise 7.1: Three dimensional plot of Laplace transform

In Fig. 7.3 the magnitude of the transform

$$\text{Eqn. (7.8):} \quad X(s) = \frac{s + 0.5}{(s + 0.5)^2 + 4\pi^2}$$

was graphed as a three-dimensional surface. In this exercise we will reproduce that figure using MATLAB, and display various cutouts of the Laplace transform surface on vertical lines $s = \sigma + j0$. The first step is to produce a set of complex values of s on a rectangular grid in the s-plane.

```
>>   [sr,si] = meshgrid([-6:0.3:6],[-15:0.5:15]);
>>   s = sr+j*si;
```

The next step is to compute the magnitude of the transform at each point on the grid. Additionally, values of magnitude that are greater than 2 will be clipped for graphing purposes.

```
>>   Xs = @(s) (s+0.5)./((s+0.5).^2+4*pi*pi);
>>   XsMag = abs(Xs(s));
>>   XsMag = XsMag.*(XsMag<=2)+2.*(XsMag>2);
```

A three-dimensional mesh plot of $|X(s)|$ can be generated with the following lines:

```
>>   mesh(sr,si,XsMag);
>>   axis([-6,6,-15,15]);
```

The script listed below produces a mesh plot complete with axis labels and color specifications.

```
1    % Script: matex_7_1a.m
2    %
3    [sr,si] = meshgrid([-6:0.3:6],[-15:0.5:15]);
4    s = sr+j*si;
5    Xs = @(s) (s+0.5)./((s+0.5).^2+4*pi*pi);   % Eqn.(7.8)
6    XsMag = abs(Xs(s));
7    XsMag = XsMag.*(XsMag<=2)+2.*(XsMag>2);
8    shading interp;              % Shading method: Interpolated
9    colormap copper;             % Specify the color map used.
10   m1 = mesh(sr,si,XsMag);
11   axis([-6,6,-15,15]);
12   % Adjust transparency of surface lines.
13   set(m1,'EdgeAlpha',0.6','FaceAlpha',0.6);
14   % Specify x,y,z axis labels.
15   xlabel('\sigma');
16   ylabel('j\omega');
17   zlabel('|X(s)|');
18   % Specify viewing angles.
19   view(gca,[23.5,38]);
```

In line 10 of the script, the handle returned by the function `mesh(..)` is assigned to the variable `m1` so that it can be used in line 13 for adjusting the transparency of the surface.

Alternatively, a contour plot of $|X(s)|$ can be produced by slightly modifying the code. The script listed below gives a bird's eye view of the magnitude of the transform by plotting points that have the same magnitude value as contours.

```
1    % Script: matex_7_1b.m
2    %
3    [sr,si] = meshgrid([-6:0.3:6],[-15:0.5:15]);
4    s = sr+j*si;
5    Xs = @(s) (s+0.5)./((s+0.5).^2+4*pi*pi);   % Eqn.(7.8)
6    XsMag = abs(Xs(s));
7    XsMag = XsMag.*(XsMag<=2)+2.*(XsMag>2);
8    shading interp;              % Shading method: Interpolated
9    colormap copper;             % Specify the color map used.
10   values = [[0:0.04:0.2],[0.3:0.1:2]];   % z value for each contour.
11   m2 = contour(sr,si,XsMag,values); grid;
12   axis([-6,6,-15,15]);
13   % Specify x,y axis labels.
14   xlabel('\sigma');
15   ylabel('j\omega');
```

The Fourier transform $X(\omega)$ is equal to the Laplace transform evaluated on the $j\omega$ axis of the s-plane, that is,

$$X(s)\Big|_{s=0+j\omega} = \mathcal{F}\{x(t)\}$$

Applying this relationship to the magnitudes of the two transforms, the magnitude of the Fourier transform is obtained by evaluating the magnitude of the Laplace transform on the $j\omega$ axis. The script listed below demonstrates this.

```
1    % Script: matex_7_1c.m
2    %
3    [sr,si] = meshgrid([-6:0.3:6],[-15:0.5:15]);
4    s = sr+j*si;
5    Xs = @(s) (s+0.5)./((s+0.5).^2+4*pi*pi);   % Eqn.(7.8)
6    XsMag = abs(Xs(s));
7    XsMag = XsMag.*(XsMag<=2)+2.*(XsMag>2);
8    % Define the trajectory s=j*omega
9    omega = [-15:0.01:15];
10   tr = j*omega;
11   % Produce a mesh plot and hold it.
12   shading interp;
13   colormap copper;
14   m1 = mesh(sr,si,XsMag);
15   hold on;
16   % Superimpose a plot of X(s) magnitude values evaluated on the
17   % trajectory using 'plot3' function.
18   m2 = plot3(real(tr),imag(tr),abs(Xs(tr)),'b-','LineWidth',1.5);
19   hold off;
20   axis([-6,6,-15,15]);
21   % Adjust transparency of surface lines.
22   set(m1,'EdgeAlpha',0.6,'FaceAlpha',0.6);
23   % Specify x,y,z axis labels.
24   xlabel('\sigma');
25   ylabel('j\omega');
26   zlabel('|X(s)|');
27   % Specify viewing angles.
28   view(gca,[23.5,38]);
```

Lines 9 and 10 create a vector of s values on the $j\omega$ axis. Line 18 graphs the values of $|X(s)|$ along the $j\omega$ axis using the plot3(..) function. It is also possible, with a few changes in the code, to cut the Laplace transform surface along the $j\omega$ axis and display the profile of the cutout. The modified script to accomplish this is listed below:

```
1    % Script: matex_7_1d.m
2    %
3    [sr,si] = meshgrid([-6:0.3:6],[-15:0.5:15]);
4    s = sr+j*si;
5    Xs = @(s) (s+0.5)./((s+0.5).^2+4*pi*pi);   % Eqn.(7.8)
6    XsMag = abs(Xs(s));
7    XsMag = XsMag.*(XsMag<=2)+2.*(XsMag>2);
8    % Define the trajectory s=j*omega
9    omega = [-15:0.01:15];
10   tr = j*omega;
11   % Produce a mesh plot and hold it.
12   shading interp;
13   colormap copper;
14   % Set the surface equal to zero in the right half of the s-plane.
15   XsMag = XsMag.*(sr<=0);
16   m1 = mesh(sr,si,XsMag);
17   hold on;
18   % Superimpose a plot of X(s) magnitude values evaluated on the
19   % trajectory using 'plot3' function.
20   m2 = plot3(real(tr),imag(tr),abs(Xs(tr)),'b-','LineWidth',1.5);
21   % Stem plot on the trajectory for a painted profile look.
22   m3 = stem3(real(tr([1:25:3000])),imag(tr([1:25:3000])),
23       abs(Xs(tr([1:25:3000]))));
24   hold off;
25   axis([-6,6,-15,15]);
26   % Adjust transparency of surface lines.
27   set(m1,'EdgeAlpha',0.6,'FaceAlpha',0.6);
```

```
28    % Adjust color of cutout profile.
29    set(m3,'Marker','none','Color',[0.01,0.74,0.25]);
30    % Specify x,y,z axis labels.
31    xlabel('\sigma');
32    ylabel('j\omega');
33    zlabel('|X(s)|');
34    % Specify viewing angles.
35    view(gca,[23.5,38]);
```

Notice how the magnitude values in the right half of the s-plane are suppressed in line 15.
Software resources:
matex_7_1a.m
matex_7_1b.m
matex_7_1c.m
matex_7_1d.m

MATLAB Exercise 7.2: Computing the Fourier transform from the Laplace transform

The Fourier transform of a signal is equal to its Laplace transform evaluated on the $j\omega$-axis of the s-plane.

$$\mathcal{F}\left\{x\left(t\right)\right\} = \left. X\left(s\right)\right|_{s=j\omega}$$

Consider the Laplace transform

$$X\left(s\right) = \frac{s+0.5}{\left(s+0.5\right)^2 + 4\pi^2}$$

The first method of computing and graphing the Fourier transform of the signal is to use an anonymous function for $X\left(s\right)$ and evaluate it on the $j\omega$-axis. The magnitude $\left|X\left(\omega\right)\right|$ is graphed using the following statements:

```
>>   Xs = @(s) (s+0.5)./((s+0.5).^2+4*pi*pi);
>>   omg = [-15:0.05:15];
>>   Xomg = Xs(j*omg);
>>   plot(omg,abs(Xomg)); grid;
```

If the phase $\angle X\left(\omega\right)$ is needed, it can be graphed using

```
>>   plot(omg,angle(Xomg)); grid;
```

The second method is to use MATLAB function freqs(..). We will begin by writing $X\left(s\right)$ in rational form with numerator and denominator polynomials ordered in descending powers of s.

$$X\left(s\right) = \frac{s+0.5}{s^2+s+39.7284}$$

Vectors "num" and "den" to hold numerator and denominator coefficients should be entered as

```
>>   num = [1,0.5];
>>   den = [1,1,39.7284];
```

Afterwards the magnitude and the phase of the Fourier transform may be computed and graphed with the statements

```
>>   omg = [-15:0.05:15];
>>   Xomg = freqs(num,den,omg);
>>   plot(omg,abs(Xomg),'r'); grid;
>>   plot(omg,angle(Xomg),'r'); grid;
```

Software resources:
matex_7_2a.m
matex_7_2b.m

MATLAB Exercise 7.3: Graphing poles and zeros

Consider a CTLTI system described by the system function

$$X(s) = \frac{s^3 - s^2 - 4s + 4}{s^4 + 7s^3 + 21s^2 + 37s + 30}$$

Poles and zeros of the system function can be graphed on the s-plane by entering coefficients of numerator and denominator polynomials as vectors and then computing the roots.

```
>>   num = [1,-1,-4,4];
>>   den = [1,7,21,37,30];
>>   z = roots(num);
>>   p = roots(den);
```

To produce the graph we need

```
>>   plot(real(z),imag(z),'o',real(p),imag(p),'x');
```

which uses "o" for a zero and "x" for a pole. The graph produced does not display the real and imaginary axes in the s-plane. A more complete pole-zero plot may be generated with the following lines:

```
>>   plot(real(z),imag(z),'o',real(p),imag(p),'x',...
         [-3,3],[0,0],'k:',[0,0],[-3,3],'k:');
>>   xlabel('\sigma');
>>   ylabel('j\omega');
```

Software resources:
matex_7_3.m

MATLAB Exercise 7.4: Residue calculations

Consider the transform

$$X_1(s) = \frac{s(s-2)}{s^4 + 9s^3 + 30s^2 + 42s + 20}$$

In expanding $X(s)$ to partial fractions, MATLAB function residue(..) can be used. Numerator and denominator polynomials are specified by means of two vectors that list the coefficients of these polynomials in the order of descending powers of s.

```
>>   num = conv([1,0],[1,-2]);
>>   den = [1,9,30,42,20];
>>   [res,poles,qt] = residue(num,den)

res =
     1.7000 - 1.9000i
```

```
       1.7000 + 1.9000i
      -4.0000
       0.6000

poles =
      -3.0000 + 1.0000i
      -3.0000 - 1.0000i
      -2.0000
      -1.0000

qt =
       []
```

Notice how the convolution function conv(..) is used for multiplying the numerator factors s and $(s-2)$ expressed through the vectors $[1,0]$ and $[1,-2]$ respectively. Returned vectors "**res**" and "**poles**" hold the residues and the poles of the partial fraction expansion. The vector "**qt**" is empty since the numerator order is less than the denominator order in this case. The order of residues and poles in these two vectors is important. We have

$$\begin{aligned} p_1 &= -3+j1\,, & k_1 &= 1.7-j1.9 \\ p_2 &= -3-j1\,, & k_2 &= 1.7+j1.9 \\ p_3 &= -2\,, & k_3 &= -4 \\ p_4 &= -1\,, & k_4 &= 0.6 \end{aligned}$$

The partial fraction expansion we seek is in the form

$$X_1(s) = \frac{(1.7-j1.9)}{s+3-j1} + \frac{(1.7+j1.9)}{s+3+j1} - \frac{4}{s+2} + \frac{0.6}{s+1}$$

Consider another transform

$$X_2(s) = \frac{3\,s^2 + 2\,s + 5}{s^2 + 4\,s + 20}$$

the residues for which can be computed through

```
>>   num = [3,2,5];
>>   den = [1,4,20];
>>   [res,poles,qt] = residue(num,den)

res =
   -5.0000 + 4.3750i
   -5.0000 - 4.3750i

poles =
   -2.0000 + 4.0000i
   -2.0000 - 4.0000i

qt =
    3
```

The vector "**qt**" has a single element, and corresponds to the quotient polynomial $Q(s) = 3$. The partial fraction expansion for $X_2(s)$ is

$$X_2(s) = 3 + \frac{(-5+j4.375)}{s+2-j4} + \frac{(-5-j4.375)}{s+2+j4}$$

Software resources:
`matex_7_4a.m`
`matex_7_4b.m`

MATLAB Exercise 7.5: Symbolic calculations for Laplace transform

It is also possible to do symbolic processing with MATLAB. As a first step we need to specify that "**s**" and "**t**" are symbolic variables rather than numeric variables. This is accomplished with the statement

```
>>  syms s t
```

Notice that there is no comma, but rather just space, separating the two symbolic variables. The next step is to define the signal $x(t)$ the Laplace transform of which we seek. For example, the signal

$$x(t) = e^{-t} \cos(2t) \, u(t)$$

is defined with the statement

```
>>  xt = exp(-t)*cos(2*t);
```

The resulting variable "**xt**" is also a symbolic variable since it utilizes "**t**" defined with the previous statement. The Laplace transform can be computed using the function `laplace(..)`.

```
>>  Xs = laplace(xt)

Xs =
(s + 1)/((s + 1)^2 + 4)
```

MATLAB returns the answer as a symbolic expression. To display the answer in a way closer to its natural form, use

```
>>  pretty(Xs)

      s + 1
   ------------
           2
   (s + 1)  + 4
```

which corresponds to

$$X(s) = \frac{s+1}{(s+1)^2 + 4}$$

Combining the statements listed above, the transform can be computed in a compact form as

```
>>  syms s t;
>>  Xs = laplace(exp(-t)*cos(2*t));
```

Some observations:

1. In using the function `laplace(..)` with just one argument we assume that the independent variable of the signal is named "**t**" and the independent variable of the transform is named "**s**". If, for some reason, the names "**v**" and "**w**" need to be used in place of "**t**" and "**s**" respectively, then an alternative syntax of the function with three arguments can be employed:

```
>>   Xs = laplace(xt,v,w)
```

2. The function `laplace(..)` computes the Laplace transform integral starting with the lower limit $t = 0$. In other words, the transform computed is the *unilateral Laplace transform* discussed in Section 7.7.

Inverse Laplace transform can be computed symbolically using the function `ilaplace(..)`. Consider the problem of finding the inverse Laplace transform of

$$X(s) = \frac{s+1}{s\,(s+2)}$$

which was solved in Example 7.22 through the use of partial fractions. The following set of statements produce the solution using symbolic processing in MATLAB:

```
>>   syms s t
>>   Xs = (s+1)/(s*(s+2));
>>   xt = ilaplace(Xs)

xt =
1/(2*exp(2*t)) + 1/2

>>   pretty(xt)

      1           1
 ---------- + -
  2 exp(2 t)    2
```

The result displayed corresponds to

$$x(t) = \frac{1}{2}\,e^{-2t}\,u(t) + \frac{1}{2}\,u(t)$$

The symbolic result obtained for the signal $x(t)$ can be graphed using the function `ezplot(..)` as follows:

```
>>   ezplot(xt,[0,5]); grid;
>>   axis([0,5,0,1.2]);
```

The graph produced is shown in Fig. 7.80.

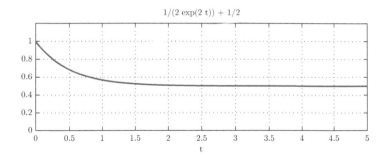

Figure 7.80 – The graph produced using function "ezplot".

Software resources:
`matex_7_5a.m`
`matex_7_5b.m`

MATLAB Exercise 7.6: Computing frequency response of a system from pole-zero layout

The problem of obtaining the frequency response of a CTLTI system from the placement of its poles and zeros in the s-domain was discussed in Section 7.5.5. In this exercise we will use MATLAB to compute the frequency response for a system characterized by the system function

$$H(s) = \frac{s-1}{s+2}$$

The system has one zero at $s = 1$ and one pole at $s = -2$. Suppose we need the frequency response of this system at the frequency $\omega = 3$ rad/s, which is equal to the system function evaluated for $s = j3$:

$$H(3) = H(s)\Big|_{s=j3} = \frac{j3-1}{j3+2}$$

Using vector notation

$$\overrightarrow{H(3)} = \frac{\overrightarrow{B}}{\overrightarrow{A}} = \frac{(\overrightarrow{j3-1})}{(\overrightarrow{j3+2})}$$

where we have defined the vectors $\overrightarrow{A}$ and $\overrightarrow{B}$ for notational convenience. The script listed below computes the vectors $\overrightarrow{A}$ and $\overrightarrow{B}$ and uses them for computing the magnitude and the phase of the system function at $\omega = 3$ rad/s:

```
1   % Script: matex_7_6a.m
2   %
3   omega = 3;
4   s = j*omega;
5   B = s-1;
6   A = s+2;
7   mag = abs(B)/abs(A);
8   phs = angle(B)-angle(A);
```

This script computes the frequency response of the system at one specific frequency. It would be more interesting if we could use the same idea to compute the frequency response of the system at a large number of frequencies so that its magnitude and phase can be graphed as functions of ω. Let us change the variable "omega" into a vector by editing line 1 of the script:

```
1   omega = [-5:0.01:5];
```

This change causes the variable "s" to become a complex vector with 1001 elements. Also, in line 5, the standard division operator "/" needs to be changed to the element-by-element division operator "./" to read

```
5   mag = abs(B)./abs(A);
```

The MATLAB script is listed below with these modifications and the addition of graphing statements:

```
1   % Script: matex_7_6b.m
2   %
3   omega = [-5:0.01:5];
4   s = j*omega;
5   B = s-1;
```

```
6    A = s+2;
7    mag = abs(B)./abs(A);
8    phs = angle(B)-angle(A);
9    clf;
10   subplot(2,1,1);
11   plot(omega,mag);
12   title('Magnitude of the frequency response');
13   xlabel('\omega (rad/s)'); grid;
14   subplot(2,1,2);
15   plot(omega,phs);
16   title('Phase of the frequency response');
17   xlabel('\omega (rad/s)'); grid;
```

Software resources:

matex_7_6a.m

matex_7_6b.m

MATLAB Exercise 7.7: Frequency response from pole-zero layout revisited

In MATLAB Exercise 7.6 we have explored a method of computing the frequency response of a CTLTI system based on the graphical interpretation of the pole-zero layout of the system function, discussed in Section 7.5.5. The idea can be generalized into the development of a MATLAB function ss_freqs(..) for computing the frequency response.

```
1    function [mag,phs] = ss_freqs(zrs,pls,gain,omega)
2      nz = length(zrs);        % Number of zeros.
3      np = length(pls);        % Number of poles.
4      nomg = length(omega);    % Number of frequency points.
5      s = j*omega;             % Get points on the imaginary axis.
6      mag = ones(1,nomg);
7      phs = zeros(1,nomg);
8      if (nz > 0),
9        for n = 1:nz
10         mag = mag.*abs(s-zrs(n));
11         phs = phs+angle(s-zrs(n));
12       end;
13     end;
14     if (np > 0),
15       for n = 1:np
16         mag = mag./abs(s-pls(n));
17         phs = phs-angle(s-pls(n));
18       end;
19     end;
20     mag = mag*gain;
21     phs = wrapToPi(phs);
```

Line 21 of the function causes phase angles to be contained in the interval $(-\pi, \pi)$. The script listed below may be used for testing ss_freqs(..) with the system function

$$H(s) = \frac{s + 0.5}{s^2 + s + 39.7284}$$

```
1    % Script: matex_7_7.m
2    %
3    num = [1,0.5];
4    den = [1,1,39.7284];
5    zrs = roots(num);       % Compute zeros.
6    pls = roots(den);       % Compute poles.
7    omg = [-15:0.05:15];    % Vector of frequencies.
```

```
8    [mag,phs] = ss_freqs(zrs,pls,1,omg);
9    clf;
10   subplot(2,1,1);
11   plot(omg,mag); grid;
12   xlabel('\omega (rad/s)');
13   subplot(2,1,2);
14   plot(omg,phs); grid;
15   xlabel('\omega (rad/s)');
```

Software resources:
ss_freqs.m
matex_7_7.m

MATLAB Exercise 7.8: **System objects**

MATLAB has some functions that create and work with *objects* to represent linear and time-invariant systems. Consider, for example, a CTLTI system described by the system function

$$H(s) = \frac{s^2 + 1}{s^3 + 5\,s^2 + 17\,s + 13}$$

An object named "sys1" representing this system can be created in MATLAB with the following lines:

```
>>   num = [1,0,1];
>>   den = [1,5,17,13];
>>   sys1 = tf(num,den)

Transfer function:
       s^2 + 1
   ---------------------
s^3 + 5 s^2 + 17 s + 13
```

The object "sys1" created with a call to the function tf(..) may be used with other functions that accept system objects. For example, a pole-zero plot can be generated through

```
>>   pzmap(sys1)
```

The impulse response of the system may be graphed using the following code:

```
>>   t = [0:0.01:5];
>>   h = impulse(sys1,t);
>>   plot(t,h); grid;
```

Similarly, the unit-step response is graphed using

```
>>   t = [0:0.01:5];
>>   y = step(sys1,t);
>>   plot(t,y); grid;
```

A system can be specified in a variety of ways. An alternative to using the numerator and the denominator coefficients is to use the zeros and the poles of the system through the function zpk(..). For example, an object "sys2" for the system function

$$H(s) = \frac{10\,s\,(s-1)}{(s+1)\,(s+2)\,(s+3)}$$

is created by

```
>>  zrs = [0,1];
>>  pls = [1,2,3];
>>  sys2 = zpk(zrs,pls,10)
```

Let the input signal to this system be

$$x(t) = \cos(4\pi t)\, u(t)$$

The response of the system may be computed and graphed with the following code:

```
>>  t = [0:0.01:3];
>>  x = cos(2*pi*2*t);
>>  [y,t] = lsim(sys2,x,t);
>>  plot(t,x,t,y);
```

Software resources:
matex_7_8a.m
matex_7_8b.m
matex_7_8c.m

MATLAB Exercise 7.9: Bode plots

Construct a Bode plot for the system with system function

$$H(s) = \frac{10\,s\,(s-1)}{(s+1)\,(s+2)\,(s+3)}$$

Solution: The easiest method is to use system objects. The following code accomplishes this task.

```
>>  zrs = [0,1];
>>  pls = [-1,-2,-3];
>>  sys = zpk(zrs,pls,10);
>>  bode(sys); grid;
```

Software resources:
matex_7_9.m

MATLAB Exercise 7.10: Solving a differential equation through Laplace transform

Two examples of using the unilateral Laplace transform for solving differential equations with specified initial conditions were given in Section 7.7. In this exercise we will explore the use of symbolic processing capabilities of MATLAB for solving similar problems. The functions laplace(..) and ilaplace(..) were explored in MATLAB Exercise 7.5. They compute the forward and the inverse Laplace transform with the assumption that the time-domain signal involved is causal. In effect they implement the unilateral variant of the Laplace transform.

Consider the RC circuit problem solved in Example 7.43. The governing differential equation is

$$\frac{dy(t)}{dt} + 4\,y(t) = 4\,x(t)$$

with the initial value of the solution specified as $y(0^+) = -2$ V. The input signal is a unit amplitude pulse with a duration of 2 seconds.

$$x(t) = u(t) - u(t-2)$$

The following script can be used for solving the problem and graphing the result:

```
1    % Script: matex_7_10.m
2    %
3    syms s t Ys
4    xt = heaviside(t)-heaviside(t-2);  % x(t)=u(t)-u(t-2)
5    Xs = laplace(xt);           % Laplace transform of x(t)
6    Y1 = s*Ys+2;                % Laplace transform of dy/dt
7    Ys = solve(Y1+4*Ys-4*Xs,Ys);    % Solve for Y(s)
8    yt = ilaplace(Ys);          % Inverse Laplace transform of Y(s)
9    ezplot(yt,[0,5]); grid;
10   axis([0,5,-2.5,1.5]);
```

In line 3 we declare three symbolic variables. The variable "**Ys**" corresponds to the yet unknown transform $Y_u(s)$. In line 4 the input signal is specified in symbolic form. Note that we make use of the symbolic function `heaviside(..)` to represent the unit-step function. The symbolic result "**Y1**" represents the Laplace transform

$$Y_1(s) = \mathcal{L}\left\{\frac{dy(t)}{dt}\right\} = s\,Y_u(s) - y(0^+) = s\,Y_u(s) + 2$$

Line 7 of the script uses the function `solve(..)` to solve the equation

$$Y_1(s) + 4\,Y_u(s) - 4\,X(s) = 0$$

for $Y_u(s)$. The graph produced by the function `ezplot(..)` on line 9 should match Fig. 7.78.
Software resources:
`matex_7_10.m`

Problems

7.1. Using the Laplace transform definition given by Eqn. (7.1) determine the transform of each signal listed below. For each transform construct a pole-zero diagram and specify the ROC.

a. $x(t) = e^{-2t}\,u(t)$

b. $x(t) = e^{-2t}\,u(t-1)$

c. $x(t) = e^{2t}\,u(-t)$

d. $x(t) = e^{2t}\,u(-t+1)$

e. $x(t) = \begin{cases} 1, & 0 < t < 1 \\ -1, & 1 < t < 2 \\ 0, & t < 0 \text{ or } t > 2 \end{cases}$

f. $x(t) = \begin{cases} 1, & 0 < t < 1 \\ -1, & t > 1 \\ 0, & t < 0 \end{cases}$

7.2. Using the Laplace transform definition given by Eqn. (7.1) determine the transform of the signal

$$x(t) = \sum_{n=0}^{\infty} e^{-anT}\,\delta(t - nT)$$

where $a > 0$ and $T > 0$. Put the transform into a closed form using the appropriate formula. Afterwards construct a pole-zero plot and indicate the ROC.

7.3. Pole-zero diagrams for four transforms are shown in Fig. P.7.3. For each, determine the ROC if it is known that the Fourier transform of $x(t)$ exists.

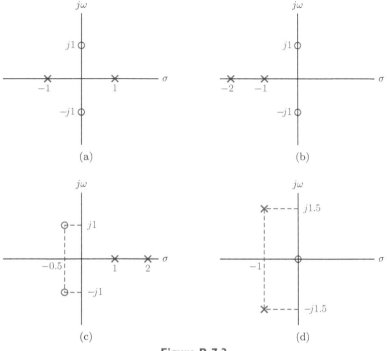

Figure P. 7.3

7.4. Consider again the four transforms with pole-zero diagrams shown in Fig. P.7.3. Determine the ROC for each of the transforms if $x(t)$ is known to be causal in each case.

7.5. Repeat Problem 7.3 if $x(t)$ is known to be anti-causal for each case.

7.6. The Laplace transform of a signal $x(t)$ is given by

$$X(s) = \frac{s-1}{s^2 - s - 2}, \qquad \text{ROC:} \quad -1 < \text{Re}\,\{s\} < 2$$

Which of the following Fourier transforms can be obtained from $X(s)$ without actually determining the signal $x(t)$? In each case, either determine the indicated Fourier transform or explain why it cannot be determined.

    a. $\mathcal{F}\{x(t)\}$
    b. $\mathcal{F}\{x(t)\,e^t\}$
    c. $\mathcal{F}\{x(t)\,e^{-3t}\}$

7.7. The Laplace transform of a signal $x(t)$ is given by

$$X(s) = \frac{1}{s^2 + 2s + 5}, \qquad \text{ROC:} \quad \text{Re}\,\{s\} > -1$$

Which of the following Fourier transforms can be obtained from $X(s)$ without actually determining the signal $x(t)$? In each case, either determine the indicated Fourier transform or explain why it cannot be determined.

    a. $\mathcal{F}\left\{x\left(t\right)\right\}$
    b. $\mathcal{F}\left\{x\left(t\right)e^{-t}\right\}$
    c. $\mathcal{F}\left\{x\left(t\right)e^{t}\right\}$
    d. $\mathcal{F}\left\{x\left(t\right)e^{3t}\right\}$

7.8. Construct a pole-zero diagram and specify the ROC for each of the transforms given below. Also, determine the Fourier transform $X\left(\omega\right)$ if it exists.

    a. $X\left(s\right) = \dfrac{s-2}{s^2+3s+2}$, $x\left(t\right)$ is causal

    b. $X\left(s\right) = \dfrac{s}{s^2-1}$, $x\left(t\right)$ is causal

    c. $X\left(s\right) = \dfrac{s+1}{s^2-4s+3}$, $x\left(t\right)$ is anti-causal

    d. $X\left(s\right) = \dfrac{s^2-s}{s^2-s-6}$, $x\left(t\right)$ is anti-causal

    e. $X\left(s\right) = \dfrac{s^2-s}{s^2+8s+15}$, $\mathcal{F}\left\{x\left(t\right)e^{4t}\right\}$ exists

7.9. Using the linearity property of the Laplace transform, determine $X\left(s\right)$ for each of the signals listed below. Also indicate the ROC in each case.

    a. $x\left(t\right) = 3\,e^{-t}\,u\left(t\right) - 5\,e^{-3t}\,u\left(t\right)$
    b. $x\left(t\right) = 3\,e^{-t}\,u\left(t\right) + 2\,e^{3t}\,u\left(-t\right)$
    c. $x\left(t\right) = \delta\left(t\right) + 2\,e^{-t}\,u\left(t\right)$
    d. $x\left(t\right) = \left(1 - e^{-t}\right)u\left(t\right)$
    e. $x\left(t\right) = \cos\left(2t\right)u\left(t\right) + 2\sin\left(3t\right)u\left(t\right)$
    f. $x\left(t\right) = e^{-2t}\cos\left(3t\right)u\left(t\right)$

7.10. Using time shifting and linearity properties of the Laplace transform as needed, determine $X\left(s\right)$ for each of the signals listed below. Also indicate the ROC in each case.

    a. $x\left(t\right) = e^{-2\left(t-1\right)}\,u\left(t-1\right)$
    b. $x\left(t\right) = e^{-2t}\,u\left(t-1\right)$
    c. $x\left(t\right) = e^{2\left(t+1\right)}\,u\left(-t-1\right)$
    d. $x\left(t\right) = e^{2t}\,u\left(-t-1\right)$
    e. $x\left(t\right) = e^{2t}\,u\left(-t+1\right)$

7.11. Determine the Laplace transform of the signal

$$x\left(t\right) = \cos\left(3t + \pi/6\right)u\left(t\right)$$

using two different methods as specified below. Show that the same result can be obtained through each.

    a.   Use a trigonometric identity to express $x\left(t\right)$ as the sum of a cosine and a sine term. Afterwards use the entries for $\cos\left(3t\right)$ and $\sin\left(3t\right)$ from the Laplace transform table.
    b.   Use Euler's formula to express $x\left(t\right)$ as a sum of complex exponentials. Afterwards use the entry for e^{at} from the Laplace transform table.

7.12. Determine the Laplace transform of each signal listed below. Specify the ROC in each case. Sketch each signal and justify the ROC based on the characteristics of each signal.

a. $x(t) = u(t) - u(t-1)$
b. $x(t) = u(t) - 2u(t-1) + u(t-2)$
c. $x(t) = u(t) - 2u(t-1)$

7.13. Using the linearity and time shifting properties, determine the Laplace transform of each signal shown graphically in Fig. P.7.13. Specify the ROC in each case.

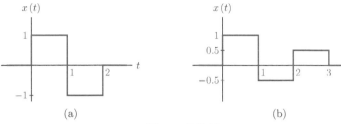

Figure P. 7.13

7.14. Express each of the signals shown in Fig. P.7.14 in terms of the unit-ramp function and its scaled and/or time shifted versions. Afterwards, determine the Laplace transform of each signal using linearity and time shifting properties, Specify the ROC in each case.

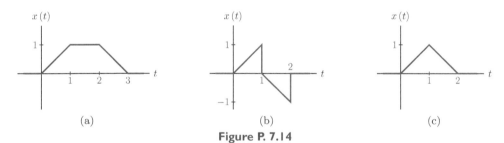

Figure P. 7.14

7.15. The signal $x(t)$ is in the form of one cycle of a sinusoidal waveform as shown in Fig. P.7.15.

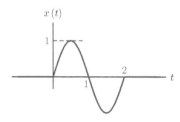

Figure P. 7.15

Determine its Laplace transform using two different methods:

a. Apply Laplace transform definition directly. Hint: Express the sinusoidal signal using Euler's formula.
b. Use the transform for $x_1(t) = \sin(\omega_0 t)\, u(t)$ and then apply linearity and time shifting properties.

7.16. The transforms given below correspond to causal signals. Using the table of Laplace transforms, determine the inverse Laplace transform of each.

Hint: For each case find a similar entry in the table and adjust its parameters to obtain the desired result.

a. $X(s) = \dfrac{1}{(s+2)^2}$

b. $X(s) = \dfrac{s+3}{(s+3)^2 + 2}$

c. $X(s) = \dfrac{1}{(s+1)^2 + 6}$

7.17. Determine the inverse Laplace transform of each function listed below using partial fraction expansion.

a. $X(s) = \dfrac{1}{s^2 + 3s + 2}$,   ROC: $\operatorname{Re}\{s\} > -1$

b. $X(s) = \dfrac{1}{s^2 + 3s + 2}$,   ROC: $\operatorname{Re}\{s\} < -2$

c. $X(s) = \dfrac{1}{s^2 + 3s + 2}$,   ROC: $-2 < \operatorname{Re}\{s\} < -1$

d. $X(s) = \dfrac{(s-1)(s-2)}{(s+1)(s+2)(s+3)}$, $x(t)$: causal

e. $X(s) = \dfrac{(s+1)(s-2)}{(s-1)(s+2)(s+3)}$, $\mathcal{F}\{x(t)\}$ exists

7.18. Use partial fraction expansion to find the inverse Laplace transform of each function listed below.

a. $X(s) = \dfrac{2s+3}{s+2}$,   ROC: $\operatorname{Re}\{s\} > -2$

b. $X(s) = \dfrac{s(s+1)}{(s+2)(s+3)}$,   ROC: $\operatorname{Re}\{s\} < -3$

c. $X(s) = \dfrac{s+5}{s^2 + 4}$,   ROC: $\operatorname{Re}\{s\} > 0$

d. $X(s) = \dfrac{s+6}{(s+1)^2 + 4}$,   ROC: $\operatorname{Re}\{s\} > 0$

e. $X(s) = \dfrac{s(s-1)}{(s+1)^2 (s+2)}$,   ROC: $\operatorname{Re}\{s\} > -1$

7.19. Given the transform

$$X(s) = \frac{s(s-3)}{(s-1)(s+1)(s+2))}$$

a. Construct a pole-zero diagram for $X(s)$.

b. List all possibilities for the ROC.

c. For each choice of the ROC listed in part (b) determine the inverse transform $x(t)$, and indicate whether it is square integrable or not.

7.20. For each Laplace transform given below, find all possible signals that might have led to that transform.

a. $X(s) = \dfrac{s+1}{(s+2)(s+3)}$

b. $X(s) = \dfrac{s+1}{(s+2)^2\,(s+3)}$

c. $X(s) = \dfrac{s\,(s-1)}{(s+3)^2}$

d. $X(s) = \dfrac{s}{(s+2)^2+9}$

7.21. Use partial fraction expansion and the time shifting property to find the inverse Laplace transforms of functions listed below. Assume each transform corresponds to a causal signal.

a. $X(s) = \dfrac{1-e^{-s}}{s+1}$

b. $X(s) = \dfrac{(1-e^{-s})\,s}{s+1}$

c. $X(s) = \dfrac{1-e^{-s}}{s\,(s+1)}$

d. $X(s) = \dfrac{1}{s}\left(1-e^{-s}+e^{-2s}-e^{-3s}\right)$

7.22. Find a system function for each CTLTI system described below by means of a differential equation.

a. $3\dfrac{dy(t)}{dt} + 2\,y(t) = 7\,x(t)$

b. $\dfrac{d^2y(t)}{dt^2} + 4\dfrac{dy(t)}{dt} + 3\,y(t) = \dfrac{dx(t)}{dt} - x(t)$

c. $\dfrac{d^2y(t)}{dt^2} + 4\,y(t) = \dfrac{d^2x(t)}{dt^2} + \dfrac{dx(t)}{dt} + 3\,x(t)$

7.23. Find a differential equation for each CTLTI system described below by means of a system function.

a. $H(s) = \dfrac{1}{s+4}$

b. $H(s) = \dfrac{s}{s+4}$

c. $H(s) = \dfrac{s+1}{s^2+5\,s+6}$

d. $H(s) = \dfrac{s\,(s-1)}{s^3+5\,s^2+8\,s+6}$

e. $H(s) = \dfrac{(s-1)\,(s+2)}{(s+1)\,(s^2+6\,s+13)}$

7.24. Let $y_u(t)$ represent the response of a CTLTI system to a unit-step input signal $x(t) = u(t)$. Determine the system function for each system the unit-step response of which is given below. Afterwards find a differential equation for each system.

a. $y_u(t) = e^{-t}\,u(t)$

b. $y_u(t) = \left(1-e^{-t}\right)u(t)$

c. $y_u(t) = \left(e^{-t}-e^{-2t}\right)u(t)$

d. $y_u(t) = \left(1-e^{-t}+2\,e^{-2t}\right)u(t)$

e. $y_u(t) = \left[1-0.3\,e^{-t}\cos(2t)\right]u(t)$

7.25. Let $y_r(t)$ represent the response of a CTLTI system to a unit-ramp input signal $x(t) = t\,u(t)$. Determine the system function for each system the unit-step response of which is given below. Afterwards find a differential equation for each system.

   a. $y_r(t) = (1 - e^{-t})\,u(t)$
   b. $y_r(t) = (e^{-t} + 2\,e^{-2t} - 4\,e^{-3t})\,u(t)$
   c. $y_r(t) = [1 - 0.3\,e^{-t}\cos(2t)]\,u(t)$

7.26. A CTLTI system is described by means of the system function

$$H(s) = \frac{s+3}{s^2 + 3s + 2}$$

Determine the response of the system to the following signals:

   a. $x(t) = e^{-0.5t}$
   b. $x(t) = e^{(-0.5+j2)t}$
   c. $x(t) = e^{j3t}$
   d. $x(t) = e^{-j3t} + e^{j3t}$
   e. $x(t) = e^{j2t} + e^{j3t}$

7.27. Show that, for a CTLTI system with a real-valued impulse response, the value of the system function at the point $s = s_0^*$ is the complex conjugate of its value at the point $s = s_0$. Given

$$H(s_0) = H_0\,e^{j\Theta_0}$$

where

$$H_0 = |H(s_0)|$$

and

$$\Theta_0 = \angle H(s_0)$$

prove that

$$H(s_0^*) = [H(s_0)]^* = H_0\,e^{-j\Theta_0}$$

Hint: Use the Laplace transform definition with the impulse response $h(t)$ and evaluate the result at $s = s_0^*$. Manipulate the resulting expression to obtain the desired proof.

7.28. A CTLTI system is described by means of the system function

$$H(s) = \frac{s(s+1)}{s^2 + 4s + 13}$$

Determine the response of the system to the following signals:

   a. $x(t) = \cos(2t)$
   b. $x(t) = e^{-0.5t}\cos(2t)$
   c. $x(t) = e^{-t}\sin(2t)$

7.29. A CTLTI system has the impulse response

$$h(t) = e^{-t}\,u(t)$$

Using Laplace transform techniques, determine the response of the system to each input signal listed below. Identify transient and steady-state components of the output signal in each case.

a. $x(t) = \cos(2t)\, u(t)$

b. $x(t) = \sin(3t)\, u(t)$

7.30. Determine the impulse response of each system specified below by means of a system function.

a. $H(s) = \dfrac{s+1}{s+3}$

b. $H(s) = \dfrac{s^2+1}{s^2+3s+2}$

c. $H(s) = \dfrac{s^2-1}{(s+2)(s^2+2s+2)}$

d. $H(s) = \dfrac{s-3}{s^2+2s+1}$

7.31. Consider again the system used in Problem 7.29. Using the Laplace transform, determine the response of the system to the signal

$$x(t) = \cos(2t)$$

which is similar to the input signal in part (a) of Problem 7.29 except it exists for all t instead of being turned on at $t=0$. Compare the output signal to that obtained in Problem 7.29(a) and comment on the result.

7.32. A CTLTI system has the impulse response

$$h(t) = e^{-t}\,[u(t) - u(t-2)]$$

Using Laplace transform techniques, determine the response of the system to the input signal $x(t) = e^{-0.5t}\cos(2t)\,u(t)$.

7.33. Construct a pole-zero diagram for each of the system functions given below.

a. $H(s) = \dfrac{s+1}{(s+2)(s+3)}$

b. $H(s) = \dfrac{s^2+1}{s^2+5s+6}$

c. $H(s) = \dfrac{s^2-1}{(s+2)(s^2+4s+13)}$

7.34. Pole-zero diagrams for four system functions are shown in Fig. P.7.34. For each, determine the the system function $H(s)$. Set $|H(0)| = 1$ for each system.

7.35. For each system function given below, sketch the magnitude and phase characteristics using the graphical method outlined in Section 7.5.5.

a. $H(s) = \dfrac{s-1}{s+2}$

b. $H(s) = \dfrac{s^2+1}{s^2+5s+6}$

c. $H(s) = \dfrac{s^2-1}{s^2+2s+5}$

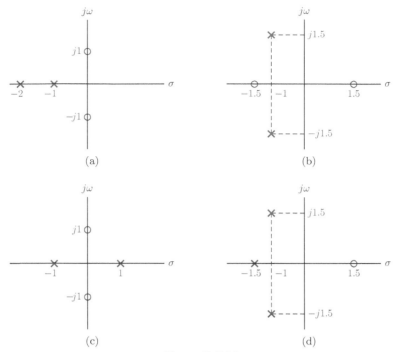

Figure P. 7.34

7.36. The system functions given below correspond to causal systems. For each system, construct a pole-zero plot and determine whether the system is stable or not.

a. $H(s) = \dfrac{s+1}{s^3 + 7s^2 + 16s + 12}$

b. $H(s) = \dfrac{s^2 - 1}{s^2 + s - 6}$

c. $H(s) = \dfrac{s^2 + 1}{s^2 + 2s + 10}$

d. $H(s) = \dfrac{s^2 + 1}{s^2 - 2s + 10}$

7.37. Determine which of the system functions given below could represent a system that is both causal and stable.

a. $H(s) = \dfrac{s-1}{s^2 + 3s + 2}$

b. $H(s) = \dfrac{s(s+1)}{s^2 + 2s + 5}$

c. $H(s) = \dfrac{s^2 + 1}{s(s^2 + 5s + 6)}$

d. $H(s) = \dfrac{s+3}{s^2 - 6s + 10}$

7.38. A causal system is described by the differential equation

$$\frac{d^2 y(t)}{dt^2} = -2\frac{dy(t)}{dt} - a\, y(t) + x(t)$$

a. Determine the system function $H(s)$.

b. Determine which values of a lead to a stable system.

7.39. Consider a first-order allpass system with system function

$$H(s) = \frac{s-2}{s+2}$$

a. Compute and sketch the phase response of this system.

b. Determine the response $y(t)$ of the system to the input signal $x(t) = e^{-t} u(t)$.

c. Find analytical expressions for the magnitude responses $X(\omega)$ and $Y(\omega)$ of the input and the output signals.

d. Sketch each magnitude spectrum.

7.40. Repeat Problem 7.39 using the second-order allpass system with the system function

$$H(s) = \frac{s^2 - 2s + 5}{s^2 + 2s + 5}$$

7.41. A CTLTI system has the system function

$$H(s) = \frac{(s+1)(s-2)}{(s+3)(s+4)}$$

a. Does the system have an inverse that is both causal and stable?

b. Express the system function in the form of the product of two system functions

$$H(s) = H_1(s) H_2(s)$$

such that $H_1(s)$ corresponds to a minimum-phase system and $H_2(s)$ corresponds to a first-order allpass system.

c. Find the inverse system $H_1^{-1}(s)$. Is it causal? Is it stable?

d. Find the overall system function of the cascade combination of $H(s)$ and $H_1^{-1}(s)$.

7.42. For each CTLTI system defined by means of a differential equation below, determine whether a causal and stable inverse exists. If the answer is yes, find a differential equation for the inverse system.

a. $\dfrac{dy(t)}{dt} + 5y(t) = 2\dfrac{dx(t)}{dt} + 3x(t)$

b. $\dfrac{d^2y(t)}{dt^2} + 7\dfrac{dy(t)}{dt} + 12y(t) = \dfrac{d^2x(t)}{dt^2} + \dfrac{dx(t)}{dt}$

c. $\dfrac{d^2y(t)}{dt^2} + 3y(t) = \dfrac{dx(t)}{dt} + 5x(t)$

d. $\dfrac{dy(t)}{dt} - 3y(t) = \dfrac{d^2x(t)}{dt^2} + 2\dfrac{dx(t)}{dt} + x(t)$

7.43. Refer to the Bode magnitude plot in Example 7.39.

a. Show that the value of the asymptote for the Bode magnitude plot is 13.98 dB at $\omega = 5$ rad/s.

b. Also show that, at $\omega = 300$ rad/s, the corner of the asymptote is at -3.52 dB.

7.44. Develop a direct-form block diagram for each system specified below by means of a system function. Assume that each system is causal and initially relaxed.

a. $H(s) = \dfrac{s+3}{s+2}$

b. $H(s) = \dfrac{s^2+1}{s^3+9\,s^2+26\,s+24}$

c. $H(s) = \dfrac{s+5}{2\,s^3+4\,s+7}$

7.45.   Develop a cascade-form block diagram for each system specified below by means of a system function. Assume that each system is causal and initially relaxed. Use first- and second-order cascade sections and ensure that all coefficients are real.

a. $H(s) = \dfrac{s^2+3\,s+2}{s^3+5\,s^2+8\,s+6}$

b. $H(s) = \dfrac{s^2+1}{s^3+9\,s^2+26\,s+24}$

c. $H(s) = \dfrac{s^2-3\,s+2}{s^2+3\,s+2}$

7.46.   Develop a parallel-form block diagram for each system specified in Problem 7.45. Assume that each system is causal and initially relaxed. Use first- and second-order parallel sections and ensure that all coefficients are real.

7.47.   Let $X_u(s)$ be the unilateral Laplace transform of the signal $x(t)$, that is

$$x(t) \overset{\mathcal{L}_u}{\longleftrightarrow} X_u(s)$$

For each of the signals listed below, write the unilateral Laplace transform $G_u(s)$ in terms of $X_u(s)$. In each case, indicate the condition for obtaining the transform directly from $X_u(s)$ without having to find $x(t)$ first.

a. $g(t) = x(t-1)$
b. $g(t) = x(t+2)$
c. $g(t) = x(2t)$
d. $g(t) = e^{-2t}\,x(t)$
e. $g(t) = t\,x(t)$

7.48.   For the RLC circuit shown in Fig. P.7.48, the initial values are

$$y(0^+) = 2\,\text{V}, \quad \text{and} \quad i(0^+) = 0.5\,\text{A}$$

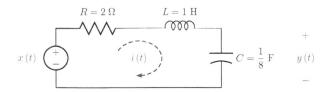

Figure P. 7.48

a.   Find a differential equation between the input $x(t)$ and the output $y(t)$.
b.   Obtain the system function $H(s)$ from the differential equation found in part (a).

c. Using unilateral Laplace transform determine the output signal if $x(t) = u(t)$.
d. Determine the output signal if $x(t) = e^{-2t} u(t)$.

7.49. For the circuit shown in Fig. P7.49, the initial values are

$$y(0^+) = 2 \text{ V}, \quad \text{and} \quad v_0(0^+) = -3 \text{ V}$$

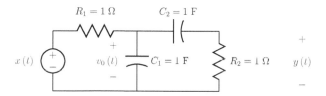

Figure P. 7.49

a. Find a differential equation between the input $x(t)$ and the output $y(t)$.
b. Obtain the system function $H(s)$ from the differential equation found in part (a).
c. Using unilateral Laplace transform determine the output signal if $x(t) = u(t)$.
d. Determine the output signal if $x(t) = e^{-t} u(t)$.

MATLAB Problems

7.50. Consider the transform

$$X(s) = \frac{s(s+2)}{(s+1)^2 + 9} \tag{7.259}$$

a. Determine the poles and zeros of $X(s)$. Manually construct a pole-zero diagram.
b. Write a MATLAB script to evaluate the magnitude of $X(s)$ at a grid of complex points in the s-plane. Use the function meshgrid(..) to generate the grid of complex points within the ranges $-5 < \sigma < 5$ and $-5 < \omega < 5$ with increments of 0.1 in each direction.
c. Use the function mesh(..) to produce a three dimensional mesh plot of $|X(s)|$.
d. Evaluate the Laplace transform for $s = j\omega$ and use the function plot3(..) to plot it over the three-dimensional mesh plot.

7.51. Refer to Problem 7.8. Using MATLAB, construct a pole-zero plot for each transform listed.

7.52. Refer to Problem 7.8. For each signal that has a Fourier transform, write a script to compute $X(\omega)$ from the Laplace transform $X(s)$ given, and to graph it in the frequency range $-10 \le \omega \le 10$ rad/s.

7.53. Each transform given below corresponds to a causal signal $x(t)$. Use symbolic mathematics capabilities of MATLAB to find the inverse Laplace transform of each. Refer to MATLAB Exercise 7.5 for an example of the solution technique.

a. $X(s) = \dfrac{1}{s^2 + 3s + 2}$

b. $X(s) = \dfrac{(s-1)(s-2)}{(s+1)(s+2)(s+3)}$

c. $X(s) = \dfrac{s(s-1)}{(s+1)^2(s+2)}$

7.54. A CTLTI system is described by means of the system function

$$H(s) = \frac{s(s+1)}{s^2 + 4s + 13}$$

Compute and graph the response of the system to the signal

$$x(t) = e^{-0.2t} u(t)$$

using the following steps:

   a.   Define a system object "**sys**" for $H(s)$.
   b.   Create a vector with samples of the input signal.
   c.   Use the function lsim(..) to compute the output signal.

7.55. Using the function residue(..) obtain a partial fraction expansion for the transform $X(s)$ given by

$$X(s) = \frac{s^4 - 9s^3 + 30s^2 - 42s + 20}{s^5 + 12s^4 + 59s^3 + 152s^2 + 200s + 96}$$

Use the results from MATLAB to write $X(s)$ in partial fraction format.

7.56. Using system objects and the function bode(..), construct Bode magnitude and phase plots for the system functions given below:

a. $H(s) = \dfrac{s+1}{s^2 + 5s + 6}$

b. $H(s) = \dfrac{s(s-1)}{s^3 + 5s^2 + 8s + 6}$

c. $H(s) = \dfrac{(s-1)(s+2)}{(s+1)(s^2 + 6s + 13)}$

7.57. Consider the second-order system function

$$H(s) = \frac{\omega_0^2}{s^2 + 2\zeta\omega_0 s + \omega_0^2}$$

Let $\omega_0 = 5$ rad/s. Using system objects and the function bode(..), construct Bode magnitude and phase plots for the system for $\zeta = 0.01,\ 0.1,\ 0.5,\ 1,\ 2$.

7.58. Refer to Problem 7.41.

a. Using system objects of MATLAB, construct objects to represent $H(s)$, $H_1(s)$ and $H_1^{-1}(s)$.

b. Compute and graph the frequency responses (magnitude and phase) for $H(s)$, $H_1(s)$, $H_1^{-1}(s)$ and $H(s)\,H_1^{-1}(s)$.

c. Compute and graph the unit-step responses of $H(s)$ and $H(s)\,H_1^{-1}(s)$. Comment on the results.

7.59. Refer to the RLC circuit in Problem 7.48. Using symbolic mathematics capabilities of MATLAB, write a script to compute and graph the response of the circuit for the input signal and the initial conditions specified in the problem statement. Use MATLAB Exercise 7.10 for an example of the solution technique.

7.60. Refer to the RLC circuit in Problem 7.49. Using symbolic mathematics capabilities of MATLAB, write a script to compute and graph the response of the circuit for the input signal and the initial conditions specified in the problem statement. Use MATLAB Exercise 7.10 for an example of the solution technique.

MATLAB Projects

7.61. Consider the second-order system analyzed in Section 7.5.10. Its system function is

$$H(s) = \frac{\omega_0^2}{s^2 + 2\zeta\omega_0 s + \omega_0^2}$$

Let $y(t)$ be the unit-step response of this system.

a. Determine the Laplace transform of the unit-step response $y(t)$. Afterwards use the final value theorem to verify that the final value of the unit-step response is unity.

b. Develop a MATLAB script to compute the unit-step response of this system. Construct a system object for specified values of parameters ω_0 and ζ. Use the function step(..) to compute the step response for $0 \le t \le 30$ s with increments of $T = 0.01$ s.

c. One performance measure of the responsiveness of a system is the *rise time* which is defined as the time it takes for the unit-step response to progress from 10 percent to 90 percent of its final value. Investigate the dependency of the rise time (t_r) on parameter ζ. Set $\omega_0 = 1$ rad/s, and use the script from part (b) repeatedly with varying values of ζ. Measure the rise time from the graph and/or the data vector for each ζ value. Roughly sketch the relationship between t_r and ζ.

7.62. Refer to Problem 7.61. Another performance measure for a system is the *settling time* which is defined as the time it takes for the unit-step response to settle within a certain percentage of its final value. Let t_s represent the 5 percent settling time, that is, the time it takes for the response to settle in the interval $0.95 < y(t) < 1.05$ and stay. Using the MATLAB script developed in Problem 7.61 investigate the dependency of the settling time on parameter ζ. Use $\omega_0 = 1$ rad/s. Roughly sketch the relationship between t_s and ζ.

Chapter 8

z-Transform for Discrete-Time Signals and Systems

Chapter Objectives

- Learn the z-transform as a more generalized version of the discrete-time Fourier transform (DTFT) studied in Chapter 5.

- Understand the convergence characteristics of the z-transform and the concept of region of convergence.

- Explore the properties of the z-transform.

- Understand the use of the z-transform for modeling DTLTI systems. Learn the z-domain system function and its use for solving signal-system interaction problems.

- Learn techniques for obtaining block diagrams for DTLTI systems based on the z-domain system function.

- Learn the use of the unilateral z-transform for solving difference equations with specified initial conditions.

8.1 Introduction

In this chapter we will consider the z-transform which, for discrete-time signals, plays the same role that the Laplace transform plays for continuous-time signals. In the process, we will observe that the development of the z-transform techniques for discrete-time signals and systems parallels the development of the Laplace transform in Chapter 7.

It was established in Chapter 7 that the Laplace transform is an extension of the Fourier transform for continuous-time signals such that the latter can be thought of as a special case, or a limited view, of the former. The relationship between the z-transform and the DTFT is similar. In Chapter 5 we have discussed the use of the DTFT for analyzing discrete-time signals. The use of the DTFT was adapted to the analysis of DTLTI systems through the concept of the system function. The z-transform is an extension of the DTFT that can be used for the same purposes.

Consider, for example, a discrete-time ramp signal $x[n] = n\,u[n]$. Its DTFT does not exist since the signal is not absolute summable. We will, however, be able to use z-transform techniques for analyzing the ramp signal. Similarly, the DTFT-based system function is only usable for stable systems since the impulse response of a system must be absolute summable for its DTFT-based system function to exist. The z-transform, on the other hand, can be used with systems that are unstable as well. The DTFT may or may not exist for a particular signal or system while the z-transform will generally exist subject to some constraints.

We will see that the DTFT is a restricted cross-section of the much more general z-transform. The DTFT of a signal, if it exists, can be obtained from its z-transform while the reverse is not generally true. In addition to analysis of signals and systems, implementation structures for discrete-time systems can be developed using z-transform based techniques. Using the unilateral variant of the z-transform we will be able to solve difference equations with non-zero initial conditions.

We begin our discussion with the basic definition of the z-transform and its application to some simple signals. In the process, the significance of the issue of convergence of the transform is highlighted. In Section 8.2 the convergence properties of the z-transform, and specifically the region of convergence concept, are studied. Section 8.3 covers the fundamental properties of the transform that are useful in computing the transforms of signals with more complicated definitions. Proofs of the significant properties of the z-transform are presented to provide further insight and experience on working with transforms. Methods for computing the inverse z-transform are discussed in Section 8.4. Application of the z-transform to the analysis of linear and time-invariant systems is the subject of Section 8.5 where the z-domain system function concept is introduced. Derivation of implementation structures for discrete-time systems based on z-domain system functions is covered in Section 8.6. In Section 8.7 we discuss the unilateral variant of the z-transform that is useful in solving difference equations with specified initial conditions.

The z-transform of a discrete-time signal $x[n]$ is defined as

$$X(z) = \sum_{n=-\infty}^{\infty} x[n]\, z^{-n} \tag{8.1}$$

where z, the independent variable of the transform, is a complex variable.

If we were to expand the summation in Eqn. (8.1) we would get an expression in the form

$$X(z) = \ldots + x[-2]\, z^2 + x[-1]\, z^1 + z[0] + x[1]\, z^{-1} + x[2]\, z^{-2} + \ldots \tag{8.2}$$

which suggests that the transform is a polynomial that contains terms with both positive and negative powers of z. Since the independent variable z is complex, the resulting transform $X(z)$ is also complex-valued in general. Notationally, the transform relationship of Eqn. (8.1) can be expressed in the compact form

$$X(z) = \mathcal{Z}\{x[n]\} \tag{8.3}$$

which is read "$X(z)$ is the z-transform of $x[n]$". An alternative way to express the same relationship is through the use of the notation

$$x[n] \overset{\mathcal{Z}}{\longleftrightarrow} X(z) \tag{8.4}$$

The derivation of the relationship between the z-transform and the DTFT is straightforward. Let the complex variable z be expressed in polar form as

$$z = r\, e^{j\Omega} \tag{8.5}$$

Since z is a complex variable, we will represent it with a point in the complex z-plane as shown in Fig. 8.1. The parameter r indicates the distance of the point z from the origin. The parameter Ω is the angle, in radians, of the line drawn from the origin to the point z, measured counter-clockwise starting from the positive real axis.

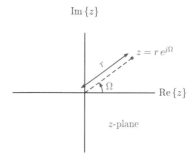

Figure 8.1 – The point $z = r\, e^{j\Omega}$ in the complex plane.

Obviously, any point in the complex z-plane can be expressed by properly choosing the parameters r and Ω. Substituting Eqn. (8.5) into the z-transform definition in Eqn. (8.1) we have

$$
\begin{aligned}
X(r, \Omega) = X(z)\Big|_{z=r\, e^{j\Omega}} \\
= \sum_{n=-\infty}^{\infty} x[n] \left(r\, e^{j\Omega} \right)^{-n} \\
= \sum_{n=-\infty}^{\infty} x[n]\, r^{-n}\, e^{-j\Omega n}
\end{aligned}
\tag{8.6}
$$

The result is a function of parameters r and Ω as well as the signal $x[n]$. We observe from Eqn. (8.6) that $X(r, \Omega)$ represents the DTFT of the signal $x[n]$ multiplied by an exponential signal r^{-n}:

$$X(r, \Omega) = X(z)\Big|_{z=r\, e^{j\Omega}} = \mathcal{F}\left\{ x[n]\, r^{-n} \right\} \tag{8.7}$$

If we choose a fixed value of $r = r_1$ in Eqn. (8.5) and allow the angle Ω to vary from 0 to 2π radians, the resulting trajectory of the complex variable z in the z-plane would be a circle with its center at the origin and with its radius equal to r_1 as shown in Fig. 8.2.

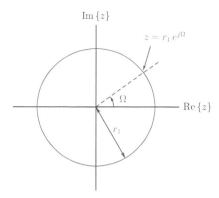

Figure 8.2 – Trajectory of $z = r_1 \, e^{j\Omega}$ on the complex plane.

This allows us to make an important observation:

Consider a circle in the z-plane with radius equal to $r = r_1$ and center at the origin. The z-transform evaluated on this circle starting at angle $\Omega = 0$ and ending at angle $\Omega = 2\pi$ is the same as the DTFT of the signal $x[n] \, r_1^{-n}$.

If the parameter r is chosen to be equal to unity, then we have $z = e^{j\Omega}$ and

$$X(z)\Big|_{z=e^{j\Omega}} = \sum_{n=-\infty}^{\infty} x[n] \, z^{-n} = \mathcal{F}\{x[n]\} \tag{8.8}$$

This is a significant observation. The z-transform of a signal $x[n]$ evaluated for $z = e^{j\Omega}$ produces the same result as the DTFT of the signal. The trajectory defined by $z = e^{j\Omega}$ in the z-plane is a circle with unit radius, centered at the origin. This circle is referred to as the *unit circle* of the z-plane. Thus we can state an important conclusion:

The DTFT of a signal $x[n]$ is equal to its z-transform evaluated at each point on the unit circle of the z-plane described by the trajectory $z = e^{j\Omega}$.

An easy way to visualize the relationship between the z-transform and the DTFT is the following: Imagine that the unit circle of the z-plane is made of a piece of string. Suppose the z-transform is computed at every point on the string. Let each point on the string be identified by the angle Ω so that the range $-\pi \leq \Omega < \pi$ covers the entire piece. If we now remove the piece of string from the z-plane and straighten it up, it becomes the Ω axis for the DTFT. The values of the z-transform computed at points on the string would be the values of one period of the DTFT.

Next we would like to develop a graphical representation of the z-transform in order to illustrate its relationship with the DTFT. Suppose the z-transform of some signal $x[n]$ is given by

$$X(z) = \frac{z \, (z - 0.7686)}{z^2 - 1.5371z + 0.9025} \tag{8.9}$$

In later sections of this chapter we will study techniques for computing the z-transform of a signal. At this point, however, our interest is in the graphical representation of the z-transform, the DTFT, and the relationship between them. Using the substitution $z = r \, e^{j\Omega}$, the transform in Eqn. (8.9) becomes

$$X\left(r,\Omega\right) = \frac{re^{j\Omega}\left(re^{j\Omega} - 0.7686\right)}{r^2 e^{j2\Omega} - 1.5371 re^{j\Omega} + 0.9025} \tag{8.10}$$

Using Eqn. (8.10) the transform at a particular point in the z-plane can be computed using r, its distance from the origin, and Ω, the angle it makes with the positive real axis. We would like to graph this transform; however, there are two issues that must first be addressed:

1. The z-transform is complex-valued. We must graph it either through its polar representation (magnitude and phase graphed separately) or its Cartesian representation (real and imaginary parts graphed separately). In this case we will choose to graph only its magnitude $|X\left(z\right)|$ as a three-dimensional surface. The magnitude of the transform can be computed by

$$|X\left(r,\Omega\right)| = \left[X\left(r,\Omega\right)X^*\left(r,\Omega\right)\right]^{1/2} \tag{8.11}$$

where $X^*\left(r,\Omega\right)$ is the complex conjugate of the transform $X\left(r,\Omega\right)$, and is computed as

$$X^*\left(r,\Omega\right) = \frac{re^{-j\Omega}\left(re^{-j\Omega} - 0.7686\right)}{r^2 e^{-j2\Omega} - 1.5371 re^{-j\Omega} + 0.9025} \tag{8.12}$$

2. Depending on the signal $x[n]$, the transform expression given by Eqn. (8.9) is valid either for $r < 0.95$ only, or for $r > 0.95$ only. This notion will elaborated upon when we discuss the "region of convergence" concept later in this chapter. In graphically illustrating the relationship between the z-transform and the DTFT at this point, we will assume that the sample transform in Eqn. (8.9) is valid for $r > 0.95$. The three-dimensional magnitude surface will be graphed for all values of r; however, we will keep in mind that it is only meaningful for values of r greater than 0.95.

The magnitude $|X\left(z\right)|$ is shown as a mesh in part (a) of Fig. 8.3. The unit circle is shown in the (x, y) plane. Also shown in part (a) of the figure is the set of values of $|X\left(z\right)|$ computed at points on the unit-circle. In part (b) of Fig. 8.3 the DTFT $X\left(\Omega\right)$ of the same signal is graphed. Notice how the unit circle of the z-plane is equivalent to the horizontal axis for the DTFT.

The z-transform defined by Eqn. (8.1) is sometimes referred to as the *bilateral* (two-sided) z-transform. A simplified variant of the transform termed the *unilateral* (one-sided) z-transform is introduced as an alternative analysis tool. The bilateral z-transform is useful for understanding signal characteristics, signal-system interaction, and fundamental characteristics of systems such as causality and stability. The unilateral z-transform is used for solving a linear constant-coefficient difference equation with specified initial conditions. We will briefly discuss the unilateral z-transform in Section 8.7, however, when we refer to z-transform without the qualifier word "bilateral" or "unilateral", we will always imply the more general bilateral z-transform as defined in Eqn. (8.1).

Interactive Demo: zt_demo1.m

The demo program "zt_demo1.m" is based on the z-transform given by Eqn. (8.9). The magnitude of the transform in question is shown in Fig. 8.3(a). The demo program computes this magnitude and graphs it as a three-dimensional mesh. It may be rotated freely for viewing from any angle by using the rotation tool in the toolbar. The magnitude of the

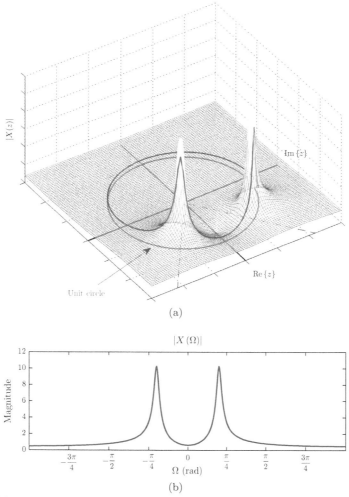

(a)

(b)

Figure 8.3 – (a) The magnitude $|X(z)|$ shown as a surface plot along with the unit circle and magnitude computed on the unit circle, (b) the magnitude of the DTFT as a function of Ω.

transform is also evaluated on a circle with radius r and graphed as a two-dimensional graph. The radius r may be varied through the use of a slider control.

Software resources:

zt_demo1.m

Software resources: See MATLAB Exercises 8.1 and 8.2.

Example 8.1: A simple z-transform example

Find the z-transform of the finite-length signal

$$x[n] = \{\ \underset{\substack{\uparrow \\ n=0}}{3.7},\ 1.3,\ -1.5,\ 3.4,\ 5.2\ \}$$

Solution: For the specified signal non-trivial samples occur for the index range $n = 0, \ldots, 4$. Therefore, the z-transform is

$$
\begin{aligned}
X(z) &= \sum_{n=-\infty}^{\infty} x[n] \, z^{-n} \\
&= x[0] + x[1] \, z^{-1} + x[2] \, z^{-2} + x[3] \, z^{-3} + x[4] \, z^{-4} \\
&= 3.7 + 1.3 \, z^{-1} - 1.5 \, z^{-2} + 3.4 \, z^{-3} + 5.2 \, z^{-4}
\end{aligned} \tag{8.13}
$$

The result is a polynomial of z^{-1}. The samples of the signal $x[n]$ become the coefficients of the corresponding powers of z^{-1}. Essentially, information about time-domain sample amplitudes of the signal is contained in the coefficients of the polynomial $X(z)$. The z-transform, therefore, is just an alternative way representing the signal $x[n]$.

The value of the transform at a specific point in the complex plane can be evaluated from the polynomial in Eqn. (8.13). For example, if we wanted to know the value of the transform at $z_1 = 1 + j2$ we would compute it as

$$
\begin{aligned}
X(1+j2) &= 3.7 + 1.3 \, (1+j2)^{-1} - 1.5 \, (1+j2)^{-2} + 3.4 \, (1+j2)^{-3} + 5.2 \, (1+j2)^{-4} \\
&= 3.7 + \frac{1.3}{(1+j2)} - \frac{1.5}{(1+j2)^2} + \frac{3.4}{(1+j2)^3} + \frac{5.2}{(1+j2)^4} \\
&= 3.7826 - j0.0259
\end{aligned}
$$

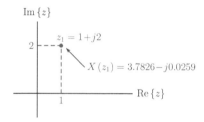

Figure 8.4 – The point z_1 in the z-plane.

The z-transform of a signal does not necessarily exist at every value of z. For example, the transform result we have found above could not be evaluated at the point $z - 0 + j0$ since the transform includes negative powers of z. In this case we conclude that the transform converges at all points in the complex z-plane with the exception of the origin.

Software resources:
ex_8_1.m

Example 8.2: **z-Transform of a non-causal signal**

Find the z-transform of the signal

$$
x[n] = \{\, 3.7,\ 1.3,\ -1.5,\ 3.4,\ 5.2 \,\}
$$
$$
\underset{\underset{n=0}{\uparrow}}{}
$$

Solution: This is essentially the same signal the z-transform of which we have computed in Example 8.1 with one difference: It has been advanced by two samples so it starts with index $n = -2$. Applying the z-transform definition given by Eqn. (8.1) we obtain

$$X\left(z\right) = \sum_{n=-\infty}^{\infty} x[n]\, z^{-n}$$

$$= x[-2]\, z^2 + x[-1]\, z^1 + x[0] + x[1]\, z^{-1} + x[2]\, z^{-2}$$

$$= 3.7\, z^2 + 1.3\, z^1 - 1.5 + 3.4\, z^{-1} + 5.2\, z^{-2}$$

As in the previous example, the transform fails to converge to a finite value at the origin of the z-plane. In addition, the transform does not converge for infinitely large values of $|z|$ because of the z^1 and z^2 terms included in $X\left(z\right)$. It converges at every point in the z-plane with the two exceptions, namely the origin and infinity.

Example 8.3: **z-Transform of the unit-impulse**

Find the z-transform of the unit-impulse signal

$$x[n] = \delta[n] = \left\{ \begin{array}{ll} 1\,, & n = 0 \\ 0\,, & n \neq 0 \end{array} \right.$$

Solution: Since the only non-trivial sample of the signal occurs at index $n = 0$, the z-transform is

$$X\left(z\right) = \mathcal{Z}\left\{\delta[n]\right\} = \sum_{n=-\infty}^{\infty} x[n]\, z^{-n} = x[0]\, z^0 = 1$$

The transform of the unit-impulse signal is constant and equal to unity. Since it does not contain any positive or negative powers of z, it converges at every point in the z-plane with no exceptions.

Example 8.4: **z-Transform of a time shifted unit-impulse**

Find the z-transform of the time shifted unit-impulse signal

$$x[n] = \delta[n - k] = \left\{ \begin{array}{ll} 1\,, & n = k \\ 0\,, & n \neq k \end{array} \right.$$

where $k \neq 0$ is a positive or negative integer.

Solution: Applying the z-transform definition we obtain

$$X\left(z\right) = \mathcal{Z}\left\{\delta[n - k]\right\} = \sum_{n=-\infty}^{\infty} x[n]\, z^{-n} = z^{-k}$$

The transform converges at every point in the z-plane with one exception:

1. If $k > 0$ then the transform does not converge at the origin $z = 0 + j0$.
2. If $k < 0$ then the transform does not converge at points with infinite radius, that is, at points for which $|z| \to \infty$.

It becomes apparent from the examples above that the z-transform must be considered in conjunction with the criteria for the convergence of the resulting polynomial $X\left(z\right)$. The collection of all points in the z-plane for which the transform converges is called the *region of convergence (ROC)*. We conclude from Examples 8.1 and 8.2 that the region of convergence for a finite-length signal is the entire z-plane with the possible exception of $z = 0 + j0$, or $|z| \to \infty$ or both.

1. The origin $z = 0 + j0$ is excluded from the region of convergence if any negative powers of z appear in the polynomial $X(z)$. Negative powers of z are associated with samples of the signal $x[n]$ with positive indices. Therefore, if the signal $x[n]$ has any non-zero valued samples with positive indices, the z-transform does not converge at the origin of the z-plane.

2. Values of z with infinite radius must be excluded from the region of convergence if the polynomial $X(z)$ includes positive powers of z. Positive powers of z are associated with samples of $x[n]$ that have negative indices. Therefore, if the signal $x[n]$ has any non-zero valued samples with negative indices, that is, if the signal is non-causal, its z-transform does not converge at $|z| \to \infty$.

For finite-length signals, the ROC can be determined based on the sample indices of the leftmost and the rightmost significant samples of the signal. If the index of the leftmost significant sample is negative, then the transform does not converge for $|z| \to \infty$. If the index of the rightmost significant sample is positive, then the transform does not converge at the origin $z = 0 + j0$.

How would we determine the region of convergence for a signal that is not finite-length? Recall that in Eqn. (8.7) we have represented the z-transform of a signal $x[n]$ as equivalent to the DTFT of the modified signal $x[n] \, r^{-n}$. As a result, the convergence of the z-transform of $x[n]$ can be linked to the convergence of the DTFT of the modified signal $x[n] \, r^{-n}$. Using Dirichlet conditions discussed in Chapter 5, this requires that $x[n] \, r^{-n}$ be absolute summable, that is,

$$\sum_{n=-\infty}^{\infty} \left| x[n] \, r^{-n} \right| < \infty \tag{8.14}$$

The condition stated in Eqn. (8.14) highlights the versatility of the z-transform over the DTFT. Recall that, for the DTFT of a signal $x[n]$ to exist, the signal has to be absolute summable. Therefore, the existence of the DTFT is a binary question; the answer to it is either yes or no. In contrast, if $x[n]$ is not absolute summable, we may still be able to find values of r for which the modified signal $x[n]r^{-n}$ is absolute summable. Therefore, the z-transform of the signal $x[n]$ may exist for some values of r. The question of existence for the z-transform is not a binary one; it is a question of which values of r allow the transform to converge.

The region of convergence concept will be discussed in detail in Section 8.2. The next few examples further highlight the need for a detailed discussion of the region of convergence.

Example 8.5: **z-transform of the unit-step signal**

Find the z-transform of the unit-step signal

$$x[n] = u[n] = \begin{cases} 1 \, , & n \geq 0 \\ 0 \, , & n < 0 \end{cases}$$

Solution: We will again apply the z-transform definition given by Eqn. (8.1).

$$X(z) = \sum_{n=-\infty}^{\infty} u[n] \, z^{-n} \tag{8.15}$$

Since $u[n] = 0$ for $n < 0$, the lower limit of the summation in Eqn. (8.15) can be changed to $n = 0$ without affecting the result:

$$X(z) = \sum_{n=0}^{\infty} u[n] \, z^{-n} \tag{8.16}$$

Furthermore, $u[n] = 1$ for $n \geq 0$, so dropping the $u[n]$ term from the summation in Eqn. (8.16) has no effect either.

$$X(z) = \sum_{n=0}^{\infty} z^{-n} \tag{8.17}$$

Eqn. (8.17) represents the sum of an infinite-length geometric series, and can be computed in closed form using the formula[1] derived in Appendix C to yield

$$X(z) = \frac{1}{1 - z^{-1}} = \frac{z}{z - 1}$$

which converges only for values of z for which

$$\left| z^{-1} \right| < 1 \qquad \Rightarrow \qquad |z| > 1$$

The ROC for the transform of the unit-step signal is the collection of points outside a circle with radius equal to unity. This region is shown shaded in Fig. 8.5. Note that the unit-circle itself is not part of the ROC since the transform does not converge for values of z on the circle.

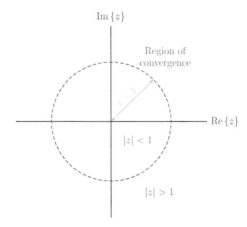

Figure 8.5 – The region of convergence for the z-transform of $x[n] = u[n]$.

Example 8.6: **z-Transform of a causal exponential signal**

Find the z-transform of the signal

$$x[n] = a^n \, u[n] = \begin{cases} a^n , & n \geq 0 \\ 0 , & n < 0 \end{cases}$$

where a is any real or complex constant.

Solution: The signal $x[n]$ is causal since $x[n] = 0$ for $n < 0$. Applying the z-transform definition given by Eqn. (8.1) we obtain

$$X(z) = \sum_{n=-\infty}^{\infty} a^n \, u[n] \, z^{-n} \tag{8.18}$$

---

[1] $\sum_{n=0}^{\infty} \alpha^n = 1/(1 - \alpha) , \quad |\alpha| < 1.$

Let us change the lower limit of the summation to $n = 0$ and drop the factor $u[n]$. Eqn. (8.18) becomes

$$X\left(z\right) = \sum_{n=0}^{\infty} a^n\, z^{-n} = \sum_{n=0}^{\infty} \left(a\, z^{-1}\right)^n$$

This is the sum of an infinite-length geometric series, and can be put into a closed form as

$$X\left(z\right) = \frac{1}{1 - a\, z^{-1}} = \frac{z}{z - a}$$

which is valid only for values of z for which

$$\left|a\, z^{-1}\right| < 1 \qquad \Rightarrow \qquad \frac{|a|}{|z|} < 1 \qquad \Rightarrow \qquad |z| > |a|$$

Four possible forms of the signal $x[n]$ are shown in Fig. 8.6a,b,c,d corresponding to possible real values of the parameter a. It should be noted that there are other possibilities; the parameter a could also have a complex value. The ROC for $X\left(z\right)$ is the region outside a circle with radius equal to $|a|$. This is shown in Fig. 8.7 for $|a| < 1$ and $|a| > 1$.

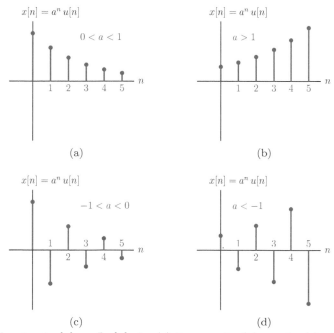

Figure 8.6 – The signal $x[n] = a^n\, u[n]$ for (a) $0 < a < 1$, (b) $a > 1$, (c) $-1 < a < 0$, and (d) $a < -1$.

For the case $|a| < 1$ the unit circle is part of the ROC whereas, for the case $|a| > 1$, it is not. Recall from the previous discussion that the DTFT is equal to the z-transform evaluated on the unit circle. Consequently, the DTFT of the signal exists only if the ROC includes the unit circle. We conclude that, for the DTFT of the signal $x[n]$ to exist, we need $|a| < 1$. This is consistent with the existence conditions discussed in Chapter 5.

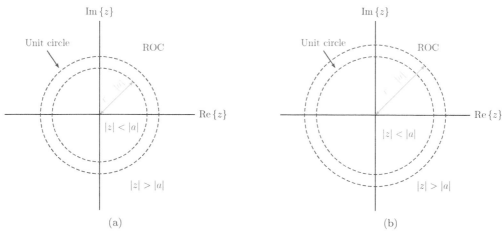

Figure 8.7 – The region of convergence for the z-transform of $x[n] = a^n\,u[n]$ for $|a| < 1$ and for $|a| > 1$.

Example 8.7: **z-Transform of an anti-causal exponential signal**

Find the z-transform of the signal

$$x[n] = -a^n\,u[-n-1] = \begin{cases} -a^n\,, & n < 0 \\ 0\,, & n \geq 0 \end{cases}$$

where a is any real or complex constant.

Solution: In this case the signal $x[n]$ is anti-causal since it is equal to zero for $n \geq 0$. Applying the z-transform definition given by Eqn. (8.1) we obtain

$$X(z) = \sum_{n=-\infty}^{\infty} -a^n\,u[-n-1]\,z^{-n} \tag{8.19}$$

Since

$$u[-n-1] = \begin{cases} 1\,, & n < 0 \\ 0\,, & n \geq 0 \end{cases}$$

changing the upper limit of the summation to $n = -1$ and dropping the factor $u[-n-1]$ would have no effect on the result. Eqn. (8.19) simplifies to

$$X(z) = -\sum_{n=-\infty}^{-1} a^n\,z^{-n} \tag{8.20}$$

Lower and upper summation limits are both negative. This can be fixed by employing a variable change $m = -n$, and the result found in Eqn. (8.20) can be written as

$$X(z) = -\sum_{m=\infty}^{1} a^{-m}\,z^{m} = -\sum_{m=1}^{\infty} \left(a^{-1}\,z\right)^{m}$$

which has the familiar sum of geometric series, however, the lower limit of the summation is $m = 1$ rather than $m = 0$ as required for the use of the closed-form formula. Let us apply another variable change to the summation, this time in the form $k = m - 1$, to get the lower limit of the summation to start at $k = 0$:

$$X(z) = - \sum_{k=0}^{\infty} \left(a^{-1} z\right)^{k+1}$$

$$= - a^{-1} z \sum_{k=0}^{\infty} \left(a^{-1} z\right)^{k} \qquad (8.21)$$

Now the summation in Eqn. (8.21) can be put into a closed form to yield

$$X(z) = a^{-1} z \left(\frac{1}{1 - a^{-1} z}\right) - \frac{z}{z - a}, \qquad \left|a^{-1}z\right| < 1 \qquad (8.22)$$

Notice how the closed-form expression found for the signal $x[n] = -a^n u[-n-1]$ in Eqn. (8.22) is identical to the one found in Example 8.6 for a different signal, namely $x[n] = a^n u[n]$. The transform expressions are identical; however, the regions in which those expressions are valid are different. The closed-form formula found in Eqn. (8.21) is valid only for values of z for which

$$\left|a^{-1} z\right| < 1 \qquad \Rightarrow \qquad \frac{|z|}{|a|} < 1 \qquad \Rightarrow \qquad |z| < |a|$$

Four possible forms of the signal $x[n]$ are shown in Fig. 8.8 corresponding to different ranges of the real-valued parameter a (keep in mind that a could also be complex). The ROC for the transform $X(z)$ found in this case is the region inside a circle with radius equal to $|a|$. This is shown in Fig. 8.9 for $|a| < 1$ and $|a| > 1$. For the case $|a| > 1$ the unit circle is part of the ROC whereas, for the case $|a| < 1$, it is not. Thus, the DTFT of the signal $x[n]$ exists only when $|a| > 1$.

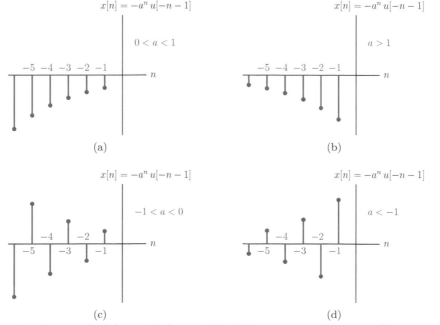

Figure 8.8 – The signal $x[n] = -a^n u[-n-1]$ for (a) $0 < a < 1$, (b) $a > 1$, (c) $-1 < a < 0$, and (d) $a < -1$.

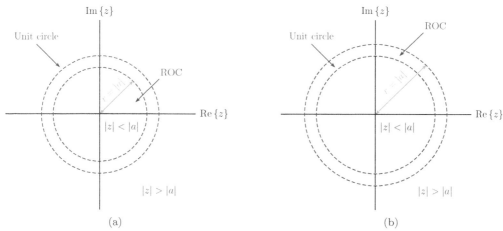

Figure 8.9 – The region of convergence for the z-transform of $x[n] = -a^n\, u[-n-1]$ for $|a| < 1$ and for $|a| > 1$.

Examples 8.6 and 8.7 demonstrate a very important concept: It is possible for two different signals to have the same transform expression for the z-transform $X(z)$. In order for us to uniquely identify which signal among the two led to a particular transform, the region of convergence must be specified along with the transform. The following two transform pairs are fundamental in determining to which of the possible signals a given transform corresponds:

$$a^n\, u[n] \overset{\mathcal{Z}}{\longleftrightarrow} \frac{z}{z-a}, \qquad \text{ROC: } |z| > |a| \tag{8.23}$$

$$-a^n\, u[-n-1] \overset{\mathcal{Z}}{\longleftrightarrow} \frac{z}{z-a}, \qquad \text{ROC: } |z| < |a| \tag{8.24}$$

Sometimes we need to deal with the inverse problem of determining the signal $x[n]$ that has a particular z-transform $X(z)$. Given a transform

$$X(z) = \frac{z}{z-a}$$

we need to know the region of convergence to determine if the signal $x[n]$ is the one in Eqn. (8.23) or the one in Eqn. (8.24). This will be very important when we work with inverse z-transform in Section 8.4 later in this chapter. In order to avoid ambiguity, we will adopt the convention that the region of convergence is an integral part of the z-transform, and must be specified explicitly, or must be implied by means of another property of the underlying signal or system, every time a transform is given.

In each of the Examples 8.5, 8.6 and 8.7 we have obtained rational functions of z for the transform $X(z)$. In the general case, a rational transform $X(z)$ is expressed in the form

$$X(z) = K\,\frac{B(z)}{A(z)} \tag{8.25}$$

where the numerator $B(z)$ and the denominator $A(z)$ are polynomials of z. The parameter K is a constant gain factor. Let the numerator polynomial be written in factored form as

$$B(z) = (z - z_1)(z - z_2)\ldots,(z - z_M) \tag{8.26}$$

and let $p_1, p_2, \ldots, p_N$ be the roots of the denominator polynomial so that

$$A(z) = (z - p_1)(z - p_2) \ldots, (z - p_N) \tag{8.27}$$

The transform can be written as

$$X(z) = K \frac{(z - z_1)(z - z_2) \ldots, (z - z_M)}{(z - p_1)(z - p_2) \ldots, (z - p_N)} \tag{8.28}$$

In Eqns. (8.26) and (8.27) the parameters M and N are the numerator order and the denominator order respectively. The larger of M and N is the order of the transform $X(z)$. The roots of the numerator polynomial are referred to as the *zeros* of the transform $X(z)$. In contrast, the roots of the denominator polynomial are the *poles*. The transform does not converge at a pole, therefore, the ROC cannot contain any poles. In the next section we will see that the boundaries of the ROC are determined by the poles of the transform.

Example 8.8: **z-Transform of a discrete-time pulse signal**

Determine the z-transform of the discrete-time pulse signal

$$x[n] = \begin{cases} 1, & 0 \le n \le N-1 \\ 0, & n < 0 \quad \text{or} \quad n > N-1 \end{cases}$$

shown in Fig. 8.10.

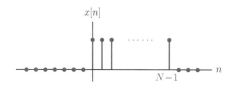

Figure 8.10 – The signal $x[n]$ for Example 8.8.

Solution: The transform in question is computed as

$$X(z) = \sum_{n=0}^{N-1} (1)\, z^{-n} = 1 + z^{-1} + \ldots + z^{-N} = \frac{1 - z^{-N}}{1 - z^{-1}} \tag{8.29}$$

In computing the closed-form result in Eqn. (8.29) we have used the finite-length geometric series sum formula derived in Appendix C. Since $x[n]$ is finite-length and causal, the transform converges at every point in the z-plane except at the origin $z = 0 + j0$. Thus we have

$$\text{ROC:} \quad |z| > 0$$

Often we find it more convenient to write the transform using non-negative powers of z. Multiplying both the numerator and the denominator polynomials in Eqn. (8.29) by bz^N leads to

$$X(z) = \frac{z^N - 1}{z^{N-1}(z-1)} \tag{8.30}$$

which is in the general factored form given by Eqn. (8.28). The appearance of the closed-form result in Eqn. (8.30) may be confusing in the way it relates to the region of convergence. It seems as though $X(z)$ might have a pole at $z = 1$. We need to realize, however, that the numerator polynomial is also equal to zero at $z = 1$:

$$B(z)\Big|_{z=1} = \left(z^N - 1\right)\Big|_{z=1} = 0$$

As a result, the numerator polynomial has a $(z-1)$ factor that effectively cancels the pole at $z=1$. We will now go through the exercise of determining the zeros and poles of the transform $X(z)$. The roots of the numerator polynomial are found by solving the equation

$$B(z) = 0 \quad \Longrightarrow \quad z^N = 1 \tag{8.31}$$

In order to determine the roots of the numerator polynomial that are not immediately obvious, we will use the fact that $e^{j2\pi k} = 1$ for all integer k, and write Eqn. (8.31) as

$$z^N = e^{j2\pi k} \tag{8.32}$$

Taking the N-th root of both sides leads to the set of roots

$$z_k = e^{j2\pi k/N} , \quad k = 0, \ldots, N-1 \tag{8.33}$$

The numerator polynomial of $X(z)$ has N roots that are all on the unit circle of the z-plane. They are equally-spaced at angles that are integer multiples of $2\pi/N$ as shown in Fig. 8.11(a). Note that in the figure we have used $N = 10$.

The roots of the denominator polynomial are found by solving

$$A(z) = 0 \quad \Rightarrow \quad z^{N-1}(z-1) = 0 \tag{8.34}$$

There are $N-1$ solutions at $p_k = 0$ and one solution at $z = 1$ as shown in Fig. 8.11(b).

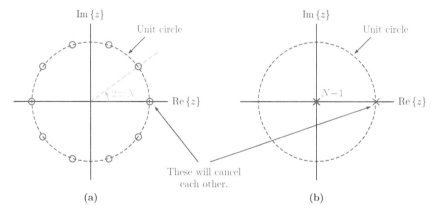

Figure 8.11 – (a) Roots of the numerator polynomial $z^N - 1$, (b) roots of the denominator polynomial $z^{N-1}(z-1)$.

The pole-zero diagram for the transform $X(z)$ can now be constructed. The factors $(z-1)$ in numerator and denominator polynomials cancel each other, therefore there is neither a zero nor a pole at $z = 1$. We are left with zeros at

$$z_k = e^{j2\pi k/N} , \quad k = 1, \ldots, N-1$$

and a total of $N-1$ poles all at

$$p_k = 0 , \quad k = 1, \ldots, N-1$$

as shown in Fig. 8.12.

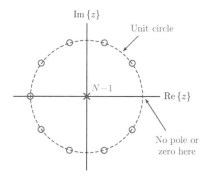

Figure 8.12 – Pole-zero diagram for $X(z)$.

In light of what we have discovered regarding the the poles of $X(z)$ it makes sense that the ROC is the entire z-plane with the exception of a singular point at the origin. Since all $N-1$ poles of the transform are at the origin, that is the only point where the transform does not converge.

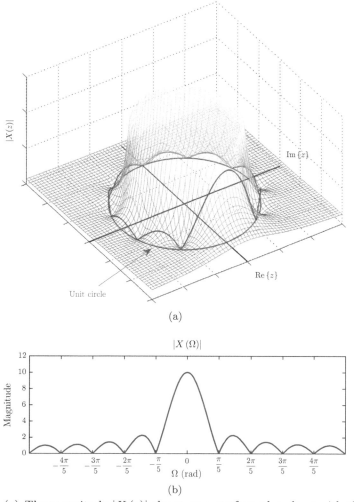

Figure 8.13 – (a) The magnitude $|X(z)|$ shown as a surface plot along with the unit circle and magnitude computed on the unit circle, (b) the magnitude of the DTFT as a function of Ω.

The magnitude of the z-transform computed in Eqn. (8.30) is shown in Fig. 8.13(a) as a surface graph for $N = 10$. Since the origin of the z-plane is excluded from the ROC, the magnitude is not computed at the origin. Also, very large magnitude values in close proximity of the origin are clipped to make the graph fit into a reasonable scale. Note how the zeros equally spaced around the unit-circle cause the magnitude surface to dip down. In addition to the surface graph, Fig. 8.13(a) also shows the unit circle of the z-plane. Magnitude values computed at points on the unit circle are marked on the surface. Fig. 8.13(b) shows the magnitude of the DTFT computed for the range of angular frequency $-\pi \leq \Omega < \pi$. It should be compared to the values marked on the surface graph that correspond to the magnitude of the z-transform evaluated on the unit circle.

To provide a slightly different perspective and to help with visualization, imagine that we take a knife and carefully cut through the surface in Fig. 8.13(a) along the perimeter of the unit-circle of the z-plane, as if the surface is a cake that we would like to fit into a cylindrical box. The profile of the cutout would match the DTFT shown in Fig. 8.13(b). This is illustrated in Fig. 8.14.

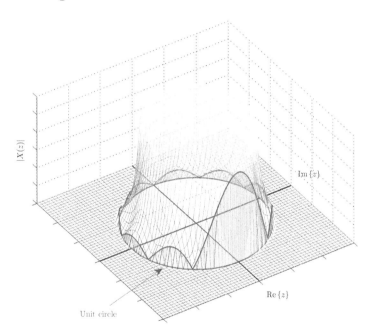

Figure 8.14 – Profile of the cutout along the perimeter of the unit circle for Example 8.8.

Interactive Demo: zt_demo2.m

The demo program "zt_demo2.m" is based on the z-transform of a discrete-time pulse analyzed in Example 8.8. The magnitude of the transform in question was shown in Fig. 8.13(a). The demo program computes this magnitude and graphs it as a three-dimensional mesh. It may be rotated freely for viewing from any angle by using the rotation tool in the toolbar. The magnitude of the transform is also evaluated on a circle with radius r and graphed as a two-dimensional graph. This corresponds to the DTFT of the modified signal $x[n]\,r^{-n}$. The radius r may be varied through the use of a slider control.
Software resources:
zt_demo2.m

Example 8.9: z-Transform of complex exponential signal

Find the z-transform of the signal

$$x[n] = e^{j\Omega_0 n} \, u[n]$$

The parameter Ω_0 is real and positive, and is in radians.

Solution: Applying the z-transform definition directly to the signal $x[n]$ leads to

$$X(z) = \sum_{n=-\infty}^{\infty} e^{j\Omega_0 n} \, u[n] \, z^{-n} \tag{8.35}$$

Since $x[n]$ is causal, the lower limit of the summation can be set to $n = 0$, and the $u[n]$ term can be dropped to yield

$$X(z) = \sum_{n=0}^{\infty} e^{j\Omega_0 n} \, z^{-n} \tag{8.36}$$

The expression in Eqn. (8.36) is the sum of an infinite-length geometric series, and can be put into closed form as

$$X(z) = \sum_{n=0}^{\infty} \left(e^{j\Omega_0} \, z^{-1} \right)^n = \frac{1}{1 - e^{j\Omega_0} \, z^{-1}} \tag{8.37}$$

The region of convergence is obtained from the convergence condition for the geometric series:

$$\left| e^{j\Omega_0} \, z^{-1} \right| < 1 \quad \Rightarrow \quad |z| > 1 \tag{8.38}$$

The transform in Eqn. (8.37) can also be written using non-negative powers of z as

$$X(z) = \frac{z}{z - e^{j\Omega_0}} \tag{8.39}$$

It has a zero at the origin and a pole at $z = e^{j\Omega_0}$. The zero and the pole of the transform as well as its ROC are shown in Fig. 8.15.

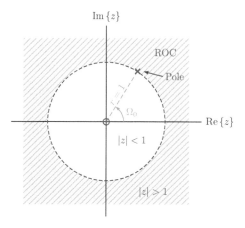

Figure 8.15 – ROC for the transform computed in Example 8.9.

Example 8.10: **A more involved z-transform example**

Determine the z-transform of the signal $x[n]$ defined by

$$x[n] = \begin{cases} (1/2)^n \,, & n \geq 0 \ \text{and even} \\ -(1/3)^n \,, & n > 0 \ \text{and odd} \\ 0 \,, & n < 0 \end{cases} \tag{8.40}$$

Solution: This example is somewhat tricky because of the particular way $x[n]$ is defined. Using the definition of the z-transform we can write

$$X(z) = \sum_{\substack{n=0 \\ n \ \text{even}}}^{\infty} \left(\frac{1}{2}\right)^n z^{-n} - \sum_{\substack{n=1 \\ n \ \text{odd}}}^{\infty} \left(\frac{1}{3}\right)^n z^{-n} \tag{8.41}$$

The first summation in Eqn. (8.41) will be carried out for only the even values of the index n, and the second summation is for only the odd values of it. To facilitate the goal of limiting the summations to even and odd index values, we will use two variable changes: For the first summation, let $n = 2m$ where m is the new integer index. This substitution will ensure that n is always an even value. Similarly, for the second summation, we will use the variable change $n = 2m + 1$ to ensure that the resulting value of n is always odd. Incorporating these variable changes into the summations, Eqn. (8.41) can be written as

$$X(z) = \sum_{m=0}^{\infty} \left(\frac{1}{2}\right)^{2m} z^{-2m} - \sum_{m=0}^{\infty} \left(\frac{1}{3}\right)^{2m+1} z^{-(2m+1)} \tag{8.42}$$

or in the equivalent form

$$X(z) = \sum_{m=0}^{\infty} \left(\frac{1}{4}\right)^m \left(z^2\right)^{-m} - \frac{1}{3} z^{-1} \sum_{m=0}^{\infty} \left(\frac{1}{9}\right)^m \left(z^2\right)^{-m}$$

$$= \sum_{m=0}^{\infty} \left(\frac{1}{4} z^{-2}\right)^m - \frac{1}{3} z^{-1} \sum_{m=0}^{\infty} \left(\frac{1}{9} z^{-2}\right)^m \tag{8.43}$$

The two summations in Eqn. (8.43) can be thought of as two transforms $X_1(z)$ and $X_2(z)$ so that

$$X(z) = X_1(z) - X_2(z)$$

The closed form expression for $X_1(z)$ and its associated ROC can be obtained as

$$X_1(z) = \sum_{m=0}^{\infty} \left(\frac{1}{4} z^{-2}\right)^m = \frac{z^2}{z^2 - \frac{1}{4}} \,, \qquad \text{ROC:} \quad \left|z^2\right| > \frac{1}{4} \tag{8.44}$$

Similarly, $X_2(z)$ is

$$X_2(z) = \frac{1}{3} z^{-1} \sum_{m=0}^{\infty} \left(\frac{1}{9} z^{-2}\right)^m = \frac{\frac{1}{3} z}{z^2 - \frac{1}{9}} \,, \qquad \text{ROC:} \quad \left|z^2\right| > \frac{1}{9} \tag{8.45}$$

Combining the closed form expressions for $X_1(z)$ and $X_2(z)$ as given by Eqns. (8.44) and (8.45) under a common denominator, we find the transform $X(z)$ and its ROC as

$$X(z) = \frac{z^4 - \frac{1}{3} z^3 - \frac{1}{9} z^2 + \frac{1}{12} z}{\left(z^2 - \frac{1}{4}\right)\left(z^2 - \frac{1}{9}\right)} \,, \qquad \text{ROC:} \quad |z| > \frac{1}{2} \tag{8.46}$$

The resulting transform has two real zeros at $z = 0$ and $z = -0.4158$ as well as a complex conjugate pair of zeros at $z = 0.3746 \pm j0.2451$. Its poles are at $z = \pm \frac{1}{2}$ and $z = \pm \frac{1}{3}$. The pole-zero diagram and the ROC for $X(z)$ are shown in Fig. 8.16.

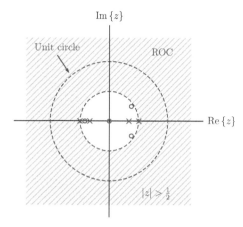

Figure 8.16 – Pole-zero diagram and the ROC for the transform computed in Example 8.10.

Example 8.11: **Transform of a downsampled signal**

The downsampling operation was briefly discussed in Chapter 6. Consider a signal $g[n]$ obtained by downsampling another signal $x[n]$ by a factor of 2:

$$g[n] = x[2n]$$

Determine the z-transform of the downsampled signal $g[n]$ in terms of the transform of the original signal $x[n]$.

Solution: Using the z-transform definition, $G(z)$ is

$$G(z) = \sum_{n=-\infty}^{\infty} g[n] z^{-n} \tag{8.47}$$

Let us substitute $g[n] = x[2n]$ into Eqn. (8.47) to obtain

$$G(z) = \sum_{n=-\infty}^{\infty} x[2n] z^{-n} \tag{8.48}$$

To resolve the summation in Eqn. (8.48) we will use the variable change $m = 2n$ and write the summation in terms of the new index m:

$$G(z) = \sum_{\substack{m=-\infty \\ m \text{ even}}}^{\infty} x[m] z^{-m/2} \tag{8.49}$$

Naturally, the summation should only have the terms for which m is even. This restriction on the values of the summation index makes it difficult for us to relate Eqn. (8.49) to the transform $X(z)$ of the original signal. To overcome this hurdle we will use a simple trick. Consider a discrete-time signal $w[n]$ with the following definition:

$$w[m] = \begin{cases} 1, & m \text{ even} \\ 0, & m \text{ odd} \end{cases}$$

The transform in Eqn. (8.49) can be written as

$$G(z) = \sum_{m=-\infty}^{\infty} x[m]\, w[m]\, z^{-m/2} \tag{8.50}$$

Notice how we removed the restriction on m in Eqn. (8.50) since the factor $w[m]$ ensures that odd-indexed terms do not contribute to the sum. One method of creating the signal $w[m]$ is through

$$w[m] = \frac{1 + (-1)^m}{2} \tag{8.51}$$

Substituting Eqn. (8.51) into Eqn. (8.50) we obtain

$$G(z) = \sum_{m=-\infty}^{\infty} x[m] \left[\frac{1 + (-1)^m}{2} \right] z^{-m/2}$$

$$= \frac{1}{2} \sum_{m=-\infty}^{\infty} x[m]\, z^{-m/2} + \frac{1}{2} \sum_{m=-\infty}^{\infty} x[m]\, (-1)^m\, z^{-m/2} \tag{8.52}$$

Recognizing that $z^{-m/2} = \left(\sqrt{z} \right)^{-m}$ and $(-1)^m\, z^{-m/2} = \left(-\sqrt{z} \right)^{-m}$, the result in Eqn. (8.52) can be rewritten as

$$G(z) = \frac{1}{2} \sum_{m=-\infty}^{\infty} x[m] \left(\sqrt{z} \right)^{-m} + \frac{1}{2} \sum_{m=-\infty}^{\infty} x[m] \left(-\sqrt{z} \right)^{-m}$$

$$= \frac{1}{2} X\left(\sqrt{z} \right) + \frac{1}{2} X\left(-\sqrt{z} \right) \tag{8.53}$$

Let the ROC for the original transform $X(z)$ be

$$r_1 < |z| < r_2$$

The ROC for $X\left(\sqrt{z} \right)$ and $X\left(-\sqrt{z} \right)$ terms in Eqn. (8.53) is

$$r_1 < \left| \pm \sqrt{z} \right| < r_2 \quad \Rightarrow \quad r_1^2 < |z| < r_2^2$$

and therefore the ROC for $G(z)$ is also

$$r_1^2 < |z| < r_2^2$$

Example 8.12: **Transform of a downsampled signal revisited**

Consider the causal exponential signal

$$x[n] = (0.9)^n\, u[n]$$

Let a new signal $g[n]$ be defined as $g[n] = x[2n]$. Find the z-transform of $g[n]$.

Solution: The transform of the signal $x[n]$ is

$$X\left(z\right) = \frac{z}{z - 0.9}\,, \qquad \text{ROC:} \quad |z| > 0.9$$

Applying the result found in Eqn. (8.53) of Example 8.11, the transform $G\left(z\right)$ of the down-sampled signal $g[n]$ is found as

$$G\left(z\right) = \frac{\frac{1}{2}\sqrt{z}}{\sqrt{z} - 0.9} + \frac{-\frac{1}{2}\sqrt{z}}{-\sqrt{z} - 0.9} \qquad (8.54)$$

Combining the two terms of Eqn. (8.54) under a common denominator we get

$$G\left(z\right) = \frac{z}{z - (0.9)^2}\,, \qquad \text{ROC:} \quad |z| > (0.9)^2 \qquad (8.55)$$

It is easy to verify the validity of the result in Eqn. (8.55) if we apply the downsampling operation in the time domain and then find the z-transform of the resulting signal $g[n]$ directly. The time domain expression for the signal $g[n]$ is

$$g[n] = x[2n] = \left(0.9^2\right)^n u[2n] = (0.81)^n u[n]$$

and its z-transform is

$$G\left(z\right) = \frac{z}{z - 0.81}\,, \qquad \text{ROC:} \quad |z| > 0.81$$

Software resources: See MATLAB Exercise 8.3.

8.2 Characteristics of the Region of Convergence

The examples we have worked on so far made it clear that the z-transform of a discrete-time signal always needs to be considered in conjunction with its region of convergence, that is, the collection of points in the z-plane for which the transform converges. In this section we will summarize and justify the fundamental characteristics of the region of convergence.

1. The ROC is circularly shaped. It is either the inside of a circle, the outside of a circle, or between two circles as shown in Fig. 8.17.

This property is easy to justify when we recall that, for $z = r\,e^{j\Omega}$, the values of the z-transform are identical to the values of the DTFT of the signal $x[n]\,r^{-n}$, as derived in Eqn. (8.7). Therefore, the convergence of the z-transform for $z = r\,e^{j\Omega}$ is equivalent to the convergence of the DTFT of the signal $x[n]\,r^{-n}$ which requires that

$$\sum_{n=-\infty}^{\infty} \left|x[n]\,r^{-n}\right| < \infty \qquad (8.56)$$

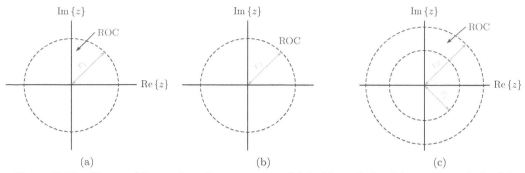

Figure 8.17 – Shape of the region of convergence: (a) inside a circle, (b) outside a circle, (c) between two circles.

Thus, the ROC depends only on r and not on Ω, explaining the circular nature of the region. Following are the possibilities for the ROC:

$$r < r_1 \ : \ \text{ Inside a circle}$$

$$r > r_2 \ : \ \text{ Outside a circle}$$

$$r_1 < r < r_2 \ : \ \text{ Between two circles}$$

2. The ROC cannot contain any poles.

By definition, poles of $X(z)$ are values of z that make the value of the transform infinitely large. For rational z-transforms, poles are the roots of the denominator polynomial. Since the transform does not converge at a pole, the ROC must naturally exclude all poles of the transform.

3. The ROC for the z-transform of a causal signal is the outside of a circle, and is expressed as

$$|z| > r_1$$

Since a causal signal does not have any significant samples for $n < 0$, its z-transform can be written as

$$X(z) - \sum_{n=0}^{\infty} x[n] z^{-n}$$

Writing z as $z = r^{j\Omega}$ and remembering that the convergence of the z-transform is equivalent to the signal $x[n] r^{-n}$ being absolute summable leads to the convergence condition

$$\sum_{n=0}^{\infty} \left| x[n] r^{-n} \right| < \infty \tag{8.57}$$

which can be expressed in the equivalent form

$$\sum_{n=0}^{\infty} |x[n]| \left(\frac{1}{|r|} \right)^n < \infty \tag{8.58}$$

If we can find a value of r for which Eqn. (8.58) is satisfied, then it is obvious that any larger value of r will satisfy Eqn. (8.58) as well. All we need to do is find the radius r_1 of the bounding circle, and the ROC is the region outside that circle.

A special case of a causal signal is one that is both causal and finite-length. Let N_1 be the largest value of the index for which the signal $x[n]$ has a non-zero value, that is

$$x[n] = 0 \quad \text{for} \quad n < 0 \text{ or } n > N_1$$

In this case the absolute summability condition in Eqn. (8.58) can be written as

$$\sum_{n=0}^{N_1} \left| x[n] \right| \left(\frac{1}{|r|} \right)^n < \infty \tag{8.59}$$

The condition in Eqn. (8.59) is satisfied for all values of r with the exception of $r = 0$. Our initial assessment that the ROC is the outside of a circle is still valid if we are willing to consider a circle with radius of $r_1 = 0$, that is, one that shrinks down to a single point at the origin. The ROC for a finite-length causal signal is therefore

$$|z| > 0 \tag{8.60}$$

4. The ROC for the z-transform of an anti-causal signal is the inside of a circle, and is expressed as

$$|z| < r_2$$

The justification of this property will be similar to that of the previous one. An anti-causal signal does not have any significant samples for $n \geq 0$, and its z-transform can be written as

$$X(z) = \sum_{n=-\infty}^{-1} x[n] \, z^{-n}$$

Using $z = r^{j\Omega}$ the condition for the convergence of the z-transform can be expressed through the equivalent condition for the absolute summability of the signal $x[n] \, r^{-n}$ as

$$\sum_{n=-\infty}^{-1} \left| x[n] \, r^{-n} \right| < \infty \tag{8.61}$$

which can be expressed in the equivalent form

$$\sum_{n=-\infty}^{-1} |x[n]| \cdot |r|^{-n} < \infty \tag{8.62}$$

Let us apply the variable change $n = -m$ to the summation in Eqn. (8.62) to write it as

$$\sum_{m=1}^{\infty} |x[-m]| \cdot |r|^{m} < \infty \tag{8.63}$$

If we can find a value of r for which Eqn. (8.63) is satisfied, then it is obvious that any smaller value of r will satisfy Eqn. (8.63) as well. If r_2 is the radius of the bounding circle, then the ROC is the region inside that circle.

A special case of an anti-causal signal is one that is both anti-causal and finite-length. Let $-N_2$ be the most negative value of the index for which the signal $x[n]$ has a non-zero value, that is

$$x[n] = 0 \quad \text{for} \quad n \geq 0 \text{ or } n < -N_2$$

In this case the absolute summability condition in Eqn. (8.62) can be written as

$$\sum_{n=-N_2}^{-1} |x[n]| \cdot |r|^{-n} < \infty \quad \Rightarrow \quad \sum_{m=1}^{N_2} |x[-m]| \cdot |r|^m < \infty \tag{8.64}$$

The condition in Eqn. (8.64) is satisfied for all values of r that are finite; the only exception would be $r \to \infty$. If we simply take the bounding circle to be one with infinite radius $r_2 = \infty$ then our conclusion that the ROC is the inside of a circle is still valid, The ROC for a finite-length anti-causal signal is therefore

$$|z| < \infty \tag{8.65}$$

5. The region of convergence for the z-transform of a signal that is neither causal nor anti-causal is a ring-shaped region between two circles, and can be expressed as

$$r_1 < |z| < r_2$$

Any signal $x[n]$ can be written as the sum of a causal signal and an anti-causal signal. Let the two signals $x_R[n]$ and $x_L[n]$ be defined in terms of the signal $x[n]$ as

$$x_R[n] = \begin{cases} x[n], & n \geq 0 \\ 0, & n < 0 \end{cases} \tag{8.66}$$

and

$$x_L[n] = \begin{cases} x[n], & n < 0 \\ 0, & n \geq 0 \end{cases} \tag{8.67}$$

so that $x_R[n]$ is causal, and $x_L[n]$ is anti-causal, and they add up to $x[n]$:

$$x[n] = x_R[n] + x_L[n] \tag{8.68}$$

Let the z-transforms of these two signals be

$$X_R(z) = \mathcal{Z}\{x_R[n]\} \quad \text{ROC: } |z| > r_1$$

$$X_L(z) = \mathcal{Z}\{x_L[n]\} \quad \text{ROC: } |z| < r_2$$

The z-transform of the signal $x[n]$ is

$$X(z) = X_R(z) + X_L(z) \tag{8.69}$$

The ROC for $X(z)$ is at least the overlap of the two regions, that is,

$$r_1 < |z| < r_2 \tag{8.70}$$

provided that $r_2 > r_1$ (otherwise there may be no overlap, and the z-transform may not exist at any point in the z-plane).

In some cases the ROC may actually be larger than the overlap in Eqn. (8.70) if the addition of the two transforms in Eqn. (8.69) results in the cancellation of a pole that sets the boundary for either $X_R(z)$ or $X_L(z)$.

As a special case, if the causal term $x_R[n]$ is of finite-length, the inner circle of the ROC may shrink down to a single point at the origin, resulting in $r_1 = 0$. Similarly, if the anti-causal term $x_L[n]$ is of finite-length, the outer circle of the ROC may grow to have infinite radius, resulting in $r_2 \to \infty$.

Example 8.13: z-Transform of two-sided exponential signal

Find the z-transform of the signal

$$x[n] = a^{|n|}$$

Solution: The specified signal exists for all values of the index n, and can be written as

$$x[n] = \begin{cases} a^n, & n \geq 0 \\ a^{-n}, & n < 0 \end{cases}$$

$$= a^n\, u[n] + a^{-n}\, u[-n-1] \tag{8.71}$$

or, equivalently as

$$x[n] = a^n\, u[n] + (1/a)^n\, u[-n-1] \tag{8.72}$$

We will think of $x[n]$ as the sum of a causal signal $x_R[n]$ and an anti-causal signal $x_L[n]$ in the form

$$x[n] = x_R[n] + x_L[n] \tag{8.73}$$

with the two components given by

$$x_R[n] = a^n\, u[n]$$
$$x_L[n] = (1/a)^n\, u[-n-1]$$

When we discuss the properties of the z-transform later in Section 8.3 we will show that the z-transform is linear, and therefore, the transform of the sum of two signals is equal to the sum of their respective transforms. Therefore $X(z)$ can be written as

$$X(z) = X_R(z) + X_L(z)$$
$$= \mathcal{Z}\{a^n\, u[n]\} + \mathcal{Z}\{(1/a)^n\, u[-n-1]\} \tag{8.74}$$

The individual transforms that make up $X(z)$ in Eqn. (8.74) can be determined by adapting the results obtained earlier in Eqns. (8.23) and (8.23) as:

$$\mathcal{Z}\{a^n\, u[n]\} = \frac{z}{z-a}, \qquad \text{ROC: } |z| > |a| \tag{8.75}$$

$$\mathcal{Z}\{(1/a)^n\, u[-n-1]\} = -\frac{z}{z-1/a}, \qquad \text{ROC: } |z| < \frac{1}{|a|} \tag{8.76}$$

The regions of convergence for $X_R(z)$ and $X_L(z)$ are shown in Fig. 8.18.

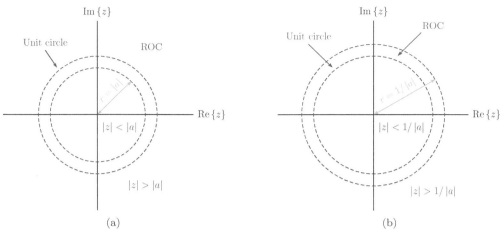

(a)   (b)

Figure 8.18 – The region of convergence for (a) $X_R(z) = \mathcal{Z}\{a^n\,u[n]\}$, (b) $X_L(z) = \mathcal{Z}\{(1/a)^n\,u[-n-1]\}$.

The transform $X(z)$ can now be computed as

$$X(z) = \frac{z}{z-a} - \frac{z}{z-1/a} = \frac{(a^2-1)\,z}{a\,z^2 - (a^2+1)\,z + a}$$

The transform has a zero at $z = 0$, and poles at $z = a$ and $z = 1/a$. For it to converge, both $X_R(z)$ and $X_L(z)$ must converge. Therefore, the ROC for $X(z)$ is the overlap of the ROCs of $X_R(z)$ and $X_L(z)$ if such an overlap exists:

$$\text{ROC:}\quad |a| < |z| < \frac{1}{|a|} \tag{8.77}$$

If $|a| < 1$, then $|a| < 1/|a|$, and the two regions shown in Fig. 8.18(a) and (b) do indeed overlap, creating a ring-shaped region between two circles with radii $|a|$ and $1/|a|$. This is shown in Fig. 8.19.

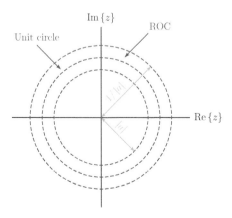

Figure 8.19 – The region of convergence for the transform $X(z)$.

If $|a| \geq 1$, then $|a| \geq 1/|a|$, and therefore the inequality in Eqn. (8.77) cannot be satisfied for any value of z. In that case, the transform found in Eqn. (8.77) does not converge at any point in the z-plane.

8.3   Properties of the z-Transform

In this section we focus on some of the important properties of the z-transform that will help us later in using the z-transform effectively for the analysis and design of discrete-time systems. The proofs of the z-transform properties are also be given. The motivation behind discussing the proofs of various properties of the z-transform is twofold:

1. The techniques used in proving various properties of the z-transform are also useful in working with z-transform problems in general.
2. The proofs provide further insight into the z-transform, and allow us to later identify opportunities for the effective use of z-transform properties in solving problems.

8.3.1   Linearity

Linearity of the z-transform:
  For any two signals $x_1[n]$ and $x_2[n]$ with their respective transforms

$$x_1[n] \overset{\mathcal{Z}}{\longleftrightarrow} X_1(z)$$

and

$$x_2[n] \overset{\mathcal{Z}}{\longleftrightarrow} X_2(z)$$

and any two constants α_1 and α_2, it can be shown that the following relationship holds:

$$\alpha_1\, x_1[n] + \alpha_2\, x_2[n] \overset{\mathcal{Z}}{\longleftrightarrow} \alpha_1\, X_1(z) + \alpha_2\, X_2(z) \tag{8.78}$$

Proof: We will prove the linearity property in a straightforward manner by using the z-transform definition given by Eqn. (8.1) with the signal $\alpha_1 x_1[n] + \alpha_2 x_2[n]$:

$$\mathcal{Z}\{\alpha_1\, x_1[n] + \alpha_2\, x_2[n]\} = \sum_{n=-\infty}^{\infty} \left(\alpha_1\, x_1[n] + \alpha_2\, x_2[n]\right) z^{-n} \tag{8.79}$$

The summation in Eqn. (8.79) can be separated into two summations as

$$\mathcal{Z}\{\alpha_1\, x_1[n] + \alpha_2\, x_2[n]\} = \sum_{n=-\infty}^{\infty} \alpha_1\, x_1[n]\, z^{-n} + \sum_{n=-\infty}^{\infty} \alpha_2\, x_2[n]\, z^{-n}$$

$$= \alpha_1 \sum_{n=-\infty}^{\infty} x_1[n]\, z^{-n} + \alpha_2 \sum_{n=-\infty}^{\infty} x_2[n]\, z^{-n}$$

$$= \alpha_1\, \mathcal{Z}\{x_1[n]\} + \alpha_2\, \mathcal{Z}\{x_2[n]\} \tag{8.80}$$

The z-transform of a weighted sum of two signals is equal to the same weighted sum of their respective transforms $X_1(z)$ and $X_2(z)$. The ROC for the resulting transform is at least the overlap of the two individual transforms, if such an overlap exists. The ROC may even be greater than the overlap of the two regions if the addition of the two transforms results in the cancellation of a pole that sets the boundary of one of the two regions.

The linearity property proven above for two signals can be generalized to any arbitrary number of signals. The z-transform of a weighted sum of any number of signals is equal to the same weighted sum of their respective transforms.

Example 8.14: **Using the linearity property of the z-transform**

Determine the z-transform of the signal

$$x[n] = 3 \left(\frac{1}{2} \right)^n u[n] - 5 \left(\frac{1}{3} \right)^n u[n]$$

Solution: A general expression for the z-transform of the causal exponential signal was found in Example 8.6 as

$$\mathcal{Z} \left\{ a^n u[n] \right\} = \frac{z}{z - a} , \qquad \text{ROC: } |z| > |a|$$

Applying this result to the exponential terms in $x[n]$ we get

$$\mathcal{Z} \left\{ \left(\frac{1}{2} \right)^n u[n] \right\} = \frac{z}{z - \frac{1}{2}} , \qquad \text{ROC: } |z| > \frac{1}{2} \tag{8.81}$$

and

$$\mathcal{Z} \left\{ \left(\tfrac{1}{3} \right)^n u[n] \right\} = \frac{z}{z - \frac{1}{3}} , \qquad \text{ROC: } |z| > \frac{1}{3} \tag{8.82}$$

Combining the results in Eqns. (8.81) and (8.82) using the linearity property we arrive at the desired result:

$$\begin{aligned} X(z) &= 3 \left(\frac{z}{z - \frac{1}{2}} \right) - 5 \left(\frac{z}{z - \frac{1}{3}} \right) \\ &= \frac{-2z \left(z - \frac{3}{4} \right)}{\left(z - \frac{1}{2} \right) \left(z - \frac{1}{3} \right)} \end{aligned} \tag{8.83}$$

The ROC for $X(z)$ is the overlap of the two regions in Eqns. (8.81) and (8.82), namely

$$|z| > \frac{1}{2}$$

The ROC for the transform $X(z)$ is shown in Fig. 8.20 along with poles and zeros of the transform.

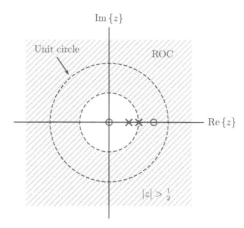

Figure 8.20 – The ROC for the transform in Eqn. (8.83).

Example 8.15: **z-Transform of a cosine signal**

Find the z-transform of the signal

$$x[n] = \cos\left(\Omega_0 n\right) u[n]$$

Solution: Using Euler's formula for the $\cos\left(\Omega_0 n\right)$ term, the signal $x[n]$ can be written as

$$\cos\left(\Omega_0 n\right) u[n] = \frac{1}{2} e^{j\Omega_0 n} u[n] + \frac{1}{2} e^{-j\Omega_0 n} u[n] \tag{8.84}$$

and its z-transform is

$$\mathcal{Z}\left\{\cos\left(\Omega_0 n\right) u[n]\right\} = \mathcal{Z}\left\{\frac{1}{2} e^{j\Omega_0 n} u[n] + \frac{1}{2} e^{-j\Omega_0 n} u[n]\right\} \tag{8.85}$$

Applying the linearity property of the z-transform, Eqn. (8.85) becomes

$$\mathcal{Z}\left\{\cos\left(\Omega_0 n\right) u[n]\right\} = \frac{1}{2} \mathcal{Z}\left\{e^{j\Omega_0 n} u[n]\right\} + \frac{1}{2} \mathcal{Z}\left\{e^{-j\Omega_0 n} u[n]\right\}$$

$$= \frac{1/2}{1 - e^{j\Omega_0} z^{-1}} + \frac{1/2}{1 - e^{-j\Omega_0} z^{-1}} \tag{8.86}$$

Combining the terms of Eqn. (8.86) under a common denominator we have

$$X\left(z\right) = \frac{1 - \cos\left(\Omega_0\right) z^{-1}}{1 - 2 \cos\left(\Omega_0\right) z^{-1} + z^{-2}} \tag{8.87}$$

Let us multiply both the numerator and the denominator by z^2 to eliminate negative powers of z:

$$X\left(z\right) = \frac{z\left[z - \cos\left(\Omega_0\right)\right]}{z^2 - 2 \cos\left(\Omega_0\right) z + 1} \tag{8.88}$$

Poles of $X\left(z\right)$ are both on the unit circle of the z-plane at $z = e^{\pm j\Omega_0}$. Since $x[n]$ is causal, the ROC is the outside of the unit circle. The unit circle itself is not included in the ROC. This is illustrated in Fig. 8.21.

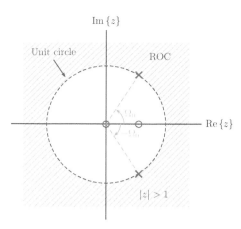

Figure 8.21 – The ROC for the transform in Eqn. (8.87).

Example 8.16: **z-Transform of a sine signal**

Find the z-transform of the signal

$$x[n] = \sin(\Omega_0 n)\, u[n]$$

Solution: This problem is quite similar to that in Example 8.15. Using Euler's formula for the signal $x[n]$ we get

$$\sin(\Omega_0 n)\, u[n] = \frac{1}{2j}\, e^{j\Omega_0 n}\, u[n] - \frac{1}{2j}\, e^{-j\Omega_0 n}\, u[n] \tag{8.89}$$

and its z-transform is

$$\begin{aligned}
\mathcal{Z}\{\sin(\Omega_0 n)\, u[n]\} &= \mathcal{Z}\left\{\frac{1}{2j}\, e^{j\Omega_0 n}\, u[n] - \frac{1}{2j}\, e^{-j\Omega_0 n}\, u[n]\right\} \\
&= \frac{1}{2j}\, \mathcal{Z}\{e^{j\Omega_0 n}\, u[n]\} - \frac{1}{2j}\, \mathcal{Z}\{e^{-j\Omega_0 n}\, u[n]\} \\
&= \frac{1/2j}{1 - e^{j\Omega_0}\, z^{-1}} - \frac{1/2j}{1 - e^{-j\Omega_0}\, z^{-1}}
\end{aligned} \tag{8.90}$$

Combining the terms of Eqn. (8.90) under a common denominator we have

$$X(z) = \frac{\sin(\Omega_0)\, z^{-1}}{1 - 2\cos(\Omega_0)\, z^{-1} + z^{-2}} \tag{8.91}$$

or, using non-negative powers of z,

$$X(z) = \frac{\sin(\Omega_0)\, z}{z^2 - 2\cos(\Omega_0)\, z + 1} \tag{8.92}$$

The ROC of the transform is $|z| > 1$ as in the previous example.

8.3.2 Time shifting

Given the transform pair

$$x[n] \overset{\mathcal{Z}}{\longleftrightarrow} X(z)$$

the following is also a valid transform pair:

$$x[n-k] \overset{\mathcal{Z}}{\longleftrightarrow} z^{-k} X(z) \tag{8.93}$$

Thus, shifting the signal $x[n]$ in the time domain by k samples corresponds to multiplication of the transform $X(z)$ by z^{-k}. The ability to express the z-transform of a time-shifted version of a signal in terms of the z-transform of the original signal will be very useful in working with difference equations.

Proof: The z-transform of $x[n-k]$ is

$$\mathcal{Z}\{x[n-k]\} = \sum_{n=-\infty}^{\infty} x[n-k]\, z^{-n} \tag{8.94}$$

Let us define a new variable $m = n - k$, and write the summation in Eqn. (8.93) in terms of this new variable:

$$\mathcal{Z}\{x[n-k]\} = \sum_{m=-\infty}^{\infty} x[m]\, z^{-(m+k)}$$

$$= z^{-k} \sum_{m=-\infty}^{\infty} x[m]\, z^{-m}$$

$$= z^{-k} X(z) \tag{8.95}$$

The region of convergence for the resulting transform $z^{-k} X(z)$ is the same as that of $X(z)$ with some possible exceptions:

1. If time-shifting the signal for $k < 0$ samples (left shift) causes some negative indexed samples to appear in $x[n-k]$, then points at $|z| \to \infty$ need to be excluded from the ROC.
2. If time-shifting the signal for $k > 0$ samples (right shift) causes some positive indexed samples to appear in $x[n-k]$, then the origin $z = 0 + j0$ of the z-plane needs to be excluded from the ROC.

Example 8.17: **z-Transform of discrete-time pulse signal revisited**

The z-transform of the discrete-time pulse signal

$$x[n] = \begin{cases} 1, & 0 \le n \le N-1 \\ 0, & n < 0 \quad \text{or} \quad n > N-1 \end{cases}$$

was determined earlier in Example 8.8. Find the same transform through the use of linearity and time-shifting properties.

Solution: The signal $x[n]$ can be expressed as the difference of a unit-step signal and a time-shifted unit-step signal, that is,

$$x[n] = u[n] - u[n - N] \tag{8.96}$$

We know from the linearity property of the z-transform that

$$X(z) = \mathcal{Z}\{u[n]\} - \mathcal{Z}\{u[n - N]\} \tag{8.97}$$

The z-transform of the unit-step function was found in Example 8.5 as

$$\mathcal{Z}\{u[n]\} = \frac{z}{z - 1}, \quad \text{ROC: } |z| > 1 \tag{8.98}$$

Using the time-shifting property of the z-transform, we find the transform of the shifted unit-step signal as

$$\mathcal{Z}\{u[n - N]\} = z^{-N}\frac{z}{z - 1}, \quad \text{ROC: } |z| > 1 \tag{8.99}$$

Adding the two transforms in Eqns. (8.98) and (8.99) yields

$$X(z) = \frac{z}{z - 1} - z^{-N}\frac{z}{z - 1} = \left(1 - z^{-N}\right)\frac{z}{z - 1} \tag{8.100}$$

The result in Eqn. (8.100) can be simplified to

$$X(z) = \frac{z^N - 1}{z^{N-1}(z - 1)} \tag{8.101}$$

which matches the earlier result found in Example 8.8. Since $x[n]$ is a finite-length causal signal, the transform converges everywhere except at the origin. Thus, the region of convergence is

$$|z| > 0$$

This is one example of the possibility mentioned earlier regarding the region of convergence: The ROC for $X(z)$ is larger than the overlap of the two regions given by Eqns. (8.98) and (8.99). The individual transforms $\mathcal{Z}\{u[n]\}$ and $\mathcal{Z}\{u[n - N]\}$ each have a pole at $z = 1$ causing the individual ROCs to be the outside of a circle with unit radius. When the two terms are added to construct $X(z)$, however, the pole at $z = 1$ is canceled, resulting in the region of convergence for $X(z)$ being larger than the overlap of the individual regions.

8.3.3 Time reversal

Given the transform pair

$$x[n] \overset{\mathcal{Z}}{\longleftrightarrow} X(z)$$

the following is also a valid transform pair:

$$x[-n] \overset{\mathcal{Z}}{\longleftrightarrow} X\left(z^{-1}\right) \tag{8.102}$$

Proof: The z-transform of the time-reversed signal $x[-n]$ is found by

$$\mathcal{Z}\left\{x[-n]\right\} = \sum_{n=-\infty}^{\infty} x[-n]\, z^{-n} \tag{8.103}$$

We will employ a variable change $m = -n$ on the summation of Eqn. (8.103) to obtain

$$\mathcal{Z}\left\{x[-n]\right\} = \sum_{m=+\infty}^{-\infty} x[m]\, z^{m} \tag{8.104}$$

The summation in Eqn. (8.104) starts at $m = +\infty$ and moves toward $m = -\infty$, adding terms from right to left. The two limits can be swapped without affecting the result since it does not matter whether we add terms from right to left or the other way around. We will also use $z^{m} = \left(z^{-1}\right)^{-m}$ to write the relationship in Eqn. (8.104) as

$$\mathcal{Z}\left\{x[-n]\right\} = \sum_{m=-\infty}^{\infty} x[m]\,\left(z^{-1}\right)^{-m} = X\left(z^{-1}\right) \tag{8.105}$$

to prove the time-reversal property.

Since we have replaced z by z^{-1} in the transform, the ROC must be adjusted for this change. Let the ROC of the original transform $X(z)$ be

$$r_1 < |z| < r_2$$

The ROC for $X\left(z^{-1}\right)$ is

$$r_1 < \left|\frac{1}{z}\right| < r_2 \qquad \Longrightarrow \qquad \frac{1}{r_2} < |z| < \frac{1}{r_1}$$

Example 8.18: **z-Transform of anti-causal exponential signal revisited**

The z-transform of the anti-causal exponential signal

$$x[n] = -a^n\, u[-n-1]$$

was found in Example 8.7 as

$$X(z) = \frac{z}{z-a}, \quad \text{ROC: } |z| < |a|$$

Obtain the same result from the z-transform of the causal exponential signal using the time-reversal and time-shifting properties.

Solution: We will start with the known transform relationship and try to convert it into the relationship that we seek, applying appropriate properties of the z-transform at each step. The causal exponential signal and its z-transform are

$$b^n\, u[n] \;\overset{z}{\longleftrightarrow}\; \frac{z}{z-b}, \quad \text{ROC: } |z| > |b| \tag{8.106}$$

where we have used the parameter b instead of a. Let us begin by applying the time-reversal operation to the signal on the left, and adjusting the transform on the right according to the time-reversal property of the z-transform:

$$b^{-n}\,u[-n] \overset{\mathcal{Z}}{\longleftrightarrow} \frac{z^{-1}}{z^{-1}-b}\,, \qquad \text{ROC:} \quad |z| < \frac{1}{|b|} \tag{8.107}$$

The transform relationship in Eqn. (8.107) can be written in the equivalent form

$$\left(\frac{1}{b}\right)^{n}\,u[-n] \overset{\mathcal{Z}}{\longleftrightarrow} \frac{-1/b}{z-1/b}\,, \qquad \text{ROC:} \quad |z| < \frac{1}{|b|} \tag{8.108}$$

We will now apply a time shift to the signal by one sample to the left through the substitution $n \to n+1$ which causes the transform to be multiplied by z, resulting in the relationship

$$\left(\frac{1}{b}\right)^{n+1}\,u[-n-1] \overset{\mathcal{Z}}{\longleftrightarrow} \frac{(-1/b)\,z}{z-1/b}\,, \qquad \text{ROC:} \quad |z| < \frac{1}{|b|} \tag{8.109}$$

Multiplying both sides of the relationship by $-b$ leads to

$$-\left(\frac{1}{b}\right)^{n}\,u[-n-1] \overset{\mathcal{Z}}{\longleftrightarrow} \frac{z}{z-1/b}\,, \qquad \text{ROC:} \quad |z| < \frac{1}{|b|} \tag{8.110}$$

Finally, choosing $b = 1/a$, we obtain the desired result:

$$-a^{n}\,u[-n-1] \overset{\mathcal{Z}}{\longleftrightarrow} \frac{z}{z-a}\,, \qquad \text{ROC:} \quad |z| < |a| \tag{8.111}$$

8.3.4   Multiplication by an exponential signal

Given the transform pair

$$x[n] \overset{\mathcal{Z}}{\longleftrightarrow} X(z)$$

the following is also a valid transform pair:

$$a^{n}\,x[n] \overset{\mathcal{Z}}{\longleftrightarrow} X(z/a) \tag{8.112}$$

Proof: The z-transform of $a^{n}\,x[n]$ is

$$\mathcal{Z}\left\{a^{n}\,x[n]\right\} = \sum_{n=-\infty}^{\infty} a^{n}\,x[n]\,z^{-n}$$

$$= \sum_{n=-\infty}^{\infty} x[n]\left(\frac{z}{a}\right)^{-n}$$

$$= X(z/a) \tag{8.113}$$

The ROC must be adjusted for the new transform variable (z/a). Let the ROC of the original transform $X(z)$ be

$$r_1 < |z| < r_2$$

The ROC for $X(z/a)$ is

$$r_1 < \left|\frac{z}{a}\right| < r_2 \qquad \Longrightarrow \qquad |a|\, r_1 < |z| < |a|\, r_2$$

Example 8.19: **Multiplication by an exponential signal**

Determine the z-transform of the signal

$$x[n] = a^n \, \cos\left(\Omega_0\right) \, u[n]$$

Assume a is real.

Solution: Let the signal $x_1[n]$ be defined as

$$x_1[n] = \cos\left(\Omega_0 n\right) \, u[n]$$

so that

$$x[n] = a^n \, x_1[n]$$

The z-transform of the signal $x_1[n]$ was found in Example 8.15 to be

$$X_1(z) = \mathcal{Z}\left\{\cos\left(\Omega_0 n\right) \, u[n]\right\} = \frac{z\left[z - \cos\left(\Omega_0\right)\right]}{z^2 - 2 \, \cos\left(\Omega_0\right) \, z + 1}$$

Multiplication of the time-domain signal by a^n causes the transform to be evaluated for $z \to z/a$, resulting in the transform relationship

$$X(z) = X_1(z/a)$$

$$= \frac{(z/a)\left[(z/a) - \cos\left(\Omega_0\right)\right]}{(z/a)^2 - 2 \, \cos\left(\Omega_0\right) \, (z/a) + 1}$$

$$= \frac{z\left[z - a \, \cos\left(\Omega_0\right)\right]}{z^2 - 2a \, \cos\left(\Omega_0\right) \, z + a^2} \tag{8.114}$$

The transform $X(z)$ has two poles at $z = a \, e^{\pm j\Omega_0}$. Its ROC is

$$\text{ROC:} \quad |z| > |a| \tag{8.115}$$

The pole-zero diagram and the ROC are shown in Fig. 8.22.

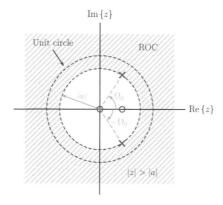

Figure 8.22 – The pole-zero diagram and the ROC for the transform in Example 8.19.

8.3.5   Differentiation in the z-domain

Given the transform pair

$$x[n] \overset{\mathcal{Z}}{\longleftrightarrow} X(z)$$

the following is also a valid transform pair:

$$n\,x[n] \overset{\mathcal{Z}}{\longleftrightarrow} (-z)\,\frac{dX(z)}{dz} \tag{8.116}$$

Proof: Let us start with the z-transform definition

$$X(z) = \sum_{n=-\infty}^{\infty} x[n]\,z^{-n}$$

Differentiating both sides with respect to z we obtain

$$\frac{d}{dz}\left[X(z)\right] = \frac{d}{dz}\left[\sum_{n=-\infty}^{\infty} x[n]\,z^{-n}\right] \tag{8.117}$$

Since summation and differentiation are both linear operators, their order can be changed: The derivative of a sum is equal to the sum of derivatives.

$$\frac{d}{dz}\left[X(z)\right] = \sum_{n=-\infty}^{\infty} \frac{d}{dz}\left[x[n]\,z^{-n}\right] \tag{8.118}$$

Carrying out the differentiation on each term inside the summation we obtain

$$\frac{d}{dz}\left[X(z)\right] = \sum_{n=-\infty}^{\infty} -n\,x[n]\,z^{-n-1}$$

$$= -z^{-1}\sum_{n=-\infty}^{\infty} n\,x[n]\,z^{-n}$$

$$= \frac{1}{(-z)}\sum_{n=-\infty}^{\infty} n\,x[n]\,z^{-n} \tag{8.119}$$

from which the z-transform of $n\,x[n]$ can be obtained as

$$\mathcal{Z}\{n\,x[n]\} = (-z)\,\frac{dX(z)}{dz} \tag{8.120}$$

Thus we prove Eqn. (8.116). The ROC for the transform of $n\,x[n]$ is the same as that of $x[n]$.

Example 8.20: **Using the differentiation property**

Determine the z-transform of the signal

$$x[n] = n\,a^n\,u[n]$$

Solution: In Example 8.6 we found the z-transform of a causal exponential signal to be

$$\mathcal{Z}\left\{a^n\,u[n]\right\} = \frac{z}{z-a}\,, \qquad \text{ROC:}\quad |z| > |a|$$

Applying the differentiation property given by Eqn. (8.116) we get

$$X\left(z\right) = (-z)\,\frac{d}{dz}\left[\frac{z}{z-a}\right]$$

$$= (-z)\,\frac{(-a)}{\left(z-a\right)^2} = \frac{az}{\left(z-a\right)^2} \tag{8.121}$$

and the ROC is

$$|z| > |a|$$

Example 8.21: z-Transform of a unit-ramp signal

Determine the z-transform of the unit-ramp signal

$$x[n] = n\,u[n]$$

Solution: The solution is straightforward if we use the result obtained in Example 8.20. The signal $n\,a^n\,u[n]$ becomes the unit-ramp signal we are considering if the parameter a is set equal to unity. Setting $a = 1$ in the transform found in Example 8.20 we have

$$X\left(z\right) = \left.\frac{az}{\left(z-a\right)^2}\right|_{a=1} = \frac{z}{\left(z-1\right)^2} \tag{8.122}$$

with the ROC

$$|z| > 1$$

Example 8.22: More on the use of the differentiation property

Determine the z-transform of the signal

$$x[n] = n\,\left(n+2\right)\,u[n]$$

Solution: The signal $x[n]$ can be written as

$$x[n] = n^2\,u[n] + 2n\,u[n] \tag{8.123}$$

The z-transform of the unit-ramp function was found in Example 8.21 through the use of the differentiation property as

$$\mathcal{Z}\left\{n\,u[n]\right\} = \frac{z}{\left(z-1\right)^2}\,, \qquad \text{ROC:}\quad |z| > 1 \tag{8.124}$$

We will now apply the differentiation property to the transform pair in Eqn. (8.124):

$$\mathcal{Z}\left\{n^2\,u[n]\right\} = (-z)\,\frac{d}{dz}\left[\frac{z}{\left(z-1\right)^2}\right]$$

$$= \frac{z\,\left(z+1\right)}{\left(z-1\right)^3}\,, \qquad \text{ROC:}\quad |z| > 1 \tag{8.125}$$

Finally, combining the results in Eqns. (8.124) and (8.125) we construct the z-transform of $x[n]$ as

$$X(z) = \mathcal{Z}\left\{n^2\,u[n]\right\} + 2\,\mathcal{Z}\left\{n\,u[n]\right\}$$

$$= \frac{z\,(z+1)}{(z-1)^3} + 2\,\frac{z}{(z-1)^2}$$

$$= \frac{3z\,(z-1/3)}{(z-1)^3}\,, \qquad \text{ROC:} \quad |z| > 1 \tag{8.126}$$

The transform found in Eqn. (8.126) has three poles at $z = 1$. Its zeros are at $z = 0$ and $z = \frac{1}{3}$. The pole-zero diagram and the ROC are shown in Fig. 8.23.

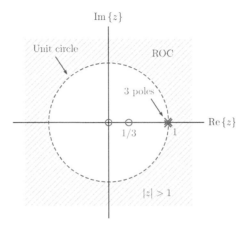

Figure 8.23 – The pole-zero diagram and the ROC for the transform in Example 8.22.

8.3.6 Convolution property

Convolution property of the z-transform is fundamental in its application to DTLTI systems. It forms the basis of the z-domain system function concept that will be explored in detail in Section 8.5.

For any two signals $x_1[n]$ and $x_2[n]$ with their respective transforms

$$x_1[n] \overset{\mathcal{Z}}{\longleftrightarrow} X_1(z)$$

$$x_2[n] \overset{\mathcal{Z}}{\longleftrightarrow} X_2(z)$$

it can be shown that the following transform relationship holds:

$$x_1[n] * x_2[n] \overset{\mathcal{Z}}{\longleftrightarrow} X_1(z)\,X_2(z) \tag{8.127}$$

Proof: We will carry out the proof by using the convolution result inside the z-transform definition. Recall that the convolution of two discrete-time signals $x_1[n]$ and $x_2[n]$ is given by

$$x_1[n] * x_2[n] = \sum_{k=-\infty}^{\infty} x_1[k]\, x_2[n-k]$$

Inserting this into the z-transform definition we obtain

$$\mathcal{Z}\{x_1[n] * x_2[n]\} = \mathcal{Z}\left\{ \sum_{k=-\infty}^{\infty} x_1[k]\, x_2[n-k] \right\}$$

$$= \sum_{n=-\infty}^{\infty} \left[\sum_{k=-\infty}^{\infty} x_1[k]\, x_2[n-k] \right] z^{-n} \tag{8.128}$$

Interchanging the order of the two summations in Eqn. (8.128) leads to

$$\mathcal{Z}\{x_1[n] * x_2[n]\} = \sum_{k=-\infty}^{\infty} \sum_{n=-\infty}^{\infty} x_1[k]\, x_2[n-k]\, z^{-n}$$

$$= \sum_{k=-\infty}^{\infty} x_1[k] \sum_{n=-\infty}^{\infty} x_2[n-k]\, z^{-n} \tag{8.129}$$

Recognizing that the inner summation on the right side of Eqn. (8.129) represents the z-transform of $x_2[n-k]$ and is therefore equal to

$$\sum_{n=-\infty}^{\infty} x_2[n-k]\, z^{-n} = \mathcal{Z}\{x_2[n-k]\} = z^{-k}\, X_2(z) \tag{8.130}$$

we obtain

$$\mathcal{Z}\{x_1[n] * x_2[n]\} = \sum_{k=-\infty}^{\infty} x_1[k]\, z^{-k}\, X_2(z)$$

$$= X_2(z) \sum_{k=-\infty}^{\infty} x_1[k]\, z^{-k}$$

$$= X_1(z)\, X_2(z) \tag{8.131}$$

As before, the ROC for the resulting transform is at least the overlap of the two individual transforms, if such an overlap exists. It may be greater than the overlap of the two regions if the multiplication of the two transforms $X_1(z)$ and $X_2(z)$ results in the cancellation of a pole that sets a boundary for either $X_1(z)$ or $X_2(z)$.

Convolution property of the z-transform is very useful in the sense that it provides an alternative to actually computing the convolution of two signals in the time-domain. Instead, we can obtain $x[n] = x_1[n] * x_2[n]$ using the procedure outlined below:

Finding convolution result through z-transform:

To compute $x[n] = x_1[n] * x_2[n]$:

1. Find the z-transforms of the two signals.

$$X_1(z) = \mathcal{Z}\{x_1[n]\}$$
$$X_2(z) = \mathcal{Z}\{x_2[n]\}$$

2. Multiply the two transforms to obtain $X(z)$.

$$X(z) = X_1(z)\, X_2(z)$$

3. Compute $x[n]$ as the inverse z-transform of $X(z)$.

$$x[n] = \mathcal{Z}^{-1}\{X(z)\}$$

Example 8.23: **Using the convolution property**

Consider two signals $x_1[n]$ and $x_2[n]$ given by

$$x_1[n] = \{\ \underset{\underset{n=0}{\uparrow}}{4}\ ,\ 3,\ 2,\ 1\ \}$$

and

$$x_2[n] = \{\ \underset{\underset{n=0}{\uparrow}}{3}\ ,\ 7,\ 4\ \}$$

Let $x[n]$ be the convolution of these two signals, i.e.,

$$x[n] = x_1[n] * x_2[n]$$

Determine $x[n]$ using z-transform techniques.

Solution: Convolution of these two signals was computed in Example 3.19 of Chapter 1 using the time-domain method. In this exercise we will use the z-transform to obtain the same result. Individual transforms of the signals $x_1[n]$ and $x_2[n]$ are

$$X_1(z) - 4 + 3\,z^{-1} + 2\,z^{-2} + z^{-3}$$

and

$$X_2(z) = 3 + 7\,z^{-1} + 4\,z^{-2}$$

Applying the convolution property, the transform of $x[n]$ can be written as the product of the two transforms:

$$
\begin{aligned}
X(z) =& X_1(z)\, X_2(z) \\
=& \left(4 + 3\,z^{-1} + 2\,z^{-2} + z^{-3}\right)\left(3 + 7\,z^{-1} + 4\,z^{-2}\right) \\
=& 12 + 37\,z^{-1} + 43\,z^{-2} + 29\,z^{-3} + 15\,z^{-4} + 4\,z^{-5} \tag{8.132}
\end{aligned}
$$

Comparing the result obtained in Eqn. (8.132) with the definition of the z-transform we conclude that

$$x[0] = 12 \quad x[1] = 37 \quad x[2] = 43 \quad x[3] = 29 \quad x[4] = 15 \quad x[5] = 4$$

Equivalently, the signal $x[n]$ can be written as

$$x[n] = \{\underset{\substack{\uparrow \\ n=0}}{12}, 37, 43, 29, 15, 4\}$$

The solution of the convolution problem above using z-transform techniques leads to an interesting result: Convolution operation can be used for multiplying two polynomials. Suppose we need to multiply two polynomials

$$A(\nu) = a_k \nu^k + a_{k-1} \nu^{k-1} + \ldots + a_1 \nu + a_0$$

and

$$B(\nu) = b_m \nu^m + b_{m-1} \nu^{m-1} + \ldots + b_1 \nu + b_0$$

and we have a convolution function available in software. To find the polynomial $C(\nu) = A(\nu) B(\nu)$, we can first create a discrete-time signal with the coefficients of each polynomial:

$$x_a[n] = \{\underset{\substack{\uparrow \\ n=0}}{a_0}, a_1, \ldots, a_{k-1}, a_k\}$$

and

$$x_b[n] = \{\underset{\substack{\uparrow \\ n=0}}{b_0}, b_1, \ldots, b_{m-1}, b_m\}$$

It is important to ensure that the polynomial coefficients are listed in ascending order of powers of ν in the signals $x_a[n]$ and $x_b[n]$, and any missing coefficients are accounted for in the form of zero-valued samples. If we now compute the convolution of the two signals as

$$x_c[n] = x_a[n] * x_b[n]$$

the resulting signal $x_c[n]$ holds the coefficients of the product polynomial $C(\nu)$, also in ascending order of powers of the independent variable. This will be demonstrated in MATLAB Exercise 8.4.

Software resources: See MATLAB Exercise 8.4.

Example 8.24: **Finding the output signal of a DTLTI system using inverse z-transform**

Consider a DTLTI system described by the impulse response

$$h[n] = (0.9)^n u[n]$$

driven by the input signal

$$x[n] = u[n] - u[n-7]$$

Compute the output signal $y[n]$ through the use of z-transform techniques. Recall that this problem was solved in Example 3.21 of Chapter 3 using the convolution operation.

Solution: The z-transform of the impulse response is

$$H\left(z\right) = \mathcal{Z}\left\{h[n]\right\} = \frac{z}{z - 0.9}, \qquad \text{ROC:} \quad |z| > 0.9$$

The z-transform of the input signal $x[n]$ is

$$X\left(z\right) = \sum_{n=0}^{6} z^{-n}$$

$$= 1 + z^{-1} + z^{-2} + z^{-3} + z^{-4} + z^{-5} + z^{-6}$$

$$= \frac{z^7 - 1}{z^6\left(z - 1\right)}$$

The finite-length geometric series form of the transform $X\left(z\right)$ will prove to be more convenient for use in this case. The z-transform of the output signal is found by multiplying the z-transforms of the input signal and the impulse response:

$$Y\left(z\right) = X\left(z\right) H\left(z\right)$$

$$= \left(1 + z^{-1} + z^{-2} + z^{-3} + z^{-4} + z^{-5} + z^{-6}\right) H\left(z\right)$$

$$= H\left(z\right) + z^{-1} H\left(z\right) + z^{-2} H\left(z\right) + z^{-3} H\left(z\right) + z^{-4} H\left(z\right)$$

$$+ z^{-5} H\left(z\right) + z^{-6} H\left(z\right)$$

The output signal $y[n]$ can now be found as the inverse z-transform of $Y\left(z\right)$ with the use of linearity and time-shifting properties of the z-transform.

$$y[n] = h[n] + h[n - 1] + h[n - 2] + h[n - 3] + h[n - 4]$$

$$+ h[n - 5] + h[n - 6]$$

$$= \left(0.9\right)^n u[n] + \left(0.9\right)^{n-1} u[n - 1] + \left(0.9\right)^{n-2} u[n - 2]$$

$$+ \left(0.9\right)^{n-3} u[n - 3] + \left(0.9\right)^{n-4} u[n - 4]$$

$$+ \left(0.9\right)^{n-5} u[n - 5] + \left(0.9\right)^{n-6} u[n - 6] \tag{8.133}$$

It can be shown that the result obtained in Eqn. (8.133) is identical to the one found in Example 3.21 when the two results are compared on a sample-by-sample basis.
Software resources:
ex_8_24.m

8.3.7 Initial value

Initial value property of the z-transform applies to causal signals only.

Given the transform pair

$$x[n] \xleftrightarrow{\mathcal{Z}} X\left(z\right)$$

where $x[n] = 0$ for $n < 0$, it can be shown that

$$x[0] = \lim_{z \to \infty} \left[X\left(z\right)\right] \tag{8.134}$$

Proof: Consider the z-transform definition given by Eqn. (8.1) applied to a causal signal:

$$X(z) = \sum_{n=0}^{\infty} x[n] z^{-n}$$

$$= x[0] + x[1] z^{-1} + x[2] z^{-2} + \dots \qquad (8.135)$$

It is obvious from Eqn. (8.135) that, as z becomes infinitely large, all terms that contain negative powers of z tend to zero, leaving behind only $x[0]$.

It is interesting to note that the limit in Eqn. (8.134) exists only for a causal signal, provided that $x[0]$ is finite. Consequently, the convergence of $\lim_{z \to \infty} [X(z)]$ can be used as a test of the causality of the signal. For a causal signal $x[n]$, the ROC of the z-transform must include infinitely large values of z. If the transform $X(z)$ is a rational function of z, the numerator order must not be greater than the denominator order.

Example 8.25: **Using the initial value property**

The z-transform of a causal signal $x[n]$ is given by

$$X(z) = \frac{3z^3 + 2z + 5}{2z^3 - 7z^2 + z - 4} \qquad (8.136)$$

Determine the initial value $x[0]$ of the signal.

Solution: Using Eqn. (8.134) we get

$$\lim_{z \to \infty} [X(z)] = \lim_{z \to \infty} \left[\frac{3z^3 + 2z + 5}{2z^3 - 7z^2 + z - 4} \right] = \lim_{z \to \infty} \left[\frac{3z^3}{2z^3} \right] = \frac{3}{2}$$

The result can be easily justified by computing the inverse z-transform of $X(z)$ using either partial fraction expansion or long division. These techniques will be explored in Section 8.4.

8.3.8 Correlation property

Cross-correlation of two discrete-time signals $x[n]$ and $y[n]$ is defined as

$$r_{xy}[m] = \sum_{n=-\infty}^{\infty} x[n] y[n-m] \qquad (8.137)$$

Given two signals $x[n]$ and $y[n]$ with their respective transforms

$$x[n] \xleftrightarrow{\mathcal{Z}} X(z)$$

$$y[n] \xleftrightarrow{\mathcal{Z}} Y(z)$$

it can be shown that

$$R_{xy}(z) = \mathcal{Z}\{r_{xy}[m]\} = X(z) Y(z^{-1}) \qquad (8.138)$$

Proof: We will first prove this property by applying the z-transform definition directly to the cross-correlation of the two signals.

$$R_{xy}(z) = \sum_{m=-\infty}^{\infty} r_{xy}[m]\, z^{-m}$$

$$= \sum_{m=-\infty}^{\infty} \left[\sum_{n=-\infty}^{\infty} x[n]\, y[n-m] \right] z^{-m} \tag{8.139}$$

By interchanging the order of the two summations in Eqn. (8.139) and rearranging the terms we get

$$R_{xy}(z) = \sum_{n=-\infty}^{\infty} x[n] \sum_{m=-\infty}^{\infty} y[n-m]\, z^{-m} \tag{8.140}$$

Applying the variable change $n - m = k$ to the inner summation in Eqn. (8.139) yields

$$R_{xy}(z) = \sum_{n=-\infty}^{\infty} x[n] \sum_{k=-\infty}^{\infty} y[k]\, z^{k-n} \tag{8.141}$$

which can be written in the equivalent form

$$R_{xy}(z) = \left[\sum_{n=-\infty}^{\infty} x[n]\, z^{-n} \right] \left[\sum_{k=-\infty}^{\infty} y[k]\, z^{k} \right]$$

$$= \left[\sum_{n=-\infty}^{\infty} x[n]\, z^{-n} \right] \left[\sum_{k=-\infty}^{\infty} y[k]\, \left(z^{-1}\right)^{-k} \right]$$

$$= X(z)\, Y\left(z^{-1}\right) \tag{8.142}$$

Alternatively, we could have proven this property with less work by using other properties of the z-transform. It is obvious from the definition of the cross-correlation in Eqn. (8.137) that $r_{xy}[n]$ is the convolution of $x[n]$ with $y[-n]$, that is,

$$r_{xy}[n] = x[n] * y[-n] \tag{8.143}$$

From the convolution property of the z-transform we know that the z-transform of $r_{xy}[n]$ must be the product of the individual transforms of $x[n]$ and $y[-n]$:

$$R_{xy}(z) = \mathcal{Z}\{r_{xy}[n]\} = \mathcal{Z}\{x[n]\}\, \mathcal{Z}\{y[-n]\} \tag{8.144}$$

In addition, we know from the time reversal property that

$$\mathcal{Z}\{y[-n]\} = Y\left(z^{-1}\right) \tag{8.145}$$

Therefore

$$R_{xy}(z) = X(z)\, Y\left(z^{-1}\right)$$

The argument for the ROC of the transform uses the same reasoning we have employed earlier: The ROC for the resulting transform $R_{xy}(z)$ is at least the overlap of the two individual transforms, if such an overlap exists. It may be greater than the overlap of the two regions if the multiplication of the two transforms $X(z)$ and $Y(z^{-1})$ results in the cancellation of a pole that sets a boundary for either $X(z)$ or $Y(z^{-1})$.

As a special case of Eqn. (8.146), the transform of the autocorrelation of a signal $x[n]$ can be found as

$$R_{xx}(z) = X(z)\,X\left(z^{-1}\right)$$

Example 8.26: **Using the correlation property**

Determine the cross-correlation of the two signals

$$x[n] = a^n\,u[n]$$

and

$$y[n] = u[n] - u[n-3]$$

using the correlation property of the z-transform.

Solution: The z-transform of $x[n]$ is

$$X(z) = \frac{z}{z-a}\,, \qquad \text{ROC:} \quad |z| > |a| \tag{8.146}$$

Next we will determine the z-transform of $y[n]$:

$$Y(z) = 1 + z^{-1} + z^{-2} \qquad \text{ROC:} \quad |z| > 0 \tag{8.147}$$

Substituting $z \to z^{-1}$ in this last result we obtain

$$Y\left(z^{-1}\right) = 1 + z + z^2 \qquad \text{ROC:} \quad |z| < \infty \tag{8.148}$$

which is the z-transform of the time-reversed signal $y[-n]$. We have modified the ROC in Eqn. (8.148) to account for the fact that the transform variable z has been reciprocated. Now we obtain $R_{xy}(z)$ by multiplying the two results in Eqns. (8.146) and (8.148):

$$
\begin{aligned}
R_{xy}(z) &= X(z)\,Y\left(z^{-1}\right) \\
&= \frac{z}{z-a}\left(1 + z + z^2\right) \\
&= \frac{z}{z-a} + z\,\frac{z}{z-a} + z^2\,\frac{z}{z-a}\,, \qquad \text{ROC:} \quad |a| < |z| < \infty
\end{aligned}
\tag{8.149}
$$

The cross-correlation $r_{xy}[n]$ is the inverse z-transform of the result in Eqn. (8.149) which, using the time-shifting property of the z-transform, can be found as

$$r_{xy}[n] = a^n\,u[n] + a^{n+1}\,u[n+1] + a^{n+2}\,u[n+2] \tag{8.150}$$

Software resources:
ex_8_26.m

Example 8.27: **Finding auto-correlation of a signal**

Using the correlation property of the z-transform, determine the auto-correlation of the signal

$$x[n] = a^n\,u[n] \tag{8.151}$$

Solution: The z-transform of the signal is

$$X\left(z\right) = \frac{z}{z-a}\,,\qquad \text{ROC:}\quad |z| > |a| \tag{8.152}$$

Substituting z^{-1} for z in Eqn. (8.152) we obtain

$$X\left(z^{-1}\right) = \frac{z^{-1}}{z^{-1}-a} = \frac{-1/a}{z-1/a}\,,\qquad \text{ROC:}\quad |z| < \frac{1}{|a|} \tag{8.153}$$

The z-transform of the auto-correlation $r_{xx}[n]$ is

$$
\begin{aligned}
R_{xx}\left(z\right) &= X\left(z\right)X\left(z^{-1}\right)\\[6pt]
&= \frac{-z}{a\,z^2 - \left(a^2+1\right)z + a}\,,\qquad \text{ROC:}\quad |a| < |z| < \frac{1}{|a|}
\end{aligned}\tag{8.154}
$$

provided that $|a| < 1$. Recall that in Example 8.13 we found the z-transform of the two-sided exponential signal to be

$$\mathcal{Z}\left\{a^{|n|}\right\} = \frac{\left(a^2-1\right)z}{a\,z^2 - \left(a^2+1\right)z + a}$$

Comparing this earlier result to what we have in Eqn. (8.154) leads to the conclusion

$$r_{xx}[n] = \left(\frac{1}{1-a^2}\right)a^{|n|} \tag{8.155}$$

8.3.9   Summation property

Given the transform pair

$$x[n] \overset{\mathcal{Z}}{\longleftrightarrow} X\left(z\right)$$

the following is also a valid transform pair:

$$\sum_{k=-\infty}^{n} x[k] \overset{\mathcal{Z}}{\longleftrightarrow} \frac{z}{z-1}X\left(z\right) \tag{8.156}$$

Proof: Let a new signal be defined as

$$w[n] = \sum_{k=-\infty}^{n} x[k] \tag{8.157}$$

Using the definition of the z transform given by Eqn. (8.1) in conjunction with $w[n]$ we get

$$W\left(z\right) = \sum_{n=-\infty}^{\infty} w[n]\,z^{-n} = \sum_{n=-\infty}^{\infty}\left[\sum_{k=-\infty}^{n} x[k]\right]z^{-n}$$

which would be difficult to put into a closed form directly. Instead, we will make use of the other properties of the z-transform discussed earlier, and employ a simple trick of writing $w[n]$ as

$$w[n] = \sum_{k=-\infty}^{n-1} x[k] + x[n]$$

$$= w[n-1] + x[n] \tag{8.158}$$

Eqn. (8.158) provides us with a difference equation between $w[n]$ and $x[n]$. Since the z-transform is linear, the transform of $w[n]$ can be written as

$$\mathcal{Z}\{w[n]\} = \mathcal{Z}\{w[n-1]\} + \mathcal{Z}\{x[n]\} \tag{8.159}$$

and using the time-shifting property we have

$$W(z) = z^1 W(z) + X(z) \tag{8.160}$$

which we can solve for $W(z)$ to obtain

$$W(z) = \frac{1}{1 - z^1} X(z) = \frac{z}{z - 1} X(z) \tag{8.161}$$

An alternative method of justifying the summation property is to think of $w[n]$ as the convolution of the signal $x[n]$ with the unit-step function. Using the convolution sum we can write

$$x[n] * u[n] = \sum_{k=-\infty}^{\infty} x[k] u[n - k] \tag{8.162}$$

Consider the term $u[n - k]$:

$$u[n - k] = \begin{cases} 1, & n - k \geq 0 \quad \Rightarrow k \leq n \\ 0, & n - k < 0 \quad \Rightarrow k > n \end{cases} \tag{8.163}$$

Therefore, dropping the $u[n - k]$ term from the summation in Eqn. (8.162) and adjusting the summation limits to compensate for it we have

$$x[n] * u[n] = \sum_{k=-\infty}^{n} x[k] = w[n] \tag{8.164}$$

Let us take the z-transform of both sides of Eqn. (8.164) using the convolution property of the z-transform:

$$W(z) = \mathcal{Z}\{x[n] * u[n]\} = \mathcal{Z}\{u[n]\} X(z) = \frac{z}{z - 1} X(z) \tag{8.165}$$

which provides us with the alternative proof we seek.

Example 8.28: **z-Transform of a unit-ramp signal revisited**

The transform of the unit-ramp signal

$$x[n] = n\, u[n]$$

was found in Example 8.21 through the use of the differentiation property. In this example we will use it as an opportunity to apply the summation property of the z-transform. The ramp signal $x[n]$ can be expressed as the running sum of a time-shifted unit-step signal as

$$x[n] = \sum_{k=-\infty}^{n} u[n-1] \tag{8.166}$$

It can easily be verified that the definition in Eqn. (8.166) produces $x[n] = 0$ for $n \leq 0$, and $x[1] = 1$, $x[2] = 2$, and so on, consistent with the ramp signal. The z-transform of the time-shifted unit-step signal is

$$\mathcal{Z}\{u[n-1]\} = z^{-1}\,\mathcal{Z}\{u[n]\} = z^{-1}\left(\frac{z}{z-1}\right) = \frac{1}{z-1} \tag{8.167}$$

Using the summation property we find the the transform $X(z)$ as

$$X(z) = \frac{z}{z-1}\,\mathcal{Z}\{u[n-1]\} = \left(\frac{z}{z-1}\right)\left(\frac{1}{z-1}\right) = \frac{z}{(z-1)^2}$$

which matches the answer found in Example 8.20.

8.4   Inverse z-Transform

Inverse z-transform is the problem of finding $x[n]$ from the knowledge of $X(z)$. Often we need to determine the signal $x[n]$ that has a specified z-transform. There are three basic techniques for computing the inverse z-transform:

1. Direct evaluation of the inversion integral
2. Partial fraction expansion technique for a rational transform
3. Expansion of the rational transform into a power series through long division

We will focus our attention on the last two methods. The inversion integral method will be briefly mentioned, but will not be considered further due to its complexity. We will find that methods 2 and 3 will be sufficient for most problems encountered in the analysis of signals and linear systems.

8.4.1   Inversion integral

Let $X(r,\Omega)$ be the function obtained by evaluating the z-transform of a signal $x[n]$ for $z = r\,e^{j\Omega}$, that is, at points on a circle with radius equal to r. We have established in Section 8.1 that $X(r,\Omega)$ is the same as the DTFT of the signal $x[n]\,r^{-n}$, that is

$$X(r,\Omega) = X(z)\Big|_{z=r\,e^{j\Omega}} = \mathcal{F}\{x[n]\,r^{-n}\} \tag{8.168}$$

Consequently, the inverse DTFT equation given by Eqn. (5.70) should yield the signal $x[n]\,r^{-n}$:

$$x[n]\,r^{-n} = \mathcal{F}^{-1}\{X(r,\Omega)\} = \frac{1}{2\pi}\int_{-\pi}^{\pi} X(r,\Omega)\,e^{j\Omega}\,d\Omega \tag{8.169}$$

Multiplying both sides of Eqn. (8.169) with r^n we obtain

$$x[n] = r^n \frac{1}{2\pi} \int_{-\pi}^{\pi} X(r, \Omega) \, e^{j\Omega} \, d\Omega$$

$$= \frac{1}{2\pi} \int_{-\pi}^{\pi} X(r, \Omega) \left(re^{j\Omega} \right)^n \, d\Omega \qquad (8.170)$$

Eqns. (8.168) through (8.170) provide us with a method for finding $x[n]$ from its z-transform:

1. Choose a value for r so that the circle with radius equal to r would be in the ROC of the transform $X(z)$.
2. For the chosen value of the radius r, determine the function $X(r, \Omega)$ by setting $z = r e^{j\Omega}$ in the transform $X(z)$:

$$X(r, \Omega) = X(z)\Big|_{z = r\, e^{j\Omega}} = \mathcal{F}\left\{ x[n]\, r^{-n} \right\} \qquad (8.171)$$

3. Evaluate the integral

$$x[n] = \frac{1}{2\pi} \int_{-\pi}^{\pi} X(r, \Omega) \left(re^{j\Omega} \right)^n \, d\Omega \qquad (8.172)$$

to find the signal $x[n]$.

The three-step procedure outlined above can be reduced to a contour integral. Since $z = re^{j\Omega}$ and r is a constant in the evaluation of the integral in Eqn. (8.172), we can write

$$dz = jr\, e^{j\Omega} \, d\Omega = jz\, d\Omega \qquad (8.173)$$

and therefore

$$d\Omega = \frac{1}{jz} \, dz \qquad (8.174)$$

In the next step we will substitute Eqn. (8.174) into Eqn. (8.172), change $re^{j\Omega}$ to z, and change $X(r, \Omega)$ to $X(z)$ to arrive at the result

$$x[n] = \frac{1}{2\pi j} \oint X(z) \, z^{n-1} \, dz \qquad (8.175)$$

where we have used the contour integral symbol $\oint$ to indicate that the integral should be evaluated by traveling counter-clockwise on a closed contour in the z-plane within the ROC of the transform. The values of the transform $X(z)$ on the closed contour are multiplied by z^{n-1} and integrated. Integration can start at an arbitrary point on the contour, but it must end at the same point. Note the similarity of the inversion integral to that given by Eqn. (7.93) for the computation of the inverse Laplace transform.

In general, direct evaluation of the contour integral is difficult when $X(z)$ is a rational function of z. An indirect method of evaluating the integral in Eqn. (8.175) is to rely on the *Cauchy residue theorem* named after Augustin-Louis Cauchy (1789-1857). In this text we will not consider the inversion integral further for computing the inverse z-transform since more practical methods exist to accomplish the same task.

8.4.2 Partial fraction expansion

It was established in Examples 8.6 and 8.7 that the z-transform of a causal exponential signal is

$$\mathcal{Z}\left\{a^n\, u[n]\right\} = \frac{z}{z-a}\,,\qquad \text{ROC:}\quad |z| > |a| \tag{8.176}$$

and the z-transform of an anti-causal exponential signal is

$$\mathcal{Z}\left\{-a^n\, u[-n-1]\right\} = \frac{z}{z-a}\,,\qquad \text{ROC:}\quad |z| < |a| \tag{8.177}$$

The two signals in Eqns. (8.176) and (8.177) lead to the same rational function for the transform, albeit with different ROCs. These two transform pairs can be used as the basis for determining the inverse z-transform of rational functions expressed using partial fractions. Consider a transform $X(z)$ given with its denominator factored out as

$$X(z) = \frac{B(z)}{(z-z_1)\,(z-z_2)\,\ldots\,(z-z_N)} \tag{8.178}$$

Provided that the order of the numerator polynomial $B(z)$ is not greater than N, expanding the transform into partial fractions in the form

$$X(z) = \frac{k_1\, z}{z-z_1} + \frac{k_2\, z}{z-z_2} + \ldots + \frac{k_N\, z}{z-z_N} \tag{8.179}$$

would allow us to use the standard forms in Eqns. (8.176) and (8.177) for finding the inverse transform of $X(z)$. Let individual terms in the partial fraction expansion be

$$X_i(z) = \frac{k_i\, z}{z-z_i} \quad \text{for}\quad i = 1,\ldots,N \tag{8.180}$$

so that

$$X(z) = X_1(z) + X_2(z) + \ldots + X_N(z) \tag{8.181}$$

The inverse transform is therefore computed as

$$x[n] = x_1[n] + x_2[n] + \ldots + x_N[n] \tag{8.182}$$

where each contributing term $x_i[n]$ represents the inverse transform of the corresponding term on the right side of Eqn. (8.181), that is,

$$x_i[n] = \mathcal{Z}^{-1}\left\{X_i(z)\right\} = \mathcal{Z}^{-1}\left\{\frac{k_i\, z}{z-z_i}\right\},\quad \text{for } i = 1,\ldots,N \tag{8.183}$$

In order to find the correct terms $x_i[n]$ for Eqn. (8.182) we need to determine for each $X_i(z)$ whether it is the transform of a causal signal as in Eqn. (8.176) or an anti-causal signal as in Eqn. (8.177). These decisions must be made by looking at the ROC for $X(z)$ and reasoning what the contribution from the ROC of each individual term $X_i(z)$ must be in order to get the overlap that we have. Since each term in the partial fraction expansion has only one pole, we will adopt the following simple rules:

Determining inverse of each partial fraction:

1. If the ROC for $X(z)$ is inside the circle that passes through the pole at z_i, then the contribution of $X_i(z)$ to the ROC is in the form $|z| < |z_i|$, and therefore the term $x_i[n]$ is an anti-causal signal in the form

$$x_i[n] = -k_i z_i^n u[-n-1] \tag{8.184}$$

2. If the ROC for $X(z)$ is outside the circle that passes through the pole at z_i, then the contribution of $X_i(z)$ to the ROC is in the form $|z| > |z_i|$, and therefore the term $x_i[n]$ is a causal signal in the form

$$x_i[n] = k_i z_i^n u[n] \tag{8.185}$$

These two rules are illustrated in Fig. 8.24. The next two examples will serve to clarify the process explained above.

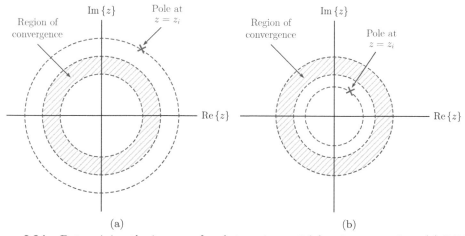

Figure 8.24 – Determining the inverse of each term in partial fraction expansion: (a) ROC is inside the circle that passes through the pole at z_i, (b) ROC is outside the circle that passes through the pole at z_i.

Example 8.29: **Expanding a rational z-transform into partial fractions**

Express the z-transform

$$X(z) = \frac{(z-1)(z+2)}{\left(z-\frac{1}{2}\right)(z-2)}$$

using partial fractions.

Solution: The first step is to divide $X(z)$ by z to obtain

$$\frac{X(z)}{z} = \frac{(z-1)(z+2)}{z\left(z-\frac{1}{2}\right)(z-2)} \tag{8.186}$$

The function $X(z)/z$ has three single poles at $z = 0, \frac{1}{2}$, and 2, and can be expressed as

$$\frac{X(z)}{z} = \frac{k_1}{z} + \frac{k_2}{z - \frac{1}{2}} + \frac{k_3}{z - 2} \tag{8.187}$$

Multiplying both sides of Eqn. (8.187) by z allows us to express $X(z)$ using familiar terms:

$$X(z) = k_1 + \frac{k_2 z}{z - \frac{1}{2}} + \frac{k_3 z}{z - 2} \tag{8.188}$$

The transition from Eqn. (8.187) to Eqn. (8.188) explains the motivation for dividing $X(z)$ by z in Eqn. (8.186) before expanding the result into partial fractions in Eqn. (8.187). Had we not divided $X(z)$ by z in Eqn. (8.187), we would not have obtained partial fractions in the standard form $z/(z - p_i)$ in Eqn. (8.188).

The residues in Eqn. (8.187) can be found by using the residue formulas derived in Appendix E:

$$k_1 = z \left. \frac{X(z)}{z} \right|_{z=0} = \left. \frac{(z-1)(z+2)}{(z-\frac{1}{2})(z-2)} \right|_{z=0} = \frac{(0-1)(0+2)}{(0-\frac{1}{2})(0-2)} = -2$$

$$k_2 = \left(z - \tfrac{1}{2}\right) \left. \frac{X(z)}{z} \right|_{z=\frac{1}{2}} = \left. \frac{(z-1)(z+2)}{z(z-2)} \right|_{z=0} = \frac{(\frac{1}{2}-1)(\frac{1}{2}+2)}{\frac{1}{2}(\frac{1}{2}-2)} = \frac{5}{3}$$

$$k_3 = (z - 2) \left. \frac{X(z)}{z} \right|_{z=2} = \left. \frac{(z-1)(z+2)}{z(z-\frac{1}{2})} \right|_{z=0} = \frac{(2-1)(2+2)}{2(2-\frac{1}{2})} = \frac{4}{3}$$

Using the residues computed, the transform $X(z)$ is

$$X(z) = -2 + \frac{\frac{5}{3}z}{z - \frac{1}{2}} + \frac{\frac{4}{3}z}{z - 2}$$

Software resources: See MATLAB Exercise 8.5.

Example 8.30: **Finding the inverse z-transform using partial fractions**

Consider again the transform $X(z)$ expanded into partial fractions in Example 8.29. The region of convergence was not specified in that example, and therefore, we have only determined the partial fraction expansion for $X(z)$.

1. How many possibilities are there for the ROC?
2. For each possible choice of ROC, determine the inverse transform.

Solution: A pole-zero diagram for $X(z)$ is shown in Fig. 8.25. We know from previous discussion that

a. There can be no poles in the ROC.
b. The boundaries of ROC must be determined by poles.

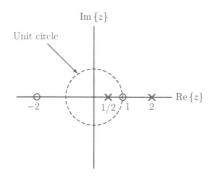

Figure 8.25 – The pole-zero diagram for $X(z)$ in Example 8.30.

We will begin by defining the three components of $X(z)$ in the partial fraction expansion given by Eqn. (8.189). Let

$$X_1(z) = -2$$

$$X_2(z) = \frac{\frac{5}{3}z}{z - \frac{1}{2}}$$

and

$$X_3(z) = \frac{\frac{4}{3}z}{z - 2}$$

The first component, $X_1(z)$, is a constant, and its ROC is always the entire z-plane. Therefore it has no effect on the overall ROC for $X(z)$. The inverse transform of $X_1(z)$ is

$$x_1[n] = \mathcal{Z}^{-1}\{-2\} = -2\,\delta[n]$$

The ROC of $X(z)$ will be determined based on the individual ROCs of the terms $X_2(z)$ and $X_3(z)$. The term $X_2(z)$ has a zero at $z = 0$, and a pole at $z = \frac{1}{2}$. Its region of convergence is either the inside or the outside of the circle with a radius of $\frac{1}{2}$. Similarly, the term $X_3(z)$ has a zero at $z = 0$, and a pole at $z = 2$. Its region of convergence is either the inside or the outside of the circle with a radius of 2. Applying the rules (a) and (b) mentioned above in conjunction with the pole-zero diagram shown in Fig. 8.25 we obtain the following possibilities for the ROC of $X(z)$:

Possibility 1: ROC: $|z| < \frac{1}{2}$
In this case both $X_2(z)$ and $X_3(z)$ must correspond to anti-causal signals, so that the overlap of individual ROCs yields the region chosen. Thus we need

$$\text{ROC for } X_2(z): \quad |z| < \frac{1}{2}$$

$$\text{ROC for } X_3(z): \quad |z| < 2$$

The resulting ROC for $X(z)$ is the overlap of the two individual ROCs, and is therefore the inside of the circle with radius $\frac{1}{2}$. This is illustrated in Fig. 8.26.

The inverse transforms of $X_2(z)$ and $X_3(z)$ are determined using the anti-causal signal given by Eqn. (8.177):

$$x_2[n] = \mathcal{Z}^{-1}\left\{\frac{\frac{5}{3}z}{z - \frac{1}{2}}\right\} = -\frac{5}{3}\left(\frac{1}{2}\right)^n u[-n-1] \qquad (8.189)$$

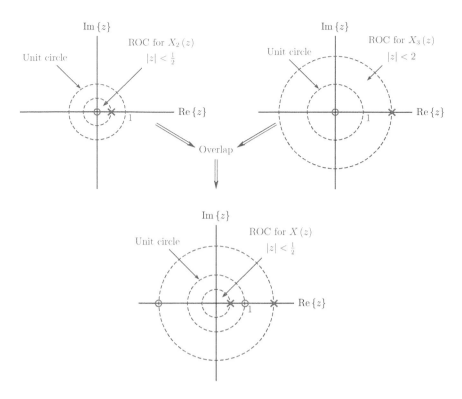

Figure 8.26 – ROCs involved in possibility 1 of Example 8.30.

$$x_3[n] = \mathcal{Z}^{-1} \left\{ \frac{\frac{4}{3} z}{z - 2} \right\} = -\frac{4}{3} (2)^n u[-n - 1] \tag{8.190}$$

Combining Eqns. (8.189), (8.189) and (8.190) we find the signal $x[n]$ to be

$$x[n] = -2\,\delta[n] - \frac{5}{3} \left(\frac{1}{2} \right)^n u[-n - 1] - \frac{4}{3} (2)^n u[-n - 1] \tag{8.191}$$

Numerical evaluation of $x[n]$ in Eqn. (8.191) for a few values of the index n results in

$$x[n] = \left\{ \dots, -53.375, -26.75, -13.5, -7, -4, \underset{\underset{n=0}{\uparrow}}{-2} \right\}$$

Possibility 2:   ROC: $|z| > 2$

This ROC is only possible as the overlap of individual ROCs if both $X_2(z)$ and $X_3(z)$ correspond to causal signals, that is,

$$\text{ROC for } X_2(z): \quad |z| > \frac{1}{2}$$

$$\text{ROC for } X_3(z): \quad |z| > 2$$

The resulting ROC for $X(z)$ in this case is the overlap of the two individual ROCs which is the outside of the circle with a radius of 2. This is illustrated in Fig. 8.27.

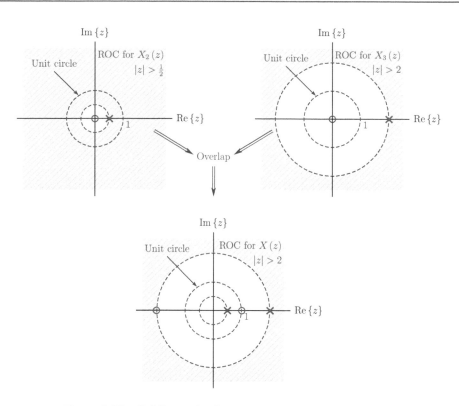

Figure 8.27 – ROCs involved in possibility 2 of Example 8.30.

In this case, the inverse transforms of both $X_2(z)$ and $X_3(z)$ are causal signals, and they are computed from Eqn. (8.176) as

$$x_2[n] = \mathcal{Z}^{-1}\left\{\frac{\frac{5}{3}z}{z - \frac{1}{2}}\right\} = \frac{5}{3}\left(\frac{1}{2}\right)^n u[n] \tag{8.192}$$

$$x_3[n] = \mathcal{Z}^{-1}\left\{\frac{\frac{4}{3}z}{z - 2}\right\} = \frac{4}{3}(2)^n u[n] \tag{8.193}$$

Using Eqns. (8.189), (8.192) and (8.193) the inverse transform for $X(z)$ is

$$x[n] = -2\,\delta[n] + \frac{5}{3}\left(\frac{1}{2}\right)^n u[n] + \frac{4}{3}(2)^n u[n] \tag{8.194}$$

In this case the first few samples of $x[n]$ are computed as

$$x[n] = \{\ \underset{\substack{\uparrow \\ n=0}}{1}\ , 3.5, 5.75, 8.2083, 10.7708, 13.3854, \ldots\}$$

Possibility 3:   ROC: $\frac{1}{2} < |z| < 2$
For this ROC to be the overlap of the two individual ROCs, $X_2(z)$ must correspond to a causal signal while $X_3(z)$ corresponds to an anti-causal signal. We need

$$\text{ROC for } X_2(z): \quad |z| > \frac{1}{2}$$

$$\text{ROC for } X_3(z): \quad |z| < 2$$

Thus the ROC for $X_2(z)$ is the outside of a circle, and the ROC for $X_3(z)$ is the inside of a circle. The overlap of the two ROCs is the region between the two circles. Fig. 8.28 illustrates this possibility.

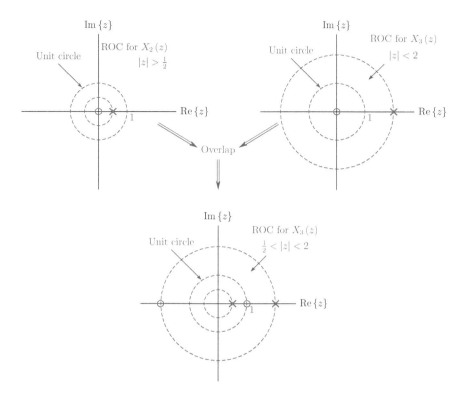

Figure 8.28 – ROCs involved in possibility 3 of Example 8.30.

Inverting each component accordingly, we obtain

$$x_2[n] = \mathcal{Z}^{-1} \left\{ \frac{\frac{5}{3} z}{z - \frac{1}{2}} \right\} = \frac{5}{3} \left(\frac{1}{2} \right)^n u[n] \tag{8.195}$$

and

$$x_3[n] = \mathcal{Z}^{-1} \left\{ \frac{\frac{4}{3} z}{z - 2} \right\} = -\frac{4}{3} (2)^n u[-n - 1] \tag{8.196}$$

Using Eqns. (8.189), (8.195) and (8.196) the inverse transform for $X(z)$ can be constructed as

$$x[n] = -2\,\delta[n] + \frac{5}{3} \left(\frac{1}{2} \right)^n u[n] - \frac{4}{3} (2)^n u[-n - 1] \tag{8.197}$$

In this case the first few samples of $x[n]$ are computed as

$$x[n] = \{ \dots, -0.1667, -0.3333, -0.6667, -0.3333, 0.8333, 0.4167, 0.2083, \dots \}$$
$$\underset{n=0}{\uparrow}$$

8.4.3   Long division

An alternative method for computing the inverse z-transform is the long division technique. Recall that the definition of the z-transform given by Eqns. (8.1) and (8.2) is essentially in the form of a power series involving powers of z and z^{-1}. The long division idea is based on converting a rational transform $X(z)$ back into its power series form, and associating the coefficients of the power series with the sample amplitudes of the signal $x[n]$. In contrast with the partial fraction expansion method discussed in the previous section, the long division method does not produce an analytical solution for the signal $x[n]$. Instead, it allows us to obtain the signal one sample at a time. Its main advantage over the partial fraction expansion method is that it is suitable for use on a computer.

Consider a rational transform in the general form

$$X(z) = \frac{B(z)}{A(z)} = \frac{b_M z^M + b_{M-1} z^{M-1} + \ldots + b_1 z + b_0}{a_N z^N + a_{N-1} z^{N-1} + \ldots + a_1 z + a_0} \tag{8.198}$$

where the M-th order denominator polynomial $B(z)$ is

$$B(z) = b_M z^M + b_{M-1} z^{M-1} + \ldots + b_1 z + b_0 \tag{8.199}$$

and the N-th order numerator polynomial $A(z)$ is

$$A(z) = a_N z^N + a_{N-1} z^{N-1} + \ldots + a_1 z + a_0 \tag{8.200}$$

Initially we will focus on the case where the signal $x[n]$ that led to the transform $X(z)$ is causal. This implies the following:

1. The ROC is in the form $|z| > r_1$.
2. The transform $X(z)$ converges at $|z| \to \infty$.
3. The order of the numerator is less then or equal to the order of the denominator, that is, $M \leq N$.

Dividing the numerator polynomial by the denominator polynomial, the transform $X(z)$ can be written in the alternative form

$$X(z) = \left(\frac{b_M}{a_N}\right) z^{-(N-M)} + \frac{\bar{B}(z)}{A(z)} \tag{8.201}$$

where $(b_M/a_N) z^{-(N-M)}$ is the quotient of the division. The remainder is a polynomial $\bar{B}(z)$ in the form

$$\bar{B}(z) = \bar{b}_{M-1} z^{M-1} + \ldots + \bar{b}_{M-N} z^{M-N} \tag{8.202}$$

Associating the expression in Eqn. (8.202) with the power series form of the z-transform we recognize that

$$x[N - M] = \frac{b_M}{a_N} \tag{8.203}$$

providing us with one sample of the signal $x[n]$. We can now take the function

$$\bar{X}(z) = \frac{\bar{B}(z)}{A(z)} \tag{8.204}$$

and repeat the process, obtaining

$$\bar{X}(z) = \left(\frac{\bar{b}_{M-1}}{a_N}\right) z^{-(N-M+1)} + \frac{\hat{B}(z)}{A(z)} \tag{8.205}$$

which produces another sample of the signal $x[n]$ as

$$x[N - M + 1] = \frac{\bar{b}_{M-1}}{a_N} \tag{8.206}$$

and a new remainder polynomial

$$\hat{B}(z) = \hat{b}_{M-2}z^{M-2} + \ldots + \hat{b}_{M-N-1}z^{M-N-1} \tag{8.207}$$

The next example will illustrate this process.

Example 8.31: **Using long division with right-sided signal**

Use the long division method to determine the first few samples of the signal $x[n]$ of Example 8.22 from its z-transform which was determined to be

$$X(z) = \frac{3z^2 - z}{(z - 1)^3}, \qquad \text{ROC:} \quad |z| > 1$$

Solution: By multiplying out the denominator, the transform $X(z)$ can be written as

$$X(z) = \frac{3z^2 - z}{z^3 - 3z^2 + 3z - 1}, \qquad \text{ROC:} \quad |z| > 1$$

The numerator polynomial is

$$B(z) = 3z^2 - z$$

and the denominator polynomial is

$$A(z) = z^3 - 3z^2 + 3z - 1$$

Dividing the numerator polynomial by the denominator polynomial we obtain an alternative form of $X(z)$ as

$$X(z) = 3z^{-1} + \frac{8z - 9 + 3z^{-1}}{z^3 - 3z^2 + 3z - 1} = 3z^{-1} + \bar{X}(z) \tag{8.208}$$

Thus we have extracted one term of the power series representation of $X(z)$, and it indicates that $x[1] = 3$. Repeating the process with $\bar{X}(z)$, we obtain

$$\bar{X}(z) = 8z^{-2} + \frac{15 - 21z^{-1} + 8z^{-2}}{z^3 - 3z^2 + 3z - 1} = 8z^{-2} + \hat{X}(z) \tag{8.209}$$

indicating that $x[2] = 8$. The results in Eqns. (8.208) and (8.209) can be combined, yielding

$$X(z) = 3z^{-1} + 8z^{-2} + \frac{15 - 21z^{-1} + 8z^{-2}}{z^3 - 3z^2 + 3z - 1}$$

Thus, each iteration through the long division operation produces one more sample of the signal $x[n]$. We are now ready to set up the long division:

$$
\begin{array}{r}
3z^{-1} + 8z^{-2} + 15z^{-3} + 24z^{-4} \\
z^3 - 3z^2 + 3z - 1 \enclose{longdiv}{3z^2 \quad -z} \\
3z^2 \ -9z \ +9 \ -3z^{-1} \\
\hline
8z \ -9 \ +3z^{-1} \\
8z \ -24 \ +24z^{-1} \ -8z^{-2} \\
\hline
15 \ -21z^{-1} \ +8z^{-2} \\
15 \ -45z^{-1} \ +45z^{-2} \ -15z^{-3} \\
\hline
24z^{-1} \ -37z^{-2} \ +15z^{-3}
\end{array}
$$

Using the quotient and the remainder of the division, the transform $X(z)$ can be written as

$$X(z) = 3z^{-1} + 8z^{-2} + 15z^{-3} + 24z^{-4} + \frac{24z^{-1} - 37z^{-2} + 15z^{-3}}{z^3 - 3z^2 + 3z - 1} \tag{8.210}$$

Comparing the result in Eqn. (8.210) with the definition of the z-transform in Eqn. (8.1) we conclude that the signal $x[n]$ is in the form

$$x[n] = \{\; \underset{\substack{\uparrow \\ n=1}}{3} \;, 8, \; 15, \; 24, \; \ldots\}$$

The sample amplitudes obtained should be in agreement with those obtained by directly evaluating the values of the signal $x[n] = n(n+2)u[n]$ that led to the transform in question.

In Example 8.31 the use of the long division technique produced a causal signal as the inverse transform of the specified function $X(z)$. That was fine since the ROC specified for the transform also supported this conclusion. What if we have a transform and associated ROC that indicate an anti-causal or a non-causal signal as the inverse transform? How would we need to modify the long division technique to produce the correct result in such a case?

Consider again a rational transform $X(z)$ with one change from Eqn. (8.198): This time we will order the terms of numerator and denominatior polynomials in ascending powers of z. The result is

$$X(z) = \frac{B(z)}{A(z)} = \frac{b_0 + b_1 z + \ldots + b_{M-1} z^{M-1} + b_M z^M}{a_0 + a_1 z + \ldots + a_{N-1} z^{N-1} + a_N z^N} \tag{8.211}$$

If we now divide the numerator polynomial by the denominator polynomial, we can write $X(z)$ as

$$X(z) = \left(\frac{b_0}{a_0}\right) + \frac{\bar{B}(z)}{A(z)} \tag{8.212}$$

The term (b_0/a_0) is the quotient of the division. The remainder is a polynomial $\bar{B}(z)$ in the form

$$\bar{B}(z) = \bar{b}_1 z + \bar{b}_2 z^2 + \ldots \tag{8.213}$$

Associating the expression in Eqn. (8.202) with the power series form of the z-transform we recognize that

$$x[0] = \frac{b_0}{a_0} \tag{8.214}$$

We can now take the function

$$\bar{X}(z) = \frac{\tilde{B}(z)}{A(z)} \tag{8.215}$$

and repeat the process, obtaining

$$\bar{X}(z) = \left(\frac{\bar{b}_1}{a_0}\right) z + \frac{\hat{B}(z)}{A(z)} \tag{8.216}$$

which produces another sample of the signal $x[n]$ as

$$x[-1] = \frac{\bar{b}_1}{a_0} \tag{8.217}$$

and a new remainder polynomial

$$\hat{B}(z) = \hat{b}_2 z^2 + \hat{b}_3 z^3 + \ldots \tag{8.218}$$

The next example will illustrate the use of long division with different types of ROCs.

Example 8.32: Using long division with specified ROC

In Example 8.29 we have used the partial fraction expansion method to determine the inverse z-transform of the rational function

$$X(z) = \frac{(z-1)(z+2)}{\left(z-\frac{1}{2}\right)(z-2)}$$

for all possible choices of the ROC. Verify the results of that example by determining the first few samples of the inverse z-transform through the use of the long division method for each possible choice of the ROC.

Solution: Multiplying out the factors of the numerator and the denominator of $X(z)$ we obtain

$$X(z) = \frac{z^2 + z - 2}{z^2 - 2.5z + 1} \tag{8.219}$$

As we have discussed in Example 8.30, there are three possible choices for the ROC:

Possibility 1: ROC: $|z| < \frac{1}{2}$
In this case the inverse transform $x[n]$ must be anti-causal. In order to obtain an anti-causal solution from the long division, we will rewrite $X(z)$ with its numerator and denominator polynomials arranged in ascending powers of z:

$$X(z) = \frac{-2 + z + z^2}{1 - 2.5z + z^2}$$

and set up the long division:

$$
\begin{array}{r}
-2 - 4z - 7z^2 - 13.5z^3 \\
\hline
1 - 2.5z + z^2 \,\big|\, -2 \;+z \;\;+z^2 \\
-2 + 5z \;\;-2z^2 \\
\hline
-4z \;\;+3z^2 \\
-4z + 10z^2 \;\;\;-4z^3 \\
\hline
-7z^2 \;\;+4z^3 \\
-7z^2 + 17.5z^3 \;\;\;\;\;-7z^4 \\
\hline
-13.5z^3 \;\;\;\;+7z^4 \\
-13.5z^3 + 33.75z^4 - 13.5z^5 \\
\hline
-26.75z^4 + 13.5z^5
\end{array}
$$

Using the quotient and the remainder of the division, the transform $X(z)$ can be written as

$$X(z) = -2 - 4z - 7z^2 - 13.5z^3 + \frac{-26.75z^4 + 13.5z^5}{1 - 2.5z + z^2} \tag{8.220}$$

Comparing the result in Eqn. (8.220) with the z-transform definition in Eqn. (8.1) we conclude that the signal $x[n]$ must be in the form

$$x[n] = \{\,\ldots,\ -13.5,\ -7,\ -4,\ -\underset{\underset{n=0}{\uparrow}}{2}\ \}$$

consistent with the answer found in Example 8.30.

Possibility 2:   ROC: $|z| > 2$
In this case the signal $x[n]$ is causal. To obtain a causal answer we will use the original form of $X(z)$ in Eqn. (8.219) with numerator and denominator polynomials arranged in descending orders of z. The long division is set up as follows:

$$
\begin{array}{r}
1 + 3.5z^{-1} + 5.75z^{-2} \\ \hline
z^2 - 2.5z + 1\ \big|\ z^2\quad +z\quad -2 \\
z^2\ -2.5z\quad +1 \\ \hline
3.5z\quad -3 \\
3.5z\ -8.75\quad +3.5z^{-1} \\ \hline
5.75\quad -3.5z^{-1} \\
5.75\ -14.375z^{-1}\ +5.75z^{-2} \\ \hline
10.875z^{-1}\ -5.75z^{-2}
\end{array}
$$

Using the results obtained so far, the transform $X(z)$ is expressed as

$$X(z) = 1 + 3.5z^{-1} + 5.75z^{-2} + \frac{10.875z^{-1} - 5.75z^2}{z^2 - 2.5z + 1}$$

The first few samples of the signal $x[n]$ are

$$x[n] = \{\ \underset{\underset{n=0}{\uparrow}}{1}\ ,\ 3.5,\ 5.75,\ \ldots\}$$

identical to the result that was found in Example 8.30.

Possibility 3:   ROC: $\frac{1}{2} < |z| < 2$
In this case the inverse transform $x[n]$ has a causal component and an anti-causal component, so it can be written in the form

$$x[n] = x_R[n] + x_L[n]$$

Accordingly, we need to partition the transform $X(z)$ into two parts: One that corresponds to the causal component $x_R[n]$, and one that corresponds to the anti-causal component $x_L[n]$. Recall that in Example 8.29 the transform $X(z)$ was expressed through partial fractions as

$$X(z) = X_1(z) + X_2(z) + X_3(z)$$

with

$$X_1(z) = -2\,, \qquad \text{ROC:}\quad \text{all } z$$

$$X_2(z) = \frac{\frac{5}{3}z}{z - \frac{1}{2}}\,, \qquad \text{ROC:}\quad |z| > \frac{1}{2}$$

and

$$X_3\left(z\right) = \frac{\frac{4}{3}\,z}{z-2}\,, \qquad \text{ROC:} \quad |z| < 2$$

It is clear that $X_2\left(z\right)$ should be included in $X_R\left(z\right)$, and $X_3\left(z\right)$ should be included in $X_L\left(z\right)$. The constant term, $X_1\left(z\right) = -2$, could be included with either function; we will choose to include it with $X_R\left(z\right)$. Consequently we have

$$X_R\left(z\right) = X_1\left(z\right) + X_2\left(z\right) = -2 + \frac{\frac{5}{3}\,z}{z-\frac{1}{2}} = \frac{-\frac{1}{3}\,z+1}{z-\frac{1}{2}}\,, \qquad \text{ROC:} \quad |z| > \frac{1}{2}$$

and

$$X_L\left(z\right) = X_3\left(z\right) = \frac{\frac{4}{3}\,z}{z-2}\,, \qquad \text{ROC:} \quad |z| < 2$$

Let us begin with $x_R[n]$. The long division for the causal component is set up using the numerator and the denominator of $X_R\left(z\right)$ arranged in descending powers of z:

$$
\begin{array}{r}
-\frac{1}{3} + \frac{5}{6}\,z^{-1} + \frac{5}{12}\,z^{-2} \\
\hline
z-\frac{1}{2}\;\big|\; -\frac{1}{3}\,z \;+1 \\
-\frac{1}{3}\,z \;+\frac{1}{6} \\
\hline
\frac{5}{6} \\
\frac{5}{6} \;-\frac{5}{12}\,z^{-1} \\
\hline
\frac{5}{12}\,z^{-1} \\
\frac{5}{12}\,z^{-1} \;-\frac{5}{24}\,z^{-2} \\
\hline
\frac{5}{24}\,z^{-2}
\end{array}
$$

We conclude that $x_R[0] = -1/3$, $x_R[1] = 5/6$ and $x_R[2] = 5/12$. Next we will set up the long division for $X_L[z]$ with numerator and denominator polynomials arranged in ascending powers of z:

$$
\begin{array}{r}
-\frac{2}{3}\,z - \frac{1}{3}\,z^2 - \frac{1}{6}\,z^3 \\
\hline
-2+z\;\big|\; \frac{4}{3}\,z \\
\frac{4}{3}\,z \;-\frac{2}{3}\,z^2 \\
\hline
\frac{2}{3}\,z^2 \\
\frac{2}{3}\,z^2 \;-\frac{1}{3}\,z^3 \\
\hline
\frac{1}{3}\,z^3 \\
\frac{1}{3}\,z^3 \;-\frac{1}{6}\,z^4 \\
\hline
\frac{1}{6}\,z^4
\end{array}
$$

Thus, the first few samples of the anti-causal component of $x[n]$ are $x_L[-1] = -2/3$, $x_L[-2] = -1/3$ and $x_L[-3] = -1/6$. Combining the results of the two long divisions we obtain $x[n]$ as

$$x[n] = \{\ldots, -\frac{1}{6}, -\frac{1}{3}, -\frac{2}{3}, \underset{\underset{n=0}{\uparrow}}{-\frac{1}{3}}, \frac{5}{6}, \frac{5}{12}, \ldots\}$$

8.5   Using the z-Transform with DTLTI Systems

In Chapter 3 we have demonstrated that the output signal of a DTLTI system is related to its input signal and its impulse response through the convolution sum. Specifically, if the impulse response of a DTLTI system is $h[n]$, and if the signal $x[n]$ is applied to the system as input, the output signal $y[n]$ is computed as

$$y[n] = x[n] * h[n] = \sum_{k=-\infty}^{\infty} x[k]\, h[n-k]$$

Based on the convolution property of the z-transform introduced by Eqn. (8.127) in Section 8.3.6, the z-transform of the output signal is equal to the product of the z-transforms of the input signal and the impulse response, that is,

$$Y(z) = X(z)\, H(z) \tag{8.221}$$

If the input signal $x[n]$ and the impulse response $h[n]$ are specified, we could find the z-transforms $X(z)$ and $H(z)$, multiply them to obtain $Y(z)$, and then determine $y[n]$ by means of inverse z-transform. This process provides us an alternative to computing the output by direct application of the convolution sum, and was demonstrated in Examples 8.23 and 8.24.

Alternatively, if the input and the output signals $x[n]$ and $y[n]$ are specified, their z-transforms can be computed, and then the z-transform of the impulse response can be determined from them. Solving Eqn. (8.221) for $H(z)$ we get

$$H(z) = \frac{Y(z)}{X(z)} \tag{8.222}$$

The function $H(z)$ is the *z-domain system function* of the system. We already know that a DTLTI system can be described fully and uniquely by means of its impulse response $h[n]$. Since the system function $H(z)$ is just the z-transform of the impulse response $h[n]$, it also represents a complete description of the DTLTI system.

8.5.1   Relating the system function to the difference equation

In Section 3.4 of Chapter 3 we have established the fact that a DTLTI system can be described by means of a constant-coefficient linear difference equation in the standard form

$$\sum_{k=0}^{N} a_k\, y[n-k] = \sum_{k=0}^{M} b_k\, x[n-k] \tag{8.223}$$

Therefore it must be possible to obtain the other two forms of description of the DTLTI system, namely the system function and the impulse response, from the knowledge of its difference equation. If we take the z-transform of both sides of Eqn. (8.223) the equality would still be valid:

$$\mathcal{Z}\left\{\sum_{k=0}^{N} a_k\, y[n-k]\right\} = \mathcal{Z}\left\{\sum_{k=0}^{M} b_k\, x[n-k]\right\} \tag{8.224}$$

Using the linearity property of the z-transform, Eqn. (8.224) can be written as

$$\sum_{k=0}^{N} a_k \, \mathcal{Z}\left\{y[n-k]\right\} = \sum_{k=0}^{M} b_k \, \mathcal{Z}\left\{x[n-k]\right\} \tag{8.225}$$

and using the time-shifting property of the z-transform leads to

$$\sum_{k=0}^{N} a_k z^{-k} \, Y(z) = \sum_{k=0}^{M} b_k z^{-k} \, X(z) \tag{8.226}$$

The transforms $X(z)$ and $Y(z)$ do not depend on the summation index k, and can therefore be factored out of the summations in Eqn. (8.226) resulting in

$$Y(z) \sum_{k=0}^{N} a_k z^{-k} = X(z) \sum_{k=0}^{M} b_k z^{-k} \tag{8.227}$$

Finally, the system function can be obtained from Eqn. (8.227) as

$$H(z) = \frac{Y(z)}{X(z)} = \frac{\displaystyle\sum_{k=0}^{M} b_k z^{-k}}{\displaystyle\sum_{k=0}^{N} a_k z^{-k}} \tag{8.228}$$

Finding the system function from the difference equation:

1. Separate the terms of the difference equation so that $y[n]$ and its time-shifted versions are on the left of the equal sign, and $x[n]$ and its time-shifted versions are on the right of the equal sign, as in Eqn. (8.223).
2. Take the z-transforms of both sides of the difference equation, and use the time-shifting property of the z-transform as in Eqn. (8.226).
3. Determine the system function as the ratio of $Y(z)$ to $X(z)$ as in Eqn. (8.228).
4. If the impulse response is needed, it can now be determined as the inverse z-transform of $H(z)$.

At this point we will make two important observations:

1. In the above development leading up to the z-domain system function we have relied on the convolution operation in Eqn. (8.221). In Eqn. (8.228) we have used the convolution property of the z-transform. We know from Chapter 3 that the convolution operation is only applicable to problems involving linear and time-invariant systems. Therefore the system function concept is meaningful only for systems that are both linear and time-invariant. This notion was introduced in earlier discussions involving system functions as well.
2. Furthermore, it was justified in Section 3.4 of Chapter 3 that a constant-coefficient difference equation corresponds to a linear and time-invariant system only if all initial conditions are set to zero.

Consequently, we conclude that, in determining the system function from the difference equation, all initial conditions must be assumed to be zero.

If we need to use z-transform based techniques to solve a difference equation subject to non-zero initial conditions, we can do that through the use of the unilateral z-transform, but not through the use of the system function. The unilateral z-transform and its use for solving difference equations will be discussed in Section 8.7.

Example 8.33: **Finding the system function from the difference equation**

Let a DTLTI system be defined by the difference equation

$$y[n] - 0.4y[n-1] + 0.89y[n-2] = x[n] - x[n-1]$$

Find the system function $H(z)$ for this system.

Solution: We will assume zero initial conditions and take the z-transforms of both sides of the difference equation to get

$$Y(z) - 0.4z^{-1}Y(z) + 0.89z^{-2}Y(z) = X(z) - z^{-1}X(z)$$

from which the system function is obtained as

$$H(z) = \frac{1 - z^{-1}}{1 - 0.4z^{-1} + 0.89z^{-2}}$$

or, using non-negative powers of z

$$H(z) = \frac{z(z-1)}{z^2 - 0.4z + 0.89}$$

We will make another important observation based on the result obtained in Example 8.33: The characteristic equation for the system considered in Example 8.33 is

$$z^2 - 0.4z + 0.89 = 0$$

and the solutions of the characteristic equation are the modes of the system as defined in Section 3.5.1 of Chapter 3. When we find the system function $H(z)$ from the difference equation we see that its denominator polynomial is identical to the characteristic polynomial. The roots of the denominator polynomial are the poles of the system function in the z-domain, and consequently, they are identical to the modes of the difference equation of the system.

Recall that in Section 3.5.1 we have reached some conclusions about the relationship between the modes of the difference equation and the natural response of the corresponding system. The same conclusions would apply to the poles of the system function. Specifically, if all poles p_k of the system are real-valued and distinct, then the natural response is in the form

$$y[n] = \sum_{k=1}^{N} c_k p_k^n$$

Complex poles appear in conjugate pairs provided that all coefficients of the system function are real-valued. A pair of complex conjugate poles

$$p_{1a} = r_1\, e^{j\Omega_1}, \quad \text{and} \quad p_{1b} = r_1\, e^{-j\Omega_1}$$

yields a response of the type

$$y_1[n] = d_1\, r_1^n\, \cos\left(\Omega_1 n + \theta_1\right)$$

Finally, a pole of multiplicity m at $z = p_1$ leads to a response in the form

$$y_1[n] = c_{11}\, p_1^n + c_{12}\, n p_1^n + \ldots + c_{1m}\, n^{m-1} p_1^n + \text{other terms}$$

Justifications for these relationships were given in Section 3.5.1 of Chapter 3 through the use of the modes of the difference equation, and will not be repeated here.

Sometimes we need to reverse the problem represented in Example 8.33 and find the difference equation from the knowledge of the system function. The next example will demonstrate this.

Example 8.34: **Finding the difference equation from the system function**

Find the difference equation of the DTLTI system defined by the system function

$$H\left(z\right) = \frac{z^2 - 5z + 6}{z^3 + 2z^2 - z - 2}$$

Solution: Let us first write $H\left(z\right)$ as a rational function of z^{-1} by multiplying both its numerator and denominator with z^{-3}:

$$H\left(z\right) = \frac{z^{-1} - 5z^{-2} + 6z^{-3}}{1 + 2z^{-1} - z^{-2} - 2z^{-3}}$$

We know that the system function is the ratio of the output transform to the input transform, that is,

$$H\left(z\right) = \frac{Y\left(z\right)}{X\left(z\right)}$$

so that we can write

$$\left(1 + 2z^{-1} - z^{-2} - 2z^{-3}\right) Y\left(z\right) = \left(z^{-1} - 5z^{-2} + 6z^{-3}\right) X\left(z\right)$$

or equivalently

$$y[n] + 2y[n-1] - y[n-2] - 2y[n-3] = x[n-1] - 5x[n-2] + 6x[n-3]$$

8.5.2   Response of a DTLTI system to complex exponential signal

An interesting interpretation of the system function concept is obtained when one considers the response of a DTLTI system to a complex exponential signal in the form

$$x[n] = z_0^n$$

where z_0 represents a point in the z-plane within the ROC of the system function. Let $h[n]$ be the impulse response of the DTLTI system under consideration. The output signal is determined through the use of the convolution sum

$$y[n] = h[n] * x[n] = \sum_{k=-\infty}^{\infty} h[k]\, x[n-k] \qquad (8.229)$$

Substituting $x[n - k] = z_0^{n-k}$ into Eqn. (8.229) and simplifying the resulting summation we obtain

$$y[n] = \sum_{k=-\infty}^{\infty} h[k] z_0^{n-k}$$

$$= z_0^n \sum_{k=-\infty}^{\infty} h[k] z_0^{-k} \tag{8.230}$$

In Eqn. (8.230) the summation corresponds to the system function $H(z)$ evaluated at the point $z = z_0$, so the response of the DTLTI system to the input signal $x[n] = z_0^n$ is

$$y[n] = z_0^n H(z_0) \tag{8.231}$$

We conclude that the response of a DTLTI system to the exponential signal $x[n] = z_0^n$ is a scaled version of the input signal. The scale factor is the value of the system function at the point $z = z_0$.

Example 8.35: **Response of a DTLTI system to a complex exponential signal**

Consider a DTLTI system with the system function

$$H(z) = \frac{z + 1}{z^2 - \frac{5}{6}z + \frac{1}{6}}$$

Find the response of the system to the exponential signal

$$x[n] = \left(0.9\, e^{j0.2\pi}\right)^n$$

Solution: The input signal is complex-valued. Using Euler's formula we can write $x[n]$ as

$$x[n] = (0.9)^n \cos(0.2\pi n) + j\,(0.9)^n \sin(0.2\pi n)$$

Real and imaginary parts of $x[n]$ are shown in Fig. 8.29. The value of the system function at $z = 0.9\, e^{j0.2\pi}$ is

$$H\left(0.9\, e^{j0.2\pi}\right) = \frac{0.9\, e^{j0.2\pi} + 1}{\left(0.9\, e^{j0.2\pi}\right)^2 - \frac{5}{6}\left(0.9\, e^{j0.2\pi}\right) + \frac{1}{6}}$$

$$= -1.0627 - j\,4.6323$$

$$= 4.7526\, e^{-j\,1.7963}$$

The output signal is found using Eqn. (8.231) as

$$y[n] = \left(0.9\, e^{j0.2\pi}\right)^n \left(4.7526\, e^{-j\,1.7963}\right)$$

$$= 4.7526\,(0.9)^n\ e^{j(0.2\pi n - 1.7963)}$$

or, in Cartesian form

$$y[n] = 4.7526\,(0.9)^n \cos(0.2\pi n - 1.7963) + j\,4.7526\,(0.9)^n \sin(0.2\pi n - 1.7963)$$

Real and imaginary parts of the output signal $y[n]$ are shown in Fig. 8.30.

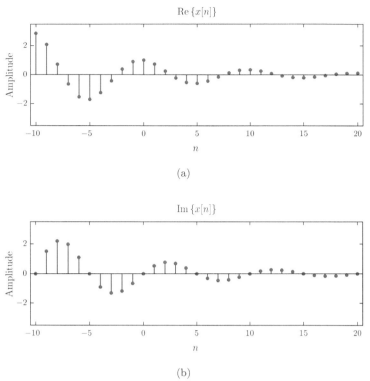

(a)

(b)

Figure 8.29 – The signal $x[n]$ for Example 8.35: (a) real part, (b) imaginary part.

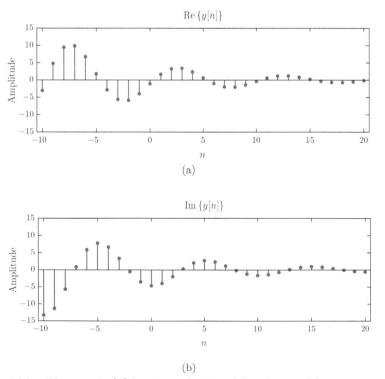

(a)

(b)

Figure 8.30 – The signal $y[n]$ for Example 8.35: (a) real part, (b) imaginary part.

Software resources:
ex_8_35a.m
ex_8_35b.m

8.5.3   Response of a DTLTI system to exponentially damped sinusoid

Next we will consider an input signal in the form of an exponentially damped sinusoid such as

$$x[n] = r_0^n \, \cos{(\Omega_0 n)} \tag{8.232}$$

In order to find the response of a DTLTI system to this signal, let us write $x[n]$ using Euler's formula:

$$x[n] = \frac{1}{2} \, r_0^n \, e^{j\Omega_0 n} + \frac{1}{2} \, r_0^n \, e^{-j\Omega_0 n} \tag{8.233}$$

Let z_0 be defined as

$$z_0 = r_0 \, e^{j\Omega_0}$$

It follows that

$$z_0^* = r_0 \, e^{-j\Omega_0}$$

and Eqn. (8.233) can be written as

$$x[n] = \frac{1}{2} \, z_0^n + \frac{1}{2} \, (z_0^*)^n \tag{8.234}$$

The response of the system is determined using the linearity of the system:

$$
\begin{aligned}
y[n] &= \mathrm{Sys} \left\{ \frac{1}{2} \, z_0^n + \frac{1}{2} \, (z_0^*)^n \right\} \\
&= \frac{1}{2} \, \mathrm{Sys} \left\{ z_0^n \right\} + \frac{1}{2} \, \mathrm{Sys} \left\{ (z_0^*)^n \right\}
\end{aligned}
\tag{8.235}
$$

We already know that the response of the system to the term z_0^n is

$$\mathrm{Sys} \left\{ z_0^n \right\} = z_0^n \, H \, (z_0) \tag{8.236}$$

and its response to the term $(z_0^*)^n$ is

$$\mathrm{Sys} \left\{ (z_0^*)^n \right\} = (z_0^*)^n \, H \, (z_0^*) \tag{8.237}$$

so that the output signal $y[n]$ can be written as

$$y[n] = \frac{1}{2} \, z_0^n \, H \, (z_0) + \frac{1}{2} \, (z_0^*)^n \, H \, (z_0^*) \tag{8.238}$$

The result obtained in Eqn. (8.238) can be further simplified. Let the value of the system function evaluated at the point $z = z_0$ be written in polar complex form as

$$H \, (z_0) = H_0 \, e^{j\Theta_0} \tag{8.239}$$

where H_0 and Θ_0 represent the magnitude and the phase of the system function at the point $z = z_0$ respectively:

$$H_0 = |H \, (z_0)| \tag{8.240}$$

and

$$\Theta_0 = \angle H(z_0) \tag{8.241}$$

It can be shown (see Problem 8.34 at the end of this chapter) that, for a system with a real-valued impulse response, the value of the system function at the point $z = z_0^*$ is the complex conjugate of its value at the point $z = z_0$, that is,

$$H(z_0^*) = [H(z_0)]^* = H_0\, e^{-j\Theta_0} \tag{8.242}$$

Using Eqns. (8.239) and (8.242) in Eqn. (8.238), the output signal $y[n]$ becomes

$$
\begin{aligned}
y[n] &= \frac{1}{2} z_0^n\, H_0\, e^{j\Theta_0} + \frac{1}{2} (z_0^*)^n\, H_0\, e^{-j\Theta_0} \\
&= \frac{1}{2} \left(r_0\, e^{j\Omega_0}\right)^n H_0\, e^{j\Theta_0} + \frac{1}{2} \left(r_0\, e^{-j\Omega_0}\right)^n H_0\, e^{-j\Theta_0} \\
&= \frac{1}{2} r_0^n\, H_0 \left[e^{j(\Omega_0 n + \Theta_0)} + e^{-j(\Omega_0 n + \Theta_0)} \right] \\
&= H_0\, r_0^n\, \cos(\Omega_0 n + \Theta_0)
\end{aligned}
\tag{8.243}
$$

Comparison of the input signal in Eqn. (8.232) and the output signal Eqn. (8.243) reveals the following:

1. The amplitude of the signal is multiplied by the magnitude of the system function evaluated at the point $z = r_0\, e^{j\Omega}$.
2. The phase of the cosine function is incremented by an amount equal to the phase of the system function evaluated at the point $z = r_0\, e^{j\Omega}$.

Example 8.36: **Response of a DTLTI system to an exponentially damped sinusoid**

Consider again the system function used in Example 8.35. Determine the response of the system to the input signal

$$x[n] = (0.9)^n\, \cos(0.2\pi n)$$

Solution: In Example 8.35 we have evaluated the system function at the point $z = 0.9\, e^{j0.2\pi}$ and have obtained

$$H\left(0.9\, e^{j0.2\pi}\right) = H_0\, e^{j\Theta_0} = 4.7526\, e^{-j\,1.7963} \tag{8.244}$$

Using Eqn. (8.243) with $H_0 = 4.7526$ and $\Theta_0 = -1.7963$ radians, we obtain the output signal as

$$y[n] = 4.7526\,(0.9)^n\, \cos(0.2\pi n - 1.7963)$$

The input and the output signals are shown in Fig. 8.31.

Software resources:
ex_8_36.m

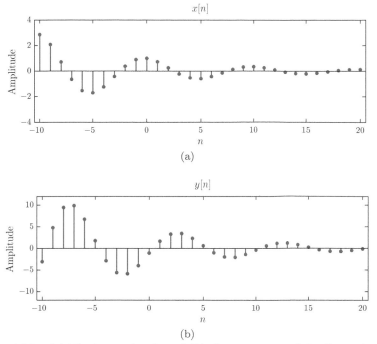

Figure 8.31 – (a) The input signal, and (b) the output signal for Example 8.36.

8.5.4   Graphical interpretation of the pole-zero plot

It was discussed earlier in this chapter that the complex variable z can be represented as a point in the z-plane. Alternatively, a complex number can be thought of as a vector in the complex plane (see Appendix A), drawn from the origin to the point of interest. Let $z = r\, e^{j\Omega}$. Fig. 8.32 illustrates the use of a vector with norm (or length) equal to r and angle equal to Ω for representing z.

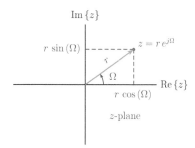

Figure 8.32 – Vector representation of the complex variable $z = r\, e^{j\Omega}$.

Using vector notation we will write

$$\overrightarrow{z} = r\, \overrightarrow{e^{j\Omega}} \tag{8.245}$$

The norm and the angle of the vector $\overrightarrow{z}$ are expressed as

$$\left| \overrightarrow{z} \right| = r \tag{8.246}$$

and

$$\angle \overrightarrow{z} = \Omega \tag{8.247}$$

Fixing Ω and changing the value of r causes the tip of the vector to move either toward or away from the origin while the direction of the vector remains unchanged. Fixing r and changing the value of Ω causes the vector to rotate while its length remains unchanged. In this case the tip of the vector stays on a circle with radius r.

Consider a fixed point z_1 in the z-plane. It will be interesting to find the vector $(\overrightarrow{z - z_1})$.

$$\overrightarrow{z - z_1} = \overrightarrow{z} - \overrightarrow{z_1}$$
$$= \overrightarrow{z} + (\overrightarrow{-z_1}) \tag{8.248}$$

Based on Eqn. (8.248) the vector $(\overrightarrow{z - z_1})$ can be found by adding the two vectors $\overrightarrow{z}$ and $(\overrightarrow{-z_1})$ using the parallelogram rule as shown in Fig. 8.33.

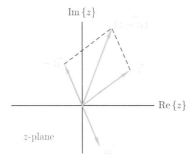

Figure 8.33 – Finding the vector representation for $(z - z_1)$.

Any vector is identified by two features: its norm (length) and its direction. The starting point of a vector is not important. Shifting a vector parallel to itself so that it starts at any desired point would be allowed as long as the norm and the direction of the vector are not changed. Accordingly, the vector $(\overrightarrow{z - z_1})$ found through the use of the parallelogram rule in Fig. 8.33 can be shifted so that it originates at the point z_1 and terminates at the point z as shown in Fig. 8.34.

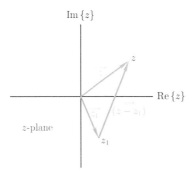

Figure 8.34 – Alternative representation of the vector for $(z - z_1)$.

The alternative representation shown in Fig. 8.34 leads us to an important conclusion:

In the z-plane, the vector $\overrightarrow{(z - z_1)}$ is drawn with an arrow that starts at the point z_1 and ends at the point z.

This conclusion will be critical in graphically interpreting the pole-zero diagram for a system function. We will begin by considering a simple DTLTI system with the impulse response

$$h[n] = a^n \, u[n] \tag{8.249}$$

and the corresponding z-domain system function

$$H(z) = \frac{z}{z - a}, \qquad \text{ROC:} \quad |z| > |a| \tag{8.250}$$

The system has a zero at $z = 0$ and a pole at $z = a$. In vector form, the system function can be written as the ratio of two vectors

$$\overrightarrow{H(z)} = \frac{\overrightarrow{z}}{\overrightarrow{(z - a)}} \tag{8.251}$$

Suppose we need to evaluate the system function at a specific point $z = z_a$ in the z-plane. Assuming that the point $z = z_a$ is within the ROC of the system function, the vector representation of the system function at the point of interest is

$$\overrightarrow{H(z_a)} = \frac{\overrightarrow{z_a}}{\overrightarrow{(z_a - a)}} \tag{8.252}$$

with its magnitude and phase computed as

$$\left| \overrightarrow{H(z_a)} \right| = \frac{\left| \overrightarrow{z_a} \right|}{\left| \overrightarrow{(z_a - a)} \right|} \tag{8.253}$$

and

$$\angle \overrightarrow{H(z_a)} = \angle \overrightarrow{z_a} - \angle \overrightarrow{(z_a - a)} \tag{8.254}$$

Example 8.37: **Evaluating $H(z)$ using vectors**

Let a DTLTI system be characterized by the system function

$$H(z) = \frac{z}{z - 0.8}, \qquad \text{ROC:} \quad |z| > 0.8$$

Discuss how the system function can be graphically evaluated at the points

a. $z_a = 1.5 + j1$
b. $z_b = -0.5 - j1$

Solution: Both points z_a and z_b are within the ROC of the system function. For $H(z_a)$ we need the vectors $\overrightarrow{z_a}$ and $\overrightarrow{(z_a - 0.8)}$ shown in Fig. 8.35.

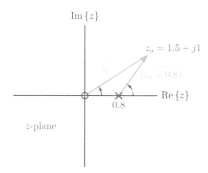

Figure 8.35 – Vectors for graphical representation of $H(z_a)$ in Example 8.37.

If the pole-zero diagram in Fig. 8.35 is drawn to scale, we could simply measure the lengths and the angles of the two vectors, and compute the magnitude and the phase of the system function at $z = z_a$ from those measurements as

$$\left|\overrightarrow{H(z_a)}\right| = \frac{\left|\overrightarrow{z_a}\right|}{\left|\overrightarrow{(z_a - 0.8)}\right|}$$

and

$$\angle \overrightarrow{H(z_a)} = \angle \overrightarrow{z_a} - \angle \overrightarrow{(z_a - 0.8)}$$

It can be shown that

$$\left|\overrightarrow{z_a}\right| = 1.8028 , \qquad\qquad \angle \overrightarrow{z_a} = 0.5880 \text{ rad} = 33.7^o$$

$$\left|\overrightarrow{(z_a - 0.8)}\right| = 1.2207 , \qquad \angle \overrightarrow{(z_a - 0.8)} = 0.9601 \text{ rad} = \ \ 55^o$$

The magnitude and the phase of the system function at $z_a = 1.5 + j1$ are

$$\left|\overrightarrow{H(1.5 + j1)}\right| = \frac{1.8028}{1.2207} = 1.4769$$

and

$$\angle \overrightarrow{H(1.5 + j1)} = 0.5880 - 0.9601 = -0.3721 \text{ rad} = -21.3^o$$

To evaluate $H(z_b)$ we need the vectors $\overrightarrow{z_b}$ and $\overrightarrow{(z_b - 0.8)}$ as shown in Fig. 8.36.

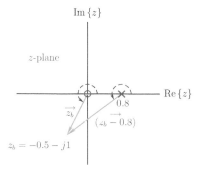

Figure 8.36 – Vectors for graphical representation of $H(z_b)$ in Example 8.37.

With these new vectors it can be shown that

$$\left|\overrightarrow{z_b}\right| = 1.1180 \,, \qquad \measuredangle \overrightarrow{z_b} = 4.2487 \text{ rad} = 243.4^o$$

$$\left|(\overrightarrow{z_b - 0.8})\right| = 1.6401 \,, \qquad \measuredangle(\overrightarrow{z_b - 0.8}) = 3.7973 \text{ rad} = 217.6^o$$

The magnitude and the phase of the system function at $z_a = 1.5 + j1$ are

$$\left|H\left(\overrightarrow{-0.5 - j1}\right)\right| = \frac{1.1180}{1.6401} = 0.6817$$

and

$$\measuredangle H\left(\overrightarrow{-0.5 - j1}\right) = 4.2487 - 3.7973 = 0.4515 \text{ rad} = 25.8^o$$

Vector representation of the factors that make up the system function $H(z)$ is particularly useful for inferring the frequency response of a system from the distribution of its z-domain poles and zeros. It was shown earlier that the DTFT-based frequency response $H(\Omega)$ of a DTLTI system can be obtained from the z-domain system function by evaluating $H(z)$ at each point on the unit circle of the z-plane. Continuing with the first-order system function given by Eqn. (8.250) we obtain

$$H(\Omega) = H(z)\Big|_{z=e^{j\Omega}} = \frac{e^{j\Omega}}{e^{j\Omega} - a} \tag{8.255}$$

In using Eqn. (8.255) we are assuming that $|a| < 1$ so that the ROC for the system function includes the unit circle, and therefore the frequency response $H(\Omega)$ of the system exists. Fig. 8.37 shows the two vectors

$$\overrightarrow{A} = \overrightarrow{e^{j\Omega}} \tag{8.256}$$

and

$$\overrightarrow{B} = \left(\overrightarrow{e^{j\Omega} - a}\right) \tag{8.257}$$

that are needed for evaluating $H(\Omega)$. The vectors originate from the zero and the pole and terminate at the point $z = e^{j\Omega}$ on the unit circle.

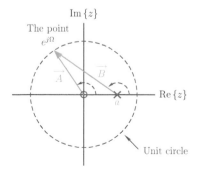

Figure 8.37 – Vectors from the zero and the pole of the system function $H(z) = z/(z-a)$ to a point on the unit circle.

As Ω is varied, the termination point moves on the unit circle, causing the lengths and the angles of the vectors to change, and therefore causing the frequency response $H(\Omega)$ to vary as a function of Ω.

Now consider a more general system function in the form

$$H(z) = K \frac{(z - z_1)(z - z_2) \ldots (z - z_M)}{(z - p_1)(z - p_2) \ldots (z - p_N)} \tag{8.258}$$

The system has M zeros at $z_1, z_2, \ldots, z_M$ and N poles at $p_1, p_2, \ldots, p_N$. Suppose we would like to evaluate the system function at the point $z = z_a$ within the ROC. Vector representation of $H(z_a)$ is

$$H(\overrightarrow{z_a}) = K \frac{(\overrightarrow{z_a - z_1})(\overrightarrow{z_a - z_2}) \ldots (\overrightarrow{z_a - z_M})}{(\overrightarrow{z_a - p_1})(\overrightarrow{z_a - p_2}) \ldots (\overrightarrow{z_a - p_N})} \tag{8.259}$$

The magnitude of the system function is

$$|H(z_a)| = K \frac{|(\overrightarrow{z_a - z_1})||(\overrightarrow{z_a - z_2})| \ldots |(\overrightarrow{z_a - z_M})|}{|(\overrightarrow{z_a - p_1})||(\overrightarrow{z_a - p_2})| \ldots |(\overrightarrow{z_a - p_N})|} \tag{8.260}$$

and its phase is

$$\angle H(z_a) = \angle(\overrightarrow{z_a - z_1}) + \angle(\overrightarrow{z_a - z_2}) + \ldots + \angle(\overrightarrow{z_a - z_M})$$
$$- \angle(\overrightarrow{z_a - p_1}) - \angle(\overrightarrow{z_a - p_2}) - \ldots - \angle(\overrightarrow{z_a - p_N}) \tag{8.261}$$

The vector-based graphical method described above is useful for understanding the correlation between pole-zero placement and system behavior. In order to build intuition in this regard, Fig. 8.38 illustrates several pole-zero layouts along with the magnitude spectrum that corresponds to each layout.

Interactive Demo: `pz_demo2.m`

The pole-zero explorer demo program "`pz_demo2.m`" allows experimentation with the placement of poles and zeros of the system function. Before using it two vectors should be created in MATLAB workspace: one containing the poles of the system and one containing its zeros. In the pole-zero explorer user interface, the "import" button is used for importing these vectors. Pole-zero layout in the z-plane is displayed along with the magnitude and the phase of the system function. The vectors from each zero and each pole to a point on the unit circle of the z-plane may optionally be displayed. Individual poles and zeros may be moved, and the effects on magnitude and phase may be observed. Complex conjugate poles and zeros move together to keep the conjugate relationship.

For example, a system function with zeros at $z_1 = 0$, $z_2 = 1$ and poles at $p_1 = 0.5 + j0.7$, $p_2 = 0.5 - j0.7$ may be studied by creating the following vectors in MATLAB workspace and importing them into the pole-zero explorer program:

$$\texttt{zrs} = [0,\ 1]$$

$$\texttt{pls} - [0.5 + j0.7,\ 0.5 - j0.7]$$

The scenarios illustrated in Fig. 8.38 can be recreated using preset layouts #1 through #5.
Software resources:
`pz_demo2.m`

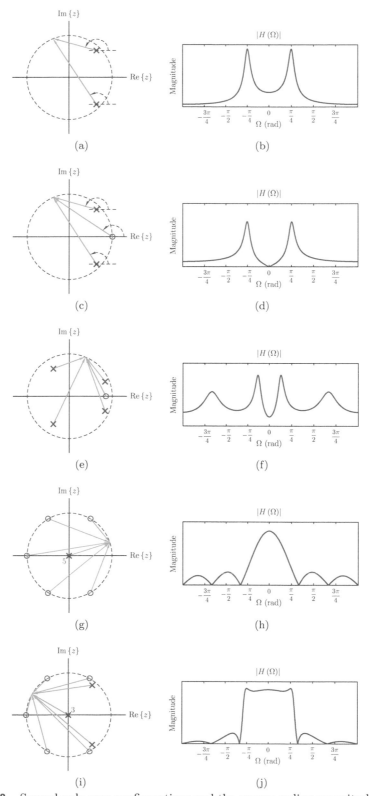

Figure 8.38 – Several pole-zero configurations and the corresponding magnitude responses.

8.5.5   System function and causality

Causality in linear and time-invariant systems was discussed in Section 3.8 of Chapter 3. For a DTLTI system to be causal, its impulse response $h[n]$ needs to be equal to zero for $n < 0$. Thus, by changing the lower limit of the summation index to $n = 0$ in the definition of the z-transform, the z-domain system function for a causal DTLTI system can be written as

$$H\left(z\right) = \sum_{n=-\infty}^{\infty} h[n]\, z^{-n} = \sum_{n=0}^{\infty} h[n]\, z^{-n} \tag{8.262}$$

As discussed in Section 8.2 of this chapter, the ROC for the system function of a causal system is the outside of a circle in the z-plane. Consequently, the system function must also converge at $|z| \to \infty$. Consider a system function in the form

$$H\left(z\right) = \frac{B\left(z\right)}{A\left(z\right)} = \frac{b_M z^M + b_{M-1} z^{M-1} + \ldots + b_1 z + b_0}{a_N z^N + a_{N-1} z^{N-1} + \ldots + a_1 z + a_0}$$

For the system described by $H\left(z\right)$ to be causal we need

$$\lim_{z \to \infty}\left[H\left(z\right)\right] = \lim_{z \to \infty}\left[\frac{b_M}{a_N}\, z^{(M-N)}\right] < \infty \tag{8.263}$$

which requires that $M - N \leq 0$ and consequently $M \leq N$. Thus we arrive at an important conclusion:

Causality condition:
   In the z-domain transfer function of a causal DTLTI system the order of the numerator must not be greater than the order of the denominator.

Note that this condition is necessary for a system to be causal, but it is not sufficient. It is also possible for a non-causal system to have a system function with $M \leq N$.

8.5.6   System function and stability

In Section 3.9 of Chapter 3 we concluded that for a DTLTI system to be stable its impulse response must be absolute summable, that is,

$$\sum_{n=-\infty}^{\infty} |h[n]| < \infty$$

Furthermore, we have established in Section 5.3.3 of Chapter 5 that the DTFT of a signal exists if the signal is absolute summable. But the DTFT of the impulse response is equal to the z-domain transfer function evaluated on the unit circle of the z-plane, that is,

$$H\left(\Omega\right) = H\left(z\right)\Big|_{z=e^{j\Omega}}$$

provided that the unit circle of the z-plane is within the ROC.

Stability condition:
   It follows that, for a DTLTI system to be stable, the ROC of its z-domain system function must include the unit circle.

What are the corresponding conditions that must be imposed on the locations of poles and zeros for stability? We will answer this question by taking three separate cases into account:

1. **Causal system:**
   The ROC for the system function of a causal system is the outside of a circle with radius r_1, and is expressed in the form

   $$|z| > r_1$$

   For the ROC to include the unit circle, we need $r_1 < 1$. Since the ROC can not have any poles in it, all the poles of the system function must be on or inside the circle with radius r_1.

   For a causal system to be stable, the system function must not have any poles on or outside the unit circle of the z-plane.

2. **Anti-causal system:**
   If the system is anti-causal, its impulse response is equal to zero for $n \geq 0$ and the ROC for the system function is the inside of a circle with radius r_2, expressed in the form

   $$|z| < r_2$$

   For the ROC to include the unit circle, we need $r_2 > 1$. Also, all the poles of the system function must reside on or outside the circle with radius r_2 since there can be no poles within the ROC.

   For an anti-causal system to be stable, the system function must not have any poles on or inside the unit circle of the z-plane.

3. **Neither causal nor anti-causal system:**
   In this case the ROC for the system function, if it exists, is the region between two circles with radii r_1 and r_2, and is expressed in the form

   $$r_1 < |z| < r_2$$

   The poles of the system function may be either

   a. On or inside the circle with radius r_1
   b. On or outside the circle with radius r_2

   and the ROC must include the unit circle.

Example 8.38: Impulse response of a stable system

A stable system is characterized by the system function

$$H(z) = \frac{z(z+1)}{(z-0.8)(z+1.2)(z-2)}$$

Determine the impulse response of the system.

Solution: The ROC for the system function is not directly stated; however, we are given enough information to deduce it. The poles of the system are at $p_1 = 0.8$, $p_2 = -1.2$ and

$p_3 = 2$. Since the system is known to be stable, its ROC must include the unit circle. The only possible choice is

$$0.8 < |z| < 1.2$$

Partial fraction expansion of $H(z)$ is

$$H(z) = -\frac{0.75\,z}{z - 0.8} - \frac{0.0312\,z}{z + 1.2} + \frac{0.7813\,z}{z - 2}$$

Based on the ROC determined above, the first partial fraction corresponds to a causal signal, and the other two correspond to anti-causal signals. The impulse response of the system is

$$h[n] = -0.75\,(0.8)^n\,u[n] + 0.0312\,(-1.2)^n\,u[-n-1] - 0.7813\,(2)^n\,u[-n-1]$$

which is shown in Fig. 8.39. We observe that $h[n]$ tends to zero as the sample index is increased in both directions, consistent with the fact that $h[n]$ must be absolute summable.

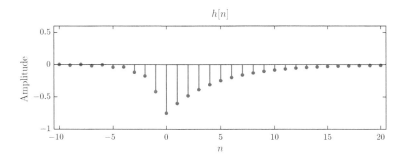

Figure 8.39 – Impulse response $h[n]$ for Example 8.38.

Software resources:
ex_8_38a.m
ex_8_38b.m

Example 8.39: **Stability of a system described by a difference equation**

A DTLTI system is characterized by the difference equation

$$y[n] = x[n-1] + 3x[n-2] + 2x[n-3] + 2.3y[n-1] - 2y[n-2] + 1.2y[n-3]$$

Comment on the stability of this system.

Solution: As in the previous example, the ROC for the system function is not directly stated. On the other hand, the difference equation of the system is written with the current output sample $y[n]$ on the left of the equal sign. All terms on the right side of the equal sign are past samples of the input and the output signals. The form of the difference equation suggests that the system is causal. When we find the system function we must choose the ROC in a way that includes infinitely large values of $|z|$.

Rearranging terms of the difference equation we obtain

$$y[n] - 2.3y[n-1] + 2y[n-2] - 1.2y[n-3] = x[n-1] + 3x[n-2] + 2x[n-3]$$

Taking the z-transform of each side yields

$$\left(1 - 2.3z^{-1} + 2z^{-2} - 1.2z^{-3}\right) Y(z) = \left(z^{-1} + 3z^{-2} + 2z^{-3}\right) X(z)$$

Finally, solving for the ratio of $Y(z)$ and $X(z)$ we find the system function as

$$H(z) = \frac{Y(z)}{X(z)} = \frac{z^{-1} + 3z^{-2} + 2z^{-3}}{1 - 2.3z^{-1} + 2z^{-2} - 1.2z^{-3}}$$

Let us multiply both the numerator and the denominator of $H(z)$ by z^3 so that we express it using powers of z:

$$H(z) = \frac{z^2 + 3z + 2}{z^3 - 2.3z^2 + 2z - 1.2}$$

Numerator and denominator polynomials can be factored to yield

$$H(z) = \frac{(z+1)(z+2)}{(z - 0.4 + j0.8)(z - 0.4 - j0.8)(z - 1.5)}$$

The pole-zero diagram for the system function is shown in Fig. 8.40. There are two zeros, one at $z = -1$ and one at $z = -2$. The system function also has a pole at $z = 1.5$ and a conjugate pair of poles at $z = 0.4 \pm j0.8$.

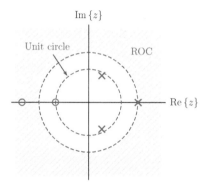

Figure 8.40 – Pole-zero diagram and ROC for the system of Example 8.39.

Since the system is known to be causal, its ROC must be the outside of a circle. The radius of the boundary circle is determined by the outermost pole of the system which happens to be at $z = 1.5$. Consequently, the ROC for the system function is

$$|z| > 1.5$$

Since the ROC does not include the unit circle, the system is unstable. This can be verified by computing the first few samples of the impulse response $h[n]$ from the system function using long division which yields

$$h[n] = \{\underset{\underset{n=1}{\uparrow}}{1}, 5.3, 12.19, 18.64, 24.85, 34.5, 52.02, 80.46, 122.42, 183.07, \ldots\}$$

The impulse response keeps growing with increasing values of the index n.
Software resources:
ex_8_39a.m
ex_8_39b.m
ex_8_39c.m

8.5.7 Allpass systems

A system the magnitude characteristic of which is constant across all frequencies is called an *allpass system*. For a system to be considered an allpass system we need

$$|H(\Omega)| = C \text{ (constant)} \qquad -\pi \le \Omega \le \pi \tag{8.264}$$

Consider a first-order DTLTI system with a complex zero at

$$z_1 = a = r\, e^{j\Omega_0} \tag{8.265}$$

and a complex pole at

$$p_1 = \frac{1}{a^*} = \frac{1}{r\, e^{-j\Omega_0}} = \frac{1}{r} e^{j\Omega_0} \tag{8.266}$$

The resulting system function is

$$H(z) = \frac{z - z_1}{z - p_1} = \frac{z - r\, e^{j\Omega_0}}{z - (1/r)\, e^{j\Omega_0}} \tag{8.267}$$

The choice made for the zero and the pole ensures that the two lie on a line that goes through the origin. Their distances from the origin are r and $1/r$ respectively. Since it is desirable to have a system that is both causal and stable, we need to choose $r > 1$ so that the resulting pole is inside the unit circle. Naturally this choice causes the zero to be outside the unit circle. This is illustrated in Fig. 8.41.

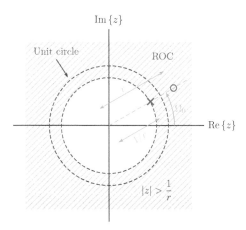

Figure 8.41 – The zero and the pole of a first-order allpass system.

A pole and a zero that have the relationship shown in Fig. 8.41 are said to be *mirror images* of each other across the unit circle. To find $H(\Omega)$ we will evaluate the z-domain system function on the unit circle of the z-plane:

$$H(\Omega) = \frac{e^{j\Omega} - r\, e^{j\Omega_0}}{e^{j\Omega} - (1/r)\, e^{j\Omega_0}} \tag{8.268}$$

For convenience we will determine the squared magnitude of $H\left(\Omega\right)$:

$$
\begin{aligned}
\left|H\left(\Omega\right)\right|^{2} &= H\left(\Omega\right) H^{*}\left(\Omega\right) \\
&= \left(\frac{e^{j\Omega} - r\,e^{j\Omega_{0}}}{e^{j\Omega} - \left(1/r\right)e^{j\Omega_{0}}}\right) \left(\frac{e^{-j\Omega} - r\,e^{-j\Omega_{0}}}{e^{-j\Omega} - \left(1/r\right)e^{-j\Omega_{0}}}\right) \\
&= \frac{1 + r^{2} - 2r\,\cos\left(\Omega - \Omega_{0}\right)}{1 + \left(1/r\right)^{2} - 2\left(1/r\right)\cos\left(\Omega - \Omega_{0}\right)} \\
&= r^{2}\quad\text{(constant)}
\end{aligned}
\tag{8.269}
$$

The magnitude of the system function is $\left|H\left(\Omega\right)\right| = r$. It can be shown (see Problem 8.40 at the end of this chapter) that the phase of the system function is

$$
\angle H\left(\Omega\right) = \tan^{-1}\left[\frac{\left(r^{2} - 1\right)\sin\left(\Omega - \Omega_{0}\right)}{2r - \left(r^{2} + 1\right)\cos\left(\Omega - \Omega_{0}\right)}\right]
\tag{8.270}
$$

The phase characteristic $\angle H\left(\Omega\right)$ is shown in Fig. 8.42 for the case of a pole and a zero both on the real axis, that is, for $\Omega_{0} = 0$. The phase is shown for $r = 2.5$, $r = 1.75$ and $r = 1.2$.

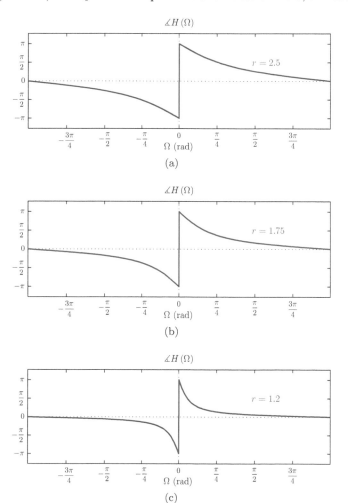

Figure 8.42 – The phase characteristic of the first-order allpass system for $\Omega_{0} = 0$ and (a) $r = 2.5$, (b) $r = 1.75$, (c) $r = 1.2$.

Since the shape of the phase response can be controlled by the choice of r while keeping the magnitude response constant, an allpass system is also referred to as a *phase-shifter*. Increased versatility in controlling the phase response can be obtained by choosing the pole and the zero to be complex-valued. This requires $\Omega_0 \neq 0$. Consider a first-order system with a pole at $p_1 = (1/r)\, e^{j\Omega_0}$ and a zero at $z_1 = r\, e^{j\Omega_0}$. Again we choose $r > 1$ to obtain a system that is both causal and stable. Fig. 8.43 illustrates the phase response $\angle H(\Omega)$ is shown for $\Omega_0 = \pi/6$ and $\Omega_0 = \pi/3$. The parameter r is fixed at $r = 1.2$ in both cases.

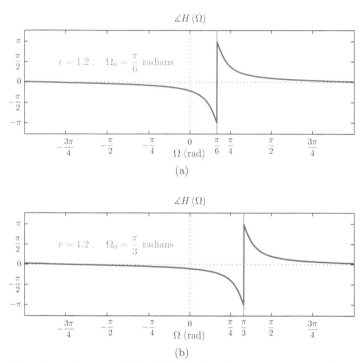

Figure 8.43 – The phase characteristic of the first-order allpass system for $r = 1.2$ and (a) $\Omega_0 = \pi/6$, (b) $\Omega_0 = \pi/3$.

If a system function with real-valued coefficients is desired, two first-order allpass sections with complex conjugate zeros and poles can be combined into a second-order allpass system. The resulting system function is

$$H(z) = \left(\frac{z - r\, e^{j\Omega_0}}{z - (1/r)\, e^{j\Omega_0}} \right) \left(\frac{z - r\, e^{-j\Omega_0}}{z - (1/r)\, e^{-j\Omega_0}} \right) \tag{8.271}$$

$$= \frac{z^2 - 2r\, \cos(\Omega_0)\, z + r^2}{z^2 - (2/r)\, \cos(\Omega_0)\, z + (1/r)^2} \tag{8.272}$$

Pole-zero diagram for a second-order allpass system with parameters r and Ω_0 is shown in Fig. 8.44.

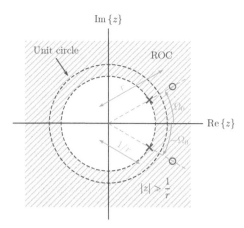

Figure 8.44 – Pole-zero diagram for second-order allpass system.

8.5.8   Inverse systems

The inverse of the system is another system which, when connected in cascade with the original system, forms an identity system. This relationship is depicted in Fig. 8.45.

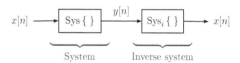

Figure 8.45 – A system and its inverse connected in cascade.

The output signal of the original system is

$$y[n] = \text{Sys}\,\{x[n]\} \tag{8.273}$$

We require the inverse system to recover the original input signal $x[n]$ from the output signal $y[n]$, therefore

$$x[n] = \text{Sys}_i\,\{y[n]\} \tag{8.274}$$

Combining Eqns. (8.273) and (8.273) yields

$$\text{Sys}_i\,\Big\{\,\text{Sys}\,\{\,x[n]\,\}\,\Big\} = x[n] \tag{8.275}$$

Let the original system under consideration be a DTLTI system with impulse response $h[n]$, and let the inverse system be also a DTLTI system with impulse response $h_i[n]$ as shown in Fig. 8.46. For the output signal of the inverse system to be identical to the input signal of the original system we need the impulse response of the cascade combination to be equal to $\delta[n]$, that is,

$$h_{eq}[n] = h[n] * h_i[n] = \delta[n] \tag{8.276}$$

or, using the convolution summation

$$h_{eq}[n] = \sum_k h[k]\,h_i[n-k] = \delta[n] \tag{8.277}$$

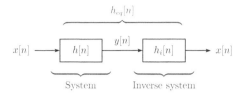

Figure 8.46 – A DTLTI system and its inverse connected in cascade.

Corresponding relationship between the system functions is found by taking the z-transform of Eqn. (8.276):

$$H_{eq}(z) = H(z)\,H_i(z) = 1 \tag{8.278}$$

Consequently, the system function of the inverse system is

$$H_i(z) = \frac{1}{H(z)} \tag{8.279}$$

Two important characteristics of the inverse system are causality and stability. We will first focus on causality. Consider again a rational system function in the standard form

$$H(z) = \frac{B(z)}{A(z)} = \frac{b_M z^M + b_{M-1} z^{M-1} + \ldots + b_1 z + b_0}{a_N z^N + a_{N-1} z^{N-1} + \ldots + a_1 z + a_0} \tag{8.280}$$

The system function for the inverse system is

$$H_i(z) = \frac{A(z)}{B(z)} = \frac{a_N z^N + a_{N-1} z^{N-1} + \ldots + a_1 z + a_0}{b_M z^M + b_{M-1} z^{M-1} + \ldots + b_1 z + b_0}$$

If the original system with system function $H(z)$ is causal then $M \le N$ as we have established in Section 8.5.5. By the same token, causality of the inverse system with system function $H_i(z)$ requires $N \le M$. Hence we need $N = M$ if both the original system and its inverse are required to be causal.

To analyze the stability of the inverse system we will find it more convenient to write the system function $H(z)$ in pole zero form. Using $M = N$ we have

$$H(z) = \frac{b_N (z - z_1)(z - z_2)\ldots(z - z_N)}{a_N (z - p_1)(z - p_2)\ldots(z - p_N)} \tag{8.281}$$

If the original system is both causal and stable, all its poles must be inside the unit circle of the z-plane (see Section 8.5.6), therefore

$$|p_k| < 1\,, \qquad k = 1, \ldots, N \tag{8.282}$$

The system function of the inverse system, written in pole-zero form, is

$$H_i(z) = \frac{a_N (z - p_1)(z - p_2)\ldots(z - p_N)}{b_N (z - z_1)(z - z_2)\ldots(z - z_N)} \tag{8.283}$$

For the inverse system to be stable, its poles must also lie inside the unit circle. The poles of the inverse system are the zeros of the original system. Therefore, for the inverse system to be stable, both zeros and poles of the original system must be inside the unit circle. In addition to Eqn. (8.282) we also need

$$|z_k| < 1\,, \qquad k = 1, \ldots, N \tag{8.284}$$

A causal DTLTI system that has all of its zeros and poles inside the unit circle is referred to as a *minimum-phase system*. A minimum-phase system and its inverse are both causal and stable.

Example 8.40: **Inverse of a system described by a difference equation**

A DTLTI system is described by a difference equation

$$y[n] = 0.1\,y[n-1] + 0.72\,y[n-2] + x[n] + 0.5\,x[n-1]$$

Determine if a causal and stable inverse can be found for this system. If yes, find a difference equation for the inverse system.

Solution: The system function is found by taking the z-transform of the difference equation and solving for the ratio of the z-transforms of the output signal and the input signal:

$$Y(z)\left[1 - 0.1\,z^{-1} - 0.72\,z^{-2}\right] = X(z)\left[1 + 0.5\,z^{-1}\right]$$

$$H(z) = \frac{1 + 0.5\,z^{-1}}{1 - 0.1\,z^{-1} - 0.72\,z^{-2}} = \frac{z\,(z+0.5)}{(z+0.8)\,(z-0.9)}$$

The zeros of the system function are at $z_1 = 0$ and $z_2 = -0.5$. Its two poles are at $p_1 = -0.8$ and $p_2 = 0.9$. Since all poles and zeros are inside the unit circle of the z-plane, $H(z)$ is a minimum phase system. The system function for the inverse system is

$$H_i(z) = \frac{(z+0.8)\,(z-0.9)}{z\,(z+0.5)} = \frac{1 - 0.1\,z^{-1} - 0.72\,z^{-2}}{1 + 0.5\,z^{-1}}$$

which is clearly causal and stable. Its corresponding difference equation is

$$y[n] = -0.5\,y[n-1] + x[n] - 0.1\,x[n-1] - 0.72\,x[n-2]$$

8.6 Implementation Structures for DTLTI Systems

Block diagram implementation of discrete-time systems was briefly discussed in Section 3.6 of Chapter 3. A method was presented for obtaining a block diagram from a linear constant-coefficient difference equation. Three types of elements were utilized; namely signal adder, constant-gain multiplier and one-sample delay. In this section we will build on the techniques presented in Section 3.6, this time utilizing the z-domain system function as the starting point.

8.6.1 Direct-form implementations

The general form of the z-domain system function for a DTLTI system is

$$H(z) = \frac{Y(z)}{X(z)} = \frac{b_0 + b_1 z^{-1} + b_2 z^{-2} + \ldots + \ldots b_M z^{-M}}{1 + a_1 z^{-1} + a_2 z^{-2} + \ldots + a_N z^{-N}} \tag{8.285}$$

For the sake of convenience we chose to write $H(z)$ in terms of negative powers of z. $X(z)$ and $Y(z)$ are the z-transforms of the input and the output signals respectively.

The method of obtaining a block diagram from a z-domain system function will be derived using a third-order system, but its generalization to higher-order system functions is quite straightforward. Consider a DTLTI system for which $M = 3$ and $N = 3$. The system function $H(z)$ is

$$H(z) = \frac{Y(z)}{X(z)} = \frac{b_0 + b_1 z^{-1} + b_2 z^{-2} + b_3 z^{-3}}{1 + a_1 z^{-1} + a_2 z^{-2} + a_3 z^{-3}} \qquad (8.286)$$

Let us use an intermediate transform $V(z)$ that corresponds to an intermediate signal $v[n]$, and express the system function as

$$H(z) = H_1(z)\, H_2(z) = \frac{Y(z)}{V(z)} \frac{V(z)}{X(z)} \qquad (8.287)$$

We will elect to associate $V(z)/X(z)$ with the numerator polynomial, that is,

$$H_1(z) = \frac{V(z)}{X(s)} = b_0 + b_1 z^{-1} + b_2 z^{-2} + b_3 z^{-3} \qquad (8.288)$$

To satisfy Eqn. (8.287), the ratio $Y(z)/V(z)$ must be associated with the denominator polynomial.

$$H_2(s) = \frac{Y(z)}{V(z)} = \frac{1}{1 + a_1 z^{-1} + a_2 z^{-2} + a_3 z^{-3}} \qquad (8.289)$$

The relationships described by Eqns. (8.286) through (8.289) are illustrated in Fig. 8.47.

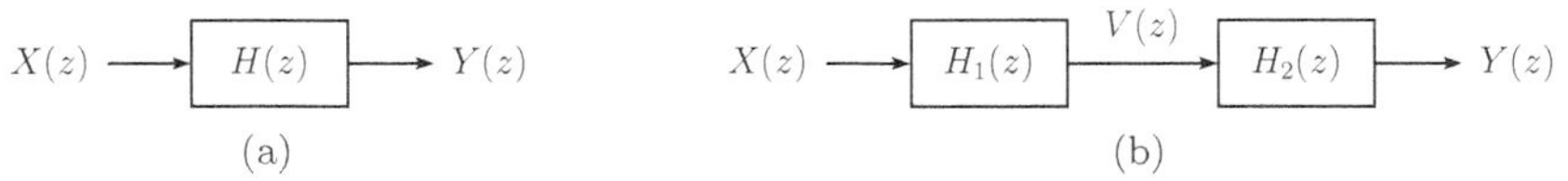

Figure 8.47 – (a) DTLTI system with system function $H(z)$, (b) cascade form using an intermediate function $V(z)$.

Solving Eqn. (8.288) for $V(z)$ results in

$$V(z) = b_0 X(z) + b_1 z^{-1} X(z) + b_2 z^{-2} X(z) + b_3 z^{-3} X(z) \qquad (8.290)$$

Taking the inverse z-transform of each side of Eqn. (8.290) and remembering that multiplication of the transform by z^{-k} corresponds to a delay of k samples in the time domain, we get

$$v[n] = b_0\, x[n] + b_1\, x[n-1] + b_2\, x[n-2] + b_3\, x[n-3] \qquad (8.291)$$

The relationship between $X(z)$ and $V(z)$ can be realized using the block diagram shown in Fig. 8.48(a). The second part of the system will be implemented by expressing the output transform $Y(z)$ in terms of the intermediate transform $V(z)$. From Eqn. (8.289) we write

$$Y(z) = V(z) - a_1 z^{-1} Y(z) - a_2 z^{-2} Y(z) - a_3 z^{-3} Y(z) \qquad (8.292)$$

The corresponding time-domain relationship is

$$y[n] = v[n] - a_1\, y[n-1] - a_2\, y[n-2] - a_3\, y[n-3] \qquad (8.293)$$

The relationship between $V(z)$ and $Y(z)$ can be realized using the block diagram shown in Fig. 8.48(b).

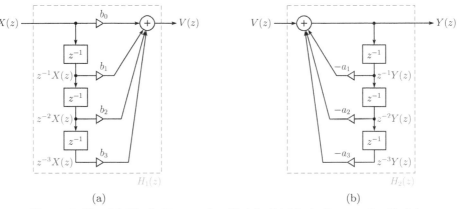

(a)   (b)

Figure 8.48 – (a) Block diagram for $H_1(z)$, (b) block diagram for $H_2(z)$.

The two block diagrams obtained for $H_1(z)$ and $H_2(z)$ can be connected in cascade as shown in Fig. 8.49 to implement the system function $H(z)$. This is the *direct-form I* realization of the system.

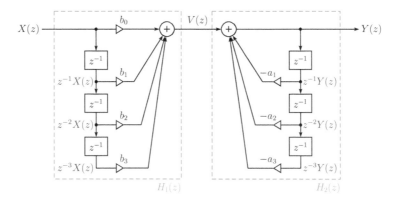

Figure 8.49 – Direct-form I realization of $H(z)$.

The block diagram in Fig. 8.49 represents the functional relationship between $X(z)$ and $Y(z)$, the z-transforms of the input and the output signals. If a diagram using time-domain quantities is desired, the conversion is easy using the following steps:

1. Replace each transform $X(z)$, $Y(z)$ and $V(z)$ with its time-domain version $x[n]$, $y[n]$ and $v[n]$.
2. Replace each block with the factor z^{-1} with one-sample time delay.

The resulting diagram is shown in Fig. 8.50.

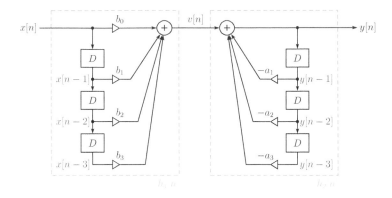

Figure 8.50 – Direct-form I realization of $H(z)$ using time-domain quantities.

Example 8.41: **Direct-form type-I implementation of a system**

Consider a causal DTLTI system described by the z-domain system function

$$H(z) = \frac{z^3 - 7z + 6}{z^4 - z^3 - 0.34\, z^2 + 0.966\, z - 0.2403}$$

Draw a direct-form I block diagram for implementing this system.

Solution: Let us multiply both the numerator and the denominator by z^{-4} to write the system function in the standard form of Eqn. (8.285):

$$H(z) = \frac{z^{-1} - 7\, z^{-3} + 6\, z^{-4}}{1 - z^{-1} - 0.34\, z^{-2} + 0.966\, z^{-3} - 0.2403\, z^{-4}}$$

We have

$$M = 4\,, \quad N = 4$$

$$b_0 = 0\,, \quad b_1 = 1\,, \quad b_2 = 0\,, \quad b_3 = -7\,, \quad b_4 = 6$$

$$a_0 = 1\,, \quad a_1 = -1;\quad a_2 = -0.34\,, \quad a_3 = 0.966\,, \quad a_4 = -0.2403$$

The block diagram is shown in Fig. 8.51.

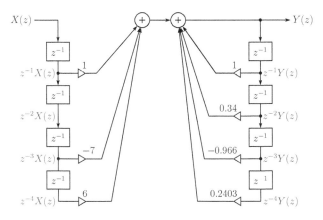

Figure 8.51 – Direct-form I block diagram for Example 8.41.

Let us consider the direct-form I diagram in Fig. 8.49 again. The subsystem with system function $H_1(z)$ takes $X(z)$ as input, and produces $V(z)$. The subsystem with system function $H_2(z)$ takes $V(z)$, the output of the first subsystem, and produces $Y(z)$ in response. Since each subsystem is linear,[2] it does not matter which one comes first in a cascade connection. An alternative way of connecting the two subsystems is shown in Fig. 8.52, and it leads to the block diagram shown in Fig. 8.53.

Figure 8.52 – Alternative placement of the two subsystems.

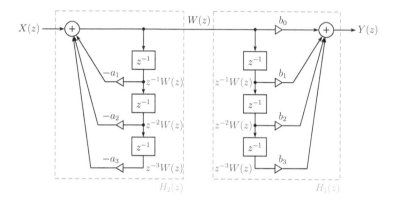

Figure 8.53 – Block diagram obtained by swapping the order of two subsystems.

The middle part of the diagram in Fig. 8.53 has two delay lines running parallel to each other, each set containing four nodes. These two sets of nodes hold identical values $W(z)$, $z^{-1}W(z)$, $z^{-2}W(z)$ and $z^{-3}W(z)$. Consequently, the two sets of nodes can be merged together to result in the diagram shown in Fig. 8.54. This is the *direct-form II* realization of the system.

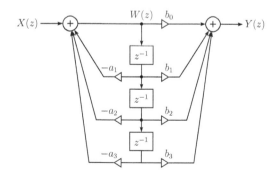

Figure 8.54 – Direct-form II realization of $H(z)$.

The block diagram in Fig. 8.54 could easily have been obtained directly from the system function in Eqn. (8.286) by inspection, using the following set of rules:

---

[2]   Remember that, for linearity, each subsystem must be initially relaxed. All initial values must be zero. In Fig. 8.50, for example, we need $x[-1] = x[-2] = x[-3] = y[-1] = y[-2] = y[-3] = 0$.

1. Begin by ordering terms of numerator and denominator polynomials in ascending powers of z^{-1}.
2. Ensure that the leading coefficient in the denominator, that is, the constant term, is equal to unity. (If this is not the case, simply scale all coefficients to satisfy this rule.)
3. Set gain factors of feed-forward branches equal to the numerator coefficients.
4. Set gain factors of feedback branches equal to the negatives of the denominator coefficients.
5. Be careful to account for any missing powers of z^{-1} in either polynomial, and treat them as terms with their coefficients equal to zero.

Example 8.42: **Direct-form II implementation of a system**

Draw a direct-form II block diagram for implementing the system used in Example 8.41.

Solution: The diagram obtained by interchanging the order of the two subsystems in Fig. 8.51 and eliminating the nodes that become redundant is shown in Fig. 8.55.

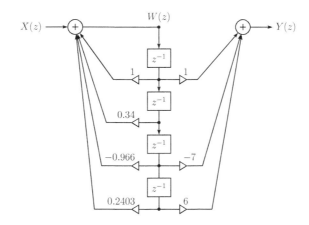

Figure 8.55 – Direct-form II block diagram for Example 8.42.

8.6.2 Cascade and parallel forms

Instead of simulating a system with the direct-form block diagram discussed in the previous section, it is also possible to express the system function as either the product or the sum of lower order sections, and base the block diagram on cascade of parallel combination of smaller diagrams. Consider a system function of order M that can be expressed in the form

$$
\begin{aligned}
H\left(z\right) =& H_1\left(z\right) H_2\left(z\right) \ldots H_M\left(z\right) \\
=& \frac{W_1\left(z\right)}{X\left(z\right)} \frac{W_2\left(z\right)}{W_1\left(z\right)} \cdots \frac{Y\left(z\right)}{W_{M-1}\left(z\right)}
\end{aligned}
\tag{8.294}
$$

One method of simulating this system would be to build a diagram for each of the subsections $H_i(s)$ using the direct-form approach discussed previously, and then to connect those sections in cascade as shown in Fig. 8.56.

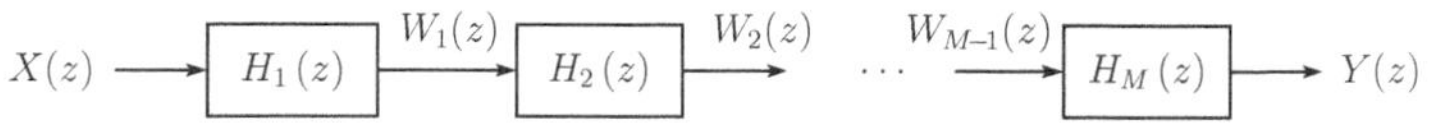

Figure 8.56 – Cascade implementation of $H(z)$.

An easy method of sectioning a system function in the style of Eqn. (8.294) is to first find the poles and the zeros of the system function, and then use them for factoring numerator and denominator polynomials. Afterwards, each section may be constructed by using one of the poles. Each zero is incorporated into one of the sections, and some sections may have constant numerators. If some of the poles and zeros are complex-valued, we may choose to keep conjugate pairs together in second-order sections to avoid the need for complex gain factors in the diagram. The next example will illustrate this process.

Example 8.43: **Cascade form block diagram**

Develop a cascade form block diagram for the DTLTI system used in Example 8.41.

Solution: The system function specified in Example 8.41 has zeros at $z_1 = -3$, $z_2 = 2$, $z_3 = 1$, and poles at $p_1 = -0.9$, $p_2 = 0.3$, $p_{3,4} = 0.8 \pm j0.5$ (see MATLAB Exercise 8.9). It can be written in factored form as

$$H(z) = \frac{(z+3)(z-1)(z-2)}{(z+0.9)(z-0.3)(z-0.8-j0.5)(z-0.8+j0.5)}$$

The system function can be broken down into cascade sections in a number of ways. Let us choose to have two second-order sections by defining (see MATLAB Exercise 8.9)

$$H_1(z) = \frac{(z+3)(z-1)}{(z+0.9)(z-0.3)} = \frac{z^2+2z-3}{z^2+0.6z-0.27}$$

and

$$H_2(z) = \frac{z-2}{(z-0.8-j0.5)(z-0.8+j0.5)} = \frac{z-2}{z^2-1.6z+0.89}$$

so that

$$H(z) = H_1(z)\,H_2(z)$$

A cascade form block diagram is obtained by using direct-form II for each of $H_1(z)$ and $H_2(z)$, and connecting the resulting diagrams as shown in Fig. 8.57.

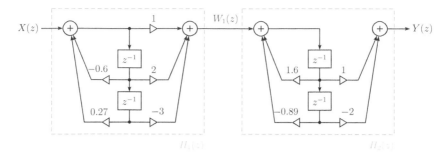

Figure 8.57 – Cascade form block diagram for Example 8.43.

It is also possible to consolidate the neighboring adders in the middle although this would cause the intermediate signal $W_1(z)$ to be lost. The resulting diagram is shown in Fig. 8.58.

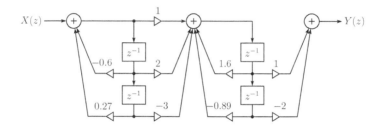

Figure 8.58 – Further simplified cascade form block diagram for Example 8.43.

Software resources: See MATLAB Exercises 8.9 and 8.10.

An alternative to the cascade form simulation diagram is a parallel form diagram which is based on writing the system function as a sum of lower order functions:

$$H(s) = \bar{H}_1(z) + \bar{H}_2(z) + \ldots + \bar{H}_M(z)$$

$$-\frac{\bar{W}_1(z)}{X(z)} + \frac{\bar{W}_2(z)}{X(z)} + \ldots \frac{\bar{W}_M(z)}{X(z)} \qquad (8.295)$$

A simulation diagram may be constructed by implementing each term in Eqn. (8.295) using the direct-form approach, and then connecting the resulting subsystems in a parallel configuration as shown in Fig. 8.59.

A rational system function $H(z)$ may be sectioned in the form of Eqn. (8.295) using partial fraction expansion. If some of the poles and zeros are complex-valued, we may choose to keep conjugate pairs together in second-order sections to avoid the need for complex gain factors in the diagram. This process will be illustrated in the next example.

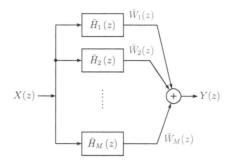

Figure 8.59 – Parallel implementation of $H(z)$.

Example 8.44: **Parallel form block diagram**

Develop a cascade form block diagram for the DTLTI system used in Example 8.41.

Solution: The system function has poles at $p_1 = -0.9$, $p_2 = 0.3$ and $p_{3,4} = 0.8 \pm j0.5$. Let us expand it into partial fractions in the form

$$H(z) = \frac{k_1}{z + 0.9} + \frac{k_2}{z - 0.3} + \frac{k_3}{z - 0.8 - j0.5} + \frac{k_4}{z - 0.8 + j0.5}$$

Note that it is not necessary to divide $H(z)$ by z before expanding it into partial fractions since we are not trying to compute the inverse transform. The residues of the partial fraction expansion are (see MATLAB Exercise 8.11)

$$k_1 = -3.0709\,, \qquad\qquad k_2 = 6.5450$$
$$k_3 = -1.2371 + j1.7480 \qquad k_4 = -1.2371 - j1.7480$$

and $H(s)$ is

$$H(z) = \frac{-3.0709}{z + 0.9} + \frac{6.5450}{z - 0.3} + \frac{-1.2371 + j1.7480}{z - 0.8 - j0.5} + \frac{-1.2371 - j1.7480}{z - 0.8 + j0.5}$$

Let us combine the first two terms to create a second-order section:

$$\bar{H}_1(z) = \frac{-3.0709}{z + 0.9} + \frac{6.5450}{z - 0.3} = \frac{3.474\,z + 6.812}{z^2 + 0.6\,z - 0.27}$$

The remaining two terms are combined into another second-order section, yielding

$$\bar{H}_2(z) = \frac{-1.2371 + j1.7480}{z - 0.8 - j0.5} + \frac{-1.2371 - j1.7480}{z - 0.8 + j0.5} = \frac{-2.474\,z + 0.2314}{z^2 - 1.6\,z + 0.89}$$

so that

$$H(z) = \bar{H}_1(z) + \bar{H}_2(z)$$

The parallel form simulation diagram is shown in Fig. 8.60.

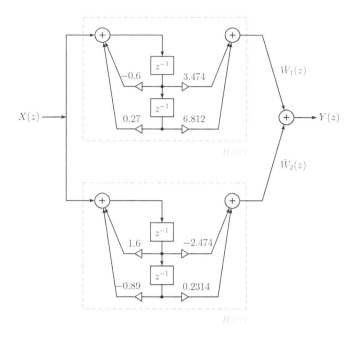

Figure 8.60 – Parallel form block diagram for Example 8.44.

Software resources: See MATLAB Exercise 8.11.

8.7   Unilateral z-Transform

It was mentioned in earlier discussion that the z-transform as defined by Eqn. (8.1) is often referred to as the bilateral z-transform.   An alternative version, known as the unilateral z-transform, is defined by

$$X_u\left(z\right)=\mathcal{Z}_u\left\{x[n]\right\}=\sum_{n=0}^{\infty}x[n]\,z^{-n} \tag{8.296}$$

We use the subscript "u" to distinguish the unilateral z-transform from its bilateral counterpart. It is clear from a comparison of Eqn. (8.296) with Eqn. (8.1) that the only difference between the two definitions is the lower index of the summation. In Eqn. (8.296) we start the summation at $n=0$ instead of $n=\infty$. If $x[n]$ is a causal signal, that is, if $x[n]=0$ for $n<0$, then the unilateral transform $X_u\left(z\right)$ becomes identical to the bilateral transform $X\left(z\right)$. On the other hand, for a signal that is $x[n]$ is not necessarily causal, the unilateral transform is essentially the bilateral transform of $x[n]$ multiplied by a unit-step function, that is,

$$\mathcal{Z}_u\left\{x[n]\right\}=\mathcal{Z}\left\{x[n]\,u[n]\right\}$$

$$=\sum_{n=-\infty}^{\infty}x[n]\,u[n]\,z^{-n}$$

Because of the way $X_u(z)$ is defined, its region of convergence is always the outside of a circle, and does not have to be explicitly stated.

One property of the unilateral z-transform that differs from its counterpart for the bilateral z-transform is the time-shifting property. Consider the time-shifted signal $x[n-1]$. In terms of the bilateral z-transform, we have proved the following relationship in earlier discussion:

$$x[n-1] \overset{\mathcal{Z}}{\longleftrightarrow} z^{-1} X(z) \tag{8.297}$$

The relationship in Eqn. (8.298) does not extend to the unilateral z-transform, that is, we cannot claim that $\mathcal{Z}_u\{x[n-1]\} = z^{-1}\,\mathcal{Z}_u\{x[n]\}$. The reason for this fundamental difference becomes obvious if we write the unilateral transform in open form:

$$\mathcal{Z}_u\{x[n]\} = \sum_{n=0}^{\infty} x[n]\,z^{-n}$$

$$= x[0] + x[1]\,z^{-1} + x[2]\,z^{-2} + x[3]\,z^{-3} + \ldots \tag{8.298}$$

The transform of the time-shifted signal is

$$\mathcal{Z}_u\{x[n-1]\} = \sum_{n=0}^{\infty} x[n-1]\,z^{-n}$$

$$= x[-1] + x[0]\,z^{-1} + x[1]\,z^{-2} + x[3]\,z^{-3} + \ldots \tag{8.299}$$

The sample $x[-1]$ is not part of $\mathcal{Z}_u\{x[n]\}$ but appears in $\mathcal{Z}_u\{x[n-1]\}$. Let us rewrite the latter by separating $x[-1]$ from the rest of the terms:

$$\mathcal{Z}_u\{x[n-1]\} = x[-1] + \sum_{n=1}^{\infty} x[n-1]\,z^{-n}$$

$$= x[-1] + z^{-1} \sum_{n=0}^{\infty} x[n]\,z^{-n}$$

$$= x[-1] + z^{-1} \left[x[0] + x[1]\,z^{-1} + x[2]\,z^{-2} + x[3]\,z^{-3} + \ldots \right] \tag{8.300}$$

Consequently, the unilateral z-transform of $x[n-1]$ can be expressed in terms of the transform of $x[n]$ as

$$\mathcal{Z}_u\{x[n-1]\} = x[-1] + z^{-1} X_u(z) \tag{8.301}$$

Through similar reasoning it can be shown that, for the signal $x[n-2]$, we have

$$\mathcal{Z}_u\{x[n-2]\} = x[-2] + x[-1]\,z^{-1} + z^{-2} X_u(z) \tag{8.302}$$

and in the general case of a signal time-shifted by $k > 0$ samples

$$\mathcal{Z}_u\{x[n-k]\} = x[-k] + x[-k+1]\,z^{-1} + \ldots + x[-1]\,z^{-k+1} + z^{-k} X_u(z)$$

$$= \sum_{n=-k}^{-1} x[n]\,z^{-n-k} + z^{-k} X_u(z) \tag{8.303}$$

If the signal is advanced rather than delayed in time, it can be shown that (see Problem 8.44 at the end of this chapter)

$$\mathcal{Z}_u\{x[n+k]\} = z^k X_u(z) - \sum_{n=0}^{k-1} x[n]\,z^{k-n} \tag{8.304}$$

for $k > 0$. The unilateral z-transform is useful in the use of z-transform techniques for solving difference equations with specified initial conditions. The next couple of examples will illustrate this.

Example 8.45: **Finding the natural response of a system through z-transform**

The homogeneous difference equation for a system is

$$y[n] - \frac{5}{6}\, y[n-1] + \frac{1}{6}\, y[n-2] = 0$$

Using z-transform techniques, determine the natural response of the system for the initial conditions

$$y[-1] = 19\,, \quad y[-2] = 53$$

Recall that this problem was solved in Example 3.12 of Chapter 3 using time-domain solution techniques.

Solution: The first step will be to compute the unilateral z-transform of both sides of the homogeneous difference equation:

$$\mathcal{Z}_u \left\{ y[n] - \frac{5}{6}\, y[n-1] + \frac{1}{6}\, y[n-2] \right\} = 0 \tag{8.305}$$

Using the linearity of the z-transform we can write

$$\mathcal{Z}_u \left\{ y[n] \right\} - \frac{5}{6}\, \mathcal{Z}_u \left\{ y[n-1] \right\} + \frac{1}{6}\, \mathcal{Z}_u \left\{ y[n-2] \right\} = 0 \tag{8.306}$$

The unilateral z-transforms of the time-shifted versions of the output signal are

$$\mathcal{Z}_u \left\{ y[n-1] \right\} = y[-1] + z^{-1}\, Y_u\,(z)$$
$$= 19 + z^{-1}\, Y_u\,(z) \tag{8.307}$$

and

$$\mathcal{Z}_u \left\{ y[n-2] \right\} = y[-2] + y[-1]\, z^{-1} + z^{-2}\, Y_u\,(z)$$
$$= 53 + 19\, z^{-1} + z^{-2}\, Y_u\,(z) \tag{8.308}$$

Substituting Eqns. (8.307) and (8.308) into Eqn. (8.306) we get

$$Y_u\,(z) - \frac{5}{6}\, \left[19 + z^{-1}\, Y_u\,(z) \right] + \frac{1}{6}\, \left[53 + 19\, z^{-1} + z^{-2}\, Y_u\,(z) \right] = 0$$

which can be solved for $Y_u\,(z)$ to yield

$$Y_u\,(z) = \frac{z\left(7z - \frac{19}{6}\right)}{z^2 - \frac{5}{6}z + \frac{1}{6}} = \frac{z\left(7z - \frac{19}{6}\right)}{\left(z - \frac{1}{2}\right)\left(z - \frac{1}{3}\right)}$$

Now we need to find $y[n]$ through inverse z-transform using partial fraction expansion. We will write $Y_u\,(z)$ as

$$Y_u\,(z) = \frac{k_1 z}{z - \frac{1}{2}} + \frac{k_2 z}{z - \frac{1}{3}}$$

The residues of the partial fraction expansion are

$$k_1 = \left(z - \frac{1}{2}\right) \left. \frac{Y_u\,(z)}{z} \right|_{z=\frac{1}{2}} = \left. \frac{\left(7z - \frac{19}{6}\right)}{\left(z - \frac{1}{3}\right)} \right|_{z=\frac{1}{2}} = \frac{7\left(\frac{1}{2}\right) - \frac{19}{6}}{\left(\frac{1}{2} - \frac{1}{3}\right)} = 2$$

and

$$k_2 = \left(z - \tfrac{1}{3}\right) \left. \frac{Y_u(z)}{z}\right|_{z=\frac{1}{3}} = \left.\frac{\left(7z - \frac{19}{6}\right)}{\left(z - \frac{1}{2}\right)}\right|_{z=\frac{1}{3}} = \frac{7\left(\frac{1}{3}\right) - \frac{19}{6}}{\left(\frac{1}{3} - \frac{1}{2}\right)} = 5$$

Thus, the partial fraction expansion for $Y_u(z)$ is

$$Y_u(z) = \frac{2z}{z - \frac{1}{2}} + \frac{5z}{z - \frac{1}{3}}$$

resulting in the homogeneous solution

$$y_h[n] = \mathcal{Z}_u^{-1}\{Y_u(z)\} = 2\left(\frac{1}{2}\right)^n u[n] + 5\left(\frac{1}{3}\right)^n u[n]$$

This result is in perfect agreement with that found in Example 3.12.

Software resources: See MATLAB Exercise 8.13.

Example 8.46: **Finding the forced response of a system through z-transform**

Consider a system defined by means of the difference equation

$$y[n] = 0.9\, y[n-1] + 0.1\, x[n]$$

Determine the response of this system for the input signal

$$x[n] = 20\,\cos(0.2\pi n)$$

if the initial value of the output is $y[-1] = 2.5$. Recall that this problem was solved in Example 3.15 of Chapter 3 using time-domain solution techniques.

Solution: Let us begin by computing the unilateral z-transform of both sides of the difference equation.

$$\mathcal{Z}_u\{y[n]\} = \mathcal{Z}_u\{0.9\, y[n-1]\} + \mathcal{Z}_u\{0.1\, x[n]\}$$
$$= 0.9\,\mathcal{Z}_u\{y[n-1]\} + 0.1\,\mathcal{Z}_u\{x[n]\} \tag{8.309}$$

In Example 8.15 we have found

$$\mathcal{Z}\{\cos(\Omega_0 n)\, u[n]\} = \frac{z\,[z - \cos(\Omega_0)]}{z^2 - 2\cos(\Omega_0)\, z + 1} \tag{8.310}$$

Adapting that result to the transform of the input signal $x[n]$ given, we get

$$\mathcal{Z}_u\{x[n]\} = \mathcal{Z}_u\{20\cos(0.2\pi n)\} = \frac{20z\,[z - \cos(0.2\pi)]}{z^2 - 2\cos(0.2\pi)\, z + 1} \tag{8.311}$$

The unilateral z-transform of the $y[n-1]$ term is

$$\mathcal{Z}_u\{y[n-1]\} = y[-1] + z^{-1} Y_u(z) = 2.5 + z^{-1} Y_u(z) \tag{8.312}$$

Substituting Eqns. (8.311) and (8.312) into Eqn. (8.309) we obtain

$$Y_u(z) = 0.9 \left[2.5 + z^{-1} Y_u(z) \right] + 0.1 \frac{20z \left[z - \cos(0.2\pi) \right]}{z^2 - 2\cos(0.2\pi) z + 1}$$

$$= 0.9 z^{-1} Y_u(z) + 2.25 + \frac{2z \left[z - \cos(0.2\pi) \right]}{z^2 - 2\cos(0.2\pi) z + 1}$$

from which the unilateral z-transform of the output signal can be obtained as

$$Y_u(z) = \frac{2z^2 \left[z - \cos(0.2\pi) \right] + 2.25 z \left(z - e^{j0.2\pi} \right) \left(z - e^{-j0.2\pi} \right)}{(z - 0.9) \left(z - e^{j0.2\pi} \right) \left(z - e^{-j0.2\pi} \right)}$$

We are now ready to determine $y[n]$ using partial fraction expansion. Let

$$Y_u(z) = \frac{k_1 z}{z - 0.9} + \frac{k_2 z}{z - e^{j0.2\pi}} + \frac{k_3 z}{z - e^{-j0.2\pi}}$$

The residues are

$$k_1 = (z - 0.9) \left. \frac{Y_u(z)}{z} \right|_{z=0.9}$$

$$= \left. \frac{2z \left[z - \cos(0.2\pi) \right] + 2.25 \left(z - e^{j0.2\pi} \right) \left(z - e^{-j0.2\pi} \right)}{\left(z - e^{j0.2\pi} \right) \left(z - e^{-j0.2\pi} \right)} \right|_{z=0.9}$$

$$= \frac{2(0.9) \left[0.9 - \cos(0.2\pi) \right] + 2.25 \left(0.9 - e^{j0.2\pi} \right) \left(0.9 - e^{-j0.2\pi} \right)}{\left(0.9 - e^{j0.2\pi} \right) \left(0.9 - e^{-j0.2\pi} \right)}$$

$$= 2.7129$$

and

$$k_2 = \left(z - e^{j0.2\pi} \right) \left. \frac{Y_u(z)}{z} \right|_{z=e^{j0.2\pi}}$$

$$= \left. \frac{2z \left[z - \cos(0.2\pi) \right] + 2.25 \left(z - e^{j0.2\pi} \right) \left(z - e^{-j0.2\pi} \right)}{(z - 0.9) \left(z - e^{-j0.2\pi} \right)} \right|_{z=e^{j0.2\pi}}$$

$$= \frac{2e^{j0.2\pi} \left[e^{j0.2\pi} - \cos(0.2\pi) \right]}{\left(e^{j0.2\pi} - 0.9 \right) \left(e^{j0.2\pi} - e^{-j0.2\pi} \right)}$$

$$= 0.7685 \quad j1.4953$$

The third residue, k_3, can be computed in a similar manner, however, we will take a shortcut and recognize the fact that complex residues of a rational transform must occur in complex conjugate pairs provided that all coefficients of the denominator polynomial are real-valued. Consequently we have

$$k_3 = k_2^* = 0.7685 + j1.4953$$

It will be more convenient to express k_2 and k_3 in polar complex form as

$$k_2 = 1.6813 \, e^{-j1.096} , \quad k_3 = 1.6813 \, e^{j1.096}$$

The forced response of the system is

$$
\begin{aligned}
y[n] =& 2.7129\,(0.9)^n\,u[n] + 1.6813\,e^{-j1.096}e^{j0.2\pi n}u[n] + 1.6813\,e^{j1.096}e^{-j0.2\pi n}u[n] \\
=& 2.7129\,(0.9)^n\,u[n] + 1.6813\,\left[e^{j(0.2\pi n - 1.096)} + e^{-j(0.2\pi n - 1.096)}\right]u[n] \\
=& 2.7129\,(0.9)^n\,u[n] + 3.3626\cos\left(0.2\pi n - 1.096\right)u[n]
\end{aligned}
$$

or equivalently

$$
y[n] = 2.7129\,(0.9)^n\,u[n] + 1.5371\cos\left(0.2\pi n\right)u[n] + 2.9907\sin\left(0.2\pi n\right)u[n]
$$

which is consistent with the answer found earlier in Example 3.15.

8.8 Further Reading

[1] U. Graf. *Applied Laplace Transforms and Z-Transforms for Scientists and Engineers: A Computational Approach Using a Mathematica Package.* Birkhäuser, 2004.

[2] D.G. Manolakis and V.K. Ingle. *Applied Digital Signal Processing: Theory and Practice.* Cambridge University Press, 2011.

[3] A.V. Oppenheim and R.W. Schafer. *Discrete-Time Signal Processing.* Prentice Hall, 2010.

[4] R.A. Schilling, R.J. Schilling, and P.D. Sandra L. Harris. *Fundamentals of Digital Signal Processing Using MATLAB.* Cengage Learning, 2010.

[5] L. Tan and J. Jiang. *Digital Signal Processing: Fundamentals and Applications.* Elsevier Science, 2013.

MATLAB Exercises

MATLAB Exercise 8.1: Three-dimensional plot of z-transform

In Fig. 8.3 the magnitude of the transform

$$
\text{Eqn. (8.9):} \qquad X\left(z\right) = \frac{z\,(z - 0.7686)}{z^2 - 1.5371\,z + 0.9025}
$$

was graphed as a three-dimensional surface. In this exercise we will reproduce that figure using MATLAB, and develop the code to evaluate and display the transform on a circular trajectory $z = r\,e^{j\Omega}$. The first step is to produce a set of complex values of z on a rectangular grid in the z-plane.

```
>>  [zr,zi] = meshgrid([-1.5:0.03:1.5],[-1.5:0.03:1.5]);
>>  z = zr+j*zi;
```

The next step is to compute the magnitude of the z-transform at each point on the grid. An anonymous function will be used for defining $X\left(z\right)$. Additionally, values of magnitude that are greater than 12 will be clipped for graphing purposes.

```
>>   Xz = @(z) z.*(z-0.7686)./(z.*z-1.5371*z+0.9025);
>>   XzMag = abs(Xz(z));
>>   XzMag = XzMag.*(XzMag<=12)+12.*(XzMag>12);
```

A three-dimensional mesh plot of $|X(z)|$ plot can be generated with the following lines:

```
>>   mesh(zr,zi,XzMag);
>>   axis([-1.5,1.5,-1.5,1.5]);
```

The script listed below produces a mesh plot complete with axis labels and color specifications.

```
1   % Script: matex_8_1a.m
2   %
3   [zr,zi] = meshgrid([-1.5:0.03:1.5],[-1.5:0.03:1.5]);
4   z = zr+j*zi;
5   Xz = @(z) z.*(z-0.7686)./(z.*z-1.5371*z+0.9025);   % Eqn.(8.9)
6   XzMag = abs(Xz(z));
7   XzMag = XzMag.*(XzMag<=12)+12.*(XzMag>12);
8   shading interp;          % Shading method: Interpolated
9   colormap copper;         % Specify the color map used.
10  m1 = mesh(zr,zi,XzMag);
11  axis([-1.5,1.5,-1.5,1.5]);
12  % Adjust transparency of surface lines.
13  set(m1,'EdgeAlpha',0.6,'FaceAlpha',0.6);
14  % Specify x,y,z axis labels.
15  xlabel('Re[z]');
16  ylabel('Im[z]');
17  zlabel('|X(z)|');
18  % Specify viewing angles.
19  view(gca,[56.5 40]);
```

In line 10 of the script, the handle returned by the function mesh(..) is assigned to the variable m1 so that it can be used in line 13 for adjusting the transparency of the surface.

The DTFT $X(\Omega)$ is equal to the z-transform evaluated on the unit circle of the z-plane, that is,

$$X(z)\Big|_{z=e^{j\Omega}} = \mathcal{F}\{x[n]\}$$

Applying this relationship to the magnitudes of the two transforms, the magnitude of the DTFT is obtained by evaluating the magnitude of the z-transform on the unit circle. The script listed below demonstrates this.

```
1   % Script: matex_8_1b.m
2   %
3   [zr,zi] = meshgrid([-1.5:0.03:1.5],[-1.5:0.03:1.5]);
4   z = zr+j*zi;
5   Xz = @(z) z.*(z-0.7686)./(z.*z-1.5371*z+0.9025);   % Eqn.(8.9)
6   XzMag = abs(Xz(z));
7   XzMag = XzMag.*(XzMag<=12)+12.*(XzMag>12);
8   % Set the trajectory.
9   r = 1;                   % Radius of the trajectory.
10  Umega = [0:0.005:1]*2*pi;
11  tr = r*exp(j*Omega);
12  % Produce a mesh plot and hold it.
13  shading interp;          % Shading method: Interpolated
14  colormap copper;         % Specify the color map used.
15  m1 = mesh(zr,zi,XzMag);
16  set(m1,'EdgeAlpha',0.6,'FaceAlpha',0.6);
```

```
17   hold on;
18   % Superimpose a plot of X(z) magnitude values  evaluated
19   % on the trajectory using 'plot3' function.
20   m2 = plot3(real(tr),imag(tr),abs(Xz(tr)),'b-','LineWidth',1.5);
21   % Show the unit-circle in red.
22   m3 = plot3(real(tr),imag(tr),zeros(size(tr)),'r-','LineWidth',1.5);
23   hold off;
24   axis([-1.5,1.5,-1.5,1.5]);
25   % Specify x,y,z axis labels.
26   xlabel('Re[z]');
27   ylabel('Im[z]');
28   zlabel('|X(z)|');
29   % Specify viewing angles.
30   view(gca,[56.5 40]);
```

Software resources:
matex_8_1a.m
matex_8_1b.m

MATLAB Exercise 8.2: **Computing the DTFT from the z-transform**

The DTFT of a signal is equal to its z-transform evaluated on the unit circle of the z-plane.

$$\mathcal{F}\{x[n]\} = X(z)\Big|_{z=e^{j\Omega}}$$

Consider the z-transform

$$X(z) = \frac{z - 0.7686}{z^2 - 1.5371\,z + 0.9025}$$

The first method of computing and graphing the DTFT of the signal is to use an anonymous function for $X(z)$ and evaluate it on the unit circle. The magnitude $|X(\Omega)|$ is graphed using the following statements:

```
>>   X = @(z) (z-0.7686)./(z.*z-1.5371*z+0.9025);
>>   Omg = [-1:0.001:1]*pi;
>>   Xdtft = X(exp(j*Omg));
>>   plot(Omg,abs(Xdtft)); grid;
```

If the phase $\angle X(\Omega)$ is needed, it can be graphed using

```
>>   plot(Omg,angle(Xdtft)); grid;
```

The second method is to use MATLAB function freqz(..). Care must be taken to ensure that the function is used with proper arguments, or erroneous results may be obtained. The function freqz(..) requires that the transform $X(z)$ be expressed using negative powers of z in the form

$$X(z) = \frac{0 + z^{-1} - 0.7686\,z^{-2}}{1 - 1.5371\,z^{-1} + 0.9025\,z^{-2}}$$

Vectors "**num**" and "**den**" to hold numerator and denominator coefficients should be entered as

```
>>   num = [0,1,-0.7686];
>>   den = [1,-1.5371,0.9025];
```

Afterwards the magnitude and the phase of the DTFT can be computed and graphed with the statements

```
>>   Omg = [-1:0.001:1]*pi;
>>   [Xdtft,Omg] = freqz(num,den,Omg);
>>   plot(Omg,abs(Xdtft)); grid;
>>   plot(Omg,angle(Xdtft)); grid;
```

Common mistake: Pay special attention to the leading 0 in the numerator coefficient vector. Had we omitted it, the function freqz(..) would have incorrectly used the transform

$$X(z) = \frac{1 - 0.7686\, z^{-1}}{1 - 1.5371\, z^{-1} + 0.9025\, z^{-2}}$$

resulting in the correct magnitude $|X(\Omega)|$ but incorrect phase $\angle X(\Omega)$. This example demonstrates the potential for errors when built-in MATLAB functions are used without carefully reading the help information.

Software resources:

matex_8_2a.m

matex_8_2b.m

MATLAB Exercise 8.3: Graphing poles and zeros

Consider a DTLTI system described by the system function

$$X(z) = \frac{z^3 - 0.5\, z + 0.5}{z^4 - 0.5\, z^3 - 0.5\, z^2 - 0.6\, z + 0.6}$$

Poles and zeros of the system function can be graphed on the z-plane by entering coefficients of numerator and denominator polynomials as vectors and then computing the roots.

```
>>   num = [1,0,-0.5,0.5];
>>   den = [1,-0.5,-0.5,-0.6,0.6];
>>   z = roots(num);
>>   p = roots(den);
```

To produce the graph we need

```
>>   plot(real(z),imag(z),'o',real(p),imag(p),'x');
```

which uses "o" for a zero and "x" for a pole. The graph produced does not display the unit circle of the z-plane, and the aspect ratio may not be correct. These issues are resolved using the code below:

```
>>   Omg = [0:0.001:1]*2*pi;
>>   c = exp(j*Omg);    % Unit circle
>>   plot(real(z),imag(z),'o',real(p),imag(p),'x',real(c),imag(c),':');
>>   axis equal
```

Alternatively, built-in function zplane(..) may be used for producing a pole-zero plot, however, caution must be used as in the previous MATLAB exercise. The function zplane(..) expects numerator and denominator polynomials to be specified in terms of negative powers of z. Using the vectors "num" and "den" as entered above and calling the function zplane(..) through

```
>>   zplane(num,den);
```

would display an incorrect pole-zero plot with an additional zero at the origin. The correct use for this problem would be

```
>>  num = [0,1,0,-0.5,0.5];
>>  den = [1,-0.5,-0.5,-0.6,0.6];
>>  zplane(num,den);
```

Software resources:
matex_8_3a.m
matex_8_3b.m

MATLAB Exercise 8.4: Using convolution function for polynomial multiplication

The function conv(..) for computing the convolution of two discrete-time signals was used in MATLAB Exercise 3.7 in Chapter 3. In Example 8.23 of this chapter we have discovered another interesting application of the conv(..) function, namely the multiplication of two polynomials. In this exercise we will apply this idea to the multiplication of the two polynomials

$$A(\nu) = 5\nu^4 + 7\nu^3 + 3\nu - 2 \tag{8.313}$$

and

$$B(\nu) = 8\nu^5 + 6\nu^4 - 3\nu^3 + \nu^2 - 4\nu + 2 \tag{8.314}$$

The problem is to determine the polynomial $C(\nu) = A(\nu) B(\nu)$. We will begin by creating a vector with the coefficients of each polynomial ordered from highest power of ν down to the constant term, being careful to account for any missing terms:

```
>>  A = [5,7,0,3,-2];        % Coefficients a_4, ..., a_0
>>  B = [8,6,-3,1,-4,2];     % Coefficients b_5, ..., b_0
```

For the convolution operation the coefficients are needed in ascending order starting with the constant term. MATLAB function fliplr(..) will be used for reversing the order.

```
>>  tmpA = fliplr(A);        % Coefficients a_0, ..., a_4
>>  tmpB = fliplr(B);        % Coefficients b_0, ..., a_5
>>  tmpC = conv(tmpA,tmpB);  % Coefficients c_0, ..., a_9
>>  C = fliplr(tmpC)         % Coefficients c_9, ..., c_0

C =
    40    86    27     8   -11   -39    23   -14    14    -4
```

The product polynomial is

$$C(\nu) = 40\nu^9 + 86\nu^8 + 27\nu^7 + 8\nu^6 - 11\nu^5 - 39\nu^4 + 23\nu^3 - 14\nu^2 + 14\nu - 4 \tag{8.315}$$

Software resources:
matex_8_4.m

MATLAB Exercise 8.5: Partial fraction expansion with MATLAB

Find a partial fraction expansion for the transform

$$X(z) = \frac{-0.2\,z^3 + 1.82\,z^2 - 3.59\,z + 3.1826}{z^4 - 1.2\,z^3 + 0.46\,z^2 + 0.452\,z - 0.5607}$$

Solution: We need to create two vectors "num" and "den" with the numerator and the denominator coefficients of $X(z)/z$ and then use the function residue(..). The statements below compute the z-domain poles and residues:

```
>>   num = [-0.2000,1.8200,-3.5900,3.1826];
>>   den = [1.0000,-1.2000,0.4600,0.4520,-0.5607,0];
>>   [r,p,k] = residue(num,den)

r =
     0.8539 + 0.0337i
     0.8539 - 0.0337i
     1.1111
     2.8571
    -5.6761

p =
     0.5000 + 0.8000i
     0.5000 - 0.8000i
     0.9000
    -0.7000
          0

k =
     []
```

Based on the results obtained, the PFE is in the form

$$\frac{X(z)}{z} = \frac{0.8539 + j0.0337}{z - 0.5 - j0.8} + \frac{0.8539 - j0.0337}{z - 0.5 + j0.8} + \frac{1.1111}{z - 0.9} + \frac{2.8571}{z + 0.7} + \frac{-5.6761}{z}$$

or, equivalently

$$X(z) = \frac{(0.8539 + j0.0337)\,z}{z - 0.5 - j0.8} + \frac{(0.8539 - j0.0337)\,z}{z - 0.5 + j0.8} + \frac{1.1111\,z}{z - 0.9} + \frac{2.8571\,z}{z + 0.7} - 5.6761$$

Software resources:
matex_8_5.m

MATLAB Exercise 8.6: Developing a function for long division

In this exercise we will develop a MATLAB function named "ss_longdiv" to compute the inverse z-transform by long division. MATLAB function deconv(..) will be used for implementing each step of the long division operation. The utility function ss_longdiv_util(..) listed below implements one step of the long division.

```
1   function [q,r] = ss_longdiv_util(num,den)
2     M = max(size(num))-1;     % The order of the numerator.
3     N = max(size(den))-1;     % The order of the denominator.
4     % Append the shorter vector with zeros.
5     if (M < N)
6       num = [num,zeros(1,N-M)];
7     elseif (M > N)
8       den = [den,zeros(1,M-N)];
9     end;
10    [q,r] = deconv(num,den);  % Find quotient and remainder.
11    r = r(2:length(r));       % Drop first element of remainder
12                              % which is 0.
13  end
```

To test this function, let us try the transform in Example 8.31.

$$X(z) = \frac{3z^2 - z}{z^3 - 3z^2 + 3z - 1}, \qquad \text{ROC:} \quad |z| > 1$$

The first sample of the inverse transform is obtained using the following:

```
>>   num = [3,-1,0];
>>   den = [1,-3,3,-1];
>>   [q,r] = ss_longdiv_util(num,den)

q =
     3

r =
     8      -9      3
```

This result should be compared to Eqn. (8.208). The next step in the long division can be performed with the statement

```
>>   [q,r] = ss_longdiv_util(r,den)

q =
     8

r =
    15     -21      8
```

and matches Eqn. (8.209). This process can be repeated as many times as desired. The function ss_longdiv(..) listed below uses the function ss_longdiv_util(..) in a loop to compute n samples of the inverse transform.

```
1    function [x] = ss_longdiv(num,den,n)
2      [q,r] = ss_longdiv_util(num,den);
3      x = q;
4      for nn = 2:n,
5        [q,r] = ss_longdiv_util(r,den);
6        x = [x,q];
7      end;
8    end
```

It can be tested using the problem in Example 8.31:

```
>>   x = ss_longdiv([3,-1,0],[1,-3,3,-1],6)

x =
     3      8      15      24      35      48
```

If a left-sided result is desired, the function ss_longdiv(..) can be still be used by simply re-ordering the coefficients in the vectors "**num**" and "**den**". Refer to the transform in Example 8.32.

$$X(z) = \frac{z^2 + z - 2}{z^2 - 2.5z + 1}, \qquad \text{ROC: } |z| < \tfrac{1}{2}$$

The left-sided solution is obtained by

```
>>   num = fliplr([1,1,-2]);
>>   den = fliplr([1,-2.5,1]);
>>   x = ss_longdiv(num,den,6)

x =
    -2.0000    -4.0000    -7.0000    -13.5000    -26.7500    -53.3750
```

Software resources:
ss_longdiv.m
ss_longdiv_util.m

MATLAB Exercise 8.7: Computing frequency response of a system from pole-zero layout

The problem of obtaining the frequency response of a DTLTI system from the placement of its poles and zeros in the z-domain was discussed in Section 8.5.4. In this exercise we will use MATLAB to compute the frequency response for a system characterized by the system function

$$H(z) = \frac{z + 0.6}{z - 0.8}$$

The system has one zero at $z = -0.6$ and one pole at $z = 0.8$. Suppose we need the frequency response of this system at the angular frequency of $\Omega = 0.3\pi$ radians, which is equal to the system function evaluated at $z = e^{j0.3\pi}$:

$$H(0.3\pi) = H(z)\Big|_{z=e^{j0.3\pi}} = \frac{e^{j0.3\pi} + 0.6}{e^{j0.3\pi} - 0.8}$$

Using vector notation, we have

$$H(\overrightarrow{0.3\pi}) = \frac{\overrightarrow{B}}{\overrightarrow{A}} = \frac{\left(e^{j0.3\pi} \overrightarrow{+ 0.6}\right)}{\left(e^{j0.3\pi} \overrightarrow{- 0.8}\right)}$$

where we have defined the vectors $\overrightarrow{A}$ and $\overrightarrow{B}$ for notational convenience. The script listed below computes the vectors $\overrightarrow{A}$ and $\overrightarrow{B}$ and uses them for computing the magnitude and the phase of the system function at $\omega = 0.3\pi$:

```
1   % Script: matex_8_7a.m
2   %
3   Omega = 0.3*pi;
4   z = exp(j*Omega);
5   B = z+0.6;
6   A = z-0.8;
7   mag = abs(B)/abs(A);
8   phs = angle(B)-angle(A);
```

This script computes the frequency response of the system at one specific frequency. It would be more interesting if we could use the same idea to compute the frequency response of the system at a large number of angular frequencies so that its magnitude and phase can be graphed as functions of Ω. Let us change the variable "Omega" into a vector by editing line 1 of the script:

```
1   Omega = [-1:0.004:1]*pi;
```

This change causes the variable "z" to become a complex vector with 501 elements. Also, in line 5, the standard division operator "/" needs to be changed to the element-by-element division operator "./" to read

```
5   mag = abs(B)./abs(A);
```

The MATLAB script is listed below with these modifications and the addition of graphing statements:

```
1   % Script: matex_8_7b.m
2   %
3   Omega = [-1:0.004:1]*pi;
4   z = exp(j*Omega);
5   B = z+0.6;
6   A = z-0.8;
7   mag = abs(B)./abs(A);
8   phs = angle(B)-angle(A);
9   clf;
10  subplot(2,1,1);
11  plot(Omega,mag);
12  title('Magnitude of the frequency response');
13  xlabel('\Omega (rad)'); grid;
14  subplot(2,1,2);
15  plot(Omega,phs);
16  title('Phase of the frequency response');
17  xlabel('\Omega (rad)'); grid;
```

Software resources:

matex_8_7a.m

matex_8_7b.m

MATLAB Exercise 8.8: Frequency response from pole-zero layout revisited

In MATLAB Exercise 8.7 we have explored a method of computing the frequency response of a DTLTI system based on the graphical interpretation of the pole-zero layout of the system function, discussed in Section 8.5.4. The idea can be generalized into the development of a MATLAB function ss_freqz(..) for computing the frequency response.

```
1   function [mag,phs] = ss_freqz(zrs,pls,gain,Omega)
2     nz = length(zrs);        % Number of zeros.
3     np = length(pls);        % Number of poles.
4     nOmg = length(Omega);    % Number of frequency points.
5     z = exp(j*Omega);        % Get points on the unit-circle.
6     mag = ones(1,nOmg);
7     phs = zeros(1,nOmg);
8     if (nz > 0),
9       for n = 1:nz
10        mag = mag.*abs(z-zrs(n));
11        phs = phs+angle(z-zrs(n));
12      end;
13    end;
14    if (np > 0),
15      for n = 1:np
16        mag = mag./abs(z-pls(n));
17        phs = phs-angle(z-pls(n));
18      end;
19    end;
20    mag = mag*gain;
21    phs = wrapToPi(phs);
```

Line 21 of the function causes phase angles to be contained in the interval $(-\pi, \pi)$. The script listed below may be used for testing ss_freqz(..) with the system function

$$H(z) = \frac{z - 0.7686}{z^2 - 1.5371\,z + 0.9025}$$

```
1   % Script: matex_8_8.m
2   %
```

```
3    num = [1,-0.7686];
4    den = [1,-1.5371,0.9025];
5    zrs = roots(num);       % Compute zeros.
6    pls = roots(den);       % Compute poles.
7    Omg = [-1:0.001:1]*pi;  % Vector of frequencies.
8    [mag,phs] = ss_freqz(zrs,pls,1,Omg);
9    clf;
10   subplot(2,1,1);
11   plot(Omg,mag); grid;
12   xlabel('\Omega (rad)');
13   subplot(2,1,2);
14   plot(Omg,phs); grid;
15   xlabel('\Omega (rad)');
```

Software resources:
ss_freqz.m
matex_8_8.m

MATLAB Exercise 8.9: Preliminary calculations for a cascade-form block diagram

In this exercise we will use MATLAB to help with the preliminary calculations for sectioning a system function into cascade subsystems to facilitate the development of a cascade-form block diagram.

Consider the system function first used in Example 8.41.

$$H(z) = \frac{z^3 - 7z + 6}{z^4 - z^3 - 0.34\,z^2 + 0.966\,z - 0.2403}$$

A cascade-form block diagram was obtained in Example 8.43. Two vectors "num" and "den" are created with the following statements to hold numerator and denominator coefficients in descending powers of z:

```
>>  num=[1,0,-7,6];
>>  den=[1,-1,-0.34,0.966,-0.2403];
```

Poles and zeros of the system function are found as

```
>>  z = roots(num)

z =
    -3.0000
     2.0000
     1.0000

>>  p = roots(den)

p =
    -0.9000
     0.8000 + 0.5000i
     0.8000 - 0.5000i
     0.3000
```

The returned vector "z" holds the three zeros of the system function that are all real-valued. The vector "p" holds the four poles. The first subsystem $H_1(z)$ will be formed using zeros #1 and #3 along with poles #1 and #4. Using the function poly(..) for polynomial construction from roots, the numerator and denominator coefficients of $H(z)$ are obtained as

```
>>   num1 = poly([z(1),z(3)])

num1 =
     1.0000      2.0000     -3.0000

>>   den1 = poly([p(1),p(4)])

den1 =
     1.0000      0.6000     -0.2700
```

corresponding to
$$H_1(z) = \frac{(z+3)(z-1)}{(z+0.9)(z-0.3)} = \frac{z^2 + 2z - 3}{z^2 + 0.6z - 0.27}$$

The remaining zero will be combined with poles #2 and #3 to yield the numerator and the denominator of $H_2(z)$:

```
>>   num2 = [1,-z(2)]

num2 =
     1.0000     -2.0000

>>   den2 = poly([p(2),p(3)])

den2 =
     1.0000     -1.6000      0.8900
```

The second subsystem is therefore
$$H_2(z) = \frac{z-2}{(z-0.8-j0.5)(z-0.8+j0.5)} = \frac{z-2}{z^2 - 1.6z + 0.89}$$

Software resources:
matex_8_9.m

MATLAB Exercise 8.10: **Preliminary calculations for a cascade-form block diagram revisited**

An alternative method of breaking up a system function into second-order cascade sections is to use MATLAB functions tf2zp(..) and zp2sos(..). Working with the same system function as in MATLAB Exercise 8.9 use the following set of statements:

```
>>   num = [0,1,0,-7,6];
>>   den = [1,-1,-0.34,0.966,-0.2403];
>>   [z,p,k] = tf2zp(num,den);
>>   [sos,G] = zp2sos(z,p,k)

sos =
          0      1.0000      3.0000      1.0000      0.6000     -0.2700
     1.0000     -3.0000      2.0000      1.0000     -1.6000      0.8900

G =
     1
```

It is important to remember that vectors "**num**" and "**den**" hold coefficients of numerator and denominator polynomials in terms of ascending negative powers of z. In general they need to be of the same length, hence requiring a zero-valued element to be added in front of the vector "**num**". Second-order sections are in the form
$$H_i(z) = \frac{b_{0i} + b_{1i}\, z^{-1} + b_{2i}\, z^{-2}}{1 + a_{1i}\, z^{-1} + a_{2i}\, z^{-2}}$$

Each row of the matrix "sos" holds coefficients of one second-order section in the order

$$b_{0i}, \ b_{1i}, \ b_{2i}, \ 1, \ a_{1i}, \ a_{2i}$$

The result obtained above corresponds to second-order sections

$$H_1(z) = \frac{z^{-1} + 3\,z^{-2}}{1 + 0.6\,z^{-1} - 0.27\,z^{-2}}$$

and

$$H_2(z) = \frac{1 - 3\,z^{-1} + 2\,z^{-2}}{1 - 1.6\,z^{-1} + 0.89\,z^{-2}}$$

Software resources:
matex_8_10.m

MATLAB Exercise 8.11: Preliminary calculations for a parallel-form block diagram

This exercise is about using MATLAB to help with the preliminary calculations for sectioning a system function into parallel subsystems to facilitate the development of a parallel-form block diagram.

Consider again the system function first used in Example 8.41.

$$H(z) = \frac{z^3 - 7z + 6}{z^4 - z^3 - 0.34\,z^2 + 0.966\,z - 0.2403}$$

A parallel-form block diagram was obtained in Example 8.44. In order to obtain the subsystems $\bar{H}_1(z)$ and $\bar{H}_2(z)$ for the parallel-form implementation we will begin by creating vectors "num" and "den" to hold numerator and denominator coefficients in descending powers of z:

```
>>   num=[1,0,-7,6];
>>   den=[1,-1,-0.34,0.966,-0.2403];
```

Residues are found using the function residue(..). Recall from the discussion in MATLAB Exercise 8.5 that there is also a function named residuez(..) specifically designed for z-domain residues. However, it utilizes a slightly different format than the conventions we have developed in Section 8.4.2 for the partial fraction expansion, and therefore will not be used here.

```
>>   [r,p,k] = residue(num,den)

r =
   -1.2371 +  1.7480i
   -1.2371 -  1.7480i
   -3.0709
    6.5450


p =
    0.8000 +  0.5000i
    0.8000 -  0.5000i
   -0.9000
    0.3000


k =
      []
```

The resulting expansion is

$$H\left(z\right) = \frac{-1.2371 + j1.7480}{z - 0.8 - j0.5} + \frac{-1.2371 - j1.7480}{z - 0.8 + j0.5} \frac{-3.0709}{z + 0.9} + \frac{6.5450}{z - 0.3}$$

We will recombine the two terms with real poles (terms #3 and #4 in the lists of poles and residues computed by MATLAB) to obtain $\bar{H}_1\left(z\right)$:

```
>>   [num1,den1] = residue(r(3:4),p(3:4),[]);
>>   H1 = tf(num1,den1,-1)

Transfer function:
 3.474 z + 6.812
------------------
z^2 + 0.6 z - 0.27

Sampling time: unspecified
```

The remaining terms (items #1 and #2) will be recombined for the second subsystem $\bar{H}_2\left(z\right)$:

```
>>   [num2,den2] = residue(r(1:2),p(1:2),[]);
>>   H2 = tf(num2,den2,-1)

Transfer function:
-2.474 z + 0.2314
------------------
z^2 - 1.6 z + 0.89

Sampling time: unspecified
```

so that

$$H\left(z\right) = \bar{H}_1\left(z\right) + \bar{H}_2\left(z\right)$$

To see if this is indeed the case, let us add the two system functions:

```
>>   H = H1+H2

Transfer function:
    z^3 - 8.882e-016 z^2 - 7 z + 6
-------------------------------------------
z^4 - z^3 - 0.34 z^2 + 0.966 z - 0.2403

Sampling time: unspecified
```

Software resources:
matex_8_11.m

MATLAB Exercise 8.12: Implementing a system using second-order sections

In Section 8.6.2 cascade and parallel implementation forms of a rational system function $H\left(z\right)$ were discussed using second-order sections.

The second-order rational system function

$$H\left(z\right) = \frac{b_0 + b_1\,z^{-1} + b_2\,z^{-2}}{1 + a_1\,z^{-1} + a_2\,z^{-2}}$$

may be used as the basis of cascade and parallel implementations, and may be implemented using the direct-form II block diagram shown in Fig. 8.61.

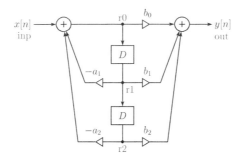

Figure 8.61 – Block diagram for second-order section.

We will develop a MATLAB function ss_iir2(..) for a generic implementation of this second-order system. The focus will be on making the function reusable within a larger block diagram implemented in either cascade or parallel form. For flexibility the function takes in one sample of the input signal at a time, and returns one sample of the output signal. The syntax of the function ss_iir2(..) is

```
[out,states] = ss_iir2(inp,coeffs,states)
```

The scalar "inp" is the current sample of the input signal. The vector "coeffs" holds the coefficients of the second-order IIR filter stage in the following order:

$$\text{coeffs} = [b_{0i}, \ b_{1i}, \ b_{2i}, \ 1, \ a_{1i}, \ a_{2i}]$$

Note that this is the same coefficient order used in MATLAB functions tf2sos, zp2sos and ss2sos. Finally, the input vector "states" holds values of the internal variables r_1 and r_2 set in the previous step:

$$\text{states} = [r_1, \ r_2]$$

The return variable "out" is the current output sample computed in response to the current input sample. The function also returns updated values of the internal variables r_1 and r_2 in the vector "states". These will be needed the for processing the next input sample.

The listing for the function ss_iir2 is given below:

```
1   function [out,states] = ss_iir2(inp,coeffs,states)
2     % Extract the filter states.
3     r1 = states(1);
4     r2 = states(2);
5     % Extract the coefficients.
6     b0 = coeffs(1);
7     b1 = coeffs(2);
8     b2 = coeffs(3);
9     a1 = coeffs(5);
10    a2 = coeffs(6);
11    % Compute the output sample.
12    r0 = inp-a1*r1-a2*r2;
13    out = b0*r0+b1*r1+b2*r2;
14    % Update the filter states:  r1<-r0,   r2<-r1
15    states(1) = r0;
16    states(2) = r1;
17  end
```

To test the function, we will use the DTLTI system used in Examples 8.43 and 8.44. The cascade form block diagram shown in Fig. 8.57 may be implemented with the following script that computes and graphs the unit-step response of the system:

```
1   % Script: matex_8_12a.m
2   %
3   states1 = [0,0];
4   states2 = [0,0];
5   coeffs1 = [1,2,-3,1,0.6,-0.27];
6   coeffs2 = [0,1,-2,1,-1.6,0.89];
7   x = ones(1,100);
8   y = zeros(1,100);
9   for n=1:100,
10    inp = x(n);
11    [w1,states1] = ss_iir2(inp,coeffs1,states1);
12    [y(n),states2] = ss_iir2(w1,coeffs2,states2);
13  end;
14  stem([0:99],y);
```

Similarly, the script listed below computes and graphs the unit-step response of the same system using the parallel form block diagram obtained in Example 8.44 and shown in Fig. 8.60.

```
1   % Script: matex_8_12b.m
2   %
3   states1 = [0,0];
4   states2 = [0,0];
5   coeffs1 = [0,3.474,6.812,1,0.6,-0.27];
6   coeffs2 = [0,-2.474,0.2314,1,-1.6,0.89];
7   x = ones(1,100);
8   y = zeros(1,100);
9   for n=1:100,
10    inp = x(n);
11    [w1,states1] = ss_iir2(inp,coeffs1,states1);
12    [w2,states2] = ss_iir2(inp,coeffs2,states2);
13    y(n) = w1+w2;
14  end;
15  stem([0:99],y);
```

Software resources:
ss_iir2.m
matex_8_12a.m
matex_8_12b.m

MATLAB Exercise 8.13: Solving a difference equation through z-transform

In Section 8.7 two examples of using the unilateral z-transform for solving difference equations with specified initial conditions were given. In this exercise we will use the symbolic processing capabilities of MATLAB to solve the problem in Example 8.45. The functions ztrans(..) and iztrans(..) are available for symbolic computation of the forward and inverse z-transform. They both assume that the time-domain signal involved is causal. In effect they implement the unilateral variant of the z-transform.

Consider the homogeneous difference equation explored in Example 8.45

$$y[n] - \frac{5}{6}\, y[n-1] + \frac{1}{6}\, y[n-2] = 0$$

with the initial conditions

$$y[-1] = 19 \ , \quad y[-2] = 53$$

Unilateral z-transforms of $y[n-1]$ and $y[n-2]$ are

$$Y_1(z) = \mathcal{Z}_u\{y[n-1]\} = 19 + z^{-1}Y_u(z)$$

and

$$Y_2(z) = \mathcal{Z}_u \{y[n-2]\} = 53 + 19\,z^{-1} + z^{-2}\,Y_u(z)$$

The following script can be used for solving the problem:

```
1    % Script: matex_8_13.m
2    %
3    syms z n Yz
4    Y1 = 19+z^(-1)*Yz;               % z-transform of y[n-1]
5    Y2 = 53+19*z^(-1)+z^(-2)*Yz;     % z-transform of y[n-2]
6    Yz = solve(Yz-5/6*Y1+1/6*Y2,Yz); % Solve for Y(z)
7    yn = iztrans(Yz)                 % Inverse z-transform of Y(z)
```

In line 3 we declare three symbolic variables. The variable "Yz" corresponds to the yet unknown transform $Y_u(z)$. Line 6 of the script uses the function solve(..) to solve the equation

$$Y_u(z) - \frac{5}{6}\,Y_1(z) + \ Y_2(z) = 0$$

for $Y_u(z)$. Line 7 generates the response

```
yn =
     2*(1/2)^n + 5*(1/3)^n
```

which is in agreement with the solution found in Example 8.45. If desired, the symbolic expression in MATLAB variable "yn" may be numerically evaluated and graphed with the following statements:

```
>>   n = [0:10];
>>   y = eval(yn);
>>   stem(n,y);
```

Software resources:
matex_8_13a.m
matex_8_13b.m

Problems

8.1.   Using the definition of the z-transform, compute $X(z)$ for the signals listed below. Write each transform using non-negative powers of z. In each case determine the poles and the zeros of the transform, and the region of convergence.

a. $x[n] = \{\ \underset{\underset{n=0}{\uparrow}}{1}\ ,1,\ 1\}$

b. $x[n] = \{\ \underset{\underset{n=0}{\uparrow}}{1}\ ,1,\ 1,\ 1,\ 1\}$

c. $x[n] = \{1,\ 1,\ \underset{\underset{n=0}{\uparrow}}{1}\ ,1,\ 1\}$

d. $x[n] = \{1,\ 1,\ 1,\ 1,\ \underset{\underset{n=0}{\uparrow}}{1}\ \}$

8.2.   For each transform $X(z)$ listed below, determine whether the DTFT of the corresponding signal $x[n]$ exists. If it does, find it.

a. $X(z) = \dfrac{z\,(z-1)}{(z+1)\,(z+2)}$,   ROC: $|z| > 2$

b. $X(z) = \dfrac{z\,(z+2)}{(z+1/2)\,(z+3/2)}$,   ROC: $\dfrac{1}{2} < |z| < \dfrac{3}{2}$

c. $X(z) = \dfrac{z^2}{z^2+5\,z+6}$,   ROC: $|z| < 2$

d. $X(z) = \dfrac{(z+1)\,(z-1)}{(z+2)\,(z-3)\,(z-4)}$,   ROC: $|z| < 2$

8.3.   Pole-zero diagrams for four transforms are shown in Fig. P.8.3. For each, determine the ROC if it is known that the DTFT of $x[n]$ exists.

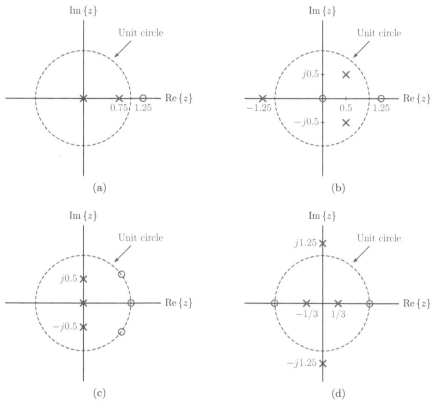

Figure P. 8.3

8.4.   Determine the z-transforms of the signals given below. Indicate the ROC for each.

a. $x[n] = \begin{cases} n\,, & n = 0,\ldots,9 \\ 0\,, & \text{otherwise} \end{cases}$

b. $x[n] = \begin{cases} n\,, & n = 0,\ldots,9 \\ 10\,, & n \geq 10 \\ 0\,, & \text{otherwise} \end{cases}$

c. $x[n] = \begin{cases} n\,, & n = 0,\ldots,9 \\ -n+20\,, & n = 10,\ldots,19 \\ 0\,, & \text{otherwise} \end{cases}$

8.5. Find the z-transform of the signal

$$x[n] = \begin{cases} 1 , & n \geq 0 \text{ and even} \\ (2/3)^n , & n > 0 \text{ and odd} \\ 0 , & n < 0 \end{cases}$$

Also indicate the ROC.

8.6. Consider the signal $g[n]$ specified as

$$g[n] = (0.9)^n \cos(0.3n) \, u[n]$$

    a. Determine the transform $G(z)$ and its ROC.
    b. Let a new signal $x[n]$ be obtained from $g[n]$ through the relationship $x[n] = g[2n]$. Determine the transform $X(z)$ and its ROC from the transform $G(z)$ using the development outlined in Example 8.11.
    c. Determine the transform $X(z)$ and its ROC by direct application of the z-transform definition to the signal $x[n]$. Compare to the result obtained in part (b).

8.7. Determine the z-transform of each signal listed below, and indicate the ROC. Use linearity, time shifting and time reversal properties of the z-transform when needed.

    a. $x[n] = \delta[n+2]$
    b. $x[n] = \delta[n-3]$
    c. $x[n] = \left(\frac{1}{2}\right)^n u[n]$
    d. $x[n] = \left(\frac{1}{2}\right)^n u[n] + \left(\frac{1}{3}\right)^n u[n]$
    e. $x[n] = \left(\frac{1}{2}\right)^{n-1} u[n-1]$
    f. $x[n] = \left(\frac{1}{2}\right)^{n-1} u[n]$
    g. $x[n] = \left(\frac{1}{2}\right)^{n+1} u[n]$
    h. $x[n] = (3)^n u[-n-1]$
    i. $x[n] = (3)^n u[-n+1]$

8.8. Consider the signal

$$x[n] = (1/2)^{|n|}$$

    a. Express $x[n]$ as the sum of a causal signal $x_R[n]$ and an anti-causal signal $x_L[n]$. Find an analytical expression for each component, and sketch each component.
    b. Determine the transforms $X_R(z)$ and $X_L(z)$ corresponding to the signals $x_R[n]$ and $x_L[n]$ respectively. Also determine the ROC for each transform.
    c. Use linearity of the z-transform to express $X(z)$ and to determine its ROC.

8.9. Use the time reversal property of the z-transform to determine $X(z)$ for the signals listed below. Indicate the ROC for each.

    a. $x[n] = u[-n]$
    b. $x[n] = u[-n-1]$
    c. $x[n] = u[-n] - u[n-5]$
    d. $x[n] = n \, u[-n]$
    e. $x[n] = \cos(\Omega_0 n) \, u[-n]$

8.10. Use the differentiation property of the z-transform to find the transform of each signal below. Also indicate the ROC for each.

a. $x[n] = (n^2 + 3n + 5) \, u[n]$
b. $x[n] = (n^2 - 5) \, u[-n - 1]$
c. $x[n] = n \cos(\Omega_0 n) \, u[n]$
d. $x[n] = (n + 1) \sin(\Omega_0 n) \, u[n]$

8.11. Transforms of several causal signals are listed below. Use the initial value property of the z transform to determine $x[0]$ for each case.

a. $X(z) = \dfrac{z^2 + z + 1}{z^2 + 3z + 2}$

b. $X(z) = \dfrac{z^2 + z + 1}{z^3 + 3z^2 + 3z + 1}$

c. $X(z) = \dfrac{z^{-1} + z^{-2} - z^{-3}}{1 + 0.7 z^{-1} + 1.2 z^{-2} - 1.5 z^{-3}}$

8.12.

a. Use the correlation property of the z-transform to determine the autocorrelation function for
$$x[n] = u[n] - u[n - 2]$$
Hint: Write $X(z)$ in polynomial form.

b. Generalize the result in part (a) to find the autocorrelation function for
$$x[n] = u[n] - u[n - N], \qquad N > 0$$
using z-transform techniques.

8.13. Use the correlation property of the z-transform to determine the cross correlation function for the signals
$$x[n] = u[n] - u[n - 3]$$
and
$$y[n] = u[n] - u[n - 5]$$
Hint: Write $X(z)$ and $Y(z)$ in polynomial form.

8.14. Prove that the cross correlation of signals $x[n]$ and $y[n]$ is equal to the convolution of $x[n]$ and $y[-n]$, that is,
$$r_{xy}[n] = x[n] * y[-n]$$

a. Using direct application of the convolution sum
b. Using correlation and time reversal properties of the z-transform

8.15. Let $x[n] = a^n u[n]$. A signal $w[n]$ is defined in terms of $x[n]$ as
$$w[n] = \sum_{k=-\infty}^{n} x[k]$$

a. Determine $X(z)$. Afterwards determine $W(z)$ from $X(z)$ using the summation property of the z-transform.
b. Determine $w[n]$ from the transform $W(z)$ using partial fraction expansion.
c. Determine $w[n]$ directly from the summation relationship using the geometric series formula and compare to the result found in part (b).

8.16. Using the summation property of the z transform prove the well-known formula

$$\sum_{k=0}^{n} k = \frac{n\,(n+1)}{2}$$

Hint: Use $x[n] = n\,u[n]$ and $w[n] = \sum_{k=-\infty}^{n} x[k]$.

8.17. Consider the z-transforms listed below along with their ROCs. For each case determine the inverse transform $x[n]$ using partial fraction expansion.

a. $X\left(z\right) = \dfrac{z}{\left(z+1\right)\left(z+2\right)}$,   ROC: $|z| < 1$

b. $X\left(z\right) = \dfrac{z+1}{\left(z+1/2\right)\left(z+2/3\right)}$,   ROC: $|z| > \dfrac{2}{3}$

c. $X\left(z\right) = \dfrac{z\left(z+1\right)}{\left(z-0.4\right)\left(z+0.7\right)}$,   ROC: $|z| > 0.7$

d. $X\left(z\right) = \dfrac{z\left(z+1\right)}{\left(z+3/4\right)\left(z-1/2\right)\left(z-3/2\right)}$,   ROC: $\dfrac{3}{4} < |z| < \dfrac{3}{2}$

e. $X\left(z\right) = \dfrac{z\left(z+1\right)}{\left(z+3/4\right)\left(z-1/2\right)\left(z-3/2\right)}$,   ROC: $\dfrac{1}{2} < |z| < \dfrac{3}{4}$

8.18. The transform $X\left(z\right)$ is given by

$$\frac{\left(z+1\right)\left(z-2\right)}{\left(z+1/2\right)\left(z-1\right)\left(z+2\right)}$$

The ROC is not specified.

a. Construct a pole-zero plot for $X\left(z\right)$ and identify all possibilities for the ROC.
b. Express $X\left(z\right)$ using partial fractions and determine the residues.
c. For each choice of the ROC, determine the inverse transform $x[n]$.

8.19. The transforms listed below have complex poles. Each is known to be the transform of a causal signal. Determine the inverse transform $x[n]$ using partial fraction expansion.

a. $X\left(z\right) = \dfrac{z^2 + 3\,z}{z^2 - 1.4\,z + 0.85}$

b. $X\left(z\right) = \dfrac{z^2}{z^2 - 1.6\,z + 1}$

c. $X\left(z\right) = \dfrac{z^2 + 3\,z}{z^2 - \sqrt{3}\,z + 1}$

8.20. The transforms listed below have multiple poles. Express each transform using partial fractions, and determine the residues.

a. $X\left(z\right) = \dfrac{z^2 + 3\,z + 2}{z^2 - 2\,z + 1}$

b. $X\left(z\right) = \dfrac{z\left(z+1\right)}{\left(z+0.9\right)^2\left(z-1\right)}$

c. $X\left(z\right) = \dfrac{z^2 + 4\,z - 7}{\left(z+0.9\right)^2\left(z-1.2\right)^2}$

8.21.   The difference equation that represents the remaining balance on a car loan was derived in Chapter 3 as

$$y[n] = (1 + c)\, y[n - 1] - x[n]$$

where c is the monthly interest rate, $x[n]$ is the payment made in month n, and $y[n]$ is the balance remaining.

a.   Treat the loan repayment system as a DTLTI system and find its system function $H(z)$.
b.   Let A and B represent the initial amount borrowed and the monthly payment amount respectively. Since we would like to treat the system as DTLTI we cannot use an initial value for the difference equation. Instead, we can view the borrowed amount as a negative payment made at $n = 0$, and monthly payments can be viewed as positive payments starting at $n = 1$. With this approach the input to the system would be

$$x[n] = -A\,\delta[n] + B\,u[n - 1]$$

Find the transform $X(z)$.
c.   Find the transform $Y(z)$ of the output signal that represents remaining balance each month.
d.   We would like to have the loan fully paid back after N payments, that is, $y[N] = 0$. Determine the amount of the monthly payment B in terms of parameters A, c and N.

8.22.   Consider again the transforms listed in Problem 8.19. Find the inverse of each transform using the long division method. In each case carry out the long division for 8 samples.

8.23.   Refer to the transform $X(z)$ in Problem 8.18 for which the ROC is not given. For each possible choice of the ROC, determine the inverse transform using long division. Carry out the long division operation to compute significant samples of $x[n]$ in the interval $-5 \leq n \leq 5$.

8.24.   Assume that the transforms listed below correspond to causal signals. Find the inverse of each transform for $n = 0, \ldots, 5$ using long division.

a. $X(z) = \dfrac{z}{(z + 1)\,(z + 2)}$

b. $X(z) = \dfrac{z + 1}{(z + 1/2)\,(z + 2/3)}$

c. $X(z) = \dfrac{z\,(z + 1)}{(z - 0.4)\,(z + 0.7)}$

8.25.   For each transform given below with its ROC, find the inverse transform using long division. For each, obtain $x[n]$ in the range $n = -4, \ldots, 4$.

a. $X(z) = \dfrac{z\,(z + 1)}{(z + 3/4)\,(z - 1/2)\,(z - 3/2)}$,   ROC: $\dfrac{3}{4} < |z| < \dfrac{3}{2}$

b. $X(z) = \dfrac{z\,(z + 1)}{(z + 3/4)\,(z - 1/2)\,(z - 3/2)}$,   ROC: $\dfrac{1}{2} < |z| < \dfrac{3}{4}$

Compare with the analytical results found in Problem 8.17 parts (d) and (e).

8.26. Find the system function $H(z)$ for each causal DTLTI system described below by means of a difference equation. Afterwards determine the impulse response of the system using partial fraction expansion.

a. $y[n] = 0.9\,y[n-1] + x[n] - x[n-1]$
b. $y[n] = 1.7\,y[n-1] - 0.72\,y[n-2] + x[n] - 2\,x[n-1]$
c. $y[n] = 1.7\,y[n-1] - 0.72\,y[n-2] + x[n] + x[n-1] + x[n-2]$
d. $y[n] = y[n-1] - 0.11\,y[n-2] - 0.07\,y[n-3] + x[n-1]$

8.27. Several causal DTLTI systems are described below by means of their system functions. Find a difference equation for each system.

a. $H(z) = \dfrac{z+1}{z^2 + 5z + 6}$

b. $H(z) = \dfrac{(z-1)^2}{(z+2/3)\,(z+1/2)\,(z-4/5)}$

c. $H(z) = \dfrac{z^2+1}{z^3 + 1.2\,z^2 - 1.8}$

8.28. A causal DTLTI system is described by the difference equation

$$y[n] = -0.1\,y[n-1] + 0.56\,y[n-2] + x[n] - 2\,x[n-1]$$

a.   Find the system function $H(z)$.
b.   Determine the impulse response $h[n]$ of the system.
c.   Using z-transform techniques, determine the unit-step response of the system.
d.   Using z-transform techniques, find the response of the system to a time-reversed unit-step signal $x[n] = u[-n]$.

8.29. Repeat Problem 8.28 for an anti-causal DTLTI system is described by the difference equation

$$y[n] = -\frac{5}{6}\,y[n+1] - \frac{1}{6}\,y[n+2] + \frac{1}{6}\,x[n+1] + \frac{1}{6}\,x[n+2]$$

8.30. A causal DTLTI system is characterized by the system function

$$H(z) = \frac{0.04\,z}{z - 0.96}$$

a.   Using z-transform techniques, determine the response of the system to the causal sinusoidal signal

$$x[n] = \sin(0.01n)\,u[n]$$

b.   Determine the response of the system to the non-causal sinusoidal signal

$$x[n] = \sin(0.01n)$$

This is the steady-state response of the system. Refer to the discussion in Section 8.5.3.
c.   Compare the responses in parts (a) and (b). Approximately how many samples does it take for the response to the causal sinusoidal signal in part (a) to be almost equal to the steady-state response found in part (b)?

8.31. The impulse response of a DTLTI system is

$$h[n] = (0.8)^n \cos(0.2\pi n)\, u[n]$$

   a. Determine the system function $H(z)$ and indicate its ROC.
   b. Draw the pole-zero plot for the system function and determine its stability.
   c. Write a difference equation for this system.
   d. Determine the unit-step response of the system using z-transform techniques.

8.32. Input and output signal pairs are listed below for several DTLTI systems. For each case, determine the system function $H(z)$ along with its ROC. Also indicate if the system considered is stable and/or causal.

   a. $x[n] = (1/2)^n u[n],\qquad y[n] = 3\,(1/2)^n u[n] + 2\,(3/4)^n u[n]$
   b. $x[n] = (1/2)^n u[n],\qquad y[n] = (1/2)^{n+3} u[n+2] + (1/2)^{n+2} u[n+1]$
   c. $x[n] = u[n],\qquad y[n] = (n-1)\,u[n-1]$
   d. $x[n] = 1.25\,\delta[n] - 0.25\,(0.8)^n u[n],\qquad y[n] = (0.8)^n u[n]$

8.33. Consider the feedback control system shown in Fig. P.8.33.

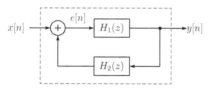

Figure P. 8.33

   a. Determine the overall system function $H(z)$ in terms of the system functions $H_1(z)$ and $H_2(z)$ of the two subsystems.
   b. Let the impulse responses of the two subsystems be $h_1[n] = u[n]$ and $h_2[n] = K\,\delta[n-1]$ where K is a constant gain parameter. Determine the system functions $H_1(z)$ and $H_2(z)$. Afterwards determine the overall system function $H(z)$.
   c. Determine the range of K for which the system is stable.
   d. Let $K = -3/2$. If the input signal is a unit step, find the signal $e[n]$ and the output signal $y[n]$.

8.34. The impulse response $h[n]$ of a DTLTI system is real-valued. Let the system function evaluated at some point $z = z_0$ be expressed in the form

$$H(z_0) = H_0\, e^{j\Theta_0}$$

where

$$H_0 = |H(z_0)|$$

and

$$\Theta_0 = \angle H(z_0)$$

Show that the value of the system function at the point $z = z_0^*$ is the complex conjugate of its value at the point $z = z_0$, that is,

$$H(z_0^*) = [H(z_0)]^* = H_0\, e^{-j\Theta_0}$$

Hint: Apply the z-transform definition to $h[n]$, evaluate the resulting expression at $z = z_0^*$, and manipulate the result to complete the proof.

8.35. A DTLTI system has the system function

$$H\left(z\right) = \frac{z^2 + 3\,z}{z^2 - 1.4\,z + 0.85}$$

Find the steady-state response of the system to each of the following signals:

a. $x[n] = (0.8)^n\, e^{j0.4\pi n}$
b. $x[n] = (0.9)^n\, \cos\left(0.3\pi n\right)$

8.36. A DTLTI system has the system function

$$H\left(z\right) = \frac{z - 0.4}{z\left(z^2 - 1.4\,z + 0.85\right)}$$

a.   Draw a pole-zero diagram to show poles and zeros of the system function as well as the unit circle of the z-plane.
b.   Using the graphical method discussed in Section 8.5.4 determine the magnitude and the phase of the system function at the angular frequency $\Omega = \pi/6$.
c.   Repeat part (b) at $\Omega = \pi/4$.

8.37.   For each function $H\left(z\right)$ given below, determine if it could be the system function of a *causal and stable* system. In each case provide the reasoning behind your answer.

a. $H\left(z\right) = \dfrac{(z - 2)\,(z + 2)}{(z + 1/2)}$

b. $H\left(z\right) = \dfrac{(z + 1)^2}{(z - 1/2)\,(z + 1/3)}$

c. $H\left(z\right) = \dfrac{z\,(z + 1)}{(z + 1/2)\,(z + 3/2)}$

8.38.   Several causal DTLTI systems are described below by their difference equations. For each system, determine the system function $H\left(z\right)$. Draw a pole-zero diagram for the system and indicate the ROC for the system function. Determine if the system is stable.

a. $y[n] = 1.5\,y[n - 1] - 0.54\,y[n - 2] + x[n] + 3\,x[n - 1]$
b. $y[n] = -0.64\,y[n - 2] + 2\,x[n]$
c. $y[n] = 0.25\,y[n - 1] - 0.125\,y[n - 2] - 0.5\,y[n - 3] + x[n]$
d. $y[n] = 0.25\,y[n - 1] - 0.5\,y[n - 2] - 0.75\,y[n - 3] + x[n] + x[n - 1]$

8.39.   Consider again the car loan repayment system discussed in Problem 8.21.

a.   Show that the loan repayment system is unstable for any practical setting. You may find it helpful to draw a pole-zero diagram.
b.   Find two examples of bounded input signals $|x[n]| < \infty$ that result in unbounded output signals.
c.   Can a bounded input signal be found that produces an output signal with a non-zero constant steady-state value? What type of a payment scheme does this represent?

8.40.   A first-order allpass filter section is characterized by the system function

$$H\left(z\right) = \frac{z - r\,e^{j\Omega_0}}{z - (1/r)\,e^{j\Omega_0}}$$

Show that the phase characteristic of the system is

$$\angle H\left(\Omega\right) = \tan^{-1}\left[\frac{\left(r^2 - 1\right)\sin\left(\Omega - \Omega_0\right)}{2r - \left(r^2 + 1\right)\cos\left(\Omega - \Omega_0\right)}\right]$$

Hint: First find $H\left(\Omega\right)$ by substituting $z = e^{j\Omega}$. Afterwards express $H\left(\Omega\right)$ in Cartesian form as

$$H\left(\Omega\right) = H_r\left(\Omega\right) + jH_i\left(\Omega\right)$$

8.41. Develop a direct-form II block diagram for each system specified below by means of a system function. Assume that each system is causal and initially relaxed.

a. $H\left(z\right) = \dfrac{z + 2}{z - 1/2}$

b. $H\left(z\right) = \dfrac{z^2 + 1}{z^3 + 0.8\,z^2 - 2.2\,z + 0.6}$

c. $H\left(z\right) = \dfrac{z^3 + 2\,z^2 - 3\,z + 4}{2\,z^3 + 0.8\,z^2 + 1.8\,z + 3.2}$

8.42. Develop a cascade-form block diagram for each system specified below by means of a system function. Assume that each system is causal and initially relaxed. Use first- and second-order cascade sections and ensure that all coefficients are real.

a. $H\left(z\right) = \dfrac{z + 1}{\left(z + 1/2\right)\left(z + 2/3\right)}$

b. $H\left(z\right) = \dfrac{z\left(z + 1\right)}{\left(z - 0.4\right)\left(z + 0.7\right)}$

c. $H\left(z\right) = \dfrac{z\left(z + 1\right)}{\left(z + 0.6\right)\left(z^2 - 1.4\,z + 0.85\right)}$

8.43. Develop a parallel-form block diagram for each system specified in Problem 8.42. Assume that each system is causal and initially relaxed. Use first and second-order parallel sections and ensure that all coefficients are real.

8.44. Let $X_u\left(z\right)$ be the unilateral z-transform of $x[n]$.

a. Using the definition of the unilateral z-transform show that the transform of $x[n+1]$ is

$$\mathcal{Z}_u\left\{x[n+1]\right\} = z\,X_u\left(z\right) - z\,x[0]$$

b. Show that the transform of $x[n+2]$ is

$$\mathcal{Z}_u\left\{x[n+2]\right\} = z^2\,X_u\left(z\right) - z^2\,x[0] - z\,x[1]$$

c. Generalize the results of parts (a) and (b), and show that for $k > 0$

$$\mathcal{Z}_u\left\{x[n+k]\right\} = z^k\,X_u\left(z\right) - \sum_{n=0}^{k-1} x[n]\,z^{k-n}$$

8.45. Using the unilateral z-transform determine the natural response of each system with the homogeneous difference equation and initial conditions given below:

a. $y[n] - 1.4y[n-1] + 0.85y[n-2] = 0$, $\quad y[-1] = 5$, $\quad$ and $\quad y[-2] = 7$

b. $y[n] - 1.6y[n-1] + 0.64y[n-2] = 0$, $\quad y[-1] = 2$, $\quad$ and $\quad y[-2] = -3$

8.46. Consider the difference equations given below with the specified input signals and initial conditions. Using the unilateral z-transform determine the solution $y[n]$ for each.

a. $y[n] - 2\,y[n-1] = x[n]\,,\quad x[n] = \cos\left(0.2\pi n\right) u[n]\,,\quad y[-1] = 5$

b. $y[n] + 0.6\,y[n-1] = x[n] + x[n-1]\,,\quad x[n] = u[n]\,,\quad y[-1] = -3$

c. $y[n] - 0.2\,y[n-1] - 0.48\,y[n-2] = x[n]\,,\quad x[n] = u[n]\,,\quad y[-1] = -1\,,\quad y[-2] = 2$

8.47. The *Fibonacci sequence* is a well-known sequence of integers. Fibonacci numbers start with the pattern

$$0, 1, 1, 2, 3, 5, 8, 13, 21, 34, 55, 89, \ldots$$

They follow the recursion relationship

$$F_{n+2} = F_{n+1} + F_n$$

with $F_0 = 0$ and $F_1 = 1$. In simplest terms, each number in the sequence is the sum of the two numbers before it. In this problem we will use difference equations and the unilateral z-transform to study some properties of the Fibonacci sequence.

a. Construct a homogeneous difference equation to represent the recurrence relationship of the Fibonacci sequence. Select $y[n] = F_{n+2}$ with initial conditions $y[-1] = F_1 = 1$ and $y[-2] = F_0 = 0$.

b. Find the solution $y[n]$ of the homogeneous difference equation using the unilateral z-transform.

c. It can be shown that, for large n, the ratio of consecutive Fibonacci numbers converges to a constant known as the *golden ratio*, that is,

$$\lim_{n \to \infty} \frac{y[n+1]}{y[n]} = \varphi$$

Using the solution $y[n]$ found in part (b) determine the value of φ.

MATLAB Problems

8.48. The signal $x[n]$ is given by

$$x[n] = (0.8)^n \left(u[n] - u[n-8]\right)$$

a. Find $X(z)$. Determine its poles and zeros. Manually construct a pole-zero diagram.

b. Write a MATLAB script to evaluate the magnitude of $X(z)$ at a grid of complex points in the z-plane. Use the function `meshgrid(..)` to generate the grid of complex points within the ranges $-1.5 < \operatorname{Re}\{z\} < 1.5$ and $-1.5 < \operatorname{Im}\{z\} < 1.5$ with increments of 0.05 in each direction.

c. Use the function `mesh(..)` to produce a three-dimensional mesh plot of $|X(z)|$.

d. Evaluate the z-transform for $z = e^{j\Omega}$ and use the function `plot3(..)` to plot it over the three-dimensional mesh plot.

8.49.

a. Repeat Problem 8.48 with the signal

$$x[n] = (0.6)^n \left(u[n] - u[n-8]\right)$$

b. Repeat Problem 8.48 with the signal

$$x[n] = (0.4)^n \ (u[n] - u[n-8])$$

8.50. Refer to the system in Problem 8.30.

a. Construct a system object for $H(z)$ using the function `zpk(..)`.
b. Compute the response of the system to the input signal

$$x[n] = \sin(0.01n) \ u[n]$$

for $n = 0, \ldots, 49$ using the function `dlsim(..)`. Compare the first few samples of the response to the result of hand calculations in Problem 8.30.

8.51. Refer to the system in Problem 8.31.

a. Write a MATLAB script to find the impulse response of the system by iteratively solving the difference equation found in Problem 8.31 part (c). Compute $h[n]$ for $n = 0, \ldots, 10$. Does it match the expected answer?
b. Modify the script written in part (a) so that the unit-step response of the system is computed by iteratively solving the difference equation. Compare the result to that found in Problem 8.31 part (d).
c. Construct a system object for $H(z)$ using the function `tf(..)`. Afterwards compute the impulse response and the unit-step response of the system using functions `dimpulse(..)` and `dstep(..)` respectively. Compare to the answers obtained in previous parts.

8.52. Refer to the DTLTI system in Problem 8.35.

a. Write a MATLAB script to compute and graph the magnitude and the phase of the frequency spectrum for the system. Your script should also mark the points critical for the two input signals used in Problem 8.35.
b. Graph the steady-state response to each of the two input signals specified.

8.53. Develop a script to compute the phase response of a first-order allpass filter with a real pole at $p = r + j0$ for parameter values $r = 0.2, 0.4, 0.6, 0.8$. Graph the four phase responses on the same coordinate system for comparison.

8.54. Develop a script to compute the phase response of a second-order allpass filter with a complex conjugate poles at

$$p_{1,2} = r \, e^{j\theta}$$

a. In your script set $\theta = \pi/6$ radians. Compute the phase response for $r = 0.4, 0.6, 0.8$, and graph the results superimposed for comparison.
b. Set $r = 0.8$. Compute the phase response for $\theta = \pi/6, \pi/4, \pi/3$ radians, and graph the results superimposed for comparison.

8.55. Refer to Problem 8.45.

a. For each difference equation given, write a script to find the solution for sample indices $n = 0, \ldots, 10$ using the iterative solution method with the specified initial conditions.
b. For each difference equation given, write a script to find the analytical solution using symbolic mathematics capabilities of MATLAB.

MATLAB Project

8.56. In Problem 1.51 of Chapter 1 the generation of DTMF signals was explored using MATLAB's trigonometric functions. In this project we will explore alternative means to sinusoidal tone generation.

In some applications we may wish to use inexpensive processors that may either have no support for trigonometric functions or may be too slow in computing them. An alternative method of computing the samples of a sinusoidal signal is to use a discrete-time system the impulse response of which is sinusoidal, and to apply an impulse signal to its input. Recall that the z-transform of a discrete-time sine signal was found in Example 8.16 as

$$\mathcal{Z}\left\{\sin\left(\Omega_i n\right) u[n]\right\} = \frac{\sin\left(\Omega_i\right) z}{z^2 - 2\cos\left(\Omega_i\right) z + 1}$$

which could be used as the basis of a second-order system as shown in Fig. P.8.56(a). DTMF signals that correspond to digits on a keypad can be generated by using two systems of this type in parallel as shown in Fig. P.8.56(b) by adjusting the coefficients properly.

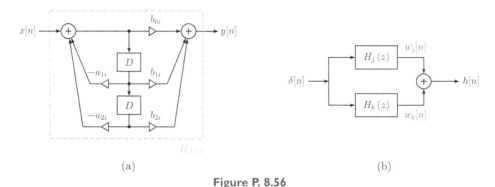

(a)   (b)

Figure P. 8.56

a. Develop a MATLAB function `ss_dtmf2(..)` to generate samples of the DTMF signal for a specified digit for a specified duration. The syntax of the function should be as follows:

```
x = ss_dtmf2(digit,fs,n)
```

The first argument "`digit`" is the digit for which the DTMF signal is to be generated. Let values digit = 0 through digit = 9 represent the corresponding keys on the keypad. Map the remaining two keys "∗" and "#" to values digit = 10 and digit = 11 respectively. Finally, the value digit = 12 should represent a pause, that is, a silent period. The arguments "`fs`" and "`n`" are the sampling rate (in Hz) and the number of samples to be produced respectively. Internally you may want to use the function `ss_iir2(..)` that was developed in MATLAB Exercise 8.12.

b. Develop a function named `ss_dtmf(..)` with the syntax

```
x = ss_dtmf(number,fs,nd,np)
```

The arguments for the function `ss_dtmf(..)` are defined as follows:

number: The phone number to be dialed, entered as a vector. For example, to dial the number 555-1212, the vector "**number**" would be entered as

```
number = [5,5,5,1,2,1,2]
```

fs: The sampling rate (in Hz) used in computing the amplitudes of the DTMF signal.

nd: Number of samples for each digit.

np: Number of samples for each pause between consecutive digits.

The function ss_dtmf(..) should use the function ss_dtmf2(..) to produce the signals for each digit (and the pauses between digits) and append them together to create the signal $x[n]$.

c. Write a script to test the function ss_dtmf(..) with the number 555-1212. Use a sampling rate of $f_s = 8000$ Hz. The duration of each digit should be 200 milliseconds with 80 millisecond pauses between digits.

d. Play back the resulting signal $x[n]$ using the sound(..) function.

Chapter 9

State-Space Analysis of Systems

Chapter Objectives

- Understand the concepts of a *state variable* and *state-space representation* of systems.

- Learn the advantages of state-space modeling over other methods of representing continuous-time and discrete-time systems.

- Learn to derive state-space models for CTLTI and DTLTI systems from the knowledge of other system descriptions such as system function, differential equation or difference equation.

- Learn how to obtain a system function, a differential equation or a difference equation from the state-space model of a system.

- Develop techniques for solving state-space models for determining the state variables and the output signal of a system.

- Explore methods of converting a state-space model for a CTLTI system to that of a DTLTI system to allow simulation of CTLTI systems on a computer.

9.1 Introduction

In previous chapters of this text we have considered three different methods for describing a linear and time-invariant system:

1. A linear constant-coefficient differential equation (or, in the case of a discrete-time system, a difference equation) involving the input and the output variables of the system in the time domain.
2. A system function $H(s)$ for a CTLTI system, or $H(z)$ for a DTLTI system, describing the frequency-domain behavior characteristics of the system.

3. An impulse response $h(t)$ for a CTLTI system, or $h[n]$ for a DTLTI system, that describes how the system responds to a continuous unit-impulse function $\delta(t)$ or a discrete-time unit-impulse function $\delta[n]$ as appropriate.

We know that these three descriptions are equivalent. For a continuous-time system, the system function $H(s)$ can be obtained from the differential equation using Laplace transform techniques, specifically using the linearity and the differentiation-in-time properties of the Laplace transform. If the system under consideration is discrete, then the z-domain transfer function $H(z)$ can be obtained from the governing difference equation in a similar manner, using the linearity and the time-shifting properties of the z-transform. Afterwards, the impulse response $h(t)$ or $h[n]$ can be determined from the appropriate system function using inverse transform techniques.

Alternatively, we may need to find the differential equation or the difference equation of a system specified through its impulse response. This can be achieved by simply reversing the order of steps described above. Thus, the three definition forms listed above for a linear and time-invariant system are interchangeable since we are able to go back and forth between them freely.

For signal-system interaction problems, that is, for problems involving the actions of a system on an input signal to produce an output signal, the solution of a general N-th order differential equation may be difficult or impossible to carry out analytically for anything other than fairly simple systems and input signals. Sometimes we can avoid the need to solve the differential equation, and attempt to solve the problem using frequency-domain techniques instead. This involves first translating the problem into the frequency domain, solving it in terms of transform-based descriptions of the signals and the systems involved, and finally translating the results back to the time domain. However, this approach only works for systems that have zero initial conditions, since the transfer function concept is valid for only CTLTI or DTLTI systems that are initially relaxed. Furthermore, if the system under consideration is nonlinear or time-varying or both, then the system function-based approach cannot be used at all, since system functions are only available for CTLTI or DTLTI systems.

In this chapter we will introduce yet another method for describing a system, namely *state-space representation* based on a set of variables called the *state variables* of the system.

Consider a continuous-time system that is originally described by means of a N-th order differential equation. A state-space model for such a system can be obtained by replacing the N-th order differential equation with N first-order differential equations that involve as many state variables and their derivatives. In general, the N first-order differential equations are coupled with each other, and must be solved together as a set rather than as individual equations.

Similarly, a discrete-time system that is described through a N-th order difference equation can be represented by N first-order difference equations using as many discrete-time state variables and their one-sample delayed versions.

As evident from the foregoing discussion, the state-space model of a system is a time-domain model rather than a frequency-domain one since it is derived from a differential equation or difference equation. State-space modeling of systems provides a number of advantages over other modeling techniques:

1. State-space modeling provides a standard method of formulating the time-domain input-output relationship of a system independent of the order of the system.

2. State-space model of a system describes the internal structure of the system, in contrast with the other description forms which simply describe the relationship between input and output signals.

3. Systems with multiple inputs and/or outputs can be handled within the standard formulation we will develop in the rest of this chapter without the need to resort to special techniques that may be dependent on the topology of the system or on the number of input or output signals.

4. State-space models for CTLTI and DTLTI systems are naturally suitable for use in computer simulation of these systems.

5. State-space modeling also lends itself to the representation of nonlinear and/or time-varying systems, a task that is not achievable through modeling with the use of the system function or the impulse response. A N-th order differential equation or difference equation can be used for representing nonlinear and/or time-varying signals; however, computer solutions for these types of systems still necessitate the use of state-space modeling techniques.

In Section 9.2 we discuss state-space models for CTLTI systems. The discussion starts with the problem of obtaining a state-space model for a system described by means of a circuit diagram, differential equation or system function. Afterwards, methods for solving state equations are given. The problem of obtaining the system function from state equations is examined. The issue of system stability is studied from a state-space perspective. Section 9.3 repeats the topics of Section 9.2 for DTLTI systems. Conversion of the state-space model of a CTLTI system into that of an approximately equivalent DTLTI system for the purpose of simulation is the subject of Section 9.4.

9.2 State-Space Modeling of Continuous-Time Systems

Consider a single-input single-output continuous-time system described by a N-th order differential equation. Let $r(t)$ and $y(t)$ be the input and the output signals of the system respectively. Note that in this chapter we will change our notation a bit, and use $r(t)$ for an input signal instead of the usual $x(t)$. This change is needed in order to reserve the use of $x_i(t)$ for state variables as it is customary. The system under consideration can be thought of as a *black box* with input signal $r(t)$, output signal $y(t)$ and N internal variables $x_i(t)$ for $i = 1, \ldots, N$ as shown in Fig. 9.1.

These internal variables are the *state variables* or, more compactly, the *states* of the system under consideration. The system can be described through a set of first-order differential equations written for each state variable:

$$\frac{dx_1(t)}{dt} = f_1\left[\, x_1(t), \ldots, x_N(t), r(t)\,\right]$$

$$\frac{dx_2(t)}{dt} = f_2\left[\, x_1(t), \ldots, x_N(t), r(t)\,\right]$$

$$\vdots$$

$$\frac{dx_N(t)}{dt} = f_N\left[\, x_1(t), \ldots, x_N(t), r(t)\,\right] \tag{9.1}$$

The first derivative of each state variable appears on the left side of each equation, and is set equal to some function $f_i[\ldots]$ of all state variables and the input $r(t)$. The equations given in Eqn. (9.1) are referred to as the *state equations* of the system. They describe how

Figure 9.1 – State-space representation of a system.

each state variable changes in the time domain in response to

1. The input signal $r(t)$
2. The current values of all state variables

In a sense, observing the time-domain behavior of the state variables provides us with information about the internal operation of the system. Once the N state equations are solved for the N state variables, the output signal can be found through the use of the *output equation*

$$y(t) = g\left[x_1(t), \dots, x_N(t), r(t)\right] \tag{9.2}$$

Eqns. (9.1) and (9.2) together form a *state-space model* for the system.

Up to this point in our discussion we focused on a system with a single input signal $r(t)$ and a single output signal $y(t)$. State-space models can be extended to systems with multiple input and/or output signals in a straightforward manner. Consider a system with K input signals $\{r_i(t); \ i = 1, \dots, K\}$ and M output signals $\{y_j(t); \ i = 1, \dots, M\}$ as shown in Fig. 9.2.

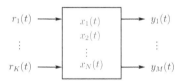

Figure 9.2 – State-space representation of a multiple-input multiple-output system.

The state equations given by Eqn. (9.1) can be modified as follows to represent this system through a state-space model:

$$\frac{dx_1(t)}{dt} = f_1\left[x_1(t), \dots, x_N(t), r_1(t), \dots, r_K(t)\right]$$

$$\frac{dx_2(t)}{dt} = f_2\left[x_1(t), \dots, x_N(t), r_1(t), \dots, r_K(t)\right]$$

$$\vdots$$

$$\frac{dx_N(t)}{dt} = f_N\left[x_1(t), \dots, x_N(t), r_1(t), \dots, r_K(t)\right] \tag{9.3}$$

Each right-side function $f_i[\dots]$ becomes a function of all input signals as well as the state variables. In addition to this change, the output equation given by Eqn. (9.2) needs to be replaced with a set of output equations for computing all output signals:

$$y_1(t) = g_1\left[x_1(t), \ldots, x_N(t), r_1(t), \ldots, r_K(t)\right]$$

$$y_2(t) = g_2\left[x_1(t), \ldots, x_N(t), r_1(t), \ldots, r_K(t)\right]$$

$$\vdots$$

$$y_M(t) = g_M\left[x_1(t), \ldots, x_N(t), r_1(t), \ldots, r_K(t)\right] \qquad (9.4)$$

Thus, the functions $g_1[\ldots], g_2[\ldots], \ldots, g_M[\ldots]$ allow the output signals to be computed from the state variables and the input signals.

In most signal-system interaction problems our main focus is on the relationship between the input and the output signals of the system. Often we are faced with one of two types of questions:

1. What is the output signal of a system in response to a specified input signal?
2. What system produces a specified output signal for a specified input signal?

The first question represents an analysis perspective, and the second question represents a design perspective. Even though both questions are posed with a single-input single-output system in mind, they can easily be adapted to multiple-input multiple-output systems. In answering either question, the role of state variables is an intermediate one. We will see in later parts of this chapter that a given system can have multiple state-space models that are equivalent in terms of the input-output relationships they define. They would differ from each other, however, in terms of how they model the internal structure of the system.

In some state-space models, the state variables $x_1(t), \ldots, x_N(t)$ chosen correspond to actual physical signals within the system. In other models they may be purely mathematical variables that do not correspond to any physical signals. If we are in a situation where we need to understand the internal workings of a system, we may want to choose state variables that correspond to actual signals within the system. If, on the other hand, our focus is solely on the input-output relationship of the system, the role of state variables is intermediate. In such a case we would choose the state variables that lead to the simplest possible structure in the state equations given by Eqn. 9.1.

9.2.1 State-space models for CTLTI systems

The formulation given in Eqns. (9.1) through (9.4) was not based on any assumptions about the system under consideration. Therefore, it will apply to systems that do not necessarily have to be linear or time-invariant. If the continuous-time system is also linear and time-invariant (CTLTI), the right-side functions $f_i[\ldots], i = 1, \ldots, N$ and $g_i[\ldots], i = 1, \ldots, M$ are linear functions of the state variables and the input signals. For a single-input single-output CTLTI system, the state-space representation is in the form

$$\frac{dx_1(t)}{dt} = a_{11}x_1(t) + a_{12}x_2(t) + \ldots + a_{1N}x_N(t) + b_1 r(t)$$

$$\frac{dx_2(t)}{dt} = a_{21}x_1(t) + a_{22}x_2(t) + \ldots + a_{2N}x_N(t) + b_2 r(t)$$

$$\vdots$$

$$\frac{dx_N(t)}{dt} = a_{N1}x_1(t) + a_{N2}x_2(t) + \ldots + a_{NN}x_N(t) + b_N r(t) \qquad (9.5)$$

and the output equation is in the form

$$y\left(t\right) = c_1\,x_1\left(t\right) + c_2\,x_2\left(t\right) + \ldots, c_N\,x_N\left(t\right) + d\,r\left(t\right) \tag{9.6}$$

Coefficients a_{ij}, b_i, c_j and d used in the state equations and the output equation are all constants independent of the time variable. Eqns. (9.5) and (9.6) can be expressed in matrix notation as follows:

$$\text{State equation:} \qquad \dot{\mathbf{x}}\left(t\right) = \mathbf{A}\,\mathbf{x}\left(t\right) + \mathbf{B}\,r\left(t\right) \tag{9.7}$$

$$\text{Output equation:} \qquad y\left(t\right) = \mathbf{C}\,\mathbf{x}\left(t\right) + d\,r\left(t\right) \tag{9.8}$$

The coefficient matrices used in Eqns. (9.7) and (9.8) are

$$\mathbf{A} = \begin{bmatrix} a_{11} & a_{12} & \cdots & a_{1N} \\ a_{21} & a_{22} & \cdots & a_{2N} \\ \vdots & & & \\ a_{N1} & a_{N2} & \cdots & a_{NN} \end{bmatrix}, \quad \mathbf{B} = \begin{bmatrix} b_1 \\ b_2 \\ \vdots \\ b_N \end{bmatrix}, \quad \mathbf{C} = \begin{bmatrix} c_1 \\ c_2 \\ \vdots \\ c_N \end{bmatrix}^T \tag{9.9}$$

The vector $\mathbf{x}\left(t\right)$ is the *state vector* that consists of the N state variables:

$$\mathbf{x}\left(t\right) = \begin{bmatrix} x_1\left(t\right) \\ x_2\left(t\right) \\ \vdots \\ x_N\left(t\right) \end{bmatrix} \tag{9.10}$$

The vector $\dot{\mathbf{x}}\left(t\right)$ contains the derivatives of the N state variables with respect to time:

$$\dot{\mathbf{x}}\left(t\right) = \begin{bmatrix} dx_1\left(t\right)/dt \\ dx_2\left(t\right)/dt \\ \vdots \\ dx_N\left(t\right)/dt \end{bmatrix} \tag{9.11}$$

Thus, a single-input single-output CTLTI system can be uniquely described by specifying the matrix $\mathbf{A}$, the vectors $\mathbf{B}$ and $\mathbf{C}$, and the scalar d. If the initial value $\mathbf{x}\left(0\right)$ of the state vector is known, the state equation given by Eqn. (9.7) can be solved for $\mathbf{x}\left(t\right)$. Once $\mathbf{x}\left(t\right)$ is determined, the output signal can be found through the use of Eqn. (9.8).

Some authors use a variant of the state equation in place of Eqn. (9.7), and express $\mathbf{x}\left(t\right)$ as follows:

$$\dot{\mathbf{x}}\left(t\right) = \mathbf{A}\,\mathbf{x}\left(t\right) + \mathbf{B}\,r\left(t\right) + \mathbf{E}\,\frac{dr\left(t\right)}{dt} \tag{9.12}$$

It can be shown (see Problem 9.2 at the end of this chapter) that this alternative form of the state equation can be converted into the standard form used in Eqn. (9.7) by the use of a linear transformation of the state vector, and therefore does not warrant separate analysis.

9.2.2 Obtaining state-space model from physical description

In this section we will present examples of obtaining a state-space model for a system starting with the physical relationships that exist between the signals within the system. The next two examples will illustrate this.

Example 9.1: **State-space model for RLC circuit**

Find a state-space model for the RLC circuit shown in Fig. 9.3 by choosing the inductor current and the capacitor voltage as the state variables.

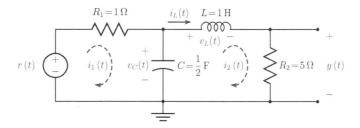

Figure 9.3 – The RLC circuit for Example 9.1.

Solution: We will begin by writing KVL for the two loops:

$$-r\left(t\right) + R_1\, i_1\left(t\right) + v_C\left(t\right) = 0 \tag{9.13}$$

$$-v_C\left(t\right) + v_L\left(t\right) + R_2\, i_2\left(t\right) = 0 \tag{9.14}$$

The voltage of an inductor is proportional to the derivative of its current:

$$v_L\left(t\right) = L\,\frac{di_2\left(t\right)}{dt} = L\,\frac{di_L\left(t\right)}{dt} \tag{9.15}$$

The current of a capacitor is proportional to the derivative of its voltage; therefore, we have

$$i_1\left(t\right) - i_2\left(t\right) = C\,\frac{dv_c\left(t\right)}{dt}$$

which can be solved for $i_1\left(t\right)$ to yield

$$i_1\left(t\right) = C\,\frac{dv_c\left(t\right)}{dt} + i_2\left(t\right) = C\,\frac{dv_c\left(t\right)}{dt} + i_L\left(t\right) \tag{9.16}$$

Substituting Eqns. (9.15) and (9.16) into Eqns. (9.13) and (9.14) and rearranging terms we obtain

$$\frac{dv_C\left(t\right)}{dt} = -\frac{1}{R_1 C}\, v_C\left(t\right) - \frac{1}{C}\, i_L\left(t\right) + \frac{1}{R_1 C}\, r\left(t\right) \tag{9.17}$$

$$\frac{di_L\left(t\right)}{dt} = \frac{1}{L}\, v_C\left(t\right) - \frac{R_2}{L}\, i_L\left(t\right) \tag{9.18}$$

The output signal is

$$y\left(t\right) = R_2\, i_L\left(t\right) \tag{9.19}$$

Let the two state variables $x_1\left(t\right)$ and $x_1\left(t\right)$ be defined as

$$x_1\left(t\right) = v_C\left(t\right) \quad \text{and} \quad x_2\left(t\right) = i_L\left(t\right)$$

Using the state variables, the state equations are

$$\frac{dx_1(t)}{dt} = -\frac{1}{R_1 C} x_1(t) - \frac{1}{C} x_2(t) + \frac{1}{R_1 C} r(t) \tag{9.20}$$

$$\frac{dx_2(t)}{dt} = \frac{1}{L} x_1(t) - \frac{R_2}{L} x_2(t) \tag{9.21}$$

and the output equation is

$$y(t) = R_2 x_2(t) \tag{9.22}$$

The state-space model in Eqns. (9.20) through (9.22) can be expressed in matrix form as

$$\dot{\mathbf{x}}(t) = \begin{bmatrix} -1/R_1 C & -1/C \\ 1/L & -R_2/L \end{bmatrix} \mathbf{x}(t) + \begin{bmatrix} 1/R_1 C \\ 0 \end{bmatrix} r(t)$$

and

$$y(t) = \begin{bmatrix} 0 & R_2 \end{bmatrix} \mathbf{x}(t) + (0) r(t)$$

With the substitution of the numerical values for circuit elements we get

$$\dot{\mathbf{x}}(t) = \begin{bmatrix} -2 & -2 \\ 1 & -5 \end{bmatrix} \mathbf{x}(t) + \begin{bmatrix} 1 \\ 0 \end{bmatrix} r(t)$$

and

$$y(t) = \begin{bmatrix} 0 & 5 \end{bmatrix} \mathbf{x}(t) + (0) r(t)$$

Example 9.2: **State-space model for series RLC circuit**

Find a state-space model for the series RLC circuit shown in Fig. 9.4.

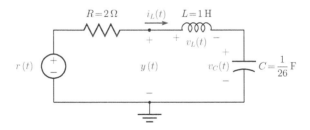

Figure 9.4 – The series RLC network for Example 9.2.

Solution: Applying KVL to the circuit, we obtain

$$r(t) = R i_L(t) + v_L(t) + v_C(t) \tag{9.23}$$

Using the fundamental properties of the capacitor and the inductor we have

$$v_L(t) = L \frac{di_L(t)}{dt} \quad \text{and} \quad i_L(t) = C \frac{dv_C(t)}{dt}$$

Using them with Eqn. (9.23) we get the two state equations:

$$\frac{di_L(t)}{dt} = -\frac{R}{L} i_L(t) - \frac{1}{L} v_C(t) + \frac{1}{L} r(t)$$

and

$$\frac{dv_C(t)}{dt} = \frac{1}{C} i_L(t)$$

The output voltage $y(t)$ is obtained as

$$y(t) = -R i_L(t) + r(t)$$

We will choose the inductor current and the capacitor voltage as the state variables of the system, so we have $x_1(t) = i_L(t)$, and $x_2(t) = v_C(t)$. The state equations are

$$\frac{dx_1(t)}{dt} = -\frac{R}{L} x_1(t) - \frac{1}{L} x_2(t) + \frac{1}{L} r(t) \tag{9.24}$$

$$\frac{dx_2(t)}{dt} = \frac{1}{C} x_1(t) \tag{9.25}$$

and the output equation is

$$y(t) = -R x_1(t) + r(t) \tag{9.26}$$

The state equations and the output equation can be written in matrix form as

$$\begin{bmatrix} dx_1(t)/dt \\ dx_2(t)/dt \end{bmatrix} = \begin{bmatrix} -R/L & -1/L \\ 1/C & 0 \end{bmatrix} \begin{bmatrix} x_1(t) \\ x_2(t) \end{bmatrix} + \begin{bmatrix} 1/L \\ 0 \end{bmatrix} r(t) \tag{9.27}$$

and

$$y(t) = \begin{bmatrix} -R & 0 \end{bmatrix} \begin{bmatrix} x_1(t) \\ x_2(t) \end{bmatrix} + (1) r(t) \tag{9.28}$$

Substituting the numerical values, Eqns. (9.27) and (9.28) become

$$\dot{\mathbf{x}}(t) = \begin{bmatrix} -2 & -1 \\ 26 & 0 \end{bmatrix} \mathbf{x}(t) + \begin{bmatrix} 1 \\ 0 \end{bmatrix} r(t)$$

and

$$y(t) = \begin{bmatrix} -2 & 0 \end{bmatrix} \mathbf{x}(t) + (1) r(t)$$

9.2.3   Obtaining state-space model from differential equation

Converting a N-th order differential equation to a state-space model requires finding N first-order differential equations that represent the same system, and an output equation that links the output signal to the state variables. The solution is not unique. In this section we will present just one method of accomplishing this task. Consider a CTLTI system described by a differential equation in the form

$$\frac{d^N y(t)}{dt^N} + a_{N-1} \frac{d^{N-1} y(t)}{dt^{N-1}} + \ldots + a_1 \frac{dy(t)}{dt} + a_0 y(t) = b_0 r(t) \tag{9.29}$$

We will rewrite the differential equation so that the highest order derivative is computed from the remaining terms.

$$\frac{d^N y(t)}{dt^N} = -a_{N-1} \frac{d^{N-1} y(t)}{dt^{N-1}} - \ldots - a_1 \frac{dy(t)}{dt} - a_0 y(t) + b_0 r(t) \tag{9.30}$$

Let us choose the state variables as follows:

$$x_1(t) = y(t)$$

$$x_2(t) = \frac{dy(t)}{dt} \implies \frac{dx_1(t)}{dt} = x_2(t)$$

$$x_3(t) = \frac{d^2y(t)}{dt^2} \implies \frac{dx_2(t)}{dt} = x_3(t)$$

$$\vdots$$

$$x_N(t) = \frac{d^{N-1}y(t)}{dt^{N-1}} \implies \frac{dx_{N-1}(t)}{dt} = x_N(t)$$

The resulting state equations are

$$\frac{dx_1(t)}{dt} = x_2(t)$$

$$\frac{dx_2(t)}{dt} = x_3(t)$$

$$\vdots$$

$$\frac{dx_{N-1}(t)}{dt} = x_N(t)$$

$$\frac{dx_N(t)}{dt} = -a_0 x_1(t) - a_1 x_2(t) - \ldots - a_{N-1} x_N(t) + b_0 r(t) \tag{9.31}$$

The last state equation is obtained by using the state variables $x_1(t)$ through $x_N(t)$ defined above in the differential equation given by Eqn. (9.30), and writing $dx_N(t)/dt$ in terms of the state variables $x_1(t)$ through $x_N(t)$ and the input signal $r(t)$. The output equation is easily found from the assignment of the first state variable:

$$y(t) = x_1(t) \tag{9.32}$$

In matrix form, the state-space model is

$$\dot{\mathbf{x}}(t) = \mathbf{A}\mathbf{x}(t) + \mathbf{B}r(t)$$
$$y(t) = \mathbf{C}\mathbf{x}(t) + dr(t)$$

with

$$\mathbf{A} = \begin{bmatrix} 0 & 1 & 0 & 0 & \ldots & 0 \\ 0 & 0 & 1 & 0 & \ldots & 0 \\ 0 & 0 & 0 & 1 & \ldots & 0 \\ \vdots & \vdots & \vdots & \vdots & \ldots & \vdots \\ 0 & 0 & 0 & 0 & \ldots & 1 \\ -a_0 & -a_1 & -a_2 & -a_3 & \ldots & -a_{N-1} \end{bmatrix}, \quad \mathbf{B} = \begin{bmatrix} 0 \\ 0 \\ 0 \\ \vdots \\ 0 \\ b_0 \end{bmatrix} \tag{9.33}$$

for the state equation, and

$$\mathbf{C} = \begin{bmatrix} 1 & 0 & 0 & \ldots & 0 & 0 \end{bmatrix}, \quad d = 0 \tag{9.34}$$

for the output equation. This form of the state-space model is called the *phase-variable canonical form*. It has special significance in the study of control systems. Pay attention to the structure of the coefficient matrix $\mathbf{A}$, repeated below in partitioned form:

$$
\mathbf{A} = \left[
\begin{array}{c:ccccc}
0 & 1 & 0 & 0 & \cdots & 0 \\
0 & 0 & 1 & 0 & \cdots & 0 \\
0 & 0 & 0 & 1 & \cdots & 0 \\
\vdots & \vdots & \vdots & \vdots & \cdots & \vdots \\
0 & 0 & 0 & 0 & \cdots & 1 \\
\hdashline
-a_0 & -a_1 & -a_2 & -a_3 & \cdots & -a_{N-1}
\end{array}
\right]
\tag{9.35}
$$

The top left partition is a $(N-1) \times 1$ vector of zeros; the top right partition is an identity matrix of order $(N-1)$; the bottom row is a vector of differential equation coefficients in reverse order and with a sign change. This is handy for obtaining the state-space model by directly inspecting the differential equation. It is important to remember, however, that the coefficient of the highest order derivative in Eqn. (9.29) must be equal to unity for this to work.

Example 9.3: **State-space model from differential equation**

Find a state-space model for a CTLTI system described by the differential equation

$$
\frac{d^3 y(t)}{dt^3} + 5 \frac{d^2 y(t)}{dt^2} + 11 \frac{dy(t)}{dt} + 15\, y(t) = 3\, r(t)
$$

Solution: Expressing the highest order derivative as a function of the other terms leads to

$$
\frac{d^3 y(t)}{dt^3} = -5 \frac{d^2 y(t)}{dt^2} - 11 \frac{dy(t)}{dt} - 15\, y(t) + 3\, r(t)
\tag{9.36}
$$

State variables can be defined as

$$
x_1(t) = y(t)
$$

$$
x_2(t) = \frac{dy(t)}{dt} \quad \Longrightarrow \quad x_2(t) = \frac{dx_1(t)}{dt}
$$

$$
x_3(t) = \frac{d^2 y(t)}{dt^2} \quad \Longrightarrow \quad x_3(t) = \frac{dx_2(t)}{dt}
$$

Recognizing that

$$
\frac{d^3 y(t)}{dt^3} = \frac{dx_3(t)}{dt}
$$

the differential equation in Eqn. (9.36) can be written as

$$
\frac{dx_3(t)}{dt} = -5\, x_3(t) - 11\, x_2(t) - 15\, x_1(t) + 3\, r(t)
$$

In matrix form, the state-space model is

$$
\dot{\mathbf{x}}(t) = \begin{bmatrix} 0 & 1 & 0 \\ 0 & 0 & 1 \\ -15 & -11 & -5 \end{bmatrix} \mathbf{x}(t) + \begin{bmatrix} 0 \\ 0 \\ 3 \end{bmatrix} r(t)
$$

$$
y(t) = \begin{bmatrix} 1 & 0 & 0 \end{bmatrix} \mathbf{x}(t)
$$

Example 9.4: **State-space model from differential equation revisited**

Find a state-space model for a CTLTI system described by the differential equation

$$\frac{d^3 y(t)}{dt^3} + 5 \frac{d^2 y(t)}{dt^2} + 11 \frac{dy(t)}{dt} + 15 y(t) = 3 r(t) + 7 \frac{dr(t)}{dt}$$

Solution: This differential equation is similar to that of Example 9.3 except for the derivative of the input signal that appears on the right side of the equal sign. While it is possible to use the alternative form of state equation suggested by Eqn. (9.12) for this system, we will choose to stay with the standard form of Eqn. (9.7). Let us first rewrite the differential equation in a rearranged form.

$$\frac{d^3 y(t)}{dt^3} - 7 \frac{dr(t)}{dt} = -5 \frac{d^2 y(t)}{dt^2} - 11 \frac{dy(t)}{dt} - 15 y(t) + 3 r(t) \tag{9.37}$$

State variables will be defined as

$$x_1(t) = y(t)$$

$$x_2(t) = \frac{dy(t)}{dt} \qquad \implies \qquad x_2(t) = \frac{dx_1(t)}{dt}$$

$$x_3(t) = \frac{d^2 y(t)}{dt^2} - 7 r(t) \qquad \implies \qquad x_3(t) = \frac{dx_2(t)}{dt} - 7 r(t)$$

We have defined the last state variable $x_3(t)$ differently from the definition in Example 9.3 so that the left side of Eqn. (9.37) is equal to the derivative of $x_3(t)$. Using these definitions, Eqn. (9.37) becomes

$$\frac{dx_3(t)}{dt} = -5 \left[x_3(t) + 7 r(t) \right] - 11 x_2(t) - 15 x_1(t) + 3 r(t)$$

In matrix form, the state-space model is

$$\dot{\mathbf{x}}(t) = \begin{bmatrix} 0 & 1 & 0 \\ 0 & 0 & 1 \\ -15 & -11 & -5 \end{bmatrix} \mathbf{x}(t) + \begin{bmatrix} 0 \\ 7 \\ -32 \end{bmatrix} r(t)$$

$$y(t) = \begin{bmatrix} 1 & 0 & 0 \end{bmatrix} \mathbf{x}(t)$$

When we compare Examples 9.3 and 9.4 we see that coefficient matrix $\mathbf{A}$ is the same in both of them, but the vector $\mathbf{B}$ is different. The inspection method for deriving the state-space model directly from the differential equation would not be as straightforward to use when we have derivatives of the input signal appearing on the right side. We will, however, generalize the relationship between the coefficients of the differential equation and the state-space model when we consider the derivation of the latter from the system function in the next section.

9.2.4 Obtaining state-space model from system function

If the CTLTI system under consideration has a system function that is simple in the sense that it has no multiple poles and all of its poles are real-valued, then a state-space model can be easily found using partial fraction expansion. This idea will be explored in the next example.

Example 9.5: **State-space model from s-domain system function**

It can be shown that the RLC circuit that was used in Example 9.1 and shown in Fig. 9.3 has the system function

$$G(s) = \frac{Y(s)}{R(s)} = \frac{10}{(s+3)(s+4)}$$

Find an alternative, but equivalent, state-space model for the system based on this system function.

Solution: Expanding the system function into partial fractions and then solving for the transform $Y(s)$ we get

$$Y(s) = \frac{10}{s+3} R(s) - \frac{10}{s+4} R(s)$$

Let us define $Z_1(s)$ and $Z_2(s)$ as

$$Z_1(s) = \frac{10}{s+3} R(s) \tag{9.38}$$

and

$$Z_2(s) = \frac{-10}{s+4} R(s) \tag{9.39}$$

so that $Y(s) = Z_1(s) + Z_2(s)$. Rearranging the terms in Eqns. (9.38) and (9.39) we get

$$s\,Z_1(s) = -3\,Z_1(s) + 10\,R(s) \tag{9.40}$$

and similarly

$$s\,Z_2(s) = -4\,Z_2(s) - 10\,R(s) \tag{9.41}$$

Taking the inverse Laplace transform of Eqns. (9.40) and (9.41) leads to the two first-order differential equations

$$\frac{dz_1(t)}{dt} = -3z_1(t) + 10\,r(t) \tag{9.42}$$

$$\frac{dz_2(t)}{dt} = -4z_2(t) - 10\,r(t) \tag{9.43}$$

Using $z_1(t)$ and $z_2(t)$ as the new state variables, the state-space model can be written as

$$\begin{bmatrix} dz_1(t)/dt \\ dz_2(t)/dt \end{bmatrix} = \begin{bmatrix} -3 & 0 \\ 0 & -4 \end{bmatrix} \begin{bmatrix} z_1(t) \\ z_2(t) \end{bmatrix} + \begin{bmatrix} 10 \\ -10 \end{bmatrix} r(t)$$

and

$$y(t) = \begin{bmatrix} 1 & 1 \end{bmatrix} \begin{bmatrix} z_1(t) \\ z_2(t) \end{bmatrix}$$

Example 9.5 was made simple by the fact that the partial fraction expansion of the system function contains only first-order terms with real coefficients. An interesting consequence of the simple technique utilized is that the state equations obtained in Eqns. (9.42) and (9.43) are uncoupled, that is, each state equation involves only one of the state variables. This corresponds to a diagonal coefficient matrix **A**. Consequently, we have the advantage of being able to solve the state equations independently of each other.

If the system function does not have the properties that allowed the simple solution of Example 9.5, finding a state-space model for a system function involves the use of a block diagram as an intermediate step. The development of this idea will be presented using a third-order system function as an example; generalization to higher-order system functions will be readily apparent. Consider the system function

$$G(s) = \frac{Y(s)}{X(s)} = \frac{b_3 s^3 + b_2 s^2 + b_1 s + b_0}{s^3 + a_2 s^2 + a_1 s + a_0} \tag{9.44}$$

Using the techniques discussed in Section 7.6.1 of Chapter 7, the simulation diagram shown in Fig. 9.5 can be obtained for $G(s)$.

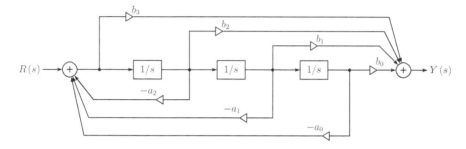

Figure 9.5 – Simulation diagram for $G(s)$ of Eqn. (9.44).

We know from integration property of the Laplace transform (see Section 7.3.8 of Chapter 7) that multiplying a transform by $1/s$ is equivalent to integrating the corresponding signal in the time domain. Therefore, a time-domain equivalent of the block diagram in Fig. 9.5 can be obtained by replacing $1/s$ components with integrators and using time-domain versions of input and output quantities. The resulting diagram is shown in Fig. 9.6.

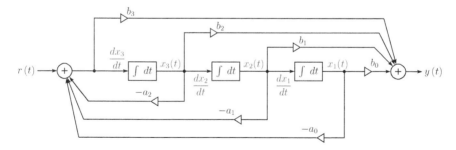

Figure 9.6 – Selecting states from the simulation diagram for $G(s)$.

A state-space model can be obtained from the diagram of Fig. 9.4 as follows:

1. Designate the output signal of each integrator as a state variable. These are marked in Fig. 9.6 as $x_1(t)$, $x_2(t)$ and $x_3(t)$.
2. The input of each integrator becomes the derivative of the corresponding state variable. These are marked in Fig. 9.6 as dx_1/dt, dx_2/dt and dx_3/dt.
3. Express the signal at the input of each integrator as a function of state variables and the input signal $r(t)$ as dictated by the diagram.

Following state equations are obtained by inspecting the diagram in Fig. 9.6:

$$\frac{dx_1(t)}{dt} = x_2(t) \tag{9.45}$$

$$\frac{dx_2(t)}{dt} = x_3(t) \tag{9.46}$$

$$\frac{dx_3(t)}{dt} = -a_0\, x_1(t) - a_1\, x_2(t) - a_0\, x_3(t) + r(t) \tag{9.47}$$

The output signal can be written as

$$y(t) = b_0\, x_1(t) + b_1\, x_2(t) + b_2\, x_3(t) + b_3\, \frac{dx_3(t)}{dt} \tag{9.48}$$

We do not like to see the derivative of a state variable on the right side of the output equation. Substituting Eqn. (9.47 into Eqn. (9.48) yields the proper form of the output equation

$$y(t) = (b_0 - b_3\, a_0)\, x_1(t) + (b_1 - b_3\, a_1)\, x_2(t) + (b_2 - b_3\, a_2)\, x_3(t) + b_3\, r(t) \tag{9.49}$$

Expressed in matrix form, the state-space model is

$$\dot{\mathbf{x}}(t) = \begin{bmatrix} 0 & 1 & 0 \\ 0 & 0 & 1 \\ -a_0 & -a_1 & -a_2 \end{bmatrix} \mathbf{x}(t) + \begin{bmatrix} 0 \\ 0 \\ 1 \end{bmatrix} r(t) \tag{9.50}$$

$$y(t) = \begin{bmatrix} (b_0 - b_3\, a_0) & (b_1 - b_3\, a_1) & (b_2 - b_3\, a_2) \end{bmatrix} \mathbf{x}(t) + b_3\, r(t) \tag{9.51}$$

The results can easily be extended to a N-th order system function in the form

$$G(s) = \frac{Y(s)}{X(s)} = \frac{b_N s^N + b_{N-1} s^{N-1} + \ldots + b_1 s + b_0}{s^N + a_{N-1} s^{N-1} + \ldots + a_1 s + a_0} \tag{9.52}$$

to produce the state-space model

$$\dot{\mathbf{x}}(t) = \begin{bmatrix} 0 & 1 & 0 & 0 & \ldots & 0 \\ 0 & 0 & 1 & 0 & \ldots & 0 \\ 0 & 0 & 0 & 1 & \ldots & 0 \\ \vdots & \vdots & \vdots & \vdots & \ldots & \vdots \\ 0 & 0 & 0 & 0 & \ldots & 1 \\ -a_0 & -a_1 & -a_2 & -a_3 & \ldots & -a_{N-1} \end{bmatrix} \mathbf{x}(t) + \begin{bmatrix} 0 \\ 0 \\ 0 \\ \vdots \\ 0 \\ 1 \end{bmatrix} r(t) \tag{9.53}$$

$$y(t) = \begin{bmatrix} (b_0 - b_N\, a_0) & (b_1 - b_N\, a_1) & \ldots & (b_{N-2} - b_N\, a_{N-2}) & (b_{N-1} - b_N\, a_{N-1}) \end{bmatrix} \mathbf{x}(t) + b_N\, r(t) \tag{9.54}$$

It is interesting to see that the structure of the coefficient matrix $\mathbf{A}$ is the same as in Eqn. (9.35). Similarly, the vector $\mathbf{B}$ is identical to that found in Section 9.2.3.

Software resources: See MATLAB Exercise 9.1.

Example 9.6: **State-space model from block diagram**

Find a state-space model for a CTLTI system described through the system function

$$G\left(s\right) = \frac{2s^3 - 26s + 24}{s^4 + 7s^3 + 21s^2 + 37s + 30}$$

Solution: A block diagram for this system was constructed in Example 7.40 of Chapter 7, and was shown in Fig. 7.69. That diagram is shown again in Fig. 9.7 with notation adapted to time domain and with the designations of the four state variables marked.

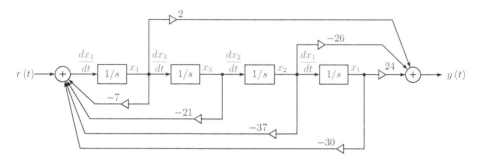

Figure 9.7 – Block diagram for Example 9.6.

The state equations are

$$\frac{dx_1\left(t\right)}{dt} = x_2\left(t\right)$$

$$\frac{dx_2\left(t\right)}{dt} = x_3\left(t\right)$$

$$\frac{dx_3\left(t\right)}{dt} = x_4\left(t\right)$$

$$\frac{dx_4\left(t\right)}{dt} = -30\,x_1\left(t\right) - 37\,x_2\left(t\right) - 21\,x_3\left(t\right) - 7\,x_4\left(t\right) + r\left(t\right)$$

and the output equation is

$$y\left(t\right) = 24\,x_1\left(t\right) - 26\,x_2\left(t\right) + 2\,x_4\left(t\right)$$

In matrix form we get

$$\dot{\mathbf{x}}\left(t\right) = \begin{bmatrix} 0 & 1 & 0 & 0 \\ 0 & 0 & 1 & 0 \\ 0 & 0 & 0 & 1 \\ -30 & -37 & -21 & -7 \end{bmatrix} \mathbf{x}\left(t\right) + \begin{bmatrix} 0 \\ 0 \\ 0 \\ 1 \end{bmatrix} r\left(t\right)$$

and

$$y\left(t\right) = \begin{bmatrix} 24 & -26 & 0 & 2 \end{bmatrix} \mathbf{x}\left(t\right)$$

Example 9.7: **State-space model from block diagram revisited**

Find a state-space model for a CTLTI system with the system function

$$G(s) = \frac{2s^3 - 26s + 24}{s^4 + 7s^3 + 21s^2 + 37s + 30}$$

using the cascade form block diagram.

Solution: A cascade form block diagram for this system was constructed in Example 7.41 of Chapter 7, and was shown in Fig. 7.72. That diagram is shown again in Fig. 9.8 with notation adapted to time domain and with the designations of the four state variables marked.

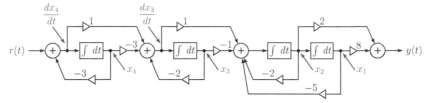

Figure 9.8 – Block diagram for Example 9.7.

Writing the input signal of each integrator in terms of the other signals in the diagram we obtain

$$\frac{dx_1(t)}{dt} = x_2(t) \tag{9.55}$$

$$\frac{dx_2(t)}{dt} = -5\,x_1(t) - 2\,x_2(t) - x_3(t) + \frac{dx_3(t)}{dt} \tag{9.56}$$

$$\frac{dx_3(t)}{dt} = -2\,x_3(t) - 3\,x_4(t) + \frac{dx_4(t)}{dt} \tag{9.57}$$

$$\frac{dx_4(t)}{dt} = -3\,x_4(t) + r(t) \tag{9.58}$$

and the output signal is

$$y(t) = 8\,x_1(t) + 2\,x_2(t)$$

State equations should not contain any derivatives on the right side of the equal sign. Using Eqn. (9.58) in Eqn. (9.57) yields

$$\frac{dx_3(t)}{dt} = -2\,x_3(t) - 6\,x_4(t) + r(t) \tag{9.59}$$

Similarly, using Eqn. (9.59) in Eqn. (9.56) we get

$$\frac{dx_2(t)}{dt} = -5\,x_1(t) - 2\,x_2(t) - 3\,x_3(t) - 6\,x_4(t) + r(t) \tag{9.60}$$

The state-space model can now be put in matrix form using Eqns. (9.55), (9.58), (9.59) and (9.60). The result is

$$\dot{\mathbf{x}}(t) = \begin{bmatrix} 0 & 1 & 0 & 0 \\ -5 & -2 & -3 & -6 \\ 0 & 0 & -2 & -6 \\ 0 & 0 & 0 & -3 \end{bmatrix} \mathbf{x}(t) + \begin{bmatrix} 0 \\ 1 \\ 1 \\ 1 \end{bmatrix} r(t)$$

and

$$y(t) = \begin{bmatrix} 8 & 2 & 0 & 0 \end{bmatrix} \mathbf{x}(t)$$

9.2.5 Alternative state-space models

In Examples 9.1 and 9.2 we have defined the state variables of each system as the capacitor voltage and the inductor current. These choices were mainly motivated by the fact that the derivatives of these two quantities readily appear in KVL equations for either circuit.

Other state variables could have been defined for these systems as well, leading to alternative state-space models. Consider a CTLTI system defined through Eqns. (9.7) and (9.8). Let $\mathbf{z}(t)$ be a length-N column vector of variables $z_1(t), \ldots, z_N(t)$, related to the original state vector $\mathbf{x}(t)$ of the CTLTI system through

$$\mathbf{x}(t) = \mathbf{P}\,\mathbf{z}(t) \tag{9.61}$$

where $\mathbf{P}$ is a constant $N \times N$ transformation matrix. The only requirement on the matrix $\mathbf{P}$ is that it be non-singular (invertible). Substituting Eqn. (9.61) into Eqn. (9.7) leads to

$$\mathbf{P}\,\dot{\mathbf{z}}(t) = \mathbf{A}\,\mathbf{P}\,\mathbf{z}(t) + \mathbf{B}\,r(t) \tag{9.62}$$

Multiplication of both sides of Eqn. (9.62) with the inverse of $\mathbf{P}$ from the left yields

$$\dot{\mathbf{z}}(t) = \mathbf{P}^{-1}\mathbf{A}\mathbf{P}\,\mathbf{z}(t) + \mathbf{P}^{-1}\mathbf{B}\,r(t) \tag{9.63}$$

Finally, using Eqn. (9.61) with Eqn. (9.8) we get

$$y(t) = \mathbf{C}\mathbf{P}\,\mathbf{z}(t) + d\,r(t) \tag{9.64}$$

Eqns. (9.63) and (9.64) provide us with an alternative state-space model for the same CTLTI system, using a new state vector $\mathbf{z}(t)$. They can be written in the standard form of Eqns. (9.7) and (9.8) as

$$\dot{\mathbf{z}}(t) = \tilde{\mathbf{A}}\,\mathbf{z}(t) + \tilde{\mathbf{B}}\,r(t) \tag{9.65}$$

$$y(t) = \tilde{\mathbf{C}}\,\mathbf{z}(t) + d\,r(t) \tag{9.66}$$

with newly defined coefficient matrices

$$\tilde{\mathbf{A}} = \mathbf{P}^{-1}\,\mathbf{A}\,\mathbf{P} \tag{9.67}$$

$$\tilde{\mathbf{B}} = \mathbf{P}^{-1}\,\mathbf{B} \tag{9.68}$$

and

$$\tilde{\mathbf{C}} = \mathbf{C}\,\mathbf{P} \tag{9.69}$$

The transformation relationship $\tilde{\mathbf{A}} = \mathbf{P}^{-1}\mathbf{A}\mathbf{P}$ is known in control system theory as a *similarity transformation*. It can be used for obtaining alternative state-space models of a system. Each choice of the transformation matrix $\mathbf{P}$ leads to a different, but equivalent, state-space model. In each of these alternative models the input-output relationship remains the same although the state variables are different. Since any non-singular matrix $\mathbf{P}$ can be used in a similarity transformation, an infinite number of different state-space models can be found for a given system.

Software resources: See MATLAB Exercise 9.2.

In the next several examples we will explore this concept further.

Example 9.8: **Alternative state-space model for series RLC circuit**

Refer to the system in Example 9.2 again. Since the system is of second order, the input-output relationship can be expressed through a second-order differential equation between $y(t)$ and $r(t)$. Find such a differential equation from the state-space model found in Example 9.2. Afterwards obtain an alternative state-space model for the same system directly from the second-order differential equation.

Solution: The state-space model found in Example 9.2 was

$$\dot{\mathbf{x}}(t) = \begin{bmatrix} -2 & -1 \\ 26 & 0 \end{bmatrix} \mathbf{x}(t) + \begin{bmatrix} 1 \\ 0 \end{bmatrix} r(t)$$

and

$$y(t) = \begin{bmatrix} -2 & 0 \end{bmatrix} \mathbf{x}(t) + r(t)$$

If we write the state equations and the output equation in open form, we obtain

$$\frac{dx_1(t)}{dt} = -2\, x_1(t) - x_2(t) + r(t) \tag{9.70}$$

$$\frac{dx_2(t)}{dt} = 26\, x_1(t) \tag{9.71}$$

$$y(t) = -2\, x_1(t) + r(t) \tag{9.72}$$

Let us differentiate Eqn. (9.72) to obtain

$$\frac{dy(t)}{dt} = -2\frac{dx_1(t)}{dt} + \frac{dr(t)}{dt} \tag{9.73}$$

Substituting Eqn. (9.70) into Eqn. (9.73)

$$\frac{dy(t)}{dt} = 4\, x_1(t) + 2\, x_2(t) - 2\, r(t) + \frac{dr(t)}{dt} \tag{9.74}$$

Differentiating Eqn. (9.74) one more time and using the result with Eqns. (9.70) and (9.71) leads to

$$\frac{d^2 y(t)}{dt^2} = 44\, x_1(t) - 4\, x_2(t) + 4\, r(t) - 2\frac{dr(t)}{dt} + \frac{d^2 r(t)}{dt^2} \tag{9.75}$$

Finally, eliminating the state variables $x_1(t)$ and $x_2(t)$ between Eqns. (9.72), (9.74) and (9.75) yields the second-order differential equation

$$\frac{d^2 y(t)}{dt^2} = -2\frac{dy(t)}{dt} - 26\, y(t) + \frac{d^2 r(t)}{dt^2} + 26\, r(t) \tag{9.76}$$

A state-space model in phase-variable canonical form can be obtained from the differential equation in Eqn. (9.76) using the technique discussed in Section 9.2.4. The result is

$$\dot{\mathbf{z}}(t) = \begin{bmatrix} 0 & 1 \\ -26 & -2 \end{bmatrix} \mathbf{z}(t) + \begin{bmatrix} 0 \\ 1 \end{bmatrix} r(t)$$

$$y(t) = \begin{bmatrix} 0 & -2 \end{bmatrix} \mathbf{z}(t) + r(t)$$

Software resources: ex_9_8.m

Example 9.9: **State variable transformation**

Show that the alternative state-space model found in Example 9.8 could have been obtained from the state-space model of Example 9.2 through the transformation

$$\mathbf{x}(t) = \mathbf{P}\,\mathbf{z}(t) = \begin{bmatrix} 0 & 1 \\ 26 & 0 \end{bmatrix} \mathbf{z}(t)$$

Solution: The original state-space model found in Example 9.1 has the coefficient matrices

$$\mathbf{A} = \begin{bmatrix} -2 & -1 \\ 26 & 0 \end{bmatrix}, \quad \mathbf{B} = \begin{bmatrix} 1 \\ 0 \end{bmatrix}, \quad \mathbf{C} = \begin{bmatrix} -2 & 0 \end{bmatrix}, \quad d = 1$$

The inverse of the transformation matrix $\mathbf{P}$ specified above is

$$\mathbf{P}^{-1} = \begin{bmatrix} 0 & 1/26 \\ 1 & 0 \end{bmatrix}$$

Applying the similarity transformation given byEqn. (9.67) to the coefficient matrix $\mathbf{A}$ yields

$$\tilde{\mathbf{A}} = \mathbf{P}^{-1}\mathbf{A}\mathbf{P} = \begin{bmatrix} 0 & 1/26 \\ 1 & 0 \end{bmatrix} \begin{bmatrix} -2 & -1 \\ 26 & 0 \end{bmatrix} \begin{bmatrix} 0 & 1 \\ 26 & 0 \end{bmatrix} = \begin{bmatrix} 0 & 1 \\ -26 & -2 \end{bmatrix}$$

which is identical to the matrix $\tilde{\mathbf{A}}$ found in Example 9.8. Vectors $\tilde{\mathbf{B}}$ and $\tilde{\mathbf{C}}$ can also be found using Eqns. (9.68) and (9.69) as

$$\tilde{\mathbf{B}} = \mathbf{P}^{-1}\,\mathbf{B} = \begin{bmatrix} 0 & 1/26 \\ 1 & 0 \end{bmatrix} \begin{bmatrix} 1 \\ 0 \end{bmatrix} = \begin{bmatrix} 0 \\ 1 \end{bmatrix}$$

and

$$\tilde{\mathbf{C}} = \mathbf{C}\,\mathbf{P} = \begin{bmatrix} -2 & 0 \end{bmatrix} \begin{bmatrix} 0 & 1 \\ 26 & 0 \end{bmatrix} = \begin{bmatrix} 0 & -2 \end{bmatrix}$$

Vectors $\tilde{\mathbf{B}}$ and $\tilde{\mathbf{C}}$ also agree with those found in Example 9.8.
Software resources:
ex_9_9.m

9.2.6   CTLTI systems with multiple inputs and/or outputs

A CTLTI system with multiple input and/or output signals can also be represented with a state-space model in the standard form of Eqns. (9.7 and (9.8). Consider a CTLTI system with K input signals $\{r_i(t)\,;\ i = 1, \ldots, K\}$ and M output signals $\{y_j(t)\,;\ i = 1, \ldots, M\}$. Let $\mathbf{r}(t)$ be the $K \times 1$ vector of input signals:

$$\mathbf{r}(t) = \begin{bmatrix} r_1(t) \\ r_2(t) \\ \vdots \\ r_K(t) \end{bmatrix} \tag{9.77}$$

Similarly, $\mathbf{y}(t)$ is the $M \times 1$ vector of output signals.

$$\mathbf{y}(t) = \begin{bmatrix} y_1(t) \\ y_2(t) \\ \vdots \\ y_M(t) \end{bmatrix} \tag{9.78}$$

With these new vectors, the state equation is in the form

$$\dot{\mathbf{x}}(t) = \mathbf{A}\,\mathbf{x}(t) + \mathbf{B}\,\mathbf{r}(t) \tag{9.79}$$

The matrix $\mathbf{A}$ is still the $N \times N$ coefficient matrix, and the vector $\mathbf{x}(t)$ is still the $N \times 1$ state vector as before. Dimensions of the matrix $\mathbf{B}$ will need to be changed, however, to make it compatible with the $K \times 1$ vector $\mathbf{r}(t)$. Therefore, $\mathbf{B}$ is now a $N \times K$ matrix. The output equation is in the form

$$\mathbf{y}(t) = \mathbf{C}\,\mathbf{x}(t) + \mathbf{D}\,\mathbf{r}(t) \tag{9.80}$$

Dimensions of $\mathbf{C}$ need to be updated as well. Since $\mathbf{y}(t)$ is a $M \times 1$ vector, $\mathbf{C}$ is a $M \times N$ matrix. Additionally, the scalar d in the output equation in Eqn. (9.8) is replaced with a $M \times K$ matrix $\mathbf{D}$.

Example 9.10: Multiple-input multiple-output system

Consider a CTLTI system with two inputs and two outputs that is described by a pair of differential equations:

$$\frac{d^2 y_1(t)}{dt^2} + 3\,\frac{dy_1(t)}{dt} + 2\,y_1(t) + y_2(t) = 3\,r_1(t) + r_2(t)$$

$$\frac{dy_2(t)}{dt} + 3\,y_2(t) + 4\,y_1(t) = r_1(t) + 4\,\frac{dr_1(t)}{dt} - 2\,r_2(t)$$

Find a state-space model for this system.

Solution: We will begin by rearranging the terms of the two differential equations to obtain

$$\frac{d^2 y_1(t)}{dt^2} = -\,3\,\frac{dy_1(t)}{dt} - 2\,y_1(t) - y_2(t) + 3\,r_1(t) + r_2(t) \tag{9.81}$$

$$\frac{dy_2(t)}{dt} - 4\,\frac{dr_1(t)}{dt} = -\,3\,y_2(t) - 4\,y_1(t) + r_1(t) - 2\,r_2(t) \tag{9.82}$$

Let us define the state variables as follows:

$$x_1(t) = y_1(t)\,, \qquad x_2(t) = \frac{dy_1(t)}{dt}\,, \qquad x_3(t) = y_2(t) - 4\,r_1(t) \tag{9.83}$$

It naturally follows that

$$\frac{dx_1(t)}{dt} = x_2(t) \tag{9.84}$$

Using the definitions of the state variables in Eqns. (9.83) and (9.84) we obtain

$$\frac{dx_2(t)}{dt} = -\,3\,x_2(t) - 2\,x_1(t) - \big[x_3(t) + 4\,r_1(t)\big] + 3\,r_1(t) + r_2(t)$$

$$= -\,2\,x_1(t) - 3\,x_2(t) - x_3(t) - r_1(t) + r_2(t) \tag{9.85}$$

and

$$\frac{dx_3(t)}{dt} = -3\left[x_3(t) + 4r_1(t)\right] - 4x_1(t) + r_1(t) - 2r_2(t)$$

$$= -4x_1(t) - 3x_3(t) - 11r_1(t) - 2r_2(t) \tag{9.86}$$

Eqns. (9.84), (9.85) and (9.86) can be put in matrix form as

$$\dot{\mathbf{x}}(t) = \begin{bmatrix} 0 & 1 & 0 \\ -2 & -3 & -1 \\ -4 & 0 & -3 \end{bmatrix} \mathbf{x}(t) + \begin{bmatrix} 0 & 0 \\ -1 & 1 \\ -11 & -2 \end{bmatrix} \mathbf{r}(t)$$

and

$$\mathbf{y}(t) = \begin{bmatrix} 1 & 0 & 0 \\ 0 & 0 & 1 \end{bmatrix} \mathbf{x}(t) + \begin{bmatrix} 0 & 0 \\ 0 & 4 \end{bmatrix} \mathbf{r}(t)$$

9.2.7 Solution of state-space model

Before deriving the general solution for the state-space model of a continuous-time system it will be useful to consider a special case in which the coefficient matrix $\mathbf{A}$ is upper-triangular, that is, all of its elements below the main diagonal are equal to zero. Example 9.11 will demonstrate that a step-by-step solution can be obtained in this case.

Example 9.11: **Solving state equations with upper-triangular coefficient matrix**

A system is described by the state-space model

$$\dot{\mathbf{x}}(t) = \begin{bmatrix} -3 & 1 & 0 \\ 0 & -2 & 5 \\ 0 & 0 & -1 \end{bmatrix} \mathbf{x}(t) + \begin{bmatrix} 0 \\ 2 \\ 1 \end{bmatrix} r(t)$$

and

$$y(t) = \begin{bmatrix} 1 & 0 & 0 \end{bmatrix} \mathbf{x}(t)$$

The initial value of the state vector is

$$\mathbf{x}(0) = \begin{bmatrix} 0 \\ 3 \\ 2 \end{bmatrix}$$

Determine the output signal in response to a unit-step input signal.

Solution: The differential equation for the state variable $x_3(t)$ does not include either of the other two state variables; therefore, it is convenient to start with it.

$$\frac{dx_3(t)}{dt} = -x_3(t) + r(t) = -x_3(t) + u(t) \;, \qquad x_3(0) = 2 \tag{9.87}$$

Let us apply the solution technique developed in Section 2.5.1 of Chapter 2. Using Eqn. (2.57) with $\alpha = 1$ and $t_0 = 0$, the solution of Eqn. (9.87) is

$$x_3(t) = e^{-t} x_3(0) + \int_0^t e^{-(t-\tau)} u(\tau)\, d\tau \;, \qquad t \geq 0$$

Using the initial value $x_3(0) = 2$ and realizing that $u(\tau) = 1$ within the integration interval we obtain

$$x_3(t) = 2\,e^{-t} + \int_0^t e^{-(t-\tau)}\,d\tau = \left(1 + e^{-t}\right)u(t) \tag{9.88}$$

Having determined $x_3(t)$, the differential equation for $x_2(t)$ is

$$\begin{aligned}
\frac{dx_2(t)}{dt} &= -2\,x_2(t) + 5\,x_3(t) + 2\,r(t) \\
&= -2\,x_2(t) + 5\left(1 + e^{-t}\right)u(t) + 2\,u(t) \\
&= -2\,x_2(t) + \left(7 + 5\,e^{-t}\right)u(t)\,, \qquad x_2(0) = 3
\end{aligned} \tag{9.89}$$

Repeating the procedure outlined above, $x_2(t)$ is determined as

$$\begin{aligned}
x_2(t) &= 3\,e^{-2t} + \int_0^t e^{-2(t-\tau)}\left(7 + 5e^{-\tau}\right)u(\tau)\,d\tau \\
&= \left(\frac{7}{2} + 5\,e^{-t} - \frac{11}{2}\,e^{-2t}\right)u(t)
\end{aligned} \tag{9.90}$$

Finally, the differential equation for $x_1(t)$ is

$$\begin{aligned}
\frac{dx_1(t)}{dt} &= -3\,x_1(t) + x_2(t) \\
&= -3\,x_1(t) + \left(\frac{7}{2} + 5\,e^{-t} - \frac{11}{2}\,e^{-2t}\right)u(t)\,, \qquad x_1(0) = 0
\end{aligned} \tag{9.91}$$

which can be solved to yield

$$x_1(t) = \left(\frac{7}{6} + \frac{5}{2}\,e^{-t} - \frac{11}{2}\,e^{-2t} + \frac{11}{6}\,e^{-3t}\right)u(t)$$

The output of the system is found using the output equation:

$$y(t) = x_1(t) = \left(\frac{7}{6} + \frac{5}{2}\,e^{-t} - \frac{11}{2}\,e^{-2t} + \frac{11}{6}\,e^{-3t}\right)u(t)$$

The solution of Example 9.11 was made simple by the fact that the coefficient matrix is upper-triangular. Starting with the last state equation and working our way up we were able to resolve the dependencies between state equations one at a time. Next, we will develop a general solution method that does not rely on an upper-triangular coefficient matrix. The solution will be obtained in two steps: First, the solution of the homogeneous state equation will be found. Next, a particular solution will be found based on the input signal $r(t)$. The complete solution of the state equation is obtained as the sum of the homogeneous solution and the particular solution. The derivation of the solution procedure should be compared to the development in Section 2.5.2 of Chapter 2 for an ordinary differential equation.

Homogeneous solution: Consider the homogeneous state equation obtained by setting $r(t) = 0$:

$$\dot{\mathbf{x}}(t) = \mathbf{A}\,\mathbf{x}(t) \tag{9.92}$$

Each of the state variables $x_i(t)$ in the state vector $\mathbf{x}(t)$ can be expressed by a Taylor series in the form

$$x_i(t) = x_i(0) + \left.\frac{dx_i(t)}{dt}\right|_{t=0} t + \left.\frac{1}{2!}\frac{d^2x_i(t)}{dt^2}\right|_{t=0} t^2 + \ldots + \left.\frac{1}{n!}\frac{d^nx_i(t)}{dt^n}\right|_{t=0} t^n + \ldots \tag{9.93}$$

Alternatively, Eqn. (9.93) can be generalized to construct a vector form of Taylor series for expressing the state vector $\mathbf{x}(t)$ as

$$\mathbf{x}(t) = \mathbf{x}(0) + \left.\frac{d\mathbf{x}(t)}{dt}\right|_{t=0} t + \frac{1}{2!}\left.\frac{d^2\mathbf{x}(t)}{dt^2}\right|_{t=0} t^2 + \ldots + \frac{1}{n!}\left.\frac{d^n\mathbf{x}(t)}{dt^n}\right|_{t=0} t^n + \ldots \qquad (9.94)$$

Each vector derivative in Eqn. (9.94) corresponds to a vector of corresponding derivatives of all state variables:

$$\left.\frac{d^n\mathbf{x}(t)}{dt^n}\right|_{t=0} = \begin{bmatrix} d^n x_1(t)/dt^n \\ d^n x_2(t)/dt^n \\ \vdots \\ d^n x_N(t)/dt^n \end{bmatrix}_{t=0} \qquad (9.95)$$

Using the homogeneous state equation, the first derivative evaluated at $t = 0$ is

$$\left.\frac{d\mathbf{x}(t)}{dt}\right|_{t=0} = \left.\mathbf{A}\,\mathbf{x}(t)\right|_{t=0} = \mathbf{A}\,\mathbf{x}(0) \qquad (9.96)$$

The second derivative of the state vector evaluated at time $t = 0$ is obtained by differentiating both sides of Eqn. (9.92) and using it in conjunction with Eqn. (9.96):

$$\left.\frac{d^2\mathbf{x}(t)}{dt^2}\right|_{t=0} = \mathbf{A}\,\left.\frac{d\mathbf{x}(t)}{dt}\right|_{t=0} = \mathbf{A}\,(\mathbf{A}\,\mathbf{x}(0)) = \mathbf{A}^2\,\mathbf{x}(0) \qquad (9.97)$$

Continuing in this fashion, the n-th derivative of the state vector evaluated at $t = 0$ is

$$\left.\frac{d^n\mathbf{x}(t)}{dt^n}\right|_{t=0} = \mathbf{A}^n\,\mathbf{x}(0) \qquad (9.98)$$

Using Eqn. (9.98), the Taylor series expansion given by Eqn. (9.94) for the state vector becomes

$$\mathbf{x}(t) = \left[\mathbf{I} + \mathbf{A}\,t + \frac{1}{2!}\,\mathbf{A}^2\,t^2 + \ldots + \frac{1}{n!}\,\mathbf{A}^n\,t^n + \ldots\right]\mathbf{x}(0) \qquad (9.99)$$

In Eqn. (9.99) the infinite series in square brackets is the matrix exponential function $e^{\mathbf{A}t}$, that is,

$$e^{\mathbf{A}t} = \mathbf{I} + \mathbf{A}\,t + \frac{1}{2!}\,\mathbf{A}^2\,t^2 + \ldots + \frac{1}{n!}\,\mathbf{A}^n\,t^n + \ldots \qquad (9.100)$$

Therefore, the homogeneous solution of the state equation is

$$\mathbf{x}(t) = e^{\mathbf{A}t}\,\mathbf{x}(0) \qquad (9.101)$$

As in the case of scalar differential equations, the homogeneous solution represents the behavior of the state variables of the system due to the initial conditions alone. In the absence of an external input signal $r(t)$ the matrix $e^{\mathbf{A}t}$ transforms the state vector at time $t = 0$ to the state vector at time t. Consequently it is referred to as the *state transition matrix*. A method for computing the state transition matrix of a state-space model will be presented in Section 9.2.8.

State transition matrix:

$$\phi(t) = e^{\mathbf{A}t} \qquad (9.102)$$

General solution: The general solution technique for the state equation will parallel the development of the general solution for a first-order differential equation in Section 2.5.1 of Chapter 2. Consider the non-homogeneous state equation

$$\dot{\mathbf{x}}(t) = \mathbf{A}\,\mathbf{x}(t) + \mathbf{B}\,r(t) \tag{9.103}$$

with the initial state of the system specified at time $t = t_0$.

We will begin with an assumed solution in the form

$$\mathbf{x}(t) = e^{\mathbf{A}t}\,\mathbf{s}(t) \tag{9.104}$$

where $\mathbf{s}(t)$ is a yet undetermined column vector of time functions $s_1(t), \ldots, s_N(t)$. Differentiation of both sides of Eqn. (9.104) with respect to time results in

$$\dot{\mathbf{x}}(t) = \mathbf{A}\,e^{\mathbf{A}t}\,\mathbf{s}(t) + e^{\mathbf{A}t}\,\dot{\mathbf{s}}(t) \tag{9.105}$$

Comparing Eqns. (9.105) and (9.103) leads us to the conclusion

$$\mathbf{B}\,r(t) = e^{\mathbf{A}t}\,\dot{\mathbf{s}}(t) \tag{9.106}$$

Solving Eqn. (9.106) for $\dot{\mathbf{s}}(t)$ yields

$$\dot{\mathbf{s}}(t) = e^{-\mathbf{A}t}\,\mathbf{B}\,r(t) \tag{9.107}$$

Integrating both sides of Eqn. (9.107) over time starting at $t = t_0$ we obtain

$$\mathbf{s}(t) = \mathbf{s}(t_0) + \int_{t_0}^{t} e^{-\mathbf{A}\tau}\,\mathbf{B}\,r(\tau)\,d\tau \tag{9.108}$$

Substituting Eqn. (9.108) into Eqn. (9.104) results in

$$\mathbf{x}(t) = e^{\mathbf{A}t}\,\mathbf{s}(t_0) + e^{\mathbf{A}t}\int_{t_0}^{t} e^{-\mathbf{A}\tau}\,\mathbf{B}\,r(\tau)\,d\tau \tag{9.109}$$

The term $\mathbf{s}(t_0)$ is obtained from Eqn. (9.104) as

$$\mathbf{s}(t_0) = e^{-\mathbf{A}t_0}\,\mathbf{x}(t_0) \tag{9.110}$$

and Eqn. (9.109) becomes

$$\mathbf{x}(t) = e^{\mathbf{A}(t-t_0)}\,\mathbf{x}(t_0) + \int_{t_0}^{t} e^{\mathbf{A}(t-\tau)}\,\mathbf{B}\,r(\tau)\,d\tau \tag{9.111}$$

The state equation

$$\dot{\mathbf{x}}(t) = \mathbf{A}\,\mathbf{x}(t) + \mathbf{B}\,r(t)\,, \qquad \mathbf{x}(t_0): \text{specified} \tag{9.112}$$

is solved as

$$\mathbf{x}(t) = e^{\mathbf{A}(t-t_0)}\,\mathbf{x}(t_0) + \int_{t_0}^{t} e^{\mathbf{A}(t-\tau)}\,\mathbf{B}\,r(\tau)\,d\tau \tag{9.113}$$

Often the initial state of a system is specified at time instant $t = 0$. In that case the general solution becomes

$$\mathbf{x}(t) = e^{\mathbf{A}t}\mathbf{x}(0) + \int_0^t e^{\mathbf{A}(t-\tau)}\mathbf{B}\,r(\tau)\,d\tau \tag{9.114}$$

9.2.8   Computation of the state transition matrix

In the previous section it became apparent that the state transition matrix plays an important role in solving the state equations. The definition of the state transition matrix $\phi(t)$ is based on the infinite series

$$\phi(t) = e^{\mathbf{A}t} = \mathbf{I} + \mathbf{A}t + \frac{1}{2!}\mathbf{A}^2 t^2 + \ldots + \frac{1}{n!}\mathbf{A}^n t^n + \ldots$$

While the series is convergent, and can be used for approximating $\phi(t)$ for specific values of t, we still need a method for determining the analytical form of $\phi(t)$. Such a method will be presented in this section.

We will begin with the homogeneous state equation given by Eqn. (9.92) and take the unilateral Laplace transform of each side of it to obtain

$$s\mathbf{X}(s) - \mathbf{x}(0) = \mathbf{A}\mathbf{X}(s) \tag{9.115}$$

The vector $\mathbf{X}(s)$ is a column vector of Laplace transforms of state variables, that is,

$$\mathbf{X}(s) = \begin{bmatrix} X_1(s) \\ X_2(s) \\ \vdots \\ X_N(s) \end{bmatrix}$$

Rearranging terms of Eqn. (9.115) we get

$$[s\mathbf{I} - \mathbf{A}]\,\mathbf{X}(s) = \mathbf{x}(0) \tag{9.116}$$

The matrix $\mathbf{I}$ in Eqn. (9.116) represents the identity matrix or order N. Its inclusion is necessary to keep the two terms in square brackets dimensionally compatible, since we could not subtract a matrix from a scalar. Solving Eqn. (9.116) for $\mathbf{X}(s)$ results in

$$\mathbf{X}(s) = [s\mathbf{I} - \mathbf{A}]^{-1}\mathbf{x}(0) \tag{9.117}$$

The state vector can be found by taking inverse Laplace transform of Eqn. (9.117) as

$$\mathbf{x}(t) = \mathcal{L}^{-1}\left\{[s\mathbf{I} - \mathbf{A}]^{-1}\right\}\mathbf{x}(0) \tag{9.118}$$

The state transition matrix $\phi(t)$ is found by comparing Eqns. (9.118) and (9.101):

$$\phi(t) = \mathcal{L}^{-1}\left\{[s\mathbf{I} - \mathbf{A}]^{-1}\right\} \tag{9.119}$$

The matrix

$$\mathbf{\Phi}(s) = [s\mathbf{I} - \mathbf{A}]^{-1} \tag{9.120}$$

is called the *resolvent matrix*, and is the Laplace transform of the state transition matrix.

In summary, computation of the state transition matrix involves the following steps:

1. Form the matrix $[s\mathbf{I} - \mathbf{A}]$.

2. Find the resolvent matrix $\boldsymbol{\Phi}(s)$ by inverting the matrix $[s\,\mathbf{I}-\mathbf{A}]$.

3. Find the state transition matrix $\phi(t)$ by computing the inverse Laplace transform of $\boldsymbol{\Phi}(s)$.

Example 9.12: **Finding the state transition matrix**

Determine the state transition matrix $\phi(t)$ for a system with the coefficient matrix

$$\mathbf{A} = \begin{bmatrix} -1 & -1 & 0 \\ 1 & 0 & -1 \\ 5 & 7 & -6 \end{bmatrix}$$

The first step is to construct the matrix $[s\,\mathbf{I}-\mathbf{A}]$:

$$[s\,\mathbf{I}-\mathbf{A}] = s\begin{bmatrix} 1 & 0 & 0 \\ 0 & 1 & 0 \\ 0 & 0 & 1 \end{bmatrix} - \begin{bmatrix} -1 & -1 & 0 \\ 1 & 0 & -1 \\ 5 & 7 & -6 \end{bmatrix}$$

$$= \begin{bmatrix} s+1 & 1 & 0 \\ -1 & s & 1 \\ -5 & -7 & s+6 \end{bmatrix}$$

In the next step we find the resolvent matrix $\boldsymbol{\Phi}(s)$:

$$\boldsymbol{\Phi}(s) = [s\,\mathbf{I}-\mathbf{A}]^{-1} = \frac{1}{\det[s\,\mathbf{I}-\mathbf{A}]}\,\mathrm{adj}\,[s\,\mathbf{I}-\mathbf{A}]$$

$$= \begin{bmatrix} \dfrac{s^2+6s+7}{(s+1)(s+2)(s+4)} & \dfrac{-s-6}{(s+1)(s+2)(s+4)} & \dfrac{1}{(s+1)(s+2)(s+4)} \\[2ex] \dfrac{1}{(s+2)(s+4)} & \dfrac{s+6}{(s+2)(s+4)} & \dfrac{-1}{(s+2)(s+4)} \\[2ex] \dfrac{5s+7}{(s+1)(s+2)(s+4)} & \dfrac{7s+2}{(s+1)(s+2)(s+4)} & \dfrac{s^2+s+1}{(s+1)(s+2)(s+4)} \end{bmatrix}$$

The state transition matrix $\phi(t)$ is found by taking the inverse Laplace term of each element of the resolvent matrix using partial fraction expansion.

$$\phi(t) = \begin{bmatrix} \left(\frac{2}{3}e^{-t}+\frac{1}{2}e^{-2t}-\frac{1}{6}e^{-4t}\right) & \left(-\frac{5}{3}e^{-t}+2e^{-2t}-\frac{1}{3}e^{-4t}\right) & \left(\frac{1}{3}e^{-t}-\frac{1}{2}e^{-2t}+\frac{1}{6}e^{-4t}\right) \\[2ex] \left(\frac{1}{2}e^{-2t}-\frac{1}{2}e^{-4t}\right) & \left(2e^{-2t}-e^{-4t}\right) & \left(-\frac{1}{2}e^{-2t}+\frac{1}{2}e^{-4t}\right) \\[2ex] \left(\frac{2}{3}e^{-t}+\frac{3}{2}e^{-2t}-\frac{13}{6}e^{-4t}\right) & \left(-\frac{5}{3}e^{-t}+6e^{-2t}-\frac{13}{3}e^{-4t}\right) & \left(\frac{1}{3}e^{-t}-\frac{3}{2}e^{-2t}+\frac{13}{6}e^{-4t}\right) \end{bmatrix}$$

Example 9.13: **Solving state equations for RLC circuit**

A state-space model for the RLC circuit in Fig. 9.3 was found in Example 9.1 as

$$\dot{\mathbf{x}}(t) = \begin{bmatrix} -2 & -2 \\ 1 & -5 \end{bmatrix}\mathbf{x}(t) + \begin{bmatrix} 1 \\ 0 \end{bmatrix}r(t)$$

$$y(t) = \begin{bmatrix} 0 & 5 \end{bmatrix}\mathbf{x}(t)$$

The initial value of the state vector is

$$\mathbf{x}(0) = \begin{bmatrix} 3 \\ -2 \end{bmatrix}$$

Compute the two state variables $x_1(t)$ and $x_2(t)$ as well as the output signal $y(t)$ in response to a unit-step input signal.

Solution: We will begin by constructing the matrix $[s\,\mathbf{I} - \mathbf{A}]$:

$$[s\,\mathbf{I} - \mathbf{A}] = s \begin{bmatrix} 1 & 0 \\ 0 & 1 \end{bmatrix} - \begin{bmatrix} -2 & -2 \\ 1 & -5 \end{bmatrix}$$

$$= \begin{bmatrix} s+2 & 2 \\ -1 & s+5 \end{bmatrix}$$

The resolvent matrix $\mathbf{\Phi}(s)$ is the inverse of $[s\,\mathbf{I} - \mathbf{A}]$:

$$\mathbf{\Phi}(s) = [s\,\mathbf{I} - \mathbf{A}]^{-1} = \begin{bmatrix} \dfrac{s+5}{(s+3)(s+4)} & \dfrac{-2}{(s+3)(s+4)} \\[2ex] \dfrac{1}{(s+3)(s+4)} & \dfrac{s+2}{(s+3)(s+4)} \end{bmatrix}$$

State transition matrix $\phi(t)$ is found by taking the inverse Laplace transform of each element of the resolvent matrix through partial fraction expansion:

$$\phi(t) = \mathcal{L}^{-1}\{\mathbf{\Phi}(s)\} = \begin{bmatrix} \left(2\,e^{-3t} - e^{-4t}\right) & \left(-2\,e^{-3t} + 2e^{-4t}\right) \\ \left(e^{-3t} - e^{-4t}\right) & \left(-e^{-3t} + 2\,e^{-4t}\right) \end{bmatrix}$$

The state vector $\mathbf{x}(t)$ can now be computed through the use of Eqn. (9.114). The first term in the solution is the homogeneous solution which is

$$\phi(t)\,\mathbf{x}(0) = \begin{bmatrix} \left(2\,e^{-3t} - e^{-4t}\right) & \left(-2\,e^{-3t} + 2e^{-4t}\right) \\ \left(e^{-3t} - e^{-4t}\right) & \left(-e^{-3t} + 2\,e^{-4t}\right) \end{bmatrix} \begin{bmatrix} 3 \\ -2 \end{bmatrix} = \begin{bmatrix} 10\,e^{-3t} - 7\,e^{-4t} \\ 5\,e^{-3t} - 7\,e^{-4t} \end{bmatrix}$$

In preparation for the computation of the second term in Eqn. (9.114), which is the particular solution, we first need to compute

$$\phi(t)\,\mathbf{B} = \begin{bmatrix} \left(2\,e^{-3t} - e^{-4t}\right) & \left(-2\,e^{-3t} + 2e^{-4t}\right) \\ \left(e^{-3t} - e^{-4t}\right) & \left(-e^{-3t} + 2\,e^{-4t}\right) \end{bmatrix} \begin{bmatrix} 1 \\ 0 \end{bmatrix} = \begin{bmatrix} 2\,e^{-3t} - e^{-4t} \\ e^{-3t} - e^{-4t} \end{bmatrix}$$

The particular solution is

$$\int_0^t \phi(t-\tau)\,\mathbf{B}\,r(\tau)\,d\tau = \int_0^t \begin{bmatrix} 2\,e^{-3(t-\tau)} - e^{-4(t-\tau)} \\ e^{-3(t-\tau)} - e^{-4(t-\tau)} \end{bmatrix} d\tau = \begin{bmatrix} \frac{5}{12} - \frac{2}{3}e^{-3t} + \frac{1}{4}e^{-4t} \\ \frac{1}{12} - \frac{1}{3}e^{-3t} + \frac{1}{4}e^{-4t} \end{bmatrix}$$

The state vector is obtained by adding the homogeneous and particular solutions:

$$\mathbf{x}(t) = \begin{bmatrix} 10\,e^{-3t} - 7\,e^{-4t} \\ 5\,e^{-3t} - 7\,e^{-4t} \end{bmatrix} + \begin{bmatrix} \frac{5}{12} - \frac{2}{3}e^{-3t} + \frac{1}{4}e^{-4t} \\ \frac{1}{12} - \frac{1}{3}e^{-3t} + \frac{1}{4}e^{-4t} \end{bmatrix} = \begin{bmatrix} \frac{5}{12} + \frac{28}{3}e^{-3t} - \frac{27}{4}e^{-4t} \\ \frac{1}{12} + \frac{14}{3}e^{-3t} - \frac{27}{4}e^{-4t} \end{bmatrix}$$

and the output signal is obtained through the use of the output equation:

$$y\left(t\right) = \left[\begin{array}{cc} 0 & 5 \end{array}\right] \left[\begin{array}{c} \frac{5}{12} + \frac{28}{3}\,e^{-3t} - \frac{27}{4}\,e^{-4t} \\ \frac{1}{12} + \frac{14}{3}\,e^{-3t} - \frac{27}{4}\,e^{-4t} \end{array}\right] = \frac{5}{12} + \frac{70}{3}\,e^{-3t} - \frac{135}{4}\,e^{-4t}$$

$$t \geq 0$$

Software resources:
ex_9_13.m

> **Software resources:** See MATLAB Exercises 9.3, 9.4 and 9.5.

9.2.9 Obtaining system function from state-space model

The problem of obtaining a state-space model of a CTLTI system from its s-domain system function was considered in Section 9.2.4. In this section we will discuss the inverse problem. Given a state-space model for a CTLTI system, determine the system function $G\left(s\right)$.

Initially we will focus on a single-input single-output system. Let the system be defined by a state-space model in the standard form of Eqns. (9.7) and (9.8). Taking the Laplace transform of both sides of the state equation yields

$$s\,\mathbf{X}\left(s\right) = \mathbf{A}\,\mathbf{X}\left(s\right) + \mathbf{B}\,R\left(s\right) \tag{9.121}$$

$\mathbf{X}\left(s\right)$ is a column vector that holds the Laplace transform of each state variable:

$$\mathbf{X}\left(s\right) = \left[\begin{array}{c} X_1\left(s\right) \\ X_2\left(s\right) \\ \vdots \\ X_N\left(s\right) \end{array}\right]$$

In taking the Laplace transform of the state equation we have assumed zero initial conditions, that is,

$$\mathbf{x}\left(0\right) = \mathbf{0} \tag{9.122}$$

since the system is CTLTI. Recall that the system function concept is only valid for a linear and time-invariant system. Rearranging terms of Eqn. (9.121) yields

$$\left(s\,\mathbf{I} - \mathbf{A}\right)\mathbf{X}\left(s\right) = \mathbf{B}\,R\left(s\right) \tag{9.123}$$

The $N \times N$ identity matrix $\mathbf{I}$ is necessary to keep the two terms in parentheses compatible. Solving Eqn. (9.124) for $\mathbf{X}\left(s\right)$ we obtain

$$\mathbf{X}\left(s\right) = \left(s\,\mathbf{I} - \mathbf{A}\right)^{-1}\mathbf{B}\,R\left(s\right) \tag{9.124}$$

Next we will take the Laplace transform of the output equation and then use Eqn. (9.124) in it to obtain

$$\begin{aligned} Y\left(s\right) &= \mathbf{C}\,\mathbf{X}\left(s\right) + d\,R\left(s\right) \\ &= \mathbf{C}\left(s\,\mathbf{I} - \mathbf{A}\right)^{-1}\mathbf{B}\,R\left(s\right) + d\,R\left(s\right) \\ &= \left[\mathbf{C}\left(s\,\mathbf{I} - \mathbf{A}\right)^{-1}\mathbf{B} + d\right]R\left(s\right) \end{aligned} \tag{9.125}$$

The system function is found from Eqn. (9.125) as

$$G(s) = \frac{Y(s)}{R(s)} = \mathbf{C}(s\mathbf{I} - \mathbf{A})^{-1}\mathbf{B} + d \qquad (9.126)$$

We recognize the term $(s\mathbf{I} - \mathbf{A})^{-1}$ in Eqn. (9.126) as the resolvent matrix $\mathbf{\Phi}(s)$. Therefore

$$G(s) = \mathbf{C}\,\mathbf{\Phi}(s)\,\mathbf{B} + d \qquad (9.127)$$

Example 9.14: **Obtaining the system function**

Determine the system function for a CTLTI system described by the state-space model

$$\dot{\mathbf{x}}(t) = \begin{bmatrix} -2 & -2 \\ 1 & -5 \end{bmatrix} \mathbf{x}(t) + \begin{bmatrix} 1 \\ 0 \end{bmatrix} r(t)$$

$$y(t) = \begin{bmatrix} 0 & 5 \end{bmatrix} \mathbf{x}(t)$$

Solution: For the specified coefficient matrix $\mathbf{A}$, the resolvent matrix was determined in Example 9.13 as

$$\mathbf{\Phi}(s) = [s\mathbf{I} - \mathbf{A}]^{-1} = \begin{bmatrix} \dfrac{s+5}{(s+3)(s+4)} & \dfrac{-2}{(s+3)(s+4)} \\ \dfrac{1}{(s+3)(s+4)} & \dfrac{s+2}{(s+3)(s+4)} \end{bmatrix}$$

Using Eqn. (9.127) the system function is

$$G(s) = \mathbf{C}\,\mathbf{\Phi}(s)\,\mathbf{B} + d$$

$$= \begin{bmatrix} 0 & 5 \end{bmatrix} \begin{bmatrix} \dfrac{s+5}{(s+3)(s+4)} & \dfrac{-2}{(s+3)(s+4)} \\ \dfrac{1}{(s+3)(s+4)} & \dfrac{s+2}{(s+3)(s+4)} \end{bmatrix} \begin{bmatrix} 1 \\ 0 \end{bmatrix}$$

$$= \frac{5}{(s+3)(s+4)}$$

Software resources: See MATLAB Exercises 9.6.

The results obtained for a single-input single-output system can easily be adapted to multiple input multiple-output systems. Consider a CTLTI system with K input signals $\{r_i(t);\ i = 1,\ldots,K\}$ and M output signals $\{y_j(t);\ i = 1,\ldots,M\}$. The matrix $\mathbf{B}$ is $(N \times K)$, and the matrix $\mathbf{C}$ is $(M \times N)$. The scalar d turns into a matrix $\mathbf{D}$ that is $(M \times K)$. As a result, Eqn. (9.127) turns into

$$\mathbf{G}(s) = \mathbf{C}\,\mathbf{\Phi}(s)\,\mathbf{B} + \mathbf{D} \qquad (9.128)$$

$\mathbf{G}(s)$ is a $(M \times K)$ matrix of system functions from each input to each output.

$$\begin{bmatrix} Y_1(s) \\ Y_2(s) \\ \vdots \\ Y_M(s) \end{bmatrix} = \begin{bmatrix} G_{11}(s) & G_{12}(s) & \cdots & G_{1K}(s) \\ G_{21}(s) & G_{22}(s) & \cdots & G_{2K}(s) \\ & \vdots & & \\ G_{M1}(s) & G_{M2}(s) & \cdots & G_{MK}(s) \end{bmatrix} \begin{bmatrix} R_1(s) \\ R_2(s) \\ \vdots \\ R_K(s) \end{bmatrix} \qquad (9.129)$$

Example 9.15: **System function for multiple-input multiple-output system**

A 2-input 2-output system is characterized by the state-space model

$$\dot{\mathbf{x}}(t) = \begin{bmatrix} 0 & 1 & 0 \\ -2 & -3 & -1 \\ -4 & 0 & -3 \end{bmatrix} \mathbf{x}(t) + \begin{bmatrix} 0 & 0 \\ -1 & 1 \\ -11 & -2 \end{bmatrix} \mathbf{r}(t)$$

$$\mathbf{y}(t) = \begin{bmatrix} 1 & 0 & 0 \\ 0 & 0 & 1 \end{bmatrix} \mathbf{x}(t) + \begin{bmatrix} 0 & 0 \\ 0 & 4 \end{bmatrix} \mathbf{r}(t)$$

Determine the system function matrix $\mathbf{G}(s)$.

Solution: We will begin by constructing the matrix $[s\,\mathbf{I} - \mathbf{A}]$.

$$[s\,\mathbf{I} - \mathbf{A}] = s \begin{bmatrix} 1 & 0 & 0 \\ 0 & 1 & 0 \\ 0 & 0 & 1 \end{bmatrix} - \begin{bmatrix} 0 & 1 & 0 \\ -2 & -3 & -1 \\ -4 & 0 & -3 \end{bmatrix}$$

$$= \begin{bmatrix} s & -1 & 0 \\ 2 & s+3 & 1 \\ 4 & 0 & s+3 \end{bmatrix} \tag{9.130}$$

The state transition matrix is found by inverting $[s\,\mathbf{I} - \mathbf{A}]$.

$$\boldsymbol{\Phi}(s) = [s\,\mathbf{I} - \mathbf{A}]^{-1} = \frac{1}{s^3 + 6s^2 + 11s + 2} \begin{bmatrix} (s+3)^2 & (s+3) & -1 \\ -2(s+1) & s(s+3) & -s \\ -4(s+3) & -4 & (s+1)(s+2) \end{bmatrix}$$

The system function matrix is

$$\mathbf{G}(s) = \mathbf{C}\,\boldsymbol{\Phi}(s)\,\mathbf{B} + \mathbf{D}$$

$$= \frac{1}{s^3 + 6s^2 + 11s + 2} \begin{bmatrix} -(s-8) & (s+5) \\ -(11s^2 + 33s + 18) & 2s(2s^2 + 11s + 19) \end{bmatrix}$$

Software resources:
ex_9_15.m

9.3 State-Space Modeling of Discrete-Time Systems

Discrete-time systems can also be modeled using state variables. Consider a discrete-time system with a single input signal $r[n]$ and a single output signal $y[n]$. Such a system can be modeled by means of the relationships between its input and output signals as well as N internal variables $x_i[n]$ for $i = 1, \ldots, N$ as shown in Fig. 9.9.

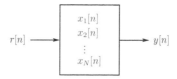

Figure 9.9 – State-space representation of a discrete-time system.

The internal variables $x_1[n], \ldots, x_N[n]$ are the state variables of the discrete-time system. Let us describe the system through a set of first-order difference equations written for each state variable:

$$x_1[n+1] = f_1\big(x_1[n], \ldots, x_N[n], r[n]\big)$$

$$x_2[n+1] = f_2\big(x_1[n], \ldots, x_N[n], r[n]\big)$$

$$\vdots$$

$$x_N[n+1] = f_N\big(x_1[n], \ldots, x_N[n], r[n]\big) \qquad (9.131)$$

The equations given in Eqn. (9.131) are the *state equations* of the system. They describe how each state variable changes in the time domain in response to

1. The input signal $r[n]$
2. The current values of all state variables

Observing the time-domain behavior of the state variables provides us with information about the internal operation of the system. Once the N state equations are solved for the N state variables, the output signal can be found through the use of the *output equation*

$$y[n] = g\big(x_1[n], \ldots, x_N[n], r[n]\big) \qquad (9.132)$$

Eqns. (9.131) and (9.132) together form a *state-space model* for the system.

Systems with multiple inputs and/or outputs can be represented with state-space models as well. Consider a discrete-time system with K input signals $\{r_i[n];\ i = 1, \ldots, K\}$ and M output signals $\{y_j[n];\ i = 1, \ldots, M\}$ as shown in Fig. 9.10.

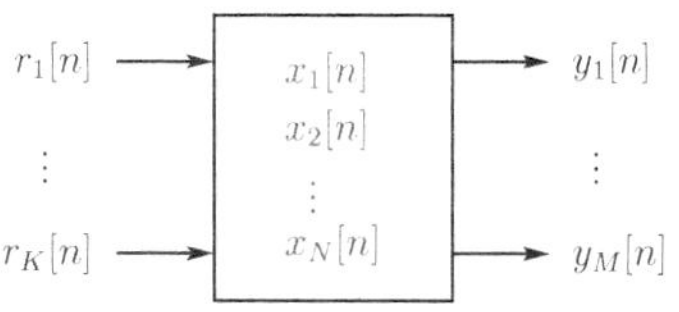

Figure 9.10 – State-space representation of a multiple-input multiple-output discrete-time system.

The state equations given by Eqn. (9.131) can be modified to apply to a multiple-input multiple-output system through a state-space model in the form

$$x_1[n+1] = f_1\big(x_1[n], \ldots, x_N[n], r_1[n], \ldots, r_K[n]\big)$$

$$x_2[n+1] = f_2\big(x_1[n], \ldots, x_N[n], r_1[n], \ldots, r_K[n]\big)$$

$$\vdots$$

$$x_N[n+1] = f_N\big(x_1[n], \ldots, x_N[n], r_1[n], \ldots, r_K[n]\big) \qquad (9.133)$$

Each right-side function $f_i(\ldots)$ includes, as arguments, all input signals as well as the state variables. In addition to this change, the output equation given by Eqn. (9.132) needs to be replaced with a set of output equations for computing all output signals:

$$y_1[n] = g_1\left(x_1[n], \ldots, x_N[n], r_1[n], \ldots, r_K[n]\right)$$

$$y_2[n] = g_2\left(x_1[n], \ldots, x_N[n], r_1[n], \ldots, r_K[n]\right)$$

$$\vdots$$

$$y_M[n] = g_M\left(x_1[n], \ldots, x_N[n], r_1[n], \ldots, r_K[n]\right) \tag{9.134}$$

9.3.1 State-space models for DTLTI systems

If the discrete-time system under consideration is also linear and time-invariant (DTLTI), the right-side functions $f_i\left(\ldots\right), i = 1, \ldots, N$ and $g_i\left(\ldots\right), i = 1, \ldots, M$ in Eqns. (9.133) and (9.134) are linear functions of the state variables and the input signals. For a single-input single-output DTLTI system, the state-space representation is in the form

$$x_1[n+1] = a_{11}\, x_1[n] + a_{12}\, x_2[n] + \ldots + a_{1N}\, x_N[n] + b_1\, r[n]$$

$$x_2[n+1] = a_{21}\, x_1[n] + a_{22}\, x_2[n] + \ldots + a_{2N}\, x_N[n] + b_2\, r[n]$$

$$\vdots$$

$$x_N[n+1] = a_{N1}\, x_1[n] + a_{N2}\, x_2[n] + \ldots + a_{NN}\, x_N[n] + b_N\, r[n] \tag{9.135}$$

and the output equation is

$$y[n] = c_1\, x_1[n] + c_2\, x_2[n] + \ldots, c_N\, x_N[n] + d\, r[n] \tag{9.136}$$

where the coefficients a_{ij}, b_i, c_j and d used in the state equations and the output equation are all constants independent of the time variable. Eqns. (9.135) and (9.136) can be expressed in matrix notation as follows:

$$\text{State equation:} \quad \mathbf{x}[n+1] = \mathbf{A}\,\mathbf{x}[n] + \mathbf{B}\, r[n] \tag{9.137}$$

$$\text{Output equation:} \quad y[n] = \mathbf{C}\,\mathbf{x}[n] + d\, r[n] \tag{9.138}$$

The coefficient matrices used in Eqns. (9.137) and (9.138) are

$$\mathbf{A} = \begin{bmatrix} a_{11} & a_{12} & \ldots & a_{1N} \\ a_{21} & a_{22} & \ldots & a_{2N} \\ & \vdots & & \\ a_{N1} & a_{N2} & \ldots & a_{NN} \end{bmatrix}, \quad \mathbf{B} = \begin{bmatrix} b_1 \\ b_2 \\ \vdots \\ b_N \end{bmatrix}, \quad \mathbf{C} = \begin{bmatrix} c_1 \\ c_2 \\ \vdots \\ c_N \end{bmatrix}^T \tag{9.139}$$

The vector $\mathbf{x}[n]$ is the *state vector* that consists of the N state variables:

$$\mathbf{x}[n] = \begin{bmatrix} x_1[n] \\ x_2[n] \\ \vdots \\ x_N[n] \end{bmatrix} \tag{9.140}$$

The vector $\mathbf{x}[n + 1]$ contains versions of the N state variables advanced by one sample:

$$\mathbf{x}[n + 1] = \begin{bmatrix} x_1[n + 1] \\ x_2[n + 1] \\ \vdots \\ x_N[n + 1] \end{bmatrix} \tag{9.141}$$

Thus, a single-input single-output DTLTI system can be uniquely described by means of the matrix $\mathbf{A}$, the vectors $\mathbf{B}$ and $\mathbf{C}$, and the scalar d. If the initial value $\mathbf{x}[0]$ of the state vector is known, the state equation given by Eqn. (9.137) can be solved for $\mathbf{x}[n]$. Once $\mathbf{x}[n]$ is determined, the output signal can be found through the use of Eqn. (9.138).

9.3.2   Obtaining state-space model from difference equation

Converting a N-th order difference equation to a state-space model requires finding N first-order difference equations that represent the same system, and an output equation that links the output signal to the state variables. As in the case of continuous-time state-space models, the solution is not unique. Consider a DTLTI system described by a difference equation in the form

$$y[n] + a_1\,y[n-1] + a_2\,y[n-2] + \ldots + a_{N-1}\,y[n-N+1] + a_N\,y[n-N] = b_0\,r[n] \tag{9.142}$$

Let us rewrite the difference equation so that $y[n]$ is computed from the remaining terms:

$$y[n] = -a_1\,y[n-1] - a_2\,y[n-2] - \ldots - a_{N-1}\,y[n-N+1] - a_N\,y[n-N] + b_0\,r[n] \tag{9.143}$$

We will choose the state variables as follows:

$$
\begin{aligned}
x_1[n] &= y[n-N] \\
x_2[n] &= y[n-N+1] &\implies&\quad x_1[n+1] = x_2[n] \\
x_3[n] &= y[n-N+2] &\implies&\quad x_2[n+1] = x_3[n] \\
&\;\;\vdots \\
x_N[n] &= y[n-1] &\implies&\quad x_{N-1}[n+1] = x_N[n]
\end{aligned}
$$

The resulting state equations are

$$
\begin{aligned}
x_1[n+1] &= x_2[n] \\
x_2[n+1] &= x_3[n] \\
&\;\;\vdots \\
x_{N-1}[n+1] &= x_N[n] \\
x_N[n+1] &= -a_N\,x_1[n] - a_{N-1}\,x_2[n] - \ldots - a_2\,x_{N-1}[n] - a_1\,x_N[n] + b_0\,r[n] \tag{9.144}
\end{aligned}
$$

The last state equation is obtained by using the state variables $x_1[n]$ through $x_N[n]$ defined above in the difference equation given by Eqn. (9.143), and writing $y[n] = x_N[n+1]$ in terms of the state variables $x_1[n]$ through $x_N[n]$ and the input signal $r[n]$. The output equation is readily obtained from the last state equation:

$$
\begin{aligned}
y[n] &= x_N[n+1] \\
&= -a_N\, x_1[n] - a_{N-1}\, x_2[n] - \ldots - a_2\, x_{N-1}[n] - a_1\, x_N[n] + b_0\, r[n]
\end{aligned}
\tag{9.145}
$$

In matrix form, the state-space model is

$$
\begin{aligned}
\mathbf{x}[n+1] &= \mathbf{A}\,\mathbf{x}[n] + \mathbf{B}\,r[n] \\
y[n] &= \mathbf{C}\,\mathbf{x}[n] + d\,r[n]
\end{aligned}
$$

with

$$
\mathbf{A} =
\begin{bmatrix}
0 & 1 & 0 & \cdots & 0 & 0 \\
0 & 0 & 1 & \cdots & 0 & 0 \\
\vdots & \vdots & \vdots & \cdots & \vdots & \vdots \\
0 & 0 & 0 & \cdots & 1 & 0 \\
0 & 0 & 0 & \cdots & 0 & 1 \\
-a_N & -a_{N-1} & -a_{N-2} & \cdots & -a_2 & -a_1
\end{bmatrix}
, \qquad
\mathbf{B} =
\begin{bmatrix}
0 \\
0 \\
0 \\
\vdots \\
0 \\
b_0
\end{bmatrix}
\tag{9.146}
$$

for the state equation, and

$$
\mathbf{C} =
\begin{bmatrix}
-a_N & -a_{N-1} & -a_{N-2} & \cdots & -a_2 & -a_1
\end{bmatrix}
, \qquad d = b_0
\tag{9.147}
$$

for the output equation. This is the familiar *phase-variable canonical form* we have seen in the analysis of continuous-time state-space models. The coefficient matrix $\mathbf{A}$ is given below in partitioned form:

$$
\mathbf{A} =
\begin{bmatrix}
0 & 1 & 0 & \cdots & 0 & 0 \\
0 & 0 & 1 & \cdots & 0 & 0 \\
\vdots & \vdots & \vdots & \cdots & \vdots & \vdots \\
0 & 0 & 0 & \cdots & 1 & 0 \\
0 & 0 & 0 & \cdots & 0 & 1 \\
-a_N & -a_{N-1} & -a_{N-2} & \cdots & -a_2 & -a_1
\end{bmatrix}
\tag{9.148}
$$

The top left partition is a $(N-1) \times 1$ vector of zeros; the top right partition is an identity matrix of order $(N-1)$; the bottom row is a vector of difference equation coefficients in reverse order and with a sign change. This is handy for obtaining the state-space model by directly inspecting the difference equation. It is important to remember, however, that the coefficient of $y[n]$ in Eqn. (9.142) must be equal to unity for this to work.

Example 9.16: **State-space model from difference equation**

Find a state-space model for a DTLTI system described by the difference equation

$$
y[n] + 1.2\, y[n-1] - 0.13\, y[n-2] - 0.36\, y[n-3] = 2\, r[n]
$$

Solution: Expressing $y[n]$ as a function of the other terms leads to

$$
y[n] = -1.2\, y[n-1] + 0.13\, y[n-2] + 0.36\, y[n-3] + 2\, r[n]
\tag{9.149}
$$

State variables can be defined as

$$x_1[n] = y[n-3]$$
$$x_2[n] = y[n-2] \implies x_1[n+1] = x_2[n]$$
$$x_3[n] = y[n-1] \implies x_2[n+1] = x_3[n]$$

Recognizing that

$$y[n] = x_3[n+1]$$

the difference equation in Eqn. (9.149) can be written as

$$x_3[n+1] = -1.2\,x_3[n] + 0.13\,x_2[n] + 0.36\,x_1[n] + 2\,r[n]$$

In matrix form, the state-space model is

$$\mathbf{x}[n+1] = \begin{bmatrix} 0 & 1 & 0 \\ 0 & 0 & 1 \\ 0.36 & 0.13 & -1.2 \end{bmatrix} \mathbf{x}[n] + \begin{bmatrix} 0 \\ 0 \\ 2 \end{bmatrix} r[n]$$

$$y[n] = \begin{bmatrix} 0.36 & 0.13 & -1.2 \end{bmatrix} \mathbf{x}[n] + 2\,r[n]$$

The inspection method for deriving the state-space model directly from the difference equation would not be as straightforward to use if delayed versions of the input signal $r[n]$ appear in the difference equation. We will, however, generalize the relationship between the coefficients of the difference equation and the state-space model when we consider the derivation of the latter from the system function in the next section.

9.3.3 Obtaining state-space model from system function

If the DTLTI system under consideration has a system function that is simple in the sense that it has no multiple poles and all of its poles are real-valued, then a state-space model can be easily found using partial fraction expansion. This idea will be explored in the next example.

Example 9.17: **State-space model from z-domain system function**

A DTLTI system has the system function

$$H(z) = \frac{Y(z)}{R(z)} = \frac{7\,z}{(z-0.6)\,(z+0.8)}$$

Find a state-space model for this system.

Solution: Expanding $H(z)$ into partial fractions we get

$$H(z) = \frac{3}{z-0.6} + \frac{4}{z+0.8}$$

Note that we did not divide $H(z)$ by z before expanding it into partial fractions, since in this case we do not need the z term in the numerator of each partial fraction. The z-transform of the output signal is

$$Y(z)\frac{3}{z-0.6}R(z) + \frac{4}{z+0.8}R(z)$$

Let us define $X_1(z)$ and $X_2(z)$ as

$$X_1(z) = \frac{3}{z - 0.6} R(z) \tag{9.150}$$

and

$$X_2(z) = \frac{4}{z + 0.8} R(z) \tag{9.151}$$

so that $Y(z) = X_1(x) + X_2(z)$. Rearranging the terms in Eqns. (9.150) and (9.151) we get

$$z X_1(z) = 0.6 X_1(z) + 3 R(z) \tag{9.152}$$

and similarly

$$z X_2(z) = -0.8 X_2(z) + 4 R(z) \tag{9.153}$$

Taking the inverse z-transform of Eqns. (9.152) and (9.153) leads to the two first-order difference equations

$$x_1[n+1] = 0.6 \, x_1[n] + 3 \, r[n] \tag{9.154}$$

$$x_2[n+1] = -0.8 \, x_2[n] + 4 \, r[n] \tag{9.155}$$

The output equation is

$$y[n] = x_1[n] + x_2[n] \tag{9.156}$$

Using $x_1[n]$ and $x_2[n]$ as the state variables, the state-space model can be written as

$$\begin{bmatrix} x_1[n+1] \\ x_2[n+1] \end{bmatrix} = \begin{bmatrix} 0.6 & 0 \\ 0 & -0.8 \end{bmatrix} \begin{bmatrix} x_1[n] \\ x_2[n] \end{bmatrix} + \begin{bmatrix} 3 \\ 4 \end{bmatrix} r[n]$$

and

$$y[n] = \begin{bmatrix} 1 & 1 \end{bmatrix} \begin{bmatrix} x_1[n] \\ x_2[n] \end{bmatrix}$$

A state-space model was easily obtained for the system in Example 9.17 owing to the fact that the system function $H(z)$ has only first-order real poles. As in the continuous-time case, the state equations obtained in Eqns. (9.154) and (9.155) are uncoupled from each other, that is, each state equation involves only one of the state variables. This leads to a state matrix $\mathbf{A}$ that is diagonal. It is therefore possible to solve the state equations independently of each other.

If the system function does not have the properties that afforded us the simplified solution of Example 9.17, finding a state-space model for a system function involves the use of a block diagram as an intermediate step.

Consider the system function

$$H(z) = \frac{Y(z)}{X(z)} = \frac{b_0 + b_1 z^{-1} + b_2 z^{-2} + b_3 z^{-3}}{1 + a_1 z^{-1} + a_2 z^{-2} + a_3 z^{-3}} \tag{9.157}$$

A direct-form II block diagram for implementing this system is shown in Fig. 9.11 (see Section 8.6 of Chapter 8 for the details of obtaining a block diagram).

We know from time shifting property of the z-transform (see Section 8.3.2 of Chapter 8) that multiplication of a transform by z^{-1} is equivalent to right shifting the corresponding signal by one sample in the time domain. A time-domain equivalent of the block diagram

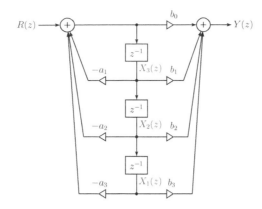

Figure 9.11 – Simulation diagram for $H(z)$ of Eqn. (9.157).

in Fig. 9.11 is obtained by replacing z^{-1} components with delays and using time-domain versions of input and output quantities. The resulting diagram is shown in Fig. 9.12.

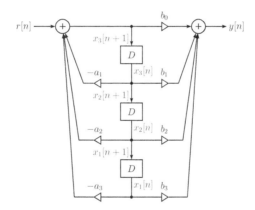

Figure 9.12 – Selecting states from the block diagram for $H(z)$.

A state-space model can be obtained from the diagram of Fig. 9.134 as follows:

1. Designate the output signal of each delay element as a state variable. These are marked in Fig. 9.12 as $x_1[n]$, $x_2[n]$ and $x_3[n]$.
2. The input of each delay element is a one sample advanced version of the corresponding state variable. These are marked in Fig. 9.12 as $x_1[n+1]$, $x_2[n+1]$ and $x_3[n+1]$.
3. Express the input signal of each delay element as a function of the state variables and the input signal as dictated by the diagram.

Following state equations are obtained by inspecting the diagram:

$$x_1[n+1] = x_2[n] \tag{9.158}$$

$$x_2[n+1] = x_3[n] \tag{9.159}$$

$$x_3[n+1] = -a_3\, x_1[n] - a_2\, x_2[n] - a_1\, x_3[n] + r[n] \tag{9.160}$$

The output signal can be written as

$$y[n] = b_3 \, x_1[n] + b_2 \, x_2[n] + b_1 \, x_3[n] + b_0 \, x_3[n+1] \tag{9.161}$$

Since we do not want the $x_3[n+1]$ term on the right side of the output equation, Eqn. (9.160) needs to be substituted into Eqn. (9.161) to obtain the proper output equation:

$$y[n] = (b_3 - b_0 \, a_3) \, x_1[n] + (b_2 - b_0 \, a_2) \, x_2[n] + (b_1 - b_0 \, a_1) \, x_3[n] + b_0 \, r[n] \tag{9.162}$$

Expressed in matrix form, the state-space model is

$$\dot{\mathbf{x}}[n+1] = \begin{bmatrix} 0 & 1 & 0 \\ 0 & 0 & 1 \\ -a_3 & -a_2 & -a_1 \end{bmatrix} \mathbf{x}[n] + \begin{bmatrix} 0 \\ 0 \\ 1 \end{bmatrix} r[n] \tag{9.163}$$

$$y[n] = \begin{bmatrix} (b_3 - b_0 \, a_3) & (b_2 - b_0 \, a_2) & (b_1 - b_0 \, a_1) \end{bmatrix} \mathbf{x}[n] + b_0 \, r[n] \tag{9.164}$$

The results can be easily extended to a N-th order system function in the form

$$H(z) = \frac{Y(z)}{X(z)} = \frac{b_0 + b_1 z^{-1} + b_2 z^{-2} + \dots + \dots b_M z^{-M}}{1 + a_1 z^{-1} + a_2 z^{-2} + \dots + a_N z^{-N}} \tag{9.165}$$

to produce the state-space model

$$\mathbf{x}[n+1] = \begin{bmatrix} 0 & 1 & 0 & 0 & \dots & 0 \\ 0 & 0 & 1 & 0 & \dots & 0 \\ 0 & 0 & 0 & 1 & \dots & 0 \\ \vdots & \vdots & \vdots & \vdots & \dots & \vdots \\ 0 & 0 & 0 & 0 & \dots & 1 \\ -a_N & -a_{N-1} & -a_{N-2} & -a_{N-3} & \dots & -a_1 \end{bmatrix} \mathbf{x}[n] + \begin{bmatrix} 0 \\ 0 \\ 0 \\ \vdots \\ 0 \\ 1 \end{bmatrix} r[n] \tag{9.166}$$

$$y[n] = \begin{bmatrix} (b_N - b_0 \, a_N) & (b_{N-1} - b_0 \, a_{N-1}) & \dots & (b_1 - b_0 \, a_1) \end{bmatrix} \mathbf{x}[n] + b_0 \, r[n] \tag{9.167}$$

The structure of the coefficient matrix $\mathbf{A}$ is the same as in Eqn. (9.146). Similarly, the vector $\mathbf{B}$ is identical to that found in Section 9.3.2.

Example 9.18: **State-space model from block diagram for DTLTI system**

Find a state-space model for the DTLTI system described through the system function

$$H(z) = \frac{z^3 - 7z + 6}{z^4 - z^3 - 0.34 \, z^2 + 0.966 \, z - 0.2403}$$

Solution: A cascade form block diagram for this system was drawn in Example 8.43 of Chapter 8 and is shown again in Fig. 9.13 with state variable assignments marked. Writing the input signal of each delay element in terms of the other signals in the diagram we obtain

$$x_1[n+1] = x_2[n] \tag{9.168}$$

$$x_2[n+1] = 0.27 \, x_1[n] - 0.6 \, x_2[n] + r[n] \tag{9.169}$$

$$x_3[n+1] = x_4[n] \tag{9.170}$$

$$x_4[n+1] = -3 \, x_1[n] + 2 \, x_2[n] + x_2[n+1] - 0.89 \, x_3[n] + 1.6 \, x_4[n] \tag{9.171}$$

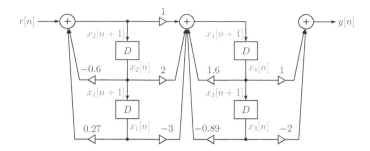

Figure 9.13 – Cascade form block diagram for Example 9.18.

Eqn. (9.171) contains the undesired term $x_2[n+1]$ on the right side of the equal sign. This needs to be resolved by substitution of Eqn. (9.169) into Eqn. (9.171):

$$x_4[n+1] = -2.73\, x_1[n] + 1.4\, x_2[n] - 0.89\, x_3[n] + 1.6\, x_4[n] + r[n] \qquad (9.172)$$

The output equation is

$$y[n] = -2\, x_3[n] + x_4[n] \qquad (9.173)$$

In matrix form, the state-space model is

$$\mathbf{x}[n+1] = \begin{bmatrix} 0 & 1 & 0 & 0 \\ 0.27 & -0.6 & 0 & 0 \\ 0 & 0 & 0 & 1 \\ -2.73 & 1.4 & -0.89 & 1.6 \end{bmatrix} \mathbf{x}[n] + \begin{bmatrix} 0 \\ 1 \\ 0 \\ 1 \end{bmatrix} r[n]$$

$$y[n] = \begin{bmatrix} 0 & 0 & -2 & 1 \end{bmatrix} \mathbf{x}[n]$$

9.3.4 Solution of state-space model

Discrete-time state equations can be solved iteratively. Consider the state-space model given by Eqns. (9.137) and (9.138). Given the initial state vector $\mathbf{x}[0]$ and the input signal $r[n]$ the iterative solution proceeds as follows:

1. Using $n = 0$, compute $\mathbf{x}[1]$, the state vector at $n + 1 = 1$ using

$$\mathbf{x}[1] = \mathbf{A}\,\mathbf{x}[0] + \mathbf{B}\,r[0]$$

2. Compute the output sample at $n = 0$ using

$$y[0] = \mathbf{C}\,\mathbf{x}[0] + d\,r[0]$$

3. Increment n by 1, and repeat steps 1 and 2.

The iterative solution method allows as many samples of the output signal to be obtained as desired. If an analytical solution is needed, however, we need to follow an alternative approach and obtain a general expression for $y[n]$, similar to the development in Section

9.2.7 for continuous-time state-space models. It is possible to find an analytical solution by obtaining the homogeneous and particular solutions and combining them. Instead, we will use the z-transform approach. Taking the unilateral z-transform of both sides of Eqn. (9.137) we obtain

$$z\,\mathbf{X}\,(z) - z\,\mathbf{x}[0] = \mathbf{A}\,\mathbf{X}\,(z) + \mathbf{B}\,R\,(z) \tag{9.174}$$

Solving Eqn. (9.174) gives

$$\mathbf{X}\,(z) = z\,[z\,\mathbf{I} - \mathbf{A}]^{-1}\,\mathbf{x}[0] + [z\,\mathbf{I} - \mathbf{A}]^{-1}\,\mathbf{B}\,R\,(z) \tag{9.175}$$

The matrix

$$\mathbf{\Phi}\,(z) = z\,[z\,\mathbf{I} - \mathbf{A}]^{-1} \tag{9.176}$$

is the z-transform version of the *resolvent matrix*. Using it in Eqn. (9.175) yields

$$\mathbf{X}\,(z) = \mathbf{\Phi}\,(z)\,\mathbf{x}\,(0) + z^{-1}\,\mathbf{\Phi}\,(z)\,\mathbf{B}\,R\,(z) \tag{9.177}$$

for the solution in the z domain. Taking the inverse z-transform of Eqn. (9.177) and recognizing that the inverse z-transform of the product $\mathbf{\Phi}\,(z)\,R\,(z)$ is the convolution of the corresponding time-domain functions leads to the solution

$$\mathbf{x}[n] = \phi[n]\,\mathbf{x}[0] + \sum_{m=0}^{n-1} \phi[n-1-m]\,\mathbf{B}\,r[m] \tag{9.178}$$

where $\phi[n]$ is the discrete-time version of the *state transition matrix* computed as

$$\phi[n] = \mathcal{Z}^{-1}\,\{\mathbf{\Phi}\,(z)\} = \mathcal{Z}^{-1}\left\{z\,[z\,\mathbf{I} - \mathbf{A}]^{-1}\right\} \tag{9.179}$$

Example 9.19: **Solution of discrete-time state-space model**

Compute the unit-step response of the discrete-time system described by the state-space model

$$\mathbf{x}[n+1] = \begin{bmatrix} 0 & -0.5 \\ 0.25 & 0.75 \end{bmatrix} \mathbf{x}[n] + \begin{bmatrix} 2 \\ 1 \end{bmatrix} r[n]$$

$$y[n] = \begin{bmatrix} 3 & 1 \end{bmatrix} \mathbf{x}[n]$$

using the initial state vector

$$\mathbf{x}[0] = \begin{bmatrix} 2 \\ 0 \end{bmatrix}$$

Solution: We will begin by forming the the matrix $[z\,\mathbf{I} - \mathbf{A}]$:

$$[z\,\mathbf{I} - \mathbf{A}] = z\begin{bmatrix} 1 & 0 \\ 0 & 1 \end{bmatrix} - \begin{bmatrix} 0 & -0.5 \\ 0.25 & 0.75 \end{bmatrix} = \begin{bmatrix} z & 0.5 \\ -0.25 & z - 0.75 \end{bmatrix}$$

The resolvent matrix is found as

$$\mathbf{\Phi}\,(z) = \frac{z}{(z - 0.5)\,(z - 0.25)}\begin{bmatrix} z - 0.75 & -0.5 \\ 0.25 & z \end{bmatrix}$$

The state transition matrix is found by taking the inverse z-transform of the resolvent matrix:

$$\phi[n] = \begin{bmatrix} -(0.5)^n\,u[n] + 2\,(0.25)^n\,u[n] & -2\,(0.5)^n\,u[n] + 2\,(0.25)^n\,u[n] \\ (0.5)^n\,u[n] - (0.25)^n\,u[n] & 2\,(0.5)^n\,u[n] - (0.25)^n\,u[n] \end{bmatrix}$$

The solution of the state equation is found by using the state transition matrix in Eqn. (9.178):

$$\mathbf{x}[n] = \begin{bmatrix} 6\,(0.5)^n\,u[n] - 4\,(0.25)^n\,u[n] \\ -6\,(0.5)^n\,u[n] + 2\,(0.25)^n\,u[n] + 4\,u[n] \end{bmatrix}$$

The output signal is

$$y[n] = \begin{bmatrix} 3 & 1 \end{bmatrix}\mathbf{x}[n]$$
$$= 12\,(0.5)^n\,u[n] - 10\,(0.25)^n\,u[n] + 4\,u[n]$$

Software resources:
ex_9_19a.m
ex_9_19b.m

9.3.5 Obtaining system function from state-space model

The z-domain system function $H(z)$ of a DTLTI system can be easily obtained from its state-space description. Consider Eqn. (9.175) derived in the process of obtaining the solution for the z-transform of the state vector:

Eqn. (9.175): $\mathbf{X}(z) = z\,[z\,\mathbf{I} - \mathbf{A}]^{-1}\,\mathbf{x}[0] + [z\,\mathbf{I} - \mathbf{A}]^{-1}\,\mathbf{B}\,R(z)$

Recall that the system function concept is valid only for a system that is both linear and time-invariant. If the system is DTLTI, then the initial value of the state vector is $\mathbf{x}[0] = \mathbf{0}$, and Eqn. (9.175) becomes

$$\mathbf{X}(z) = [z\,\mathbf{I} - \mathbf{A}]^{-1}\,\mathbf{B}\,R(z) \tag{9.180}$$

Using $\mathbf{X}(z)$ in the output equation leads to

$$Y(z) = \mathbf{C}\,\mathbf{X}(z) + d\,R(z)$$
$$= \mathbf{C}\,[z\,\mathbf{I} - \mathbf{A}]^{-1}\,\mathbf{B}\,R(z) + d\,R(z) \tag{9.181}$$

from which the system function follows as

$$H(z) = \frac{Y(z)}{R(z)} = \mathbf{C}\,[z\,\mathbf{I} - \mathbf{A}]^{-1}\,\mathbf{B} + d$$

9.4 Discretization of Continuous-Time State-Space Model

As discussed in Section 9.3.4 of this chapter, discrete-time state-space models have the added advantage that they can be solved iteratively one step at a time. On the other hand, a continuous-time state-space model cannot be solved iteratively unless it is first converted to an approximate discrete-time model. There are times when we may want to simulate an analog system on a digital computer. In this section we will discuss methods for *discretizing* a continuous-time state-space model. Only single-input-single-output systems will discussed although extension to systems with multiple inputs and/or outputs is straightforward. Let an analog system be described by the equations

$$\dot{\mathbf{x}}_{\mathbf{a}}(t) = \mathbf{A}\,\mathbf{x}_{\mathbf{a}}(t) + \mathbf{B}\,r_a(t) \tag{9.182}$$

$$y_a(t) = \mathbf{C}\,\mathbf{x}_{\mathbf{a}}(t) + d\,r_a(t) \tag{9.183}$$

In Eqns. (9.182) and (9.182) the subscript "a" is used to indicate analog variables. The discretization problem is stated as follows:

Given the state-space model for the analog system, find a discrete-time system with the state-space model

$$\mathbf{x}[n+1] = \bar{\mathbf{A}}\,\mathbf{x}[n] + \bar{\mathbf{B}}\,r[n] \tag{9.184}$$

$$y[n] = \bar{\mathbf{C}}\,\mathbf{x}[n] + \bar{d}\,r[n] \tag{9.185}$$

so that, for a sampling interval T_s we have

$$\mathbf{x}[n] \approx \mathbf{x_a}\,(nT_s) \qquad \text{and} \qquad y[n] \approx y_a\,(nT_s) \tag{9.186}$$

when $r[n] = r_a\,(nT_s)$. The value of T_s must be chosen to satisfy the sampling criteria (see Chapter 6).

The solution for the continuous-time state-space model was given in Section 9.2.7 by Eqn. (9.113) which is repeated here:

$$\text{Eqn. (9.113):} \qquad \mathbf{x_a}\,(t) = e^{\mathbf{A}\,(t-t_0)}\,\mathbf{x_a}\,(t_0) + \int_{t_0}^{t} e^{\mathbf{A}\,(t-\tau)}\,\mathbf{B}\,r_a\,(\tau)\,d\tau \tag{9.187}$$

Eqn. (9.113) starts with the solution at time t_0 and returns the solution at time t. By selecting the limits of the integral to be $t_0 = nT_s$ and $t = (n+1)T_s$, Eqn. (9.113) becomes

$$\mathbf{x_a}\,((n+1)\,T_s) = e^{\mathbf{A}T_s}\,\mathbf{x_a}\,(nT_s) + \int_{nT_s}^{(n+1)T_s} e^{\mathbf{A}\,((n+1)T_s-\tau)}\,\mathbf{B}\,r_a\,(\tau)\,d\tau$$

If T_s is small enough the term $r_a\,(\tau)$ may be assumed to be constant within the span of the integral, that is,

$$r_a\,(\tau) \approx r_a\,(nT_s) \qquad \text{for} \quad nT_s \le \tau \le (n+1)\,T_s \tag{9.188}$$

With this assumption Eqn. (9.188) becomes

$$\mathbf{x_a}\,((n+1)\,T_s) \approx e^{\mathbf{A}T_s}\,\mathbf{x_a}\,(nT_s) + \left[\int_{nT_s}^{(n+1)T_s} e^{\mathbf{A}\,((n+1)T_s-\tau)}\,d\tau\right]\,\mathbf{B}\,r_a\,(nT_s) \tag{9.189}$$

Using the variable change $\lambda = (n+1)\,T_s - \tau$ on the integral in Eqn. (9.189) we obtain

$$\mathbf{x_a}\,((n+1)\,T_s) \approx e^{\mathbf{A}T_s}\,\mathbf{x_a}\,(nT_s) + \left[\int_{0}^{T_s} e^{\mathbf{A}\lambda}\,d\lambda\right]\,\mathbf{B}\,r_a\,(nT_s)$$

$$= e^{\mathbf{A}T_s}\,\mathbf{x_a}\,(nT_s) + \mathbf{A}^{-1}\left[e^{\mathbf{A}T_s} - \mathbf{I}\right]\,\mathbf{B}\,r_a\,(nT_s) \tag{9.190}$$

Finally, substitution of $r_a\,(nT_s) = r[n]$ and $\mathbf{x_a}\,(nT_s) = \mathbf{x}[n]$ yields

$$\mathbf{x}[n+1] = e^{\mathbf{A}T_s}\,\mathbf{x}[n] + \mathbf{A}^{-1}\left[e^{\mathbf{A}T_s} - \mathbf{I}\right]\,\mathbf{B}\,r[n]$$

The output equation is simply

$$y[n] = \mathbf{C}\,\mathbf{x}[n] + d\,r[n] \tag{9.191}$$

In summary, the continuous-time system described by Eqns. (9.182) and (9.183) can be approximated with the discrete-time system in Eqns. (9.184) and (9.185) by choosing the coefficient matrices as follows:

Discretization of state-space model:

$$\bar{\mathbf{A}} = e^{\mathbf{A}T_s}, \qquad \bar{\mathbf{B}} = \mathbf{A}^{-1}\left[e^{\mathbf{A}T_s} - \mathbf{I}\right]\mathbf{B}, \qquad \bar{\mathbf{C}} = \mathbf{C}, \qquad \bar{d} = d \qquad (9.192)$$

Example 9.20: **Discretization of state-space model for RLC circuit**

A state-space model for the RLC circuit in Fig. 9.3 was found in Example 9.1 as

$$\dot{\mathbf{x}}(t) = \begin{bmatrix} -2 & -2 \\ 1 & -5 \end{bmatrix}\mathbf{x}(t) + \begin{bmatrix} 1 \\ 0 \end{bmatrix}r(t)$$

$$y(t) = \begin{bmatrix} 0 & 5 \end{bmatrix}\mathbf{x}(t)$$

The state transition matrix corresponding to this model was found in Example 9.13 as

$$e^{\mathbf{A}t} = \phi(t) = \begin{bmatrix} \left(2\,e^{-3t} - e^{-4t}\right) & \left(-2\,e^{-3t} + 2e^{-4t}\right) \\ \left(e^{-3t} - e^{-4t}\right) & \left(-e^{-3t} + 2\,e^{-4t}\right) \end{bmatrix}$$

Using a step size of $T_s = 0.1$ s, find a discrete-time state-space model to approximate the continuous-time model given above.

Solution: Let us begin by evaluating the state transition matrix at $t = T_s = 0.1$ s.

$$e^{\mathbf{A}T_s} = \phi(0.1) = \begin{bmatrix} \left(2\,e^{-0.3} - e^{-0.4}\right) & \left(-2\,e^{-0.3} + 2e^{-0.4}\right) \\ \left(e^{-0.3} - e^{-0.4}\right) & \left(-e^{-0.3} + 2\,e^{-0.4}\right) \end{bmatrix} = \begin{bmatrix} 0.8113 & -0.1410 \\ 0.0705 & 0.5998 \end{bmatrix}$$

The state matrix of the discrete-time model is

$$\bar{\mathbf{A}} = e^{\mathbf{A}T_s} = \begin{bmatrix} 0.8113 & -0.1410 \\ 0.0705 & 0.5998 \end{bmatrix}$$

The inverse of state matrix $\mathbf{A}$ is

$$\mathbf{A}^{-1} = \begin{bmatrix} -0.4167 & 0.1667 \\ -0.0833 & -0.1667 \end{bmatrix}$$

which can be used for computing the vector $\bar{\mathbf{B}}$ as

$$\bar{\mathbf{B}} = \mathbf{A}^{-1}\left[e^{\mathbf{A}T_s} - \mathbf{I}\right]\mathbf{B}$$

$$= \begin{bmatrix} -0.4167 & 0.1667 \\ -0.0833 & -0.1667 \end{bmatrix}\left(\begin{bmatrix} 0.8113 & -0.1410 \\ 0.0705 & 0.5998 \end{bmatrix} - \begin{bmatrix} 1 & 0 \\ 0 & 1 \end{bmatrix}\right)\begin{bmatrix} 1 \\ 0 \end{bmatrix} = \begin{bmatrix} -0.2547 \\ 0.0952 \end{bmatrix}$$

Therefore, the discretized state-space model for the RLC circuit in question is

$$\mathbf{x}[n+1] = \begin{bmatrix} 0.8113 & -0.1410 \\ 0.0705 & 0.5998 \end{bmatrix}\mathbf{x}[n] + \begin{bmatrix} -0.2547 \\ 0.0952 \end{bmatrix}r[n]$$

$$y[n] = \begin{bmatrix} 0 & 5 \end{bmatrix}\mathbf{x}[n]$$

Software resources:
ex_9_20.m

Software resources: See MATLAB Exercise 9.8.

The only potential drawback for the discretization technique derived above is that it requires inversion of the matrix $\mathbf{A}$ and computation of the state transition matrix. Simpler approximations are also available. An example of a simpler technique is the *Euler method* which was applied to a first-order scalar differential equation in MATLAB Exercise 2.4 in Chapter 2.

Let us evaluate the continuous-time state equation given by 9.182 at $t = nT_s$:

$$\dot{\mathbf{x}}_{\mathbf{a}}(t)\Big|_{t=nT_s} = \mathbf{A}\,\mathbf{x}_{\mathbf{a}}(nT_s) + \mathbf{B}\,r_a(nT_s) \tag{9.193}$$

The derivative on the left side of Eqn. (9.193) can be approximated using the first difference

$$\dot{\mathbf{x}}_{\mathbf{a}}(t)\Big|_{t=nT_s} \approx \frac{1}{T_s}\left[\mathbf{x}_{\mathbf{a}}((n+1)T_s) - \mathbf{x}_{\mathbf{a}}(nT_s)\right] \tag{9.194}$$

Substituting Eqn. (9.194) into Eqn. (9.193) and rearranging terms gives

$$\mathbf{x}((n+1)T_s) \approx [\mathbf{I} + \mathbf{A}\,T_s]\,\mathbf{x}(nT_s) + T_s\,\mathbf{B}\,r(nT_s) \tag{9.195}$$

Through substitutions $r_a(nT_s) = r[n]$ and $\mathbf{x}_{\mathbf{a}}(nT_s) = \mathbf{x}[n]$ we obtain

$$\mathbf{x}[n+1] = [\mathbf{I} + \mathbf{A}\,T_s]\,\mathbf{x}[n] + T_s\,\mathbf{B}\,r[n] \tag{9.196}$$

The output equation is

$$y[n] = \mathbf{C}\,\mathbf{x}[n] + d\,r[n] \tag{9.197}$$

as before.

The results can be summarized as follows:

Discretization using Euler method:

$$\bar{\mathbf{A}} = \mathbf{I} + \mathbf{A}\,T_s\,, \qquad \bar{\mathbf{B}} = T_s\,\mathbf{B}\,, \qquad \bar{\mathbf{C}} = \mathbf{C}\,, \qquad \bar{d} = d \tag{9.198}$$

Since this is a first-order approximation method, the step size (sampling interval) T_s chosen should be significantly small compared to the time constant.

Software resources: See MATLAB Exercise 9.9.

9.5   Further Reading

[1] R.C. Dorf and R.H. Bishop. *Modern Control Systems*. Prentice Hall, 2011.

[2] B. Friedland. *Control System Design: An Introduction to State-Space Methods*. Dover Publications, 2012.

[3] K. Ogata. *Modern Control Engineering*. Instrumentation and Controls Series. Prentice Hall, 2010.

[4] L.A. Zadeh and C.A. Desoer. *Linear System Theory: The State Space Approach*. Dover Civil and Mechanical Engineering Series. Dover Publications, 2008.

MATLAB Exercises

MATLAB Exercise 9.1: Obtaining state-space model from system function

MATLAB function `tf2ss(..)` facilitates conversion of a rational system function in the s domain to a state-space model. Consider a CTLTI system with the system function

$$H\left(s\right) = \frac{s^2 + s - 2}{s^2 + 6\,s + 5}$$

A state-space model for the system is obtained as

```
>>   num = [1,1,-2];
>>   den = [1,6,5];
>>   [A,B,C,d] = tf2ss(num,den)

A =
     -6      -5
      1       0

B =
      1
      0

C =
     -5      -7

d =
      1
```

and corresponds to

$$\dot{\mathbf{x}}\left(t\right) = \begin{bmatrix} -6 & -5 \\ 1 & 0 \end{bmatrix} \mathbf{x}\left(t\right) + \begin{bmatrix} 1 \\ 0 \end{bmatrix} r\left(t\right)$$

$$y\left(t\right) = \begin{bmatrix} -5 & -7 \end{bmatrix} \mathbf{x}\left(t\right) + r\left(t\right)$$

This state-space model is different from what would be obtained using the technique developed in Section 9.2.4 Eqns. (9.53) and 9.54), but it is equivalent. As an exercise we will develop our own version of the function `tf2ss(..)` to produce the result obtained in Section 9.2.4, and name it "ss_t2s". The listing for the function `ss_t2s` is given below:

```
1    function [A,B,C,d] = ss_t2s(num,den)
2      nden = length(den);    % Size of denominator vector
3      N = nden - 1;          % Order of system
4      % If 'num' has fewer elements than 'den', put zeros in front.
5      while (length(num)<nden)
6        num = [0,num];
7      end;
8      % Divide all coefficients by leading denominator coefficient
9      num = num/den(1);
```

```
10    den = den/den(1);
11    num = fliplr(num);    % Reorder numerator coefficients
12    den = fliplr(den);    % Reorder denominator coefficients
13    A = [zeros(N-1,1),eye(N-1);-den(1:N)];
14    B = [zeros(N-1,1);1];
15    C = num(1:N)-num(N+1)*den(1:N);
16    d = num(N+1);
17  end
```

After line 12 of the code is executed, the vectors "**num**" and "**den**" are structured as follows:

$$\text{num} = [a_0, a_1, \dots, a_{N-1}, 1] \tag{9.199}$$
$$\text{den} = [b_0, b_1, \dots, b_{N-1}, b_N] \tag{9.200}$$

The construction of state-space model in lines 13 through 16 follows the development in Section 9.2.4 Eqns. (9.53) and (9.54). Using the function "**ss_t2s**" another state-space model for the same system is obtained as

```
>>   num = [1,1,-2];
>>   den = [1,6,5];
>>   [A,B,C,d] = ss_t2s(num,den)

A =
        0      1
       -5     -6

B =
        0
        1

C =
       -7     -5

d =
        1
```

and corresponds to

$$\dot{\mathbf{z}}(t) = \begin{bmatrix} 0 & 1 \\ -5 & -6 \end{bmatrix} \mathbf{z}(t) + \begin{bmatrix} 0 \\ 1 \end{bmatrix} r(t)$$

$$y(t) = \begin{bmatrix} -7 & -5 \end{bmatrix} \mathbf{z}(t) + r(t)$$

It can easily be shown (see Problem 9.9 at the end of this chapter) that this state-space model is related to the one found using MATLAB function tf2ss(..) by the transformation

$$\mathbf{x}(t) = \begin{bmatrix} 0 & 1 \\ 1 & 0 \end{bmatrix} \mathbf{z}(t)$$

Software resources:
matex_9_1a.m
matex_9_1b.m
ss_t2s.m

MATLAB Exercise 9.2: Diagonalizing the state matrix

The state-space description of a system is not unique. It was discussed in Section 9.2.5 that, for a given state-space model that uses the state vector $\mathbf{x}(t)$, alternative models can be obtained through similarity transformations. An alternative form using the new state vector $\mathbf{z}(t)$ is given by

$$\dot{\mathbf{z}}(t) = \tilde{\mathbf{A}}\mathbf{z}(t) + \tilde{\mathbf{B}}r(t)$$
$$y(t) = \tilde{\mathbf{C}}\mathbf{z}(t) + dr(t)$$

Sometimes it is desirable to find a state-space model with a diagonal coefficient matrix so that the resulting first-order differential equations are uncoupled from each other. Let λ_1, λ_2, ..., λ_N be the eigenvalues of the matrix $\mathbf{A}$ with corresponding eigenvectors $\mathbf{v_1}$, $\mathbf{v_2}$, ..., $\mathbf{v_N}$. It can be shown that, if $\mathbf{A}$ has distinct eigenvalues, a transformation matrix $\mathbf{P}$ constructed as

$$\mathbf{P} = [\mathbf{v_1} \mid \mathbf{v_2} \mid \ldots \mid \mathbf{v_N}]$$

leads to a state-space model with $\bar{\mathbf{A}} = \mathbf{P}^{-1}\mathbf{A}\mathbf{P}$ a diagonal matrix. MATLAB function `eig(..)` can be used to obtain both the eigenvalues and the eigenvectors.

Let the state-space model of a system be

$$\dot{\mathbf{x}}(t) = \begin{bmatrix} 6.5 & 2 & 1.5 \\ -10.5 & -2 & -1.5 \\ 5.5 & 2 & 2.5 \end{bmatrix} \mathbf{x}(t) + \begin{bmatrix} 1 \\ -2 \\ 0 \end{bmatrix} r(t)$$

$$y(t) = \begin{bmatrix} -1 & -1 & -2 \end{bmatrix} \mathbf{x}(t)$$

The state-space model may be entered into MATLAB and an alternative model with a diagonal state matrix may be computed through the following:

```
>>  A = [6.5,2,1.5;-10.5,-2,-1.5;5.5,2,2.5];
>>  B = [1;,-2;0];
>>  C = [-1,-1,-2];
>>  [P,E] = eig(A);
>>  A_bar = inv(P)*A*P

A_bar =
    4.0000   -0.0000   -0.0000
    0.0000    1.0000    0.0000
    0.0000    0.0000    2.0000

>>  B_bar = inv(P)*B

B_bar =
   -0.0000
    2.9580
   -4.9749

>>  C_bar = C*P
C_bar =

    0.4082    0.3381   -0.0000
```

The results lead to a new state-space model as follows:

$$\dot{\mathbf{z}}(t) = \begin{bmatrix} 4 & 0 & 0 \\ 0 & 1 & 0 \\ 0 & 0 & 2 \end{bmatrix} \mathbf{z}(t) + \begin{bmatrix} 0 \\ 2.9580 \\ -4.9749 \end{bmatrix} r(t)$$

$$y(t) = \begin{bmatrix} 0.4082 & 0.3381 & 0 \end{bmatrix} \mathbf{z}(t) + r(t)$$

Software resources:
matex_9_2.m

MATLAB Exercise 9.3: Computation of the state transition matrix

In this exercise we will explore various approaches to the problem of computing the state transition matrix through the use of MATLAB. The definition of the state transition matrix was given by Eqn. (9.102) which is repeated here:

$$\text{Eqn. (9.102):} \quad \phi(t) = e^{\mathbf{A}t}$$

An infinite series form of the state transition matrix was also given, and is repeated below:

$$\text{Eqn. (9.100):} \quad e^{\mathbf{A}t} = \mathbf{I} + \mathbf{A}\,t + \frac{1}{2!}\,\mathbf{A}^2\,t^2 + \ldots + \frac{1}{n!}\,\mathbf{A}^n\,t^n + \ldots$$

Consider the coefficient matrix $\mathbf{A}$ that was used in Example 9.12:

$$\mathbf{A} = \begin{bmatrix} -1 & -1 & 0 \\ 1 & 0 & -1 \\ 5 & 7 & -6 \end{bmatrix}$$

It is possible to compute the complex exponential function $e^{\mathbf{A}t}$ at one specific time instant using the MATLAB function expm(..). The following segment of code computes $\phi(0.2)$, the value of the state transition matrix at time $t = 0.2$:

```
>>   A = [-1,-1,0; 1,0,-1; 5,7,-6];
>>   expm(A*0.2)

ans =
    0.8061    -0.1737     0.0126
    0.1105     0.8913    -0.1105
    0.5778     0.7103     0.2410
```

It is important to remember to use the function expm(..) and not the function exp(..) for this purpose. The latter function computes the exponential function of each element of the matrix which is not the same as computing $e^{\mathbf{A}t}$.

It is also possible to approximate the state transition matrix by using the first few terms of the infinite series for it. The infinite series can be written in nested form as

$$\phi(t) = e^{\mathbf{A}t} = \mathbf{I} + \mathbf{A}\,t\left[\mathbf{I} + \frac{1}{2}\,\mathbf{A}\,t\left[\mathbf{I} + \frac{1}{3}\,\mathbf{A}\,t\left[\mathbf{I} + \frac{1}{4}\,\mathbf{A}\,t\left[\mathbf{I} + \ldots\ldots\ldots\right]\right]\right]\right] \qquad (9.201)$$

The following code segment computes the first five terms of the series for $\phi(0.2)$:

```
>>   id = eye(3);
>>   At = A*0.2;
>>   id+At*(id+At/2*(id+At/3*(id+At/4)))

ans =
    0.8057    -0.1743     0.0130
    0.1093     0.8891    -0.1093
    0.5727     0.7003     0.2461
```

Even though we have used only the first five terms, the result obtained is quite close to that obtained through the use of the function expm(..). The accuracy of the approximation can be improved by including more terms.

Both code segments listed above compute the state transition matrix at one specific time instant. If elements of the state transition matrix are needed as functions of time over a specified interval, the function expm(..) can be used within a loop. The script listed below computes and graphs $\phi_{12}(t)$, the row-1 column-2 element of the state transition matrix, in the time interval $0 \leq t \leq 4$ sec.

```
1   % Script: matex_9_3a.m
2   %
3   t = [0:0.01:4];     % Set a time vector "t"
4   % Create an empty vector "phi12" to fill later
5   phi12 =[];
6   % Compute the STM for each element of vector "t"
7   for i=1:length(t),
8       time = t(i);              % Pick a time instant
9       stm = expm(A*time);       % Compute the STM at time instant t(i)
10      phi12 = [phi12,stm(1,2)]; % Append element (1,2) to vector "phi12"
11  end;
12  % Graph element (1,2) of the STM
13  plot(t,phi12);
14  grid;
15  title('Element 1,2 of the state transition matrix');
16  xlabel('t (sec)');
```

The graph produced is shown in Fig. 9.14.

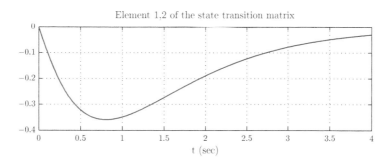

Figure 9.14 – Graph obtained in MATLAB Exercise 9.3.

This result should be compared to the analytical expression obtained in Example 9.12 for the element $\phi_{12}(t)$:

$$\phi_{12}(t) = -\frac{5}{3}e^{-t} + 2e^{-2t} - \frac{1}{3}e^{-4t}$$

The following code segment computes and graphs the analytical form of $\phi_{12}(t)$ for comparison:

```
>>  phi12a = -5/3*exp(-t)+2*exp(-2*t)-1/3*exp(-4*t);
>>  plot(t,phi12a);
```

Software resources:
matex_9_3a.m
matex_9_3b.m

MATLAB Exercise 9.4: Solving the homogeneous state equation

Consider the homogeneous state equation

$$\dot{\mathbf{x}}(t) = \begin{bmatrix} 0 & 5 \\ -2 & -2 \end{bmatrix} \mathbf{x}(t)$$

Let the initial value of the state vector be

$$\mathbf{x}(0) = \begin{bmatrix} 1 \\ -2 \end{bmatrix}$$

Using MATLAB, compute and graph the state transition matrix $\phi(t)$ in the interval $0 \leq t \leq 4$ s. Afterwards, compute and graph the state vector $\mathbf{x}(t)$ in the same time interval.

Solution: The solution of the homogeneous state equation is computed as

$$\mathbf{x}(t) = \phi(t)\,\mathbf{x}(0) \qquad t \geq 0$$

The state transition matrix is the form

$$\phi(t) = \begin{bmatrix} \phi_{11}(t) & \phi_{12}(t) \\ \phi_{21}(t) & \phi_{22}(t) \end{bmatrix}$$

Using the state transition matrix with the initial value of the state vector we obtain

$$\mathbf{x}(t) = \begin{bmatrix} \phi_{11}(t) & \phi_{12}(t) \\ \phi_{21}(t) & \phi_{22}(t) \end{bmatrix} \begin{bmatrix} 1 \\ -2 \end{bmatrix} = \begin{bmatrix} \phi_{11}(t) - 2\,\phi_{12}(t) \\ \phi_{21}(t) - 2\,\phi_{22}(t) \end{bmatrix}$$

The function expm(..) was first used in MATLAB Exercise 9.3 for computing the state transition matrix at for specific value of t. In the script listed below it is used in a loop structure for computing the state transition matrix at a set of time instants. The elements of the state transition matrix are computed and graphed. The result is shown in Fig. 9.15.

```
1   % Script: matex_9_4a.m
2   %
3   A = [0,5;-2,-2];        % State matrix
4   t = [0:0.01:4];         % Set a time vector 't'
5   % Create empty vectors 'phi11', 'phi12', etc. to fill later
6   phi11 =[];
7   phi12 =[];
8   phi21 =[];
9   phi22 =[];
10  % Compute the STM for each element of vector 't'
11  for i=1:length(t),
12      time = t(i);            % Pick a time instant
13      stm = expm(A*time);     % Compute the STM at time instant t(i)
14      phi11 = [phi11,stm(1,1)];   % Append element (1,1) to vector 'phi11'
15      phi12 = [phi12,stm(1,2)];   % Append element (1,2) to vector 'phi12'
16      phi21 = [phi21,stm(2,1)];   % Append element (2,1) to vector 'phi21'
17      phi22 = [phi22,stm(2,2)];   % Append element (2,2) to vector 'phi22'
18  end;
19  % Graph elements of the STM
20  clf;
21  subplot(2,2,1)
22  plot(t,phi11); grid;
23  title('\phi_{11}(t)');
24  subplot(2,2,2)
25  plot(t,phi12); grid;
```

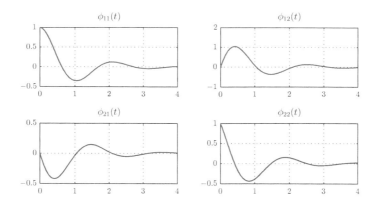

Figure 9.15 – Elements of state transition matrix for MATLAB Exercise 9.4.

```
26    title('\phi_{12}(t)');
27    subplot(2,2,3)
28    plot(t,phi21); grid;
29    title('\phi_{21}(t)');
30    subplot(2,2,4)
31    plot(t,phi22); grid;
32    title('\phi_{22}(t)');
```

The script listed below computes and graphs the elements of the state vector. It relies on the previous script to be run first, so that vectors "t", "phi11", "phi12", "phi21" and "phi22" are already in MATLAB workspace.

```
1     % Script: matex_9_4b.m
2     %
3     % Compute elements of the state vector
4     x1 = phi11-2*phi12;
5     x2 = phi21-2*phi22;
6     % Graph elements of the state vector
7     clf;
8     subplot(2,1,1)
9     plot(t,x1); grid;
10    title('x_{1}(t)');
11    subplot(2,1,2)
12    plot(t,x2); grid;
13    title('x_{2}(t)');
```

The graph produced by the script `matex_9_4b.m` is shown in Fig. 9.16.

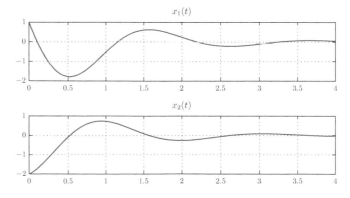

Figure 9.16 – Elements of state vector for MATLAB Exercise 9.4.

Software resources:
```
matex_9_4a.m
matex_9_4b.m
```

MATLAB Exercise 9.5: Symbolic computation of the state transition matrix

It is also possible to compute the state transition matrix using symbolic mathematics capabilities of MATLAB. This is especially useful for checking solutions to homework problems. Consider the system a state-space model of which was given in Example 9.13. The coefficient matrix is

$$\mathbf{A} = \begin{bmatrix} -2 & -2 \\ 1 & -5 \end{bmatrix}$$

The state transition matrix $\mathbf{phi}\,(t)$ can be computed as a symbolic expression using the following lines of code:

```
>>   A = [-2,-2; 1,-5];
>>   t = sym('t');    % Define symbolic variable.
>>   expm(A*t)        % Compute the state transition matrix.

ans =

[ 2/exp(3*t) - 1/exp(4*t), 2/exp(4*t) - 2/exp(3*t)]
[ 1/exp(3*t) - 1/exp(4*t), 2/exp(4*t) - 1/exp(3*t)]
```

The symbolic answer produced should match the result found in Example 9.13.
   An alternative approach would be to

1. Obtain the resolvent matrix $\boldsymbol{\Phi}\,(s)$.
2. Obtain the state transition matrix as the inverse Laplace transform of the resolvent matrix.

Mathematically we have

$$\text{Eqn. (9.120):} \qquad \boldsymbol{\Phi}\,(s) = [s\,\mathbf{I} - \mathbf{A}]^{-1} \tag{9.202}$$

$$\text{Eqn. (9.119):} \qquad \phi\,(t) = \mathcal{L}^{-1}\left\{[s\,\mathbf{I} - \mathbf{A}]^{-1}\right\} \tag{9.203}$$

The following lines of code implement this approach using the symbolic mathematics capabilities of MATLAB:

```
>>   s = sym('s');       % Define symbolic variable.
>>   tmp = s*eye(2)-A;
>>   rsm = inv(tmp)      % Resolvent matrix.

rsm =

[ (s + 5)/(s^2 + 7*s + 12),       -2/(s^2 + 7*s + 12)]
[       1/(s^2 + 7*s + 12), (s + 2)/(s^2 + 7*s + 12)]

>>   stm = ilaplace(rsm)  % State transition matrix.

stm =

[ 2/exp(3*t) - 1/exp(4*t), 2/exp(4*t) - 2/exp(3*t)]
[ 1/exp(3*t) - 1/exp(4*t), 2/exp(4*t) - 1/exp(3*t)]
```

Software resources:
matex_9_5a.m
matex_9_5b.m

MATLAB Exercise 9.6: Obtaining system function from continuous-time state-space model

In Example 9.14 we have obtained the system function of a second-order CTLTI system from its state-space model. Verify the solution using symbolic mathematics capabilities of MATLAB.

Solution: Symbolic computation of the state transition matrix was discussed in MATLAB Exercise 9.5. Extending the technique used in that exercise, the system function can be found with the following statements:

```
>>   A = [-2,-2;1,-5];
>>   B = [1;0];
>>   C = [0,5];
>>   D = [0];
>>   s = sym('s');
>>   tmp = s*eye(2)-A

tmp =

[ s + 2,       2]
[    -1,  s + 5]

>>   rsm = inv(tmp)

rsm =

[ (s + 5)/(s^2 + 7*s + 12),        -2/(s^2 + 7*s + 12)]
[         1/(s^2 + 7*s + 12), (s + 2)/(s^2 + 7*s + 12)]

>>   G = C*rsm*B+D

G =

5/(s^2 + 7*s + 12)
```

The system function obtained matches the result in Example 9.14.

Software resources:
matex_9_6.m

MATLAB Exercise 9.7: Obtaining system function from discrete-time state-space model

Determine the system function for a DTLTI sytem described by the state-space model

$$\mathbf{x}[n+1] = \begin{bmatrix} 0 & -0.5 \\ 0.25 & 0.75 \end{bmatrix} \mathbf{x}[n] + \begin{bmatrix} 2 \\ 1 \end{bmatrix} r[n]$$

$$y[n] = \begin{bmatrix} 3 & 1 \end{bmatrix} \mathbf{x}[n]$$

Solution: The system function $H(z)$ can be found with the following statements:

```
>>   A = [0,-0.5;0.25,0.75];
>>   B = [2;1];
>>   C = [3,1];
>>   D = [0];
```

```
>>   z = sym('z');
>>   tmp = z*eye(2)-A;
>>   rsm = z*inv(tmp)

rsm =

[ (2*z*(4*z - 3))/(8*z^2 - 6*z + 1),   -(4*z)/(8*z^2 - 6*z + 1)]
[               (2*z)/(8*z^2 - 6*z + 1), (8*z^2)/(8*z^2 - 6*z + 1)]

>>   H = C*rsm*B+D

H =

(8*z^2)/(8*z^2-6*z+1) - (8*z)/(8*z^2-6*z+1) + (12*z*(4*z-3))/(8*z^2-6*z+1)
```

Software resources:
matex_9_7.m

MATLAB Exercise 9.8: Discretization of state-space model

A state-space model for the RLC circuit in Fig. 9.3 was found in Example 9.1 as

$$\dot{\mathbf{x}}_{\mathbf{a}}(t) = \begin{bmatrix} -2 & -2 \\ 1 & -5 \end{bmatrix} \mathbf{x}_{\mathbf{a}}(t) + \begin{bmatrix} 1 \\ 0 \end{bmatrix} r_a(t)$$

$$y_a(t) = \begin{bmatrix} 0 & 5 \end{bmatrix} \mathbf{x}_{\mathbf{a}}(t)$$

The unit-step response of the circuit was determined in Example 9.13 to be

$$y_a(t) = \frac{5}{12} + \frac{70}{3} e^{-3t} - \frac{135}{4} e^{-4t}, \qquad t \geq 0$$

subject to the initial state vector $\mathbf{x}_{\mathbf{a}}(0) = [3 \ -2]^T$. Convert the state-space description to discrete-time, and obtain a numerical solution for the unit-step response using a sampling interval of $T_s = 0.1$ s.

Solution: The discrete-time state-space model can be computed with the following statements that are based on Eqn. (9.192):

```
>>   A = [-2,-2;1,-5];
>>   B = [1;0];
>>   C = [0,5];
>>   d = 0;
>>   Ts = 0.1;
>>   A_bar = expm(A*Ts)

A_bar =
      0.8113    -0.1410
      0.0705     0.5998

>>   B_bar = inv(A)*(A_bar-eye(2))*B

B_bar =
      0.0904
      0.0040

>>   C_bar = C

C_bar =
      0         5
```

```
>>   d_bar = d

d_bar =
       0
```

Once coefficient matrix and vectors "A_bar", "B_bar", "C_bar", and "d_bar"are created, the output signal can be approximated by iteratively solving the discrete-time state-space model. The script listed below computes the iterative solution for $n = 0, \ldots, 30$. It also graphs the approximate solution found (red dots) superimposed with the analytical solution found in Example 9.13.

```
1    % Script: matex_9_8.m
2    %
3    Ts = 0.1;
4    A_bar = expm(A*Ts);
5    B_bar = inv(A)*(A_bar-eye(2))*B;
6    C_bar = C;
7    d_bar = d;
8    xn = [3;-2];          % Initial value of state vector
9    n = [0:30];           % Vector of indices
10   yn = [];              % Empty vector to start
11   for nn=0:30,
12     xnp1 = A_bar*xn+B_bar;   % 'xnp1' represents x[n+1]
13     yn = [yn,C*xn];          % Append to vector 'yn'
14     xn = xnp1;               % New becomes old for next iteration
15   end;
16   % Graph correct vs. approximate solution.
17   t = [0:0.01:3];
18   ya = 5/12+70/3*exp(-3*t)-135/4*exp(-4*t);   % From Example 9-13
19   plot(t,ya,'b-',n*Ts,yn,'ro'); grid;
20   title('y_{a}(t) and y[n]');
21   xlabel('t (sec)');
```

The graph is shown in Fig. 9.17.

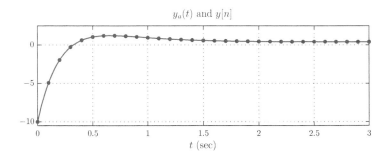

Figure 9.17 – Graph obtained in MATLAB Exercise 9.8.

Software resources:

matex_9_8.m

MATLAB Exercise 9.9: Discretization using Euler method

Consider again the state-space model for the RLC circuit of Fig. 9.3 which was used in MATLAB Exercise 9.9. Convert the state-space description to discrete-time using Euler's method., and obtain a numerical solution for the unit-step response using a sampling interval of $T_s = 0.1$ s. The initial state vector is $\mathbf{x_a}(0) = [3 \;\; -2]^T$.

Solution: The discrete-time state-space model can be computed with the following statements that are based on Eqn. (9.198):

```
>>  A = [-2,-2;1,-5];
>>  B = [1;0];
>>  C = [0,5];
>>  d = 0;
>>  Ts = 0.1;
>>  A_bar = eye(2)+A*Ts

A_bar =
    0.8000    -0.2000
    0.1000     0.5000

>>  B_bar = B*Ts;

B_bar =
    0.1000
         0

>>  C_bar = C

C_bar =
    0     5

>>  d_bar = d

d_bar =
    0
```

The script listed below computes the iterative solution for $n = 0, \ldots, 30$. It also graphs the approximate solution found (red dots) superimposed with the analytical solution found in Example 9.13.

```
1    % Script: matex_9_9.m
2    %
3    Ts = 0.1;
4    A_bar = eye(2)+A*Ts;
5    B_bar = B*Ts;
6    C_bar = C;
7    d_bar = d;
8    xn = [3;-2];              % Initial value of state vector
9    n = [0:30];               % Vector of indices
10   yn = [];                  % Empty vector to start
11   for nn=0:30,
12      xnp1 = A_bar*xn+B_bar; % 'xnp1' represents x[n+1]
13      yn = [yn,C*xn];        % Append to vector 'yn'
14      xn = xnp1;             % New becomes old for next iteration
15   end;
16   % Graph correct vs. approximate solution.
17   t = [0:0.01:3];
18   ya = 5/12+70/3*exp(-3*t)-135/4*exp(-4*t);  % From Example 9-13
19   plot(t,ya,'b-',n*Ts,yn,'ro'); grid;
20   title('y_{a}(t) and y[n]');
21   xlabel('t (sec)');
```

The graph is shown in Fig. 9.18.

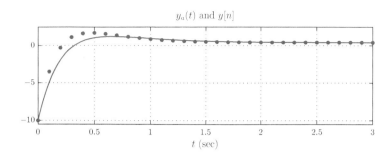

Figure 9.18 – Graph obtained in MATLAB Exercise 9.9.

The quality of approximation is certainly not as good as the one in MATLAB Exercise 9.8, but it can be improved by reducing T_s.
Software resources:
`matex_9_9.m`

Problems

9.1. A continuous-time system is described by the following set of differential equations.

$$\frac{dx_1\,(t)}{dt} = 2\,x_1\,(t) + 2\,x_2\,(t) - x_3\,(t) + r_2\,(t)$$

$$\frac{dx_2\,(t)}{dt} = -3\,x_1\,(t) + 2\,x_3\,(t) - r_2\,(t)$$

$$\frac{dx_3\,(t)}{dt} = x_1\,(t) + x_2\,(t) - 2\,x_3\,(t) + r_1\,(t) + r_2\,(t)$$

$$y_1\,(t) = x_1\,(t) + x_3\,(t) + r_1\,(t)$$

$$y_2\,(t) = x_2\,(t) + x_3\,(t) + r_1\,(t)$$

Write the state-space model of the system in matrix form.

9.2. Consider the alternative form of the matrix state equation given by Eqn. (9.12):

$$\dot{\mathbf{x}}\,(t) = \mathbf{A}\,\mathbf{x}\,(t) + \mathbf{B}\,r\,(t) + \mathbf{E}\,\frac{dr\,(t)}{dt}$$

a. Show that this alternative form can be converted to the standard form given by Eqn. (9.7) by defining a new state vector

$$\mathbf{z} = \mathbf{x} - \mathbf{F}\,r\,(t)$$

Express the vector $\mathbf{F}$ in terms of $\mathbf{A}$, $\mathbf{B}$ and $\mathbf{E}$.

b. The state-space model for a CTLTI system is

$$\dot{\mathbf{x}}\,(t) = \begin{bmatrix} -1 & 3 \\ -2 & -2 \end{bmatrix} \mathbf{x}\,(t) + \begin{bmatrix} 0 \\ 1 \end{bmatrix} r\,(t) + \begin{bmatrix} 2 \\ 1 \end{bmatrix} \frac{dr\,(t)}{dt}$$

Convert the state-space model into the standard form by defining a new state vector $\mathbf{z}\,(t)$.

9.3.   Find a state-space model for the circuit shown in Fig. P.9.3. Choose the capacitor voltages $v_{C1}(t)$ and $v_{C2}(t)$ as the state variables $x_1(t)$ and $x_2(t)$ respectively.

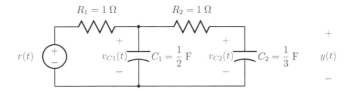

Figure P. 9.3

9.4.   Find a state-space model for the RLC circuit shown in Fig. P.9.4. Choose the capacitor voltage $v_C(t)$ and the inductor current $i_L(t)$ as the state variables $x_1(t)$ and $x_2(t)$ respectively.

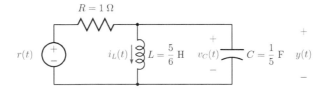

Figure P. 9.4

9.5.   Using the approach taken in Examples 9.3 and 9.4 find a state-space model for each CTLTI system described by a differential equation below:

a. $\dfrac{d^2 y(t)}{dt^2} + 3\,\dfrac{dy(t)}{dt} + 2\,y(t) = 2\,r(t)$

b. $\dfrac{d^2 y(t)}{dt^2} + 4\,\dfrac{dy(t)}{dt} + 3\,y(t) = 3\,r(t)$

c. $\dfrac{d^2 y(t)}{dt^2} - y(t) = r(t)$

d. $\dfrac{d^3 y(t)}{dt^3} + 6\,\dfrac{d^2 y(t)}{dt^2} + 11\,\dfrac{dy(t)}{dt} + 6\,y(t) = r(t)$

e. $\dfrac{d^3 y(t)}{dt^3} + 6\,\dfrac{d^2 y(t)}{dt^2} + 11\,\dfrac{dy(t)}{dt} + 6\,y(t) = r(t) + 2\,\dfrac{dr(t)}{dt}$

9.6.   Find a state-space model for each CTLTI system described below by means of a system function. Use the partial fraction expansion of the system function so that the resulting state-space model has a diagonal coefficient matrix $\mathbf{A}$.

a. $G(s) = \dfrac{2}{(s+1)(s+2)}$

b. $G(s) = \dfrac{10\,s + 1}{(s+2)(s+4)}$

c. $G(s) = \dfrac{(s-1)(s+4)}{(s+1)(s+2)(s+3)}$

9.7. In Example 9.4 a state-space model was obtained using an ad hoc method for the CTLTI system defined by the differential equation

$$\frac{d^3 y(t)}{dt^3} + 5\frac{d^2 y(t)}{dt^2} + 11\frac{dy(t)}{dt} + 15\, y(t) = 3\, r(t) + 7\frac{dr(t)}{dt}$$

Obtain a state-space model for the system using a more formal approach by

    a. First finding a system function $G(s)$ for the system.
    b. Deriving a state space model from the system function.

9.8. A CTLTI system is described by the state-space model

$$\dot{\mathbf{x}}(t) = \begin{bmatrix} -4 & 3 \\ -1 & 0 \end{bmatrix} \mathbf{x}(t) + \begin{bmatrix} 2 \\ 1 \end{bmatrix} r(t)$$

$$y(t) = \begin{bmatrix} 3 & 1 \end{bmatrix} \mathbf{x}(t) + (2)\, r(t)$$

    a. Find a transformation matrix in the form

$$\mathbf{Px} = \begin{bmatrix} 1 & a \\ b & 1 \end{bmatrix}$$

    such that the transformation

$$\mathbf{x}(t) = \mathbf{P}\,\mathbf{z}(t)$$

    converts the state-space model into Jordan canonical form.
    b. Write the state-space model in terms of the new state vector $\mathbf{z}(t)$.

9.9. Refer to MATLAB Exercise 9.1 in which two different state-space models were obtained for the same system. The first model was

$$\dot{\mathbf{x}}(t) = \begin{bmatrix} -6 & -5 \\ 1 & 0 \end{bmatrix} \mathbf{x}(t) + \begin{bmatrix} 1 \\ 0 \end{bmatrix} r(t)$$

$$y(t) = \begin{bmatrix} -5 & -7 \end{bmatrix} \mathbf{x}(t) + r(t)$$

and the second model was

$$\dot{\mathbf{z}}(t) = \begin{bmatrix} 0 & 1 \\ -5 & -6 \end{bmatrix} \mathbf{z}(t) + \begin{bmatrix} 0 \\ 1 \end{bmatrix} r(t)$$

$$y(t) = \begin{bmatrix} -7 & -5 \end{bmatrix} \mathbf{z}(t) + r(t)$$

Show that the second state-space model can be obtained from the first by means of the similarity transformation

$$\mathbf{x}(t) = \begin{bmatrix} 0 & 1 \\ 1 & 0 \end{bmatrix} \mathbf{z}(t)$$

9.10. Refer to Example 9.1. The state-space model obtained for the RLC circuit in Fig. 9.3 was

$$\dot{\mathbf{x}}(t) = \begin{bmatrix} -2 & -2 \\ 1 & -5 \end{bmatrix} \mathbf{x}(t) + \begin{bmatrix} 1 \\ 0 \end{bmatrix} r(t)$$

$$y(t) = \begin{bmatrix} 0 & 5 \end{bmatrix} \mathbf{x}(t) + (0)\, r(t)$$

a. Write the state equations in open form. Express $dy(t)/dt$ and $d^2y(t)/dt^2$ in terms of the state variables $x_1(t)$ and $x_2(t)$ and the input signal $r(t)$.
b. Find a second-order differential equation between $y(t)$ and $r(t)$ by eliminating the state variables between the equations for $y(t)$, $dy(t)/dt$ and $d^2y(t)/dt^2$.
c. Using the second-order differential equation obtained in part (b), find an alternative state-space model in phase-variable canonical form.

9.11. Refer to Problem 9.10. Find a similarity transformation

$$\mathbf{x}(t) = \mathbf{P}\,\mathbf{z}(t)$$

that would transform the original state-space model to the alternative model in phase-variable canonical form that was found.

9.12. Refer again to Example 9.1 in which a state-space model for the RLC circuit in Fig. 9.3 was found to be

$$\dot{\mathbf{x}}(t) = \begin{bmatrix} -2 & -2 \\ 1 & -5 \end{bmatrix} \mathbf{x}(t) + \begin{bmatrix} 1 \\ 0 \end{bmatrix} r(t)$$

$$y(t) = \begin{bmatrix} 0 & 5 \end{bmatrix} \mathbf{x}(t) + (0)\, r(t)$$

In this problem we will explore an alternative approach for finding a differential equation between the input and the output signals of the system.

a. Write the state equations in open form. Take the Laplace transform of each equation (two state equations and one output equation) by making use of linearity and time differentiation properties of the Laplace transform.
b. Eliminate $X_1(s)$ and $X_2(s)$ to find an equation between $Y(s)$ and $R(s)$. Determine the system function $G(s)$.
c. Find a differential equation for the system from the system function found in part (b).

9.13. A CTLTI system with two inputs and two outputs is described by the following pair of differential equations:

$$\frac{dy_1(t)}{dt} = -2\,y_1(t) - y_2(t) + r_1(t) + r_2(t)$$

$$\frac{dy_2(t)}{dt} = -3\,y_2(t) + y_1(t) + \frac{dr_1(t)}{dt}$$

Following the approach used in Example 9.10 find a state-space model for this system.

9.14. A CTLTI system with two inputs and two outputs is described by the state-space model

$$\dot{\mathbf{x}}(t) = \begin{bmatrix} -1 & 2 \\ -3 & -6 \end{bmatrix} \mathbf{x}(t) + \begin{bmatrix} 1 & 2 \\ 0 & -1 \end{bmatrix} \mathbf{r}(t)$$

$$\mathbf{y}(t) = \begin{bmatrix} -1 & -3 \\ 2 & 0 \end{bmatrix} \mathbf{x}(t) + \begin{bmatrix} 0 & 1 \\ 4 & -1 \end{bmatrix} \mathbf{r}(t)$$

a. Using the appropriate similarity transformation, find an equivalent state-space model for this system with a diagonal state matrix.
b. Using the state-space model found, express output transforms $Y_1(s)$ and $Y_2(s)$ in terms of input transforms $R_1(s)$ and $R_2(s)$.

c. Using the results found in part (b), write the input-output relationships of the system in the s-domain as

$$\begin{bmatrix} Y_1(s) \\ Y_2(s) \end{bmatrix} = \begin{bmatrix} G_{11}(s) & G_{12}(s) \\ G_{21}(s) & G_{22}(s) \end{bmatrix} \begin{bmatrix} R_1(s) \\ R_2(s) \end{bmatrix}$$

9.15. Refer to Example 9.11 in which the state variables and the output signal of a third-order system were determined in response to unit-step input signal and subject to specified initial conditions. Verify that the solutions found for each of the state variables and the output signal satisfy the state-space model and the initial conditions.

9.16. A continuous-time system is described by the state-space model

$$\dot{\mathbf{x}}(t) = \begin{bmatrix} -3 & 2 \\ 0 & -1 \end{bmatrix} \mathbf{x}(t) + \begin{bmatrix} 0 \\ 1 \end{bmatrix} r(t)$$

$$y(t) = \begin{bmatrix} 1 & 1 \end{bmatrix} \mathbf{x}(t) + r(t)$$

The initial value of the state vector is

$$\mathbf{x}(0) = \begin{bmatrix} -2 \\ 3 \end{bmatrix}$$

Taking advantage of the upper-triangular property of the state matrix determine the output signal in response to a unit-step input signal. Follow the approach taken in Example 9.11.

9.17. Refer to the system in Problem 9.16.

a. Using the appropriate similarity transformation, obtain an equivalent state-space model for the system in which the state matrix is diagonal.

b. Use the alternative state-space model with the diagonal state matrix to determine the output signal in response to a unit-step input signal.

9.18. Consider the homogeneous state-space model

$$\dot{\mathbf{x}}(t) = \begin{bmatrix} -3 & 2 \\ 0 & -1 \end{bmatrix} \mathbf{x}(t)$$

a. Determine the state transition matrix $\phi(t)$.

b. Using the initial state vector

$$\mathbf{x}(0) = \begin{bmatrix} -2 \\ 3 \end{bmatrix}$$

and the state transition matrix found in part (a), find $\mathbf{x}(2)$, the value of the state vector at time instant $t = 2$.

c. Find $\mathbf{x}(5)$.

9.19. The general solution of the continuous-time state-space model can also be obtained using Laplace transform techniques. Let $\mathbf{X}(s)$ be the vector that holds the Laplace transform of each state variable.

a. Take the unilateral Laplace transform of both sides of the state equation given by Eqn. (9.7) taking $\mathbf{x}(0)$, the initial value of the state vector, into account.

b. Solve for $\mathbf{X}(s)$ in terms of the Laplace transform $R(s)$ of the input signal.

c. Take the Laplace transform of the output equation given by Eqn. (9.8) and substitute the solution found for $\mathbf{X}(s)$ to obtain the relationship between $Y(s)$ and $R(s)$.

d. Write the time-domain relationship between the output signal $y(t)$ and the input signal $r(t)$ remembering that

$$\phi(t) = e^{\mathbf{A}t} = \mathcal{L}^{-1}\left\{[s\mathbf{I} - \mathbf{A}]^{-1}\right\}$$

and that the product of two Laplace transforms corresponds to convolution of the corresponding signals in the time domain.

9.20. Refer to Example 9.13. Verify that the solutions found for the state variables and the output signal satisfy the state-space model and the specified initial conditions.

9.21. A state-space model for the RLC circuit of Fig. 9.4 was found in Example 9.2 as

$$\dot{\mathbf{x}}(t) = \begin{bmatrix} -2 & -1 \\ 26 & 0 \end{bmatrix} \mathbf{x}(t) + \begin{bmatrix} 1 \\ 0 \end{bmatrix} r(t)$$

$$y(t) = \begin{bmatrix} -2 & 0 \end{bmatrix} \mathbf{x}(t) + (1)\, r(t)$$

Find the system function $G(s) = Y(s)/R(s)$.

9.22. Find a state-space model for each DTLTI system described below by means of a difference equation.

a. $y[n] - 0.9\, y[n-1] = r[n]$
b. $y[n] - 1.7\, y[n-1] + 0.72\, y[n-2] = 3\, r[n]$
c. $y[n] - y[n-1] + 0.11\, y[n-2] + 0.07\, y[n-3] = r[n]$

9.23. System functions for several DTLTI systems are given below. Using the technique outlined in Example 9.17 that is based on partial fraction expansion, find a state-space model for each system. Ensure that the matrix $\mathbf{A}$ is diagonal in each case.

a. $H(z) = \dfrac{z+1}{(z+1/2)\,(z+2/3)}$

b. $H(z) = \dfrac{z\,(z+1)}{(z-0.4)\,(z+0.7)}$

c. $H(z) = \dfrac{z\,(z+1)}{(z+3/4)\,(z-1/2)\,(z-3/2)}$

9.24. Consider the system functions given in Problem 9.23. For each system draw a direct-form II block diagram and select the state variables. Then find a state-space model based on the block diagram.

9.25. Consider the DTLTI systems described below by means of their difference equations. The direct method of finding a state-space model explained in Example 9.16 would be difficult to use in this case, due to the inclusion of delayed input terms $r[n-1]$, $r[n-2]$, etc. Find a state-space model for each system in two steps: First find the system function $H(z)$, and then find a state-space model in phase-variable canonical form by inspecting the system function.

a. $y[n] - 0.9\, y[n-1] = r[n] + r[n-1]$
b. $y[n] - 1.7\, y[n-1] + 0.72\, y[n-2] = 3\, r[n] + 2\, r[n-2]$
c. $y[n] - y[n-1] + 0.11\, y[n-2] + 0.07\, y[n-3] = r[n] - r[n-1] + r[n-2]$

9.26. A discrete-time system has the state-space model

$$\mathbf{x}[n+1] = \begin{bmatrix} 0 & -0.5 \\ 0.25 & 0.75 \end{bmatrix} \mathbf{x}[n] + \begin{bmatrix} 2 \\ 1 \end{bmatrix} r[n]$$

$$y[n] = \begin{bmatrix} 3 & 1 \end{bmatrix} \mathbf{x}[n]$$

The initial value of the state vector is

$$\mathbf{x}[0] = \begin{bmatrix} 2 \\ 0 \end{bmatrix}$$

The input to the system is a unit-step function, that is, $r[n] = u[n]$. Iteratively solve for the output signal $y[n]$ for $n = 0, \ldots, 3$.

9.27. A discrete-time system is described by the state-space model

$$\mathbf{x}[n+1] = \begin{bmatrix} -0.1 & -0.7 \\ -0.8 & 0 \end{bmatrix} \mathbf{x}[n] + \begin{bmatrix} 3 \\ 1 \end{bmatrix} r[n]$$

$$y[n] = \begin{bmatrix} 2 & -1 \end{bmatrix} \mathbf{x}[n]$$

  a. Find the resolvent matrix $\mathbf{\Phi}(z)$.
  b. Find the state transition matrix $\phi[n]$.
  c. Compute the unit-step response of the system using the initial state vector

$$\mathbf{x}[0] = \begin{bmatrix} 2 \\ 0 \end{bmatrix}$$

9.28. A discrete-time system is described by the state-space model

$$\mathbf{x}[n+1] = \begin{bmatrix} -0.1 & -0.7 \\ -0.8 & 0 \end{bmatrix} \mathbf{x}[n] + \begin{bmatrix} 3 \\ 1 \end{bmatrix} r[n]$$

$$y[n] = \begin{bmatrix} 2 & -1 \end{bmatrix} \mathbf{x}[n]$$

Find the system function $H(z)$.

MATLAB Problems

9.29. A DTLTI system has the system function

$$H(z) = \frac{z^3 - 7z + 6}{z^4 - 0.2\, z^3 - 0.93\, z^2 + 0.198\, z + 0.1296}$$

Develop a MATLAB script to do the following:

  a. Use function tf2ss(..) to find a state-space model.
  b. Use function eig(..) to find a transformation matrix $\mathbf{P}$ that converts the state-space model found in part (a) to one with a diagonal state matrix.
  c. Obtain an alternative state-space model using the transformation matrix $\mathbf{P}$ found in part (b).

Caution: The use of the function tf2ss(..) with discrete-time system functions requires some care. Write the numerator and the denominator of the system function as polynomials of z^{-1}, and then set up vectors for numerator and denominator coefficients for use with the function tf2ss(..).

9.30. Refer to the system described in Problem 9.26. Write a script to compute and graph the unit-step response of the system for $n = 0, \ldots, 99$ using the iterative method.

9.31. Refer to the system in Problem 9.27. Write a script to determine the unit-step response of the system using symbolic mathematics functions of MATLAB. Verify the analytical solution found in Problem 9.27.

9.32. Refer to the system in Problem 9.28. Write a script to determine the system function $H(z)$ using symbolic mathematics functions of MATLAB. Verify that it matches the solution found manually in Problem 9.28.

9.33. Repeat MATLAB Exercise 9.9 using a sampling interval of $T = 0.02$ s. Simulate the system for $0 \le t \le 3$ s. Graph the approximated output signal against the analytical solution found in Example 9.13.

9.34. Consider the system a state-space model for which was found in Example 9.3.

    a. Find a discrete-time state-space model to approximate this system using the technique outlined in Eqns. (9.187) through (9.192). Use a sampling interval of $T_s = 0.1$ s.

    b. Using the discretized state-space model found, compute an approximation to the unit-step response of the system. Assume zero initial conditions.

9.35. Consider the system a state-space model for which was found in Example 9.4.

    a. Find a discrete-time state-space model to approximate this system using Euler's method with the sampling interval of $T_s = 0.1$ s.

    b. Using the discretized state-space model found, compute an approximation to the unit-step response of the system. Assume zero initial conditions.

Chapter 10

Analysis and Design of Filters

Chapter Objectives

- Develop the concept of analog and discrete-time filters.

- Learn ideal filter behavior and the conditions for distortionless transmission.

- Learn design specifications for realizable filters.

- Explore methods for designing realizable analog filters using Butterworth and Chebyshev approximation formulas.

- Study design methods for obtaining IIR discrete-time filters from analog prototype filters.

- Study design methods for obtaining FIR discrete-time filters.

10.1   Introduction

In many signal processing applications the need arises to change the strength, or the relative significance, of various frequency components in a given signal. Sometimes we may need to eliminate certain frequency components in a signal; at other times we may need to boost the strength of a range of frequencies over others. This act of changing the relative amplitudes of frequency components in a signal is referred to as *filtering*, and the system that facilitates this is referred to as a *filter*.

In Chapters 4 and 7 we have developed techniques for transform-domain analysis of CTLTI systems. Specifically, the relationship between the input and the output signals of a CTLTI system is expressed in the transform domain as

$$Y(\omega) = H(\omega) X(\omega) \tag{10.1}$$

in terms of the Fourier transforms of the signals involved, or as

$$Y(s) = H(s) X(s) \tag{10.2}$$

using the Laplace transforms of the signals. In either case, the system function $H(\omega)$ or $H(s)$ serves as a multiplier function that shapes the spectrum of the input signal to create an output signal.

In a general sense any CTLTI system can be seen as a filter. Consider, for example, a communication channel the purpose of which is to transmit an information bearing signal from one point to another. Assuming that the particular channel we are considering can be reasonably modeled as a CTLTI system, we have the model shown in Fig. 10.1.

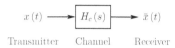

$$x(t) \longrightarrow \boxed{H_c(s)} \longrightarrow \bar{x}(t)$$

Transmitter   Channel   Receiver

Figure 10.1 – CTLTI model for a communication channel.

Ideally we would like the signal at the output of the channel, $\bar{x}(t)$, to be identical to the signal transmitted, $x(t)$. This is generally not the case, however, and the signal $\bar{x}(t)$ is a *distorted* version of $x(t)$. Therefore the channel acts as a filter the action of which is an undesired one. In order to compensate for the distortion effect of the channel, a second system may be connected in cascade with the channel as shown in Fig. 10.2. This second system is referred to as an *equalizer*. Its purpose is to reduce the distortion effect caused by the channel (or filter) so that the output $y(t)$ is a better representation of the original signal $x(t)$.

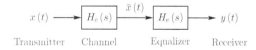

$$x(t) \longrightarrow \boxed{H_c(s)} \xrightarrow{\bar{x}(t)} \boxed{H_e(s)} \longrightarrow y(t)$$

Transmitter   Channel   Equalizer   Receiver

Figure 10.2 – Cascade connection of the channel with an equalizer.

If we want the output $y(t)$ to be identical to the original input $x(t)$, it follows from Fig. 10.2 that the cascade combination of the channel and the equalizer must form an identity system, that is,

$$H_c(s)\, H_e(s) = 1 \tag{10.3}$$

This requires the system function of the equalizer to be

$$H_e(s) = \frac{1}{H_c(s)} \tag{10.4}$$

In this case $H_c(s)$ represents a filter (channel) with undesired effects on the signal transmitted through it. In contrast, $H_e(s)$ represents a filter that we may design for the purpose of eliminating, or at least reducing, the undesired effects of the channel.

10.2   Distortionless Transmission

A system driven by an input signal $x(t)$ produces an output signal $y(t)$ that is, in general, different in shape from the input signal. Thus the system modifies the signal as it transmits it from its input to its output. This modification may be a desired effect of the system or an undesired side effect. Regardless, it is referred to as *distortion*.

Consider a CTLTI system with system function $H\left(s\right)$ driven by an input signal $x\left(t\right)$. The input and the output signals are related to each other through the convolution relationship

$$y\left(t\right) = h\left(t\right) * x\left(t\right) = \int_{-\infty}^{\infty} h\left(\lambda\right) x\left(t - \lambda\right) d\lambda \tag{10.5}$$

The frequency spectra of input and output signals are related to each other by

$$Y\left(\omega\right) = H\left(\omega\right) X\left(\omega\right) \tag{10.6}$$

The type of distortion caused by a CTLTI system is *linear distortion*, and it exhibits itself in two distinct forms, namely *amplitude distortion* and *phase distortion*. Let us write the magnitude spectrum of the output signal from Eqn. (10.6):

$$\left|Y\left(\omega\right)\right| = \left|H\left(\omega\right)\right| \left|X\left(\omega\right)\right| \tag{10.7}$$

Considering that the input signal $x\left(t\right)$ is a specific mixture of sinusoidal components at various frequencies, Eqn. (10.7) describes how the system modifies the amplitude of each of these sinusoidal components. The scale factor applied by the system to a sinusoid at a particular frequency may be different from the scale factor applied to a sinusoid at another frequency. This non-uniform scaling of sinusoidal components of the input signal leads to *amplitude distortion*.

The phase spectrum of the output signal is found from Eqn. (10.6) as

$$\angle Y\left(\omega\right) = \angle H\left(\omega\right) + \angle X\left(\omega\right) \tag{10.8}$$

The system adds an offset to the phase of each sinusoidal component of the input signal. Phase-shifting a sinusoidal component of the input signal causes a corresponding time shift (delay). If signal components are delayed by differing amounts, their relative alignment in time is disturbed, causing the output signal to be different from the input signal. This effect is known as *phase distortion*.

What type of a system function would be needed in order to obtain a *distortionless* version of the input signal at the output of the system? As a starting point for discussion let us assume that what we mean by distortionless transmission is the output signal being identical to the input signal, that is,

$$y\left(t\right) = x\left(t\right) \tag{10.9}$$

This clearly requires that we have

$$Y\left(\omega\right) = X\left(\omega\right) \tag{10.10}$$

which in turn requires

$$H\left(\omega\right) = 1 , \quad \text{all } \omega \tag{10.11}$$

and

$$h\left(t\right) = \delta\left(t\right) \tag{10.12}$$

Thus, we need the impulse response of the system to be an impulse. While Eqns. (10.11) and (10.12) represent a mathematically satisfying result, the corresponding system is physically unrealizable. No physical system is capable of responding to an input signal instantaneously, without a time delay between the input and the output. In order to come up with a realizable result, we will relax our definition of distortionless transmission, and accept an output signal in the form

$$y\left(t\right) = K\,x\left(t - t_d\right) \tag{10.13}$$

where K is a constant scale factor and t_d is a constant time-delay. Thus, $y\left(t\right)$ in Eqn. (10.13) is a scaled and delayed version of the input signal $x\left(t\right)$. Taking the Fourier transform of both sides of Eqn. (10.13) yields

$$Y\left(\omega\right) = K\,e^{-j\omega t_d}\,X\left(\omega\right) \tag{10.14}$$

and the system function needed for distortionless transmission is

$$H\left(\omega\right) = K\,e^{-j\omega t_d} \tag{10.15}$$

Magnitude and phase characteristics of the system function are

$$\left|H\left(\omega\right)\right| = K \tag{10.16}$$

and

$$\Theta\left(\omega\right) = \angle H\left(\omega\right) = -\omega t_d \tag{10.17}$$

A distortion-free system has a magnitude that is constant, and a phase that is a linear function of frequency as shown in Fig. 10.3.

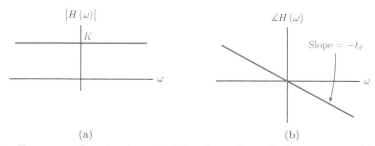

(a)   (b)

Figure 10.3 – Frequency-domain characteristics for a distortionless system: (a) magnitude, (b) phase.

Example 10.1: **Distortion caused by the RC circuit**

Consider again the simple RC circuit that was used in several examples in Chapters 2, 4 and 7.

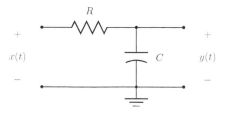

Figure 10.4 – RC circuit for Example 10.1.

In some cases the RC circuit can be used as a simple model for a wired electrical connection between two points. Such a model would provide a reasonably good approximation to the characteristics of the wired connection provided that the distance between the two points being connected is not too long. (For long distance connections the resistance and the capacitance would have to be modeled as distributed parameters rather than being concentrated at one point each.) The system function for the RC circuit is

$$H(\omega) = \frac{1}{1 + j\omega RC}$$

Clearly this is not a distortionless system as defined by Eqn. (10.15), however, under certain circumstances it can be considered near distortion-free. The magnitude of the system function is

$$\left| H(\omega) \right| = \frac{1}{\sqrt{1 + (\omega/\omega_c)^2}} \tag{10.18}$$

with the 3-dB cutoff frequency defined as

$$\omega_c \triangleq \frac{1}{RC} \tag{10.19}$$

This magnitude characteristic is shown in Fig. 10.5(a). The phase of the system function is

$$\Theta(\omega) = \angle H(\omega) = -\tan^{-1}\left(\frac{\omega}{\omega_c}\right) \tag{10.20}$$

and the corresponding time delay is

$$t_d(\omega) = -\frac{\Theta(\omega)}{\omega} = \frac{1}{\omega}\tan^{-1}\left(\frac{\omega}{\omega_c}\right) \tag{10.21}$$

The phase characteristic is shown in Fig. 10.5(b).

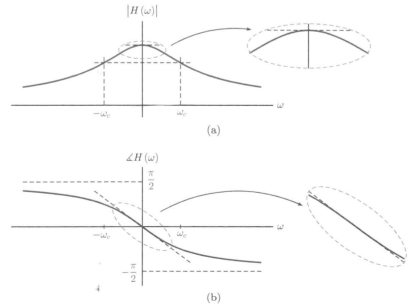

(a)

(b)

Figure 10.5 – Magnitude and phase characterictics of the RC circuit of Example 10.1 at low frequencies.

The magnitude characteristic of the system seems to be relatively flat for frequencies in the vicinity of $\omega = 0$. Similarly, the phase characteristic is close to a straight line, and the time delay is relatively constant for low frequencies. Thus, if we limit the use of the system to input signals that contain only the frequencies $\omega << \omega_c$, we could consider the system as near distortion-free.

What is the frequency range (in terms of ω_c) in which one can consider the RC circuit approximately distortionless? Of course, the answer depends on the amount of variation in magnitude and/or time delay that we consider negligible. Since both the magnitude and the time-delay characteristics exhibit their peak values at $\omega = 0$, and both are symmetric around that point, we could simply look for a frequency at which the characteristic drops down by a certain percentage. Let ω_0 be the frequency at which the magnitude of the system function drops below the peak value by p percent. Using Eqn. (10.18) and realizing that the maximum magnitude is $H(0) = 1$, we can write

$$\frac{1}{\sqrt{1 + (\omega_0/\omega_c)^2}} \leq \frac{100 - p}{100} \tag{10.22}$$

Solving this inequality for ω_0 yields

$$\omega_0 \leq \omega_c \sqrt{\left(\frac{100}{100 - p}\right)^2 - 1} \tag{10.23}$$

For the frequencies $|\omega| \leq 0.1425\,\omega_c$, the variation in the magnitude of the system function will be 1 percent or less. If we are willing to tolerate up to 5 percent magnitude variation, then the usable frequency range is $|\omega| \leq 0.3287\,\omega_c$.

Next consider the phase characteristic given by Eqn. (10.20). Differentiating it with respect to frequency, we obtain

$$\frac{d\Theta}{d\omega} = -\frac{1}{\omega_c}\left[\frac{1}{1 + (\omega/\omega_c)^2}\right] \tag{10.24}$$

and the slope of the phase characteristic at frequency $\omega = 0$ is

$$\left.\frac{d\Theta}{d\omega}\right|_{\omega=0} = -\frac{1}{\omega_c} \tag{10.25}$$

The line that is tangent to the phase characteristic at the frequency $\omega = 0$ is also shown in Fig. 10.5(b) for comparison. The phase remains fairly close to the tangent line for low frequencies. Similar analysis can be performed for the time-delay characteristic of the system; however, because of the form of Eqn. (10.21), arriving at an analytical result is tedious. Instead of trying to duplicate the methodology used in Eqns. (10.22) and (10.23), let us just take the frequency ranges obtained for specified magnitude variations above, and determine the corresponding variation in time delay. It can easily be shown that

$$t_d(0) = 0.1592/f_c \tag{10.26}$$

$$t_d(0.1425f_c) = 0.1581/f_c \tag{10.27}$$

$$t_d(0.3287f_c) = 0.1538/f_c \tag{10.28}$$

If we accept 1 percent variation in magnitude, the corresponding variation in time delay is 0.67 percent. For a 5 percent magnitude variation, the time delay varies by 3.38 percent.

Software resources:
ex_10_1.m

Interactive Demo: dist_demo.m

The demo program "dist_demo" is modeled after Example 10.1 and Fig. 10.5. Magnitude, phase and time-delay characteristics of the RC circuit are graphed. Circuit parameters may be varied using slider controls. Allowed deviation of the magnitude response from a constant value is also specified. A dashed rectangle showing the deviation from ideal behavior is shown on each graph. Zoom-in feature may be used on each graph to get a closer look at the critical values.

1. Begin with $R = 1$ MΩ and $C = 1$ μF. Let the tolerance limit be 1 percent. Observe the frequency range in which the magnitude stays within 1 percent of the dc value. Also observe deviations from ideal behavior for the phase and the time delay, and compare them to theoretical values.
2. Gradually reduce the value of R down to 0.5 MΩ and observe the changes in the frequency range.
3. While keeping $R = 0.5$ MΩ increase the tolerance limit to 2, 3, 4 and 5 percent, and observe the changes in the frequency interval.

Software resources:
dist_demo.m

10.3   Ideal Filters

The conditions for a CTLTI system to be distortionless were identified in Eqns. (10.16) and (10.17) of the previous section, and were graphically illustrated in Fig. 10.3. We observe that, in order to be truly distortionless, a system must have infinite bandwidth. In other words, it must be able to respond to signals containing an infinite range of frequencies. No physically realizable system is capable of doing this.

Furthermore, signals encountered in engineering applications have limited bandwidth, that is, they contain a limited range of frequencies. Consider such a signal $x(t)$. A CTLTI system that satisfies Eqns. (10.16) and (10.17) not necessarily at every frequency, but at all frequencies contained in the signal $x(t)$, would still facilitate distortionless transmission of this signal. Ideal frequency-selective filters are systems that allow distortionless transmission of signal components in a particular frequency range while eliminating all other signal components.

One type is an *ideal lowpass filter* that is characterized by the system function

$$H_{LP}(\omega) = \Pi\left(\frac{\omega}{2\omega_0}\right) e^{-j\omega t_d}$$

$$= \begin{cases} e^{-j\omega t_d}, & |\omega| \leq \omega_0 \\ 0, & |\omega| > \omega_0 \end{cases} \tag{10.29}$$

Any signal that contains only the frequencies in the range $|\omega| \leq \omega_0$ is transmitted through this system without any distortion of its magnitude or phase characteristics. The magnitude and the phase of the system function $H_{LP}(\omega)$ of an ideal lowpass filter are graphed in Fig. 10.6.

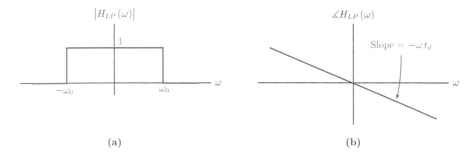

Figure 10.6 – The frequency spectrum of the ideal lowpass filter: (a) magnitude, (b) phase.

The magnitude of the system function is constant for frequencies in the range $-\omega_0 \leq \omega \leq \omega_0$. The phase of the system function is a straight line that passes through the origin, with the slope $-\omega\, t_d$. Signals that do not contain any frequencies outside the range $|\omega| \leq \omega_0$ would be transmitted through this system without experiencing any distortion. The frequency ω_0 in Eqn. (10.29) is the *cutoff frequency* of the ideal lowpass filter. The parameter t_d indicates the time delay caused by the filter.

Some signals are bandpass in nature, that is, they contain frequencies in a bandpass frequency range $\omega_1 \leq |\omega| < \omega_2$. Examples of these signals are encountered often in the study of communication systems. The system function of an *ideal bandpass filter* that would allow distortionless transmission of the appropriate bandpass signal is given by

$$H_{BP}(\omega) = \begin{cases} e^{-j\omega\, t_d}, & \omega_1 \leq |\omega| \leq \omega_2 \\ 0, & \text{otherwise} \end{cases} \qquad (10.30)$$

and is illustrated in Fig. 10.7.

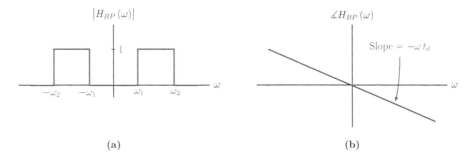

Figure 10.7 – The frequency spectrum of the ideal bandpass filter: (a) magnitude, (b) phase.

Similarly, ideal highpass and band-reject filters can also be defined. An *ideal highpass filter* transmits, without distortion, signals that contain only frequencies greater than a minimum value, that is, $|\omega| \geq \omega_0$. An *ideal band-reject filter*, on the other hand, facilitates distortionless transmission of signals that contain only those frequencies that are outside a specified range. Frequencies that satisfy either $|\omega| \leq \omega_1$ or $|\omega| \geq \omega_2$ are transmitted; all other frequencies are suppressed. Figs. 10.8 and 10.9 illustrate the system functions of ideal highpass and band-reject filters respectively.

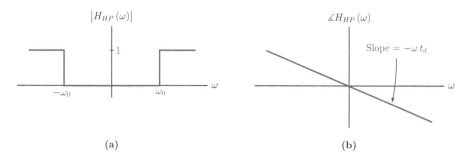

Figure 10.8 – The frequency spectrum of the ideal highpass filter: (a) magnitude, (b) phase.

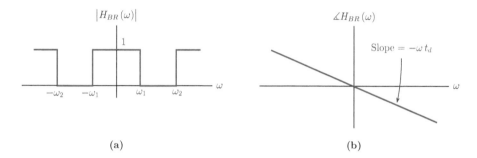

Figure 10.9 – The frequency spectrum of the ideal band-reject filter: (a) magnitude, (b) phase.

The ideal filters discussed above are not practically realizable. In order to see why this is the case, let us determine the impulse response of an ideal filter. For the ideal lowpass filter of Eqn. (10.29), the impulse response can be easily computed from the system function through inverse Fourier transform. Using the inverse Fourier transform relationship given by Eqn. (4.126) on the ideal lowpass filter system function in Eqn. (10.29) we obtain

$$h_{LP}(t) = \mathcal{F}^{-1}\{H_{LP}(\omega)\} = \frac{\omega_0}{\pi} \operatorname{sinc}\left(\frac{\omega_0}{\pi}(t - t_d)\right) \qquad (10.31)$$

For convenience, let us express $h_{LP}(t)$ using f_0 instead of ω_0:

$$h_{LP}(t) = 2f_0 \operatorname{sinc}(2f_0(t - t_d)) \qquad (10.32)$$

The resulting impulse response is shown in Fig. 10.10.

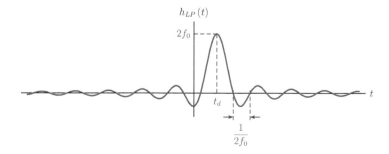

Figure 10.10 – Impulse response of the ideal lowpass filter.

A few observations are in order regarding the impulse response of the ideal lowpass filter:

1. The impulse response $h_{LP}(t)$ given by Eqn. (10.32) is in the form of a sinc function with a peak amplitude of $2f_0$. This peak occurs at the time instant $t = t_d$.
2. The zero crossings of the impulse response are uniformly spaced. Consecutive zero crossings are $1/(2f_0)$ apart from each other.
3. The impulse response exists for all time instants including $t < 0$. Thus, the ideal lowpass filter is a non-causal system, and is therefore physically unrealizable.
4. Furthermore, no amount of additional time delay would make the impulse response in Eqn. (10.32) causal.

It is also interesting and instructive to derive the impulse response of an ideal bandpass filter the spectrum of which is shown in Fig. 10.7. We could compute the impulse response as the inverse Fourier transform of the system function in Eqn. (10.30), using the approach that was employed in finding the impulse response of the ideal lowpass filter. Instead we will take a simpler approach and express the ideal bandpass filter spectrum in terms of frequency shifts applied to the ideal lowpass filter spectrum. In order to accomplish this, we will initially consider an ideal lowpass filter with zero phase and no time delay, i.e., $t_d = 0$. A comparison of Figs. 10.6 and 10.7 reveals that the ideal bandpass filter spectrum with zero phase can be written as

$$H_{BP}(\omega) = H_{LP}(\omega - \omega_b) + H_{LP}(\omega + \omega_b)$$

$$= \Pi\left(\frac{\omega - \omega_b}{2\omega_0}\right) + \Pi\left(\frac{\omega + \omega_b}{2\omega_0}\right) \tag{10.33}$$

with parameter ω_b set to the midpoint frequency of the passband

$$\omega_b = \frac{\omega_2 + \omega_1}{2} \tag{10.34}$$

and the cutoff frequency of the lowpass filter chosen as

$$\omega_0 = \frac{\omega_2 - \omega_1}{2} \tag{10.35}$$

The construction of the bandpass filter spectrum from the lowpass filter spectrum is illustrated in Fig. 10.11.

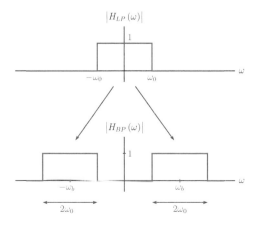

Figure 10.11 – Obtaining the spectrum of the ideal bandpass filter from the spectrum of the ideal lowpass filter.

Using the frequency-shifting property of the Fourier transform, we are now ready to write the impulse response of the ideal bandpass filter as

$$
\begin{aligned}
h_{BP}(t) &= h_{LP}(t)\, e^{j2\pi f_b t} + h_{LP}(t)\, e^{-j2\pi f_b t} \\
&= 2\, h_{LP}(t)\, \cos\left(2\pi f_b t\right) \\
&= 4f_0\, \operatorname{sinc}\left(2f_0 t\right)\, \cos\left(2\pi f_b t\right)
\end{aligned}
\tag{10.36}
$$

For convenience we have used frequencies in Hertz rather than the radian frequencies in Eqn. (10.36). This impulse response is for an ideal bandpass filter with no time delay. It may be generalized by adding a linear phase term and the corresponding constant time delay to obtain

$$
h_{BP}(t) = 4f_0\, \operatorname{sinc}\left[2f_0\left(t - t_d\right)\right]\, \cos\left[2\pi f_b\left(t - t_d\right)\right]
\tag{10.37}
$$

Finally, by substituting Eqns. (10.34) and (10.35) into Eqn. (10.37) we get

$$
h_{BP}(t) = 2\left(f_2 - f_1\right)\, \operatorname{sinc}\left[\left(f_2 - f_1\right)\left(t - t_d\right)\right]\, \cos\left[\pi\left(f_2 + f_1\right)\left(t - t_d\right)\right]
\tag{10.38}
$$

which is shown graphically in Fig. 10.12.

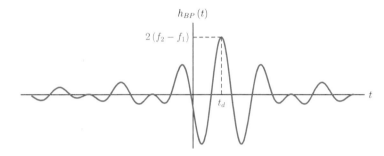

Figure 10.12 – Impulse response of the ideal bandpass filter.

Interactive Demo: `ilpf_demo.m`

The demo program "`ilpf_demo1.m`" illustrates the impulse response of the ideal lowpass filter as defined in Eqn. (10.32), and shown in Fig. 10.10. The magnitude and the phase of the filter transfer function $H_{LP}(f)$ are displayed along with the impulse response $h_{LP}(t)$. The cutoff frequency f_0 and the time delay t_d may be varied using slider controls.

1. With the time delay set to $t_d = 0$ vary the cutoff frequency and observe its effects on the impulse response.
2. With the cutoff frequency fixed at $f_0 = 100$ Hz vary the time delay and observe its effects on the impulse response.

Software resources:
`ilpf_demo.m`

Interactive Demo: `ibpf_demo.m`

This demo program illustrates the impulse response of the ideal bandpass filter the system function of which was described by Eqn. (10.38) and Fig. 10.12. The magnitude and the phase of the filter transfer function $H_{BP}(f)$ are displayed along with the impulse response

$h_{BP}(t)$. Slider controls allow adjustments to parameters f_0 and f_b as defined by Eqns. (10.34) and (10.35).

1. With the time delay set to $t_d = 0$ and $f_b = 100$ Hz vary f_0 and observe its effects on the impulse response.
2. With $t_d = 0$ and $f_0 = 200$ Hz vary f_b and observe its effects on the impulse response.
3. With $f_0 = 100$ Hz and $f_b = 150$ Hz vary the time delay and observe its effects on the impulse response.

Software resources:
ibpf_demo.m

10.4   Design of Analog Filters

In this section we will focus on designing analog filters to meet a set of user specifications. We have seen in Section 10.3 that filters with ideal characteristics are not realizable since they are non-causal. In practical applications of filtering we seek realizable approximations to ideal filter behavior. The requirements on the frequency spectrum of the filter must be relaxed in order to obtain realizable filters.

In the design of analog filters, specifications for the desired filter are usually given in terms of a set of tolerance limits for the magnitude response $|H(\omega)|$. Specification diagrams for the four frequency-selective filter types, namely lowpass, highpass, bandpass and band-reject filters, are shown in Fig. 10.13.

Consider the lowpass filter specifications illustrated by Fig. 10.13(a). The frequency interval $0 < \omega < \omega_1$ is referred to as the *passband* of the filter. Within the passband the magnitude of the filter spectrum is required to remain in the range $1 - \Delta_1 \leq |H(\omega)| \leq 1$. For frequencies greater than ω_2, that is, in the *stopband* of the filter, the magnitude is required to stay below Δ_2. The parameters Δ_1 and Δ_2 are called the *passband tolerance* and the *stopband tolerance* respectively.

The frequency range $\omega_1 < \omega < \omega_2$ between the passband and the stopband is referred to as the *transition band* where there are no constraints on the magnitude behavior of the filter. The inclusion of a transition band along with the tolerance limits on passband and stopband behavior should allow us to come up with practically realizable designs. It should be noted that the impulse response $h(t)$ is required to be real-valued. Recall from the discussion in Section 4.3.5 of Chapter 4 that if the impulse response is real then the system function is conjugate symmetric, and therefore the magnitude of the system function is an even function of ω. This is the reason why the specification diagrams in Fig. 10.13 display only the positive frequencies.

Specifications for a highpass filter, shown in Fig. 10.13(b) mirror those of the lowpass filter. Bandpass and band-reject filter specifications, shown in Fig. 10.13(c),(d) have two transition bands each.

Tolerance limits for a desired filter can also be given on a decibel (dB) scale as shown in Fig. 10.14.

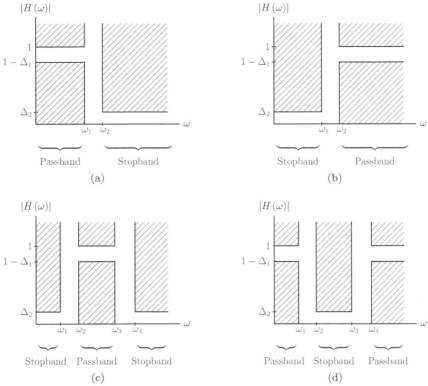

Figure 10.13 – Specification diagrams for frequency selective filters: (a) lowpass, (b) highpass, (c) bandpass, (d) band-reject.

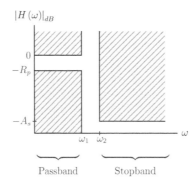

Figure 10.14 – Decibel tolerance specifications for analog lowpass filter.

The maximum value of the magnitude response is set as 0 dB. The maximum allowed dB *passband ripple* R_p and the minimum required dB *stopband attenuation* A_s are related to the parameters Δ_1 and Δ_2 as

$$R_p = 20 \log_{10} \left(\frac{1}{1 - \Delta_1} \right) \tag{10.39}$$

and

$$A_s = 20 \log_{10} \left(\frac{1}{\Delta_2} \right) \tag{10.40}$$

Given the specifications of the desired filter in the form of one of the specification diagrams in Fig. 10.13, the design problem can be stated as follows:

Analog filter design problem:
    Find the system function $H(s)$ of a filter the magnitude response of which remains in the allowed area of the specification diagram.

In general, the analog filter design problem begins with the design of a lowpass prototype filter regardless of the type of filter that is ultimately needed. If a highpass, bandpass or band-reject filter is desired, it is obtained subsequently by means of a *frequency transformation* to be applied to the lowpass prototype filter. This approach is motivated by the availability of well established methods for the design of analog lowpass filters.

A number of approximation formulas exist in the literature for specifying the squared-magnitude of a realizable lowpass filter spectrum. Some of the better known formulas will be mentioned here.

Butterworth lowpass filters are characterized by the squared-magnitude function

$$|H(\omega)|^2 = \frac{1}{1 + (\omega/\omega_c)^{2N}} \tag{10.41}$$

where the parameters N and ω_c are the filter-order and the cutoff frequency respectively.

As an alternative to the Butterworth approximation formula, the *Chebyshev type-I* approximation formula for the squared-magnitude function of a lowpass filter is

$$|H(\omega)|^2 = \frac{1}{1 + \varepsilon^2 C_N^2(\omega/\omega_1)} \tag{10.42}$$

where ε is a positive constant. The frequency ω_1 is the passband edge frequency. The function $C_N(\nu)$ in the denominator represents the *Chebyshev polynomial* of order N.

Chebyshev type-II approximation formula for the squared-magnitude function is similar to the Chebyshev type-I formula.

$$|H(\omega)|^2 = \frac{\varepsilon^2 C_N^2(\omega_2/\omega)}{1 + \varepsilon^2 C_N^2(\omega_2/\omega)} \tag{10.43}$$

The parameter ω_2 is the *stopband edge frequency*.

Finally, an *elliptic* lowpass filter is characterized by the squared-magnitude function

$$|H(\omega)|^2 = \frac{1}{1 + \varepsilon^2 \psi_N^2(\omega/\omega_1)} \tag{10.44}$$

The parameter ε is a positive constant. The function $\psi_N(\nu)$ is called a *Chebyshev rational function*, and is defined in terms of *Jacobi elliptic functions*.

In the rest of this section we will provide a brief introduction to analog filter design methods. A thorough treatment of analog filter design would be well beyond the scope of this text. There are a number of excellent texts on the subject.

10.4.1   Butterworth lowpass filters

Consider again the Butterworth squared-magnitude function given by Eqn. (10.41). At the frequency $\omega = \omega_c$ the magnitude is equal to $1/\sqrt{2}$. Since this approximately corresponds to the -3 dB point on a logarithmic scale, the parameter ω_c is also referred to as the $3\text{-}dB$ *cutoff frequency* of the Butterworth filter. Furthermore, it can easily be shown that the first $2N-1$ derivatives of the magnitude spectrum $|H(\omega)|$ are equal to zero at $\omega=0$, that is,

$$\frac{d^n}{d\omega^n}\left\{|H(\omega)|\right\}\bigg|_{\omega=0} = 0\,, \qquad n=1,\ldots,2N-1 \tag{10.45}$$

The magnitude characteristic $|H(\omega)|$ for a Butterworth filter is said to be *maximally flat*. Magnitude spectra for Butterworth filters of various orders are shown in Fig. 10.15.

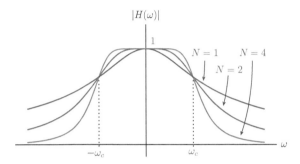

Figure 10.15 – Magnitude spectra for Butterworth lowpass filters or order $N=1$, $N=2$, and $N=4$.

We know that the squared-magnitude function can be expressed as the product of the system function and its complex conjugate, that is

$$|H(\omega)|^2 = H(\omega)\,H^*(\omega) \tag{10.46}$$

In addition, since the impulse response $h(t)$ of the filter is real-valued, its system function exhibits conjugate symmetry (see Section 4.3.5 of Chapter 4):

$$H^*(\omega) = H(-\omega) \tag{10.47}$$

Combining Eqns. (10.46) and (10.47) yields

$$|H(\omega)|^2 = H(\omega)\,H(-\omega) \tag{10.48}$$

Using the s-domain system function $H(s)$, the problem of designing a Butterworth filter can be stated as follows: Given the values of the two filter parameters ω_c and N, find the s-domain system function $H(s)$ for which the squared-magnitude of the function $H(\omega)$ matches the right side of Eqn. (10.41).

For a system function $H(s)$ with real coefficients, it can be shown that

$$|H(\omega)|^2 = H(s)\,H(-s)\bigg|_{s=j\omega} \tag{10.49}$$

The procedure in Eqn. (10.49) needs to be reversed to find the system function $H(s)$. Substituting s for $j\omega$, or equivalently, $-s^2$ for ω^2 in Eqn. (10.41) we obtain

$$H(s)\,H(-s) = \frac{1}{1+\left(-s^2/\omega_c^2\right)^N} \tag{10.50}$$

If $H(s)$ has a pole in the left half s-plane at $p_1 = \sigma_1 + j\omega_1$, then $H(-s)$ has a corresponding pole at $\bar{p}_1 = -\sigma_1 - j\omega_1$. Thus the poles of $H(s)$ are mirror images of the poles of $H(-s)$ with respect to the origin. In order to extract $H(s)$ from the product $H(s)H(-s)$, we need to find the poles of this product, and separate the two sets of poles that belong to $H(s)$ and $H(-s)$ respectively. The cases of even and odd filter-orders N need to be treated separately.

Case 1: N is odd

The poles of the product $H(s)H(-s)$ are the values of s that satisfy the equation

$$s^{2N} - \omega_c^{2N} = 0 \tag{10.51}$$

which can be written in the alternative form

$$s^{2N} = \omega_c^{2N} e^{j2\pi k} \tag{10.52}$$

where we have used the fact that $e^{j2\pi k} = 1$ for all integer k. Poles of $H(s)H(-s)$ are

$$p_k = \omega_c\, e^{j\pi k/N}, \qquad k = 0, \ldots, 2N - 1 \tag{10.53}$$

A few observations are in order:

1. All the poles of the product $H(s)H(-s)$ are located on a circle with radius equal to ω_c.
2. Furthermore, the poles are equally spaced on the circle, and the angle between any two adjacent poles is π/N radians.
3. Since $H(s)$ and $H(-s)$ have only real coefficients, all complex poles appear in conjugate pairs.

Fig. 10.16 depicts the situation for $\omega_c = 2$ rad/s, and $N = 5$.

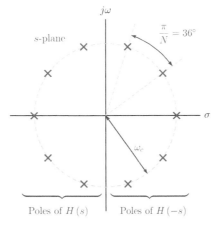

Figure 10.16 – Poles of the product $H(s)H(-s)$ for Butterworth lowpass filter of order $N = 5$ with 3-dB cutoff frequency $\omega_c = 2$ rad/s.

We are interested in obtaining a filter that is both causal and stable. It is therefore necessary to use the poles in the left half of the s-plane for $H(s)$. The remaining poles, the ones in the right half of the s-plane, must be associated with $H(-s)$. The system function $H(s)$ for the Butterworth lowpass filter is constructed in the form

$$H(s) = \frac{\omega_c^N}{\prod_k (s - p_k)}, \qquad \text{for all } k \text{ that satisfy} \quad \frac{\pi}{2} < \frac{k\pi}{N} < \frac{3\pi}{2} \qquad (10.54)$$

Case 2: N is even
The poles of the product $H(s)H(-s)$ are the solutions of the equation

$$s^{2N} + \omega_c^{2N} = 0 \qquad (10.55)$$

which can be written as

$$s^{2N} = \omega_c^{2N} e^{j\pi(2k+1)} \qquad (10.56)$$

In this case, the general solution for the poles is

$$p_k = \omega_c e^{j\pi(2k+1)/2N}, \qquad \text{for} \quad k = 0, \ldots, 2N - 1 \qquad (10.57)$$

For even filter orders, poles of the magnitude-squared system function are still equally spaced on a circle. Compared to odd filter orders, the only difference in this case is that no poles appear on the real axis. The system function $H(s)$ can be constructed in the same way as it was done for odd values of N.

Poles of $H(s)H(-s)$ for the Butterworth lowpass filter:

$$p_k = \begin{cases} \omega_c\, e^{jk\pi/N}, & k = 0, \ldots, 2N - 1 & \text{if } N \text{ is odd} \\ \omega_c\, e^{j(2k+1)\pi/2N}, & k = 0, \ldots, 2N - 1 & \text{if } N \text{ is even} \end{cases}$$

Interactive Demo: `btw_demo.m`

The demo program "`btw_demo.m`" illustrates the magnitude characteristic of a Butterworth lowpass filter the squared version of which is given by Eqn. (10.41). The desired 3-dB cutoff frequency f_c and filter order N may be specified. Observe the following:

1. The edges of the magnitude characteristic become sharper as the filter order is increased.
2. At the frequency $f = f_c$ the magnitude is equal to $1/\sqrt{2}$ times its peak value independent of the filter order.

Software resources:
`btw_demo.m`

Example 10.2:   **Butterworth filter design**

Determine the system function $H(s)$ of a third-order Butterworth lowpass filter with a 3-dB cutoff frequency of $\omega_c = 20\pi$ rad/s.

Solution: Using the Butterworth approximation formula in Eqn. (10.41), the squared-magnitude function of the desired filter can be written as

$$|H(\omega)|^2 = \frac{1}{1 + (\omega/20\pi)^6} \tag{10.58}$$

Substituting $\omega^2 \rightarrow -s^2$ in Eqn. (10.58) we obtain

$$H(s)\,H(-s) = \frac{1}{1 + (-s^2/400\pi^2)^3}$$

or equivalently

$$H(s)\,H(-s) = \frac{(20\pi)^6}{(20\pi)^6 - s^6}$$

The poles of the function $H(s)\,H(-s)$ are those values of s that satisfy

$$s^6 = (20\pi)^6$$

Solutions of Eqn. (10.59) are

$$p_k = (20\pi)\,e^{j2\pi k/6}\,, \qquad k = 0,\ldots,5$$

as shown in Fig. 10.17.

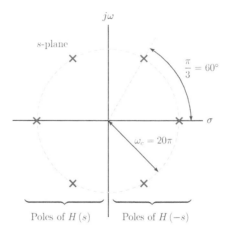

Figure 10.17 – Poles of $H(s)\,H(-s)$ for the third-order Butterworth lowpass filter of Example 10.2.

We are particularly interested in the poles that lie in the left half of the s-plane since those are the poles that are associated with $H(s)$. By inspection of Fig. 10.17 it is clear that the poles of $H(s)$ are

$$p_2 = (20\pi)\,e^{j2\pi/3}$$

$$p_3 = (20\pi)\,e^{j\pi} = -20\pi$$

$$p_4 = (20\pi)\,e^{j4\pi/3} = (20\pi)\,e^{-j2\pi/3}$$

We are now ready to construct the transfer function by using these poles:

$$H\left(s\right) = \frac{\left(20\pi\right)^3}{\left(s - p_2\right)\left(s - p_3\right)\left(s - p_4\right)}$$

$$= \frac{\left(20\pi\right)^3}{\left(s + 20\pi\right)\left(s^2 + 20\pi\,s + 400\pi^2\right)} \qquad (10.59)$$

In order to compute the magnitude and the phase characteristics of the system we need to substitute $s = j\omega$ into Eqn. (10.59) to obtain

$$H\left(\omega\right) = \frac{\left(20\pi\right)^3}{\left(j\omega + 20\pi\right)\left(-\omega^2 + j\,20\pi\omega + 400\pi^2\right)}$$

The magnitude of the system function is

$$\left|H\left(\omega\right)\right| = \frac{\left(20\pi\right)^3}{\sqrt{\omega^2 + \left(20\pi\right)^2}\,\sqrt{\left(-\omega^2 + 400\pi^2\right)^2 + \left(20\pi\omega\right)^2}}$$

and the phase of the system function is

$$\angle H\left(\omega\right) = -\tan^{-1}\left(\frac{\omega}{20\pi}\right) - \tan^{-1}\left(\frac{20\pi\omega}{-\omega^2 + 400\pi^2}\right)$$

Magnitude and phase responses of the designed filter are shown in Fig. 10.18.

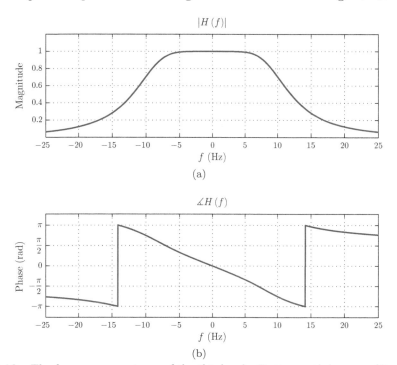

Figure 10.18 – The frequency spectrum of the third-order Butterworth lowpass filter designed in Example 10.2: (a) magnitude, (b) phase.

Software resources:
ex_10_2.m

Software resources: See MATLAB Exercise 10.1.

Obtaining N and ω_c for the Butterworth lowpass filter

The desired behavior of a lowpass filter to be designed is usually specified in terms of the two critical frequencies ω_1 and ω_2 as well as the decibel tolerance values R_p and A_s as shown in Fig. 10.14. The corresponding parameter values for ω_c and N need to be determined from the provided set of specifications so that the filter can be designed. If $R_p = 3$ dB, then we have $\omega_c = \omega_1$. Otherwise the requirements at the two critical frequencies can be expressed as inequalities.

At $\omega = \omega_1$ we require

$$-10 \log_{10} \left[1 + \left(\frac{\omega_1}{\omega_c} \right)^{2N} \right] \geq -R_p \tag{10.60}$$

Similarly at $\omega = \omega_2$ we require

$$-10 \log_{10} \left[1 + \left(\frac{\omega_2}{\omega_c} \right)^{2N} \right] \leq -A_s \tag{10.61}$$

Rearranging Eqns. (10.60) and (10.61) yields

$$\left(\frac{\omega_1}{\omega_c} \right)^{2N} \leq 10^{R_p/10} - 1 \tag{10.62}$$

and

$$\left(\frac{\omega_2}{\omega_c} \right)^{2N} \geq 10^{A_s/10} - 1 \tag{10.63}$$

Dividing both sides of Eqn. (10.62) by the corresponding terms of Eqn. (10.63) we obtain

$$\left(\frac{\omega_1}{\omega_2} \right)^{2N} \leq \frac{10^{R_p/10} - 1}{10^{A_s/10} - 1} \tag{10.64}$$

which can be solved for the filter order N that is needed:

$$N \geq \frac{\log_{10} \sqrt{\left(10^{A_s/10} - 1 \right) / \left(10^{R_p/10} - 1 \right)}}{\log_{10} \left(\omega_2/\omega_1 \right)} \tag{10.65}$$

The right side of the inequality in Eqn. (10.65) may or may not be an integer. If it is not, then the value of N needs to be rounded up. The excess tolerance created by rounding N up to the next integer can be used in either the passband or the stopband.

In summary, the parameters of the Butterworth lowpass filter are computed as follows:

Given dB tolerances R_p and A_s and critical frequencies ω_1 and ω_2:

$$N = \frac{\log_{10} \sqrt{\left(10^{A_s/10} - 1 \right) / \left(10^{R_p/10} - 1 \right)}}{\log_{10} \left(\omega_2/\omega_1 \right)} \qquad \Longrightarrow \qquad \text{Round up to next integer}$$

After determining N, compute ω_c from one of the following:

$$\left(\frac{\omega_1}{\omega_c}\right)^{2N} = 10^{R_p/10} - 1 \qquad \text{or} \qquad \left(\frac{\omega_2}{\omega_c}\right)^{2N} = 10^{A_s/10} - 1$$

10.4.2 Chebyshev lowpass filters

The squared-magnitude function for a Chebyshev lowpass filter is given by

$$|H(\omega)|^2 = \frac{1}{1 + \varepsilon^2 C_N^2(\omega/\omega_1)} \tag{10.66}$$

Parameter ε is a positive constant, N is the order of the filter, and ω_1 is the passband edge frequency in rad/s (see Fig. 10.13(a)). The term $C_N(\nu)$ represents the *Chebyshev polynomial* of order N. Chebyshev filters of this type are sometimes called *Chebyshev type-I* filters. A variant of the squared-magnitude response in Eqn. (10.66) will be introduced in the next section as the *inverse Chebyshev filter* or the *Chebyshev type-II filter*.

Once the values of the parameters ε, N and ω_1 are specified, the design procedure proceeds as follows:

1. Find $C_N(\nu)$, the Chebyshev polynomial of order N.
2. Substitute $\nu = \omega/\omega_1$ in the polynomial $C_N(\nu)$.
3. Form the Chebyshev type I squared-magnitude function as specified in Eqn. (10.66).
4. Find the poles of the system function $H(s)$ from the squared-magnitude function. As in the case of the Butterworth lowpass filter, this involves substituting $j\omega = s$ into the squared-magnitude function, finding the roots of the denominator polynomial, and separating those poles into two sets, one to be associated with $H(s)$ and the other to be associated with $H(s)$.

Before we carry out the steps outlined above, it will be useful to review the properties of Chebyshev polynomials.

Chebyshev polynomials

The Chebyshev polynomial of order N is defined as

$$C_N(\nu) = \cos\left(N \cos^{-1}(\nu)\right) \tag{10.67}$$

A better approach for understanding the definition in Eqn. (10.67) would be to split it into two equations as

$$\nu = \cos(\theta) \tag{10.68}$$

and

$$C_N(\nu) = \cos(N\theta) \tag{10.69}$$

The Chebyshev polynomial of a specified order N can be obtained by

1. Using trigonometric identities to express $\cos(N\theta)$ as a function of $\cos(\theta)$.
2. Replacing each $\cos(\theta)$ term with ν.

The first two polynomials are easy to obtain from this definition:

$$C_0\left(\nu\right) = 1$$

$$C_1\left(\nu\right) = \nu$$

To obtain $C_2\left(\nu\right)$ let us write $\cos\left(2\theta\right)$ in terms of $\cos\left(\theta\right)$ using a trigonometric identity:

$$\cos\left(2\theta\right) = 2\cos^2\left(\theta\right) - 1$$

Thus, the second-order Chebyshev polynomial is

$$C_2\left(\nu\right) = 2\nu^2 - 1$$

Higher-order Chebyshev polynomials may be obtained by continuing in this fashion. As the order increases, however, the procedure outlined above becomes increasingly tedious. Fortunately, it is possible to derive a recursive formula to facilitate the derivation of higher-order Chebyshev polynomals.

Let us write $\cos\left(\left(N+1\right)\theta\right)$ and $\cos\left(\left(N-1\right)\theta\right)$ using trigonometric identities:

$$\cos\left(\left(N+1\right)\theta\right) = \cos\left(\theta\right)\cos\left(N\theta\right) - \sin\left(\theta\right)\sin\left(N\theta\right)$$

$$\cos\left(\left(N-1\right)\theta\right) = \cos\left(\theta\right)\cos\left(N\theta\right) + \sin\left(\theta\right)\sin\left(N\theta\right)$$

Adding the two equations results in

$$\cos\left(\left(N+1\right)\theta\right) + \cos\left(\left(N-1\right)\theta\right) = 2\cos\left(\theta\right)\cos\left(N\theta\right)$$

which can be rearranged to produce the recursive formula we seek:

$$\cos\left(\left(N+1\right)\theta\right) = 2\cos\left(\theta\right)\cos\left(N\theta\right) - \cos\left(\left(N-1\right)\theta\right)$$

Thus, the recursive formula for obtaining Chebyshev polynomials is

$$C_{N+1}\left(\nu\right) = 2\nu\,C_N\left(\nu\right) - C_{N-1}\left(\nu\right) \tag{10.70}$$

The recursive formula in Eqn. (10.70) allows any order Chebyshev polynomial to be found provided that the polynomials for the previous two orders are known. With the knowledge of $C_0\left(\nu\right) = 1$ and $C_1\left(\nu\right) = \nu$, the polynomial $C_2\left(\nu\right)$ can be found as

$$
\begin{aligned}
C_2\left(\nu\right) &= 2\,\nu\,C_1\left(\nu\right) - C_0\left(\nu\right) \\
&= 2\nu^2 - 1
\end{aligned}
$$

Similarly $C_3\left(\nu\right)$ is found as

$$
\begin{aligned}
C_3\left(\nu\right) &= 2\nu\,C_2\left(\nu\right) - C_1\left(\nu\right) \\
&= 2\nu\left(2\nu^2 - 1\right) - \nu \\
&= 4\nu^3 - 3\nu
\end{aligned}
$$

The behaviors of the first few orders of Chebyshev polynomials are illustrated in Fig. 10.19.

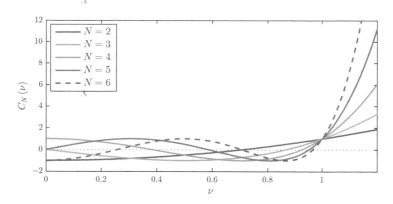

Figure 10.19 – Some Chebyshev polynomials.

A few interesting characteristics of Chebyshev polynomials can be readily observed:

1. All Chebyshev polynomials pass through the point $(1,1)$, that is, $C_N(1) = 1$ for all N.
2. For all Chebyshev polynomials, if $|\nu| \leq 1$ then $|C_N(\nu)| \leq 1$.
3. For $|\nu| > 1$ all Chebyshev polynomials grow monotonically without bound.
4. At $\nu = 0$ the behavior of Chebyshev polynomials depends on the order N. $C_N(0) = 0$ if N is odd, and $C_N(0) = \pm 1$ if N is even.

Fig. 10.20 shows the squared Chebyshev polynomials for the first few orders N.

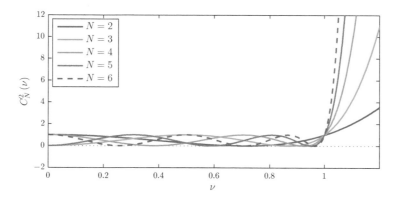

Figure 10.20 – Squared Chebyshev polynomials.

Squared Chebyshev polynomials $C_N^2(\nu)$ oscillate in the amplitude range $[0,1]$ for $|\nu| \leq 1$ and $N \geq 2$. For $|\nu| > 1$ they grow monotonically without bound.

Software resources: See MATLAB Exercise 10.2.

The squared magnitude function given by Eqn. (10.66) is shown in Fig. 10.21 for several values of N using $\varepsilon = 0.4$ and $\omega_1 = 1$ rad/s.

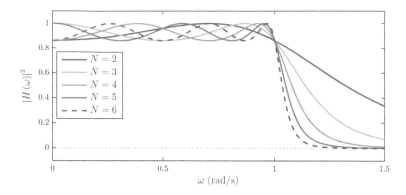

Figure 10.21 – Squared magnitude for Chebyshev filters with $\omega_1 = 1$ rad/s.

Poles for the Chebyshev lowpass filter

In constructing the system function for a Chebyshev lowpass filter we will use an approach similar to that taken with a Butterworth lowpass filter in Section 10.4.1. First we will determine the poles of the product $H(s) H(-s)$. Recall that the product $H(s) H(-s)$ is obtained by starting with the squared magnitude function $|H(\omega)|^2$ and replacing $j\omega$ with s:

$$H(s) H(-s) = \frac{1}{1 + \varepsilon^2 C_N^2 \left(\dfrac{s}{j\omega_1} \right)} \tag{10.71}$$

The poles p_k of $H(s) H(-s)$ are the solutions of the equation

$$1 + \varepsilon^2 C_N^2 \left(\frac{p_k}{j\omega_1} \right) = 0 \tag{10.72}$$

for $k = 0, \ldots, 2N - 1$.

Let us define

$$\nu_k = \frac{s}{j\omega_1} \tag{10.73}$$

so that

$$1 + \varepsilon^2 C_N^2 (\nu_k) = 0 \tag{10.74}$$

Using the definition of the Chebyshev polynomial given by Eqn. (10.69) we can rewrite Eqn. (10.74) as

$$1 + \varepsilon^2 \cos^2 (N\theta_k) = 0 \tag{10.75}$$

where $\nu_k = \cos(\theta_k)$. Solving Eqn. (10.75) for $\cos(N\theta_k)$ yields

$$\cos(N\theta_k) = \pm \frac{j}{\varepsilon} \tag{10.76}$$

Let $\theta_k = \alpha_k + j\beta_k$ with α_k and β_k both as real parameters. Eqn. (10.76) becomes

$$\cos(N\alpha_k + jN\beta_k) = \pm \frac{j}{\varepsilon} \tag{10.77}$$

which, using the appropriate trigonometric identity, can be written as

$$\cos\left(N\alpha_k\right)\cos\left(jN\beta_k\right) - \sin\left(N\alpha_k\right)\sin\left(jN\beta_k\right) = \pm\frac{j}{\varepsilon} \tag{10.78}$$

Recognizing that $\cos\left(jN\beta_k\right) = \cosh\left(N\beta_k\right)$ and $\sin\left(jN\beta_k\right) = j\sinh\left(N\beta_k\right)$, we obtain

$$\cos\left(N\alpha_k\right)\cosh\left(N\beta_k\right) - j\sin\left(N\alpha_k\right)\sinh\left(N\beta_k\right) = \pm\frac{j}{\varepsilon} \tag{10.79}$$

Equating real and imaginary parts of both sides of Eqn. (10.79) yields

$$\cos\left(N\alpha_k\right)\cosh\left(N\beta_k\right) = 0 \tag{10.80}$$

and

$$\sin\left(N\alpha_k\right)\sinh\left(N\beta_k\right) = \pm\frac{1}{\varepsilon} \tag{10.81}$$

To satisfy Eqn. (10.80) the cosine term must be set equal to zero, leading to

$$\cos\left(N\alpha_k\right) = 0 \quad\Longrightarrow\quad \alpha_k = \frac{(2k+1)\,\pi}{2N}\,, \quad k = 0,\ldots,2N-1 \tag{10.82}$$

Using this value of α_k in Eqn. (10.81) results in

$$\sin\left(N\alpha_k\right) = \pm 1 \tag{10.83}$$

and

$$\beta_k = \frac{\sinh^{-1}\left(1/\varepsilon\right)}{N} \tag{10.84}$$

The poles of the product $H\left(s\right)H\left(-s\right)$ can now be determined. Using Eqn. (10.73)

$$p_k = j\omega_1\,\nu_k = j\omega_1\cos\left(\theta_k\right)\,, \quad k = 0,\ldots,2N-1 \tag{10.85}$$

Using the values of α_k and β_k found, the poles of $H\left(s\right)H\left(-s\right)$ are

$$p_k = j\omega_1\left[\cos\left(\alpha_k\right)\cosh\left(\beta_k\right) - j\sin\left(\alpha_k\right)\sinh\left(\beta_k\right)\right] \tag{10.86}$$

for $k = 0,\ldots,2N-1$. It can be shown that the poles given by Eqn. (10.86) are on an elliptical trajectory. The ones in the left half s-plane are associated with $H\left(s\right)$ in order to obtain a causal and stable filter.

Poles of $H\left(s\right)H\left(-s\right)$ for the Chebyshev type-I lowpass filter:

$$\text{Using:} \quad \alpha_k = \frac{(2k+1)\,\pi}{2N} \quad \text{and} \quad \beta_k = \frac{\sinh^{-1}\left(1/\varepsilon\right)}{N}$$

$$p_k = j\omega_1\left[\cos\left(\alpha_k\right)\cosh\left(\beta_k\right) - j\sin\left(\alpha_k\right)\sinh\left(\beta_k\right)\right]\,, \quad k = 0,\ldots,2N-1$$

Interactive Demo: cheb1_demo.m

This demo program illustrates the magnitude response of the Chebyshev type-I lowpass filter, the squared version of which is given by Eqn. (10.42). The cutoff frequency f_1, the parameter ε and the filter order N may be specified as desired. Observe the following:

1. The edges of the magnitude characteristic become sharper as the filter order is increased.
2. At the frequency $f = f_1$ the magnitude is equal to

$$|H(f_1)| = \frac{1}{\sqrt{1 + \varepsilon^2}}$$

Software resources:
cheb1_demo.m

Example 10.3: **Design of a Chebyshev lowpass filter**

Design a third-order Chebyshev type-I analog lowpass filter with a passband edge frequency of $\omega_1 = 1$ rad/s and $\varepsilon = 0.4$. Afterwards compute and graph the magnitude response of the designed filter.

Solution: Using Eqn. (10.82) the parameter α_k is found as

$$\alpha_k = \frac{(2k+1)\,\pi}{6}, \quad k = 0, \ldots, 5$$

and the parameter β_k is obtained from Eqn. (10.84) as

$$\beta_k = \frac{\sinh^{-1}(1/0.4)}{3} = 0.5491$$

Poles of $H(s)H(-s)$ are found through the use of Eqn. (10.86):

$$p_k = j\left[\cos\left(\frac{(2k+1)\,\pi}{6}\right)\cosh(0.5491) - j\sin\left(\frac{(2k+1)\,\pi}{6}\right)\sinh(0.5491)\right],$$

$$k = 0, \ldots, 5 \qquad (10.87)$$

Evaluating Eqn. (10.87) for $k = 0, \ldots, 5$ yields the following poles:

$$
\begin{aligned}
p_0 &= -0.2885 + j\,1 \\
p_1 &= 0.5771 \\
p_2 &= 0.2885 - j\,1 \\
p_3 &= -0.2885 - j\,1 \\
p_4 &= -0.5771 \\
p_5 &= -0.2885 + j\,1
\end{aligned}
$$

The first three poles, namely p_0, p_1 and p_2 are in the right half s-plane; therefore they belong to $H(-s)$. The system function $H(s)$ should be constructed using the remaining three poles.

$$H\left(s\right) = \frac{0.6250}{\left(s + 0.2885 + j\,1\right)\left(s + 0.5771\right)\left(s + 0.2885 - j\,1\right)}$$

$$= \frac{0.6250}{s^3 + 1.1542\,s^2 + 1.4161\,s + 0.6250}$$

The numerator of $H\left(s\right)$ was adjusted to achieve $\left|H\left(0\right)\right| = 0$. The magnitude and the phase of $H\left(s\right)$ are shown in Fig. 10.22.

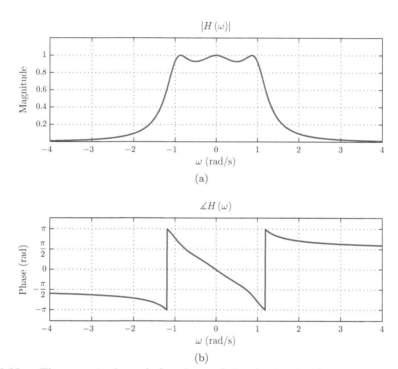

(a)

(b)

Figure 10.22 – The magnitude and the phase of the third-order Chebyshev lowpass filter designed in Example 10.3.

Software resources:
ex_10_3.m

Software resources:	See MATLAB Exercise 10.3.

Obtaining N and ε for the Chebyshev lowpass filter

In some design problems it is desired to find the lowest-order filter that satisfies a set of design specifications. The desired behavior of the lowpass filter is typically specified in terms of the two critical frequencies ω_1 and ω_2 as well as the dB tolerance values R_p and A_s for the passband and the stopband respectively.

At the frequency $\omega = \omega_1$ we need

$$20 \log_{10} \left[\frac{|H(\omega_1)|}{|H(\omega)|_{\max}} \right] = -R_p \tag{10.88}$$

Recall that $C_N(1) = 1$ for all Chebyshev polynomials. Therefore Eqn. (10.88) becomes

$$10 \log_{10} \left(\frac{1}{1 + \varepsilon^2} \right) = -R_p \tag{10.89}$$

which can be solved for ε to yield

$$\varepsilon = \sqrt{10^{R_p/10} - 1} \tag{10.90}$$

In the next step we need to determine the filter order N. At the stopband edge frequency we have the inequality

$$20 \log_{10} |H(\omega_2)| \leq -A_s \tag{10.91}$$

In order to simplify the notation let us define $\omega_0 = \omega_2/\omega_1$. Eqn. (10.91 can be expressed as

$$10 \log_{10} \left(\frac{1}{1 + \varepsilon^2 C_N^2(\omega_0)} \right) \leq -A_s \tag{10.92}$$

In the worst case Eqn. (10.92) can be taken with equality and solved for $C_N(\omega_0)$ to yield

$$C_N(\omega_0) = \sqrt{\frac{10^{A_s/10} - 1}{10^{R_p/10} - 1}} \tag{10.93}$$

in which we have used the value of ε found in Eqn. (10.90). In order to solve Eqn. (10.93) for N we will define some intermediate variables. Let

$$F = \sqrt{\frac{10^{A_s/10} - 1}{10^{R_p/10} - 1}} \tag{10.94}$$

and

$$\cos(\theta_0) = \omega_0 \tag{10.95}$$

The equation that needs to be solved for N is

$$\cos(N\theta_0) = F \tag{10.96}$$

Since θ_0 is complex (remember that $\omega_0 > 1$) we will express it in Cartesian complex form as $\theta_0 = \alpha_0 + j\beta_0$, and write (10.96) as

$$\cos(N\alpha_0)\cosh(N\beta_0) - j\sin(N\alpha_0)\sinh(N\beta_0) = F \tag{10.97}$$

Since F is real, the left side of Eqn. (10.97) must be real as well. Therefore we must have

$$\alpha_0 = 0 \qquad \Longrightarrow \qquad \theta_0 = j\beta_0 \tag{10.98}$$

and

$$\cosh(N\beta_0) = F \tag{10.99}$$

Using Eqn. (10.95) we can write

$$\beta_0 = -j\,\theta_0 = \cosh^{-1}(\omega_0) \qquad (10.100)$$

Solving Eqns. (10.99) and (10.100) for the filter order N we obtain

$$N = \frac{\cosh^{-1}(F)}{\cosh^{-1}(\omega_0)} \qquad (10.101)$$

The value of N obtained through Eqn. (10.101) may not necessarily be an integer, and must be rounded up to the next integer. The excess tolerance obtained by rounding N up causes the stopband inequality given by Eqn. (10.92) to improve. If it is desired to use the excess tolerance for improving the passband response instead, then ε may be recalculated from Eqn. (10.92) using the rounded-up value of N.

In summary, the parameters of a Chebyshev type-I lowpass filter are computed as follows:

Given dB tolerances R_p and A_s and critical frequencies ω_1 and ω_2:

$$\omega_0 = \frac{\omega_2}{\omega_1}\,, \qquad \text{and} \qquad F = \sqrt{\frac{10^{A_s/10} - 1}{10^{R_p/10} - 1}}$$

$$N = \frac{\cosh^{-1}(F)}{\cosh^{-1}(\omega_0)} \quad \Longrightarrow \quad \text{Round up to next integer}$$

Compute ε from one of the following:

$$10\log_{10}\left(\frac{1}{1+\varepsilon^2}\right) = -R_p \qquad \text{or} \qquad 10\log_{10}\left(\frac{1}{1+\varepsilon^2 C_N^2(\omega_0)}\right) = -A_s$$

Software resources: See MATLAB Exercise 10.4.

10.4.3   Inverse Chebyshev lowpass filters

The squared-magnitude function for an *inverse Chebyshev* lowpass filter, also referred to as a *Chebyshev type-II* lowpass filter, is

$$|H(\omega)|^2 = \frac{\varepsilon^2 C_N(\omega_2/\omega)}{1+\epsilon^2 C_N^2(\omega_2/\omega)} \qquad (10.102)$$

As in the case of Chebyshev type-I filters, ε is a positive constant and $C_N(\nu)$ represents the Chebyshev polynomial of order N. The parameter ω_2 is the stopband edge frequency (see Fig. 10.13(a)). Magnitude responses of several Chebyshev type-II filters are shown in Fig. 10.23. The magnitude response of the Chebyshev type-II filter is smooth in the passband and has equiripple behavior in the stopband.

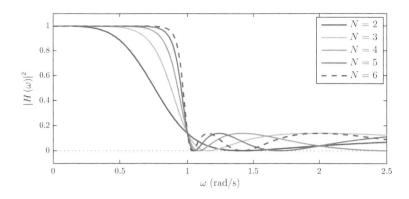

Figure 10.23 – Squared magnitude for Chebyshev filters with $\omega_2 = 1$ rad/s.

Poles and zeros of the inverse Chebyshev filter

The denominator of the squared magnitude function $|H(\omega)|^2$ in Eqn. (10.102) is very similar to that of the type-I Chebyshev filter squared magnitude response given by Eqn. (10.66). Consequently, most of the results obtained in the previous section through the derivation of the poles of Chebyshev type-I lowpass filter will be usable. For an inverse Chebyshev filter the poles p_k of the product $H(s)H(-s)$ are the solutions of the equation

$$1 + \varepsilon^2 C_N^2 \left(\frac{j\omega_2}{p_k} \right) = 0 \tag{10.103}$$

for $k = 0, \ldots, 2N - 1$.

Let

$$\nu_k = \frac{j\omega_2}{p_k} \tag{10.104}$$

so that Eqn. (10.103) becomes

$$1 + \varepsilon^2 C_N^2 (\nu_k) = 0 \tag{10.105}$$

This result is identical to that obtained in Eqn. (10.74) for computing the poles of the Chebyshev type-I lowpass filter, therefore, the solution obtained for p_k may be used here as well. We have

$$\nu_k = \cos(\theta_k) \tag{10.106}$$

with

$$\alpha_k = \text{Re}\{\theta_k\} = \frac{(2k+1)\pi}{2N} \tag{10.107}$$

and

$$\beta_k = \text{Im}\{\theta_k\} = \frac{\sinh^{-1}(1/\varepsilon)}{N} \tag{10.108}$$

The poles p_k are determined from α_k and β_k as

$$p_k = \frac{j\omega_2}{\nu_k} = \frac{j\omega_2}{\cos(\alpha_k + j\beta_k)}$$

$$= \frac{j\omega_2}{\cos(\alpha_k)\cosh(\beta_k) - j\sin(\alpha_k)\sinh(\beta_k)} \tag{10.109}$$

As discussed before, the poles in the left half s-plane are associated with $H(s)$ in order to obtain a causal and stable filter. The zeros of $H(s)H(s)$ are found by solving

$$C_N^2 \left(\frac{j\omega_2}{z_k} \right) = 0 \tag{10.110}$$

It can be shown that the zeros of the inverse Chebyshev lowpass filter are

$$z_k = \frac{\pm j\omega_2}{\cos \left(\dfrac{(2k-1)\,\pi}{2N} \right)}, \qquad k = 1, \ldots, K \tag{10.111}$$

where the upper limit K is computed as

$$K = \begin{cases} (N-1)/2\,, & N \text{ odd} \\ N/2\,, & N \text{ even} \end{cases} \tag{10.112}$$

A summary of the results is given below:

Poles of $H(s)\,H(-s)$ for the Chebyshev type-II lowpass filter:

$$\text{Using:} \qquad \alpha_k = \frac{(2k+1)\,\pi}{2N} \qquad \text{and} \qquad \beta_k = \frac{\sinh^{-1}(1/\varepsilon)}{N}$$

$$p_k = \frac{j\omega_2}{\cos(\alpha_k)\cosh(\beta_k) - j\sin(\alpha_k)\sinh(\beta_k)}, \qquad k = 0, \ldots, 2N-1$$

Zeros of $H(s)$ for the Chebyshev type-II lowpass filter:

$$z_k = \frac{\pm j\omega_2}{\cos \left(\dfrac{(2k-1)\,\pi}{2N} \right)}, \qquad k = 1, \ldots, K$$

The upper limit K is computed as

$$K = \begin{cases} (N-1)/2\,, & N \text{ odd} \\ N/2\,, & N \text{ even} \end{cases}$$

Interactive Demo: cheb2_demo.m

This demo program illustrates the magnitude response of the Chebyshev type-II lowpass filter, the squared version of which is given by Eqn. (10.43). The stopband edge frequency f_2, the parameter ϵ and the filter order N may be specified as desired. Observe the following:

1. The edges of the magnitude characteristic become sharper as the filter order is increased.

2. At the frequency $f = f_2$ the magnitude is equal to

$$|H(f_2)| = \sqrt{\frac{\varepsilon^2}{1 + \varepsilon^2}}$$

Software resources:
cheb2_demo.m

Software resources: See MATLAB Exercise 10.5.

Obtaining N and ε for the inverse Chebyshev lowpass filter

The inverse Chebyshev lowpass filter is uniquely described by the choice of parameters ω_2, ε and N. If the desired filter is described by the specification diagram of Fig. 10.14 then we have the set of parameters R_p, A_s, ω_1 and ω_2 from which ε and N must be determined.

The first step is to choose the parameter ε so that the magnitude response of the filter fills the allowed area completely in the stopband.

Evaluating the squared magnitude response given by Eqn. (10.102) at the stopband edge frequency $\omega = \omega_2$ and remembering that $C_N(1) = 1$ for all Chebyshev polynomials we obtain

$$10\log_{10}\left(\frac{\varepsilon^2}{1+\varepsilon^2}\right) = -A_s \tag{10.113}$$

which can be solved for ε to yield

$$\varepsilon = \frac{1}{\sqrt{10^{A_s/10}-1}} \tag{10.114}$$

Next the filter order N needs to be determined. At the passband edge frequency we have the following inequality:

$$20\log_{10}|H(j\omega_1)| \geq -R_p \tag{10.115}$$

Let us define $\omega_0 = \omega_2/\omega_1$ to simplify the notation. Eqn. (10.115) can be written as

$$10\log_{10}\left[\frac{\varepsilon^2 C_N^2(\omega_0)}{1+\varepsilon^2 C_N^2(\omega_0)}\right] \geq -R_p \tag{10.116}$$

In the worst case Eqn. (10.116) can be taken with equality and solved for $C_N(\omega_0)$ to yield

$$C_N(\omega_0) = \sqrt{\frac{10^{A_s/10}-1}{10^{R_p/10}-1}} \tag{10.117}$$

after the substitution of the result found in Eqn. (10.114). Incidentally this is identical to Eqn. (10.93); therefore, its solution must be identical as well. Using the definition of parameter F as given by Eqn. (10.94) the minimum filter order that satisfies the specifications is found as

$$N = \frac{\cosh^{-1}(F)}{\cosh^{-1}(\omega_0)} \tag{10.118}$$

If Eqn. (10.118) does not yield an integer, it must be rounded up to the next integer. The excess tolerance created by the rounding up of N can be used in either frequency band by solving Eqn. (10.113) or Eqn. (10.116) for ε.

In summary, the parameters of a Chebyshev type-II lowpass filter are computed as follows:

Given dB tolerances R_p and A_s and critical frequencies ω_1 and ω_2:

$$\omega_0 = \frac{\omega_2}{\omega_1}, \qquad \text{and} \qquad F = \sqrt{\frac{10^{A_s/10} - 1}{10^{R_p/10} - 1}}$$

$$N = \frac{\cosh^{-1}(F)}{\cosh^{-1}(\omega_0)} \implies \text{Round up to next integer}$$

Compute ε from one of the following:

$$\varepsilon = \frac{1}{\sqrt{10^{A_s/10} - 1}} \qquad \text{or} \qquad \varepsilon = \frac{1}{\sqrt{C_N^2(\omega_0)\left(10^{R_p/10} - 1\right)}}$$

10.4.4 Analog filter transformations

In preceding sections we have discussed the design of analog lowpass filters. Design formulas for Butterworth, Chebyshev and elliptic squared-magnitude functions were presented only for lowpass filters. If a different filter type such as a highpass, bandpass or band-reject filter is needed, it is obtained from a lowpass filter by means of an *analog filter transformation*.

Let $G(s)$ be the system function of an analog lowpass filter, and let $H(\lambda)$ represent the new filter to be obtained from it. For the new filter we use λ as the Laplace transform variable. $H(\lambda)$ is obtained from $G(s)$ through a transformation such that

$$H(\lambda) = G(s)\Big|_{s=f(\lambda)} \tag{10.119}$$

The function $s = f(\lambda)$ is the transformation that converts the lowpass filter into the type of filter desired.

Lowpass to highpass transformation

Consider the specification diagrams shown in Fig. 10.24 for a lowpass filter with magnitude response $|G(\omega)|$ and a highpass filter with magnitude response $|H(\omega)|$.

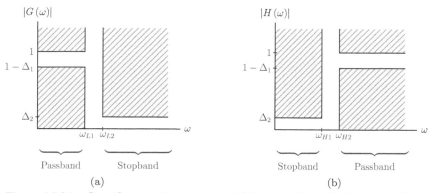

Figure 10.24 – Specification diagrams for (a) lowpass filter, (b) highpass filter.

It is desired to obtain the highpass filter system function $H(\lambda)$ from the lowpass filter system function $G(s)$. The transformation

$$\frac{s}{\omega_{L1}} = \frac{\omega_{H2}}{\lambda} \quad\Longrightarrow\quad s = \frac{\omega_{L1}\,\omega_{H2}}{\lambda} = \frac{\omega_0}{\lambda} \tag{10.120}$$

can be used for this purpose The magnitudes of the two filters are identical at their respective passband edge frequencies, that is,

$$|H(\omega_{H2})| = |G(\omega_{L1})| \tag{10.121}$$

The stopband edges of the two filters are related by

$$\omega_{L2}\,\omega_{H1} = \omega_{L1}\,\omega_{H2} \tag{10.122}$$

so that we have

$$|H(\omega_{H1})| = |G(\omega_{L2})| \tag{10.123}$$

Example 10.4: **Design of a Chebyshev highpass filter**

Recall that a third-order Chebyshev type-I analog lowpass filter was designed in Example 10.3 with a cutoff frequency of $\omega_1 = 1$ rad/s and $\varepsilon = 0.4$. The system function of the filter was found to be

$$G(s) = \frac{0.6250}{s^3 + 1.1542\,s^2 + 1.4161\,s + 0.6250}$$

Convert this filter to a highpass filter with a critical frequency of $\omega_{H2} = 3$ rad/s. Afterwards compute and graph the magnitude response of the designed filter.

Solution: The critical frequency of the lowpass filter is $\omega_{L1} = 1$ rad/s; therefore, we will use the transformation

$$s = \frac{3}{\lambda}$$

which leads to the system function

$$H(\lambda) = \frac{0.6250}{\left(\dfrac{3}{\lambda}\right)^3 + 1.1542\left(\dfrac{3}{\lambda}\right)^2 + 1.4161\left(\dfrac{3}{\lambda}\right) + 0.6250}$$

$$= \frac{0.6250\,\lambda^3}{0.6250\,\lambda^3 + 4.2482\,\lambda^2 + 10.3875\,\lambda + 27}$$

The magnitude and the phase of $H(\lambda)$ are shown in Fig. 10.25.

Software resources:

ex_10_4.m

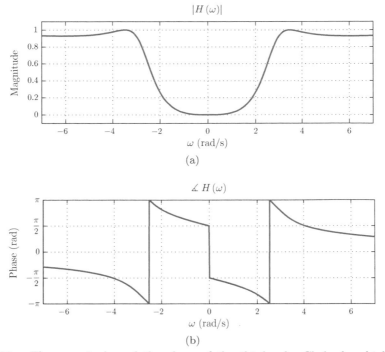

$$|H(\omega)|$$

(a)

$$\angle H(\omega)$$

(b)

Figure 10.25 – The magnitude and the phase of the third-order Chebyshev highpass filter designed in Example 10.4.

Software resources: See MATLAB Exercise 10.6.

Lowpass to bandpass transformation

Specification diagrams in Fig. 10.26 show a lowpass filter with magnitude response $|G(\omega)|$ and a bandpass filter with magnitude response $|H(\omega)|$.

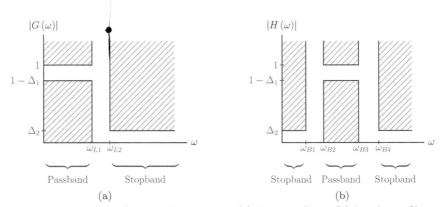

Figure 10.26 – Specification diagrams for (a) lowpass filter, (b) bandpass filter.

It is desired to obtain the bandpass filter system function $H(\lambda)$ from the lowpass filter system function $G(s)$. The transformation

$$s = \frac{\lambda^2 + \omega_0^2}{B\lambda} \tag{10.124}$$

can be used for this purpose. We require

$$|H(\omega_{B2})| = |G(-\omega_{L1})| \tag{10.125}$$

and

$$|H(\omega_{B3})| = |G(\omega_{L1})| \tag{10.126}$$

It can be shown that (see Problem 10.10) these requirements are met if the parameters ω_0 and B are chosen as

$$\omega_0 = \sqrt{\omega_{B2}\,\omega_{B3}} \qquad \text{and} \qquad B = \frac{\omega_{B3} - \omega_{B2}}{\omega_{L1}} \tag{10.127}$$

The parameter ω_0 is the *geometric mean* of the passband edge frequencies of the bandpass filter. The parameter B is the ratio of the bandwidth of the bandpass filter to the bandwidth of the lowpass filter.

Lowpass to band-reject transformation

Specification diagrams in Fig. 10.27 show a lowpass filter with magnitude response $|G(\omega)|$ and a band-reject filter with magnitude response $|H(\omega)|$.

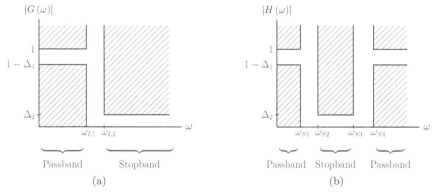

Figure 10.27 – Specification diagrams for (a) lowpass filter, (b) band-reject filter.

In order to obtain the band-reject filter system function $H(\lambda)$ from the lowpass filter system function $G(s)$, the transformation to be used is in the form

$$s = \frac{B\lambda}{\lambda^2 + \omega_0^2} \tag{10.128}$$

For a lowpass filter to be converted into a highpass filter we require

$$|H(\omega_{S4})| = |G(-\omega_{L1})| \tag{10.129}$$

and

$$|H(\omega_{S1})| = |G(\omega_{L1})| \tag{10.130}$$

It can be shown that (see Problem 10.12) these requirements are met if the parameters ω_0 and B are chosen as

$$\omega_0 = \sqrt{\omega_{S1}\,\omega_{S4}} \qquad \text{and} \qquad B = (\omega_{S4} - \omega_{S1})\,\omega_{L1} \tag{10.131}$$

10.5 Design of Digital Filters

In this section we will briefly discuss the design of discrete-time filters. Discrete-time filters are viewed under two broad categories: *infinite impulse response (IIR)* filters, and *finite impulse response (FIR)* filters. The system function of an IIR filter has both poles and zeros. Consequently the impulse response of the filter is of infinite length. It should be noted, however, that the impulse response of a stable IIR filter must also be absolute summable, and must therefore decay over time. In contrast, the behavior of an FIR filter is controlled only by the placement of the zeros of its system function. For causal FIR filters all of the poles are at the origin, and they do not contribute to the magnitude characteristics of the filter.

For a given set of specifications, an IIR filter is generally more efficient than a comparable FIR filter. On the other hand, FIR filters are always stable. Additionally, a linear phase characteristic is possible with FIR filters whereas causal and stable IIR filters cannot have linear phase. The significance of a linear phase characteristic is that the time delay is constant for all frequencies. This is desirable in some applications, and requires the use of an FIR filter.

10.5.1 Design of IIR filters

Consider again the specification diagram for an analog lowpass filter shown in Fig. 10.13(a). It can be adapted for the system function of a discrete-time filter as shown in Fig. 10.28.

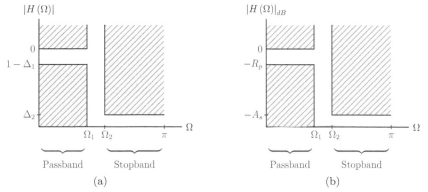

Figure 10.28 – Tolerance specifications for a discrete-time lowpass filter.

The magnitude specifications of the discrete-time filter are given in the angular frequency range $0 \leq \Omega < \pi$. Since the impulse response of the filter to be designed is real, the magnitude is an even function of Ω, and therefore the part of the magnitude response in the range $-\pi \leq \Omega < 0$ is obtained as the mirror image of what is shown in Fig. 10.28. Furthermore, the magnitude response is 2π-periodic.

Specification diagrams for other filter types (highpass, bandpass or band-reject) can be adapted to discrete-time filters in a similar manner.

The most common method of designing IIR filters is to start with an appropriate analog filter, and to convert its system function to the system function of a discrete-time filter by means of some transformation. Designing a discrete-time filter by this approach involves a three step procedure:

1. The specifications of the desired discrete-time filter are converted to the specifications of an appropriate analog filter that can be used as a prototype. Let the desired discrete-time filter be specified through critical frequencies Ω_1 and Ω_2 along with tolerance values Δ_1 and Δ_2. Analog prototype filter parameters ω_1 and ω_2 need to be determined. (If the filter type is bandpass or band-reject, two additional frequencies need to be specified.)

2. An analog prototype filter that satisfies the design specifications in step 1 is designed. Its system function $G(s)$ is constructed.

3. The analog prototype filter is converted to a discrete-time filter by means of a transformation. Specifically, a z-domain system function $H(z)$ is obtained from the analog prototype system function $G(s)$. The designed discrete-time filter can be analyzed using the techniques discussed in Chapters 5 and 8, and implemented using the block diagram methods discussed in Chapter 8.

Step 2 of the design procedure, namely the design of an analog prototype, was discussed in Section 10.4. The problem of converting an analog system function to a discrete-time system function (step 3) will be discussed next, followed by the conversion of specifications (step 1). The reason for this particular order is because step 1 is dependent on which method is to be used in step 3.

Analog to discrete-time conversion

Step 3 of the design procedure discussed above requires a discrete-time system function to be obtained from the analog system function. The goal is to come up with a system function $H(z)$ the characteristics of which resemble those of the analog system function $G(s)$ in some way. The goal is somewhat loosely defined; there is no clear definition of when two filters resemble each other. There is also the question of which characteristics of the two filters (magnitude, phase, time-delay, impulse response, etc.) exhibit similarity. In most cases analog prototype filters are designed based on their magnitude responses. Therefore, our goal will be to obtain a reasonable degree of similarity between the magnitude characteristics of the two filters.

The main challenge is the fact that the system function $H(\Omega)$ is 2π-periodic whereas the analog prototype filter system function $G(\omega)$ is not. Consequently, an exact match between the system functions is not possible. Multiple methods exist for obtaining an approximate match. Two of these methods, namely impulse invariance and bilinear transformation, will be discussed in the remainder of this section.

Impulse invariance

A possible method of obtaining a discrete-time filter from an analog prototype is to ensure that the impulse response is preserved in the conversion process. Specifically, the discrete-time filter is chosen such that its impulse response is equal to a scaled and sampled version of the impulse response of the analog prototype:

$$h[n] = T\, g(nT) \tag{10.132}$$

The relationship in Eqn. (10.132) is the sampling relationship discussed in Chapter 6. The reason for the scale factor T will become obvious when we consider the magnitude spectrum of the resulting discrete-time filter.

Let the system function for a causal analog prototype filter be $G(s)$ which can be expanded into partial fractions in the form

$$G(s) = \sum_{i=1}^{N} \frac{k_i}{s - p_i} \tag{10.133}$$

where p_i are the poles, and k_i are the corresponding residues. The impulse response is

$$g(t) = \sum_{i=1}^{N} k_i\, e^{p_i t}\, u(t) \tag{10.134}$$

Under impulse-invariant design the impulse response of the discrete-time system is required to be

$$h[n] = T\, g(nT) = \sum_{i=1}^{N} T\, k_i\, e^{p_i nT}\, u[n] \tag{10.135}$$

The system function of the discrete-time filter is found by taking the z transform of $h[n]$ as

$$H(z) = \sum_{i=1}^{N} \frac{T\, k_i\, z}{z - e^{p_i T}} \tag{10.136}$$

Comparing Eqns. (10.133) and (10.136) we conclude that $H(z)$ can be obtained directly from the partial fraction expansion of $G(z)$ without the intermediate steps of computing $g(t)$ and sampling it.

An important issue in the conversion of an analog filter to a discrete-time filter is the requirement that a stable discrete-time filter be obtained from a stable analog filter. The poles of the analog prototype filter are at

$$p_i = \sigma_i + j\omega_i\,, \qquad i = 1, \ldots, N \tag{10.137}$$

The corresponding poles of the discrete-time filter obtained through impulse invariance are at

$$\bar{p}_i = e^{p_i T} = e^{\sigma_i T}\, e^{j\omega_i T}\,, \qquad i = 1, \ldots, N \tag{10.138}$$

For a casual and stable analog prototype filter the poles are in the left half s-plane, and therefore $\sigma_i < 0$ for $i = 1, \ldots, N$. This implies that

$$|\bar{p}_i| = e^{\sigma_i T} < 1\,, \qquad i = 1, \ldots, N \tag{10.139}$$

We are assured that a casual and stable analog prototype filter leads to a casual and stable discrete-time filter when the impulse invariance technique is used.

Finally we need to understand the relationship between the magnitude responses of the analog prototype filter and the resulting discrete-time filter. This step is easy owing to the sampling relationship between $g(t)$ and $h[n]$ given by Eqn. (10.132). Using Eqn. (6.25) from Chapter 6 we obtain

$$H(\Omega) = \sum_{k=-\infty}^{\infty} G\left(\frac{\Omega - 2\pi k}{T}\right) \tag{10.140}$$

In order to avoid aliasing, the analog prototype filter must be bandlimited such that

$$G(\omega) = 0\,, \qquad |\omega| \geq \frac{\pi}{T} \tag{10.141}$$

so that

$$H\left(\Omega\right) = G\left(\frac{\Omega}{T}\right) \qquad -\pi \le \omega < \pi \qquad (10.142)$$

and the spectrum of the resulting discrete-time filter is a frequency-scaled version of the analog prototype filter spectrum in the range $-\pi \le < \omega < \pi$. Nevertheless, the bandlimiting condition given by Eqn. (10.141) is often not satisfied by the analog prototype filter (none of the analog designs considered in Section 10.4 are perfectly bandlimited). Therefore, the spectrum of the resulting discrete-time filter is an aliased version of the spectrum of the analog prototype. This is illustrated in Fig. 10.29.

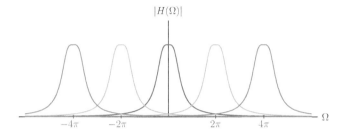

Figure 10.29 – Aliasing effect in impulse-invariant design.

Based on Eqn. (10.140) the value of the system function $H\left(\Omega\right)$ at a specific frequency $\Omega = \Omega_0$ is determined by contributions from the analog prototype system function $G\left(\omega\right)$ at frequencies

$$\omega_k = \frac{\Omega_0 - 2\pi k}{T}, \qquad \text{all } k \qquad (10.143)$$

Due to the aliasing of the spectrum, impulse-invariant designs are only useful for lowpass filters for which aliasing can be kept at acceptable levels.

Interactive Demo: `impinv_demo.m`

This demo program illustrates the mapping of the $j\omega$-axis of the s-plane to the unit circle of the z-plane for impulse invariant design. Magnitude response of an analog prototype filter with system function

$$G\left(s\right) = \frac{2}{s+2}$$

is shown along with the magnitude response of discrete-time filter derived from it using the impulse invariance technique. The sampling interval T may be specified using a slider control. For a selected value of Ω, the values of ω at which the analog prototype contributes to the discrete-time filter are shown.
Software resources:
`impinv_demo.m`

Example 10.5: **Impulse invariant design of a lowpass filter**

Consider the third-order Chebyshev type-II analog lowpass filter designed in Example 10.3. Convert this filter to a discrete-time filter using the impulse invariance technique with $T = 0.2$ s. Afterwards compute and graph the magnitude response of the discrete-time filter.

Solution: The system function of the analog prototype filter designed in Example 10.3 was

$$G\left(s\right) = \frac{0.6250}{s^3 + 1.1542\,s^2 + 1.4161\,s + 0.6250}$$

which can be written in partial fraction form as

$$G\left(s\right) = \frac{-0.2885 - j\,0.0833}{s + 0.2886 - j\,1} + \frac{-0.2885 + j\,0.0833}{s + 0.2886 + j\,1} + \frac{0.5771}{s + 0.5771}$$

Using Eqn. (10.136) the discrete-time filter system function is written in partial fraction form as

$$H\left(z\right) = \frac{\left(-0.0577 - j\,0.0167\right)z}{z - 0.9251 - j\,0.1875} + \frac{\left(-0.0577 + j\,0.0167\right)z}{z - 0.9251 + j\,0.1875} + \frac{0.1154\,z}{z - 0.8910}$$

The closed form expression for $H\left(z\right)$ is

$$H\left(z\right) = \frac{0.0023\,z^2 + 0.0021\,z}{z^3 - 2.7412\,z^2 + 2.5395\,z - 0.7939}$$

The magnitude and the phase of $H\left(z\right)$ are shown in Fig. 10.30.

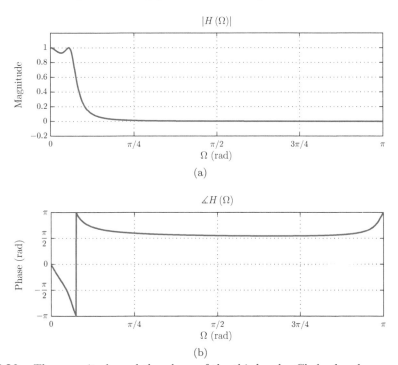

Figure 10.30 – The magnitude and the phase of the third-order Chebyshev lowpass IIR filter designed in Example 10.5.

Note that aliasing can be kept at a negligible level with the choice of T, and the analog frequency $\omega_1 = 1$ rad/s corresponds to the discrete-time frequency $\Omega = 0.2$ radians.

Software resources:

ex_10_5a.m

ex_10_5b.m

Software resources: See MATLAB Exercise 10.8.

Example 10.5 demonstrates that, if the analog prototype filter system function $G(s)$ is fixed, then the sampling interval should be chosen as small as possible to minimize the effect of aliasing. However, a typical IIR filter design problem begins with the critical frequencies Ω_1 and Ω_2 of the discrete-time filter. Let the corresponding critical frequencies of the analog prototype filter to be designed be ω_1 and ω_2 respectively. Recall from the sampling relationship that the analog frequencies and the discrete-time frequencies are related by $\omega = \Omega/T$. If the sampling interval is T, the sampling rate is $\omega_s = 2\pi/T$, and we obtain

$$\frac{\omega_1}{\omega_s} = \frac{\Omega_1}{2\pi} = \text{constant} \tag{10.144}$$

and

$$\frac{\omega_2}{\omega_s} = \frac{\Omega_2}{2\pi} = \text{constant} \tag{10.145}$$

For specified discrete-time critical frequencies, increasing the sampling rate ω_s does not improve the aliasing condition since the bandwidth of the signal $g(t)$ to be sampled also increases proportionally. Therefore the choice of T is irrelevant and we often use $T = 1$ s for simplicity. Example 10.6 will illustrate this.

Example 10.6: **Impulse invariant design specifications**

A lowpass IIR filter is to be designed with the following specifications:

$$\Omega_1 = 0.2\pi\,, \qquad \Omega_2 = 0.25\pi\,, \qquad R_p = 1 \text{ dB}\,, \qquad A_s = 30 \text{ dB}$$

Impulse invariance technique is to be used for converting an analog prototype filter to a discrete-time filter. Determine the specifications of the analog prototype if the sampling interval is to be

  a. $T = 1$ s

  b. $T = 2$ s

Solution:

  a. Using $T = 1$ s, the critical frequencies of the analog prototype filter are

$$\omega_1 = \frac{\Omega_1}{T} = \frac{\Omega_1}{1} = 0.2\pi\,, \qquad \omega_2 = \frac{\Omega_2}{T} = \frac{\Omega_2}{1} = 0.25\pi$$

The dB tolerance limits are unchanged:

$$R_p = 1 \text{ dB}\,, \qquad A_s = 30 \text{ dB}$$

Let the system function for the analog prototype filter be $G_1(\omega)$ yielding an impulse response $g_1(t)$. The impulse response of the discrete-time filter is

$$h_1[n] = g_1(nT) = g_1(n)$$

  b. With $T = 2$ s, the critical frequencies of the analog prototype filter are

$$\omega_1 = \frac{\Omega_1}{T} = \frac{\Omega_1}{2} = 0.1\pi\,, \qquad \omega_2 = \frac{\Omega_2}{T} = \frac{\Omega_2}{2} = 0.125\pi$$

The dB tolerance limits are unchanged as before. Let the system function for the analog prototype filter be $G_2(\omega)$ so that its impulse response is $g_2(t)$. The impulse response of the discrete-time filter is obtained by sampling $g_2(t)$ every 2 seconds, that is,

$$h_2[n] = g_2(nT) = g_2(2n)$$

We have thus obtained two discrete-time filters with the two choices of the sampling interval T. What is the relationship between these two filters? Let us realize that $G_1(0.2\pi) = G_2(0.1\pi)$ and $G_1(0.25\pi) = G_2(0.125\pi)$. Generalizing these relationships we have

$$G_1(\omega) = G_2\left(\frac{\omega}{2}\right)$$

Based on the scaling property of the Fourier transform (see Section 4.3.5 of Chapter 4) this implies that

$$g_1(n) = g_2(2n)$$

and therefore

$$h_1[n] = h_2[n]$$

The two IIR filters designed are identical and independent from the choice of T.

Bilinear transformation

The impulse invariance technique has a severe shortcoming that prevents its adoption as a general method applicable to the design of all filter types. The aliasing effect that results from sampling the impulse response limits its usefulness to lowpass filters only. An alternative technique known as *bilinear transformation* avoids aliasing and overcomes this limitation.

In order to develop the idea we will work with a first-order system; however, the technique developed will be applicable to higher-order systems as well. Consider a first-order analog prototype filter described by the system function

$$G(s) = \frac{Y_a(s)}{X_a(s)} = \frac{a}{s - b} \tag{10.146}$$

The differential equation for the filter is

$$\frac{dy_a(t)}{dt} = b\,y_a(t) + a\,x_a(t) \tag{10.147}$$

Let us define an intermediate variable $w_a(t)$ as

$$w_a(t) = \frac{dy_a(t)}{dt}$$

so that Eqn. (10.147) becomes

$$w_a(t) = b\,y_a(t) + a\,x_a(t) \tag{10.148}$$

Since the goal is to obtain a discrete-time filter, we will write the discrete-time version of Eqn. (10.148) as

$$w[n] = b\,y[n] + a\,x[n] \tag{10.149}$$

where we have defined $w[n] = w_a(nT)$, $y[n] = y_a(nT)$ and $x[n] = x_a(nT)$. The equivalent relationship between $y_a(t)$ and $w_a(t)$ is

$$y_a(t) = \int_{t_0}^{t} w_a(t)\, dt + y_a(t_0) \tag{10.150}$$

Letting $t_0 = (n-1)T$ and $t = nT$, a discrete-time approximation to $y_a(t)$ can be obtained using the trapezoidal approximation method (see Problem 3.35 at the end of Chapter 3):

$$y[n] = \frac{T}{2}\left(w[n] + w[n-1]\right) + y[n-1] \tag{10.151}$$

Substituting Eqn. (10.149) into Eqn. (10.151) the difference equation between $x[n]$ and $y[n]$ is found as

$$y[n] = \frac{T}{2}\left(b\,y[n] + a\,x[n] + b\,y[n-1] + a\,x[n-1]\right) + y[n-1] \tag{10.152}$$

which can be simplified to

$$\left(1 - b\frac{T}{2}\right)y[n] = \left(1 + b\frac{T}{2}\right)y[n-1] + a\frac{T}{2}x[n] + a\frac{T}{2}x[n-1] \tag{10.153}$$

The system function of the corresponding discrete-time filter is obtained from Eqn. (10.153) as

$$H(z) = \frac{a}{\dfrac{2}{T}\dfrac{1 - z^{-1}}{1 + z^{-1}} - b} \tag{10.154}$$

By comparing Eqns. (10.146) and (10.154) we conclude that the transformation between the s-plane and the z-plane is given by

$$s = \frac{2}{T}\frac{1 - z^{-1}}{1 + z^{-1}} \tag{10.155}$$

This is referred to as the *bilinear transformation*. Solving Eqn. (10.155) for z yields

$$z = \frac{2 + sT}{2 - sT} \tag{10.156}$$

as the inverse of bilinear transformation. Since the relationship between s and z is invertible, bilinear transformation represents a one-to-one mapping between the s-plane and the z-plane. For each point in the s-plane, there is one and only one point in the z-plane, and vice versa. Contrast this with the impulse invariance technique that maps multiple frequencies on the $j\omega$-axis of the s-plane to a single point on the unit circle of the z-plane, resulting in aliasing.

In order to understand how the $j\omega$-axis of the s-plane is mapped to the z-plane we will evaluate Eqn. (10.156) for $s = j\omega$:

$$z = \frac{2 + j\omega T}{2 - j\omega T} \tag{10.157}$$

The magnitude of the expression is

$$|z| = \frac{\sqrt{4 + \omega^2 T^2}}{\sqrt{4 + \omega^2 T^2}} = 1 \tag{10.158}$$

Since the magnitude of z is equal to unity for all ω, the $j\omega$-axis of the s-plane is mapped onto the unit circle of the z-plane. Furthermore, the entire $j\omega$-axis of the s-plane is mapped onto the unit circle of the z-plane only once, and there is no aliasing. Bilinear transform does not suffer from the limitations of the impulse invariance technique, and can be used for designing all four frequency-selective filter types. Setting $z = e^{j\Omega}$ in Eqn. (10.158) we have

$$ e^{j\Omega} = \frac{\sqrt{4 + \omega^2 T^2}}{\sqrt{4 + \omega^2 T^2}} = 1 \tag{10.159} $$

and

$$ \Omega = \angle\, e^{j\Omega} = 2\, \tan^{-1}\left(\frac{\omega T}{2}\right) \tag{10.160} $$

Eqn. (10.160) gives the relationship between analog frequencies ω and discrete-time frequencies Ω. It is shown graphically in Fig. 10.31.

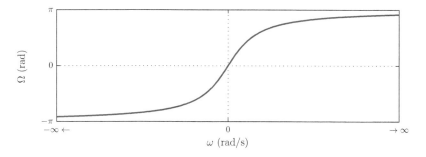

Figure 10.31 – Mapping of frequencies in bilinear transformation.

It is evident from Fig. 10.31 that the mapping of frequencies is highly nonlinear. For low frequencies the relationship between ω and Ω is almost linear, however, for high frequencies we see that significant changes in ω trigger minuscule changes in Ω. This behavior is necessary if we want to map an infinite range of analog frequencies to a 2π-radian range of discrete-time frequencies. Fig. 10.32 illustrates the mapping from the $j\omega$-axis of the s-plane and the unit circle of the z-plane.

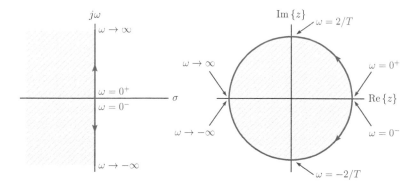

Figure 10.32 – Mapping $j\omega$-axis of the s-plane to the unit circle of the z-plane.

The $j\omega$-axis of the s-plane is mapped onto the unit circle of the z-plane. Each point on the $j\omega$-axis of the s-plane maps to one and only one point on the unit circle of the z-plane.

Furthermore, each point in the left half s-plane maps to one and only one point inside the unit circle of the z-plane. This ensures that a stable analog prototype filter leads to a stable discrete-time filter under bilinear transformation.

Interactive Demo: biln_demo.m

This demo program illustrates the mapping of the $j\omega$-axis of the s-plane to the unit circle of the z-plane for bilinear transformation. Magnitude response of an analog prototype filter

$$G(s) = \frac{2}{s+2}$$

is shown along with the magnitude response of discrete-time filter derived from it using bilinear transformation. The sampling interval T may be specified using a slider control. For a selected value of ω, the corresponding value of Ω is shown.

Software resources:
biln_demo.m

Example 10.7: **Applying bilinear transformation**

A second-order analog filter has the system function

$$G(s) = \frac{2}{(s+1)(s+2)}$$

Convert this filter to a discrete-time filter using bilinear transformation with $T = 2$ s. Afterwards compute and graph the magnitude response of the discrete-time filter.

Solution: Substituting

$$s = \frac{2}{T}\frac{1-z^{-1}}{1+z^{-1}} = \frac{1-z^{-1}}{1+z^{-1}} \tag{10.161}$$

we obtain

$$H(z) = \frac{2}{\left(\frac{1-z^{-1}}{1+z^{-1}}+1\right)\left(\frac{1-z^{-1}}{1+z^{-1}}+2\right)}$$

which can be simplified to

$$H(z) = \frac{z^2+2z+1}{z(3z+1)}$$

The magnitude and the phase of $H(z)$ are shown in Fig. 10.33.
Software resources:
ex_10_7.m

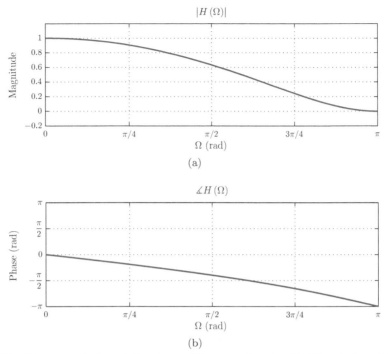

Figure 10.33 – The magnitude and the phase of the second-order IIR filter designed in Example 10.7.

Obtaining analog prototype specifications

As discussed earlier, the first step in designing IIR filters is to convert the specifications of the desired discrete-time filter to the corresponding specifications of an appropriate analog prototype filter. For lowpass and highpass filters two edge frequencies Ω_1 and Ω_2 are given along with the tolerance values Δ_1 and Δ_2, or their decibel equivalents R_p and A_s. The critical frequencies for the discrete-time IIR filter must be translated to the critical frequencies of the analog prototype based on which method is to be used in step 3 of the design process. If impulse invariance will be used in converting the analog prototype to a discrete-time filter, critical frequencies of the analog prototype are computed as

$$\omega_1 = \frac{\Omega_1}{T} \quad \text{and} \quad \omega_2 = \frac{\Omega_2}{T} \tag{10.162}$$

since the conversion process is based on sampling the impulse response of the analog prototype filter. If, on the other hand, bilinear transformation is to be used, then the distortion of the frequency axis must be taken into account. In this case the critical frequencies for the analog prototype filter are computed as

$$\omega_1 = \frac{2}{T} \tan\left(\frac{\Omega_1}{2}\right) \quad \text{and} \quad \omega_2 = \frac{2}{T} \tan\left(\frac{\Omega_2}{2}\right) \tag{10.163}$$

This is called *prewarping*. Recall that, when bilinear transformation is applied to the analog prototype filter in step 3 of the design process, the frequency axis is distorted as described by Eqn. (10.160). Prewarping counteracts this distortion at the critical frequencies.

Example 10.8: IIR filter design using bilinear transformation

Using bilinear transformation design a Butterworth lowpass filter with the following specifications:

$$\Omega_1 = 0.2\pi\,, \qquad \Omega_2 = 0.36\pi\,, \qquad R_p = 2\ \text{dB}\,, \qquad A_s = 20\ \text{dB}$$

Solution: We will use $T = 1$ s. The critical frequencies of the analog prototype filter are found using the prewarping equation applied to Ω_1 and Ω_2:

$$\omega_1 = 2\tan(0.2\pi/2) = 0.6498\ \text{rad}$$

$$\omega_2 = 2\tan(0.36\pi/2) = 1.2692\ \text{rad}$$

Next the filter order N needs to be determined using Eqn. (10.65).

$$N \geq \frac{\log_{10}\sqrt{\left(10^{2/10}-1\right)/\left(10^{20/10}-1\right)}}{\log_{10}\left(0.6498/1.2692\right)} = 3.8326$$

The filter order needs to be chosen as $N = 4$. The 3-dB cutoff frequency is found by setting the magnitude at the stopband edge to $-A_s$ dB by solving

$$\left(\frac{\omega_2}{\omega_c}\right)^{2N} = 10^{A_s/10} - 1$$

so that

$$\left(\frac{1.2692}{\omega_c}\right)^{8} = 10^{20/10} - 1 = 99$$

which yields $\omega_c = 0.7146$ rad/s. The analog prototype filter can now be designed using Butterworth lowpass filter design technique described in Section 10.4.1 with the values of N and ω_c found. The system function is

$$G(s) = \frac{0.2608}{s^4 + 1.8675\,s^3 + 1.7437\,s^2 + 0.9537\,s + 0.2608}$$

The system function for the discrete-time filter is found through bilinear transformation using the replacement

$$s \rightarrow 2\,\frac{1 - z^{-1}}{1 + z^{-1}}$$

which results in

$$H(z) = \frac{0.0065\,z^4 + 0.0260\,z^3 + 0.0390\,z^2 + 0.0260\,z + 0.0065}{z^4 - 2.2209\,z^3 + 2.0861\,z^2 - 0.9204\,z + 0.1594}$$

The dB magnitude of the resulting filter is shown in Fig. 10.34 with the tolerance limits.

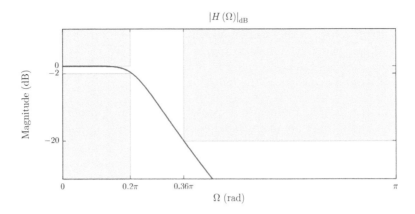

Figure 10.34 – dB Magnitude for the filter designed in Example 10.8.

Software resources:
ex_10_8a.m
ex_10_8b.m

Software resources: See MATLAB Exercises 10.9, 10.10, and 10.11.

10.5.2   Design of FIR filters

A length-N FIR filter is completely characterized by its finite-length impulse response $h[n]$ for $n = 0, \ldots, N - 1$. The system function for such a filter is computed as

$$
\begin{aligned}
H(z) &= \sum_{n=0}^{N-1} h[n]\, z^{-n} \\
&= h[0] + h[1]\, z^{-1} + \ldots + h[N-1]\, z^{-(N-1)}
\end{aligned}
\tag{10.164}
$$

The system function can also be written using non-negative powers of z in the form

$$
H(z) = \frac{h[0]\, z^{N-1} + h[1]\, z^{N-2} + \ldots + h[N-2]\, z + h[N-1]}{z^{N-1}}
\tag{10.165}
$$

which leads us to the following conclusions:

1. Length-N FIR filter has a system function that is of order $N - 1$.
2. The system function $H(z)$ has $N - 1$ zeros and as many poles.
3. The placement of zeros of the filter is determined by the sample amplitudes of the impulse response $h[n]$.
4. In contrast, all poles of the length-N FIR filter are at the origin, independent of the impulse response. Therefore, FIR filters are inherently stable.
5. The magnitude response of the filter is determined only by the locations of the zeros since the poles at the origin do not contribute to the magnitude response (see Section 8.5.4).

The sample amplitudes of the impulse response $h[n]$ for $n = 0, \ldots, N-1$ are often referred to as the *filter coefficients*. In this section we will present two approaches to the problem of designing FIR filters: one very simplistic approach that is nevertheless instructive, and one elegant approach that is based on computer optimization. In a general sense, the design procedure for FIR filters consists of the following steps:

1. Start with the desired frequency response. Select the appropriate length N of the filter. This choice may be an educated guess, or may rely on empirical formulas.
2. Choose a design method that attempts to minimize, in some way, the difference between the desired frequency response and the actual frequency response that results.
3. Determine the filter coefficients $h[n]$ using the design method chosen.
4. Compute the system function and decide if it is satisfactory. If not, repeat the process with a different value of N and/or a different design method.

It was discussed earlier in this chapter that one of the main reasons for preferring FIR filters over IIR filters is the possibility of a linear phase characteristic. Linear phase is desirable since it leads to a time-delay characteristic that is constant independent of frequency. As far as real-time implementations of IIR and FIR filters are concerned, we have seen that IIR filters are mathematically more efficient and computationally less demanding of hardware resources compared to FIR filters. If linear phase is not a significant concern in a particular application, then an IIR filter may be preferred. If linear phase is a requirement, on the other hand, an FIR filter must be chosen even though its implementation may be more costly. Therefore, in the discussion of FIR design we will focus on linear-phase FIR filters.

Conditions for linear phase

It can be shown that a length-N FIR filter with impulse response $h[n]$ has a linear phase characteristic if

$$h[n] = \mp h[N - 1 - n] \qquad \text{for } n = 0, \ldots, K \tag{10.166}$$

where the upper limit K is set as follows:

$$K = \begin{cases} N/2 - 1, & N \text{ even} \\ (N-1)/2, & N \text{ odd} \end{cases} \tag{10.167}$$

The condition given by Eqns. (10.166) and (10.167) leads to four possible scenarios:

1. N : even, and $h[n] = h[N - 1 - n]$ for $N = 0, \ldots, N/2 - 1$
2. N : even, and $h[n] = -h[N - 1 - n]$ for $N = 0, \ldots, N/2 - 1$
3. N : odd, and $h[n] = h[N - 1 - n]$ for $N = 0, \ldots, (N-1)/2$
4. N : odd, and $h[n] = -h[N - 1 - n]$ for $N = 0, \ldots, (N-1)/2$

In addition, it can be shown that the constant time delay for a length-N linear-phase FIR filter is

$$n_d = \frac{N-1}{2} \tag{10.168}$$

for all four possibilities. It is interesting to see that the time-delay n_d is not an integer when N is even.

A complete proof of linear-phase conditions must address each of the four possibilities listed above (see Problems 10.17 and 10.19).

Example 10.9: **Phase of FIR filter with symmetric impulse response**

A length-5 FIR filter has the impulse response

$$h[n] = \{ \underset{\substack{\uparrow \\ n=0}}{3} , 2, 1, 2, 3 \}$$

Show that the phase characteristic of $H(\Omega)$ is a linear function of Ω.

Solution: The DTFT of the impulse response is

$$H(\Omega) = 3 + 2\,e^{-j\Omega} + e^{-j2\Omega} + 2\,e^{-j3\Omega} + 3\,e^{-j4\Omega}$$

Let us factor out $e^{-j2\Omega}$ and write the result as

$$H(\Omega) = \left[3\,e^{j2\Omega} + 2\,e^{j\Omega} + 1 + 2\,e^{-j\Omega} + 3\,e^{-j2\Omega} \right] e^{-j2\Omega} \tag{10.169}$$

The expression in square brackets contains symmetric exponentials. Using Euler's formula, it becomes

$$\begin{aligned} A(\Omega) &= 3\,e^{j2\Omega} + 2\,e^{j\Omega} + 1 + 2\,e^{-j\Omega} + 3\,e^{-j2\Omega} \\ &= 1 + \cos(\Omega) + \frac{3}{2}\cos(2\Omega) \end{aligned} \tag{10.170}$$

which is purely real. Using Eqn. (10.170) in Eqn. (10.169) we obtain

$$H(\Omega) = A(\Omega)\, e^{-j2\Omega}$$

The phase response of the filter is

$$\Theta(\Omega) = -j2\Omega$$

corresponding to a time delay of $n_d = 2$ samples. The magnitude and the phase of the system function are shown in Fig. 10.35.

It should be noted that the phase characteristic is linear in spite of the vertical jumps in Fig. 10.35(b). These vertical sections are due to the fact that $A(\Omega)$ is negative for some frequencies. Therefore, the magnitude $|H(\Omega)|$ is

$$|H(\Omega)| = |A(\Omega)|$$

and the phase is

$$\angle H(\Omega) = \begin{cases} \Theta(\Omega) , & \text{if } A(\Omega) \geq 0 \\ \Theta(\Omega) \mp \pi , & \text{if } A(\Omega) < 0 \end{cases}$$

What matters for linear phase is that all non-vertical sections of the phase characteristic have the same slope value.

Software resources:
ex_10_9.m

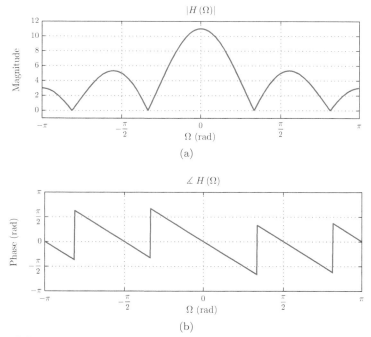

(a)

(b)

Figure 10.35 – Magnitude and the phase of the FIR filter used in Example 10.9.

We are now ready to discuss FIR filter design. First the Fourier series design method will be discussed. The use of *window functions* to overcome the fundamental problems encountered will be explored. Afterwards we will briefly discuss the Parks McClellan technique.

Fourier series design of FIR filters

An ideal lowpass discrete-time filter with a cutoff frequency Ω_c has the system function

$$H_d(\Omega) = \begin{cases} 1, & |\Omega| < \Omega_c \\ 0, & \Omega_c < |\Omega| < \pi \end{cases} \tag{10.171}$$

as shown in Fig. 10.36. The filter has zero phase and zero time delay.

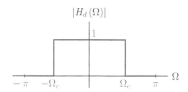

Figure 10.36 – Ideal discrete-time lowpass filter frequency response.

The impulse response of the ideal discrete-time lowpass filter is found by taking the inverse DTFT of $H_d(\Omega)$ given by Eqn. (10.171).

$$h_d[n] = \frac{1}{2\pi} \int_{-\pi}^{\pi} H_d(\Omega)\, e^{j\Omega n}\, d\Omega$$

$$= \frac{1}{2\pi} \int_{-\Omega_c}^{\Omega_c} (1)\, e^{j\Omega n}\, d\Omega = \frac{\Omega_c}{\pi}\, \text{sinc}\left(\frac{\Omega_c n}{\pi}\right) \tag{10.172}$$

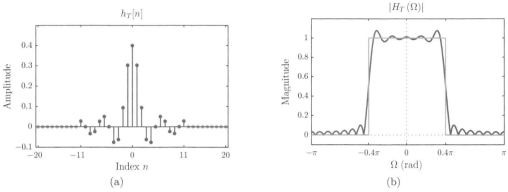

Figure 10.38 – (a) The truncated impulse response $h_d[n]$ for $M = 11$, and (b) its magnitude spectrum.

The result in Eqn. (10.172) is shown in Fig. 10.37 for $\Omega_c = 0.4\pi$.

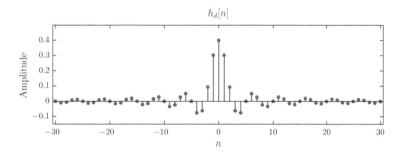

Figure 10.37 – Impulse response of the ideal lowpass filter with $\Omega_c = 0.4\pi$.

The result obtained in Eqn. (10.172) for $h_d[n]$ is valid for all N. Therefore, $h_d[n]$ infinitely long, and cannot be the impulse response of an FIR filter. On the other hand, sample amplitudes of $h_d[n]$ get smaller as the sample index is increased in both directions. A finite-length impulse response can be obtained by truncating $h_d[n]$ as follows:

$$h_T[n] = \begin{cases} h_d[n], & -M \geq n \geq M \\ 0, & \text{otherwise} \end{cases} \tag{10.173}$$

The truncated impulse response has $2M+1$ samples for $n = -M, \ldots, M$. Truncation of the ideal impulse response causes the spectrum of the filter to deviate from the ideal spectrum. The system function for the resulting filter is

$$H_T(\Omega) - \int_{n=-\infty}^{\infty} h_T[n]\, c^{-j\Omega n} = \int_{n=-M}^{M} h_d[n]\, e^{-j\Omega n} \tag{10.174}$$

An example is shown in Fig. 10.38. The truncated impulse response with $M = 11$ is shown in Fig. 10.38(a), and the magnitude spectrum of the resulting length-23 filter is shown in Fig. 10.38(b).

The truncated impulse response $h_T[n]$ is still non-causal owing to the fact that the filter has zero phase. In order to obtain a causal filter, a delay of M samples must be incorporated into the impulse response, resulting in

$$h[n] = h_T[n - M], \qquad n = 0, \ldots, 2M \tag{10.175}$$

Thus, $h[n]$ corresponds to a causal FIR filter. Using the time shifting property of the DTFT (see Section 5.3.5 of Chapter 5), the system function $H(\Omega)$ of the causal filter is related to $H_T(\Omega)$ by

$$H(\Omega) = H_T(\Omega)\,e^{-j\Omega M} \qquad (10.176)$$

The addition of the M-sample delay only affects the phase of the system function and not its magnitude. Since $h[n]$ is both causal and finite-length, it is the impulse response of a valid FIR filter.

It is worth observing that the impulse response $h_d[n]$ given by Eqn. (10.172) is an even function of n. Even symmetry is preserved in $h_T[n]$ obtained in Eqn. (10.173) by truncating $h_d[n]$. Finally, when $h[n]$ is obtained by time shifting $h_T[n]$ by M samples, the symmetry necessary for linear phase is preserved. Therefore, any filter designed using this technique will have linear phase.

Example 10.10: **Fourier series design**

Using the Fourier series method, design a length-15 FIR lowpass filter to approximate an ideal lowpass filter with $\Omega_c = 0.3\pi$ rad.

Solution: The impulse response of the ideal lowpass filter is

$$h_d[n] = \frac{0.3\pi}{\pi}\ \text{sinc}\left(\frac{0.3\pi n}{\pi}\right) = 0.3\ \text{sinc}\,(0.3n)$$

Since $N = 2M + 1 = 15$ we have $M = 7$. The truncated impulse response is

$$h_T[n] = \begin{cases} 0.3\ \text{sinc}\,(0.3n)\,, & -7 \le n \le 7 \\ 0\,, & \text{otherwise} \end{cases}$$

The impulse response of the FIR filter is

$$h[n] = h_T[n-7] = 0.3\ \text{sinc}\,(0.3\,(n-7))\,, \qquad n = 0,\dots,14$$

which can be evaluated as

$$h[n] = \{\,0.014,\ -0.031,\ -0.064,\ -0.047,\ 0.033,\ 0.151,\ 0.258,\ 0.3,\ 0.258,\ 0.151,$$
$$\underset{n=0}{\uparrow}$$

$$0.032,\ -0.047,\ -0.064,\ -0.031,\ 0.014\,\}$$

The magnitude response of the designed filter is shown in Fig. 10.39.

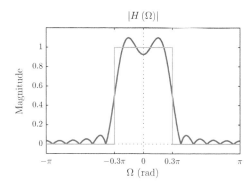

Figure 10.39 – Magnitude response of the FIR filter designed in Example 10.10.

Software resources:
ex_10_10a.m
ex_10_10b.m

A by-product of truncating the impulse response in Eqn. (10.173) is the oscillatory behavior of the frequency response that is particularly evident around the cutoff frequency Ω_c. This effect was observed as the *Gibbs phenomenon* in earlier chapters. An alternative way of expressing the truncation of $h_d[n]$ given by Eqn. (10.173) is

$$h_T[n] = h_d[n]\, w[n] \qquad (10.177)$$

where $w[n]$ is a *rectangular window* function defined as

$$w[n] = \begin{cases} 1\,, & -M \le n \le M \\ 0\,, & \text{otherwise} \end{cases} \qquad (10.178)$$

Based on the multiplication property of the DTFT (see Section 5.3.5 of Chapter 5), the spectrum of $h_T[n]$ is the convolution of the spectra of $h_d[n]$ and $w[n]$.

$$H_T\left(\Omega\right) = \frac{1}{2\pi} \int_{-\pi}^{\pi} H_d\left(\lambda\right) W\left(\Omega - \lambda\right) d\lambda \qquad (10.179)$$

which is illustrated in Fig. 10.40.

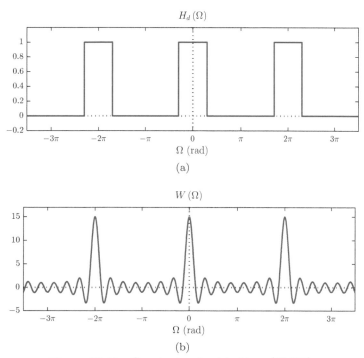

$$H_d\left(\Omega\right)$$

(a)

$$W\left(\Omega\right)$$

(b)

Figure 10.40 – Spectra involved in Eqn. (10.179).

As the spectra in Fig. 10.40 reveal, the reason behind the oscillatory behavior of $H_T\left(\Omega\right)$ especially around $\Omega = \Omega_c$ is the shape of the spectrum $W\left(\Omega\right)$. The high-frequency content of $W\left(\Omega\right)$, on the other hand, is mainly due to the abrupt transition of the rectangular window from unit amplitude to zero amplitude at its two edges. The solution is to use an alternative window function in Eqn. (10.177), one that smoothly tapers down to zero at its edges, in place of the rectangular window. The chosen window function must also be an even function of n in order to keep the symmetry of $h_T[n]$ for linear phase. A large number of window functions exist in the literature. A few of them are listed below:

Triangular (Bartlett) window:

$$w[n] = 1 - \frac{|n|}{M} , \qquad -M \leq n \leq M \tag{10.180}$$

Hamming window:

$$w[n] = 0.54 - 0.46 \cos\left(\frac{\pi\,(n+M)}{M}\right) , \qquad -M \leq n \leq M \tag{10.181}$$

Hanning window:

$$w[n] = 0.5 - 0.5 \cos\left(\frac{\pi\,(n+M)}{M}\right) , \qquad -M \leq n \leq M \tag{10.182}$$

Blackman window:

$$w[n] = 0.42 - 0.5 \cos\left(\frac{\pi\,(n+M)}{M}\right) + 0.08 \cos\left(\frac{2\pi\,(n+M)}{M}\right) , \qquad -M \leq n \leq M \tag{10.183}$$

The four window functions listed are shown in Fig. 10.41 for $M = 12$.

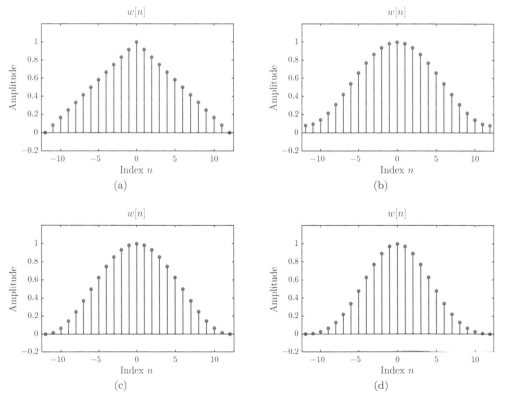

Figure 10.41 – Window functions: (a) triangular (Bartlett) window, (b) Hamming window, (c) Hanning window, (d) Blackman window.

Fig 10.42 shows the decibel magnitude spectra for rectangular, Hamming and Blackman windows for comparison.

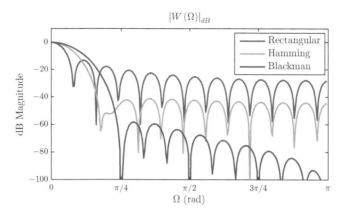

$$|W\left(\Omega\right)|_{dB}$$

Figure 10.42 – Decibel magnitude responses of various window functions.

It is seen that the rectangular window has stronger side lobes than the other two window functions. Hamming window provides side lobes that are at least 40 dB below the main lobe, and the side lobes for the Blackman window are at least 60 dB below the main lobe. The downside to side lobe suppression is the widening of the main lobe.

We are now ready to summarize the Fourier series design method for linear-phase FIR filters:

1. Determine the ideal impulse response $h_d[n]$ as the inverse DTFT of the ideal spectrum being approximated.
2. Multiply the ideal impulse response with the selected length-$(2M+1)$ window function to obtain a finite-length impulse response $h_T[n] = h_d[n]\,w[n]$.
3. Obtain the causal FIR impulse response by time shifting $h_T[n]$, that is, $h[n] = h_T[n - M]$.

Example 10.11: **Fourier series design using window functions**

Redesign the filter of Example 10.10 using Hamming and Blackman windows.

Solution: Using a window function the truncated impulse response is

$$h_T[n] = 0.3\,\text{sinc}\left(0.3n\right)w[n]\,, \qquad -7 \leq n \leq 7$$

and the impulse response of the FIR filter is

$$h[n] = h_T[n-7] = 0.3\,\text{sinc}\left(0.3\left(n-7\right)\right)w[n-7]\,, \qquad n = 0,\dots,14 \qquad (10.184)$$

For the Hamming window, Eqn. (10.184) yields

$$h[n] = \{\,0.006,\ -0.017,\ -0.042,\ -0.036,\ 0.028,\ 0.142,\ 0.254,\ 0.3,\ 0.254,\ 0.142,$$
$$\underset{\underset{n=0}{\uparrow}}{}$$
$$0.028,\ -0.036,\ -0.042,\ -0.017,\ 0.006\,\}$$

When the Blackman window is used, we get

$$h[n] = \{\, 0.003, \, -0.011, \, -0.031, \, -0.030, \, 0.025, \, 0.135, \, 0.250, \, 0.3, \, 0.250, \, 0.135,$$

$$\underset{\substack{\uparrow \\ n=0}}{}$$

$$0.025, \, -0.030, \, -0.031, \, -0.011, \, 0.003 \,\}$$

Magnitude responses of the two filters are shown in Fig. 10.43.

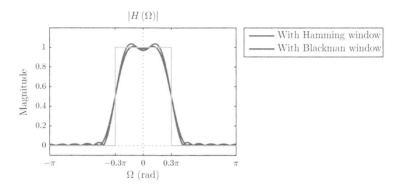

Figure 10.43 – Magnitude responses of the FIR filters designed in Example 10.10.

Software resources:
ex_10_11a.m
ex_10_11b.m

Filter types other than lowpass

In the discussion above we have concentrated on the design of lowpass FIR filters. Other types of filters can also be designed; the only modification that is needed is in determining the ideal impulse response $h_d[n]$. Expressions for the ideal impulse responses of highpass, bandpass and band-reject filters can be derived from Eqn. (10.172) as follows:

Highpass:

$$H_d\left(\Omega\right) = \begin{cases} 1\,, & \Omega_c < |\Omega| < \pi \\ 0\,, & |\Omega| < \Omega_c \end{cases} \qquad h_d[n] = \delta[n] - \frac{\Omega_c}{\pi}\,\text{sinc}\left(\frac{\Omega_c n}{\pi}\right)$$

Bandpass:

$$H_d\left(\Omega\right) = \begin{cases} 1\,, & \Omega_1 < |\Omega| < \Omega_2 \\ 0\,, & |\Omega| < \Omega_1 \\ & \text{or } \Omega_2 < |\Omega| < \pi \end{cases} \qquad h_d[n] = \frac{\Omega_2}{\pi}\,\text{sinc}\left(\frac{\Omega_2 n}{\pi}\right) - \frac{\Omega_1}{n}\,\text{sinc}\left(\frac{\Omega_1 n}{\pi}\right)$$

Band-reject:

$$H_d\left(\Omega\right) = \begin{cases} 1\,, & |\Omega| < \Omega_1 \\ & \text{or } \Omega_2 < |\Omega| < \pi \\ 0\,, & \Omega_1 < |\Omega| < \Omega_2 \end{cases} \qquad h_d[n] = \delta[n] - \frac{\Omega_2}{\pi}\,\text{sinc}\left(\frac{\Omega_2 n}{\pi}\right) + \frac{\Omega_1}{\pi}\,\text{sinc}\left(\frac{\Omega_1 n}{\pi}\right)$$

Software resources: See MATLAB Exercise 10.12.

Parks-McClellan technique for FIR filter design

FIR filter design technique due to Parks and McClellan is also referred to as the *Remez exchange method*. It is based on the idea of finding a polynomial approximation to the desired frequency response by minimizing the largest approximation error.

The Fourier series design method discussed in the previous section, when used with a rectangular window, provides the optimum solution in the sense of mean-squared error. If the error between the desired frequency response and the actual frequency response is $E(\Omega)$, then the length-N FIR filter designed using the Fourier series method leads to the mean-squared error

$$\mathrm{MSE} = \frac{1}{2\pi} \int_{-\pi}^{\pi} |E(\Omega)|^2 \, d\Omega \qquad (10.185)$$

that is the smallest possible with N coefficients. On the other hand, we have seen that the magnitude response of the filter suffers from oscillatory behavior near discontinuities of the spectrum.

Better approximations can be obtained if we minimize the maximum value of the error rather than its mean-squared value. Parks-McClellan algorithm begins with an initial guess of the set of frequencies for the extrema of $E(\Omega)$. The locations of the extrema are then iteratively shifted until the maximum deviation from the ideal frequency response is made as small as possible.

Computational details of the Parks-McClellan algorithm are outside the scope of this text and may be found in references. We will provide examples of using MATLAB for designing Parks-McClellan filters in the MATLAB Exercises section.

Software resources: See MATLAB Exercise 10.13.

10.6   Further Reading

[1] A.V. Oppenheim and R.W. Schafer. *Discrete-Time Signal Processing*. Prentice Hall, 2010.

[2] R. Schaumann and M.E. Van Valkenburg. *Design of Analog Filters*. Oxford University Press, 2010.

[3] R.A. Schilling, R.J. Schilling, and P.D. Sandra L. Harris. *Fundamentals of Digital Signal Processing Using MATLAB*. Cengage Learning, 2010.

[4] L. Tan and J. Jiang. *Digital Signal Processing: Fundamentals and Applications*. Elsevier Science, 2013.

[5] F. Taylor. *Digital Filters: Principles and Applications with MATLAB*. IEEE Series on Digital and Mobile Communication. Wiley, 2011.

[6] L.D. Thede. *Practical Analog and Digital Filter Design*. Artech House, 2005.

[7] L. Wanhammar. *Analog Filters Using MATLAB*. Springer London, Limited, 2009.

MATLAB Exercises

MATLAB Exercise 10.1: Butterworth analog filter design

MATLAB function `butter(..)` may be used for designing analog Butterworth filters. For example, a fourth-order Butterworth lowpass filter with a 3-dB cutoff frequency of $\omega_c = 3$ rad/s is designed as follows:

```
>> [num,den] = butter(4,3,'s')

num =
          0          0          0          0    81.0000

den =
     1.0000     7.8394    30.7279    70.5544    81.0000
```

The third argument 's' is needed for specifying that an *analog* filter is desired since the function `butter(..)` can be used for designing both analog and discrete-time Butterworth filters. The vectors "**num**" and "**den**" returned by the function indicate an s-domain system function

$$H\left(s\right) = \frac{81}{s^4 + 7.8394\,s^3 + 30.7279\,s^2 + 70.5544\,s + 81}$$

for the designed filter. Magnitude and phase characteristics of the designed filter may be graphed with the following statements:

```
>>   omg = [-10:0.01:10];
>>   H = freqs(num,den,omg);
>>   plot(omg,abs(H));
>>   plot(omg,angle(H));
```

Software resources:
`matex_10_1a.m`
`matex_10_1b.m`

MATLAB Exercise 10.2: Chebyshev polynomials

In this exercise we will develop and test a function for computing the coefficients of the Chebyshev polynomial of a specified order N. The N-th order Chebyshev polynomial $C_N\left(\nu\right)$ can be computed from the order of the two lower order Chebyshev polynomials through the use of the recursive formula given by Eqn. (10.70) and repeated below.

$$\text{Eqn. (10.70):} \qquad C_{N+1}\left(\nu\right) = 2\nu\,C_N\left(\nu\right) - C_{N-1}\left(\nu\right)$$

The function `ss_chebpol(..)` listed below.

```
1    function [c] = ss_chebpol(N)
2      if (N==0),
3        c = [1];
4      elseif (N==1),
5        c = [1,0];
6      else
7        cnm1 = [1];
8        cn = [1,0];
9        for i = 2:N,
10         c = 2*conv([1,0],cn)-[0,0,cnm1];
11         cnm1 = cn;
```

```
12            cn = c;
13         end;
14      end;
```

The loop between lines 9 and 13 implements the recursive formula. The script listed below uses the function `ss_chebpol(..)` along with the function `polyval(..)` to evaluate and graph the Chebyshev polynomials for $N = 2, 4, 6$.

```
1    % Script: matex_10_2.m
2    %
3    nu = [0:0.005:1.2];
4    c2 = ss_chebpol(2);
5    p2 = polyval(c2,nu);
6    c4 = ss_chebpol(4);
7    p4 = polyval(c4,nu);
8    c6 = ss_chebpol(6);
9    p6 = polyval(c6,nu);
10   plot(nu,p2,nu,p4,nu,p6); grid;
11   axis([0,1.2,-2,12]);
12   legend('N=2','N=4','N=6','Location','NorthWest');
13   title('Chebyshev polynomials for N=2,4,6');
14   xlabel('\nu');
15   ylabel('C_N(\nu)');
```

The graph produced is shown in Fig. 10.44.

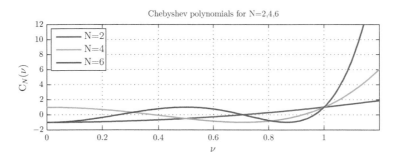

Figure 10.44 – MATLAB graph of Chebyshev polynomials for $N = 2, 4, 6$.

Software resources:
matex_10_2a.m
matex_10_2b.m

MATLAB Exercise 10.3: Chebyshev type-I analog filter design

In Example 10.3 a Chebyshev type-I lowpass filter with $N = 3$ and $\varepsilon = 0.4$ was designed. In this exercise we will verify the results using function `cheby1(..)`. The second argument for the function `cheby1(..)` is the decibel passband tolerance R_p which is computed as

$$R_p = 10 \log_{10} \left(1 + \varepsilon^2\right)$$

The desired filter can be obtained with the following statements:

```
>>   N = 3;
>>   omg1 = 1;
>>   epsilon = 0.4;
```

```
>>   Rp = 10*log10(1+epsilon^2);
>>   [num,den] = cheby1(N,Rp,omg1,'s')

num =
          0            0            0        0.6250

den =
     1.0000       1.1542       1.4161       0.6250
```

Again the third argument 's' is needed for specifying that an *analog* filter is desired since the function cheby1(..) can be used for designing both analog and discrete-time Chebyshev type-I filters. The vectors "**num**" and "**den**" returned by the function indicate an *s*-domain system function

$$H\left(s\right)=\frac{0.6250}{s^3+1.1542\,s^2+1.4161\,s+0.6250}$$

for the designed filter. Magnitude and phase characteristics of the designed filter may be graphed with the following statements:

```
>>   omg = [-4:0.01:4];
>>   H = freqs(num,den,omg);
>>   plot(omg,abs(H));
>>   plot(omg,angle(H));
```

Software resources:
matex_10_3a.m
matex_10_3b.m

MATLAB Exercise 10.4: Determining Chebyshev analog filter parameters

Design a Chebyshev type-I lowpass filter with the set of specifications

$$\omega_1=3\ \text{rad/s}\,,\qquad \omega_2=4\ \text{rad/s}\,,\qquad R_p=1\ \text{dB}\,,\qquad A_s=30\ \text{dB}$$

Solution: The design can be carried out using MATLAB functions cheb1ord(..) and cheby1(..) as follows:

```
>>   omg1 = 3;
>>   omg2 = 4;
>>   Rp = 1;
>>   As = 30;
>>   N = cheb1ord(omg1,omg2,Rp,As,'s')

N =
      7

[num,den] = cheby1(N,Rp,omg1,'s')

num =
       0       0       0       0       0       0       0    67.156


den =
    1.000    2.769   19.585   38.577   109.961   133.315   155.766   67.156
```

The results correspond to a seventh-order system function

$$H\left(s\right)=\frac{67.156}{s^7+2.769\,s^6+19.585\,s^5+38.577\,s^4+109.961\,s^3+133.315\,s^2+155.766\,s+67.156}$$

Magnitude characteristics of the designed filter may be graphed with the following statements:

```
>>   omg = [-10:0.01:10];
>>   H = freqs(num,den,omg);
>>   plot(omg,abs(H)); grid;
>>   axis([-10,10,-0.1,1.1]);
>>   title('|H(\omega)|');
>>   xlabel('\omega (rad/s)');
>>   ylabel('Magnitude');
```

The graph produced is shown in Fig. 10.45.

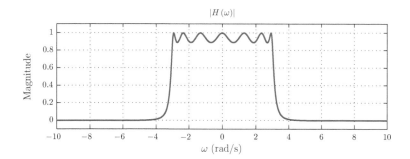

Figure 10.45 – Magnitude response of the filter designed in MATLAB Exercise 10.4.

Software resources:
`matex_10_4a.m`
`matex_10_4b.m`

MATLAB Exercise 10.5: **Chebyshev type-II analog filter design**

Design a Chebyshev type-II lowpass filter with the set of specifications

$$\omega_1 = 3 \text{ rad/s}, \qquad \omega_2 = 4 \text{ rad/s}, \qquad R_p = 1 \text{ dB}, \qquad A_s = 30 \text{ dB}$$

Solution: The design can be carried out using MATLAB functions `cheb2ord(..)` and `cheby2(..)` as follows:

```
>>   omg1 = 3;
>>   omg2 = 4;
>>   Rp = 1;
>>   As = 30;
>>   N = cheb2ord(omg1,omg2,Rp,As,'s');
>>   [num,den] = cheby2(N,As,omg2,'s');
```

The results correspond to a seventh-order system function

$$H(s) = \frac{0.89\,s^6 + 113.39\,s^4 + 3628.57\,s^2 + 33175.48}{s^7 + 18.09\,s^6 + 163.21\,s^5 + 959.29\,s^4 + 3933.07\,s^3 + 11877.01\,s^2 + 23394.27\,s + 33175.48}$$

Magnitude characteristics of the designed filter may be graphed with the following statements:

```
>>    omg = [-10:0.01:10];
>>    H = freqs(num,den,omg);
>>    plot(omg,abs(H)); grid;
>>    axis([-10,10,-0.1,1.1]);
>>    title('|H(\omega)|');
>>    xlabel('\omega (rad/s)');
>>    ylabel('Magnitude');
```

The graph produced is shown in Fig. 10.46.

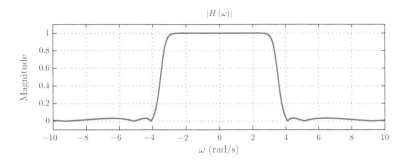

Figure 10.46 – Magnitude response of the filter designed in MATLAB Exercise 10.5.

Software resources:
matex_10_5a.m
matex_10_5b.m

MATLAB Exercise 10.6: Lowpass to highpass filter transformation

Recall that a Chebyshev type-I analog lowpass filter was designed in MATLAB Exercise 10.3. Using it as the starting point, obtain a highpass filter with a passband edge at $\omega_{H2} = 3$ rad/s.

Solution: The function lp2hp(..) can be used for this purpose. The filter of MATLAB Exercise 10.3 is redesigned and then converted to a highpass filter with the following set of statements:

```
>>    Rp = 10*log10(1+0.4^2);
>>    [num,den] = cheby1(3,Rp,1,'s');
>>    [numh,denh] = lp2hp(num,den,3)

numh =
      1.0000    -0.0000    -0.0000     0.0000

denh =
      1.0000     6.7971    16.6201    43.2000
```

Magnitude and phase characteristics of the highpass filter may be graphed with the following statements:

```
>>    omg = [-7:0.01:7];
>>    H = freqs(numh,denh,omg);
>>    plot(omg,abs(H)); grid;
>>    plot(omg,angle(H)); grid;
```

Software resources:
matex_10_6a.m
matex_10_6b.m

MATLAB Exercise 10.7: Lowpass to bandpass and lowpass to band-reject transformations

Using the Chebyshev filter designed in MATLAB Exercise 10.3 as a basis, design the following two filters:

a. A bandpass filter with the critical frequencies

$$\omega_{B2} = 1.5 \text{ rad/s} \qquad \omega_{B3} = 2.5 \text{ rad/s} \tag{10.186}$$

b. A band-reject filter with the critical frequencies

$$\omega_{B1} = 1 \text{ rad/s} \qquad \omega_{B2} = 3.5 \text{ rad/s} \tag{10.187}$$

Solution:

a. The statements listed below repeat the lowpass filter design of MATLAB Exercise 10.3 and then convert it to a bandpass filter through the use of the function `lp2bp(..)`. The magnitude response of the resulting bandpass filter is graphed.

```
>> omgL1 = 1;
>> epsilon = 0.4;
>> Rp = 10*log10(1+epsilon^2);
>> [num,den] = cheby1(3,Rp,omgL1,'s');
>> omgB2 = 1.5;
>> omgB3 = 2.5;
>> omg0 = sqrt(omgB2*omgB3);
>> B = (omgB3-omgB2)/omgL1;
>> [numb,denb] = lp2bp(num,den,omg0,B)

numb =
    0.6250    0.0000   -0.0000   -0.0000

denb =
    1.0000    1.1542   12.6661    9.2813   47.4977   16.2305   52.7344

>> omg = [-10:0.01:10];
>> H = freqs(numb,denb,omg);
>> plot(omg,abs(H)); grid;
>> title('|H(\omega)|');
>> xlabel('\omega  (rad/s)');
>> ylabel('Magnitude');
```

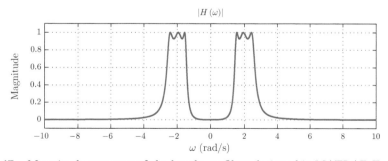

Figure 10.47 – Magnitude response of the bandpass filter designed in MATLAB Exercise 10.7.

b. The statements listed below repeat the lowpass filter design of MATLAB Exercise 10.3 and then convert it to a band-reject filter through the use of the function `lp2bs(..)` and graph the magnitude response of the resulting band-reject filter.

```
>>  omgL1 = 1;
>>  epsilon = 0.4;
>>  Rp = 10*log10(1+epsilon^2);
>>  [num,den] = cheby1(3,Rp,omgL1,'s');
>>  omgS1 = 1;
>>  omgS4 = 3.5;
>>  omg0 = sqrt(omgS1*omgS4);
>>  B = (omgS4-omgS1)*omgL1;
>>  [nums,dens] = lp2bs(num,den,omg0,B)

nums =
     1.0000    -0.0000    10.5000    -0.0000    36.7500    -0.0000    42.8750

dens =
     1.0000    16.9927   114.3754   793.9487   400.3140   208.1602    42.8750

>>  omg = [-10:0.01:10];
>>  H = freqs(nums,dens,omg);
>>  plot(omg,abs(H)); grid;
>>  title('|H(\omega)|');
>>  xlabel('\omega  (rad/s)');
>>  ylabel('Magnitude');
```

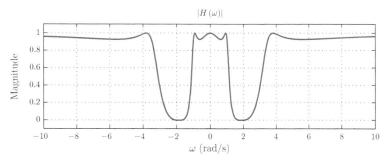

Figure 10.48 – Magnitude response of the band-reject filter designed in MATLAB Exercise 10.7.

Software resources:
matex_10_7a.m
matex_10_7b.m

MATLAB Exercise 10.8: Impulse-invariant design

In Example 10.5 impulse invariance technique was used for obtaining a discrete-time filter from a Chebyshev type-I analog prototype with system function

$$G\left(s\right) = \frac{0.6250}{s^3 + 1.1542\,s^2 + 1.4161\,s + 0.6250}$$

MATLAB function impinvar(..) can be used to accomplish this task as follows:

```
>>  T = 0.2;
>>  num = [0.6250],
>>  den = [1,1.1542,1.4161,0.6250];
>>  [numz,denz] = impinvar(num,den,1/T)

numz =

   -0.0000     0.0023     0.0021
```

```
denz =

   1.0000    -2.7412    2.5395    -0.7939
```

The results obtained correspond to the z-domain system function

$$H(z) = \frac{0.0023\, z^2 + 0.0021\, z}{z^3 - 2.7412\, z^2 + 2.5395\, z - 0.7939}$$

which matches the answer found in Example 10.5. The magnitude and the phase of the system function may be graphed with the following statements:

```
>>   Omg = [0:0.01:1]*pi;
>>   H = freqz(numz,denz,Omg);
>>   plot(Omg,abs(H));
>>   plot(Omg,angle(H));
```

Software resources:
matex_10_8a.m
matex_10_8b.m

MATLAB Exercise 10.9: IIR filter design using bilinear transformation

Consider again the IIR lowpass filter design problem solved in Example 10.8. In this exercise we will rely on MATLAB to solve it. Recall that the specifications for the filter to be designed were given as follows:

$$\Omega_1 = 0.2\pi\,, \qquad \Omega_2 = 0.36\pi\,, \qquad R_p = 2 \text{ dB}\,, \qquad A_s = 20 \text{ dB}$$

Even though the design problem can be solved quickly with two function calls (see MATLAB Exercise 10.10 for this alternative approach), we will opt to solve it using the three steps in Section 10.5.1. This will allow checking the intermediate results obtained in Example 10.8. The first step is to find the specifications for the analog prototype filter. Critical frequencies ω_1 and ω_2 for the analog prototype are found using the prewarping formula. Afterwards N and ω_c are determined using the function buttord(..).

```
>>   T = 1;
>>   Rp = 2;
>>   As = 20;
>>   omg1 = 2/T*tan(0.2*pi/2);
>>   omg2 = 2/T*tan(0.36*pi/2);
>>   [N,omgc] = buttord(omg1,omg2,Rp,As,'s')

N =
     4

omgc =
     0.7146
```

The last argument of the function buttord(..) signifies that we seek the order and cutoff frequency for an *analog* Butterworth filter. Having determined N and ω_c, the analog prototype filter can be designed using the function butter(..).

```
>>   [num,den] = butter(N,omgc,'s')

num =
          0          0          0          0     0.2608

den =
     1.0000     1.8675     1.7437     0.9537     0.2608
```

Again the last argument specifies that we are designing an analog filter. The vectors "num" and "den" correspond to the analog prototype system function

$$G(s) = \frac{0.2608}{s^4 + 1.8675\,s^3 + 1.7437\,s^2 + 0.9537\,s + 0.2608}$$

If desired, magnitude and phase characteristics of the analog prototype filter may be graphed with the following lines:

```
>>   omg = [0:0.01:5];
>>   G = freqs(num,den,omg);
>>   plot(omg,abs(G)); grid;
>>   plot(omg,angle(G)); grid;
```

In step 3 the analog prototype filter is converted to a discrete-time filter using bilinear transformation. MATLAB function bilinear(..) may be used for this purpose.

```
>>   [numz,denz] = bilinear(num,den,1/T)

numz =
     0.0065     0.0260     0.0390     0.0260     0.0065

denz =
     1.0000    -2.2209     2.0861    -0.9204     0.1594
```

The system function for the discrete-time system is

$$H(z) = \frac{0.0065\,z^4 + 0.0260\,z^3 + 0.0390\,z^2 + 0.0260\,z + 0.0065}{z^4 - 2.2209\,z^3 + 2.0861\,z^2 - 0.9204\,z + 0.1594}$$

If desired, magnitude and phase characteristics of the discrete-time filter may be graphed with the following lines:

```
>>   Omg = [0:0.01:1]*pi;
>>   H = freqz(numz,denz,Omg);
>>   plot(Omg,abs(H));
>>   plot(Omg,angle(H));
```

Software resources:
matex 10_9.m

MATLAB Exercise 10.10: IIR filter design using bilinear transformation revisited

The Butterworth IIR filter of MATLAB Exercise 10.9 can be designed quickly by using the discrete-time filter design capabilities of the functions buttord(..) and butter(..) as follows:

```
>>  T = 1;
>>  Omg1 = 0.2*pi;
>>  Omg2 = 0.36*pi;
>>  Rp = 2;
>>  As = 20;
>>  [N,Omgc] = buttord(Omg1/pi,Omg2/pi,Rp,As);
>>  [numz,denz] = butter(N,Omgc)

numz =
    0.0065    0.0260    0.0390    0.0260    0.0065

denz =
    1.0000   -2.2209    2.0861   -0.9204    0.1594
```

Note that we leave out the last argument 's' to obtain discrete-time solutions directly. The result obtained is identical to that of MATLAB Exercise 10.9.

Software resources:

matex_10_10.m

MATLAB Exercise 10.11: **A complete IIR filter design example**

A Chebyshev type-I IIR bandpass filter is to be designed with the following set of specifications:

$$\Omega_1 = 0.2\pi \, , \qquad \Omega_2 = 0.26\pi \, , \qquad \Omega_3 = 0.48\pi \, , \qquad \Omega_4 = 0.54\pi$$
$$R_p = 1 \text{ dB} \, , \qquad A_s = \quad 25 \text{ dB}$$

The specification diagram is shown in Fig. 10.49.

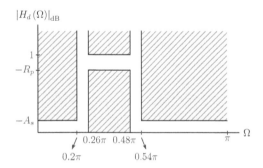

Figure 10.49 – Specification diagram for MATLAB Exercise 10.11.

Solution: The normalized edge frequencies are

$$6F_1 = \frac{\Omega_1}{2\pi} = 0.1; , \qquad F_2 = \frac{\Omega_2}{2\pi} = 0.13; , \qquad F_3 = \frac{\Omega_3}{2\pi} = 0.24; , \qquad F_4 = \frac{\Omega_4}{2\pi} = 0.27$$

Our first step will be to use the function cheb1ord(..) to determine the minimum filter order that allows the requirements to be satisfied. In this case the cheb1ord(..) will be used without the 's' parameter (see MATLAB Exercise 10.4) since we want to compute the filter order from IIR digital filter specifications and not from those of the analog prototype. The following code computes N:

```
>>  Rp = 1;
>>  As = 25;
```

```
>>  F1 = 0.1;
>>  F2 = 0.13;
>>  F3 = 0.24;
>>  F4 = 0.27;
>>  N = cheb1ord([2*F2,2*F3],[2*F1,2*F4],Rp,As)

N =
    5
```

The result indicates that the analog prototype is of the fifth order. Consequently, the bandpass IIR filter will be of order 10.

It is important to pay attention to the way the function cheb1ord(..) is used. The first vector in the argument list contains the two passband edge frequencies *normalized in MATLAB sense*, and the second vector in the argument list contains the two stopband edge frequencies also *normalized in MATLAB sense*. Unfortunately MATLAB defines normalized frequencies differently from the common practice. In our conventions the angular frequency $\Omega = \pi$ radians corresponds to the normalized frequency $F = 0.5$. In MATLAB conventions it corresponds to $\bar{F} = 1$. Therefore, in specifying a discrete-time filter to MATLAB through normalized frequencies we always use twice the conventional normalized frequency values.

Once N is found, the filter is designed using the cheby1(..). Its magnitude spectrum is computed using the function freqz(..).

```
>>  [numz,denz] = cheby1(N,Rp,[2*F2,2*F3]);
>>  Omg = [-1:0.001:1]*pi;
>>  H = freqz(numz,denz,Omg);
>>  plot(Omg,abs(H)); grid;
>>  axis([-pi,pi,-0.1,1.1]);
>>  title('|H(\Omega)|');
>>  xlabel('\Omega (rad)');
>>  ylabel('Magnitude');
```

The magnitude response of the filter obtained is shown in Fig. 10.50.

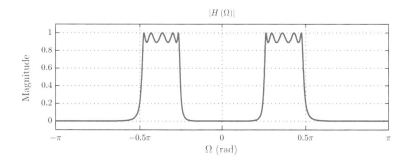

Figure 10.50 – Magnitude response of the filter designed in MATLAB Exercise 10.11.

Software resources:
matex_10_11a.m
matex_10_11b.m

MATLAB Exercise 10.12: **FIR filter design using Fourier series method**

Using the Fourier series design method with a triangular (Bartlett) window, design a 24th-order FIR bandpass filter with passband edge frequencies $\Omega_1 = 0.4\pi$ and $\Omega_2 = 0.7\pi$ as shown in Fig. 10.51.

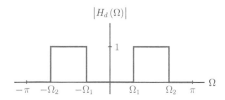

Figure 10.51 – Desired FIR magnitude response for MATLAB Exercise 10.12.

Solution: The order of the FIR filter is $N - 1 = 24$. Normalized passband edge frequencies are

$$6F_1 = \frac{\Omega_1}{2\pi} = 0.2 \qquad \text{and} \qquad F_2 = \frac{\Omega_2}{2\pi} = 0.35$$

The following two statements create a length-25 Bartlett window and then use it for designing the bandpass filter required.

```
>>  wn = bartlett(25);
>>  hn = fir1(24,[0.4,0.7],wn)
```

Some important details need to be highlighted. The function `bartlett(..)` uses the filter length N (this applies to other window generation functions such as `hamming(..)`, `hann(..)`, and `blackman(..)` as well). On the other hand, the design function `fir1(..)` uses the filter order which is $N - 1$. The second argument to the function `fir1(..)` is a vector of two normalized edge frequencies which results in a bandpass filter being designed. (A lowpass filter results if only one edge frequency is used.) The frequencies are *normalized in MATLAB sense* which means they must be twice our normalized frequencies.

The magnitude spectrum may be computed and graphed for the bandpass filter with impulse response in vector "**hn**" using the following statements.

```
>>  Omg = [-256:255]/256*pi;
>>  H = fftshift(fft(hn,512));
>>  Omgd = [-1,-0.7,-0.7,-0.4,-0.4,0.4,0.4,0.7,0.7,1]*pi;
>>  Hd = [0,0,1,1,0,0,1,1,0,0];
>>  plot(Omg,abs(H),Omgd,Hd); grid;
>>  axis([-pi,pi,-0.1,1.1]);
>>  title('|H(\Omega)|');
>>  xlabel('\Omega (rad)');
>>  ylabel('Magnitude');
```

The magnitude response is shown in Fig. 10.52.

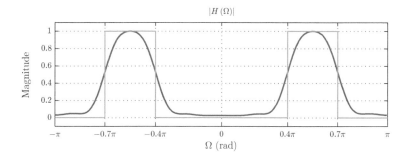

Figure 10.52 – Magnitude response of the filter designed in MATLAB Exercise 10.12.

Software resources:
matex_10_12a.m
matex_10_12b.m

MATLAB Exercise 10.13: **FIR filter design using Parks-McClellan technique**

In this exercise we will explore the use of MATLAB for designing FIR filters with the Parks-McClellan algorithm. The algorithm works better if the design specifications allow for transition bands between passbands and stopbands. Consider the desired magnitude response shown in Fig. 10.53 for a bandpass filter. Only the right side of the spectrum is shown for the angular frequency range $0 \leq \Omega \leq \pi$.

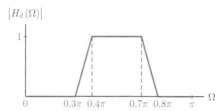

Figure 10.53 – Desired FIR magnitude response for MATLAB Exercise 10.13.

In using MATLAB function firpm(..) for Parks-McClellan FIR filter design we need to specify two vectors. The first vector lists the corner frequencies of the desired magnitude behavior shown in Fig. 10.53, *normalized in MATLAB sense*, and in ascending order. The first normalized frequency must be 0, and the last normalized frequency must be 1. The second vector lists the magnitude values at the corner frequencies listed in the first vector. The statements listed below create the two vectors which will be named 'frq" and "mag".

```
>>   frq = [0,0.3,0.4,0.7,0.8,1];
>>   mag = [0,0,1,1,0,0];
```

An easy method of checking the correctness of the two vectors 'frq" and "mag" is to use them as arguments to the function plot(..) as

```
>>   plot(pi*frq,mag);
```

The graph produced should be identical to the desired magnitude behavior in Fig. 10.53. The FIR bandpass filter is designed, and its magnitude spectrum is graphed through the following statements:

```
>>   hn = firpm(24,frq,mag)
>>   Omg = [0:255]/256*pi;
>>   H = fft(hn,512);
>>   H = H(1:256);
>>   plot(Omg,abs(H),pi*frq,mag);
>>   axis([0,pi,-0.1,1.1]);
>>   title('|H(\Omega)|');
>>   xlabel('\Omega (rad)');
>>   ylabel('Magnitude');
```

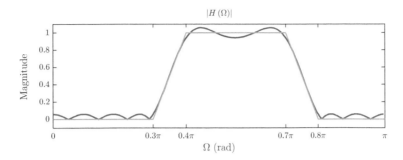

Figure 10.54 – Magnitude response of the filter designed in MATLAB Exercise 10.13.

It is also possible to use weight factors in the design. The following statement leads to a design that emphasizes better accuracy in the passband in exchange for some degradation in stopband performance.

```
>>  hn = firpm(24,frq,mag,[1,10,1])
```

Finally, increasing the filter order improves the performance further. A 44th-order FIR filter is obtained by

```
>>  hn = firpm(44,frq,mag,[1,10,1])
```

Magnitude characteristics of the last two designs are shown in Fig. 10.55.

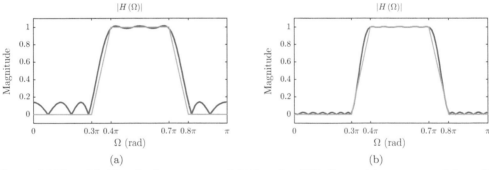

Figure 10.55 – (a) Magnitude response of 24th-order FIR filter with uneven weights, (b) magnitude response of 44th-order FIR filter.

Software resources:
matex_10_13a.m
matex_10_13b.m
matex_10_13c.m

Problems

10.1.   Consider the RC circuit used in Example 10.1 and shown in Fig. 10.4. Let the parameter values be $R = 1 \text{ k}\Omega$ and $C = 1 \text{ µF}$.

    a. Find the frequency range in Hertz in which the magnitude of the system function exhibits 1 percent or less deviation from its value at $\omega = 0$.

b. For the frequency range found in part (a) determine how much the phase characteristic deviates from a straight line. Determine the endpoints of the phase characteristic in the range of interest, and compare them to the endpoints of a straight line with the appropriate slope.

c. Find the variation in the time delay over the frequency range established in part (a).

d. Repeat parts (a) through (c) if 2 percent variation in magnitude is allowed.

10.2. Consider a second-order lowpass filter with the system function

$$H\left(s\right) = \frac{1}{s^2 + \sqrt{2}\,s + 1}$$

a. Find the frequency range in Hz in which the magnitude of the system function exhibits 1 percent or less deviation from its value at $\omega = 0$.

b. For the frequency range found in part (a) determine how much the phase characteristic deviates from a straight line. Determine the endpoints of the phase characteristic in the range of interest, and compare to the endpoints of a straight line with the appropriate slope.

c. Find the variation in the time delay over the frequency range established in part (a).

d. Repeat parts (a) through (c) if 2 percent variation in magnitude is allowed.

10.3. The magnitude spectrum of an ideal lowpass filter is shown in Fig. P.10.3. The time delay is to be $t_d = 0.1$ s for all frequencies.

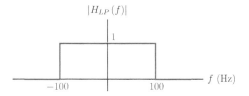

Figure P. 10.3

a. Write a mathematical expression for the system function $H_{LP}\left(f\right)$.

b. Determine the impulse response $h_{LP}\left(t\right)$.

c. Sketch the impulse response found in part (b).

10.4.   The magnitude spectrum of an ideal bandpass filter is shown in Fig. P.10.4. The time delay is to be $t_d = 0.3$ s for all frequencies.

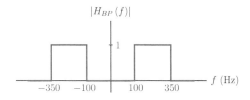

Figure P. 10.4

a. Write a mathematical expression for the system function $H_{BP}\left(f\right)$.

b. Determine the impulse response $h_{BP}\left(t\right)$.

c. Sketch the impulse response found in part (b).

10.5. Design a fourth-order Butterworth analog lowpass filter with 3-dB cutoff frequency of $\omega_c = 2$ rad/s. Find its poles and construct the system function $H(s)$.

10.6. A Butterworth analog lowpass filter is to be designed with the following specifications:

$$\omega_1 = 3 \text{ rad/s} \qquad \omega_2 = 4.5 \text{ rad/s} \qquad R_p = 1 \text{ dB} \qquad A_s = 30 \text{ dB}$$

Using the method presented in Section 10.4.1 determine the filter order N and the 3-dB cutoff frequency ω_c. Do not actually design the filter.

10.7. Consider the problem of determining Butterworth analog lowpass filter parameters from design specifications. First N is determined from Eqn. (10.65). In general it is not an integer, and must be rounded up to the nearest integer. This rounding up causes the tolerance specifications to be exceeded in the designed filter.

- If Eqn. (10.62) is used for computing ω_c, then specifications are met exactly at the passband edge $\omega = \omega_1$ and exceeded at the stopband edge $\omega = \omega_2$.

- If Eqn. (10.63) is used for computing ω_c, then specifications are met exactly at the stopband edge $\omega = \omega_2$ and exceeded at the passband edge $\omega = \omega_1$.

An alternative would be to divide the excess tolerance equally (in a dB sense) between the passband and the stopband. Derive the general solution for ω_c to accomplish this.

10.8. Design a fourth-order Chebyshev type-I analog filter with a passband edge frequency of $\omega_1 = 2$ rad/s and $\varepsilon = 0.3$. Find the poles of the filter and then construct the system function $H(s)$.

10.9. Design a fourth-order Chebyshev type-II analog filter with a stopband edge frequency of $\omega_2 = 2$ rad/s and $\varepsilon = 0.3$. Find the poles of the filter and then construct the system function $H(s)$.

10.10. An analog lowpass filter is needed with the following specifications:

$$\omega_1 = 2 \text{ rad/s} \qquad \omega_2 = 3.5 \text{ rad/s} \qquad R_p = 1 \text{ dB} \qquad A_s = 20 \text{ dB}$$

a. Determine N and ω_c for a Butterworth filter that meets the requirements. Find the system function for the corresponding filter.
b. Determine N and ε for a Chebyshev type-I filter that meets the requirements. Find the system function for the corresponding filter.
c. Determine N and ε for a Chebyshev type-II filter that meets the requirements. Find the system function for the corresponding filter.

10.11. Consider the lowpass to bandpass filter transformation given by Eqn. (10.124). Show that the parameters ω_0 and B must be chosen as given by Eqn. (10.127) in order to satisfy the conditions given by Eqns. (10.125) and (10.126).
Hint: Let $s = j\omega$ and $\lambda = jp$. Write the relationship between the radian frequencies ω and p and sketch ω as a function of p. Impose the conditions of Eqns. (10.125) and (10.126) and solve for the unknown parameters.

10.12. Consider the lowpass to band-reject filter transformation given by Eqn. (10.128). Show that the parameters ω_0 and B must be chosen as given by Eqn. (10.131) in order to satisfy the conditions given by Eqns. (10.129) and (10.130).
Hint: Let $s = j\omega$ and $\lambda = jp$. Write the relationship between the radian frequencies ω and p and sketch ω as a function of p. Impose the conditions of Eqns. (10.129) and (10.130) and solve for the unknown parameters.

10.13. Consider a lowpass filter with the system function

$$G\left(s\right) = \frac{2}{s+2}$$

This filter is to be converted to a highpass filter with system function $H\left(\lambda\right)$ by means of the transformation $s = \omega_0/\lambda$.

a. Determine ω_0 so that the frequency 2 rad/s for the lowpass filter corresponds to the frequency 5 rad/s for the highpass filter.

b. Using the value of ω_0 found in part (a), determine the frequency for the highpass filter that corresponds to the frequency 6 rad/s for the lowpass filter.

c. Determine the system function $H\left(\lambda\right)$ for the highpass filter. Evaluate its magnitude at the frequency $\omega = 5$ rad/s and verify that it is equal to the magnitude of the lowpass filter at $\omega = 2$ rad/s.

10.14. A Butterworth analog highpass filter is to be designed with the following specifications:

$$\omega_{H1} = 2 \text{ rad/s} \qquad \omega_{H2} = 5 \text{ rad/s} \qquad R_p = 1 \text{ dB} \qquad A_s = 30 \text{ dB}$$

a. Find the corresponding specifications ω_{L1}, ω_{L2}, R_p and A_s for an appropriate lowpass prototype filter.

b. Determine the parameters N and ω_c for the lowpass prototype filter.

c. Design the lowpass prototype filter. Find its poles, and construct its system function $G\left(s\right)$.

d. Convert the lowpass prototype filter to a highpass filter using the appropriate transformation. Determine the system function $H\left(\lambda\right)$.

e. Evaluate the decibel magnitude of the highpass filter at the frequencies ω_{H1} and ω_{H2}. Verify that the designed filter meets the requirements.

10.15. An analog filter with the system function

$$G\left(s\right) = \frac{2}{\left(s+1\right)\left(s+2\right)}$$

is to be used as a prototype for the design of a discrete-time filter through the impulse invariance technique.

a. Find the impulse response $g\left(t\right)$ of the analog prototype filter.

b. Obtain the impulse response of the discrete-time filter as a sampled and scaled version of the analog prototype impulse response, that is,

$$h[n] = T\, g\left(nT\right)$$

Use the sampling interval $T = 0.5$ s.

c. Determine the system function $H\left(z\right)$ of the discrete-time filter.

10.16. An analog filter with the system function

$$G\left(s\right) = \frac{2}{\left(s+1\right)\left(s+2\right)}$$

is to be used as a prototype for the design of a discrete-time filter through bilinear transformation. Using $T = 0.5$ s, convert the analog prototype system function $G\left(s\right)$ to a discrete-time filter system function $H\left(z\right)$.

10.17. The impulse response of an FIR filter is

$$h[n] , \qquad n = 0, \dots, N - 1$$

where N, the length of the FIR filter, is even.

a. Show that, if
$$h[n] = h[N - 1 - n] \text{ for } n = 0, \dots, N/2 - 1$$

then the system function $H(\Omega)$ is in the form

$$H(\Omega) = A(\Omega) \, e^{-j\Omega(N-1)/2}$$

where $A(\Omega)$ is some function that is purely real.

b. Show that, if
$$h[n] = -h[N - 1 - n] \text{ for } n = 0, \dots, N/2 - 1$$

then the system function $H(\Omega)$ is in the form

$$H(\Omega) = B(\Omega) \, e^{-j\Omega(N-1)/2} e^{j\pi/2}$$

where $B(\Omega)$ is some function that is purely real.

10.18. Repeat the proofs in Problem 10.17 for the case where N is odd.

10.19. Using the Fourier series design method with a Hamming window, design a length-19 FIR filter to approximate the ideal bandpass magnitude characteristic shown in Fig. P.10.19.

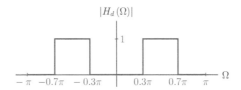

Figure P. 10.19

10.20. In the use of Fourier series design method for FIR filters we have focused on designing filters with an odd number of coefficients. This was necessitated by step 2 of the design process where the ideal impulse response $h_d[n]$ is truncated to the interval $-M \le n \le M$ resulting in filter length $N = 2M + 1$, always an odd value.

It is possible to modify the method slightly to allow the design of both odd-length and even-length filters. The key is to anticipate the correct time delay due to the linear phase characteristics of the final design, and make it part of the specifications. Let the ideal spectrum to be approximated be given by

$$H_d(\Omega) = \begin{cases} e^{-j\Omega(N-1)/2} , & |\Omega| \le \Omega_c \\ 0 , & \text{otherwise} \end{cases}$$

Show that, by computing the ideal impulse response as the inverse DTFT of $H_d(\Omega)$ and truncating the result to retain samples in the interval $n = 0, \dots, N - 1$, a linear-phase FIR filter can be obtained for both odd and even values of N.

10.21. Refer to Problem 10.20. If a window function is to be used with the modified design technique described in Problem 10.20, its sample amplitudes need to be specified for the index range $n = 0, \ldots, N - 1$. In addition, window samples must have the symmetry property

$$w[n] = w[N - 1 - n], \qquad n = 0, \ldots, N - 1$$

Modify the window function definitions given in Eqns. (10.180) through (10.183) to fit this form.

MATLAB Problems

10.22.   Consider again the RC circuit of Problem 10.1. Develop a MATLAB script to graph magnitude, phase and time delay characteristics of the circuit in a specified range of frequency $-f_0 < f < f_0$.

   a. Using the script developed, graph the characteristics for the frequency range found in part (a) of Problem 10.1.
   b. Repeat for for the frequency range found in part (d) of Problem 10.1.

10.23.   Consider the second-order lowpass filter of Problem 10.2. Develop a MATLAB script to graph magnitude, phase and time delay characteristics of the system in a specified range of frequency $-f_0 < f < f_0$.

   a. Using the script developed, graph the characteristics for the frequency range found in part (a) of Problem 10.2.
   b. Repeat for for the frequency range found in part (d) of Problem 10.2.

10.24.

   a. Develop a MATLAB function `ss_ilp(..)` to compute the impulse response of an ideal lowpass filter with the system function

$$H(\omega) = \begin{cases} 1\, e^{-j\omega t_d}, & |\omega| \leq \omega_0 \\ 0, & \text{otherwise} \end{cases}$$

   The syntax of your function should be

```
h = ss_ilp(omg0,td,t)
```

   where "`omg0`" and "`td`" are parameters of the ideal lowpass filter, and "`t`" is the vector of time instants at which the impulse response if to be evaluated.
   b. Write a script to test the function `ss_ilp(..)` for several sets of parameter values.

10.25.

   a. Develop a MATLAB function `ss_ibp(..)` to compute the impulse response of an ideal bandpass filter with the system function

$$H(\omega) = \begin{cases} 1\, e^{-j\omega t_d}, & \omega_1 \leq |\omega| \leq \omega_2 \\ 0, & \text{otherwise} \end{cases}$$

   The syntax of your function should be

```
h = ss_ibp(omg1,omg2,td,t)
```

where "omg1", "omg2" and "td" are parameters of the ideal lowpass filter, and "t" is the vector of time instants at which the impulse response if to be evaluated. The function ss_ibp(..) should internally utilize the function ss_ilp(..) developed in Problem 10.24.

 b. Write a script to test the function ss_ibp(..) for several sets of parameter values.

10.26. Refer to the Butterworth analog lowpass filter design considered in Problem 10.5. Design the same filter using MATLAB. Graph the magnitude and the phase characteristics of the designed filter.

10.27. Refer to Problem 10.6. Using the values of N and ω_c found in Problem 10.6 and the MATLAB function butter(..) design the filter. Compute and graph the dB magnitude response of the filter and verify that it satisfies the design specifications.

10.28.

 a. Design the filter specified in Problem 10.8 using MATLAB.
 b. Compute and graph the decibel magnitude response and verify that it meets the requirements of the design.
 c. Graph the pole-zero diagram of the system function.

10.29.

 a. Design the filter specified in Problem 10.9 using MATLAB.
 b. Compute and graph the decibel magnitude response and verify that it meets the requirements of the design.
 c. Graph the pole-zero diagram of the system function.

10.30. Refer to Problem 10.10.

 a. Repeat each design (Butterworth, Chebyshev type-I and Chebyshev type-II) using appropriate MATLAB functions. Compare to the results of hand calculations obtained in Problem 10.10.
 b. Compute and graph the magnitude response of each design.
 c. Compute and graph the decibel magnitude responses of all three designs on the same coordinate system.

10.31. Repeat the Butterworth analog highpass design in Problem 10.14 using MATLAB functions buttord(..), butter(..) and lp2hp(..). Afterwards compute and graph the decibel magnitude of the highpass filter and check the decibel tolerance values at the two critical frequencies.

10.32. Refer to the discrete-time filter designed in Problem 10.15 using the impulse invariance method with the sampling interval $T = 0.5$ s.

 a. Using the function impinvar(..) repeat the design process of Problem 10.15. Compare the result obtained for $H(z)$ to the one found in Problem 10.15.
 b. Using MATLAB, compute and graph the magnitude response of the analog prototype filter in the frequency interval $-\pi/T \leq \omega \leq \pi/T$.
 c. Compute and graph the magnitude response of the discrete-time filter in the angular frequency interval $-\pi \leq \Omega \leq \pi$.
 d. Suppose that the discrete-time filter is used for processing a continuous-time signal as shown in Fig. P.10.32.

Figure P. 10.32

Determine the relationship between the frequencies ω and Ω. Based on that relationship, compute and graph the overall magnitude response of the system in Fig. P.10.32 simultaneously with the magnitude response of the analog prototype filter for comparison. In other words, compute and graph $|Y_a(\omega)/X_a(\omega)|$ and $|G(\omega)|$.

10.33. Repeat Problem 10.32 using a sampling interval of $T = 0.25$ s instead of $T = 0.5$ s. Comment on how this change affects the overall performance of the analog system in Fig. P.10.32 that utilizes a discrete-time filter.

10.34. Refer to the discrete-time filter designed in Problem 10.16 using the bilinear transformation method with the sampling interval $T = 0.5$ s.

   a. Using the function `bilinear(..)` repeat the design process of Problem 10.16. Compare the result obtained for $H(z)$ to the one found in Problem 10.16.
   b. Using MATLAB, compute and graph the magnitude response of the analog prototype filter in the frequency interval $-\pi/T \le \omega \le \pi/T$.
   c. Compute and graph the magnitude response of the discrete-time filter in the angular frequency interval $-\pi \le \Omega \le \pi$.
   d. Suppose that the discrete-time filter is used for processing a continuous-time signal as shown in Fig. P.10.32. Compute and graph the overall magnitude response of the system in Fig. P.10.32 simultaneously with the magnitude response of the analog prototype filter for comparison. In other words, compute and graph $|Y_a(\omega)/X_a(\omega)|$ and $|G(\omega)|$.

10.35. Refer to the FIR filter specified in Problem 10.19.

   a. Write a script to compute and graph the magnitude response of the length-15 filter designed manually in Problem 10.19.
   b. Write a script to design a length-45 filter to meet the same specifications. Compute and graph the decibel magnitude characteristics of the length-15 and length-45 designs in a superimposed fashion for comparison.
   c. Repeat the length-45 design using a Blackman window instead of a Hamming window. Compute and graph the decibel magnitude characteristics of the two length-45 designs together and compare.

10.36.

   a. Develop a MATLAB function `ss_fir1(..)` to design a FIR lowpass filter of any order N using the technique explored in Problem 10.20. The syntax of the function should be

```
hd = ss_fir1(Omgc,N)
```

where "`Omgc`" is the cutoff frequency in radians, and "`N`" is the filter length. The returned vector "`hd`" holds samples of the impulse response. No windowing is done within this function; a window function may be applied to the vector "`hd`" separately if desired.

b. Test the function developed in part (a) by using it to design a length-36 lowpass FIR filter with angular cutoff frequency $\Omega_c = 0.4\pi$ radians. Compute and graph the magnitude response of the designed filter.

MATLAB Projects

10.37. FIR filters obtained through the use of the Fourier series design method suffer from Gibbs oscillations especially around the discontinuities of the desired magnitude behavior. We have seen that using a window function in the design process alleviates this problem. Another solution is to change the desired filter spectrum to a more realistic one. Consider the desired lowpass filter magnitude spectrum shown in Fig. P.10.37(a). Two frequencies Ω_1 and Ω_2 define the edges of the passband and the stopband. Between the two frequencies is a transition band. If this magnitude characteristic is sampled at the DFT frequencies we obtain

$$|H[k]| = \left| H_d \left(\frac{2\pi k}{N} \right) \right|$$

as shown in Fig. P.10.37(b).

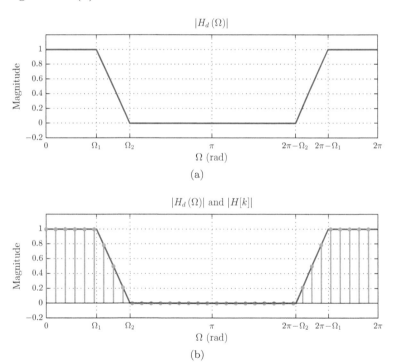

(a)

(b)

Figure P. 10.37

With the addition of a linear phase term, $H[k]$ is found as

$$H[k] = |H[k]|\, e^{-j2\pi k M/N}\,, \qquad k = 0, \ldots, N-1$$

where M is the delay of the filter computed as $M = (N-1)/2$. The impulse response of the FIR filter is found as the inverse DFT of $H[k]$:

$$h[n] = \text{DFT}^{-1}\{H[k]\}$$

This is the basis of the design method known as *frequency-sampling design*.

a. Using the idea outlined, develop a MATLAB function `ss_fir2(..)` with the following syntax:

```
h = ss_fir2(Omg,mag,N)
```

The vector "`Omg`" holds the corner frequencies of the desired magnitude spectrum starting with 0 and ending with 2π. For example, for a lowpass filter with edge frequencies $\Omega_1 = 0.3\pi$ and $\Omega_2 = 0.5\pi$ the vector "`Omg`" would have the elements

$$\text{Omg} = [0,\ 0.3\pi,\ 0.5\pi,\ 1.5\pi,\ 1.7\pi,\ 2\pi]$$

The vector "`mag`" holds desired magnitude values at the corner frequencies in the vector "`Omg`". For the example above, the vector "`mag`" would have the elements

$$\text{mag} = [1,\ 1,\ 0,\ 0,\ 1,\ 1]$$

As a simple test, the statement

```
>> plot(Omg,mag)
```

should graph the desired magnitude in the interval $0 \le \Omega \le 2\pi$.

b. Use the function `ss_fir2(..)` to design a length-35 lowpass FIR linear-phase filter with the critical frequencies $\Omega_1 = 0.2\pi$ and $\Omega_2 = 0.3\pi$. Compute and graph the magnitude response of the designed filter.

10.38.   Refer to the function `ss_iir2(..)` developed in MATLAB Exercise 8.12 in Chapter 8 for implementing a second-order section with system function

$$H(z) = \frac{b_0 + b_1\,z^{-1} + b_2\,z^{-2}}{1 + a_1\,z^{-1} + a_2\,z^{-2}}$$

One of the input arguments of the function `ss_iir2(..)` is a vector of filter coefficients in a prescribed order:

$$\text{coeffs} = [b_{0i},\ b_{1i},\ b_{2i},\ 1,\ a_{1i},\ a_{2i}]$$

GUI-based MATLAB program `fdatool` has the capability to export the coefficients of a designed filter to the workspace as a "SOS matrix" which corresponds to an implementation of the filter in the form of second-order cascade sections. Each row of the "SOS matrix" contains the coefficients of one second-order section in the same order given above.

a. Develop a MATLAB function `ss_iirsos(..)` for implementing an IIR filter designed using the program `fdatool(..)`. Its syntax should be

```
out = ss_iirsos(inp,sos,gains)
```

The matrix "sos" is the coefficient matrix exported from `fdatool(..)` using the menu selections *file, export*. The vector "**gains**" is the vector of gain factors for each second-order section. It is also exported from `fdatool(..)`. The vector "**inp**" holds samples of the input signal. The computed output signal is returned in vector "**out**". The function `ss_iirsos(..)` should internally utilize the function `ss_iir2(..)` for implementing the filter in cascade.

b. Use `fdatool(..)` to design an elliptic bandpass filter with passband edges at $\Omega_2 = 0.4\pi$, $\Omega_3 = 0.7\pi$ and stopband edges at $\Omega_1 = 0.3\pi$, $\Omega_4 = 0.8\pi$. The passband ripple must not exceed 1 dB, and the stopband attenuation must be at least 30 dB. Export the coefficients and gain factors of the designed filter to MATLAB workspace.

c. Use the function `ss_iirsos(..)` to compute the unit-step response of the designed filter. Compare to the step response computed by the program `fdatool(..)`.

Chapter 11

Amplitude Modulation

Chapter Objectives

- Learn the significance of and the need for modulation in analog and digital communication systems.

- Understand amplitude modulation as one method of embedding an information bearing signal into a high frequency sinusoidal carrier.

- Explore time- and frequency-domain characteristics of amplitude modulated signals. Understand variants of amplitude modulation (DSB and SSB).

- Study techniques for obtaining modulated signals as well as extracting information bearing signals from modulated signals.

11.1 Introduction

The main purpose of any communication system is to facilitate the transfer of an information bearing signal from one point in space to another, with an acceptable level of quality.

Consider, for example, two people communicating through their cell phones. The sound of the person who acts as information source is detected by the microphone on his or her phone, converted to an electrical signal for processing within the phone, and eventually converted to an electromagnetic signal which is transmitted from the antenna of the phone. This electromagnetic signal is received by the antennas within the cell phone network, transmitted from cell to cell as needed, possibly travels through various types of connections such fiber optics and satellite links, and finally reaches the antenna of the phone that is designated as its destination. Within the receiving phone, the electromagnetic signal is converted back to an electrical signal, and is eventually converted to sound.

Thus, any communication system has a transmitter and a receiver. The collection of all systems between the transmitter and the receiver is referred to as the *communication channel*. In general, a channel has two undesirable effects on the signals that are transmitted through it:

1. It distorts them.
2. It contaminates them with noise.

As a result, the electrical signal reproduced in the receiving phone and the sound reproduced through its speaker are only approximations of the original signals transmitted. In the design of communication systems we take steps to ensure that the signals at the receiving end are reasonably good representations of the signals transmitted.

In this chapter we will present an introduction to modulation which is one of the steps taken in the design of a communication system to ensure reliable transmission of signals. In doing so, we will utilize the analysis tools and transforms developed in previous chapters for linear systems.

Modulation is the process of embedding an information bearing signal, referred to as the *message signal*, in another signal known as the *carrier*. The carrier by itself does not contain any information, yet it may have other features that make it desirable. Some examples of these desirable features are: 1) suitability for transmission through a particular medium, 2) better performance in the presence of noise and interference, 3) suitability to more efficient utilization of resources, and so on.

Fig. 11.1 illustrates the general structure of a very simple analog communication system.

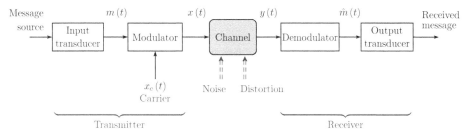

Figure 11.1 – Structure of a simple analog communication system.

The overall operation of the communication system can be summarized as follows:

1. The role of the input transducer is to convert the source message from its native format to the form of an electrical signal $m(t)$, referred to as the *message signal*. Some examples of input transducers are microphone, video camera, pressure gauge, temperature sensor, etc.

2. The modulator embeds the message signal $m(t)$ into the carrier signal $x_c(t)$ to create the modulated carrier $x(t)$ which is then transmitted through the channel.

3. The output signal of the channel is a distorted version of $x(t)$ which is also contaminated with noise. In simple terms we have

$$y(t) = \text{Sys}\{x(t)\} + w(t)$$

where $\text{Sys}\{\ldots\}$ represents the effect of the channel on the modulated signal, and $w(t)$ represents the additive channel noise.[1]

4. At the receiver side of the system the demodulator reverses the action of the modulator, and extracts an approximate version $\hat{m}(t)$ of the original message signal $m(t)$ from the received modulated signal $y(t)$. The demodulator may or may not need a local copy of the carrier signal $x_c(t)$.

---

[1]   This is actually an oversimplification. In an actual communication channel, the signal is not necessarily distorted first and contaminated by noise afterwards; distortion and contamination take place simultaneously in a distributed fashion, resulting in a much more difficult model for analysis.

5. Finally, the output transducer converts the electrical signal $\hat{m}(t)$ to its native format. Examples of output transducers are speaker, video display, etc.

11.2   The Need for Modulation

We need to address the question of why a modulator and a demodulator are needed in a communication system. Using a modulated carrier for transmitting a message has a number of benefits over transmitting the message signal $m(t)$ directly through the channel without modulation:

1. **Compatibility of the signal with the channel:**
   Modulation gives us the ability to produce a signal that is compatible with the characteristics of the channel that will transmit it.
   Most message signals we encounter in our daily lives involve relatively low frequencies, referred to as the *baseband* frequencies. For example, voice signals used in telephone communications occupy the frequency range from about 300 Hz to about 3600 Hz. High fidelity music signals are typically in the frequency range from about 20 Hz to about 20 kHz. Similarly, baseband video signals such as those encountered on a TV monitor occupy roughly the frequency range from 0 to about 4.2 MHz. Signals such as these can be transmitted using wire links or optical fiber connections, but they are not suitable for over-the-air transmission due to their large wavelengths. The antennas required for over-the-air transmission of these signals in their original form would have to be unreasonably long. Modulating a high frequency carrier with these message signals makes over-the-air broadcast possible while keeping required antenna lengths at practical levels.

2. **Reduced susceptibility to noise and interference:**
   Modulation of a carrier with the message signal results in a shift of the range of signal frequencies. This frequency shift can be used in a way that allows us to position the frequency spectrum of the signal where noise and interference conditions are more favorable compared to the conditions at baseband frequencies.
   Additionally, some modulation techniques allow performance improvements in terms of noise susceptibility provided that increased bandwidth is available.

3. **Multiplexing:**
   Modulation allows for more efficient use of available resources by multiplexing different signals in the same medium.
   In sinusoidal modulation, the frequency shift that results from modulation can also be used for mixing multiple messages in one medium, with each message signal modulating a carrier at a different frequency. These modulated carrier signals can be combined and transmitted together. They can later be separated from each other provided that the frequency ranges occupied by different carriers do not overlap with each other. This effect is known as *frequency division multiplexing (FDM)*. On radio and television, FDM forms the basis of simultaneous broadcast of multiple stations in the same air space.
   In digital pulse modulation, the gaps between the successive pulses of a carrier can be used for accommodating messages from multiple sources, a concept known as *time division multiplexing (TDM)*. Multiple digital information sources can utilize the same channel simultaneously, by taking turns in sending data.

11.3   Types of Modulation

The type of modulation used in a communication system depends on

1. The type of carrier used
2. The particular parameter of the carrier that is utilized for embedding the message signal into it

We will focus on two commonly used types of carrier signals, namely a sinusoidal carrier and a pulse train. Consider the sinusoidal carrier signal:

$$x_c(t) = A_c \cos\left(2\pi f_c t + \phi_c\right) \tag{11.1}$$

The carrier described by Eqn. (11.1) has three parameters, namely the amplitude A_c, the frequency f_c and the phase ϕ_c as shown in Fig. 11.2.

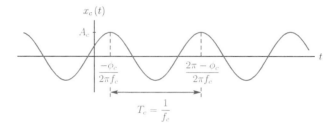

Figure 11.2 – Sinusoidal carrier signal.

If the values of these three parameters are fixed, then the carrier clearly has no information value since its future behavior is completely predictable. Such a carrier signal is referred to as a *blank carrier* or *unmodulated carrier*. If, on the other hand, we allow one of these parameters to vary in time, then the resulting signal will be a more interesting one, and certainly not as predictable. Furthermore, if the time variations of the parameter in question are somehow related to an information bearing message signal $m(t)$, then these variations can be tracked at the receiving end of the communication system, and the message signal $m(t)$ can be extracted from the variations of carrier parameters. This is the essence of sinusoidal modulation. The modulated carrier can be used as a vehicle for transmitting the message signal $m(t)$ from one point to another. Different modulation schemes can be devised, depending on which of the three parameters is varied in time:

1. In *amplitude modulation*, the amplitude A_c is varied as a function of the message signal $m(t)$.
2. In *angle modulation*, the angle $2\pi f_c t + \phi_c$ of the sinusoidal carrier is varied as a function of the message signal $m(t)$. This can be achieved by varying either the phase or the frequency, and leads to two variants of angle modulation known as *phase modulation* and *frequency modulation* respectively.

An alternative to the sinusoidal carrier described in Eqn. (11.1) is a pulse train in the form

$$x_c(t) = \sum_{k=-\infty}^{\infty} A_c \, \Pi\left(\frac{t - kT_0}{\tau}\right) \tag{11.2}$$

which is shown in Fig. 11.3.

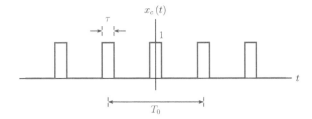

Figure 11.3 – Pulse train as a carrier signal.

If the pulse train is used for analog modulation, it presents three opportunities for a message signal to be embedded in it.

1. The amplitudes of pulses can be varied as a function of the message signal, resulting in *pulse-amplitude modulation (PAM)*.
2. The duration of each pulse can be varied as a function of the message signal. This leads to *pulse-width modulation (PWM)*.
3. Finally, the position of each pulse within its allocated period can be varied. The resulting modulation scheme is known as *pulse-position modulation (PPM)*.

The pulse train can also be used as the basis of digital modulation. The first step is to modulate the amplitude of the pulse train with the message signal using pulse-amplitude modulation. Afterwards the amplitude of each pulse is rounded up or down to be associated with one of the discrete amplitude levels from a predetermined set, an operation known as *quantization*. Finally, each quantized amplitude is *encoded* so that it can be represented using a set of binary digits (bits). The combination of pulse-amplitude modulation, quantization and encoding is referred to as *pulse-code modulation (PCM)*.

11.4   Amplitude Modulation

Consider the unmodulated sinusoidal carrier given by Eqn. (11.1). Its amplitude A_c is constant. One method of incorporating the message signal $m(t)$ into the carrier is to allow the amplitude of the carrier vary in a way that is related to $m(t)$. Let us replace the constant amplitude term A_c with $[A_c + m(t)]$ to obtain

$$\begin{aligned} x_{AM}(t) &= [A_c + m(t)] \cos(2\pi f_c t) \\ &= A_c \cos(2\pi f_c t) + m(t) \cos(2\pi f_c t) \end{aligned} \tag{11.3}$$

The result is an *amplitude-modulated (AM)* carrier. To set the terminology straight we will list the roles of the signals involved in Eqn. (11.3) as follows:

$$\begin{aligned} m(t) &: \quad \text{Modulating signal} \\ x_c(t) &: \quad \text{Unmodulated (blank) carrier} \\ x_{AM}(t) &: \quad \text{Modulated carrier} \end{aligned}$$

The parameter f_c is referred to as the *carrier frequency*. Note that the phase term ϕ_c in Eqn. (11.1) is not important for this discussion, and we will simplify the formulation of amplitude modulation without losing any generality by assuming that $\phi_c = 0$. One example of an AM signal is depicted in Fig. 11.4.

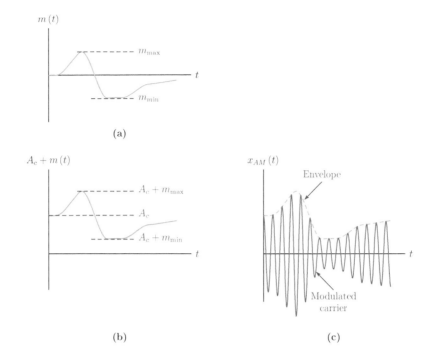

Figure 11.4 – Waveforms involved in amplitude modulation: (a) the message signal, (b) the message signal with offset A_c, and (c) the amplitude-modulated carrier and its envelope.

In this particular example we have assumed that the amplitudes of the message signal are constrained to remain in the interval $m_{\min} \leq m(t) \leq m_{\max}$. Further, we will assume that the time average of the signal $m(t)$ is zero. The new amplitude function $[A_c + m(t)]$ will remain in the range

$$A_c + m_{\min} \leq [A_c + m(t)] \leq A_c + m_{\max}$$

Furthermore, the relationship between A_c and $m(t)$ is chosen such that $[A_c + m(t)] \geq 0$ for all t. As a result, the envelope of the modulated carrier, that is, the waveform obtained by connecting the positive peaks of $x_{AM}(t)$, matches $[A_c + m(t)]$. This will be important as we consider ways of recovering the message signal $m(t)$ from the AM signal $x_{AM}(t)$.

In contrast, consider the example in Fig. 11.5 where $[A_c + m(t)]$ is allowed to be negative for some values of t. In this case the envelope of the AM waveform does not equal $[A_c + m(t)]$, but rather its absolute value $|A_c + m(t)|$.

Combining the two cases considered above we obtain a general expression for the envelope of the AM signal as

$$x_e(t) = |A_c + m(t)| \tag{11.4}$$

If $[A_c + m(t)] \geq 0$ for all t, then the envelope is a replica of the message signal $m(t)$ subject to a constant offset. Otherwise, the envelope is a distorted version of the message signal. In order to preserve the shape of the message signal in the envelope of the AM signal, we need

$$A_c + m_{\min} \geq 0 \tag{11.5}$$

Let us define the modulation index as

$$\mu = \frac{|m_{\min}|}{A_c} \tag{11.6}$$

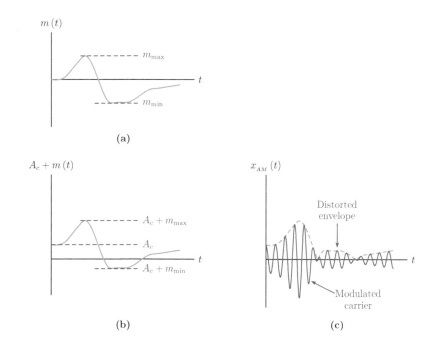

Figure 11.5 – Waveforms involved in amplitude modulation when $[A_c + m\,(t)] \geq 0$ is not satisfied for all t: (a) the message signal, (b) the message signal with offset A_c, and (c) the amplitude-modulated carrier and its distorted envelope.

which can be expressed either as a fraction or as a percentage. For the envelope to display the correct shape of the message signal, we need $0 \leq \mu \leq 1$. An amplitude modulated carrier with a modulation index $\mu < 1$ is said to be *under-modulated*. A modulation index of $\mu = 1$ leads to the *critically-modulated* case. Both of these will allow a simple system known as an *envelope detector* to be used in recovering the message signal $m\,(t)$ from the AM signal $x_{AM}\,(t)$. On the other hand, if $\mu > 1$, the carrier is said to be *over-modulated*. In this case the envelope is a distorted version of the message signal, and the envelope detector is no longer a viable option for the recovery of $m\,(t)$. A more complicated synchronous detector will be needed to extract the message.

Interactive Demo: am_demo1

This demo uses the signals in Figs. 11.4 and 11.5. The message signal and the blank carrier are shown on the left side of the graphical user interface. The main graph on the right side displays the modulated carrier computed as in Eqn. (11.3), as well as its envelope given by Eqn. (11.4). A slider control is provided for allowing the modulation index (refer to Eqn. (11.6)) to be varied in the range $0.4 \leq \mu \leq 1.8$. Alternatively, the desired value of the modulation index may be typed into the edit field (the same range restrictions still apply). Note that, in the implementation of this demo, the peak amplitudes $m_{\min}$ and $m_{\max}$ of the message signal are fixed. Change of modulation index μ is facilitated by changing the parameter A_c as dictated by Eqn. (11.6).

1. Observe the changes in the AM signal and its envelope as the modulation index μ is varied.
2. Pay attention to the envelope crossover that begins just as the modulation index exceeds 1.

Software resources:
am_demo1.m

Example 11.1: Tone modulation

Sometimes it is insightful to analyze a modulation technique by using a message signal that has a simple analytical description. One particular message signal that can be used in this way is a single tone. Let the modulating signal be

$$m(t) = A_m \cos(2\pi f_m t)$$

Substituting $m(t)$ in Eqn. (11.3) the corresponding AM waveform is

$$x_{AM}(t) = [A_c + A_m \cos(2\pi f_m t)] \cos(2\pi f_c t) \qquad (11.7)$$

which can also be written in the form

$$x_{AM}(t) = A_c [1 + \mu \cos(2\pi f_m t)] \cos(2\pi f_c t) \qquad (11.8)$$

with the modulation index defined as $\mu = A_m/A_c$. The envelope of the modulated carrier is

$$x_e(t) = \left| A_c [1 + \mu \cos(2\pi f_m t)] \right| \qquad (11.9)$$

The tone-modulated carrier is shown in Fig. 11.6 for various values of the modulation index. Notice how the envelope is distorted when $\mu > 1$.

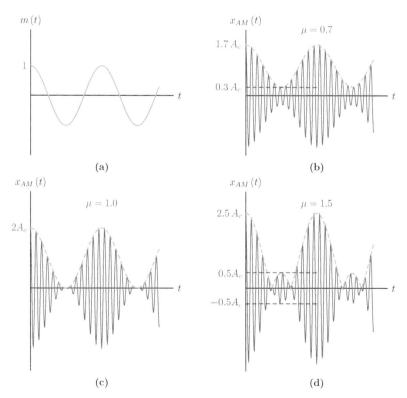

Figure 11.6 – Amplitude modulation using a single-tone message signal in Example 11.1: (a) the message signal, (b) under-modulated AM signal with $\mu = 0.7$, (c) critically-modulated AM signal with $\mu = 1$, (d) over-modulated AM signal with $\mu = 1.5$.

Software resources:
ex_11_1.m

Interactive Demo: am_demo2

The demo program "am_demo2.m" illustrates amplitude modulation of a carrier by a single tone as discussed in Example 11.1. Three sliders are provided for varying the message frequency f_m, the carrier frequency f_c, and the modulation index μ. Sliders allow f_m to be changed from 0.5 to 3 kHz, and f_c to be changed from 10 to 25 kHz. Modulation index may be varied between 0.4 and 1.8.

1. Observe the effect of each change on the AM signal and its envelope.
2. See how the envelope crosses over the horizontal axis when the modulation index becomes greater than 1.
3. Pay attention to the phase reversal of the carrier at the points of envelope crossover. The easiest method would be to use the "zoom" button in the toolbar at the top of the program window and zoom into the area where envelope crossover occurs.

Software resources:
am_demo2.m

Software resources: See MATLAB Exercise 11.1.

11.4.1 Frequency spectrum of the AM signal

Consider again the expression for an amplitude modulated sinusoidal carrier given by Eqn. (11.3), repeated here for convenience:

Eqn. (11.3): $x_{AM}(t) = A_c \cos(2\pi f_c t) + m(t) \cos(2\pi f_c t)$

Taking the Fourier transforms of both sides, we obtain the frequency spectrum of the AM signal as

$$X_{AM}(f) = \frac{1}{2} A_c \delta(f + f_c) + \frac{1}{2} A_c \delta(f - f_c)$$

$$+ \frac{1}{2} M(f + f_c) + \frac{1}{2} M(f - f_c) \tag{11.10}$$

where we made use of the modulation theorem of the Fourier transform discussed in Section 4.3.5. The first two terms in Eqn. (11.10) are due to the carrier frequency. The term $\delta(f + f_c)$ represents an impulse function shifted to the frequency $f = -f_c$. Similarly, $\delta(f - f_c)$ is an impulse function shifted to the frequency $f = f_c$. The last two terms in Eqn. (11.10) contain shifted versions of the message spectrum. The amount of frequency shift is controlled by the carrier frequency. The term $M(f + f_c)$ is a left-shifted version of the message spectrum, and $M(f - f_c)$ a right-shifted one. This is depicted in Fig. 11.7.

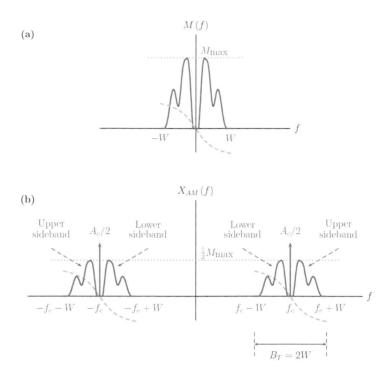

Figure 11.7 – (a) Frequency spectrum $M(f)$ of the message signal; magnitude (solid) and phase (dashed), (b) frequency spectrum $X_{AM}(f)$ of the amplitude-modulated signal; magnitude (solid) and phase (dashed).

Note that the shape of the message spectrum $M(f)$ in Fig. 11.7 is chosen arbitrarily for illustration purposes only; different message signals will have different looking spectra. For positive frequencies, the part of the AM spectrum that occupies frequencies greater than the carrier frequency f_c is referred to as the *upper sideband*. Conversely, the part of the spectrum that occupies frequencies less than f_c is referred to as the *lower sideband*. For the negative side of the frequency axis, these definitions are mirrored. Another interesting point to observe is that, for positive frequencies, lower and upper sidebands occupy the frequency range from $f_c - W$ to $f_c + W$. Thus, if the spectrum of the message signal has a one-sided bandwidth of W Hz, then the spectrum of the amplitude-modulated signal has a one-sided bandwidth of $B_T = 2W$ Hz, that is, twice the message bandwidth. B_T is called the *transmission bandwidth* of the AM signal.

This illuminates one of the shortcomings of amplitude modulation: It is wasteful of bandwidth. A message signal which occupies a bandwidth of W Hz in its baseband form requires twice that bandwidth when it is embedded into the amplitude of a carrier. We will see in Section 11.6 that *single sideband modulation* provides a solution to the problem of wasting bandwidth.

The relationship between the carrier frequency f_c and the message bandwidth W is also important. To avoid overlaps between spectral components when the message spectrum is shifted right and left, we need $f_c > W$, that is, we need the carrier frequency to be greater than the highest frequency in the message spectrum. Going one step further, if the envelope of the AM signal is required to be a good-quality representation of the message signal, then we need $f_c \gg W$, that is, we need the carrier frequency to be significantly greater than the message bandwidth. In practical applications, f_c is often several magnitudes greater than W.

The demo program "am_demo3.m" illustrates the frequency spectrum of the AM signal, and is based on Fig. 11.7. The sample message spectrum picked for this demo has a bandwidth of 1 kHz and a peak magnitude of $\max\{|M(f)|\} = 1$. Amplitude parameter for the carrier is set as $A_c = 2$. As a result, the modulation index is fixed. A slider control allows the carrier frequency to be varied in the range $2\,\text{kHz} \le f_c \le 9\,\text{kHz}$. Move the slider to vary the carrier frequency f_c, and the observe the effect on the spectrum of the AM signal. Pay attention to the shifting of the spectrum when f_c is changed.

Software resources:
am_demo3.m

Example 11.2: **Frequency spectrum of the tone-modulated carrier**

Compute the frequency spectrum of the tone-modulated AM carrier of Example 11.1 in terms of the carrier frequency f_c, the message frequency f_m, the modulation index μ and the amplitude A_c.

Solution: Recall that the tone-modulated signal is in the form

$$x_{AM}(t) = A_c\left[1 + \mu\,\cos(2\pi f_m t)\right]\cos(2\pi f_c t)$$

which can also be written as

$$x_{AM}(t) = A_c\cos(2\pi f_c t) + \mu A_c\cos(2\pi f_m t)\cos(2\pi f_c t) \tag{11.11}$$

Applying the appropriate trigonometric identity[2] to the second term of Eqn. (11.11) we obtain

$$x_{AM}(t) = A_c\cos(2\pi f_c t) + \frac{\mu A_c}{2}\cos(2\pi[f_c + f_m]\,t)$$

$$+ \frac{\mu A_c}{2}\cos(2\pi[f_c - f_m]\,t) \tag{11.12}$$

Eqn. (11.12) contains three sinusoidal terms at frequencies f_c, $f_c + f_m$ and $f_c - f_m$ respectively. Individual Fourier transforms of the three terms are

$$A_c\cos(2\pi f_c t) \xleftrightarrow{\mathcal{F}} \frac{A_c}{2}\delta(f + f_c) + \frac{A_c}{2}\delta(f - f_c)$$

$$\frac{\mu A_c}{2}\cos(2\pi[f_c + f_m]\,t) \xleftrightarrow{\mathcal{F}} \frac{\mu A_c}{4}\delta(f + [f_c + f_m]) + \frac{\mu A_c}{4}\delta(f - [f_c + f_m])$$

$$\frac{\mu A_c}{2}\cos(2\pi[f_c - f_m]\,t) \xleftrightarrow{\mathcal{F}} \frac{\mu A_c}{4}\delta(f + [f_c - f_m]) + \frac{\mu A_c}{4}\delta(f - [f_c - f_m])$$

Thus, the spectrum of the tone-modulated AM signal is obtained as

$$X_{AM}(f) = \frac{A_c}{2}\delta(f + f_c) + \frac{A_c}{2}\delta(f - f_c)$$

$$+ \frac{\mu A_c}{4}\delta(f + [f_c + f_m]) + \frac{\mu A_c}{4}\delta(f - [f_c + f_m])$$

$$+ \frac{\mu A_c}{4}\delta(f + [f_c - f_m]) + \frac{\mu A_c}{4}\delta(f - [f_c - f_m]) \tag{11.13}$$

---

[2] $\cos(a)\cos(b) = \frac{1}{2}\cos(a + b) + \frac{1}{2}\cos(a - b)$.

Frequency spectra for the single-tone message signal and the amplitude-modulated signal are shown in Fig. 11.8.

(a)

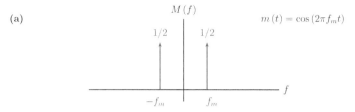

(b)

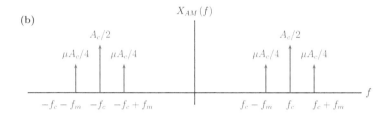

Figure 11.8 – (a) Frequency spectrum $M(f)$ of the single-tone message signal, and (b) frequency spectrum of the tone-modulated AM signal of Example 11.2.

The first two terms in Eqn. (11.13) are due to the carrier alone. The remaining four terms are the sidebands due to the message signal, and they are around positive and negative carrier frequencies with displacements of $\pm f_m$ in either direction.

Interactive Demo: am_demo4

This demo illustrates the effects of various parameters on the frequency spectrum of the tone-modulated AM signal, and is based on Fig. 11.8 and Eqn. (11.13) of Example 11.2. Use the slider controls to vary the message frequency f_m, the carrier frequency f_c, and the modulation index μ. Sliders allow f_m to be changed from 0.5 to 3 kHz, and f_c to be changed from 10 to 25 kHz. Modulation index may be varied between 0.4 and 1.8. The amplitude is fixed at $A_c = 1$. Vary the parameters, and observe

1. The frequency shifting of the sidebands of the spectrum as the carrier frequency is changed
2. How the locations of impulses representing sidebands are related to the message frequency f_m and the carrier frequency f_c
3. The area under each impulse as the modulation index is changed

Software resources:
am_demo4.m

Example 11.3: **AM spectrum for a multi-tone modulating signal**

Determine the frequency spectrum of the AM signal produced by modulating a carrier at frequency $f_c = 1$ kHz with the two-tone signal

$$m(t) = 3 \cos(2\pi(100)t - \pi/6) + 2 \sin(2\pi(200)t)$$

Use the modulation index $\mu = 0.7$.

Solution: First we need to determine the message spectrum. In order to get the phase relationship between the two terms correctly, the sine function in $m(t)$ needs to be converted to a cosine function. Using the appropriate trigonometric identity,[3] $m(t)$ can be written as

$$m(t) = 3\cos\left(2\pi(100)t - \pi/6\right) + 2\cos\left(2\pi(200)t - \pi/2\right) \tag{11.14}$$

Using Euler's formula, Eqn. (11.14) becomes

$$m(t) = \frac{3}{2}e^{j(2\pi(100)t - \pi/6)} + \frac{3}{2}e^{-j(2\pi(100)t - \pi/6)}$$
$$+ e^{j(2\pi(200)t - \pi/2)} + e^{-j(2\pi(200)t - \pi/2)}$$

which yields the spectrum shown in Fig. 11.9.

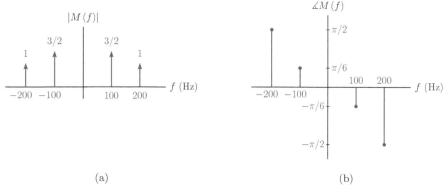

(a)　　　　　　　　　　　　　　　　　　　　　(b)

Figure 11.9 – Frequency spectrum $M(f)$ of the two-tone message signal of Example 11.3: (a) magnitude, (b) phase.

The next step is to determine the carrier amplitude A_c that will yield the desired modulation index value of $\mu = 0.7$. This, in turn, requires the minimum value of the message signal which is found from Eqn. (11.14) to be

$$m_{\min} = -3.2536$$

and occurs at time instant $t = 4.28$ ms as shown in Fig. 11.10.

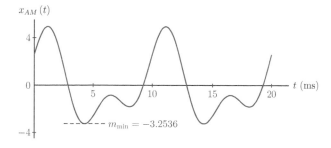

Figure 11.10 – The two-tone message signal of Example 11.3.

---

[3] $\sin(a) = \cos(a - \pi/2)$.

Using Eqn. (11.6) the carrier parameter A_c is found as

$$A_c = \frac{|m_{\min}|}{\mu} = \frac{|-3.2536|}{0.7} = 4.648$$

Thus, the AM signal is

$$x_{AM}(t) = [A_c + m(t)] \cos(2\pi f_c t)$$
$$= [4.648 + 3 \cos(2\pi(100)t - \pi/6) + 2 \cos(2\pi(200)t - \pi/2)] \cos(2\pi(1000)t)$$

and is shown in Fig. 11.11(a). The frequency spectrum of the AM signal can be easily found using Eqn. (11.10) along with the message spectrum in Fig. 11.9. The magnitude and the phase spectra of the AM signal are shown in Fig. 11.11(b) and (c).

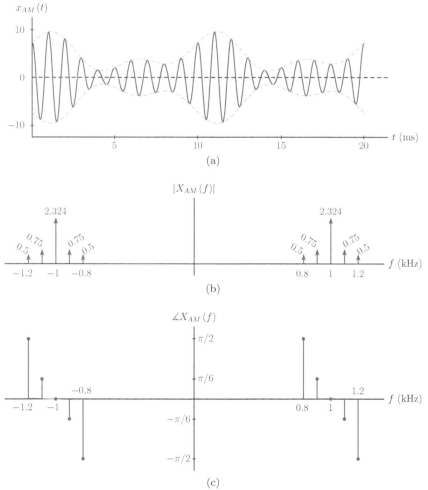

Figure 11.11 – (a) The AM signal for Example 11.3, (b) the magnitude of the frequency spectrum, (c) the phase of the frequency spectrum.

Software resources:
ex_11_3.m

Software resources: See MATLAB Exercise 11.2.

Example 11.4: **Phasor analysis of the tone-modulated AM signal**

The mathematical expression for a carrier with frequency f_c that is amplitude-modulated by a single tone with frequency f_m was derived in Eqn. (11.8) as

$$\text{Eqn. (11.8):} \qquad x_{AM}(t) = A_c \left[1 + \mu \cos(2\pi f_m t)\right] \cos(2\pi f_c t)$$

which, using trigonometric identities, can be written as

$$x_{AM}(t) = A_c \cos(2\pi f_c t) + \frac{\mu A_c}{2} \cos(2\pi (f_c + f_m)) + \frac{\mu A_c}{2} \cos(2\pi (f_c - f_m)) \qquad (11.15)$$

The first term in Eqn. (11.15) is the carrier term. The remaining two terms are the upper sideband and lower sideband terms terms respectively as we have seen in Fig. 11.8. Let

$$x_{\text{car}}(t) \triangleq A_c \cos(2\pi f_c t) \qquad (11.16)$$

$$x_{\text{usb}}(t) \triangleq \frac{\mu A_c}{2} \cos(2\pi (f_c + f_m) t) \qquad (11.17)$$

and

$$x_{\text{lsb}}(t) \triangleq \frac{\mu A_c}{2} \cos(2\pi (f_c - f_m) t) \qquad (11.18)$$

so that the amplitude modulated signal can be written as

$$x_{AM}(t) = x_{\text{car}}(t) + x_{\text{usb}}(t) + x_{\text{lsb}}(t) \qquad (11.19)$$

In Section 1.3.7 of Chapter 1 we have discussed means of representing sinusoidal signals with phasors. The techniques developed in that section can be applied to the three terms of the AM signal in Eqns. (11.16) through (11.18) to obtain

$$x_{\text{car}}(t) = \text{Re}\left\{A_c\, e^{j2\pi f_c t}\right\} \qquad (11.20)$$

$$x_{\text{usb}}(t) = \text{Re}\left\{\frac{\mu A_c}{2} e^{j2\pi (f_c + f_m)t}\right\} \qquad (11.21)$$

$$x_{\text{lsb}}(t) = \text{Re}\left\{\frac{\mu A_c}{2} e^{-j2\pi (f_c - f_m)t}\right\} \qquad (11.22)$$

If we wanted to apply the phasor notation of Section 1.3.7 in a strict sense, we would need to express Eqns. (11.20), (11.21) and (11.22) as

$$\boldsymbol{X}_{\text{car}} = A_c \angle 0° \qquad \Longrightarrow \qquad x_{\text{car}}(t) = \text{Re}\left\{\boldsymbol{X}_{\text{car}}\, e^{j2\pi f_c t}\right\} \qquad (11.23)$$

$$\boldsymbol{X}_{\text{usb}} = \frac{\mu A_c}{2} \angle 0° \qquad \Longrightarrow \qquad x_{\text{usb}}(t) = \text{Re}\left\{\boldsymbol{X}_{\text{usb}}\, e^{j2\pi (f_c + f_m)t}\right\} \qquad (11.24)$$

$$\boldsymbol{X}_{\text{lsb}} = \frac{\mu A_c}{2} \angle 0° \qquad \Longrightarrow \qquad x_{\text{lsb}}(t) = \text{Re}\left\{\boldsymbol{X}_{\text{lsb}}\, e^{j2\pi (f_c - f_m)t}\right\} \qquad (11.25)$$

The phasor $\boldsymbol{X}_{\text{car}} = A_c \angle 0°$ rotates counterclockwise at the rate of f_c Hz, that is, f_c revolutions per second. The phasors $\boldsymbol{X}_{\text{usb}} = (\mu A_c/2) \angle 0°$ and $\boldsymbol{X}_{\text{lsb}} = (\mu A_c/2) \angle 0°$ rotate at the rates of $(f_c + f_m)$ and $(f_c - f_m)$ Hz respectively. It is impractical to work with three phasors when each rotates at a different speed. Instead, we will take a slightly unconventional approach and rewrite Eqns. (11.24) and (11.25) as

$$x_{\text{usb}}(t) = \text{Re}\left\{\boldsymbol{X}_{\text{usb}}\, e^{j2\pi f_m t}\, e^{j2\pi f_c t}\right\} = \text{Re}\left\{\tilde{\boldsymbol{X}}_{\text{usb}}\, e^{j2\pi f_c t}\right\} \tag{11.26}$$

and

$$x_{\text{lsb}}(t) = \text{Re}\left\{\boldsymbol{X}_{\text{lsb}}\, e^{-j2\pi f_m t}\, e^{j2\pi f_c t}\right\} = \text{Re}\left\{\tilde{\boldsymbol{X}}_{\text{lsb}}\, e^{j2\pi f_c t}\right\}$$

The modified phasors

$$\tilde{\boldsymbol{X}}_{\text{usb}} = (\mu A_c/2)\; \measuredangle 2\pi f_m t \tag{11.27}$$

and

$$\tilde{\boldsymbol{X}}_{\text{lsb}} = (\mu A_c/2)\; \measuredangle -2\pi f_m t \tag{11.28}$$

have the same base rotation rate of f_c Hz as the phasor $\boldsymbol{X}_{\text{car}}$. This allows us to add the three phasors together as vectors. In addition to the base rotation at the rate of f_c Hz, each of the modified phasors $\tilde{\boldsymbol{X}}_{\text{usb}}$ and $\tilde{\boldsymbol{X}}_{\text{lsb}}$ has a secondary rotation at the rate of $\pm f_m$ Hz relative to the carrier phasor $\boldsymbol{X}_{\text{car}}$. This is illustrated in Fig. 11.12 for a modulation index of $\mu < 1$.

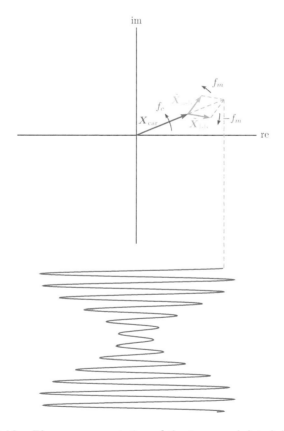

Figure 11.12 – Phasor representation of the tone-modulated AM signal.

The phasor $\boldsymbol{X}_{\text{car}}$ is drawn starting at the origin. The phasors $\tilde{\boldsymbol{X}}_{\text{usb}}$ and $\tilde{\boldsymbol{X}}_{\text{lsb}}$ are shown attached to the tip of the phasor $\boldsymbol{X}_{\text{car}}$, and they rotate around the tip of $\boldsymbol{X}_{\text{car}}$ with speeds of $\pm f_m$. The entire system of three phasors rotates around the origin at a speed of f_c. The AM signal can be written as

$$x_{AM}(t) = \mathrm{Re}\left\{\left(\boldsymbol{X}_{\mathrm{car}} + \tilde{\boldsymbol{X}}_{\mathrm{usb}} + \tilde{\boldsymbol{X}}_{\mathrm{lsb}}\right) e^{j2\pi f_c t}\right\} \tag{11.29}$$

One interesting use of the phasor representation derived above is finding the envelope of the AM signal. In the time domain, we know that the envelope of the AM signal is obtained by connecting the positive peaks of the high-frequency carrier. The equivalent of this action in phasor domain would be to freeze the base rotation (around the origin) of the system of three phasors at a 0-degree angle, and allow only the slower rotations of the phasors $\tilde{\boldsymbol{X}}_{\mathrm{usb}}$ and $\tilde{\boldsymbol{X}}_{\mathrm{lsb}}$ at the rates $\pm f_m$ Hz around the tip of $\boldsymbol{X}_{\mathrm{car}}$. The use of this idea for determining the envelope is shown in Fig. 11.13.

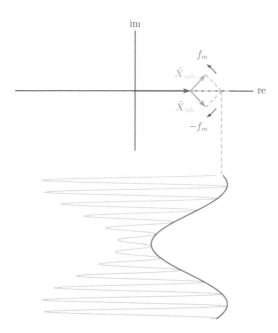

Figure 11.13 – Using phasor representation to determine the envelope of the AM signal.

The resulting envelope is

$$x_e(t) = \mathrm{Re}\left\{A_c + \frac{\mu A_c}{2} e^{j2\pi f_m t} + \frac{\mu A_c}{2} e^{-j2\pi f_m t}\right\}$$

$$= A_c + \mu A_c \cos(2\pi f_m t) \tag{11.30}$$

which is consistent with the expression found earlier for $\mu < 1$.

Interactive Demo: am_demo5

This demo illustrates the concepts discussed in Example 11.4. The system of three phasors $\boldsymbol{X}_{\mathrm{car}}$, $\tilde{\boldsymbol{X}}_{\mathrm{usb}}$ and $\tilde{\boldsymbol{X}}_{\mathrm{lsb}}$ that make up the phasor description of the AM signal is shown in the upper right part of the graphical user interface. The carrier frequency f_c, the single-tone message frequency f_m and the modulation index μ of the AM signal may each be controlled using the appropriate slider control. The time variable may be advanced using a slider control, and the rotation of the system of phasors may be observed. Pay particular attention to

1. The base rotation around the origin, and the slower rotation of the two sideband phasors $\tilde{\boldsymbol{X}}_{\text{usb}}$ and $\tilde{\boldsymbol{X}}_{\text{lsb}}$ around the tip of the carrier phasor $\boldsymbol{X}_{\text{car}}$
2. How the real part of the vector sum of three phasors corresponds to the amplitude of the AM signal $x_{AM}(t)$ shown in the graph at the bottom

Software resources:
am_demo5.m

Interactive Demo: am_demo6

This demo shows how the system of three phasors discussed in Example 11.4 relates to the envelope of the tone-modulated AM signal. In order to see just the envelope without the carrier, the base rotation of the system of phasors is stopped, that is, the carrier phasor $\boldsymbol{X}_{\text{car}}$ lays on the horizontal axis without rotating. The sideband phasors $\tilde{\boldsymbol{X}}_{\text{usb}}$ and $\tilde{\boldsymbol{X}}_{\text{lsb}}$ rotate with the frequency f_m. In this case, the real part of the vector sum of three phasors corresponds to the envelope of the AM signal.

Software resources:
am_demo6.m

11.4.2   Power balance and modulation efficiency

In the AM signal of Eqn. (11.3), the first term $A_c \cos(2\pi f_c t)$, referred to as the *carrier component*, is independent of the message signal. The sole purpose of transmitting the carrier component along with the message-dependent second term is to ensure that the envelope of the resulting signal resembles the message signal. From a power perspective, the transmission of the carrier component represents a waste of power. In contrast, the second term in Eqn. (11.3) includes the message signal $m(t)$, and thus represents the useful component of the AM signal. It would be interesting to find what percentage of the power in the transmitted AM signal is useful, and what percentage is wasted. The normalized average power of the AM signal is

$$P_x = \left\langle x_{AM}^2(t) \right\rangle = \left\langle [A_c + m(t)]^2 \cos^2(2\pi f_c t) \right\rangle$$

where the $\langle \ldots \rangle$ operator indicates time averaging as we have discussed in Section 1.3.5 of Chapter 1. Using the appropriate trigonometric identity,[4] Eqn. (11.31) can be written as

$$P_x = \left\langle \frac{1}{2} [A_c + m(t)]^2 \right\rangle + \left\langle \frac{1}{2} [A_c + m(t)]^2 \cos(4\pi f_c t) \right\rangle \tag{11.31}$$

---

[4] $\cos^2(a) = \frac{1}{2}[1 + \cos(2a)]$.

Assuming that the carrier frequency is significantly greater than the bandwidth of the message signal, the second term in Eqn. (11.31) averages to zero, and we have

$$P_x = \left\langle \frac{1}{2} \left[A_c + m\,(t) \right]^2 \right\rangle$$

$$= \left\langle \frac{1}{2} A_c^2 + A_c\, m\,(t) + \frac{1}{2}\, m^2\,(t) \right\rangle$$

$$= \frac{1}{2}\, A_c^2 + A_c\, \langle m\,(t) \rangle + \frac{1}{2}\, \langle m^2\,(t) \rangle \tag{11.32}$$

For most message signals $\langle m\,(t) \rangle = 0$, and Eqn. (11.32) reduces to

$$P_x = \frac{1}{2}\, A_c^2 + \frac{1}{2}\, \langle m^2\,(t) \rangle = P_c + P_m$$

The first term in Eqn. (11.33), $P_c = \frac{1}{2} A_c^2$, is the power in the carrier component, that is, the wasted power. In contrast, the second term $P_m = \frac{1}{2}\langle m^2\,(t) \rangle$ is the power in the sidebands, or the useful power. We will define the modulation efficiency as the percentage of the useful power within the total power of the AM signal:

$$\eta = \frac{\text{Power in sidebands}}{\text{Total power}} = \frac{P_m}{P_c + P_m} = \frac{\langle m^2\,(t) \rangle}{A_c^2 + \langle m^2\,(t) \rangle} \tag{11.33}$$

As expected, modulation efficiency depends on the significance of A_c^2 next to $\langle m^2\,(t) \rangle$. For practical message signals, lower and upper bounds of the allowable amplitude range are often symmetric, that is, $|m_{\min}| = |m_{\max}|$. For the case $\mu \leq 1$, this would lead to

$$|m_{\min}|, |m_{\max}| \leq A_c \tag{11.34}$$

As a result, the numerator term is bound by

$$\langle m^2\,(t) \rangle \leq A_c^2 \tag{11.35}$$

with the equality achieved only when the message signal is in the form of a square-wave type signal with amplitudes of $\pm A_c$ only. Thus, the maximum theoretical efficiency that can be attained with amplitude modulation (when the envelope is preserved) is $\eta = 0.5$, or 50 percent. Actual efficiency values obtained in practice will be significantly lower.

In summary, amplitude modulation is wasteful of power. This is the second major shortcoming of amplitude modulation besides the issue of inefficient bandwidth utilization discussed in the previous section.

Example 11.5: **Efficiency in tone modulation**

Determine the modulation efficiency for the tone-modulated AM signal of Example 11.1.

Solution: Recall that the modulating signal was given by

$$m\,(t) = \mu A_c\, \cos\,(2\pi f_m t)$$

Thus, the average normalized power of the message signal is

$$\langle m^2\,(t) \rangle = \langle \mu^2 A_c^2\, \cos^2\,(2\pi f_m t) \rangle = \frac{\mu^2 A_c^2}{2}$$

Using this result in Eqn. (11.33) we find the modulation efficiency of a tone-modulated AM signal to be

$$\eta = \frac{\mu^2}{2 + \mu^2}$$

In the case of tone-modulation, the maximum efficiency that can be achieved while preserving the signal envelope is $\eta = 0.333$, or 33.3 percent for the critically-modulated case $\mu = 1$. For smaller values of μ the efficiency will be even lower.

Example 11.6: **Multi-tone modulating signal revisited**

Determine the power efficiency of the AM signal of Example 11.3.

Solution: Recall that the AM signal obtained in Example 11.3 was

$$x_{AM}(t) = [4.648 + 3\cos(2\pi(100)t - \pi/6) + 2\cos(2\pi(200)t - \pi/2)]\cos(2\pi(1000)t)$$

The average normalized power in the message signal is

$$\langle m^2(t) \rangle = \langle 9\cos^2(2\pi(100)t - \pi/6) \rangle + \langle 4\sin^2(2\pi(200)t) \rangle$$
$$= \frac{9}{2} + \frac{4}{2} = 6.5$$

The average normalized power of the AM signal including the carrier is

$$\langle x_{AM}^2(t) \rangle = \langle A_c^2\cos^2(2\pi(1000)t) \rangle + \langle m^2(t) \rangle$$
$$= \frac{A_c^2}{2} + \langle m^2(t) \rangle$$
$$= \frac{(4.648)^2}{2} + 6.5 = 17.3$$

Power efficiency is found as the ratio of the two results:

$$\eta = \frac{6.5}{17.3} = 0.3756 = 37.6 \text{ percent}$$

11.4.3   Generation of AM signals

Consider again the basic form of the AM signal given by Eqn. (11.3). Theoretically it is possible to produce this signal by using a *product modulator* as shown in Fig. 11.14.

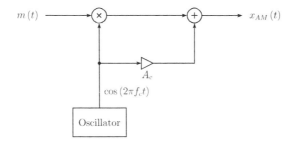

Figure 11.14 – Product modulator block diagram for generating AM signal.

The system uses an analog multiplier, an analog adder, and an amplifier. An oscillator generates the sinusoidal carrier signal $\cos(2\pi f_c t)$. The message signal and the carrier signal are first multiplied to obtain the second term of Eqn. (11.3). Afterwards, a properly scaled version of the carrier signal is added into the mix to complete the AM signal. The adder and the amplifier in Fig. 11.14 can be implemented easily; however, multiplication of two analog signals in time is an operation that is only practical at relatively low power levels. Radio broadcast stations produce AM signals at power levels typically in the 10 to 50 kW range. Implementation of analog multipliers is impractical at such power levels, hence we need to look into alternative means of generating high-power AM signals. We will consider two of these alternatives, namely a switching modulator, and a square-law modulator.

Switching modulator

Consider the arrangement shown in Fig. 11.15 where the carrier signal $x_c(t) = B_c \cos(2\pi f_c t)$ and the message signal $m(t)$ are added together, and their sum is applied to a diode rectifier that consists of a diode and a resistor.

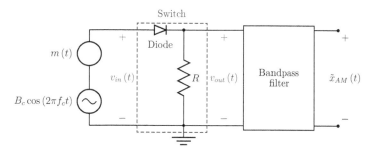

Figure 11.15 – Switching modulator circuit.

Ideally, the role of the diode rectifier would be to replace negative signal amplitudes with zero while keeping positive amplitudes unchanged. It is assumed that the amplitude of the sum signal is large enough for the diode to be modeled as an ideal switch. Furthermore, we will ensure that the amplitude of the carrier term within the sum is significantly greater than the message term, that is, $B_c \gg m(t)$, so that the switching action of the diode and resistor combination is controlled largely by the carrier term alone. With these assumptions, the combination of the diode and the load resistor can be modeled as an ideal switch with the characteristic as shown in Fig. 11.16

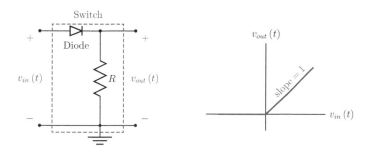

Figure 11.16 – Diode-based switch and its input-output characteristic.

The input to the switching circuit with the diode and the load resistor is

$$v_{in}(t) = B_c \cos(2\pi f_c t) + m(t) \tag{11.36}$$

and the output voltage of the switching circuit is

$$v_{out}(t) = \begin{cases} v_{in}(t), & \text{if } x_c(t) \geq 0 \\ 0, & \text{otherwise} \end{cases} \tag{11.37}$$

The assumption $B_c \gg m(t)$ ensures that the switching action in Eqn. (11.37) is controlled by $x_c(t)$, and the effect of the message signal $m(t)$ on switching is negligibly small. Mathematically, the switching relationship in Eqn. (11.37) is equivalent to multiplying $v_i(t)$ with a hypothetical pulse train:

$$v_{out}(t) = v_{in}(t)\, p(t) \tag{11.38}$$

This relationship is depicted in Fig. 11.17.

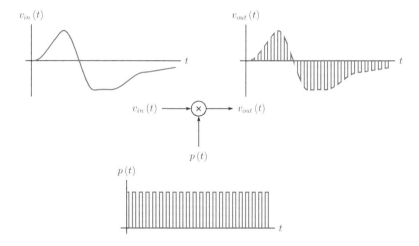

Figure 11.17 – Modeling the switching circuit through multiplication by a pulse train.

The pulse train $p(t)$ is periodic with the same period $T_c = 1/f_c$ as the carrier, and it has a duty cycle of $d = 0.5$. (The assumption $B_c \gg m(t)$ that we made earlier was needed for this purpose.) The sinusoidal carrier is shown in Fig. 11.17 with dashed lines to illustrate its relationship to the pulse train $p(t)$. An alternative yet equivalent perspective is given in Fig. 11.18 where the relationship in Eqn. (11.38) is implemented in the form of an electronic switch controlled by the pulse train $p(t)$.

For the duration of each pulse, that is, when $p(t) = 1$, the switch is in the top position, connecting the signal $v_{in}(t)$ to the input of the bandpass filter. Between pulses, the switch is in the lower position, grounding the input of the bandpass filter. The TFS representation of the pulse train $p(t)$ in Eqn. (11.38) is (see Section 4.2.2 of Chapter 4)

$$p(t) = a_0 + \sum_{k=1}^{\infty} a_k \cos(2\pi k f_c t) \tag{11.39}$$

Note that the fundamental frequency in Eqn. (11.39) is the same as the carrier frequency f_c. The coefficients of the Fourier series expansion are found as

$$a_0 = \frac{1}{2} \tag{11.40}$$

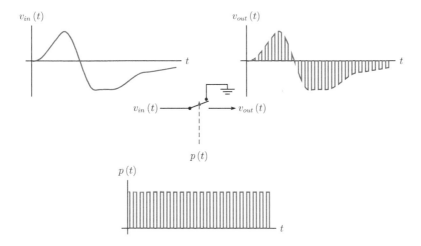

Figure 11.18 – Switching modulator circuit using a switch controlled by a pulse train.

and

$$a_k = \frac{2}{\pi n} \sin\left(\frac{\pi n}{2}\right), \quad k = 1, 2, \ldots$$

$$= \left\{ \frac{2}{\pi}, \, 0, \, -\frac{2}{3\pi}, \, 0, \, \frac{2}{5\pi}, \ldots \right\}, \quad k = 1, 2, \ldots \tag{11.41}$$

Using Eqns. (11.39) through (11.41) along with Eqn. (11.37), we can write Eqn. (11.38) as

$$v_{out}(t) = [B_c \cos(2\pi f_c t) + m(t)]$$

$$\times \left[\frac{1}{2} + \frac{2}{\pi} \cos(2\pi f_c t) - \frac{2}{3\pi} \cos(6\pi f_c t) + \frac{2}{5\pi} \cos(10\pi f_c t) - \ldots \right] \tag{11.42}$$

or as

$$v_{out}(t) = \frac{B_c}{\pi} + \frac{1}{2} m(t) + \frac{B_c}{2} \cos(2\pi f_c t) + \frac{2}{\pi} m(t) \cos(2\pi f_c t)$$

$$+ \{\text{high frequency terms}\} \tag{11.43}$$

The purpose of the bandpass filter in Fig. 11.15 becomes apparent from Eqn. (11.43). The first two terms are at low frequencies: B_c/π is a constant, and $\frac{1}{2} m(t)$ has frequencies extending up to W Hz. The next two terms correspond to an AM signal, and they are the terms we need to keep so that we have

$$\tilde{x}_{AM}(t) = \frac{B_c}{2} \cos(2\pi f_c t) + \frac{2}{\pi} m(t) \cos(2\pi f_c t) \tag{11.44}$$

The remaining terms in Eqn. (11.42) are at higher frequencies. The bandpass filter is needed to eliminate the undesired low-frequency and high-frequency terms. It needs to be designed such that the content in the frequency range

$$f_c - W \le f \le f_c + W$$

is kept, and the frequencies outside that range are eliminated. Also, in order to avoid any overlap between the spectrum of the AM signal and the spectra of low-frequency terms, we need $f_c > 2W$. This is not a very restricting requirement since, in most practical cases, $f_c >> W$. In Eqn. (11.44), the parameter B_c can be adjusted to obtain the desired modulation index. Comparing Eqn. (11.44) with Eqn. (11.8) it is obvious that $A_c = B_c/2$, and the modulation index can be determined by

$$\mu = \frac{(2/\pi)\,\max\left\{|m\,(t)\,|\right\}}{B_c/2} = \frac{4\,\max\left\{|m\,(t)\,|\right\}}{\pi B_c} \tag{11.45}$$

Interactive Demo: am_demo7

This demo program implements a simulation of the switching modulator discussed above. Implementation is based on Eqns. (11.36) through (11.45), using a single-tone modulating signal $m\,(t) = \cos\,(2\pi f_m t)$. The parameters B_c, f_c and f_m are adjustable through the use of slider controls in the interface. The undesired terms of Eqn. (11.43) are filtered out using a fifth-order Butterworth bandpass filter. The passband edge frequencies f_1 and f_2 of this filter (refer to MATLAB Example 11.3) are controlled indirectly by means of a center frequency f_b and a passband width f_v so that the lower edge of the passband is at

$$f_1 = f_b - \frac{1}{2}\,f_v$$

and its higher edge is at

$$f_2 = f_b + \frac{1}{2}\,f_v$$

It should be understood that, by "edge frequency" for a Butterworth filter, we are referring to a frequency where the magnitude of the system function decreases from its peak value by 3 dB.

   The modulation index that results from this process is given by Eqn. (11.45). For the single-tone modulating signal used in this case $\max\{|m\,(t)|\} = 1$, and Eqn. (11.45) reduces to

$$\mu = \frac{4}{\pi B_c}$$

which is displayed as the parameter B_c is varied. The signals $v_{in}\,(t)$, $v_{out}\,(t)$, and $\tilde{x}_{AM}\,(t)$ are computed and graphed. Additionally, the ideal envelope

$$x_e\,(t) = \frac{B_c}{2}\,[1 + \mu\,m\,(t)]$$

is displayed superimposed with $\tilde{x}_{AM}\,(t)$ for comparison. Vary the simulation parameters and observe the following:

1. The relationship between the modulation index μ and the carrier amplitude B_c

2. The summing of the carrier and the message signals to obtain the signal $v_{in}\,(t)$

3. The relationship between the signals $v_{in}\,(t)$ and $v_{out}\,(t)$

4. The time lag between the ideal envelope and the synthesized AM signal that is due to the Butterworth bandpass filter

5. The effect of the edge frequencies of the Butterworth filter on the quality of the synthesized signal and on the time lag

6. The relationship between f_c, f_m, and the bandpass filter edge frequencies as dictated by Eqn. (11.43)

Software resources:
am_demo7.m

Software resources: See MATLAB Exercises 11.3 and 11.4.

Square-law modulator

The basis of a square-law modulator is a nonlinear device with a strong second-order non-linearity. Consider an ideal second-order nonlinear device with the following input-output characteristic:

$$v_{out}\left(t\right) = a\,v_{in}\left(t\right) + b\,v_{in}^2\left(t\right) \tag{11.46}$$

In practice, a diode or a transistor operating in small-signal mode would produce a characteristic that is fairly close to Eqn. (11.46). As in the case of the switching modulator, let the input voltage to this nonlinear device be the sum of the carrier and the message signal:

$$v_{in}\left(t\right) = B_c\,\cos\left(2\pi f_c t\right) + m\left(t\right) \tag{11.47}$$

Using Eqn. (11.47) in Eqn. (11.46), the output voltage of the nonlinear device is found to be

$$v_{out}\left(t\right) = a\,\left[B_c\,\cos\left(2\pi f_c t\right) + m\left(t\right)\right] + b\,\left[B_c\,\cos\left(2\pi f_c t\right) + m\left(t\right)\right]^2 \tag{11.48}$$

which can be written as

$$\begin{aligned} v_{out}\left(t\right) =&\, a\,B_c\,\cos\left(2\pi f_c t\right) + a\,m\left(t\right) + b\,B_c^2\,\cos^2\left(2\pi f_c t\right) \\ &+ b\,m^2\left(t\right) + 2b\,B_c\,m\left(t\right)\,\cos\left(2\pi f_c t\right) \end{aligned} \tag{11.49}$$

In Eqn. (11.49) we recognize the first and the last terms as the ones that we need for an amplitude modulated carrier. The three terms in between are the terms that we will need to eliminate. Let us rearrange Eqn. (11.49) by combining the useful terms together to obtain

$$\begin{aligned} v_{out}\left(t\right) =&\, a\,B_c\,\left[1 + \frac{2b}{a}\,m(t)\right]\,\cos\left(2\pi f_c t\right) \\ &+ a\,m\left(t\right) + b\,B_c^2\,\cos^2\left(2\pi f_c t\right) + b\,m^2\left(t\right) + 2b\,B_c\,m\left(t\right)\,\cos\left(2\pi f_c t\right) \end{aligned} \tag{11.50}$$

It is obvious that the first term is the AM signal with $A_c = aB_c$ and $\mu = 2b/a$. The remaining terms need to be eliminated to yield an amplitude-modulated signal

$$\tilde{x}_{AM}\left(t\right) = aB_c\,\left[1 + \frac{2b}{a}\,m\left(t\right)\right]\,\cos\left(2\pi f_c t\right) \tag{11.51}$$

In order to find a way to eliminate the undesired terms, we need to understand what range of frequencies each term occupies in the frequency spectrum of $v_{out}\left(t\right)$. Assuming that the message signal $m\left(t\right)$ has a bandwidth of W, frequency ranges of individual terms of Eqn. (11.50) are detailed in Table 11.1.

Index	Term	Frequencies
1	$a\,B_c\,\left[1 + (2b/a)\,m\left(t\right)\right]\cos\left(2\pi f_c t\right)$	$f_c - W \leq f \leq f_c + W$
2	$a\,m\left(t\right)$	$-W \leq f \leq W$
3	$b\,B_c^2\,\cos^2\left(2\pi f_c t\right)$	$f = 0$, and $f = 2f_c$
4	$b\,m^2\left(t\right)$	$-2W \leq f \leq 2W$

Table 11.1 – Terms in Eqn. (11.50) and the frequencies they occupy.

The frequency spectrum of $v_{out}(t)$ is shown in Fig. 11.19. The terms of the spectrum are numbered based on their indices in Table 11.1.

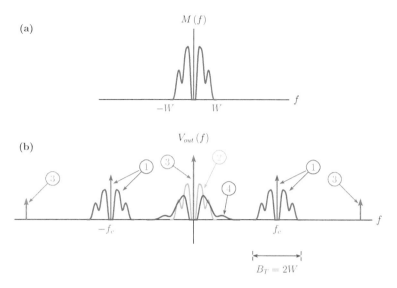

(a)

(b)

Figure 11.19 – Frequency spectrum of the signal v_{out} of Eqn. (11.50.

An inspection of the spectrum in Fig. 11.19 reveals that the amplitude modulated carrier may be isolated and the undesired terms may be removed by processing the signal $v_{out}(t)$ through a bandpass filter. Thus, the square-law modulator is completed by connecting an appropriate bandpass filter to the output of the nonlinear device. A block diagram for it is shown in Fig. 11.20.

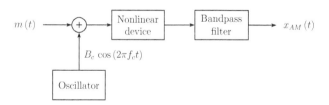

Figure 11.20 – Square-law modulator.

Interactive Demo: am_demo8

This demo provides a simulation of the square-law modulator considered in Fig. 11.20, Eqns. (11.46) through (11.50). As in the demo program "am_demo7.m", a single-tone modulating signal $m(t) = \cos(2\pi f_m t)$ is used. Parameters encountered in square-law modulator discussion above, namely f_c, f_m, a and b, may be adjusted using the slider controls provided. The signals $v_{in}(t)$ and $v_{out}(t)$ are computed by direct application of Eqns. (11.47) and (11.48) respectively. Afterwards, the signal $\tilde{x}_{AM}(t)$ is computed by processing the signal $v_{out}(t)$ through a fifth-order Butterworth bandpass filter with user-specified critical frequencies f_1 and f_2 which are again specified indirectly by means of parameters f_b and f_v as

$$f_1 = f_b - \frac{1}{2} f_v$$

and its higher edge is at

$$f_2 = f_b + \frac{1}{2} f_v$$

The effect of the parameter B_c on the modulator output signal $\hat{x}_{AM}(t)$ is relatively insignif-
icant; therefore this parameter is fixed as $B_c = 1$. The three signals $v_{in}(t)$, $v_{out}(t)$ and
$\tilde{x}_{AM}(t)$ are graphed. The modulation index for the synthesized signal is given by $\mu = 2b/a$,
and is displayed as the parameters a and b are varied. The ideal envelope

$$x_e(t) = aB_c [1 + \mu\, m(t)]$$

is displayed superimposed with $\tilde{x}_{AM}(t)$ for comparison. Vary the simulation parameters
and observe the following:

1. The relationship between the modulation index μ and the parameters a and b of the
   nonlinear characteristic
2. The relationship between the signals $v_{in}(t)$ and $v_{out}(t)$
3. The time lag between the ideal envelope and the synthesized AM signal that is due
   to the Butterworth bandpass filter
4. The effect of the edge frequencies of the Butterworth filter on the quality of the
   synthesized signal and on the time lag
5. The relationship between f_c, f_m, and the bandpass filter edge frequencies as dictated
   by Table (11.1)

Software resources:
am_demo8.m

Software resources: See MATLAB Exercises 11.5 and 11.6.

11.4.4   Demodulation of AM signals

The term *demodulation* refers to the act of recovering the message signal $m(t)$ as closely
to its original form as possible from the modulated waveform $x_{AM}(t)$. It was illustrated
in Figs. 11.4 and 11.5 that, as long as the modulation index is less than 100 percent, the
envelope of the modulated carrier resembles the message signal. In that case the simplest
means of demodulation is by using an envelope detector.

Envelope detector
Consider the simple circuit that consists of a diode, a capacitor and a resistor as shown in
Fig. 11.21.

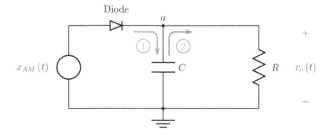

Figure 11.21 – Diode-based envelope detector circuit.

The modulated carrier is applied to this circuit in the form of a voltage signal. The output voltage approximately tracks the envelope of the AM waveform as we will illustrate. Two cases need to be considered:

Case 1: $x_{AM}(t) \geq v_c(t)$

> At time instants when the amplitude of the AM waveform is greater than or equal to the capacitor voltage, the diode of the envelope detector circuit is forward biased, and it conducts current. If we assume an ideal diode with no voltage drop between its terminals, the capacitor charges quickly to maintain $v_c(t) = x_{AM}(t)$. This corresponds to current path labeled (1) in Fig. 11.21.

Case 2: $x_{AM}(t) < v_c(t)$

> At time instants when the amplitude of the AM waveform is less than the capacitor voltage, the diode becomes reverse biased, and therefore no current flows through it. The capacitor that was charged to the peak value of the carrier must now discharge through the resistor. The discharge current follows the path labeled (2) in Fig. 11.21. Let the peak value of the carrier be $v_c(t_0) = V_0$ just before the capacitor begins its discharge at time $t = t_0$. We know that the current of the capacitor is related to its voltage by $i_c(t) = C\,dv_c(t)/dt$. Writing Kirchhoff's current law at the node labeled "a", we have

$$C\frac{dv_c(t)}{dt} + \frac{v_c(t)}{R} = 0 \tag{11.52}$$

which leads to the differential equation that governs the discharge of the capacitor:

$$\frac{dv_c(t)}{dt} = -\frac{1}{RC}v_c(t) \tag{11.53}$$

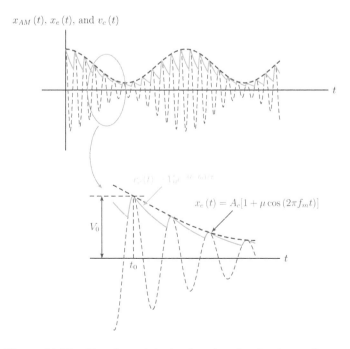

Figure 11.22 – Envelope detector input and output waveforms.

subject to the initial condition $v_c(t_0) = V_0$. The solution of this differential equation is in the form

$$v_c(t) = V_0 e^{-(t-t_0)/RC} \qquad (11.54)$$

where V_0 is the initial value at the beginning of the discharge cycle, that is, the peak value of the carrier at the point it switches from case 1 to case 2. This behavior of the envelope detector is illustrated in Fig. 11.22.

The time constant of the discharge is $\tau = RC$. It is important to note that the rate of decay for the exponential solution in Eqn. (11.54) is critical for the proper operation of the envelope detector. Thus, the time constant must be chosen carefully. If $v_c(t)$ decays too rapidly, the resulting envelope will look quite busy with too much high-frequency detail. If, on the other hand, $v_c(t)$ decays too slowly, then output will have sections where it deviates significantly from the correct envelope behavior as illustrated in Fig. 11.23.

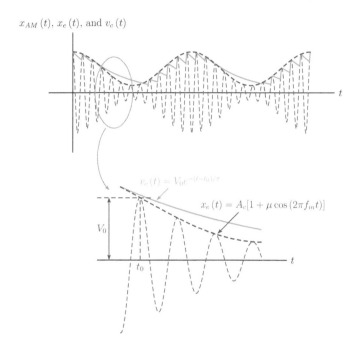

Figure 11.23 – Envelope detector with inappropriate choice of time constant.

We will address the problem of finding a proper value of τ in Example 11.7.

Example 11.7: **Envelope detector for tone-modulated AM signal**

Consider the envelope detector of Fig. 11.21 with a tone-modulated AM waveform as the input signal. Determine the range of the time constant τ needed to properly recover the single-tone message signal from the modulated waveform.

Solution: The envelope detector fills the gaps between the peaks of the carrier with decaying exponential voltage segments due to the discharging behavior of the RC circuit. In this process, the critical parameter is the time constant τ of the discharge. It was illustrated in Fig. 11.23 that, if the time constant is too large, the envelope detector output voltage fails to track the actual envelope. This happens because the slope of the actual envelope is more negative than the slope of the exponential discharge. Therefore, it would make sense to compare the slopes of the actual envelope and the exponential discharge to come up with the necessary condition that must be imposed on the choice of τ. We will begin with the actual envelope which, for the tone-modulated case, is

$$x_e(t) = A_c \left[1 + \mu \cos(2\pi f_m t)\right] \tag{11.55}$$

with the assumption that $\mu < 1$. The slope of the envelope is

$$\frac{dx_e(t)}{dt} = -2\pi f_m A_c \mu \sin(2\pi f_m t) \tag{11.56}$$

Let us pick a time instant $t = t_0$ at one of the peaks of the carrier as shown in Fig. 11.22. At that instant, the slope of the envelope is

$$\left.\frac{dx_e(t)}{dt}\right|_{t=t_0} = -2\pi f_m A_c \mu \sin(2\pi f_m t_0) \tag{11.57}$$

The envelope detector output right after the time instant $t = t_0$ is

$$\begin{aligned} v_c(t) &= x_e(t_0)\, e^{-(t-t_0)/\tau} \\ &= A_c \left[1 + \mu \cos(2\pi f_m t_0)\right] e^{-(t-t_0)/\tau} \end{aligned} \tag{11.58}$$

and the slope of the envelope detector output is

$$\frac{dv_c(t)}{dt} = -\frac{A_c}{\tau} \left[1 + \mu \cos(2\pi f_m t_0)\right] e^{-(t-t_0)/\tau} \tag{11.59}$$

Evaluating this slope at $t = t_0$ where it is maximum, we obtain

$$\left.\frac{dv_c(t)}{dt}\right|_{t=t_0} = -\frac{A_c}{\tau} \left[1 + \mu \cos(2\pi f_m t_0)\right] \tag{11.60}$$

Now we are ready to state the condition for the envelope detector output to correctly track the envelope of the AM signal. We need the slope in Eqn. (11.60) to be more negative than the value in Eqn. (11.57):

$$\left.\frac{dv_c(t)}{dt}\right|_{t=t_0} \leq \left.\frac{dx_e(t)}{dt}\right|_{t=t_0} \tag{11.61}$$

Let $\alpha \triangleq 2\pi f_m t_0$ to simplify the notation, and Eqn. (11.61) becomes

$$-\frac{1}{\tau}\left[1 + \mu \cos(\alpha)\right] \leq -2\pi f_m \mu \sin(\alpha) \tag{11.62}$$

or equivalently

$$\tau \leq \frac{1}{2\pi f_m}\left(\frac{1 + \mu \cos(\alpha)}{\mu \sin(\alpha)}\right) \tag{11.63}$$

It is important for the inequality in Eqn. (11.63) to be satisfied for all values of α, including when the right side has its smallest possible value. To determine the minimum value of the expression in parentheses, we need to differentiate it with respect to α and set the result equal to 0. This yields $\cos(\alpha) = -\mu$ and, consequently, $\sin(\alpha) = \sqrt{1 - \mu^2}$. Substituting these into Eqn. (11.63) we obtain

$$\tau \leq \frac{1}{2\pi f_m}\left(\frac{\sqrt{1 - \mu^2}}{\mu}\right) \tag{11.64}$$

The time constant $\tau = RC$ should satisfy the inequality in Eqn. (11.64) in order to avoid the loss of tracking for the envelope detector. The result we have derived is for the case when the modulating signal is a single tone. If the modulating signal has a range of frequencies, then Eqn. (11.64) can still be used heuristically, and the inequality should be satisfied for the highest frequency in the message signal, i.e.,

$$\tau \leq \frac{1}{2\pi\left[\max\left(f_m\right)\right]}\left(\frac{\sqrt{1-\mu^2}}{\mu}\right) \tag{11.65}$$

Interactive Demo: `am_demo9`

This demo is based on Example 11.7. It simulates an envelope detector used on an AM signal modulated by a single-tone message signal. Carrier frequency f_c, message frequency f_m, modulation index μ, and envelope detector time constant τ may be adjusted by using the slider controls provided. The program computes and graphs the AM waveform, the correct envelope, and the envelope detector output for the parameter values specified. The minimum time constant needed for proper detection of the envelope is computed using Eqn. (11.65), and is displayed.

1. Observe the dependency of the maximum time constant on the modulation index μ and the message frequency f_m.
2. Explain why the maximum time constant is not dependent on the carrier frequency f_c.
3. Push the time constant beyond its maximum value, and observe the envelope detector output. At what part of the signal does the envelope detector stop tracking the correct envelope?

Software resources:
`am_demo9.m`

Coherent demodulation

The simple demodulation technique discussed in the previous section only works if the modulation index is less than 100 percent. For $\mu > 1$ we have seen that the envelope of the modulated carrier is a distorted version of the message signal, and is therefore not useful in the demodulation process. The alternative is to use *coherent demodulation* which requires the availability of a local carrier signal synchronized with the original carrier signal that was used in generating $x_{AM}(t)$. Consider the block diagram shown in Fig. 11.24.

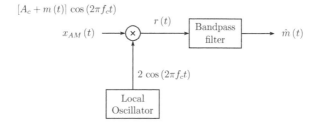

Figure 11.24 – Coherent demodulation of an AM signal.

The product $r(t)$ at the output of the multiplier is

$$r(t) = 2\,x_{AM}(t)\cos(2\pi f_c t)$$
$$= 2\left[A_c + m(t)\right]\cos^2(2\pi f_c t) \qquad (11.66)$$

Using the appropriate trigonometric identity Eqn. (11.66) can be written as

$$r(t) = A_c + m(t) + \left[A_c + m(t)\right]\cos(4\pi f_c t) \qquad (11.67)$$

Using the sample message spectrum $M(f)$ and the corresponding AM signal spectrum $X_{AM}(f)$ shown in Fig. 11.7 the spectrum of the intermediate signal $r(t)$ is as shown in Fig. 11.25.

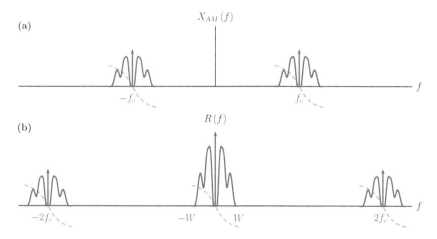

Figure 11.25 – Frequency spectrum $R(f)$ of the intermediate signal in the coherent demodulator shown in Fig. 11.24.

It is clear from Fig. 11.25 that, in order to recover the message spectrum $M(f)$ from $R(f)$ the high frequency component around $2f_c$ and the dc component should be removed (we are assuming that the message signal $m(t)$ does not have a dc component). That is the role of the bandpass filter.

11.5   Double-Sideband Suppressed Carrier Modulation

In Section 11.4.2 we have seen that one of the shortcomings of amplitude modulation is the amount of power wasted on transmitting the carrier along with the two sidebands. Modulation efficiency given by Eqn. (11.33) is theoretically limited to 50 percent, and is much lower in practice. Less than half the transmitted power is used on information bearing signals, and the rest of the power is wasted for transmitting the carrier.

One method of fixing this shortcoming is to simply drop the power-wasting term $A_c \cos(2\pi f_c t)$ from the AM waveform expression in Eqn. (11.3). This results in the signal

$$x_{DSB}(t) = m(t)\cos(2\pi f_c t) \qquad (11.68)$$

which is referred to as the *double-sideband suppressed-carrier (DSB-SC)* modulated signal. In the remainder of this text, we will refer to this signal simply as the DSB modulated signal. Fig. 11.26 illustrates the DSB signal in the time domain.

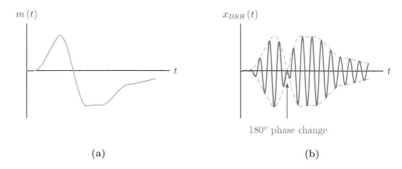

Figure 11.26 – Waveforms involved in DSB modulation: (a) the message signal, (b) the DSB modulated carrier.

The power efficiency of the DSB modulated signal is 100 percent since no power is being wasted on any non-information-bearing signal components. On the other hand, a comparison of Fig. 11.26 to Fig. 11.4 makes it clear that the envelope is lost in the case of DSB modulation, so we can no longer rely on the simplicity of an envelope detector for recovering the message signal from the modulated signal. A demodulation technique with higher complexity must be employed instead. Specifically, at the receiver end of the communication system, we will need a carrier the frequency and the phase of which are in perfect synchronization with the original carrier used in creating the DSB modulated signal (coherent demodulation). This is the price paid for improved power efficiency.

Interactive Demo: dsb_demo1

This demo uses the signals in Fig. 11.26. The message signal and the blank carrier are shown on the left side of the graphical user interface. On the right side of the graphical user interface, the DSB modulated signal is shown as computed by Eqn. (11.68). A slider control and an edit control are provided for changing the carrier frequency f_c.

1. Observe the changes in the DSB signal as the carrier frequency f_c is varied.
2. Notice that the envelope no longer resembles the message signal $m(t)$.
3. Pay attention to the 180-degree phase reversal of the carrier at the point of envelope crossover.

Software resources:
dsb_demo1.m

Example 11.8: **Tone modulation with DSB**

Determine and graph the DSB-SC signal when the modulating signal is a single tone in the form

$$m(t) = A_m \cos(2\pi f_m t)$$

Solution: Using the specified $m(t)$ the resulting DSB modulated signal is

$$x_{DSB}(t) = A_m \cos(2\pi f_m t) \cos(2\pi f_c t) \tag{11.69}$$

which is shown in Fig. 11.27.

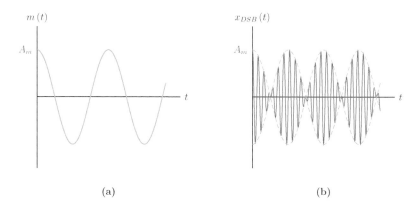

Figure 11.27 – DSB-SC modulation using a single-tone message signal in Example 11.8: (a) message signal, (b) DSB-SC signal.

The positive envelope of the DSB signal is

$$x_e\left(t\right) = \left|m\left(t\right)\right| = \left|A_m \, cos\left(2\pi f_m t\right)\right| \tag{11.70}$$

As expected, it no longer resembles the message signal $m\left(t\right)$.

Software resources:
ex_11_8.m

Interactive Demo: dsb_demo2

This demo is based on Example 11.8 and Fig. 11.27. The DSB modulated signal is graphed with the single-tone modulating signal as given by Eqn. (11.69). Slider controls are provided for changing the values of the message amplitude A_m, the single-tone message frequency f_m, and the carrier frequency f_c. Observe the following:

1. The relationship between the message signal $m(t)$ and the positive envelope $x_e(t)$ of the DSB signal
2. The effects of changing f_c and f_m
3. The 180-degree phase change that occurs in the carrier when the envelope crosses over the horizontal axis

Software resources:
dsb_demo2.m

11.5.1   Frequency spectrum of the DSB-SC signal

The mathematical expression for the double-sideband suppressed-carrier signal was given by Eqn. (11.68). Taking the Fourier transform of both sides of Eqn. (11.68) and using the modulation property discussed in Section 4.3.5 yields

$$X_{DSB}\left(f\right) = \frac{1}{2} M\left(f + f_c\right) + \frac{1}{2} M\left(f - f_c\right) \tag{11.71}$$

which is illustrated in Fig. 11.28.

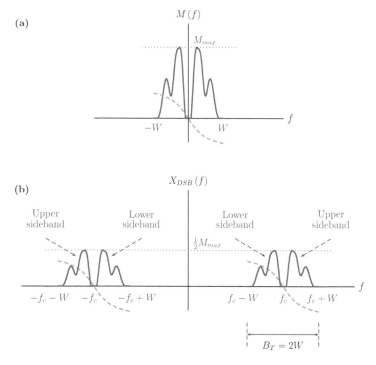

Figure 11.28 – (a) Frequency spectrum $M(f)$ of the message signal; magnitude (solid) and phase (dashed), (b) frequency spectrum $X_{DSB}(f)$ of the double-sideband suppressed carrier; magnitude (solid) and phase (dashed).

A quick comparison of Fig. 11.28 and Eqn. (11.71) with their counterparts for amplitude modulation in Section 11.4.1, namely Fig. 11.7 and Eqn. (11.10), reveals that the only difference is the absence of the impulses at $\pm f_c$ from the DSB-SC spectrum. Upper and lower sidebands are present as before. The transmission bandwidth, $B_T = 2W$, is also unchanged.

Interactive Demo: dsb_demo3

This demo follows Fig. 11.28 and Eqn. (11.71). The spectrum of a sample message signal, and the spectrum of the corresponding DSB-SC signal are shown. The sample message spectrum picked for this demo has a bandwidth of 1 kHz and a peak magnitude of $\max\{|M(f)|\} = 1$. A slider control is provided for adjusting the carrier frequency f_c in the range $2\,\text{kHz} \leq f_c \leq 9\,\text{kHz}$.

1. Observe how the sidebands move inward or outward as the carrier frequency is changed.
2. Compare the spectrum of the DSB signal to that of the AM signal in Fig. 11.7 and MATLAB demo am_demo3.m.

Software resources:
dsb_demo3.m

Example 11.9: **Frequency spectrum of the tone-modulated DSB-SC signal**

Determine and graph the frequency spectrum of the tone-modulated DSB-SC signal.

Solution: The expression for a tone-modulated DSB signal was given in Eqn. (11.69). Using the appropriate trigonometric identity it can be written as

$$x_{DSB}(t) = \frac{A_m}{2}\cos(2\pi[f_c + f_m]t) + \frac{A_m}{2}\cos(2\pi[f_c - f_m]t) \qquad (11.72)$$

Eqn. (11.72) contains two cosine terms at frequencies $f_c + f_m$ and $f_c - f_m$. For the tone-modulated case these two cosine terms constitute the upper and the lower sideband respectively. Individual Fourier transforms of the two terms in Eqn. (11.72) are

$$\frac{A_m}{2}\cos(2\pi[f_c + f_m]t) \quad\overset{\mathcal{F}}{\longleftrightarrow}\quad \frac{A_m}{4}\delta(f + [f_c + f_m]) + \frac{A_m}{4}\delta(f - [f_c + f_m])$$

$$\frac{A_m}{2}\cos(2\pi[f_c - f_m]t) \quad\overset{\mathcal{F}}{\longleftrightarrow}\quad \frac{A_m}{4}\delta(f + [f_c - f_m]) + \frac{A_m}{4}\delta(f - [f_c - f_m])$$

and the spectrum of the tone-modulated DSB signal is obtained as

$$X_{DSB}(f) = \frac{A_m}{4}\delta(f + [f_c + f_m]) + \frac{A_m}{4}\delta(f - [f_c + f_m])$$

$$+ \frac{A_m}{4}\delta(f + [f_c - f_m]) + \frac{A_m}{4}\delta(f - [f_c - f_m]) \qquad (11.73)$$

Frequency spectra for the single-tone message signal and the amplitude-modulated signal of Eqn. (11.73) are shown in Fig. 11.29.

(a)

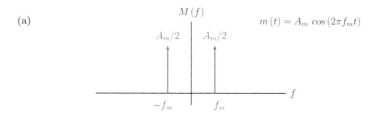

(b)

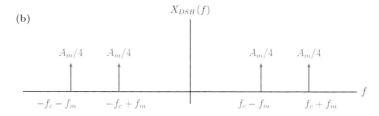

Figure 11.29 – (a) Frequency spectrum $M(f)$ of the single-tone message signal, and (b) frequency spectrum of the tone-modulated DSB-SC signal of Example 11.9.

The DSB spectrum for a single tone modulating signal is similar to the spectrum of the AM carrier with the same modulating signal which was shown in Fig. 11.8. The main difference is the lack of impulses at frequencies $\pm f_c$.

Interactive Demo: dsb_demo4

This demo illustrates the concepts explored in Example 11.9 and Fig. 11.29. The two graphs show the spectrum of the single-tone message signal $m(t) = A_m \cos(2\pi f_m t)$ and the spectrum of the corresponding DSB-SC signal. Slider controls allow adjustments to the message amplitude A_m, the message frequency f_m and the carrier frequency f_c.

1. Pay attention to the frequency shift of the sidebands of the spectrum as the carrier frequency is changed.
2. Observe how the sideband impulses move toward each other or away from each other as the message frequency is changed.
3. Observe the values of the impulses as the message amplitude is changed.

Software resources:
dsb_demo4.m

11.6 Single-Sideband Modulation

One of the shortcomings of amplitude modulation, namely the power efficiency, was addressed in Section 11.5. Dropping the carrier term is the method of improving the power efficiency at the expense of giving up the envelope that resembles the message signal, and the opportunity for simple non-coherent detection afforded by it. Another shortcoming of amplitude modulation is the doubling of the bandwidth. For a message signal with bandwidth W Hz, the transmission bandwidth of both the AM and the DSB-SC waveforms is $B_T = 2W$ as is evident from Figs. 11.7 and 11.28.

Consider a message signal $m(t)$ with the frequency spectrum $M(f)$ shown in Fig. 11.30.

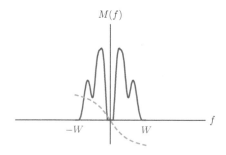

Figure 11.30 – Frequency spectrum $M(f)$ for a sample message signal $m(t)$.

The message signal $m(t)$ is real-valued. Based on the discussion in Section 4.3.5, its frequency spectrum $M(f)$ exhibits conjugate symmetry:

$$M^*(f) = M(-f) \tag{11.74}$$

As a result, the magnitude of the spectrum $M(f)$ is an even function of frequency, and the phase of it is an odd function of frequency

$$|M(f)| = |M(-f)| \tag{11.75}$$

and

$$\angle M(f) = -\angle M(-f) \tag{11.76}$$

Let us split the message spectrum into two parts defined as

$$M_n(f) = \begin{cases} M(f), & f < 0 \\ 0, & f > 0 \end{cases} \tag{11.77}$$

and

$$M_p(f) = \begin{cases} 0, & f > 0 \\ M(f), & f > 0 \end{cases} \tag{11.78}$$

so that $M(f) = M_n(f) + M_p(f)$. Clearly, $M_n(f)$ represents the part of the message spectrum for negative frequencies, that is, the left side. Conversely, $M_p(f)$ represents the right side. This is illustrated in Fig. 11.31.

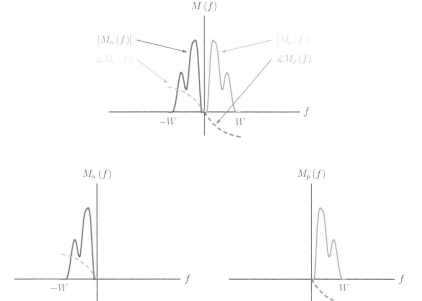

Figure 11.31 – Splitting the message spectrum $M(f)$ into its left and right halves.

Based on the symmetry properties outlined in Eqns. (11.74) through (11.76), one of the two components of the spectrum, either $M_n(f)$ or $M_p(f)$, makes the other term redundant. For example, if we know $M_p(f)$, we can obtain $M_n(f)$ as

$$M_n(f) = M_p^*(-f) \tag{11.79}$$

As a result, the transmission of a modulated signal with just one of the sidebands in it is sufficient for the receiver to get the message signal. This is referred to as *single-sideband (SSB)* modulation.

Recall that the spectrum of the DSB-SC signal was given by Eqn. (11.71), and includes both lower and upper sidebands. The DSB spectrum can be rewritten using the two components of $M(f)$ as

$$\begin{aligned} X_{DSB}(f) &= \frac{1}{2}M(f + f_c) + \frac{1}{2}M(f - f_c) \\ &= \frac{1}{2}M_n(f + f_c) + \frac{1}{2}M_p(f + f_c) + \frac{1}{2}M_n(f - f_c) + \frac{1}{2}M_p(f - f_c) \end{aligned} \tag{11.80}$$

A single-sideband spectrum can be formed by retaining either the upper sidebands or the lower sidebands. If the upper sidebands are retained, the resulting spectrum is

$$X_{SSB,U}(f) = \frac{1}{2} M_n(f + f_c) + \frac{1}{2} M_p(f - f_c) \qquad (11.81)$$

We will refer to $X_{SSB,U}(f)$ as the *upper-sideband SSB spectrum*. Alternatively, we can retain the lower sidebands to obtain

$$X_{SSB,L}(f) = \frac{1}{2} M_p(f + f_c) + \frac{1}{2} M_n(f - f_c) \qquad (11.82)$$

which is the *lower-sideband SSB spectrum*. Fig. 11.32 illustrates the results obtained in Eqns. (11.80) through (11.82).

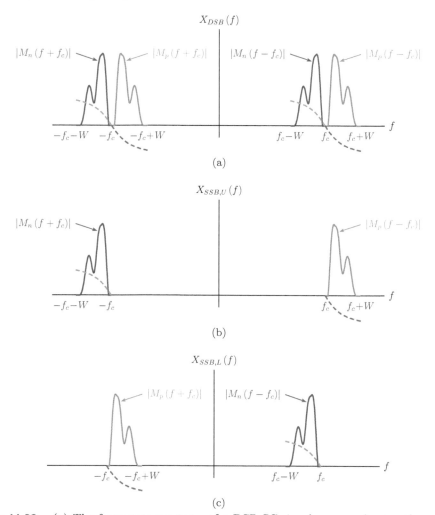

Figure 11.32 – (a) The frequency spectrum of a DSB-SC signal expressed using the spectral components $M_n(f)$ and $M_p(f)$, (b) the upper-sideband SSB spectrum $X_{SSB,U}(f)$, (c) the lower-sideband SSB spectrum $X_{SSB,L}(f)$.

The time-domain signals that correspond to the two SSB spectra of Eqns. (11.81) and (11.82) would be

$$x_{SSB,U}(t) = \mathcal{F}^{-1}\{X_{SSB,U}(f)\} \qquad (11.83)$$

and

$$x_{SSB,L}(t) = \mathcal{F}^{-1}\{X_{SSB,L}(f)\} \tag{11.84}$$

Theoretically, the SSB signals $x_{SSB,U}(t)$ and $x_{SSB,L}(t)$ can be obtained from the DSB signal $x_{DSB}(t)$ through sideband filtering as illustrated in Fig. 11.33.

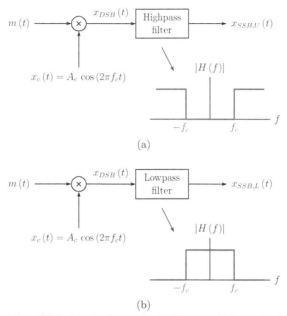

(a)

(b)

Figure 11.33 – Obtaining SSB signals from the DSB signal through sideband filtering: (a) block diagram for obtaining the upper-sideband SSB signal, and (b) block diagram for obtaining the lower-sideband SSB signal.

The block diagrams in Fig. 11.33 assume that the filters used have ideal characteristics. This approach is not used in practical implementation of SSB modulation since ideal filters are not available and would be near impossible to approximate since the allowed transition band would be extremely narrow relative to the carrier frequency.

One question that we have not yet considered up to this point is: *What does the time-domain SSB signal look like?* We will use Eqns. (11.81) and (11.82) to derive time-domain expressions for the lower-sideband and upper-sideband SSB signals. The message signal $m(t)$ can be written in the form

$$m(t) = m_n(t) + m_p(t) \tag{11.85}$$

where $m_n(t)$ and $m_p(t)$ are the time-domain signals that correspond to the two components of the spectrum, $M_n(f)$ and $M_p(f)$, respectively:

$$m_n(t) \overset{\mathcal{F}}{\longleftrightarrow} M_n(f) \tag{11.86}$$

$$m_p(t) \overset{\mathcal{F}}{\longleftrightarrow} M_p(f) \tag{11.87}$$

Using the frequency-shifting property of the Fourier transform, time-domain equivalents of Eqns. (11.81) and (11.82) can be derived as

$$x_{SSB,U}(t) = \frac{1}{2}m_n(t)\,e^{-j2\pi f_c t} + \frac{1}{2}m_p(t)\,e^{j2\pi f_c t} \tag{11.88}$$

and

$$x_{SSB,L}\left(t\right) = \frac{1}{2}\, m_p\left(t\right) e^{-j2\pi f_c t} + \frac{1}{2}\, m_n\left(t\right) e^{j2\pi f_c t} \tag{11.89}$$

Let us first consider $m_p\left(t\right)$ and its Fourier transform $M_p\left(f\right)$. The transform $M_p\left(f\right)$ is nonzero only for positive frequencies, and is not conjugate symmetric. As a result, we expect the time-domain signal $m_p\left(t\right)$ to be complex-valued. One way of relating the transform $M_p\left(f\right)$ to the message spectrum $M\left(f\right)$ is through the use of the signum function:

$$\begin{aligned}M_p\left(f\right) &= M\left(f\right)\left[\frac{1}{2} + \frac{1}{2}\, \mathrm{sgn}\left(f\right)\right] \\ &= \frac{1}{2}\, M\left(f\right) + \frac{1}{2}\, \mathrm{sgn}\left(f\right) M\left(f\right)\end{aligned} \tag{11.90}$$

Consider a signal $\hat{m}\left(t\right)$ derived from $m\left(t\right)$ and $M\left(f\right)$ through the Fourier transform relationship

$$\hat{m}\left(t\right) = \mathcal{F}^{-1}\left\{-j\,\mathrm{sgn}\left(f\right) M\left(f\right)\right\} \tag{11.91}$$

The Fourier transform of $\hat{m}\left(t\right)$ is

$$\hat{M}\left(f\right) = \mathcal{F}\left\{\hat{m}\left(t\right)\right\} = -j\,\mathrm{sgn}\left(f\right) M\left(f\right) \tag{11.92}$$

The signal $\hat{m}\left(t\right)$ is known as the *Hilbert transform* of the signal $m\left(t\right)$. It can be modeled as the output signal of a *Hilbert transform filter* with system function

$$H\left(f\right) = -j\,\mathrm{sgn}\left(f\right) \tag{11.93}$$

driven by the message signal $m\left(t\right)$ as illustrated in Fig. 11.34.

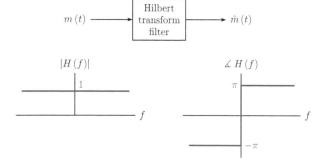

Figure 11.34 – Hilbert transform $\hat{m}\left(t\right)$ of the signal $m\left(t\right)$ modeled as the output of a Hilbert transform filter.

Substituting Eqn. (11.92) into Eqn. (11.90) we obtain

$$M_p\left(f\right) = \frac{1}{2}\, M\left(f\right) + j\,\frac{1}{2}\,\hat{M}\left(f\right) \tag{11.94}$$

and

$$m_p\left(t\right) = \frac{1}{2}\, m\left(t\right) + j\,\frac{1}{2}\,\hat{m}\left(t\right) \tag{11.95}$$

Mathematical expressions for the left half of the spectrum and the corresponding time-domain signal can also be obtained through similar derivations:

$$M_n\left(f\right) = \frac{1}{2}\, M\left(f\right) - j\,\frac{1}{2}\,\hat{M}\left(f\right) \tag{11.96}$$

and

$$m_n\left(t\right) = \frac{1}{2}\,m\left(t\right) - j\,\frac{1}{2}\,\hat{m}\left(t\right) \tag{11.97}$$

We are now ready to write the time-domain expressions for the upper-sideband SSB and lower-sideband SSB signals in terms of the message signal. Substituting Eqns. (11.95) and (11.97) into Eqn. (11.88) yields the the expression for the upper-sideband SSB signal as

$$x_{SSB,U}\left(t\right) = \frac{1}{2}\left[\frac{1}{2}\,m\left(t\right) - j\,\frac{1}{2}\,\hat{m}\left(t\right)\right]e^{-j2\pi f_c t} + \frac{1}{2}\left[\frac{1}{2}\,m\left(t\right) + j\,\frac{1}{2}\,\hat{m}\left(t\right)\right]e^{j2\pi f_c t} \tag{11.98}$$

which can be simplified to

$$\begin{aligned} x_{SSB,U}\left(t\right) &= \frac{1}{4}\,m\left(t\right)\left[e^{j2\pi f_c t} + e^{-j2\pi f_c t}\right] + \frac{1}{4}\,\hat{m}\left(t\right)\left[j\,e^{j2\pi f_c t} - j\,e^{-j2\pi f_c t}\right]\\ &= \frac{1}{2}\,m\left(t\right)\cos\left(2\pi f_c t\right) - \frac{1}{2}\,\hat{m}\left(t\right)\sin\left(2\pi f_c t\right) \end{aligned} \tag{11.99}$$

Similarly, the lower-sideband SSB signal can be found by substituting Eqns. (11.95) and (11.97) into Eqn. (11.89), and simplifying the result.

$$x_{SSB,L}\left(t\right) = \frac{1}{2}\,m\left(t\right)\cos\left(2\pi f_c t\right) + \frac{1}{2}\,\hat{m}\left(t\right)\sin\left(2\pi f_c t\right) \tag{11.100}$$

11.7   Further Reading

[1] L.W. Couch. *Digital and Analog Communication Systems*. Always Learning. Prentice Hall, 2013.

[2] B.P. Lathi and Z. Ding. *Modern Digital and Analog Communication Systems*. The Oxford Series in Electrical and Computer Engineering. Oxford University Press, 2009.

[3] J.G. Proakis, M. Salehi, and G. Bauch. *Contemporary Communication Systems Using MATLAB, 3rd ed.* Cengage Learning, 2011.

MATLAB Exercises

MATLAB Exercise 11.1: Compute and graph the AM signal

Develop a MATLAB script for generating samples of the tone-modulated AM signal given by Eqn. (11.8), using parameter values $f_c = 25$ kHz, $f_m = 2$ kHz, $A_c = 2$, and $\mu = 0.7$. Graph the resulting signal.

Solution: An important issue is the time increment used in evaluating amplitude values of the AM signal. Even though our final graph will mimic the behavior of an analog signal, it is obvious that any signal we generate with MATLAB is discrete by nature. In essence, we will be sampling the analog AM signal of Eqn. (11.8) to produce a discrete-time signal, and then graphing it to look like an analog signal. The time increment used in the calculation of AM signal values is the same as the sampling interval $T_s = 1/f_s$ discussed in Chapter 6. It needs to be chosen sufficiently small compared to the period of the carrier so that the resulting graph is an accurate representation of the AM signal. Nyquist sampling theorem states that the sampling rate f_s needs to be at least twice the highest signal frequency. In this example, however, we will go well beyond that requirement so that the resulting analog graph looks smooth. For the parameter values chosen, the highest frequency in the signal $x_{AM}(t)$ is $f_c + f_m = 27$ kHz. We will choose the time increment as $T_s = 10^{-6}$ seconds, corresponding to a sampling rate of $f_s = 1$ MHz. MATLAB code is listed below.

```
1    % Script: matex_11_1.m
2    %
3    Ac = 2;
4    Mu = 0.7;
5    fc = 25000;
6    fm = 2000;
7    t = [0:1e-6:1e-3];          % Vector of time instants.
8    % Compute the AM signal.
9    x_am = Ac*(1+Mu*cos(2*pi*fm*t)).*cos(2*pi*fc*t);   % Eqn. (11.8)
10   % Compute the envelope.
11   x_env = abs(Ac*(1+Mu*cos(2*pi*fm*t)));             % Eqn. (11.9)
12   % Graph the AM signal and the envelope.
13   clf;
14   plot(t,x_am,'-',t,x_env,'m--');
15   grid;
16   title('Tone modulated AM signal and its envelope');
17   xlabel('Time (sec)');
18   ylabel('Amplitude');
19   legend('AM signal','Envelope');
```

The graph that is generated by the script `matex_11_1.m` is shown in Fig. 11.35.

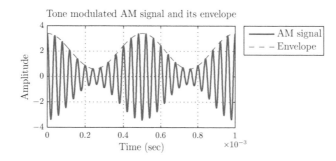

Figure 11.35 – MATLAB graph for single-tone modulated signal.

Software resources:
`matex_11_1.m`

MATLAB Exercise 11.2: EFS spectrum of the tone-modulated AM signal

Refer to the tone-modulated AM signal graphed in MATLAB Exercise 11.1. Determine its fundamental period. Afterwards compute and graph its EFS spectrum.

Solution: First we need to determine the periodicity of the signal. With parameter values $f_c = 25$ kHz and $f_m = 2$ kHz, the frequencies contained in the AM signal are $f_c = 25$ kHz, $f_c - f_m = 23$ kHz and $f_c + f_m = 27$ kHz. The fundamental frequency of the signal $x_{AM}(t)$ is $f_0 = 1$ kHz corresponding to a fundamental period of $T_0 = 1$ ms (see the discussion in Section 1.3.4 of Chapter 1). The script listed below computes samples of exactly one period of the tone-modulated AM signal and then uses the function `ss_efsapprox(..)` developed in MATLAB Exercise 5.13 of Chapter 5 to approximate the EFS line spectrum of the signal.

```
1    % Script: matex_11_2.m
2    %
3    Ac = 2;
4    Mu = 0.7;
5    fc = 25000;
6    fm = 2000;
7    t = [0:999]*1e-6;
8    x_am = Ac*(1+Mu*cos(2*pi*fm*t)).*cos(2*pi*fc*t);
9    clf;
10   k=[-30:30];
11   ck = ss_efsapprox(x_am,k);
12   stem(k,ck);
13   title('EFS coefficients for x_{AM}(t)');
14   xlabel('Index k');
```

The graph that is generated by the script `matex_11_2.m` is shown in Fig. 11.36.

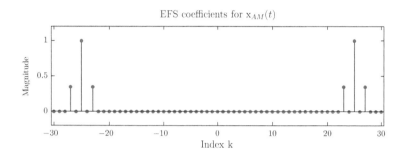

Figure 11.36 – MATLAB graph for line spectrum of tone-modulated AM signal.

Software resources:
`matex_11_2.m`

MATLAB Exercise 11.3: Function to simulate a switching modulator

In this example, we will develop a MATLAB function to simulate the behavior of the switching modulator discussed. In MATLAB, the switching behavior in Eqn. (11.37) can be implemented with the following statements:

```
>>    control = (carrier >=0);
>>    v_out = v_in.*control;
```

The two statements may also be combined as

```
>>    v_out = v_in.*(carrier >=0);
```

The next step is to use a bandpass filter for cleaning up the undesired terms. Function `butter(..)` will be used for designing a suitable bandpass filter for this purpose (see MATLAB Exercises 10.1, 10.9 and 10.10 in Chapter 10).

```
>>    [numz,denz] = butter(N,Wn,'bandpass');
```

Once a suitable bandpass filter is designed, function `filter(..)` is used for processing the switched signal through it. The listing for the function `ss_switchmod(..)` is given below. The input arguments to the function are msg: Vector holding samples of the message signal.

Bc: Amplitude of the carrier to be added to the message signal.

fc: Carrier frequency in Hz.

Ts: Time increment in seconds, used for the message signal in vector "msg", and also in computing samples of the synthesized AM signal.

f1: Lower edge frequency in Hz for the bandpass filter.

f2: Upper edge frequency in Hz for the bandpass filter.

The function returns a vector holding the samples of the synthesized AM waveform.

```
1   function x_am = ss_switchmod(msg,Bc,fc,Ts,f1,f2)
2     nSamp = length(msg);          % Number of samples in "msg"
3     t = [0:nSamp-1]*Ts;           % Vector of time instants
4     carrier = Bc*cos(2*pi*fc*t);
5     % Compute input to the diode switch.
6     v_in = carrier+msg;           % Eqn. (11.36)
7     % Simulate the diode switch.
8     v_out = v_in.*(carrier>=0);   % Eqn. (11.37)
9     % Design the bandpass filter.
10    [numz,denz] = butter(5,[2*f1*Ts,2*f2*Ts],'bandpass');
11    % Process switch output through bandpass filter.
12    x_am = filter(numz,denz,v_out);
```

It is important to understand the relationship between analog frequencies in Hz and the normalized frequencies used in bandpass filter design. Given the frequency interval $f_1 < f < f_2$ Hz that we would like to retain and a sampling rate of f_s used in the simulation, the normalized edge frequencies of the bandpass filter are

$$F_1 = \frac{f_1}{f_s} \qquad \text{and} \qquad F_2 = \frac{f_2}{f_s}$$

Since MATLAB defines normalized frequencies differently (see Chapter 10), the values used in the call to function butter(..) are $2F_1 = 2f_1/f_s$ and $2F_2 = 2f_2/f_s$.

Software resources:

ss_switchmod.m

MATLAB Exercise 11.4: **Testing the switching modulator**

In this exercise a tone-modulated AM waveform with message frequency $f_m = 2$ kHz, carrier frequency $f_c = 25$ kHz, and modulation index $\mu = 0.5$ will be generated in order to test the function ss_switchmod(..) developed in MATLAB Exercise 11.3. Let the sampling interval be $T_s = 1$ μs corresponding to a sampling rate of $f_s = 1$ MHz.

For the bandpass filter, a passband between $f_1 = 15$ kHz and $f_2 = 35$ kHz is sufficient. MATLAB script to test the switching modulator function is listed below.

```
1   % File: matex_11_4
2   %
3   t = [0:999]*1e-6;          % Vector of time instants
4   msg = cos(2*pi*2000*t);    % Single-tone message signal
5   Bc = 4/(0.5*pi);           % Achieve a modulation index of 0.5
6   x_am = ss_switchmod(msg,Bc,25000,1e-6,15000,35000);
7   % Compute correct envelope for comparison purposes.
8   x_env = Bc/2*(1+0.5*msg);
9   % Graph the synthesized AM signal and correct envelope.
10  clf;
11  plot(t,x_am,t,x_env);
12  grid;
```

```
13    title('AM signal synthesis using a switching modulator');
14    xlabel('t (sec)');
15    ylabel('Amplitude');
16    legend('Synthesized signal','Correct envelope');
```

The AM signal produced is shown in Fig. 11.37 superimposed with the ideal envelope of the single-tone modulated signal.

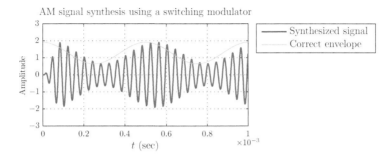

Figure 11.37 – MATLAB graph for switching modulator simulation.

A comparison of the synthesized AM waveform and the correct envelope in Fig. 11.37 indicates a lag in the AM waveform. This is due to the delay characteristics of the fifth-order Butterworth bandpass filter used. This delay could be increased or decreased by changing the filter order or the passband edge frequencies.

Software resources:

matex_11_4.m

MATLAB Exercise 11.5: Function to simulate a square-law modulator

The listing for a MATLAB function ss_sqlawmod(..) to implement a square-law modulator is given below. The input arguments to the function are

msg:	Vector holding samples of the message signal.
Bc:	Amplitude of the carrier to be added to the message signal.
fc:	Carrier frequency in Hz.
Ts:	Time increment in seconds, used for the message signal in vector "msg", and also in computing samples of the synthesized AM signal.
a:	The parameter a in the input-output characteristic of the nonlinear device in Eqn. (11.46).
b:	The parameter b in the input-output characteristic of the nonlinear device in Eqn. (11.46).
f1:	Lower edge frequency in Hz for the bandpass filter.
f2:	Upper edge frequency in Hz for the bandpass filter.

```
1    function x_am = ss_sqlawmod(msg,Bc,fc,Ts,a,b,f1,f2)
2
3      nSamp = length(msg);        % Number of samples in "msg"
4      t = [0:nSamp-1]*Ts;         % Vector of time instants
5      carrier = Bc*cos(2*pi*fc*t);
6      % Compute input to the nonlinear device.
7      v_in = carrier+msg;         % Eqn. (11.47)
8      % Simulate the nonlinear device switch.
9      v_out = a*v_in+b*v_in.*v_in; % Eqn. (11.46)
```

```
10    % Design the bandpass filter.
11    [num,den] = butter(5,[2*f1*Ts,2*f2*Ts],'bandpass');
12    % Process v_out through bandpass filter.
13    x_am = filter(num,den,v_out);
```

Software resources:
ss_sqlawmod.m

MATLAB Exercise 11.6: Testing the square-law modulator

In this exercise we test the function ss_sqlawmod(..) developed in MATLAB Exercise 11.5. A tone-modulated AM waveform with message frequency $f_m = 2$ kHz, carrier frequency $f_c = 25$ kHz, and modulation index $\mu = 0.5$ is used for testing. Based on Eqn. (11.51) the modulation index produced by the square-law modulator using the nonlinear characterictic of Eqn. (11.46) is $\mu = 2b/a$. Choosing $a = 1$ and $b = 0.25$ allows us to obtain $\mu = 0.5$. The sampling interval used in the generation of various signals is $T_s = 1$ μs corresponding to a sampling rate of $f_s = 1$ MHz. MATLAB script to test the square-law modulator is listed below.

```
1     % File: matex_11_6.m
2     %
3     t = [0:999]*1e-6;          % Vector of time instants
4     msg = cos(2*pi*2000*t);    % Single-tone message signal
5     % Set parameter value for Ac=1 and modulation index of 0.5
6     Bc = 1;
7     a = 1;
8     b = 0.25;
9     x_am = ss_sqlawmod(msg,Bc,25000,1e-6,1,0.25,15000,35000);
10    % Compute correct envelope for comparison purposes.
11    x_env = 1+0.5*msg;
12    % Graph the synthesized AM signal and correct envelope.
13    clf;
14    plot(t,x_am,t,x_env);
15    grid;
16    title('AM signal synthesis using a square-law modulator');
17    xlabel('t (sec)');
18    ylabel('Amplitude');
19    legend('Synthesized signal','Correct envelope');
```

The AM signal produced is shown in Fig. 11.38 superimposed with the ideal envelope of the single-tone modulated signal.

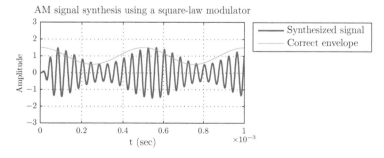

Figure 11.38 – MATLAB graph for square-law modulator simulation.

As in MATLAB Exercise 11.4, comparison of the synthesized AM waveform and the correct envelope in Fig. 11.38 indicates a lag in the AM waveform due to the delay characteristics of the fifth-order Butterworth bandpass filter used.

Software resources:

matex_11_6.m

MATLAB Exercise 11.7: Function to simulate envelope detector

In this exercise we develop a MATLAB function to simulate the actions of the envelope detector described in Section 11.4.4. Refer to the circuit in Fig. 11.21. To mimic the behavior of the envelope detector, two cases need to be considered, namely 1) $x_{AM}(t) \geq v_c(t)$, and 2) $x_{AM}(t) < v_c(t)$. The code for the function ss_envdet(..) is given below. The input arguments to the function are

> x_am: Vector holding samples of the AM signal.
>
> Ts: Time increment used in computing the samples of the AM signal.
>
> tau: Time constant for the RC circuit of the envelope detector.

```
1    function x_env = ss_envdet(x_am,Ts,tau)
2      x_am = x_am.*(x_am>0);     % Rectify the AM signal
3      % Create a vector to hold samples of the envelope.
4      x_env = zeros(size(x_am));
5      nSamp = length(x_am);      % Number of samples in "x_am"
6      out = -1;
7      % The loop to compute the output.
8      for i=1:nSamp,
9        inp = x_am(i);
10       if (inp>=out)     % Case 1: x_am(t) > vc(t)
11         out = inp;
12       else              % Case 2: x_am(t) < vc(t)
13         out = out*(1-Ts/tau);
14       end;
15       x_env(i) = out;
16     end;
```

Note that, for case 2, we need to solve the differential equation given in Eqn. (11.53) using a numerical approximation technique. Evaluating the differential equation at $t = t_0$ we have

$$\left. \frac{dv_c}{dt} \right|_{t=t_0} = -\frac{1}{\tau} v_c(t_0) \tag{11.101}$$

Replacing the derivative on the left side of Eqn. (11.101) with a finite difference approximation at time $t = t_0$ we obtain

$$\frac{v_c(t_0 + T_s) - v_c(t_0)}{T_s} \approx -\frac{1}{\tau} v_c(t_0) \tag{11.102}$$

Solving Eqn. (11.102) for $v_c(t_0 + T_s)$ yields

$$v_c(t_0 + T_s) \approx v_c(t_0) \left[1 - \frac{T_s}{\tau} \right] \tag{11.103}$$

This result allows us to approximate the output voltage at time $t = t_0 + T_s$ from the knowledge of its value at time $t = t_0$, and can be used iteratively to approximate the output at time instants $t_0 + 2T_s$, $t_0 + 3T_s$, and so on. It is a special case of what is known as Euler's

method for numerically solving a first-order differential equation one time step at a time (see MATLAB Exercise 2.4 in Chapter 2).
Software resources:
`ss_envdet.m`

MATLAB Exercise 11.8: Testing the envelope detector function

In this exercise the envelope detector function `ss_envdet(..)` developed in MATLAB Example 11.7 will be tested. For the sake of simplicity, a single-tone modulated AM signal will be used for testing. The single-tone message frequency is $f_m = 1.5$ and the carrier frequency is $f_c = 10$ kHz. The modulation index is $\mu = 0.6$. Let the time constant of the envelope detector circuit be chosen as $\tau = 0.1$ ms. The MATLAB script that generates an AM waveform with the prescribed parameter values and then simulates an envelope detector to demodulate it is listed below:

```
1    % File: matex_11_8a.m
2    %
3    time = [1:999]*1e-6;        % Vector of time instants
4    fc = 10000;                 % Carrier frequency
5    fm = 1500;                  % Single-tone message frequency
6    tau = 0.0001;               % Time constant for envelope detector
7    mu = 0.6;                   % Modulation index
8    % Compute the AM waveform.
9    x_AM = (1+mu*cos(2*pi*fm*time)).*cos(2*pi*fc*time);
10   % Compute the ideal envelope.
11   x_env = abs(1+mu*cos(2*pi*fm*time));
12   % Compute the envelope detector output.
13   x_det = ss_envdet(x_AM,1e-6,tau);
14   % Graph the AM waveform, its envelope, and the detector output.
15   plot(1000*time,x_AM,1000*time,x_env,1000*time,x_det);
16   title('Envelope detector simulation');
17   ylabel('Amplitude');
18   xlabel('t (ms)');
19   legend('AM waveform','Ideal envelope','Detector output');
```

The MATLAB graph that is produced by the script listed above is shown in Fig. 11.39.

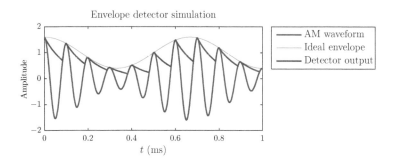

Figure 11.39 – MATLAB graph for envelope detector simulation with $f_c = 10$ kHz, $f_m = 1.5$ kHz, $\mu = 0.6$, and $\tau = 0.1$ ms.

It can be seen that our choice of parameter values above produces a properly detected envelope in Fig. 11.39. (Compare the detector output signal to the correct envelope of the AM waveform.) Let us increase the time constant to $\tau = 0.3$ ms and observe the resulting envelope detector output by simply modifying one line of the code:

```
6    tau = 0.0003;
```

The new value of the time constant leads to the graph shown in Fig. 11.40.

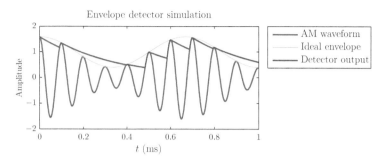

Figure 11.40 – MATLAB graph for envelope detector simulation with $f_c = 10$ kHz, $f_m = 1.5$ kHz, $\mu = 0.6$, and $\tau = 0.3$ ms.

Observe how the envelope detector loses track of the envelope as a result of a time constant that is too large.

Software resources:
matex_11_8a.m
matex_11_8b.m

Problems

11.1. Consider the single-tone modulated AM signal $x_{AM}(t)$ given by Eqn. (11.7). For each set of parameters listed below, determine the missing parameter value.

a. $A_m = 2$, $\mu = 0.7$, $A_c = ?$
b. $A_m = 3$, $\mu = 0.85$, $A_c = ?$
c. $A_c = 2$, $\mu = 0.9$, $A_m = ?$
d. $A_c = 3$, $A_m = 2.5$, $\mu = ?$

11.2. Manually sketch each single-tone modulated AM signal, the parameters of which are given below. Indicate critical values on each axis.

a. $A_c = 3$, $\mu = 0.6$, $f_c = 10$ Hz , $f_m = 2$ Hz
b. $A_c = 3$, $\mu = 1.3$, $f_c = 10$ Hz , $f_m = 2$ Hz
c. $A_c = 5$, $A_m = 4$, $f_c = 25$ Hz , $f_m = 8$ Hz

11.3. A sinusoidal carrier is in the form

$$x_c(t) = A_c \cos(20\pi t)$$

For each modulating signal listed below, the modulation index is required to be $\mu = 0.7$. Determine the parameter A_c in each case.

a. $m(t) = \cos(4\pi t) + 2\cos(6\pi t)$
b. $m(t) = \cos(4\pi t) + 3\sin(6\pi t)$
c. $m(t) = 3\sin(4\pi t) + 2\cos(6\pi t - \pi/3)$

11.4.   Refer to the tone-modulated AM signals, the parameters of which are listed below. For each, draw the phasors $\boldsymbol{X}_{\text{car}}$, $\tilde{\boldsymbol{X}}_{\text{usb}}$ and $\tilde{\boldsymbol{X}}_{\text{lsb}}$ at the time instant indicated.

a.	$A_c = 3$,	$\mu = 0.6$,	$f_c = 10$ Hz,	$f_m = 2$ Hz,	$t = 0.35$ s
b.	$A_c = 3$,	$\mu = 1.3$,	$f_c = 10$ Hz,	$f_m = 2$ Hz,	$t = 0.6$ s
c.	$A_c = 5$,	$A_m = 4$,	$f_c = 25$ Hz,	$f_m = 8$ Hz,	$t = 150$ ms

11.5.   Refer to the tone-modulated AM signals the parameters of which are listed in Problem 11.1. Compute and graph the frequency spectrum of each signal.

11.6.   Refer to Problem 11.3. For each modulating signal determine and sketch the frequency spectrum of $x_{AM}(t)$. Afterwards determine the modulation efficiency η. (Use the value of A_c computed in Problem 11.2 for each case.)

11.7.   The frequency spectrum $M(f)$ of a message signal $m(t)$ is shown in Fig. P.11.7.

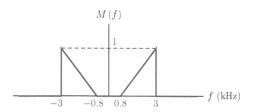

Figure P. 11.7

The signal $m(t)$ is used as input to the AM modulator shown in Fig. 11.14. The carrier amplitude is $A_c = 1.5$ and the carrier frequency is $f_c = 10$ kHz. Sketch the frequency spectrum of the signal $x_{AM}(t)$.

11.8.   Refer to the message signal $m(t)$ the spectrum of which is shown in Fig. P.11.7. Assume that this signal is used as input to the AM modulator shown in Fig. 11.20. The carrier amplitude is $B_c = 1.5$ and the carrier frequency is $f_c = 10$ kHz. The nonlinear device has the input-output characteristic

$$v_{out}(t) = 2\,v_{in}(t) + 0.8\,v_{in}^2(t)$$

  a.   Determine and sketch the frequency spectrum of the signal at the output of the nonlinear device.
  b.   Specify the appropriate bandpass filter to produce the AM signal $x_{AM}(t)$.

11.9.   A single-tone modulated signal is shown in Fig. P.11.9.

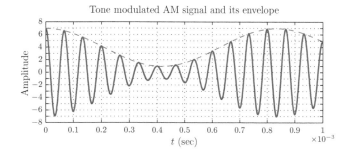

Figure P. 11.9

Estimate the following from the figure:

   a. Carrier frequency f_c
   b. Single-tone message frequency f_m
   c. Modulation index
   d. Efficiency of modulation

11.10. Refer to Problem 11.3. For each modulating signal determine the efficiency of modulation using the value of A_c computed in Problem 11.3.

11.11. Consider the switching modulator shown in Fig. 11.15. Let the message signal be the single tone

$$m\left(t\right) = 3\,\cos\left(200\pi t\right)$$

The carrier amplitude is $B_c = 5$, and the carrier frequency is $f_c = 800$ Hz.

   a. Sketch the signal $v_{in}\left(t\right)$.
   b. Assuming the diode is ideal, sketch the signal $v_{out}\left(t\right)$.
   c. Determine the magnitude characteristic of an ideal bandpass filter that would be appropriate for use in this switching modulator.

11.12. Consider the *balanced modulator* block diagram shown in Fig. P.11.12 which can be used for producing DSB with suppressed carrier without the need for filtering.

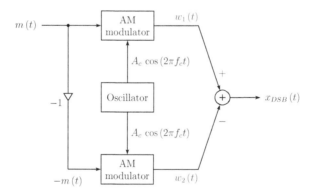

Figure P. 11.12

Determine the intermediate signals $w_1\left(t\right)$ and $w_2\left(t\right)$. Show that the output signal is the proper DSB-modulated signal with the carrier suppressed.

11.13. Refer to the message signal $m\left(t\right)$ the frequency spectrum $M\left(f\right)$ of which is shown in Fig. P.11.7. A sinusoidal carrier with amplitude $A_c = 1.5$ and the carrier frequency $f_c = 10$ kHz is to be used in conjunction with this signal.

   a. If the signal $m\left(t\right)$ is used for modulating a sinusoidal carrier using DSB, sketch the frequency spectrum of the signal $x_{DSB}\left(t\right)$.
   b. Repeat part (a) using SSB with the upper sideband retained.
   c. Repeat part (a) using SSB with the lower sideband retained.

11.14. A single-tone modulated AM signal given by

$$x_{AM}\left(t\right) = \left[4 + 3.2\,\cos\left(200\pi t\right)\right]\cos\left(2000\pi t\right)$$

is applied to the envelope detector shown in Fig. 11.21. Determine an appropriate time constant for the envelope detector so that the single-tone message signal is recovered correctly. Based on the time constant found, pick practical values for R and C.

11.15. Refer to the coherent demodulator shown in Fig. P.11.15. Assume that the local oscillator signal is not perfectly synchronized with the original carrier, but rather has a phase error θ as shown.

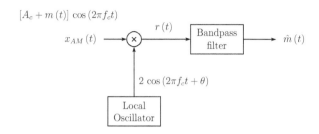

Figure P. 11.15

Determine the signal $r(t)$ and the bandpass filter output $\hat{m}(t)$. Explain the effect of the phase error on these signals.

11.16. Refer to the coherent demodulator shown in Fig. P.11.16. Assume that the local oscillator signal is not perfectly synchronized with the original carrier, but rather has a frequency error so that the local carrier frequency is $f_c + \Delta f$.

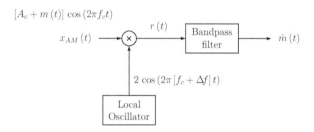

Figure P. 11.16

Determine the signal $r(t)$ and the bandpass filter output $\hat{m}(t)$. Explain the effect of the frequency error on these signals.

MATLAB Problems

11.17. Refer to Problem 11.2. For each set of parameters given, develop a MATLAB script to compute and graph the corresponding AM signal. In each case choose an appropriate sampling interval and produce a graph that spans at least three periods of the modulating tone.

11.18. Refer to Problem 11.3. For each modulating signal write a script to compute and graph $x_{AM}(t)$. Choose the sampling interval appropriately, and graph the corresponding signal in a time interval sufficient to show relevant details.

11.19.   Refer to Problem 11.3. Determine the fundamental frequency and the fundamental period for each multi-tone-modulated AM signal. Write a script for each case to produce 1024 samples of exactly one period. Compute and graph the approximate EFS spectrum using the function `ss_efsapprox(..)`.

11.20.   Refer to Problem 11.11. Write a MATLAB script to

a.   Compute and graph the signal $v_{in}(t)$.
b.   Compute and sketch the signal $v_{out}(t)$ assuming that the diode is ideal.
c.   Use the function `ss_switchmod(..)` developed in MATLAB Exercise 11.3 to compute the output signal $x_{AM}(t)$. Graph the result.

11.21.   Using the `ss_switchmod(..)` as the starting point, develop a new MATLAB function `ss_switchmod2(..)` to simulate a switching modulator that uses the more realistic diode model shown in Fig. P.11.21 instead of the ideal diode model.

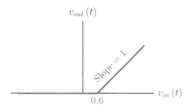

Figure P. 11.21

Afterwards rework Problem 11.20 using the function `ss_switchmod2(..)` instead of `ss_switchmod(..)`. Compare the results.

11.22.   Refer to Problem 11.14. Write a MATLAB script to generate samples of the signal $x_{AM}(t)$ and then to simulate the envelope detector using the function `ss_envdet(..)` developed in MATLAB Exercise 11.7. First, use the value of τ found in Problem 11.14. Afterwards repeat with half the value found and with twice the value found. Graph the envelope detector output signal for each case and compare the results.

MATLAB Projects

11.23.   A mystery AM signal has been saved into the data file "`ch11data.mat`". It can be loaded into MATLAB workspace with the statement

```
>>   load ch11data.mat
```

creating two vectors named "`t`" and "`xAM`" in the workspace.

a.   Graph the signal $x_{AM}(t)$.
b.   Determine the sampling rate used in creating the vector "`xAM`" by inspecting the elements of the vector "`t`". Compute the FFT of the signal, and graph the spectrum using actual frequencies in Hz based on the sampling rate used.
c.   Inspect the spectrum graph obtained in part (b). What seems to be the carrier frequency? How many tones are in the modulating signal $m(t)$, and what are their frequencies?

d. Using function `cheby1(..)` design a Chebyshev type-I band-reject filter to remove the carrier frequency. Adjust the passband cutoff frequencies so that the sideband frequencies are kept and the carrier frequency is removed. A ninth-order filter with a passband ripple of 0.5 dB may be an appropriate starting point. Compute and graph the magnitude response of the designed filter using the function `freqz(..)` Adjust design parameters and repeat the design as necessary.

e. Use the function `filter(..)` to apply the AM signal to the Chebyshev type-I filter designed in part (d). Create a vector "x2" with the samples of the Chebyshev filter output signal. Graph the output signal.

f. Compute and graph the frequency spectrum of the output signal using the FFT. Is the carrier frequency gone?

g. Estimate the power in the original AM signal and in the signal at the output of the Chebyshev band-reject filter. (Hint: Scalar product operator can be used in MATLAB with some scaling afterwards to estimate the power in a signal.) Using these two power estimates, compute an estimate of modulation efficiency.

11.24.

a. Develop a MATLAB function `ss_coherent(..)` to simulate the coherent demodulator shown in Fig. 11.24. The syntax of the function should be as follows:

```
>>   mhat = ss_coherent(xAM,Ts,fc,theta,deltaf)
```

The input arguments to the function are

`xAM:`	Vector holding samples of the AM signal.
`Ts:`	Sampling intetrval used for signal in vector "`xAM`".
`fc:`	Carrier frequency in Hz.
`theta:`	Phase error of the local carrier (in radians).
`deltaf:`	Frequency error of the local carrier (in Hz).

The vector "`mhat`" holds samples of the output signal. Internally, implement the bandpass filter as the cascade of two filters: A fifth-order Butterworth lowpass filter with the appropriate cutoff frequency, and a sixth-order Chebyshev type-I highpass filter to remove frequencies less than 2 percent of the carrier frequency.

b. Test the function `ss_coherent(..)` with the signal

$$x_{AM}(t) = [4 + 3.2 \cos(200\pi t)] \cos(2000\pi t)$$

using a sampling rate of $f_s = 25$ kHz. Set phase and frequency errors equal to zero. Generate samples of the AM signal for $t = 0, \ldots, 100$ ms. Compute and graph the simulated output signal $\hat{m}(t)$.

c. Repeat part (b) using a phase error of $\theta = \pi/10$ radians.

d. Repeat part (b) using a frequency error of $\Delta f = 60$ Hz.

11.25. A SSB-modulated signal with lower sideband retained was given by Eqn. (11.100) and is repeated here:

Eqn. (11.100): $\quad x_{SSB,L}(t) = \dfrac{1}{2} m(t) \cos(2\pi f_c t) + \dfrac{1}{2} \hat{m}(t) \sin(2\pi f_c t) \qquad (11.104)$

The term $\hat{m}(t)$ represents the Hilbert transform of $m(t)$

a. Using the `fdatool` program in MATLAB, design a length-55 FIR filter to approximate an ideal Hilbert transform filter. Export the impulse response of the designed filter to a data file.

b. Let the message signal be
$$m(t) = 2 \cos(200\pi t)$$

and use the carrier frequency $f_c = 1000$ Hz. Write a script to compute the signal $x_{SSB,L}(t)$ as given above. Your script should

1. Load the length-55 Hilbert transform filter impulse from data file.
2. Generate samples of $m(t)$ using a sampling rate of $f_c = 25$ kHz.
3. Process $m(t)$ through the FIR filter designed to obtain an approximation to the Hilbert transform $\hat{m}(t)$.
4. Compute an approximation to $x_{SSB,L}(t)$ as specified. Hint: Keep in mind that the length-55 FIR filter has a delay of $(55 - 1)/2 = 27$ samples, so it produces an approximation to a delayed version of $\hat{m}(t)$.

Appendix A

Complex Numbers and Euler's Formula

A.1 Introduction

Complex numbers allow us to solve equations for which no real solution can be found. Consider, for example, the equation

$$x^2 + 9 = 0 \qquad\qquad (A.1)$$

which cannot be satisfied for any real number. By introducing an imaginary unit[1] $j = \sqrt{-1}$ so that $j^2 = -1$, Eqn. (A.1) can be solved to yield

$$x = \mp j3$$

Similarly, the equation

$$(x + 2)^2 + 9 = 0 \qquad\qquad (A.2)$$

has the solutions

$$x = -2 \mp j3 \qquad\qquad (A.3)$$

A general complex number is in the form

$$x = a + jb \qquad\qquad (A.4)$$

where a and b are real numbers. The values a and b are referred to as the *real part* and *imaginary part* of the complex number x respectively. Following notation is used for real and imaginary parts of a complex number:

$$a = \operatorname{Re}\{x\}$$
$$b = \operatorname{Im}\{x\}$$

---

[1] The imaginary unit was first introduced by Italian mathematician Gerolamo Cardano as he worked on the solutions of cubic and quartic equations, although its utility was not fully understood and appreciated until the works of Leonhard Euler and Carl Friedrich Gauss almost two centuries later.

For two complex numbers to be equal, both their real parts and imaginary parts must be equal. Given two complex numbers $x_1 = a + jb$ and $x_2 = c + jd$, the equality $x_1 = x_2$ implies that $a = c$ and $b = d$.

It is often convenient to graphically represent a complex number as a point in the *complex plane* constructed using a horizontal axis corresponding to the real part and a vertical axis corresponding to the imaginary part as depicted in Fig. A.1.

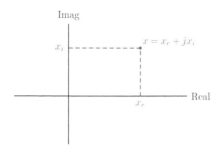

Figure A.1 – Complex number $x = x_r + jx_i$ shown as a point in the complex plane.

If the imaginary part of a complex number is equal to zero, the resulting number is said to be *purely real*. An example is $x = 3 + j0$. Similarly, a complex number the real part of which is equal to zero is said to be *purely imaginary*. An example of a purely imaginary number is $x = 0 + j5$. Purely real and purely imaginary numbers are simply special cases of the more general class of complex numbers.

Another method of graphical representation of a complex number can be found through the use of a vector as shown in Fig. A.2. Vector representation of complex numbers is quite useful since vector operations can be used for manipulating complex numbers.

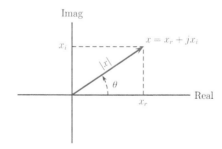

Figure A.2 – Complex number $x = x_r + jx_i$ shown as a vector.

A complex number written as the sum of its real and imaginary parts such as $x = x_r + jx_i$ is said to be in *Cartesian form*. An alternative form referred to as a *polar form* is derived from the vector representation of a complex number by using the norm $|x|$ and the angle θ of the vector instead of the real and imaginary parts x_r and x_i. The two forms of the complex number x are:

$$x = x_r + jx_i , \quad \text{Cartesian form}$$

$$x = |x|\, e^{j\theta} , \quad \text{Polar form}$$

The norm $|x|$ is the distance of the point representing the complex number from the origin. The angle θ is the angle measured counterclockwise starting with the positive real axis.

A complex number can easily be converted from Cartesian form to polar form and vice versa. Through simple geometric relationships it can be shown that

$$|x| = \sqrt{x_r^2 + x_i^2} \tag{A.5}$$

and

$$\theta = \tan^{-1}\left(\frac{x_i}{x_r}\right) \tag{A.6}$$

Alternately, real and imaginary parts can be obtained from the norm and the angle through

$$x_r = |x|\cos(\theta) \tag{A.7}$$

and

$$x_i = |x|\sin(\theta) \tag{A.8}$$

A.2   Arithmetic with Complex Numbers

Definitions of arithmetic operators are easily extended to apply to complex numbers. In Section A.2.1 addition and subtraction operators will be discussed. Multiplication and division of complex numbers will be discussed in Section A.2.2.

A.2.1   Addition and subtraction

Consider two complex numbers $x_1 = a + jb$ and $x_2 = c + jd$. Addition of these two complex numbers is carried out as follows:

$$\begin{aligned} y &= x_1 + x_2 \\ &= (a + jb) + (c + jd) \\ &= (a + c) + j(b + d) \end{aligned} \tag{A.9}$$

Real and imaginary parts of the result can be written separately as

$$\begin{aligned} \operatorname{Re}\{y\} &= \operatorname{Re}\{x_1\} + \operatorname{Re}\{x_2\} \\ \operatorname{Im}\{y\} &= \operatorname{Im}\{x_1\} + \operatorname{Im}\{x_2\} \end{aligned}$$

Subtraction of complex numbers is similar to addition. The difference $z = x_1 - x_2$ is computed as follows:

$$\begin{aligned} z &= x_1 - x_2 \\ &= (a + jb) - (c + jd) \\ &= (a - c) + j(b - d) \end{aligned} \tag{A.10}$$

Real and imaginary parts of the result can be written separately as

$$\begin{aligned} \operatorname{Re}\{z\} &= \operatorname{Re}\{x_1\} - \operatorname{Re}\{x_2\} \\ \operatorname{Im}\{z\} &= \operatorname{Im}\{x_1\} - \operatorname{Im}\{x_2\} \end{aligned}$$

Addition and subtraction operators for complex numbers are analogous for addition and subtraction of vectors using the parallelogram rule. This is illustrated in Fig. A.3.

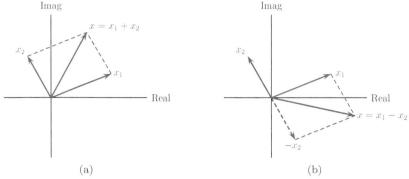

(a)   (b)

Figure A.3 – Addition and subtraction operators with complex vectors.

If the complex numbers to be added are expressed in polar form, they must first be converted to Cartesian form before they can be added. Consider two complex numbers in polar form given as

$$x_1 = |x_1| \, e^{j\theta_1}, \quad \text{and} \quad x_2 = |x_2| \, e^{j\theta_2} \tag{A.11}$$

The sum z of these complex numbers is found as

$$
\begin{aligned}
z &= x_1 + x_2 \\
&= |x_1| \, \cos{(\theta_1)} + j \, |x_1| \, \sin{(\theta_1)} + |x_2| \, \cos{(\theta_2)} + j \, |x_2| \, \sin{(\theta_2)} \\
&= \Big[|x_1| \, \cos{(\theta_1)} + |x_2| \, \cos{(\theta_2)} \Big] + j \, \Big[|x_1| \, \sin{(\theta_1)} + |x_2| \, \sin{(\theta_2)} \Big]
\end{aligned}
\tag{A.12}
$$

which can be put into polar form as

$$z = |z| \, e^{j\theta_z} \tag{A.13}$$

where

$$|z| = |x_1|^2 + |x_2|^2 + 2 \, |x_1| \, |x_2| \, \cos{(\theta_1 - \theta_2)} \tag{A.14}$$

and

$$\theta_z = \tan^{-1}\left(\frac{|x_1| \, \sin{(\theta_1)} + |x_2| \, \sin{(\theta_2)}}{|x_1| \, \cos{(\theta_1)} + |x_2| \, \cos{(\theta_2)}} \right) \tag{A.15}$$

A.2.2   Multiplication and division

Again consider two complex numbers $x_1 = a + jb$ and $x_2 = c + jd$. The product of these two complex numbers is computed as follows:

$$
\begin{aligned}
y &= x_1 \, x_2 \\
&= (a + jb)\,(c + jd) \\
&= ac + j \, ad + j \, bc + j^2 \, bd \\
&= (ac - bd) + j \, (ad + bc)
\end{aligned}
\tag{A.16}
$$

Real and imaginary parts of the result can be written separately as

$$
\begin{aligned}
\text{Re}\,\{y\} &= \text{Re}\,\{x_1\} \, \text{Re}\,\{x_2\} - \text{Im}\,\{x_1\} \, \text{Im}\,\{x_2\} \\
\text{Im}\,\{y\} &= \text{Re}\,\{x_1\} \, \text{Im}\,\{x_2\} + \text{Im}\,\{x_1\} \, \text{Re}\,\{x_2\}
\end{aligned}
$$

Product of a complex number and its own complex conjugate is equal to the squared norm of the complex number:

$$\begin{aligned}
x_1\, x_1^* &= (a + jb)\, (a + jb)^* \\
&= (a + jb)\, (a - jb) \\
&= a^2 + b^2 = |x_1|^2
\end{aligned}$$

Division of complex numbers is slightly more involved. Consider the complex number z defined as

$$z = \frac{x_1}{x_2} = \frac{(a + jb)}{(c + jd)} \tag{A.17}$$

The expression in Eqn. (A.17) can be simplified by multiplying both the numerator and the denominator by the complex conjugate of the denominator:

$$z = \frac{x_1}{x_2} = \frac{(a + jb)}{(c + jd)} \frac{(c - jd)}{(c - jd)} \tag{A.18}$$

A.3   Euler's Formula

A complex exponential function can be expressed in the form

$$e^{jx} = \cos(x) + j\,\sin(x) \tag{A.19}$$

This relationship is known as *Euler's formula*, and will be used extensively in working with signals, linear systems and various transforms. If the sign of x changes in Eqn. (A.20), we get

$$e^{-jx} = \cos(-x) + j\,\sin(-x) = \cos(x) - j\,\sin(x) \tag{A.20}$$

Using Eqns. (A.19) and (A.2), trigonometric functions $\cos(x)$ and $\sin(x)$ may be expressed in terms of complex exponential functions as

$$\cos(x) = \frac{e^{jx} + e^{-jx}}{2} \tag{A.21}$$

and

$$\sin(x) = \frac{e^{jx} - e^{-jx}}{2j} \tag{A.22}$$

Appendix B

Mathematical Relations

B.1 Trigonometric Identities

$$\cos(a \pm b) = \cos(a)\cos(b) \mp \sin(a)\sin(b) \tag{B.1}$$

$$\sin(a \pm b) = \sin(a)\cos(b) \pm \cos(a)\sin(b) \tag{B.2}$$

$$\tan(a \pm b) = \frac{\tan(a) \pm \tan(b)}{1 \mp \tan(a)\tan(b)} \tag{B.3}$$

$$\cos(a)\cos(b) = \tfrac{1}{2}\cos(a+b) + \tfrac{1}{2}\cos(a-b) \tag{B.4}$$

$$\sin(a)\sin(b) = \tfrac{1}{2}\cos(a-b) - \tfrac{1}{2}\cos(a+b) \tag{B.5}$$

$$\sin(a)\cos(b) = \tfrac{1}{2}\sin(a+b) + \tfrac{1}{2}\sin(a-b) \tag{B.6}$$

$$\cos(a) + \cos(b) = 2\cos\left(\frac{a+b}{2}\right)\cos\left(\frac{a-b}{2}\right) \tag{B.7}$$

$$\cos(a) - \cos(b) = -2\sin\left(\frac{a+b}{2}\right)\sin\left(\frac{a-b}{2}\right) \tag{B.8}$$

$$\sin(a) + \sin(b) = 2\sin\left(\frac{a+b}{2}\right)\cos\left(\frac{a-b}{2}\right) \tag{B.9}$$

$$\sin(a) - \sin(b) = 2\sin\left(\frac{a-b}{2}\right)\cos\left(\frac{a+b}{2}\right) \tag{B.10}$$

$$\cos(2a) = \cos^2(a) - \sin^2(a) \tag{B.11}$$

$$\sin(2a) = 2\sin(a)\cos(a) \tag{B.12}$$

$$\tan(2a) = \frac{2\tan(a)}{1 - \tan^2(a)} \tag{B.13}$$

$$\cos^2(a) = \tfrac{1}{2} + \tfrac{1}{2}\cos(2a) \tag{B.14}$$

$$\sin^2(a) = \tfrac{1}{2} - \tfrac{1}{2}\cos(2a) \tag{B.15}$$

B.2 Indefinite Integrals

$$\int x\,e^{ax}\,dx = \frac{ax-1}{a^2}\,e^{ax} \tag{B.16}$$

$$\int \frac{1}{x^2 + a^2}\,dx = \frac{1}{a}\,\tan^{-1}\left(\frac{x}{a}\right) \tag{B.17}$$

$$\int \frac{x}{x^2 + a^2}\,dx = \frac{1}{2}\,\ln\left(x^2 + a^2\right) \tag{B.18}$$

$$\int \frac{x^2}{x^2 + a^2}\,dx = x - a\,\tan^{-1}\left(\frac{x}{a}\right) \tag{B.19}$$

$$\int x\,\cos(ax)\,dx = \frac{1}{a^2}\left[\cos(ax) + ax\,\sin(ax)\right] \tag{B.20}$$

$$\int x\,\sin(ax)\,dx = \frac{1}{a^2}\left[\sin(ax) - ax\,\cos(ax)\right] \tag{B.21}$$

$$\int e^{ax}\,\cos(bx)\,dx = \frac{e^{ax}}{a^2 + b^2}\left[a\,\cos(bx) + b\,\sin(bx)\right] \tag{B.22}$$

$$\int e^{ax}\,\sin(bx)\,dx = \frac{e^{ax}}{a^2 + b^2}\left[a\,\sin(bx) - b\,\cos(bx)\right] \tag{B.23}$$

$$\int (a + bx)^n\,dx = \frac{(a + bx)^{n+1}}{b\,(n+1)},\qquad n > 0 \tag{B.24}$$

$$\int \frac{1}{(a + bx)^n}\,dx = \frac{-1}{b\,(n-1)\,(a + bx)^{n-1}},\qquad n > 1 \tag{B.25}$$

B.3 Laplace Transform Pairs

Signal	Transform	ROC
$\delta(t)$	1	all s
$u(t)$	$\dfrac{1}{s}$	$\mathrm{Re}\{s\} > 0$
$u(-t)$	$-\dfrac{1}{s}$	$\mathrm{Re}\{s\} < 0$

Signal	Transform	ROC
$e^{at} u(t)$	$\dfrac{1}{s-a}$	$\mathrm{Re}\{s\} > a$
$-e^{at} u(-t)$	$\dfrac{1}{s-a}$	$\mathrm{Re}\{s\} < a$
$e^{j\omega_0 t} u(t)$	$\dfrac{1}{s-j\omega_0}$	$\mathrm{Re}\{s\} > 0$
$e^{-|t|}$	$\dfrac{-2}{s^2-1}$	$-1 < \mathrm{Re}\{s\} < 1$
$\Pi\left(\dfrac{t-\tau/2}{\tau}\right)$	$\dfrac{1-e^{-s\tau}}{s}$	$\mathrm{Re}\{s\} > -\infty$
$\cos(\omega_0 t) u(t)$	$\dfrac{s}{s^2+\omega_0^2}$	$\mathrm{Re}\{s\} > 0$
$\sin(\omega_0 t) u(t)$	$\dfrac{\omega_0}{s^2+\omega_0^2}$	$\mathrm{Re}\{s\} > 0$
$e^{at}\cos(\omega_0 t) u(t)$	$\dfrac{s-a}{(s-a)^2+\omega_0^2}$	$\mathrm{Re}\{s\} > a$
$e^{at}\sin(\omega_0 t) u(t)$	$\dfrac{\omega_0}{(s-a)^2+\omega_0^2}$	$\mathrm{Re}\{s\} > a$

B.4 z-Transform Pairs

Signal	Transform	ROC
$\delta[n]$	1	all z
$u[n]$	$\dfrac{z}{z-1}$	$|z| > 1$
$u[-n]$	$\dfrac{-z}{z-1}$	$|z| < 1$
$a^n u[n]$	$\dfrac{z}{z-a}$	$|z| > |a|$
$-a^n u[-n-1]$	$\dfrac{z}{z-a}$	$|z| < |a|$
$\cos(\Omega_0 n) u[n]$	$\dfrac{z[z-\cos(\Omega_0)]}{z^2-2\cos(\Omega_0)z+1}$	$|z| > 1$
$\sin(\Omega_0 n) u[n]$	$\dfrac{\sin(\Omega_0)z}{z^2-2\cos(\Omega_0)z+1}$	$|z| > 1$
$a^n\cos(\Omega_0 n) u[n]$	$\dfrac{z[z-a\cos(\Omega_0)]}{z^2-2a\cos(\Omega_0)z+a^2}$	$|z| > |a|$
$a^n\sin(\Omega_0 n) u[n]$	$\dfrac{a\sin(\Omega_0)z}{z^2-2a\cos(\Omega_0)z+a^2}$	$|z| > |a|$
$n a^n u[n]$	$\dfrac{az}{(z-a)^2}$	$|z| > |a|$

Appendix C

Closed Forms for Sums of Geometric Series

Summations of geometric series appear often in problems involving various transforms and in the analysis of linear systems. In this appendix we give derivations of closed-form formulas for infinite- and finite-length geometric series.

C.1   Infinite-Length Geometric Series

Consider the sum of infinite-length geometric series in the form

$$P = \sum_{n=0}^{\infty} a^n \tag{C.1}$$

For the summation in Eqn. (C.1) to converge we must have $|a| < 1$. Assuming that is the case, let us write Eqn. (C.1) in open form:

$$P = 1 + a + a^2 + a^3 + \ldots \tag{C.2}$$

Subtracting unity from both sides of Eqn. (C.2) we obtain

$$P - 1 = a + a^2 + a^3 + \ldots \tag{C.3}$$

in which the terms on the right side of the equal sign have a common factor a. Factoring it out leads to

$$\begin{aligned} P - 1 &= a\left(1 + a + a^2 + a^3 + \ldots\right) \\ &= a\,P \end{aligned} \tag{C.4}$$

which can be solved for P to yield

$$P = \frac{1}{1-a}, \qquad |a| < 1 \tag{C.5}$$

C.2   Finite-Length Geometric Series

Consider the finite-length sum of a geometric series in the form

$$Q = \sum_{n=0}^{L} a^n \tag{C.6}$$

Unlike P of the previous section, the convergence of Q does not require $|a| < 1$. The only requirement is that $|a| < \infty$. Let us write Q in open form:

$$Q = 1 + a + \ldots + a^L \tag{C.7}$$

Subtracting unity from both sides of Eqn. (C.7) leads to

$$Q - 1 = a + \ldots + a^L \tag{C.8}$$

As before, the terms on the right side of the equal sign have a common factor a which can be factored out to yield

$$\begin{aligned} Q - 1 &= a\left(1 + \ldots + a^{L-1}\right) \\ &= a\left(Q - a^L\right) \end{aligned} \tag{C.9}$$

which can be solved for Q:

$$Q = \frac{1 - a^{L+1}}{1 - a} \tag{C.10}$$

Consistency check: If L is increased, in the limit Q approaches P provided that $|a| < 1$, that is

$$\lim_{L \to \infty} [Q] = P \quad \text{if} \quad |a| < 1 \tag{C.11}$$

C.3   Finite-Length Geometric Series (Alternative Form)

Consider an alternative form of the finite-length geometric series sum in which the lower limit is not equal to zero.

$$Q = \sum_{n=L_1}^{L_2} a^n \tag{C.12}$$

In order to obtain a closed form formula we will apply the variable change $m = n - L_1$ with which Eqn. (C.12) becomes

$$Q = \sum_{m=0}^{L_2 - L_1} a^{m + L_1} \tag{C.13}$$

which can be simplified as

$$Q = a^{L_1} \sum_{m=0}^{L_2 - L_1} a^m \tag{C.14}$$

The summation on the right side of Eqn. (C.14) is in the standard form of Eqn. (C.6), therefore

$$Q = a^{L_1} \left(\frac{1 - a^{L_2 - L_1 + 1}}{1 - a} \right) = \frac{a^{L_1} - a^{L_2 + 1}}{1 - a} \tag{C.15}$$

Consistency check: For $L_1 = 0$ and $L_2 = L$, Eqn. (C.15) reduces to Eqn. (C.10).

Appendix D

Orthogonality of Basis Functions

Orthogonality properties are used extensively in the development of Fourier series representations for the analysis of continuous-time and discrete-time signals. The use of an orthogonal set of basis functions ensures that the relationship between a periodic signal and its Fourier series representation is one-to-one, that is, each periodic signal has a unique set of Fourier series coefficients, and each set of coefficients corresponds to a unique signal. In this appendix we will summarize several forms of orthogonality properties and carry out their proofs.

D.1 Orthogonality for Trigonometric Fourier Series

Consider the two real-valued basis function sets defined as

$$\phi_k(t) = \cos(k\omega_0 t) \; ; \quad k = 1, \ldots, \infty \tag{D.1}$$

and

$$\psi_k(t) = \sin(k\omega_0 t) \; ; \quad k = 1, \ldots, \infty \tag{D.2}$$

where ω_0 is the fundamental frequency in rad/s. The period that corresponds to the fundamental frequency ω_0 is

$$T_0 = \frac{2\pi}{\omega_0}$$

It can be shown that each of the two sets in Eqns. (D.1) and (D.2) is orthogonal within itself, that is,

$$\int_0^{T_0} \phi_k(t)\,\phi_m(t)\,dt = \begin{cases} T_0/2, & k = m \\ 0, & k \neq m \end{cases} \tag{D.3}$$

and

$$\int_0^{T_0} \psi_k(t)\,\psi_m(t)\,dt = \begin{cases} T_0/2, & k = m \\ 0, & k \neq m \end{cases} \tag{D.4}$$

Furthermore, the two sets are orthogonal to each other:

$$\int_0^{T_0} \phi_k(t)\, \psi_m(t)\, dt = 0\,, \quad \text{for any } k, m \tag{D.5}$$

Writing Eqns. (D.3), (D.4) and (D.5) using the definitions of the basis function sets in Eqns. (D.1) and (D.2), we arrive at the orthogonality properties

$$\int_0^{T_0} \cos(k\omega_0 t)\, \cos(m\omega_0 t)\, dt = \begin{cases} T_0/2\,, & k = m \\ 0\,, & k \neq m \end{cases} \tag{D.6}$$

$$\int_0^{T_0} \sin(k\omega_0 t)\, \sin(m\omega_0 t)\, dt = \begin{cases} T_0/2\,, & k = m \\ 0\,, & k \neq m \end{cases} \tag{D.7}$$

$$\int_0^{T_0} \cos(k\omega_0 t)\, \sin(m\omega_0 t)\, dt = 0\,, \quad \text{for any } k, m \tag{D.8}$$

Proof:
Using the trigonometric identity in Eqn. (B.4), Eqn. (D.6) becomes

$$\int_0^{T_0} \cos(k\omega_0 t)\, \cos(m\omega_0 t)\, dt = \frac{1}{2}\int_0^{T_0} \cos((k+m)\,\omega_0 t)\, dt$$

$$+ \frac{1}{2}\int_0^{T_0} \cos((k-m)\,\omega_0 t)\, dt \tag{D.9}$$

Let us first assume $k \neq m$. The integrand of the first integral on the right side is periodic with period

$$\frac{2\pi}{(k+m)\,\omega_0} = \frac{T_0}{(k+m)}$$

The integrand $\cos((k+m)\,\omega_0 t)$ has $(k+m)$ full periods in the interval from 0 to T_0. Therefore the result of the first integral is zero. Similarly the integrand of the second integral on the right side of Eqn. (D.9) is periodic with period

$$\frac{2\pi}{|k-m|\,\omega_0} = \frac{T_0}{|k-m|}$$

The integrand $\cos((k-m)\,\omega_0 t)$ has $|k-m|$ full periods in the interval from 0 to T_0. Therefore the result of the second integral is also zero.

If $k = m$ then Eqn. (D.9) becomes

$$\int_0^{T_0} \cos(k\omega_0 t)\, \cos(m\omega_0 t)\, dt = \int_0^{T_0} \cos^2(k\omega_0 t)\, dt$$

$$= \frac{1}{2}\int_0^{T_0} \cos(2k\,\omega_0 t)\, dt + \frac{1}{2}\int_0^{T_0} dt$$

$$= \frac{T_0}{2} \tag{D.10}$$

completing the proof of Eqn. (D.6). Eqns. (D.7) and (D.8) can be proven similarly, using the trigonometric identities given by (B.5) and (B.6) and recognizing integrals that span an integer number of periods of sine and cosine functions.

D.2   Orthogonality for Exponential Fourier Series

Consider the set of complex periodic basis functions

$$\phi_k\left(t\right) = e^{jk\omega_0 t} \; ; \quad k = -\infty, \ldots, \infty \tag{D.11}$$

where the parameter ω_0 is the fundamental frequency in rad/s as in the case of the trigonometric set of basis functions of the previous section, and $T_0 = 2\pi/\omega_0$ is the corresponding period. It can be shown that this basis function set in Eqn. (D.11) is orthogonal in the sense

$$\int_0^{T_0} \phi_k\left(t\right) \phi_m^*\left(t\right), dt = \left\{ \begin{array}{ll} T_0 \, , & k = m \\ 0 \, , & k \neq m \end{array} \right. \tag{D.12}$$

The second term in the integrand is conjugated due to the fact that we are working with a complex set of basis functions. Writing Eqn. (D.12) in open form using the definition of the basis function set in Eqns. (D.11) we arrive at the orthogonality property

$$\int_0^{T_0} e^{jk\omega_0 t} \, e^{-jm\omega_0 t} \, dt = \left\{ \begin{array}{ll} T_0 \, , & k = m \\ 0 \, , & k \neq m \end{array} \right. \tag{D.13}$$

Proof:
Let us combine the exponential terms in the integral of Eqn. (D.13) and then apply Euler's formula to write it as

$$\int_0^{T_0} e^{jk\omega_0 t} \, e^{-jm\omega_0 t} \, dt = \int_0^{T_0} e^{j(k-m)\omega_0 t} \, dt$$

$$= \int_0^{T_0} \cos\left(\left(k-m\right)\omega_0 t\right) \, dt + j \int_0^{T_0} \sin\left(\left(k-m\right)\omega_0 t\right) \, dt \tag{D.14}$$

If $k \neq m$, both integrands on the right side of Eqn. (D.14) are periodic with period

$$\frac{2\pi}{|k-m|\,\omega_0} = \frac{T_0}{|k-m|}$$

Both integrands $\cos\left(\left(k-m\right)\omega_0 t\right)$ and $\sin\left(\left(k-m\right)\omega_0 t\right)$ have exactly $|k-m|$ full periods in the interval from 0 to T_0. Therefore the result of each integral is zero.

If $k = m$ we have

$$\int_0^{T_0} e^{jk\omega_0 t} \, e^{-jm\omega_0 t} \, dt = \int_0^{T_0} e^{j(0)\omega_0 t} \, dt = \int_0^{T_0} dt = T_0 \tag{D.15}$$

which completes the proof for Eqn. (D.13).

D.3   Orthogonality for Discrete-Time Fourier Series

Consider the set of discrete-time complex periodic basis functions

$$w_N^k = e^{-j\frac{2\pi}{N}k} \; ; \quad k = 0, \ldots, N-1 \tag{D.16}$$

Parameters N and k are both integers. It can be shown that this basis function set in Eqn. (D.16) is orthogonal in the sense

$$\sum_{n=0}^{N-1} e^{j(2\pi/N)kn} e^{-j(2\pi/N)mn} = \begin{cases} 0 \,, & k \neq m \\ N \,, & k = m \end{cases} \tag{D.17}$$

Proof:

The summation in Eqn. (D.17) can be put into a closed form using the finite-length geometric series formula in Eqn. (C.10) to obtain

$$\sum_{n=0}^{N-1} e^{j(2\pi/N)kn} e^{-j(2\pi/N)mn} = \sum_{n=0}^{N-1} e^{j(2\pi/N)\,(k-m)n}$$

$$= \frac{1 - e^{j(2\pi/N)\,(k-m)N}}{1 - e^{j(2\pi/N)\,(k-m)}}$$

which can be simplified as

$$\sum_{n=0}^{N-1} e^{j(2\pi/N)kn} e^{-j(2\pi/N)mn} = \frac{1 - e^{j\,2\pi\,(k-m)}}{1 - e^{(j2\pi/N)\,(k-m)}} \tag{D.18}$$

Let us begin with the case $k \neq m$. The numerator of the fraction on the right side of Eqn. (D.18) is equal to zero. Since both integers k and m are in the interval $k, m = 0, \ldots, N-1$ their absolute difference $|k - m|$ is in the interval $n = 1, \ldots, N-1$. Therefore the denominator of the fraction is non-zero, and the result is zero, proving the first part of Eqn. (D.17).

If $k = m$ then the denominator of the fraction in Eqn. (D.18) also becomes zero, and L'Hospital's rule must be used leading to

$$\sum_{n=0}^{N-1} e^{j(2\pi/N)kn} e^{-j(2\pi/N)mn} = N \,, \qquad \text{for } k = m \tag{D.19}$$

Appendix E

Partial Fraction Expansion

The technique of partial fraction expansion is very useful in the study of signals and systems, particularly in finding the time-domain signals that correspond to inverse Laplace transforms and inverse z-transforms of rational functions, that is, a ratio of two polynomials. It is based on the idea of writing a rational function, say of s or z, as a linear combination of simpler rational functions the inverse transforms of which can be easily determined. Since both the Laplace transform and the z-transform are linear transforms, writing the rational function $X(z)$ or $X(z)$ as a linear combination of simpler functions and finding the inverse transform of each of those simpler functions allows us to construct the inverse transform of the original function in a straightforward manner. The main problem is in determining which simpler functions should be used in the expansion, and how much weight each should have.

In this appendix we will present the details of the partial fraction expansion technique first from the perspective of the Laplace transform for continuous-time signals, and then from the perspective of the z-transform for discrete-time signals.

E.1 Partial Fraction Expansion for Continuous-Time Signals and Systems

As an example, consider a rational function $X(s)$ with two poles, expressed in the form

$$X(s) = \frac{B(s)}{(s-a)(s-b)} = \frac{k_1}{s-a} + \frac{k_2}{s-b} \tag{E.1}$$

The rational function $X(s)$ is expressed as a weighted sum of partial fractions

$$\frac{1}{s-a}$$

and

$$\frac{1}{s-b}$$

with weights k_1 and k_2 respectively. The weights k_1 and k_2 are called the *residues* of the partial fraction expansion. Before we address the issue of determining the values of the residues, a question that comes to mind is: What limitations, if any, should be placed on the numerator polynomial $B(s)$ for the expansion in Eqn. (E.1) to be valid? In order to find the answer to this question we will combine the two partial fractions under a common denominator to obtain

$$\frac{k_1}{s-a} + \frac{k_2}{s-b} = \frac{(k_1 + k_2) s - (k_1 b + k_2 a)}{(s-a)(s-b)} \tag{E.2}$$

It is obvious from Eqn. (E.2) that the numerator polynomial $B(s)$ may not be higher than first-order since no s^2 term or higher-order terms can be obtained in the process of combining the two partial fractions under a common denominator. Consider a more general case of a rational function with P poles in the form

$$X(s) = \frac{B(s)}{(s-s_1)(s-s_2)\ldots(s-s_P)}$$

$$= \frac{k_1}{s-s_1} + \frac{k_2}{s-s_2} + \ldots + \frac{k_P}{s-s_P} \tag{E.3}$$

If the P terms in the partial fraction expansion of Eqn. (E.3) are combined under a common denominator, the highest-order numerator term that could be obtained is s^{P-1}.

As a general rule, for the partial fraction expansion in the form of Eqn. (E.3) to be possible, the numerator order must be less than the denominator order.

If this is not the case, some additional steps need to be taken before partial fraction expansion can be used. Suppose $X(s)$ is a rational function in which the numerator order is greater than or equal to the denominator order. Let

$$X(s) = \frac{b_Q s^Q + b_{Q-1} s^{Q-1} + \ldots + b_1 s + b_0}{a_P s^P + a_{P-1} s^{P-1} + \ldots + a_1 s + a_0} \tag{E.4}$$

with $Q \geq P$. We can use long division on numerator and denominator polynomials to write $X(s)$ in the modified form

$$X(s) = C(s) + \frac{\bar{b}_{P-1} s^{P-1} + \bar{b}_{P-2} s^{P-2} + \ldots + \bar{b}_1 s + \bar{b}_0}{a_P s^P + a_{P-1} s^{P-1} + \ldots + a_1 s + a_0} \tag{E.5}$$

where $C(s)$ is a polynomial of s. Now the new function

$$\bar{X}(s) = X(s) - C(s) = \frac{\bar{b}_{P-1} s^{P-1} + \bar{b}_{P-2} s^{P-2} + \ldots + \bar{b}_1 s + \bar{b}_0}{a_P s^P + a_{P-1} s^{P-1} + \ldots + a_1 s + a_0} \tag{E.6}$$

has a numerator that is of lower order than its denominator, and can therefore be expanded into partial fractions. Example E.1 will illustrate this.

Example E.1:

Consider the function

$$X(s) = \frac{(s+1)(s+2)}{(s+5)(s+6)}$$

Since both the numerator and the denominator are of second order, the function cannot be expanded into partial fractions directly. In other words, an attempt to express $X(s)$ as

$$X(s) \overset{?}{=} \frac{k_1}{s+5} + \frac{k_2}{s+6}$$

would fail since the s^2 term in the numerator cannot be matched by combining the two partial fractions under a common denominator. Instead, we will use long division to extract a polynomial $C(s)$ from $X(s)$ so that the difference $X(s) - C(s)$ can be expanded into partial fractions. Let us first multiply out the factors in $X(s)$ to write it as

$$X(s) = \frac{s^2 + 3s + 2}{s^2 + 11s + 30}$$

Now we can divide the numerator polynomial by the denominator polynomial using long division to obtain

$$
\begin{array}{r}
1 \\
s^2 \quad +11s \quad +30 \enclose{longdiv}{ s^2 \quad +3s \quad +2} \\
\underline{s^2 \quad +11s \quad +30} \\
-8s \quad -28
\end{array}
\tag{E.7}
$$

Thus, the function $X(s)$ can be written in the equivalent form

$$X(s) = 1 + \frac{-8s - 28}{s^2 + 11s + 30} \tag{E.8}$$

Let

$$\bar{X}(s) = X(s) - 1 = \frac{-8s - 28}{s^2 + 11s + 30} \tag{E.9}$$

We can now expand the function $\bar{X}(s)$ into partial fractions as

$$\bar{X}(s) = \frac{k_1}{s+5} + \frac{k_2}{s+6} \tag{E.10}$$

and write $X(s)$ as

$$X(s) = 1 + \bar{X}(s) = 1 + \frac{k_1}{s+5} + \frac{k_2}{s+6} \tag{E.11}$$

Example E.2:

Consider the function

$$X(s) = \frac{(s+1)(s+2)(s+3)}{(s+5)(s+6)}$$

the factors of which can be multiplied to yield

$$X(s) = \frac{s^3 + 6s^2 + 11s + 6}{s^2 + 11s + 30}$$

Using long division on $X(s)$ we obtain

$$
\begin{array}{r}
s - 5 \\
s^2 \quad +11s \quad +30 \quad \big| \quad s^3 \quad +6s^2 \quad +11s \quad +6
\end{array}
$$

$$
\begin{array}{r}
s^3 \quad +11s^2 \quad +30s
\end{array}
$$

$$
\begin{array}{r}
-5s^2 \quad -19s \quad +6
\end{array}
$$

$$
\begin{array}{r}
-5s^2 \quad -55s \quad -150
\end{array}
$$

$$
\begin{array}{r}
36s \quad +156
\end{array}
$$

$$(E.12)$$

Thus, the function $X(s)$ can be written in the equivalent form

$$
X(s) = s - 5 + \bar{X}(s) = s - 5 + \frac{36s + 156}{s^2 + 11s + 30} \tag{E.13}
$$

and can be expressed using partial fractions as

$$
X(s) = s - 5 + \frac{k_1}{s + 5} + \frac{k_2}{s + 6} \tag{E.14}
$$

Now that we know how to deal with a function that has a numerator of equal or greater order than its denominator, we will focus our attention on determining the residues. For the purpose of determining the residues, we will assume that the order of the numerator polynomial is less than that of the denominator polynomial. In addition, we will initially consider a denominator polynomial with only simple roots. Thus, the function $X(s)$ is in the form

$$
X(s) = \frac{B(s)}{(s - s_1)(s - s_2) \ldots (s - s_P)} \tag{E.15}
$$

to be expanded into partial fractions in the form

$$
X(s) = \frac{k_1}{s - s_1} + \frac{k_2}{s - s_2} + \ldots + \frac{k_P}{s - s_P} \tag{E.16}
$$

Let us multiply both sides of Eqn. (E.16) by $(s - s_1)$ to obtain

$$
(s - s_1) X(s) = k_1 + \frac{k_2 (s - s_1)}{s - s_2} + \ldots + \frac{k_P (s - s_1)}{s - s_P} \tag{E.17}
$$

If we now set $s = s_1$ on both sides of Eqn. (E.17) we get

$$
(s - s_1) X(s) \Big|_{s=s_1} = k_1 \Big|_{s=s_1} + \frac{k_2 (s - s_1)}{s - s_2} \Big|_{s=s_1} + \ldots + \frac{k_P (s - s_1)}{s - s_P} \Big|_{s=s_1}
$$

$$
= k_1 \tag{E.18}
$$

Thus, the residue of the pole at $s = s_1$ can be found by multiplying $X(s)$ with the factor $(s - s_1)$ and then setting $s = s_1$. Using the form of $X(s)$ in Eqn. (E.15) the residue k_1 is found as

$$
(s - s_1) X(s) \Big|_{s=s_1} = \frac{B(s_1)}{(s_1 - s_2) \ldots (s_1 - s_P)} \tag{E.19}
$$

which amounts to canceling the $(s - s_1)$ term in the denominator, and evaluating what is left for $s = s_1$. Generalizing this result, the residue of the n-th pole is found as

$$
k_n = (s - s_n) X(s) \Big|_{s=s_n} \qquad n = 1, \ldots, P \tag{E.20}
$$

Example E.3:

Find a partial fraction expansion for the rational function

$$X\left(s\right) = \frac{\left(s+1\right)\left(s+2\right)}{\left(s+3\right)\left(s+4\right)\left(s+5\right)}$$

Solution: The partial fraction expansion we are looking for is in the form

$$X\left(s\right) = \frac{k_1}{s+3} + \frac{k_2}{s+4} + \frac{k_3}{s+5} \tag{E.21}$$

We will use Eqn. (E.20) to determine the residues. The first residue k_1 is obtained as

$$k_1 = \left.\left(s+3\right) X\left(s\right)\right|_{s=-3}$$

$$= \left.\frac{\left(s+1\right)\left(s+2\right)}{\left(s+4\right)\left(s+5\right)}\right|_{s=-3}$$

$$= \frac{\left(-3+1\right)\left(-3+2\right)}{\left(-3+4\right)\left(-3+5\right)} = 1$$

The remaining two residues k_2 and k_3 are obtained similarly:

$$k_2 = \left.\left(s+4\right) X\left(s\right)\right|_{s=-4}$$

$$= \left.\frac{\left(s+1\right)\left(s+2\right)}{\left(s+3\right)\left(s+5\right)}\right|_{s=-4}$$

$$= \frac{\left(-4+1\right)\left(-4+2\right)}{\left(-4+3\right)\left(-4+5\right)} = -6$$

$$k_3 = \left.\left(s+5\right) X\left(s\right)\right|_{s=-5}$$

$$= \left.\frac{\left(s+1\right)\left(s+2\right)}{\left(s+3\right)\left(s+4\right)}\right|_{s=-5}$$

$$= \frac{\left(-5+1\right)\left(-5+2\right)}{\left(-5+3\right)\left(-5+4\right)} = 6$$

Thus, the partial fraction expansion for $X\left(z\right)$ is

$$X\left(s\right) = \frac{1}{s+3} - \frac{6}{s+4} + \frac{6}{s+5}$$

Next we will consider a rational function the denominator of which has repeated roots. Let $X\left(s\right)$ be a rational function with a pole of order r at $s = s_1$.

$$X\left(s\right) = \frac{B\left(s\right)}{\left(s-s_1\right)^r\left(s-s_2\right)} \tag{E.22}$$

The partial fraction expansion for $X\left(s\right)$ needs to be in the form

$$X\left(s\right) = \frac{k_{1,1}}{s-s_1} + \frac{k_{1,2}}{\left(s-s_1\right)^2} + \ldots + \frac{k_{1,r}}{\left(s-s_1\right)^r} + \frac{k_2}{s-s_2} \tag{E.23}$$

For the multiple pole at $s = s_1$ as many terms as the multiplicity of the pole are needed so that combining the terms on the right side of Eqn. (E.23) can yield a result that matches the function $X(s)$ given by Eqn. (E.22).

The residue k_2 of the single pole at $s = s_2$ can easily be determined as discussed above. We will focus our attention on determining the residues of the r-th order pole at $s = s_1$. Let us begin by multiplying both sides of Eqn. (E.23) by $(s - s_1)^r$ to obtain

$$(s - s_1)^r X(s) = k_{1,1} (s - s_1)^{r-1} + k_{1,2} (s - s_1)^{r-2} + \ldots + k_{1,r} + \frac{k_2 (s - s_1)^r}{s - s_2} \quad \text{(E.24)}$$

If we now set $s = s_1$ on both sides of E.24) we would get

$$(s - s_1)^r X(s) \Big|_{s=s_1} = k_{1,1} (s - s_1)^{r-1} \Big|_{s=s_1} + k_{1,2} (s - s_1)^{r-2} \Big|_{s=s_1}$$

$$+ \ldots + k_{1,r} \Big|_{s=s_1} + \frac{k_2 (s - s_1)^r}{s - s_2} \Big|_{s=s_1}$$

$$= k_{1,r} \quad \text{(E.25)}$$

It can be shown that the other residues for the pole at $s = s_1$ can be found as follows:

$$k_{1,r-1} = \frac{d}{ds} \left[(s - s_1)^r X(s) \right] \Big|_{s=s_1}$$

$$k_{1,r-2} = \frac{1}{2} \frac{d^2}{ds^2} \left[(s - s_1)^r X(s) \right] \Big|_{s=s_1}$$

$$\vdots$$

$$k_{1,2} = \frac{1}{(r-2)!} \frac{d^{r-2}}{ds^{r-2}} \left[(s - s_1)^r X(s) \right] \Big|_{s=s_1}$$

$$k_{1,1} = \frac{1}{(r-1)!} \frac{d^{r-1}}{ds^{r-1}} \left[(s - s_1)^r X(s) \right] \Big|_{s=s_1}$$

Generalizing these results, the residues of a pole of multiplicity r are calculated using

$$k_{1,n} = \frac{1}{(r-n)!} \frac{d^{r-n}}{ds^{r-n}} \left[(s - s_1)^r X(s) \right] \Big|_{s=s_1} \qquad n = 1, \ldots, r \quad \text{(E.26)}$$

In using the general formula in Eqn. (E.26) for $n = r$ we need to remember that $0! = 1$ and

$$\frac{d^0}{ds^0} \left[(s - s_1)^r X(s) \right] = (s - s_1)^r X(s) \quad \text{(E.27)}$$

Example E.4:

Find a partial fraction expansion for the rational function

$$X(s) = \frac{s}{(s + 1)^3 (s + 2)}$$

Solution: The partial fraction expansion for $X(s)$ is in the form

$$X(s) = \frac{k_{1,1}}{s + 1} + \frac{k_{1,2}}{(s + 1)^2} + \frac{k_{1,3}}{(s + 1)^3} + \frac{k_2}{s + 2} \quad \text{(E.28)}$$

The residue of the single pole at $s = -2$ is easily found using the technique developed earlier:

$$k_2 = (s + 2) X (s)\Big|_{s=-2} = \frac{s}{(s+1)^3}\Big|_{s=-2} = \frac{-2}{(-2+1)^3} = 2$$

We will use Eqn. (E.26) to find the residues of the third-order pole at $s = -1$.

$$k_{1,3} = (s + 1)^3 X (s)\Big|_{s=-1} = \frac{s}{(s+2)}\Big|_{s=-1} = \frac{-1}{(-1+2)} = -1$$

$$k_{1,2} = \frac{d}{ds}\left[(s + 1)^3 X (s)\right]\Big|_{s=-1} = \frac{d}{ds}\left[\frac{s}{s+2}\right]\Big|_{s=-1} = \frac{2}{(s+2)^2}\Big|_{s=-1} = 2$$

$$k_{1,1} = \frac{1}{2!}\frac{d^2}{ds^2}\left[(s + 1)^3 X (s)\right]\Big|_{s=-1} = \frac{1}{2!}\frac{d^2}{ds^2}\left[\frac{s}{s+2}\right]\Big|_{s=-1} = \frac{-2}{(s+2)^3}\Big|_{s=-1} = -2$$

Thus, the partial fraction expansion for $X (z)$ is

$$X (s) = -\frac{2}{s+1} + \frac{2}{(s+1)^2} - \frac{1}{(s+1)^3} + \frac{2}{s+2}$$

E.2 Partial Fraction Expansion for Discrete-Time Signals and Systems

The residues of the partial fraction expansion for rational z-transforms are obtained using the same techniques employed in the previous section. One subtle difference is that, when working with the inverse z-transform, we often need the partial fraction expansion in the form

$$\begin{aligned}X (z) &= \frac{B (z)}{(z - z_1) (z - z_2) \ldots (z - z_P)}\\ &= \frac{k_1 z}{z - z_1} + \frac{k_2 z}{z - z_2} + \ldots + \frac{k_P z}{z - z_P}\end{aligned} \tag{E.29}$$

In order to use the residue formulas established in the previous section and still obtain the expansion in the form of Eqn. (E.29) with a z factor in each term, we simply expand $X (z) / z$ into partial fractions:

$$\begin{aligned}\frac{X (z)}{z} &= \frac{B (z)}{(z - z_1) (z - z_2) \ldots (z - z_P) z}\\ &= \frac{k_1}{z - z_1} + \frac{k_2}{z - z_2} + \ldots + \frac{k_P}{z - z_P} + \frac{k_{P+1}}{z}\end{aligned} \tag{E.30}$$

The last term in Eqn. (E.30) may or may not be needed depending on whether the numerator polynomial $B (z)$ has a root at $z = 0$ or not. Once the residues of the expansion in Eqn. (E.30) are found, we can revert to the form in Eqn. (E.29) by multiplying both sides with z.

Appendix F

Review of Matrix Algebra

In this section we present some basic definitions for matrix algebra.

Matrix

A matrix is a collection of real or complex numbers arranged to form a rectangular grid:

$$\mathbf{A} = \begin{bmatrix} a_{11} & a_{12} & \cdots & a_{1M} \\ a_{21} & a_{22} & \cdots & a_{2M} \\ \vdots & & & \\ a_{N1} & a_{N2} & \cdots & a_{NM} \end{bmatrix} \tag{F.1}$$

A total of NM numbers arranged into N rows and M columns as shown in Eqn. (F.1) form a $N \times M$ matrix. The numbers that make up the matrix are referred to as the *elements* of the matrix. Two subscripts are used to specify the placement of an element in the matrix. The element at the in row-i and column-j is denoted by a_{ij}.

Vector

A vector is a special matrix with only one row or only one column. A matrix with only one row (a $1 \times N$ matrix) is called a *row vector* of length N. A matrix with only one column (a $M \times 1$ matrix) is called a *column vector* of length M.

Square matrix

A *square matrix* is one in which the number of rows is equal to the number of columns. An example is

$$\mathbf{A} = \begin{bmatrix} 5 & 2 & 0 & 4 \\ 3 & -1 & 0 & 0 \\ 0 & 1 & 1 & 3 \\ -7 & 4 & 3 & 5 \end{bmatrix} \tag{F.2}$$

Diagonal matrix

A *diagonal matrix* is a square matrix in which all elements that are not on the main diagonal are equal to zero. The main diagonal of a square matrix is the diagonal from the top left corner to the bottom right corner. Elements on the main diagonal have identical row and column indices such as $a_{11}, a_{22}, \ldots, a_{NN}$. An example is

$$\mathbf{A} = \begin{bmatrix} 5 & 0 & 0 & 0 \\ 0 & -1 & 0 & 0 \\ 0 & 0 & 1 & 0 \\ 0 & 0 & 0 & 4 \end{bmatrix} \tag{F.3}$$

Identity matrix

An *identity matrix* is a diagonal matrix in which all diagonal elements are equal to unity. For example, the identity matrix of order 3 is

$$\mathbf{I_3} = \begin{bmatrix} 1 & 0 & 0 \\ 0 & 1 & 0 \\ 0 & 0 & 1 \end{bmatrix} \tag{F.4}$$

Equality of two matrices

Two matrices $\mathbf{A}$ and $\mathbf{B}$ are equal to each other if they have the same dimensions and if all corresponding elements are equal, that is

$$\mathbf{A} = \mathbf{B} \quad \Longrightarrow \quad a_{ij} = b_{ij} \quad \text{for } i = 1, \ldots, M \text{ and } j = 1, \ldots, N \tag{F.5}$$

Trace of a square matrix

The *trace* of a square matrix is defined as the arithmetic sum of all of its elements on the main diagonal.

$$\text{Trace}(\mathbf{A}) = \sum_{i=1}^{N} a_{ii} \tag{F.6}$$

Transpose of a matrix

The *transpose* of a matrix $\mathbf{A}$ is denoted by $\mathbf{A}^T$, and is obtained by interchanging rows and columns so that row-i of $\mathbf{A}$ becomes column-i of $\mathbf{A}^T$. For example, the transpose of the matrix $\mathbf{A}$ in Eqn. (F.2) is

$$\mathbf{A}^T = \begin{bmatrix} 5 & 3 & 0 & -7 \\ 2 & -1 & 1 & 4 \\ 0 & 0 & 1 & 3 \\ 4 & 0 & 3 & 5 \end{bmatrix} \tag{F.7}$$

Determinant of a square matrix

Each square matrix has a scalar value associated with it referred to as the *determinant*. Methods for computing determinants can be found in most texts on linear algebra. As an example, the determinant of the matrix

$$\mathbf{A} = \begin{bmatrix} 5 & 2 \\ 3 & -1 \end{bmatrix} \tag{F.8}$$

is computed as

$$|\mathbf{A}| = (5)(-1) - (2)(3) = -11 \tag{F.9}$$

Minors of a square square matrix

The row-i column-j minor m_{ij} of a square matrix $\mathbf{A}$ is defined as the determinant of the submatrix obtained by deleting row-i and column-j from the matrix $\mathbf{A}$. Consider, for example the square matrix

$$\mathbf{A} = \begin{bmatrix} 5 & 2 & 4 \\ 3 & -1 & 2 \\ -1 & 0 & 3 \end{bmatrix} \tag{F.10}$$

The minor m_{23} is found by deleting row-2 and column-3 and computing the determinant of the matrix left behind as

$$m_{23} = \begin{vmatrix} 5 & 2 \\ -1 & 0 \end{vmatrix} = 2 \tag{F.11}$$

Cofactors of a square square matrix

The row-i column-j cofactor Δ_{ij} of a square matrix $\mathbf{A}$ is defined as

$$\Delta_{ij} = (-1)^{i+j} \, m_{ij} \tag{F.12}$$

Using the matrix in Eqn. (F.10) as an example, the cofactor Δ_{23} is

$$\Delta_{23} = (-1)^{2+3} \, m_{23} = -2 \tag{F.13}$$

Adjoint of a square square matrix

The adjoint of a square matrix $\mathbf{A}$ is found by first transposing $\mathbf{A}$ and then replacing each element by its cofactor. For example, the adjoint of the matrix $\mathbf{A}$ in Eqn. (F.10) is

$$\text{Adj}\,(\mathbf{A}) = \begin{bmatrix} -3 & -6 & 8 \\ -11 & 19 & 2 \\ -1 & -2 & -11 \end{bmatrix} \tag{F.14}$$

Scaling a matrix

Scaling a matrix by a constant scale factor means multiplying each element of the matrix with that scale factor.

Addition of two matrices

Addition of two matrices is only defined for matrices with identical dimensions. The sum of two matrices $\mathbf{A}$ and $\mathbf{B}$ is the matrix $\mathbf{C}$ in which each element is equal to the sum of corresponding elements of $\mathbf{A}$ and $\mathbf{B}$.

$$\mathbf{C} = \mathbf{A} + \mathbf{B} \quad \implies \quad c_{ij} = a_{ij} + b_{ij} \quad \text{for } i = 1, \ldots, M \text{ and } j = 1, \ldots, N \tag{F.15}$$

Scalar product of two vectors

The scalar product of a $1 \times N$ row vector $\mathbf{x}$ and a $N \times 1$ column vector $\mathbf{y}$ is defined as

$$\mathbf{x} \cdot \mathbf{y} = \sum_{i=1}^{N} x_i \, y_i \tag{F.16}$$

Multiplication of two matrices

The product of an $M \times K$ matrix $\mathbf{A}$ and a $K \times N$ matrix $\mathbf{B}$ is a $M \times N$ matrix $\mathbf{C}$ the elements of which are computed as

$$c_{ij} = \sum_{k=1}^{K} a_{ik} \, b_{kj} \tag{F.17}$$

In other words, the row-i column-j element of matrix $\mathbf{C}$ is the scalar product of row-i of of matrix $\mathbf{A}$ and column-j of matrix $\mathbf{B}$.

Inversion of a matrix

The inverse of a square matrix $\mathbf{A}$ denoted by $\mathbf{A}^{-1}$ is also a square matrix. It that satisfies the equation

$$\mathbf{A}^{-1}\,\mathbf{A} = \mathbf{A}\,\mathbf{A}^{-1} = \mathbf{I} \tag{F.18}$$

It is computed by scaling the adjoint matrix by the reciprocal of the determinant, that is,

$$\mathbf{A}^{-1} = \frac{\mathrm{Adj}\,(\mathbf{A})}{|\mathbf{A}|} \tag{F.19}$$

The inverse does not exist if the determinant of $\mathbf{A}$ is equal to zero. Such a matrix is called a *singular* matrix.

Eigenvalues and eigenvectors

The scalar λ and the $N \times 1$ vector $\mathbf{v}$ are called an *eigenvalue* and an *eigenvector* of the $N \times N$ matrix $\mathbf{A}$ respectively if they satisfy the equation

$$\mathbf{A}\,\mathbf{v} = \lambda\,\mathbf{v} \tag{F.20}$$

An $N \times N$ matrix $\mathbf{A}$ has N eigenvalues and associated eigenvectors although they are not necessarily unique.

Index